Praise for *Financial Data Analytics*

Really interesting, and an impressive masterpiece! *Financial Data Analytics* contains a rich amount of material, with original research findings in almost every chapter; many parts of the book will even be directly helpful for my own teaching in business school. In view of its dedication towards data-driven analytical tools genuinely needed in financial problems, I believe that it is the very book that defines the scope of financial data analytics.

—Alain Bensoussan, Fellow of AMS, IEEE, and SIAM; President of *INRIA* (1984–1996); President of *CNES (Centre National d'Etudes Spatiales)* (1996–2003); Chairman of *ESA Council (European Space Agency)* (1999–2002); Former Member of Advisory Board, *Mathematical Finance*; Lars Magnus Ericsson Chair Professor of Management, *Naveen Jindal School of Management, University of Texas at Dallas*

Financial Data Analytics provides a timely and thorough exploration of crucial topics in contemporary data science, specifically tailored for a quantitative finance audience. It skillfully balances introductory and advanced concepts, seamlessly integrating mathematical foundations with detailed coding examples. Designed to appeal to a broad audience, the content accommodates varying levels of familiarity with the mathematical and computational aspects of quantitative finance. The authors' adept presentation of complex ideas, coupled with practical applications, renders *Financial Data Analytics* an invaluable resource for both novice and seasoned professionals alike.

—KC Gary Chan, Fellow of ASA and IMS; President (Western North American Region), *International Biometrics Society* (2022); Professor, Department of Statistics, *University of Washington*

This book presents a wide coverage of state-of-the-art topics in data analytics, which are crucial in our current era of big data. Its organic blend of mathematical derivations of the theory and practical applications in FinTech and InsurTech via tailor-made implementable Python and R codes is exceptional. To given but a single example, based on my own research interests, the novel CIBer is a very interesting and original new tool. It is a joy to read *Financial Data Analytics*, a book that cannot be missed on the bookshelf of any researcher or student interested in this topic.

—Jan Dhaene, Full Professor, Director of Master of Science in Financial and Actuarial Engineering, and Head of Actuarial Research Group, Department of Accountancy, Finance and Insurance, Faculty of Business and Economics, *KU Leuven*; Head of Division "Actuariële Toepassingen voor Verzekerings-ondernemingen en Pensioenfondsbeheer", *KU Leuven Research and Development*; Member of *Institute of Actuaries of Belgium*, Member and Vice-chair of Actuarial Education Network, *International Actuarial Association*

Financial Data Analytics is a fantastic book that offers rare gifts to industry practitioners. The important theories are brought to life through R and Python program codes, developed by the authors for the book, and easily adaptable for industry use. The book has comprehensive coverage of state-of-the-art techniques for every need. I like reading the practical applications, which help develop intuition for the more complicated methodologies, and surely someone can make a handsome profit implementing them. A super read, and a must-have for professionals if numbers rule your world.

—Kaiser Fung, Bestselling author, *Numbers Rule Your World* and *Numbersense*; Founding Director, MSc programme in Applied Analytics, *Columbia University*; Founder, *Principal Analytics Prep*

Financial Data Analytics is an exceptional book that integrates mathematics, practical examples, and real-life scenarios. With its focus on real datasets and practical programming codes in Python and R, the book offers a comprehensive exploration of various topics. It presents novel research findings and provides valuable insights for researchers, practitioners, and actuarial students. The book strikes a balance between foundational concepts and advanced techniques, making it an invaluable reference for professionals in the field. Additionally, its relevance extends to actuarial students preparing for their professional examinations. By redefining the landscape of financial data analytics in FinTech and InsurTech, this book establishes itself as a trusted guide in the industry.

—Simon Lam, Fellow of SOA, CFA, and FRM; President of *The Actuarial Society of Hong Kong* (2018, 2023); Deputy CEO & General Manager, *Munich Re (Hong Kong)*

The book will certainly play an impactful role in the advancement of financial analytics and should be on the bookshelf of every serious student of the topic.

—Wai Keung Li, Fellow of ASA and IMS; Emeritus Professor, *The University of Hong Kong*; Dean, *Faculty of Liberal Arts and Social Sciences, The Education University of Hong Kong*

Financial Data Analytics is a masterfully written book that encompasses a wide spectrum of statistical models and algorithms, with a special emphasis on financial and insurance applications. Drawing upon their multidisciplinary background and extensive research experience, as well as their close connection with the industry, the authors skillfully explain the theoretical underpinnings of both conventional and contemporary statistical methods that are truly relevant to the industry (including but not limited to regression learning, classification trees, neural networks, as well as the specification and assessment of these models), and amply illustrate the practical applications of these methods in various disciplines, by an abundance of real financial and insurance data, and using both Python and R. The dual focus on theory and applications, together with the discussion on recent advancements of the fields, makes the book one of a kind, even field-defining, among books on similar topics, and an ideal resource for anyone interested in understanding and implementing statistical models in this era of big data, as well as for students preparing for professional examinations on data analytics, such as the SRM, PA and ATPA exams of the Society of Actuaries.

—Ambrose Lo, Fellow of SOA, Chartered Enterprise Risk Analyst; Author of *ACTEX Study Manual for SOA Exam SRM*, *ACTEX Study Manual for SOA Exam PA*, and *ACTEX Study Manual for SOA Exam ATPA*

It is a tome!

—Suresh P. Sethi, Fellow of INFORMS, IEEE, POMS, and SIAM; Eugene McDermott Chair Professor of Operations Management, *Naveen Jindal School of Management, University of Texas at Dallas*

Financial Data Analytics is an encyclopedic documentation of in-depth and extensive statistical analysis in finance and beyond. It provides an end-to-end systematic approach to academics and practitioners with theories, tremendous examples and data, and algorithms with coding that are readily applicable in real life. The book encompasses four dimensions of coverage—theoretical framework to application and coding, distributional characteristics to data diagnosis and simulation, learning, and lastly coverage of both qualitative and quantitative data. The book consolidates classical knowledge with the most contemporary research in all subjects. *Financial Data Analytics* is one comprehensive biblical handbook for academic researchers, financial practitioners, and graduate students for both methodologies and applications. The book also lays a systematic framework for future extension and enrichment for financial data analytics.

—Nai-pan Tang, Former Chief Risk Officer and Member of Executive Committee, Hang Seng Bank; Former Deputy CEO and Chief Risk Officer, Shanghai Commercial Bank Ltd.; Former Director of the Board, Deputy CEO, Alternative CEO, Chief Risk Officer, and Vice Chairman of Asset Management, China CITIC Bank International; Director, The Hong Kong Institute of Bankers (2019–2021); Professor of Practice, Department of Finance, Chinese University of Hong Kong

Financial Data Analytics is a very impressive work with extensive coverage and many interesting topics. It stands out among similar books by the unique blend of detailed mathematical derivations, practical examples, real-life datasets, and readily available programme codes in Python and R. It also contains many novel results from the authors' recent research. Whether you are researchers on related fields, practitioners in the financial industry, or students preparing for a few exams at The Society of Actuaries or The Institute and Faculty of Actuaries, *Financial Data Analytics* will certainly be a valuable reference book.

—Hailiang Yang, Associate of SOA and Honorary Fellow of IFoA; Editor of *Insurance: Mathematics and Economics*; Professor, *Department of Financial and Actuarial Mathematics, Xi'an Jiaotong–Liverpool University*

Throughout my career at JP Morgan Chase and then CITIC Securities, I have participated in and witnessed how various data analytics tools revolutionized financial industry. *Financial Data Analytics* is a perfect example echoing this trend, by discussing a wide spectrum of modern tools in data analytics, from both a theoretical viewpoint and a practical aspect, with readily implementable programme codes in both Python and R for real-life examples. This combination is so unique amongst the few books on data analytics; it not only reflects the broad range of theoretical knowledge of the authors, but also demonstrates their close ties with the finance and insurance industries. I actually had the opportunity to testify some profit-making strategies mentioned in the book, and their performance was genuinely impressive. This book is far more than an academic monograph for scholars; it is certainly an illuminating guide for practitioners to explore their own alchemy of finance.

—Wei Zhou, Executive Director of Equity Derivatives Quantitative Research, *JP Morgan Chase* (2016–2021); Executive Director and Head of Quantitative Modelling, *CITIC Securities*

Financial Data Analytics

Founded in 1807, John Wiley & Sons is the oldest independent publishing company in the United States. With offices in North America, Europe, Australia and Asia, Wiley is globally committed to developing and marketing print and electronic products and services for our customers' professional and personal knowledge and understanding.

The Wiley Finance series contains books written specifically for finance and investment professionals as well as sophisticated individual investors and their financial advisors. Book topics range from portfolio management to e-commerce, risk management, financial engineering, valuation and financial instrument analysis, as well as much more.

For a list of available titles, visit our Web site at www.WileyFinance.com.

Financial Data Analytics

with Machine Learning, Optimization and Statistics

SAM CHEN
Hang Seng University of Hong Kong

KA CHUN CHEUNG
University of Hong Kong

PHILLIP YAM
Chinese University of Hong Kong

with programme codes by Kaiser Fan

WILEY

This edition first published 2025

Registered Office(s)
John Wiley & Sons, Inc., 111 River Street, Hoboken, NJ 07030, USA
John Wiley & Sons Ltd, The Atrium, Southern Gate, Chichester, West Sussex, PO19 8SQ, UK

Editorial Office
The Atrium, Southern Gate, Chichester, West Sussex, PO19 8SQ, UK

For details of our global editorial offices, customer services, and more information about Wiley products visit us at www.wiley.com.
Wiley also publishes its books in a variety of electronic formats and by print-on- demand. Some content that appears in standard print versions of this book may not be available in other formats. Designations used by companies to distinguish their products are often claimed as trademarks. All brand names and product names used in this book are trade names, service marks, trademarks or registered trademarks of their respective owners. The publisher is not associated with any product or vendor mentioned in this book.

Library of Congress Cataloging-in-Publication Data Is Available:

ISBN 9781119863373 (Cloth)
ISBN 9781119863380 (ePDF)
ISBN 9781119863397 (ePub)
ISBN 9781119863403 (oBook)

Cover Design: Wiley
Cover Image: © da-kuk/Getty Images

SKY10084189_091024

To our parents and families

Contents

About the Authors

Yongzhao Chen (Sam) received his BSc in Actuarial Science with first class honours and PhD in Actuarial Science from The University of Hong Kong. He is currently an Assistant Professor at the Department of Mathematics, Statistics and Insurance of the Hang Seng University of Hong Kong. His research interests include actuarial science, especially credibility theory, and data analytics.

Ka Chun Cheung received his BSc in Actuarial Science with first class honours and PhD from The University of Hong Kong. He was the Director of the Actuarial Science Programme, and is currently Head and full Professor at the Department of Statistics and Actuarial Science in School of Computing and Data Science, The University of Hong Kong. He is an Associate of the Society of Actuaries and an elected member of the International Statistical Institute. He is serving on the editorial boards of Insurance, Mathematics and Economics and Journal of Industrial and Management Optimization. His current research interests include various topics in actuarial science, including optimal reinsurance, stochastic orders, dependence structures, and extreme value theory.

Phillip Yam received his BSc in Actuarial Science with first class honours and MPhil from The University of Hong Kong. Supported by the two scholarships awarded by the Croucher Foundation (Hong Kong), he obtained an MASt (Master of Advanced Study) degree, Part III of the Mathematical Tripos, with Distinction in Mathematics from University of Cambridge and a DPhil in Mathematics from University of Oxford. During his postgraduate studies, he was awarded with the E. M. Burnett Prize in Mathematics from University of Cambridge, and the junior research fellowship from The Erwin Schrödinger International Institute for Mathematics and Physics of University of Vienna.

Phillip is currently the Co-Director of the Interdisciplinary Major Programme in Quantitative Finance and Risk Management Science, and a full Professor at the Department of Statistics of The Chinese University of Hong Kong (CUHK). He is also Assistant Dean (Education) of CUHK Faculty of Science, and Fellow of the Centre for Promoting Science Education in the Faculty. He has been appointed as a research fellow in the Hausdorff Research Institute for Mathematics at the University of Bonn and a Visiting Professor in both the Department of Statistics at Columbia University in the City of New York and Naveen Jindal of Management at University of Texas at Dallas. He has published about a hundred journal articles in actuarial science, applied mathematics, data analytics, engineering, financial mathematics, operations management, and statistics, and has also been serving in editorial boards of several journals in these fields. Together with Alain Bensoussan and Jens Frehse, he wrote the first monograph on mean field games and mean field type control theory. His research project with the title "Comonotone-independence Bayes Classifier (CIBer)" was awarded a Silver Medal in the 48th International Exhibition of Inventions Geneva in 2023. Besides academia, he has provided consulting services for various financial institutions and insurance companies, and established close connections in these industries; many of his students also work in international investment banking and insurance companies.

Kaiser Fan received his BSc in Risk Management Science with first class honours and MPhil from The Chinese University of Hong Kong under the guidance of Professor Phillip Yam. As a data scientist, his research interests include data analytics, and machine learning especially in deep learning. He contributes to the programming and many examples and illustrations in the book.

Foreword

To the memory of

Tze Leung Lai (1945–2023)

Late Ray Lyman Wilbur Professor of Statistics,
Stanford University

We were saddened to hear about the sudden passing away of Professor Tze Leung Lai. We would like to thank him for his care for the younger generation, including ourselves, as well as all of his valuable guidance in the past two decades, since Ka Chun's and Phillip's senior-year undergraduate and master studies at The University of Hong Kong; we all learned a lot from him, both indirectly or directly. He was certainly a renowned scholar. Due to the pandemic, we could not make the trip to visit him in person; we were hoping to meet him again last summer after the pandemic eventually came to an end, only to learn that he departed too soon. During the book writing process, we sent a draft version of the book to him. He was glad of what we had achieved and also graciously offered to write a foreword for this book, which can no longer become a reality now. However, this foreword is always reserved for him. We thank you again for your generous offer, Professor Lai; thank you, our mentor, may you rest in peace.

Winter, 2023

Preface

In the field of finance, nothing is more important than gaining profits, and any innovation that draws people's attention must lead to at least the same level of profit as the existing methods; indeed, this has always been the main driving force of updates to relevant curricula over time. For instance, with the development of option pricing and portfolio selection in 1970s, the financial training from mid-1980s to 2000s heavily involved Itô's stochastic calculus and partial differential equations. On the other hand, volatility models such as GARCH were proposed by Robert Engle in the early 1980s for a better estimation of parameters facilitating derivative valuation and portfolio management, and various academic curricula quickly followed suit by placing more emphasis on financial econometrics. Quantitative analysis of game theory, particularly the numerical algorithm for discovering equilibrium points, gained more importance and attention in academia after John Nash won the Nobel Memorial Prize in Economic Sciences in 1994; the trend continues today with further sophistication and generalization towards the context of mean field games in the last dozen years. In the 2000s, as behavioural finance was gaining increasing attention in society, people wanted to learn more methods in the realm of experimental behavioural economics and finance, especially on how the market makes use of statistical methods to understand the impact of different human behaviour and devise advertising strategies accordingly, which explains why case studies and primitive statistics have been prevalent in the financial classes in recent decades.

Recently, attention has been diverted towards AI. The revolutionary developments in machine learning and deep learning have brought new elements of data analytics into finance, particularly including the heated areas of *InsurTech*, *FinTech* and *RegTech*. To catch up with the trend, curriculum designs should be revised to cover financial or business data analytics in a comprehensive manner, and statistics is certainly at the core of them; this is precisely why we wanted to write this book. Among the few books in this field, involving the use of standard statistics in financial analysis with either Python or **R**, and statistical applications in financial engineering, focus is usually put on the possible financial applications of conventional statistical tools, yet some practical problems may require tools beyond traditional statistics, and we aim to address a few relevant issues in this book. Another important motivation for us is certainly the positive feedback from students regarding our teaching materials in the past decade, which we have consolidated as the foundation of this book.

This book investigates contemporary practical techniques of financial data analytics that are specific for real-life scenarios and leave room for a high profit-making potential, with 15 chapters in total covering a wide range of important and frontier topics in this field. For example, we shall explore data analytics in investment strategy, financial forensics, and the immediate use of deep learning in finance. We also critically discuss the pros and cons of machine learning tools. While we raise caution against potential pitfalls of new approaches like deep learning, we also propose a novel feature engineering scheme as part of CIBer (see Chapter 12) to overcome limitations of existing methods regarding input features, which also achieves a promising classification performance. Examples are provided throughout the whole book, in which we focus on a few typical datasets from real-life financial markets to facilitate intuitive comparisons among models, allowing readers to form their own judgements on their pros and cons, and hence apply suitable data analysis methods to their own datasets. Executable detailed programme codes in Python and **R** are also readily available with corresponding examples. Practitioners including quants and fund managers can gain insights into the latest developments of data analytics from the book, and help to formulate effective investment strategies or to facilitate better product designs. This up-to-date knowledge may further help them conduct novel applied research in different business disciplines. It is also our hope that senior-year students and postgraduates can deepen their understanding on this field and find the book useful for their future academic research. Meanwhile, the contents of this book also cover a large part of syllabi of modules from different public professional examinations on predictive analytics, including but not limited to the Statistics for Risk Modeling (SRM) Exam and Predictive Analytics (PA) Exam of the *Society of Actuaries*, making it a suitable main or supplementary reading for these professional examinations.

To benefit the most from this book, it is advisable that readers have a solid background in probability and statistics, linear algebra, and advanced calculus, preferably at the sophomore level. For some parts of the book, some basic knowledge in real and complex analysis would enhance a full understanding of them. Especially, some sections marked by asterisks necessitate a higher level of mathematical understanding and may be omitted during the initial reading. Anyhow, for the convenience of the readers, some of the relevant basic knowledge is reviewed in Chapter 1. Acquaintance with programming languages such as Python or **R** is also instrumental. To make the book more self-contained, a quick overview of these two programming languages is provided in Chapter 2. In addition, readers interested in more sophisticated investment strategies and derivative pricing also need a rudimentary background in Itô's stochastic calculus and continuous martingale theory, which is unavoidable given the technical nature of the subject matter.

In writing this book, our multidisciplinary background proved helpful; we all had diverse training in actuarial science, economics and finance, mathematics, probability and statistics, and our research also involves the application of data analytics in diverse applied areas. We have established long-term research collaborations with industry practitioners, and many of our undergraduate and postgraduate students, friends, and colleagues are also working in world-leading companies in finance and insurance sectors. Growing up in the traditional global financial hub of Hong Kong has also equipped us with practical financial knowledge that benefits our pragmatic research, while we are not bounded by the routine methods in solving both research and practical problems. In addition, our graduate student Kaiser Fan also made a unique contribution by implementing most programme codes in the examples in a tailor-made and illuminating fashion.

With the publication of this book, we welcome valuable feedback and comments from readers. Due to limitations in both scope and time, we could not delve into all topics in detail, and we apologize for any missing information. While we drew inspiration from a wide range of literature, and we tried to cite all of them, some may still have unintentionally slipped our mind over time, and we sincerely apologize for any oversights. We also benefited from courses on financial data analytics or equivalents, including teaching materials and course design, in renowned universities in Asia, Australia, Europe and North America.

While we believe our book offers a valuable collection of tools in financial data analytics, we deliberately left out blockchains, as they have shifted towards being an internet and network security concern rather than an analytical tool for financial information. Meanwhile, we have an upcoming book on deep learning with some applications in finance, to which we briefly hint in the closing chapter of this book. Hopefully readers will enjoy the current book and stay tuned for the upcoming release.

Sam Chen, Ka Chun Cheung, and Phillip Yam
Hong Kong, December 2023

Acknowledgements

First and foremost, we would like to thank Wiley for their kind consideration of our book, as well as their assistance and patience in different stages of preparation, so that we could have more time to polish this highly original book.

This book would also be impossible without the valuable teaching materials and datasets generously provided by Pui Lam Leung, and we would like to express our sincere gratitude to him. Moreover, as the soul figure of the whole book, Phillip Yam took the initiative to design its framework, and led the writing of all chapters using his solid mathematical knowledge and deep financial insights; the other authors express their sincere gratitude for his dedicated effort and unique role played.

Throughout the writing of this book over the last five years (mostly during the COVID-19 pandemic), we received much positive feedback and constructive suggestions from experts in both academia and industry after their careful reading of our book, including but not limited to Alain Bensoussan, Jan Dhaene, Tze Leung Lai, Wai Keung Li, Hailiang Yang; as well as Kaiser Fung, Wah Tung Lau, Kam Pui Wat, John Alexander Wright, and Wei Zhou. Our special thanks go to all of them, particularly Professor Lai; he will be remembered for long.

The assistance on typesetting and proofreading by other members in our research team, namely Pok Him Cheng, Adrian Patrick Kennedy, Junyou Li, Kenneth Tsz Hin Ng, Zhengyao Sun, and Jing Zhang, is also critically important to us. We truly appreciate their efforts, without which our book cannot be completed within a limited period of time at the last stage.

We also acknowledge funding support from various research and teaching grants. Yongzhao Chen (Sam) was partially supported by UGC/FDS14/P03/23 with the project title "*InsurTech: Risk Classification and Premium Calibration with Data Analytics*"; Ka Chun Cheung was partially supported by HKGRF-17303721 with the project title "*A contemporary study of optimal reinsurance arrangements*"; Phillip Yam was partially supported by HKGRF-14300319 with the project title "*Shape-constrained Inference: Testing for Monotonicity*", HKGRF-14301321 with the project title "*General Theory for Infinite Dimensional Stochastic Control: Mean Field and Some Classical Problems*", HKGRF-14300123 with the project title "*Well-posedness of Some Poisson-driven Mean Field Learning Models and their Applications*", CUHK Teaching Development and Language Enhancement Grant (TDLEG) for the 2022–25 Triennium with the project title "*Computational Thinking (CT) as a Problem-solving Skill – A Multidisciplinary Virtual Learning Package*", CUHK Teaching Development and Language Enhancement Grant (TDLEG) for the 2019–22 Triennium: Funding Scheme for Engaging Postgraduate Students in Teaching and Teaching Development with the project title "*Supporting Statistics Research Postgraduates to Teach Quantitative Data Analysis to Postgraduate Students without*

Statistics Background – Phase I", and CUHK Teaching Development and Language Enhancement Grant (TDLEG) for the 2022–25 Triennium: Funding Scheme for Engaging Postgraduate Students in Teaching and Teaching Development with project title "*Supporting Statistics Research Postgraduates to Teach Quantitative Data Analysis to Postgraduate Students without Statistics Background – Phase II*". The work described in this book was supported by a grant from the Germany/Hong Kong Joint Research Scheme sponsored by the Research Grants Council of Hong Kong and the German Academic Exchange Service of Germany (Reference No. G-CUHK411/23).

Last but certainly not least, we give heartful thanks to our families for their unwavering support. The writing of this book occupied much of our spare time that we could have spent with our families. We are immensely grateful for their understanding and support throughout. We can never thank them enough, and we love them from the bottom of our hearts.

Introduction

"We know the past but cannot control it. We control the future but cannot know it."

—Claude Shannon [18]

DEVELOPMENT OF FINANCIAL DATA ANALYTICS

Financial data analytics can be described as the application of statistical models and algorithms to learn from the data in financial markets and investment portfolios in a timely manner, and then using the knowledge gained to control their future performance trends and make predictions accordingly. This recently emerging term highlights the importance of incorporating real-world information, especially those from extremal events, in the era of big data. The use of computational intelligence, statistical rules and algorithms allows businesses to make informed decisions and predictions. Furthermore, under an increasing emphasis on the concept of risk management, market practitioners can better measure the econometric patterns and trends using the large datasets available.

While this term sounds quite new, the practice itself has actually been present throughout mankind's history, although in different forms. Even in ancient times, our ancestors already engaged in commercial trading and recorded data related to their daily lives; indeed, economic transactions in the form of exchange among goods, services, and commodities such as crops, livestock, and textiles, were common among ancient civilizations. They also developed some methods to systematically record data as well as to track and document their economic activities, though they are certainly less mature or complex than modern ones. Take the *Babylonians* for example: clay tablets have been discovered from *Mesopotamia*, which contain records of various business transactions, contracts, and trade agreements. While we cannot assert that the tablets were specifically dedicated to financial data, these archaeological findings do provide insights into early business activities and trade. In fact, the Babylonians are also well-known for their advanced mathematical and astronomical knowledge, and they might be the first to develop accounting and record-keeping to track trade, debts, and other business transactions. These represent the initial attempts of humans to systematically record and organize economic information; see more discussions in [11].

As the notions of contracts and stocks emerged, financial activities saw further developments in the medieval era. For example, the first forward contracts were traded on the *Dojima Rice Exchange* in 1697 in Japan, as the *Samurai* were paid in rice and needed a stable method of converting rice to currency; such contracts were beneficial to agricultural producers, who could use them to hedge against future price changes. While people paid more attention to the patterns behind the data, their investment decisions depended mainly on such market sentiments; this also contributed to the financial bubbles during this period. The *Tulip Mania* in the Netherlands during the 1630s is often regarded as one of the most famous examples of a speculative bubble in financial history. During this crisis, the prices of tulip bulbs skyrocketed to extraordinarily high levels before eventually collapsing. Since then, similar bubbles have appeared from time to time, including the well-known *South Sea Company Bubble* in 1720, when many investors were lured to invest in the stock but ended up with a huge loss. The most famous name among these unfortunate stockholders is arguably *Sir Isaac Newton*; despite being an excellent mathematician, his losses in this investment exceeded £20,000, which is comparable with around $20 million nowadays. Allegedly, Newton commented on this incident that he could "*calculate the motions of the heavenly bodies, but not the madness of people*"; one can find many more interesting stories in [3].

In the same era, attempts at data analytics were also made in the field of insurance; in particular, the renowned English astronomer and mathematician, *Edmond Halley*, studied the demographic records in *Breslau* (now *Wrocław* in Poland) and created life tables accordingly to estimate life expectancies and survival probabilities for different age groups, which now play a crucial role in pricing annuities and insurance policies [7]. Compared with the *Ulpian*'s life table, developed in ancient *Rome* without reference to valid data sources, Halley's data-driven studies in the late 17th Century truly laid the foundation for the development of actuarial science as a quantitative subject; also see [22].

Moving on to the mid 19th century, the modern version of commodity and futures markets developed from forward contracts began to take shape in metropolitan areas like *New York* and *London*. The formation of organized exchanges brought several benefits to market participants. It enhanced market liquidity by bringing together a larger number of buyers and sellers, facilitating efficient price discovery. Standardized contracts and rules were introduced, ensuring uniformity and transparency in trading practices. However, during this period, these markets relied mainly on raw business data rather than a large-scale systematic study or model fitting for prediction purposes.

The next wave of advance in finance, particularly financial modelling and option pricing, was initiated at the turn of the 20th century by the French mathematician *Louis Bachelier*. His groundbreaking thesis, "*Théorie de la spéculation*" [1], introduced the use of the *arithmetic Brownian motion* (ABM) to model security prices. Although Bachelier did not have access to reliable market data or sophisticated data analysis techniques, he attempted to replicate real market data as closely as possible. He used a volatile model to capture the movement of prices, yet it was not the best representation of the observed market dynamics. The importance of his work was later acknowledged by the renowned economist *Paul Samuelson*, who further improved the model for asset prices by using geometric Brownian motion (GBM) instead of ABM, as GBM can better capture the long-term growth trend observed in financial markets by the drift term; see [17]. Further along this line of development, in 1973, economists *Fischer Black* and *Myron Scholes* proposed the game-changing

Black–Scholes option pricing model [2] by incorporating the no-arbitrage principle, with contributions from *Robert Merton* extending its application to other financial matters [15]. Scholes and Merton were later awarded the Nobel Memorial Prize in Economic Sciences in 1997 for their fundamental works. The Black–Scholes formula revolutionized the options market by providing a mathematical framework for pricing options and understanding their sensitivities to various factors, and in turn boosted the need for various statistical methods to estimate the underlying model parameters, particularly the volatilities.

Earlier in the same century, the prevailing classical economic theories established in the 19th century still proved effective and useful until the *Great Depression* in the 1930s, which then prompted a re-evaluation. *John Maynard Keynes*, an influential economist during that period, patched the theory by introducing Keynesian economics, which emphasized government intervention to stimulate economic growth [9]. While econometrics, the application of statistical methods to economic data, was not as advanced during Keynes's time, his ideas did provide the ground for future developments in macroeconomic analysis and guidance for policy-making.

In the 1960s, the *Capital Asset Pricing Model* (CAPM) proposed by *Harry Markowitz* [14] marked a significant breakthrough in the branch of financial economics. CAPM provided a framework for understanding the equilibrium relation between the return and risk of an asset in an efficient market. Markowitz's work emphasized portfolio optimization and diversification by showing that investors are only compensated for non-diversifiable or "systematic" risk under CAPM. He introduced the concept of efficient portfolios, aiming to maximize returns for a given level of risk or minimize risk for a given level of returns. Quadratic programming techniques were first used to look for optimal portfolios, considering the expected returns and covariances of different assets. *William Sharpe* further enriched this model by introducing the Sharpe ratio [19], also known as the price of risk, which measures the excess return of an asset relative to its volatility. It has since become an industry-standard metric for comparing the risk-adjusted returns of different investments.

Following Markowitz's and Sharpe's works, scholars like *Eugene Fama* and *Kenneth French* conducted empirical tests of the CAPM and expanded upon it. They further introduced multifactor models [5] that incorporated additional factors such as company size and book-to-market ratio. These models aimed to explain variations in stock returns: for example, smaller companies exhibited greater than expected returns over the long term, and challenged the CAPM's ability to fully capture market dynamics. Various techniques such as principal component analysis (PCA) have also been employed to further improve multifactor models that capture additional sources of risk and return beyond the CAPM framework. *Robert Shiller* also contributed to the topic by exploring the role of investor behaviour in asset pricing. He emphasized the impact of psychological factors on market prices, challenging the notion of fully rational markets taken in the framework of CAPM; for example, see [20] and its following articles. More discussions can also be found in [12]. These factors may be helpful in eventually providing an explanation for the outburst of financial bubbles in history, such as the aforementioned Tulip Mania and the South Sea Company bubble.

After the development of the CAPM, *Robert Engle* made notable contributions in the early 1980s by introducing the *Generalized Autoregressive Conditional Heteroskedasticity* (GARCH) model [4]. This model aimed at capturing the phenomenon of volatility clustering, namely the tendency that fluctuates at a similar scale to occur collectively, and account for time-varying volatility in stock returns. The GARCH model extended the traditional autoregressive models by incorporating the concept of conditional heteroskedasticity, which allows the volatility of returns to vary over time; and he was later awarded the Nobel Memorial Prize in Economic Sciences in 2003 for his contribution to "*methods of analysing economic time series with time-varying volatility*".

Subsequently, one prominent area where data analysis has made significant contributions is the study of detecting statistical arbitrage. This approach involves leveraging complex models and analysing dependencies, mostly related to the use of copulas, between various financial instruments to identify and exploit trading opportunities. Investment bankers have successfully utilized model dependence structures to implement profitable strategies, although the effectiveness of such approaches can be subject to market conditions and unforeseen events, as exemplified by the blow-up of a large hedge fund, *Long-Term Capital Management* (LTCM), in 1998 and the credit crunch in 2008; see [13] for further discussions.

Credit risk assessment and modelling is another area where data analysis has proven essential. Using copulae and analysing vast amounts of data on credit-related variables, researchers and practitioners have developed mathematical models to understand the relationship between securities and quantify credit risks. This was pioneered by *Abe Sklar* [21], and utilized by *David Li* in [10] on primitive credit derivative pricing. In particular, the copula technique is used to model the dependence structure between securities to enable a more accurate description of the joint behaviour of multiple risk factors.

Macroeconomic data analysis has also been instrumental in understanding and forecasting economic trends. *Lars Hansen*, the 2013 Nobel Laureate in Economic Sciences, has advocated the generalized method of moments [8], and other techniques, such as regression analysis, unit root, factor analysis, and nowcasting, have also been widely used to study macroeconomic indicators and their impact on financial markets.

Furthermore, data analysis techniques such as *Benford's Law* [16] and *Zipf's Law* [6] have been employed to detect potential accounting flaws and irregularities. These statistical phenomena serve as a tool for auditors and investigators to identify anomalies in business data, thus contributing to the integrity and transparency of financial reporting.

To this end, in the contemporary context, financial data analytics is essentially about providing tools and understanding for individuals and businesses in the financial industry, with a focus on achieving detailed results and optimization in various aspects in finance, such as profitability and risk reduction.

The aforementioned problem investigation methods over the history share a common philosophy: they establish a connection between financial goals and variables with data available through mathematical models. This approach effectively reduces the dimensionality of the problem from infinite to finite; to this end, only few (possibly very massive) parameters need to be estimated by statistical methods. Looking ahead, with the recent emergence of machine learning, new possibilities are also emerging for financial analysis; for instance, some (gated) recurrent neural networks such as the *long-short term memory* (LSTM) model have proven relevant to the prediction of security prices and identification of market anomalies. In general, deep learning takes a different approach by employing deep neural networks to replace those mathematical models already mentioned. The challenge with deep neural networks lies in their black-box nature, necessitating the exploration of concise mathematical descriptions to understand their inner workings. This quest for a comprehensive understanding is in the context of explainable AI, which will be a pivotal focus in the upcoming decade. In particular, we eagerly look forward to new developments along this direction in the realm of financial data analytics in the foreseeable future.

ORGANIZATION OF THE BOOK

The book is divided into three distinct parts, each focusing on a distinct aspect of the subject matter.

Part I of the book serves as a foundation, covering essential knowledge in mathematics, statistics, and programming. Chapters 1 and 2 review some basic mathematical concepts and programming techniques essential for understanding the subsequent chapters. Topics covered include matrix analysis, probability theory, elementary Bayesian statistics, as well as a brief introduction to Python and **R**, demonstrated by applications in CAPM with ANOVA diagnosis. Chapter 3 introduces statistical diagnosis techniques used to analyse and assess distributions of financial data. Chapter 4 delves into the application of data analytics in financial forensics, and explores techniques for detecting fraudulent activities, including anomaly detection and forensic accounting. Chapter 5 introduces some typical methods for sample generation and variance reduction techniques in simulations for numerical finance, with a focus on the well-known Monte Carlo method illustrated by the calculation of some Greek letters of options. Chapter 6 introduces several useful numerical algorithms for parameter estimation in model inference, iterative optimization, and generation of complex probability distributions. Chapter 7 explores the modelling and analysis of security data in continuous time, covering different hierarchical time series models, and also introduces some profitable trading strategies with an analysis of calendar effects via the goodness index (Kelly formula). The concept of risk measures in finance is examined in Chapter 8. It covers popular risk measures in accordance with the regulations of the Basel Accord, such as Value-at-Risk and Expected Shortfall, as well as some advanced modelling methodologies such as *extreme value theory* and use of copulae for dependence structures.

As we move on from Part I to Parts II and III of the book, we shift our focus from fundamental concepts and data cleansing techniques to the exploration of various linear and nonlinear models commonly employed in financial data analytics. Part II introduces more advanced modelling techniques, providing insights into linear models, dimension reduction, and classification methods. We kick off Part II by introducing *Principal Component Analysis* (PCA) for dimension reduction and feature extraction in Chapter 9. Collaborative filtering techniques for the Recommender system are also discussed. Several commonly used models in regression learning, such as polynomial regression, logistic regression, Poisson regression, and principal component regression, as well as practical considerations in model selection and diagnosis, are discussed in Chapter 10. Chapter 11 explores the use of linear classifiers in finance, including perceptrons and support vector machines. Versatile techniques such as stochastic gradient descent method and quadratic programming are also covered.

Part III shares the same focus as Part II, but the emphasis is on the more enriched non-linear models, which play an equally crucial role as linear ones in effectively analysing real-world financial data. Chapter 12 introduces Bayesian modelling techniques; in particular, several Bayes-type classifiers will be discussed, including an original and efficient method recently developed by the authors, called the *Comonotone-Independence Bayes classifier* (CIBer).[1] Chapter 13 focuses on tree-based methods for classification and regression analysis, such as *Classification and Regression Trees* (CART) and *Random Forests*; the concepts of various *entropies* in information theory are also introduced. Chapter 14 introduces the cluster analysis for unsupervised learning when labels are not available. We introduce two methods, which are two sides of the same coin, namely K-means that separates the samples into distinct classes, and K-nearest neighbours (K-NN) that classifies a new observation given the existing separately identifiable classes. As a wrap-up of our book, Chapter 15 sheds light on the applications of Deep Neural Networks (DNNs) in finance. It discusses several topics such as deep learning architectures, multilayer perceptrons (MLP), and the mechanism of error backpropagation. We also provide a quick introduction to primitive (gated) recurrent neural networks such as LSTM, together with its potential use for financial prediction.

Last but not least, for the ease of the readers, the Python and **R** codes illustrated in this book, the datasets adopted, as well as future updates, are readily accessible through GitHub via the link https://github.com/kaiser1999/Financial-Data-Analytics/; readers can also be redirected to the link via the QR code in Figure 1.

FIGURE 1 GitHub repository for this book.

[1] This tool won a Silver Medal in the 48th International Exhibition of Inventions Geneva in April 2023.

REFERENCES

1. Bachelier, L. (1900). Théorie de la spéculation. In *Annales scientifiques de l'École normale supérieure* (vol. 17), pp. 21–86.
2. Black, F., and Scholes, M. (1973). The pricing of options and corporate liabilities. *Journal of Political Economy*, 81(3), 637–654.
3. Cooper, G. (2008). *The Origin of Financial Crises*. Vintage.
4. Engle, R.F. (1982). Autoregressive conditional heteroscedasticity with estimates of the variance of United Kingdom inflation. *Econometrica*, 50(4), 987–1007.
5. Fama, E.F., and French, K.R. (1993). Common risk factors in the returns on stocks and bonds. *Journal of Financial Economics*, 33(1), 3–56.
6. Gabaix, X. (1999). Zipf's law for cities: an explanation. *The Quarterly Journal of Economics*, 114(3), 739–767.
7. Halley, E. (1874). An estimate of the degrees of the mortality of mankind, drawn from curious tables of the Births and Funerals at the City of Breslaw; with an attempt to ascertain the price of annuities upon lives. *Journal of the Institute of Actuaries*, 18(4), 251–262.
8. Hansen, L.P. (1982). Large sample properties of generalized method of moments estimators. *Econometrica*, 50(4), 1029–1054.
9. Keynes, J.M. (1937). The general theory of employment. *The Quarterly Journal of Economics*, 51(2), 209–223.
10. Li, D.X. (2000). On default correlation: a copula function approach. *Journal of Fixed Income*, 9(4), 43–54.
11. Lo, A.W., and Hasanhodzic, J. (2010). *The Evolution of Technical Analysis: Financial Prediction from Babylonian Tablets to Bloomberg Terminals*, Vol. 96. Wiley.
12. Lo, A.W., and Mackinlay, A.C. (1999). *A Non-Random Walk Down Wall Street*. Princeton University Press.
13. Lowenstein, R. (2000). *When Genius Failed: the Rise and Fall of Long-Term Capital Management*. Random House.
14. Markowitz, H.M. (1991). Foundations of portfolio theory. *The Journal of Finance*, 46(2), 469–477.
15. Merton, R.C. (1990). *Continuous-Time Finance*. Blackwell.
16. Nigrini, M.J. (2012). *Benford's Law: Applications for Forensic Accounting, Auditing, and Fraud Detection*. Wiley.
17. Samuelson, P.A. (1983). *Foundations of Economic Analysis*, Vol. 197, No. 1. Harvard University Press.
18. Shannon, C.E. (1959). Coding theorems for a discrete source with a fidelity criterion. *IRE National Convention Record*, 7(4), 142–163.
19. Sharpe, W.F. (1994). The Sharpe ratio. *The Journal of Portfolio Management*, 21(1), 49–58.
20. Shiller, R.J. (1981). Do stock prices move too much to be justified by subsequent changes in dividends? *The American Economic Review*, 71(3), 421–436.
21. Sklar, A. (1973). Random variables, joint distribution functions, and copulas. *Kybernetika*, 9(6), 449–460.
22. Turnbull, C. (2017). *A History of British Actuarial Thought*. Springer International Publishing.

PART One

Data Cleansing and Analytical Models

CHAPTER 1

Mathematical and Statistical Preliminaries

This chapter provides a quick review of basic advanced calculus, linear algebra and probability and statistics, serving as background knowledge for the material in later chapters. Readers familiar with the relevant terminologies are encouraged to start their journey from Chapter 3 onward, and then turn back to this chapter as a reference from time to time.

With the rapid advance in computing technology and the enrichment of data storage, we often collect a huge amount of information and observations with a large number of feature variables. A dataset with p numbers of variables and n numbers of observations can be arranged in the form of an $n \times p$ data matrix:

$$X = \begin{pmatrix} x_{11} & \cdots & x_{1p} \\ \vdots & \ddots & \vdots \\ x_{n1} & \cdots & x_{np} \end{pmatrix} = \begin{pmatrix} \boldsymbol{x}_1^\top \\ \vdots \\ \boldsymbol{x}_n^\top \end{pmatrix},$$

where the j-th row $\boldsymbol{x}_j^\top$, for $j = 1, \ldots, n$, represents the j-th observation, the transpose of the column vector $\boldsymbol{x}_j$. Data scientists believe that there exists a data generating mechanism so that each sample of multivariate observation is randomly drawn from a common but often unknown joint probability distribution. Statistically speaking, the data matrix X is the realization of the following random matrix:

$$\mathrm{X} = \begin{pmatrix} \mathrm{x}_{11} & \cdots & \mathrm{x}_{1p} \\ \vdots & \ddots & \vdots \\ \mathrm{x}_{n1} & \cdots & \mathrm{x}_{np} \end{pmatrix} = \begin{pmatrix} \mathbf{x}_1^\top \\ \vdots \\ \mathbf{x}_n^\top \end{pmatrix}, \tag{1.1}$$

where the row vectors $\mathbf{x}_1^\top, \ldots, \mathbf{x}_n^\top$ are independent *random (row) vectors* following a common joint distribution. Throughout this book, we shall use roman type letters, such as x and X, to denote random variables, and use italic letters, such as x and X, to represent the corresponding observed values of sample data. In addition, we denote vectors by bold lowercase letters, e.g. $\boldsymbol{x}$ and $\mathbf{x}$ represent deterministic and random vectors, respectively, and denote matrices by bold uppercase letters, e.g. $\boldsymbol{X}$ and $\mathbf{X}$ represent deterministic and random matrices, respectively. Due to the frequent use of vectors, matrices and their random counterparts in this book, the traditional notations in probability and statistics may cause some ambiguity. To this point our notations follow the convention in [9]; the same practice is also adopted in the companion book [4] of the present one. Also, for the notation of natural logarithm, we shall follow the notation "ln" throughout the book, while we may also use "log" and "$\log_e$" interchangeably depending on the context to avoid any ambiguity.

1.1 RANDOM VECTOR

From (1.1), we learn that a $p \times 1$ random (column) vector, denoted by $\mathbf{x} = (\mathrm{x}_1, \ldots, \mathrm{x}_p)^\top$, is just a vertical stack of p univariate random variables. The mean of $\mathbf{x}$ is the $p \times 1$ vector

$$\mathbb{E}(\mathbf{x}) := \begin{pmatrix} \mathbb{E}(\mathrm{x}_1) \\ \vdots \\ \mathbb{E}(\mathrm{x}_p) \end{pmatrix} = \begin{pmatrix} \mu_1 \\ \vdots \\ \mu_p \end{pmatrix} =: \boldsymbol{\mu},$$

the variance-covariance matrix (or covariance matrix) of $\mathbf{x}$ is the following $p \times p$ (symmetric positive definite) matrix:

$$\mathrm{Var}(\mathbf{x}) = \mathbb{E}\left((\mathbf{x} - \mathbb{E}(\mathbf{x}))(\mathbf{x} - \mathbb{E}(\mathbf{x}))^\top\right) =: \boldsymbol{\Sigma} = \begin{pmatrix} \sigma_{11} & \cdots & \sigma_{1p} \\ \vdots & \ddots & \vdots \\ \sigma_{p1} & \cdots & \sigma_{pp} \end{pmatrix} =: (\sigma_{ij})_{i,j=1,\ldots,p},$$

or (σ_{ij}) in shortened form if there is no cause for ambiguity; the correlation matrix of $\mathbf{x}$ is another symmetric and positive definite $p \times p$ matrix:

$$\mathrm{Corr}(\mathbf{x}) =: \boldsymbol{R} = \begin{pmatrix} 1 & \rho_{12} & \cdots & \rho_{1p} \\ \rho_{21} & 1 & \cdots & \vdots \\ \vdots & \vdots & \ddots & \vdots \\ \rho_{p1} & \cdots & \cdots & 1 \end{pmatrix} =: (\rho_{ij})_{i,j=1,\ldots,p},$$

where $\rho_{ij} := \frac{\sigma_{ij}}{\sqrt{\sigma_{ii}}\sqrt{\sigma_{jj}}}$ is the correlation coefficient between x_i and x_j. The following are some important properties that we shall use later:

1. If $\boldsymbol{A}$ is an $r \times p$ deterministic matrix, then $\mathbf{y} = \boldsymbol{A}\mathbf{x}$ is an $r \times 1$ random vector with mean $\mathbb{E}(\mathbf{y}) = \boldsymbol{A}\boldsymbol{\mu}$ and covariance matrix $\mathrm{Var}(\mathbf{y}) = \boldsymbol{A}\boldsymbol{\Sigma}\boldsymbol{A}^\top$; moreover, for another $r \times 1$ random vector $\mathbf{z} = \boldsymbol{B}\mathbf{x}$, where $\boldsymbol{B}$ is another $r \times p$ constant matrix, the covariance of $\mathbf{y}$ and $\mathbf{z}$ is $\mathrm{Cov}(\boldsymbol{A}\mathbf{x}, \boldsymbol{B}\mathbf{x}) = \boldsymbol{A}\boldsymbol{\Sigma}\boldsymbol{B}^\top$.

2. The correlation matrix $R = (\rho_{ij})$ of $\mathbf{x}$ can be computed from its covariance matrix $\boldsymbol{\Sigma} = (\sigma_{ij})$ by the formula $R = D^{-1/2}\boldsymbol{\Sigma}D^{-1/2}$, where D is the diagonal matrix denoted by $\text{diag}(\sigma_{11}, \ldots, \sigma_{pp})$, and $D^{-1/2} := \text{diag}(1/\sqrt{\sigma_{11}}, \ldots, 1/\sqrt{\sigma_{pp}})$.

1.1.1 Random Sample, Sample Mean and Covariance Matrix

In many real-life applications, the underlying probability density function (or cumulative distribution function) is unknown and requires estimation. We are particularly interested in estimating the unknown mean vector $\boldsymbol{\mu} = (\mu_1, \ldots, \mu_p)^\top$ and covariance matrix $\boldsymbol{\Sigma} = (\sigma_{ij})$. For a given data matrix X, it is customary to estimate $\boldsymbol{\mu}$ and $\boldsymbol{\Sigma}$ by using the sample mean vector $\overline{x} := (\overline{x}_1, \ldots, \overline{x}_p)^\top$ and the sample covariance matrix $S := (s_{ij})$, respectively, where

$$\overline{x}_i = \sum_{k=1}^{n} \frac{x_{ki}}{n} \text{ and } s_{ij} = \frac{1}{n-1}\sum_{k=1}^{n}(x_{ki} - \overline{x}_i)(x_{kj} - \overline{x}_j) = \frac{1}{n-1}\left(\sum_{k=1}^{n} x_{ki}x_{kj} - n\overline{x}_i\overline{x}_j\right).$$

The next proposition demonstrates that these estimation methods enjoy the well-known desirable property known as *unbiasedness.*

Proposition 1.1 *(Unbiasedness of the Sample Mean Vector and Covariance Matrix)* Let $\mathbf{x}_1, \ldots, \mathbf{x}_n$ be a random sample from a common joint probability distribution which has the mean vector $\boldsymbol{\mu}$ and covariance matrix $\boldsymbol{\Sigma}$. Then the $p \times 1$ sample mean vector

$$\overline{\mathbf{x}} = \frac{1}{n}\sum_{i=1}^{n} \mathbf{x}_i$$

is an unbiased estimator of $\boldsymbol{\mu}$, that is, $\mathbb{E}(\overline{\mathbf{x}}) = \boldsymbol{\mu}$; besides, the $p \times p$ covariance matrix of $\overline{\mathbf{x}}$ is given by

$$\text{Var}(\overline{\mathbf{x}}) = \frac{1}{n}\boldsymbol{\Sigma}.$$

Moreover, the $p \times p$ sample covariance matrix

$$\mathrm{S} = \frac{1}{n-1}\sum_{i=1}^{n}(\mathbf{x}_i - \overline{\mathbf{x}})(\mathbf{x}_i - \overline{\mathbf{x}})^\top$$

is an unbiased estimator of $\boldsymbol{\Sigma}$, that is, $\mathbb{E}(\mathrm{S}) = \boldsymbol{\Sigma}$.

Proof.

$$\mathbb{E}(\overline{\mathbf{x}}) = \mathbb{E}\left(\frac{1}{n}\sum_{i=1}^{n}\mathbf{x}_i\right) = \frac{1}{n}\sum_{i=1}^{n}\mathbb{E}(\mathbf{x}_i) = \frac{1}{n}\cdot n \cdot \boldsymbol{\mu} = \boldsymbol{\mu}.$$

Note that

$$\mathrm{Var}(\overline{\mathbf{x}}) = \mathbb{E}\left((\overline{\mathbf{x}}-\boldsymbol{\mu})(\overline{\mathbf{x}}-\boldsymbol{\mu})^{\top}\right) = \mathbb{E}\left(\left(\frac{1}{n}\sum_{j=1}^{n}(\mathbf{x}_j-\boldsymbol{\mu})\right)\left(\frac{1}{n}\sum_{l=1}^{n}(\mathbf{x}_l-\boldsymbol{\mu})\right)^{\top}\right)$$
$$= \frac{1}{n^2}\sum_{j=1}^{n}\sum_{l=1}^{n}\mathbb{E}\left((\mathbf{x}_j-\boldsymbol{\mu})(\mathbf{x}_l-\boldsymbol{\mu})^{\top}\right).$$

For $j \neq l$, each entry in $\mathbb{E}\left((\mathbf{x}_j-\boldsymbol{\mu})(\mathbf{x}_l-\boldsymbol{\mu})^{\top}\right)$ vanishes because each entry is the covariance between a component of $\mathbf{x}_j$ and that of $\mathbf{x}_l$, and they are independent by the definition of random sample. Therefore,

$$\mathrm{Var}(\overline{\mathbf{x}}) = \frac{1}{n^2}\sum_{j=1}^{n}\mathbb{E}\left((\mathbf{x}_j-\boldsymbol{\mu})(\mathbf{x}_j-\boldsymbol{\mu})^{\top}\right).$$

Since $\boldsymbol{\Sigma} = \mathbb{E}\left((\mathbf{x}_j-\boldsymbol{\mu})(\mathbf{x}_j-\boldsymbol{\mu})^{\top}\right)$ is the common population covariance matrix for each $\mathbf{x}_j$, we have

$$\mathrm{Var}(\overline{\mathbf{x}}) = \frac{1}{n^2}\sum_{j=1}^{n}\mathbb{E}\left((\mathbf{x}_j-\boldsymbol{\mu})(\mathbf{x}_j-\boldsymbol{\mu})^{\top}\right) = \frac{1}{n^2}\cdot n\cdot\boldsymbol{\Sigma} = \frac{1}{n}\boldsymbol{\Sigma}.$$

Next, we consider the $p \times p$ matrix:

$$\mathbf{A} := \sum_{i=1}^{n}(\mathbf{x}_i-\overline{\mathbf{x}})(\mathbf{x}_i-\overline{\mathbf{x}})^{\top} = \sum_{i=1}^{n}\mathbf{x}_i\mathbf{x}_i^{\top} - n\overline{\mathbf{x}}\,\overline{\mathbf{x}}^{\top}, \tag{1.2}$$

which represents the matrix of sums of squares and cross products. Note that

$$\mathbb{E}\left(\mathbf{x}_i\mathbf{x}_i^{\top}\right) = \boldsymbol{\Sigma} + \boldsymbol{\mu}\boldsymbol{\mu}^{\top} \text{ and } \mathbb{E}\left(\overline{\mathbf{x}}\,\overline{\mathbf{x}}^{\top}\right) = \frac{1}{n}\boldsymbol{\Sigma} + \boldsymbol{\mu}\boldsymbol{\mu}^{\top}.$$

Hence,

$$\mathbb{E}(\mathbf{A}) = \sum_{i=1}^{n}\mathbb{E}\left(\mathbf{x}_i\mathbf{x}_i^{\top}\right) - n\mathbb{E}\left(\overline{\mathbf{x}}\,\overline{\mathbf{x}}^{\top}\right) = n\boldsymbol{\Sigma} + n\boldsymbol{\mu}\boldsymbol{\mu}^{\top} - n\left(\frac{1}{n}\boldsymbol{\Sigma} + \boldsymbol{\mu}\boldsymbol{\mu}^{\top}\right)$$
$$= n\boldsymbol{\Sigma} + n\boldsymbol{\mu}\boldsymbol{\mu}^{\top} - \boldsymbol{\Sigma} - n\boldsymbol{\mu}\boldsymbol{\mu}^{\top} = (n-1)\boldsymbol{\Sigma}.$$

It then follows that

$$\mathbb{E}(\mathbf{S}) = \mathbb{E}\left(\frac{1}{n-1}\mathbf{A}\right) = \boldsymbol{\Sigma}.$$

□

Example 1.1 ***(Blue Chips on the Hong Kong Stock Exchange)*** The data file `fin-ratio.csv`[1] contains financial ratios of 680 securities listed on the main board of the Hong Kong Stock Exchange in 2002. There are six financial variables in order, namely, **Earning Yield** (EV), **Cash Flow to Price** (CFTP), **logarithm of Market Value** (ln_MV), **Dividend Yield** (DY), **Book to Market Equity** (BTME), and **Debt to Equity Ratio** (DTE). The last column **HSI** is a binary variable which takes value one if the company is a Blue Chip or zero otherwise. Among these companies, there were 32 Blue Chips which were the *Hang Seng Index Constituent Stocks* at that time. For the time being we ignore this HSI variable and compute the sample mean vector, sample covariance matrix, and sample correlation matrix using Python with the codes shown in Programme 1.1; the **R** codes are shown in Programme 1.2. For those who are not yet very familiar with Python and **R**, please refer to Chapter 2 for a quick introduction.

```
1  import pandas as pd
2  import numpy as np
3  np.set_printoptions(precision=4) # Control display into 4 digits
4
5  HSI_2002 = pd.read_csv("fin-ratio.csv")
6  print(HSI_2002.columns)
7
8  X_2002 = HSI_2002.drop(columns="HSI").values # A 680x6 data matrix
9  mu_2002 = np.mean(X_2002, axis=0) # Mean vector
10 print(mu_2002)
11
12 S_2002 = np.cov(X_2002, rowvar=False) # Covariance matrix
13 print(S_2002)
14
15 R_2002 = np.corrcoef(X_2002, rowvar=False) # Correlation matrix
16 print(R_2002)
```

```
1  ['EY' 'CFTP' 'ln MV' 'DY' 'BTME' 'DTE' 'HSI']
2  [-0.6502 -0.2339  6.2668  2.4962  1.9083  0.7097]
3  [[18.4979  2.909   1.1602  1.9204  1.4781  0.338 ]
4   [ 2.909   3.6931  0.7663  1.2371  1.8228  0.3288]
5   [ 1.1602  0.7663  2.7439  0.9721 -0.7734 -0.0741]
6   [ 1.9204  1.2371  0.9721 13.8716 -0.2575  0.1582]
7   [ 1.4781  1.8228 -0.7734 -0.2575 68.3082  1.9618]
8   [ 0.338   0.3288 -0.0741  0.1582  1.9618 12.9929]]
9  [[ 1.      0.352   0.1628  0.1199  0.0416  0.0218]
10  [ 0.352   1.      0.2407  0.1728  0.1148  0.0475]
11  [ 0.1628  0.2407  1.      0.1576 -0.0565 -0.0124]
12  [ 0.1199  0.1728  0.1576  1.     -0.0084  0.0118]
13  [ 0.0416  0.1148 -0.0565 -0.0084  1.      0.0659]
14  [ 0.0218  0.0475 -0.0124  0.0118  0.0659  1.    ]]
```

Programme 1.1 Computations of sample mean vector, sample covariance matrix, and sample correlation matrix of `fin-ratio.csv` in Python.

[1] This dataset is available at https://github.com/kaiser1999/Financial-Data-Analytics.

```
1  > options(digits=4) # Control display into 4 digits
2  >
3  > HSI_2002 <- read.csv("fin-ratio.csv")
4  > names(HSI_2002)
5  [1] "EY"     "CFTP"   "ln_MV" "DY"     "BTME"  "DTE"    "HSI"
6  >
7  > X_2002 <- HSI_2002[,1:6] # A 680x6 data matrix
8  > (mu_2002 <- apply(X_2002, 2, mean)) # Mean vector
9       EY    CFTP   ln_MV      DY    BTME     DTE
10 -0.6502 -0.2339  6.2668  2.4962  1.9083  0.7097
11 >
12 > (S_2002 <- var(X_2002)) # Covariance matrix
13            EY   CFTP     ln_MV      DY    BTME      DTE
14 EY     18.498 2.9090   1.16019  1.9204  1.4781  0.33795
15 CFTP    2.909 3.6931   0.76630  1.2371  1.8228  0.32879
16 ln_MV   1.160 0.7663   2.74394  0.9721 -0.7734 -0.07413
17 DY      1.920 1.2371   0.97207 13.8716 -0.2575  0.15815
18 BTME    1.478 1.8228  -0.77342 -0.2575 68.3082  1.96177
19 DTE     0.338 0.3288  -0.07413  0.1582  1.9618 12.99291
20 >
21 > (R_2002 <- cor(X_2002)) # Correlation matrix
22             EY    CFTP    ln_MV        DY       BTME      DTE
23 EY     1.00000 0.35195  0.16285  0.119884  0.041583  0.02180
24 CFTP   0.35195 1.00000  0.24072  0.172848  0.114767  0.04746
25 ln_MV  0.16285 0.24072  1.00000  0.157561 -0.056493 -0.01242
26 DY     0.11988 0.17285  0.15756  1.000000 -0.008366  0.01178
27 BTME   0.04158 0.11477 -0.05649 -0.008366  1.000000  0.06585
28 DTE    0.02180 0.04746 -0.01242  0.011780  0.065850  1.00000
```

Programme 1.2 Computations of sample mean vector, sample covariance matrix, and sample correlation matrix of `fin-ratio.csv` in R.

1.2 MATRIX THEORY

Before we embark on our journey, let us first recall some basic notations and concepts in introductory matrix theory. Readers may refer to [11, 18, 19] for more detailed introductions to matrix theory at an elementary level; more advanced topics are discussed in [8, 20, 21], including various algorithms for matrix solvers and techniques related to singular value decomposition.

(I) Determinant

Let A and B be two $p \times p$ square matrices and let $|A|$ denote the *determinant* of A. Basic properties include the following:

1. $|A^\top| = |A|$.
2. $|\alpha A| = \alpha^p |A|$, for any scalar $\alpha \in \mathbb{R}$.
3. $|AB| = |A| \cdot |B|$. [Note that $|A+B| \neq |A| + |B|$ in general.]

(II) Inverse

Let A be a $p \times p$ *nonsingular* matrix; that is, its inverse A^{-1} exists, such that $AA^{-1} = A^{-1}A = I_p$, where I_p is the $p \times p$ identity matrix. We have the following properties:

4. $(A^{-1})^\top = (A^\top)^{-1}$.
5. $(AB)^{-1} = B^{-1}A^{-1}$.
6. $|A^{-1}| = 1/|A|$.
7. A is an *orthogonal* matrix if and only if $A^{-1} = A^\top$.
8. If $D = \text{diag}(d_1, \ldots, d_p)$ with $d_i \neq 0$ for $i = 1, \ldots, p$, then $D^{-1} = \text{diag}(d_1^{-1}, \ldots, d_p^{-1})$.

(III) Trace

Given a $p \times p$ matrix $A = (a_{ij})$, the trace of A, denoted by tr(A), is the sum of the diagonal entries of A:

$$\text{tr}(A) = \sum_{i=1}^{p} a_{ii}.$$

The trace of a matrix satisfies the following properties:

9. $\text{tr}(A) = \text{tr}(A^\top)$.
10. $\text{tr}(\alpha A + \beta B) = \alpha\text{tr}(A) + \beta\text{tr}(B)$, for $\alpha, \beta \in \mathbb{R}$.
11. Cyclic invariance: $\text{tr}(ABC) = \text{tr}(CAB) = \text{tr}(BCA)$. In general, for any k numbers of $p \times p$ matrices $A_1, \ldots, A_k$, we have:

$$\text{tr}(A_1A_2 \cdots A_k) = \text{tr}(A_iA_{i+1} \cdots A_kA_1 \cdots A_{i-1}), \text{ for any } i = 2, \ldots, k.$$

The derivative of the trace of a matrix with respect to itself satisfies the following properties [5]. Using the notation $\nabla_A f := \left(\frac{\partial}{\partial a_{ij}} f\right)$ for the *gradient* of f with respect to A (we sometimes also use $\frac{\partial f}{\partial A}$ instead), we have

12. $\nabla_A \text{tr}(AB) = B^\top$.
13. $\nabla_A \text{tr}(ABA^\top C) = CAB + C^\top AB^\top$.
14. $\nabla_{A^\top} \text{tr}(ABA^\top C) = B^\top A^\top C + BA^\top C$.

We prove only the first statement, while the second and third statements can follow directly from the chain rule. Recall that the components of A and B are denoted as (a_{ij}) and (b_{ij}), respectively. Since

$$\text{tr}(AB) = \sum_{l=1}^{p} \sum_{k=1}^{p} a_{lk} b_{kl},$$

we then have, for any $i, j = 1, \ldots, p$,

$$\frac{\partial \text{tr}(\boldsymbol{AB})}{\partial a_{ij}} = \frac{\partial}{\partial a_{ij}} \sum_j \sum_k a_{lk} b_{kl} = b_{ji},$$

which confirms the assertion of Property 12.

(IV) Positive definite matrix

A $p \times p$ symmetric matrix $\boldsymbol{A}$ is called *positive definite* (usually denoted by $\boldsymbol{A} \succ \boldsymbol{0}$) if for any nonzero $p \times 1$ vector $\boldsymbol{x}$, $\boldsymbol{x}^\top \boldsymbol{A}\boldsymbol{x} > 0$; if $\boldsymbol{x}^\top \boldsymbol{A}\boldsymbol{x} = 0$ for some nonzero $p \times 1$ vector $\boldsymbol{x}$, while $\boldsymbol{x}^\top \boldsymbol{A}\boldsymbol{x} \geq 0$ still holds for any $\boldsymbol{x} \in \mathbb{R}^p$, we say that $\boldsymbol{A}$ is *positive semi-definite*, denoted as $\boldsymbol{A} \succeq \boldsymbol{0}$, where $\boldsymbol{0}$ denotes a zero vector.

15. If $\boldsymbol{A} \succ \boldsymbol{0}$, then $\boldsymbol{A}^{-1} \succ \boldsymbol{0}$.
16. For any $n \times p$ matrix $\boldsymbol{B}$, $\boldsymbol{B}^\top \boldsymbol{B} \succeq \boldsymbol{0}$; indeed, for any nonzero $p \times 1$ vector $\boldsymbol{x}$, let $\boldsymbol{y} = \boldsymbol{B}\boldsymbol{x}$, which is an $n \times 1$ vector, then $\boldsymbol{x}^\top \boldsymbol{B}^\top \boldsymbol{B}\boldsymbol{x} = \boldsymbol{y}^\top \boldsymbol{y} = \sum_{i=1}^n y_i^2 \geq 0$.
17. The sample covariance matrix is positive semi-definite. It suffices to show that the SSCP (Sum of Squares and Cross Products) matrix $\boldsymbol{S} \succeq \boldsymbol{0}$. Note that

$$\boldsymbol{S} = \sum_{i=1}^n (\boldsymbol{x}_i - \overline{\boldsymbol{x}})(\boldsymbol{x}_i - \overline{\boldsymbol{x}})^\top = (\boldsymbol{X} - \boldsymbol{1}_n \overline{\boldsymbol{x}}^\top)^\top (\boldsymbol{X} - \boldsymbol{1}_n \overline{\boldsymbol{x}}^\top) \succeq \boldsymbol{0},$$

by using Property 16, where $\boldsymbol{1}_n$ denotes an n-dimensional vector whose all entries are 1.

(V) Eigenvalues and eigenvectors

For any $p \times p$ matrix $\boldsymbol{A}$, the determinant $|\boldsymbol{A} - \lambda \boldsymbol{I}_p|$ is a polynomial in λ of degree p. The p roots of the polynomial equation $|\boldsymbol{A} - \lambda \boldsymbol{I}_p| = 0$, denoted by $\lambda_1, \ldots, \lambda_p$ (counting multiplicity), are called the *eigenvalues* (or *latent roots*) of $\boldsymbol{A}$. As a result, each $\boldsymbol{A} - \lambda_i \boldsymbol{I}_p$ is a singular matrix, hence there exists a nonzero vector $\boldsymbol{h}_i$ such that $(\boldsymbol{A} - \lambda_i \boldsymbol{I}_p)\boldsymbol{h}_i = 0$ (or equivalently $\boldsymbol{A}\boldsymbol{h}_i = \lambda_i \boldsymbol{h}_i$); $\boldsymbol{h}_i$ is called an *eigenvector* of $\boldsymbol{A}$ corresponding to λ_i. If $\boldsymbol{h}_i$ is taken to have a unit length, i.e. $\boldsymbol{h}_i^\top \boldsymbol{h}_i = 1$, then it is called a *normalized (unit) eigenvector* of $\boldsymbol{A}$.

Some important properties of eigenvalues and eigenvectors are as follows:

18. If $\boldsymbol{A}$ is a real symmetric matrix, then all its eigenvalues are real.
19. If $\boldsymbol{A} \succeq \boldsymbol{0}$, then all eigenvalues of $\boldsymbol{A}$ are non-negative.
20. The eigenvalues, counting multiplicity, of $\boldsymbol{A}$ and $\boldsymbol{A}^\top$, are the same.
21. If $\boldsymbol{A}$ and $\boldsymbol{B}$ are $p \times p$ nonsingular matrices, then the eigenvalues of $\boldsymbol{AB}$ and $\boldsymbol{BA}$ are equal.
22. If $\lambda_1, \ldots, \lambda_p$ are non-zero eigenvalues of a nonsingular matrix $\boldsymbol{A}$, then $\lambda_1^{-1}, \ldots, \lambda_p^{-1}$ are the eigenvalues of $\boldsymbol{A}^{-1}$.
23. If $\boldsymbol{A}$ is a real symmetric matrix, and λ_i and λ_j are two distinct eigenvalues of $\boldsymbol{A}$, then the corresponding eigenvectors $\boldsymbol{h}_i$ and $\boldsymbol{h}_j$ are orthogonal; indeed, by definition, $\boldsymbol{A}\boldsymbol{h}_i = \lambda_i \boldsymbol{h}_i$ and $\boldsymbol{A}\boldsymbol{h}_j = \lambda_j \boldsymbol{h}_j$, and therefore $\boldsymbol{h}_j^\top \boldsymbol{A}\boldsymbol{h}_i = \lambda_i \boldsymbol{h}_j^\top \boldsymbol{h}_i$ and $\boldsymbol{h}_i^\top \boldsymbol{A}\boldsymbol{h}_j = \lambda_j \boldsymbol{h}_i^\top \boldsymbol{h}_j$. By the symmetry of $\boldsymbol{A}$, $\boldsymbol{h}_j^\top \boldsymbol{A}\boldsymbol{h}_i = \boldsymbol{h}_i^\top \boldsymbol{A}\boldsymbol{h}_j$, hence $(\lambda_i - \lambda_j)\boldsymbol{h}_j^\top \boldsymbol{h}_i = 0$. As $\lambda_i \neq \lambda_j$, we must have $\boldsymbol{h}_j^\top \boldsymbol{h}_i = 0$.

24. If A is a $p \times p$ real symmetric matrix, then it has p orthogonal eigenvectors. Let H be a $p \times p$ matrix whose columns are normalized eigenvectors $h_1, \ldots, h_p$ of A, i.e., $H = (h_1, \ldots, h_p)$, then H is orthogonal, and

$$H^\top AH = D = \text{diag}(\lambda_1, \ldots, \lambda_p) \quad (\textit{Diagonalization} \text{ of } A),$$

or equivalently,

$$A = HDH^\top (\textit{Spectral decomposition} \text{ of } A).$$

To see this, one can simply write $AH = (Ah_1, \ldots, Ah_p) = (\lambda_1 h_1, \ldots, \lambda_p h_p) = HD$, therefore $H^\top AH = D$.

25. The trace of a matrix A is the sum of all its eigenvalues and the determinant of A is the product of all its eigenvalues, i.e., $\text{tr}(A) = \sum_{i=1}^p \lambda_i$ and $|A| = \prod_{i=1}^p \lambda_i$, respectively.

(VI) Symmetric square root of a positive semi-definite matrix

Let $A \geq \mathbf{0}$ be a $p \times p$ real symmetric matrix with non-negative eigenvalues $\lambda_1, \ldots, \lambda_p$ and their corresponding eigenvectors $h_1, \ldots, h_p$. As above, define the matrix $H = (h_1, \ldots, h_p), D = \text{diag}(\lambda_1, \ldots, \lambda_p)$ and $D^{1/2} := \text{diag}(\sqrt{\lambda_1}, \ldots, \sqrt{\lambda_p})$. Recall that $A = HDH^\top$. The *symmetric square root* of A is given by $A^{1/2} := HD^{1/2}H^\top$; indeed, a simple calculation gives $A^{1/2}A^{1/2} = HD^{1/2}H^\top HD^{1/2}H^\top = HDH^\top = A$.

All the above matrix concepts can be implemented in both **R** and Python by using built-in matrix functions. Let us continue with the blue chips dataset in Example 1.1. The Python code `S_2002.T` returns the transpose of S from the 2002 dataset, `@` computes matrix multiplication, and so on. All the codings for other related computations can be found in Programme 1.3 for Python (resp. Programme 1.4 for **R**).

```
1  print(np.linalg.det(np.linalg.inv(S_2002))) # |A^-1| = 1/|A|
2  print(1/np.linalg.det(S_2002))
3
4  eig_val_2002, H_2002 = np.linalg.eig(S_2002)
5  print(eig_val_2002)
6  print(H_2002)
7
8  print(np.round(H_2002.T @ H_2002, 3)) # H^T H = I
9  print(H_2002[:,0].T @ H_2002[:,1]) # h_1^T h_2 = 0
10 print(np.round(H_2002.T @ S_2002 @ H_2002, 3)) # H^T AH = D
11 D_2002 = np.diag(eig_val_2002)
12 print(D_2002)
13
14 sqrt_S_2002 = H_2002 @ np.sqrt(D_2002) @ H_2002.T # A^1/2 = HD^1/2 H^T
15 print(sqrt_S_2002 @ sqrt_S_2002)
16 print(S_2002)
```

```
1  5.706216005035965e-07
2  5.706216005035965e-07
3  [68.4874 19.9176  3.3412  2.2574 13.2054 12.8986]
4  [[-0.0311 -0.9118  0.1822  0.0364 -0.3644  0.0161]
5   [-0.0295 -0.1914 -0.819  -0.5398  0.0184  0.0059]
6   [ 0.0109 -0.091  -0.5321  0.8403  0.0438 -0.0201]
7   [ 0.003  -0.3462  0.1122 -0.0186  0.912  -0.1884]
8   [-0.9984  0.0338  0.0126  0.0233  0.0076 -0.0365]
9   [-0.0357 -0.0509  0.013   0.0172  0.1822  0.9811]]
10 [[ 1.  0.  0.  0. -0.  0.]
11  [ 0.  1.  0. -0. -0.  0.]
12  [ 0.  0.  1. -0.  0.  0.]
13  [ 0. -0. -0.  1.  0. -0.]
14  [-0. -0.  0.  0.  1. -0.]
15  [ 0.  0.  0. -0. -0.  1.]]
16 5.637851296924623e-17
17 [[68.487  0.      0.      0.     -0.      0.    ]
18  [ 0.    19.918  0.      0.     -0.      0.    ]
19  [ 0.     0.      3.341 -0.      0.      0.    ]
20  [ 0.     0.     -0.      2.257 -0.     -0.    ]
21  [-0.    -0.      0.     -0.     13.205 -0.    ]
22  [ 0.     0.      0.     -0.     -0.     12.899]]
23 [[68.4874  0.       0.       0.       0.       0.     ]
24  [ 0.     19.9176  0.       0.       0.       0.     ]
25  [ 0.      0.      3.3412  0.       0.       0.     ]
26  [ 0.      0.      0.       2.2574  0.       0.     ]
27  [ 0.      0.      0.       0.      13.2054  0.     ]
28  [ 0.      0.      0.       0.       0.      12.8986]]
29 [[18.4979  2.909   1.1602  1.9204  1.4781  0.338 ]
30  [ 2.909   3.6931  0.7663  1.2371  1.8228  0.3288]
31  [ 1.1602  0.7663  2.7439  0.9721 -0.7734 -0.0741]
32  [ 1.9204  1.2371  0.9721 13.8716 -0.2575  0.1582]
33  [ 1.4781  1.8228 -0.7734 -0.2575 68.3082  1.9618]
34  [ 0.338   0.3288 -0.0741  0.1582  1.9618 12.9929]]
35 [[18.4979  2.909   1.1602  1.9204  1.4781  0.338 ]
36  [ 2.909   3.6931  0.7663  1.2371  1.8228  0.3288]
37  [ 1.1602  0.7663  2.7439  0.9721 -0.7734 -0.0741]
38  [ 1.9204  1.2371  0.9721 13.8716 -0.2575  0.1582]
39  [ 1.4781  1.8228 -0.7734 -0.2575 68.3082  1.9618]
40  [ 0.338   0.3288 -0.0741  0.1582  1.9618 12.9929]]
```

Programme 1.3 Illustrative examples of Matrix Theory in Python using the dataset `fin-ratio.csv`.

```
1  > det(solve(S_2002))  # |A^-1| = 1/|A|
2  [1] 5.706e-07
3  > 1 / det(S_2002)
4  [1] 5.706e-07
5  > eig_2002 <- eigen(S_2002)
6  > eig_2002$values
7  [1] 68.487 19.918 13.205 12.899  3.341  2.257
8  > (H_2002 <- eig_2002$vector)
9              [,1]     [,2]      [,3]      [,4]     [,5]     [,6]
10 [1,]  0.031107  0.91185  0.364402  0.016136  0.18218 -0.03637
11 [2,]  0.029477  0.19142 -0.018447  0.005915 -0.81898  0.53980
12 [3,] -0.010938  0.09104 -0.043829 -0.020125 -0.53213 -0.84030
13 [4,] -0.003038  0.34621 -0.911976 -0.188393  0.11223  0.01856
14 [5,]  0.998380 -0.03383 -0.007556 -0.036516  0.01255 -0.02334
15 [6,]  0.035663  0.05094 -0.182190  0.981058  0.01304 -0.01720
16 > round(t(H_2002)%*%H_2002, 3)  # H^T H = I
17      [,1] [,2] [,3] [,4] [,5] [,6]
18 [1,]    1    0    0    0    0    0
19 [2,]    0    1    0    0    0    0
20 [3,]    0    0    1    0    0    0
21 [4,]    0    0    0    1    0    0
22 [5,]    0    0    0    0    1    0
23 [6,]    0    0    0    0    0    1
24 > t(H_2002[,1])%*%H_2002[,2]  # h_1^T h_2 = 0
25 [,1]
26 [1,] -7.589e-17
27 > round(t(H_2002)%*%S_2002%*%H_2002, 3)  # H^T AH = D
28       [,1]  [,2]  [,3] [,4]  [,5]  [,6]
29 [1,] 68.49  0.00  0.00  0.0 0.000 0.000
30 [2,]  0.00 19.92  0.00  0.0 0.000 0.000
31 [3,]  0.00  0.00 13.21  0.0 0.000 0.000
32 [4,]  0.00  0.00  0.00 12.9 0.000 0.000
33 [5,]  0.00  0.00  0.00  0.0 3.341 0.000
34 [6,]  0.00  0.00  0.00  0.0 0.000 2.257
35 > (D_2002 <- diag(eig_2002$values))
36       [,1]  [,2]  [,3] [,4]  [,5]  [,6]
37 [1,] 68.49  0.00  0.00  0.0 0.000 0.000
38 [2,]  0.00 19.92  0.00  0.0 0.000 0.000
39 [3,]  0.00  0.00 13.21  0.0 0.000 0.000
40 [4,]  0.00  0.00  0.00 12.9 0.000 0.000
41 [5,]  0.00  0.00  0.00  0.0 3.341 0.000
42 [6,]  0.00  0.00  0.00  0.0 0.000 2.257
43 > sqrt_S_2002 <- H_2002%*%sqrt(D_2002)%*%t(H_2002)  # A^1/2 = HD^1/2 H^T
44 > sqrt_S_2002%*%sqrt_S_2002
45        [,1]   [,2]    [,3]    [,4]    [,5]     [,6]
46 [1,] 18.498 2.9090 1.16019  1.9204  1.4781  0.33795
47 [2,]  2.909 3.6931 0.76630  1.2371  1.8228  0.32879
48 [3,]  1.160 0.7663 2.74394  0.9721 -0.7734 -0.07413
49 [4,]  1.920 1.2371 0.97207 13.8716 -0.2575  0.15815
```

```
50 [5,]  1.478 1.8228 -0.77342 -0.2575 68.3082  1.96177
51 [6,]  0.338 0.3288 -0.07413  0.1582  1.9618 12.99291
52 > S_2002
53             EY   CFTP     ln_MV      DY    BTME       DTE
54 EY      18.498 2.9090  1.16019  1.9204  1.4781   0.33795
55 CFTP     2.909 3.6931  0.76630  1.2371  1.8228   0.32879
56 ln_MV    1.160 0.7663  2.74394  0.9721 -0.7734  -0.07413
57 DY       1.920 1.2371  0.97207 13.8716 -0.2575   0.15815
58 BTME     1.478 1.8228 -0.77342 -0.2575 68.3082   1.96177
59 DTE      0.338 0.3288 -0.07413  0.1582  1.9618  12.99291
```

Programme 1.4 Illustrative examples of Matrix Theory in **R** using the dataset `fin-ratio.csv`.

(VII) Algorithm for Cholesky Decomposition

The *Cholesky decomposition* of a $p \times p$ positive definite matrix A is to find a $p \times p$ upper triangular matrix $\mathbf{C}$ such that $\mathbf{C}^{\mathsf{T}}\mathbf{C} = A$. Let $A = (a_{ij})$ and $\mathbf{C} = (c_{ij})$ where $c_{ij} = 0$ whenever $i > j$. The following steps describe how to find $\mathbf{C}$ from a given matrix A:

Step 1. Set $c_{11} = \sqrt{a_{11}}$, and $c_{1j} = a_{1j}/c_{11}$ for $j = 2, \ldots, p$.
Step 2. For $i = 2, \ldots, p$, set

$$c_{ii} = \sqrt{a_{ii} - \sum_{k=1}^{i-1} c_{ki}^2}, \text{ and } c_{ij} = \frac{1}{c_{ii}}\left(a_{ij} - \sum_{k=1}^{i-1} c_{ki}c_{kj}\right), \text{ for } j = i+1, \ldots, p.$$

To see why this algorithm can give a valid $\mathbf{C}$, note that the (i, j) element of $\mathbf{C}^{\mathsf{T}}\mathbf{C} = \sum_{k=1}^{p} c_{ki}c_{kj} = a_{ij}$. When $i = 1$ and $j = 1$, $c_{11}^2 = a_{11}$, and so we simply take $c_{11} = \sqrt{a_{11}}$. When $i = 1$ and $j = 2, \ldots, p$, $a_{1j} = \sum_{k=1}^{p} c_{k1}c_{kj} = c_{11}c_{1j}$ since $c_{k1} = 0$ for $k = 2, \ldots, p$, therefore $c_{1j} = a_{1j}/c_{11}$. For $i = 2, \ldots, p$, $a_{ii} = \sum_{k=1}^{p} c_{ki}^2 = \sum_{k=1}^{i} c_{ki}^2 = \sum_{k=1}^{i-1} c_{ki}^2 + c_{ii}^2$ since $c_{ij} = 0$ for $i > j$, we then take

$$c_{ii} = \sqrt{a_{ii} - \sum_{k=1}^{i-1} c_{ki}^2}.$$

Finally, for $j = i + 1, \ldots, p$,

$$a_{ij} = \sum_{k=1}^{i} c_{ki}c_{kj} = c_{ii}c_{ij} + \sum_{k=1}^{i-1} c_{ki}c_{kj} \Rightarrow c_{ij} = \frac{1}{c_{ii}}\left(a_{ij} - \sum_{k=1}^{i-1} c_{ki}c_{kj}\right).$$

1.3 VECTORS AND MATRIX NORMS

The notion of a norm is to measure the length of different abstract objects. Here we recall some useful results and commonly used norms for vectors and matrices.

(I) Vector norm
We consider the space $\mathbb{R}^p$, such that each $x \in \mathbb{R}^p$ is a p-dimensional vector. A function $\phi : \mathbb{R}^p \to \mathbb{R}$ is a **norm** for the Euclidean space $\mathbb{R}^p$ if it satisfies:

1. positivity: $\phi(x) \geq 0$ for all $x \in \mathbb{R}^p$, and $\phi(x) = 0 \iff x = \mathbf{0}$;
2. triangle inequality: for any $x, y \in \mathbb{R}^p$, $\phi(x + y) \leq \phi(x) + \phi(y)$;
3. absolute homogeneity: for any $x \in \mathbb{R}^p$ and scalar $a \in \mathbb{R}$, $\phi(ax) = |a|\phi(x)$.

One of the most commonly encountered norms is the q-norm: for $x = (x_1, \ldots, x_p)^\top \in \mathbb{R}^p$ and $q \geq 1$, the q-norm is defined by:

$$\|x\|_q := \left(\sum_{i=1}^{p} |x_i|^q \right)^{\frac{1}{q}}.$$

The case when $q = \infty$ is defined in a different way; indeed, the ∞-norm, properly called sup-norm, is defined as

$$\|x\|_\infty := \max_{1 \leq i \leq p} |x_i|.$$

When $q = 0$, we define the "0-norm" by the number of non-zero entries in x. The 0-norm is not a norm as it fails to satisfy the homogeneity condition.

(II) Matrix norm
The concept of a norm can be extended to the space $\mathbb{R}^{p\times k}$ of $p \times k$ matrices. A function $\phi : \mathbb{R}^{p\times k} \to \mathbb{R}$ is a **norm** for the space $\mathbb{R}^{p\times k}$ if it satisfies

1. positivity: $\phi(A) \geq 0$ for all $A \in \mathbb{R}^{p\times k}$, and $\phi(A) = 0 \iff A = \mathbf{0}$;
2. triangle inequality: for any $A, B \in \mathbb{R}^{p\times k}$, $\phi(A + B) \leq \phi(A) + \phi(B)$;
3. homogeneity: for any $A \in \mathbb{R}^{p\times k}$ and scalar $a \in \mathbb{R}$, $\phi(aA) = |a|\phi(A)$.

In addition, if $p = k$ and ϕ satisfies

4. sub-multiplicativity: $\phi(AB) \leq \phi(A)\phi(B)$ for any $A \in \mathbb{R}^{p\times k}$,

then ϕ is said to be a **sub-multiplicative norm.**

There are several common examples of matrix norms. In particular, the q-norm for square matrices is commonly used. A matrix norm $\|\cdot\|$ on $\mathbb{R}^{p\times p}$ is said to be *compatible* with a vector norm $\|\cdot\|_v$ on $\mathbb{R}^p$ if for any $A \in \mathbb{R}^{p\times p}$ and $x \in \mathbb{R}^p$, it holds that

$$\|Ax\|_v \leq \|A\| \cdot \|x\|_v.$$

Let $\|\cdot\|_q$ be the vector q-norm on $\mathbb{R}^p$. Then the q-norm $\|\cdot\|$ on the space of $p \times p$ square matrices is defined as

$$\|A\| = \max_{x \neq 0} \frac{\|Ax\|_q}{\|x\|_q} = \max_{\|x\|_q = 1} \|Ax\|_q, \quad A \in \mathbb{R}^{p \times p}. \tag{1.3}$$

It is clear that this definition verifies the properties of a matrix norm. In addition, it is obvious that the q-norms for square matrices and vectors are compatible. The definition can easily be generalized to $p \times k$ matrices, by considering the norm $\|Ax\|_q$ in (1.3) as the q-norm of Ax in $\mathbb{R}^p$ for any $x \in \mathbb{R}^k$ which possesses the corresponding q-norm in $\mathbb{R}^k$. In the following, without the risk of causing any ambiguity, we shall denote the q-norms for both vectors and matrices by $\|\cdot\|_q$. Below are several important special cases of the q-norm for $p \times k$ matrices:

1. 1-norm:

$$\|A\|_1 = \max_{j=1,\ldots,k} \left(\sum_{i=1}^{p} |a_{ij}| \right).$$

2. 2-norm (a.k.a. the *spectral norm*):

$$\|A\|_2 = \sqrt{\lambda_{\max}(A^\top A)},$$

where $\lambda_{\max}(A^\top A)$ is the maximum eigenvalue of the matrix $A^\top A$.

3. ∞-norm/sup-norm:

$$\|A\|_\infty = \max_{i=1,\ldots,p} \left(\sum_{j=1}^{k} |a_{ij}| \right).$$

Apart from the q-norm, another commonly encountered matrix norm in machine learning and deep learning is the *Frobenius norm*, which is defined as

$$\|A\|_F := \left(\sum_{i=1}^{p} \sum_{j=1}^{k} |a_{ij}|^2 \right)^{\frac{1}{2}} = \sqrt{\operatorname{tr}(A^\top A)}.$$

1.4 COMMON PROBABILITY DISTRIBUTIONS

In this section, we shall review some basic probability distributions learnt in elementary statistical courses. Readers may refer to [10, 13] for more a comprehensive treatment of this topic, and [1, 14, 17] for a more advanced discussion, and on tackling block matrices in particular. These distributions are important and serve as the building blocks for constructing and deriving the null distributions of many test statistics. Throughout the book, "iid" stands for "*independent and identically distributed*", referring to an underlying family of random variables.

1.4.1 Univariate Distributions

To begin with, we review the definitions and related important properties of several commonly used univariate distributions.

1. **Normal distribution: $\mathcal{N}(\mu, \sigma^2)$.** A real random variable x is said to follow a normal distribution with a mean μ and variance σ^2, denoted as $\mathrm{x} \sim \mathcal{N}(\mu, \sigma^2)$, if its *probability density function* (pdf) takes the form:

$$f(x) = \frac{1}{\sqrt{2\pi\sigma^2}} \exp\left\{-\frac{(x-\mu)^2}{2\sigma^2}\right\}, \quad x \in \mathbb{R}. \tag{1.4}$$

2. **Chi-square distribution: χ^2_ν.** A non-negative random variable x is said to follow a chi-square distribution with degrees of freedom $\nu \in \mathbb{N}$, denoted by $\mathrm{x} \sim \chi^2_\nu$, if its pdf takes the form:

$$f(x) = \frac{1}{2^{\nu/2}\Gamma(\nu/2)} x^{\nu/2-1} e^{-x/2}, \quad x > 0, \tag{1.5}$$

   where $\Gamma(\cdot)$ is the well-known gamma function:

$$\Gamma(z) = \int_0^\infty t^{z-1} e^{-t}\, dt, \quad \text{for} \quad z > 0.$$

   If $\mathrm{z}_1, \ldots, \mathrm{z}_n$ are iid $\mathcal{N}(0, 1)$ random variables, then one can easily show that $\mathrm{x} = \mathrm{z}_1^2 + \cdots + \mathrm{z}_n^2 \sim \chi^2_n$. Furthermore, if $\mathrm{y}_1, \ldots, \mathrm{y}_n$ are iid $\mathcal{N}(\mu, \sigma^2)$, we also have the following distribution of the sample variance s_y^2:

$$\frac{(n-1)\mathrm{s}_\mathrm{y}^2}{\sigma^2} = \sum_{i=1}^n \frac{(\mathrm{y}_i - \bar{\mathrm{y}})^2}{\sigma^2} \sim \chi^2_{n-1};$$

   this result is just a special case of Proposition 1.2, which we will meet shortly.

3. **Student's t-distribution: t_ν.** If $\mathrm{z} \sim \mathcal{N}(0, 1)$ is independent of $\mathrm{x} \sim \chi^2_\nu$, then $\mathrm{t} := \frac{\mathrm{z}}{\sqrt{\mathrm{x}/\nu}}$ is said to follow a t-distribution t_ν with ν degrees of freedom. As a result, if $\mathrm{y}_1, \ldots, \mathrm{y}_n$ are iid $\mathcal{N}(\mu, \sigma^2)$, then

$$\frac{\sqrt{n}(\bar{\mathrm{y}} - \mu)}{\mathrm{s}_\mathrm{y}} \sim t_{n-1};$$

   indeed, since $\bar{\mathrm{y}} \sim \mathcal{N}(\mu, \frac{\sigma^2}{n})$, we have $\mathrm{z} := \frac{\sqrt{n}(\bar{\mathrm{y}}-\mu)}{\sigma} \sim \mathcal{N}(0, 1)$, and $\mathrm{x} := \frac{(n-1)\mathrm{s}_\mathrm{y}^2}{\sigma^2} \sim \chi^2_{n-1}$, furthermore $\bar{\mathrm{y}}$ is independent of s_y^2 in accordance with Proposition 1.2. Therefore,

$$\mathrm{t} = \frac{\mathrm{z}}{\sqrt{\mathrm{x}/(n-1)}} = \frac{\sqrt{n}(\bar{\mathrm{y}} - \mu)}{\mathrm{s}_\mathrm{y}} \sim t_{n-1}.$$

4. **Fisher's F-distribution: $F_{u,v}$.** If $\mathrm{x} \sim \chi_u^2$ and $\mathrm{y} \sim \chi_v^2$ are independent, then $\frac{\mathrm{x}/u}{\mathrm{y}/v} \sim F_{u,v}$, an F-distribution with parameters u and v in order. Furthermore, if $\mathrm{x}_1, \ldots, \mathrm{x}_m$ are iid $\mathcal{N}(\mu_1, \sigma_1^2)$ and are independent of $\mathrm{y}_1, \ldots, \mathrm{y}_n$ being iid $\mathcal{N}(\mu_2, \sigma_2^2)$, then $\frac{\mathrm{s}_\mathrm{x}^2/\sigma_1^2}{\mathrm{s}_\mathrm{y}^2/\sigma_2^2} = \frac{\mathrm{s}_\mathrm{x}^2\sigma_2^2}{\sigma_1^2\mathrm{s}_\mathrm{y}^2} \sim F_{m-1,n-1}$; indeed, since $\frac{(m-1)\mathrm{s}_\mathrm{x}^2}{\sigma_1^2} \sim \chi_{m-1}^2$ and $\frac{(n-1)\mathrm{s}_\mathrm{y}^2}{\sigma_2^2} \sim \chi_{n-1}^2$ are independent,

$$\frac{\mathrm{s}_\mathrm{x}^2/\sigma_1^2}{\mathrm{s}_\mathrm{y}^2/\sigma_2^2} = \frac{\mathrm{s}_\mathrm{x}^2\sigma_2^2}{\sigma_1^2\mathrm{s}_\mathrm{y}^2} \sim F_{m-1,n-1}.$$

   In particular, if $\mathrm{t} \sim t_\nu$, then $\mathrm{t}^2 \sim F_{1,\nu}$.
5. **Sampling distributions of $\overline{\mathrm{x}}$ and s_x^2 (a special case of Proposition 1.2):** if $\mathrm{x}_1, \ldots, \mathrm{x}_n$ are iid $\mathcal{N}(\mu, \sigma^2)$, then $\overline{\mathrm{x}} \sim \mathcal{N}(\mu, \sigma^2/n)$ and is independent of $(n-1)\mathrm{s}_\mathrm{x}^2/\sigma^2 \sim \chi_{n-1}^2$.

1.4.2 The Multivariate Normal Distribution

The random vector $\mathbf{x} = (\mathrm{x}_1, \ldots, \mathrm{x}_p)^\mathsf{T}$ is said to follow a p-variate normal distribution with a mean vector $\boldsymbol{\mu}$ and covariance matrix $\boldsymbol{\Sigma}$,[2] denoted by $\mathbf{x} \sim \mathcal{N}_p(\boldsymbol{\mu}, \boldsymbol{\Sigma})$, if its joint probability density function is given by

$$f(\boldsymbol{x}) = \frac{1}{(2\pi)^{p/2}|\boldsymbol{\Sigma}|^{1/2}} \exp\left\{-\frac{1}{2}(\boldsymbol{x}-\boldsymbol{\mu})^\mathsf{T}\boldsymbol{\Sigma}^{-1}(\boldsymbol{x}-\boldsymbol{\mu})\right\}, \quad \boldsymbol{x} \in \mathbb{R}^p. \tag{1.6}$$

The following are some important properties of the multivariate normal distribution:

1. Consider $\mathbf{x} \sim \mathcal{N}_p(\boldsymbol{\mu}, \boldsymbol{\Sigma})$, then for a deterministic $\boldsymbol{B} \in \mathbb{R}^{q\times p}$ and $\boldsymbol{b} \in \mathbb{R}^q$, we have $\mathrm{y} = \boldsymbol{B}\mathbf{x} + \boldsymbol{b} \sim \mathcal{N}_q(\boldsymbol{B}\boldsymbol{\mu} + \boldsymbol{b}, \boldsymbol{B}\boldsymbol{\Sigma}\boldsymbol{B}^\mathsf{T})$.
2. Let $\mathbf{x}_1, \ldots, \mathbf{x}_n \overset{\text{iid}}{\sim} \mathcal{N}_p(\boldsymbol{\mu}, \boldsymbol{\Sigma})$. Then $\overline{\mathbf{x}} \sim \mathcal{N}_p(\boldsymbol{\mu}, (1/n)\boldsymbol{\Sigma})$.
3. Let $\mathbf{x}_1, \ldots, \mathbf{x}_n \overset{\text{iid}}{\sim} \mathcal{N}_p(\boldsymbol{\mu}, \boldsymbol{\Sigma})$. Then $n(\overline{\mathbf{x}} - \boldsymbol{\mu})^\mathsf{T}\boldsymbol{\Sigma}^{-1}(\overline{\mathbf{x}} - \boldsymbol{\mu}) \sim \chi_p^2$. To see this, let $\mathrm{y} = \sqrt{n}\boldsymbol{\Sigma}^{-1/2}(\overline{\mathbf{x}} - \boldsymbol{\mu})$. By Property 1, $\mathrm{y} \sim \mathcal{N}_p(\mathbf{0}, \boldsymbol{I}_p)$. Therefore, $n(\overline{\mathbf{x}} - \boldsymbol{\mu})^\mathsf{T}\boldsymbol{\Sigma}^{-1}(\overline{\mathbf{x}} - \boldsymbol{\mu}) = \mathrm{y}^\mathsf{T}\mathrm{y} = \sum_{i=1}^p \mathrm{y}_i^2 \sim \chi_p^2$ in light of Property 2 in Subsection 1.4.1.
4. (The *Central Limit Theorem*). Let $\mathbf{x}_1, \ldots, \mathbf{x}_n$ be p-dimensional iid random vectors from an arbitrary distribution with a finite mean $\boldsymbol{\mu}$ and finite covariance matrix $\boldsymbol{\Sigma}$. Then $\sqrt{n}(\overline{\mathbf{x}} - \boldsymbol{\mu}) \sim \mathcal{N}_p(\mathbf{0}, \boldsymbol{\Sigma})$ approximately for a large enough sample size n (relative to p).

   Consequently, by using Slutsky's theorem [15], we further have $n(\overline{\mathbf{x}} - \boldsymbol{\mu})^\mathsf{T}$ $\mathrm{S}^{-1}(\overline{\mathbf{x}} - \boldsymbol{\mu}) \sim \chi_p^2$ approximately for a large sample size n (relative to p).

[2] Here, we only consider the nondegenerate case where $\boldsymbol{\Sigma}$ is positive definite.

5. Let $\mathbf{x}_1, \ldots, \mathbf{x}_n$ be iid $\mathcal{N}_p(\boldsymbol{\mu}, \boldsymbol{\Sigma})$ random vectors. Then the squared generalized distance (a.k.a. the *Mahalanobis distance*) $\mathrm{d}_i^2 = (\mathbf{x}_i - \overline{\mathbf{x}})^\top \mathrm{S}^{-1}(\mathbf{x}_i - \overline{\mathbf{x}}) \sim \chi_p^2$ approximately for a large value of n (relative to p). This is a direct consequence of an application of *Slutsky's theorem* by replacing $\boldsymbol{\mu}$ by $\overline{x}$ and $\boldsymbol{\Sigma}$ by S in Property 4 above.

In simple linear regression, we often use the marginal and conditional distributions of the bivariate normal distribution; the corresponding results are readily extended to the multivariate case for their application in the context of multiple linear regression.

Let $\mathbf{x} \sim \mathcal{N}_p(\boldsymbol{\mu}, \boldsymbol{\Sigma})$. Partition $\mathbf{x}$, $\boldsymbol{\mu}$ and $\boldsymbol{\Sigma}$ into:

$$\mathbf{x} = \begin{pmatrix} \mathbf{x}_1 \\ \hline \mathbf{x}_2 \end{pmatrix}, \quad \boldsymbol{\mu} = \begin{pmatrix} \boldsymbol{\mu}_1 \\ \hline \boldsymbol{\mu}_2 \end{pmatrix}, \quad \boldsymbol{\Sigma} = \left(\begin{array}{c|c} \boldsymbol{\Sigma}_{11} & \boldsymbol{\Sigma}_{12} \\ \hline \boldsymbol{\Sigma}_{21} & \boldsymbol{\Sigma}_{22} \end{array}\right),$$

where $\mathbf{x}_1$, $\boldsymbol{\mu}_1$ are $q \times 1$ vectors; $\mathbf{x}_2$, $\boldsymbol{\mu}_2$ are $(p-q) \times 1$ vectors, and $\boldsymbol{\Sigma}_{11}, \boldsymbol{\Sigma}_{12}, \boldsymbol{\Sigma}_{21}$, and $\boldsymbol{\Sigma}_{22}$ are respectively $q \times q$, $q \times (p-q)$, $(p-q) \times q$, and $(p-q) \times (p-q)$ matrices, for some $q < p$. Note that $\boldsymbol{\Sigma}_{21} = \boldsymbol{\Sigma}_{12}^\top$. We summarize below some useful properties about the marginal and conditional distributions of the multivariate normal distribution.

1. If $\mathbf{x} \sim \mathcal{N}_p(\boldsymbol{\mu}, \boldsymbol{\Sigma})$, then the marginal distribution of any q-component subset of $\mathbf{x}$ is q-variate normal. Without loss of generality, it suffices to show that $\mathbf{x}_1 \sim \mathcal{N}_q(\boldsymbol{\mu}_1, \boldsymbol{\Sigma}_{11})$. Let $B = [I_q \,\vdots\, \mathbf{0}]$ be a $q \times p$ matrix. By Property 1 in Subsection 1.4.2, $\mathbf{x}_1 = B\mathbf{x} \sim \mathcal{N}_q(B\boldsymbol{\mu}, B\boldsymbol{\Sigma}B^\top) = \mathcal{N}_q(\boldsymbol{\mu}_1, \boldsymbol{\Sigma}_{11})$.
2. $\mathbf{x}_1$ and $\mathbf{x}_2$ are independent if and only if $\boldsymbol{\Sigma}_{12} = \mathbf{0}$.
3. $\mathbf{x}_1 - \boldsymbol{\Sigma}_{12}\boldsymbol{\Sigma}_{22}^{-1}\mathbf{x}_2 \sim \mathcal{N}_q(\boldsymbol{\mu}_1 - \boldsymbol{\Sigma}_{12}\boldsymbol{\Sigma}_{22}^{-1}\boldsymbol{\mu}_2, \boldsymbol{\Sigma}_{11\cdot 2})$ and is independent of $\mathbf{x}_2$, where $\boldsymbol{\Sigma}_{11\cdot 2} := \boldsymbol{\Sigma}_{11} - \boldsymbol{\Sigma}_{12}\boldsymbol{\Sigma}_{22}^{-1}\boldsymbol{\Sigma}_{21}$. To show this, we first let $C = \begin{pmatrix} I_q & -\boldsymbol{\Sigma}_{12}\boldsymbol{\Sigma}_{22}^{-1} \\ \mathbf{0} & I_{p-q} \end{pmatrix}$.

   Then $C\mathbf{x} = \begin{pmatrix} \mathbf{x}_1 - \boldsymbol{\Sigma}_{12}\boldsymbol{\Sigma}_{22}^{-1}\mathbf{x}_2 \\ \mathbf{x}_2 \end{pmatrix}$ is p-variate normal with the mean $C\boldsymbol{\mu} = \begin{pmatrix} \boldsymbol{\mu}_1 - \boldsymbol{\Sigma}_{12}\boldsymbol{\Sigma}_{22}^{-1}\boldsymbol{\mu}_2 \\ \boldsymbol{\mu}_2 \end{pmatrix}$ and covariance matrix

$$C\boldsymbol{\Sigma}C^\top = \begin{pmatrix} I_q & -\boldsymbol{\Sigma}_{12}\boldsymbol{\Sigma}_{22}^{-1} \\ \mathbf{0} & I_{p-q} \end{pmatrix} \begin{pmatrix} \boldsymbol{\Sigma}_{11} & \boldsymbol{\Sigma}_{12} \\ \boldsymbol{\Sigma}_{21} & \boldsymbol{\Sigma}_{22} \end{pmatrix} \begin{pmatrix} I_q & \mathbf{0} \\ -\boldsymbol{\Sigma}_{22}^{-1}\boldsymbol{\Sigma}_{21} & I_{p-q} \end{pmatrix}$$

$$= \begin{pmatrix} \boldsymbol{\Sigma}_{11\cdot 2} & \mathbf{0} \\ \mathbf{0} & \boldsymbol{\Sigma}_{22} \end{pmatrix}.$$

4. The conditional distribution of $\mathbf{x}_1$ given $\mathbf{x}_2 = x_2$ is $\mathcal{N}_q(\boldsymbol{\mu}_1 + \boldsymbol{\Sigma}_{12}\boldsymbol{\Sigma}_{22}^{-1}(x_2 - \boldsymbol{\mu}_2), \boldsymbol{\Sigma}_{11\cdot 2})$. This result follows immediately from Property 3 of this subsection by setting $\mathbf{x}_2 = x_2$.

1.4.3 Wishart Distribution

The *Wishart distribution* is a multivariate extension of the chi-square distribution. Recall Property 2 of Subsection 1.4.1, namely if $x_1, \dots, x_n$ are iid distributed as $\mathcal{N}(\mu, \sigma^2)$, then

$$\sum_{i=1}^{n} \frac{(x_i - \overline{x})^2}{\sigma^2} = \frac{(n-1)s^2}{\sigma^2} \sim \chi^2_{n-1}.$$

We now consider iid multivariate normal random vectors $\mathbf{z}_i \sim \mathcal{N}_p(\mathbf{0}, \boldsymbol{\Sigma})$, $i = 1, \dots, n$, with $n > p$. Then, the $p \times p$ random matrix

$$\mathbf{A} := \sum_{i=1}^{n} \mathbf{z}_i \mathbf{z}_i^{\top}$$

is said to follow the Wishart distribution with degrees of freedom n and scaling matrix $\boldsymbol{\Sigma}$, denoted as $W_p(\boldsymbol{\Sigma}, n)$. The density function of $\mathbf{A}$ is given by:

$$w(a \mid \boldsymbol{\Sigma}, n) = \frac{|a|^{(n-p-1)/2} \exp\left\{-\frac{1}{2}\mathrm{tr}(\boldsymbol{\Sigma}^{-1} a)\right\}}{2^{pn/2} |\boldsymbol{\Sigma}|^{n/2} \Gamma_p\left(\frac{n}{2}\right)}, \quad a \in \mathbb{R}^{p \times p}, \tag{1.7}$$

where Γ_p is the multivariate gamma function:

$$\Gamma_p(t) = \pi^{\frac{p(p-1)}{4}} \prod_{i=1}^{p} \Gamma\left(t - \frac{1}{2}(i-1)\right). \tag{1.8}$$

We remark that if $n < p$, $\mathbf{A}$ does not admit a density; nevertheless, in that case, we still denote its distribution by $W_p(\boldsymbol{\Sigma}, n)$; see more detailed discussion in [7].

Let $\mathbf{x}_i \sim \mathcal{N}_p(\boldsymbol{\mu}, \boldsymbol{\Sigma})$, $i = 1, \dots, n$, be iid multivariate normal random vectors. The following important result gives the sampling distribution of the sample covariance matrix S.

Proposition 1.2 The sample mean $\overline{\mathbf{x}} = \frac{1}{n} \sum_{i=1}^{n} \mathbf{x}_i$ is independent of the sample covariance matrix $\mathbf{S} = \frac{1}{n-1} \sum_{i=1}^{n} (\mathbf{x}_i - \overline{\mathbf{x}})(\mathbf{x}_i - \overline{\mathbf{x}})^{\top}$, and $\mathbf{S} \sim W_p\left(\frac{1}{n-1}\boldsymbol{\Sigma}, n-1\right)$.

Proof. Consider the SSCP matrix $\mathbf{A} = (n-1)\mathbf{S} = \sum_{i=1}^{n} (\mathbf{x}_i - \overline{\mathbf{x}})(\mathbf{x}_i - \overline{\mathbf{x}})^{\top} = \sum_{i=1}^{n} \mathbf{x}_i \mathbf{x}_i^{\top} - n\overline{\mathbf{x}}\,\overline{\mathbf{x}}^{\top}$. Let $B = (b_{ij})$ be an $n \times n$ orthogonal matrix with the last row being

$$b_{nj} = \frac{1}{\sqrt{n}}, \quad j = 1, \dots, n.$$

Also let $\mathbf{z}_i := \sum_{j=1}^n b_{ij}\mathbf{x}_j$, $i = 1, \ldots, n$. The orthogonality of B gives $\sum_{i=1}^n b_{ij}b_{ik} = \delta_{jk}$, where the *Dirac delta* δ_{jk} equals 1 if $j = k$, or 0 otherwise. Then, we have

$$\begin{aligned}\sum_{i=1}^n \mathbf{z}_i\mathbf{z}_i^\top &= \sum_{i=1}^n \left(\sum_{j=1}^n b_{ij}\mathbf{x}_j\right)\left(\sum_{k=1}^n b_{ik}\mathbf{x}_k\right)^\top \\ &= \sum_{j=1}^n \sum_{k=1}^n \left(\sum_{i=1}^n b_{ij}b_{ik}\right)\mathbf{x}_j\mathbf{x}_k^\top \\ &= \sum_{j=1}^n \mathbf{x}_j\mathbf{x}_j^\top.\end{aligned}$$

By definition, as $\mathbf{z}_n = \sum_{j=1}^n b_{nj}\mathbf{x}_j = \frac{1}{\sqrt{n}}\sum_{j=1}^n \mathbf{x}_j = \sqrt{n}\bar{\mathbf{x}}$, we can rewrite

$$\mathbf{A} = \sum_{i=1}^n \mathbf{x}_i\mathbf{x}_i^\top - n\bar{\mathbf{x}}\bar{\mathbf{x}}^\top = \sum_{i=1}^n \mathbf{z}_i\mathbf{z}_i^\top - \mathbf{z}_n\mathbf{z}_n^\top = \sum_{i=1}^{n-1} \mathbf{z}_i\mathbf{z}_i^\top. \tag{1.9}$$

Note that $\mathbf{z}_i$'s, for $i = 1, \ldots, n-1$, are jointly multivariate normally distributed because they are linear combinations of the multivariate normal random vectors $\mathbf{x}_j$'s. To establish their independence, it suffices to check the vanishing of their pairwise covariance matrix. Also note that $\mathbb{E}(\mathbf{z}_n) = \mathbb{E}(\sqrt{n}\bar{\mathbf{x}}) = \sqrt{n}\boldsymbol{\mu}$. For any $i \neq n$, the orthogonality of B and the definition of $\sqrt{n} \cdot b_{nj} = 1$, for all $j = 1, 2, \ldots, n$, together give

$$\mathbb{E}(\mathbf{z}_i) = \sum_{j=1}^n b_{ij}\mathbb{E}(\mathbf{x}_j) = \sum_{j=1}^n b_{ij}\boldsymbol{\mu} = \left(\sum_{j=1}^n b_{ij}b_{nj}\right)(\sqrt{n}\boldsymbol{\mu}) = \mathbf{0}.$$

Finally, by the independence of $\mathbf{x}_i$'s and the orthogonality of B, we have

$$\begin{aligned}\mathrm{Cov}(\mathbf{z}_i, \mathbf{z}_j) &= \mathbb{E}\left(\left(\sum_{k=1}^n b_{ik}(\mathbf{x}_k - \boldsymbol{\mu})\right)\left(\sum_{l=1}^n b_{jl}(\mathbf{x}_l - \boldsymbol{\mu})^\top\right)\right) \\ &= \sum_{k=1}^n \sum_{l=1}^n b_{ik}b_{jl}\mathbb{E}\left((\mathbf{x}_k - \boldsymbol{\mu})(\mathbf{x}_l - \boldsymbol{\mu})^\top\right) \\ &= \sum_{k=1}^n \sum_{l=1}^n b_{ik}b_{jl}(\delta_{kl}\boldsymbol{\Sigma}) = \sum_{k=1}^n b_{ik}b_{jk}\boldsymbol{\Sigma} = \delta_{ij}\boldsymbol{\Sigma},\end{aligned}$$

and so $\mathbf{z}_j \sim \mathcal{N}_p(\mathbf{0}, \boldsymbol{\Sigma})$, for $j = 1, 2, \ldots, n-1$, and the $\mathbf{z}_i$'s are mutually independent. By virtue of (1.9) and the very definition of the Wishart distribution, we can then see that $\mathbf{A} \sim W_p(\boldsymbol{\Sigma}, n-1)$, and therefore $\mathbf{S} \sim W_p\left(\frac{1}{n-1}\boldsymbol{\Sigma}, n-1\right)$. The independence of $\mathbf{S}$ and $\bar{\mathbf{x}}$ is inherited from the independence of $\mathbf{A}$ and $\mathbf{z}_n$, also due to (1.9). □

1.5 INTRODUCTORY BAYESIAN STATISTICS

A Bayesian approach to the estimation of parameters is a probabilistic method in which an optimal decision rule is taken, in contrast to the usual maximum likelihood estimation (MLE) method. Compared with MLE, whose resulting $\hat{\theta}$ maximizes the probability of obtaining the given sample of training dataset, the Bayesian approach evaluates the conditional probability of θ given the sample taking various values instead; see Figure 1.1 for an illustration of the comparison. Indeed, for *Bayesian inference*, the training stage updates our prior belief on the probabilities of θ using the input data, resulting in posterior probabilities as shown in Figure 1.1b.

The theoretical foundation of Bayesian inference is based on an extended form of *Bayes' theorem*: Let $A_1, \dots, A_n$ be mutually exclusive events and $\cup_{i=1}^{n} A_i =: A$. Then, for any event $B \subseteq A$, we have

$$\mathbb{P}(A_{(j_1,\dots,j_k)}|B) = \frac{\mathbb{P}(B|A_{(j_1,\dots,j_k)})\mathbb{P}(A_{(j_1,\dots,j_k)})}{\sum_{i=1}^{n} \mathbb{P}(B|A_i)\mathbb{P}(A_i)} = \sum_{\alpha=1}^{k} \frac{\mathbb{P}(B|A_{j_\alpha})\mathbb{P}(A_{j_\alpha})}{\sum_{i=1}^{n} \mathbb{P}(B|A_i)\mathbb{P}(A_i)}, \tag{1.10}$$

where $A_{(j_1,\dots,j_k)} = \cup_{\alpha=1}^{k} A_{j\alpha}$.

Example 1.2 John undertook a diagnostic test for heart disease. It is known that 1% of the whole population suffers from the disease. Suppose that 95% of patients (having disease) would be diagnosed by the test as positive, i.e., the accuracy of the test is 95%. On the other hand, 95% of healthy individuals would be correctly diagnosed as negative. Unfortunately, John's test result was positive.

Let A be the event of John having heart disease and B be the event of having a positive test result. Then we have:

$$\mathbb{P}(B|A) = 0.95 = \mathbb{P}(\overline{B}|\overline{A}).$$

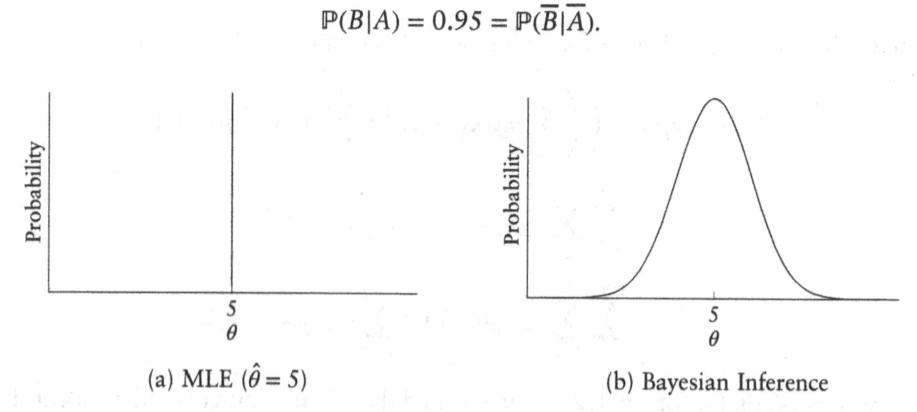

FIGURE 1.1 Comparison between MLE (a) and Bayesian inference (b). MLE draws the conclusion $\hat{\theta} = 5$, while Bayesian inference deduces that the conditional probability of $\theta = 5$ given the sample is the highest, yet it does not eliminate other possible values as the true parameter value of θ.

Prior to the test, we know $\mathbb{P}(A) = 0.01$. Given the positive result, the conditional probability of John having the disease becomes

$$\mathbb{P}(A|B) = \frac{0.95 \times 0.01}{0.95 \times 0.01 + 0.05 \times 0.99} \approx 0.16. \tag{1.11}$$

Therefore, even though this diagnostic test has seemingly high accuracy, the probability that John really suffered from the disease given the positive diagnosis result is fairly low. Actually, as we shall elaborate in Section 10.6, once diagnosed as positive, John can only be classified as either true positive (TP) or false positive (FP), and the quantity (1.11) is exactly the precision of the diagnosis, defined as:

$$\text{Precision} := \frac{\text{TP}}{\text{TP} + \text{FP}} = \mathbb{P}(A|B).$$

(I) Normal Distribution with Known Variance

We here present a special case of Bayesian inference for the mean parameter of a normal distribution $\mathcal{N}(\mu, \sigma^2)$; in particular, we first assume that the variance σ^2 is known and only make inference on the unknown mean μ. The case when σ is unknown is discussed at the end of this section.

Let $\mathbf{x} = \{\mathrm{x}_i\}_{i=1}^n$ be a random sample from $\mathcal{N}(\mu, \sigma^2)$, then given μ, the density of x_i is:

$$f_{\mathrm{x}_i}(x_i|\mu) = \frac{1}{\sqrt{2\pi\sigma^2}} \exp\left(-\frac{(x_i - \mu)^2}{2\sigma^2}\right), \quad \text{for } x_i \in \mathbb{R}, \tag{1.12}$$

and the conditional joint density of $\mathbf{x}$ given μ is:

$$f(\mathbf{x}|\mu) = \prod_{i=1}^{n} f_{\mathrm{x}_i}(x_i|\mu). \tag{1.13}$$

In the Bayesian framework, the knowledge of μ is considered uncertain, and its uncertainty is described by a (subjective) probability distribution. By Bayes' theorem, the conditional density of μ given that the random sample $\mathbf{x} = x$ is:

$$f_{\mu|\mathbf{x}}(\mu|x) = \frac{f(x|\mu)\pi(\mu)}{\int_{-\infty}^{\infty} f(x|v)\pi(v)dv}, \tag{1.14}$$

where $\pi(\mu)$ is known as the *prior* distribution of μ, and $f(\mu|x)$ is known as the *posterior* distribution of μ given the sample. Without any prior knowledge, we usually prefer the prior to be uniform over $\mathbb{R}$ so that μ takes any possible real value with equal probability before training, while noting that such a distribution is ill-posed as it cannot be integrated to 1. Alternatively, we may assume the prior distribution to be normal with mean μ_0 and variance σ_0^2, i.e., $\mu \sim \mathcal{N}(\mu_0, \sigma_0^2)$; indeed, it will resemble a uniform distribution over $\mathbb{R}$ as $\sigma_0^2 \to \infty$, as shown in Figure 1.2, which justifies this choice of prior.

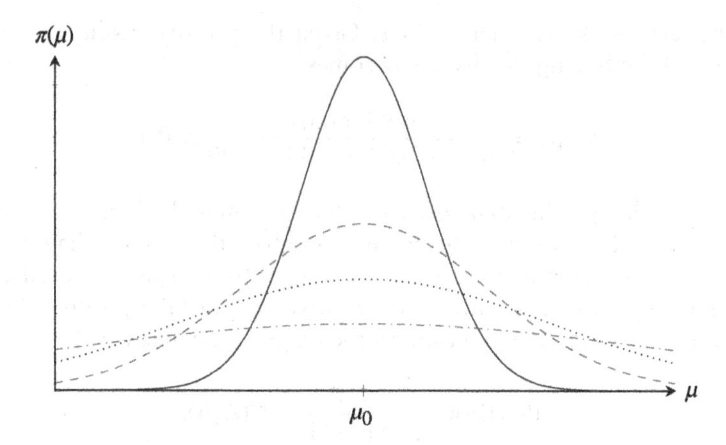

FIGURE 1.2 Density of $\mathcal{N}(\mu_0, \sigma_0^2)$ with $\sigma_0^2 = 1$ (red solid), 5 (orange dashed), 10 (blue dotted) and 20 (green dot-dashed).

We next derive the posterior distribution $f_{\mu|\mathbf{x}}(\mu|\mathbf{x})$. Let

$$C := \int_{-\infty}^{\infty} f(\mathbf{x}|\nu)\pi(\nu)d\nu$$

be the normalizing constant, then the posterior can be rewritten as:

$$f_{\mu|\mathbf{x}}(\mu|\mathbf{x}) = \frac{1}{C} f(\mathbf{x}|\mu)\pi(\mu), \tag{1.15}$$

which integrates to 1. For notational convenience, we further denote $\beta := \frac{1}{\sigma^2}$ and $\beta_0 := \frac{1}{\sigma_0^2}$; formally, such reciprocals of the variance of a normal distribution are called the *precision*. Since the prior distribution is $\mathcal{N}(\mu_0, \sigma_0^2)$, we have:

$$f_{\mu|\mathbf{x}}(\mu|\mathbf{x}) \propto \exp\left(-\frac{\beta}{2} \sum_{i=1}^{n} (x_i - \mu)^2 - \frac{\beta_0}{2}(\mu - \mu_0)^2 \right), \tag{1.16}$$

where the proportionality constant is independent of μ. Then, we expand the exponent in (1.16) as:

$$\begin{aligned}
&-\frac{\beta}{2} \sum_{i=1}^{n} (x_i - \mu)^2 - \frac{\beta_0}{2}(\mu - \mu_0)^2 \\
&= -\frac{n\beta + \beta_0}{2}\mu^2 + \left(\beta \sum_{i=1}^{n} x_i + \beta_0 \mu_0 \right) \mu + k_1 \\
&= -\frac{n\beta + \beta_0}{2} \left(\mu - \frac{\beta \sum_{i=1}^{n} x_i + \beta_0 \mu_0}{n\beta + \beta_0} \right)^2 + k_2,
\end{aligned} \tag{1.17}$$

where the constants k_1 and k_2 are independent of μ. By writing

$$\beta_n := n\beta + \beta_0 \quad \text{and} \quad \mu_n := \frac{\beta \sum_{i=1}^{n} x_i + \beta_0 \mu_0}{n\beta + \beta_0}, \tag{1.18}$$

and substituting (1.17) and (1.18) into (1.16), we see that

$$f_{\mu|\mathbf{x}}(\mu|\boldsymbol{x}) \propto \exp\left(-\frac{\beta_n}{2}(\mu - \mu_n)^2\right). \tag{1.19}$$

Hence, we conclude that the posterior distribution of μ is still a normal distribution but with a mean μ_n and variance β_n^{-1}, i.e., $\mu|\mathbf{x} = \boldsymbol{x} \sim \mathcal{N}(\mu_n, \beta_n^{-1})$.

As mentioned before, to mimic a uniform prior, we simply let $\sigma_0^2 \to \infty$. Then in the limiting case, $\beta_0 = 0$ and $\mu_n = \frac{1}{n}\sum_{i=1}^{n} x_i$, i.e., the sample mean $\overline{x}_n$ of the training data; this result is consistent with the maximum likelihood estimator of μ. On the other hand, as $\beta_0 \to 0$, the variance of the posterior distribution tends to

$$\beta_n^{-1} = \frac{1}{n\beta} = \frac{\sigma^2}{n},$$

which decreases in magnitude with n; graphically, the bell-shaped curve in Figure 1.3 would degenerate at μ_n.

For a small sample size n, we have a rather low confidence that the true value of μ is equal to μ_n, in the sense that the posterior variance β_n^{-1} is not too small. This confidence increases with n, and eventually becomes certain as $n \to \infty$, by then $\beta_n^{-1} \to 0$ which further implies $\mathbb{P}(|\mu - \mu_n| > \epsilon|\mathbf{x} = \boldsymbol{x}) \to 0$ for any $\epsilon > 0$ in light of *Chebyshev's inequality*. This is also consistent with the conclusion drawn by using MLE. In this way, we may consider *Bayesian inference* as supplementing MLE.

Secondly, from (1.18), the posterior mean μ_n is indeed the weighted average of the sample mean $\overline{x}_n$ and the prior mean μ_0, with their respective weights proportional

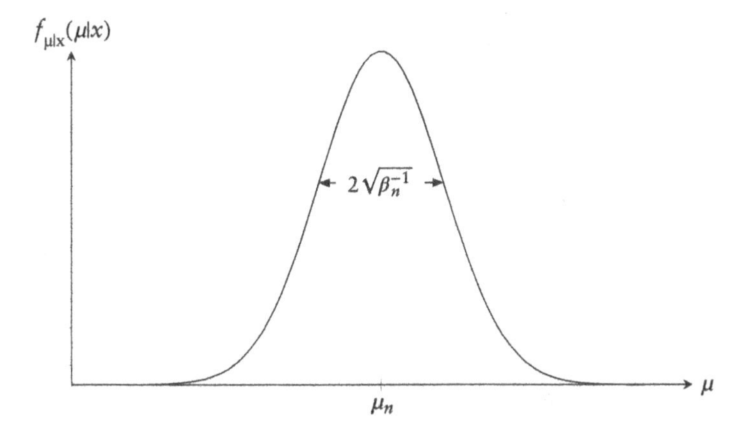

FIGURE 1.3 The posterior distribution $f_{\mu|\mathbf{x}}(\mu|\boldsymbol{x})$.

to $n\beta$ and β_0: the weight for $\overline{x}_n$ in $\mu_n = \frac{n\beta\overline{x}_n+\beta_0\mu_0}{n\beta+\beta_0}$, namely $\frac{n\beta}{n\beta+\beta_0}$, is known as the *credibility factor* in actuarial science. For any fixed value of σ_0^2 (or equivalently, β_0), μ_n always depends increasingly more on $\overline{x}_n$ than μ_0 when n is large, meaning that for a sufficiently large sample, the choice of prior does not significantly affect the posterior estimate of μ.

Let us illustrate this claim using a concrete example. We generate a random sample of size n from $\mathcal{N}(2, 1)$, and perform Bayesian inference on the mean $\mu = 2$ with a prior distribution of $\mathcal{N}(-1, 1)$ by pretending that we did not know the mean. Figure 1.4 shows the posterior distributions of μ, and the posterior mean and variance are denoted by μ_n and σ_n^2, respectively. We can observe that the posterior distribution of the mean detaches from the prior distribution and gets closer to the true value of 2 as n increases.

(II) Mathematical connection between MLE and Bayesian inference

We next provide another mathematical perspective to justify the connection between Bayesian inference and MLE in general. Recall the posterior distribution formula (1.15):

$$f_{\mu|x}(\mu|x) = \frac{1}{C} f(x|\mu)\pi(\mu).$$

For the common choice of the prior distribution $\pi(\mu)$ as a constant function independent of μ (which can be made well-posed by restricting it to compact domain), we

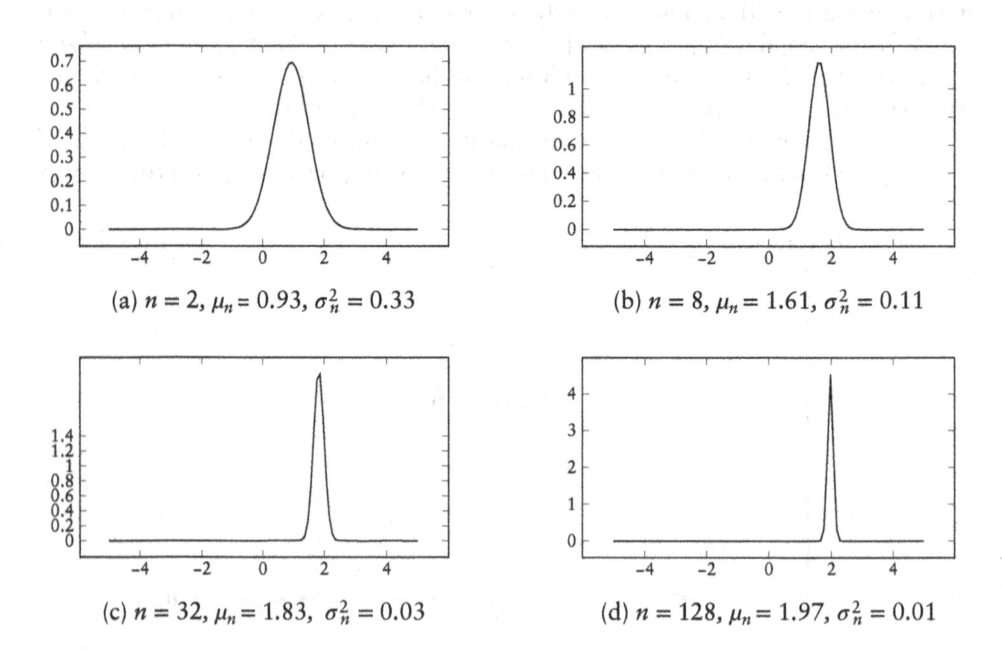

FIGURE 1.4 Posterior distributions of μ with different values of n.

have $f(\mu|x) \propto f(x|\mu)$, and the optimal point μ maximizes both the posterior distribution $f(\mu|x)$ and the likelihood function $f(x|\mu)$ simultaneously, which implies agreement between MLE and Bayesian inference. A similar interpretation can be drawn if the likelihood function $f(x|\mu)$ "dominates" the prior distribution $\pi(\mu)$ especially when the domain for μ is bounded. This observation serves as another justification for considering Bayesian inference to be in some sense an extension of MLE.

(III) Inference on the Sampling Distribution

Besides the parameters of the distribution of **x**, sometimes we are also interested in estimating the distribution itself. Following the previous example, with the knowledge of μ and σ^2, the distribution of x is $\mathcal{N}(\mu, \sigma^2)$. However, the actual value of μ is unknown, hence we have to make use of the posterior distribution $f_{\mu|\mathbf{x}}(\mu|x)$ to describe the distribution of x, under the assumption that σ^2 is known.

By virtue of Bayes' theorem, the predictive distribution of x_{n+1} given the sample $x_1, \ldots, x_n$ is simply the conditional distribution of the feature observation x_{n+1} given the sample, so it is the weighted average of the conditional distribution $\mathcal{N}(\mu, \sigma^2)$ of **x** with weights $f_{\mu|\mathbf{x}}(\mu|x)$, i.e.,

$$\hat{f}_{\mathrm{x}}(x) := f_{\mathrm{x}_{n+1}|\mathbf{x}}(x|x) = \int_{-\infty}^{\infty} f_{\mu|\mathbf{x}}(\mu|x)\, f_{\mathcal{N}}(x|\mu, \sigma^2)\, d\mu, \tag{1.20}$$

where $f_{\mathcal{N}}$ is the normal density with a mean μ and variance σ^2. Recalling that $\mu|\mathbf{x} = x \sim \mathcal{N}(\mu_n, \beta_n^{-1})$, (1.20) becomes:

$$\hat{f}_{\mathrm{x}}(x) \propto \int_{-\infty}^{\infty} \exp\left(-\frac{\beta_n}{2}(\mu - \mu_n)^2 - \frac{\beta}{2}(x - \mu)^2\right) d\mu, \tag{1.21}$$

where the exponent of the integrand can be rearranged as:

$$\begin{aligned}
&-\frac{\beta_n}{2}(\mu - \mu_n)^2 - \frac{\beta}{2}(x - \mu)^2 \\
&= -\frac{\beta_n + \beta}{2}\mu^2 + (\beta_n\mu_n + \beta x)\mu - \frac{\beta_n\mu_n^2 + \beta x^2}{2} \\
&= -\frac{\beta_n + \beta}{2}\left(\mu - \frac{\beta_n\mu_n + \beta x}{\beta_n + \beta}\right)^2 + \frac{(\beta_n\mu_n + \beta x)^2}{2(\beta_n + \beta)} - \frac{\beta_n\mu_n^2 + \beta x^2}{2}.
\end{aligned} \tag{1.22}$$

Therefore, we have

$$\begin{aligned}
\hat{f}_{\mathrm{x}}(x) \propto \exp&\left(\frac{(\beta_n\mu_n + \beta x)^2}{2(\beta_n + \beta)} - \frac{\beta_n\mu_n^2 + \beta x^2}{2}\right) \\
&\cdot \int_{-\infty}^{\infty} \exp\left(-\frac{\beta_n + \beta}{2}\left(\mu - \frac{\beta_n\mu_n + \beta x}{\beta_n + \beta}\right)^2\right) d\mu.
\end{aligned} \tag{1.23}$$

Note that the exponent of the first term on the right-hand side can be rewritten as:

$$\frac{(\beta_n\mu_n + \beta x)^2}{2(\beta_n + \beta)} - \frac{(\beta_n\mu_n^2 + \beta x^2)}{2} = -\frac{(x-\mu_n)^2}{2(\beta^{-1} + \beta_n^{-1})} + c, \tag{1.24}$$

where the constant c is independent of x, while the integrand in the second term is just proportional to a normal density with mean $\frac{\beta_n\mu_n+\beta x}{\beta_n+\beta}$ and variance $\frac{1}{\beta_n+\beta}$, hence

$$\int_{-\infty}^{\infty} \exp\left(-\frac{\beta_n+\beta}{2}\left(\mu - \frac{\beta_n\mu_n+\beta x}{\beta_n+\beta}\right)^2\right) d\mu = \sqrt{\frac{2\pi}{\beta_n+\beta}}.$$

Altogether this yields

$$\hat{f}_{\mathrm{x}}(x) \propto \exp\left(-\frac{(x-\mu_n)^2}{2(\beta^{-1}+\beta_n^{-1})}\right),$$

which implies that the predictive distribution is $\mathcal{N}(\mu_n, \beta^{-1}+\beta_n^{-1})$.

We observe that an additional factor β_n^{-1} appears in the variance of this predictive distribution compared with its original conditional distribution $\mathcal{N}(\mu, \sigma^2)$ given (μ, σ^2); this reflects the additional uncertainty regarding the convergence of μ_n to μ. Recall that β_n^{-1} is the posterior variance of μ given $\mathbf{x} = x$, which decreases to zero with the increase of the sample size n, and so does the variance of this predictive distribution. That means as $n \to \infty$, $\beta_n^{-1} \to 0$, and the variance of x reduces to $\beta^{-1} = \sigma^2$, which echoes the convergence of μ_n towards the true value of μ. Altogether this implies that the predictive distribution of x_{n+1} converges to the true distribution as $n \to \infty$.

As an illustration, we revisit the example in Figure 1.4, where the samples are drawn from $\mathcal{N}(2, 1)$, and we perform Bayesian inference on the mean with the prior distribution $\mathcal{N}(-1, 1)$. In Figure 1.5, we plot the predictive distribution of x_{n+1} for various values of sample size n, namely $\mathcal{N}(\mu_n, s_n^2 := \beta^{-1} + \beta_n^{-1})$ (in solid lines), with the true distribution $\mathcal{N}(2, 1)$ (in dashed line), and clearly the former converges to the latter as n increases.

(IV) Normal Distribution with Unknown Variance

In Bayesian inference, we can choose virtually any arbitrary distribution as the prior $\pi(\cdot)$. However, in practice, we usually prefer to choose $\pi(\mu)$ that could simplify the expression of the posterior distribution (1.15). One common choice can be warranted by the concept of *conjugate prior* for the likelihood function, so that the prior distribution itself and its corresponding posterior one belong to the same family of distribution. For instance, the prior distribution of $\mu \sim \mathcal{N}(\mu_0, \sigma_0^2)$ introduced earlier in the section is a conjugate prior for the likelihood function of a normally distributed sample, since the posterior $\mu|\mathbf{x} = x \sim \mathcal{N}(\mu_n, \beta_n^{-1})$ is also a normal distribution.

Let us illustrate the technique by studying the extended case when both parameters μ and σ^2 of the normal distribution are unknown. For notational simplicity,

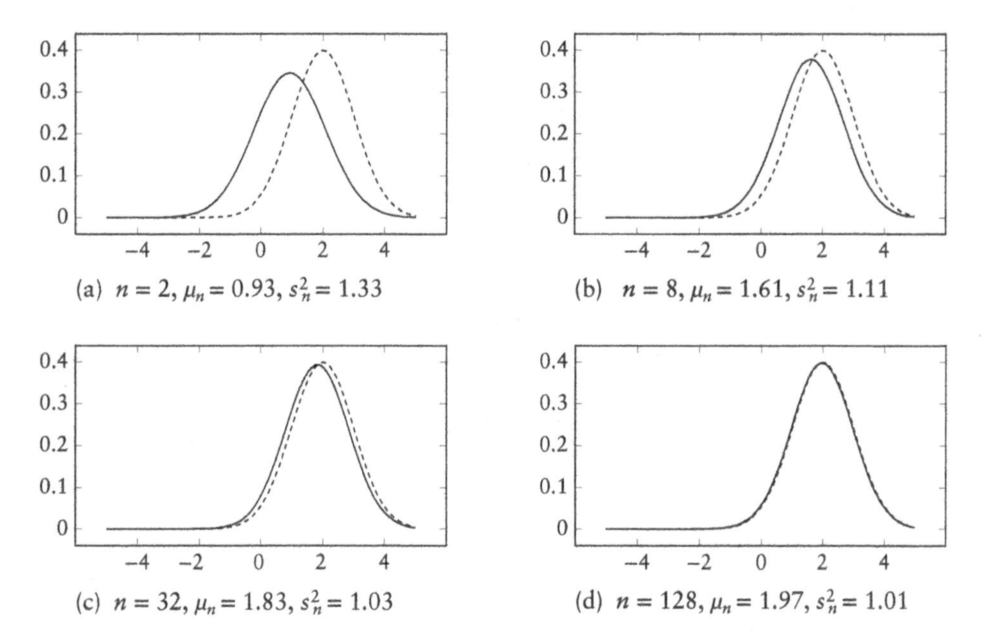

(a) $n = 2, \mu_n = 0.93, s_n^2 = 1.33$ (b) $n = 8, \mu_n = 1.61, s_n^2 = 1.11$

(c) $n = 32, \mu_n = 1.83, s_n^2 = 1.03$ (d) $n = 128, \mu_n = 1.97, s_n^2 = 1.01$

FIGURE 1.5 Predictive distributions of x_{n+1} for various values of n via Bayesian inference.

we use the precision $\lambda = \sigma^{-2}$ instead of variance, then we have $\mathrm{x}|(\upmu = \mu, \uplambda = \lambda) \sim \mathcal{N}(\mu, \lambda^{-1})$. With observations $\boldsymbol{x} = (x_1, \ldots, x_n)$, the likelihood function reads:

$$f(\boldsymbol{x}|\mu, \lambda) = \prod_{i=1}^{n} f_{\mathrm{x}|\upmu,\uplambda}(x_i|\mu, \lambda), \tag{1.25}$$

where $f_{\mathrm{x}|\upmu,\uplambda}(x_i|\mu, \lambda) = \sqrt{\frac{\lambda}{2\pi}} \exp\left(-\frac{\lambda(x_i-\mu)^2}{2}\right)$.

We next consider the *normal-gamma distribution*, denoted by NormalGamma $\left(\mu_0, \beta_0, \frac{\beta_0}{2} + 1, b\right)$, with the following joint probability density function:

$$\pi(\mu, \lambda) = \frac{b^{\frac{\beta_0}{2}+1}}{\Gamma\left(\frac{\beta_0}{2}+1\right)\sqrt{2\pi}} \lambda^{\frac{\beta_0}{2}} e^{-b\lambda} \sqrt{\beta_0 \lambda} \exp\left(-\frac{\beta_0 \lambda (\mu - \mu_0)^2}{2}\right). \tag{1.26}$$

It is actually the joint distribution of $(\upmu, \uplambda)$, where $\upmu|\uplambda = \lambda \sim \mathcal{N}(\mu_0, \frac{1}{\beta_0 \lambda})$ and $\uplambda \sim \Gamma\left(\frac{\beta_0}{2} + 1, b\right)$ with the density proportional to $\lambda^{\frac{\beta_0}{2}} e^{-b\lambda}$.[3] For the rest of this section, we aim to show that a normal-gamma distribution is a conjugate prior for the normal

[3] Equivalently, the variance σ^2 is said to follow an *inverse gamma* distribution.

likelihood by deriving the posterior distribution; also see [2, 6]. By (1.25) and (1.26), the posterior density is given by:

$$f(\mu, \lambda|\boldsymbol{x}) \propto (\beta_0\lambda)^{\frac{1}{2}} \lambda^{\frac{\beta_0+n}{2}} e^{-b\lambda} \exp\left(-\frac{\lambda}{2}\sum_{i=1}^{n}(x_i-\mu)^2 - \frac{\beta_0\lambda}{2}(\mu-\mu_0)^2\right).$$

Recall that

$$\sum_{i=1}^{n}(x_i-\mu)^2 = \sum_{i=1}^{n}(x_i-\overline{x}_n)^2 + n(\overline{x}_n-\mu)^2 =: nV + n(\overline{x}_n-\mu)^2,$$

where $V = \frac{\sum_{i=1}^{n}(x_i-\overline{x}_n)^2}{n}$. Hence, we can rewrite:

$$\begin{aligned} f(\mu, \lambda|\boldsymbol{x}) &\propto \lambda^{\frac{\beta_0+n+1}{2}} e^{-b\lambda} \exp\left(-\frac{\beta_0\lambda}{2}(\mu-\mu_0)^2 - \frac{\lambda}{2}(nV + n(\overline{x}_n-\mu)^2)\right) \\ &= \lambda^{\frac{\beta_0+n+1}{2}} \exp\left(-\lambda\left(b+\frac{nV}{2}\right)\right) \cdot \exp\left(-\frac{\beta_0\lambda}{2}(\mu-\mu_0)^2 - \frac{n\lambda}{2}(\overline{x}_n-\mu)^2\right). \end{aligned} \tag{1.27}$$

Upon rearrangement, we note that:

$$\begin{aligned} &\frac{\beta_0\lambda}{2}(\mu-\mu_0)^2 + \frac{n\lambda}{2}(\overline{x}_n-\mu)^2 \\ &= \frac{(\beta_0+n)\lambda}{2}\left(\mu^2 - 2\mu\frac{\beta_0\mu_0+n\overline{x}_n}{\beta_0+n} + \left(\frac{\beta_0\mu_0+n\overline{x}_n}{\beta_0+n}\right)^2\right) \\ &\quad - \frac{(\beta_0+n)\lambda}{2}\left(\frac{\beta_0\mu_0+n\overline{x}_n}{\beta_0+n}\right)^2 + \frac{\beta_0\lambda}{2}\mu_0^2 + \frac{n\lambda}{2}\overline{x}_n^2 \\ &= \frac{(\beta_0+n)\lambda}{2}\left(\mu - \frac{\beta_0\mu_0+n\overline{x}_n}{\beta_0+n}\right)^2 + \frac{n\beta_0\lambda}{2(\beta_0+n)}(\overline{x}_n-\mu_0)^2. \end{aligned} \tag{1.28}$$

Then substituting (1.28) into (1.27) yields:

$$\begin{aligned} f(\mu, \lambda|\boldsymbol{x}) &\propto \lambda^{\frac{\beta_0+n}{2}} \exp\left(-\lambda\left(b+\frac{nV}{2}+\frac{n\beta_0}{2(\beta_0+n)}(\overline{x}_n-\mu_0)^2\right)\right) \\ &\quad \cdot \sqrt{(\beta_0+n)\lambda}\exp\left(-\frac{(\beta_0+n)\lambda}{2}\left(\mu-\frac{\beta_0\mu_0+n\overline{x}_n}{\beta_0+n}\right)^2\right), \end{aligned} \tag{1.29}$$

which is the joint probability density of a NormalGamma($\mu^*, \beta^*, \frac{\beta^*}{2}+1, b^*$) distribution, with

$$\mu^* := \frac{\beta_0\mu_0+n\overline{x}_n}{\beta_0+n}, \quad \beta^* := \beta_0+n, \quad b^* := b+\frac{nV}{2}+\frac{n\beta_0}{2(\beta_0+n)}(\overline{x}_n-\mu_0)^2;$$

in other words, $\mu|(\lambda, \mathbf{x}) = (\lambda, x) \sim \mathcal{N}\left(\mu^*, \frac{1}{\beta^*\lambda}\right)$ and $\lambda|\mathbf{x} = x \sim \Gamma\left(\frac{\beta^*}{2} + 1, b^*\right)$. This confirms that the NormalGamma$(\mu_0, \beta_0, \frac{\beta_0}{2} + 1, b)$ distribution is indeed a conjugate prior for the likelihood function.

In addition, we provide the predictive distribution of x_{n+1} by using:

$$\hat{f}_{\mathrm{x}}(x) = \int_0^{\infty} \int_{-\infty}^{\infty} f(\mu, \lambda|\mathbf{x}) f_{\mathrm{x}|\mu,\lambda}(x|\mu, \lambda) d\mu d\lambda.$$

Recall that the conditional density of x_{n+1} given μ, λ takes the following form:

$$f(x|\mu, \lambda) \propto \lambda^{\frac{1}{2}} \exp\left(-\frac{\lambda}{2}(x - \mu)^2\right), \tag{1.30}$$

while by (1.29), the posterior density is:

$$f(\mu, \lambda|\mathbf{x}) \propto \lambda^{\frac{\beta^*}{2}} e^{-b^*\lambda} \sqrt{\beta^*\lambda} \exp\left(-\frac{\beta^*\lambda}{2}(\mu - \mu^*)^2\right). \tag{1.31}$$

We then integrate the product of (1.30) and (1.31), first with respect to μ and then with respect to λ. By some standard yet tedious calculations, we get:

$$\hat{f}_{\mathrm{x}}(x) \propto \left(\frac{1}{b^* + \frac{\beta^*}{2(\beta^*+1)}(x - \mu^*)^2}\right)^{\frac{\beta^*+1}{2}+1},$$

which is the predictive probability density of a location-scale t-distribution; in other words, predictively $\mathrm{x}_{n+1} = \mu^* + \tau \mathrm{t}$, where $\mathrm{t} \sim t_\nu$ with $\nu = \beta^* + 2$, and $\tau = \sqrt{\frac{2b^*}{\beta^*}}$.

The model above can also be extended to high-dimensional settings as follows. Assume that given the p-dimensional mean vector $\boldsymbol{\mu}$ and the $p \times p$ covariance matrix $\boldsymbol{\Sigma}$, the random vector $\mathbf{x} = (\mathrm{x}_1, \ldots, \mathrm{x}_p)^\top$ follows a multivariate normal distribution. Via the use of the *precision matrix* $\mathrm{P} = \boldsymbol{\Sigma}^{-1}$, we have $\mathbf{x} \mid (\boldsymbol{\mu} = \mu, \mathrm{P} = P) \sim \mathcal{N}_p(\mu, P^{-1})$. It can be analogously shown that the conjugate prior in this scenario becomes the multivariate extension of the normal-gamma distribution, namely the *normal-Wishart distribution*, denoted by $\mathrm{NW}(\mu_0, \lambda, \boldsymbol{\Psi}, \nu)$, which is the joint distribution of $(\boldsymbol{\mu}, \mathrm{P})$ with $\boldsymbol{\mu} \mid \mathrm{P} = P \sim \mathcal{N}_p(\mu_0, (\lambda P)^{-1})$ and $\mathrm{P} \sim W_p(\boldsymbol{\Psi}, \nu)$; also see [13, 16] for more discussion.

We conclude this section with a remark on the practical use of normal-Wishart distributions in actuarial science. In credibility theory, they have been widely adopted in the formulation of hierarchical models, such as models for severity per claim, which helps to determine the pure premium by multiplying the estimated mean annual claim frequency and the estimated mean severity per claim to estimate the mean of the annual aggregate loss. In addition, these models can also facilitate an efficient estimation of the variance of each individual, leading to a justifiable pricing of the risk premium for the policyholders; see [12] for example. Meanwhile, in the rise of the trendy *evolutionary credibility theory*, which allows for time-varying mean and variance, the incorporation of temporal structures is inevitable for enhancing the practical value of the normal-Wishart model. A possible resolution has been proposed in a more recent study [3] under a semi-parametric evolutionary setting, allowing simultaneous estimation of both mean and variance of the policyholder at each time point based on his/her past claim history; this formulation includes the class of static normal-Wishart models above as an interesting special case, and can hence be regarded as its viable extension with the presence of a temporal dependence structure. We anticipate more research developments to emerge along this direction.

REFERENCES

1. Anderson, T.W. (1962). *An Introduction to Multivariate Statistical Analysis*. Wiley.
2. Carlin, B., and Louis, T. (2008). *Bayesian Methods for Data Analysis* (3rd ed.). Chapman & Hall/CRC.
3. Chen, Y., Cheung, K.C., Choi, H.M.C., and Yam, S.C.P. (2020). Evolutionary credibility risk premium. *Insurance: Mathematics and Economics*, 93, 216–229.
4. Chen, Y., Fan, N.S., and Yam, S.C.P. (2024+). *Statistical Deep Learning with Python and R*. Preprint.
5. Deisenroth, M.P., Faisal, A.A., and Ong, C.S. (2020). *Mathematics for Machine Learning*. Cambridge University Press.
6. Gelman, A., Carlin, J.B., Stern, H.S., Dunson, D.B., Vehtari, A., and Rubin, D.B. (2013). *Bayesian Data Analysis* (3rd ed.). Chapman & Hall/CRC.
7. Ghosh, M., and Sinha, B.K. (2002). A simple derivation of the Wishart distribution. *The American Statistician*, 56(2), 100–101.
8. Golub, G.H., and Van Loan, C.F. (2013). *Matrix Computations*. JHU Press.
9. Goodfellow, I., Bengio, Y., and Courville, A. (2016). *Deep Learning*. MIT Press.
10. Hogg, R.V., McKean, J.W., and Craig, A.T. (2019). *Introduction to Mathematical Statistics*. Pearson.
11. Horn, R.A., and Johnson, C.R. (2012). *Matrix Analysis*. Cambridge University Press.
12. Jing, B.Y., Li, Z., Pan, G., and Zhou, W. (2016). On SURE-type double shrinkage estimation. *Journal of the American Statistical Association*, 111(516), 1696–1704.
13. Johnson, R.A., and Wichern, D.W. (2002). *Applied Multivariate Statistical Analysis*. Prentice Hall.
14. Lai, T.L., and Xing, H. (2008). *Statistical Models and Methods for Financial Markets*. Springer.
15. McLachlan, G.J. (1999). Mahalanobis distance. *Resonance*, 4(6), 20–26.
16. Morrison, D.F. (1990). *Multivariate Statistical Methods*. McGraw-Hill.
17. Rencher, A.C., and Christensen, W.F. (2012) *Methods of Multivariate Analysis* (3rd ed.). Wiley.
18. Strang, G. (2006). *Linear Algebra and its Applications*. Thomson, Brooks/Cole.
19. Strang, G. (2019). *Linear Algebra and Learning from Data*. Wellesley-Cambridge Press.
20. Watkins, D.S. (2004). *Fundamentals of Matrix Computations*. Wiley.
21. Zhang, X.D. (2017). *Matrix Analysis and Applications*. Cambridge University Press.

CHAPTER 2

Introduction to Python and R

2.1 WHAT IS PYTHON?

Python is used *very* widely as a scripting language. For instance, *Pixar*[1] uses it in all their productions, making computer-animated movies such as *Toy Story 3*. Python bears some resemblance to Java, but generally is cleaner and easier to read. The name "Python" came from a British comedy group called *Monty Python*[2], reflecting the goal of Python's developers – keep the programme funny.

In Python, we can enjoy the following features:

1. No need to declare the types of variables. For example, we write `x = 1` in Python compared to `int x = 1` in `C++`. You can first write `x = 1` (an integer), then `x = "statistics"` (a string) in the later parts of the same programme;
2. The environment, in which you work, lets you type bits of code and allows you to see what happens without an intermediate compiling step, making experimentation and testing very simple;
3. It is very easy to read, and uses English keywords and natural-sounding syntax, for instance,

   Example

```
1 x = 1
2 if x > 0:
3     print("x is positive")
4     print("Time to begin with machine learning!")
```

4. Instead of using curly braces to delimit code blocks, Python uses (for instance) 4 whitespace indentations (in fact any number of indentations is allowed as long as the practice is consistent throughout the programme). Correctly nesting your code now has a semantic meaning.

[1] Pixar Animation Studios, a subsidiary of *The Walt Disney Company*. Company website: https://www.pixar.com/.

[2] "General Python FAQ". Python v2.7.3 documentation. Docs.python.org. Retrieved 4 June 2020.

Python is similar to Java in the sense that it also handles your memory management: it allocates memory for different uses, and has the capacity to free up that memory once it is no longer needed. In the rest of this book, the Python coding shown is consistent with version 3.9 of Python.

2.2 WHAT IS R?

R is a free and integrated programming language for statistical computing and graphical display. It is widely used among probabilists, statisticians and data scientists. Some elementary statistical tests and functions are already built in **R**, such that no extra libraries have to be called out. The name **R**[3] partially came after the first names of the first two **R** creators: Ross Ihaka and Robert Gentleman; another reason is to acknowledge **S**, a common programming language used in statistics that precedes **R**.

2.3 PACKAGE MANAGEMENT IN PYTHON AND R

Python and **R** are two very popular programming languages for machine learning and deep learning developers, and there are hundreds of packages and libraries available online.

In Python, one cannot install the packages directly in the console. Using *Anaconda*[4], one of the most popular data science platforms, we first locate *Anaconda Prompt* in the computer machine. If you are using the Microsoft Windows computer system, it can be searched for in the *Windows Search Bar*. Then, we install the packages in the *Anaconda Prompt* by typing `pip install tensorflow` as in Figure 2.1.

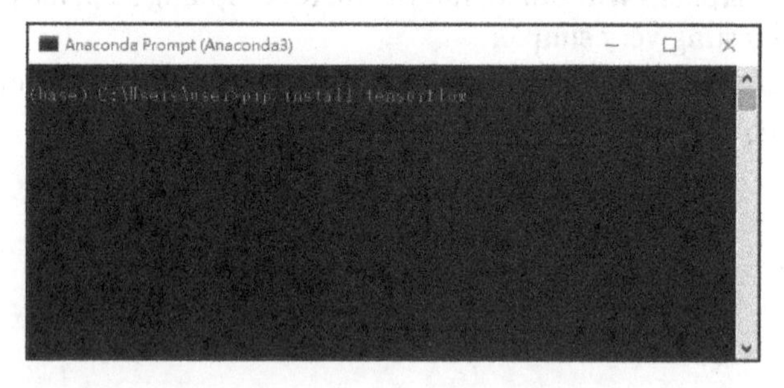

FIGURE 2.1 Typing "pip install tensorflow" in *Anaconda prompt.*

[3] Kurt Hornik. The **R** FAQ: Why **R**? https://cran.r-project.org/doc/manuals/R-FAQ.pdf. (accessed 12 January 2021)

[4] *Anaconda* is a distribution mainly for Python and **R**, it consists of different *Integrated Development Environments* (IDEs), including *Spyder*, *Jupyter Notebook*, *RStudio*, *Visual Studio Code*. We can imagine *Anaconda* as a computer system between Microsoft Windows, Mac OS, or Linux, and the IDEs as the web browsers, let say *Chrome*, *Microsoft Edge*, or *Safari*; these computer systems provide us a platform to search online, in which we can choose different web browsers. We can install *Anaconda* from https://www.anaconda.com/.

Here `pip` is used to install the Python related packages. There is another command `conda`, which is used to install packages written in any languages available in *Anaconda*. After the installation, we can import the package in the console by typing:

```
1 import tensorflow as tf
```

Here `tf` is used as a self-defined, short-handed notation for `tensorflow`. The tensorflow package is widely used for deep learning. It has an advanced version that utilises the full power of *Graphics Processing Unit* (GPU), which is called the `tensorflow-gpu`. It significantly reduces the training time of a deep learning model, but it requires a decent GPU and additional installations of software *Compute Unified Device Architecture* (CUDA) and *CUDA Deep Neural Network library* (cuDNN). Moreover, the importation of `tensorflow-gpu` is achieved by the command "`import tensorflow as tf`"; GPU will then automatically be used instead of CPU. We can check our CPU and GPU by:

```
1 from tensorflow.python.client import device_lib
2 print(device_lib.list_local_devices())
```

```
1  [name: "/device:CPU:0"
2  device_type: "CPU"
3  memory_limit: 268435456
4  locality {
5  }
6  incarnation: 14792078817090895925
7  , name: "/device:GPU:0"
8  device_type: "GPU"
9  memory_limit: 7020285133
10 locality {
11     bus_id: 1
12     links {
13     }
14 }
15 incarnation: 13190594769463994O8
16 physical_device_desc: "device: 0, name: GeForce RTX 2070 SUPER,
       pci bus id: 0000:01:00.0, compute capability: 7.5"
17 ]
```

Here "`GeForce RTX 2070 SUPER`" is the GPU used and the reference index for this GPU is 0. If there exists another GPU in use, the reference index for the second GPU is 1, and so on. The memory limits for CPU and GPU are shown in bytes, and we see that, for instance, the limit of the present GPU is 30 times more than that of CPU.

Meanwhile, In **R**, the installation and the importation of packages are done directly in the console:

```
1 > install.packages("e1071")
2 > library(e1071)
```

Here e1071 is one of the packages for building a *support vector machine* (SVM) model in R. This SVM model will be discussed in detail in Section 11.2.

2.4 BASIC OPERATIONS IN PYTHON AND R

In this section, some basic operations in Python and R which are essential for statistical computing are briefly described. For more comprehensive treatments of these two languages, we refer the reader to [2, 4, 5, 6]. The dataset fin-ratio.csv (see Example 1.1) is used as an illustration for the first and the second items:

(I) **Read and Write csv/txt Files:**

csv (Comma Separated Values) files or txt (TeXT) files are two common file types for data storage. In Python, we need the pandas (deriving its name from PANel DAta, *i.e.* observations over multiple time periods for the same object) library:

```
1 import pandas as pd
2 
3 # Read csv in Python
4 data = pd.read_csv('fin-ratio.csv')
5 # Write csv in Python
6 data.to_csv("fin-ratio_new.csv")
```

Programme 2.1 Read and write csv files in Python.

Note that data must be a pandas.DataFrame object in order to use the to_csv() function. In R, there is no need to import any libraries for this action.

```
1 > # Read csv in R
2 > data <- read.csv("fin-ratio.csv")
3 > # Write csv in R
4 > write.csv(data, "fin-ratio_new.csv")
```

Programme 2.2 Read and write csv files in R.

(II) **Computations of Sample Mean, Sample Variance, and Sample Covariance Matrix:**

In Python, the pandas library have built-in functions for sample means, sample variances, and sample covariances.

```
1 # Assign data to x except the label y = HSI stock or not
2 x = data.drop(data.columns[-1], axis=1)
3 # data.columns[-1] = HSI, axis=1 means column (axis=0 for row)
4 # Compute sample means and sample variances
5 print(x.mean())
6 print(x.var())
7 
8 # Compute sample covariance matrix
9 print(x.cov())
```

```
1  EY        -0.650240
2  CFTP      -0.233896
3  ln_MV      6.266807
4  DY         2.496174
5  BTME       1.908263
6  DTE        0.709732
7  dtype: float64
8  EY        18.497907
9  CFTP       3.693061
10 ln_MV      2.743936
11 DY        13.871563
12 BTME      68.308197
13 DTE       12.992907
14 dtype: float64
15                EY      CFTP     ln_MV         DY       BTME        DTE
16 EY      18.497907  2.908964  1.160189   1.920377   1.478128   0.337953
17 CFTP     2.908964  3.693061  0.766300   1.237147   1.822839   0.328791
18 ln_MV    1.160189  0.766300  2.743936   0.972071  -0.773423  -0.074132
19 DY       1.920377  1.237147  0.972071  13.871563  -0.257534   0.158153
20 BTME     1.478128  1.822839 -0.773423  -0.257534  68.308197   1.961765
21 DTE      0.337953  0.328791 -0.074132   0.158153   1.961765  12.992907
```

Programme 2.3 Basic statistics in Python.

In R, there is no need to import any libraries for this action.

```
1  > # Assign data to x except the label y = HSI stock or not
2  > x <- data[-ncol(data)]
3  > # Compute sample means and sample variances for each column
4  > apply(x, MARGIN=2, FUN=mean)
5          EY       CFTP      ln_MV         DY       BTME        DTE
6  -0.6502403 -0.2338956  6.2668068  2.4961735  1.9082626  0.7097322
7  > apply(x, MARGIN=2, FUN=var)
8         EY      CFTP     ln_MV        DY      BTME       DTE
9  18.497907  3.693061  2.743936 13.871563 68.308197 12.992907
10 >
11 > # Compute sample covariance matrix
12 > cov(x)
13               EY      CFTP       ln_MV         DY       BTME         DTE
14 EY     18.497907 2.9089644  1.16018856  1.9203766  1.4781279  0.33795296
15 CFTP    2.908964 3.6930613  0.76629950  1.2371466  1.8228390  0.32879079
16 ln_MV   1.160189 0.7662995  2.74393616  0.9720714 -0.7734227 -0.07413221
17 DY      1.920377 1.2371466  0.97207145 13.8715626 -0.2575337  0.15815280
18 BTME    1.478128 1.8228390 -0.77342270 -0.2575337 68.3081966  1.96176520
19 DTE     0.337953 0.3287908 -0.07413221  0.1581528  1.9617652 12.99290723
```

Programme 2.4 Basic statistics in R.

Here the `apply()` function "applies" a function specified by the parameter `FUN` by returning a vector along the dimension as specified by the parameter `MARGIN`, in which the argument `MARGIN=2` indicates the function is applied on the second dimension, *i.e.* along the *column* direction.

(III) **Building a Mathematical Function:**

In Python, we need to import the `math` library or the `numpy` (stands for NUMerical PYthon) library for the `exp` operation for taking exponents.

```
1 import numpy as np
2
3 def MyFun(x):
4     return np.exp(x)
```

Programme 2.5 A simple exponential function `MyFun()` saved as `My_Function.py` in Python.

Within the same folder, we can call back our `MyFun()` function in another file with the following command:

```
1 from My_Function import MyFun
2
3 MyFun(2)
```

```
1 7.38905609893065                    # = e^2
```

Programme 2.6 Calling the function `MyFun()` saved in Programme 2.5 in Python.

In **R**, there is no need to import any libraries for the `exp` operation for taking exponents. Nevertheless, let us build a simple function as follows:

```
1 > MyFun <- function(x){
2 +     return (exp(x))
3 + }
```

Programme 2.7 A simple exponential function `MyFun()` saved as `My_Function.R` in **R**.

Similarly, within the same folder, we can call back our `MyFun()` function in another file with the following command:

```
1 > source("My_Function.R")
2 >
3 > MyFun(2)
4 [1] 7.389056                    # = e^2
```

Programme 2.8 Calling the function `MyFun()` saved in Programme 2.7 in **R**.

For both Python and **R**, due to the limited space for display, the printed results can only show a few significant figures, while in the calculations themselves, more significant figures are used. Actually, Python uses more digits in various calculations than **R**.

(IV) **Recursion Operation – For-Loop:**

In Python, any loop iteration starts from zero:

```
1 # For loop in Python
2 for i in range(5):
3     print(i)
```

```
0
1
2
3
4
```

Programme 2.9 A simple "for-loop" in Python.

However, in **R**, any loop iteration starts from 1:

```
> # For loop in R
> for (i in 1:5){
+     print(i)
+ }
[1] 1
[1] 2
[1] 3
[1] 4
[1] 5
```

Programme 2.10 A simple "for-loop" in **R**.

Similarly, all indices in Python start at 0 while indices in **R** start at 1 as seen above. Therefore, additional attention has to be taken when translating indexing between Python and **R**.

(V) **Help Function:**

In Python, you can call `help(function)`. Moreover, in *Spyder*, one of the most popular programming platforms for the Python language, one can just place the cursor over the function to see documentation (See Figure 2.2). In **R**, you can always call `help(function)` or `?function` for helpful documentation about the function. For example, `help(mean)` or `?mean` provides the documentation of the mean function.

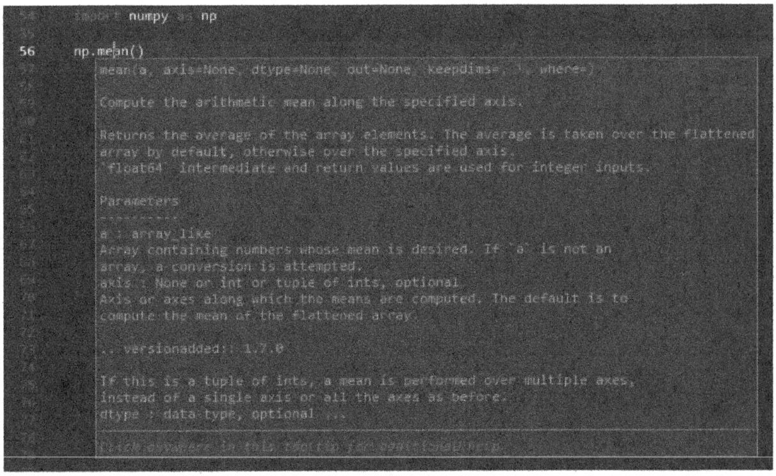

FIGURE 2.2 Function documentation can be shown when the cursor is placed over that function.

(VI) Matrix Computation:

In Python, we need to import the numpy library for matrix operations as shown below:

```
1  import numpy as np
2
3  A = np.array([1, 3, 5, 2, 6, 4, 2, 3, 1]).reshape(3,3)
4  B = np.array([3, 1, 2, 4, 2, 8, 1, 3, 1]).reshape(3,3)
5  print(A)
6  print(B)
7
8  # Matrix Multiplication
9  print(np.dot(A, B))
10 print(A @ B)
11
12 # Hadamard Product
13 print(np.multiply(A, B))
14 print(A * B)
15
16 # Inverse
17 print(np.linalg.inv(A))          # LINear ALGebra
```

```
1  [[1 3 5]
2   [2 6 4]
3   [2 3 1]]
4
5  [[3 1 2]
6   [4 2 8]
7   [1 3 1]]
8
9  [[20 22 31]
10  [34 26 56]
11  [19 11 29]]
12
13 [[20 22 31]
14  [34 26 56]
15  [19 11 29]]
16
17 [[ 3  3 10]
18  [ 8 12 32]
19  [ 2  9  1]]
20
21 [[ 3  3 10]
22  [ 8 12 32]
23  [ 2  9  1]]
24
25 [[ 0.33333333 -0.66666667  1.        ]
26  [-0.33333333  0.5        -0.33333333]
27  [ 0.33333333 -0.16666667  0.        ]]
```

Programme 2.11 Simple matrix operations in Python (line breaks are added for easy identification; the similar practice is adopted throughout the book where appropriate).

In R, we do not need to import any libraries to conduct basic matrix manipulation. Some simple illustrations are shown below.

```
1  > (A <- matrix(c(1, 3, 5, 2, 6, 4, 2, 3, 1), nrow=3, byrow=T))
2       [,1] [,2] [,3]
3  [1,]    1    3    5
4  [2,]    2    6    4
5  [3,]    2    3    1
6  > (B <- matrix(c(3, 1, 2, 4, 2, 8, 1, 3, 1), nrow=3, byrow=T))
7       [,1] [,2] [,3]
8  [1,]    3    1    2
9  [2,]    4    2    8
10 [3,]    1    3    1
11 >
12 > A%*%B                                  # Matrix Multiplication
13      [,1] [,2] [,3]
14 [1,]   20   22   31
15 [2,]   34   26   56
16 [3,]   19   11   29
17 > A * B                                    # Hadamard Product
18      [,1] [,2] [,3]
19 [1,]    3    3   10
20 [2,]    8   12   32
21 [3,]    2    9    1
22 > solve(A)                                 # Inverse
23            [,1]       [,2]       [,3]
24 [1,]  0.3333333 -0.6666667  1.0000000
25 [2,] -0.3333333  0.5000000 -0.3333333
26 [3,]  0.3333333 -0.1666667  0.0000000
```

Programme 2.12 Simple matrix operations in R.

2.5 ONE-WAY ANOVA AND TUKEY'S HSD FOR STOCK MARKET INDICES

One-way analysis of variance (ANOVA) is used to compare the means of $|C| := |\{c_1, \ldots, c_K\}| \geq 2$ categorical and independent groups of data with only one continuous feature (independent) variable[5]. It requires the assumption that all observations are normally and independently distributed with a common variance. Statistically, let $\mu_1, \ldots, \mu_{|C|}$ be the respective population means of these $|C|$ categorical groups governed by the only continuous feature variable $\mathrm{x}^{(1)}$. The null hypothesis of one-way ANOVA is:

$$H_0 : \mu_1 = \mu_2 = \cdots = \mu_{|C|} = \mu,$$

[5] For two or more continuous independent variables, this ANOVA analysis is extended to one-way *multivariate analysis of variance* (MANOVA); see more details in [7].

where, as always, we reject the null hypothesis by claiming that the population means are not all equal if the p-value of the test statistics is smaller than 0.05. However, if the null hypothesis is rejected, one-way ANOVA cannot provide us with much knowledge on which population mean(s) μ_k('s) is (are) significantly different from the others. [8] proposed the Tukey's HSD (*Honest Significant Difference*) which considers all possible pairs of population means.

Suppose that for $k = 1, \ldots, |C|$, we draw an iid sample of size N_k from the k-th group following a normal distribution with the mean μ_k and the variance σ^2 that is common across groups, while we also assume mutual independence among groups. The pooled variance of the combined sample is defined as $S_p^2 := \frac{\sum_{k=1}^{|C|}(N_k-1)S_k^2}{\sum_{k=1}^{|C|}(N_k-1)}$, where S_k^2 is the sample variance of the group k. To test the new null hypothesis that two chosen groups i, j have the same means, Tukey's method uses the following t-statistic for this pairwise comparison:

$$q_{\text{Tukey},ij} = \frac{\overline{x}_i^{(1)} - \overline{x}_j^{(1)}}{S_p\sqrt{\frac{1}{N_i} + \frac{1}{N_j}}}, \quad \text{for all } i \neq j, \tag{2.1}$$

where $\overline{x}_i^{(1)}$ and $\overline{x}_j^{(1)}$ are the sample means of the groups i and j, respectively.

In the following, for any pair $i \neq j$, the statistic (2.1) is shown to follow a t-distribution under H_0. By the iid normal nature of the sample within a group, for each group $k = 1, \ldots, |C|$, the sample mean follows a normal distribution, i.e. $\overline{\mathrm{x}}_k^{(1)} \sim \mathcal{N}(\mu, \frac{\sigma^2}{N_k})$, and the scaled sample variance follows a chi-square distribution, i.e., $\frac{(N_k-1)S_k^2}{\sigma^2} \sim \chi^2_{N_k-1}$. The independence of these two statistics is warranted by the normality assumption. Furthermore, by the independence of samples across groups, the difference between the respective sample means of the groups i and j is also normally distributed:

$$\overline{\mathrm{x}}_i^{(1)} - \overline{\mathrm{x}}_j^{(1)} \sim \mathcal{N}\left(0, \sigma^2\left(\frac{1}{N_i} + \frac{1}{N_j}\right)\right);$$

as for the pooled variance, its distribution can be obtained by the additive property of independent chi-square distributions, where $N := \sum_{k=1}^{|C|} N_k$ denotes the total sample size:

$$\frac{(N - |C|)S_p^2}{\sigma^2} = \sum_{k=1}^{|C|} \frac{(N_k - 1)S_k^2}{\sigma^2} \sim \chi^2_{N-|C|}.$$

The independence between the sample mean and sample variance in each group further indicates the independence of $\overline{\mathrm{x}}_i^{(1)} - \overline{\mathrm{x}}_j^{(1)}$ and S_p^2. Altogether this allows us to construct the following statistic that follows a t-distribution with $N - |C|$ degrees of freedom under H_0:

$$\mathrm{t}_{ij} := \frac{\frac{\overline{\mathrm{x}}_i^{(1)} - \overline{\mathrm{x}}_j^{(1)}}{\sigma\sqrt{\frac{1}{N_i}+\frac{1}{N_j}}}}{\sqrt{\frac{(N-|C|)S_p^2}{\sigma^2}\frac{1}{N-|C|}}} = \frac{\overline{\mathrm{x}}_i^{(1)} - \overline{\mathrm{x}}_j^{(1)}}{\mathrm{S}_p\sqrt{\frac{1}{N_i}+\frac{1}{N_j}}},$$

which is the same form as (2.1), thus establishing our desired claim.

As a remark, the pooled variance S_p^2 can also be rearranged as follows:

$$S_p^2 = \frac{\sum_{k=1}^{|C|}(N_k - 1)S_k^2}{\sum_{k=1}^{|C|}(N_k - 1)} = \frac{\sum_{k=1}^{|C|}(N_k - 1)\cdot\frac{1}{N_k-1}\sum_{\substack{1\le n\le N\\ c(x_n^{(1)})=k}}(x_n^{(1)} - \overline{x}_k^{(1)})^2}{N - |C|}$$

$$= \frac{\sum_{k=1}^{|C|}\sum_{\substack{1\le n\le N\\ c(x_n^{(1)})=k}}(x_n^{(1)} - \overline{x}_k^{(1)})^2}{N - |C|},$$

where $c(x_n^{(1)})$ denotes the corresponding group index of observation $x_n^{(1)}, n = 1, \ldots, N$. In particular, the numerator is sometimes called the *within-group sum of squared errors* (within-group SSE):

$$\mathrm{SSE}_{\mathrm{within}} := \sum_{k=1}^{|C|}\sum_{\substack{1\le n\le N\\ c(x_n^{(1)})=k}}(x_n^{(1)} - \overline{x}_k^{(1)})^2.$$

Then Tukey's HSD statistic can also be written as:

$$q_{\mathrm{Tukey},ij} = \frac{\overline{x}_i^{(1)} - \overline{x}_j^{(1)}}{\sqrt{\left(\frac{1}{N_i}+\frac{1}{N_j}\right)\frac{\mathrm{SSE}_{\mathrm{within}}}{N-|C|}}}, \quad \text{for all } i \neq j. \tag{2.2}$$

The common practice for performing the Tukey HSD test is on one chosen pair of groups at a time, which we shall follow in our discussions below. In contrast, the test can also be conducted for all pairs with a total of $\binom{|C|}{2}$ Tukey's HSD statistics in a simultaneous manner. To this end, we carry out simultaneous pairwise comparisons with the corresponding t-distributions to conclude if a difference between any two group means is significantly different from 0. Moreover, recalling that each t_{ij} has a common t-distribution with $N - |C|$ degrees of freedom and they are still dependent on each other as a group, then the upper bound of the p-value of the simultaneous test can be obtained using the celebrated *Bonferroni inequality*:

$$p = \mathbb{P}\left(\bigcup_{i,j}\{|\mathrm{t}_{ij}| > q_{ij}\}\right) < \sum_{i,j}\mathbb{P}(|\mathrm{t}_{ij}| > q_{ij}) =: \sum_{i,j}p_{ij},$$

where p_{ij} is the p-value of the single pairwise comparison of the groups i and j. Then we reject the null hypothesis of simultaneous equality of pairwise means across groups at a significant level α if $\sum_{i,j} p_{ij} < \alpha$.

Next, we shall illustrate the applications of one-way ANOVA and Tukey HSD for three stock market indices, namely the S&P500, HSI, and FTSE via both Python and R; note that for this particular example, we shall prioritize on elaborating the implementation in R, supplemented by their Python counterparts. The data files `SPX.csv`, `HSI.csv`, and `FTSE.csv` are the adjusted closing prices of S&P500, HSI, and FTSE respectively, dating from 4 January 1984 to 26 April 2021. In particular, two time periods of data will be analyzed, namely (i) from 2005 to 2007; and (ii) from 2018 to 2020.

```
1  > SPX <- read.csv("SPX.csv", header=TRUE, row.names=1)
2  > HSI <- read.csv("HSI.csv", header=TRUE, row.names=1)
3  > FTSE <- read.csv("FTSE.csv", header=TRUE, row.names=1)
4  >
5  > SPX$Year <- format(as.Date(rownames(SPX), format="%d/%m/%Y"), "%Y")
6  > HSI$Year <- format(as.Date(rownames(HSI), format="%d/%m/%Y"), "%Y")
7  > FTSE$Year <- format(as.Date(rownames(FTSE), format="%d/%m/%Y"), "%Y")
8  >
9  > SPX$log_return <- c(NA, diff(log(SPX$Price)))
10 > HSI$log_return <- c(NA, diff(log(HSI$Price)))
11 > FTSE$log_return <- c(NA, diff(log(FTSE$Price)))
12 >
13 > # Remove first day of data for log return
14 > SPX <- SPX[-1,]
15 > HSI <- HSI[-1,]
16 > FTSE <- FTSE[-1,]
17 >
18 > Remove_Outlier <- function(index, outlier_factor=1.5){
19 +     q25 <- quantile(index$log_return, probs=.25, na.rm=FALSE)
20 +     q75 <- quantile(index$log_return, probs=.75, na.rm=FALSE)
21 +     iqr <- q75 - q25  #Inter-quartile range
22 +     lower_bound <- q25 - outlier_factor*iqr
23 +     upper_bound <- q75 + outlier_factor*iqr
24 +     pos <- index$log_return>lower_bound & index$log_return<upper_bound
25 +     return (index[pos,])
26 + }
27 >
28 > # Hyperparameter
29 > outlier_removal <- TRUE
30 > if (outlier_removal){
31 +     SPX <- Remove_Outlier(SPX)
32 +     HSI <- Remove_Outlier(HSI)
33 +     FTSE <- Remove_Outlier(FTSE)
34 + }
```

Programme 2.13 Cleansing of datasets of stock market indices for one-way ANOVA and Tukey HSD in R.

For Programme 2.13 in **R**, `as.Date()` function converts the date character string from the index name of the dataframe into a datetime object, in which the argument `format="%d/%m/%Y"` indicates that the input string has a date format of (two digits) / month (two digits) / year (four digits). After that, `format()` is immediately applied with an argument `"%Y"` indicating that only the year is returned, and this is augmented into the corresponding stock market index dataframe in a new column named as `Year`. The logarithmic return, i.e., the logarithm of the ratio of consecutive prices, is computed by the `log()` function, then taking the difference of consecutive terms via the `diff()` function. Note that the logarithmic return on the first day (4 January 1984) is unavailable because we do not have the adjusted closing price on the previous day. Therefore, we supplement at the beginning of the list of logarithmic return figures with `NA`, and we place it in the new column named `log_return`. We then remove the data on the very first day by using `[-1,]`.

For the `Remove_Outlier()` function (defined in **R** in Programme 2.13 or in Python in Programme 2.14), we used the convention that any data points which are lower than the lower quartile minus 1.5 times the inter-quartile range or greater than the upper quartile plus 1.5 times the inter-quartile range are regarded as outliers. If each stock market index follows a standard normal distribution, then the 25% (resp. 75%) quantile represents a quantile value of −0.6745 (resp. 0.6745) under $\mathcal{N}(0,1)$. With a multiplier of 1.5 times of the interquartile range as the `outlier_factor`, the `lower_bound` (resp. `upper_bound`) represents the quantile value $-0.6745 + 1.5 \cdot (0.6745 - (-0.6745)) = -2.6960$ (resp. 2.6960). Therefore, under $\mathcal{N}(0,1)$, we are selecting the middle portion of $\Phi(2.6960) - \Phi(-2.6960) = 0.9930$, where $\Phi(\cdot)$ is the *cumulative distribution function* (cdf) of a standard normal distribution. Therefore, if the stock market index follows a heavier-tailed distribution than $\mathcal{N}(0,1)$, then the outliers account for roughly more than 1% of the dataset.

The equivalent code in Python is shown below:

```
1  import pandas as pd
2  import numpy as np
3  import matplotlib.pyplot as plt
4
5  plt.rc('font', size=20); plt.rc('axes', titlesize=30, labelsize=30)
6  plt.rc('xtick', labelsize=25); plt.rc('ytick', labelsize=25)
7
8  SPX = pd.read_csv("SPX.csv", index_col=0)
9  HSI = pd.read_csv("HSI.csv", index_col=0)
10 FTSE = pd.read_csv("FTSE.csv", index_col=0)
11
12 SPX["Year"] = pd.to_datetime(SPX.index.values).year
13 HSI["Year"] = pd.to_datetime(HSI.index.values).year
14 FTSE["Year"] = pd.to_datetime(FTSE.index.values).year
15
16 SPX["log_return"] = np.log(SPX.Price) - np.log(SPX.Price.shift(1))
17 HSI["log_return"] = np.log(HSI.Price) - np.log(HSI.Price.shift(1))
18 FTSE["log_return"] = np.log(FTSE.Price) - np.log(FTSE.Price.shift(1))
19
20 SPX, HSI, FTSE = SPX.iloc[1:], HSI.iloc[1:], FTSE.iloc[1:]
21
```

```
22 def Remove_Outlier(index, outlier_factor=1.5):
23     q25 = np.quantile(index.log_return, q=0.25)
24     q75 = np.quantile(index.log_return, q=0.75)
25     iqr = q75 - q25    # Inter-quartile range
26         lower_bound = q25 - outlier_factor*iqr
27     upper_bound = q75 + outlier_factor*iqr
28     boolean = (lower_bound<index.log_return) & \
29         (index.log_return<upper_bound)
30     return index[boolean]
31
32 # Hyperparameter
33 outlier_removal = True
34 if outlier_removal:
35     SPX = Remove_Outlier(SPX)
36     HSI = Remove_Outlier(HSI)
37     FTSE = Remove_Outlier(FTSE)
```

Programme 2.14 Cleansing of datasets of stock market indices for one-way ANOVA and Tukey HSD in Python.

Now, we extract the data from 2005 to 2007 and plot a histogram for the logarithmic returns of each cleansed stock market index. Programme 2.15 contains the **R** code; Programme 2.16 the Python.

```
35 > start_year <- 2005
36 > end_year (<- 2007
37 > year_name <- paste0(" Year: ", start_year, "-", end_year)
38 > Chosen_Year <- start_year:end_year
39 > Chosen_SPX <- SPX[SPX$Year%in%Chosen_Year,]
40 > Chosen_HSI <- HSI[HSI$Year%in%Chosen_Year,]
41 > Chosen_FTSE <- FTSE[FTSE$Year%in%Chosen_Year,]
42 >
43 > par(cex.lab=2.5, cex.axis=2, cex.main=2.5, mar=c(5,5,4,4))
44 > hist(Chosen_SPX$log_return, breaks=20, xlab="SPX",
45 +      main=paste0("SPX daily log-return.", year_name))
46 > hist(Chosen_HSI$log_return, breaks=20, xlab="HSI",
47 +      main=paste0("HSI daily log-return.", year_name))
48 > hist(Chosen_FTSE$log_return, breaks=20, xlab="FTSE",
49 +      main=paste0("FTSE daily log-return.", year_name))
```

Programme 2.15 Continuation of Programme 2.13: plotting histograms for stock market indices from 2005 to 2007 in **R**.

Through various statistical tests (such as the Kolmogorov–Smirnov test) to be introduced in Chapter 3, we shall see in general that the logarithmic returns of stock market indices barely follow a normal distribution, except the central portion during fair weather days. Nevertheless, Figure 2.3 and Figure 2.5, produced via **R** and Python respectively, show that the histograms of the daily logarithmic returns for all three stock market indices having a bell-shaped distribution. Despite the lack of normality, as long as the underlying distribution is symmetric enough and light-tailed, the celebrated Central Limit Theorem can still warrant that the corresponding sample mean of log-returns of each index closely follows a normal distribution even with

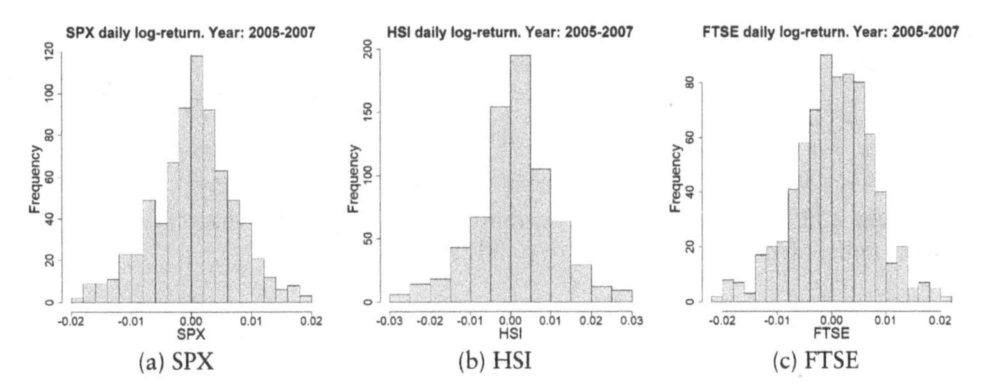

(a) SPX (b) HSI (c) FTSE

FIGURE 2.3 Histograms of the daily logarithmic returns from 2005 to 2007 in **R**.

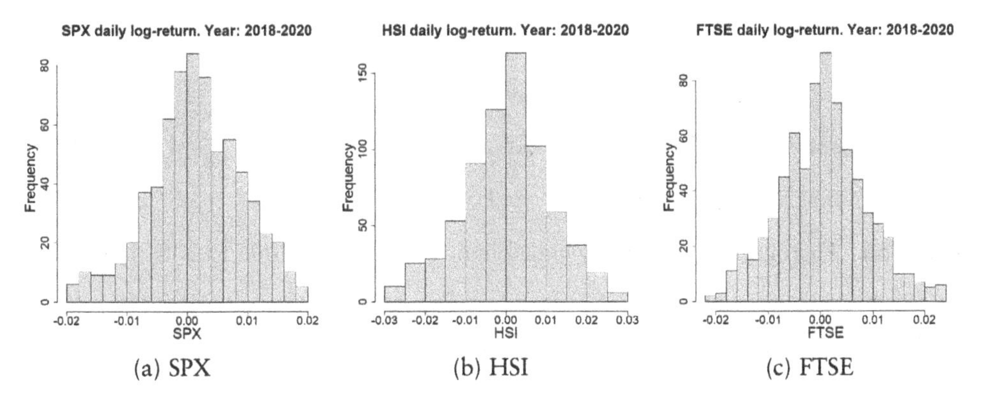

(a) SPX (b) HSI (c) FTSE

FIGURE 2.4 Histograms of the daily logarithmic returns from 2018 to 2020 in **R**.

a small sample size, thanks to the Berry–Esseen bound [1, 3]. Therefore, the results of one-way ANOVA and Tukey HSD tests are still robust enough to cover the present comparative study.

Figure 2.4 shows the results (from **R**) of replicating Programme 2.15, but this time over a different time period, 2018 to 2020, via `start_year <- 2018` and `end_year <- 2020`.

We have the same illustration via Python codes in Programme 2.16 with histograms shown in Figure 2.5 for the period 2005 to 2007 and Figure 2.6 for 2018 to 2020, respectively.

```
38 start_year = 2005    # also for 2018
39 end_year = 2007      # also for 2020
40 year_name = " Year: " + str(start_year) + "-" + str(end_year)
41 Chosen_Year = range(start_year, end_year + 1)
42 Chosen_SPX = SPX[SPX.Year.isin(Chosen_Year)]
43 Chosen_HSI = HSI[HSI.Year.isin(Chosen_Year)]
44 Chosen_FTSE = FTSE[FTSE.Year.isin(Chosen_Year)]
```

```
45
46 for idx, Chosen_idx in zip(["SPX", "HSI", "FTSE"],
47                             [Chosen_SPX,Chosen_HSI,Chosen_FTSE]):
48     fig, ax = plt.subplots(1, 1, figsize=(10,10))
49     ax.set_title(f"{idx} daily log-return." + year_name)
50     ax.hist(Chosen_idx.log_return, bins="sturges", ec='black'
```

Programme 2.16 Continuation of Programme 2.14: plotting histograms for stock market indices from 2005 to 2007 (resp. 2018 to 2020) in Python.

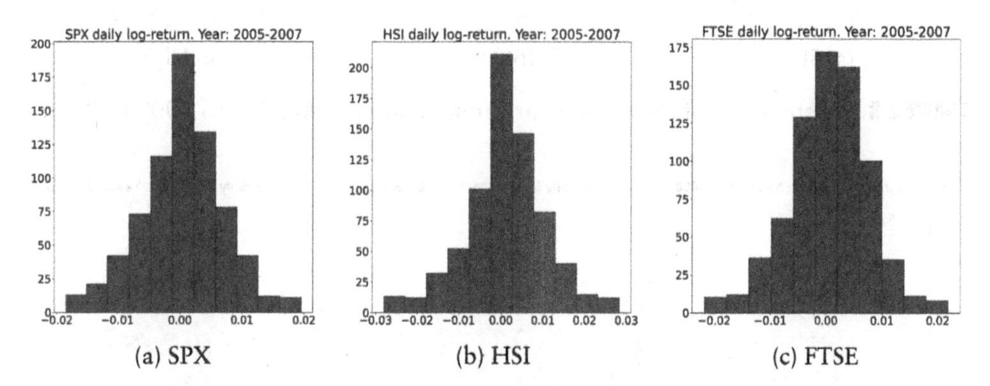

(a) SPX (b) HSI (c) FTSE

FIGURE 2.5 Histograms of the daily logarithmic returns from 2005 to 2007 in Python.

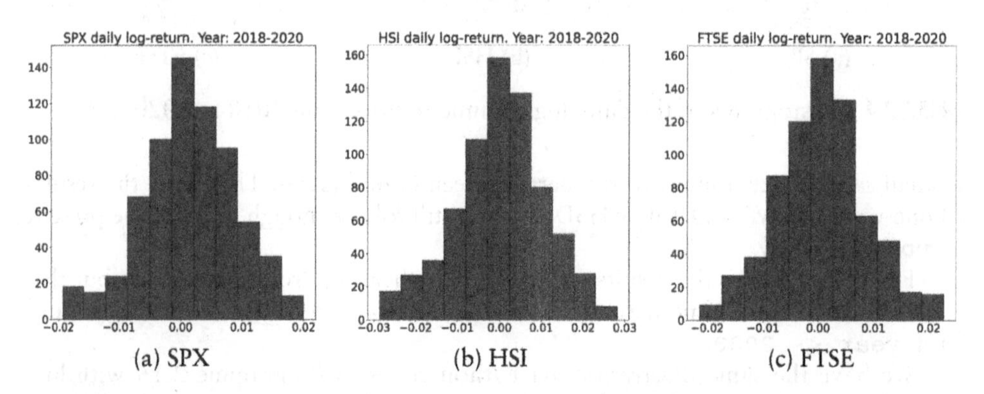

(a) SPX (b) HSI (c) FTSE

FIGURE 2.6 Histograms of the daily logarithmic returns from 2018 to 2020 in Python.

Next, in Programme 2.17, we perform both one-way ANOVA and Tukey HSD tests in **R** for the period 2005 to 2007. The `aov()` function is used for performing one-way ANOVA with (i) a group variable `Index` which consists of three categories `SPX`, `HSI`, and `FTSE`; and (ii) a numerical response variable `log_return` for the daily logarithmic returns of the corresponding market index. The relatively large

p-value (0.152 > 0.05) of the ANOVA test indicates that we fail to reject the null hypothesis of the equality of all means of μ_{SPX}, μ_{FTSE} and μ_{HSI} over the period 2005 to 2007. This claim is further supported by the large p-values of the Tukey HSD tests on all three possible pairs of stock market indices via the `TukeyHSD()` function in R, as there are three stock market indices, and there are $\binom{3}{2} = 3$ possible pairs. More graphical illustrations can be seen in Figure 2.7.

```
50  > boxplot(Chosen_SPX$log_return, Chosen_HSI$log_return,
51  +         Chosen_FTSE$log_return, names=c("SPX", "HSI", "FTSE"),
52  +         xlab="Index", ylab="Daily log-return", frame=FALSE,
53  +         col=c("#00AFBB", "#E7B800", "#FC4E07"), boxwex=0.75,
54  +         main=paste0("Boxplot.", year_name))
55  >
56  > Chosen_SPX$Index <- "SPX"
57  > Chosen_HSI$Index <- "HSI"
58  > Chosen_FTSE$Index <- "FTSE"
59  > AllIndex <- rbind(Chosen_SPX, Chosen_HSI, Chosen_FTSE)
60  > AllIndex$Index <- factor(AllIndex$Index, c("SPX", "HSI", "FTSE"))
61  >
62  > library("gplots")
63  > plotmeans(log_return~Index, data=AllIndex,
64  +           xlab="Index", ylab="Daily log-return",
65  +           main=paste0("Mean Plot with 95% CI.", year_name))
66  >
67  > anova.test <- aov(log_return~Index, data=AllIndex)
68  > summary(anova.test)
69               Df  Sum Sq   Mean Sq F value Pr(>F)
70  Index         2 0.00023 1.166e-04   1.887  0.152
71  Residuals  2184 0.13495 6.179e-05
72  > # Reject the null hypothesis?
73  > summary(anova.test)[[1]][[1, "Pr(>F)"]] < 0.05
74  [1] FALSE
75  >
76  > (tukey.test <- TukeyHSD(anova.test))
77    Tukey multiple comparisons of means
78      95% family-wise confidence level
79
80  Fit: aov(formula = log_return ~ Index, data = AllIndex)
81
82  $Index
83                    diff           lwr          upr     p adj
84  HSI-SPX   0.0007591526 -0.0002091937 0.0017274989 0.1573120
85  FTSE-SPX  0.0001519578 -0.0008093795 0.0011132951 0.9270513
86  FTSE-HSI -0.0006071949 -0.0015745675 0.0003601777 0.3045921
87
88  > plot(tukey.test)
```

Programme 2.17 Continuation of Programme 2.15: One-way ANOVA and Tukey HSD for three stock market indices from 2005 to 2007 in R.

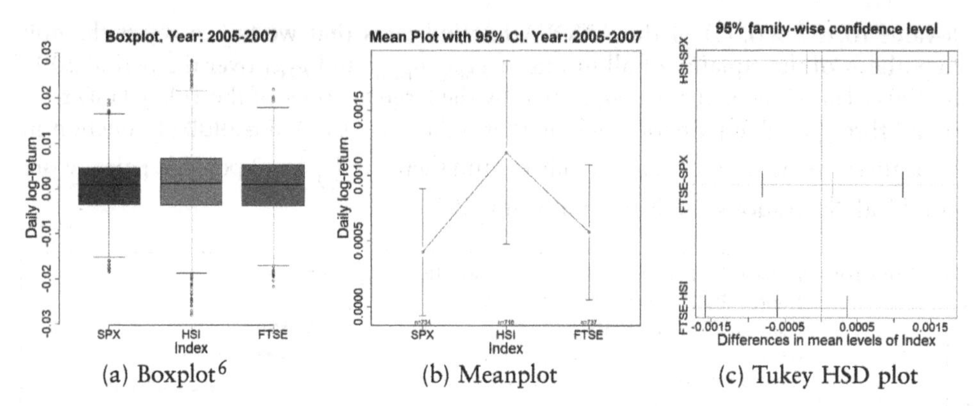

(a) Boxplot[6] (b) Meanplot (c) Tukey HSD plot

FIGURE 2.7 Boxplot, meanplot, and Tukey HSD plot for three stock market indices from 2005 to 2007 in **R**.

However, over the period 2018 to 2020, using Programme 2.18 (written in **R**), the small p-value ($0.019 < 0.05$) from the ANOVA test indicates that we should reject the null hypothesis on the equality of all means, meaning that the corresponding population means μ_{SPX}, μ_{FTSE} and μ_{HSI} are not all equal. We further adopt the Tukey HSD tests to the three possible pairs of stock market indices to test which pair(s) of means are unequal. The corresponding small p-value ($0.0168939 \ll 0.05$) of the Tukey HSD test for `HSI` against `SPX` indicates that the population means for `HSI` and `SPX` are unlikely the same, i.e. $\mu_{\text{HSI}} \neq \mu_{\text{SPX}}$. Meanwhile, for the Tukey HSD test for `FTSE` against `HSI`, the large p-value ($0.7091423 \gg 0.05$) indicates that we do not reject the equality of the means $\mu_{\text{FTSE}} = \mu_{\text{HSI}}$; the analogous conclusion of $\mu_{\text{FTSE}} = \mu_{\text{SPX}}$ can also be drawn from the Tukey HSD test for `FTSE` against `SPX`, yet based on its relatively small p-value of 0.1226088 (> 0.05), it is more likely that the two means may be unequal. Indeed, the discrepancy between US and Hong Kong markets may be explained by the economic conflict between US and China throughout that period, which brought along uncertainty and concern on various prospects of the economic growth in Asia-Pacific region, including but not limited to tariff levels and trade negotiations. As a result, Chinese companies with a significant exposure to the US market were most severely affected, and the overall performance of the Hang Seng Index, where these companies played a crucial role, deviated inevitably from that of the S&P500 Index. We can also use the Sharpe ratio to study any possible differences between the performances of financial markets in different countries, for instance, see [10, 11].

[6] This name was coined by *John Tukey* possibly because of the box between the first and third quartiles in the plot; also see [9].

```
89                                   ⋮
90  > summary(anova.test)
91                   Df  Sum Sq   Mean Sq F value Pr(>F)
92  Index             2 0.00063 3.156e-04   3.972  0.019 *
93  Residuals      2117 0.16824 7.947e-05
94  ---
95  Signif. codes:  0 '***' 0.001 '**' 0.01 '*' 0.05 '.' 0.1 ' ' 1
96  > # Reject the null hypothesis?
97  > summary(anova.test)[[1]][[1, "Pr(>F)"]] < 0.05
98  [1] TRUE
99  >
100 > (tukey.test <- TukeyHSD(anova.test))
101   Tukey multiple comparisons of means
102     95% family-wise confidence level
103
104 Fit: aov(formula = log_return ~ Index, data = AllIndex)
105
106 $Index
107                    diff           lwr           upr     p adj
108 HSI-SPX  -0.0013055645 -0.0024218836 -0.0001892454 0.0168939
109 FTSE-SPX -0.0009336742 -0.0020511337  0.0001837853 0.1226088
110 FTSE-HSI  0.0003718903 -0.0007319837  0.0014757644 0.7091423
111
112 > plot(tukey.test)
```

Programme 2.18 Continuation of Programme 2.15: One-way ANOVA and Tukey HSD for three stock market indices from 2018 to 2020 in R.

From Figure 2.8b, the mean of the daily logarithmic returns for `SPX` is significantly higher than the corresponding means of `HSI`, which is in agreement with the result obtained from the Tukey HSD tests that $\mu_{\text{HSI}} \neq \mu_{\text{SPX}}$. Certainly, there are limitations to our present study. For instance, we assume the stationarity of both the

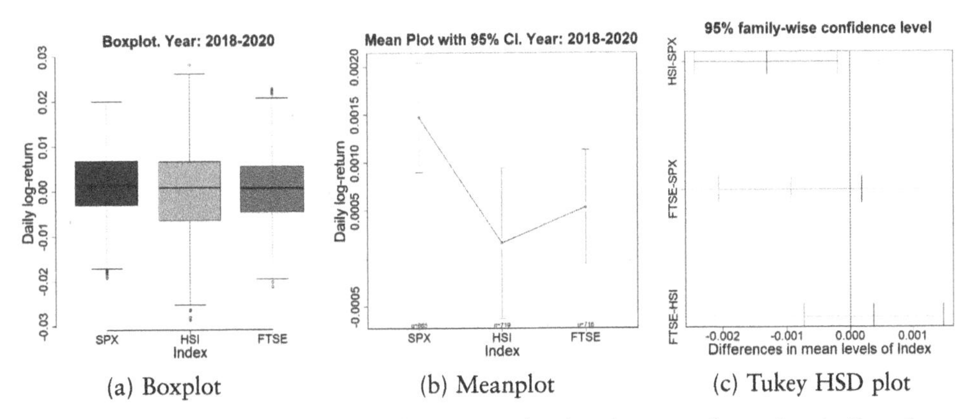

(a) Boxplot (b) Meanplot (c) Tukey HSD plot

FIGURE 2.8 Boxplot, meanplot, Tukey HSD plot for three stock market indices from 2018 to 2020 in R.

means and variance of log-returns over not too short periods. Since the economic environments over those periods were rather stable, one can argue this assumption is legitimate. Nevertheless, we shall introduce more sophisticated models allowing for more realistic assumptions later in Chapters 3, 7 and 8.

We also have the same tests, written in Python, shown in Programme 2.19 for the period 2005 to 2007 and Programme 2.20 for the period 2018 to 2020. The resulting plots are illustrated in Figures 2.9 and 2.10, respectively.

```
51  import seaborn as sns
52  import scipy.stats as stats
53  from statsmodels.stats.multicomp import MultiComparison
54
55  fig, ax = plt.subplots(1, 1, figsize=(11,10))
56  plt.subplots_adjust(wspace=0.4)       # horizontal spacing between plots
57  ax.set_title("Boxplot." + year_name)
58  ax.set_xlabel("Location")
59  ax.set_ylabel("Daily log-return")
60  ax.boxplot([Chosen_SPX.log_return, Chosen_HSI.log_return,
61              Chosen_FTSE.log_return],
62             widths=0.7, labels=["SPX", "HSI", "FTSE"])
63
64  Chosen_SPX.loc[:, "Index"] = "SPX"
65  Chosen_HSI.loc[:, "Index"] = "HSI"
66  Chosen_FTSE.loc[:, "Index"] = "FTSE"
67  AllIndex = pd.concat([Chosen_SPX, Chosen_HSI, Chosen_FTSE])
68
69  fig, ax = plt.subplots(1, 1, figsize=(11,10))
70  sns.pointplot(x='Index', y="log_return", data=AllIndex, ci=95, ax=ax)
71  ax.set_title("Mean Plot with 95% CI." + year_name)
72  ax.set_xlabel("Index")
73  ax.set_ylabel("Daily log-return")
74
75  import statsmodels.api as sm
76  from statsmodels.formula.api import ols
77
78  index_lm = ols('log_return ~ Index', data=AllIndex).fit()
79  print(sm.stats.anova_lm(index_lm, typ=2))
80
81  comp = MultiComparison(data=AllIndex.log_return, groups=AllIndex.Index,
82                         group_order=["SPX", "HSI", "FTSE"])
83  TurkeyHSD_result = comp.tukeyhsd()
84  print(TurkeyHSD_result)
85
86  fig, ax = plt.subplots(1, 1, figsize=(11,10))
87  TurkeyHSD_result.plot_simultaneous(figsize=(11, 10), ax=ax)
88  ax.set_xlabel("Daily log-return")
89  ax.set_ylabel("Index")
```

```
1                  sum_sq       df          F     PR(>F)
2  Index        0.000233       2.0   1.886774   0.151807
3  Residual     0.134945    2184.0        NaN        NaN
4  Multiple Comparison of Means - Tukey HSD, FWER=0.05
5  ====================================================
6  group1 group2 meandiff p-adj   lower  upper  reject
7  ----------------------------------------------------
8     SPX    HSI   0.0008 0.1573 -0.0002 0.0017  False
9     SPX   FTSE   0.0002 0.9271 -0.0008 0.0011  False
10    HSI   FTSE  -0.0006 0.3046 -0.0016 0.0004  False
11 ----------------------------------------------------
```

Programme 2.19 Continuation of Programme 2.16: One-way ANOVA and Tukey HSD for three stock market indices from 2005 to 2007 in Python.

```
1                  sum_sq       df          F     PR(>F)
2  Index        0.000631       2.0   3.971531   0.018985
3  Residual     0.168242    2117.0        NaN        NaN
4  Multiple Comparison of Means - Tukey HSD, FWER=0.05
5  =====================================================
6  group1 group2 meandiff p-adj   lower   upper  reject
7  -----------------------------------------------------
8     SPX    HSI  -0.0013 0.0169 -0.0024 -0.0002   True
9     SPX   FTSE  -0.0009 0.1226 -0.0021  0.0002  False
10    HSI   FTSE   0.0004 0.7091 -0.0007  0.0015  False
11 -----------------------------------------------------
```

Programme 2.20 Continuation of Programme 2.16: One-way ANOVA and Tukey HSD for three stock market indices from 2018 to 2020 in Python.

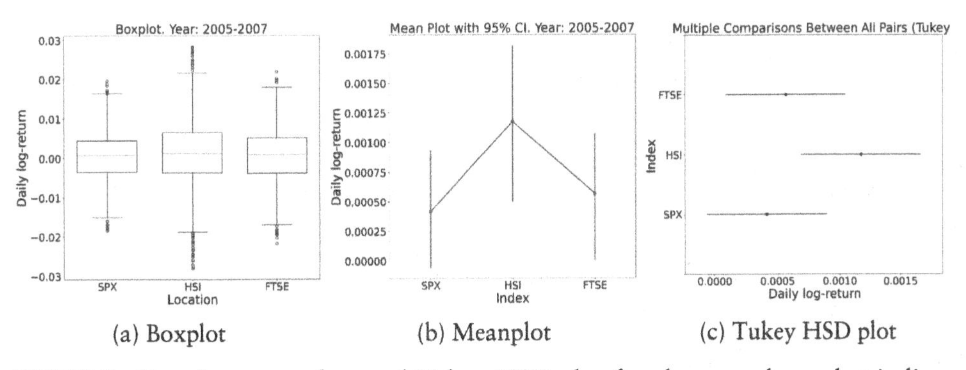

(a) Boxplot (b) Meanplot (c) Tukey HSD plot

FIGURE 2.9 Boxplot, meanplot, and Tukey HSD plot for three stock market indices from 2005 to 2007 in Python.

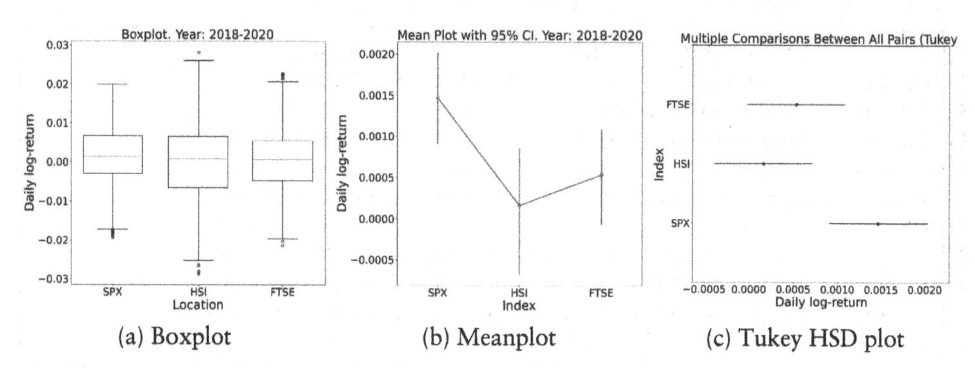

(a) Boxplot (b) Meanplot (c) Tukey HSD plot

FIGURE 2.10 Boxplot, meanplot, and Tukey HSD plot for three stock market indices from 2018 to 2020 in Python.

Lastly, in Programmes 2.21 and 2.22 via **R** (resp. Programmes 2.23 and 2.24 via Python), we replicate the result of `tukey.test` in **R** (resp. `tukeyhsd` in Python) based on (2.2). We first compute the individual sample means μ_{SPX}, μ_{FTSE}, μ_{HSI}, the individual sum of squared errors SSE_{SPX}, SSE_{FTSE}, SSE_{HSI}, and the total degrees of freedom ($N - |C| = N_{\text{SPX}} + N_{\text{FTSE}} + N_{\text{HSI}} - 3 = 2,184$ for the period 2005 to 2007 and $N - |C| = 2,117$ for the period 2018 to 2020). Then, we compute the within-group mean squared error defined as $\text{MSE}_{\text{within}} = \text{SSE}_{\text{within}}/(N - |C|) = (\text{SSE}_{\text{SPX}} + \text{SSE}_{\text{FTSE}} + \text{SSE}_{\text{HSI}})/(N - |C|)$. In both **R** and Python, $\text{MSE}_{\text{within}}$ is recorded in the summary table of the one-way ANOVA; in **R** for Programme 2.17 for the period 2005 to 2007 (resp. Programme 2.18 for the period 2018 to 2020), it is stated directly in the `Mean Sq` of `Residuals`; in Python for Programme 2.19 for the period 2005 to 2007 (resp. Programme 2.20 for the period 2018 to 2020), it can be computed by dividing the `sum_sq` by `df` of `Residual`. The `ptukey()` function in **R** (resp. the `psturng()` function in Python) computes the probability $\mathbb{P}(q_{\text{Tukey},ij} > q)$ of the Tukey HSD test given (i) the quantile value `q`; (ii) the number of groups `nmeans` in **R** (resp. `r` in Python); and the degrees of freedom `df` in **R** (resp. `v` in Python).

```
113 > mu_SPX <- mean(Chosen_SPX$log_return)
114 > mu_HSI <- mean(Chosen_HSI$log_return)
115 > mu_FTSE <- mean(Chosen_FTSE$log_return)
116 >
117 > SSE_SPX <- sum((Chosen_SPX$log_return - mu_SPX)^2)
118 > SSE_HSI <- sum((Chosen_HSI$log_return - mu_HSI)^2)
119 > SSE_FTSE <- sum((Chosen_FTSE$log_return - mu_FTSE)^2)
120 >
121 > n_SPX <- length(Chosen_SPX$log_return)
122 > n_HSI <- length(Chosen_HSI$log_return)
123 > n_FTSE <- length(Chosen_FTSE$log_return)
124 >
125 > (df <- length(AllIndex$log_return) - 3)
126 [1] 2184
127 > (Within_group_MSE <- (SSE_SPX + SSE_HSI + SSE_FTSE) / df)
128 [1] 6.178807e-05
129 >
```

```
130 > SPX_HSI_SE_ANOVA <- sqrt(Within_group_MSE * (1/n_SPX + 1/n_HSI))
131 > FTSE_SPX_SE_ANOVA <- sqrt(Within_group_MSE * (1/n_FTSE + 1/n_SPX))
132 > FTSE_HSI_SE_ANOVA <- sqrt(Within_group_MSE * (1/n_FTSE + 1/n_HSI))
133 >
134 > # q_Tukey = sqrt(2) t
135 > SPX_vs_HSI <- abs(mu_SPX - mu_HSI) / SPX_HSI_SE_ANOVA * sqrt(2)
136 > FTSE_vs_SPX <- abs(mu_FTSE - mu_SPX) / FTSE_SPX_SE_ANOVA * sqrt(2)
137 > FTSE_vs_HSI <- abs(mu_FTSE - mu_HSI) / FTSE_HSI_SE_ANOVA * sqrt(2)
138 >
139 > ptukey(q=SPX_vs_HSI, nmeans=3, df=df, lower.tail=FALSE)
140 [1] 0.157312
141 > ptukey(q=FTSE_vs_SPX, nmeans=3, df=df, lower.tail=FALSE)
142 [1] 0.9270513
143 > ptukey(q=FTSE_vs_HSI, nmeans=3, df=df, lower.tail=FALSE)
144 [1] 0.3045921
```

Programme 2.21 Continuation of Programme 2.17: Replication of the Tukey HSD tests for the period 2005 to 2007 in **R**.

```
145                              ⋮
146 > (df <- length(AllIndex$log_return) -3)
147 [1] 2117
148 > (Within_group_MSE <- (SSE_SPX + SSE_HSI + SSE_FTSE) / df)
149 [1] 7.947169e-05
150                              ⋮
151 > ptukey(q=SPX_vs_HSI, nmeans=3, df=df, lower.tail=FALSE)
152 [1] 0.01689388
153 > ptukey(q=FTSE_vs_SPX, nmeans=3, df=df, lower.tail=FALSE)
154 [1] 0.1226088
155 > ptukey(q=FTSE_vs_HSI, nmeans=3, df=df, lower.tail=FALSE)
156 [1] 0.7091423
```

Programme 2.22 Continuation of Programme 2.18: Replication of the Tukey HSD tests for the period 2018 to 2020 in **R**.

```
90  mu_SPX = np.mean(Chosen_SPX.log_return)
91  mu_HSI = np.mean(Chosen_HSI.log_return)
92  mu_FTSE = np.mean(Chosen_FTSE.log_return)
93
94  SSE_SPX = np.sum((Chosen_SPX.log_return - mu_SPX)**2)
95  SSE_HSI = np.sum((Chosen_HSI.log_return - mu_HSI)**2)
96  SSE_FTSE = np.sum((Chosen_FTSE.log_return - mu_FTSE)**2)
97
98  n_SPX = len(Chosen_SPX.log_return)
99  n_HSI = len(Chosen_HSI.log_return)
100 n_FTSE = len(Chosen_FTSE.log_return)
101
102 df = len(AllIndex.log_return) - 3 + 1 - 1        # df: degrees of freedom
103 print(df)
104
105 Within_group_MSE = (SSE_SPX + SSE_HSI + SSE_FTSE) / df
106 print(Within_group_MSE)
```

```
107
108 SPX_HSI_SE_ANOVA = np.sqrt(Within_group_MSE * (1/n_SPX + 1/n_HSI))
109 FTSE_SPX_SE_ANOVA = np.sqrt(Within_group_MSE * (1/n_FTSE + 1/n_SPX))
110 FTSE_HSI_SE_ANOVA = np.sqrt(Within_group_MSE * (1/n_FTSE + 1/n_HSI))
111
112 # q_Tukey = sqrt(2) t
113 SPX_vs_HSI = np.abs(mu_SPX - mu_HSI) / SPX_HSI_SE_ANOVA * np.sqrt(2)
114 FTSE_vs_SPX = np.abs(mu_FTSE - mu_SPX) / FTSE_SPX_SE_ANOVA * np.sqrt(2)
115 FTSE_vs_HSI = np.abs(mu_FTSE - mu_HSI) / FTSE_HSI_SE_ANOVA * np.sqrt(2)
116
117 from statsmodels.stats.libqsturng import psturng
118
119 print(psturng(q=SPX_vs_HSI, r=3, v=df))
120 print(psturng(q=FTSE_vs_SPX, r=3, v=df))
121 print(psturng(q=FTSE_vs_HSI, r=3, v=df))
```

```
1 2184
2 6.178807223949821e-05
3 [0.15748396]
4 0.9
5 [0.30513622]
```

Programme 2.23 Continuation of Programme 2.19: Replication of the Tukey HSD tests for the period 2005 to 2007 in Python.

```
1 2117
2 7.947169047304725e-05
3 [0.01689736]
4 [0.12277521]
5 [0.69161587]
```

Programme 2.24 Continuation of Programme 2.19: Replication of the Tukey HSD tests for the period 2018 to 2020 in Python.

Notice that the p-values of the Tukey HSD tests computed from `tukeyhsd` and `qsturng`, both provided by the `statsmodels` library in Python, are slightly different, no matter which time period (2005–2007 or 2018–2020). The difference in values is due to the inability of the solver `fminbound` used in `psturng` to locate the minimizer accurately. We therefore recommend readers to use the `tukeyhsd` function in Python directly for performing the Tukey HSD tests in practice.

REFERENCES

1. Berry, A.C. (1941). The accuracy of the Gaussian approximation to the sum of independent variates. *Transactions of the American Mathematical Society*, 49(1), 122–136.
2. Crawley, M.J. (2012). *The R Book*. Wiley.
3. Esseen, C.G. (1942). On the Liapunov limit error in the theory of probability. *Arkiv för matematik, astronomi och fysik*, 28, 1–19.

4. Kabacoff, R.I. (2022). *R in Action: Data Analysis and Graphics with R* (3rd ed.). Manning Publications Company.
5. Lutz, M. (2013). *Learning Python: Powerful Object-Oriented Programming*. O'Reilly Media, Inc.
6. McKinney, W. (2022). *Python for Data Analysis* (3rd ed.). O'Reilly Media, Inc.
7. Timm, N.H. (Ed.). (2002). *Applied Multivariate Analysis*. Springer.
8. Tukey, J.W. (1949). Comparing individual means in the analysis of variance. *Biometrics*, 5(2), 99–114.
9. Tukey, J.W. (1972). Some graphic and semigraphic displays. *Statistical Papers in Honor of George W. Snedecor*, 5, 293–316.
10. Wong, W.K., Wright, J.A., Yam, S.C.P., and Yung, S.P. (2012). A mixed Sharpe ratio. *Risk and Decision Analysis*, 3(1–2), 37–65.
11. Wright, J.A., Yam, S.C.P., and Yung, S.P. (2014). A test for the equality of multiple Sharpe ratios. *The Journal of Risk*, 16(4), 3.

CHAPTER 3

Statistical Diagnostics of Financial Data

The simple version of *Black–Scholes–Merton* model assumes that the returns of the stock on different days are iid normal, but in reality, with real securities data, we see that the normality assumption is usually invalid because of the nature of the fatter-tailed distribution of the returns. In the latter part of this chapter, we shall replace the normal distribution with other fatter-tailed distributions for the logarithm of the ratio of prices, and then investigate the validity of these new models.

3.1 NORMALITY ASSUMPTION FOR RELATIVE STOCK PRICE CHANGES

The basic assumption of the Black–Scholes–Merton model [1] is that the percentage change in the stock price over a short period of time of length δt is normally distributed; that means, approximately,

$$\delta S/S \sim \mathcal{N}(\mu\delta t, \sigma^2\delta t), \tag{3.1}$$

where δS is the change in the stock price S over the short time interval $(t, t+\delta t)$, $\mu\delta t$ and $\sigma\sqrt{\delta t}$ are the mean and standard deviation of the percentage change, respectively. Here, μ is called *drift rate* and σ stands for the *volatility*. Usually the stock price is observed over a fixed time interval (let's say the closing price is observed daily, so that $\delta t = 1/252$ as there are normally around 252 trading days in a year). For n consecutive daily stock prices $S_1, \ldots, S_n$, we define the daily arithmetic return as:

$$u_i := (S_i - S_{i-1})/S_{i-1} \sim \mathcal{N}(\mu\tau, \sigma^2\tau), \quad i = 2, \ldots, n, \tag{3.2}$$

where $\tau = 1/252$. To see whether this modeling is appropriate in practice, we can use some graphical methods to test for normality. Let us illustrate this by the following example.

Example 3.1 The file `stock_1999_2002.csv` contains adjusted daily closing prices of the stocks HSBC (0005), CLP (0002) and Cheung Kong (0001) from 1 January 1999 to 31 December 2002. Let us first read in these stock prices and compute the corresponding u_i for each stock according to (3.2); we also centralize them by subtracting $\hat{\mu}\tau =: \overline{\mu}$. We then produce a time series plot, a histogram, and a normal quantile-quantile plot (normal Q-Q plot) for u_i through Programme 3.1 via Python (resp. Programme 3.2 via **R**). Often, $\hat{\mu}\tau$ is so small that we may take it as zero whence testing for normality.

```
1  import pandas as pd
2  import numpy as np
3  import matplotlib.pyplot as plt
4  from scipy import stats
5  import statsmodels.api as sm
6
7  plt.rc('font', size=20); plt.rc('axes', titlesize=20, labelsize=20)
8  plt.rc('xtick', labelsize=20); plt.rc('ytick', labelsize=20)
9
10 d = pd.read_csv("stock_1999_2002.csv", index_col=0)
11 u = np.diff(d, axis=0) / d.iloc[:-1, :] # Arithmetic return
12
13 fig, ax = plt.subplots(1, 1, figsize=(15, 20))
14 axes = d.plot(subplots=True, layout=(3,1), ax=ax, fontsize=25)
15 for a in axes: a[0].legend(fontsize=35)
16
17 fig, ax = plt.subplots(1, 1, figsize=(15, 20))
18 axes = u.plot(subplots=True, layout=(3,1), ax=ax, fontsize=25)
19 for a in axes: a[0].legend(fontsize=35)
20
21 col = ["blue", "orange", "green"]
22 for i, comp in enumerate(u.columns):
23     axs[i,0].hist(u[comp], color=col[i], ec='black', bins="sturges")
24     axs[i,0].set_title(f"Histogram of {comp} Return")
25     sm.qqplot(u[comp], dist=stats.norm, ax=axs[i,1], line="q",
26               markerfacecolor=col[i], markeredgewidth=0)
27     axs[i,1].set_title(f"Normal Q-Q Plot of {comp} Return")
```

Programme 3.1 Example 3.1: Graphical statistics of the prices and returns of HSBC, CLP, and CK via Python.

```
1  > d <- read.csv("stock_1999_2002.csv", row.names=1)
2  > d <- as.ts(d)
3  > u <- (lag(d) - d) / d
4  > colnames(u) <- colnames(d)
5  >
6  > library(zoo)
7  > par(cex.lab=2, cex.axis=2, cex.main=3)
8  > plot(zoo(d), plot.type="multiple", col=c("blue", "orange", "green"))
9  > plot(zoo(u), plot.type="multiple", col=c("blue", "orange", "green"))
10 >
11 > par(mfrow=c(3,2), mar=c(5,5,4,4), cex.lab=2, cex.axis=2, cex.main=2)
```

```
12 > # if the dist is normal, the plot should close to this line
13 > # histogram; qq-normal plot; add a line for reference
14 > hist(u[,"HSBC"]); qqnorm(u[,"HSBC"]); qqline(u[,"HSBC"])
15 > hist(u[,"CLP"]); qqnorm(u[,"CLP"]); qqline(u[,"CLP"])
16 > hist(u[,"CK"]); qqnorm(u[,"CK"]); qqline(u[,"CK"])
```

Programme 3.2 Example 3.1: Graphical statistics of the prices and returns of HSBC, CLP, and CK via R.

From Figure 3.1 via Python (resp. Figure 3.3 via **R**), these time series plots show the sample u_i's fluctuating around zero. The volatility is small over certain periods and large over others, appearing in clusters. Obviously, the volatility varies with time. We shall discuss how to model this time-changing volatility in Chapter 7, but first let us focus on the plausibility of the normality assumption for u_i.

According to (3.2), u_i is expected to be normally distributed. To check the normality assumption, the most basic method is through a graphical approach with histograms. However, a histogram only gives us a general idea about the distribution. A more sophisticated graphical method is through the use of the normal Q-Q plot. To this end, we order the sample u_i in ascending order of magnitude denoted by $u_{(i)}$, resulting in $u_{(1)} \leq u_{(2)} \leq \ldots \leq u_{(n)}$. The normal Q-Q plot is the plot of these ordered $u_{(i)}$'s against $q_{(i)}$'s, the i-th quantile of the standard normal distribution, where

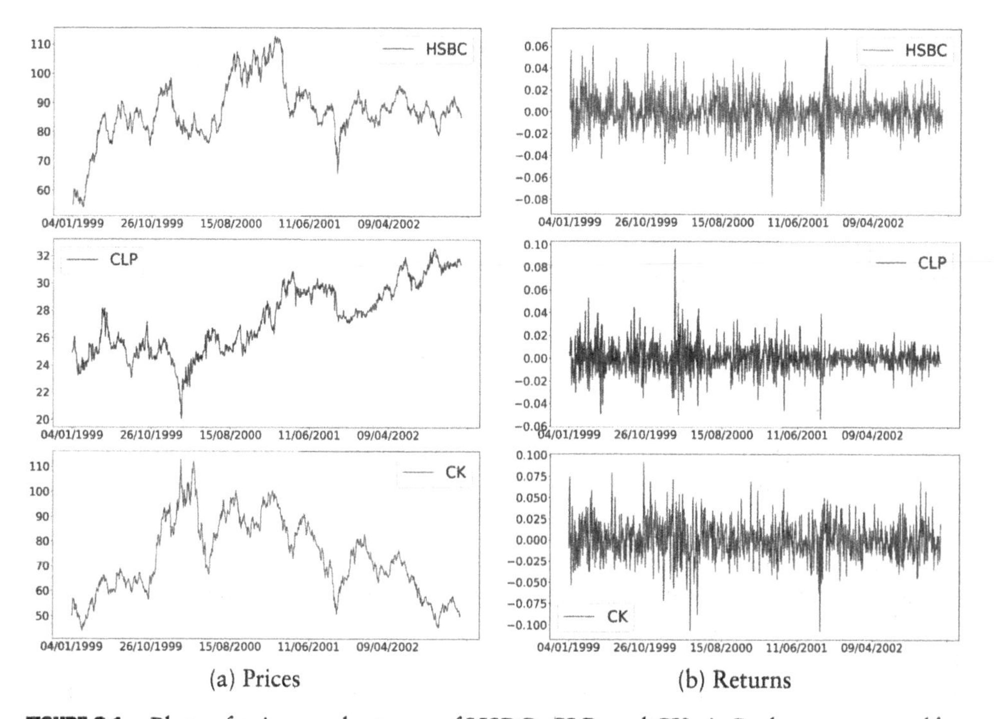

(a) Prices (b) Returns

FIGURE 3.1 Plots of prices and returns of HSBC, CLP, and CK via Python, generated by Programme 3.1.

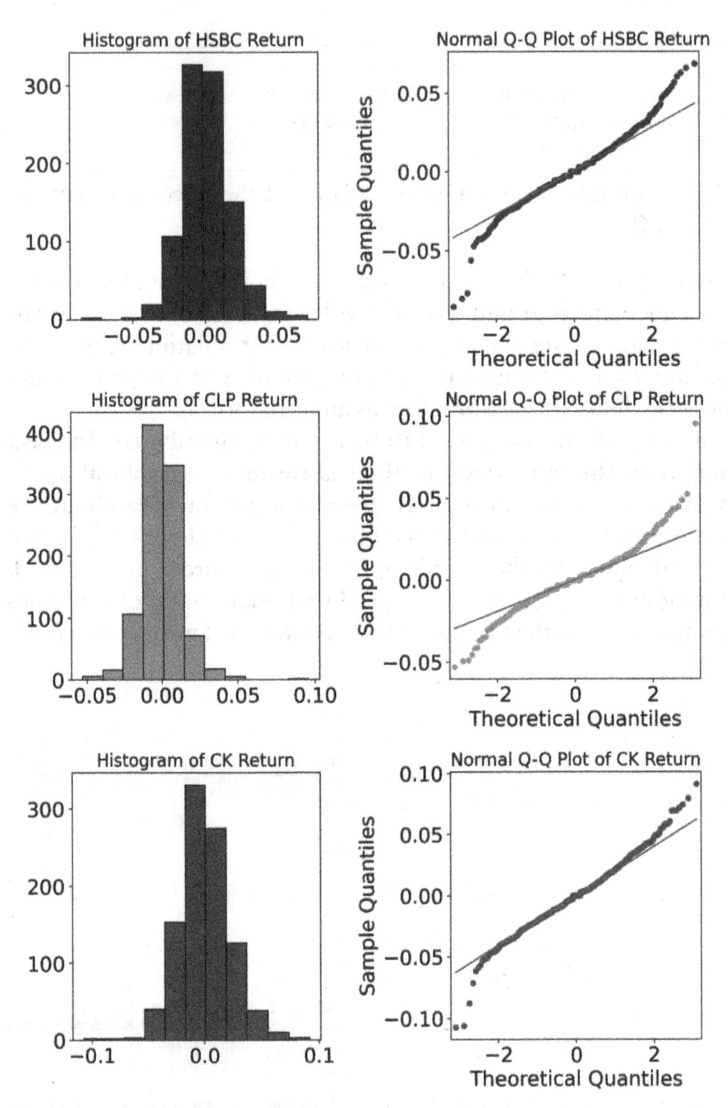

FIGURE 3.2 Histograms and Normal Q-Q plots of the returns of HSBC, CLP, and CK via Python, generated by Programme 3.1.

$q_{(i)}$ is the value that satisfies $\mathbb{P}(z < q_{(i)}) = (i - 0.5)/n$ (with this 0.5 as a continuity adjustment) with z being the standard normal random variable. Conceptually, the right-hand side should be i/n, yet the probability of $(i - 0.5)/n$ is used to avoid the difficulty encountered when $i = n$ as there will likely be some other observations larger than this present largest $u_{(n)}$. If the samples u_i's are normally distributed, this plot should be close to a straight line making, an angle of $\pi/4$ with the horizontal axis. The sample plots illustrate that the proportional changes are usually heavily tailed in favour of large gains and losses, and a mild positive skewness is also noted from

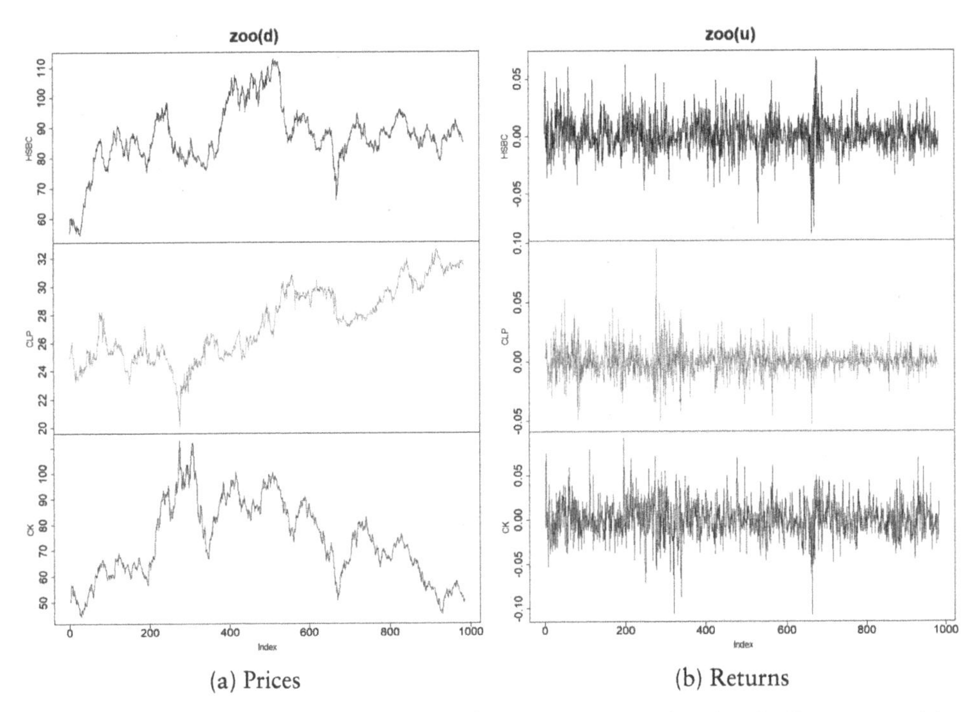

(a) Prices (b) Returns

FIGURE 3.3 Plots of prices and returns of HSBC, CLP, and CK's via **R**, generated by Programme 3.2.

the histograms. In particular, the normal Q-Q plots of the samples u_i's show that the distribution of each sequence of u_i's have a fatter tail than the normal distribution. More seriously, the *Value-at-Risk* (V@R) will be under-estimated if the normal model is adopted; see more discussions in Chapter 8.

The normal Q-Q plot is a graphical method for testing univariate normality, and indeed is very effective. That said, there are other formal statistical tests for testing normality; for instance, this Q-Q plot can be officially formalized as a *Shapiro–Wilk* (SW) test [15], which is a formal way of testing whether a sample is normally distributed. The corresponding test statistic is $W := \frac{(\sum_{i=1}^{n} a_i x_{(i)})^2}{\sum_{i=1}^{n} (\bar{x}-x_i)^2}$, where $x_{(i)}$'s are the ordered sample values of $x_1, \ldots, x_n$, and the row vector $\boldsymbol{a} = \boldsymbol{m}^\top V^{-1}/c$, where the Euclidean 2-norm $c = \|\boldsymbol{m}^\top V^{-1}\|_2 = (\boldsymbol{m}^\top V^{-1} V^{-1} \boldsymbol{m})^{1/2}$, and the vector $\boldsymbol{m}$ is the column vector of the expected values of the order statistics of iid random variables sampled from the standard normal distribution; finally, V is the covariance matrix of those normal order statistics; see more discussion in Appendix *3.A.1. In Programme 3.3 via Python (resp. Programme 3.4 via **R**), we illustrate this test's application to the daily arithmetic returns of these stocks. However, a shortcoming of using the Shapiro–Wilk test is the finding of the exact theoretical values of $\boldsymbol{m}$ and V, which is not so immediate. And there are other simpler normality tests available.

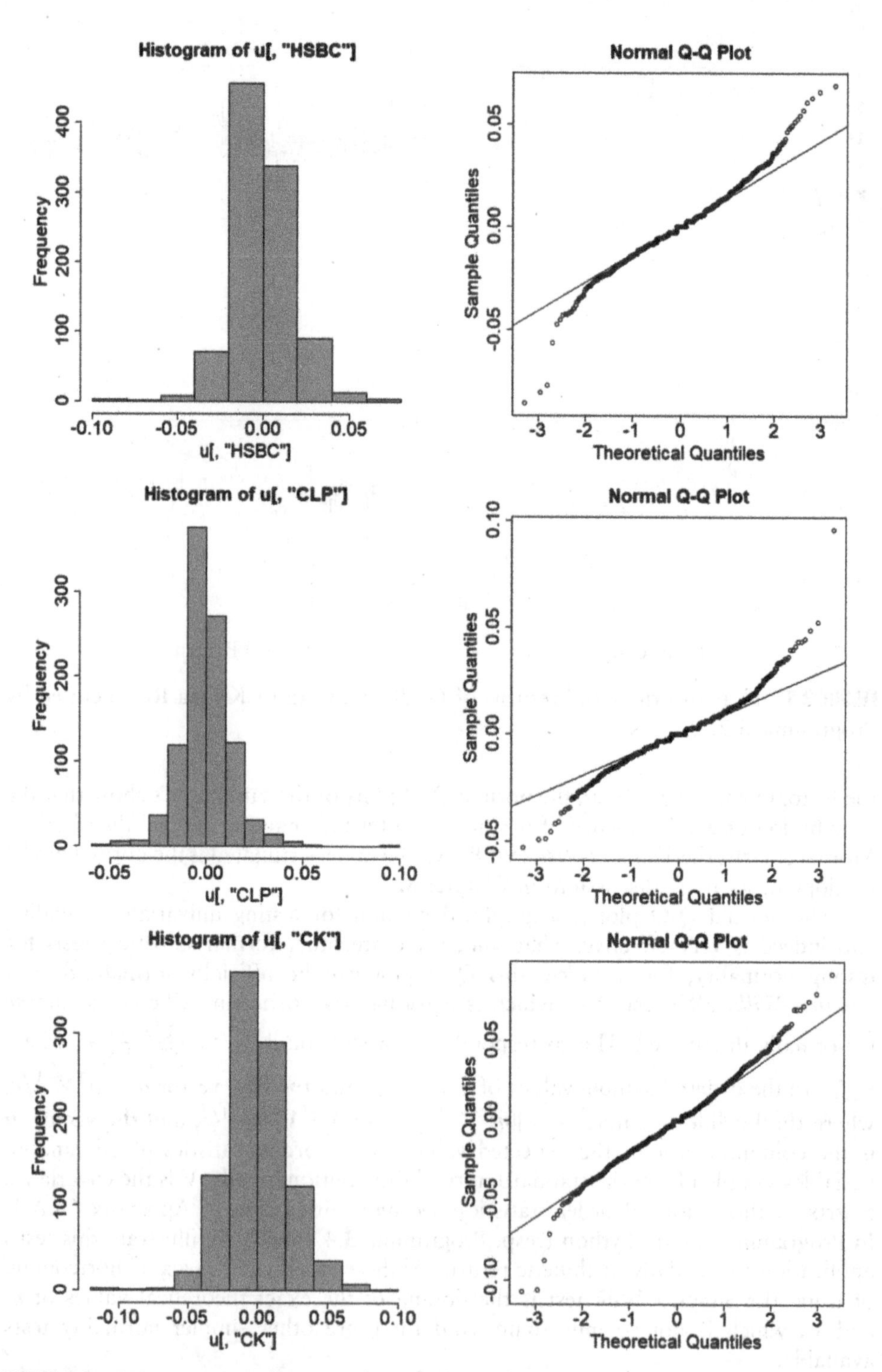

FIGURE 3.4 Histograms and Normal Q-Q plots of returns of HSBC, CLP, and CK via R, generated by Programme 3.2.

```
1  from scipy import stats
2
3  print(stats.shapiro(u.HSBC))
4  print(stats.shapiro(u.CLP))
5  print(stats.shapiro(u.CK))
```

```
1  ShapiroResult(statistic=0.9731917977333069, pvalue=1.711250065261627e-12)
2  ShapiroResult(statistic=0.9575617909431458, pvalue=2.8419345043810865e-16)
3  ShapiroResult(statistic=0.9860042929649353, pvalue=4.3795928661438666e-08)
```

Programme 3.3 Shapiro–Wilk test for the normality assumption for the returns of HSBC, CLP, and CK via Python.

```
1  > shapiro.test(u[,"HSBC"])
2      Shapiro-Wilk normality test
3  data:  u[, "HSBC"]
4  W = 0.97319, p-value = 1.71e-12
5  > shapiro.test(u[,"CLP"])
6      Shapiro-Wilk normality test
7  data:  u[, "CLP"]
8  W = 0.95756, p-value = 2.84e-16
9  > shapiro.test(u[,"CK"])
10     Shapiro-Wilk normality test
11 data:  u[, "CK"]
12 W = 0.986, p-value = 4.37e-08
```

Programme 3.4 Shapiro–Wilk test for the normality assumption for returns of HSBC, CLP, and CK via R.

The *Kolmogorov–Smirnov* (KS) test can be used to test whether a random sample is coming from a specific distribution. The test statistic is $D_n := \sup_{x \in \mathbb{R}} |F_n(x) - F(x)|$, where $F_n(x)$ and $F(x)$ are respectively the empirical distribution function, based on a random sample of size n, and theoretical distribution function. Note that the distribution of $\sqrt{n}D_n$ converges to that of the absolute maximum value of a standardized *Brownian bridge*[1] on the time interval $[0, 1]$ as n goes to infinity; see [17] for details. More discussions on this test statistic are referred to Appendix 3.A.2. The corresponding programmes in Python and in **R** are shown in Programmes 3.5 and 3.6, respectively.

```
1  print(stats.kstest(u.HSBC, stats.norm.cdf, method="asymp",
2                     args=(np.mean(u.HSBC), np.std(u.HSBC, ddof=1))))
3  print(stats.kstest(u.CLP, stats.norm.cdf, method="asymp",
4                     args=(np.mean(u.CLP), np.std(u.CLP, ddof=1))))
5  print(stats.kstest(u.CK, stats.norm.cdf, method="asymp",
6                     args=(np.mean(u.CK), np.std(u.CK, ddof=1))))
```

[1] The distribution of a Brownian bridge $\{B_t\}_{t\in[0,T]}$ is the conditional distribution of a standard Brownian motion $\{W_t\}$ given that $W_T = 0$.

```
1 KstestResult(statistic=0.06382812407395322, pvalue=0.0006645538367210273,
      statistic_location=0.005763688760806957, statistic_sign=1)
2 KstestResult(statistic=0.07405502536812492, pvalue=4.154699368674999e-05,
      statistic_location=0.0033783783783804945, statistic_sign=1)
3 KstestResult(statistic=0.04710872371545094, pvalue=0.025479917030897994,
      statistic_location=0.005524861878452851, statistic_sign=1)
```

Programme 3.5 Kolmogorov–Smirnov test for the normality assumption for the returns of HSBC, CLP, and CK via Python.

```
1  > u1 <- u[,"HSBC"]; u2 <- u[,"CLP"]; u3 <- u[,"CK"]
2      # Extract the column of returns of each stock
3  > u1 <- u[,"HSBC"]; u2 <- u[,"CLP"]; u3 <- u[,"CK"]
4  > ks.test(u1, pnorm, mean=mean(u1), sd=sd(u1))
5      Asymptotic one-sample Kolmogorov-Smirnov test
6  data:  u1
7  D = 0.063828, p-value = 0.0006646
8  alternative hypothesis: two-sided
9  > ks.test(u2, pnorm, mean=mean(u2), sd=sd(u2))
10     Asymptotic one-sample Kolmogorov-Smirnov test
11 data:  u2
12 D = 0.074055, p-value = 4.155e-05
13 alternative hypothesis: two-sided
14 > ks.test(u3, pnorm, mean=mean(u3), sd=sd(u3))
15     Asymptotic one-sample Kolmogorov-Smirnov test
16 data:  u3
17 D = 0.047109, p-value = 0.02548
18 alternative hypothesis: two-sided
```

Programme 3.6 Kolmogorov–Smirnov test for the normality assumption for the returns of HSBC, CLP, and CK via **R**.

Another commonly used normality test is the *Jarque–Bera* (JB) test. It is based on the fact that the skewness = 0 and the kurtosis = 3 for the standard normal distribution. The JB test statistic is $n(\hat{\zeta}_1^2/6 + \hat{\zeta}_2^2/24)$ which asymptotically follows χ_2^2, the chi-square distribution of degrees of freedom 2, where $\hat{\zeta}_1 := \frac{\sum_{i=1}^n (x_i-\bar{x})^3}{[\sum_{i=1}^n (x_i-\bar{x})^2]^{3/2}}$ and $\hat{\zeta}_2 := \frac{\sum_{i=1}^n (x_i-\bar{x})^4}{[\sum_{i=1}^n (x_i-\bar{x})^2]^2} - 3$ are respectively the standardized sample skewness (the theoretical counterpart being $\text{Skew}(x) := \mathbb{E}(((x-\mu)/\sigma)^3)$) and standardized sample excess kurtosis (the theoretical counterpart being $\text{EKurt}(x) := \mathbb{E}(((x-\mu)/\sigma)^4) - 3$). The `jarque_bera()` function from `stats` library in Python and the `jarque.bera.test()` function from `tseries` library in **R** can both be used to compute the JB statistics. Meanwhile, for the sake of illustration, we also write the following function `JB_test()` in Python in Programme 3.7 (resp. the function `JB.test()` in **R** in Programme 3.8) to perform this JB test.

```
1  def JB_test(u):
2      z = u - np.mean(u)  # Remove mean
3      n = len(z) # Sample size
4      s = np.std(z) # Population standard deviation
5      sk = sum(z**3) / (n*s**3) # Skewness
6      ku = sum(z**4) / (n*s**4) - 3 # Excess Kurtosis
7      JB = n * (sk**2/6 + ku**2/24) # JB test statistics
8      p = 1 - stats.chi2.cdf(JB, 2) # chi-square p-value
9      return ({"JB-test": JB, "p-value": p})
10
11 print(stats.jarque_bera(u.HSBC))
12 print(JB_test(u.HSBC))
13
14 print(stats.jarque_bera(u.CLP))
15 print(JB_test(u.CLP))
16
17 print(stats.jarque_bera(u.CK))
18 print(JB_test(u.CK))
```

```
1  SignificanceResult(statistic=250.92302711005993, pvalue
       =3.256544160517786e-55)
2  {'JB-test': 250.92302711005883, 'p-value': 0.0}
3  SignificanceResult(statistic=831.0302232280023, pvalue
       =3.500095374289533e-181)
4  {'JB-test': 831.0302232280412, 'p-value': 0.0}
5  SignificanceResult(statistic=92.53310618800974, pvalue
       =8.066614349465839e-21)
6  {'JB-test': 92.53310618800832, 'p-value': 0.0}
```

Programme 3.7 Jarque–Bera test for the normality assumption for the returns of HSBC, CLP, and CK via Python.

```
1  > library("tseries")
2  >
3  > JB.test <- function(u){
4  +     z <- u - mean(u)                # Remove mean
5  +     n <- length(z)                  # Sample size
6  +     s <- sd(z)*sqrt((n-1)/n)        # Population standard deviation
7  +     sk <- sum(z^3)/(n*s^3)          # Skewness
8  +     ku <- sum(z^4)/(n*s^4) - 3      # Excess Kurtosis
9  +     JB <- n * (sk^2/6 + ku^2/24)    # JB test statistics
10 +     p <- 1 - pchisq(JB, 2)          # chi-square p-value
11 +     list("JB_stat"=JB, "p_value"=p)
12 + }
13 >
14 > JB.test(u[,"HSBC"])
15 $JB_stat
16 [1] 250.923
17
```

```
18  $p_value
19  [1] 0
20
21  > jarque.bera.test(u[,"HSBC"])
22      Jarque Bera Test
23  data:  u[, "HSBC"]
24  X-squared = 250.92, df = 2, p-value < 2.2e-16
25  >
26  > JB.test(u[,"CLP"])
27  $JB_stat
28  [1] 831.0302
29
30  $p_value
31  [1] 0
32
33  > jarque.bera.test(u[,"CLP"])
34      Jarque Bera Test
35  data:  u[, "CLP"]
36  X-squared = 831.03, df = 2, p-value < 2.2e-16
37  >
38  > JB.test(u[,"CK"])
39  $JB_stat
40  [1] 92.53311
41
42  $p_value
43  [1] 0
44
45  > jarque.bera.test(u[,"CK"])
46      Jarque Bera Test
47  data:  u[, "CK"]
48  X-squared = 92.533, df = 2, p-value < 2.2e-16
```

Programme 3.8 Jarque–Bera test for the normality assumption for the returns of HSBC, CLP, and CK via **R**.

The p-values for u_1, u_2 and u_3 from both the KS test and the JB test are small, thus we conclude the normality assumptions for them are not valid, which is consistent with the observation from the normal Q-Q plots.

3.2 STUDENT'S t_ν-DISTRIBUTION FOR STOCK PRICE CHANGES

To cope with the heavy tails commonly observed among securities, one may use a scaled Student's t-distribution, denoted by t_ν, with ν degrees of freedom and a scaling parameter σ to model relative price changes u_i's. The density of the t_ν-distribution is given by:

$$T \sim f(t;\nu) = \frac{\Gamma((\nu+1)/2)}{\Gamma(\nu/2)\sqrt{\nu\pi\sigma^2}}\left(1+\frac{t^2}{\nu\sigma^2}\right)^{-(\nu+1)/2}, \quad \text{for } \nu > 2,$$

where its first four moments are: (i) mean: $\mu = \mathbb{E}(T) = 0$; (ii) variance: $\sigma^2 = \mathbb{E}(T^2) = \sigma^2\nu/(\nu - 2)$; (iii) standardized skewness: $\zeta_1 = \mathbb{E}(T^3)/\sigma^3 = 0$; (iv) standardized excess kurtosis: $\zeta_2 = \mathbb{E}(T^4)/\sigma^4 - 3 = 6/(\nu - 4)$. More details for the t-distribution as well as its common use in statistics can be found in standard textbooks such as [2, 4, 7, 16].

The tail of the t_ν-distribution is heavier than that of any normal distribution. In principle, it becomes normal when $\nu \to \infty$ and t_ν-distributions are commonly used to model security data with fat tails. For common securities, the value of ν ranges from 3 to 6 as pointed out in [11]; specifically, Harry Markowitz noted that a degree of freedom around 4 yields the best estimate and is consistent with the observations from S&P500 stock returns. We can lay down t_ν Q-Q plots to see whether the fittings look better. First, we need to estimate the degrees of freedom ν. One simple way is to match the sample excess kurtosis $\hat{\zeta}_2$ computed from the sample u_i's to the theoretical value of $6/(\nu - 4)$, which gives an estimate of $\nu = 6/\hat{\zeta}_2 + 4$. The p-th quantile of the t_ν-distribution is obtained by setting $\mathbb{P}\{T \leq t_\nu(p)\} = p$, denoted by $q_p := t_\nu(p)$. In Programmes 3.9 (in Python) and 3.10 (in **R**), we can respectively use the `t_QQ_plot()` and `t.QQ.plot()` functions to produce the t_ν Q-Q plots.

```
1  def t_QQ_plot(u, color="blue", comp="", ax=None):
2          z = u - np.mean(u)   # Remove mean
3          su = np.sort(z) # Sort z
4          n = len(z) # Sample size
5          s = np.std(z) # Population standard deviation
6          ku = sum(z**4) / (n*s**4) - 3 # Excess Kurtosis
7          nu = 6/ku + 4 # degrees of freedom
8          i = (np.arange(1, n+1)-0.5)/n # a vector of percentiles
9          q = stats.t.ppf(i, nu) # percentile points from t(v)
10
11         b, w = np.linalg.lstsq(np.vstack([np.ones(n),q]).T,su,rcond=None)[0]
12         ax.scatter(q, su, color=color)
13         ax.plot(q, w*q + b, color="red")
14         ax.set_title(f"Self-defined t Q-Q Plot of {comp} Return")
15         return nu
16
17 fig, axs = plt.subplots(3, 2, figsize=(10,15))
18 df_HSBC = t_QQ_plot(u.HSBC, color="blue", comp="HSBC", ax=axs[0,0])
19 stats.probplot(u.HSBC, dist="t", sparams=df_HSBC, plot=axs[0,1])
20 axs[0,1].set_title("t Q-Q Plot of HSBC Return")
21 axs[0,1].get_lines()[0].set_color('blue')
22
23 df_CLP = t_QQ_plot(u.CLP, color="orange", comp="CLP", ax=axs[1,0])
24 stats.probplot(u.CLP, dist="t", sparams=df_CLP, plot=axs[1,1])
25 axs[1,1].set_title("t Q-Q Plot of CLP Return")
26 axs[1,1].get_lines()[0].set_color('orange')
27
28 df_CK = t_QQ_plot(u.CK, color="green", comp="CK", ax=axs[2,0])
29 stats.probplot(u.CK, dist="t", sparams=df_CK, plot=axs[2,1])
30 axs[2,1].set_title("t Q-Q Plot of CK Return")
31 axs[2,1].get_lines()[0].set_color('green')
32 fig.tight_layout()
```

Programme 3.9 Student's t Q-Q plots for the returns of HSBC, CLP, and CK via Python.

From Figures 3.5 and 3.6, we note that the t_v-distribution fits the fat tails better than the normal distribution. Again, we can use the Kolmogorov–Smirnov (KS) test to test whether u_1, u_2 and u_3 follow a t_v-distribution via Programmes 3.11 (in Python) and 3.12 (in **R**).

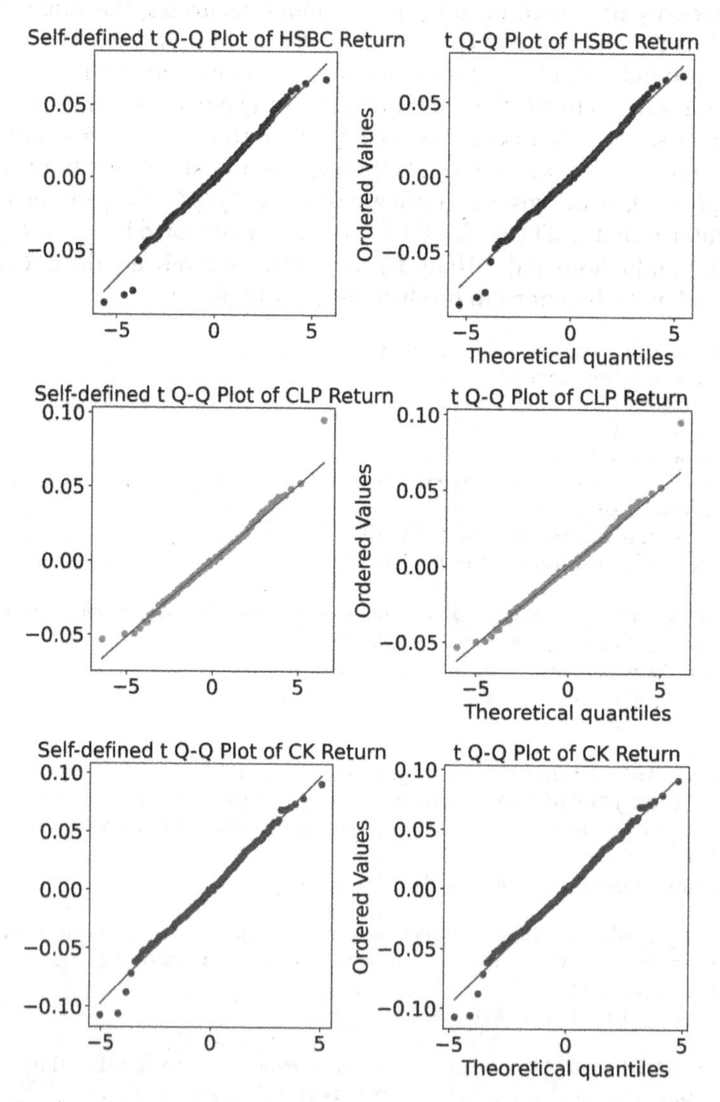

FIGURE 3.5 Student's t Q-Q plots for the returns of HSBC, CLP, and CK via Python, generated by Programme 3.9.

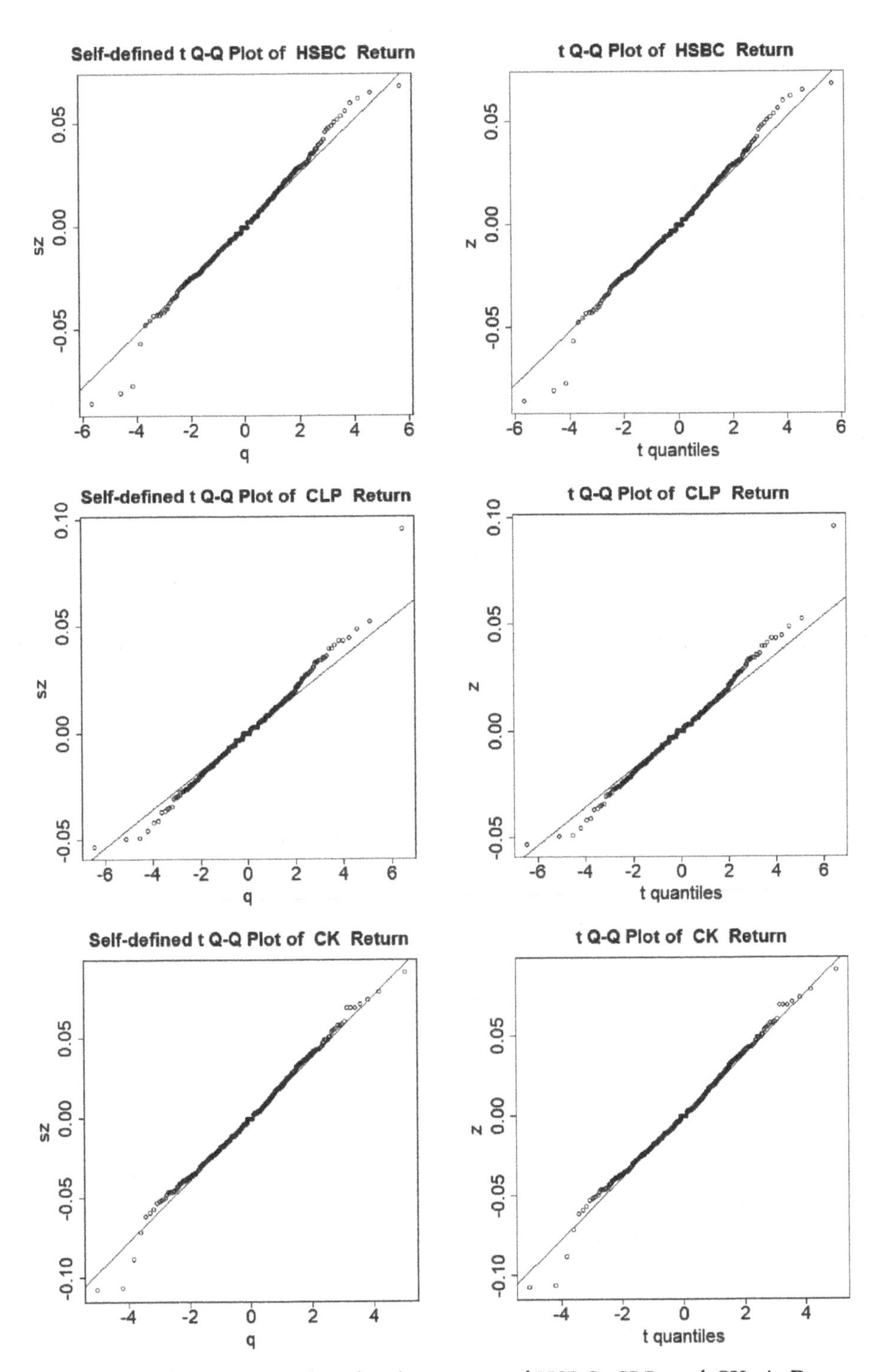

FIGURE 3.6 Student's *t* Q-Q plots for the returns of HSBC, CLP, and CK via **R**, generated by Programme 3.10.

```
1  > library("car")
2  >
3  > t.QQ.plot <- function(u, comp=""){
4  +     z <- u - mean(u)                 # Remove mean
5  +     sz <- sort(z)                    # sort z
6  +     n <- length(z)                   # sample size
7  +     s <- sd(z)*sqrt((n-1)/n)         # Population standard deviation
8  +     ku <- sum(z^4)/(n*s^4) - 3       # Excess kurtosis
9  +     nu <- 6/ku + 4 # Degrees of freedom
10 +     i <- ((1:n)-0.5)/n               # create a vector of percentiles
11 +     q <- qt(i, nu)                   # percentile points from t(v)
12 +
13 +     plot(q, sz, main=paste("Self-defined t Q-Q Plot of ", comp, " Return"))
14 +     qqline(sz, distribution=function(p) qt(p, df=nu), probs=c(0.25, 0.75))
15 +     qqPlot(z, distribution="t", df=nu, envelope=FALSE, line="quartiles",
16 +            col.lines="black", lwd=1, cex=1, grid=FALSE, id=FALSE,
17 +            main=paste("t Q-Q Plot of ", comp, " Return"))
18 +     nu
19 + }
20 >
21 > par(mfrow=c(3,2), mar=c(5,5,4,4), cex.lab=2, cex.axis=2, cex.main=2)
22 > df_HSBC <- t.QQ.plot(u[,"HSBC"], comp="HSBC")
23 > df_CLP <- t.QQ.plot(u[,"CLP"], comp="CLP")
24 > df_CK <- t.QQ.plot(u[,"CK"], comp="CK")
```

Programme 3.10 Student's *t* Q-Q plots for the returns of HSBC, CLP, and CK via R.

```
1 print([df_HSBC, df_CLP, df_CK])
2
3 t_HSBC = u.HSBC/np.std(u.HSBC, ddof=1)*np.sqrt(df_HSBC/(df_HSBC-2))
4 print(stats.kstest(t_HSBC,stats.t.cdf,args=(df_HSBC,),method="asymp"))
5 t_CLP = u.CLP/np.std(u.CLP, ddof=1)*np.sqrt(df_CLP/(df_CLP-2))
6 print(stats.kstest(t_CLP, stats.t.cdf, args=(df_CLP,), method="asymp"))
7 t_CK = u.CK/np.std(u.CK, ddof=1)*np.sqrt(df_CK/(df_CK-2))
8 print(stats.kstest(t_CK, stats.t.cdf, args=(df_CK,), method="asymp"))
```

```
1 [6.424557788359385, 5.357740314185319, 7.999967780163342]
2 KstestResult(statistic=0.04730417090539163, pvalue=0.0245721259927215,
      statistic_location=0.0, statistic_sign=1)
3 KstestResult(statistic=0.06459816887080366, pvalue=0.0005471321107018978,
      statistic_location=0.0, statistic_sign=-1)
4 KstestResult(statistic=0.04018311291963372, pvalue=0.08362674752852832,
      statistic_location=0.0, statistic_sign=1)
```

Programme 3.11 Kolmogorov–Smirnov test for the Student's *t* assumption for the returns of HSBC, CLP, and CK via Python.

```
1  > t_HSBC <- u[,"HSBC"]/sd(u[,"HSBC"])*sqrt(df_HSBC/(df_HSBC-2))
2  > ks.test(t_HSBC, pt, df_HSBC)
3      Asymptotic one-sample Kolmogorov-Smirnov test
4  data:  t_HSBC
5  D = 0.047304, p-value = 0.02457
6  alternative hypothesis: two-sided
7  > t_CLP <- u[,"CLP"]/sd(u[,"CLP"])*sqrt(df_CLP/(df_CLP-2))
8  > ks.test(t_CLP, pt, df_CLP)
9      Asymptotic one-sample Kolmogorov-Smirnov test
10 data:  t_CLP
11 D = 0.064598, p-value = 0.0005471
12 alternative hypothesis: two-sided
13 > t_CK <- u[,"CK"]/sd(u[,"CK"])*sqrt(df_CK/(df_CK-2))
14 > ks.test(t_CK, pt, df_CK)
15     Asymptotic one-sample Kolmogorov-Smirnov test
16 data:  t_CK
17 D = 0.040183, p-value = 0.08363
18 alternative hypothesis: two-sided
```

Programme 3.12 Kolmogorov–Smirnov test for the Student's t assumption for the returns of HSBC, CLP, and CK via **R**.

Again the p-values of HSBC and CLP are small, yet much bigger than those from fitting the normal distribution. We still conclude u_1 and u_2 (returns of HSBC and CLP) do not completely follow a t_ν-distribution. However, since the p-value of CK is greater than 0.05, we fail to reject the null hypothesis, and it is viable to claim u_3 follows a t_ν-distribution. Significant deviation from the t_ν-distribution occurs at the very extreme negative values of relative price changes, as the extreme return happens more likely during market downturns.

3.3 TESTING FOR MULTIVARIATE NORMALITY

The normal Q-Q plots in the previous section are only for checking normality in a univariate setting. We may want to test whether the p-dimensional daily returns of a portfolio of p stocks jointly follow a multivariate normal distribution, that is,

$$\mathbf{u}_i = \begin{pmatrix} \mathrm{u}_{i1} \\ \vdots \\ \mathrm{u}_{ip} \end{pmatrix} = \begin{pmatrix} \delta \mathrm{S}_{i1}/\mathrm{S}_{i1} \\ \vdots \\ \delta \mathrm{S}_{ip}/\mathrm{S}_{ip} \end{pmatrix} \sim \mathcal{N}_p \left[\begin{pmatrix} \mu_1 \\ \vdots \\ \mu_p \end{pmatrix} \delta t, \delta t \cdot \boldsymbol{\Sigma} \right], \quad \text{for} \quad i = 1, \ldots, n.$$

We can extend the idea of normal Q-Q plots to test multivariate normality by using the notion of the squared generalized distance (*Mahalanobis distance* [12]):

$$d_i^2 = (\boldsymbol{u}_i - \overline{\boldsymbol{u}})^\top S^{-1} (\boldsymbol{u}_i - \overline{\boldsymbol{u}}) \quad \text{for} \quad i = 1, \ldots, n, \tag{3.3}$$

where $\overline{\boldsymbol{u}}$ and S are the sample unbiased estimates of $\boldsymbol{\mu} \cdot \delta t$ and $\delta t \cdot \boldsymbol{\Sigma}$ respectively. Note that d_i is a number representing the distance from an individual observation $\boldsymbol{u}_i$

to the central tendency indicated by the sample mean, and d_i^2 is sometimes called the *squared generalized distance* of u_i. It can be shown that if $\mathbf{u}_i$ has a multivariate normal distribution, then for a large n, d_i^2 will approximately follow a chi-square distribution with p degrees of freedom; also recall Property 6 in Section 1.4.2. Therefore, we can generate a chi-square Q-Q plot for those Mahalanobis distances as follows:

1. Order the d_i^2 in ascending order in magnitude, and denote them by $d_{(i)}^2$ in order;
2. Compute $q_p(i)$, the i-th quantile of the chi-square distribution with p degrees of freedom, that means $\mathbb{P}(\chi_p^2 < q_p(i)) = (i - 0.5)/n$, after the continuity correction;
3. Plot $d_{(i)}^2$ against $q_p(i)$. If the points are close to a straight line at an angle of $\pi/4$ against the horizontal axis, the original u_i's are likely to come from a multivariate normal distribution.

Let us illustrate the approach of using a chi-square Q-Q plot for $\mathbf{u} = (\mathrm{u}_1, \mathrm{u}_2, \mathrm{u}_3)^\top$ to see whether it has a tri-variate normal distribution; refer to Programmes 3.13 via Python and 3.14 via **R**. First, we compute the sample mean vector and the sample variance-covariance matrix S of the daily returns using the most recent 180 days. Then we compute the inverse of S, the squared generalized distances and the quantiles of the chi-square distribution with degrees of freedom 3. Note that Python does not have a built-in function for finding the inverse of a matrix, yet we can use the function `inv()` in `numpy.linalg` to do it. From the Q-Q plots in Figure 3.7, the distribution of the return vector in the past 180 days, stored as `u_180`, is not significantly different from a multivariate normal distribution, except for the tail part. We can again use the KS test to test whether $\mathrm{d}_i^2 \sim \chi_p^2$.

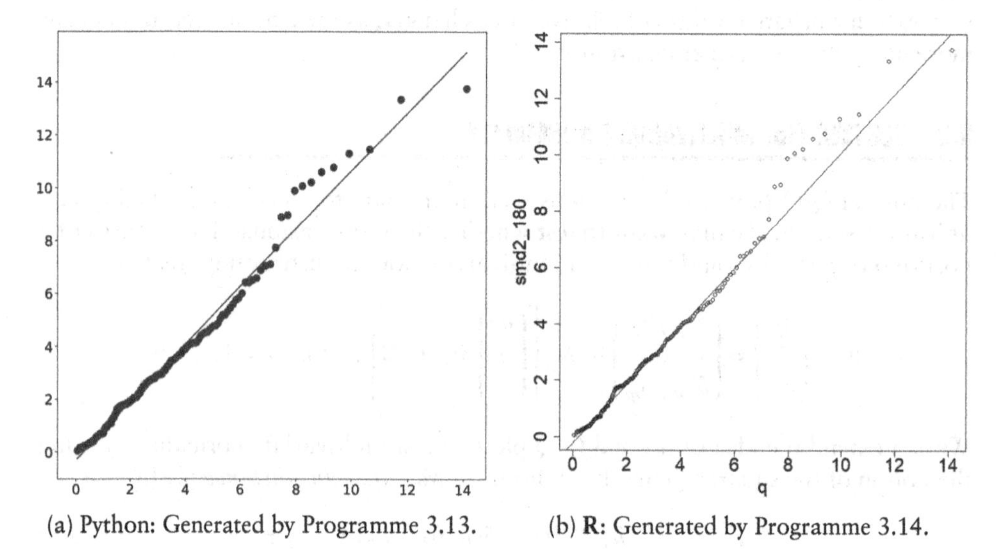

(a) Python: Generated by Programme 3.13. (b) **R**: Generated by Programme 3.14.

FIGURE 3.7 Chi-square Q-Q plots for the returns of HSBC, CLP, and CK.

```
1  n = 180
2  u_180 = u.iloc[len(u)-n:, :]
3  mu_180 = np.mean(u_180, axis=0)
4  S_180 = np.cov(u_180, rowvar=False, ddof=1)
5
6  z_180 = (u_180 - mu_180).values.reshape(n, -1)
7  md2_180 = np.sum((z_180 @ np.linalg.inv(S_180)) * z_180, axis=1)
8  smd2_180 = np.sort(md2_180)
9  i = (np.arange(1, n+1)-0.5)/n
10 q = stats.chi2.ppf(i, 3)
11
12 fig = plt.figure(figsize=(10, 10))
13 b, w = np.linalg.lstsq(np.vstack([np.ones(n), q]).T,
14                        smd2_180, rcond=None)[0]
15 plt.scatter(q, smd2_180, color="blue", s=100)
16 plt.plot(q, w*q + b, color="blue")
17
18 print(stats.kstest(md2_180, stats.chi2.cdf, args=(3,), method="asymp"))
```

```
1  KstestResult(statistic=0.08087630304409366, pvalue=0.1896718485816787,
       statistic_location=0.714597606388447, statistic_sign=1)
```

Programme 3.13 Chi-square Q-Q plots for the returns of HSBC, CLP, and CK via Python.

```
1  > n <- 180
2  > u_180 <- tail(u, n)
3  > mu_180 <- apply(u_180, 2, mean)
4  > S_180 <- cov(u_180)
5  >
6  > z_180 <- sweep(u_180, 2, mu_180)
7  > md2_180 <- rowSums((z_180%*%solve(S_180)) * z_180)
8  > smd2_180 <- sort(md2_180)      # sort md2 in ascendingly
9  > i <- ((1:n)-0.5)/n           # create percentile vector
10 > q <- qchisq(i,3)        # compute quantiles
11 >
12 > par(mfrow=c(1,1))
13 > qqplot(q, smd2_180, main="") # Q-Q chi-square plot
14 > qqline(smd2_180, distribution=function(p) qchisq(p, df=3))
15 >
16 > ks.test(smd2_180, pchisq, 3)
17         Asymptotic one-sample Kolmogorov-Smirnov test
18 data:  smd2_180
19 D = 0.080876, p-value = 0.1897
20 alternative hypothesis: two-sided
```

Programme 3.14 Chi-square Q-Q plots for the returns of HSBC, CLP, and CK via R.

Note that as the p-value is larger than 0.05, we may not reject the null hypothesis that the vector of d_i's stored in `md2_180` follows a chi-square distribution, or $\mathrm{u} = (\mathrm{u}_1, \mathrm{u}_2, \mathrm{u}_3)^T$ has an almost tri-variate normal distribution. In Chapter 8, we shall illustrate a better fit by using copulae. More advanced methods of testing for normality can be found in [18].

3.4 SAMPLE CORRELATION MATRIX

When we consider more than one random variable, correlations between any two random variables serve as important parameters. Let us compute the correlation matrix of `u_180` by Programmes 3.15 in Python and 3.16 in R.

```
1 print(np.corrcoef(u_180, rowvar=False, ddof=1))
```

```
1 [[ 1.          -0.01344783  0.61284746]
2  [-0.01344783  1.           0.11466795]
3  [ 0.61284746  0.11466795  1.        ]]
```

Programme 3.15 Sample correlation matrix for the returns over the most recent 180 days of HSBC, CLP, and CK via Python.

```
1 > cor(u_180)
2             HSBC         CLP        CK
3 HSBC  1.00000000 -0.01344783 0.6128475
4 CLP  -0.01344783  1.00000000 0.1146680
5 CK    0.61284746  0.11466795 1.0000000
```

Programme 3.16 Sample correlation matrix for the returns over the most recent 180 days HSBC, CLP, and CK via R.

For Python, we can generate the scatter plot matrix of these daily returns by using the function `pandas.plotting.scatter_matrix()`. This can be considered as a graphical illustration of the correlation matrix.

From the output of Programmes 3.15 and 3.16, generally speaking we see that the pairs of `u1` and `u2` as well as `u2` and `u3` are mostly uncorrelated; but `u1` and `u3` are positively correlated. The sample correlation matrix can be computed from the sample covariance matrix S by using the relations $r_{ij} = \frac{s_{ij}}{\sqrt{s_{ii}s_{jj}}}$, for $i, j = 1, \ldots, p$. Pictorial illustrations of the pairwise correlations are provided in Figure 3.8 and Figure 3.9 via Programme 3.17 (Python) and Programme 3.18 (R) respectively.

```
1  fig = plt.figure(figsize=(10, 10))
2  hist_kwds = {'color':'blue','bins':'sturges','ec':'black','alpha':0.8}
3  axes = pd.plotting.scatter_matrix(u, figsize=(10,10), color="blue",
4                                    hist_kwds=hist_kwds)
5  new_labels = [round(float(i.get_text()), 2) for i in
6                axes[0,0].get_yticklabels()]
7  axes[0,0].set_yticklabels(new_labels)
```

Programme 3.17 Correlation analysis on the returns of HSBC, CLP, and CK via Python.

```
1  > pairs(u_180)
```

Programme 3.18 Correlation analysis on the returns of HSBC, CLP, and CK via R.

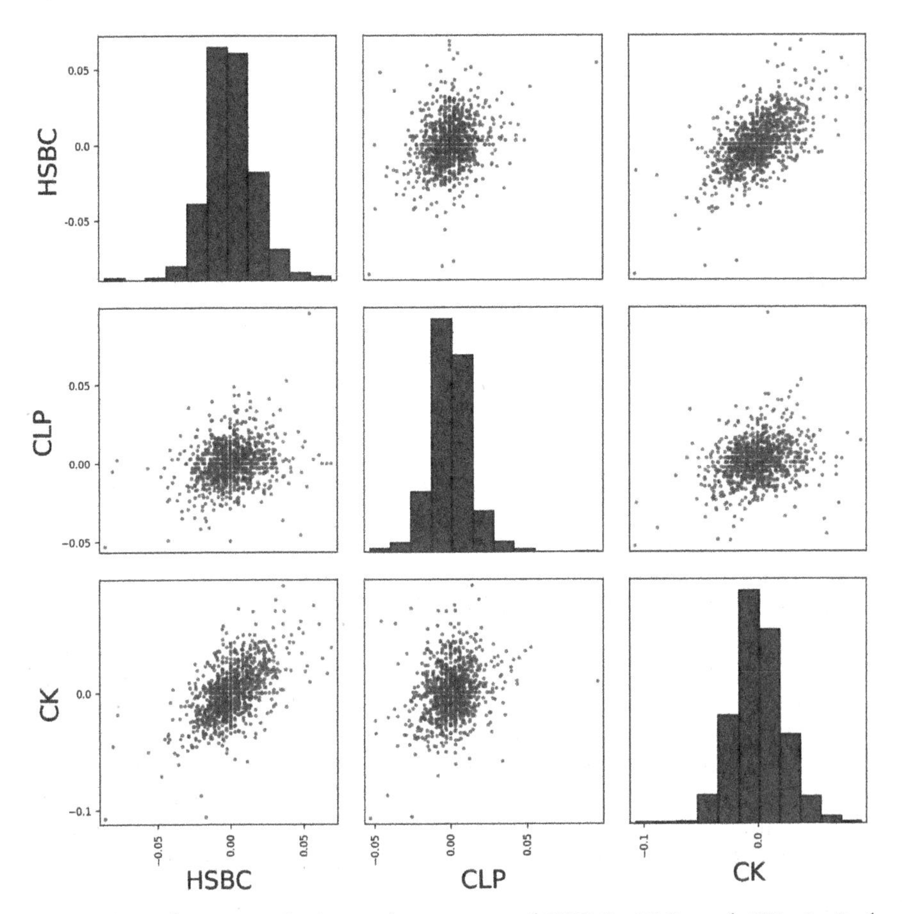

FIGURE 3.8 Correlation analysis on the returns of HSBC, CLP, and CK via Python, generated by Programme 3.17.

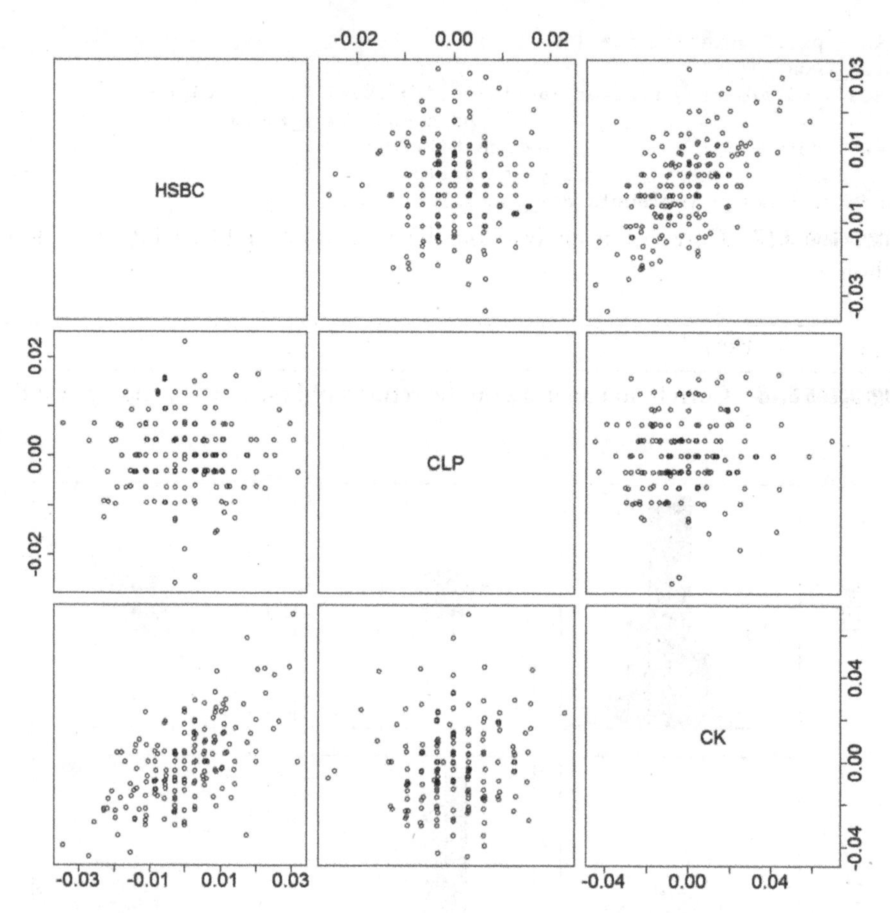

FIGURE 3.9 Correlation analysis on the returns of HSBC, CLP, and CK via R.

3.5 EMPIRICAL PROPERTIES OF STOCK PRICES

Without a doubt, modeling the price of a stock is one of the most active research areas in *Quantitative Finance* and *Econometrics*. In the last few sections, we demonstrated some primitive approaches of modeling stock prices. However, in reality stock prices are affected by many factors such as interest rates, and various economic and political factors; see more discussion in [3, 10] for instance. Despite the complexity, there are some empirical findings and conventions about stock prices that are widely accepted in the industry. Here is a list of the most notable ones; a more comprehensive account can be found in [3].

1. Instead of modeling a stock price itself, we usually model the (daily) returns of the stock. There are two types of returns:
   (i) logarithmic return $\tilde{u}_i = \ln(S_i) - \ln(S_{i-1})$, and
   (ii) arithmetic return $u_i = (S_i - S_{i-1})/S_{i-1} = S_i/S_{i-1} - 1$.

   These two returns are close in value if u_i is small. This can be easily seen by the fact that $\tilde{u}_i = \ln(S_i/S_{i-1}) = \ln(1 + u_i) \approx u_i$. Although the logarithmic return is additive, we usually use arithmetic returns in practice, for simplicity.
2. If the time interval is short, say a day, $\delta t = 1/252 = 0.004$, out of 252 trading days per year on average, then we may assume the mean of u_i to be zero. This reduces the complexity of the model, although this is less of a concern with high computing performance these days.
3. There is little autocorrelation in u_i, i.e., $\mathrm{Corr}(\mathrm{u}_{i+k}, \mathrm{u}_i) \approx 0$ for $k = 1, 2, \ldots$ In other words, returns are almost impossible to be "linearly" predicted from their own history.
4. Although the Black–Scholes–Merton model assumes that the distribution is normal, its distribution is generally believed to have fatter tails than those of the normal in many circumstances. Fatter tails imply a higher probability of large losses (or profit) than that offered by the normal distribution; see [5] for more details on the modelling in this regard. This is particularly important when one calculates Value-at-Risks (V@Rs); more details can be seen in Chapter 8.
5. The standard deviation of u_i is large compared with the mean of u_i. Empirical findings show that the standard deviation of daily returns usually lies between 0.95% and 3.15% during normal times, but was between 2% and 8% (for the US market) during the first half of 2020 – the pandemic year!
6. This standard deviation certainly varies with time and depends on many unknown and uncontrollable factors, and so the Black–Scholes–Merton model is not so reliable for long-term modelling; more sophisticated time series models will be introduced in Chapter 7.
7. Since the sample mean of u_i is taken to be zero, the variance of u_i can be estimated by the sample mean of the squared returns u_i^2's. For these squared returns, some positive autocorrelation exists, meaning $\mathrm{Corr}(\mathrm{u}_{i+k}^2, \mathrm{u}_i^2) > 0$ for small values of k.
8. A generic stochastic model for the return is $\mathrm{u}_i = \mu_i + \sigma_i \mathrm{z}_i$, where z_i's are iid according to a distribution with a mean 0 and variance 1. A special case of this is $\mathrm{u}_i = \mu\delta t + \sigma\sqrt{\delta t}\mathrm{z}_i$, where z_i's follow iid $\mathcal{N}(0, 1)$, which is the celebrated Black–Scholes–Merton model.

To understand the difference in statistical properties between the original time series S_i's of stock prices and the relative returns u_i's, we generate the respective histogram and the normal Q-Q plot for S_i, S_{i+1} vs S_i and u_{i+1} vs u_i in Figures 3.10 in Python and 3.11 in **R** via Programmes 3.19 and 3.20 respectively.

```
1  fig, axs = plt.subplots(4, 3, figsize=(15,20))
2  col = ["blue", "orange", "green"]
3  for i, comp in enumerate(d.columns):
4      axs[0,i].hist(d[comp], color=col[i], ec='black', bins="sturges")
5      axs[0,i].set_title(f"Histogram of {comp} Price")
6      stats.probplot(d[comp], dist="norm", plot=axs[1,i])
7      axs[1,i].set_title(f"Normal Q-Q Plot of {comp} Price")
8      axs[1,i].get_lines()[0].set_color(col[i])
9      pd.plotting.lag_plot(d[comp], lag=1, ax=axs[2,i], c=col[i])
10     axs[2,i].set_xlabel(f"{comp}(t)")
11     axs[2,i].set_ylabel(f"{comp}(t+1)")
12     axs[2,i].set_title(f"1-day Lagged Plot of {comp} Price")
13     pd.plotting.lag_plot(u[comp], lag=1, ax=axs[3,i], c=col[i])
14     axs[3,i].set_xlabel(f"{comp}(t)")
15     axs[3,i].set_ylabel(f"{comp}(t+1)")
16     axs[3,i].set_title(f"1-day Lagged Plot of {comp} Return")
```

Programme 3.19 Illustrating some empirical properties of the prices and returns of HSBC, CLP, and CK via Python.

```
1  > par(mfrow=c(4,3), mar=c(5,5,4,4))
2  >
3  > hist(d[,"HSBC"]); hist(d[,"CLP"]); hist(d[,"CK"])
4  >
5  > qqnorm(d[,"HSBC"]); qqline(d[,"HSBC"])
6  > qqnorm(d[,"CLP"]); qqline(d[,"CLP"])
7  > qqnorm(d[,"CK"]); qqline(d[,"CK"])
8  >
9  > plot(d[,"HSBC"], lag(d[,"HSBC"]))
10 > plot(d[,"CLP"], lag(d[,"CLP"]))
11 > plot(d[,"CK"], lag(d[,"CK"]))
12 >
13 > plot(u[,"HSBC"], lag(u[,"HSBC"]))
14 > plot(u[,"CLP"], lag(u[,"CLP"]))
15 > plot(u[,"CK"], lag(u[,"CK"]))
```

Programme 3.20 Illustrating empirical properties of the prices and returns of HSBC, CLP, and CK via R.

Compared with the plots of u_i in Section 3.1, it is obvious that the distributions of the S_i's are non-normal. More importantly, strong autocorrelations exist in S_i's while there is almost none in u_i's. These plots show that $\{S_i\}$ has strong autocorrelation (of lag 1) while $\{u_i\}$ has a small or no autocorrelation, which confirms the claim made in item 3. We can also plot the *autocorrelation function* (*acf*) of lag k for sample sequence $\{a_1, \ldots, a_n\}$:

$$r_k := \frac{\sum_{i=1}^{n-k}(a_i - \overline{a})(a_{i+k} - \overline{a})}{\sum_{i=1}^{n}(a_i - \overline{a})^2}, \text{ for } k = 1, \ldots, K,$$

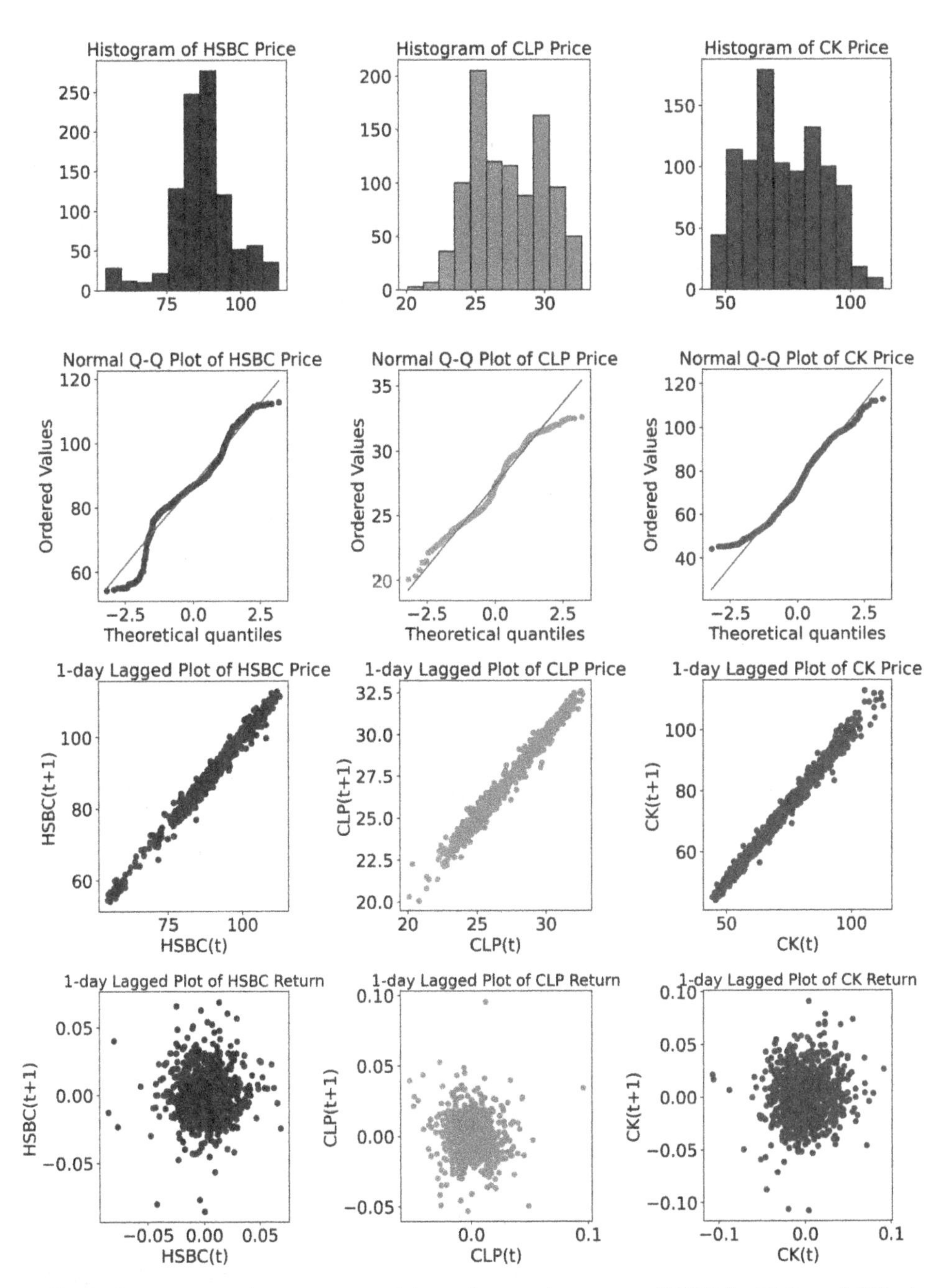

FIGURE 3.10 Empirical properties of the prices and returns of HSBC, CLP, and CK via Python, generated by Programme 3.19.

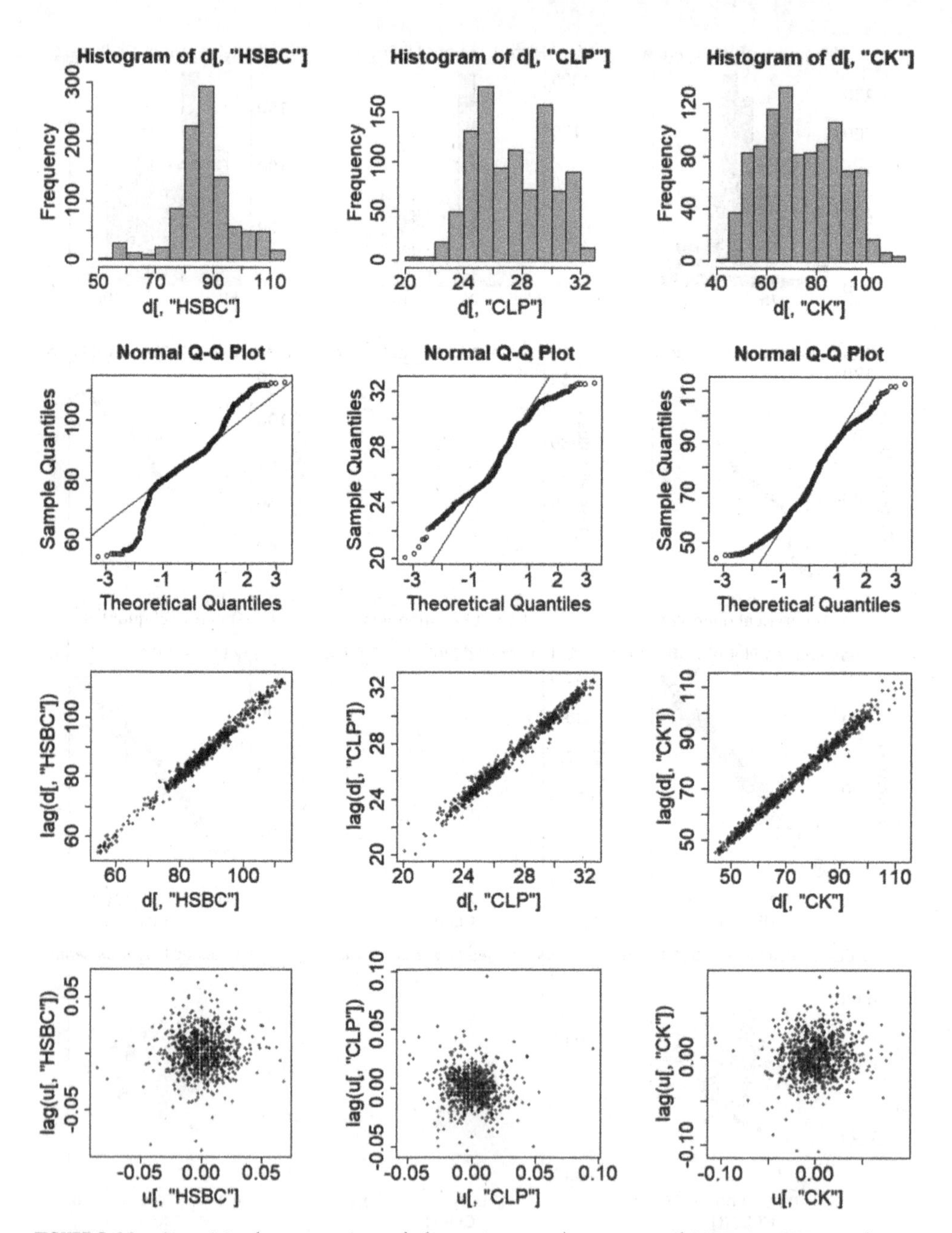

FIGURE 3.11 Empirical properties of the prices and returns of HSBC, CLP, and CK via R, generated by Programme 3.20.

where $\overline{a}$ is the sample mean of a_i's; we shall discuss this concept in more detail in Chapter 7. Figures 3.12 in Python and 3.13 in **R**, generated respectively from Programmes 3.21 and 3.22, also show that a strong autocorrelation exists in S_i, but there is little or no autocorrelation in u_i, and some autocorrelation in $\{u_i^2\}$ that means there could be a dependence structure at the volatility level; more details will be discussed in the GARCH modelings of Chapter 7.

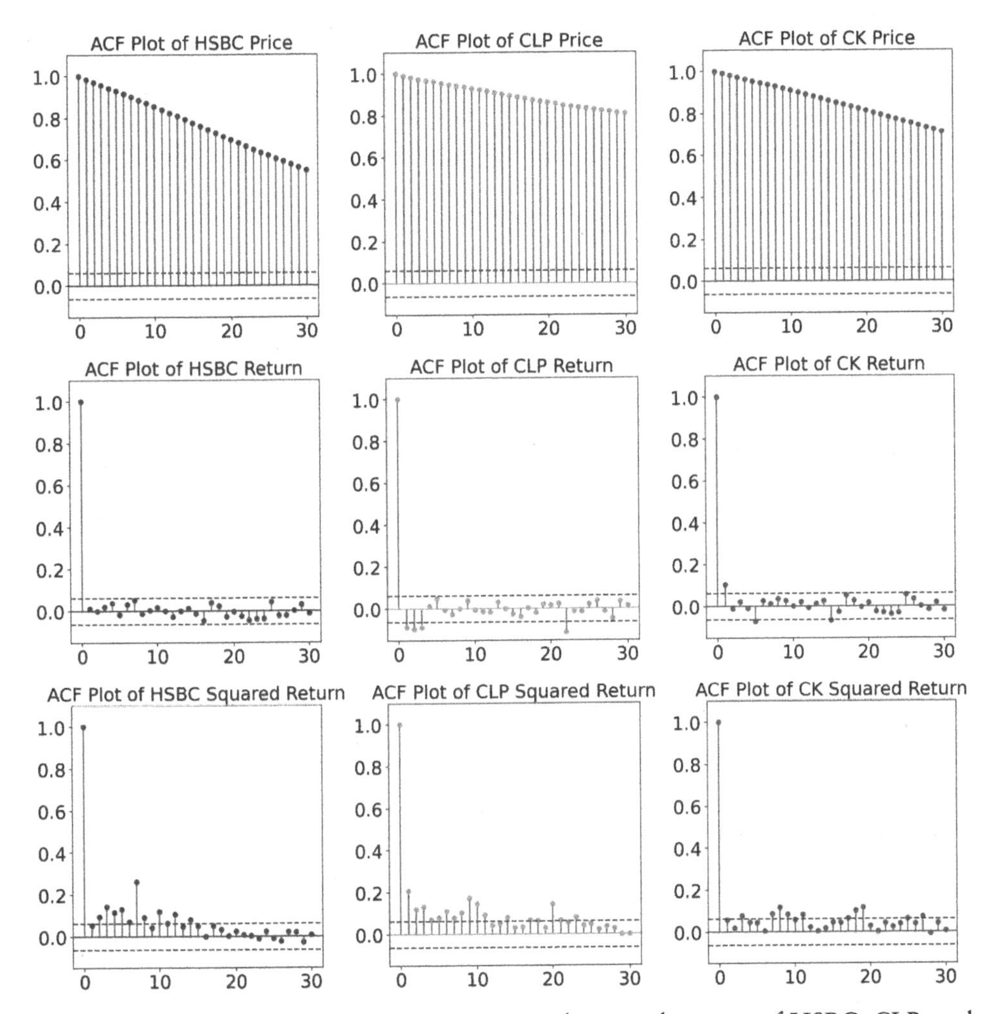

FIGURE 3.12 ACF plots of the prices, returns and squared returns of HSBC, CLP, and CK via Python, generated by Programme 3.21.

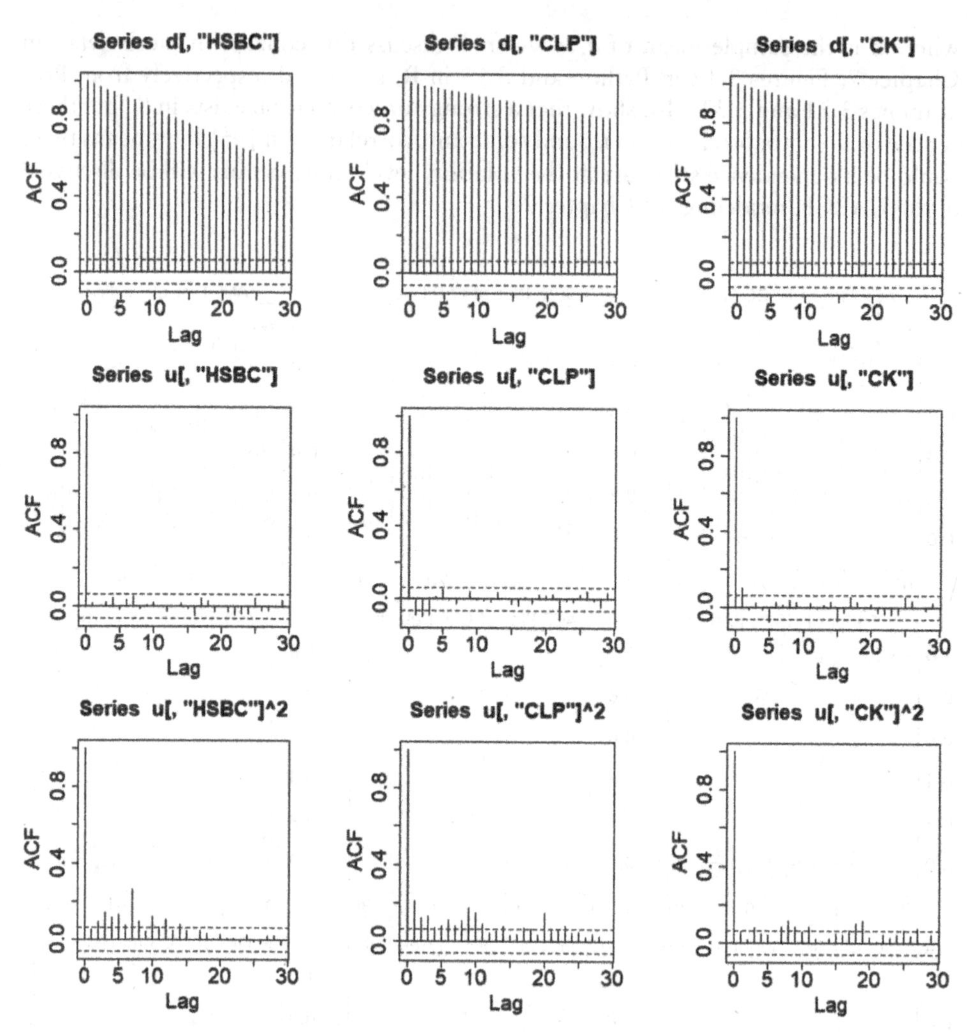

FIGURE 3.18 ACF plots of the prices, returns and squared returns of HSBC, CLP, and CK via **R**, generated by Programme 3.22.

```
1  from statsmodels.graphics.tsaplots import plot_acf
2  fig, axs = plt.subplots(3, 3, figsize=(15,15))
3  alpha = 0.05
4  c_i = stats.norm.ppf(1-alpha/2) / np.sqrt(len(d))
5
6  for i, comp in enumerate(d.columns):
7      plot_acf(d[comp], alpha=None, c=col[i],
8               title=f"ACF Plot of {comp} Price", ax=axs[0,i])
9      axs[0,i].set_ylim((-0.15, 1.1))
10     axs[0,i].axhline(c_i, linestyle="--", c="purple")
11     axs[0,i].axhline(-c_i, linestyle="--", c="purple")
12     plot_acf(u[comp], alpha=None, c=col[i],
13              title=f"ACF Plot of {comp} Return", ax=axs[1,i])
```

```
14        axs[1,i].set_ylim((-0.15, 1.1))
15        axs[1,i].axhline(c_i, linestyle="--", c="purple")
16        axs[1,i].axhline(-c_i, linestyle="--", c="purple")
17        plot_acf(u[comp]**2, alpha=None, c=col[i],
18                 title=f"ACF Plot of {comp} Squared Return", ax=axs[2,i])
19        axs[2,i].set_ylim((-0.15, 1.1))
20        axs[2,i].axhline(c_i, linestyle="--", c="purple")
21        axs[2,i].axhline(-c_i, linestyle="--", c="purple")
```

Programme 3.21 ACF plots of the prices, returns and squared returns of HSBC, CLP, and CK via Python.

```
1  > par(mfrow=c(3,3), mar=c(5,5,4,4))
2  >
3  > acf(d[,"HSBC"]); acf(d[,"CLP"]); acf(d[,"CK"])
4  > acf(u[,"HSBC"]); acf(u[,"CLP"]); acf(u[,"CK"])
5  > acf(u[,"HSBC"]^2); acf(u[,"CLP"]^2); acf(u[,"CK"]^2)
```

Programme 3.22 ACF plots of the prices, returns and squared returns of HSBC, CLP, and CK via R.

3.A APPENDIX

*3.A.1 Shapiro–Wilk Statistic

Very often, we test whether the hypothesis that the underlying distribution of a random sample is normal is true or not. To this end, we first rank the sample in ascending order $y_1 \leq \cdots \leq y_n$. If the hypothesis of normality is true, we can write $y_i = \mu + \sigma z_{(i)}$ for $i = 1, \ldots, n$, where $z_{(1)}, \ldots, z_{(n)}$ is an *ordered* random sample of n iid standard normal random variables. Here, μ and σ are the unknown population mean and standard deviation. We denote by $\boldsymbol{m} = \mathbb{E}(\mathbf{z})$ and $V = \mathrm{Var}(\mathbf{z})$ the mean vector and covariance matrix, respectively, of these n normal order statistics $\mathbf{z} = (z_{(1)}, \ldots, z_{(n)})^\top$. And $\mathbf{1}_n$ is an n-dimensional vector having all components equal to 1, $A := (\mathbf{1}_n, \boldsymbol{m})$ is an $n \times 2$ matrix, and $\boldsymbol{\theta} := (\mu, \sigma)^\top \in \mathbb{R}^2$. We can have the following decomposition into the linear model part and error term:

$$\mathbf{y} = \mu \mathbf{1}_n + \sigma \boldsymbol{m} + \sigma(\mathbf{z} - \boldsymbol{m}) = A\boldsymbol{\theta} + \boldsymbol{\epsilon}, \tag{3.4}$$

where $\boldsymbol{\epsilon} := \sigma(\mathbf{z} - \boldsymbol{m})$ is the error vector whose mean and covariance are respectively:

$$\mathbb{E}(\boldsymbol{\epsilon}) = \sigma(\mathbb{E}(\mathbf{z}) - \mathbf{m}) = \mathbf{0} \qquad \text{and} \qquad \mathrm{Cov}(\boldsymbol{\epsilon}) = \sigma^2 V.$$

As in [9], the weighted least square estimate for $\boldsymbol{\theta}$ of model (3.4), which minimizes the loss function

$$\|(V^{-1/2}A)\boldsymbol{\theta} - (V^{-1/2}\boldsymbol{y})\|_2^2, \tag{3.5}$$

is given by

$$\begin{aligned}\hat{\boldsymbol{\theta}} &= [(V^{-1/2}A)^\top (V^{-1/2}A)]^{-1}(V^{-1/2}A)^\top (V^{-1/2}\boldsymbol{y}) \\ &= (A^\top V^{-1}A)^{-1}A^\top V^{-1}\boldsymbol{y}\end{aligned}$$

$$
\begin{aligned}
&= \begin{pmatrix} \mathbf{1}_n^\top V^{-1}\mathbf{1}_n & \mathbf{1}_n^\top V^{-1}\boldsymbol{m} \\ \mathbf{1}_n^\top V^{-1}\boldsymbol{m} & \boldsymbol{m}^\top V^{-1}\boldsymbol{m} \end{pmatrix}^{-1} \begin{pmatrix} \mathbf{1}_n^\top \\ \boldsymbol{m}^\top \end{pmatrix} V^{-1}\boldsymbol{y} \\
&= C \begin{pmatrix} \boldsymbol{m}^\top V^{-1}\boldsymbol{m} & -\mathbf{1}_n^\top V^{-1}\boldsymbol{m} \\ -\mathbf{1}_n^\top V^{-1}\boldsymbol{m} & \mathbf{1}_n^\top V^{-1}\mathbf{1}_n \end{pmatrix} \begin{pmatrix} \mathbf{1}_n^\top \\ \boldsymbol{m}^\top \end{pmatrix} V^{-1}\boldsymbol{y} \\
&= C \begin{pmatrix} \boldsymbol{m}^\top V^{-1}\boldsymbol{m}\mathbf{1}_n^\top - \mathbf{1}_n^\top V^{-1}\boldsymbol{m}\boldsymbol{m}^\top \\ -\mathbf{1}_n^\top V^{-1}\boldsymbol{m}\mathbf{1}_n^\top + \mathbf{1}_n^\top V^{-1}\mathbf{1}_n\boldsymbol{m}^\top \end{pmatrix} V^{-1}\boldsymbol{y}, \qquad (3.6)
\end{aligned}
$$

where

$$C := \frac{1}{\mathbf{1}_n^\top V^{-1}\mathbf{1}_n \cdot \boldsymbol{m}^\top V^{-1}\boldsymbol{m} - (\mathbf{1}_n^\top V^{-1}\boldsymbol{m})^2}.$$

Since $z_1, \ldots, z_n$ follow a symmetric distribution, the two sets of ordered observations $z_{(1)}, \ldots, z_{(n)}$ and $-z_{(n)}, \ldots, -z_{(1)}$ have exactly the same joint distribution; we define $J := \begin{pmatrix} 0 & \cdots & 1 \\ \vdots & \ddots & \vdots \\ 1 & \cdots & 0 \end{pmatrix}$ with all anti-diagonal entries being 1 and all other entries zero, then y and $-J\mathrm{y}$ have the same joint distribution, and hence $\boldsymbol{m} = -J\boldsymbol{m}$ and $V = JVJ$. Together with the fact that $J^{-1} = J$, we have

$$\mathbf{1}_n^\top V^{-1}\boldsymbol{m} = \mathbf{1}_n^\top J V^{-1} J(-J\boldsymbol{m}) = -\mathbf{1}_n^\top V^{-1}\boldsymbol{m},$$

and thus $\mathbf{1}_n^\top V^{-1}\boldsymbol{m} = 0$, with which we can simplify (3.6) as:

$$\hat{\theta} = \frac{1}{\mathbf{1}_n^\top V^{-1}\mathbf{1}_n \cdot \boldsymbol{m}^\top V^{-1}\boldsymbol{m}} \begin{pmatrix} \boldsymbol{m}^\top V^{-1}\boldsymbol{m} \cdot \mathbf{1}_n^\top V^{-1}\boldsymbol{y} \\ \mathbf{1}_n^\top V^{-1}\mathbf{1}_n \cdot \boldsymbol{m}^\top V^{-1}\boldsymbol{y} \end{pmatrix},$$

hence the weighted least square estimates for μ and σ are, respectively:

$$
\begin{aligned}
\hat{\mu} &= \frac{(\boldsymbol{m}^\top V^{-1}\boldsymbol{m})(\mathbf{1}_n^\top V^{-1}\boldsymbol{y})}{(\mathbf{1}_n^\top V^{-1}\mathbf{1}_n)(\boldsymbol{m}^\top V^{-1}\boldsymbol{m})} = \frac{\mathbf{1}_n^\top V^{-1}\boldsymbol{y}}{\mathbf{1}_n^\top V^{-1}\mathbf{1}_n}; \\
\hat{\sigma} &= \frac{(\mathbf{1}_n^\top V^{-1}\mathbf{1}_n)(\boldsymbol{m}^\top V^{-1}\boldsymbol{y})}{(\mathbf{1}_n^\top V^{-1}\mathbf{1}_n)(\boldsymbol{m}^\top V^{-1}\boldsymbol{m})} = \frac{\boldsymbol{m}^\top V^{-1}\boldsymbol{y}}{\boldsymbol{m}^\top V^{-1}\boldsymbol{m}}, \qquad (3.7)
\end{aligned}
$$

where the last equality follows from using the fact that $\mathbf{1}_n^\top V^{-1}\boldsymbol{m} = 0$.

The key idea of the Shapiro–Wilk test is that when one is regressing the ordered sample $y_1, \ldots, y_n$ against the corresponding expected normal order statistics $\boldsymbol{m}$ under the null hypothesis, a linear fit can be expected if the sample really comes from a normal distribution. The Shapiro–Wilk test statistic W, motivated by the R^2-statistic in simple linear regression, determines how good this linear fit is. Recall that R^2 [21] for a simple linear regression of $y_i = \beta_0 + \beta_1 x_i + \epsilon_i$, for $i = 1, \ldots, n$, is given by

$$R^2 = \frac{\sum_{i=1}^n (\hat{y}_i - \bar{y})^2}{\sum_{i=1}^n (y_i - \bar{y})^2},$$

where $\hat{y}_i$ is the fitted value of this simple linear model for the actual observation i. Note that the numerator can be rewritten as

$$\begin{aligned}\sum_{i=1}^{n}(\hat{y}_i-\overline{y})^2 &= \sum_{i=1}^{n}(\hat{\beta}_0+\hat{\beta}_1 x_i-\overline{y})^2=\sum_{i=1}^{n}(\overline{y}-\hat{\beta}_1\overline{x}+\hat{\beta}_1 x_i-\overline{y})^2\\ &=\hat{\beta}_1^2\sum_{i=1}^{n}(x_i-\overline{x})^2=\frac{\left(\sum_{i=1}^{n}(x_i-\overline{x})(y_i-\overline{y})\right)^2}{\sum_{i=1}^{n}(x_i-\overline{x})^2}\\ &=(\boldsymbol{\alpha}^\top \boldsymbol{y})^2, \end{aligned} \tag{3.8}$$

where the second last equality holds by recalling $\hat{\beta}_1=\frac{\sum_{i=1}^{n}(x_i-\overline{x})(y_i-\overline{y})}{\sum_{j=1}^{n}(x_j-\overline{x})^2}$, and $\boldsymbol{\alpha}:=(\alpha_1,\dots,\alpha_n)^\top$ such that $\alpha_i:=\frac{x_i-\overline{x}}{\sqrt{\sum_{j=1}^{n}(x_j-\overline{x})^2}}$ for $i=1,\dots,n$; clearly, $\|\boldsymbol{\alpha}\|_2^2=\sum_{i=1}^{n}\alpha_i^2=1$. Similar to this R^2 in simple linear regression, we aim to construct the Shapiro–Wilk test statistic as a quotient of the sample variance estimate of residuals and the total sum of squares. Note that the total sum of squares is still $s^2=\sum_{i=1}^{n}(y_i-\overline{y})^2$, and by (3.7), the residual sum of squares can be taken as $\hat{\sigma}^2=\left(\frac{\boldsymbol{m}^\top \boldsymbol{V}^{-1}}{\boldsymbol{m}^\top \boldsymbol{V}^{-1}\boldsymbol{m}}\boldsymbol{y}\right)^2$, which is in a similar form as (3.8), but the coefficient vector of $\frac{\boldsymbol{m}^\top \boldsymbol{V}^{-1}}{\boldsymbol{m}^\top \boldsymbol{V}^{-1}\boldsymbol{m}}$ may not have a unit 2-norm like $\boldsymbol{\alpha}$. Therefore, the Shapiro–Wilk test statistic is proposed in the following form: for some $c>0$,

$$W=\frac{(c\cdot\hat{\sigma})^2}{s^2}=\frac{1}{s^2}\left(\frac{c\boldsymbol{m}^\top \boldsymbol{V}^{-1}}{\boldsymbol{m}^\top \boldsymbol{V}^{-1}\boldsymbol{m}}\boldsymbol{y}\right)^2, \tag{3.9}$$

such that $\left\|\frac{c\boldsymbol{m}^\top \boldsymbol{V}^{-1}}{\boldsymbol{m}^\top \boldsymbol{V}^{-1}\boldsymbol{m}}\right\|_2^2=c^2\frac{\|\boldsymbol{m}^\top \boldsymbol{V}^{-1}\|_2^2}{(\boldsymbol{m}^\top \boldsymbol{V}^{-1}\boldsymbol{m})^2}=1$, just like $\boldsymbol{\alpha}$, which yields

$$c=\frac{\boldsymbol{m}^\top \boldsymbol{V}^{-1}\boldsymbol{m}}{\sqrt{\boldsymbol{m}^\top \boldsymbol{V}^{-1}\boldsymbol{V}^{-1}\boldsymbol{m}}}. \tag{3.10}$$

Defining $\boldsymbol{a}:=c\left(\frac{\boldsymbol{m}^\top \boldsymbol{V}^{-1}}{\boldsymbol{m}^\top \boldsymbol{V}^{-1}\boldsymbol{m}}\right)^\top=\frac{\boldsymbol{V}^{-1}\boldsymbol{m}}{\sqrt{\boldsymbol{m}^\top \boldsymbol{V}^{-1}\boldsymbol{V}^{-1}\boldsymbol{m}}}$, we then conclude with the following definite form of the Shapiro–Wilk statistic:

$$W=\frac{(\boldsymbol{a}^\top \boldsymbol{y})^2}{s^2}=\frac{\left(\sum_{i=1}^{n}a_i y_i\right)^2}{\sum_{i=1}^{n}(y_i-\overline{y})^2}. \tag{3.11}$$

Finally, after this standardization, we claim that W has the same property as R^2 on determining goodness-of-fit, namely the closer in value to 1, the better the linear fit is, which further implies the normal nature of the underlying sample. Under the null hypothesis, if the population distribution is normal, then $\boldsymbol{y}$ has a linear relation with $\boldsymbol{m}$, which leads to a stronger alignment between $\boldsymbol{y}$ and $\boldsymbol{a}$ and so a larger value of W.

On the other hand, since $\mathbf{1}_n^T a = 0$ by recalling that $\mathbf{1}_n^T V^{-1} m = 0$ as mentioned before, and $||a||_2 = 1$, by the *Cauchy–Schwarz inequality*,

$$W = \frac{\left(\sum_{i=1}^n a_i y_i\right)^2}{\sum_{i=1}^n (y_i - \overline{y})^2} = \frac{\left(\sum_{i=1}^n a_i (y_i - \overline{y})\right)^2}{\sum_{i=1}^n (y_i - \overline{y})^2} \leq \frac{\sum_{i=1}^n a_i^2 \sum_{i=1}^n (y_i - \overline{y})^2}{\sum_{i=1}^n (y_i - \overline{y})^2} = 1,$$

which tells us the largest possible value for W is one. As a result, if W is much smaller than 1, we are more confident to reject the null hypothesis of the normal nature of the underlying sample. For more discussion on the statistical properties of the test, such as its *consistency*, we refer readers to [8, 13, 18]. Meanwhile, some variants of the Shapiro–Wilk test are also available in various references including [14, 18].

3.A.2 Kolmogorov–Smirnov Test

Suppose that $x_{(1)} < x_{(2)} < \cdots < x_{(n)}$ are n real-valued observations, arranged in the ascending order, of $x_1, \ldots, x_n$ drawn from an unknown distribution F of continuous type on $\mathbb{R}$. For simplicity, it is assumed that all observations are distinct. The empirical distribution function is defined by:

$$\hat{F}_n(x) = \begin{cases} 0, & x < x_{(1)}; \\ \frac{k}{n}, & x_{(k)} \leq x < x_{(k+1)},\ k = 1, 2, \ldots, n-1; \\ 1, & x_{(n)} \leq x. \end{cases}$$

It has a jump of size $1/n$ occurring at each observed value x_i, and $B_x := \hat{F}_n(x)$, the value of the empirical distribution function at x, can be thought of as a random variable that can only take values of $0, 1/n, 2/n, \ldots$, or 1. It is clear that

$$\mathbb{P}(nB_x = k) = \mathbb{P}\left(B_x = \frac{k}{n}\right) = \binom{n}{k}(F(x))^k(1 - F(x))^{n-k}, \quad k = 0, 1, 2, \ldots, n,$$

which implies that $\mathbb{E}(nB_x) = nF(x)$ and $\mathrm{Var}(nB_x) = n(F(x))(1 - F(x))$. As a result, $\mathbb{E}(\hat{F}_n(x)) = \mathbb{E}(B_x) = F(x)$, and $\mathrm{Var}(\hat{F}_n(x)) = \mathrm{Var}(B_x) = \frac{F(x)(1-F(x))}{n}$. By Chebyshev's inequality, $\hat{F}_n(x)$ approaches its mean $F(x)$ as n approaches to infinity. In fact, we also have the following stronger result of uniform convergence.

Theorem 3.1 *(Glivenko–Cantelli* [6, 19, 20]*)* $\hat{F}_n(x)$ converges to $F(x)$ uniformly in x on $\mathbb{R}$ as $n \to \infty$.

The celebrated Kolmogorov–Smirnov statistic for a random sample following a distribution F_0 is defined as:

$$D_n := \sup_x |\hat{F}_n(x) - F_0(x)|,$$

which means the largest possible, or more precisely, the least upper bound, of all pointwise deviations $|\hat{F}_n(x) - F_0(x)|$ for any $x \in \mathbb{R}$. The exact distribution of D_n does not depend on the underlying distribution function $F_0(x)$, since $U := F_0(X)$ has a uniform distribution $U(0,1)$ whenever $X \sim F_0$; while under the null hypothesis $H_0 : F(x) \equiv F_0(x)$,

$$D_n = \sup_{0<F_0(x)<1} \left| \frac{1}{n} \sum_{i=1}^{n} \mathbb{1}_{\{F_0(x_i) \le F_0(x)\}} - F_0(x) \right| = \sup_{0<u<1} \left| \frac{1}{n} \sum_{i=1}^{n} \mathbb{1}_{\{u_i \le u\}} - u \right|,$$

where $u_i := F_0(x_i)$, $i = 1, \ldots, n$ are sample from $U(0,1)$. Therefore D_n is a distribution-free statistic. When testing the null hypothesis $H_0 : F(x) \equiv F_0(x)$ against all possible alternatives $H_1 : F(x) \not\equiv F_0(x)$, for some specified distribution function F_0, the null hypothesis H_0 is rejected if the observed value of D_n is greater than some chosen critical value. To form a confidence band based on a sample of size n at the critical level α, one can use the equation $\mathbb{P}(D_n \ge d_\alpha) = \alpha$ such that

$$\begin{aligned} 1 - \alpha &= \mathbb{P}(\sup_x |\hat{F}_n(x) - F(x)| \le d_\alpha) \\ &= \mathbb{P}(|\hat{F}_n(x) - F(x)| \le d_\alpha, \text{ for all } x) \\ &= \mathbb{P}(\hat{F}_n(x) - d_\alpha \le F(x) \le \hat{F}_n(x) + d_\alpha, \text{ for all } x). \end{aligned}$$

Define $\hat{F}_L(x) := \max(\hat{F}_n(x) - d_\alpha, 0)$ and $\hat{F}_U(x) := \min(\hat{F}_n(x) + d_\alpha, 1)$ for $x \in \mathbb{R}$. The two step functions $\hat{F}_L(x)$ and $\hat{F}_U(x)$ together yield a $100(1-\alpha)\%$ confidence band for the underlying distribution function $F(x)$.

REFERENCES

1. Black, F., and Scholes, M. (1972). The pricing of options and corporate liabilities. *The Journal of Political Economy*, 81(3), 637–654.
2. DeGroot, M.H., and Schervish, M.J. (2011). *Probability and Statistics* (4th ed.). Pearson.
3. Engle, R.F., Bali, T.G., and Murray, S. (2016). *Empirical Asset Pricing the Cross Section of Stock Returns*. Wiley.
4. Hogg, R., Tanis, E., and Zimmerman, D. (2013). *Probability and Statistical Inference* (9th ed.). Pearson.
5. Jondeau, E., Poon, S.H., and Rockinger, M. (2007). *Financial Modeling Under Non-Gaussian Distributions*. Springer.
6. Kallenberg, O. (1997). *Foundations of Modern Probability*. Springer.
7. Lehmann, E.L., and Romano, J.P. (2022). *Testing Statistical Hypothesis* (4th ed.). Springer.
8. Leslie, J.R., Stephens, M.A., and Fotopoulos, S. (1986). Asymptotic distribution of the Shapiro–Wilk *W* for testing for normality. *Annals of Statistics*, 14, 1497–1506.
9. Lloyd, E.H. (1952). Least-squares estimation of location and scale parameters using order statistics. *Biometrika*, 39(1/2), 88–95.
10. Lo, A.W., and Mackinlay, A.C. (1999). *A Non-Random Walk Down Wall Street*. Princeton University Press.

11. Markowitz, H.M., and Usmen, M. (1996). The likelihood of various stock market return distributions, Part 2: Empirical results. *Journal of Risk and Uncertainty*, 13, 221–247.
12. McLachlan, G.J. (1999). Mahalanobis distance. *Resonance*, 4(6), 20–26.
13. Sarkadi, K. (1981), On the consistency of some goodness of fit tests. *Proceedings of the Sixth Conference on Probability Theory, Brasov, 1979*, pp. 195–204.
14. Shapiro, S.S., and Francia, R.S. (1972). An approximate analysis of variance test for normality. *Journal of the American Statistical Association*, 67, 215–216.
15. Shapiro, S.S., and Wilk, M.B. (1965). An analysis of variance test for normality (complete samples). *Biometrika*, 52(3/4), 591–611.
16. Shorack, G.R. (2017). *Probability for Statisticians*. Springer.
17. Shorack, G.R., and Wellner, J.A. (2009). *Empirical Processes with Applications to Statistics*. Society for Industrial and Applied Mathematics.
18. Thode, H.C. (2002). *Testing for Normality*. CRC Press.
19. Van der Vaart, A.W. (2000). *Asymptotic Statistics*. Cambridge University Press.
20. Van der Vaart, A.W., and Wellner, J.A. (1996). *Weak Convergence and Empirical Processes with Applications to Statistics*. Springer.
21. Weisberg, S. (2014). *Applied Linear Regression*. Wiley.

CHAPTER 4

Financial Forensics

Banks and financial corporations are often vulnerable to fraudulent activities due to some misbehaviour of employees or procedures not followed correctly; criminals might take advantage of some basic accounting loopholes to commit illegal financial activities like money laundering. Financial forensics is an important tool for detecting such behaviour. In the era of big data, there is a vast amount of information to be processed, and it becomes nearly impossible for humans, no matter how large the working team is, to look deeply in every case worthy of investigation. As examples, Benford's Law and Zipf's Law can be adopted to quickly locate any potential areas that are prone to fraudulent activity. For real-world naturally arising numbers, Benford's Law tells us the distribution of the first few leading-digits in financial statements, whereas Zipf's Law tells us the rank-frequency distribution in macroeconomic data and social sciences.

4.1 BENFORD'S LAW

In 1881, American mathematician *Simon Newcomb* (1835–1909) made an offbeat observation about books of logarithms. He noticed that earlier pages of the books are more smudged and worn much more than later pages; those numbers with smaller leading digits were being looked up more often for calculations than those with larger ones, and he formulated the empirical law that the proportion of numbers starting with the digit d is not uniform, instead it is governed by $\log_{10}(1 + 1/d)$, for $d = 1, 2, \ldots, 9$. This phenomenon was mentioned in an article in *American Journal of Mathematics* in 1881 yet at that time he could not provide a rigorous mathematical justification for his own claim.

Later in 1938, physicist *Frank Benford* made the same observation as Newcomb. He compiled a table of 20,229 numbers from a wide variety of datasets of areas of rivers, death rates, baseball statistics, numbers in magazine articles and many others. His results were largely consistent with the pattern previously observed by Newcomb. This counterintuitive law of nature, rather than the uniform one, has been named after him.

Undoubtedly, many sets of numbers do not follow Benford's law, such as random numbers and numbers drawn from other statistical distributions. In 1996,

Theodore Hill, from the Georgia Institute of Technology, showed that if the sampling distributions are selected at random such that samples are drawn from each of these distributions, i.e. drawing samples from an uncertain distribution, the leading-digit frequencies of the combined sample would eventually converge to Benford's law, even though this may not be true for individual samples taken from a fixed distribution.

An alternative way to understand this empirical Benford's law is to see the collection of all numerical figures as base-independent, and the same law is applicable to any other base systems different from base-10. We expect the same law should remain valid even after a change of measurement unit, which is achieved by a multiplication with some scaling number. This property is called *scaling invariance*. For instance, 100 meters can be transformed to 0.1 kilometers or $10,000$ centimeters, but the leading digit 1 remains unchanged regardless of the convention of measurement used. In 1961, the mathematician *Roger Pinkham* of Rutgers University proved that scaling invariance does imply Benford's law. As an illustration, consider a set of numbers from the $U[1,10)$ distribution so that the expected frequency of each leading digit d is 1/9. We change the unit by multiplying with a factor 2. Table 4.1 shows the resulting leading digits of the newly obtained numbers and their corresponding expected frequencies (probabilities). It clearly shows that the uniform nature of the distribution is lost after this simple scaling, with an expectation that $5/9$ of the resulting numbers have 1 as their leading digit. Hence, the uniform distribution cannot be scale-invariant.

TABLE 4.1 The expected frequencies of the leading digits of randomly selected numbers after scaling by a factor 2.

Original Interval	Expected Frequencies before ×2	Leading Digit after ×2
$[1,1.5)$	$\frac{1}{2}\times\frac{1}{9}$	2
$[1.5,2)$	$\frac{1}{2}\times\frac{1}{9}$	3
$[2,2.5)$	$\frac{1}{2}\times\frac{1}{9}$	4
$[2.5,3)$	$\frac{1}{2}\times\frac{1}{9}$	5
$[3,3.5)$	$\frac{1}{2}\times\frac{1}{9}$	6
$[3.5,4)$	$\frac{1}{2}\times\frac{1}{9}$	7
$[4,4.5)$	$\frac{1}{2}\times\frac{1}{9}$	8
$[4.5,5)$	$\frac{1}{2}\times\frac{1}{9}$	9
$[5,10)$	$\frac{5}{9}$	1

4.2 SCALING INVARIANCE AND BENFORD'S LAW

Consider a σ-finite measure[1] μ on $[1, \infty)$ with a (continuous) density f_μ:

$$\mu(a, b) := \int_a^b f_\mu(x)\, dx, \qquad 1 \le a < b < \infty,$$

it is said to be scale-invariant if for any $\alpha > 1$,

$$\mu(\alpha a, \alpha b) = \mu(a, b), \qquad \text{for any} \quad 1 < a < b < \infty.$$

In particular, we can deduce the following property of f_μ:

$$\mu(\alpha a, \alpha b) = \int_{\alpha a}^{\alpha b} f_\mu(x)\, dx = \int_a^b f_\mu(\alpha y) \cdot \alpha\, dy = \mu(a, b) = \int_a^b f_\mu(y)\, dy,$$

via a simple change-of-variables $x = \alpha y$. Since a and b are arbitrary,

$$f_\mu(\alpha y) \cdot \alpha = f_\mu(y), \qquad \text{for a.e. } y \in [1, \infty).$$

Further, if f_μ is continuous, we can conclude that

$$f_\mu(\alpha y) \cdot \alpha = f_\mu(y), \qquad \text{for all } y \in [1, \infty). \tag{4.1}$$

Next, we define another σ-finite measure ν on $[0, \infty)$ such that, for some $\theta > 1$ and $a, b \in [0, \infty)$,

$$\nu(a, b) = \mu(\theta^a, \theta^b) = \int_{\theta^a}^{\theta^b} f_\mu(x)\, dx = \int_a^b f_\mu(\theta^z) \cdot \theta^z \ln\theta\, dz =: \int_a^b f_\nu(z)\, dz,$$

where we introduce the variable substitution $x = \theta^z$. Note that, for any $\varepsilon > 0$, by using (4.1),

$$\begin{aligned} f_\nu(z + \varepsilon) &= f_\mu(\theta^{z+\varepsilon}) \cdot \theta^{z+\varepsilon} \ln\theta = f_\mu(\theta^z \cdot \theta^\varepsilon) \cdot \theta^z \cdot \theta^\varepsilon \ln\theta \\ &= \theta^{-\varepsilon} f_\mu(\theta^z) \cdot \theta^z \cdot \theta^\varepsilon \ln\theta = f_\mu(\theta^z) \cdot \theta^z \ln\theta = f_\nu(z). \end{aligned}$$

Hence, f_ν is translational invariant since both z and $\varepsilon > 0$ are arbitrary, and so $f_\nu(z) = 1/C$ for some $C > 0$, meaning that ν is clearly a uniform (σ-finite) measure on $[0, \infty)$. By working backwards, we can also deduce that

$$f_\mu(x) = \frac{1}{\ln\theta} \cdot \frac{1}{Cx}, \qquad \text{for } x \ge 1. \tag{4.2}$$

[1] A σ-finite measure μ means that for any compact subset K of $\mathbb{R}$ now, $\mu(K)$ is finite; this allows the possibility that the μ-measure of the totality can be infinite.

We next assign a scale-invariant σ-finite measure on $[1, \infty)$, with the aim of explaining the phenomenon of the distribution of the leading digit. We first rewrite the number $x > 1$ in its scientific notation $x = \overline{x} \times 10^k$, where $1 \leq \overline{x} < 10$ is called the *mantissa* of x for some $k \in \mathbb{Z}_+$. The leading digit, denoted by $d_1(x)$, is simply the integral part of $\overline{x}$, that is, $d_1(x) = \lfloor \overline{x} \rfloor$. Pick a point x at random with a probability density proportional to $f_\mu(x)$ given by (4.2) on $[1, M]$; we denote the corresponding density by $f_{\mathrm{x}}^{(M)}(x) = \frac{1}{x} \Big/ \int_1^M \frac{1}{s}\, ds = \frac{1}{x \ln M}$. By the very definition, the leading digit of a number after scaling, with whatever factor $\alpha > 1$, remains in $[1, 10)$, then what sort of probability distribution does this leading digit $d_1(\mathrm{x})$ follow if $\mathrm{x} \in [1, M]$ follows $f_{\mathrm{x}}^{(M)}(x)$ when M is very large? Define the set $\mathcal{K}_M := \{0, \ldots, \lfloor \log_{10} M \rfloor\}$. We first compute, for $n = 1, 2, \ldots, 9$, the following probability:

$$
\begin{aligned}
&\mu(\mathrm{x} \in [1, M], d_1(\mathrm{x}) = n) \\
&= \mu(\mathrm{x} \in [1, M], n \leq \overline{\mathrm{x}} < n + 1) \\
&= \mu(\mathrm{x} \in [1, M], n \times 10^k \leq \mathrm{x} < (n+1) \times 10^k, \text{for some } k \in \mathcal{K}_M) \\
&= \nu(\mathrm{z} \in [0, \log_{10} M], \log_{10} n + k \leq \mathrm{z} < \log_{10}(n+1) + k, \text{for some } k \in \mathcal{K}_M), \text{for } \theta = 10 \\
&= \sum_{k \in \mathcal{K}_M} \nu(\mathrm{z} \in [0, \log_{10} M], \log_{10} n + k \leq \mathrm{z} < \log_{10}(n+1) + k) \\
&= \sum_{k \in \mathcal{K}_M \setminus \{\lfloor \log_{10} M \rfloor\}} \left(\frac{\log_{10}(n+1) + k}{C} - \frac{\log_{10} n + k}{C} \right) \\
&\qquad + \nu(\mathrm{z} \in (\lfloor \log_{10} M \rfloor, \log_{10} M], \log_{10} n + \lfloor \log_{10} M \rfloor \leq \mathrm{z} < \log_{10}(n+1) + \lfloor \log_{10} M \rfloor) \\
&=: (\lfloor \log_{10} M \rfloor - 1 + 1) \cdot \frac{\log_{10}(n+1) - \log_{10} n}{C} + \nu_{n,\text{remainder}}.
\end{aligned}
$$

Here, x is bounded above by $M > 1$ such that the summation over $k \in \mathcal{K}_M$ only ranges from 0 to $\lfloor \log_{10} M \rfloor - 1$, i.e. the sum is taken over all $\lfloor \log_{10} M \rfloor$ numbers of blue empty boxes in Figure 4.1, and there is an additional remainder term corresponding to when z falls in the interval $(\lfloor \log_{10} M \rfloor, \log_{10} M]$, i.e., the red shaded box in Figure 4.1. These blue empty boxes represent the interval $[\log_{10} n + k, \log_{10}(n+1) + k)$ for $k = 0, \ldots, \lfloor \log_{10} M \rfloor - 1$. Notice that any possible red shaded box lying in $(\lfloor \log_{10} M \rfloor, \lfloor \log_{10} M \rfloor + 1)$ does not contribute to the sum, and the term $\nu_{n,\text{remainder}}$ may be smaller in size than the blue empty boxes.

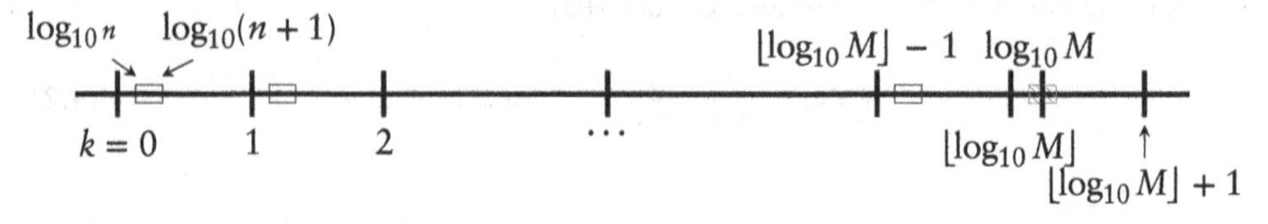

FIGURE 4.1 Summation over $k \in \mathcal{K}_M$ under the ν measure.

We next consider the following quotient which is the probability of $d_1(\mathrm{x}) = n$ under the density $f_{\mathrm{x}}^{(M)}$:

$$\frac{\mu(\mathrm{x} \in [1, M], d_1(\mathrm{x}) = n)}{\mu(\mathrm{x} \in [1, M])} = \lfloor \log_{10} M \rfloor \cdot \frac{\log_{10}(n+1) - \log_{10} n}{\log_{10} M} + \frac{v_{n,\text{remainder}}}{\log_{10} M}$$

$$\longrightarrow \log_{10}(n+1) - \log_{10} n = \log_{10}\left(1 + \frac{1}{n}\right), \quad \text{as } M \to \infty.$$

Moreover, we can also analyze the frequency of first two digits. Given the expression $x = \overline{x_{1,2}} \times 10^k$, where $10 \leq \overline{x_{1,2}} < 100$ and $k \in \mathcal{K}_{2,M} := \{0, \ldots, \lfloor \log_{10} M \rfloor - 1\}$, the first two digits are $d_1(x)$ and $d_2(x)$, respectively, such that $d_1(x) = \lfloor \overline{x_{1,2}}/10 \rfloor \in \{1, \ldots, 9\}$ and $d_2(x) = \lfloor \overline{x_{1,2}} - 10 \times d_1(x) \rfloor \in \{0, \ldots, 9\}$. Denote $\overline{n_1 n_2} := 10 \times n_1 + n_2$, for $n_1 = 1, \ldots, 9$ and $n_2 = 0, \ldots, 9$. We then compute:

$$\begin{aligned}
&\mu(\mathrm{x} \in [10, M], d_1(\mathrm{x}) = n_1, d_2(\mathrm{x}) = n_2) \\
&= \mu(\mathrm{x} \in [10, M], \overline{n_1 n_2} \leq \overline{\mathrm{x}_{1,2}} < \overline{n_1 n_2} + 1) \\
&= \mu(\mathrm{x} \in [10, M], \overline{n_1 n_2} \times 10^k \leq \mathrm{x} < (\overline{n_1 n_2} + 1) \times 10^k, \text{for some } k \in \mathcal{K}_{2,M}) \\
&= v(\mathrm{z} \in [1, \log_{10} M], \log_{10}\overline{n_1 n_2} + k \leq \mathrm{z} < \log_{10}(\overline{n_1 n_2} + 1) + k, \text{for some } k \in \mathcal{K}_{2,M}), \\
&\quad \text{for } \theta = 10 \\
&= \sum_{k \in \mathcal{K}_{2,M}} v(\mathrm{z} \in [1, \log_{10} M], \log_{10}\overline{n_1 n_2} + k \leq \mathrm{z} < \log_{10}(\overline{n_1 n_2} + 1) + k) \\
&= (\lfloor \log_{10} M \rfloor - 1 - 1 + 1) \cdot \frac{\log_{10}(\overline{n_1 n_2} + 1) - \log_{10}\overline{n_1 n_2}}{C} + v_{1,2,\text{remainder}},
\end{aligned}$$

where the remainder $v_{1,2,\text{remainder}}$ is similar to that for the first digit case in that $v_{1,2,\text{remainder}}/\log_{10} M$ converges to 0 as $M \to \infty$.

Again, we consider the quotient which stands for the probability that $d_1(\mathrm{x}) = n_1$ and $d_2(\mathrm{x}) = n_2$ under $f_{\mathrm{x}}^{(M)}$. By using a similar argument as the first digit case,

$$\frac{\mu(\mathrm{x} \in [10, M], d_1(\mathrm{x}) = n_1, d_2(\mathrm{x}) = n_2)}{\mu(10, M)} \longrightarrow \log_{10}\left(1 + \frac{1}{\overline{n_1 n_2}}\right),$$

as $M \to \infty$, for $n_1 = 1, \ldots, 9$ and $n_2 = 0, \ldots, 9$.

Therefore, the limiting marginal pmf, as M goes to ∞, of the second digit is

$$\mathbb{P}(d_2(\mathrm{x}) = n_2) = \sum_{n_1=1}^{9} \log_{10}\left(1 + \frac{1}{\overline{n_1 n_2}}\right), \qquad \text{for} \quad n_2 = 0, \ldots, 9.$$

Furthermore, we can now ask a question: given the first leading digit $d_1(x)$ is n_1, what is the probability that the second digit $d_2(x)$ is n_2? The conditional probability, for $n_2 = 0, \dots, 9$, is then given by

$$\mathbb{P}(d_2(x) = n_2 \mid d_1(x) = n_1) = \log_{10}\left(1 + \frac{1}{\overline{n_1 n_2}}\right) \Big/ \log_{10}\left(1 + \frac{1}{n_1}\right).$$

For example, for $n_1 = 1, n_1 = 9$, those probabilities simplify to

$$\log_2\left(1 + \frac{1}{\overline{1n_2}}\right) \quad \text{and} \quad \frac{\log_{10}\left(1 + \frac{1}{\overline{9n_2}}\right)}{1 - \log_{10} 9},$$

respectively.

4.3 BENFORD'S LAW IN BUSINESS REPORTS

The use of Benford's law as a method of fraud detection has become prevalent in recent years, motivated by the ever increasing amount of data available for checking by regulators, for which automated methods of fraud detection are needed in order to scan through these data effectively. This is of particular importance since investigations by regulators are often very costly, in terms of both time and money, and it is more favourable to have a computer-based checking to clear the fog so that they only have to focus on the most suspicious transactions. Therefore, Benford's law can serve as a very useful empirical law to screen data for any possible irregularities. If irregularities are detected, it may not be due to fraudulent activities, and a more comprehensive investigation is initiated to ascertain their cause. Among the earliest research, which proposed utilizing the leading digit frequencies and Benford's law to serve as a baseline for fraud detection, was [49], wherein the author indicated that socio-economic or financial data is often drafted with a bias to support critical policy plannings by manual intervention. He further suggested that a common way of creating fictitious data tends to distribute the digits involved uniformly owing to human intuition and prejudice on natural symmetry. In contrast, most realistic socio-economic or financial data will obey scale-invariance with respect to different measurement units, which ultimately leads the digits to follow Benford's law approximately. The scholar *Mark J. Nigrini*, from West Virginia University, has coined the name of using Benford's law for accounting fraud detection as "digit analysis", which is a special quantitative method in "forensic auditing". In particular, he describes:

> "*Benford's Law provides auditors with the expected digit frequencies in tabulated data. By examining the digit and the number frequencies, auditors can gain data insights that might be missed using traditional analytical procedures and sampling methods. The digit and number patterns could point to number invention, systematic frauds, data errors, or biases in the data. Research is currently underway on advanced tests to detect anomalies in data subsets.*"

Furthermore, Nigrini [31] proposed a "five-digit test" as a formal and systematic framework to quickly screen out any fraudulent numbers present in public filings, so that companies with the most anomalous data are targeted for another deeper level of expert auditing. The "five-digit test" examines the validity of the following digits against Benford's law, with the respective expected frequencies listed aside:

1. The first leading digit; also see Figure 4.2a:

$$\mathbb{P}(d_1(x) = n) = \log_{10}\left(1 + \frac{1}{n}\right), \qquad \text{for } n = 1, \ldots, 9; \tag{4.3}$$

2. The second leading digit; also see Figure 4.2b:

$$\mathbb{P}(d_2(x) = n_2) = \sum_{n_1=1}^{9} \log_{10}\left(1 + \frac{1}{n_1 n_2}\right), \qquad \text{for } n_2 = 0, 1, \ldots, 9; \tag{4.4}$$

3. The first and second leading digits:

$$\mathbb{P}(d_1(x) = n_1, d_2(x) = n_2) = \log_{10}\left(1 + \frac{1}{n_1 n_2}\right), \tag{4.5}$$

for $n_1 = 1, \ldots, 9$, $n_2 = 0, 1, \ldots, 9$;

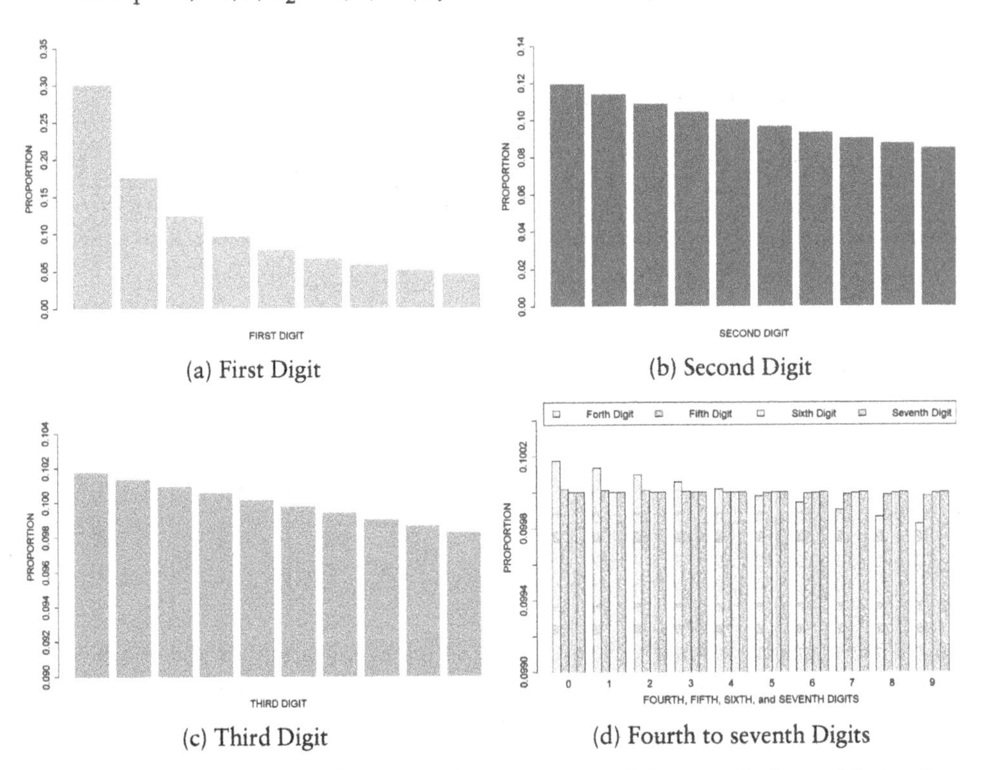

(a) First Digit

(b) Second Digit

(c) Third Digit

(d) Fourth to seventh Digits

FIGURE 4.2 Histograms for the expected proportions of the set of all possible leading digits in accordance with Benford's law.

4. The first, second and third leading digits:

$$\mathbb{P}(d_1(\mathrm{x}) = n_1, d_2(\mathrm{x}) = n_2, d_3(\mathrm{x}) = n_3) = \log_{10}\left(1 + \frac{1}{\overline{n_1 n_2 n_3}}\right), \quad (4.6)$$

for $n_1 = 1, \dots, 9$ and $n_2, n_3 = 0, 1, \dots, 9$;

5. The last two digits, i.e., the fourth digit onwards: the frequencies of the latter digit should almost be uniform; also see Figure 4.2d.

To see the reason behind this almost-uniformity for the latter digits, let us consider the $(\ell + 1)$-th one. By using similar arguments to those for the first two leading digits, we can see that the probabilities of $d_{\ell+1}(\mathrm{x}) = 0, \dots, 9$ are respectively:

$$\mathbb{P}(d_{\ell+1}(\mathrm{x}) = n_{\ell+1}) = \sum_{n_1=1}^{9} \sum_{n_2=0}^{9} \cdots \sum_{n_\ell=0}^{9} \log_{10}\left(1 + \frac{1}{\overline{n_1 n_2 \cdots n_\ell n_{\ell+1}}}\right),$$

where $\overline{n_1 \cdots n_{\ell+1}} := n_1 \times 10^{\ell} + n_2 \times 10^{\ell-1} + \cdots + n_\ell \times 10 + n_{\ell+1}$. We can also see that these probabilities are decreasing in $n_{\ell+1}$, hence

$$\begin{aligned}
0 &< \sup_{i>j\in\{0,\dots,9\}} \left(\mathbb{P}(d_{\ell+1}(\mathrm{x}) = i) - \mathbb{P}(d_{\ell+1}(\mathrm{x}) = j)\right) \\
&\leq \mathbb{P}(d_{\ell+1}(\mathrm{x}) = 0) - \mathbb{P}(d_{\ell+1}(\mathrm{x}) = 9) \\
&= \sum_{n_1=1}^{9} \sum_{n_2=0}^{9} \cdots \sum_{n_\ell=0}^{9} \left[\log_{10}\left(1 + \frac{1}{\overline{n_1 n_2 \cdots n_\ell 0}}\right) - \log_{10}\left(1 + \frac{1}{\overline{n_1 n_2 \cdots n_\ell 9}}\right)\right] \\
&\leq \sum_{n_1=1}^{9} \sum_{n_2=0}^{9} \cdots \sum_{n_\ell=0}^{9} 1 \cdot \left(\frac{1}{\overline{n_1 n_2 \cdots n_\ell 0}} - \frac{1}{\overline{n_1 n_2 \cdots n_\ell 9}}\right) \\
&\leq \sum_{n_1=1}^{9} \sum_{n_2=0}^{9} \cdots \sum_{n_\ell=0}^{9} \frac{9}{(\overline{n_1 n_2 \cdots n_\ell 0})^2} \\
&= \frac{9}{10^2}\left[\frac{1}{(\underbrace{10\cdots00}_{\ell-1 \text{ zeros}})^2} + \frac{1}{(10\cdots01)^2} + \cdots + \frac{1}{(\underbrace{99\cdots99}_{\ell \text{ nines}})^2}\right],
\end{aligned}$$

where the second inequality is due to a simple application of the mean value theorem. As a truncated tailed sum of the convergent Euler's series $\pi^2/6 = 1 + 1/2^2 + 1/3^2 + \cdots$, we clearly see that the last expression must converge to zero as ℓ goes to infinity. Therefore, for large ℓ, all $\mathbb{P}(d_\ell(\mathrm{x}) = i)$'s, for $i = 1, \dots, 9$, are essentially the same numerically.

Next we describe the five-digit test procedure. Firstly, tests against the first and second leading digits are regarded as the high level ones for screening out whether or not a company's public filing is reasonable. For those companies having financial reports containing the first and second digits whose frequencies deviate significantly from Benford's law, the first two and the first three digits are further tested in order to narrow down the list of companies to be targeted for auditing. Further, if these two tests cannot be passed, we further test for the last two digits so as to detect any fabricated and rounded numbers; readers with interest can also refer to [31] for a more extensive discussion and in-depth guidelines on Benford's law in forensic accounting.

In [31], there is an empirical analysis about the personal spending of the former president of the United States *Bill Clinton*. Since his presidency in 1993, taxes paid by both Bill and *Hillary Clinton* were repeatedly under investigation. For instance, in [40] and [48] the authors used line-item[2] analysis to assess whether the requested tax deductions by the Clintons were legitimate. By 1994, the Clintons had made their entire tax returns over the period of 1977 to 1992 public so as to address public concern[3]. In [31], Nigrini studied what he called "manipulatable items", that is ones for which the taxpayer could make blatant adjustments, as these were the numerical figures most likely subject to fraudulent modifications; a detailed discussion of incorporating accounting concepts in relation to taxable returns can be found in [31]. After manually identifying these potentially manipulatable numbers over these many years, Nigrini suspected 380 manipulatable figures with a total value of approximately US$3.4 million; based on these numerical figures, Nigrini concluded that Bill Clinton's income and deduction numbers barely conform with Benford's law by using the common goodness-of-fit chi-square test [35] (also see Appendix 4.1), suggesting the stated figures could be authentic, albeit possibly containing rounding errors.

Lastly, in [10], two tables (Tables 4.2 and 4.3) are constructed, which summarize when analysis via Benford's law is appropriate or not. For instance, numbers could be easily influenced by human thought as some numbers are considered (un)lucky; ATM withdrawals tend to be biased towards a certain symmetry. Both these cases do not follow Benford's law [32]. In particular, machine assigned numbers, like cheque numbers or purchase order numbers, are more likely to follow a uniform distribution on $\{1, 2, \ldots, 9\}$. Moreover, [10] also indicated that sellers tend to round-down the prices to raise psychological attention, for example the selling price of $2.00 is likely to be rounded down to $1.99 in order to attract more customers, causing an excess of last digit 9s at the expense of having 0s. In contrast, [4] suggested that managers tend to round-up net incomes that were just below the psychological boundaries.

[2] Line-item in accounting is an item of revenue or expenditure in a budget or other financial statements or reports.

[3] Note that tax returns of other former presidents, along with nominees, can be found online at https://www.taxnotes.com/presidential-tax-returns.

TABLE 4.2 When Benford analysis is likely useful?

Likely Useful	Examples
Figures resulting from multiplication, division, or raising to integer powers of numbers each of which is sampled from an unknown (uncertain) distribution [3]	Accounts receivable (= number sold × price) Accounts payable (= number bought × price)
Data arisen from transaction flows	Disbursements, sales, expenses
Big financial data	Full year's transactions
Set of numerical figures with a sample mean being greater than its sample median with a positive skewness [50]	Most financial reports of different accounting items

TABLE 4.3 When Benford analysis is not likely useful?

Not Likely Useful	Examples
Collection of assigned numbers	Cheque numbers, invoice numbers, zip codes
Numbers subject to behavioral influence of human such as firm-specific numbers	Prices that with psychological preferences ($1.99 vs $2.00), ATM withdrawals, typical refunds
Set of numerical figures with a notable portion of missed transaction records	Bribes, thefts, kickbacks, contract rigging

4.3.1 Benford's Law in US Census Data (2010)

Here we outline an interesting example of Nigrini's testing procedure [31] on US census data of the year 2010. This dataset contains population sizes of 3,142 counties in the United States, and the purpose is to investigate the quality of the data via testing whether these altogether conform with Benford's law by using Nigrini's "five-digit test". To begin, we load the census data in Programme 4.1 via Python (resp. Programme 4.2 via R).

```
1 import pandas as pd
2
3 df_census = pd.read_csv("census_2000_2010.csv")
4 print(df_census.head(5))
5 census_2010 = df_census["pop.2010"]
```

```
1     fips            name     area  pop.2000  pop.2010
2 0   1001  Autauga County   594.44   43671.0     54571
3 1   1003  Baldwin County  1589.78  140415.0    182265
4 2   1005  Barbour County   884.88   29038.0     27457
5 3   1007     Bibb County   622.58   20826.0     22915
6 4   1009   Blount County   644.78   51024.0     57322
```

Programme 4.1 Load the population census dataset in Python. (Data source: United States Census Bureau Data Repository)

```
1  > library(benford.analysis)        # US Census (2010) dataset
2  > library(stringr)                 # Pad leading zero
3  > data(census.2000_2010)           # load census dataset
4  >
5  > head(census.2000_2010, 5)        # print the first 5 rows
6    fips           name      area pop.2000 pop.2010
7  1 1001 Autauga County  594.44    43671    54571
8  2 1003 Baldwin County 1589.78   140415   182265
9  3 1005 Barbour County  884.88    29038    27457
10 4 1007    Bibb County  622.58    20826    22915
11 5 1009  Blount County  644.78    51024    57322
12 > census_2010 <- census.2000_2010$pop.2010
```

Programme 4.2 Load the population census dataset in **R**. (Data source: United States Census Bureau Data Repository)

Let K denote the total number of possible digits in a chosen test, e.g. $K = 9$ and $K = 10$ for the first digit test and the second digit test, respectively. Other than the test based on *Mean Absolute Deviation* (MAD) proposed in [31], we here introduce two common tests for conformity of the dataset with Benford's law by computing the absolute deviation between the actual proportion AP_i and the expected (theoretical) proportion EP_i, for $i = 1, \ldots, K$; also see Appendix 4.A.1 for an introduction to goodness-of-fit tests.

1. *Folded Z-statistic* for each possible digit:

$$Z_{i,\text{folded}} = \frac{|\text{AP}_i - \text{EP}_i|}{\sqrt{\text{EP}_i(1 - \text{EP}_i)/n}} \xrightarrow{d} \mathcal{N}^+\left(\sqrt{\frac{2}{\pi}}, 1 - \frac{2}{\pi}\right), \quad \text{for } i = 1, \ldots, K, \tag{4.7}$$

   where $\mathcal{N}^+$ denotes the *folded normal distribution*.

2. *Chi-square Goodness-of-fit Test Statistics* for all possible digits at once:

$$\sum_{i=1}^{K} \frac{(n\text{AP}_i - n\text{EP}_i)^2}{n\text{EP}_i} \xrightarrow{d} \chi^2_{K-1}. \tag{4.8}$$

The density of a folded normal distribution on $[0, \infty)$, denoted by $\mathcal{N}^+(\mu, \sigma^2)$, is given by:

$$f(x) = \frac{1}{\sqrt{2\pi\sigma^2}} \exp\left(-\frac{(x-\mu)^2}{2\sigma^2}\right) + \frac{1}{\sqrt{2\pi\sigma^2}} \exp\left(-\frac{(x+\mu)^2}{2\sigma^2}\right), \qquad 0 \le x < \infty.$$

In light of the Central Limit theorem, since $\sqrt{n}(\text{AP}_i - \text{EP}_i)/\sqrt{\text{EP}_i(1 - \text{EP}_i)} \xrightarrow{d} \mathcal{N}(0, 1)$, the folded normal density can be computed in Python with `2*norm.pdf(x)` and in R with `2*dnorm(x)`, respectively. In the chi-square goodness-of-fit test, Table 4.4 is a list of suggested p-value critical levels for the levels of conformity with Benford's law.

TABLE 4.4 Suggested "five-digit test" *p*-value criteria for the chi-square goodness-of-fit test.

Range	Conformity
0.1 to 1	Close
0.05 to 0.1	Acceptable
0.01 to 0.05	Marginally acceptable
0 to 0.01	Not acceptable

```
1  import numpy as np
2  import matplotlib.pyplot as plt
3  from scipy.stats import norm, chi2
4
5  def Benford_Analysis(Count, EP, DIGITS, x_label=""):
6      # Fill in missing digits with zero
7      missing_digit=np.setdiff1d(DIGITS,np.array(Count.index,dtype=int))
8      if len(missing_digit) > 0:
9          Count = pd.concat([Count, pd.Series(0, index=missing_digit)])
10         Count = Count.sort_index()
11
12     # Remove any additional digit in Count not in the group DIGITS
13     Count = Count[DIGITS]
14
15     AP = Count / Count.sum()
16     N = Count.sum()
17     folded_z = np.abs(AP - EP) / np.sqrt(EP * (1 - EP) / N)
18     p_val = 2 * norm.pdf(folded_z)  # p-values for folded z-scores
19     print(DIGITS[p_val < 0.05])    # Rejected digits by Benford's law
20
21     # Build a bar chart for each digit
22     colors = np.repeat("burlywood", len(DIGITS))
23     colors[p_val < 0.05] = "red"
24     bar_name = [str(d).zfill(len(str(DIGITS[-1]))) for d in DIGITS]
25     fig, ax = plt.subplots(figsize=(10, 7))
26     ax.bar(bar_name,AP,color=colors,edgecolor='none',label="Actual")
27     ax.plot(bar_name,EP,color="blue",linewidth=2,label="Benford's Law")
28     ax.set_xlabel(x_label, fontsize=20)
29     ax.set_ylabel("PROPORTION", fontsize=20)
30     ax.tick_params(axis='both', which='major', labelsize=18)
31     ax.xaxis.set_major_locator(plt.MaxNLocator(10))
32
33     handles, _ = ax.get_legend_handles_labels()
34     labels = ["Actual", "Rejected", "Benford's Law"]
35     handles = [plt.Rectangle((0,0),1,1, color="burlywood"),
36                plt.Rectangle((0,0),1,1, color="red"),
37                handles[0]]
38     plt.legend(handles, labels, loc="upper right")
39
40     # Return the p-value of chi-square goodness-of-fit test statistics
41     return chi2.pdf(np.sum(N*(AP - EP)**2 / EP), len(DIGITS)-1)
```

Programme 4.3 Function for computing folded *Z*-statistics and chi-square goodness-of-fit test statistics of the US population census dataset in Python.

```
1  > par(cex.lab=1.5, cex.axis=1.5, cex.main=2)
2  > Benford_Analysis <- function(Count, EP, DIGITS, x_label=""){
3  +   # Fill in missing digits with zero
4  +   if(length(setdiff(DIGITS, as.numeric(names(Count)))) > 0) {
5  +     missing_digit <- setdiff(DIGITS, as.numeric(names(Count)))
6  +     Count[as.character(missing_digit)] <- 0
7  +     Count <- Count[order(as.numeric(names(Count)))]
8  +   }
9  +
10 +   # Remove any additional digit in Count not in the group DIGITS
11 +   if(length(setdiff(as.numeric(names(Count)), DIGITS)) > 0) {
12 +     Count <- Count[as.character(DIGITS)]
13 +   }
14 +
15 +   AP <- as.numeric(Count/sum(Count))
16 +   N <- sum(Count)
17 +   folded_z <- abs(AP - EP)/sqrt(EP*(1-EP)/N)
18 +   p_val <- 2*dnorm(folded_z)      # p-values for folded z-scores
19 +   print(DIGITS[p_val < 0.05])     # Rejected digits by Benford's law
20 +
21 +   # Build a bar chart for each digit
22 +   col <- rep("burlywood1", length(DIGITS))
23 +   col[p_val < 0.05] <- "red"
24 +   bar_name <- str_pad(DIGITS, nchar(tail(DIGITS, 1)), pad="0")
25 +   census.barplot <- barplot(AP,names.arg=bar_name,col=col,border=NA,
26 +                             ylab="PROPORTION", xlab=x_label)
27 +   lines(census.barplot, EP, col="blue", lwd=2)
28 +   legend("topright", ncol=1, c("Actual", "Rejected", "Benford's Law"),
29 +          lty=c(0,0,1), lwd=c(0,0,2), fill=c("burlywood1", "red", 0),
30 +          border=NA, col=c(0, 0, "blue"), cex=1.5)
31 +
32 +   # Return the p-value of chi-square goodness-of-fit test statistics
33 +   return(dchisq(sum(N*(AP - EP)^2/EP), length(DIGITS)-1))
34 + }
```

Programme 4.4 Function for computing folded Z-statistics and chi-square goodness-of-fit test statistics of the US population census dataset in **R**.

In Programme 4.3 via Python (resp. Programme 4.4 via **R**), the `Benford_Analysis()` function is defined in the same fashion with four input arguments, namely: `Count` as the Python `pandas` dataframe (resp. **R** `table`) of actual frequencies for each digit to be inputted; `EP` for the expected proportion for each digit of the chosen test calculated in accordance with Benford's law; `DIGITS` for the set of the possible leading digits of interest for a test; and `x_label` for the name of the chosen test printed in the x-axis of the bar chart. Moreover, this `Benford_Analysis()` function also exports a frequency plot that shows how the actual digit frequencies deviate from those expected ones according to Benford's law (depicted by the blue broken line). In particular, the digits which violate Benford's law (having a small p-value (< 0.05) value using the folded Z-statistics) will be colored in red. Lastly, the p-value of the chi-square goodness-of-fit test for all digits at once is also returned.

As an illustration for the Nigrini "five-digit test", we shall perform all five tests without regard to their individual conformity with Benford's law. For each of these five Nigrini's digit tests, we first extract the digit frequencies of interest from our dataset, and then construct a Python `pandas` dataframe (resp. **R** `table`) that contains the corresponding actual frequencies observed in the numerical figures of the dataset along with the expected proportions (EP) as predicted by Benford's law in accordance with (4.3), (4.4), (4.5) and (4.6).

(I) First digit test:

```
1 FT_1 = [int(str(d)[0]) for d in census_2010]              # Get first digit
2 Count_1 = pd.Series(FT_1).value_counts().sort_index()# A summary table
3 DIGITS_1 = np.arange(1, 10)                         # Possible first digit values
4 EP_1 = np.log10(1 + 1 / DIGITS_1)                   # Expected proportion (4.3)
5 print(Benford_Analysis(Count_1, EP_1, DIGITS_1, "FIRST DIGIT"))
```

```
1 [5]
2 0.06153655621418872
```

Programme 4.5 First digit test for the US population census (2010) dataset in Python.

```
1 > FT_1 <- substr(census_2010, 1, 1)              # Get first digit
2 > Count_1 <- table(as.numeric(FT_1))             # A summary table
3 > DIGITS_1 <- 1:9                         # Possible first digit values
4 > EP_1 <- log10(1 + 1/DIGITS_1)                  # Expected proportion (4.3)
5 > Benford_Analysis(Count_1, EP_1, DIGITS_1, "FIRST DIGIT")
6 [1] 5
7 [1] 0.06153656
```

Programme 4.6 First digit test for the US population census (2010) dataset in R.

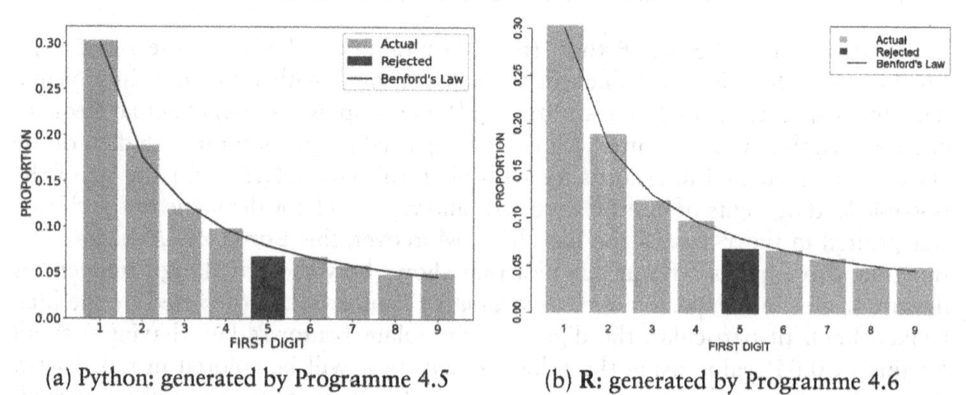

(a) Python: generated by Programme 4.5 (b) **R**: generated by Programme 4.6

FIGURE 4.8 Actual first digit frequencies vs those expected ones given by Benford's law for the US population census (2010) dataset.

(II) Second digit test:

```
1  FT_2 = [int(str(d)[1]) for d in census_2010]          # Get second digit
2  Count_2 = pd.Series(FT_2).value_counts().sort_index()# A summary table
3  DIGITS_2 = np.arange(10)                        # Possible second digit values
4  EP_2 = np.zeros_like(DIGITS_2, dtype=float) # Expected proportion (4.4)
5  for i in range(1, 10):
6      EP_2 += np.log10(1 + 1 / np.arange(10 * i, 10 * i + 10))
7
8  print(Benford_Analysis(Count_2, EP_2, DIGITS_2, "SECOND DIGIT"))
```

```
1  []
2  0.10399271634363873
```

Programme 4.7 Second digit test for the US population census (2010) dataset in Python.

```
1  > FT_2 <- substr(census_2010, 2, 2)          # Get second digit
2  > Count_2 <- table(as.numeric(FT_2))         # A summary table
3  > DIGITS_2 <- 0:9                    # Possible second digit values
4  > EP_2 <- rep(0, length(DIGITS_2))           # Expected proportion (4.4)
5  > for (i in 1:9){
6  +     EP_2 <- EP_2 + log10(1 + 1/((10 * i + 0):(10 * i + 9)))
7  + }
8  > Benford_Analysis(Count_2, EP_2, DIGITS_2, "SECOND DIGIT")
9  integer(0)
10 [1] 0.1039927
```

Programme 4.8 Second digit test for the US population census (2010) dataset in R.

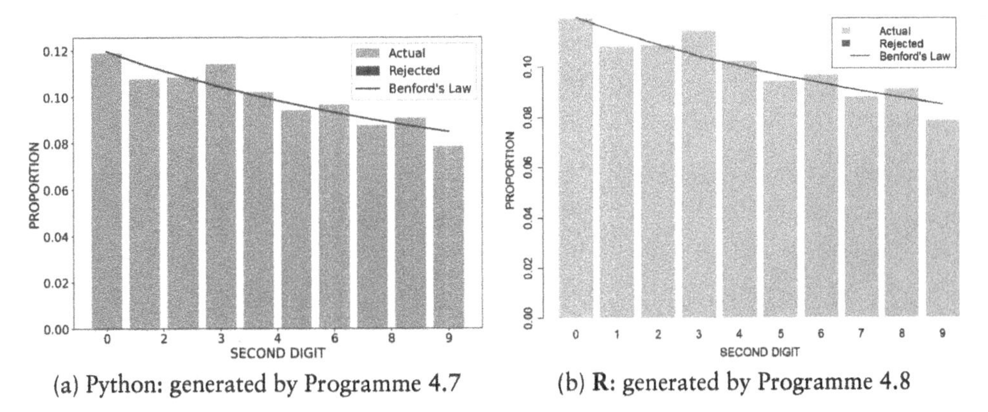

(a) Python: generated by Programme 4.7 (b) R: generated by Programme 4.8

FIGURE 4.4 Actual second digit frequencies vs those expected given by Benford's law for the US population census (2010) dataset.

(III) First-two digit test:

```
1 FT_3 = [int(str(d)[:2]) for d in census_2010]     # Get first-two digits
2 Count_3 = pd.Series(FT_3).value_counts().sort_index()# A summary table
3 DIGITS_3 = np.arange(10, 100)            # Possible first-two digits values
4 EP_3 = np.log10(1 + 1 / DIGITS_3)      # Expected proportion (4.5)
5 print(Benford_Analysis(Count_3, EP_3, DIGITS_3, "FIRST-TWO DIGITS"))
```

```
1 []
2 0.02865213051561239
```

Programme 4.9 First-two digit test for the US population census (2010) dataset in Python.

```
1 > FT_3 <- substr(census_2010, 1, 2)              # Get first-two digits
2 > Count_3 <- table(as.numeric(FT_3))             # A summary table
3 > DIGITS_3 <- 10:99                  # Possible first-two digit values
4 > EP_3 <- log10(1 + 1/DIGITS_3)                  # Expected proportion (4.5)
5 > Benford_Analysis(Count_3, EP_3, DIGITS_3, "FIRST-TWO DIGITS")
6 integer(0)
7 [1] 0.02865213
```

Programme 4.10 First-two digit test for the US population census (2010) dataset in **R**.

(IV) First-three digit test:

```
1 FT_4 = [int(str(d)[:3]) for d in census_2010] # Get first-three digits
2 Count_4 = pd.Series(FT_4).value_counts().sort_index()# A summary table
3 DIGITS_4 = np.arange(100, 1000)    # Possible first-three digits values
4 EP_4 = np.log10(1 + 1 / DIGITS_4)               # Expected proportion (4.6)
5 print(Benford_Analysis(Count_4, EP_4, DIGITS_4, "FIRST-THREE DIGITS"))
```

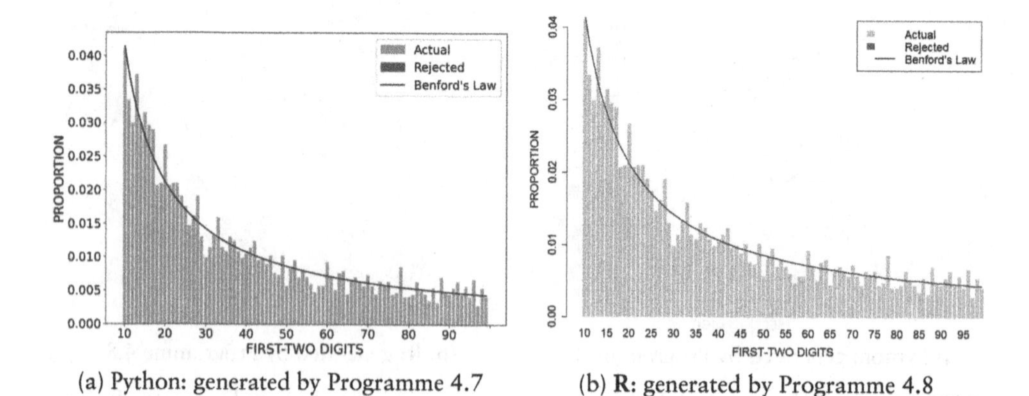

(a) Python: generated by Programme 4.7 (b) **R**: generated by Programme 4.8

FIGURE 4.5 Actual first-two digit frequencies vs those expected ones given by Benford's law for the US population census (2010) dataset.

```
1 [138 139 200 223 233 282 330 331 371 477 522 600 630 670 754 783 785
2  836 869 887 925 963 987 998]
3 0.007937533523065457
```

Programme 4.11 First-three digit test for the US population census (2010) dataset in Python.

```
1 > FT_4 <- substr(census_2010, 1, 3)          # Get first-three digits
2 > Count_4 <- table(as.numeric(FT_4))          # A summary table
3 > DIGITS_4 <- 100:999            # Possible first-three digits values
4 > EP_4 <- log10(1 + 1/DIGITS_4)               # Expected proportion (4.6)
5 > Benford_Analysis(Count_4, EP_4, DIGITS_4, "FIRST-THREE DIGITS")
6  [1] 138 139 200 223 233 282 330 331 371 477 522 600 630 670 754 783 785
7 [18] 836 869 887 925 963 987 998
8 [1] 0.007937534
```

Programme 4.12 First-three digit test for the US population census (2010) dataset in R.

(V) Last digit test (Uniformity: at least four-digit figures):

```
1 print(pd.Series([len(str(d)) for d in census_2010]
2                  ).value_counts().sort_index()) # Table for digit length
3 
4 census_4_digit = [d for d in census_2010 if len(str(d)) >= 4]
5 FT_5 = [int(str(d)[-2:]) for d in census_4_digit] # Get last-two digits
6 Count_5 = pd.Series(FT_5).value_counts().sort_index()# A summary table
7 DIGITS_5 = np.arange(100)                 # Possible last-two digits values
8 EP_5 = np.ones_like(DIGITS_5) / len(DIGITS_5)      # Expected proportion
9 print(Benford_Analysis(Count_5, EP_5, DIGITS_5, "LAST-TWO DIGITS"))
```

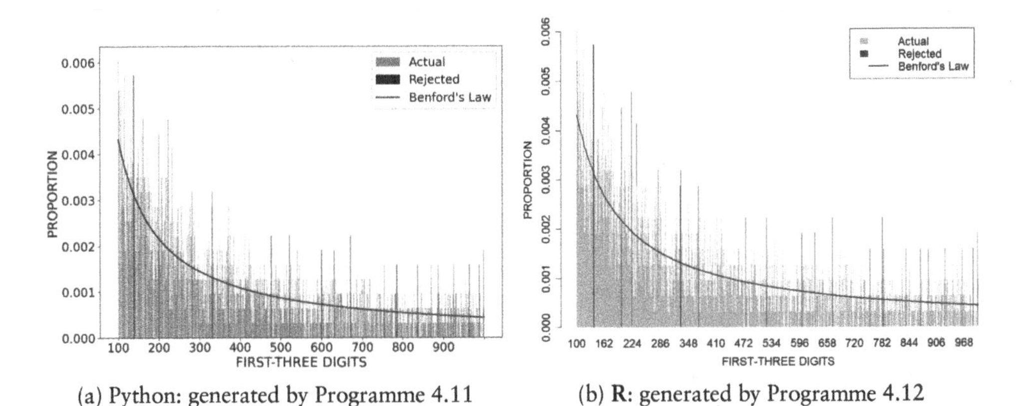

(a) Python: generated by Programme 4.11 (b) R: generated by Programme 4.12

FIGURE 4.6 Actual first-three digit frequencies vs those expected ones given by Benford's law for the US population census (2010) dataset.

```
1 2        2
2 3       33
3 4      663
4 5     1867
5 6      539
6 7       39
7 Name: count, dtype: int64
8 [31 34 38 42 54]
9 0.004605253453075102
```

Programme 4.13 Last-two digit test for the US population census (2010) dataset in Python.

```
1  > table(nchar(census_2010))                          # Table for digit length
2     2    3    4    5    6    7
3     2   33  663 1867  539   39
4  > census_4_digit <- census_2010[nchar(census_2010) >= 4]
5  > FT_5 <- substr(census_4_digit, nchar(census_4_digit)-2+1,
6  +                nchar(census_4_digit))              # Get last-two digits
7  > Count_5 <- table(as.numeric(FT_5))                 # A summary table
8  > DIGITS_5 <- 0:99                      # Possible last-two digits values
9  > EP_5 <- rep(1/length(DIGITS_5), length(DIGITS_5)) # EP
10 > Benford_Analysis(Count_5, EP_5, DIGITS_5, "LAST-TWO DIGITS")
11 [1] 31 34 38 42 54
12 [1] 0.004605253
```

Programme 4.14 Last-two digit test for the US population census (2010) dataset in R.

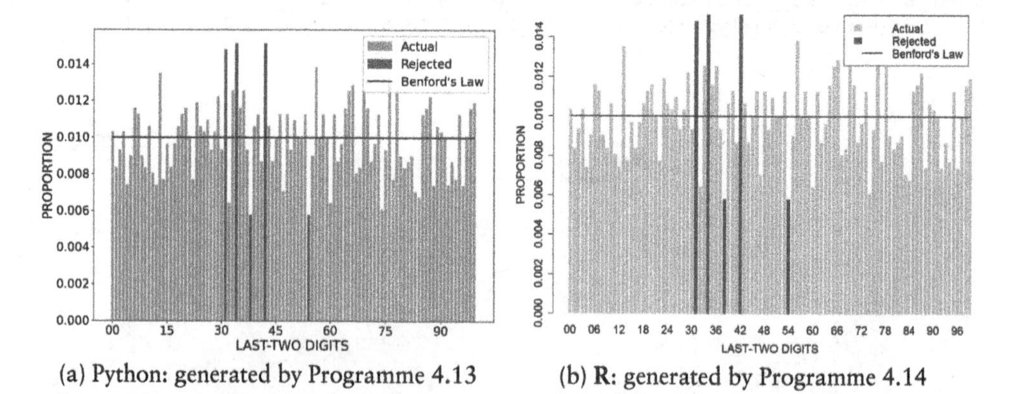

(a) Python: generated by Programme 4.13 (b) R: generated by Programme 4.14

FIGURE 4.7 Actual last-two digit frequencies vs those expected ones given by Benford's law for the US population census (2010) dataset.

In conclusion, Table 4.5 shows a summary of the "five-digit test" for the US census dataset in 2010. For the first (resp. second) digit test, its p-value of 0.061537 (resp. 0.103993) of the chi-square goodness-of-fit test concludes acceptable (resp. close) conformity of the dataset with Benford's law; indeed, we clearly see an astonishingly

TABLE 4.5 A summary table of Nigrini's "five-digit test" for US census data (2010).

Digit test	Violation(s)	p-value of χ^2	Conformity
First	5	0.061537	Acceptable
Second	None	0.103993	Close
First-Two	None	0.028652	Marginally Acceptable
First-Three	138, 139, 200, 223, 233, 282, 330, 331, 371, 477, 522, 600, 630, 670, 754, 783, 785, 836, 869, 887, 925, 963, 987, 998	0.007938	Not acceptable
Last-Two	31, 34, 38, 42, 54	0.004605	Not acceptable

high degree of alignment between the actual and expected proportions in Figure 4.3 (resp. Figure 4.4). In particular, for the first digit test, the small p-value of the folded Z test for the first digit of 5 shows rejection of the Benford's law, and this is further highlighted by the bar with a deeper color in red in Figure 4.3. However, for the second digit test, no digit is rejected by the test using the folded Z-statistics.

Although the first digit and second digit tests conclude there is at least an acceptable conformity with Benford's law, for the purpose of illustration, we also continue the discussion on the remaining three digit tests. For the first-two digit test (also see Figure 4.5), its p-value of 0.028652 of the chi-square goodness-of-fit test indicates a marginally acceptable conformity of the dataset with Benford's law. Interestingly, no digit in the first-two digit test is rejected by the test using folded Z-statistics. For the tests on first three and last two digits (also see Figures 4.6 and 4.7, respectively), their small respective p-values of 0.007938 and 0.004605 of the corresponding chi-squared goodness-of-fit tests suggest a rejection to Benford's Law: particularly, five numbers listed at the bottom for the last-two test case and 24 numbers listed in the first-three test case are rejected by the test using folded Z-statistics.

4.4 BENFORD'S LAW IN GROWTH FIGURES

4.4.1 For Macroeconomic and Social Science Data

We here discuss how a sequence of temporal data, arisen in finance and econometrics, with an exponential growth (or decay) can naturally conform with Benford's law, at least approximately, and we defer its rigorous mathematical proof until Subsection *4.4.3. Let $X(t)$ denote the nominal gross domestic product (GDP) of a country at time t, and model the evolution of this process in a discrete-time setting as:

$$X(t+1) = (1 + G_t)X(t), \quad \text{for } t = 1, 2, \ldots, \tag{4.9}$$

where G_t's are assumed as iid random variables such that each G_t represents the growth rate of the overall economic activity for the period from t to $t+1$. One criticism of this simple model might be that the growth rate G_t is not completely independent of the present level of GDP $X(t)$ or the timing of t, since more developed economies and aging populations tend to grow slower; however, so long as the overall time period does not span several decades, the approximation of using iid assumption is still reasonably good and Benford's law likely holds. A formal statement of this claim is based on Theorem 4.1, together with its proof, and the discussion thereafter in Subsection *4.4.3. Likewise, the same kind of dynamics modeling can explain the emergence of Benford's law in stock prices, company revenues and profits, units of product sold and so on. More discussions can be found in [25].

Over the years, many research studies have shown that Benford's law can detect economic fraudulent data from government authorities. For instance, [36] found that among eurozone countries, Greek economic data deviated the most from Benford's law, and this deviation persisted across leading digits and first few leading digits combinations. The authors also indicated that this possibility of manipulating data was later confirmed by the European Commission.

Benford's law is also very useful for monitoring the fairness of elections, so as to detect any electoral fraud through screening various official election statistics. That said, this detection is complicated by the respective privacy protection rules. Recent examples include the 2009 Iranian presidential election [38] and the presidential and parliamentary elections in Turkey over 2017–2018 [21]. Unfortunately, the efficacy of Benford's law in election monitoring has not yet been rigorously established [8] due to the lack of a prominent and definite explanation for why the phenomenon may emerge in most election data.

4.4.2 COVID-19 Figures in Healthcare Management

A recent work [20] identified the emergence of Benford's law in the early stages of the epidemic of the contemporary outbreak of COVID-19. Let $\mathrm{I}(t)$ and $\mathrm{S}(t)$ denote the respective numbers of infected and susceptible individuals in a population. Then the common susceptible-infected-susceptible compartmental epidemiological model [1] assumes that, for $t = 0, 1, 2, \ldots,$

$$\mathrm{I}(t+1) = \mathrm{I}(t) + \left(\theta + \epsilon^I_{t+1}\right)\mathrm{I}(t) - \left(\delta + \epsilon^R_{t+1}\right)\mathrm{I}(t);$$
$$\mathrm{S}(t+1) = \mathrm{S}(t) - \left(\theta + \epsilon^I_{t+1}\right)\mathrm{I}(t) + \left(\delta + \epsilon^R_{t+1}\right)\mathrm{I}(t),$$

where $\theta > 0$ and $\delta > 0$ are the infectiousness and recovery rates, respectively, and ϵ^I_{t+1} and ϵ^R_{t+1} are iid random noise terms corresponding to these two rates. Especially in the early stages of the epidemic, the number of infections is negligible compared to the total population size M, i.e., $\mathrm{I}(t) \ll M$. Then the above model reduces to, for $t = 0, 1, 2, \ldots,$

$$\mathrm{I}(t+1) = \left(1 + \theta - \delta + \epsilon^I_{t+1} - \epsilon^R_{t+1}\right)\mathrm{I}(t) =: (1 + \xi_{t+1})\mathrm{I}(t), \tag{4.10}$$

where ξ_t's are iid random variables. And the model (4.10) can be rewritten as

$$\mathrm{I}(t+1) = \mathrm{I}(0)\prod_{s=1}^{t+1}(1 + \xi_s),$$

whose form is the reason why the Benford's law emerges. As mentioned previously, the formal statement of this claim is made in Theorem 4.2, and its proof is laid down in Subsection *4.4.3. In particular, evidence of Benford's law is illustrated by using various case numbers for COVID-19 across many different countries. Figure 4.8 shows the leading digit frequencies of daily COVID-19 case figures and Table 4.6 shows the p-values of the chi-square goodness-of-fit tests in nine countries over a period of $T = 30, 40$ days:

$$\sum_{d\in\{1,\cdots,9\}} \frac{(x_{T,d} - E_{T,d})^2}{E_{T,d}} \overset{approximately}{\sim} \chi^2_{9-1},$$

where $x_{T,d}$ is the actual frequency and $E_{T,d} = T \cdot \log_{10}(1 + 1/d)$ is the expected Benford's frequency of the leading digit d over a period of T days.

Clearly, from Figure 4.8, with $T = 30$, numbers corresponding to Canada, Germany, and USA do not follow Benford's law. This observation is further supported by the small p-values (< 0.05) in their chi-square goodness-of-fit tests shown in Table 4.6. A possible explanation is that the start day of counting is too early for these three countries, in which the cumulative growths of figures appear not too geometric in nature. Finally, China's data over a period of $T = 40$ days also appears to not follow Benford's law. A plausible explanation is that the Chinese central government introduced lockdowns at a very early stage to badly affected cities in Hubei province, and as a result those ξ_t's could not evolve freely.

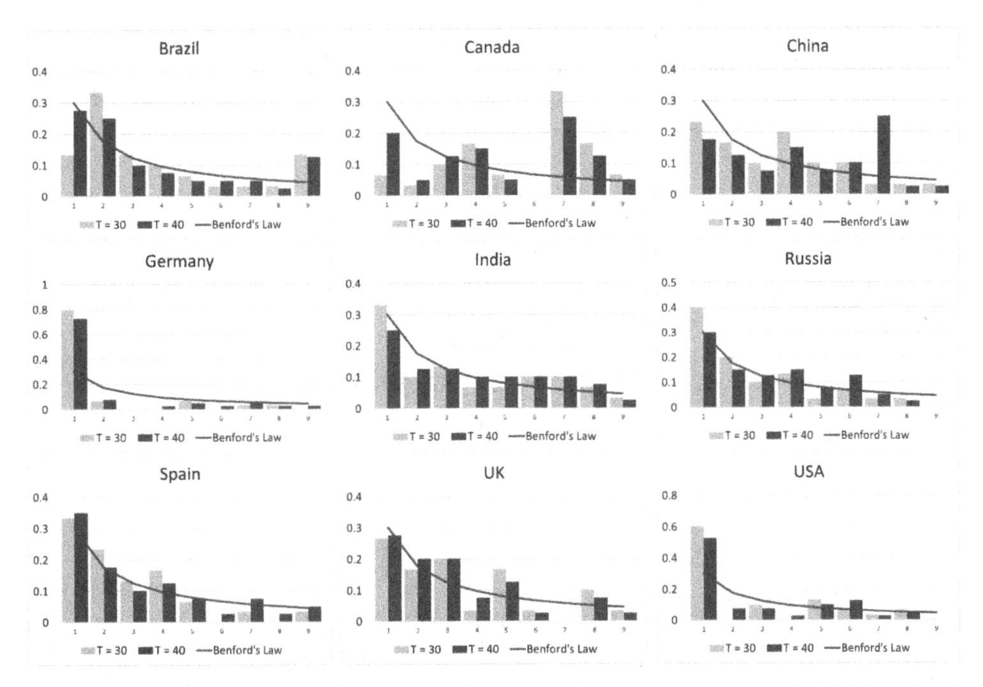

FIGURE 4.8 Leading digit frequencies of nine countries over a period of 30 and 40 days.

TABLE 4.6 *p*-values of chi-square goodness-of-fit tests of nine countries over a period of 30 and 40 days, respectively. The bolded figures in red are below the 1% significant level.

Country	$T = 30$	$T = 40$
Brazil	0.107468087	0.395583595
Canada	< 0.005	< 0.005
China	0.739600837	**0.000106815**
Germany	< 0.005	< 0.005
India	0.92827951	0.870120886
Russia	0.82875403	0.675233371
Spain	0.624441849	0.954165215
UK	0.31099018	0.486382207
USA	**0.009652637**	0.032997104

*4.4.3 Proof by Fourier Analysis for the Emergence of Benford's Law in Temporal Data

As we have seen, there can be more than one explanation behind the emergence of Benford's law for leading digits. In particular, there is an alternative interpretation as a "central limit" theorem for the mantissas of random variables, but under multiplication [3]. Following the idea of [3], we here introduce how Benford's law can be established via *Fourier analysis* when the random growth rates of the figures over different periods are independent and follow a common distribution. For example, if a city has an initial population P_0 and grows with a rate of r_i in the *i*-th year, then its population in the *n*-th year is given by

$$P_n = P_0(1 + r_1)(1 + r_2)\dots(1 + r_n), \tag{4.11}$$

which is clearly a random variable by its own right.

Given two numerical figures $x = \overline{x} \times 10^{k_x}$ and $y = \overline{y} \times 10^{k_y}$ are multiplied, the mantissa $\overline{(x \cdot y)}$ of the resulting product xy is:

$$\overline{(x \cdot y)} = \begin{cases} \overline{x} \cdot \overline{y}, & \text{if } 1 \le \overline{x} \cdot \overline{y} \le 10; \\ \frac{\overline{x} \cdot \overline{y}}{10}, & \text{if } 10 \le \overline{x} \cdot \overline{y} \le 100. \end{cases}$$

Equivalently, we must have

$$\log_{10}(\overline{(x \cdot y)}) = \log_{10}\overline{x} + \log_{10}\overline{y} \pmod 1, \tag{4.12}$$

where (mod 1) stands for the modular arithmetic on $[0, 1)$, dealing with the fraction parts of numbers. As a result, the limiting distribution of the logarithm $\overline{z}$ of a random variable $z = \prod_{i=1}^{n} x_i$ is determined by the limiting distribution of the sum of random variables $\sum_{i=1}^{n} \log_{10}\overline{x}_i \pmod 1$. In the rest of this section, we shall prove the following version of the central limit theorem on a torus by means of Fourier analysis. Its immediate consequence will give the emergence of Benford's law for time series data governed by the dynamics given by (4.9) or (4.11).

Theorem 4.1 *(Central Limit Theorem on the torus T^1)* Let $x_1, \dots, x_k$ be independent and identically distributed continuous random variables on $[0, 1]$ with a common probability density function $f \in L^2[0, 1]$, also denote $w = \sum_{i=1}^{k} x_i \pmod 1$. Then, as $k \to \infty$, w approaches a uniform distribution on $[0, 1]$.

To see Theorem 4.1 can warrant the emergence of Benford's law, for if y is a random variable on $\mathbb{R}_+ = [0, \infty)$ such that $\log_{10} y \pmod 1$ follows a uniform distribution on $[0, 1]$, we can show that $d_1(y)$ must follow a Benford's law. Indeed, for $n = 1, 2, \dots, 9$,

$$\begin{aligned}
\mathbb{P}(d_1(y) = n) &= \mathbb{P}(n \leq d_1(y) < n + 1) \\
&= \mathbb{P}(n \times 10^k \leq y < (n + 1) \times 10^k, \text{ for some } k \in \mathbb{Z}_+) \\
&= \mathbb{P}(\log_{10} n + k \leq \log_{10} y < \log_{10}(n + 1) + k, \text{ for some } k \in \mathbb{Z}_+) \\
&= \mathbb{P}(\log_{10} n \leq \log_{10} y \pmod 1 < \log_{10}(n + 1)) \\
&= \log_{10}(n + 1) - \log_{10} n = \log_{10}\left(1 + \frac{1}{n}\right),
\end{aligned}$$

where the second last equality follows by the definition of a uniform distribution on $[0, 1]$.

On the other hand, to prove Theorem 4.1, for the sake of convenience, under 1-modular arithmetics, we assume that the density function $f(t)$ is periodic with period 1 and so, if defined on $\mathbb{R}_+$,

$$\int_a^{a+1} f(t)\,dt = 1 \text{ for all possible } a \in \mathbb{R}_+.$$

Once we set f as a periodic function on $[0, 1]$, we can talk about its *Fourier coefficients*, each of which is defined by

$$\hat{f}(n) = \int_0^1 f(t)e^{-2\pi int}\,dt, \text{ for any } n \in \mathbb{Z}.$$

Moreover, as $f(t)$ is real-valued, $\hat{f}(-n) = \overline{\hat{f}(n)}$. According to the celebrated *Parseval's theorem* [34, 39], whenever $f \in L^2[0, 1]$, i.e., f is square-integrable with $\int_0^1 f^2(x)dx < \infty$, the infinite sum of $\sum_{n=-\infty}^{\infty} f(n)e^{2\pi int}$ (almost everywhere) converges to $f(t)$ under the L^2 norm, and the convergence is pointwise at all points t at which $f(t)$ is continuous [45]. For the purpose of computing probabilities, since by Cauchy–Schwarz inequality $\|g\|_1 \leq \|g\|_2$ on $[0,1]$ for any $g \in L^2[0,1]$, so L^2-convergence implies L^1-convergence in $L^2[0,1]$. Furthermore, the mapping $\mathcal{F} : f \in L^2[0,1] \mapsto \{\hat{f}(n)\}_{n\in\mathbb{Z}}$ is a Hilbert space isomorphism in light of *Parseval's identity* [34, 39]:

$$\int_0^1 f^2(t)\,dt = \sum_{n=-\infty}^{\infty} |\hat{f}(n)|^2.$$

Let us consider some elementary properties of these Fourier coefficients before proving the main Theorem 4.1.

Lemma 4.1 Assume that $f \in L^2[0,1]$ is a probability density function. Then,

1. $\hat{f}(0) = 1$;
2. $|\hat{f}(n)| < 1$ whenever $n \neq 0$; indeed, $\sup_{n\neq 0}|\hat{f}(n)| < 1$.

Proof. The first claim is obvious since $\int_0^1 f(t)\,dt = 1$. For a fixed n and a continuous density f, we first note that for any $\theta \in [0,1]$:

$$\left|\int_0^1 f(t)e^{-2\pi int}\,dt\right| = \left|e^{2\pi in\theta}\int_0^1 f(t)e^{-2\pi int}\,dt\right| = \left|\int_0^1 f(t)e^{-2\pi in(t-\theta)}\,dt\right|,$$

with this in mind, without loss of generality as this can always be furnished by an arbitrary shifting, we can assume that $f(0) = f(1) > 0$ and $t = 0$ is a point of continuity. Then,

$$\int_0^1 f(t)e^{-2\pi int}\,dt = \int_0^\delta f(t)e^{-2\pi int}\,dt + \int_{1-\delta}^1 f(t)e^{-2\pi int}\,dt + \int_\delta^{1-\delta} f(t)e^{-2\pi int}\,dt.$$

Hence,

$$\begin{aligned}\left|\int_0^1 f(t)e^{-2\pi int}\,dt\right| &\leq \int_\delta^{1-\delta} |f(t)|\,dt + \left|\int_0^\delta f(t)e^{-2\pi int}\,dt + \int_0^\delta f(1-t)e^{2\pi int}\,dt\right| \\ &= \int_\delta^{1-\delta} f(t)\,dt + \left|\int_0^\delta \left(f(t)e^{-2\pi int} + f(1-t)e^{2\pi int}\right)dt\right|.\end{aligned}$$

For the second term of the last expression,

$$\begin{aligned}&\int_0^\delta \left((f(t)e^{-2\pi int} + f(1-t)e^{2\pi int}\right)\,dt \\ &= \int_0^\delta \left((f(t) + f(1-t))\cos(2\pi nt) + i(f(1-t) - f(t))\sin(2\pi nt)\right)\,dt,\end{aligned}$$

while for the norm of the integrand, we have:

$$\begin{aligned}
&\left|\left(f(t)+f(1-t)\right)\cos(2\pi nt)+i\left(f(1-t)-f(t)\right)\sin(2\pi nt)\right|^2\\
&=\left(f(t)+f(1-t)\right)^2\cos^2(2\pi nt)+\left(f(1-t)-f(t)\right)^2\sin^2(2\pi nt)\\
&=f^2(t)+f^2(1-t)+2f(t)f(1-t)\cos(4\pi nt)\\
&<f^2(t)+f^2(1-t)+2f(t)f(1-t)=\left(f(t)+f(1-t)\right)^2,
\end{aligned}$$

where the strict inequality holds for all irrational $t > 0$. Therefore, for small enough δ, such that $f(t) > 0$, for all $t \in [0, \delta)$, we have the strict inequality:

$$\begin{aligned}
&\left|\int_0^\delta \left(f(t)e^{-2\pi int}+f(1-t)e^{2\pi int}\right)dt\right|\\
&\leq \int_0^\delta \left|\left(f(t)+f(1-t)\right)\cos(2\pi nt)+i\left(f(1-t)-f(t)\right)\sin(2\pi nt)\right| dt\\
&<\int_0^\delta \left(f(t)+f(1-t)\right)dt.
\end{aligned}$$

Combining with the above, we then obtain, for all $n \neq 0$,

$$\left|\int_0^1 f(t)e^{-2\pi int}\,dt\right| < \int_\delta^{1-\delta} f(t)\,dt + \int_0^\delta \left(f(t)+f(1-t)\right)dt = \int_0^1 f(t)\,dt = 1.$$

To establish the last claim about the boundedness of the supremum, note that by Parseval's identity, $|\hat{f}(n)| \to 0$ as $|n| \to \infty$, therefore, all $|\hat{f}(n)|$'s except finitely many of them are less than $1/2$, while for those finitely many $|\hat{f}(n)|$'s which are not less than $1/2$, each of them is still strictly less than 1, and so the claim follows. □

Consider $z = x + y \pmod 1$, where x and y are continuous random variables on $[0, 1)$ with respective density functions f and g. Then z is also a continuous random variable, and it has a density h such that

$$h(t) = \int_0^1 f(u)g(t-u)\,du = \left(f * g\right)(t),$$

which is the usual convolution of f and g on a bounded range. The Fourier coefficient $\hat{h}(n)$, for $n \in \mathbb{Z}$, can then be computed as the product of $\hat{f}(n)$ and $\hat{g}(n)$:

Lemma 4.2 Let f, g and h be the continuous probability density functions of x, y and $z = x + y \pmod 1$, respectively, where x and y are continuous random variables on $[0, 1]$. Then, $\hat{h}(n) = \hat{f}(n)\hat{g}(n)$.

Proof. Note that both g and $e^{-2\pi int}$ are periodic with a period 1, we then have, by using Fubini's theorem for these legitimate functions f, g and $e^{-2\pi int}$,

$$\begin{aligned}\hat{h}(n) &= \int_0^1 h(t)e^{-2\pi int}\,dt \\ &= \int_0^1 \left(\int_0^1 f(u)g(t-u)\,du\right) e^{-2\pi int}\,dt \\ &= \int_0^1 \int_0^1 f(u)g(t-u)e^{-2\pi in(t-u)}e^{-2\pi inu}\,dudt \\ &= \int_0^1 f(u)e^{-2\pi inu}\left(\int_0^1 g(t-u)e^{-2\pi in(t-u)}\,dt\right) du = \hat{f}(n)\hat{g}(n).\end{aligned}$$

□

Proof of Theorem 4.1: Let $x_1, \ldots, x_k$ be iid continuous random variables on $[0,1)$ with a common density $f \in L^2[0,1]$ and consider $w := \sum_{i=1}^{k} x_i \pmod 1$. Then it is clear that w is also a continuous random variable on $[0,1)$ and its density function h, by Lemma 4.2, has Fourier coefficients $\hat{h}(n) = \left(\hat{f}(n)\right)^k$ for $n \in \mathbb{Z}$. In order to show that w approaches to a uniform distribution on $[0,1)$, it suffices to show that $h \to 1$ in the L^2-norm on $L^2[0,1]$ as $k \to \infty$. Indeed, by Parseval's identity:

$$\begin{aligned}\|h-1\|_2^2 &= \int_0^1 (h(t)-1)^2\,dt = \sum_{n\neq 0} |\hat{h}(n)|^2 = \sum_{n\neq 0} |\hat{f}(n)|^{2k} \\ &= \sum_{n\neq 0} |\hat{f}(n)|^2 |\hat{f}(n)|^{2k-2} \leq \left(\sup_{n\neq 0} |\hat{f}(n)|\right)^{2k-2} \cdot \sum_{n\neq 0} |\hat{f}(n)|^2.\end{aligned}$$

By Lemma 4.1 and $f \in L^2[0,1]$, we have:

$$\sup_{n\neq 0} |\hat{f}(n)| < 1 \quad \text{and} \quad \sum_{n\neq 0} |\hat{f}(n)|^2 = \int_0^1 f^2(t)\,dt < \infty,$$

therefore it is clear that $\|h-1\|_2^2 \to 0$ as $k \to \infty$.

Theorem 4.1 can be generalized to one involving quotients and powers of random variables; indeed, similar to (4.12), it is easy to see that

$$\log_{10}(\overline{(x/y)}) = \log_{10}(\overline{x}) - \log_{10}(\overline{y}) \pmod 1 \quad \text{and} \quad \log_{10}(\overline{x^n}) = n\log_{10}(\overline{x}) \pmod 1.$$

The following lemma is similar to Lemma 4.2 and its proof is immediate.

Lemma 4.3 Let x be a continuous random variable on $[0, 1)$ with a probability density $f \in L^2[0, 1]$, and let $z = mx \pmod 1$, for a fixed $m \in \mathbb{Z}$. Then z is continuous and has density h such that

1. $\hat{h}(n) = \hat{f}(mn)$ for all $n \in \mathbb{Z}$;
2. if $m = -1$, we simply have $\hat{h}(n) = \hat{f}(-n) = \overline{\hat{f}(n)}$.

The following statement is a variant of Theorem 4.1, in which we allow random discount in addition to random geometric growth, that is to say, we allow those r_i's in (4.11) to also take negative values. Thus, these explain the emergence of Benford's law for COVID-19 figures governed by the dynamics of (4.10).

Theorem 4.2 Let $x_1, \dots, x_k$ be independent and identically distributed continuous random variables on $[0, 1)$ with a common probability density $f \in L^2[0, 1]$, also let $v := x_1 + \dots + x_j - x_{j+1} - \dots - x_k \pmod 1$. Then, v also approaches a uniform distribution on $[0, 1)$ as $k \to \infty$.

Proof. Let h be the density of v. As before, it suffices to show $\|h - 1\|_2^2 \to 0$ as $k \to \infty$; to this end, using Lemmas 4.2 and 4.3, we have

$$\|h(t) - 1\|_2^2 = \sum_{n \neq 0} |\hat{h}(n)|^2 = \sum_{n \neq 0} |\hat{f}(n)|^{2j} \cdot |\overline{\hat{f}(n)}|^{2k-2j} = \sum_{n \neq 0} |\hat{f}(n)|^{2k},$$

which converges to 0 as $k \to \infty$ by the last few lines of arguments used in the proof for Theorem 4.1. □

4.5 ZIPF'S LAW

Zipf's law is another empirical law that has been observed to naturally occur in datasets from both social and physical sciences. The law is named after the linguist *George Kingsley Zipf* who noticed that the frequency of a given word in a corpus of natural language is inversely proportional to its rank in the corresponding frequency table [51, 52]. Mathematically, given an ordered dataset $z_1 > z_2 > \dots > z_n$, in which each datum z_r is labeled by its own rank (r), Zipf's law is said to hold if the log-log plot of $\ln(z_r)$ and $\ln(r)$ is approximately linear. More precisely, Zipf's law asserts that

$$\ln(r) = \alpha - \beta \ln(z_r) + \varepsilon_r \tag{4.13}$$

for $r = 1, 2, \dots, n$, where $\alpha, \beta > 0$, and these ε_r's are independent error terms. This linear regression relationship is also referred to as the rank-size rule. Some authors, such as [14], use Zipf's law to refer specifically to phenomena following power law distributions, of which this rank-size rule is a direct consequence. This claim can be proved by using the theory of order statistics under the power law distribution. In particular, for a power law distribution Z with the survival function $\mathbb{P}(Z > z) \sim \frac{e^{\tilde{\alpha}}}{z^{\beta}}$, we take the logarithm of both sides to obtain $\ln \mathbb{P}(Z > z) \approx \tilde{\alpha} - \beta \ln z$; on the other hand, let $q_{(\gamma)}$ be the quantile at the upper tail level γ, i.e., $\mathbb{P}(Z > q_{(\gamma)}) = \gamma$, then we have $\ln \gamma \approx \tilde{\alpha} - \beta \ln q_{(\gamma)}$. Therefore, for the sample observation, since $\mathbb{P}(Z > z)|_{z=z_r} \approx \frac{r}{n}$, we then have $\ln \frac{r}{n} \approx \tilde{\alpha} - \beta \ln z_r$ or $\ln r \approx (\tilde{\alpha} + \ln n) - \beta \ln z_r$.

Zipf's law has been observed in different areas ranging from pure sciences to social sciences, for example the ranking of city sizes by population found in [14]. In fact, that city rankings have been demonstrated to obey, approximately, Zipf's law is particularly interesting given the uncontrolled nature of city growths. This observation holds in the United States throughout the 20th Century [23, 24], and for most other countries [37]; indeed, a study by Moura and Ribeiro [29] found that Brazilian cities also obey the law over a period of at least 50 years. On the other hand, the emergence of Zipf's law in epidemiological data is of significance as it provides a practical means of flagging data for further analysis in *healthcare management*. To see why Zipf's law emerges naturally in epidemiology, we first present a brief summary of the explanation in the context of city population ranking below.

Let S_t^i denote the normalized size of city i at time t among a collection of n cities, and this total number of cities is fixed and finite. Moreover, the normalized city size is simply city's population divided by the total population in all n cities, and the initial distribution of population to cities is arbitrary. It is assumed that the normalized city sizes, at least over a certain range, evolve randomly over time. More precisely, assume that in the upper tail the evolution of the i-th city size is in the form

$$S_{t+1}^i = \rho_{t+1}^i S_t^i, \quad i = 1, 2, \ldots, n, \tag{4.14}$$

where $0 < \rho_{t+1}^i$'s are iid random variables with the common non-constant density $f(\rho)$. Define $G_t(x) := \mathbb{P}(S_t^i > x)$ and note that

$$G_{t+1}(x) = \mathbb{P}(\rho_{t+1}^i S_t^i > x) = \mathbb{P}(S_t^i > x/\rho_{t+1}^i) = \mathbb{E}_f(G_t(x/\rho_{t+1}^i)),$$

where the last equality follows by the *tower property* of expectation. Hence, the continuous equilibrium state survival distribution $G = \lim_{t\to\infty} G_t$, if it exists, must satisfy

$$G(x) = \int_0^{\infty} G\left(\frac{x}{\rho}\right) f(\rho) d\rho, \quad \text{for all } x > 0. \tag{4.15}$$

The works [14, 15] note that possible ways of ensuring the existence of an equilibrium state distribution are to add a small constant to (4.14) to prevent cities from getting too small, or a lower bound for city sizes enforced by a reflective barrier. More importantly, according to [15], these adjustments typically do not affect the power law in the upper tail, for which only the growth rate matters. One then easily verifies that a solution in the upper tail is $G(x) \propto x^{-\beta}$, for some $\beta > 0$, provided that $\mathbb{E}((\rho_t^i)^{\beta}) = 1$. To see that it is the only solution, for if

$G_1(x)$ and $G_2(x)$ are two distinct solutions satisfying (4.15), then $|G_1(x) - G_2(x)| = |\int_0^\infty (G_1(\frac{x}{\rho}) - G_2(\frac{x}{\rho}))f(\rho)d\rho| \leq \int_0^\infty |G_1(\frac{x}{\rho}) - G_2(\frac{x}{\rho})| f(\rho)d\rho$, which further implies that, as long as the support $\text{supp}(f) = \mathbb{R}_+$, $\max_x |G_1(x) - G_2(x)| < \int_0^\infty f(\rho)d\rho \cdot \max_x | G_1(x) - G_2(x)| = 1 \cdot \max_x |(G_1(x) - G_2(x)|$, henceforth $G_1 \equiv G_2$. Furthermore, [14] observes that if the city sizes are normalized, the average normalized city size is constant, and this requires that $\mathbb{E}(\rho_t^i) = 1$; so it follows that Zipf's law holds in the upper tail with $\beta = 1$. This serves as the proposed explanation for the emergence of Zipf's law for city sizes. In addition, [15] points out, however, that if small cities grow much faster than larger ones, then $\beta > 1$.

As discussed above, the rank-size relationship (4.13) is then a consequence of the power law distribution

$$\mathbb{P}(S_\infty^i > x) = G(x) \propto x^{-\beta}. \tag{4.16}$$

It is easy to see that if (4.13) holds for normalized z_r, it clearly holds for unnormalized z_r by absorbing the normalizing constant into α of (4.13).

Few empirical phenomena obey power law distributions for all values of x; instead, power laws typically only hold for values greater than some minimum value $x_{\min}$, which must be estimated. In that case, Zipf's law remains valid for large enough values of z_r. In particular, the rank-size relationship can be seen in the size rankings of naturally occurring cities [19], gross domestic products [6], world cities [6], and English text frequencies [28].

4.6 ZIPF'S LAW AND COVID-19 FIGURES

The emergence of Zipf's law in epidemiological data is of significance as it provides a practical means of flagging data for further analysis in healthcare management. Investigations are generally costly, and they should be initiated only if there is a good reason to suspect data of being fraudulent. Central governments compile data from many different sources, such as local governments or directly from healthcare faculties, and so there is a clear need to automate and streamline fraud detection procedures. Furthermore, the use of Zipf's law for fraud detection is strengthened by the central government's monopoly on civil information; if the government does not publish the regional breakdown of the data in question, it would be difficult for local authorities to falsify data in accordance with Zipf's law. By monitoring the data over time, and not publishing them in detail, the central government can ascertain the expected Zipfian pattern in the data, upon which an automated fraud detection system can be developed.

After explaining the emergence of Zipf's law in city population rankings in the previous discussion, we now turn to the topic of why it should emerge within the early stages of an epidemic. Consider a country comprised of a collection of regions such as states or provinces If a common disease is simultaneously imported to each region and begins to spread locally, let $I_i(t)$ be the normalized number of infections in the region i, for $i = 1, 2, \ldots, n$; in other words, $I_i(t)$ is the number of infected individuals in the region i divided by the total number of infected nationally over n regions.

Considering that the disease is identical across each region, with a given level of infectiousness, and supposing that there is homogeneity among regions in terms of culture, population density, and development, assume that each region shares a common level of infectiousness $\theta > 0$ and a common recovery rate $\delta > 0$. We then consider a model of the form, for $i = 1, 2, \ldots, n$,

$$\mathrm{I}_i(t+1) = \mathrm{I}_i(t) + (\theta + \epsilon^I_{t+1,i})\mathrm{I}_i(t) - (\delta + \epsilon^R_{t+1,i})\mathrm{I}_i(t), \tag{4.17}$$

where $(\epsilon^I_{t+1,i}, \epsilon^R_{t+1,i})$'s are appropriately defined iid random noise vectors such that $1 + \theta + \epsilon^I_{t+1,i} - \delta - \epsilon^R_{t+1,i} \geq 0$ a.s.; alternatively (4.17) can be rewritten as

$$\mathrm{I}_i(t+1) = \gamma^i_{t+1}\mathrm{I}_i(t), \tag{4.18}$$

where $0 < \gamma^i_{t+1}$'s are iid random variables. Based on our previous discussions in Section 4.5, whenever there is a $\beta > 0$ such that

$$\mathbb{E}((1 + \theta - \delta + \epsilon^I_{t+1,i} - \epsilon^R_{t+1,i})^\beta) = 1,$$

then the equilibrium state distribution of infections within each region will be given by (4.16). Since $\mathrm{I}_i(t)$'s are assumed to be normalized, $\mathbb{E}(\gamma^i_{t+1}) = 1$, and so we must have

$$\mathbb{E}(\theta - \delta + \epsilon^I_{t+1,i} - \epsilon^R_{t+1,i}) = 0, \tag{4.19}$$

which suggests the emergence of Zipf's law with $\beta = 1$. We should emphasize that Condition (4.19) does not mean that each number of infections does not increase; since we are considering the normalized numbers of infections, this condition only refers to the number of infections in the region i relative to the rest of the whole country.

There are three obvious areas where the above construction might fall short, and they must be considered when determining the cause of anomalous data. Firstly, if the regions are very different from one another in terms of culture, population density, demographics, or development, then clearly the assumption of common infectiousness and recovery parameters is questionable, and this likely leads to deviations from Zipf's law. Therefore, if there is a handful of regions that deviate significantly from the nationally predicted rank-size relationship, the first course of action is to check whether these regions differ meaningfully from the rest of the country. This includes whether the policy response of the local authorities differed from that of the central government, or if the response followed a very different timeline. Secondly, the assumption of the iid nature of γ^i_t's, which implies the independence of the number of infections, may not be absolutely correct in practice. In particular, centralized and decentralized preventative measures could impose an effect on the dynamics of disease spread, and the likelihood of such measures being implemented clearly increases with disease prevalence in a region. Finally, the assumption that the disease is imported to all regions simultaneously can be susceptible, as in reality, the disease emerges in larger transport hubs long before the rural regions. The timeline of infections should therefore be examined to check whether cases in anomalous regions could emerge much earlier or later than the rest of the country. More discussions on this topic can be traced in our work [20].

For the estimation of $x_{\min}$, we recall the notion of Kolmogorov–Smirnov (KS) distance as introduced in Section 3.1, and then follow the approach proposed in [5]. Consider the truncated KS distance beyond $x_{\min}$ between the empirical and theoretical power law distributions defined by:

$$K(x_{\min}) := \max_{x \geq x_{\min}} \left|\hat{F}(x) - F(x)\right|, \tag{4.20}$$

where $\hat{F}(x)$ is the empirical distribution of the data above $x_{\min}$, and $F(x)$ is the power law distribution of best fit for data above $x_{\min}$; also see [7]. To estimate $x_{\min}$, [5] suggests the estimate $\hat{x}_{\min}$ to be the minimizer of $K(x_{\min})$, which can be numerically obtained via the **R** package `poweRlaw`.

Having obtained an estimate $\hat{x}_{\min}$, the two most common approaches to estimating the parameter β in (4.13) are ordinary least squares (OLS) and Hill's estimator. As noted in [16], while OLS is the most common method employed in the empirical literature, it tends to underestimate the power exponent. An alternative estimator known as Hill's estimator was proposed by [17], and is defined by

$$\hat{\beta}_{\text{Hills}} := \frac{n_{\text{tail}} - 1}{\sum_{i=1}^{n_{\text{tail}}-1}(\ln(z_i) - \ln(z_{n_{\text{tail}}}))}. \tag{4.21}$$

A test for whether the underlying distribution of the data above this estimated minimal threshold $\hat{x}_{\min}$ is a power law distribution with the estimated tail index $\hat{\beta}$ is proposed in [5]; again, this can be implemented using the package `poweRlaw`. This test is motivated by the notion of *bootstrapping* (see alternative discussion in Chapter 8), in the sense that the distribution of the test statistic, namely the truncated KS distance $K(\hat{x}_{\min})$, is approximated by the empirical distribution of the truncated KS distances of the bootstrapped samples from the fitted distribution.

Specifically, let n_{tail} denote the number of data points greater than $\hat{x}_{\min}$ in the observed dataset of size n, whose truncated distance is denoted by $K^*(\hat{x}_{\min})$. We generate a total of M (usually ranging from 1,000 to 10,000) bootstrapped samples, each of size n, where every sample point is generated independently in the following manner: with a probability of $\frac{n_{\text{tail}}}{n}$, the point is generated from the power law distribution with the scaling parameter $\hat{\beta}$ and the support $x \geq \hat{x}_{\min}$; otherwise, with a probability of $1 - \frac{n_{\text{tail}}}{n}$, the point is drawn uniformly from the portion below $\hat{x}_{\min}$ in the original observed dataset. We then calculate the truncated KS distance for the m-th bootstrapped sample, denoted by $K_m(\hat{x}_{\min})$, for $m = 1, \ldots, M$, and the p-value of this test is defined as the empirical survival function of the bootstrapped truncated KS distances at $K^*(\hat{x}_{\min})$:

$$p\text{-value} := \frac{|\{m : K_m(\hat{x}_{\min}) > K^*(\hat{x}_{\min})\}|}{M}.$$

Intuitively, a small p-value below a pre-specified significance level α indicates that the truncated KS distance of the observed dataset is larger than most of those in the bootstrapped samples, leading to the rejection of the null hypothesis at the significance level α that the upper tail of the dataset follows the fitted power law. Also see [5] for detailed discussions on the test.

In what follows, we consider an empirical analysis of the regional COVID-19 case figures as of 11 May 2020 for 14 countries[4], and the results are shown in Table 4.7. In particular, the figures in (Mainland) China without those from Hubei are also investigated so as to address the possible effect from the epicenter. We shall briefly summarize the noteworthy findings below; interested readers may refer to [20] for more discussions.

We focus on the column R^2 in Table 4.7, which is the R^2 statistic for the regression model (4.13). For all countries except China and Russia, R^2 exceeds 0.9, testifying to the presence of Zipf's law for the regional distribution of COVID-19 cases in a country.

Among the only exceptions, the low R^2 for China (without Hubei) and Russia can be justified by a change in the power law exponent, as shown in the log-log plots of Figure 4.9. This is because the estimation procedure for $x_{\min}$ as outlined in [5] locates the point where the second power law ends, but not the one where the power law exponent first changes, resulting in inaccurate results. Such change points

TABLE 4.7 Regional COVID-19 cases by country extracted from [20].

Country	$\hat{\alpha}_{\text{OLS}}$	$\hat{\beta}_{\text{OLS}}$	R^2	$\hat{\beta}_{\text{Hills}}$	$\hat{x}_{\min}$	n_{tail}	p-value
Brazil	9.0833	0.8249	0.9713	0.8104	1290	22	0.79
Canada	3.9787	0.3471	0.9563	0.3448	120	9	0.94
China	6.225	0.6164	0.8601	0.6997	127	27	0.04
China excl. Hubei	8.0087	0.9371	0.8325	0.8098	127	26	0
Germany	8.2871	0.7332	0.9365	0.8463	2556	13	0.76
India	7.5975	0.7313	0.9405	0.7176	696	15	0.13
Italy	8.775	0.7695	0.9846	0.8233	2572	15	1
Malaysia	6.191	0.7676	0.9364	0.7533	89	14	0.6
Mexico	8.4837	0.9223	0.9764	1.1451	582	15	0.94
Romania	10.6855	1.3611	0.9015	1.8568	379	16	0.43
Russia	10.1011	0.9645	0.846	1.4677	772	54	0.05
Spain	10.3289	0.9137	0.9236	1.2207	9291	8	0.65
Sweden	8.7594	0.9781	0.9209	1.4245	805	11	0.73
UK	23.0209	2.2431	0.9592	2.6948	11468	8	0.98
USA	11.2446	0.8665	0.9688	0.8212	6150	36	0.81

[4] These include Brazilian states [26], Canadian provinces [11], Chinese provinces [47], German states [22], Indian states and union territories [9], regions of Italy [42], states and federal territories of Malaysia [30], Mexican states [27], Romanian counties [43], Russian federal subjects [44], autonomous communities of Spain [13], Swedish counties [33], regions of the United Kingdom [46], and states and territories of the United States [12].

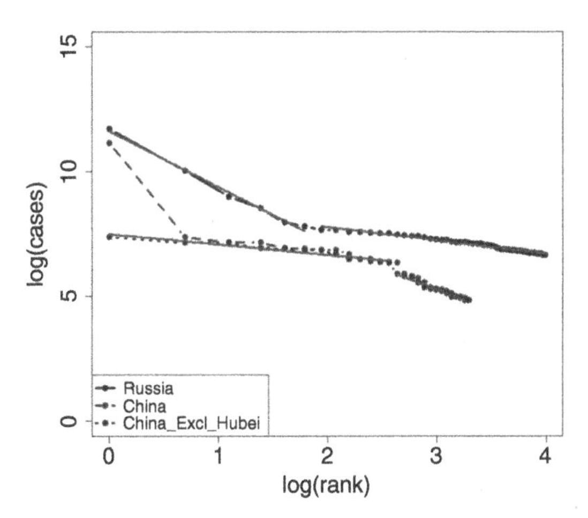

FIGURE 4.9 Log-log plots of confirmed COVID-19 cases in Russia (black solid), China with Hubei (green dashed), and China without Hubei (blue dotted). Note the change in power law indicated by the change in slope.

might be explained if the regions can be separated into two homogeneous subgroups, such as urban and rural regions, so that each of them possesses similar within-group growth trajectories. In particular, further analysis on the Russia data shows that the first power law applies to major developed cities and seaports in the European part of the country, and a similar analysis for China (without Hubei) indicates the presence of the first power law mostly in coastal provinces, which are more developed. As a result, the change points can be justified in relation to the level of development and geographical location of the place, which in turn explains the relatively low R^2 values.

On the other hand, when Hubei is included in the analysis for China, then the low R^2 can no longer be explained by a change point; there were far more cases in Hubei relative to the remaining provinces. However, this discrepancy might be explained by the entirely different growth trajectories following Chinese government intervention. The virus spread uncontrollably in Hubei for three weeks in January 2020, before the extraordinary quarantine measures undertaken by the government came into force just before the Chinese New Year, which effectively halted the transmission of the virus to other provinces via the *Chunyun* (a.k.a. *Spring Festival Travel Rush*).

The applicability of Zipf's law can also be assessed visually via log-log plots. Figures 4.9 and 4.10 present log-log plots of regional COVID-19 cases for the countries considered. Again, the plots are roughly linear for all countries except China and Russia, which supports Zipf's law in general. The estimates of β for the countries using either OLS or Hill estimation are mostly around 1, echoing the theoretical arguments based on the construction by [14] and our discussions in Section 4.5, which predicts a power law exponent of $\beta = 1$; yet there are still some deviations in which some estimates vary quite substantially and notably different from 1. For instance, Canada has an OLS estimate of β as small as 0.34, while that for the United Kingdom is as large as 2.24. This difference is caused by the fact that the Canadian cases are highly concentrated in the provinces of Quebec and Ontario, which together account for

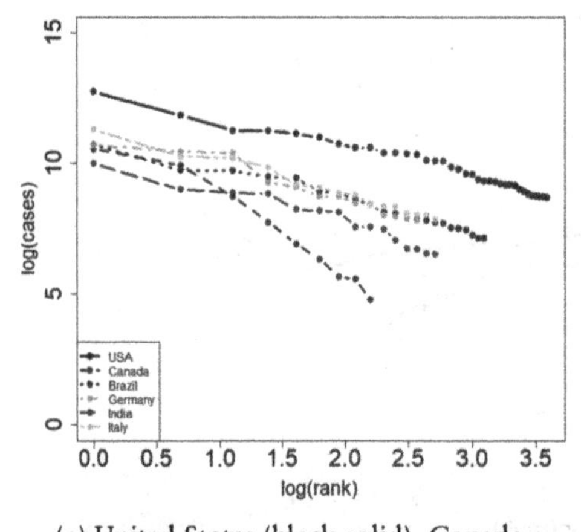

(a) United States (black solid), Canada (red dashed), Brazil (blue dotted), Germany (grey dash-dot), India (green long-dashed), and Italy (yellow double dashed)

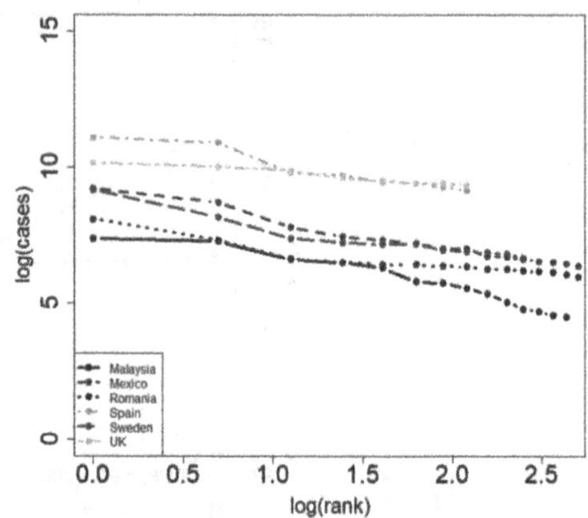

(b) Malaysia (black solid), Mexico (red dashed), Romania (blue dotted), Spain (grey dash-dot), Sweden (green long-dashed), and the United Kingdom (yellow double dashed)

FIGURE 4.10 Log-log plots of confirmed COVID-19 cases in each country.

84% of COVID-19 cases in Canada. On the other hand, the two largest regions in the United Kingdom accounted for only 30% of all cases.

Besides the examples illustrated in this chapter, Benford's law and Zipf's law still have various potential applications in a wide range of disciplines. Nevertheless, data collection is always the first and foremost step of such analyses. In the event that the data are heavy-tailed, they may be handled better by utilizing other statistical tools, e.g., *Taylor's law* of fluctuation for heavy-tailed data recently studied in [2]. We shall leave these explorations for future research.

4.A APPENDIX

4.A.1 Chi-Square Goodness-of-Fit Test

Let $x_1, \ldots, x_n$ be iid categorical variables taking K ($\ll n$) different possible values labelled as $A_1, \ldots, A_K$. Denote by $p_{(K)} = (p_1, \ldots, p_K)^\top$ where $p_j = \mathbb{P}(x = A_j)$ such that $\sum_{j=1}^{K} p_j = 1$. The maximum likelihood estimator $\hat{p}_{(K)} := (\hat{p}_1, \ldots, \hat{p}_K)^\top$ of $p_{(K)}$ can be obtained by maximizing the log-likelihood function

$$\ell(p_{(K)}) = \ln \prod_{i=1}^{n} \prod_{j=1}^{K} p_j^{y_{ij}} = \sum_{i=1}^{n} \sum_{j=1}^{K} y_{ij} \ln p_j \tag{4.22}$$

subject to the constraint of $p_1 + \cdots + p_K = 1$, where $y_{ij} := \mathbb{1}_{\{x_i = A_j\}}$. In order to solve this constrained optimization problem, we consider the associated *Lagrangian* with the multiplier λ:

$$\mathcal{L}(p_{(K)}, \lambda) := \sum_{i=1}^{n} \sum_{j=1}^{K} y_{ij} \ln p_j + \lambda \left(1 - \sum_{j=1}^{K} p_j \right).$$

Differentiating the Lagrangian with respect to p_j yields:

$$\frac{\partial}{\partial p_j} \mathcal{L}(p_{(K)}, \lambda) = \sum_{i=1}^{n} \frac{y_{ij}}{p_j} - \lambda, \quad \text{for } j = 1, \ldots, k,$$

and equating each of these with zero, we obtain $\lambda = \sum_{i=1}^{n} \frac{y_{ij}}{p_j}$, hence

$$\hat{p}_j = \sum_{i=1}^{n} \frac{y_{ij}}{\lambda}.$$

Since summing all $\hat{p}_j$'s is always equal to 1, we have

$$1 = \sum_{j=1}^{K} \hat{p}_j = \sum_{j=1}^{K} \sum_{i=1}^{n} \frac{y_{ij}}{\lambda},$$

which gives $\lambda = n$. Therefore, the maximum likelihood estimators $\hat{p}_j$'s are given by:

$$\hat{p}_j = \frac{1}{n} \sum_{i=1}^{n} y_{ij} = \frac{1}{n} \sum_{i=1}^{n} \mathbb{1}_{\{x_i = A_j\}}, \quad \text{for } j = 1, \ldots, K,$$

as expected.

Next, we first compute the expectation and covariance of $\mathbf{y} = (y_1, \ldots, y_K)^\top$, where $y_j = \mathbb{1}_{\{x = A_j\}}$:

1. **Expectation:** Since each y_j is a Bernoulli random variable with the success probability equal to p_j, we certainly have $\mathbb{E}(\mathbf{y}) = \mathbf{p}_{(K)}$.
2. **Covariance:** The covariance of $y_u \sim \text{Ber}(p_u)$ and $y_v \sim \text{Ber}(p_v)$ equals

$$\text{Cov}(y_u, y_v) = \mathbb{E}(y_u y_v) - \mathbb{E}(y_u)\mathbb{E}(y_v) = \begin{cases} p_u(1 - p_u), & \text{if } u = v, \\ -p_u p_v, & \text{if } u \neq v. \end{cases}$$

Or equivalently, in matrix form:

$$\boldsymbol{\Sigma}_{(K)} := \text{Cov}(\mathbf{y}) = \text{diag}(\mathbf{p}_{(K)}) - \mathbf{p}_{(K)} \mathbf{p}_{(K)}^\top. \tag{4.23}$$

3. **Correlation:** The correlation of $y_u \sim \text{Ber}(p_u)$ and $y_v \sim \text{Ber}(p_v)$ is:

$$\rho_{ij} := \text{Corr}(y_u, y_v) = \frac{\text{Cov}(y_u, y_v)}{\sqrt{\text{Var}(y_u)\text{Var}(y_v)}} = \begin{cases} 1, & \text{if } u = v, \\ -\sqrt{\frac{p_u p_v}{(1-p_u)(1-p_v)}}, & \text{if } u \neq v. \end{cases} \tag{4.24}$$

Before proceeding to the discussion of the chi-square goodness-of-fit test, we first introduce an important matrix inverse formula:

Proposition 4.1 *(Sherman–Morrison formula* [18, 41]*)* Suppose that A is an $n \times n$ invertible square matrix, and $\boldsymbol{u}, \boldsymbol{v}$ are n-dimensional vectors. The inverse $(A + \boldsymbol{u}\boldsymbol{v}^\top)^{-1}$ exists if and only if $1 + \boldsymbol{v}^\top A^{-1}\boldsymbol{u} \neq 0$, and it is given by:

$$(A + \boldsymbol{u}\boldsymbol{v}^\top)^{-1} = A^{-1} - \frac{A^{-1}\boldsymbol{u}\boldsymbol{v}^\top A^{-1}}{1 + \boldsymbol{v}^\top A^{-1}\boldsymbol{u}}. \tag{4.25}$$

Proof. Suppose that the inverse $(A + \boldsymbol{u}\boldsymbol{v}^\top)^{-1}$ exists, which is equivalent to the solvability of the system $(A + \boldsymbol{u}\boldsymbol{v}^\top)\boldsymbol{x} = \boldsymbol{b}$, for every $\boldsymbol{b} \in \mathbb{R}^n$. If $\boldsymbol{x}$ is a solution of this system, then

$$\boldsymbol{x} = A^{-1}\boldsymbol{b} - A^{-1}\boldsymbol{u}\boldsymbol{v}^\top\boldsymbol{x}. \tag{4.26}$$

Putting $r = \boldsymbol{v}^\top\boldsymbol{x} \in \mathbb{R}$, we have $r = \boldsymbol{v}^\top(A^{-1}\boldsymbol{b} - A^{-1}\boldsymbol{u}r)$, hence

$$r(1 + \boldsymbol{v}^\top A^{-1}\boldsymbol{u}) = \boldsymbol{v}^\top A^{-1}\boldsymbol{b}.$$

As $\boldsymbol{b}$ is arbitrary, $1 + \boldsymbol{v}^\top A^{-1}\boldsymbol{u} \neq 0$. Conversely, one can directly check that

$$(A + \boldsymbol{u}\boldsymbol{v}^\top)\left(A^{-1} - \frac{A^{-1}\boldsymbol{u}\boldsymbol{v}^\top A^{-1}}{1 + \boldsymbol{v}^\top A^{-1}\boldsymbol{u}}\right) = I_n,$$

thus (4.25) holds true. □

Proposition 4.2 Define the error vector by $\mathrm{e}_{(K)} := \hat{\mathrm{p}}_{(K)} - p_{(K)}$. Then as sample size n goes to infinity, the Pearson's goodness-of-fit test [35] statistic

$$\sum_{i=1}^{K} \frac{(n\hat{\mathrm{p}}_i - np_i)^2}{np_i} = n\mathrm{e}_{(K-1)}^\top \boldsymbol{\Sigma}_{(K-1)}^{-1} \mathrm{e}_{(K-1)} \xrightarrow{d} \chi^2_{K-1}.$$

Proof. By the Sherman–Morrison formula in Proposition 4.1, we first observe that

$$\begin{aligned}
\boldsymbol{\Sigma}_{(K-1)}^{-1} &= (\text{diag}(\boldsymbol{p}_{(K-1)}) - \boldsymbol{p}_{(K-1)}\boldsymbol{p}_{(K-1)}^{\top})^{-1} \\
&= \text{diag}(\boldsymbol{p}_{(K-1)})^{-1} - \frac{\text{diag}(\boldsymbol{p}_{(K-1)})^{-1}(-\boldsymbol{p}_{(K-1)})\boldsymbol{p}_{(K-1)}^{\top}\text{diag}(\boldsymbol{p}_{(K-1)})^{-1}}{1 + \boldsymbol{p}_{(K-1)}^{\top}\text{diag}(\boldsymbol{p}_{(K-1)})^{-1}(-\boldsymbol{p}_{(K-1)})} \\
&= \text{diag}(\boldsymbol{p}_{(K-1)})^{-1} + \frac{\mathbf{1}_{K-1}\mathbf{1}_{K-1}^{\top}}{1 - \mathbf{1}_{K-1}^{\top}\boldsymbol{p}_{(K-1)}} \\
&= \text{diag}(\boldsymbol{p}_{(K-1)})^{-1} + \frac{1}{p_K}\mathbf{1}_{K-1}\mathbf{1}_{K-1}^{\top}.
\end{aligned}$$

Therefore, the Pearson's goodness-of-fit test statistic can be rewritten as:

$$\begin{aligned}
&\sum_{j=1}^{K} \frac{(n\hat{p}_j - np_j)^2}{np_j} \\
&= n\sum_{j=1}^{K-1} \frac{(\hat{p}_j - p_j)^2}{p_j} + n\frac{(\hat{p}_K - p_K)^2}{p_K} \\
&= n\mathrm{e}_{(K-1)}^{\top}\text{diag}(\boldsymbol{p}_{(K-1)})^{-1}\mathrm{e}_{(K-1)} + n\frac{(\hat{p}_K - p_K)^2}{p_K} \\
&= n\mathrm{e}_{(K-1)}^{\top}\text{diag}(\boldsymbol{p}_{(K-1)})^{-1}\mathrm{e}_{(K-1)} + n\frac{(1 - \mathbf{1}_{K-1}^{\top}\hat{\mathrm{p}}_{(K-1)} - 1 + \mathbf{1}_{K-1}^{\top}\boldsymbol{p}_{(K-1)})^2}{p_K} \\
&= n\mathrm{e}_{(K-1)}^{\top}\text{diag}(\boldsymbol{p}_{(K-1)})^{-1}\mathrm{e}_{(K-1)} + n\frac{(\hat{\mathrm{p}}_{(K-1)} - \boldsymbol{p}_{(K-1)})^{\top}\mathbf{1}_{K-1}\mathbf{1}_{K-1}^{\top}(\hat{\mathrm{p}}_{(K-1)} - \boldsymbol{p}_{(K-1)})}{p_K} \\
&= n\mathrm{e}_{(K-1)}^{\top}\left(\text{diag}(\boldsymbol{p}_{(K-1)})^{-1} + \frac{1}{p_K}\mathbf{1}_{K-1}\mathbf{1}_{K-1}^{\top}\right)\mathrm{e}_{(K-1)} \\
&= n\mathrm{e}_{(K-1)}^{\top}\boldsymbol{\Sigma}_{(K-1)}^{-1}\mathrm{e}_{(K-1)}.
\end{aligned}$$

Furthermore, using (4.23), we then have

$$\text{Cov}(\mathrm{e}_{(K-1)}) = \text{Cov}(\hat{\mathrm{p}}_{(K-1)}) = \frac{1}{n}\left(\text{diag}(\boldsymbol{p}_{(K-1)}) - \boldsymbol{p}_{(K-1)}\boldsymbol{p}_{(K-1)}^{\top}\right) = \frac{1}{n}\boldsymbol{\Sigma}_{(K-1)}.$$

By the multivariate Central Limit Theorem, $\sqrt{n}\mathrm{e}_{(K-1)} \xrightarrow{d} \mathcal{N}_{K-1}(\mathbf{0}_{K-1}, \boldsymbol{\Sigma}_{(K-1)})$ and, from Subsection 1.4.2, the Pearson's goodness-of-fit test statistic approximately follows χ^2_{K-1} as $n \to \infty$. □

REFERENCES

1. Brauer, F., Van den Driessche, P., and Wu, J. (Eds.). (2008). *Mathematical Epidemiology* (Vol. 1945, pp. 3–17). Springer.
2. Brown, M., Cohen, J., Tang, C.F., and Yam, S.C.P. (2021). Taylor's law of fluctuation scaling for semivariances and higher moments of heavy-tailed data. *Proceedings of the National Academy of Sciences of the United States of America*, 118(46), e2108031118.
3. Boyle, J. (1994). An application of Fourier series to the most significant digit problem. *The American Mathematical Monthly*, 101(9), 879–886.
4. Carslaw, C.A. (1988). Anomalies in income numbers: evidence of goal oriented behavior. *Accounting Review*, 63(2), 321–327.
5. Clauset, A., Shalizi, C.R., and Newman, M.E. (2009). Power-law distributions in empirical data. *SIAM Review*, 51(4), 661–703.
6. Cristelli, M., Batty, M., and Pietronero, L. (2012). There is more than a power law in Zipf. *Scientific Reports*, 2(1), 1–7.
7. Daniel, W.W. (1978). *Applied Nonparametric Statistics*. Houghton Mifflin.
8. Deckert, J., Myagkov, M., and Ordeshook, P.C. (2011). Benford's law and the detection of election fraud. *Political Analysis*, 19(3), 245–268.
9. Diwanji, S. (2020). Coronavirus cases in India by state. *Statistica* (accessed 12 May 2020).
10. Durtschi, C., Hillison, W., and Pacini, C. (2004). The effective use of Benford's law to assist in detecting fraud in accounting data. *Journal of Forensic Accounting*, 5(1), 17–34.
11. Elflein, J. (2020). Canada: COVID-19 cases by province. *Statistica* (accessed 12 May 2020).
12. Elflein, J. (2020). U.S. COVID-19 cases by state. *Statistica* (accessed 12 May 2020).
13. Forte, F. (2020). Spain: Coronavirus cases by region. *Statistica* (accessed 12 May 2020).
14. Gabaix, X. (1999). Zipf's law for cities: an explanation. *The Quarterly Journal of Economics*, 114(3), 739–767.
15. Gabaix, X. (2009). Power laws in economics and finance. *Annual Review of Economics*, 1(1), 255–294.
16. Gabaix, X., and Ioannides, Y.M. (2004). The evolution of city size distributions. In *Handbook of Regional and Urban Economics* (Vol. 4, pp. 2341–2378). Elsevier.
17. Hill, B.M. (1975). A simple general approach to inference about the tail of a distribution. *The Annals of Statistics*, 3(5), 1163–1174.
18. Horn, R.A., and Johnson, C.R. (2012). *Matrix Analysis*. Cambridge University Press.
19. Jiang, B., Yin, J., and Liu, Q. (2015). Zipf's law for all the natural cities around the world. *International Journal of Geographical Information Science*, 29(3), 498–522.
20. Kennedy, A.P., and Yam, S.C.P. (2020). On the authenticity of COVID-19 case figures. *PLoS ONE*, 15(12), e0243123.
21. Klimek, P., Jiménez, R., Hidalgo, M., Hinteregger, A., and Thurner, S. (2018). Forensic analysis of Turkish elections in 2017–2018. *PLoS ONE*, 13(10), e0204975.
22. Koptyug, E. (2020). Coronavirus (COVID-19) case numbers in Germany by state 2020. *Statistica* (accessed 12 May 2020).
23. Krugman, P. (1996). Confronting the mystery of urban hierarchy. *Journal of the Japanese and International Economies*, 10(4), 399–418.
24. Krugman, P. (1996). *The Self Organizing Economy*. Wiley Blackwell.
25. Miller, S.J. (Ed.). (2015). *Benford's Law*. Princeton University Press.
26. Montanez, A. (2020). Brazil: COVID-19 cases by state. *Statistica* (accessed 12 May 2020).
27. Montanez, A. (2020). Mexico: COVID-19 cases by state. *Statistica* (accessed 12 May 2020).
28. Moreno-Sánchez, I., Font-Clos, F., and Corral, Á. (2016). Large-scale analysis of Zipf's law in English texts. *PLoS ONE*, 11(1), e0147073.

29. Moura Jr, N.J., and Ribeiro, M.B. (2006). Zipf law for Brazilian cities. *Physica A: Statistical Mechanics and its Applications*, 367, 441–448.
30. Müller, J. (2020). Malaysia: COVID-19 cases by state 2020. *Statistica* (accessed 12 May 2020).
31. Nigrini, M.J. (2012). *Benford's Law: Applications for Forensic Accounting, Auditing, and Fraud Detection*. Wiley.
32. Nigrini, M.J., and Mittermaier, L.J. (1997). The use of Benford's law as an aid in analytical procedures. *Auditing*, 16(2), 52–67.
33. Norrestad, F. (2020). Sweden: Coronavirus cases by region. *Statistica* (accessed 12 May 2020).
34. Parseval, M.A. (1806). Mémoire sur les séries et sur l'intégration complète d'une équation aux différences partielles linéaires du second ordre, à coefficients constants. *Mém. prés. par divers savants, Acad. des Sciences, Paris*, 1(1), 638–648.
35. Pearson, K. (1900). X. On the criterion that a given system of deviations from the probable in the case of a correlated system of variables is such that it can be reasonably supposed to have arisen from random sampling. *The London, Edinburgh, and Dublin Philosophical Magazine and Journal of Science*, 50(302), 157–175.
36. Rauch, B., Göchsche, M., and Brähler, G. (2011). Fact and fiction in EU-governmental economic data. *German Economic Review*, 12(3), 243–255.
37. Rosen, K.T., and Resnick, M. (1980). The size distribution of cities: an examination of the Pareto law and primacy. *Journal of Urban Economics*, 8(2), 165–186.
38. Roukema, B.F. (2014). A first-digit anomaly in the 2009 Iranian presidential election. *Journal of Applied Statistics*, 41(1), 164–199.
39. Rudin, W. (1976). *Principles of Mathematical Analysis*. McGraw-Hill.
40. Sheppard, L.A. (1994). Tax Notes audits the Clintons. *Tax Notes*, 18–23.
41. Sherman, J., and Morrison, W.J. (1950). Adjustment of an inverse matrix corresponding to a change in one element of a given matrix. *The Annals of Mathematical Statistics*, 21(1), 124–127.
42. Statistica Research Department. (2020). Italy: Coronavirus cases by region. *Statistica* (accessed 12 May 2020).
43. Statistica Research Department. (2020). Romania: Confirmed COVID-19 cases 2020. *Statistica* (accessed 12 May 2020).
44. Statistica Research Department. (2020). Russia: Coronavirus status. *Statistica* (accessed 12 May 2020).
45. Stein, E.M., and Shakarchi, R. (2011). *Fourier Analysis: An Introduction*. Princeton University Press.
46. Stewart, C. (2020). UK: Regional coronavirus cases. *Statistica* (accessed 12 May 2020).
47. Thomala, L. (2020). Greater China: Coronavirus statistics by region. *Statistica* (accessed 12 May 2020).
48. Tritch, T., and Sprouse, M. (1994). Money audits the Clintons. *Money Magazine*, 84–98.
49. Varian, H. (1972). Benford's law (letters to the editor). *The American Statistician*, 26(3), 62–65.
50. Wallace, W.A. (2002). Assessing the quality of data used for benchmarking and decision-making. *The Journal of Government Financial Management*, 51(3), 16.
51. Zipf, G.K. (1935). *The Psycho-Biology of Language: An Introduction to Dynamic Philology*. MIT Press.
52. Zipf, G.K. (1949). *Human Behavior and the Principle of Least Effort: An Introduction to Human Ecology*. Addison-Wesley Press.

CHAPTER 5

Numerical Finance

This chapter mainly introduces the use of simulations for the pricing of financial derivatives. Different basic simulation techniques will be introduced first. Regarding the origins of simulation, the idea started with the building of atomic bombs in the *Manhattan Project* [8] in 1942 during World War II. Scientists, including *Albert Einstein*, *J. Robert Oppenheimer* and *John von Neumann*, were the first to investigate the phenomenon of nuclear fission. When the nucleus of an atom is bombarded by a neutron, it will decompose into two lighter nuclei and release some more neutrons, emitting a huge amount of energy as a byproduct. These newly separated-out neutrons will further collide with other nuclei, and henceforth trigger a chain reaction. Physicists found a good use of statistics to predict and describe the aggregated behavior of the collective atom movements by treating every single atom as a random object whose moving trajectory is independent of those of the others.

5.1 FUNDAMENTALS OF SIMULATION

One hurdle faced in the implementation of modelling atomic behavior is the generation of genuine random numbers between 0 and 1; indeed, true randomness cannot be achieved by routine calculating machines. A *pseudorandom number generator* is used to mimic the truly random phenomenon of generating seemingly unordered values, and does so through a deterministic mechanism related to *chaotic dynamical systems*.

In *Chaos Theory*, the apparent random and irregular occurrence of states of a dynamical system are actually led by deterministic initial conditions. This kind of dynamical generating mechanism is very sensitive to the initial environment; indeed, even a tiny disturbance in the initial condition, after a short time, can cause the output state of the dynamical system to differ significantly from the original state. Due to this apparent unpredictability, a source of randomness can be obtained.

For pseudorandom number generators, the general idea is to choose an arbitrary initial seed every time, and then use the same dynamical system to generate a long sequence of its future states. We then dispose of the first portion of the numbers generated, and take the tail sequence of numbers as the random sample observations. Mathematically, this procedure of random number generation is justifiable, as discussed below.

One possible dynamical system used is the *linear congruential generator* which is one of the most common pseudorandom number generators. The generating algorithm is simple, and is defined by:

$$x_{n+1} \equiv (ax_n + c)\,(\mathrm{mod}\, m), \quad \text{for } n = 0, 1, \ldots,$$

where x_0 is a known initial "seed", the integer m is usually selected to be a large prime number, and c is chosen to be small ($0 < c < m$) in comparison with m. However, the value of $(ax_k + c)(\mathrm{mod}\, m)$ is very sensitive to a slight change of x_k; meanwhile, any tiny difference in the value of $ax_k + c$ can lead to a substantial deviation in the later sequential values. After a long run, the sequence of numbers x_n's will notably be uncorrelated with the original seed x_0 or any other x_k's obtained in the previous iteration steps. We can then take the fraction x_k/m as a sample observed value of the uniform random variable on $[0, 1]$. This procedure is justified in accordance with the following well-known result in *Ergodic Theory* [5, 19]: for any $0 < a < b < 1$,

$$\lim_{n\to\infty} \frac{1}{n} \sum_{j=0}^{n-1} \mathbb{1}_{[a,b]}\left(\frac{x_j}{m}\right) = b - a,$$

which ensures that for an arbitrary initial seed x_0, the relative frequency of those later values of x_k/m lying in an interval of $(a, b) \subset (0, 1)$ converges to $b - a$ as the number of iterations n tends to infinity. Programme codes in Python and R for the linear congruential generator are shown in Programmes 5.1 and 5.2, respectively; its comparisons with the built-in random number generators in Python and R are also provided in Figures 5.1a and 5.1b, respectively.

```
1  import numpy as np
2  import matplotlib.pyplot as plt
3  np.random.seed(4002)
4
5  plt.rc('font', size=20); plt.rc('axes', titlesize=23, labelsize=23)
6  plt.rc('xtick', labelsize=17); plt.rc('ytick', labelsize=17)
7
8  def pse_uni_gen(a=16807, c=1, m=2**31-1, seed=123456789,
9                  size=1, burn_in=1000):
10     x = [(a*seed + c) % m]
11     for i in range(1, size + burn_in):
12         x.append((a*x[-1] + c) % m)
13     return np.array(x[burn_in:]) / m
14
15 def pse_uniform_gen(lower=0, upper=1, seed=123456789,
16                     size=1, burn_in=1000):
17     U = pse_uni_gen(seed=seed, size=size, burn_in=burn_in)
18     return lower + (upper - lower)*U
19
20 pse_sample = pse_uniform_gen(lower=0, upper=1, size=10000)
21 built_in_sample = np.random.rand(10000)
22 plt.figure(figsize=(8,6))
```

```
23  plt.hist(pse_sample, ec='black', bins="sturges",
24          alpha=0.7, label="Pseudo Generator", color="gray")
25  plt.hist(built_in_sample, ec='black', bins="sturges",
26          alpha=0.7, label="Numpy Generator", color="red")
27  plt.legend(loc='upper right')
```

Programme 5.1 Pseudo Uniform Generator in Python.

```
1  > set.seed(4002)
2  >
3  > pse_uni_gen <- function(a=16807, c=1, m=(2^31)-1, seed=123456789,
4  +                          size=1, burn_in=1000){
5  +     x <- (a*seed + c) %% m
6  +     for (i in 1:(size+burn_in-1)){
7  +         x <- c(x, (a*x[length(x)] + c) %% m)
8  +     }
9  +     return (x[(burn_in+1):length(x)] / m)
10 + }
11 >
12 > pse_uniform_gen <- function(lower=0, upper=1, seed=123456789,
13 +                              size=1, burn_in=1000){
14 +     U <- pse_uni_gen(seed=seed, size=size, burn_in=burn_in)
15 +     return (lower+(upper - lower)*U)
16 + }
17 >
18 > pse_sample <- pse_uniform_gen(size=10000)
19 > built_in_sample <- runif(10000)
20 > hist(pse_sample, col=rgb(red=.5,green=.5,blue=.5,alpha=0.5),
21 +      main="Histogram")
22 > hist(built_in_sample, col=rgb(red=1.0,green=0,blue=0,alpha=0.5),
23 +      add=TRUE)
24 > legend('topright',legend=c("Pseudo Generator","R Generator"),
25 +        fill=c(rgb(red=.5,green=.5,blue=.5,alpha=0.5),
26 +               rgb(red=1.0,green=0,blue=0,alpha=0.5)), cex=1.5)
```

Programme 5.2 Pseudo Uniform Generator in **R**.

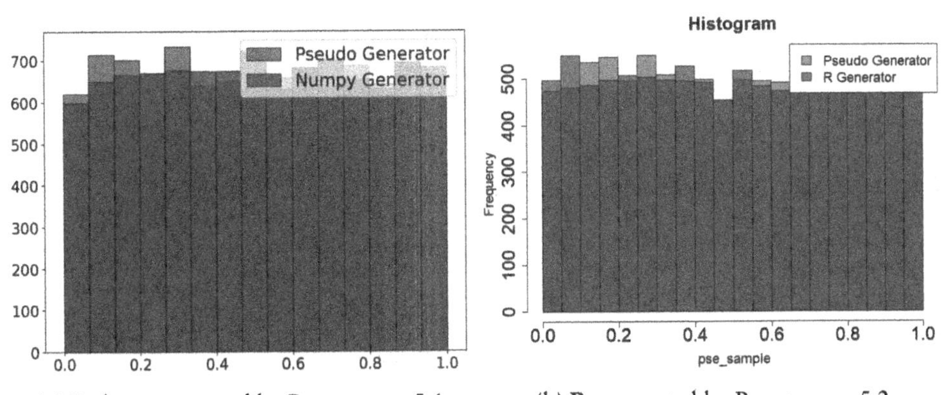

(a) Python: generated by Programme 5.1

(b) **R**: generated by Programme 5.2

FIGURE 5.1 Histograms of the samples from pseudo generators for the uniform distribution on $[0, 1]$.

Next, we discuss some common simulation techniques for non-uniform distributions; some standard references of this topic include [4, 15, 16].

Let $f_{\mathrm{x}}(x)$ and $F_{\mathrm{x}}(x)$ be the probability density function and the cumulative distribution function of a random variable x respectively. Suppose that we can generate a sample out from a uniform distribution on $[0, 1]$. Then we can simulate x using the knowledge of the explicit expression of the inverse of F_{x}; indeed, if $\mathrm{u} \sim U(0, 1)$, i.e. $\mathbb{P}(\mathrm{u} \leq u) = u$, for $u \in (0, 1)$, then

$$\tilde{\mathrm{x}} := F_{\mathrm{x}}^{-1}(\mathrm{u}) = \min\{t : F_{\mathrm{x}}(t) \geq \mathrm{u}\} \sim F_{\mathrm{x}}.$$

To see this, we note that $\mathbb{P}(F_{\mathrm{x}}^{-1}(\mathrm{u}) \leq x) = \mathbb{P}(\mathrm{u} \leq F_{\mathrm{x}}(x)) = F_{\mathrm{x}}(x)$. Therefore, if one can generate uniform samples, then in principle, if $F_{\mathrm{x}}^{-1}(\cdot)$ can be explicitly spelt out, we can generate a random sample from F_{x}. This simulation approach is said to be via *inverse function transform*. Conversely, we also have $F_{\mathrm{x}}(\mathrm{x}) \sim U(0, 1)$ if $\mathrm{x} \sim F_{\mathrm{x}}$ and F_{x} is continuous; its proof is simple, and we leave it as an exercise for readers. By applying this inverse transform technique, we can now simulate random variables of other distributions such as the common exponential distribution; see the programme codes for Python in Programme 5.3 and for **R** in Programme 5.4. The corresponding histograms are shown in Figures 5.2a and 5.2b, respectively.

```
1  import numpy as np
2  import matplotlib.pyplot as plt
3  from scipy.stats import expon
4  np.random.seed(4002)
5
6  def pseudo_exp_gen(lamb, seed=123456789, size=1, burn_in=1000):
7      U = pse_uniform_gen(seed=seed, size=size, burn_in=burn_in)
8      X = -(1/lamb)*np.log(1-U)
9      return X
10
11 pse_sample = pseudo_exp_gen(lamb=1, size=10000)
12 built_in_sample = np.random.exponential(1, size=10000)
13 plt.figure(figsize=(8,6))
14 _, bins, _ = plt.hist(pse_sample, alpha=0.7, label="Pseudo Generator",
15                       color="gray", ec='black', bins="sturges")
16 plt.hist(built_in_sample, alpha=0.7, label="Numpy Generator",
17          color="red", ec='black', bins=bins)
18 x = np.linspace(expon.ppf(0.01), expon.ppf(0.99), 10000)
19 plt.plot(x, expon.pdf(x)*10000)
20 plt.legend(loc='upper right')
```

Programme 5.3 Pseudo Exponential Generator in Python.

```
1  > set.seed(4012)
2  > par(cex.lab=1.5, cex.axis=2, cex.main=2)
3  >
4  > pse_exp_gen <- function(lamb, seed=123456789,
5  +                         size=1, burn_in=1000){
6  +     U <- pse_uni_gen(seed=seed, size=size, burn_in=burn_in)
```

```
7  +     X <- -(1/lamb)*log(1-U)
8  +     return (X)
9  + }
10 >
11 > pse_sample <- pse_exp_gen(lamb=1, size=10000)
12 > built_in_sample <- rexp(10000, 1)
13 > hist(pse_sample, col=rgb(red=.5,green=.5,blue=.5,alpha=0.5),
14 +      main="Histogram")
15 > hist(built_in_sample,col=rgb(red=1.0,green=0,blue=0,alpha=0.5),
16 +      add=TRUE)
17 > curve(10000*dexp(x, rate=1, log=FALSE), add=TRUE, col='black',
18 +       lwd=2.5)
19 > legend('topright',legend=c("Pseudo Generator","R Generator"),
20 +        fill=c(rgb(red=.5,green=.5,blue=.5,alpha=0.5),
21 +               rgb(red=1.0,green=0,blue=0,alpha=0.5)), cex=1.5)
```

Programme 5.4 Pseudo Exponential Generator in **R**.

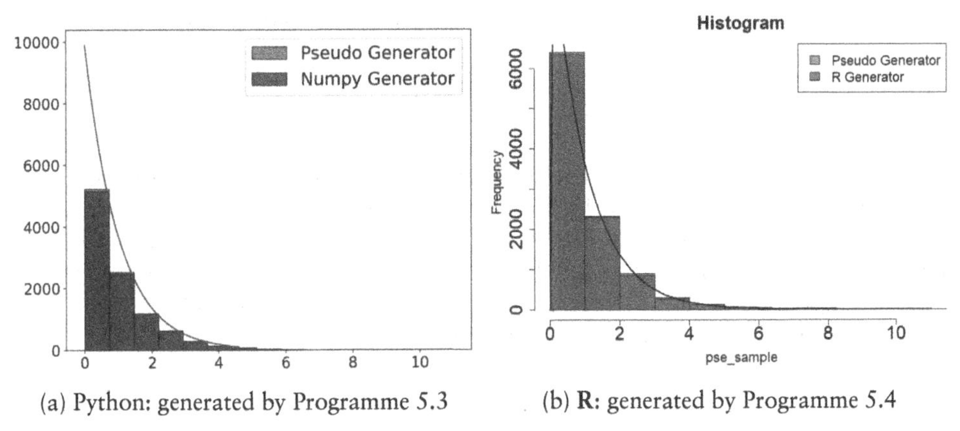

(a) Python: generated by Programme 5.3 (b) **R**: generated by Programme 5.4

FIGURE 5.2 Histograms of the samples from pseudo generators for the exponential distribution.

Although the *inverse function transform* method looks fundamental for general distribution simulations, an immediate limitation to this approach is that very often the inverse F_x^{-1} cannot be expressed explicitly, and without this the programme coding cannot be implemented conveniently. This is particularly the case for the normal distribution, as the inverse of a normal cumulative distribution function cannot be written as a sum of finitely many elementary functions. Fortunately, an alternative approach called the *Box–Muller transformation* can be adopted, allowing a pair of standard normal random variables to be easily be generated. Consider the following transformation, where u_1 and u_2 are independent uniformly distributed random variables on $[0, 1]$:

$$z_1 := \sqrt{-2\ln u_1}\cos(2\pi u_2) \quad \text{and} \quad z_2 := \sqrt{-2\ln u_1}\sin(2\pi u_2),$$

then z_1 and z_2 follow $\mathcal{N}(0,1)$ independently. To see this, we first express u_1 and u_2 in terms of z_1 and z_2 as follows:

$$\begin{cases} u_1 = \exp\left\{-\frac{z_1^2+z_2^2}{2}\right\} =: \exp\left\{\frac{-q}{2}\right\}, \\ u_2 = \frac{1}{2\pi}\arctan\left\{\frac{z_2}{z_1}\right\}, \quad \text{if } z_1 \neq 0, \end{cases}$$

with the Jacobian J of (u_1, u_2) with respect to (z_1, z_2) given by:

$$J = \begin{vmatrix} -z_1 \exp\left\{-\frac{q}{2}\right\} & -z_2 \exp\left\{-\frac{q}{2}\right\} \\ \frac{-z_2}{2\pi(z_1^2+z_2^2)} & \frac{z_1}{2\pi(z_1^2+z_2^2)} \end{vmatrix} = -\frac{1}{2\pi}\exp\left\{-\frac{q}{2}\right\}.$$

Since the joint probability density function of u_1 and u_2 is simply:

$$f(u_1, u_2) = 1, \quad u_1, u_2 \in (0,1),$$

it follows that the joint probability density function of z_1 and z_2 should be:

$$g(z_1, z_2) = \left|\frac{-1}{2\pi}\exp\left\{-\frac{q}{2}\right\}\right| \cdot (1) = \frac{1}{2\pi}\exp\left\{-\frac{z_1^2+z_2^2}{2}\right\}, \quad z_1, z_2 \in \mathbb{R}.$$

Note that there is some difficulty with the well-posedness of the inverse transformation, particularly when $z_1 = 0$; fortunately, this singularity only occurs with a zero probability measure, hence this causes no issues in the actual implementation. Again, programme codes in the Python and R environments are included in Programmes 5.5 and 5.6 respectively; while their respective comparisons with the built-in normal distribution generators are illustrated in Figures 5.3a and 5.3b.

```
1  import numpy as np
2  import matplotlib.pyplot as plt
3  np.random.seed(4002)
4
5  def pseudo_normal_gen(mu=0.0, sigma=1.0, seed=123456789,
6                        size=1, burn_in=1000):
7      U = pse_uniform_gen(seed=seed, size=2*size, burn_in=burn_in)
8      U1, U2 = U[:size], U[size:]
9      Z0 = np.sqrt(-2*np.log(U1))*np.cos(2*np.pi * U2)
10     Z1 = np.sqrt(-2*np.log(U1))*np.sin(2*np.pi * U2)
11     return Z0*sigma + mu
12
13 pse_sample = pseudo_normal_gen(mu=0, sigma=1, size=10000)
14 built_in_sample = np.random.normal(0, 1, size=10000)
15 plt.figure(figsize=(8,6))
16 _, bins, _ = plt.hist(pse_sample, alpha=0.7, label="Pseudo Generator",
17                       color="gray", ec='black', bins="sturges")
18 plt.hist(built_in_sample, alpha=0.7, label="Numpy Generator",
19          color="red", ec='black', bins=bins)
20 plt.legend()
```

Programme 5.5 Pseudo Normal Generator via Box-Muller transformation in Python.

```
1  > set.seed(4002)
2  >
3  > pse_norm_gen <- function(mu=0.0, sigma=1.0, seed=123456789,
4  +                          size=1, burn_in=1000){
5  +     U <-  pse_uni_gen(seed=seed, size=2*size, burn_in=burn_in)
6  +     U1 <- U[1:size]
7  +     U2 <- U[(size+1):(2*size)]
8  +     Z0 <- sqrt(-2*log(U1))*cos(2*pi * U2)
9  +     Z1 <- sqrt(-2*log(U1))*sin(2*pi * U2)
10 +     return (Z0*sigma + mu)
11 + }
12 > pse_sample <- pse_norm_gen(mu=0, sigma=1, size=10000)
13 > built_in_sample <- rnorm(10000)
14 > hist(pse_sample, col=rgb(red=.5,green=.5,blue=.5,alpha=0.5),
15 +      main="Histogram")
16 > hist(built_in_sample, col=rgb(red=1.0,green=0,blue=0,alpha=0.5),
17 +      add=TRUE)
18 > legend('topright',legend=c("Pseudo Generator","R Generator"),
19 +        fill=c(rgb(red=.5,green=.5,blue=.5,alpha=0.5),
20 +               rgb(red=1.0,green=0,blue=0,alpha=0.5)), cex=1.5)
```

Programme 5.6 Pseudo Normal Generator via Box-Muller transformation in R.

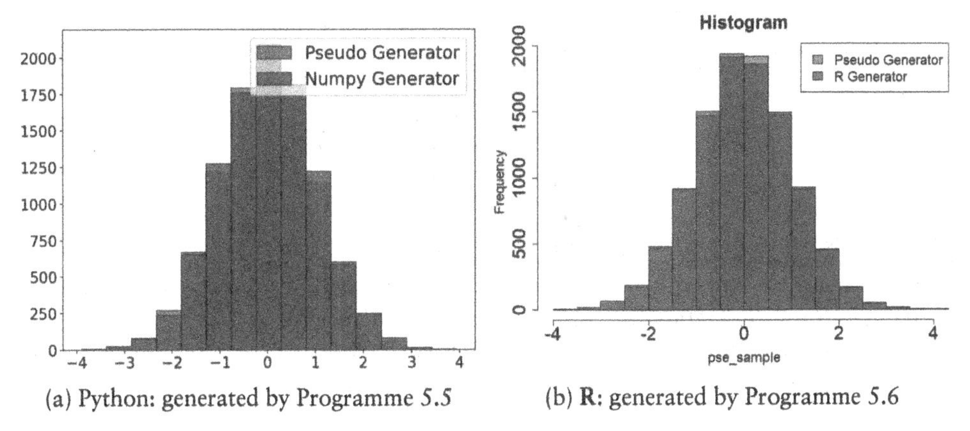

(a) Python: generated by Programme 5.5 (b) **R**: generated by Programme 5.6

FIGURE 5.3 Histograms of the samples from pseudo generators for normal distribution.

In the more general case beyond normal distribution, the method of *rejection sampling*, also known as the *acceptance-rejection method*, is commonly adopted; this method essentially dates back to the needle experiment by French mathematician *Comte de Buffon* in estimating the circumference-diameter ratio π. We here only elaborate the case of univariate distribution, while the multivariate case is analogous.

To generate samples from a random variable x with a density $f(x)$ over its support $\mathcal{X}_f$, we can propose a candidate density $g(x)$ with a support $\mathcal{X}_g$ such that $\mathcal{X}_f \subset \mathcal{X}_g$, from which it is much easier to draw samples, that "envelopes" $f(x)$; more precisely, as long as there exists a finite constant $M > 1$ such that $f(x) \le Mg(x)$ for all $x \in \mathcal{X}_f$. The following algorithm explains the general procedure of the rejection sampling:

1. Generate $\mathrm{x}_i \sim g(x)$;
2. Generate $\mathrm{u}_i \sim U(0,1)$, and accept x_i if $\mathrm{u}_i \le \frac{f(\mathrm{x}_i)}{Mg(\mathrm{x}_i)}$, otherwise reject x_i;
3. Repeat steps 1) and 2) for N times, and only keep those accepted x_i's as the generated sample.

To justify the validity of this approach, we consider the following sample average based on the accepted random observations generated from the algorithm, particularly for any test function h:

$$\frac{\sum_{i=1}^{N} h(\mathrm{x}_i)\mathbb{1}_{\left\{\mathrm{u}_i \le \frac{f(\mathrm{x}_i)}{Mg(\mathrm{x}_i)}\right\}}}{\sum_{i=1}^{N} \mathbb{1}_{\left\{\mathrm{u}_i \le \frac{f(\mathrm{x}_i)}{Mg(\mathrm{x}_i)}\right\}}} = \frac{\frac{1}{N}\sum_{i=1}^{N} h(\mathrm{x}_i)\mathbb{1}_{\left\{\mathrm{u}_i \le \frac{f(\mathrm{x}_i)}{Mg(\mathrm{x}_i)}\right\}}}{\frac{1}{N}\sum_{i=1}^{N} \mathbb{1}_{\left\{\mathrm{u}_i \le \frac{f(\mathrm{x}_i)}{Mg(\mathrm{x}_i)}\right\}}}.$$

By the strong law of large numbers, as $N \to \infty$, it converges to:

$$\begin{aligned}\frac{\mathbb{E}\left(h(\mathrm{x})\mathbb{1}_{\left\{\mathrm{u} \le \frac{f(\mathrm{x})}{Mg(\mathrm{x})}\right\}}\right)}{\mathbb{P}\left(\mathrm{u} \le \frac{f(\mathrm{x})}{Mg(\mathrm{x})}\right)} &= \frac{\int_{\mathcal{X}_g} h(x)\frac{f(x)}{Mg(x)}g(x)dx}{\mathbb{E}\left(\mathbb{P}\left(\mathrm{u} \le \frac{f(\mathrm{x})}{Mg(\mathrm{x})}\middle|\mathrm{x}\right)\right)} = \frac{\frac{1}{M}\int_{\mathcal{X}_f} h(x)f(x)dx}{\mathbb{E}\left(\frac{f(\mathrm{x})}{Mg(\mathrm{x})}\right)} \\ &= \frac{\frac{1}{M}\int_{\mathcal{X}_f} h(x)f(x)dx}{\frac{1}{M}\int_{\mathcal{X}_g} \frac{f(x)}{g(x)}g(x)dx} = \int_{\mathcal{X}_f} h(x)f(x)dx,\end{aligned}$$

which holds by noting the very definition of the density f, namely $\int_{\mathcal{X}_f} f(x)dx = 1$ and $f(x) = 0$ whenever $x \in \mathcal{X}_g \backslash \mathcal{X}_f$. In other words, the aforementioned sample average converges to the theoretical quantity of $\mathbb{E}(h(\mathrm{x}))$, and hence the algorithm is indeed valid. As an illustration, for instance, readers may refer to Subsection *6.4.2 for an application of this sampling method in the context of *stochastic volatility* models.

5.2 VARIANCE REDUCTION TECHNIQUE

Most often, after knowing how to simulate a sequence of random variables, the next immediate application is to estimate various quantities of interest, or simply some statistics, of a more complicated probability distribution. For instance, we are often interested in estimating the underlying unknown theoretical mean $\mu_{\mathrm{x}} := \mathbb{E}(\mathrm{x})$, and one may think of using the *Monte Carlo* method. As pointed out in Section 1.4, one direct approach is to use the sample mean

$$\widehat{\mu_{\mathrm{x}}} = \frac{1}{n}\sum_{i=1}^{n} \mathrm{x}_i,$$

as the estimator, where $x_1, \ldots, x_n$ is a random sample simulated from the distribution of x. The accuracy of this estimator can be assessed by examining its variance. As discussed in Subsection 1.4.2, it is given by:

$$\mathrm{Var}(\widehat{\mu_x}) = \mathrm{Var}\left(\frac{1}{n}\sum_{i=1}^{n} x_i\right) = \frac{1}{n^2}\sum_{i=1}^{n}\mathrm{Var}(x_i) = \frac{1}{n}\mathrm{Var}(x).$$

Clearly, from the expression above, one common way to obtain a more accurate estimate is to increase the number of simulations n, as this reduces the variance of the sample mean estimator. Another way is to cleverly find an alternative random variable z with the same theoretical mean as x but a smaller variance, i.e. $\mu_z := \mathbb{E}(z) = \mathbb{E}(x) = \mu_x$ and $\mathrm{Var}(z) \le \mathrm{Var}(x)$, so that $\mathrm{Var}(\widehat{\mu_z}) = \frac{1}{n}\mathrm{Var}(z) \le \frac{1}{n}\mathrm{Var}(x) = \mathrm{Var}(\widehat{\mu_x})$, and we may then instead use $\widehat{\mu_z} = \frac{1}{n}\sum_{i=1}^{n} z_i$ as an estimator of μ_x to achieve a higher precision. Later in this section, we shall describe two common approaches for looking for a more effective z.

Example 5.1 Suppose that we would like to estimate the integral $\mathcal{I} := \int_0^1 x^3 (1-x)^{2.5}\,dx$ with $n = 10,000$ generated samples. We observe that $\mathcal{I}$ can be expressed as $\mathbb{E}(u^3(1-u)^{2.5})$ where $u \sim U(0,1)$. Thus, we can first simulate $u_1, \ldots, u_{10,000}$, and then compute the following estimate:

$$\hat{\theta} = \frac{1}{n}\sum_{i=1}^{n} u_i^3(1-u_i)^{2.5}.$$

Indeed, this integral is related to the **beta distribution** with a pdf:

$$f(x) = \frac{1}{B(\alpha,\beta)}x^{\alpha-1}(1-x)^{\beta-1}, \quad \text{for } 0 < x < 1,$$

where $\alpha, \beta > 0$ and

$$B(\alpha,\beta) = \frac{\Gamma(\alpha)\Gamma(\beta)}{\Gamma(\alpha+\beta)}.$$

It is clear that $\mathcal{I} = B(4, 3.5)$. In Programme 5.7 (resp. Programme 5.8) with Python (resp. **R**) codes, we repeat the above simulation procedure 1,000 times and return the mean and the (sample) variance of the estimates obtained.

```
1  import numpy as np
2  from scipy.special import beta
3  np.random.seed(4002)
4
5  n, psi = 10000, []
6  for i in range(1000):
7      u = np.random.rand(n)
8      psi.append(np.sum(u**3 * (1-u)**2.5) / n)
9
10 print(np.mean(psi))
11 print(beta(4, 3.5))
12 print(np.var(psi, ddof=1))
```

```
1 0.010656565769388606
2 0.010656010656010658
3 6.94092512239374e-09
```

Programme 5.7 Example 5.1: Estimation of B(4, 3.5) in Python.

```
1  > set.seed(4002)
2  >
3  > n <- 10000
4  > psi <- c()
5  > for (i in 1:1000){
6  +     u <- runif(n)
7  +     psi <- c(psi, sum(u^3*(1-u)^2.5)/n)
8  + }
9  > mean(psi)
10 [1] 0.01065083
11 > beta(4, 3.5)
12 [1] 0.01065601
13 > var(psi)
14 [1] 6.500005e-09
```

Programme 5.8 Example 5.1: Estimation of B(4, 3.5) in R.

5.2.1 Antithetic Variables

Let z be a random variable such that $\mathbb{E}(z) = \mathbb{E}(x)$ and $\text{Var}(z) = \text{Var}(x)$. Then, for two independent random samples $x_1, \ldots, x_{\frac{n}{2}}$ and $z_1, \ldots, z_{\frac{n}{2}}$, the estimator:

$$\psi_A := \frac{1}{n/2} \sum_{i=1}^{n/2} \frac{x_i + z_i}{2} \tag{5.1}$$

has a variance of

$$\begin{aligned} \text{Var}(\psi_A) &= \frac{1}{n^2} \sum_{i=1}^{n/2} \text{Var}(x_i + z_i) \\ &= \frac{1}{n^2} \sum_{i=1}^{n/2} (\text{Var}(x_i) + \text{Var}(z_i) + 2\text{Cov}(x_i, z_i)) \\ &= \frac{1}{n}(\text{Var}(x_1) + \text{Cov}(x_1, z_1)). \end{aligned}$$

Notice that, as long as one can find a random variable z that is negatively correlated with x, i.e. $\text{Cov}(x_1, z_1) < 0$, we can then have $\text{Var}(\psi_A) < \text{Var}(x)/n = \text{Var}(\widehat{\mu_x})$. One easy way to generate the pair (x, z) that satisfies all the above requirements is to take $(x, z) := (F_x^{-1}(u), F_x^{-1}(1 - u))$, where u is a uniform random variable on [0, 1]. For example, when $x \sim U(0, 1)$, we just take $(x, z) = (u, 1 - u)$, where $u \sim U(0, 1)$; when $x \sim \mathcal{N}(\mu, \sigma^2)$, we may take $(x, z) = (x, 2\mu - x)$. These pairs of variables (x, z) are called *antithetic*. The following theorem shows that antithetic variables are indeed negatively correlated.

Proposition 5.1 Given an increasing function $h(\theta, \gamma)$ in $\theta \in (0, 1)$ with a parameter γ, if $u \sim U(0, 1)$, γ_1 and γ_2 are independent random variables, then

$$\mathrm{Cov}(h(u, \gamma_1), h(1 - u, \gamma_2)) \leq 0.$$

Then $x := h(u, \gamma_1)$ and $z := h(1 - u, \gamma_2)$ form a pair of antithetic variables.

Proof. We first prove a special case: if the deterministic functions $g_1(\theta)$ and $g_2(\theta)$ are increasing in $\theta \in \mathbb{R}$, then for any random variable x, the variables $g_1(x)$ and $g_2(x)$ are positively correlated. To this end, let $\tilde{x}$ be another random variable which is independent of x yet has the same distribution as that of x. As both g_1 and g_2 are increasing, clearly $(g_1(x) - g_1(\tilde{x}))(g_2(x) - g_2(\tilde{x})) \geq 0$ for any $x, \tilde{x} \in \mathbb{R}$, and so

$$\mathbb{E}((g_1(x) - g_1(\tilde{x}))(g_2(x) - g_2(\tilde{x}))) \geq 0.$$

Expanding the terms gives

$$\mathbb{E}(g_1(x)g_2(x)) + \mathbb{E}(g_1(\tilde{x})g_2(\tilde{x})) \geq \mathbb{E}(g_1(\tilde{x})g_2(x)) + \mathbb{E}(g_1(x)g_2(\tilde{x})).$$

As x and $\tilde{x}$ are iid, this inequality reduces to $2\mathbb{E}(g_1(x)g_2(x)) \geq 2\mathbb{E}(g_1(x))\mathbb{E}(g_2(x))$, which immediately implies that $\mathrm{Cov}(g_1(x), g_2(x)) \geq 0$. Now, let $x := h(u, \gamma_1)$ and $z := h(1 - u, \gamma_2)$. For any fixed γ, $h(\theta, \gamma)$ increases in θ, so both $\mathbb{E}(h(u, \gamma_1)|u = u)$ and $-\mathbb{E}(h(1 - u, \gamma_2)|u = u)$ are increasing in u. By the first assertion, we just established that

$$\mathrm{Cov}(\mathbb{E}(x|u), -\mathbb{E}(z|u)) = \mathrm{Cov}(\mathbb{E}(h(u, \gamma_1)|u), -\mathbb{E}(h(1 - u, \gamma_2)|u)) \geq 0.$$

On the other hand, since γ_1 and γ_2 are independent themselves, and they are also independent of u, $h(u, \gamma_1)$ and $h(1 - u, \gamma_2)$ are still conditionally independent given u, thus the conditional covariance given u is

$$\mathrm{Cov}(x, z|u) = \mathrm{Cov}(h(u, \gamma_1), h(1 - u, \gamma_2)|u) = 0.$$

Finally, by the tower property,

$$\mathrm{Cov}(x, z) = \mathbb{E}(\mathrm{Cov}(x, z|u)) + \mathrm{Cov}(\mathbb{E}(x|u), \mathbb{E}(z|u)) \leq 0.$$

□

For applications of this approach, readers may refer to [11, 15, 16] for more discussions. Meanwhile, we shall continue to use the target integral of Example 5.1 for the sake of consistent presentation and comparison. Example 5.2 serves as a counter-example that a mis-use of this approach may lead to a worse performance in terms of the accuracy of estimation. While many other textbooks focus on the merits such as the convenient implementation of Proposition 5.1, Example 5.2 suggests that extra caution should be exercised in actual practice.

Example 5.2 ***(Continuation of Example 5.1)*** We first simulate $u_1, \ldots, u_{5,000}$ and then compute the antithetic samples with $v_i = 1 - u_i$, for $i = 1, \ldots, 5,000$. The estimate is then given as in (5.1):

$$\hat{\theta} = \frac{1}{n/2} \sum_{i=1}^{n/2} \frac{u_i^3(1-u_i)^{2.5} + v_i^3(1-v_i)^{2.5}}{2},$$

where we take $x_i := u_i^3(1-u_i)^{2.5}$ and $z_i := v_i^3(1-v_i)^{2.5}$. Again, we repeat the simulation procedure for 1,000 times, and in Programme 5.9 via Python (resp. Programme 5.10 via **R**), we also report the mean and the (sample) variance of the estimates obtained in Python (resp. **R**). Clearly, the function $f(w) = w^3(1-w)^{2.5}$ is not monotonic, and the numerical value of the sample (bootstrapped) variance is larger than that in Example 5.1. Altogether this illustrates that the sufficient condition of Proposition 5.1 is in some sense also necessary.

```
1  import numpy as np
2  np.random.seed(4002)
3
4  n, psi_A = 10000, []
5  for i in range(1000):
6      u = np.random.rand(np.int(n/2))
7      v = 1 - u
8      psi_A.append(np.sum(u**3 * (1-u)**2.5 + v**3 * (1-v)**2.5)/n)
9
10 print(np.mean(psi_A))
11 print(np.var(psi_A, ddof=1))
```

```
1  0.010657115014611696
2  1.2923542794361985e-08
```

Programme 5.9 Example 5.2: Antithetic variables in Python.

```
1  > set.seed(4002)
2  >
3  > n <- 10000
4  > psi_A <- c()
5  > for (i in 1:1000){
6  +     u <- runif(n/2)
7  +     v <- 1 - u
8  +     psi_A <- c(psi_A, sum(u^3*(1-u)^2.5 + v^3*(1-v)^2.5)/n)
9  + }
```

```
10 > mean(psi_A)
11 [1] 0.01065189
12 > var(psi_A)
13 [1] 1.294975e-08
```

Programme 5.10 Example 5.2: Antithetic variables in R.

5.2.2 Control Variates

Let y be a random variable with a known mean $\mathbb{E}(y) = \mu_y$. Instead of conditioning, we can refine the random variable x directly, so as to reduce the variance of the estimator. For any constant $b \in \mathbb{R}$, the random variable $z := x + b(y - \mu_y)$ is clearly an unbiased estimator of μ_x. For the *control variate technique*, we aim to pick a b such that $\mathrm{Var}(z) < \mathrm{Var}(x)$. To this end, note that

$$\mathrm{Var}(z) = \mathrm{Var}(x + b(y - \mu_y)) = \mathrm{Var}(x) + b^2\mathrm{Var}(y) + 2b\mathrm{Cov}(x, y). \tag{5.2}$$

Differentiating (5.2) with respect to b and equating this derivative to zero yields that

$$\frac{d\mathrm{Var}(z)}{db} = 2b\mathrm{Var}(y) + 2\mathrm{Cov}(x, y) = 0 \quad \text{at} \quad b^* := -\frac{\mathrm{Cov}(x, y)}{\mathrm{Var}(y)}.$$

Substituting b^* back into (5.2) gives

$$\mathrm{Var}(z)|_{b=b^*} = \mathrm{Var}(x) + \frac{\mathrm{Cov}(x, y)^2}{\mathrm{Var}(y)} - \frac{2\mathrm{Cov}(x, y)^2}{\mathrm{Var}(y)} = \mathrm{Var}(x)(1 - \mathrm{Corr}(x, y)^2);$$

together with the fact that $0 \leq \mathrm{Corr}(x, y)^2 \leq 1$, we clearly have $\mathrm{Var}(z)\Big|_{b=b^*} \leq \mathrm{Var}(x)$. In other words, as long as x and y are correlated, variance reduction can be achieved by this method, and the corresponding estimator of μ_x is given by

$$\begin{aligned}\psi_C &:= \frac{1}{n}\sum_{i=1}^{n} z_i = \frac{1}{n}\sum_{i=1}^{n}\left(x_i - \frac{\mathrm{Cov}(x, y)}{\mathrm{Var}(y)}(y_i - \mu_y)\right) \\ &= \overline{x} - \frac{\mathrm{Cov}(x, y)}{\mathrm{Var}(y)}(\overline{y} - \mu_y),\end{aligned}$$

where (x_i, y_i) are iid copies of (x, y). In practice, the value of $b^* = -\frac{\mathrm{Cov}(x,y)}{\mathrm{Var}(y)}$ is not known *a priori*, but it can be estimated using the simulated sample (x_i, y_i) by using the sample correlation coefficient,

$$\widehat{b^*} = -\frac{\sum_{i=1}^{n}(x_i - \overline{x})(y_i - \overline{y})}{\sum_{i=1}^{n}(y_i - \overline{y})^2},$$

and the estimator ψ_C becomes $\overline{x} + \widehat{b^*}(\overline{y} - \mu_y)$.

Example 5.3 *(Continuation of Example 5.1)* We choose $y := (1-u)^{2.5}$ as the control variate, whose mean can be computed explicitly:

$$\mu_y = \int_0^1 (1-x)^{2.5}\,dx = -\int_0^1 (1-x)^{2.5}\,d(1-x) = \frac{1}{3.5} = \frac{2}{7}.$$

As before, we repeat the simulation procedure for 1,000 times, and in Programme 5.11 (resp. Programme 5.12), we also report the mean and the (sample) variance of the estimates obtained in Python (resp. R). We can clearly see that the sample (bootstrapped) variance is smaller than that obtained in Example 5.1.

```
1  import numpy as np
2  np.random.seed(4002)
3
4  n, psi_C = 10000, []
5  mu_y = 2/7
6  for i in range(1000):
7      u = np.random.rand(n)
8      x, y = u**3 * (1-u)**2.5, (1-u)**2.5
9      # Using sample variance and sample covariance
10     s_cov = np.cov(x, y, ddof=1)[0][1]
11     s_var = np.var(y, ddof=1)
12     psi_C.append(np.mean(x) - s_cov/s_var * (np.mean(y) - mu_y))
13
14 print(np.mean(psi_C))
15 print(np.var(psi_C, ddof=1))
```

```
1  0.010658497273011184
2  5.573765313817774e-09
```

Programme 5.11 Example 5.3: Control variate in Python.

```
1  > set.seed(4002)
2  >
3  > n <- 10000
4  > psi_C <- c()
5  > mu_y <- 2/7
6  > for (i in 1:1000){
7  +     u <- runif(n)
8  +     x <- u^3*(1-u)^2.5
9  +     y <- (1-u)^2.5
10 +     psi_C <- c(psi_C, mean(x) - cov(x, y)/var(y)*(mean(y) - mu_y))
11 + }
12 > mean(psi_C)
13 [1] 0.01065137
14 > var(psi_C)
15 [1] 5.357306e-09
```

Programme 5.12 Example 5.3: Control variate in R.

5.2.2.1 Multiple Control Variates We next consider the high-dimensional setting: let y be a p-dimensional random vector with a known expectation $\boldsymbol{\mu}_{\mathrm{y}} \in \mathbb{R}^p$. For any $\boldsymbol{b} \in \mathbb{R}^p$, the random variable $\mathrm{z} := \mathrm{x} + \boldsymbol{b}^\top(\mathrm{y} - \boldsymbol{\mu}_{\mathrm{y}})$ is an unbiased estimator of the unknown $\theta := \mathbb{E}(\mathrm{x})$. By similar arguments to the above, we consider the variance of this estimator:

$$\mathrm{Var}(\mathrm{z}) = \mathrm{Var}(\mathrm{x} + \boldsymbol{b}^\top(\mathrm{y} - \boldsymbol{\mu}_{\mathrm{y}})) = \mathrm{Var}(\mathrm{x}) + \boldsymbol{b}^\top\mathrm{Var}(\mathrm{y})\boldsymbol{b} + 2\boldsymbol{b}^\top\mathrm{Cov}(\mathrm{x}, \mathrm{y}), \tag{5.3}$$

where we have used the short-handed notation

$$\mathrm{Cov}(\mathrm{x}, \mathrm{y}) := (\mathrm{Cov}(\mathrm{x}, \mathrm{y}_1), \dots, \mathrm{Cov}(\mathrm{x}, \mathrm{y}_p))^\top.$$

Differentiating (5.3) with respect to $\boldsymbol{b}$ and setting it to zero vector yields

$$\nabla_{\boldsymbol{b}}\mathrm{Var}(\mathrm{z}) = 2\mathrm{Var}(\mathrm{y})\boldsymbol{b} + 2\mathrm{Cov}(\mathrm{x}, \mathrm{y}) = \mathbf{0} \;\text{ at }\; \boldsymbol{b}^* := -\mathrm{Var}(\mathrm{y})^{-1}\mathrm{Cov}(\mathrm{x}, \mathrm{y}).$$

Substituting $\boldsymbol{b}^*$ into (5.3), we obtain:

$$\begin{aligned}\mathrm{Var}(\mathrm{z}) &= \mathrm{Var}(\mathrm{x}) + (\mathrm{Var}(\mathrm{y})^{-1}\mathrm{Cov}(\mathrm{x}, \mathrm{y}))^\top\mathrm{Cov}(\mathrm{x}, \mathrm{y}) - 2(\mathrm{Var}(\mathrm{y})^{-1}\mathrm{Cov}(\mathrm{x}, \mathrm{y}))^\top\mathrm{Cov}(\mathrm{x}, \mathrm{y}) \\ &= \mathrm{Var}(\mathrm{x})\left(1 - \frac{\mathrm{Cov}(\mathrm{x}, \mathrm{y})^\top\mathrm{Var}(\mathrm{y})^{-1}\mathrm{Cov}(\mathrm{x}, \mathrm{y})}{\mathrm{Var}(\mathrm{x})}\right) =: \mathrm{Var}(\mathrm{x})(1 - R^2).\end{aligned}$$

This motivates us to construct the following estimator:

$$\begin{aligned}\Psi_{MC} &:= \frac{1}{n}\sum_{i=1}^{n} \mathrm{z}_i \\ &= \frac{1}{n}\sum_{i=1}^{n}\left(\mathrm{x}_i + \boldsymbol{b}^{*\top}(\mathrm{y}_i - \boldsymbol{\mu}_{\mathrm{y}})\right) = \overline{\mathrm{x}} - \mathrm{Cov}(\mathrm{x}, \mathrm{y})^\top\mathrm{Var}(\mathrm{y})^{-1}(\overline{\mathrm{y}} - \boldsymbol{\mu}_{\mathrm{y}}),\end{aligned}$$

where $(\mathrm{x}_i, \mathrm{y}_i)$ are iid copies of (x, y).

Example 5.4 *(Continuation of Example 5.1)* We choose $\mathrm{y}_1 := (1 - \mathrm{u})^{2.5}$ and $\mathrm{y}_2 := \mathrm{u}^3$ as two control variates. The means of them are, respectively,

$$\mathbb{E}(\mathrm{y}_1) = \int_0^1 (1 - x)^{2.5}\, dx = \frac{2}{7} \quad \text{and} \quad \mathbb{E}(\mathrm{y}_2) = \int_0^1 x^3\, dx = \frac{1}{4}.$$

We repeat the simulation for 1,000 times and report the mean and the (sample) variance of the estimate obtained in Programme 5.13 in Python (resp. Programme 5.14 in R). From the outputs, there is a further variance reduction by incorporating one more variate (i.e. y_2) to the one considered in Example 5.3.

```python
import numpy as np
np.random.seed(4002)

n, psi_MC = 10000, []
mu_y = np.array([2/7, 1/4])
for i in range(1000):
    u = np.random.rand(n)
    x, y = u**3 * (1-u)**2.5, np.vstack([(1-u)**2.5, u**3])
    # Using sample variance and sample covariance
    s_cov = np.cov(x, y, ddof=1)[0,1:]
    diff_y = np.mean(y, axis=1) - mu_y
    inv_s_var = np.linalg.inv(np.cov(y, ddof=1))
    psi_MC.append(np.mean(x) - s_cov @ inv_s_var @ diff_y)

print(np.mean(psi_MC))
print(np.var(psi_MC, ddof=1))
```

```
0.010656542910182573
6.819311631463463e-10
```

Programme 5.13 Example 5.4: Multiple control variates in Python.

```r
> set.seed(4002)
>
> n <- 10000
> psi_MC <- c()
> mu_y <- c(2/7, 1/4)
> for (i in 1:1000){
+     u <- runif(n)
+     x <- u^3*(1-u)^2.5
+     y <- cbind((1-u)^2.5, u^3)
+     inv_S <- solve(var(y))
+     diff_y <- apply(y, 2, mean) - mu_y
+     psi_MC <- c(psi_MC, mean(x) - cov(x, y)%*%inv_S%*%diff_y)
+ }
> mean(psi_MC)
[1] 0.01065427
> var(psi_MC)
[1] 6.842187e-10
```

Programme 5.14 Example 5.4: Multiple control variates in R.

5.2.3 Stratified Sampling

Let y = h(x) where x is a random variable with the density function $p(x)$ on the domain $\mathcal{X}$ of x. We now want to estimate the unknown parameter:

$$\theta = \mathbb{E}(\mathrm{y}) = \mathbb{E}(h(\mathrm{x})) = \int_{\mathcal{X}} h(x)p(x)\,dx\,.$$

Instead of estimating θ by the estimator

$$\hat{\theta} = \frac{1}{n}\sum_{j=1}^{n} h(\mathrm{x}_j), \tag{5.4}$$

with an arbitrarily chosen random sample $\mathrm{x}_1, \ldots, \mathrm{x}_n \sim p$, the idea of *stratified sampling* is to first partition the domain $\mathcal{X}$ into several separate regions, and then take a random sample from each region, after assigning some suitable weights, to finally estimate $\mathbb{E}(h(\mathrm{x}))$. More precisely, suppose that $\mathcal{X}$ is partitioned as:

$$\mathcal{X} = \bigcup_{j=1}^{J} \mathcal{X}_j, \quad \text{where} \quad \mathcal{X}_i \cap \mathcal{X}_j = \emptyset \text{ whenever } i \neq j.$$

Let $\omega_j := \mathbb{P}(\mathrm{x} \in \mathcal{X}_j) > 0$. Assuming that the values of ω_j's are all known for $j = 1, 2, \ldots, J$, then the conditional density $p_j(x)$ of x given $\mathrm{x} \in \mathcal{X}_j$ is simply

$$p_j(x) = \frac{1}{\omega_j}\mathbb{1}_{\{x\in\mathcal{X}_j\}}p(x), \text{ for all } x \in \mathcal{X}.$$

For each $j = 1, \ldots, J$, we pick a random sample x_{ij}'s of size n_j from $\mathcal{X}_j$, with $\mathrm{x}_{ij} \sim p_j$ for $i = 1, \ldots, n_j$. Then the *stratified estimator* of θ is given by

$$\psi_S := \sum_{j=1}^{J} \overline{w}_j \left(\frac{1}{n_j}\sum_{i=1}^{n_j} h(\mathrm{x}_{ij}) \right) = \sum_{j=1}^{J}\sum_{i=1}^{n_j} \frac{\overline{w}_j}{n_j} h(\mathrm{x}_{ij}), \tag{5.5}$$

for some weights $\overline{w}_j$, $j = 1, \ldots, J$. The main purpose here is to determine the weights $\overline{\omega}_j$ in an alternative way other than $\omega_j = \mathbb{P}(\mathrm{x} \in \mathcal{X}_j)$ so that variance can still be reduced. To this end, we let $\mu_j := \mathbb{E}(h(\mathrm{x})|\mathrm{x} \in \mathcal{X}_j)$, $\sigma_j^2 := \mathrm{Var}(h(\mathrm{x})|\mathrm{x} \in \mathcal{X}_j)$, $\mu := \mathbb{E}(h(\mathrm{x}))$, $\sigma^2 := \mathrm{Var}(h(\mathrm{x}))$, and $n := n_1 + \cdots + n_J$. By the very nature of the stratified sampling mechanism, the mean and variance of the estimator ψ_S are given by:

$$\mathbb{E}(\psi_S) = \sum_{j=1}^{J} \overline{w}_j \left(\frac{1}{n_j}\sum_{i=1}^{n_j} \mathbb{E}(h(\mathrm{x}_{ij})|\mathrm{x}_{ij} \in \mathcal{X}_j) \right) = \sum_{j=1}^{J} \overline{w}_j \mu_j,$$

and

$$\mathrm{Var}(\psi_S) = \sum_{j=1}^{J}\sum_{i=1}^{n_j} \frac{\overline{w}_j^2}{n_j^2}\mathrm{Var}(h(\mathrm{x}_{ij})|\mathrm{x}_{ij} \in \mathcal{X}_j) = \sum_{j=1}^{J} \frac{\overline{w}_j^2}{n_j}\sigma_j^2,$$

respectively. On the other hand, reverting to estimator $\hat{\theta}$, by the *law of total variance*, the variance of $\hat{\theta}$ equals:

$$\mathrm{Var}\left(\frac{1}{n}\sum_{i=1}^{n} h(\mathrm{x}_i) \right) = \frac{1}{n}\mathrm{Var}(h(\mathrm{x})) = \frac{1}{n}(\mathbb{E}(\mathrm{Var}(h(\mathrm{x})|\mathrm{z})) + \mathrm{Var}(\mathbb{E}(h(\mathrm{x})|\mathrm{z}))),$$

where we take z to be the discrete random variable such that $z = j$ whenever $x \in \mathcal{X}_j$, for $j = 1, \ldots, J$. Then,

$$\mathbb{E}(\mathrm{Var}(h(x)|z)) = \sum_{j=1}^{J} \mathbb{P}(z = j)\,\mathrm{Var}(h(x)|x \in \mathcal{X}_j) = \sum_{j=1}^{J} \omega_j \sigma_j^2,$$

$$\begin{aligned}\mathrm{Var}(\mathbb{E}(h(x)|z)) &= \mathbb{E}\left[(\mathbb{E}(h(x)|z) - \mathbb{E}(\mathbb{E}(h(x)|z)))^2\right] = \mathbb{E}((\mathbb{E}(h(x)|z) - \mu)^2) \\ &= \sum_{j=1}^{J} \omega_j (\mu_j - \mu)^2,\end{aligned}$$

as a result,

$$\mathrm{Var}\left(\frac{1}{n}\sum_{i=1}^{n} h(x_i)\right) = \frac{1}{n}\sum_{j=1}^{J} \omega_j(\sigma_j^2 + (\mu_j - \mu)^2).$$

In order to achieve $\mathrm{Var}(\psi_S) \leq \mathrm{Var}(\hat{\theta})$, from the results computed above we require that:

$$\sum_{j=1}^{J} \frac{\overline{w}_j^2}{n_j}\sigma_j^2 \leq \sum_{j=1}^{J} \frac{\omega_j}{n}(\sigma_j^2 + (\mu_j - \mu)^2),$$

which can be fulfilled by choosing n_j and $\overline{w}_j$ so that $\overline{w}_j^2 = \omega_j \cdot n_j/n$ for each $j = 1, \ldots, J$. For if n is sufficiently large, we may pick samples with $n_j = \lfloor n\omega_j \rfloor$, then $\overline{\omega}_j \approx \omega_j$, so we can have both $\mathrm{Var}(\psi_S) \leq \mathrm{Var}(\hat{\theta})$ and $\mathbb{E}(\psi_S) = \sum_{j=1}^{J} \overline{w}_j \mu_j = \mu$ asymptotically. As long as $\mu_j \neq \mu$ for at least one j, this stratified estimator ψ_S should have a smaller variance than that of the usual Monte Carlo estimator $\hat{\theta}$ in (5.4).

Example 5.5 ***(Continuation of Example 5.1)*** With $J = 10$, we repeat the simulation 1,000 times and report the mean and the (sample) variance of the estimate obtained in Programme 5.15 in Python (resp. Programme 5.16 in **R**).

```
1  import numpy as np
2  np.random.seed(4002)
3
4  n, psi_S, J = 10000, [], 10
5  for i in range(1000):
6      u_j = np.random.rand(np.int(n/J), J)
7      u_j = (u_j + np.arange(10).reshape(1, -1))/J
8      psi_S.append(np.sum(u_j**3 * (1-u_j)**2.5) / n)
9
10 print(np.mean(psi_S))
11 print(np.var(psi_S, ddof=1))
```

```
1 0.01065677096019911
2 2.3711840587199497e-10
```

Programme 5.15 Example 5.5: Stratified sampling in Python.

```
1  > set.seed(4002)
2  >
3  > n <- 10000
4  > psi_S <- c()
5  > J <- 10
6  > for (i in 1:1000){
7  +     u_j <- matrix(runif(n), ncol=J)
8  +     u_j <- sweep(u_j, 2, 0:(J-1), "+")/J
9  +     psi_S <- c(psi_S, sum(u_j^3*(1-u_j)^2.5)/n)
10 + }
11 > mean(psi_S)
12 [1] 0.01065621
13 > var(psi_S)
14 [1] 2.267159e-10
```

Programme 5.16 Example 5.5: Stratified sampling in R.

To summarize, let us compare the sample means and sample (bootstrapped) variances of various estimates obtained by the different methods shown in the previous examples. From Table 5.1 in Python (resp. Table 5.2 in R), stratified sampling provides the smallest sample variance among all estimations; while the antithetic variable gives the largest, even larger than that of the plain Monte Carlo method; as a counter-example, recall that the function of interest $w^3(1-w)^{2.5}$, for $w \in [0,1]$, does not satisfy the monotone condition. When comparing the single control variate technique with the multiple counterpart, the latter results in a smaller variance as expected. However, one should notice that choosing appropriate control variates might be challenging in an actual implementation.

TABLE 5.1 Sample means and sample variances of various variance reduction techniques in Python.

B(4, 3.5) = 0.01065601	Mean	Variance
Monte Carlo	0.01065657	6.940925e-09
Antithetic variable	0.01065712	1.292354e-08
Control variate	0.01065850	5.573765e-09
Multiple control variate	0.01065654	6.819312e-10
Stratified sampling	0.01065677	2.371184e-10

TABLE 5.2 Sample means and sample variances of various variance reduction techniques in R; also note the differences in values obtained comparing with those in Table 5.1.

B(4, 3.5) = 0.01065601	Mean	Variance
Monte Carlo	0.01065083	6.500005e-09
Antithetic variable	0.01065189	1.294975e-08
Control variate	0.01065137	5.357306e-09
Multiple control variate	0.01065427	6.842187e-10
Stratified sampling	0.01065621	2.267159e-10

5.3 A REVIEW OF FINANCIAL CALCULUS AND DERIVATIVE PRICING

To facilitate a better understanding of the rest of this chapter, a rudimentary knowledge of financial derivatives and basic stochastic calculus cannot be avoided. Excellent accounts of this subject area can be found in standard textbooks [3, 7, 9, 17, 18, 20] amongst others.

Definition 5.1 A stochastic process $\{w_t\}$, for $t \geq 0$, is the standard one-dimensional *Wiener process* (Brownian motion) *if and only if* the following conditions hold:

1. $w_0 = 0$ with a probability 1;
2. almost all sample paths are continuous with probability 1;
3. $\{w_t\}$ has stationary and independent increments;
4. the increment $w_t - w_s$ has the normal distribution with mean 0 and variance $t - s$ for any $0 \leq s < t$.

The very existence of a standard Wiener process is one of the cornerstone results in *Stochastic Analysis*. Next, we list some key properties and provide a proof for the heuristic that $\delta w \approx \sqrt{\delta t}$ and $dw \approx \sqrt{dt}$.

Lemma 5.1 *(Finiteness of quadratic variation)* Let $0 = t_0^n < t_1^n < \cdots < t_n^n = T$, where $t_i^n = \frac{iT}{n}$, be a partition of the interval $[0, T]$ into n equal parts, and denote $\delta_i^n w := w_{t_{i+1}^n} - w_{t_i^n}$ as the corresponding increment of the Wiener process w_t over $[t_i^n, t_{i+1}^n]$. We have

$$\lim_{n\to\infty} \mathbb{E}\left(\left[\sum_{i=0}^{n-1} (\delta_i^n w)^2 - T\right]^2\right) = 0;$$

in other words,

$$\lim_{n\to\infty} \sum_{i=0}^{n-1} (\delta_i^n w)^2 = T \text{ in } L^2.$$

The left-hand side expression is called the *quadratic variation* of w. on $[0, T]$, and we say it is equal to T in the L^2 sense.

Proof. From Definition 5.1, we know that $\delta_i^n \mathrm{w} \sim \mathcal{N}(0, T/n)$, hence

$$\mathbb{E}(\delta_i^n \mathrm{w}) = 0, \quad \mathbb{E}((\delta_i^n \mathrm{w})^2) = \frac{T}{n}, \quad \mathbb{E}((\delta_i^n \mathrm{w})^4) = \frac{3T^2}{n^2}.$$

Also for $i \neq j$, $(\delta_i^n \mathrm{w})^2$ and $(\delta_j^n \mathrm{w})^2$ are independent, and so

$$\begin{aligned}\mathbb{E}\left(\left[(\delta_i^n \mathrm{w})^2 - \frac{T}{n}\right]\left[(\delta_j^n \mathrm{w})^2 - \frac{T}{n}\right]\right) &= \mathbb{E}\left((\delta_i^n \mathrm{w})^2 - \frac{T}{n}\right)\mathbb{E}\left((\delta_j^n \mathrm{w})^2 - \frac{T}{n}\right)\\ &= \left(\frac{T}{n} - \frac{T}{n}\right)\left(\frac{T}{n} - \frac{T}{n}\right) = 0.\end{aligned}$$

It follows that

$$\begin{aligned}&\mathbb{E}\left(\left[\sum_{i=0}^{n-1} (\delta_i^n \mathrm{w})^2 - T\right]^2\right)\\ &= \mathbb{E}\left(\left[\sum_{i=0}^{n-1} \left((\delta_i^n \mathrm{w})^2 - \frac{T}{n}\right)\right]^2\right)\\ &= \mathbb{E}\left(\sum_{i=0}^{n-1} \left((\delta_i^n \mathrm{w})^2 - \frac{T}{n}\right)^2\right) + \mathbb{E}\left(\sum_{i\neq j} \left[(\delta_i^n \mathrm{w})^2 - \frac{T}{n}\right]\left[(\delta_j^n \mathrm{w})^2 - \frac{T}{n}\right]\right)\\ &= \sum_{i=0}^{n-1} \left[\mathbb{E}((\delta_i^n \mathrm{w})^4) - \frac{2T}{n}\mathbb{E}((\delta_i^n \mathrm{w})^2) + \frac{T^2}{n^2}\right]\\ &= \sum_{i=0}^{n-1} \left[\frac{3T^2}{n^2} - \frac{2T^2}{n^2} + \frac{T^2}{n^2}\right] = \frac{2T^2}{n} \to 0,\end{aligned}$$

as $n \to \infty$. □

Recall from (3.1) that the change in the stock price over a short period of time of length δt under the Black–Scholes–Merton model is approximately $\delta \mathrm{S}_t = \mu \mathrm{S}_t \delta t + \sigma \mathrm{S}_t \delta \mathrm{w}_t$, where $\delta \mathrm{w}_t \sim \mathcal{N}(0, \delta t)$. In financial mathematics, as a limit, the infinitesimal form of the stock price is often modeled to be governed by a *Geometric Brownian Motion*:

$$d\mathrm{S}_t = \mu \mathrm{S}_t \, dt + \sigma \mathrm{S}_t \, d\mathrm{w}_t, \qquad t \in [0, T], \tag{5.6}$$

where S_t is the stock price at time t, μ is the constant drift term, σ is the constant volatility, and $\{\mathrm{w}_t\}$ is the standard Brownian motion with $\mathrm{w}_t \sim \mathcal{N}(0, t)$ for any $t \geq 0$. A fundamental theorem of *Stochastic Analysis* states that the sample path $\{\mathrm{S}_t : t \geq 0\}$ generated by (5.6) is (almost surely) nowhere differentiable, thus we cannot use the usual Riemann integration to interpret the infinitesimal dynamics of the stock price evolution. Indeed (5.6) is a typical example of a class of differential equations called *stochastic differential equations* (SDEs). To solve it, we need a chain rule of differentiation for the calculus of this class of random functions: the famous *Itô's lemma*. This chain rule is different from the ordinary Newton–Leibniz one as we are now dealing with random functions.

Theorem 5.1 ***(One-dimensional Itô's formula, heuristic version)*** Let x_t be a diffusion process satisfying

$$d\mathrm{x}_t = a(t, \mathrm{x}_t)dt + b(t, \mathrm{x}_t)d\mathrm{w}_t,$$

where a and b are smooth enough functions in their temporal and spatial variables. Consider a function $F(t, x)$ which is continuously differentiable in t and twice continuously differentiable in x. Then $F(t, \mathrm{x}_t)$ satisfies the following stochastic differential equation:

$$\begin{aligned} dF(t, \mathrm{x}_t) = &\left[\frac{\partial F}{\partial t}(t, \mathrm{x}_t) + a(t, \mathrm{x}_t)\frac{\partial F}{\partial x}(t, \mathrm{x}_t) + \frac{b^2(t, \mathrm{x}_t)}{2}\frac{\partial^2 F}{\partial x^2}(t, \mathrm{x}_t)\right] dt \\ &+ b(t, \mathrm{x}_t)\frac{\partial F}{\partial x}(t, \mathrm{x}_t)d\mathrm{w}_t. \end{aligned} \tag{5.7}$$

Proof. We consider δF first and then pass it to its limit. From $\delta \mathrm{x}_t \approx a_t\delta t + b_t\delta \mathrm{w}_t$, where $a_t := a(t, \mathrm{x}_t)$ and $b_t := b(t, \mathrm{x}_t)$, and heuristically $\delta \mathrm{w}_t \approx \sqrt{\delta t}$, we can substitute them into the Taylor expansion of F to obtain

$$\begin{aligned} \delta F &= \frac{\partial F}{\partial t}\delta t + \frac{\partial F}{\partial x}\delta \mathrm{x}_t + \frac{1}{2}\frac{\partial^2 F}{\partial x^2}(\delta \mathrm{x}_t)^2 + \frac{1}{2}\frac{\partial^2 F}{\partial t^2}(\delta t)^2 + \frac{\partial^2 F}{\partial x \partial t}(\delta \mathrm{x}_t)(\delta t) + O_p(\delta t^{3/2}) \\ &= \frac{\partial F}{\partial t}\delta t + \frac{\partial F}{\partial x}(a_t\delta t + b_t\delta \mathrm{w}_t) + \frac{1}{2}\frac{\partial^2 F}{\partial x^2}(a_t^2\delta t^2 + 2a_t b_t \delta t \delta \mathrm{w}_t + b_t^2\delta \mathrm{w}_t^2) + O_p(\delta t^{3/2}) \\ &= \frac{\partial F}{\partial t}\delta t + \frac{\partial F}{\partial x}(a_t\delta t + b_t\delta \mathrm{w}_t) + \frac{1}{2}\frac{\partial^2 F}{\partial x^2}(b_t^2\delta \mathrm{w}_t^2) + O_p(\delta t^{3/2}) \\ &= \left[\frac{\partial F}{\partial t} + a_t\frac{\partial F}{\partial x} + \frac{b_t^2}{2}\frac{\partial^2 F}{\partial x^2}\right]\delta t + b_t\frac{\partial F}{\partial x}\delta \mathrm{w}_t + O_p(\delta t^{3/2}). \end{aligned}$$

The higher-order terms of $O_p(\delta_t^{3/2})$ refer to terms that contain δt of power strictly greater than 1. Now if we take everything to the limit (to the infinitesimal), the higher-order terms go away and (5.7) follows. □

With Itô's lemma in hand, we are now ready to solve the SDE in (5.6). Consider an *Ansatz* $\mathrm{x}_t := S_0 e^{\alpha t + \beta \mathrm{w}_t}$ with constants α and β. By Itô's lemma,

$$\begin{aligned} d\mathrm{x}_t = d(S_0 e^{\alpha t+\beta \mathrm{w}_t}) &= \left(\alpha S_0 e^{\alpha t+\beta \mathrm{w}_t} + \frac{\beta^2}{2} S_0 e^{\alpha t+\beta \mathrm{w}_t}\right) dt + \beta S_0 e^{\alpha t+\beta \mathrm{w}_t} d\mathrm{w}_t \\ &= \left(\alpha + \frac{\beta^2}{2}\right) \mathrm{x}_t dt + \beta \mathrm{x}_t d\mathrm{w}_t. \end{aligned}$$

So $x_t := S_0 e^{\alpha t + \beta w_t}$ is a solution to the following SDE:

$$\begin{cases} dx_t = \left(\alpha + \frac{\beta^2}{2}\right) x_t dt + \beta x_t dw_t, \\ x_0 = S_0. \end{cases}$$

Comparing this with the stock price equation, the solution of (5.6) is

$$S_t = S_0 \exp\left\{ \left(\mu - \frac{\sigma^2}{2}\right) t + \sigma w_t \right\}. \tag{5.8}$$

Noting that $w_t \sim \mathcal{N}(0, t)$ from Theorem 5.1, we can generate the time-t stock price S_t with the standard normal random sample $z_t \sim \mathcal{N}(0, 1)$ and compute (5.8) with $w_t = \sqrt{t} z_t$.

Moreover, as another application of Itô's lemma, we here give a quick review of derivative pricing. Let $V(S_t, t)$ denote the price of a European option at time t with the underlying stock price S_t described by the geometric Brownian motion as mentioned above. Then $V(S_t, t)$ satisfies the following *Black–Scholes pricing equation*:

$$\frac{\partial V(S_t, t)}{\partial t} + rS \frac{\partial V(S_t, t)}{\partial S} + \frac{1}{2}\sigma^2 S_t^2 \frac{\partial^2 V(S_t, t)}{\partial S^2} = rV(S_t, t).$$

To derive this, let Π_t be the time-t value of the hedging portfolio of selling one option and holding M_t units of stocks such that $\Pi_t = -V(S_t, t) + M_t S_t$. Applying Itô's lemma to $V(S_t, t)$ and substituting the stock price SDE $dS_t = \mu S_t dt + \sigma S_t dw_t$, we have

$$dV(S_t, t) = \left(\frac{\partial V(S_t, t)}{\partial S} \mu S_t + \frac{\partial V(S_t, t)}{\partial t} + \frac{1}{2}\sigma^2 S_t^2 \frac{\partial^2 V(S_t, t)}{\partial S^2} \right) dt + \frac{\partial V(S_t, t)}{\partial S} \sigma S_t dw_t,$$

with which we can also obtain the following SDE for Π_t,

$$\begin{aligned} d\Pi_t &= -dV(S_t, t) + M_t dS_t, \text{ or equivalently,} \\ &= -\left(\frac{\partial V(S_t, t)}{\partial t} + \left(\frac{\partial V(S_t, t)}{\partial S} - M_t \right) \mu S_t + \frac{1}{2}\sigma^2 S_t^2 \frac{\partial^2 V(S_t, t)}{\partial S^2} \right) dt \\ &\quad + \sigma S_t \left(M_t - \frac{\partial V(S_t, t)}{\partial S} \right) dw_t. \end{aligned}$$

If we pick M_t to be $\partial V(S_t, t)/\partial S$, then $d\Pi_t$ will have no terms that involve dw_t and it must be riskless over this infinitesimal period. Thus, by the no-arbitrage assumption, $d\Pi_t$ must grow at the risk-free rate such that $d\Pi_t = r\Pi_t dt$. Equating the two expressions for $d\Pi_t$, we obtain

$$-\left(\frac{\partial V(S_t, t)}{\partial t} + \frac{1}{2}\sigma^2 S_t^2 \frac{\partial^2 V(S_t, t)}{\partial S^2} \right) = r\left(-V(S_t, t) + \frac{\partial V(S_t, t)}{\partial S} S_t \right),$$

which after rearrangement reduces to the Black–Scholes pricing equation.

Next, we consider a basket of p assets with dynamics:

$$d\mathbf{S}_t = \mu \odot \mathbf{S}_t dt + \mathbf{S}_t \odot \Sigma^{1/2} d\mathbf{w}_t \quad \text{or}$$

$$d\begin{bmatrix} S_{t,1} \\ \vdots \\ S_{t,p} \end{bmatrix} = \begin{bmatrix} \mu_1 \\ \vdots \\ \mu_p \end{bmatrix} \odot \begin{bmatrix} S_{t,1} \\ \vdots \\ S_{t,p} \end{bmatrix} dt + \begin{bmatrix} S_{t,1} \\ \vdots \\ S_{t,p} \end{bmatrix} \odot \begin{bmatrix} \sigma_{11} & \cdots & \sigma_{1p} \\ \vdots & \ddots & \vdots \\ \sigma_{p1} & \cdots & \sigma_{pp} \end{bmatrix}^{1/2} d\begin{bmatrix} w_{t,1} \\ \vdots \\ w_{t,p} \end{bmatrix},$$

where $w_{t,i}$'s are p independent one-dimensional Wiener processes such that formally, $\text{Var}(dw_{t,i}) = dt$, for $i = 1, \ldots, p$, and $\text{Cov}(dw_{t,i}, dw_{t,j}) = 0$, for $i \neq j \in \{1, \ldots, p\}$, and $\odot$ denotes the *Hadamard product*. The problem of simulating their stock price dynamics (approximately) over a tiny time interval $[t, t + \delta t]$ then lies in generating random samples from a p-variate normal distribution with a mean vector $\delta t\boldsymbol{\mu}$ and a covariance matrix $\delta t\boldsymbol{\Sigma} = \delta t(\sigma_{ij})_{i,j=1,\ldots,p}$. To this end, we shall make use of the Cholesky decomposition of $\boldsymbol{\Sigma} = \mathrm{C}^\top \mathrm{C}$ for some upper triangular matrix C; recall Section 1.2 for more details. More precisely, we aim to generate p-variate normal random vectors with a mean vector $\mathbf{0}_p$ and a covariance matrix $\boldsymbol{\Sigma}$ through p independent standard normal random numbers:

1. First, we generate p independent standard normal random numbers, $\mathbf{z}_t = (z_{t,1}, \ldots, z_{t,p})^\top$, where $z_{t,i}$'s are iid $\mathcal{N}(0, 1)$ for $i = 1, 2, \ldots, p$;
2. Compute $\sqrt{\delta t}\mathrm{C}^\top \mathbf{z}_t \sim \mathcal{N}_p(\mathbf{0}_p, \delta t\boldsymbol{\Sigma})$.

Then, the following recursive formula computes the approximated simulated joint sample paths of the p stocks:

$$\mathbf{S}_{t+\delta t} = \mathbf{S}_t \odot (\mathbf{1}_p + \delta t\boldsymbol{\mu} + \sqrt{\delta t}\mathrm{C}^\top \mathbf{z}_t). \tag{5.9}$$

We shall illustrate the use of the recursive relation of (5.9) to simulate the stock prices of HSBC, CLP and CK via Python in Programme 5.17 (via **R** in Programme 5.18). Since the time unit for both $\boldsymbol{\mu}$ and $\boldsymbol{\Sigma}$ is one day, we simply take $\delta t = 1$. At the start of each programme, leaving room for setting the initial (random) seed in the simulation ensures that the same set of pseudo-random numbers is generated each time. This is useful as it allows us to check and debug our programmes.

```
1  np.random.seed(4012)
2  mu_180 = np.mean(u_180)
3  S_180 = np.cov(u_180, rowvar=False)
4  C_180 = np.linalg.cholesky(S_180)
5  s0 = d.iloc[-1,:]
6  s_pred = []
7  for i in range(90):
8      z = np.random.randn(3)
9      v = mu_180 + C_180.T @ z
10     s1 = s0 * (1 + v)
```

```
11        s_pred.append(s1.values)
12        s0 = s1
13
14  df_pred = pd.DataFrame(np.array(s_pred),
15                         columns=d.columns.values + "_pred")
16  df_pred.index = np.arange(len(d)+1, len(d)+90+1)
17  d.index = np.arange(len(d))
18
19  col = ["blue", "orange", "green", "pink", "brown", "red"]
20  pd.merge(d, df_pred, how='outer', left_index=True, right_index=True
21           ).plot(figsize=(10, 7), color=col, style=["-", "--", "."]*2
22                  ).legend(loc='upper right', fontsize=14, ncol=2)
```

Programme 5.17 Continuation of Programme 3.13: Stock price simulation for HSBC, CLP, and CK via Python.

```
1  > set.seed(4002)
2  > mu_180 <- apply(u_180, 2, mean)
3  > S_180 <- cov(u_180)
4  > C_180 <- chol(S_180)  # Cholesky decomposition of Sigma
5  > s0 <- tail(d, 1)      # set s0 to the most recent price
6  > s_pred <- c()
7  > for (i in 1:90) {
8  +     z <- rnorm(3)
9  +     v <- mu_180 + t(C_180)%*%z
10 +     s1 <- s0 * (1 + t(v)) # new stock price
11 +     s_pred <- rbind(s_pred, s1)
12 +     s0 <- s1    # update s0
13 + }
14 >
15 > s_pred <- ts(s_pred, start=nrow(d)+1)
16 > data <- ts.union(d, s_pred)
17 > par(mfrow=c(1,1))
18 >
19 > col <- c("blue", "orange", "green", "pink", "brown", "red")
20 > plot(data, plot.type="s", col=col, lwd=3, lty=1:6)
21 > legend("topright", col=col, lwd=3, lty=1:6, ncol=2, cex=1.5,
22 +        legend=c(colnames(d), paste0(colnames(d), "_pred")))
```

Programme 5.18 Continuation of Programme 3.14: Stock price simulation for HSBC, CLP, and CK via R.

In Programme 5.17, the command line for the random seed generator is placed outside the "for-loop", so as to ensure the pseudo-randomly generated numbers are under the same random seed. For the far-end terms of the generated sequence, they can be regarded as iid samples in accordance with the ergodic theory mentioned in Section 5.1. If we instead put the command inside the "for-loop", it will generate a new seed every time when the algorithm gets back into the "for-loop", and the numbers generated will end up not possessing the pseudo-independence structure.

With these simulated stock prices, we can calibrate the price of the portfolio or that of relevant derivative products. Here we simulated only one sample path of these stock prices as shown in Figures 5.4 and 5.5. In principle, we can simulate as many paths as we want and build up the corresponding (empirical) loss distribution of a portfolio from which the V@R can be calibrated.

For instance, in accordance with no-arbitrage pricing theory [3, 20], the price V_0 of a financial derivative with a payoff $H_T = H_T(\omega)$ is given by $V_0 = \mathbb{E}^{\mathbb{Q}}(e^{-rT}H_T)$,

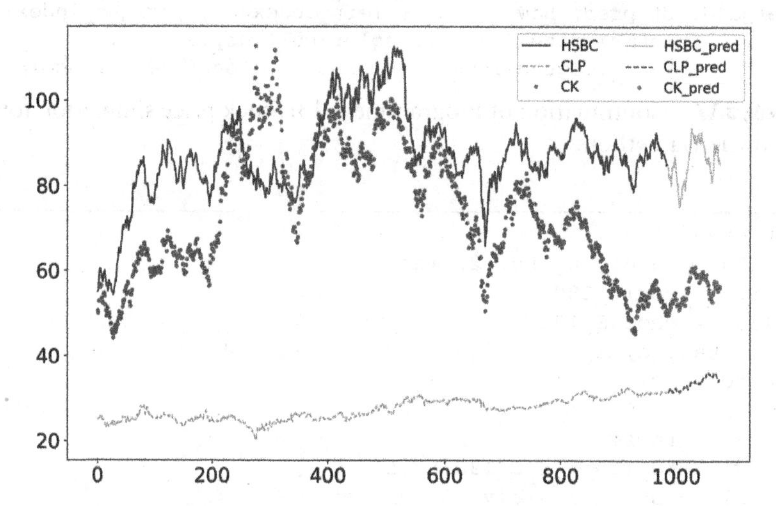

FIGURE 5.4 Actual and simulated stock prices for HSBC (solid), CLP (dashed), and CK (dotted) via Python, generated by Programme 5.17.

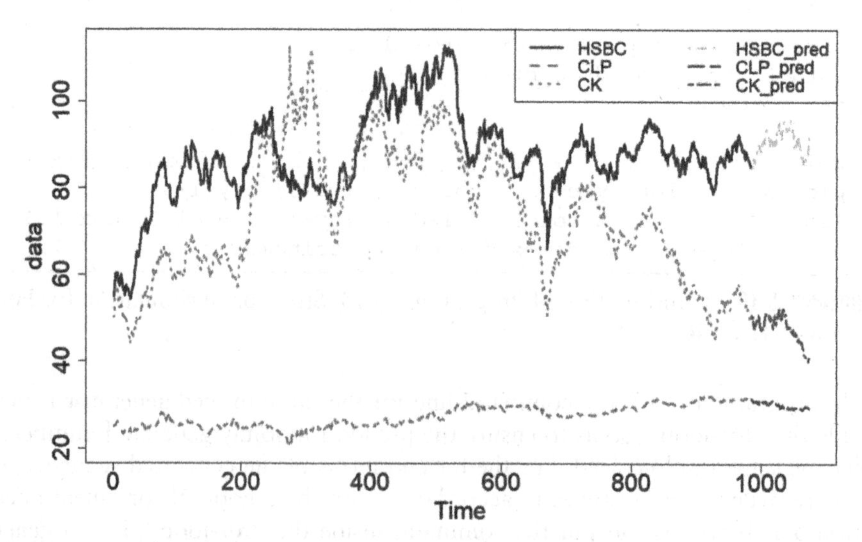

FIGURE 5.5 Actual and simulated stock prices for HSBC (solid), CLP (dashed), and CK (dotted) via R, generated by Programme 5.18.

where Q is the risk-neutral probability measure [3, 20], r is the risk-free interest rate, T is the time to maturity of the derivative, and $\boldsymbol{\omega}$ represents all the random terminal information about the underlying securities in H_T. In general, the expression for V_0 is mostly intractable. However, if $\boldsymbol{\omega}$ can be easily computationally generated, the Monte Carlo method provides an alternate approximate solution:

$$\hat{V}_0(n) = \frac{1}{n}\sum_{i=1}^{n} e^{-rT} H_T(\omega_i),$$

where ω_i, for $i = 1, \ldots, n$, are different terminal sample points.

Example 5.6 Under the Black–Scholes model and the risk-neutral probability measure Q, a European call option with strike price K has a payoff C_T given by:

$$\mathrm{C}_T = \max(\mathrm{S}_T - K, 0) := (\mathrm{S}_T - K)_+ = \left(S_0 \exp\left[\left(r - \frac{\sigma^2}{2}\right)T + \sigma\sqrt{T}\mathrm{z}\right] - K\right)_+,$$

where $\mathrm{z} \sim \mathcal{N}(0, 1)$. By taking the expectation, we can see that the fair price of the call option at time 0 is given by

$$\mathrm{C}_0 := \mathbb{E}^Q(\mathrm{C}_T) = S_0\Phi(d_+) - e^{-rT}K\Phi(d_-),$$

where Φ is the cdf of the standard normal distribution, $d_+ := \frac{\ln(S_0/K)+(r+\sigma^2/2)T}{\sigma\sqrt{T}}$, and $d_- := d_+ - \sigma\sqrt{T}$. In the meantime, we can easily approximate the price of the European call option with the following procedure:

1. Generate $\mathrm{z}_1, \ldots, \mathrm{z}_n \overset{iid}{\sim} \mathcal{N}(0, 1)$;
2. Compute $\mathrm{S}_T^{(i)} = S_0 \exp\left[\left(r - \frac{\sigma^2}{2}\right)T + \sigma\sqrt{T}\mathrm{z}_i\right]$ for $i = 1, \ldots, n$;
3. Also compute $\mathrm{C}_T^{(i)} = e^{-rT}(\mathrm{S}_T^{(i)} - K)_+$ for $i = 1, \ldots, n$;
4. $\hat{\mathrm{C}}_{T,n} = \frac{1}{n}\sum_{i=1}^{n} \mathrm{C}_T^{(i)}$.

This estimator is unbiased, i.e. $\mathbb{E}^Q(\hat{\mathrm{C}}_{T,n}) = \mathrm{C}_0$, and it is certainly consistent, i.e. $\lim_{n\to\infty} \hat{\mathrm{C}}_{T,n} = \mathrm{C}_0$.

To see the convergence rate of this proposed Monte Carlo estimator, let $\mathrm{x}_1, \ldots, \mathrm{x}_n$ be n iid random variables drawn from a common distribution with a mean μ and a finite variance σ^2. The sample mean of these x_i's has a theoretical mean and variance given by $\mathbb{E}\left(\frac{1}{n}\sum_{i=1}^{n} \mathrm{x}_i\right) = \mu$ and $\mathrm{Var}\left(\frac{1}{n}\sum_{i=1}^{n} \mathrm{x}_i\right) = \frac{\sigma^2}{n}$, and a simple application of the

Central Limit Theorem gives $\frac{\frac{1}{n}\sum_{i=1}^{n} x_i - \mu}{\sigma/\sqrt{n}} \xrightarrow{d} \mathcal{N}(0,1)$. Hence, for the call option in particular, due to the finiteness of the variance of a log-normal distribution, $\mathrm{Var}(C_T^{(1)}) =: \sigma_{C_T}^2 < \infty$[1], we then have

$$\frac{\hat{C}_{T,n} - C_0}{\sigma_{C_T}/\sqrt{n}} \xrightarrow{d} \mathcal{N}(0,1). \tag{5.10}$$

Therefore, the error $|\hat{C}_{T,n} - C_0|$ has the order of $O_p\left(\frac{1}{\sqrt{n}}\right)$. In other words, to reduce the error by half, we need to perform 4 times as many simulations. Furthermore, from (5.10), we can construct a confidence interval for our estimate. In particular, we have:

$$\mathbb{P}\left(-z_{1-\frac{\alpha}{2}} \le \frac{\hat{C}_{T,n} - C_0}{\sigma_{C_T}/\sqrt{n}} \le z_{1-\frac{\alpha}{2}}\right) \approx 1 - \alpha,$$

where $z_{1-\frac{\alpha}{2}}$ is the upper tail percentile of the standard normal distribution such that $\Phi(z_{1-\frac{\alpha}{2}}) = 1 - \frac{\alpha}{2}$. Therefore, the (approximate) $100(1-\alpha)\%$ confidence interval of C_0 is

$$\left[\hat{C}_{T,n} - z_{1-\frac{\alpha}{2}}\frac{\sigma_{C_T}}{\sqrt{n}}, \hat{C}_{T,n} + z_{1-\frac{\alpha}{2}}\frac{\sigma_{C_T}}{\sqrt{n}}\right].$$

If we want $\mathbb{P}(|\hat{C}_{T,n} - C_0| \le \varepsilon) \approx 1 - \alpha$ for some $\varepsilon > 0$, we choose sample size n such that $z_{1-\frac{\alpha}{2}}\frac{\sigma_{C_T}}{\sqrt{n}} = \varepsilon$, which is equivalent to $n = \frac{\sigma_{C_T}^2}{\varepsilon^2} z_{1-\frac{\alpha}{2}}^2$.

Next, let us consider another class of stock price dynamics models, called *local volatility* models, which are more general than the Black–Scholes model and had been commonly used before the 2007–2008 Credit Crunch. With a constant interest rate under the risk-neutral probability measure, the underlying stock price $\{S_t\}$ of a financial derivative is assumed to follow the stochastic differential equation (SDE):

$$dS_t = rS_t\,dt + \sigma(S_t)S_t\,dw_t, \quad t \in [0,T].$$

If the payoff of the derivative is path-dependent, its pricing via the Monte Carlo method requires us to simulate paths of the stock price evolutions instead of just simulating the terminal stock price value of S_T as in Example 5.6. To simulate the i-th evolution path $S_t^{(i)}$, which ought to be continuous, we discretize the interval $[0,T]$ into M sub-intervals of equal length and consider

$$\hat{S}_{m+1}^{(i)} = \hat{S}_m^{(i)}(1 + r\delta t + \sigma(\hat{S}_m^{(i)})\sqrt{\delta t}\cdot z_m^{(i)}), \qquad \text{for } m = 0,\ldots,M-1, \tag{5.11}$$

[1] The explicit form of $\sigma_{C_T}^2$ can be quite complicated. One may simply use the bootstrapped sample variance as an estimate for most practical considerations.

with the initial condition $\hat{S}_0 = S_0$, where these $z_m^{(i)}$'s $\overset{iid}{\sim} \mathcal{N}(0,1)$ and $\delta t = T/M$. We estimate $S^{(i)}_{m\delta t}$ by $\hat{S}^{(i)}_m$, and for any $u \in [m\delta t, (m+1)\delta t]$, $S^{(i)}_u \approx (1-u)\hat{S}^{(i)}_m + u\hat{S}^{(i)}_{m+1}$.

Generally, if x_t satisfies a more generic SDE $dx_t = a(x_t)\,dt + b(x_t)\,dw_t$ with an initial value $x_0 = x$, we can simulate paths of x_t using the following *discretization schemes* [10]: for $m = 0, 1, \ldots, M-1$ and $\delta t = \frac{T}{M}$,

1. *Euler–Maruyama Scheme*:

$$\hat{x}_{(m+1)\delta t} = \hat{x}_{m\delta t} + a(\hat{x}_{m\delta t})\delta t + b(\hat{x}_{m\delta t})\sqrt{\delta t}\cdot z_m, \quad z_m\text{'s} \overset{iid}{\sim} \mathcal{N}(0,1),$$

which can be seen by direct discretization of the SDE. The simulation method (5.11) is just a special case of this scheme.

2. *Milstein Scheme*:

$$\begin{aligned}\hat{x}_{(m+1)\delta t} &= \hat{x}_{m\delta t} + a(\hat{x}_{m\delta t})\delta t + b(\hat{x}_{m\delta t})\sqrt{\delta t}\cdot z_m \\ &\quad + b(\hat{x}_{m\delta t})\frac{\partial}{\partial x}b(\hat{x}_{m\delta t})\frac{\delta t}{2}(z_m^2 - 1), \quad z_m\text{'s} \overset{iid}{\sim} \mathcal{N}(0,1),\end{aligned}$$

The Milstein Scheme technically has a superior convergence behavior (in both the weak and strong convergence sense) than the Euler–Maruyama Scheme. A heuristic derivation of this is as follows: consider the generic SDE $dx_t = a(x_t)dt + b(x_t)dw_t$. We can rewrite it in an integral form as:

$$\begin{aligned}x_{t+\delta t} - x_t &= \int_t^{t+\delta t} a(x_u)du + \int_t^{t+\delta t} b(x_u)dw_u \\ &= \int_t^{t+\delta t}\left(a(x_t) + \int_t^{u_1}\frac{\partial a}{\partial x}(x_{u_2})dx_{u_2}\right)du_1 \\ &\quad + \int_t^{t+\delta t}\left(b(x_t) + \int_t^{u_1}\frac{\partial b}{\partial x}(x_{u_2})dx_{u_2}\right)dw_{u_1} \\ &= a(x_t)\cdot\delta t + \underbrace{\int_t^{t+\delta t}\int_t^{u_1}\frac{\partial a}{\partial x}(x_{u_2})\big(a(x_{u_2})du_2 + b(x_{u_2})dw_{u_2}\big)du_1}_{\text{of order } O_p((\delta t)^{3/2}) \text{ which can be neglected}} \\ &\quad + b(x_t)\delta w + \int_t^{t+\delta t}\int_t^{u_1}\frac{\partial b}{\partial x}(x_{u_2})\big(a(x_{u_2})du_2 + b(x_{u_2})dw_{u_2}\big)dw_{u_1}.\end{aligned}$$

In particular, we have

$$\begin{aligned}
&\int_t^{t+\delta t}\int_t^{u_1}\frac{\partial b}{\partial x}(\mathrm{x}_{u_2})\left(a(\mathrm{x}_{u_2})du_2+b(\mathrm{x}_{u_2})d\mathrm{w}_{u_2}\right)d\mathrm{w}_{u_1}\\
&=O_p((\delta t)^{3/2})+\int_t^{t+\delta t}\int_t^{u_1}b(\mathrm{x}_{u_2})\frac{\partial b}{\partial x}(\mathrm{x}_{u_2})d\mathrm{w}_{u_2}d\mathrm{w}_{u_1}\\
&=O_p((\delta t)^{3/2})+b(\mathrm{x}_t)\frac{\partial b}{\partial x}(\mathrm{x}_t)\int_t^{t+\delta t}\int_t^{u_1}d\mathrm{w}_{u_2}d\mathrm{w}_{u_1}.
\end{aligned}$$

By Itô's lemma, the double integral

$$\int_t^{t+\delta t}\int_t^{u_1}d\mathrm{w}_{u_2}d\mathrm{w}_{u_1}=\int_t^{t+\delta t}(\mathrm{w}_{u_1}-\mathrm{w}_t)d\mathrm{w}_{u_1}=\frac{(\mathrm{w}_{t+\delta t}-\mathrm{w}_t)^2}{2}-\frac{1}{2}\delta t.$$

Finally, with $\delta\mathrm{w}\overset{iid}{\sim}\sqrt{\delta t}\cdot\mathrm{z}$ where z is standard normal, we have

$$\mathrm{x}_{t+\delta t}-\mathrm{x}_t=a(\mathrm{x}_t)\cdot\delta t+b(\mathrm{x}_t)\sqrt{\delta t}\cdot\mathrm{z}+b(\mathrm{x}_t)\frac{\partial b}{\partial x}(\mathrm{x}_t)\frac{\delta t}{2}(\mathrm{z}^2-1)+O_p((\delta t)^{3/2}).$$

Next, we discuss the pricing of a continuous arithmetic Asian call option whose payoff is equal to $(A(T)-K)_+$ with K as the strike price and $A(T)=\frac{1}{T}\int_0^T S(t)dt$, the average price of the underlying stock over the period $[0,T]$. By using either of the Euler–Maruyama or Milstein schemes, we can first generate n sample paths $\{\hat{\mathrm{S}}_m^{(i)}\}_{m=0}^M$, for $i=1,\ldots,n$. Similar to Example 5.6, due to the fundamental theorem of asset pricing, the arbitrage-free price of this Asian option is given by the risk-neutral expectation of the discounted terminal payoff. Therefore the time-zero current price of the Asian call can be estimated by:

$$\frac{1}{n}\sum_{i=1}^{n}e^{-rT}\left(\frac{1}{M+1}\sum_{m=0}^{M}\hat{\mathrm{S}}_m^{(i)}-K\right)_+. \tag{5.12}$$

On the other hand, unlike Example 5.6, there is no elementary closed-form pricing formula for arithmetic Asian call options, even under the Black–Scholes framework, due to the difficulty in determining the distribution of the sum of lognormal random variables. Researchers and practitioners have tried various approaches to tackle this problem, for instance via numerical PDEs or infinite series. In practice, for large enough values of M and n, the result of (5.12) calculated via the standard Monte Carlo approach, and utilizing the exact expression of S_t, is usually regarded as the theoretical price.

Example 5.7 Consider an Asian call option with $S_0 = 10, K = 8, r = 0.05, \sigma = 0.3$, and $T = 1$. Under the Black–Scholes framework, $\sigma(\hat{S}_m^{(i)}) \equiv \sigma$ in (5.11) for $m = 0, \ldots, M-1$ and $i = 1, \ldots, n = 100,000$. We use the two discretization schemes discussed above to estimate the current price of this Asian call option by using different values of $M \in \{10, 100, 1000\}$. We shall first lay down the code for the simulation of the Asian call option in Programme 5.19 via Python and in Programme 5.20 via **R**. Note that for the sake of fair comparison between the two schemes in this example, we also compute an "exact" price for every candidate value of M via the standard Monte Carlo approach, as a benchmark of comparison. Readers may verify on their own, by choosing larger values of M and n in the programmes, that the "exact" theoretical price of the continuous arithmetic Asian option in question lies within the range of 2.19 ± 0.01.

```
1  import numpy as np
2  import pandas as pd
3
4  # Initialize variables
5  S_0 = 10; K = 8; r = 0.05; sigma = 0.3; T = 1
6  M_lst = [1e1, 1e2, 1e3]; n = 1e5
7
8  def Sim_Asian(n, M, S_0, K, r, sigma, T, theta):
9      delta_t = T / M
10
11     def Euler(z, S_t, theta):
12         return S_t*(1 + r*delta_t) + sigma*S_t**(theta/2)*np.sqrt(delta_t)*z
13
14     def Milstein(z, S_t, theta):
15         return (Euler(z, S_t, theta) +
16                 sigma**2*theta/2*S_t**(theta-1)*delta_t/2*(z**2-1))
17
18     def Exact(z, S_t, theta=2):
19         return S_t*np.exp((r - sigma**2/2)*delta_t + sigma*np.sqrt(delta_t)*z)
20
21     # Asian_call
22     S_Eul = S_0; avg_Eul = S_0 / (M+1)
23     S_Mil = S_0; avg_Mil = S_0 / (M+1)
24     if (theta == 2):
25         S_Ext = S_0; avg_Ext = S_0 / (M+1)
26     else:
27         S_Ext = 0; avg_Ext = 0
28
29     for m in range(int(M)):
30         z = np.random.normal(size=int(n))
31         S_Eul = Euler(z, S_Eul, theta)
32         avg_Eul += S_Eul/(M+1)
33
34         S_Mil = Milstein(z, S_Mil, theta)
35         avg_Mil += S_Mil/(M+1)
36
```

```
37          if (theta == 2):
38              S_Ext = Exact(z, S_Ext, theta)
39              avg_Ext += S_Ext/(M+1)
40
41      return {"Eul":{"price":S_Eul,"payoff":np.exp(-r*T)*np.maximum(avg_Eul-K,0)},
42              "Mil":{"price":S_Mil,"payoff":np.exp(-r*T)*np.maximum(avg_Mil-K,0)},
43              "Ext":{"price":S_Ext,"payoff":np.exp(-r*T)*np.maximum(avg_Ext-K,0)}}
```

Programme 5.19 Simulating the Asian call price in Example 5.7 via Python.

```
1  > # Initialize variables
2  > S_0 <- 10; K <- 8; r <- 0.05; sigma <- 0.3; T <- 1
3  > M_lst <- c(1e1, 1e2, 1e3); n <- 1e5
4  >
5  > Sim_Asian <- function(n, M, S_0, K, r, sigma, T, theta){
6  +   delta_t <- T / M
7  +
8  +   Euler <- function(z, S_t, theta){
9  +     S_t*(1 + r*delta_t) + sigma*S_t^(theta/2)*sqrt(delta_t)*z
10 +   }
11 +
12 +   Milstein <- function(z, S_t, theta){
13 +     (Euler(z, S_t, theta) +
14 +        sigma^2*theta/2*S_t^(theta-1)*delta_t/2*(z^2-1))
15 +   }
16 +
17 +   Exact <- function(z, S_t, theta=2){
18 +     S_t*exp((r - sigma^2/2)*delta_t + sigma*sqrt(delta_t)*z)
19 +   }
20 +
21 +   # Asian_call
22 +   S_Eul <- S_0; avg_Eul <- S_0 / (M+1)
23 +   S_Mil <- S_0; avg_Mil <- S_0 / (M+1)
24 +   if (theta == 2) {
25 +     S_Ext <- S_0; avg_Ext <- S_0 / (M+1)
26 +   } else {
27 +     S_Ext <- NA; avg_Ext <- NA
28 +   }
29 +
30 +   for (m in 1:M){
31 +     z <- rnorm(n)
32 +     S_Eul <- Euler(z, S_Eul, theta)
33 +     avg_Eul <- avg_Eul + S_Eul/(M+1)
34 +
35 +     S_Mil <- Milstein(z, S_Mil, theta)
36 +     avg_Mil <- avg_Mil + S_Mil/(M+1)
37 +
38 +     if (theta == 2){
```

```
39 +        S_Ext <- Exact(z, S_Ext, theta)
40 +        avg_Ext <- avg_Ext + S_Ext/(M+1)
41 +      }
42 +    }
43 +
44 +    list(Eul=list(price=S_Eul, payoff=exp(-r*T)*pmax(avg_Eul-K, 0)),
45 +         Mil=list(price=S_Mil, payoff=exp(-r*T)*pmax(avg_Mil-K, 0)),
46 +         Ext=list(price=S_Ext, payoff=exp(-r*T)*pmax(avg_Ext-K, 0)))
47 + }
```

Programme 5.20 Simulating the Asian call price in Example 5.7 via **R**.

To compare the performance of the two discretization methods, the histograms of $\hat{S}_M^{(i)} - S_M^{(i)}$, for $i = 1, \ldots, 10^5$, are considered, in which each $\hat{S}_M^{(i)}$ is simulated by the corresponding numerical scheme and each $S_M^{(i)}$ is generated as in Example 5.6. The results are reported in Programme 5.19 via Python (resp. Programme 5.20 via **R**).

```
1  import matplotlib.pyplot as plt
2
3  plt.rc('font', size=20); plt.rc('axes', titlesize=30, labelsize=30)
4  plt.rc('xtick', labelsize=25); plt.rc('ytick', labelsize=25)
5
6  BS_results = pd.DataFrame({"M": M_lst, "n": n, "Asian_Eul": 0.0,
7                             "Asian_Mil": 0.0, "Asian_Ext": 0.0})
8  for i in range(len(M_lst)):
9      M = M_lst[i]
10     np.random.seed(4002)
11     Asian_BS = Sim_Asian(n, M, S_0, K, r, sigma, T, 2)
12     BS_results.iloc[i, -3] = np.mean(Asian_BS["Eul"]["payoff"])
13     BS_results.iloc[i, -2] = np.mean(Asian_BS["Mil"]["payoff"])
14     BS_results.iloc[i, -1] = np.mean(Asian_BS["Ext"]["payoff"])
15
16     Eul_diff = Asian_BS["Ext"]["price"] - Asian_BS["Eul"]["price"]
17     Mil_diff = Asian_BS["Ext"]["price"] - Asian_BS["Mil"]["price"]
18     print(np.ptp(Eul_diff), np.ptp(Mil_diff), np.ptp(Eul_diff)/np.ptp(Mil_diff))
19
20     fig, axes = plt.subplots(ncols=2, figsize=(16,7))
21     axes[0].hist(Eul_diff, alpha=0.7, color="blue", ec='black', bins="sturges")
22     axes[0].set_title("Euler Scheme")
23     axes[1].hist(Mil_diff, alpha=0.7, color="red", ec='black', bins="sturges")
24     axes[1].set_title("Milstein Scheme")
25
26 print(BS_results)
```

```
1  3.3256636798751913 0.5123649808989157 6.490809879395934
2  0.8619869210188611 0.029918976226060323 28.81070911337015
3  0.320405147120308 0.0030512221022078734 105.00879201434135
4        M          n  Asian_Eul  Asian_Mil  Asian_Ext
5  0    10.0  100000.0   2.189853   2.183738   2.185303
6  1   100.0  100000.0   2.193288   2.192745   2.192906
7  2  1000.0  100000.0   2.190134   2.190106   2.190122
```

Programme 5.21 Estimating the Asian call price in Example 5.7 via Python.

```
1  > BS_results <- data.frame(M=M_lst, n=n, Asian_Eul=0,
2  +                           Asian_Mil=0, Asian_Ext=0)
3  > par(mfrow=c(1,2))
4  > for (i in 1:length(M_lst)){
5  +     M <- M_lst[i]
6  +     set.seed(4002)
7  +     Asian_BS <- Sim_Asian(n, M, S_0, K, r, sigma, T, 2)
8  +     BS_results$Asian_Eul[i] <- mean(Asian_BS$Eul$payoff)
9  +     BS_results$Asian_Mil[i] <- mean(Asian_BS$Mil$payoff)
10 +     BS_results$Asian_Ext[i] <- mean(Asian_BS$Ext$payoff)
11 +
12 +     Eul_diff <- Asian_BS$Ext$price - Asian_BS$Eul$price
13 +     Mil_diff <- Asian_BS$Ext$price - Asian_BS$Mil$price
14 +     print(paste(diff(range(Eul_diff)), diff(range(Mil_diff)),
15 +                 diff(range(Eul_diff))/diff(range(Mil_diff))))
16 +     hist(Eul_diff, xlim=c(-max(abs(Eul_diff)), max(abs(Eul_diff))),
17 +          xlab="Error", main="Euler Scheme")
18 +     hist(Mil_diff, xlim=c(-max(abs(Mil_diff)), max(abs(Mil_diff))),
19 +          xlab="Error", main="Milstein Scheme")
20 + }
21 [1] "3.05029742248599 0.481686137929922 6.33254142540793"
22 [1] "0.947418972094894 0.0425023425464754 22.2909824572354"
23 [1] "0.339418163200872 0.00329830822820831 102.906744826953"
24 > BS_results
25      M     n Asian_Eul Asian_Mil Asian_Ext
26 1   10 1e+05  2.186955  2.181032  2.182584
27 2  100 1e+05  2.191797  2.191197  2.191351
28 3 1000 1e+05  2.193214  2.193114  2.193129
```

Programme 5.22 Estimating the Asian call price in Example 5.7 via **R**.

From Figures 5.6 via Python and 5.7 via **R**, we see that the range of all the histograms of the Milstein scheme for $M = 10, 100, 1000$ are much smaller in size than that of the Euler–Maruyama scheme, and the ratio of the two ranges of Euler–Maruyama to Milstein gets larger in value (from around 6.5 to about 105) as M increases from 10 to 1,000. As predicted by the theory, when the number of discretization steps M increases, the Milstein scheme provides a much better approximation of the Asian call price than the Euler–Maruyama scheme. Table 5.3 summarizes the approximated Asian call prices using Python and **R**, and all approximated values converge to the true value of about 2.19.

As a generalization of the Black–Scholes model, the class of *constant elasticity of variance* (CEV) models is also prevalent in the financial industries, at least before the 2007–2008 Credit Crunch, for various purposes such as the pricing of equities and commodities. First proposed by *John Cox* in [6], these models can capture random volatility and the leverage effect in a more flexible manner. In particular, by connecting the volatility of the underlying asset with its price level, the resulting *implied volatility smile*[2] is similar to the volatility smile curves observed in practice. Let us illustrate the pricing of an Asian call option under this class of CEV models in the following Example 5.8.

[2] This curve, depicting the implied volatility against exercise price of a financial option, gets its name since it is usually convex and resembles a smile facing us.

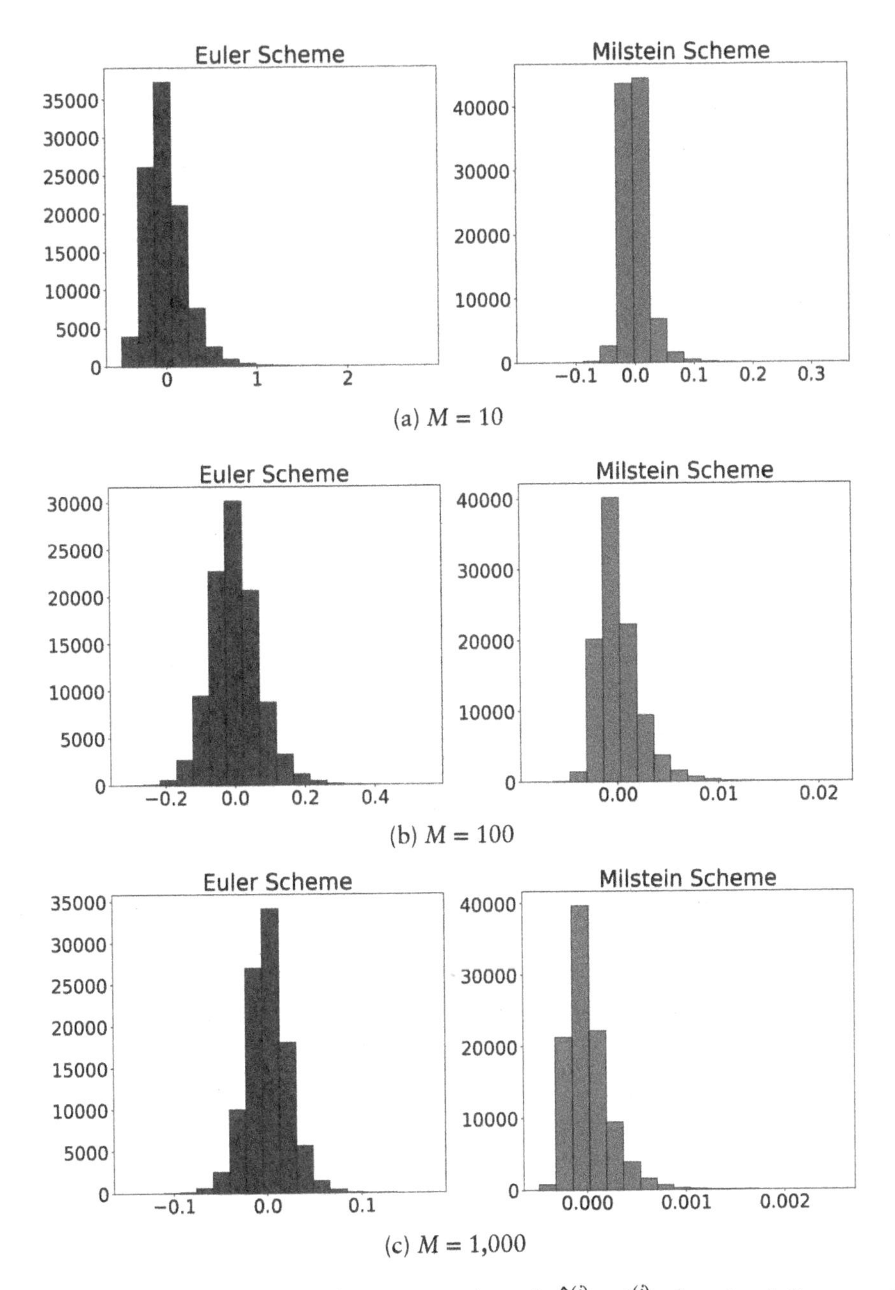

FIGURE 5.6 The histograms of the sample of $\hat{S}_M^{(i)} - S_M^{(i)}$ for $i = 1, 2, \ldots, n$ for Euler–Maruyama (left) and Milstein (right) schemes at the terminal time T via Python, generated in Programme 5.21.

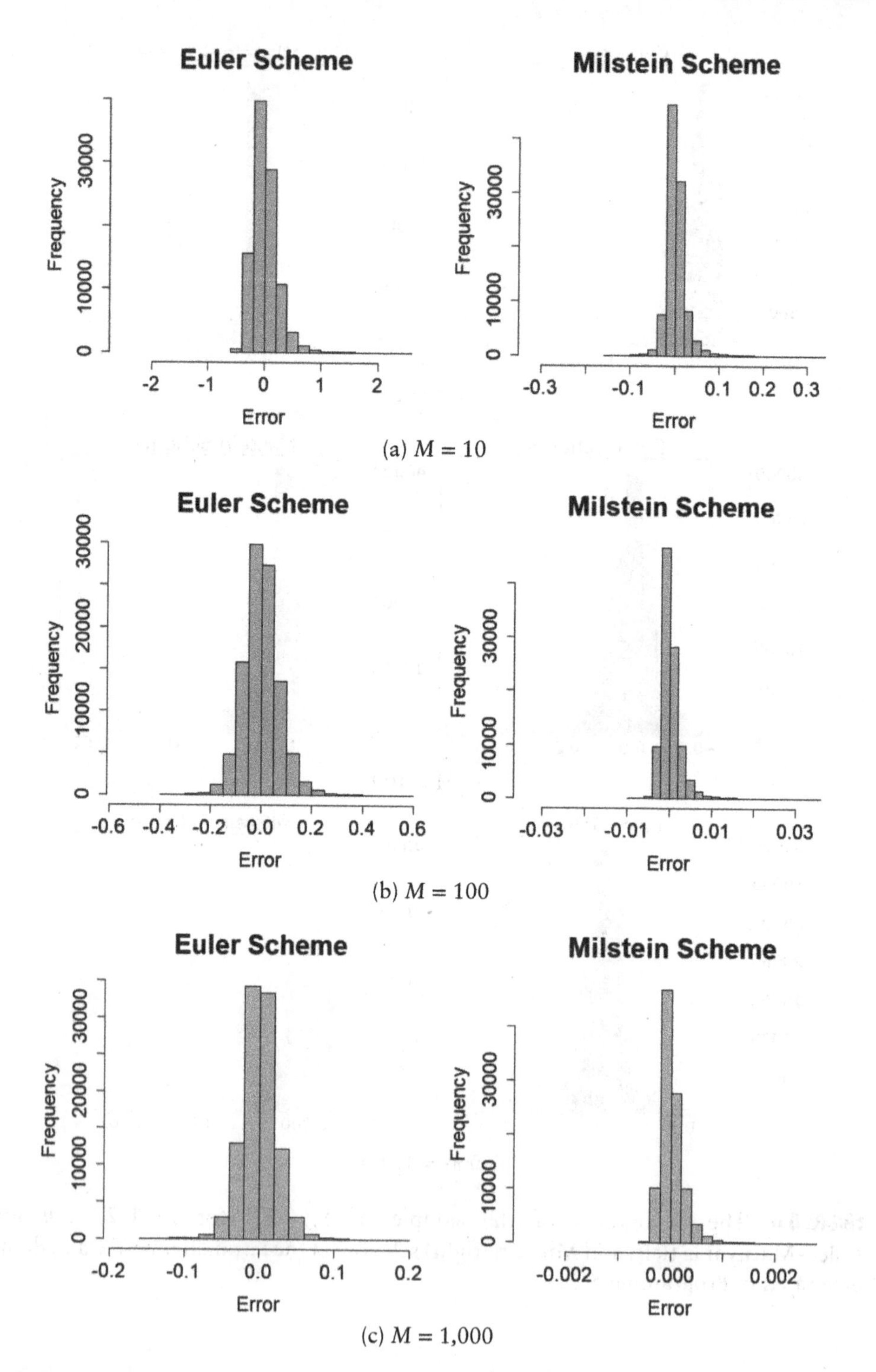

FIGURE 5.7 The histograms of the sample of $\hat{S}_M^{(i)} - S_M^{(i)}$ for $i = 1, 2, \ldots, n$ for Euler–Maruyama (left) and Milstein (right) schemes at the terminal time T via **R**, generated in Programme 5.22.

TABLE 5.3 The estimated price of Asian call option in Example 5.7 in Python and R.

	Python			R		
M	10	100	1,000	10	100	1,000
Euler	2.189853	2.193288	2.190134	2.186955	2.191797	2.193214
Milstein	2.183738	2.192745	2.190106	2.181032	2.191197	2.193114
Exact	2.185303	2.192906	2.190122	2.182584	2.191351	2.193129

Example 5.8 ***(Continuation of Example 5.7)*** Suppose that the stock price dynamics now follows a CEV model:

$$dS_t = \mu S_t dt + \sigma S_t^{\theta/2} dw_t, \tag{5.13}$$

where θ is a parameter that controls the elasticity of variance. In particular, when $\theta = 2$, it reduces to the Black–Scholes framework in Example 5.7. Here, we inspect $\theta = 1$ and 1.8. Since the SDEs of the CEV model under $\theta = 1$ and 1.8 do not possess an analytical solution, the simulated Asian call price, by using the standard Monte Carlo method (which is exactly the Euler–Maruyama scheme in this case) with $n = 10^6$ and $M = 10^4$, is selected as the benchmark that is expected to be very close to the true price. Moreover, the control variate technique, first discussed in Section 5.2, with the terminal stock price $\hat{S}_M$ adopted as the control variable is also inspected. For example, let $\hat{C}_{1,\text{Asian}}, \ldots, \hat{C}_{n,\text{Asian}}$ be the simulated Asian call prices, namely $\hat{C}_{i,\text{Asian}} = e^{-rT}\left(\frac{1}{M+1}\sum_{m=0}^{M}\hat{S}_m^{(i)} - K\right)_+$, and $\hat{S}_M^{(1)}, \ldots, \hat{S}_M^{(n)}$ be the corresponding terminal stock prices at time T, then the estimator of Asian call price is given by $\overline{C}_{n,\text{Asian}} + \hat{b}^*(\overline{S}_M - \mu_S)$, where $\overline{C}_{n,\text{Asian}}$ is the sample mean of $\hat{C}_{1,\text{Asian}}, \ldots, \hat{C}_{n,\text{Asian}}$, $\overline{S}_M$ is the sample mean of $\hat{S}_M^{(1)}, \ldots, \hat{S}_M^{(n)}$, $\mu_S = \mathbb{E}(S_T)$, and $\hat{b}^* = -\frac{\sum_{i=1}^{n}(\hat{C}_{i,\text{Asian}} - \overline{C}_{n,\text{Asian}})(\hat{S}_M^{(i)} - \overline{S}_M)}{\sum_{i=1}^{n}(\hat{S}_M^{(i)} - \overline{S}_M)^2}$; notice that for the cases of $\theta = 1$ and $\theta = 1.8$, μ_S is taken as the sample mean of the simulated paths for generating the corresponding benchmark prices mentioned before. The results are reported in Programmes 5.23 for $\theta = 1$ and 5.24 for $\theta = 1.8$ via Python (resp. Programme 5.25 for $\theta = 1$ and Programme 5.26 for $\theta = 1.8$ via **R**).

```
1   from tqdm import trange
2   pd.set_option('display.max_columns', None)  # print all columns
3
4   def control_variate(x, y, mu_y):
5       s_cov = np.cov(x, y, ddof=1)[0][1]
6       s_var = np.var(y, ddof=1)
7       return np.mean(x) - s_cov/s_var * (np.mean(y) - mu_y)
8
9   theta = 1; epochs = 100
10  CEV_results = pd.DataFrame({"M": M_lst, "n": n, "theta": theta,
11                              "mu_Eul": 0.0, "mu_Mil": 0.0,
12                              "mu_Eul_cv": 0.0, "mu_Mil_cv": 0.0,
```

```
13                                "sd_Eul": 0.0, "sd_Mil": 0.0,
14                                "sd_Eul_cv": 0.0, "sd_Mil_cv": 0.0,})
15 np.random.seed(4002)
16 benchmark = Sim_Asian(1e6, 1e4, S_0, K, r, sigma, T, theta)
17 Asian_Eul, Asian_Mil = np.empty(epochs), np.empty(epochs)
18 Asian_Eul_cv, Asian_Mil_cv = np.empty(epochs), np.empty(epochs)
19 mu_y = np.mean(benchmark["Eul"]["price"])
20 for i in range(len(M_lst)):
21     M = M_lst[i]
22     np.random.seed(4002)
23     for j in trange(epochs, desc=f"M = {M}"):
24         Asian_theta = Sim_Asian(n, M, S_0, K, r, sigma, T, theta)
25 
26         Asian_Eul[j] = np.mean(Asian_theta["Eul"]["payoff"])
27         Asian_Mil[j] = np.mean(Asian_theta["Mil"]["payoff"])
28         Asian_Eul_cv[j] = control_variate(Asian_theta["Eul"]["payoff"],
29                                           Asian_theta["Eul"]["price"],
30                                           mu_y)
31         Asian_Mil_cv[j] = control_variate(Asian_theta["Mil"]["payoff"],
32                                           Asian_theta["Mil"]["price"],
33                                           mu_y)
34 
35     CEV_results.iloc[i, -8:] = [np.mean(Asian_Eul), np.mean(Asian_Mil),
36                                 np.mean(Asian_Eul_cv),
37                                 np.mean(Asian_Mil_cv),
38                                 np.std(Asian_Eul, ddof=1),
39                                 np.std(Asian_Mil, ddof=1),
40                                 np.std(Asian_Eul_cv, ddof=1),
41                                 np.std(Asian_Mil_cv, ddof=1)]
42 
43 print(CEV_results)
44 print(np.mean(benchmark["Eul"]["payoff"]),
45       np.mean(benchmark["Mil"]["payoff"]))
```

```
 1 M = 10.0: 100% 100/100 [00:8<00:00,  8.05it/s]
 2 M = 100.0: 100% 100/100 [00:30<00:00,  3.28it/s]
 3 M = 1000.0: 100% 100/100 [04:56<00:00,  2.97s/it]
 4         M         n  theta    mu_Eul    mu_Mil  mu_Eul_cv  mu_Mil_cv  \
 5 0    10.0  100000.0      1  2.144034  2.144033   2.143991   2.143988
 6 1   100.0  100000.0      1  2.144421  2.144421   2.143762   2.143762
 7 2  1000.0  100000.0      1  2.144137  2.144137   2.143750   2.143750
 8 
 9      sd_Eul    sd_Mil  sd_Eul_cv  sd_Mil_cv
10 0  0.001796  0.001795   0.000804   0.000804
11 1  0.001870  0.001870   0.000696   0.000696
12 2  0.001789  0.001789   0.000875   0.000875
13 2.1438965096390334 2.143896201773467
```

Programme 5.23 Estimating the Asian call price in Example 5.8 under a CEV model with $\theta = 1$ via Python.

```
1  M = 10.0: 100% 100/100 [00:10<00:00,  9.67it/s]
2  M = 100.0: 100% 100/100 [01:42<00:00,  1.02s/it]
3  M = 1000.0: 100% 100/100 [17:04<00:00, 10.24s/it]
4          M         n  theta    mu_Eul    mu_Mil  mu_Eul_cv  mu_Mil_cv  \
5  0    10.0  100000.0    1.8  2.159759  2.157439   2.158437   2.156093
6  1   100.0  100000.0    1.8  2.160748  2.160493   2.158690   2.158438
7  2  1000.0  100000.0    1.8  2.160075  2.160047   2.158784   2.158759
8
9       sd_Eul    sd_Mil  sd_Eul_cv  sd_Mil_cv
10 0  0.004514  0.004527   0.002058   0.002055
11 1  0.004566  0.004567   0.001744   0.001745
12 2  0.004324  0.004323   0.002181   0.002182
13 2.1591996467773225 2.159192913776342
```

Programme 5.24 Estimating the Asian call price in Example 5.8 under a CEV model with $\theta = 1.8$ via Python.

```
1  > control_variate <- function(x, y, mu_y){
2  +   mean(x) - cov(x, y)/var(y) * (mean(y) - mu_y)
3  + }
4  >
5  > theta <- 1; epochs <- 100
6  > CEV_results <- data.frame(M=M_lst, n=n, theta=theta,
7  +                           mu_Eul=0,mu_Mil=0,mu_Eul_cv=0,mu_mil_cv=0,
8  +                           sd_Eul=0,sd_Mil=0,sd_Eul_cv=0,sd_mil_cv=0)
9  > set.seed(4002)
10 > benchmark <- Sim_Asian(1e6, 1e4, S_0, K, r, sigma, T, theta)
11 > Asian_Eul <- array(NA, epochs); Asian_Mil <- array(NA, epochs)
12 > Asian_Eul_cv <- array(NA, epochs); Asian_Mil_cv <- array(NA, epochs)
13 >
14 > mu_y <- mean(benchmark$Eul$price)
15 > for (i in 1:length(M_lst)){
16 +   M <- M_lst[i]
17 +   set.seed(4002)
18 +   prog_bar <- txtProgressBar(min=0, max=epochs, width=50, style=3)
19 +   for (j in 1:epochs){
20 +     Asian_theta <- Sim_Asian(n, M, S_0, K, r, sigma, T, theta)
21 +
22 +     Asian_Eul[j] <- mean(Asian_theta$Eul$payoff)
23 +     Asian_Mil[j] <- mean(Asian_theta$Mil$payoff)
24 +     Asian_Eul_cv[j] <- control_variate(Asian_theta$Eul$payoff,
25 +                                        Asian_theta$Eul$price, mu_y)
26 +     Asian_Mil_cv[j] <- control_variate(Asian_theta$Mil$payoff,
27 +                                        Asian_theta$Mil$price, mu_y)
28 +     setTxtProgressBar(prog_bar, j); cat(paste0(" M = ", M))
29 +   }
30 +   CEV_results[i,4:11] <- c(mean(Asian_Eul), mean(Asian_Mil),
31 +                            mean(Asian_Eul_cv), mean(Asian_Mil_cv),
32 +                            sd(Asian_Eul), sd(Asian_Mil),
33 +                            sd(Asian_Eul_cv), sd(Asian_Mil_cv))
34 + }
35   |==================================================| 100%>  M = 1000
36 > CEV_results
```

```
37        M      n theta    mu_Eul    mu_Mil mu_Eul_cv mu_mil_cv        sd_Eul
38 1     10 1e+05     1 2.143928 2.143925  2.144634  2.144633 0.001779274
39 2    100 1e+05     1 2.144030 2.144029  2.144262  2.144262 0.001861063
40 3   1000 1e+05     1 2.144338 2.144338  2.144409  2.144409 0.001771763
41          sd_Mil     sd_Eul_cv     sd_mil_cv
42 1 0.001777727 0.0007630969 0.0007624736
43 2 0.001861233 0.0008836217 0.0008834729
44 3 0.001771597 0.0009108815 0.0009107896
45 > mean(benchmark$Eul$payoff); mean(benchmark$Mil$payoff)
46 [1] 2.144733
47 [1] 2.144733
```

Programme 5.25 Estimating the Asian call price in Example 5.8 under a CEV model with $\theta = 1$ via R.

```
1  > theta <- 1.8; epochs <- 100
2                                   ⋮
3  > CEV_results
4        M      n theta    mu_Eul    mu_Mil mu_Eul_cv mu_mil_cv        sd_Eul
5  1    10 1e+05   1.8 2.159599 2.157248  2.160410  2.158076 0.004412953
6  2   100 1e+05   1.8 2.159802 2.159548  2.160314  2.160065 0.004652656
7  3  1000 1e+05   1.8 2.160523 2.160498  2.160673  2.160649 0.004461119
8            sd_Mil    sd_Eul_cv    sd_mil_cv
9  1 0.004411812 0.001941221 0.001937783
10 2 0.004656519 0.002186274 0.002187519
11 3 0.004458810 0.002299323 0.002298143
12 > mean(benchmark$Eul$payoff); mean(benchmark$Mil$payoff)
13 [1] 2.16169
14 [1] 2.161687
```

Programme 5.26 Estimating the Asian call price in Example 5.8 under a CEV model with $\theta = 1.8$ via R.

For $\theta = 1$, from the results in Programme 5.23 via Python and Programme 5.25 via **R**, as the number of discretization steps M increases, the corresponding approximated Asian call prices for both schemes get closer to the benchmark. Moreover, with the aid of $\hat{S}_M$ as the control variate, the resulting sample average of the corresponding estimator gets even closer to the benchmark, and the standard deviation of the estimator is reduced by half. To obtain a better estimation, we can replace $\hat{S}_M$ by the Asian call payoff of a continuous geometric Wiener process (corresponding to $\theta = 2$). However, it is beyond the scope of this book and we shall leave it as an exercise for readers. Similar results are obtained in Programme 5.24 via Python and Programme 5.26 via **R** for $\theta = 1.8$. Tables 5.4 and 5.5 summarize the estimated option prices and the standard deviations of the corresponding estimator under the CEV model with $\theta = 1$ and 1.8, respectively, obtained from Python and **R**, and the estimated values attained converge to a value around 2.14 for $\theta = 1$ and 2.16 for $\theta = 1.8$, respectively.

TABLE 5.4 The estimated price (top) and standard deviation (bottom) of 100 simulations of the Asian call option under Euler–Maruyama and Milstein schemes with and without S_T as the control variate (CV) in Example 5.8 under a CEV model with $\theta = 1$ in Python and **R**.

$\theta = 1$		Python			R	
M	10	100	1,000	10	100	1,000
Euler	2.144034 0.001796	2.144421 0.001870	2.144137 0.001789	2.143928 0.001779	2.144030 0.001861	2.144338 0.001772
Milstein	2.144033 0.001795	2.144421 0.001870	2.144137 0.001789	2.143925 0.001778	2.144029 0.001861	2.144338 0.001772
Euler (CV)	2.143991 0.000804	2.143762 0.000696	2.143750 0.000875	2.144634 0.000763	2.144262 0.000884	2.144409 0.000911
Milstein (CV)	2.143988 0.000804	2.143762 0.000696	2.143750 0.000875	2.144633 0.000762	2.144262 0.000883	2.144409 0.000911

TABLE 5.5 The estimated price (top) and standard deviation (bottom) of 100 simulations of the Asian call option under Euler–Maruyama and Milstein schemes with and without S_T as the control variate (CV) in Example 5.8 under a CEV model with $\theta = 1.8$ in Python and **R**.

$\theta = 1.8$		Python			R	
M	10	100	1,000	10	100	1,000
Euler	2.159759 0.004514	2.160748 0.004566	2.160075 0.004324	2.159599 0.004413	2.159802 0.004653	2.160523 0.004461
Milstein	2.157439 0.004527	2.160493 0.004567	2.160047 0.004323	2.157248 0.004412	2.159548 0.004657	2.160498 0.004459
Euler (CV)	2.158437 0.002058	2.158690 0.001744	2.158784 0.002181	2.160410 0.001941	2.160314 0.002186	2.160673 0.002299
Milstein (CV)	2.156093 0.002055	2.158438 0.001745	2.158759 0.002182	2.158076 0.001938	2.160065 0.002188	2.160649 0.002298

*5.4 GREEKS AND THEIR APPROXIMATIONS

Various sensitivities of derivatives (or portfolios) prices V's with respect to different model parameters, such as time t, risk-free interest rate r, current asset price s and volatility σ, are called *Greek letters* or simply *Greeks* for short: see Table 5.6 for some common examples.

For example, a fund manager may want to trade a portfolio that is insensitive to the (small) change of the current price of the underlying asset, that is, a portfolio with a value V such that $\Delta = \frac{\partial V}{\partial S} = 0$. A portfolio with such a property is called a *delta neutral portfolio*. For instance, a *delta-neutral portfolio* can be formed by selling one European option and holding $\Delta = \frac{\partial V(S_t,t)}{\partial S}$ amount of stocks, and this hedging portfolio will remain *riskless* throughout the period under consideration.

TABLE 5.8 Commonly used Greeks. Note that vega is not the name of any Greek letter; it gets the name due to its resemblance to the Latin letter V.

Delta	Gamma	Theta	Rho	Vega
$\Delta = \frac{\partial V}{\partial S}$	$\Gamma = \frac{\partial^2 V}{\partial S^2}$	$\Theta = \frac{\partial V}{\partial t}$	$\rho = \frac{\partial V}{\partial r}$	$\mathcal{V} = \frac{\partial V}{\partial \sigma}$

Furthermore, the *delta* of a European call option is expected to be positive as the price of a call option increases with the current price of the underlying asset. When $S_t \ll K$, i.e. when the call option is far out-of-the-money, call options are mostly worthless, and they have deltas close to 0. On the other hand, managers can construct a portfolio with a positive delta so as to bet on the greater increase in the price of the underlying asset. More specifically, to calculate the Greeks of a European call option under the Black–Scholes framework, we first recall their celebrated formula (also see Example 5.6): at time t, its price is

$$C_t = S_t\Phi(d_+) - e^{-r(T-t)}K\Phi(d_-),$$

where $\Phi(\cdot)$ is the cdf of the standard normal distribution, $d_+ := \frac{\ln \frac{S_t}{K} + \left(r + \frac{\sigma^2}{2}\right)(T-t)}{\sigma\sqrt{T-t}}$ and $d_- := d_+ - \sigma\sqrt{T-t}$. We can compute various Greeks of a European call option as follows:

1. *Delta*:

$$\begin{aligned}\frac{\partial C_t}{\partial S_t} &= \Phi(d_+) + S_t\frac{\partial\Phi(d_+)}{\partial d_+}\cdot\frac{\partial d_+}{\partial S_t} - e^{-r(T-t)}K\frac{\partial\Phi(d_-)}{\partial d_-}\cdot\frac{\partial d_-}{\partial S_t}\\ &= \Phi(d_+) + \frac{1}{S_t\sigma\sqrt{T-t}}[S_t\phi(d_+) - Ke^{-r(T-t)}\phi(d_-)]\\ &= \Phi(d_+) > 0, \end{aligned} \tag{5.14}$$

where $\phi(\cdot)$ is the corresponding pdf of the standard normal distribution.

2. *Gamma*:

$$\frac{\partial^2 C_t}{\partial S_t^2} = \frac{\partial}{\partial S_t}\left(\frac{\partial C_t}{\partial S_t}\right) = \frac{\partial\Phi(d_+)}{\partial d_+}\cdot\frac{\partial d_+}{\partial S_t} = \frac{\phi(d_+)}{S_t\sigma\sqrt{T-t}} > 0. \tag{5.15}$$

3. *Theta*:

$$\begin{aligned}\frac{\partial C_t}{\partial t} &= S_t\frac{\partial\Phi(d_+)}{\partial d_+}\cdot\frac{\partial d_+}{\partial t} - re^{-r(T-t)}K\Phi(d_-) - e^{-r(T-t)}K\frac{\partial\Phi(d_-)}{\partial d_-}\cdot\frac{\partial d_-}{\partial t}\\ &= -rKe^{-r(T-t)}\Phi(d_-) - \frac{\sigma S_t\phi(d_+)}{2\sqrt{T-t}} < 0. \end{aligned} \tag{5.16}$$

4. *Rho*:

$$\frac{\partial C_t}{\partial r} = S_t \frac{\partial \Phi(d_+)}{\partial d_+} \cdot \frac{\partial d_+}{\partial r} - Ke^{-r(T-t)} \frac{\partial \Phi(d_-)}{\partial d_-} \cdot \frac{\partial d_-}{\partial r} + (T-t)Ke^{-r(T-t)}\Phi(d_-)$$
$$= (T-t)Ke^{-r(T-t)}\Phi(d_-) > 0. \tag{5.17}$$

5. *Vega*:

$$\frac{\partial C_t}{\partial \sigma} = S_t \frac{\partial \Phi(d_+)}{\partial d_+} \cdot \frac{\partial d_+}{\partial \sigma} - Ke^{-r(T-t)} \frac{\partial \Phi(d_-)}{\partial d_-} \cdot \frac{\partial d_-}{\partial \sigma} = S_t\phi(d_+)\sqrt{T-t} > 0. \tag{5.18}$$

In general, there are no immediate tractable expressions for Greeks of most generic financial derivatives. In such cases, using a numerical approximation via the Monte Carlo method serves as a resolution. For simplicity, take the current time as 0, and let $y(\theta)$ be the discounted payoff of a European-type derivative which will expire at terminal time T, and let $\alpha(\theta) = \mathbb{E}(y(\theta))$ be the current price of the derivative, where the expectation $\mathbb{E}$ is taken with respect to the risk-neutral probability measure, and θ represents a particular parameter of interest, such as the current value of the underlying asset or its volatility. We aim to calibrate the sensitivity of the derivative price with respect to θ, by estimating the derivative $\alpha'(\theta) = \frac{\partial \alpha}{\partial \theta}$ via different simulation methods. We shall introduce three common approaches, namely (i) finite-difference approximation; (ii) pathwise differentiation; and (ii) the likelihood ratio method. We refer to [14, 20] for more discussions.

(I) Finite-difference approximation

By differentiability, we clearly have $\alpha'(\theta) \approx \frac{\alpha(\theta+h)-\alpha(\theta)}{h}$ for small enough $h > 0$, or equivalently

$$\alpha'(\theta) \approx \frac{\mathbb{E}(y(\theta+h)) - \mathbb{E}(y(\theta))}{h},$$

then we can approximate $\mathbb{E}(y(\theta))$ and $\mathbb{E}(y(\theta+h))$ using the Monte Carlo method by first generating n iid samples of each of $y(\theta)$ and $y(\theta+h)$, and then take the sample averages to obtain $\overline{y}_n(\theta)$ and $\overline{y}_n(\theta+h)$, respectively. We also define the *forward difference estimator* as

$$\hat{\Delta} := \frac{\overline{y}_n(\theta+h) - \overline{y}_n(\theta)}{h},$$

then we write

$$\begin{aligned}
\mathbb{E}(\hat{\Delta}) &= \frac{\alpha(\theta+h) - \alpha(\theta)}{h} \\
&= \frac{\alpha(\theta) + h\alpha'(\theta) + \frac{1}{2}h^2\alpha''(\theta) + o(h^2) - \alpha(\theta)}{h} \\
&= \alpha'(\theta) + \frac{h}{2}\alpha''(\theta) + o(h),
\end{aligned}$$

from which we can see that the bias of the forward difference estimator is equal to $\frac{h}{2}\alpha''(\theta) + o(h)$, which depends on the curvature of the derivative price at the parametric value θ. On the other hand, we can achieve an alternative estimator with a smaller bias by using the *central difference estimator*:

$$\hat{\Delta} := \frac{\overline{y}_n(\theta + h) - \overline{y}_n(\theta - h)}{2h},$$

whose theoretical mean is given by

$$\begin{aligned}\mathbb{E}(\hat{\Delta}) &= \frac{\alpha(\theta + h) - \alpha(\theta - h)}{2h} \\ &= \frac{\alpha(\theta) + h\alpha'(\theta) + \frac{1}{2}h^2\alpha''(\theta) + o(h^2) - (\alpha(\theta) - h\alpha'(\theta) + \frac{1}{2}h^2\alpha''(\theta) + o(h^2))}{2h} \\ &= \alpha'(\theta) + o(h),\end{aligned}$$

and we note its bias is of order $o(h)$ only, which is smaller than that from using the forward difference estimator.

Example 5.9 ***(Under a CEV model, delta and vega with finite differences)*** Under the CEV framework with $\theta = 2, 1.8$, and 1, consider a European call option with $S_0 = 10, K = 8, \sigma = 0.3, r = 0.05$, and $T = 1$. Its delta and vega are computed using forward and central differencing in Programme 5.27 via Python (resp. Programme 5.28 via **R**) by using (5.14) and (5.18), respectively, and the numerical scheme used is the Milstein one.

```
1  import numpy as np
2  import pandas as pd
3
4  # Initialize variables
5  S_0 = 10; K = 8; r = 0.05; sigma = 0.3; T = 1
6
7  def Sim_greek(h_d, h_v, n, M, S_0, K, r, sigma, T, theta, seed=4002):
8      delta_t = T / M
9      def Milstein(z, S_t, r, sigma):
10         return (S_t*(1+r*delta_t) +
11                 sigma*S_t**(theta/2)*np.sqrt(delta_t)*z +
12                 sigma**2*theta/2*S_t**(theta-1)*delta_t/2*(z**2-1))
13
14     def Euro_call(S_t, K, r, sigma, T):
15         np.random.seed(seed)
16         for m in range(M):
17             z = np.random.randn(n)
18             S_t = Milstein(z, S_t, r, sigma)
19         return np.mean(np.exp(-r*T) * np.maximum(S_t - K, 0))
20
21     Y = Euro_call(S_0, K, r, sigma, T)
22
```

```
23      # Estimate delta using forward and central difference method
24      Y_S0_neg = Euro_call(S_0-h_d, K, r, sigma, T)
25      Y_S0_pos = Euro_call(S_0+h_d, K, r, sigma, T)
26
27      # Estimate vega using forward and central difference method
28      Y_sig_neg = Euro_call(S_0, K, r, sigma-h_v, T)
29      Y_sig_pos = Euro_call(S_0, K, r, sigma+h_v, T)
30
31      return {'delta': {'forward': (Y_S0_pos - Y)/h_d,
32                        'central': (Y_S0_pos - Y_S0_neg)/(2*h_d)},
33              'vega': {'forward': (Y_sig_pos - Y)/h_v,
34                       'central': (Y_sig_pos - Y_sig_neg)/(2*h_v)}}
```

Programme 5.27 Forward and central differencing for estimating delta and vega of the European call option in Example 5.9 via Python.

```
1  > # Initialize variables
2  > S_0 <- 10; K <- 8; r <- 0.05; sigma <- 0.3; T <- 1
3  >
4  > Sim_greek <- function(h_d, h_v, n, M, S_0, K, r, sigma, T,
5  +                       theta, seed=4002){
6  +   delta_t <- T / M
7  +   Milstein <- function(z, S_t, r, sigma){
8  +     (S_t*(1 + r*delta_t) + sigma*S_t^(theta/2)*sqrt(delta_t)*z +
9  +        sigma^2*theta/2*S_t^(theta-1)*delta_t/2*(z^2-1))
10 +   }
11 +
12 +   Euro_call <- function(S_t, K, r, sigma, T){
13 +     set.seed(seed)
14 +     for (m in 1:M){
15 +       z <- rnorm(n)
16 +       S_t <- Milstein(z, S_t, r, sigma)
17 +     }
18 +     mean(exp(-r*T) * pmax(S_t - K, 0))
19 +   }
20 +
21 +   Y <- Euro_call(S_0, K, r, sigma, T)
22 +
23 +   # Estimate delta using forward and central difference method
24 +   Y_S0_neg <- Euro_call(S_0-h_d, K, r, sigma, T)
25 +   Y_S0_pos <- Euro_call(S_0+h_d, K, r, sigma, T)
26 +
27 +   # Estimate vega using forward and central difference method
28 +   Y_sig_neg <- Euro_call(S_0, K, r, sigma-h_v, T)
29 +   Y_sig_pos <- Euro_call(S_0, K, r, sigma+h_v, T)
30 +
31 +   list(delta=list(forward=(Y_S0_pos - Y)/h_d,
32 +                   central=(Y_S0_pos - Y_S0_neg)/(2*h_d)),
33 +        vega=list(forward=(Y_sig_pos - Y)/h_v,
34 +                  central=(Y_sig_pos - Y_sig_neg)/(2*h_v)))
35 + }
```

Programme 5.28 Forward and central differencing for estimating delta and vega of the European call option in Example 5.9 via R.

Next we estimate the delta and vega of the European call, under CEV model with $\theta = 2, 1.8$, and 1, by using the methods of forward and central differences by choosing $h = 0.5, 0.45, \ldots, 0.1, 0.05$ for delta, and $0.05, 0.045, \ldots, 0.01, 0.005$ for vega. We implement the respective executions of Programmes 5.27 and 5.28 in Programme 5.29 via Python and Programme 5.30 via **R**.

```
1  from scipy.stats import norm
2  from tqdm import trange
3  
4  # Compute the exact delta and vega for theta=2 (BS)
5  d_plus = (np.log(S_0/K) + (r+sigma**2/2)*T) / (sigma*np.sqrt(T))
6  delta = norm.cdf(d_plus)
7  vega = S_0 * np.sqrt(T) * norm.pdf(d_plus)
8  print("When theta=2, the exact delta and vega are:")
9  print(delta)
10 print(vega)
11 
12 h_delta = np.arange(0.5, 0.05 - 0.001, -0.05)
13 h_vega = np.arange(0.05, 0.005 - 0.0001, -0.005)
14 n, M = int(1e6), int(1e4)
15 theta_lst = [2, 1.8, 1]
16 
17 for theta in theta_lst:
18     delta_finite = pd.DataFrame({'h': h_delta, 'theta': theta,
19                                  'delta_forward': 0.0,
20                                  'delta_central': 0.0})
21     vega_finite = pd.DataFrame({'h': h_vega, 'theta': theta,
22                                  'vega_forward':0.0,'vega_central':0.0})
23     for i in trange(len(h_delta)):
24         h_d, h_v = h_delta[i], h_vega[i]
25         Euro_CEV=Sim_greek(h_d, h_v, n, M, S_0, K, r, sigma, T, theta)
26         delta_finite.iloc[i, -2:] = Euro_CEV['delta'].values()
27         vega_finite.iloc[i, -2:] = Euro_CEV['vega'].values()
28 
29     print(delta_finite)
30     print(vega_finite)
```

```
1  When theta=2, the exact delta and vega are:
2  0.8555365179482681
3  2.273542597558268
4  100% 10/10 [9:13:51<00:00, 3323.10s/it]
5        h  theta  delta_forward  delta_central
6  0  0.50      2       0.872601       0.853643
7  1  0.45      2       0.870961       0.853918
8  2  0.40      2       0.869294       0.854162
9  3  0.35      2       0.867600       0.854376
10 4  0.30      2       0.865877       0.854565
11 5  0.25      2       0.864125       0.854724
12 6  0.20      2       0.862343       0.854844
13 7  0.15      2       0.860537       0.854931
14 8  0.10      2       0.858707       0.854985
15 9  0.05      2       0.856878       0.855027
16        h  theta  vega_forward  vega_central
17 0  0.050      2      2.383822      2.230370
18 1  0.045      2      2.371816      2.234120
```

```
19 2  0.040     2      2.359503      2.237442
20 3  0.035     2      2.346864      2.240385
21 4  0.030     2      2.333844      2.242906
22 5  0.025     2      2.320525      2.245068
23 6  0.020     2      2.307028      2.246933
24 7  0.015     2      2.293288      2.248442
25 8  0.010     2      2.279327      2.249651
26 9  0.005     2      2.264988      2.250272
27 100% 10/10 [15:31:28<00:00, 5588.85s/it]
28        h  theta  delta_forward  delta_central
29 0  0.50    1.8       0.910004       0.890567
30 1  0.45    1.8       0.908407       0.890940
31 2  0.40    1.8       0.906783       0.891274
32 3  0.35    1.8       0.905124       0.891568
33 4  0.30    1.8       0.903429       0.891823
34 5  0.25    1.8       0.901707       0.892043
35 6  0.20    1.8       0.899962       0.892233
36 7  0.15    1.8       0.898179       0.892381
37 8  0.10    1.8       0.896364       0.892492
38 9  0.05    1.8       0.894495       0.892562
39         h  theta  vega_forward  vega_central
40 0  0.050    1.8      1.583429      1.428944
41 1  0.045    1.8      1.570701      1.431849
42 2  0.040    1.8      1.557725      1.434472
43 3  0.035    1.8      1.544481      1.436751
44 4  0.030    1.8      1.530969      1.438738
45 5  0.025    1.8      1.517174      1.440400
46 6  0.020    1.8      1.503018      1.441631
47 7  0.015    1.8      1.488689      1.442606
48 8  0.010    1.8      1.474007      1.443228
49 9  0.005    1.8      1.458945      1.443577
50 100% 10/10 [8:33:48<00:00, 3082.85s/it]
51        h  theta  delta_forward  delta_central
52 0  0.50      1       0.998536       0.995710
53 1  0.45      1       0.998444       0.995966
54 2  0.40      1       0.998341       0.996189
55 3  0.35      1       0.998228       0.996380
56 4  0.30      1       0.998103       0.996548
57 5  0.25      1       0.997963       0.996687
58 6  0.20      1       0.997808       0.996800
59 7  0.15      1       0.997639       0.996890
60 8  0.10      1       0.997446       0.996950
61 9  0.05      1       0.997233       0.996983
62         h  theta  vega_forward  vega_central
63 0  0.050      1      0.045896      0.028328
64 1  0.045      1      0.043400      0.027551
65 2  0.040      1      0.040940      0.026830
66 3  0.035      1      0.038555      0.026187
67 4  0.030      1      0.036247      0.025623
68 5  0.025      1      0.033997      0.025142
69 6  0.020      1      0.031829      0.024736
70 7  0.015      1      0.029768      0.024423
71 8  0.010      1      0.027795      0.024226
72 9  0.005      1      0.025929      0.024116
```

Programme 5.29 Delta and vega under a CEV model with $\theta = 2, 1.8, 1$ in Example 5.9 estimated by Programme 5.27 via Python.

```
1  > # Compute the exact delta and vega for theta=2 (BS)
2  > d_plus <- (log(S_0/K) + (r+sigma^2/2)*T) / (sigma*sqrt(T))
3  > (exact_delta <- pnorm(d_plus))
4  [1] 0.8555365
5  > (exact_vega <- S_0 * sqrt(T) * dnorm(d_plus))
6  [1] 2.273543
7  >
8  > h_delta <- seq(0.5, 0.05, by=-0.05)
9  > h_vega <- seq(0.05, 0.005, by=-0.005)
10 > n <- 1e6; M <- 1e4
11 > theta_lst <- c(2, 1.8, 1)
12 >
13 > for (theta in theta_lst){
14 +   delta_finite <- data.frame(h=h_delta, theta=theta,
15 +                              delta_forward=0, delta_central=0)
16 +   vega_finite <- data.frame(h=h_vega, theta=theta,
17 +                             vega_forward=0, vega_central=0)
18 +   prog_bar <- txtProgressBar(min=0, max=length(h_delta), width=50,
19 +                              style=3)
20 +   for (i in 1:length(h_delta)) {
21 +     h_d <- h_delta[i]; h_v <- h_vega[i]
22 +     Euro_CEV <- Sim_greek(h_d, h_v, n, M, S_0, K, r, sigma, T,
23 +                           theta=theta)
24 +     delta_finite$delta_forward[i] <- Euro_CEV$delta$forward
25 +     delta_finite$delta_central[i] <- Euro_CEV$delta$central
26 +     vega_finite$vega_forward[i] <- Euro_CEV$vega$forward
27 +     vega_finite$vega_central[i] <- Euro_CEV$vega$central
28 +     setTxtProgressBar(prog_bar, i)
29 +   }
30 +   cat("\n")
31 +   print(delta_finite)
32 +   print(vega_finite)
33 + }
34   |==================================================| 100%
35       h theta delta_forward delta_central
36 1  0.50     2     0.8733182     0.8542754
37 2  0.45     2     0.8716856     0.8545515
38 3  0.40     2     0.8700280     0.8547959
39 4  0.35     2     0.8683560     0.8550182
40 5  0.30     2     0.8666531     0.8552091
41 6  0.25     2     0.8649138     0.8553712
42 7  0.20     2     0.8631497     0.8555153
43 8  0.15     2     0.8613602     0.8556317
44 9  0.10     2     0.8595393     0.8557208
45 10 0.05     2     0.8576865     0.8557784
46        h theta vega_forward vega_central
47 1  0.050     2     2.411100     2.255973
48 2  0.045     2     2.399072     2.259671
49 3  0.040     2     2.386731     2.262980
50 4  0.035     2     2.374104     2.265920
51 5  0.030     2     2.361143     2.268382
52 6  0.025     2     2.347756     2.270414
```

```
53 | 7  0.020     2     2.333852     2.272035
54 | 8  0.015     2     2.319609     2.273226
55 | 9  0.010     2     2.304932     2.274032
56 | 10 0.005     2     2.289894     2.274441
57 |   |==========================================| 100%
58 |       h theta delta_forward delta_central
59 | 1  0.50    1.8      0.9103602      0.8909566
60 | 2  0.45    1.8      0.9087710      0.8913262
61 | 3  0.40    1.8      0.9071471      0.8916541
62 | 4  0.35    1.8      0.9054860      0.8919460
63 | 5  0.30    1.8      0.9037901      0.8922029
64 | 6  0.25    1.8      0.9020538      0.8924104
65 | 7  0.20    1.8      0.9002801      0.8925793
66 | 8  0.15    1.8      0.8984693      0.8927126
67 | 9  0.10    1.8      0.8966335      0.8928160
68 | 10 0.05    1.8      0.8947876      0.8928859
69 |        h theta vega_forward vega_central
70 | 1  0.050    1.8      1.601889      1.447216
71 | 2  0.045    1.8      1.589190      1.450110
72 | 3  0.040    1.8      1.576159      1.452677
73 | 4  0.035    1.8      1.562818      1.454928
74 | 5  0.030    1.8      1.549324      1.456956
75 | 6  0.025    1.8      1.535552      1.458695
76 | 7  0.020    1.8      1.521498      1.460143
77 | 8  0.015    1.8      1.507047      1.461244
78 | 9  0.010    1.8      1.492318      1.461971
79 | 10 0.005    1.8      1.477429      1.462322
80 |   |==========================================| 100%
81 |       h theta delta_forward delta_central
82 | 1  0.50      1      0.9985591      0.9956954
83 | 2  0.45      1      0.9984610      0.9959559
84 | 3  0.40      1      0.9983533      0.9961849
85 | 4  0.35      1      0.9982344      0.9963828
86 | 5  0.30      1      0.9981044      0.9965517
87 | 6  0.25      1      0.9979649      0.9966945
88 | 7  0.20      1      0.9978105      0.9968086
89 | 8  0.15      1      0.9976420      0.9968967
90 | 9  0.10      1      0.9974624      0.9969636
91 | 10 0.05      1      0.9972552      0.9970053
92 |        h theta vega_forward vega_central
93 | 1  0.050      1    0.05128461    0.03352689
94 | 2  0.045      1    0.04870554    0.03274183
95 | 3  0.040      1    0.04620229    0.03203062
96 | 4  0.035      1    0.04376974    0.03139444
97 | 5  0.030      1    0.04143237    0.03084050
98 | 6  0.025      1    0.03919174    0.03035746
99 | 7  0.020      1    0.03703987    0.02996309
100| 8  0.015      1    0.03497519    0.02964453
101| 9  0.010      1    0.03295957    0.02936410
102| 10 0.005      1    0.03103239    0.02921822
```

Programme 5.30 Delta and vega under a CEV model with $\theta = 2, 1.8, 1$ in Example 5.9 estimated by Programme 5.28 via R.

Programmes 5.29 via Python and 5.30 via **R** report the delta and vega estimates of the European call option under a CEV model with different parameter θ using the forward and central difference methods. When $\theta = 2$, equivalent to the classical Black–Scholes model, the delta and vega estimates of both methods converge to their theoretical values of 0.8555 and 2.2735 respectively when h gets smaller. One observes that the central difference method provides more accurate estimates compared to the forward difference method; the estimation performance of vega is inferior to that of delta no matter whether Python or **R** is used. Note that $\sigma(t) = \sigma S_t$ for $\theta = 2$ and $\sigma(t) = \sigma S_t^{0.9}$ for $\theta = 1.8$, the corresponding dynamics of S_t for $\theta < 2$ is less volatile against its current price level in comparison to that for $\theta = 2$, leading σ to be relatively less important in the option price, and so we anticipate a relatively smaller value of vega when θ gets smaller. In contrast, the initial stock price S_0, in the absence of larger volatility, becomes more critical to the option price, and so we expect delta to have a greater value as θ gets smaller. In particular, for $\theta = 1.8$, the delta and vega estimates converge to 0.893 and 1.4, respectively; for $\theta = 1$, the delta and vega estimates converge to 0.997 and 0.024, respectively.

(II) Pathwise Differentiation

The interchangeability assumption states that $\alpha'(\theta) = \frac{d}{d\theta}\mathbb{E}(y(\theta)) = \mathbb{E}\left(\frac{dy(\theta)}{d\theta}\right)$. This can be valid when the following condition holds:

$$|y(\theta_1) - y(\theta_2)| < C|\theta_1 - \theta_2|, \text{ for all } \theta_1 \text{ and } \theta_2, \text{ almost surely, for some } C > 0.$$

Indeed, this condition warrants the interchange of differentiation and the expectations taking in accordance with the *Lebesgue Dominated Convergence Theorem*.

Suppose that the payoff of a derivative is given by $y = f(S_T)$. To illustrate how the *pathwise differentiation approach* works, we consider the delta of this derivative under the Black–Scholes framework, which is the price sensitivity with respect to the current stock price S_0. Based on the interchangeability assumption, we consider the derivative of the random payoff with respect to the current price:

$$\frac{\partial y}{\partial S_0} = f'(S_T)\frac{\partial S_T}{\partial S_0} = f'(S_T)\exp\left[\left(r - \frac{\sigma^2}{2}\right)T + \sigma w_T\right] = f'(S_T)\frac{S_T}{S_0}.$$

Therefore, we can express the delta as

$$\Delta = \mathbb{E}\left(f'(S_T)\frac{S_T}{S_0}\right), \tag{5.19}$$

which allows us to the use the Monte Carlo method to estimate it. For example, recall the payoff of a European call option is $f(S_T) = e^{-rT}(S_T - K)_+$, hence $f'(S_T) = e^{-rT}\mathbb{1}_{\{S_T>K\}}$. Formula (5.19) then reduces to

$$\Delta = \mathbb{E}\left(f'(S_T)\frac{S_T}{S_0}\right) = \mathbb{E}\left[\mathbb{1}_{\{S_T>K\}}\exp\left(-\frac{\sigma^2}{2}T + \sigma w_T\right)\right].$$

As another example, we consider the vega of the European-type derivative:

$$\frac{\partial \mathrm{y}}{\partial \sigma} = f'(\mathrm{S}_T)\frac{\partial \mathrm{S}_T}{\partial \sigma} = f'(\mathrm{S}_T)\mathrm{S}_T(\mathrm{w}_T - \sigma T).$$

Therefore, we have

$$\mathcal{V} = \mathbb{E}(f'(\mathrm{S}_T)\mathrm{S}_T(\mathrm{w}_T - \sigma T));$$

and so the vega of a European call option can be written as:

$$\mathcal{V} = e^{-rT}\mathbb{E}(\mathbb{1}_{\{S_T>K\}}\mathrm{S}_T(\mathrm{w}_T - \sigma T)).$$

In Example 5.9, under the CEV model setting, we proposed the approximation of delta and vega of a European call through the finite differencing method. Let us now illustrate how to realize the pathwise differentiation method for this non-trivial model. We approximate the Greeks through the Monte Carlo simulation of the corresponding SDE system, which becomes two-dimensional in nature. To see why. for any parameter γ, differentiating (5.13) with respect to γ gives

$$d\frac{\partial \mathrm{S}_t}{\partial \gamma} = \left(\frac{\partial r}{\partial \gamma} + r\cdot\frac{\partial \mathrm{S}_t}{\partial \gamma}\right)dt + \left(\frac{\partial \sigma}{\partial \gamma}\mathrm{S}_t^{\theta/2} + \sigma\cdot\frac{\theta}{2}\mathrm{S}_t^{\theta/2-1}\frac{\partial \mathrm{S}_t}{\partial \gamma}\right)d\mathrm{w}_t;$$

or $$d\mathrm{y}_t = \left(\frac{\partial r}{\partial \gamma} + r\mathrm{y}_t\right)dt + \left(\frac{\partial \sigma}{\partial \gamma}\mathrm{S}_t^{\theta/2} + \sigma\cdot\frac{\theta}{2}\mathrm{S}_t^{\theta/2-1}\mathrm{y}_t\right)d\mathrm{w}_t, \text{for } 0 \le t \le T,$$

where we define $\mathrm{y}_t := \frac{\partial S_t}{\partial \gamma}$. Therefore, the Greeks of the European call option with respect to γ can be obtained through $\mathbb{E}(f'(\mathrm{S}_T)\mathrm{y}_T) = \mathbb{E}(e^{-rT}\mathbb{1}_{\{S_T>K\}}\mathrm{y}_T)$ by the Monte Carlo simulation of the following system of SDEs:

$$\begin{cases} d\mathrm{y}_t = \left(\frac{\partial r}{\partial \gamma} + r\mathrm{y}_t\right)dt + \left(\frac{\partial \sigma}{\partial \gamma}\mathrm{S}_t^{\theta/2} + \sigma\cdot\frac{\theta}{2}\mathrm{S}_t^{\theta/2-1}\mathrm{y}_t\right)d\mathrm{w}_t; \\ d\mathrm{S}_t = r\mathrm{S}_t dt + \sigma\mathrm{S}_t^{\theta/2}d\mathrm{w}_t, \end{cases}$$

with initial conditions $y_0 = \frac{\partial S_0}{\partial \gamma}$ and S_0. For instance, considering the following Greeks:

1. Delta with the parameter $\gamma = S_0$:

$$\begin{cases} d\mathrm{y}_t = r\mathrm{y}_t dt + \sigma\cdot\frac{\theta}{2}\mathrm{S}_t^{\theta/2-1}\mathrm{y}_t d\mathrm{w}_t; \\ d\mathrm{S}_t = r\mathrm{S}_t dt + \sigma\mathrm{S}_t^{\theta/2}d\mathrm{w}_t, \end{cases} \tag{5.20}$$

with initial conditions $y_0 = \frac{\partial S_0}{\partial S_0} = 1$ and S_0;

2. Vega with the parameter $\gamma = \sigma$:

$$\begin{cases} dy_t = ry_t dt + \left(S_t^{\theta/2} + \sigma \cdot \frac{\theta}{2} S_t^{\theta/2-1} y_t\right) dw_t; \\ dS_t = rS_t dt + \sigma S_t^{\theta/2} dw_t, \end{cases} \tag{5.21}$$

with initial conditions $y_0 = \frac{\partial S_0}{\partial \sigma} = 0$ and S_0.

Example 5.10 ***(Continuation of the CEV model in Example 5.9, now via pathwise differentiation)*** Under the CEV framework with $\theta = 2, 1.8$, and 1, consider a European call option with $S_0 = 10, K = 8, \sigma = 0.3, r = 0.05$, and $T = 1$. We now use the standard Euler–Maruyama scheme for the SDE systems obtained through the pathwise differentiations for numerical calibrations for delta and vega. The implementation is carried out in Programme 5.31 via Python (resp. Programme 5.32 via **R**) by using (5.20) and (5.21), respectively.

```
1  def greek_pathwise(n, M, S_0, K, r, sigma, T, theta, seed=4002):
2      delta_t = T / M
3      y_t = 1; v_t = 0; S_t = S_0
4      np.random.seed(seed)
5      for m in range(int(M)):
6          dw_t = np.sqrt(delta_t) * np.random.normal(size=int(n))
7          y_t += r*y_t*delta_t + sigma*theta/2*S_t**(theta/2-1)*y_t*dw_t
8          v_t += (r*v_t*delta_t +
9                  S_t**(theta/2-1)*(S_t + sigma*theta/2*v_t)*dw_t)
10         S_t += r*S_t*delta_t + sigma*S_t**(theta/2)*dw_t
11
12     return {"delta": np.mean(np.exp(-r*T)*(S_t > K) * y_t),
13             "vega": np.mean(np.exp(-r*T)*(S_t > K) * v_t)}
14
15 n_size = np.arange(500, 100000+1, 500); M = 1e4
16 theta = 2
17 delta_pathwise, vega_pathwise = [], []
18 # Estimate delta and vega with different n
19 for i in range(len(n_size)):
20     Euro_CEV = greek_pathwise(n_size[i], M, S_0, K, r, sigma, T, theta)
21     delta_pathwise.append(Euro_CEV["delta"])
22     vega_pathwise.append(Euro_CEV["vega"])
23
24 plt.rc('xtick', labelsize=15); plt.rc('ytick', labelsize=15)
25 plt.figure(figsize=(10, 7))
26 plt.plot(n_size, delta_pathwise, "ro-", markersize=8,
27          linewidth=2, label="pathwise differentiation")
28 plt.axhline(y=exact_delta, color="blue", linewidth=2, label="exact",
29             linestyle="dotted")
30 plt.xlabel("number of paths")
31 plt.ylabel("delta")
32 plt.legend(fontsize=17)
33
34 plt.figure(figsize=(10, 7))
```

```
35 plt.plot(n_size, vega_pathwise, "ro-", markersize=8, linewidth=2,
36          label="pathwise differentiation")
37 plt.axhline(y=exact_vega, color="blue", linewidth=2, label="exact",
38             linestyle="dotted")
39 plt.xlabel("number of paths")
40 plt.ylabel("vega")
41 plt.legend(fontsize=17)
```

Programme 5.31 Using pathwise differentiation method to approximate delta and vega of the European call under CEV model in Example 5.10 via Python.

```
1  > greek_pathwise <- function(n,M,S_0,K,r,sigma,T,theta,seed=4002){
2  +   delta_t <- T / M
3  +   y_t <- 1; v_t <- 0; S_t <- S_0
4  +   set.seed(seed)
5  +   for (m in 1:M){
6  +     dw_t <- sqrt(delta_t) * rnorm(n)
7  +     y_t <- (y_t + r*y_t*delta_t
8  +             + sigma*theta/2*S_t^(theta/2-1)*y_t*dw_t)
9  +     v_t <- (v_t + r*v_t*delta_t
10 +             + S_t^(theta/2-1)*(S_t + sigma*theta/2*v_t)*dw_t)
11 +     S_t <- S_t + r*S_t*delta_t + sigma*S_t^(theta/2)*dw_t
12 +   }
13 +
14 +   list(delta=mean(exp(-r*T)*(S_t > K) * y_t),
15 +        vega=mean(exp(-r*T)*(S_t > K) * v_t))
16 + }
17 >
18 > n_size <- seq(500, 100000, by=500); M <- 1e3
19 >
20 > theta <- 2
21 > delta_pathwise <- c(); vega_pathwise <- c()
22 > # Estimate delta with different n
23 > for (i in 1:length(n_size)){
24 +   Euro_CEV <- greek_pathwise(n_size[i], M, S_0, K, r, sigma, T, theta)
25 +   delta_pathwise[i] <- Euro_CEV$delta
26 +   vega_pathwise[i] <- Euro_CEV$vega
27 + }
28 >
29 > par(cex.lab=1.5, cex.axis=1.5, cex.main=2.5, mar=c(5,5,4,4))
30 > plot(n_size, delta_pathwise, type="o", col="red",
31 +      xlab="number of paths", ylab="delta", main="", lwd=2, pch=21)
32 > abline(h=exact_delta, col="blue", lwd=2, lty=3)
33 > legend("topright", col=c("red", "blue"), lwd=2, cex=1.5, lty=c(1,3),
34 +        legend=c("pathwise differentiation", "exact"), pch=c(21, NA))
35 >
36 > plot(n_size, vega_pathwise, type="o", col="red",
37 +      xlab="number of paths", ylab="vega", main="", lwd=2, pch=21)
38 > abline(h=exact_vega, col="blue", lwd=2, lty=3)
39 > legend("topright", col=c("red", "blue"), lwd=2, cex=1.5, lty=c(1,3),
40 +        legend=c("pathwise differentiation", "exact"), pch=c(21, NA))
```

Programme 5.32 Using pathwise differentiation method to approximate delta and vega of the European call under a CEV model in Example 5.10 via R.

From Figures 5.8 by Python and 5.9 by **R**, the approximations of both delta and vega under the Black–Scholes framework with $\theta = 2$ respectively converge to the true values of delta and vega as the number of iterations n increases.

Similarly, we can observe similar convergence results for $\theta = 1.8$ in Figure 5.10 by Python (resp. Figure 5.11 by **R**) and for $\theta = 1$ in Figure 5.12 by Python (resp. Figure 5.13 by **R**). We leave it as an exercise for readers to write down the Milstein scheme for the SDE system obtained through pathwise differentiation method; as expected, the corresponding convergence is even faster.

Suppose that we are only interested in approximating one Greek $\frac{\partial f_\gamma(S_T)}{\partial \gamma}$ of a derivative with the payoff $f_\gamma(S_T)$. For the finite differencing method, we have to simulate exactly two out of three different SDE dynamics simultaneously, depending on whether forward or central differencing is used, namely: the three dynamics for the stock price S_t with $\gamma - h$, γ, and $\gamma + h$ for some desired small enough $h > 0$. For the pathwise differentiation method, two dynamics still have to be simulated simultaneously, namely the dynamics for $\frac{\partial S_T}{\partial \gamma}$ and S_t. In general, pathwise differentiation is

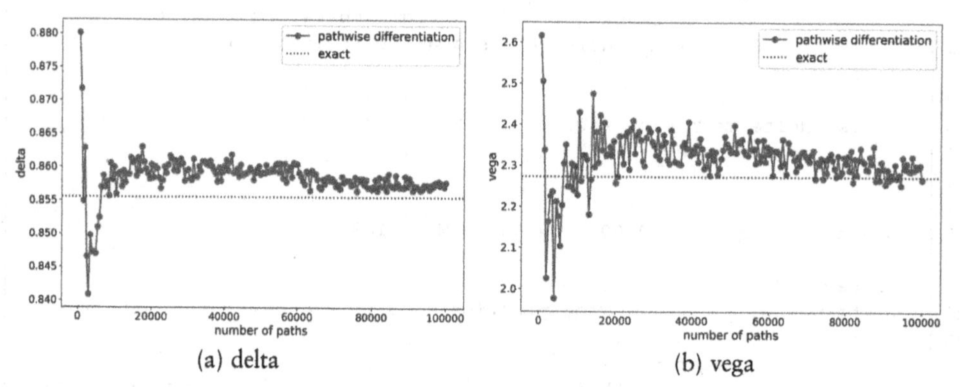

(a) delta (b) vega

FIGURE 5.8 Convergence of the pathwise differentiation method under a CEV model with $\theta = 2$ via Python, generated by Programme 5.31; the blue dotted line represents the exact value.

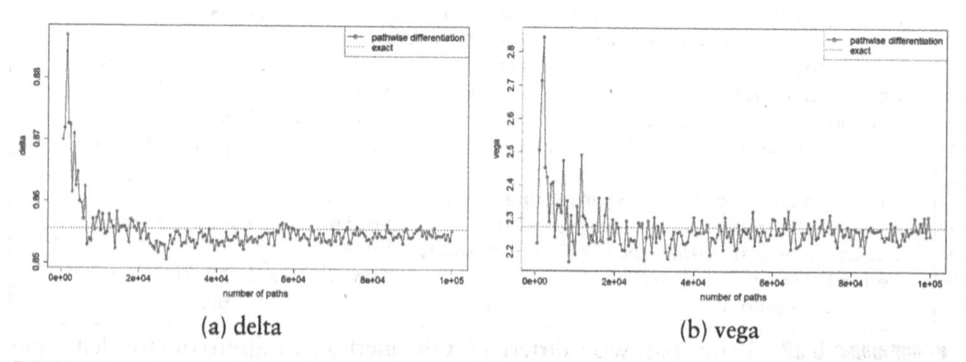

(a) delta (b) vega

FIGURE 5.9 Convergence of the pathwise differentiation method under a CEV model with $\theta = 2$ via **R**, generated by Programme 5.32; the blue dotted line represents the exact value.

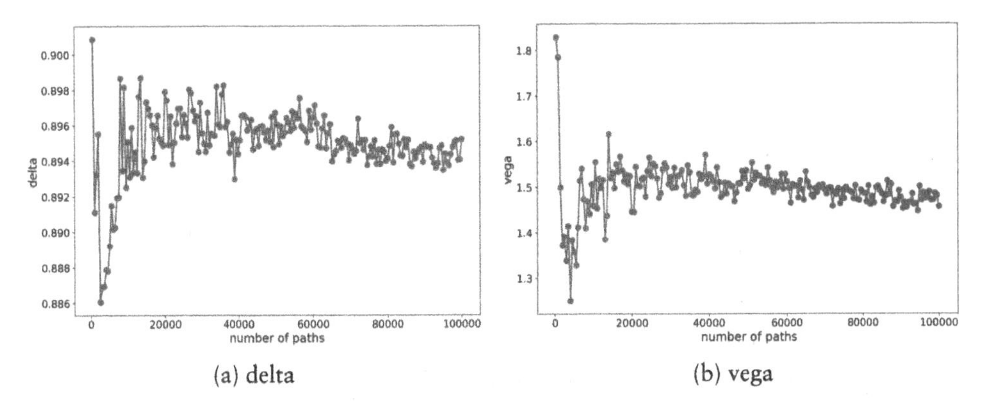

(a) delta (b) vega

FIGURE 5.10 Convergence of the pathwise differentiation method under a CEV model with $\theta = 1.8$ via Python, generated by Programme 5.31.

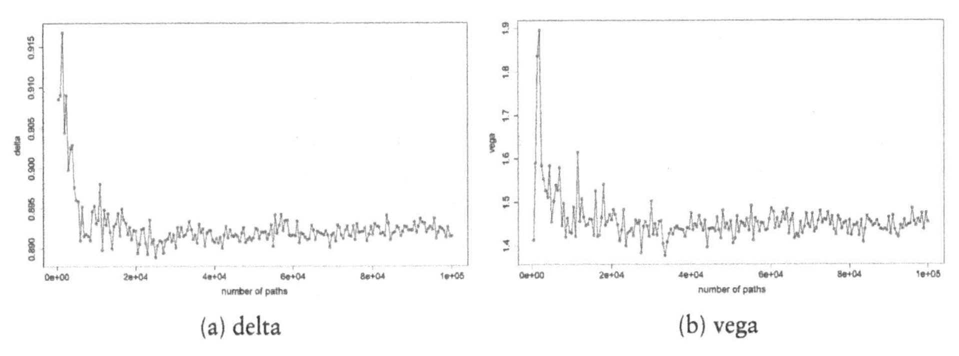

(a) delta (b) vega

FIGURE 5.11 Convergence of the pathwise differentiation method under a CEV model with $\theta = 1.8$ via **R**, generated by Programme 5.32.

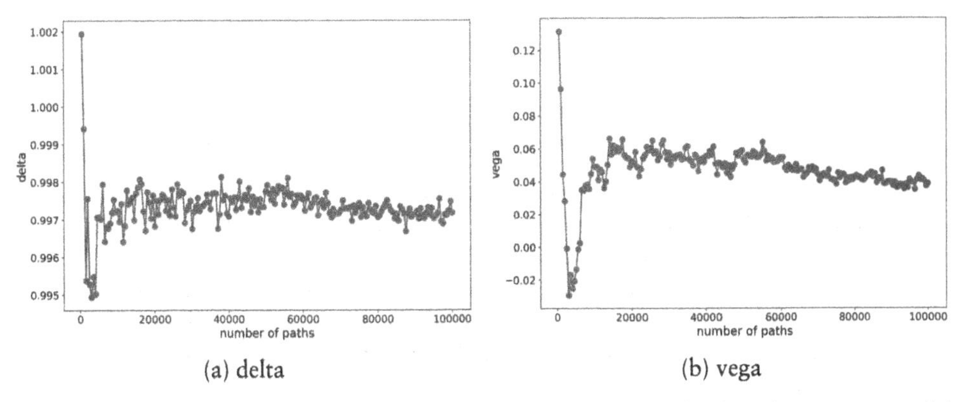

(a) delta (b) vega

FIGURE 5.12 Convergence of the pathwise differentiation method under a CEV model with $\theta = 1$ via Python, generated by Programme 5.31.

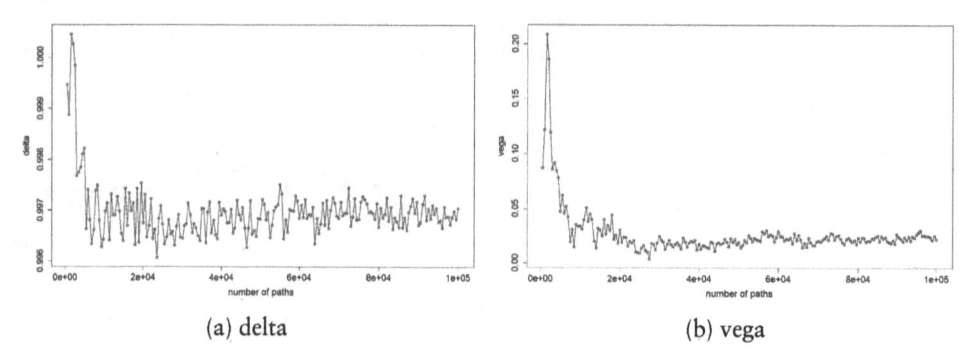

(a) delta (b) vega

FIGURE 5.13 Convergence of the pathwise differentiation method under a CEV model with $\theta = 1$ via R, generated by Programme 5.32.

generally faster and accurate enough, while it is at least as memory-efficient as the finite differencing method.

(III) Likelihood Ratio Method

Let the density function of S_T be g_θ, where θ is the parameter of interest, and for the moment, take it for granted that we know this expression as a function of θ. Then the expected discounted payoff can be written as:

$$\mathbb{E}(y) = \int_{-\infty}^{\infty} f(x) g_\theta(x)\, dx.$$

Moreover, if the discounted payoff function $f(x)$ is continuous almost surely and $g_\theta(x)$ is differentiable in θ, then the sensitivity of the expected discounted payoff with respect to θ can be computed as

$$\frac{\partial \mathbb{E}(y)}{\partial \theta} = \int_{-\infty}^{\infty} f(x) \frac{\partial g_\theta(x)}{\partial \theta}\, dx = \mathbb{E}\left(f(S_T) \frac{\partial \ln g_\theta(S_T)}{\partial \theta} \right). \tag{5.22}$$

Like the method of pathwise differentiation, we express the price sensitivity as an expectation, which can then be estimated by the Monte Carlo method. The *likelihood ratio method* involves the differentiation of the density function of S_T with respect to θ, rather than the differentiation of the payoff function as is the case with the method of pathwise differentiation.

As an immediate illustration, under the Black–Scholes framework,

$$\mathbb{P}(S_T \leq x) = \mathbb{P}\left(S_0 e^{\left(r - \frac{\sigma^2}{2}\right)T + \sigma\sqrt{T}z} \leq x \right) = \int_{-\infty}^{h(x)} \frac{1}{\sqrt{2\pi}} e^{-\frac{z^2}{2}}\, dz,$$

where $z \sim \mathcal{N}(0,1)$ and $h(x) := \frac{1}{\sigma\sqrt{T}}\left(\ln\frac{x}{S_0} - \left(r - \frac{\sigma^2}{2}\right)T\right)$. Therefore, $g_{S_0}(x)$, the density function of S_T, is given by

$$g_{S_0}(x) = \frac{d}{dx}\mathbb{P}(S_T \le x) = \frac{1}{\sqrt{2\pi}}e^{-\frac{h^2(x)}{2}} \cdot \frac{1}{x\sigma\sqrt{T}},$$

and so

$$\frac{\partial \ln g_{S_0}(x)}{\partial S_0} = \frac{d}{dS_0}\left(-\frac{h^2(x)}{2}\right) = -h(x)\frac{d}{dS_0}h(x) = h(x)\frac{1}{\sigma\sqrt{T}S_0}.$$

Since

$$h(S_T) = \frac{1}{\sigma\sqrt{T}}\left(\ln\frac{S_T}{S_0} - \left(r - \frac{\sigma^2}{2}\right)T\right) = \frac{w_T}{\sqrt{T}}, \tag{5.23}$$

it follows from (5.22) that the delta of this European-type derivative is $\mathbb{E}\left(f(S_T)\frac{w_T}{S_0\sigma T}\right)$, which can then be estimated by the Monte Carlo method. By computing the derivative of $\ln g_\theta(x)$ with respect to another model parameter θ, one can also approximate other Greek letters. For instance, considering the case of vega under the Black–Scholes framework, we first have:

$$\begin{aligned}
\frac{\partial \ln g_\sigma(x)}{\partial\sigma} &= \frac{d}{d\sigma}\left(-\frac{h^2(x)}{2}\right) + \frac{d}{d\sigma}(-\ln\sigma) = -h(x)\frac{d}{d\sigma}h(x) - \frac{1}{\sigma} \\
&= -h(x)\left(-\frac{1}{\sigma^2\sqrt{T}}\ln\frac{x}{S_0} + \frac{1}{\sigma^2\sqrt{T}}\left(r - \frac{\sigma^2}{2}\right)T - \frac{1}{\sigma\sqrt{T}}(-\sigma T)\right) - \frac{1}{\sigma} \\
&= \frac{h(x)}{\sigma^2\sqrt{T}}\left(\ln\frac{x}{S_0} - \left(r - \frac{\sigma^2}{2}\right)T - \sigma^2 T\right) - \frac{1}{\sigma}.
\end{aligned}$$

By (5.23), it follows that

$$\frac{\partial \ln g_\sigma(S_T)}{\partial\sigma} = \frac{w_T}{\sigma^2 T}(\sigma w_T - \sigma^2 T) - \frac{1}{\sigma} = \frac{1}{\sigma}\left(\frac{w_T^2}{T} - 1\right) - w_T,$$

and thus the vega of this European-type derivative is $\mathbb{E}\left(f(S_T)\left(\frac{1}{\sigma}\left(\frac{w_T^2}{T} - 1\right) - w_T\right)\right)$.

Example 5.11 ***(Continuation of Example 5.9 for the Black–Scholes model, $\theta = 2$ only, with pathwise differentiation and likelihood ratio methods)*** With the same setting as in Example 5.9, we use the likelihood ratio method to estimate the delta and vega of the European call option under the Black–Scholes framework. Additionally, we compare the results with those obtained by the pathwise differentiation method. The estimates obtained from both methods are plotted against the number of simulations implemented in Programme 5.33 via Python (resp. Programme 5.34 via **R**).

```
1  np.random.seed(4012)
2  n_size = np.arange(500, 100000+1, 500)
3  delta_pathwise, vega_pathwise = [], []
4  delta_likelihood, vega_likelihood = [], []
5  mu = (r-sigma**2/2)*T; sd = sigma*np.sqrt(T)
6
7  # Estimate delta and vega with different n
8  for i in range(len(n_size)):
9      # Generate the Black-Scholes sample
10     w_T = np.sqrt(T) * np.random.normal(size=int(n_size[i]))
11     S_T = S_0 * np.exp(mu + sigma * w_T)
12
13     d_payoff = np.exp(-r * T) * (S_T > K)
14     # Estimate delta and vega using pathwise differentiation method
15     delta_pathwise.append(np.mean(d_payoff * S_T/S_0))
16     vega_pathwise.append(np.mean(d_payoff * S_T*(w_T - sigma*T)))
17
18     payoff = np.exp(-r * T) * np.maximum(S_T - K, 0)
19     # Estimate delta and vega using likelihood ratio method
20     delta_likelihood.append(np.mean(payoff * w_T/(S_0*sigma*T)))
21     vega_likelihood.append(np.mean(payoff * ((w_T**2/T-1)/sigma-w_T)))
22
23 import matplotlib.pyplot as plt
24
25 # Plot the graph of estimated delta and exact delta
26 plt.figure(figsize=(10, 7))
27 plt.plot(n_size, delta_pathwise, "ro-", markersize=8,
28          linewidth=2, label="pathwise differentiation")
29 plt.plot(n_size, delta_likelihood, "ko-", markersize=8, linewidth=2,
30          label="likelihood ratio")
31 plt.axhline(y=exact_delta, color="blue", linewidth=2, label="exact")
32 plt.ylim((0.75, 0.95))
33 plt.xlabel("number of paths")
34 plt.ylabel("delta")
35 plt.legend(fontsize=17)
36
37 # Plot the graph of estimated vega and exact vega
38 plt.figure(figsize=(10, 7))
39 plt.plot(n_size, vega_pathwise, "ro-", markersize=8, linewidth=2,
40          label="pathwise differentiation")
41 plt.plot(n_size, vega_likelihood, "ko--", markersize=8, linewidth=2,
42          label="likelihood ratio")
43 plt.axhline(y=exact_vega, color="blue", linewidth=2, label="exact",
44             linestyle="dotted")
45 plt.ylim((0, 3.6))
46 plt.xlabel("number of paths")
47 plt.ylabel("vega")
48 plt.legend(fontsize=17)
```

Programme 5.33 The Black–Scholes delta and vega in Example 5.9 estimated by both pathwise differentiation and likelihood ratio methods via Python.

```
1  > set.seed(4002)
2  > n_size <- seq(500, 100000, by=500)
3  > delta_pathwise <- c(); vega_pathwise <- c()
4  > delta_likelihood <- c(); vega_likelihood <- c()
5  > mu <- (r-sigma^2/2)*T
6  >
7  > # Estimate delta and vega with different n
8  > for (i in 1:length(n_size)) {
9  +   # Generate the Black-Scholes sample
10 +   w_T <- sqrt(T) * rnorm(n_size[i])
11 +   S_T <- S_0 * exp(mu + sigma * w_T)
12 +
13 +   d_payoff <- exp(-r * T) * (S_T > K)
14 +   # Estimate delta and vega using pathwise differentiation method
15 +   delta_pathwise[i] <- mean(d_payoff * S_T/S_0)
16 +   vega_pathwise[i] <- mean(d_payoff * S_T*(w_T - sigma*T))
17 +
18 +   payoff <- exp(-r * T) * pmax(S_T - K, 0)
19 +   # Estimate delta and vega using likelihood ratio method
20 +   delta_likelihood[i] <- mean(payoff * w_T/(S_0*sigma*T))
21 +   vega_likelihood[i] <- mean(payoff * ((w_T^2/T - 1)/sigma - w_T))
22 + }
23 >
24 > # Plot the graph of estimated delta and exact delta
25 > plot(n_size, delta_pathwise, type="o", col="red", ylim=c(0.8, 0.93),
26 +      xlab="number of paths",ylab="delta",main="",lwd=2,pch=21,lty=1)
27 > lines(n_size, delta_likelihood, type="o", col="black", lwd=2, pch=21,
28 +       lty=2)lty=2)
29 > abline(h=exact_delta, col="blue", lwd=2, lty=3)
30 > legend("topright",col=c("red","black","blue"),lwd=2,cex=1.5,lty=1:3,
31 +        legend=c("pathwise differentiation", "likelihood ratio",
32 +                 "exact"), pch=c(21, 21, NA))
33 >
34 > # Plot the graph of estimated vega and exact vega
35 > plot(n_size, vega_pathwise, type="o", col="red", ylim=c(0.8, 4.1),
36 +      xlab="number of paths",ylab="vega",main="",lwd=2,pch=21,lty=1)
37 > lines(n_size,vega_likelihood,type="o",col="black",lwd=2,pch=21,lty=2)
38 > abline(h=exact_vega, col="blue", lwd=2, lty=3)
39 > legend("topright",col=c("red","black","blue"),lwd=2,cex=1.5,lty=1:3,
40 +        legend=c("pathwise differentiation", "likelihood ratio",
41 +                 "exact"), pch=c(21, 21, NA))
```

Programme 5.34 The Black–Scholes delta and vega in Example 5.9 estimated by pathwise differentiation and likelihood ratio methods via **R**.

Figures 5.14 (via Python) and 5.15 (via **R**) illustrate the respective convergences of the estimators by using pathwise differentiation and likelihood ratio methods. We observe that both methods gradually approach the blue dotted line, leveled at the true value, as the number of paths (simulations) increases, which validates the pathwise differentiation and likelihood ratio methods. However, the approximation of both Greeks in the pathwise differentiation method demonstrates a superior convergence performance compared to the likelihood ratio method.

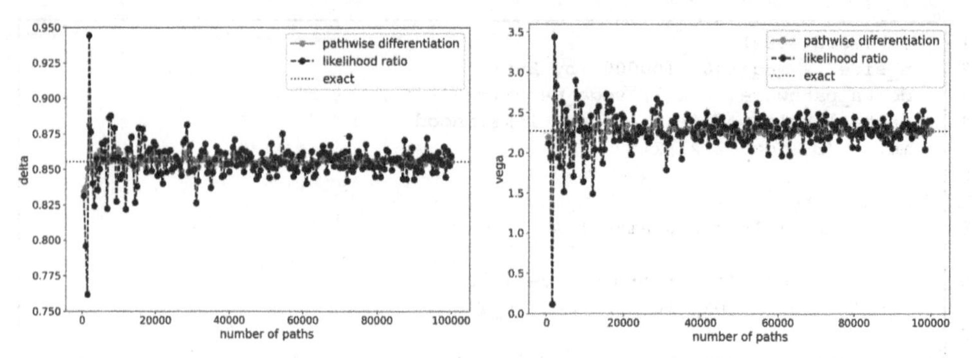

FIGURE 5.14 Comparison of convergence of the pathwise differentiation and likelihood ratio methods via Python, generated by Programme 5.33.

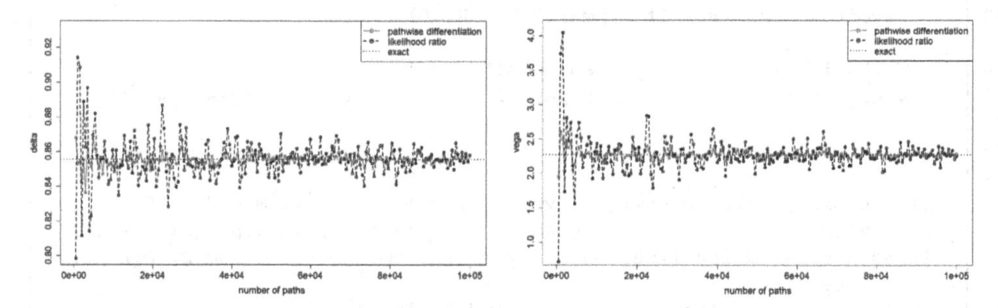

FIGURE 5.15 Comparison of convergence of the pathwise differentiation and likelihood ratio methods via **R**, generated by Programme 5.34.

In order to apply this likelihood ratio method, one needs the explicit expression of the sensitivity, with respect to the parameter of interest, of the density of the terminal state. Usually, this may not be immediately achievable in practice. Nevertheless, most security prices are modeled by SDEs, and by using *Malliavin Calculus* [2, 13], one can still write the density as a function of expectation functionals involving the so-called *Malliavin derivatives* of these SDEs. On this basis, the likelihood ratio method can be used on more occasions. That said, from the last illustration, the pathwise differentiation method shows better performance. Interestingly, its philosophy is closely related to the concept of *Jacobian flow* and consequently it intimately connects with Malliavin's differentiability of SDEs.

Last but not least, Malliavin calculus can help on expressing the hedging portfolio weights in various optimal portfolio selection problems via the celebrated *Clark–Ocone formula*, which again involves Malliavin derivatives but of the terminal payoffs this time. Due to our page limit, we cannot cover this wonderful topic of stochastic analysis in this book, and we invite readers to check out the details in [1, 12] and the references therein.

REFERENCES

1. Alós, E., and Lorite, D.G. (2021). *Malliavin Calculus in Finance: Theory and Practice.* CRC Press.
2. Bally, V. (2003). An elementary introduction to Malliavin calculus (Doctoral dissertation, INRIA).
3. Baxter, M., and Rennie, A. (1996). *Financial Calculus: An Introduction to Derivative Pricing.* Cambridge University Press.
4. Charpentier, A. (Ed.). (2014).*Computational Actuarial Science with R.* CRC Press.
5. Cornfeld, I.P., Fomin, S.V., and Sinai, Y.G.E. (2012). *Ergodic Theory.* Springer Science & Business Media.
6. Cox, J.C. (1996). The constant elasticity of variance option pricing model. *Journal of Portfolio Management*, 23, 15–17.
7. Etheridge, A. (2002). *A Course in Financial Calculus.* Cambridge University Press.
8. Gosling, F.G. (1999). *The Manhattan Project: Making the Atomic Bomb.* Diane Publishing.
9. Kennedy, D. (2016). *Stochastic Financial Models.* CRC Press.
10. Kloeden, P.E., and Platen, E. (1989). A survey of numerical methods for stochastic differential equations. *Stochastic Hydrology and Hydraulics*, 3, 155–178.
11. Korn, R., Korn, E., and Kroisandt, G. (2010). *Monte Carlo Methods and Models in Finance and Insurance.* CRC Press.
12. Malliavin, P. (2006). *Stochastic Calculus of Variations in Mathematical Finance.* Springer.
13. Nualart, D., and Nualart, E. (2018). *Introduction to Malliavin Calculus.* Cambridge University Press.
14. Oosterlee, C.W., and Grzelak, L.A. (2020). *Mathematical Modeling and Computation in Finance: With Exercises and Python and Matlab Computer Codes.* World Scientific.
15. Robert, C.P., and Casella, G. (1999). *Monte Carlo Statistical Methods.* Springer.
16. Ross, M.R. (2022). *Simulation* (6th ed.). Academic Press.
17. Shreve, S.E. (2004). *Stochastic Calculus for Finance II: Continuous-Time Models.* Springer.
18. Steele, J.M. (2001). *Stochastic Calculus and Financial Applications.* Springer.
19. Walters, P. (2000). *An Introduction to Ergodic Theory.* Springer Science & Business Media.
20. Wilmott, P., Howison, S., and Dewynne, J. (1995). *The Mathematics of Financial Derivatives: A Student Introduction.* Cambridge University Press.

CHAPTER 6

Approximation for Model Inference

The process of fitting models is often accomplished by minimizing the sum of squares (as in regression analysis) or minimizing cross-entropy (as in classification method). These are examples of the *maximum likelihood approach* to fitting. This chapter provides a comprehensive explanation of the maximum likelihood approach and introduces several methods of approximation for the purpose of model inference.

The examples provided in this chapter highlight the practical applications and importance of the algorithms to be introduced. For instance, by leveraging the maximum likelihood approach to fitting within numerical algorithms, it becomes possible to implement handwriting number recognition systems that accurately identify handwritten numbers commonly found on cheques and many business form filings. Estimating quantiles using numerical methods is essential for risk assessment and portfolio management, such as *Value-at-Risk* calibration (to be introduced in Chapter 8). After providing a comprehensive and self-contained introduction to *Markov Chain theory*, we also introduce, with an illustrative example, the very useful *Markov Chain Monte Carlo* (MCMC) method for sampling from complex probability distributions.

6.1 EM ALGORITHM

The *Expectation–Maximization Algorithm* (or *EM Algorithm* for short) is a parametric learning method first introduced in [5] for tackling parametric estimations, especially with unobservable variates. Instead of looking for an explicit analytic solution, the EM algorithm finds a maximum point of a log-likelihood function via an iterative procedure that can easily be carried out by modern computers. An in-depth account of the general theory of the EM algorithm and its various extensions is given by the classical text [22].

More precisely, let $x_1, \ldots, x_n$ be an observable p-dimensional random sample of size n from a population with a common "mixed joint density" function $f(x; \boldsymbol{\theta})$, where $\boldsymbol{\theta}$ is a k-dimensional vector of parameters. Note that if all components of the random observation $\mathbf{x}$ are continuous (resp. discrete), then f is just the joint density function (resp. joint probability mass function) as usual, yet this is not always the case. For if some components of $\mathbf{x}$ are continuous while the others are discrete (or even categorical), f is taken as a mixed joint density function. As an example, let $\mathbf{x} = (\mathrm{x}_1, \mathrm{x}_2)^\top$ be a 2-dimensional random vector and $\boldsymbol{\theta} = (\theta_1, \theta_2, \theta_3)^\top$ be a 3-dimensional vector of positive parameters, such that $\mathrm{x}_1 \sim \mathrm{Poi}(\theta_1)$ and $\mathrm{x}_2 = \sum_{i=1}^{\mathrm{x}_1} \mathrm{G}_i$, where $\mathrm{G}_i \overset{\text{iid}}{\sim}$ $\mathrm{Gamma}(\theta_2, \theta_3)$ with the pdf $\frac{\theta_3^{\theta_2}}{\Gamma(\theta_2)} x^{\theta_2-1} e^{-\theta_3 x}$, for $i = 1, 2, \ldots$. Then the density of x_2 given $\mathrm{x}_1 = x_1$ is that of $\mathrm{Gamma}(x_1\theta_2, \theta_3)$ in light of the additivity property of gamma random variables:

$$f(x_2|x_1;\theta) = \frac{\theta_3^{x_1\theta_2}}{\Gamma(x_1\theta_2)} x_2^{x_1\theta_2-1} e^{-\theta_3 x_2},$$

and the "mixed joint density" of $\mathbf{x}$ is given by

$$f(x_1, x_2; \theta) = e^{-\theta_1} \frac{\theta_1^{x_1}}{x_1!} \cdot \frac{\theta_3^{x_1\theta_2}}{\Gamma(x_1\theta_2)} x_2^{x_1\theta_2-1} e^{-\theta_3 x_2},$$

for $x_1 = 0, 1, \ldots$, and $x_2 \in \mathbb{R}^+$.

If $x_1, \ldots, x_n$ are the random observed values, then the likelihood function takes the form

$$L(\theta; x_1, \ldots, x_n) := \prod_{i=1}^{n} f(x_i; \theta),$$

and the maximum likelihood estimate $\hat{\theta}_1$ for the parameter θ is given by

$$\hat{\theta}_1 = \arg\max_{\theta} \ln L(\theta; x_1, \ldots, x_n) = \arg\max_{\theta} \sum_{i=1}^{n} \ln f(x_i; \theta).$$

However, in practice the random feature vector $\mathbf{x}$ may not be fully observable, say $\mathbf{x} := \begin{pmatrix} \mathrm{y} \\ \mathbf{z} \end{pmatrix}$, where y is the observable part, and $\mathbf{z}$ is the unobservable part whose components are called *latent variables*. In this case, we denote the observable random samples (without latent variables) by $y_1, \ldots, y_n$, and using the shorthand notation for multiple integrals, the maximum likelihood estimate $\hat{\theta}_2$ of θ is now given by

$$\hat{\theta}_2 = \arg\max_{\theta} \ln L(\theta; y_1, \ldots, y_n) = \arg\max_{\theta} \sum_{i=1}^{n} \ln \int_{\mathrm{supp}(\mathbf{z})} f(y_i, z; \theta) dz, \qquad (6.1)$$

where supp(z) is the range or support of $\mathbf{z}$. Here, $f(y;\theta) := \int_{\mathrm{supp}(\mathbf{z})} f(y, z; \theta) dz$ is the marginal mixed joint density of y, which boils down to a summation for discrete components of $\mathbf{z}$, otherwise an ordinary integration is meant, $L(\theta; y_1, \ldots, y_n)$ is the marginal likelihood of the observations $y_1, \ldots, y_n$. Note that the maximization (6.1) is usually difficult since the marginal likelihood is often inconvenient to obtain. For instance, if the dimension of $\mathbf{z}$ grows exponentially in the sample size n, then the exact calculation will be extremely difficult. The EM algorithm is a prevalent approach to tackle this issue.

6.1.1 General Theory and Algorithm

Without causing much ambiguity, we shall omit the subscripts of the joint density functions in this subsection, as they can be well-identified from the arguments; this means that while $f_{\mathbf{y}}(y;\theta)$ denoted the marginal mixed joint density of $\mathbf{y}$ before, we write it simply as $f(y;\theta)$. Similarly, $f_{\mathbf{z}|\mathbf{y}}(z|y;\theta)$, the conditional mixed joint density of $\mathbf{z}$ given $\mathbf{y}=y$, is now written as $f(z|y;\theta)$, etc.. The EM algorithm consists of repeating the following two steps: (i) an expectation (E) taking step and (ii) a maximization (M) finding step. These are elaborated in Algorithm 6.1 below, where ϵ is a pre-specified threshold.

Algorithm 6.1 EM algorithm

1. Initialize $\theta^{(0)}$ (chosen randomly);
2. Repeat the following steps 2a) and 2b): for $t = 1, 2, \ldots,$

   2a). **E-Step:** Calculate

$$Q(\theta \mid \theta^{(t)}) := \sum_{i=1}^{n} \mathbb{E}\left(\ln f(y_i, \mathbf{z}; \theta)\middle| y_i; \theta^{(t)}\right)$$

$$= \sum_{i=1}^{n} \int_{\text{supp}(\mathbf{z})} f(z|y_i;\theta^{(t)}) \ln f(y_i, z; \theta) dz. \tag{6.2}$$

   2b). **M-Step:** Maximize $Q(\theta \mid \theta^{(t)})$ over θ and set

$$\theta^{(t+1)} := \arg\max_{\theta} Q(\theta \mid \theta^{(t)}). \tag{6.3}$$

3. Stop the iteration if $\left\|Q(\theta^{(t+1)} \mid \theta^{(t)}) - Q(\theta^{(t)} \mid \theta^{(t)})\right\| < \varepsilon$, or $\left\|\theta^{(t+1)} - \theta^{(t)}\right\| < \varepsilon$.

Here we describe the idea behind the EM algorithm. At each iteration, we hope to obtain an increased value of $\ln L(\theta; y_1, \ldots, y_n) = \sum_{i=1}^{n} \ln f(y_i;\theta)$; we want to see that $\ln L(\theta^{(t)}; y_1, \ldots, y_n)$ is monotonically increasing in t. Indeed, conditioning tells us $f(y, z; \theta) = f(z|y;\theta) f(y;\theta)$, which gives $\ln f(y;\theta) = \ln f(y, z; \theta) - \ln f(z|y;\theta)$. Define

$$R(\theta|\theta^{(t)}) := \sum_{i=1}^{n} \int_{\text{supp}(\mathbf{z})} f(z|y_i;\theta^{(t)}) \ln f(z|y_i;\theta) dz. \tag{6.4}$$

Compared with $Q(\theta|\theta^{(t)})$, $R(\theta|\theta^{(t)})$ admits a similar form, except that the conditional mixed joint density of $\mathbf{z}$ given $\mathbf{y} = y$ replaces the mixed joint density of $(\mathbf{y}, \mathbf{z})$ in the integrand of $Q(\theta|\theta^{(t)})$. Then, using (6.2), we have

$$\begin{aligned}
&Q(\theta|\theta^{(t)}) - R(\theta|\theta^{(t)}) \\
&= \sum_{i=1}^{n} \int_{\mathrm{supp}(\mathbf{z})} (f(z|y_i;\theta^{(t)})(\ln f(y_i, z;\theta) - \ln f(z|y_i;\theta)))\, dz \\
&= \sum_{i=1}^{n} \int_{\mathrm{supp}(\mathbf{z})} f(z|y_i;\theta^{(t)}) \ln f(y_i;\theta) dz \\
&= \sum_{i=1}^{n} (\ln f(y_i;\theta)\cdot 1) = \ln L(\theta; y_1,\ldots,y_n). \qquad (6.5)
\end{aligned}$$

Substituting $\theta = \theta^{(t)}$ and $\theta = \theta^{(t+1)}$, respectively into (6.5), and then considering their difference, we obtain

$$\begin{aligned}
&\ln L(\theta^{(t+1)}; y_1,\ldots,y_n) - \ln L(\theta^{(t)}; y_1,\ldots,y_n) \\
&= \left(Q(\theta^{(t+1)}|\theta^{(t)}) - Q(\theta^{(t)}|\theta^{(t)})\right) - \left(R(\theta^{(t+1)}|\theta^{(t)}) - R(\theta^{(t)}|\theta^{(t)})\right). \qquad (6.6)
\end{aligned}$$

Clearly, the very definition of $\theta^{(t+1)}$ as a maximum point in (6.3) guarantees that

$$Q(\theta^{(t+1)}|\theta^{(t)}) - Q(\theta^{(t)}|\theta^{(t)}) \geq 0. \qquad (6.7)$$

On the other hand, we also have

$$\begin{aligned}
R(\theta^{(t+1)}|\theta^{(t)}) - R(\theta^{(t)}|\theta^{(t)}) &= \sum_{i=1}^{n} \int_{\mathrm{supp}(\mathbf{z})} f(z|y_i;\theta^{(t)}) \ln\left(\frac{f(z|y_i;\theta^{(t+1)})}{f(z|y_i;\theta^{(t)})}\right) dz \\
&\leq \sum_{i=1}^{n} \ln\left(\int_{\mathrm{supp}(\mathbf{z})} f(z|y_i;\theta^{(t)}) \cdot \frac{f(z|y_i;\theta^{(t+1)})}{f(z|y_i;\theta^{(t)})} dz\right) \\
&= \sum_{i=1}^{n} \ln\left(\int_{\mathrm{supp}(\mathbf{z})} f\left(z|y_i;\theta^{(t+1)}\right) dz\right) = 0,
\end{aligned}$$

where the first inequality follows from *Jensen's inequality*. Hence, together with (6.7), we conclude that

$$\ln L(\theta^{(t+1)}; y_1,\ldots,y_n) - \ln L(\theta^{(t)}; y_1,\ldots,y_n) \geq 0,$$

which asserts that the EM algorithm does provide progressively improving approximants.

Note that in most commonly encountered practical circumstances, the log-likelihood $\ln L(\theta; y_1,\ldots,y_n)$ is bounded, and so the monotone nature of the EM algorithm ensures the convergence of $\theta^{(t)}$. However, it is sensitive to the initial choice

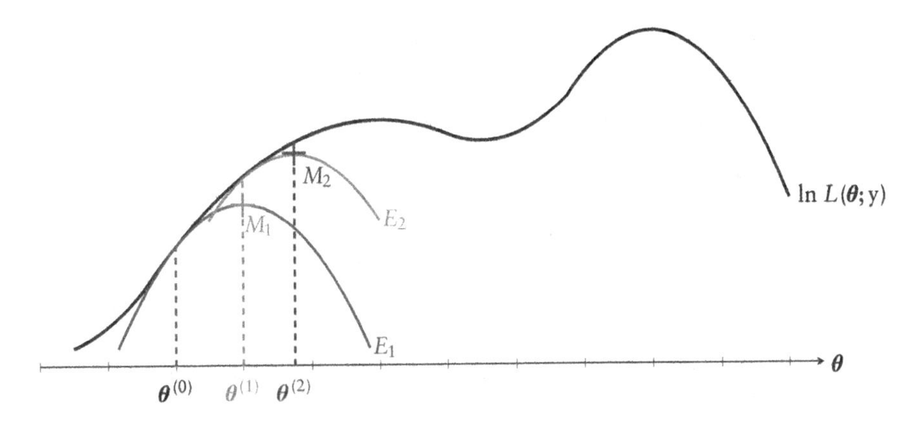

FIGURE 6.1 Illustration of the EM algorithm after the first two iterations.

of $\boldsymbol{\theta}^{(0)}$, and therefore the EM algorithm only asserts convergence to a local maxima of (6.1), not necessarily to the ultimate maximum likelihood estimate; see Figure 6.1. Therefore, we should try more initial choices of $\boldsymbol{\theta}^{(0)}$ and look for the largest possible ultimate maxima.

6.1.2 Application to Handwriting Recognition

In this subsection, we discuss an immediate application of the EM algorithm to the unsupervised learning of recognizing handwritten Arabic numbers. This can help us easily identify handwritten numerical figures found on cheques and business documents. The training dataset is a collection of labeled black-and-white images of handwritten digits from MNIST[1]; see Figure 6.2 for some sample images from the dataset. The goal here is to use the EM algorithm to correctly identify a given handwritten digit with a number from 0 to 9. Eventually, we summarize in a single artificial image that represents each of the ten digits by averaging over all the relevant classified observations in the training dataset; see Figure 6.3 for an illustration of the digit 2. The idea can also be employed to generate artificial faces from multiple images of real human faces which show similar "characteristics".

For each handwritten digit, we use 1 to represent a black pixel and 0 for white. The array of pixels is rearranged in a linear order from top down to form a feature vector $\boldsymbol{x} = (x_1, x_2, \ldots, x_K)^\top$, where K is the total number of pixels, i.e. the size of the rectangular image. For $m = 0, 1, \ldots, 9$, let S_m be the subset of images of the training

[1] The *Modified National Institute of Standards and Technology* database, MNIST database for short, is revised (modified) from the original *National Institute of Standards and Technology* (NIST) database (https://www.nist.gov/srd/nist-special-database-19) by *Yann LeCun*. MNIST consists of 60,000 training images and 10,000 testing images of handwritten digits by high school students and employees of the *United States Census Bureau*, and is commonly used in training image processing systems. The database can be retrieved at http://yann.lecun.com/exdb/mnist/.

FIGURE 6.2 A sample of handwritten digits from MNIST.

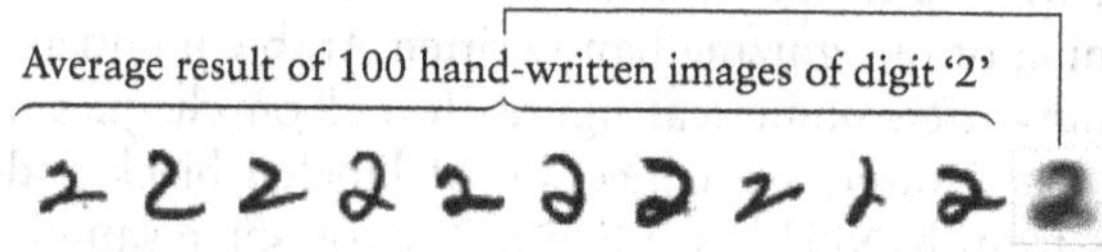

Average result of 100 hand-written images of digit '2'

FIGURE 6.3 An artificial image (right) as the average of handwritten images (left) of the digit 2.

dataset that contains the images of the digit m. Then the total sample is a disjoint union $S = \bigsqcup_{m=0}^{9} S_m$ with total sample size n, i.e. $n = \sum_{m=0}^{9} |S_m|$. The mean vector $\overline{\mu}_m$ for the digit m is defined by

$$\overline{\mu}_m := \frac{1}{|S_m|} \sum_{i=1}^{n} x_i \mathbb{1}_{\{x_i \in S_m\}}, \tag{6.8}$$

where $x_i := (x_{i1}, \ldots, x_{iK})^\top$ is the feature vector, or pixel vector, of the i-th image, for $i = 1, \ldots, n$. Each entry of $\overline{\mu}_m := (\overline{\mu}_{m1}, \ldots, \overline{\mu}_{mK})^\top$ lies in the closed interval $[0, 1]$. In Figure 6.3, the image on the right is obtained by averaging a sample of 100 handwritten digits labeled as "2". One can identify the shape of the digit 2 from this artificial image without much difficulty, even though the image is fairly blurred.

Generally speaking, suppose the training dataset with size n consists of M different types of handwritten "digits", to allow for the possibility that M is not necessarily 10. With the presence of multiple digits in the dataset, we wish to determine the most "representative" image for each digit. At the same time, we will also try to estimate the prior probability that the representative image of a particular digit is shown. To this end, we consider M different parameter vectors $\boldsymbol{\mu}_m$, $m = 1, \ldots, M$; here m stands as an index instead of the actual value of the particular chosen digit. Denote by π_m, $m = 1, \ldots, M$, the prior probability of selecting $\boldsymbol{\mu}_m$. Naturally these should sum to 1. In the following, we shall denote the collection of parameters by $\boldsymbol{\theta} := (\boldsymbol{\mu}_1, \ldots, \boldsymbol{\mu}_M, \pi_1, \ldots, \pi_M)$, with the constraint $\sum_{m=1}^{M} \pi_m = 1$. The pixel vector $\boldsymbol{x}$ of a particular image is the observable part, and the parameter vector $\mathbf{z}$, as a hyper-random vector, chosen to generate this image, is unknown. Here, $\mathbf{z}$ acts as a hidden/latent vector, and it can only take M (vector) values of $\boldsymbol{\mu}_1, \ldots, \boldsymbol{\mu}_M$. We consider components of $\boldsymbol{\mu}_m$ as the success probabilities for the respective components of Bernoulli random vector $\mathrm{i}_m = (\mathrm{i}_{m1}, \ldots, \mathrm{i}_{mK})^\top$, where i_{mk} stands for the color (1 for black, 0 for white) of the k-th pixel in the same image if the m-th type of digit is drawn. All the pixel vectors are assumed to be iid in the present modelling framework[2]. Note that the components of the random vector are assumed to be (conditionally) independent since the images are chosen fairly arbitrarily. For example, some deficits may occur in the writing process as in Figure 6.4, wherein a few drops of ink or an abrupt halt in the middle of writing have occurred. Therefore, it still sounds plausible to assume, even in the same image, that pixels are colored independently of each other.

With this in mind, given that $\mathbf{z} = \boldsymbol{\mu}_m = (\mu_{m1}, \ldots, \mu_{mK})^\top$, the conditional probability of generating an image with the observed pixel vector $\boldsymbol{x}$ is:

$$f(\boldsymbol{x}|\mathbf{z} = \boldsymbol{\mu}_m; \boldsymbol{\theta}) = \prod_{k=1}^{K} (\mu_{mk})^{x_k} (1 - \mu_{mk})^{1-x_k}. \tag{6.9}$$

Thus, the marginal probability of generating this image is, by the law of total probability,

$$f(\boldsymbol{x}; \boldsymbol{\theta}) = \sum_{m=1}^{M} \pi_m f(\boldsymbol{x}|\mathbf{z} = \boldsymbol{\mu}_m; \boldsymbol{\theta}).$$

(a) a digit of 2 with a few drops of ink

(b) a digit of 6 with an unexpected halt in the middle of writting

FIGURE 6.4 Examples of deficits in handwriting images.

[2] Recall the aphorism added by *George Box* (see [2]): "*All models are wrong, but some of them are useful.*"

As a whole, the likelihood of obtaining the training dataset is precisely given by:

$$L(\theta) = \prod_{i=1}^{n} f(x_i; \theta) = \prod_{i=1}^{n} \left(\sum_{m=1}^{M} \pi_m f(x_i | z = \mu_m; \theta) \right). \tag{6.10}$$

This model is an example of a mixture of Bernoulli distributions. However, the presence of both a product and summation in (6.10) renders an even more complicated expression after taking logarithm. Deriving explicit forms of partial derivatives of the resulting log-likelihood function is unfeasible. Therefore, instead of handling the intractable expression (6.10) analytically, we propose to solve for the optimal generator θ by the EM algorithm as follows:

1. Initialize $\theta^{(0)} = (\mu_1^{(0)}, \ldots, \mu_M^{(0)}, \pi_1^{(0)}, \ldots, \pi_M^{(0)})$, with the constraint that $\sum_{i=1}^{M} \pi_m^{(0)} = 1$.
2. For the t-th iteration:

2a). **E-step:** Compute the posterior probabilities:

$$\begin{aligned} p_{mi}^{(t)} &= \mathbb{P}(\text{the } m\text{-th parameter vector is used} \mid x_i \text{ is realized}) \\ &= \frac{\pi_m^{(t)} f(x_i | z = \mu_m^{(t)}; \theta^{(t)})}{\sum_{r=1}^{M} \pi_r^{(t)} f(x_i | z = \mu_r^{(t)}; \theta^{(t)})}, \quad i = 1, \ldots, n, \text{ and } m = 1, \ldots, M. \end{aligned} \tag{6.11}$$

Based on these computed posterior probabilities, we then compute the expected log-likelihood (6.2) (note that some integrations are approximately replaced by summations for discrete counterparts):

$$\begin{aligned} &Q(\theta \mid \theta^{(t)}) \\ &= \sum_{m=1}^{M} \sum_{i=1}^{n} p_{mi}^{(t)} \ln \left(\pi_m f(x_i | z = \mu_m; \theta) \right) \\ &= \sum_{m=1}^{M} \sum_{i=1}^{n} p_{mi}^{(t)} \left\{ \ln \pi_m + \sum_{k=1}^{K} \left(x_{ik} \ln \mu_{mk} + (1 - x_{ik}) \ln(1 - \mu_{mk}) \right) \right\}. \end{aligned} \tag{6.12}$$

2b). **M-step:** To maximize Q with respect to μ_m and π_m, for $m = 1, \ldots, M$, subject to the constraint $\sum_{m=1}^{M} \pi_m = 1$, we have to consider the following Lagrangian (function):

$$Q_\lambda = Q + \lambda \left(\sum_{m=1}^{M} \pi_m - 1 \right).$$

By differentiating Q_λ with respect to μ_{mk} for $m = 1, \ldots, M$, $k = 1, \ldots, K$, and setting the derivative to zero, we then have

$$\frac{\partial Q_\lambda}{\partial \mu_{mk}} = \sum_{i=1}^{n} p_{mi}^{(t)} \left(\frac{x_{ik}}{\mu_{mk}} - \frac{1 - x_{ik}}{1 - \mu_{mk}} \right) = 0,$$

which implies that

$$\mu_{mk} = \frac{\sum_{i=1}^{n} p_{mi}^{(t)} x_{ik}}{\sum_{i=1}^{n} p_{mi}^{(t)}}. \tag{6.13}$$

On the other hand, by differentiating Q_λ with respect to π_m and setting the derivative to zero, we obtain

$$\frac{\partial Q_\lambda}{\partial \pi_m} = \sum_{i=1}^{n} \frac{p_{mi}^{(t)}}{\pi_m} + \lambda = 0,$$

which implies that

$$-\lambda \pi_m = \sum_{i=1}^{n} p_{mi}^{(t)}. \tag{6.14}$$

Summing over $m = 1, 2, \ldots, M$ yields

$$-\lambda = -\lambda \sum_{m=1}^{M} \pi_m = \sum_{m=1}^{M} \sum_{i=1}^{n} p_{mi}^{(t)} = \sum_{i=1}^{n} \left(\sum_{m=1}^{M} p_{mi}^{(t)} \right) = \sum_{i=1}^{n} 1 = n.$$

Substituting this last result back to (6.14) yields

$$\pi_m = \frac{\sum_{i=1}^{n} p_{mi}^{(t)}}{n}, \text{for } m = 1, \ldots, M. \tag{6.15}$$

Therefore, we update $\theta^{(t)}$ to $\theta^{(t+1)}$ according to (6.13) and (6.15) as

$$\mu_{mk}^{(t+1)} = \frac{\sum_{i=1}^{n} p_{mi}^{(t)} x_{ik}}{\sum_{i=1}^{n} p_{mi}^{(t)}} \quad \text{and} \quad \pi_m^{(t+1)} = \frac{\sum_{i=1}^{n} p_{mi}^{(t)}}{n}, \tag{6.16}$$

for $m = 1, \ldots, M, k = 1, \ldots, K$, where

$$p_{mi}^{(t)} = \frac{\pi_m^{(t)} \prod_{k=1}^{K} \left(\left(\mu_{mk}^{(t)} \right)^{x_{ik}} \left(1 - \mu_{mk}^{(t)} \right)^{1-x_{ik}} \right)}{\sum_{r=1}^{M} \pi_r^{(t)} \prod_{k=1}^{K} \left(\left(\mu_{rk}^{(t)} \right)^{x_{ik}} \left(1 - \mu_{rk}^{(t)} \right)^{1-x_{ik}} \right)}. \tag{6.17}$$

3. Repeat Steps 2a) and b) until $||\theta^{(t+1)} - \theta^{(t)}||_2 < \epsilon$ or $|Q(\theta^{(t+1)}|\theta^{(t)}) - Q(\theta^{(t)}|\theta^{(t)})| < \epsilon$ for some pre-specified threshold ϵ.

In the rest of this subsection, we illustrate the implementation of the EM algorithm for the hand-written digits from MNIST. For **R**, we first load the MNIST dataset from the `keras`[3] library, then we split the dataset into a training dataset with 60,000 samples and a test dataset with 10,000 samples. This splitting of the training and test datasets is preassigned by the `Keras` library. Each sample has a dimension of 28×28 pixels, and each pixel has value ranging from 0 to 255[4]. However, since $\boldsymbol{\mu} \in [0,1]^K$ and $x_j \in [0,255]$, computing the conditional probability using (6.9) gives extremely small values. Thus, we narrow down the range of x_j by a zero-one function: if the pixel is colored with a non-zero value, we simply set x_j to 1; otherwise, x_j is taken as 0. This transformation for the test dataset is not data snooping, we are just making a greater contrast for the feature variables, which is a legitimate data cleansing procedure. All the above procedures are implemented in **R** via Programme 6.1.

```
1  > library(keras)                          # MNIST dataset
2  >
3  > # Load MNIST dataset
4  > mnist <- dataset_mnist()
5  > X_train <- mnist$train$x                 # x: feature variables
6  > y_train <- mnist$train$y                 # y: labels
7  > X_test <- mnist$test$x
8  > y_test <- mnist$test$y
9  >
10 > # Limit the pixel value between 0 and 1 to avoid computational explode
11 > X_train[X_train!=0] <- 1
12 > X_test[X_test!=0] <- 1
13 >
14 > # Set index
15 > M <- length(unique(y_train))   # 10: numbers from 0 to 9
16 > img_dim <- dim(X_train)[2:3]   # 28x28 pixels
17 > N_train <- length(y_train)     # 60000 training samples
18 > N_test <- length(y_test)       # 10000 test samples
```

Programme 6.1 Initial settings for EM in **R**.

Here, the `unique()` function picks out all the distinguishable elements, in this case $\{0, 1, \ldots, 9\}$. The dimension of the training dataset is `60000, 28, 28`, where 60,000 is the total number of samples in the training dataset, and the two 28's are the length and width of each image datum respectively. Next, we build a function to compute the expected log-likelihood $Q(\boldsymbol{\theta} \mid \boldsymbol{\theta}^{(t)})$ according to (6.12); refer to Programme 6.2 for an implementation of this in **R**.

[3] According to the *Keras* website https://keras.io/about/, the name *Keras* comes from the Greek word "horn". It is a reference to a literary image from ancient Greek and Latin literature. In Python, *Keras* is a deep learning (e.g. ANN, CNN, LSTM) API (Application Programming Interface) running on top of the machine learning platform *TensorFlow*.

[4] Each pixel of a grayscale image has a value ranging from 0 to 255, whose value represents the amount of light intensity.

```
1  > # post: posterior probability; prior: π; and mu: μ
2  > loglike_func <- function(post, prior, mu){
3  +   loglike <- 0
4  +   mu[mu < 1e-323] <- 1e-323
5  +   mu[mu > 1 - 1e-323] <- 1 - 1e-323
6  +   for (m in 1:M){
7  +     for (n in 1:N_train){
8  +       log_sums <- sum(X_train[n,,]*log(mu[m,,])
9  +                       + (1-X_train[n,,])*log(1-mu[m,,]))
10 +       loglike <- loglike + post[n,m]*(log(prior[m]) + log_sums)
11 +     }
12 +   }
13 +   return(loglike)
14 + }
```

Programme 6.2 Expected log-likelihood function Q in R.

Referring to lines 4 and 5, we adjust the values of mu to avoid **R** returning `Inf` when computing $\ln 0$; instead it will return a very negative value. Then, in Step 1, we initialize $\boldsymbol{\pi}$ using uniform random numbers. Although the information in label y should not be used in unsupervised learning, we here initialize $\boldsymbol{\mu}$ using (6.8) as an illustration; see Programme 6.3 for coding this in **R**.

```
1  > # Initialize the algorithm with prior and mu
2  > prior <- runif(M)
3  > prior <- prior/sum(prior)          # Normalize pi such that it sums to 1
4  > mu <- array(data=0, dim=c(M, img_dim))
5  > for (m in 1:M){
6  +   digit <- X_train[which(y_train==(m-1)),,]
7  +   mu[m,,] <- colMeans(digit)        # mu dimension: 10x28x28
8  + }
```

Programme 6.3 Initializing $\boldsymbol{\pi}$ and $\boldsymbol{\mu}$ in **R**.

In Programme 6.4 via **R**, the model is trained based on Step 2 using (6.16) and (6.17) with 10 iterations (see Figure 6.5a for the choice of iterations as it is quite stable around this chosen value of 10).

```
1  > # Record prior and likelihood for each iterations
2  > iter <- 10
3  > prior_record <- array(data=0, dim=c(M, iter))
4  > loglike_record <- array(data=0, dim=c(iter, 1))
5  >
6  > for (it in 1:iter){
7  +   # Around 50secs for each iteration
8  +   print(paste("no. iteration:", it))
9  +   # Compute posterior probabilities using (6.17)
10 +   posterior <- array(data=0, dim=c(N_train, M))
11 +   nominator <- array(data=0, dim=c(M, 1))
12 +   for (n in 1:N_train){
13 +     for (m in 1:M){
```

```
14 +       binm <- (mu[m,,]^X_train[n,,])*(1-mu[m,,])^(1-X_train[n,,])
15 +       nominator[m] <- prior[m]*prod(binm)
16 +     }
17 +     posterior[n,] <- nominator/sum(nominator)
18 +   }
19 +   loglike_record[it] <- loglike_func(posterior, prior, mu)
20 +
21 +   # Update μ and π using (6.16)
22 +   for (m in 1:M){
23 +     px <- array(data=0, dim=img_dim)
24 +     for (n in 1:N_train){
25 +       px <- px + posterior[n,m]*X_train[n,,]
26 +     }
27 +     mu[m,,] <- px/sum(posterior[,m])
28 +     prior[m] <- sum(posterior[,m])/N_train
29 +   }
30 +   prior <- prior/sum(prior)
31 +   prior_record[,it] <- prior
32 + } # Iterate for iter=10 times with the following output
33 [1] "no. iteration: 1"
34 [1] "no. iteration: 2"
35 [1] "no. iteration: 3"
36 [1] "no. iteration: 5"
37 [1] "no. iteration: 6"
38 [1] "no. iteration: 7"
39 [1] "no. iteration: 8"
40 [1] "no. iteration: 9"
41 [1] "no. iteration: 10"
```

Programme 6.4 Training stage in R.

In the testing stage, to give a predicted label of a test datum $\boldsymbol{x}_{\text{test}}$, we first compute the posterior probabilities

$$\mathbb{P}(\text{the } m\text{-th parameter vector is used} \mid \boldsymbol{x}_{\text{test}} \text{ is realized}), \text{ for } m = 1, \ldots, M,$$

each of which is similar to (6.11) except that the parameter $\boldsymbol{\theta}$ used is the MLE obtained in the training stage. See Programme 6.5 for an implementation in R. Then the predicted label m is the one corresponding to the largest posterior probability.

```
1 > y_hat <- array(data=0, dim=c(N_test, 1))
2 > for (n in 1:N_test){
3 +   y_prob <- array(data=0, dim=c(M, 1))
4 +   for (m in 1:M){
5 +     binm <- (mu[m,,]^X_test[n,,]) * (1-mu[m,,])^(1-X_test[n,,])
6 +     y_prob[m] <- prior[m] * prod(binm)
7 +   }
8 +   y_hat[n] <- which.max(y_prob) - 1
9 + }
```

Programme 6.5 Test stage in R.

Note that we omit the division of the marginal probability after Line 7, since we are just required to obtain the index position m which maximizes the posterior probability.

Finally, in Programme 6.6 via **R**, the *confusion matrix* (*cross tabulation table*) is used to illustrate the overall performance of the model. We also calculate its accuracy and plot the log-likelihoods for all 10 iterations (see Figure 6.5a).

```
> # Table of predictions and prediction accuracy
> table(y_test, y_hat)                                  # Confusion matrix
      y_hat
y_test    0    1    2    3    4    5    6    7    8    9
     0  742    0    2   68    6  143   10    1    8    0
     1    0 1051    3    5    0   43    6    0   17   10
     2   12   10  828   59   30   26   30    9   26    2
     3    5   24   60  741    6   54    8   10   68   34
     4    0    3    3    2  447  152   17    2   26  330
     5   20    6    4  331   34  319   12    5  135   26
     6   16   23   22    8   18   78  790    0    1    2
     7    0   19   10    4   48    8    0  721   68  150
     8   15   16    7  233   26   69    8    8  539   53
     9    7   10    2    9  218   25    0   53   79  606
> sum(diag(table(y_test, y_hat)))/N_test     # Accuracy
[1] 0.6784
>
> # Plot for the log-likelihood for all 10 iterations
> plot(loglike_record, ylab="Log-likelihood", xlab="iterations")
>
> par(mfrow=c(3,4), mar=c(2,2,2,2))
> for (m in 1:M) image(t(apply(mu[m,,], 2, rev)))
```

Programme 6.6 Results in R.

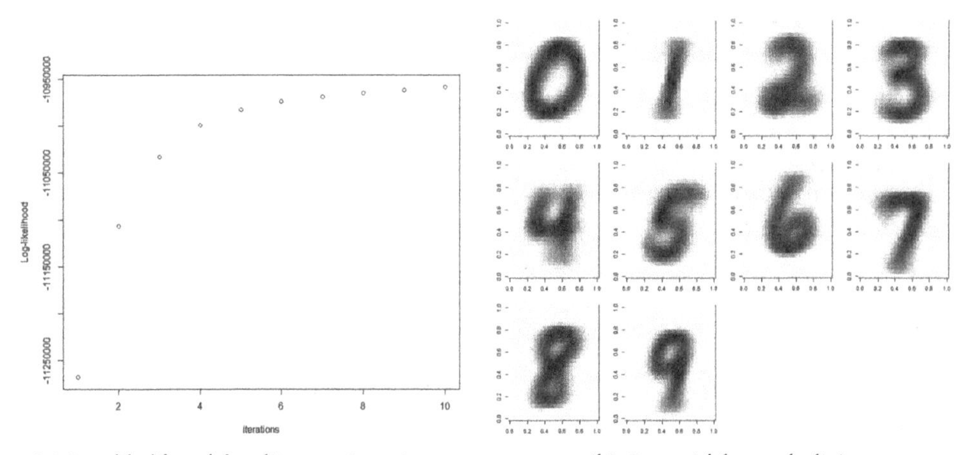

(a) Log-likelihood for different iterations.

(b) Centroid for each digit.

FIGURE 6.5 MNIST with Bernoulli Mixture Model, generated by Programme 6.6 in R.

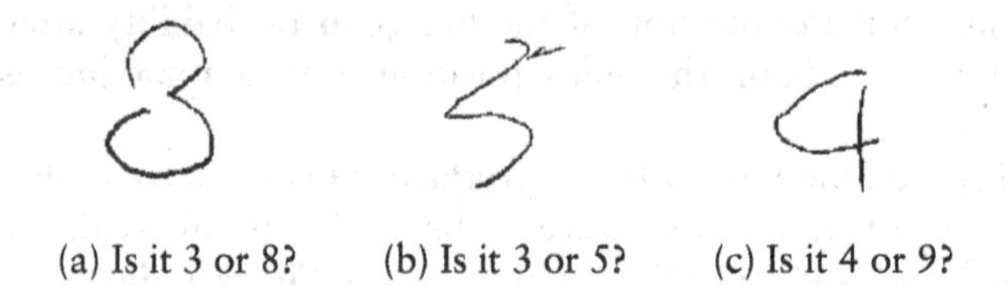

(a) Is it 3 or 8? (b) Is it 3 or 5? (c) Is it 4 or 9?

FIGURE 6.6 Some examples of confusing handwritten digits.

The accuracy of the model is around 68% and from Figure 6.5a, the log-likelihood becomes stable after the eighth iteration, indicating that only a few iterations are enough for the model to learn. Finally, the average handwritten images of each digit are plotted in Figure 6.5b. From the confusion matrix, we observe that the model performs less favourably in classifying numbers between 3 (`y_hat`) and 8 (`y_test`), 3 and 5, 4 and 9, 5 and 0, and 9 and 4. Indeed, sometimes even us humans struggle to recognise the true number, as Figure 6.6 demonstrates.

In the following, we provide the Python implementation of the entire procedure above.

```
import numpy as np
import tensorflow as tf
from tqdm import trange
from sklearn.metrics import confusion_matrix
import matplotlib.pyplot as plt

((X_train, y_train),
 (X_test, y_test)) = tf.keras.datasets.mnist.load_data()
# Limit the pixel value to avoid computational explode
X_train[np.where(X_train!=0)] = 1
X_test[np.where(X_test!=0)] = 1
M = len(np.unique(y_train))                  # 10: numbers from 0 to 9
img_dim = X_train.shape[1:]                  # 28x28 pixels
N_train = len(X_train)                       # 60000 training samples
N_test = len(X_test)                         # 10000 test samples

def loglike_func(post, prior, mu):
    loglike = 0
    mu = np.clip(mu, a_min=1e-323, a_max=1 - 1e-323)
    for m in range(M):
        # X_train / log_sums: (n, img1, img2); mu: (M, img1, img2)
        log_sums = np.sum(X_train*np.log(mu[m]) +
                          (1-X_train)*np.log(1-mu[m]), axis=(1,2))
        loglike += np.sum(post[:,m]*(np.log(prior[m])+log_sums))

    return loglike
```

```
27
28 np.random.seed(4002)
29 prior = np.random.rand(M)
30 prior /= np.sum(prior)
31 mu = np.empty((M,) + img_dim)
32 for m in range(M):
33     digit = X_train[np.where(y_train == m)[0]]
34     mu[m] = np.mean(digit, axis=0)
```

Programme 6.7 Initial settings for EM in Python.

Programme 6.8 illustrates the coding for the training stage of the EM algorithm in Python.

```
1  epochs = 10
2  prior_record = np.empty((M, epochs))
3  loglike_record = np.empty((epochs))
4  t = trange(epochs, desc="log likelihood: 00000")
5  for ep in t:
6      # Compute posterior probabilities
7      nominator = np.zeros((M, N_train))
8      for m in range(M):
9          binm = mu[m]**X_train * (1 - mu[m])**(1 - X_train)
10         nominator[m] = prior[m] * np.prod(binm, axis=(1,2))
11
12     # denominator = sum nominator
13     posterior = nominator.T / np.sum(nominator, axis=0).reshape(-1, 1)
14     loglike_record[ep] = loglike_func(posterior, prior, mu)
15     t.set_description(f"log likelihood:{np.round(loglike_record[ep])}")
16
17     # Update mu and prior
18     for m in range(M):
19         px = np.sum(posterior[:,m].reshape((-1, 1, 1))*X_train, axis=0)
20         mu[m] = px / np.sum(posterior[:,m])
21         prior[m] = np.sum(posterior[:,m]) / N_train
22
23     prior = prior/np.sum(prior)
24     prior_record[:, ep] = prior.T
```

Programme 6.8 Training stage in Python.

Lastly, we shall make predictions on the test dataset in Programme 6.9 (via Python). The results match those obtained via **R** in Programme 6.6.

```
1 y_prob = np.zeros((M, len(X_test)))
2 for m in range(M):
3     binm = mu[m]**X_test * (1 - mu[m])**(1 - X_test)
4     y_prob[m, :] = prior[m] * np.prod(binm, axis=(1,2))
```

```
5
6  y_hat = np.array(y_prob).argmax(axis=0)
7
8  # Table of predictions and prediction accuracy
9  tab = confusion_matrix(y_hat, y_test)           # Confusion matrix
10 print(tab)
11
12 # No need to inter-change labels cuz mu is first initialized by class
13 print(np.sum(tab.diagonal()) / len(y_test))     # Accuracy
14
15 # Plot for the log-likelihood for all 10 iterations
16 plt.figure(figsize=(10, 8))
17 plt.plot(np.arange(len(model.loglike_record))+1,
18          model.loglike_record, 'x-', linewidth=3)
19 plt.xlabel("Iterations", fontsize=15)
20 plt.ylabel("Log-likelihood", fontsize=15)
21 plt.title("BMM EM Algoritm", fontsize=15)
22
23 fig = plt.figure(figsize=(16, 12))
24 for m in range(M):
25     y = fig.add_subplot(3, 4, m+1)              # rows, columns, index
26     y.imshow(mu[m,:,:], aspect='auto', cmap="gray_r")
```

```
1  log likelihood:-10957705.0: 100% 10/10 [02:01<00:00, 12.17s/it]
2  [[ 742    0   12    5    0   20   15    0   14    7]
3   [   0 1049   10   24    2    6   23   19   16    9]
4   [   2    3  828   60    4    4   22    9    7    2]
5   [  68    5   59  742    1  331    8    4  235    9]
6   [   6    0   30    6  446   33   18   47   27  216]
7   [ 143   38   26   55  148  320   79    8   68   25]
8   [  10    6   30    8   17   12  790    0    8    0]
9   [   1    0    9    9    1    5    0  721    7   54]
10  [   8   24   26   67   31  136    1   73  539   80]
11  [   0   10    2   34  332   25    2  147   53  607]]
12 0.6784
```

Programme 6.8 Results in Python.

6.2 MM ALGORITHM

The *MM algorithm* is a general technique to minimize or maximize a function iteratively. When one is performing minimization, "MM" stands for "Majorize-Minimization"; when one is performing maximization, "MM" stands for "Minorize-Maximization". The main idea of the MM algorithm is to optimize a *surrogate function* that minorizes or majorizes the objective function at each iteration. Suppose that $f : \mathbb{R}^n \to \mathbb{R}$ is the objective function we want to minimize. MM algorithm produces a sequence $\{x^{(t)}\}$ iteratively as follows: at the t-th iteration,

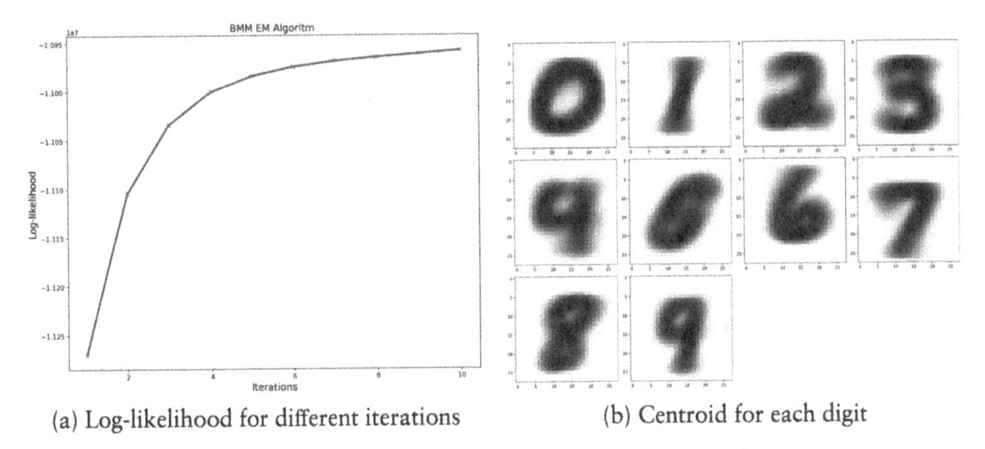

(a) Log-likelihood for different iterations (b) Centroid for each digit

FIGURE 6.7 MNIST with Bernoulli Mixture Model, generated by Programme 6.9 in Python.

a surrogate function $\boldsymbol{x} \mapsto g(\boldsymbol{x} \mid \boldsymbol{x}^{(t)})$ that satisfies

$$\begin{cases} g(\boldsymbol{x} \mid \boldsymbol{x}^{(t)}) \geq f(\boldsymbol{x}) & \text{for all } \boldsymbol{x}, \\ g(\boldsymbol{x}^{(t)} \mid \boldsymbol{x}^{(t)}) = f(\boldsymbol{x}^{(t)}), \end{cases}$$

is first found. Such surrogate function $g(\cdot \mid \boldsymbol{x}^{(t)})$ majorizes the objective function f and it "touches" f at the current iterate $\boldsymbol{x}^{(t)}$. We then minimize this surrogate function and label the minimizer $\boldsymbol{x}^{(t+1)}$. The sequence $\{\boldsymbol{x}^{(t)}\}$ produced by this algorithm has the desirable descent property that

$$f(\boldsymbol{x}^{(t+1)}) \leq f(\boldsymbol{x}^{(t)}) \quad \text{for all } t,$$

which follows from the defining properties of the surrogate function $g(\cdot \mid \boldsymbol{x}^{(t)})$:

$$f(\boldsymbol{x}^{(t+1)}) \leq g(\boldsymbol{x}^{(t+1)} \mid \boldsymbol{x}^{(t)}) \leq g(\boldsymbol{x}^{(t)} \mid \boldsymbol{x}^{(t)}) = f(\boldsymbol{x}^{(t)}).$$

When one is trying to *maximize* the objective function, a surrogate function that *minorizes* the objective function and touches it at the current iterate has to be constructed; the next iterate is then the *maximum* point of this surrogate function. Without loss of generality, we shall focus mainly on the minimization problem.

It is known that if the objective function f is strictly convex, then under some mild conditions the MM algorithm will converge to the unique minimum point (if it exists) no matter the initial seed $\boldsymbol{x}_0$ chosen. If convexity is lacking, but all stationary points are isolated, then the MM algorithm will converge to one of them, depending on the position of the initial guess. The performance and the rate of convergence of the MM algorithm depend on how well the surrogate function $g(\cdot \mid \boldsymbol{x}^{(t)})$

approximates f, and how easy the minimization of the surrogate function is. In what follows, we highlight several commonly used approaches to construct appropriate surrogate functions. More detailed discussions can be found in [18, 19, 20] and the references therein.

Firstly, suppose that the objective function $f : \mathbb{R}_+^p \to \mathbb{R}$ has the form $f(x) := h(c^\top x)$ for some $c \in \mathbb{R}_+^p$ and a convex function $h : \mathbb{R}_+ \to \mathbb{R}$. By Jensen's inequality, given the current iterate $x^{(t)} \in \mathbb{R}_+^p$,

$$f(x) = h(c^\top x) = h\left(\sum_{i=1}^{n} \frac{c_i x_i^{(t)}}{c^\top x^{(t)}} \frac{c^\top x^{(t)}}{x_i^{(t)}} x_i\right) \le \sum_{i=1}^{n} \frac{c_i x_i^{(t)}}{c^\top x^{(t)}} h\left(\frac{c^\top x^{(t)}}{x_i^{(t)}} x_i\right),$$

provided that $c^\top x^{(t)} \neq 0$. If we define $g(x \mid x^{(t)})$ to be the function on the right most, then $f(x) \le g(x \mid x^{(t)})$ and the two functions are equal at $x = x^{(t)}$. Thus $g(\cdot \mid x^{(t)})$ is a surrogate function of f at $x^{(t)}$. This surrogate function has the important property of being simply a weighted sum of one-dimensional convex functions so that the components of x are separated. This makes the minimization step much easier to perform. Secondly, a similar way to construct a separable surrogate function in the absence of the positivity of c or x was proposed in [4]:

$$\begin{aligned} f(x) = h(c^\top x) &= h\left(\sum_{i=1}^{n} \alpha_i \left(\frac{c_i}{\alpha_i}(x_i - x_i^{(t)}) + c^\top x^{(t)}\right)\right) \\ &\le \sum_{i=1}^{n} \alpha_i h\left(\frac{c_i}{\alpha_i}(x_i - x_i^{(t)}) + c^\top x^{(t)}\right) =: g(x \mid x^{(t)}), \end{aligned}$$

making use of Jensen's inequality again. Here, the α_i's are all positive numbers which sum to 1, and $\alpha_i > 0$ whenever $c_i = 0$.

Thirdly, if the objective function $f : \mathbb{R}^p \to \mathbb{R}$ is twice differentiable and the Hessian matrix $\nabla_x \nabla_x^\top f(x)$ is uniformly bounded from above (in the sense that there is a symmetric matrix B such that $B - \nabla_x \nabla_x^\top f(x)$ is still positive semidefinite for any x), then we can construct a surrogate function based on the second-order Taylor expansion of f at the current iterate $x^{(t)}$:

$$\begin{aligned} f(x) &= f(x^{(t)}) + \nabla_x f(x^{(t)})(x - x^{(t)}) + \frac{1}{2}(x - x^{(t)})^\top \nabla_x \nabla_x^\top f(z)(x - x^{(t)}) \\ &\le f(x^{(t)}) + \nabla_x f(x^{(t)})(x - x^{(t)}) + \frac{1}{2}(x - x^{(t)})^\top B (x - x^{(t)}) \\ &=: g(x \mid x^{(t)}). \end{aligned}$$

The surrogate function $g(x \mid x^{(t)})$ is quadratic in x, hence its minimization is practically easy. Indeed, the next iterate $x^{(t+1)}$ is simply the solution of the equation

$$\nabla_x g(x \mid x^{(t)}) = \nabla_x f(x^{(t)}) + (x - x^{(t)})^\top B = 0,$$

which is a system of linear equations that can be solved efficiently by many well-known algorithms. In the one-dimensional case, we have the explicit formula $x^{(t+1)} = x^{(t)} - f'(x^{(t)})/B$, provided $B \neq 0$.

Fourthly, quadratic surrogate functions are used quite often because the coefficients can be obtained by solving the tangency condition rather easily and finding the minimum point of a quadratic function is straightforward. The next example, taken from [12], uses a quadratic surrogate function to estimate quantiles of a distribution. As such, it has direct relevance for calculating Value-at-Risk, an important measure in risk management.

Example. Suppose that we are given n numbers $x_1, \ldots, x_n$, which are realizations of a certain unknown distribution F. Our objective is to estimate $F^{-1}(q)$, the q-quantile of F, by using the empirical q-quantile μ_q, which is a number that satisfies the condition

$$\frac{\sum_{i=1}^{n} \mathbb{1}_{\{x_i \le \mu_q\}}}{n} \ge q \quad \text{and} \quad \frac{\sum_{i=1}^{n} \mathbb{1}_{\{x_i \ge \mu_q\}}}{n} \ge 1 - q. \tag{6.18}$$

Using this definition to find μ_q relies on sorting $x_1, \ldots, x_n$. For instance, to find the empirical median $\mu_{0.5}$, we first sort the x_i's in ascending order $x_{(1)} \le \cdots \le x_{(n)}$. If n is odd, then $\mu_{0.5}$ is simply $x_{(\frac{n+1}{2})}$, the number exactly in the middle of the sorted sequence. Similarly, if n is even, the $\mu_{0.5}$ can be any number in the interval $[x_{(\frac{n}{2})}, x_{(\frac{n}{2}+1)}]$.

An alternative method to find μ_q, rather than dealing with time-consuming sorting, is to note that it minimizes the function

$$f_q(\mu) := \sum_{i=1}^{n} \rho_q(x_i - \mu), \quad \text{where } \rho_q(r) := qr - r\mathbb{1}_{\{r<0\}},$$

which can be seen directly from (6.18). We can minimize f_q using MM algorithm by constructing a suitable surrogate function. To this end, we first construct a surrogate function for the component function ρ_q at an arbitrary non-zero point, say r^*. As ρ_q is V-shaped, an obvious choice would be a convex quadratic function that touches ρ_q at both $|r^*|$ and $-|r^*|$. If we write the surrogate function as $\tilde{g}(r \mid r^*) = ar^2 + br + c$, the tangency consideration leads to the following system of equations:

$$\begin{cases} a|r^*|^2 + b|r^*| + c &= q|r^*|, \\ 2a|r^*| + b &= q; \\ a|r^*|^2 - b|r^*| + c &= -(q-1)|r^*|, \\ -2a|r^*| + b &= q - 1. \end{cases}$$

Solving this system yields $a = \frac{1}{4|r^*|}, b = q - \frac{1}{2}, c = \frac{|r^*|}{4}$, and so the quadratic surrogate function is given by

$$\tilde{g}(r \mid r^*) = \frac{1}{4|r^*|} r^2 + \left(q - \frac{1}{2}\right) r + \frac{|r^*|}{4} = \frac{1}{4}\left[\frac{r^2}{|r^*|} + (4q-2)r + |r^*|\right].$$

Since $\tilde{g}(r \mid r^*) \geq \rho_q(r)$ for all r and $\tilde{g}(r^* \mid r^*) = \rho_q(r^*)$, if we define

$$g(\mu \mid \mu^{(t)}) := \sum_{i=1}^{n} \tilde{g}(x_i - \mu \mid x_i - \mu^{(t)}),$$

then we have

$$g(\mu \mid \mu^{(t)}) \geq \sum_{i=1}^{n} \rho_q(x_i - \mu) = f_q(\mu),$$

$$\text{and} \qquad g(\mu^{(t)} \mid \mu^{(t)}) = \sum_{i=1}^{n} \rho_q(x_i - \mu^{(t)}) = f_q(\mu^{(t)}).$$

Thus $g(\mu \mid \mu^{(t)})$ is a surrogate function, and the next iterate $\mu^{(t+1)}$ can be obtained by minimizing $g(\mu \mid \mu^{(t)})$ over μ. Since the surrogate function is convex, its minimizer $\mu^{(t+1)}$ solves the first order condition

$$\frac{\partial}{\partial \mu} g(\mu \mid \mu^{(t)}) = \frac{1}{4} \sum_{i=1}^{n} \left[\frac{2(x_i - \mu)}{|x_i - \mu^{(t)}|} + (4q - 2) \right] = 0.$$

This yields

$$\mu^{(t+1)} = \frac{\sum_{i=1}^{n} w_i^{(t)} x_i + (2q - 1)n}{\sum_{i=1}^{n} w_i^{(t)}}, \quad \text{where } w_i^{(t)} = \frac{1}{|x_i - \mu^{(t)}|}.$$

An interesting feature of this algorithm is that it does not require the numbers $x_1, \ldots, x_n$ to be sorted. We refer to [12] for more discussions on this topic, such as its convergence and further applications.

In Section 8.3, we shall estimate Value-at-Risk by bootstrapping historical losses. Before then, let us implement the MM algorithm for finding this as a quantile in Programme 6.10 via Python (resp. Programme 6.11 via **R**). Readers may come back to this example after reading the first few sections of Chapter 8.

```
1  import numpy as np
2
3  def MM_quantile(x, q, tol=1e-18, maxit=1e4):
4    n = len(x)
5    mu_old = np.mean(x)
6    for i in range(int(maxit)):
7      w = 1/np.abs(x - mu_old)
8      mu_new = (sum(w * x) + (2*q - 1)*n)/sum(w)
9      if np.isnan(mu_new): return mu_old
10     if abs(mu_new - mu_old) < tol: return mu_new
11     mu_old = mu_new
12
13   return mu_new
```

Programme 6.10 Finding quantiles by using MM algorithm in Python.

```
1  > MM_quantile <- function(x, q, tol=1e-18, maxit=1e4){
2  +   n <- length(x)
3  +   mu_old <- mean(x)
4  +   for (i in 1:maxit){
5  +     w <- 1/abs(x - mu_old)
6  +     mu_new <- (sum(w * x) + (2*q - 1)*n)/sum(w)
7  +     if (is.na(mu_new)) return (mu_old)
8  +     if (abs(mu_new - mu_old) < tol) return (mu_new)
9  +     mu_old <- mu_new
10 +   }
11 +   return (mu_new)
12 + }
```

Programme 6.11 Finding quantiles by using MM algorithm in R.

Next, in Programme 6.12 in Python (resp. Programme 6.13 in **R**), we use the `MM_quantile()` function to compute the 1-day 99% Value-at-Risk with the losses obtained from Programme 8.1 in Python (resp. Programme 8.2 in **R**).

```
1 VaR_sim = MM_quantile(loss, 0.99) # 1-day 99% V@R
2 print(VaR_sim)
```

```
1 3570.8713433394432
```

Programme 6.12 Continuation of Programmes 6.10 and 8.1: Bootstrapping 1-day 99% Value-at-Risk of a portfolio consisting of HSBC, CLP, and CK with MM algorithm in Python.

```
1 > (VaR_sim <- MM_quantile(loss, 0.99)) # 1-day 99% V@R
2 [1] 3570.871
```

Programme 6.13 Continuation of Programmes 6.11 and 8.2: Bootstrapping 1-day 99% Value-at-Risk of a portfolio consisting of HSBC, CLP, and CK with MM algorithm in **R**.

We note here that since $nq = 983 \times 0.99 = 973.17$ is not an integer, the 0.99 quantile may be any number in the interval $(x_{(\lfloor 973.17 \rfloor)}, x_{(\lfloor 973.17 \rfloor+1)}] = (x_{(973)}, x_{(974)}] = (3,538.054, 3,570.871]$. Under MM algorithm, the quantile is chosen to be the right-hand boundary of 3,570.871. Indeed, when $\mu^{(t)}$ approaches $3,570.871$, $w_{(974)}^{(t)}$ becomes extremely large, causing $\mu^{(t+1)} \approx x_{(974)} = 3,570.871$. In comparison, as we shall discuss in detail in Section 8.3 of Chapter 8, the bootstrapping method results in an estimated 0.99 quantile of 3,543.961, which is very close to the linear interpolation for $x_{(973.17)}$ using $x_{(973)}$ and $x_{(974)}$, i.e. $x_{(973)} + 0.17(x_{(974)} - x_{(973)}) = 3,543.633$.

More detailed discussions on the MM algorithm can be found in [18, 19, 20] and the references therein.

*6.3 A SHORT COURSE ON THE THEORY OF MARKOV CHAINS

In this section, we provide a concise course on the theory of Markov chains based on which some MCMC algorithms can be built. In particular, we shall see how the network structure of the directed graph of a Markov chain determines its possession of a stationary distribution. For a more comprehensive account of the theory of Markov chains, we refer to [9, 16, 21, 23, 25]. A *Markov Chain* on a finite or countable state space is a discrete time stochastic process $\{\mathbf{x}_n\}_{n\in\mathbb{N}_0}$ with a state space either a finite set $S = \{s_1, \dots, s_K\}$ or a countable set $S = \{s_1, s_2, \dots\}$, such that the *transition probability* of the process moving from state s_i at time n to state s_j over the next time step has the following property:

$$\begin{aligned}&\mathbb{P}(\mathbf{x}_{n+1} = s_j \mid \mathbf{x}_0 = s_{i_0}, \dots, \mathbf{x}_{n-1} = s_{i_{n-1}}, \mathbf{x}_n = s_i)\\&= \mathbb{P}(\mathbf{x}_{n+1} = s_j \mid \mathbf{x}_n = s_i) =: p_{i,j}^{n,n+1},\end{aligned}$$

that is, it only depends on the last state visited, but not the whole past history. Further, if this transition probability does not depend on n, that is,

$$\mathbb{P}(\mathbf{x}_{n+1} = s_j \mid \mathbf{x}_n = s_i) = p_{i,j}^{n,n+1} = p_{i,j}, \quad \forall n \geq 0,$$

the corresponding Markov chain is said to be *stationary*.

If $\{\mathbf{x}_n\}$ is a stationary Markov chain with K states, the transition probability matrix is defined as

$$P = \begin{array}{c} (i,j) \\ \mathbf{x}_n = 1 \\ \mathbf{x}_n = 2 \\ \vdots \\ \mathbf{x}_n = K \end{array} \begin{array}{c} \begin{array}{cccc} \mathbf{x}_{n+1} = 1 & \mathbf{x}_{n+1} = 2 & \cdots & \mathbf{x}_{n+1} = K \end{array} \\ \begin{pmatrix} p_{1,1} & p_{1,2} & \cdots & p_{1,K} \\ p_{2,1} & p_{2,2} & \cdots & p_{2,K} \\ \vdots & \vdots & \ddots & \vdots \\ p_{K,1} & p_{K,2} & \cdots & p_{K,K} \end{pmatrix} \end{array},$$

where the i-th row is the conditional probability distribution of $\mathbf{x}_{n+1}$ given that $\mathbf{x}_n = i$. In particular, all entries of the transition probability matrix $\boldsymbol{P}$ are non-negative and each row sums to 1.

For a stationary Markov chain $\{\mathbf{x}_n\}$ with a transition probability matrix $\boldsymbol{P} = (p_{i,j})$, we define the n-step transition probabilities as

$$p_{i,j}^{(n)} = \mathbb{P}(\mathbf{x}_n = s_j \mid \mathbf{x}_0 = s_i).$$

Obviously, $p_{i,j}^{(0)} = 1$ if $i = j$, or 0 otherwise. For $n \geq 1$, $p_{i,j}^{(n)}$ can be computed recursively by the *Chapman–Kolmogorov equations*:

$$p_{i,j}^{(n)} = \sum_{k=0}^{\infty} p_{i,k} p_{k,j}^{(n-1)},$$

which can be proven as follows:

$$
\begin{aligned}
p_{i,j}^{(n)} &= \mathbb{P}(\mathbf{x}_n = s_j \mid \mathbf{x}_0 = s_i) \\
&= \sum_{k=0}^{\infty} \mathbb{P}(\mathbf{x}_n = s_j, \mathbf{x}_1 = s_k \mid \mathbf{x}_0 = s_i) \\
&= \sum_{k=0}^{\infty} \mathbb{P}(\mathbf{x}_n = s_j \mid \mathbf{x}_1 = s_k, \mathbf{x}_0 = s_i)\mathbb{P}(\mathbf{x}_1 = s_k \mid \mathbf{x}_0 = s_i) \\
&= \sum_{k=0}^{\infty} p_{k,j}^{(n-1)} p_{i,k}.
\end{aligned}
$$

If we denote by $P^{(n)}$ the n-step transition matrix so that its (i,j)-entry is $p_{i,j}^{(n)}$, then the Chapman–Kolmogorov equations imply that

$$
P^{(n)} = P \cdot P \cdot \cdots \cdot P = P^n. \tag{6.19}
$$

Often in statistical contexts, the state space $\mathcal{S}$ of a Markov chain is uncountable, say $\mathcal{S} = \mathbb{R}^p$ for example. It is implicitly assumed that $\mathcal{S}$ is endowed with a suitable topology and an appropriate σ-field, together with a measure on the latter. In the case of $\mathcal{S} = \mathbb{R}^p$, these correspond to the topology of *Borel sets*, the *Borel σ-field* on which the *Lebesgue measure* is defined. When the state space is uncountable, the transition probability cannot be expressed in a matrix form. Instead, the Markov property can be formulated by making use of the *transition operator* $\mathcal{P}_n : C_b(\mathcal{S}) \to C_b(\mathcal{S})$ defined by

$$
\mathcal{P}_n f(s) := \mathbb{E}[f(\mathbf{x}_{n+1}) | \mathbf{x}_n = s], \quad \text{for any } f \in C_b(\mathcal{S}),\ s \in \mathcal{S},
$$

where $C_b(\mathcal{S})$ denotes the collection of all bounded and continuous real-valued functions defined on $\mathcal{S}$. Without wading too deep into unnecessary technical details, we implicitly assume sufficient regularity of the transition density (conditional probability given $\mathbf{x}_n$) of the pair $(\mathbf{x}_n, \mathbf{x}_{n+1})$ from $\mathbf{x}_n$ to $\mathbf{x}_{n+1}$. With the transition operator, the *Markov property* can be formulated as

$$
\mathbb{P}\left(\mathbf{x}_{n+1} \in A_{n+1} | \mathbf{x}_n = s, \mathbf{x}_{n-1} \in A_{n-1}, \ldots, \mathbf{x}_1 \in A_1\right) = \mathcal{P}_n \mathbb{1}_{A_{n+1}}(s),
$$

for any $s \in \mathcal{S}$ and measurable subsets $A_i \subset \mathcal{S}$ for $i = 1, \ldots, n+1$. In what follows, even for a continuum state space, we confine our discussions to *time-homogeneous* Markov chains so that their corresponding $\mathcal{P}_n$'s are independent of n. Henceforth, we omit the subscript n and simply denote the transition operator by $\mathcal{P}$. Similar to the case with a finite state space, we can define the n-step transition operator $\mathcal{P}^{(n)} : C_b(\mathcal{S}) \to C_b(\mathcal{S})$ by

$$
\mathcal{P}^{(n)} f(s) = \mathbb{E}[f(\mathbf{x}_n) | \mathbf{x}_0 = s], \quad f \in C_b(\mathcal{S}),\ s \in \mathcal{S}.
$$

Of course, $\mathcal{P}^{(0)}$ is simply the identity operator. By the tower property,

$$\mathcal{P}^{(n)}f(s) = \mathbb{E}[\mathbb{E}[f(\mathbf{x}_n)|\mathbf{x}_{n-1}]|\mathbf{x}_0 = s] = \mathbb{E}[\mathcal{P}f(\mathbf{x}_{n-1})|\mathbf{x}_0 = s] = \mathcal{P}^{(n-1)}\circ\mathcal{P}f(s).$$

Applying this relation recursively, we see the Chapman–Kolmogorov equations in this setting become:

$$\mathcal{P}^{(n)} = \mathcal{P}^n, \quad n = 1, 2, \ldots. \tag{6.20}$$

*6.3.1 Regularity, Irreducibility, and Aperiodicity

A Markov chain with a finite state space and a transition probability matrix $\boldsymbol{P}$ is said to be *regular* if all entries in $\boldsymbol{P}^n$ are strictly positive for some integer $n \geq 0$. We aim to show that if the Markov chain $\{\mathbf{x}_n\}$ is regular and has a finite state space $\mathcal{S} = \{s_1, s_2, \ldots, s_K\}$, then the limiting distribution, $(\pi_1, \ldots, \pi_K)$ defined by

$$\pi_j = \lim_{n\to\infty} \mathbb{P}(\mathbf{x}_n = s_j \mid \mathbf{x}_0 = s_i) = \lim_{n\to\infty} p_{i,j}^{(n)}, \quad j = 1, \ldots, K,$$

exists, and these limits do not depend on the choice of the initial state i.

State s_j is said to be *accessible* from state s_i, denoted by $s_i \to s_j$, if $p_{i,j}^{(n)} > 0$ for some integer $n \geq 0$. Two states s_i and s_j are said to *communicate* (denoted by $s_i \leftrightarrow s_j$) if they are accessible from each other. It is straightforward to establish that *communication* is an *equivalence relation*: (i) $s_i \leftrightarrow s_i$ (reflexivity); (ii) if $s_i \leftrightarrow s_j$, then $s_j \leftrightarrow s_i$ (symmetry); and (iii) if $s_i \leftrightarrow s_j$ and $s_j \leftrightarrow s_k$, then $s_i \leftrightarrow s_k$ (transitivity). As a result, the states of a Markov chain can be partitioned into disjoint classes of communicating states so that two states s_i and s_j belong to the same class if and only if $s_i \leftrightarrow s_j$. If the whole state space forms one single communicating class, that is, every state communicates with each other, then the Markov chain is said to be *irreducible*. A sufficient (but not necessary) condition for a Markov chain to be regular is that it is irreducible.

Next, we define the *period* of state i of a Markov chain as follows:

$$d(i) = \begin{cases} \gcd\{n \geq 1 : p_{i,i}^{(n)} > 0\}, & \text{if } p_{i,i}^{(n)} > 0 \text{ for some } n \geq 1; \\ 0, & \text{if } p_{i,i}^{(n)} = 0, \quad \forall n \geq 1. \end{cases}$$

where gcd stands for the greatest common divisor of the set of numbers. Here are several properties of periods:

1. If $s_i \leftrightarrow s_j$, then $d(i) = d(j)$. i.e., the period is constant in each equivalence class.
   Proof. If $s_i \leftrightarrow s_j$, we have $p_{i,j}^{(k)} > 0$ for some k and $p_{j,i}^{(m)} > 0$ for some m. Consider

$$p_{i,i}^{(k+m)} = \sum_{r=0}^{\infty} p_{i,r}^{(k)} p_{r,i}^{(m)} \geq p_{i,j}^{(k)} p_{j,i}^{(m)} > 0.$$

By definition, $d(i)$ is the greatest common divisor of all periods n where $p_{i,i}^{(n)} > 0$, hence $k+m$ must be a multiple of $d(i)$. Since $s_i \leftrightarrow s_j$, there exists a path from s_j back to s_j, i.e. $p_{j,j}^{(l)} > 0$ for some l. Consider the path $s_i \to s_j \to s_j \to s_i$,

$$p_{i,i}^{(k+l+m)} = \sum_{r=0}^{\infty} \sum_{s=0}^{\infty} p_{i,r}^{(k)} p_{r,s}^{(l)} p_{s,i}^{(m)} \geq p_{i,j}^{(k)} p_{j,j}^{(l)} p_{j,i}^{(m)} > 0$$

Therefore, $k+l+m$ is also a multiple of $d(i)$. Since $k+m$ is a multiple of $d(i)$, l is also a multiple of $d(i)$. This is true for any l such that $p_{j,j}^{(l)} > 0$, which means $d(j) \geq d(i)$ by the definition of gcd. Finally, by reflexivity, one also has $d(i) \geq d(j)$. We therefore conclude that $d(i) = d(j)$.

2. If state s_j has a period $d(j)$, then there is an integer N depending on s_j such that for all integer $n \geq N$, $p_{j,j}^{(nd(j))} > 0$. i.e., a return to state s_j can occur at all sufficiently large multiples of $d(j)$. The proof of this claim is essentially same as that for the following Lemma 6.1, and so we omit it here.
3. Further, if $p_{j,i}^{(m)} > 0$, then $p_{j,i}^{(m+nd(i))} > 0$ for sufficiently large n.

A finite state Markov chain is said to be *aperiodic* if every state has period 1. It is easy to see that if a Markov chain is regular so that all entries of P^n are strictly positive for some n, then so do $P^{n+1}, P^{n+2}, \ldots$, hence the Markov chain is aperiodic. In what follows, we prove that the converse is also true: an irreducible and aperiodic Markov chain with a finite state space is regular. We first present a lemma, whose proof is taken from [1].

Lemma 6.1 Let $\mathcal{A} = \{a_1, a_2, \ldots\}$ be a set of positive integers which is

1. nonlattice: $\gcd(a_1, a_2, \ldots) = 1$; and
2. closed under addition: if $a, a' \in \mathcal{A}$, then $a + a' \in \mathcal{A}$.

Then there exists an integer N such that $n \in \mathcal{A}$ for all $n \geq N$.

If the first condition in this lemma is replaced by $\gcd(a_1, a_2, \ldots) = d$, then one can show, by considering the set $\tilde{\mathcal{A}} := \{ad^{-1} : a \in \mathcal{A}\}$, that there exists an integer N such that $nd \in \mathcal{A}$ for all $n \geq N$.

Proof of Lemma 6.1. Let $\tilde{\mathcal{A}}$ be the set of all finite linear combinations $c_1 a_1 + \cdots + c_n a_n$ with $a_i \in \mathcal{A}$ and coefficients $c_i \in \mathbb{Z}$, and $m^* = c_1^* a_1^* + \cdots + c_n^* a_n^*$ be the smallest possible positive element in $\tilde{\mathcal{A}}$. If $m^* > 1$, then there must exist at least one element, say $\tilde{a}$, in $\mathcal{A}$ that is not divided by m^* because $\mathcal{A}$ is nonlattice. In this case, the remainder when $\tilde{a}$ is divided by m^* is still a linear combination of $a_1^*, \ldots, a_n^*$ together with $\tilde{a}$ and it is still a positive element in $\tilde{\mathcal{A}}$, yet it is smaller than m^*, contradicting the minimum nature of m^*. Therefore, $m^* = 1$. Now, we define $M := |c_1^*|a_1^* + \cdots + |c_n^*|a_n^*$, and take

$N := M^2$. For any $n \geq N$, we write $n = qM + r$ with a quotient $q \geq M$ and a remainder $0 \leq r < M$. Then

$$n = qM + r = qM + rm^* = \sum_{i=1}^{n}(q|c_i^*| + rc_i^*)a_i^*.$$

As $q \geq M > r$, $q|c_i^*| + rc_i^* \geq 0$, so $n \in \mathcal{A}$. □

Theorem 6.1 Suppose that $\{\mathbf{x}_n\}$ is an aperiodic Markov Chain with a state space $\mathcal{S} = \{s_1, \ldots, s_K\}$ and a transition matrix $\boldsymbol{P}$. Then there exists an integer N such that

$$p_{i,i}^{(n)} > 0, \qquad \forall i = 1, 2, \ldots, K \quad \text{and} \quad \forall n \geq N.$$

Proof. For $s_i \in \mathcal{S}$, let $\mathcal{A}_i = \{n \geq 1 : p_{i,i}^{(n)} > 0\}$, that is, $\mathcal{A}_i$ is the set of possible return times to state s_i starting from s_i. As the Markov chain is assumed to be aperiodic, $\mathcal{A}_i$ is nonlattice. Furthermore, if $a, a' \in \mathcal{A}_i$, then

$$\mathbb{P}(\mathbf{x}_a = s_i \mid \mathbf{x}_0 = s_i) > 0 \quad \text{and} \quad \mathbb{P}(\mathbf{x}_{a+a'} = s_i \mid \mathbf{x}_a = s_i) > 0.$$

Thus,

$$\begin{aligned}\mathbb{P}(\mathbf{x}_{a+a'} = s_i \mid \mathbf{x}_0 = s_i) &\geq \mathbb{P}(\mathbf{x}_{a+a'} = s_i, \mathbf{x}_a = s_i \mid \mathbf{x}_0 = s_i) \\ &= \mathbb{P}(\mathbf{x}_{a+a'} = s_i \mid \mathbf{x}_a = s_i)\mathbb{P}(\mathbf{x}_a = s_i \mid \mathbf{x}_0 = s_i) > 0.\end{aligned}$$

This shows that $a + a' \in \mathcal{A}_i$ and so $\mathcal{A}_i$ is closed under addition. By Lemma 6.1, there exists an integer n_i such that $p_{i,i}^{(n)} > 0$ for all $n \geq n_i$. The theorem is then fully established by taking $N := \max\{n_1, \ldots, n_K\}$. □

Corollary 6.1 Suppose that $\{\mathbf{x}_n\}$ is an irreducible and aperiodic Markov chain with a state space $\mathcal{S} = \{s_1, \ldots, s_K\}$ and a transition matrix $\boldsymbol{P}$. Then there exists an integer M such that

$$p_{i,j}^{(n)} > 0, \qquad \forall i, j = 1, 2, \ldots, K \quad \text{and} \quad \forall n \geq M.$$

In other words, the Markov chain is regular.

Proof. By Theorem 6.1, there exists an integer N such that $p_{i,i}^{(n)} > 0$ for all i and for all $n \geq N$. Consider any two states $s_i, s_j \in \mathcal{S}$. As the chain is irreducible, we can find some integer $n_{i,j} \geq 0$ such that $p_{i,j}^{(n_{i,j})} > 0$. Let $M_{i,j} = N + n_{i,j}$. For any $m \geq M_{i,j}$, we have

$$\begin{aligned}\mathbb{P}(\mathbf{x}_m = s_j \mid \mathbf{x}_0 = s_i) &\geq \mathbb{P}(\mathbf{x}_{m-n_{i,j}} = s_i, \mathbf{x}_m = s_j \mid \mathbf{x}_0 = s_i) \\ &= \mathbb{P}(\mathbf{x}_m = s_j \mid \mathbf{x}_{m-n_{i,j}} = s_i)\mathbb{P}(\mathbf{x}_{m-n_{i,j}} = s_i \mid \mathbf{x}_0 = s_i) > 0.\end{aligned}$$

Therefore, $p_{i,j}^{(m)} > 0$ for all $m \geq M_{i,j}$. The corollary is proven by taking

$$M := \max\{M_{1,1}, M_{1,2}, \ldots, M_{1,K}, M_{2,1}, \ldots, M_{K,K}\}. \qquad \square$$

Let us now discuss the analogous results for Markov chains with an uncountable state space. Let $C_b^+(\mathcal{S})$ be the collection of all non-negative, bounded and continuous real-valued functions defined on $\mathcal{S}$. The Markov chain $\{\mathbf{x}_n\}$ is said to be *regular* if there exists $n \in \mathbb{N}$ such that, for $f \in C_b^+(\mathcal{S})$ that is not identically zero, we have $\mathcal{P}^n f(s) > 0$ for all $s \in \mathcal{S}$. Given any two states $s_1, s_2 \in \mathcal{S}$, we say that s_2 is *accessible* from s_1, (again) denoted by $s_1 \to s_2$, if there exists $n \in \mathbb{N}$ such that $\mathcal{P}^n f(s_1) > 0$ for any $f \in C_b^+(\mathcal{S})$ with $f(s_2) > 0$. In this case, we say that s_2 is accessible in n steps from s_1. Again, the two states are said to *communicate*, denoted by $s_1 \leftrightarrow s_2$, if $s_1 \to s_2$ and $s_2 \to s_1$. Similar to the case of a finite or countable state space, communication of states forms an equivalence relation, the proof of which can be found in Appendix *6.A.1. A Markov chain is said to be *irreducible* if every state communicates with each other.

To simplify the overall treatment, we impose the following technical assumptions.

Assumption 6.1

(i) $\mathcal{S}$ is a compact subset of $\mathbb{R}^p$ for some $p \in \mathbb{N}$.
(ii) For any $n \geq 0$, the transition density of $(\mathbf{x}_n, \mathbf{x}_{n+1})$ defined on $\mathcal{S} \times \mathcal{S}$ is jointly absolutely continuous.

Remark 6.1 The same argument of the proofs in this subsection can be extended to a non-compact continuous state space $\mathcal{S}$, in particular, when the transition density has a light tail.

The period of a state $s \in \mathcal{S}$ can be defined analogously as follows. For any $s \in \mathcal{S}$, let $I_{\mathcal{P}}(s) := \{n : \mathcal{P}^n f(s) > 0 \text{ for any } f \in C_b^+(\mathcal{S}), f(s) > 0\} \subset \mathbb{N}_0 := \{0, 1, 2, \ldots\}$. The period $d(s)$ of s is defined as:

$$d(s) := \begin{cases} \gcd\left(I_{\mathcal{P}}(s)\right), & \text{if } I_{\mathcal{P}}(s) \neq \emptyset; \\ 0, & \text{otherwise.} \end{cases}$$

An uncountable state Markov chain $\{\mathbf{x}_n\}$ is said to be *aperiodic* if $d(s) = 1$ for all $s \in \mathcal{S}$.

The following proposition summarizes some properties of periods of a Markov chain with an uncountable state space. It shares the same structure as those for finite state Markov chains, and its proof can be found in Appendix *6.A.2.

Proposition 6.1

(i) If $s_1 \leftrightarrow s_2$, then $d(s_1) = d(s_2)$.
(ii) Under Assumption 6.1, for every $s \in \mathcal{S}$, there exists $N_s \in \mathbb{N}$ depending on s such that $\mathcal{P}^{nd(s)} f(s) > 0$ for any $f \geq 0$ with $f(s) > 0$ whenever $n \geq N_s$.

(iii) If s_1 can access s_2 in m steps, then s_1 can also access s_2 in $m + nd(s_1)$ steps for any sufficiently large $n \geq 0$.

The following results manifest that an aperiodic and irreducible Markov chain is regular; they are analogous to Theorem 6.1 and Corollary 6.1 in the finite state space case. Their proofs are provided in Appendix *6.A.3.

Theorem 6.2 Let $\{\mathrm{x}_n\}$ be an aperiodic Markov chain with an uncountable state space. Under Assumption 6.1, there exists $N \in \mathbb{N}$ such that for any $n \geq N$ and any $s \in S$, it holds that $\mathcal{P}^n f(s) > 0$ for any $f \in C_b^+(S)$ with $f(s) > 0$.

Corollary 6.2 Let $\{\mathrm{x}_n\}$ be an aperiodic and irreducible Markov chain with an uncountable state space. Under Assumption 6.1, there exists $N \in \mathbb{N}$ such that for any $n \geq N$ and any $s_1, s_2 \in S$, it holds that $\mathcal{P}^n f(s_1) > 0$ for any $f \in C_b^+(S)$ with $f(s_2) > 0$.

*6.3.2 Stationary Markov Chains

A *stationary distribution* of a Markov chain with a transition probability matrix $\boldsymbol{P}$ and a finite state space $S = \{s_1, \ldots, s_K\}$ is a probability distribution $\boldsymbol{\pi} = (\pi_1, \ldots, \pi_K)$ that satisfies $\boldsymbol{\pi P} = \boldsymbol{\pi}$, that is, $\sum_{i=1}^K \pi_i p_{i,j} = \pi_j$ for $j = 1, \ldots, K$. The term "stationary" comes from the property that if a Markov chain distributes according to a stationary distribution initially, then the chain will follow the same distribution at all future time points.

By definition, when the limiting distribution $(\pi_1, \ldots, \pi_K)$, defined by

$$\pi_j = \lim_{n\to\infty} \mathbb{P}(\mathrm{x}_n = s_j \mid \mathrm{x}_0 = s_i) = \lim_{n\to\infty} p_{i,j}^{(n)},$$

exists, it is always a stationary distribution. However, the converse is not true. For instance, for the periodic Markov chain with the following transition probability matrix:

$$\boldsymbol{P} = \begin{pmatrix} 0 & 1 \\ 1 & 0 \end{pmatrix},$$

there cannot be a limiting distribution; on the other hand, $\boldsymbol{\pi} = (\frac{1}{2}, \frac{1}{2})$ is clearly its stationary distribution.

A fundamental theorem in Markov chain theory is that every irreducible and aperiodic Markov chain with a finite state space has a unique stationary distribution. To establish this result, we need the following lemma.

Lemma 6.2 Suppose that $\{\mathrm{x}_n\}$ is an irreducible and aperiodic Markov chain with a finite state space $S = \{s_1, \ldots, s_K\}$ and a transition matrix $\boldsymbol{P}$. For any two states $s_i, s_j \in S$, let $\mathrm{T}_{i,j} := \min\{n \geq 1, \mathrm{x}_n = s_j \mid \mathrm{x}_0 = s_i\}$. Then the mean hitting time $\tau_{i,j} := \mathbb{E}(\mathrm{T}_{i,j})$ is finite. In particular, $\mathbb{P}(\mathrm{T}_{i,j} < \infty) = 1$.

Any irreducible Markov chains possessing all finite mean hitting times between any two states are said to be *positive recurrent*.

Proof. By Corollary 6.1, we can find an integer M such that $p_{i,j}^{(m)} > 0$ for every pair of i, j whenever $m \geq M$. Set $\alpha := \min_{i,j}\{p_{i,j}^{(M)}\} > 0$. Fix any two states s_i and s_j, and suppose that the chain starts at s_i. Then

$$\mathbb{P}(\mathrm{T}_{i,j} > M) \leq \mathbb{P}(\mathbf{x}_M \neq s_j \mid \mathbf{x}_0 = s_i) \leq 1 - \alpha.$$

Furthermore, given everything that has happened up to time M, we have a conditional probability at least α of hitting state s_j at the time $2M$ no matter where the chain is at time M, so that

$$\begin{aligned}\mathbb{P}(\mathrm{T}_{i,j} > 2M) &= \mathbb{P}(\mathrm{T}_{i,j} > M)\mathbb{P}(\mathrm{T}_{i,j} > 2M \mid \mathrm{T}_{i,j} > M)\\ &\leq \mathbb{P}(\mathrm{T}_{i,j} > M)\mathbb{P}(\mathbf{x}_{2M} \neq s_j \mid \mathbf{x}_0 = s_i, \mathrm{T}_{i,j} > M) \leq (1-\alpha)^2.\end{aligned}$$

Iterating this argument, we see that for any ℓ,

$$\mathbb{P}(\mathrm{T}_{i,j} > \ell M) \leq (1-\alpha)^{\ell} \to 0, \qquad \text{as} \quad \ell \to \infty.$$

Therefore, $\mathbb{P}(\mathrm{T}_{i,j} = \infty) = 0$. Moreover, we have

$$\begin{aligned}\mathbb{E}(\mathrm{T}_{i,j}) &= \sum_{n=0}^{\infty} \mathbb{P}(\mathrm{T}_{i,j} > n) = \sum_{\ell=0}^{\infty} \sum_{n=\ell M}^{(\ell+1)M-1} \mathbb{P}(\mathrm{T}_{i,j} > n)\\ &\leq \sum_{\ell=0}^{\infty} \sum_{n=\ell M}^{(\ell+1)M-1} \mathbb{P}(\mathrm{T}_{i,j} > \ell M) \leq M \sum_{\ell=0}^{\infty} (1-\alpha)^{\ell}\\ &= \frac{M}{\alpha} < \infty.\end{aligned}$$

This finishes the proof of this lemma. □

Theorem 6.3 ***(Existence of Stationary Distribution)*** Every irreducible and aperiodic Markov chain with a finite state space has at least one stationary distribution.

Proof. For $i = 1, \ldots, K$, let

$$\rho_i := \mathbb{E}\left\{\sum_{n=0}^{\infty} \mathbb{1}_{\{\mathbf{x}_n = s_i, \mathrm{T}_{1,1} > n\}}\right\} = \sum_{n=0}^{\infty} \mathbb{P}(\mathbf{x}_n = s_i, \mathrm{T}_{1,1} > n)$$

be the expected number of visits to state s_i up to the time $\mathrm{T}_{1,1} - 1$. By Lemma 6.2, $\mathbb{E}(\mathrm{T}_{1,1}) = \tau_{1,1} < \infty$, and clearly $\rho_i < \tau_{1,1}$, which implies $\rho_i < \infty$. Consider the following distribution:

$$\pi = (\pi_1, \ldots, \pi_K) := \left(\frac{\rho_1}{\tau_{1,1}}, \frac{\rho_2}{\tau_{1,1}}, \ldots, \frac{\rho_K}{\tau_{1,1}}\right).$$

We claim that this is indeed a stationary distribution for the Markov chain. To this end, we first consider the case where $j \neq 1$. Then

$$\begin{aligned}
\pi_j = \frac{\rho_j}{\tau_{1,1}} &= \frac{1}{\tau_{1,1}} \sum_{n=0}^{\infty} \mathbb{P}(\mathbf{x}_n = s_j, \mathrm{T}_{1,1} > n) \\
&= \frac{1}{\tau_{1,1}} \sum_{n=1}^{\infty} \mathbb{P}(\mathbf{x}_n = s_j, \mathrm{T}_{1,1} > n) \\
&= \frac{1}{\tau_{1,1}} \sum_{n=1}^{\infty} \mathbb{P}(\mathbf{x}_n = s_j, \mathrm{T}_{1,1} > n-1) \\
&= \frac{1}{\tau_{1,1}} \sum_{n=1}^{\infty} \sum_{i=1}^{K} \mathbb{P}(\mathbf{x}_n = s_j, \mathbf{x}_{n-1} = s_i, \mathrm{T}_{1,1} > n-1) \\
&= \frac{1}{\tau_{1,1}} \sum_{n=1}^{\infty} \sum_{i=1}^{K} \mathbb{P}(\mathbf{x}_{n-1} = s_i, \mathrm{T}_{1,1} > n-1)\mathbb{P}(\mathbf{x}_n = s_j \mid \mathbf{x}_{n-1} = s_i) \\
&= \frac{1}{\tau_{1,1}} \sum_{n=1}^{\infty} \sum_{i=1}^{K} p_{i,j}\mathbb{P}(\mathbf{x}_{n-1} = s_i, \mathrm{T}_{1,1} > n-1) \\
&= \frac{1}{\tau_{1,1}} \sum_{i=1}^{K} p_{i,j} \sum_{m=0}^{\infty} \mathbb{P}(\mathbf{x}_m = s_i, \mathrm{T}_{1,1} > m) \\
&= \frac{\sum_{i=1}^{K} \rho_i p_{i,j}}{\tau_{1,1}} = \sum_{i=1}^{K} \pi_i p_{i,j}.
\end{aligned}$$

For the case of $j = 1$, by definition $\rho_1 = \mathbb{P}(\mathbf{x}_0 = s_1, \mathrm{T}_{1,1} > 0) + 0 + \cdots + 0 = 1$, and so

$$\begin{aligned}
\rho_1 = 1 &= \mathbb{P}(\mathrm{T}_{1,1} < \infty) \\
&= \sum_{n=1}^{\infty} \mathbb{P}(\mathrm{T}_{1,1} = n) = \sum_{n=1}^{\infty} \mathbb{P}(\mathrm{T}_{1,1} > n-1, \mathbf{x}_n = s_1) \\
&= \sum_{n=1}^{\infty} \sum_{i=1}^{K} \mathbb{P}(\mathbf{x}_{n-1} = s_i, \mathrm{T}_{1,1} > n-1, \mathbf{x}_n = s_1) \\
&= \sum_{n=1}^{\infty} \sum_{i=1}^{K} \mathbb{P}(\mathbf{x}_n = s_1 \mid \mathbf{x}_{n-1} = s_i)\mathbb{P}(\mathbf{x}_{n-1} = s_i, \mathrm{T}_{1,1} > n-1) \\
&= \sum_{n=1}^{\infty} \sum_{i=1}^{K} p_{i,1}\mathbb{P}(\mathbf{x}_{n-1} = s_i, \mathrm{T}_{1,1} > n-1)
\end{aligned}$$

$$= \sum_{i=1}^{K} p_{i,1} \sum_{m=0}^{\infty} \mathbb{P}(\mathbf{x}_m = s_i, \mathrm{T}_{1,1} > m)$$

$$= \sum_{i=1}^{K} \rho_i p_{i,1}.$$

Hence $\pi_1 = \frac{\rho_1}{\tau_{1,1}} = \sum_{i=1}^{K} \frac{\rho_i p_{i,1}}{\tau_{i,1}} = \sum_{i=1}^{K} \pi_i p_{i,1}$, as desired. □

Suppose that $\boldsymbol{\nu}^{(1)} = (\nu_1^{(1)}, \ldots, \nu_k^{(1)})$ and $\boldsymbol{\nu}^{(2)} = (\nu_1^{(2)}, \ldots, \nu_k^{(2)})$ are probability distributions on the finite state space $\mathcal{S} = \{s_1, \ldots, s_K\}$. The *total variation distance* between $\boldsymbol{\nu}^{(1)}$ and $\boldsymbol{\nu}^{(2)}$ is defined as

$$d_{\mathrm{TV}}(\boldsymbol{\nu}^{(1)}, \boldsymbol{\nu}^{(2)}) = \frac{1}{2} \sum_{i=1}^{K} \left| \nu_i^{(1)} - \nu_i^{(2)} \right|.$$

The constant $\frac{1}{2}$ is included to make the total variation distance d_{TV} take values between 0 and 1. If $d_{\mathrm{TV}}(\boldsymbol{\nu}^{(1)}, \boldsymbol{\nu}^{(2)}) = 0$, then $\boldsymbol{\nu}^{(1)} = \boldsymbol{\nu}^{(2)}$. For the other extreme case where $d_{\mathrm{TV}}(\boldsymbol{\nu}^{(1)}, \boldsymbol{\nu}^{(2)}) = 1$, $\boldsymbol{\nu}^{(1)}$ and $\boldsymbol{\nu}^{(2)}$ are singular to each other in the sense that $\mathcal{S}$ can be partitioned into two disjoint subsets $\mathcal{S}'$ and $\mathcal{S}''$ such that $\boldsymbol{\nu}^{(1)}$ puts all of its probability mass on $\mathcal{S}'$ while $\boldsymbol{\nu}^{(2)}$ puts all of its probability mass on $\mathcal{S}''$. If $\boldsymbol{\nu}^{(1)}, \boldsymbol{\nu}^{(2)}, \ldots,$ and $\boldsymbol{\nu}$ are probability distributions on $\mathcal{S}$, then we say that $\boldsymbol{\nu}^{(n)}$ converges to $\boldsymbol{\nu}$ in total variation as $n \to \infty$ if

$$\lim_{n\to\infty} d_{\mathrm{TV}}(\boldsymbol{\nu}^{(n)}, \boldsymbol{\nu}) = 0.$$

The total variation distance also has the natural interpretation of being the maximal absolute difference between the probabilities that the two distributions assign to any one event:

$$d_{\mathrm{TV}}(\boldsymbol{\nu}^{(1)}, \boldsymbol{\nu}^{(2)}) = \max_{A\subseteq\mathcal{S}} |\boldsymbol{\nu}^{(1)}(A) - \boldsymbol{\nu}^{(2)}(A)|. \tag{6.21}$$

Theorem 6.4 *(Markov Chain Convergence Theorem: Finite State Space)* Let $\{\mathbf{x}_n\}$ be an irreducible and aperiodic Markov chain with a finite state space $\mathcal{S} = \{s_1, \ldots, s_K\}$, a transition matrix $\boldsymbol{P}$, and an arbitrary initial distribution $\boldsymbol{\mu}^{(0)}$. Then for any distribution $\boldsymbol{\pi}$ which is stationary for $\boldsymbol{P}$, we have

$$\boldsymbol{\mu}^{(n)} \overset{\mathrm{TV}}{\to} \boldsymbol{\pi},$$

where $\boldsymbol{\mu}^{(n)}$ is the distribution of $\mathbf{x}_n$ initialized by $\boldsymbol{\mu}^{(0)}$.

Proof. When studying the behavior of $\boldsymbol{\mu}^{(n)}$, we may assume that $(\mathbf{x}_0, \mathbf{x}_1, \ldots)$ has been obtained by the simulation method:

$$\mathbf{x}_0 = \psi_{\boldsymbol{\mu}^{(0)}}(\mathrm{u}_0) \sim \boldsymbol{\mu}^{(0)}, \quad \mathbf{x}_n = \phi(\mathbf{x}_{n-1}, \mathrm{u}_n) \sim \mathbb{P}(\mathbf{x}_n = \cdot|\mathbf{x}_{n-1}), \quad n = 1, 2, \ldots,$$

where $\psi_{\mu^{(0)}}$ is a valid initiation function[5] for $\mu^{(0)}$, ϕ is a valid update function for $\boldsymbol{P}$, and $(\mathrm{u}_0, \mathrm{u}_1, \ldots)$ is an iid sequence of uniform random variables on $[0,1]$; note that with a suitable measurable choice of $\psi_{\mu^{(0)}}$ and ϕ, we need not take these uniform random elements from the hypercube $[0,1]^p$. We then introduce a second Markov chain $\{\mathbf{x}'_n\}$ by letting $\psi_{\boldsymbol{\pi}}$ be a valid initiation function for the distribution $\boldsymbol{\pi}$, also letting $(\mathrm{u}'_0, \mathrm{u}'_1, \ldots)$ be another iid sequence (independent of $(\mathrm{u}_0, \mathrm{u}_1, \ldots)$) of uniform random variables on $[0,1]$, and setting

$$\mathbf{x}'_0 = \psi_{\boldsymbol{\pi}}(\mathrm{u}'_0) \sim \boldsymbol{\pi}, \quad \mathbf{x}'_n = \phi(\mathbf{x}'_{n-1}, \mathrm{u}'_n) \sim \mathbb{P}(\mathbf{x}_n = \cdot | \mathbf{x}_{n-1}) \equiv \boldsymbol{\pi}, \quad n = 1, 2, , \ldots.$$

Since $\boldsymbol{\pi}$ is a stationary distribution, we have that $\mathbf{x}'_n \sim \boldsymbol{\pi}$ for all n. Also, the chains $(\mathbf{x}_0, \mathbf{x}_1, \ldots)$ and $(\mathbf{x}'_0, \mathbf{x}'_1, \ldots)$ are independent of each other in accordance with the assumption that the sequences of random variables $(\mathrm{u}_0, \mathrm{u}_1, \ldots)$ and $(\mathrm{u}'_0, \mathrm{u}'_1, \ldots)$ are independent of each other.

A key step in the proof is to show that, with probability 1, the two chains will meet, meaning that $\mathbf{x}_n = \mathbf{x}'_n$ for some n. To show this, define the first meeting time $\mathrm{T} = \min\{n : \mathbf{x}_n = \mathbf{x}'_n\}$, with the convention that $\mathrm{T} = \infty$ if the two chains never meet. Since the Markov chain $\{\mathbf{x}_n\}$ is irreducible and aperiodic, we can find, using Corollary 6.1, an integer M such that $p_{i,j}^{(m)} > 0$ for all i, j and $m \geq M$. Setting $\alpha := \min_{i,j} p_{i,j}^{(M)} > 0$, we have

$$\begin{aligned}
&\mathbb{P}(\mathrm{T} \leq M) \geq \mathbb{P}(\mathbf{x}_M = \mathbf{x}'_M) \\
&\geq \mathbb{P}(\mathbf{x}_M = s_1, \mathbf{x}'_M = s_1) \\
&= \mathbb{P}(\mathbf{x}_M = s_1)\mathbb{P}(\mathbf{x}'_M = s_1) \\
&= \left(\sum_{i=1}^{K} \mathbb{P}(\mathbf{x}_M = s_1, \mathbf{x}_0 = s_i)\right) \cdot \left(\sum_{i=1}^{K} \mathbb{P}(\mathbf{x}'_M = s_1, \mathbf{x}'_0 = s_i)\right) \\
&= \left(\sum_{i=1}^{K} \mathbb{P}(\mathbf{x}_M = s_1 \mid \mathbf{x}_0 = s_i)\mathbb{P}(\mathbf{x}_0 = s_i)\right) \\
&\quad \cdot \left(\sum_{i=1}^{K} \mathbb{P}(\mathbf{x}'_M = s_1 \mid \mathbf{x}'_0 = s_i)\mathbb{P}(\mathbf{x}'_0 = s_i)\right) \\
&\geq \left(\alpha \sum_{i=1}^{K} \mathbb{P}(\mathbf{x}_0 = s_i)\right) \cdot \left(\alpha \sum_{i=1}^{K} \mathbb{P}(\mathbf{x}'_0 = s_i)\right) = \alpha^2.
\end{aligned}$$

Hence, we have $\mathbb{P}(\mathrm{T} > M) \leq 1 - \alpha^2$. Similarly, given everything that has happened up to time M, we have a conditional probability of at least α^2 of having $\mathbf{x}_{2M} = \mathbf{x}'_{2M} = s_1$

[5] This is actually the inverse distribution function introduced in Section 5.1 of Chapter 5, in which various simulation techniques are discussed.

no matter where the chains are at time M, so that

$$\mathbb{P}(\mathbf{x}_{2M} \neq \mathbf{x}'_{2M} \mid \mathrm{T} > M) \leq 1 - \alpha^2.$$

Hence,

$$\begin{aligned}\mathbb{P}(\mathrm{T} > 2M) &= \mathbb{P}(\mathrm{T} > M)\mathbb{P}(\mathrm{T} > 2M \mid \mathrm{T} > M)\\ &\leq (1-\alpha^2)\mathbb{P}(\mathrm{T} > 2M \mid \mathrm{T} > M)\\ &\leq (1-\alpha^2)\mathbb{P}(\mathbf{x}_{2M} \neq \mathbf{x}'_{2M} \mid \mathrm{T} > M)\\ &\leq (1-\alpha^2)^2.\end{aligned}$$

By iterating this argument, we arrive at, for any ℓ,

$$\mathbb{P}(\mathrm{T} > \ell M) \leq (1-\alpha^2)^{\ell} \to 0, \qquad \text{as} \quad \ell \to \infty, \tag{6.22}$$

and so $\mathbb{P}(\mathrm{T} = \infty) = 0$. In other words, the two chains will meet eventually with probability 1.

Using the *coupling* concept, the next step of the proof is to construct a third Markov chain $\{\mathbf{x}''_n\}$ by setting $\mathbf{x}''_0 = \mathbf{x}_0$ (so that $\mathbf{x}''_0 \sim \boldsymbol{\mu}^{(0)}$), and for each $n \geq 0$,

$$\mathbf{x}''_{n+1} = \begin{cases} \phi(\mathbf{x}''_n, \mathbf{u}_{n+1}), & \text{if } \mathbf{x}''_n \neq \mathbf{x}'_n;\\ \phi(\mathbf{x}''_n, \mathbf{u}'_{n+1}), & \text{if } \mathbf{x}''_n = \mathbf{x}'_n.\end{cases}$$

In other words, the chain $\{\mathbf{x}''_n\}$ evolves exactly like the chain $\{\mathbf{x}_n\}$ until the time T when it first meets the chain $\{\mathbf{x}'_n\}$. It then switches to evolving like the chain $\{\mathbf{x}'_n\}$. It is important to realize that the chain $\{\mathbf{x}''_n\}$ has the same transition matrix $\boldsymbol{P}$, since at each update, the update function is exposed to a new independent uniform random variable, while each $\mathbf{u}_{n+1}$ and $\mathbf{u}'_{n+1}$ by assumption are independent to all previous uniform random variables with the same distribution.

As $\mathbf{x}''_0 \sim \boldsymbol{\mu}^{(0)}$, it follows that $\mathbf{x}''_n \sim \boldsymbol{\mu}^{(n)}$ for all n. For any $i \in \{1, \dots, K\}$, we get

$$\begin{aligned}\mu_i^{(n)} - \pi_i &= \mathbb{P}(\mathbf{x}''_n = s_i) - \mathbb{P}(\mathbf{x}'_n = s_i)\\ &\leq \mathbb{P}(\mathbf{x}''_n = s_i, \mathbf{x}'_n \neq s_i)\\ &\leq \mathbb{P}(\mathbf{x}''_n \neq \mathbf{x}'_n)\\ &= \mathbb{P}(\mathrm{T} > n) \to 0, \qquad \text{as } n \to \infty.\end{aligned} \tag{6.23}$$

Using the same argument (with the roles of $\mathbf{x}''_n$ and $\mathbf{x}'_n$ interchanged), we also see that

$$\pi_i - \mu_i^{(n)} \leq \mathbb{P}(\mathrm{T} > n) \to 0, \qquad \text{as} \quad n \to \infty. \tag{6.24}$$

Combining (6.23) and (6.24), we obtain that

$$\lim_{n\to\infty} |\mu_i^{(n)} - \pi_i| = 0, \quad \text{for every } i = 1, 2, \dots, K,$$

and so

$$\lim_{n\to\infty} d_{\mathrm{TV}}(\boldsymbol{\mu}^{(n)}, \boldsymbol{\pi}) = \lim_{n\to\infty} \left(\frac{1}{2} \sum_{i=1}^{K} |\mu_i^{(n)} - \pi_i| \right) = 0.$$

This completes the proof. □

Theorem 6.5 *(Uniqueness of the Stationary Distribution)* Any irreducible and aperiodic Markov chain with a finite state space has exactly one stationary distribution.

Proof. Let $\{\mathbf{x}_n\}$ be an irreducible and aperiodic Markov chain with a finite state space. By Theorem 6.3, there exists at least one stationary distribution, so we only need to show that there is at most one stationary distribution. Let $\boldsymbol{\pi}$ and $\boldsymbol{\pi}'$ be two stationary distributions for the Markov chain. Suppose that the Markov chain starts with an initial distribution $\boldsymbol{\mu}^{(0)} = \boldsymbol{\pi}'$. Then, by the stationarity assumption of $\boldsymbol{\pi}'$, $\boldsymbol{\mu}^{(n)} = \boldsymbol{\pi}'$ for all n. On the other hand, by Theorem 6.4,

$$0 = \lim_{n\to\infty} d_{\mathrm{TV}}(\boldsymbol{\mu}^{(n)}, \boldsymbol{\pi}) = \lim_{n\to\infty} d_{\mathrm{TV}}(\boldsymbol{\pi}', \boldsymbol{\pi}) = d_{\mathrm{TV}}(\boldsymbol{\pi}', \boldsymbol{\pi}).$$

This implies that $\boldsymbol{\pi} = \boldsymbol{\pi}'$. □

We summarize Theorems 6.3 and 6.4: if a Markov chain is irreducible and aperiodic with a finite state space, then it has a unique stationary distribution $\boldsymbol{\pi}$, and the distribution $\boldsymbol{\mu}^{(n)}$ of the chain at time n approaches $\boldsymbol{\pi}$ as $n \to \infty$, regardless of the initial distribution $\boldsymbol{\mu}^{(0)}$. By referring to (6.22), (6.23) and (6.24), we see this convergence is mostly exponentially fast in n; also see *Doeblin's theorem*.

In the remainder of this subsection, we discuss the case when the state space of the Markov chain is uncountable. In this circumstance, a probability measure $\boldsymbol{\pi}$ is said to be *stationary* if it is invariant with respect to the transition operator $\mathcal{P}$: for any $f \in C_b(S)$, the integral $\boldsymbol{\pi}(f) := \int_S f(s)\boldsymbol{\pi}(ds)$ satisfies $\boldsymbol{\pi}(\mathcal{P}f) = \boldsymbol{\pi}(f)$. The following Theorem 6.6 is an analogy of Theorem 6.3, which asserts that any irreducible and aperiodic Markov chain with an uncountable state space admits a stationary distribution. Its proof is relegated to Appendix *6.A.4.

Theorem 6.6 *(Existence of Stationary Distribution with an Uncountable State Space)* Under Assumption 6.1, every irreducible and aperiodic Markov chain with an uncountable state space admits a stationary distribution.

Now, let us provide Theorem 6.7, which serves an analogy of Theorem 6.4. It states that the stationary distribution of an irreducible and aperiodic Markov chain with an uncountable state space is unique. We recall that the total variation distance for general state spaces can still be defined as in (6.21) for discrete state space. The proof of Theorem 6.7, which makes use of the idea of "*almost-coupling*" to adapt the proof of Theorem 6.4 for the uncountable state spaces case, can be found in Appendix *6.A.5.

Theorem 6.7 *(Markov Chain Convergence Theorem: Continuum State Space)* Let $\{\mathbf{x}_n\}$ be an irreducible, aperiodic Markov chain with an uncountable state space $\mathcal{S}$ such that Assumption 6.1 is fulfilled. For $n \in \mathbb{N}_0$, let $\boldsymbol{\mu}^{(n)}$ be the distribution of $\mathbf{x}_n$ (i.e., $\boldsymbol{\mu}^{(n)}(A) = \mathbb{P}(\mathbf{x}_n \in A)$ for any measurable subset A of $\mathcal{S}$) with an initial distribution $\boldsymbol{\mu}^{(0)}$. Then $\boldsymbol{\mu}^{(n)} \to \boldsymbol{\pi}$ in total variation distance for any stationary distribution $\boldsymbol{\pi}$ of $\{\mathbf{x}_n\}$, i.e., $\lim_{n\to\infty} d(\boldsymbol{\mu}^{(n)}, \boldsymbol{\pi}) = 0$.

*6.3.3 Reversible Markov Chain

On a finite state space $\mathcal{S} = \{s_1, \dots, s_K\}$, a probability distribution $\boldsymbol{\pi} = (\pi_1, \dots, \pi_K)$ is said to be *(time-)reversible* for the Markov chain with a transition probability matrix $\boldsymbol{P}$ if for any $i, j = 1, \dots, K$, we have

$$\pi_i p_{i,j} = \pi_j p_{j,i}. \tag{6.25}$$

Theorem 6.8 Let $\{\mathbf{x}_n\}$ be a Markov chain with a finite state space $\mathcal{S} = \{s_1, \dots, s_K\}$ and a transition matrix $\boldsymbol{P}$. If $\boldsymbol{\pi}$ is a reversible distribution for the chain, then it is also a stationary distribution for the chain.

Proof. For any $j = 1, 2, \dots, K$, by using the reversibility assumption, we have

$$\pi_j = \pi_j \sum_{i=1}^{K} p_{j,i} = \sum_{i=1}^{K} \pi_j p_{j,i} = \sum_{i=1}^{K} \pi_i p_{i,j}.$$

Therefore, the distribution $\boldsymbol{\pi}$ satisfies $\boldsymbol{\pi} = \boldsymbol{\pi} \boldsymbol{P}$, making it stationary. □

When $\mathcal{S}$ is uncountable, the reversibility of a Markov chain can be defined as follows. Let $\mu(\cdot|s)$ be the transition density of the Markov chain $\{\mathbf{x}_n\}$ with an initial state $s \in \mathcal{S}$. A distribution with a density function $\boldsymbol{\pi}$ is said to be *time-reversible* for $\{\mathbf{x}_n\}$ if for any $s_1, s_2 \in \mathcal{S}$,

$$\boldsymbol{\pi}(s_1)\mu(s_2|s_1) = \boldsymbol{\pi}(s_2)\mu(s_1|s_2). \tag{6.26}$$

The following statement is an analogy of Theorem 6.8; its proof is given in Appendix *6.A.6.

Theorem 6.9 For a Markov chain $\{\mathbf{x}_n\}$ with an uncountable state space and a transition density $\mu(\cdot|*)$ so that Assumption 6.1 holds, any time-reversible distribution $\boldsymbol{\pi}$ which satisfies (6.26) is stationary.

*6.4 MARKOV CHAIN MONTE CARLO

As explained in Chapter 5, sometimes it is convenient to approximate $\mathbb{E}(g(\mathbf{x}))$ by taking the empirical average of $\sum_{i=1}^{n} \frac{g(x_i)}{n}$, where $x_1, x_2, \ldots, x_n$ are n independent realizations of a distribution π of a random variable $\mathbf{x}$. This (vanilla) Monte Carlo method is only applicable if the form of π is known in a rather explicit manner. To remedy this limitation, the *Markov Chain Monte Carlo* (MCMC) algorithm has been proposed. Its idea is to mechanically construct an irreducible and aperiodic (most often time-reversible) Markov chain $\{\mathbf{x}_n\}$ with π as the only stationary distribution. To this end, given a target distribution π on a discrete state space, we first construct a transition probability matrix P satisfying reversibility condition (6.25) such that the corresponding Markov chain is also irreducible and aperiodic. Then starting from an arbitrary initial state, we run the Markov chain with this transition probability matrix P and collect the chain values $\{x_n\}$. The limiting distribution of the chain is then given by π in accordance with Theorem 6.4. Furthermore, by the *martingale convergence theorem* [15, 31], we can estimate $\mathbb{E}(g(\mathbf{x}))$ from

$$\frac{1}{M} \sum_{t=T+1}^{T+M} g(x_t),$$

for a sufficiently large T. The time duration before T is called the *burn-in period*, and all samples prior to time T will be discarded. The MCMC algorithm finds great use in statistics, especially in Bayesian inference (to be introduced in Chapter 12). One can also refer to [7, 8, 10, 26] for more details. In the following subsections, we shall introduce some common systematic algorithms to generate Markov chain samples. We shall discuss these algorithms for stationary distributions on both discrete and continuum state spaces.

*6.4.1 Metropolis–Hastings Algorithm

A time-reversible Markov chain whose stationary distribution is some desired π, such as the posterior distribution of parameters given the observations of some feature vectors, can be constructed systematically by the celebrated *Metropolis–Hastings algorithm*. For discrete state spaces, we start by proposing (guessing) transition probabilities $p_{i,j}$, which are formally called *proposal transition probabilities*. If the $p_{i,j}$'s already satisfy time reversibility condition (6.25), the corresponding Markov chain is the one we want. Otherwise, either $\pi_i p_{i,j} > \pi_j p_{j,i}$ or $\pi_i p_{i,j} < \pi_j p_{j,i}$ for some pairs of i and j. Without loss of generality, suppose that $\mathbb{P}(\mathbf{x}_n = s_j, \mathbf{x}_{n-1} = s_i) = \pi_i p_{i,j} > \pi_j p_{j,i} = \mathbb{P}(\mathbf{x}_n = s_i, \mathbf{x}_{n-1} = s_j)$, meaning that there is a higher probability of moving from s_i to s_j than from s_j to s_i. We then introduce a probability $\alpha_{i,j}$ to reduce the probability of moving from state s_i to state s_j such that $\pi_i p_{i,j} \alpha_{i,j} = \pi_j p_{j,i}$, hence

$$\alpha_{i,j} = \frac{\pi_j p_{j,i}}{\pi_i p_{i,j}}.$$

On the other hand, we do not want to reduce the likelihood of moving from state s_j to state s_i, hence we set $\alpha_{j,i} = 1$. These considerations motivate us to define the *acceptance probability*: for any $i \neq j$,

$$\alpha_{i,j} := \min\left\{\frac{\pi_j p_{j,i}}{\pi_i p_{i,j}}, 1\right\}, \tag{6.27}$$

and the probabilities

$$p'_{i,j} := p_{i,j}\alpha_{i,j} \text{ for any } j \neq i, \quad \text{and} \quad p'_{i,i} = p_{i,i} + \sum_{k\neq i} p_{i,k}(1 - \alpha_{i,k}), \tag{6.28}$$

which clearly satisfy $\sum_{j=1}^{K} p'_{i,j} = 1$. By construction, $\boldsymbol{\pi}$ is indeed the reversible (and hence stationary) distribution for the Markov chain with the transition probability matrix $\boldsymbol{P}' = (p'_{i,j})$, i.e. $\pi_i p'_{i,j} = \pi_j p'_{j,i}$ for all i and j. To see why this is true, by symmetry it suffices to consider the case $\pi_i p_{i,j} \leq \pi_j p_{j,i}$. By definition, $\alpha_{i,j} = \min\left\{\frac{\pi_j p_{j,i}}{\pi_i p_{i,j}}, 1\right\} = 1$ and $\alpha_{j,i} = \min\left\{\frac{\pi_i p_{i,j}}{\pi_j p_{j,i}}, 1\right\} = \frac{\pi_i p_{i,j}}{\pi_j p_{j,i}}$, therefore $\pi_i p'_{i,j} = \pi_i p_{i,j}\alpha_{i,j} = \pi_i p_{i,j} = \pi_j p_{j,i}\frac{\pi_i p_{i,j}}{\pi_j p_{j,i}} = \pi_j p_{j,i}\alpha_{j,i} = \pi_j p'_{j,i}$.

The transition probability matrix $\boldsymbol{P}'$ corresponds to the following transition mechanism: given that $\mathbf{x}_n = s_i$, a proposal state s_j for the next move is drawn in accordance with the i-th row of $\boldsymbol{P}$, the transition matrix of a conveniently chosen Markov chain, so that $\alpha_{i,j}$ can also be easily computed. We then accept this proposal and set $\mathbf{x}_{n+1} = s_j$ with the probability of $\alpha_{i,j}$; otherwise, we set $\mathbf{x}_{n+1} = s_i$. Recall Theorem 6.4, this Markov chain over a countable state space, with transition probability matrix $\boldsymbol{P}'$ being irreducible and aperiodic, the limiting distribution of $\mathbf{x}_n$ uniquely exists and is given by the stationary distribution $\boldsymbol{\pi}$, as requested.

In the case of an uncountable state space S, even for a non-compact space such as $\mathbb{R}^p$ (with appropriate light tail requirements), the expression of the acceptance probability is analogous to (6.27). Let $\boldsymbol{\pi}(\cdot) : S \to \mathbb{R}_+$ be the target probability density, and $q(\cdot|\cdot) : S \times S \to \mathbb{R}_+$ be a pre-specified jointly continuous and strictly positive transition density (kernel). Then the acceptance probability takes the following form:

$$\alpha_{\boldsymbol{u},\boldsymbol{v}} = \min\left\{\frac{\pi(\boldsymbol{v})q(\boldsymbol{u} \mid \boldsymbol{v})}{\pi(\boldsymbol{u})q(\boldsymbol{v} \mid \boldsymbol{u})}, 1\right\}, \quad \text{for any } \boldsymbol{u}, \boldsymbol{v} \in S. \tag{6.29}$$

Similar to (6.28), using (6.29), we can construct a reversible Markov chain with a transition density function:

$$\mu(\boldsymbol{v}|\boldsymbol{u}) := \alpha_{\boldsymbol{u},\boldsymbol{v}}q(\boldsymbol{v}|\boldsymbol{u}) + \delta_{\boldsymbol{u}}(\boldsymbol{v})\int_S (1 - \alpha_{\boldsymbol{u},\boldsymbol{y}})q(\boldsymbol{y}|\boldsymbol{u})d\boldsymbol{y}, \tag{6.30}$$

where $\delta_{\boldsymbol{u}}(\cdot)$ is the Dirac delta function with a point mass at $\boldsymbol{u}$. Clearly,

$$\int_S \mu(\boldsymbol{v}|\boldsymbol{u})d\boldsymbol{v} = \int_S \alpha_{\boldsymbol{u},\boldsymbol{v}}q(\boldsymbol{v}|\boldsymbol{u})d\boldsymbol{v} + 1 \cdot \int_S (1 - \alpha_{\boldsymbol{u},\boldsymbol{y}}) \cdot q(\boldsymbol{y}, \boldsymbol{u})d\boldsymbol{y} = 1.$$

Following the ideas in Theorems 6.7 and 6.9, one can still establish that the Markov chain with a transition density μ, being a mixture of absolutely continuous and singular parts, can still converge to the stationary distribution π. In particular, the transition density μ constructed in (6.30) carries a point mass along the diagonal $\{(u, v) : u = v\}$ of $S \times S$ because of the presence of the Dirac delta function in the second summand of (6.30), thereby making μ discontinuous on the diagonal and hence apparently violating Assumption 6.1(ii). Nevertheless, the presence of the first term on the right of (6.30), which is jointly continuous in u and v (technically called the core part of *Harris recurrence* [6, 15, 27]), guarantees that the conclusion of Theorem 6.7 remains valid. The proof of this extension essentially involves the same adaptation as the proof of Theorem 6.7, but we do not go into details here due to the many technicalities involved. Interested readers with a thorough understanding of Theorem 6.7 can try to establish the proof of this extension as a mini-project, while we shall also certainly try to provide a definite answer to this issue in our future publications.

Next, we consider an application of the Metropolis–Hastings algorithm in Bayesian inference, where it helps to update the posterior distribution of parameters given the data. The procedure can be stated as follows: let $x_1, \ldots, x_n$ be n independent realizations from the same distribution with r unknown parameters $\boldsymbol{\theta} = (\theta_1, \ldots, \theta_r)^\top \in \mathbb{R}^r$ which initially jointly follow a prior distribution $\pi(\boldsymbol{\theta})$ that will be revised to the posterior distribution $\pi(\boldsymbol{\theta}|x_1, \ldots, x_n)$ once more data are collected. Our objective is to construct an appropriate Markov chain $\{\boldsymbol{\theta}^{(t)} = (\theta_1^{(t)}, \ldots, \theta_r^{(t)})^\top\}$ whose stationary distribution is the posterior distribution $\pi(\boldsymbol{\theta} \mid x_1, \ldots, x_n) \propto \pi(\boldsymbol{\theta})L(x_1, \ldots, x_n \mid \boldsymbol{\theta})$, where L is the likelihood function of the sample.

Algorithm 6.2 states the general procedure of applying the Metropolis–Hastings algorithm in this context, in the continuum state space setting. Note that we actually do not need to know the explicit form of the posterior $\pi(\boldsymbol{\theta}|x_1, \ldots, x_n)$. We specify a multidimensional proposal transition density $q(\boldsymbol{\theta} \mid \boldsymbol{\vartheta}) : \mathbb{R}^r \times \mathbb{R}^r \to \mathbb{R}_+$. As mentioned above, given a sufficiently large burn-in time T, an extension of Theorem 6.7 for (6.30) (or Theorem 6.4 when the parameter space is discrete) tells us that the distribution of $\boldsymbol{\theta}^{(t)}$ is close to the joint posterior distribution of $\pi(\boldsymbol{\theta} \mid x_1, \ldots, x_n)$ for $t \geq T$. Thus, we simulate the Markov chain $\boldsymbol{\theta}^{(T+1)}, \boldsymbol{\theta}^{(T+2)}, \ldots$, and the empirical distribution of these subsequent chain values will converge to the desired posterior distribution of $\pi(\boldsymbol{\theta} \mid x_1, \ldots, x_n)$.

Let us illustrate the use of this algorithm in the parameter estimation of the coefficient vector $\boldsymbol{\beta}$ in binary logistic regression, and compare it with the commonly adopted maximum likelihood approach via the *Iteratively Reweighted Least Squares* (IRLS) method which will be discussed in detail in Section 10.4. From the Bayesian perspective, we generate a sequence of $\boldsymbol{\beta}$'s through its proposal transition density, and the MH estimate can be the mean, median, or mode statistic of the generated empirical distribution. According to Algorithm 6.2, we have to define two probability

Algorithm 6.2 Metropolis–Hastings algorithm for posterior distribution

1: Initialize $\theta^{(0)} = (\theta_1^{(0)}, \theta_2^{(0)}, \ldots, \theta_r^{(0)})^\top$ with random values;
2: **for** $t = 0, 1, \ldots, T$ **do**
3: Generate $\tilde{\theta}^{(t+1)}$ from $q(\theta \mid \theta^{(t)})$;
4: Generate u from $U(0, 1)$, and calculate

$$\alpha = \min\left\{ \frac{\pi(\tilde{\theta}^{(t+1)})L(x_1, \ldots, x_n \mid \tilde{\theta}^{(t+1)})q(\theta^{(t)} \mid \tilde{\theta}^{(t+1)})}{\pi(\theta^{(t)})L(x_1, \ldots, x_n \mid \theta^{(t)})q(\tilde{\theta}^{(t+1)} \mid \theta^{(t)})}, 1 \right\}.$$

5: Set

$$\theta^{(t+1)} = \begin{cases} \tilde{\theta}^{(t+1)}, & \text{if } u < \alpha; \\ \theta^{(t)}, & \text{otherwise.} \end{cases}$$

6: **end for**

density functions, namely the prior π and proposal transition density q. In actual practice, we often take the natural logarithm of the densities to avoid numerical underflow. Recall that by Bayes' theorem, the posterior distribution $\pi(\beta|X, y)$ is proportional to the product of the likelihood function and the prior density, where we denote $X = (x_1, \ldots, x_n)$. Thus $\ln \pi(\beta|X, y)$ "=" $\ln L(\beta) + \ln \pi(\beta)$ up to a constant, where $\ln L(\beta)$ is given in (10.14), and $\pi(\beta)$ is the prior distribution of β. Note that $\pi(\beta)$ is often chosen to be non-informative so that the MH algorithm is not biased to any specific features of the prior distribution. For logistic regression, for the prior distribution, we consider a multivariate normal distribution, with a mean vector $\mathbf{0}$ and large componentwise variances with independent components, such as $\beta \sim \mathcal{N}_p(\mathbf{0}, 100^2 I_p)$. The proposal transition density q is often chosen to be related to (more technically, to be *conjugate* to) the prior distribution, and its parameter should depend on the current iterate $\beta^{(t)}$. In light of this, we adopt another multivariate normal distribution as the proposal transition distribution with mean vector $\beta^{(t)}$ and covariance matrix $\nu(X^\top X)^{-1}$, where $\nu > 0$ is a small positive hyperparameter. By the symmetric expressions of normal distributions, we see that $q(\beta^{(t+1)}|\beta^{(t)}) = q(\beta^{(t)}|\beta^{(t+1)})$, hence the acceptance probability can be simplified a bit. It becomes:

$$\begin{aligned} \alpha &= \min\left\{ \exp\left(\ln \pi(\beta^{(t+1)}|X, y) - \ln \pi(\beta^{(t)}|X, y) \right), 1 \right\} \\ &= \min\left\{ \exp\left(\ln L(\beta^{(t+1)}) + \ln \pi(\beta^{(t+1)}) - \left(\ln L(\beta^{(t)}) + \ln \pi(\beta^{(t)}) \right) \right), 1 \right\}. \end{aligned}$$

First we present the implementation in Python, starting with the construction of the Metropolis–Hastings algorithm in Programme 6.14.

```
1  import numpy as np
2  from scipy.stats import multivariate_normal
3  from tqdm import trange
4  import matplotlib.pyplot as plt
5
6  plt.rc('font', size=20); plt.rc('axes', titlesize=30, labelsize=30)
7  plt.rc('xtick', labelsize=25); plt.rc('ytick', labelsize=25)
8
9  class MCMC_logistic:
10     def __init__(self, num_it=5e4, burn_in=1e5, nu=0.5, seed=4002):
11         self.num_it = int(num_it)
12         self.burn_in = int(burn_in)
13         self.nu = nu
14         self.seed = seed
15         self.proposal_rvs = multivariate_normal.rvs
16
17     def sigmoid(self, beta, X, y=None):
18         exp_eta = np.exp(X @ beta)
19         return (exp_eta / (1 + exp_eta)).reshape(y.shape)
20
21     def log_prior(self, beta):              # ln Pi(beta)
22         # large cov to be a non-informative prior
23         return np.sum(multivariate_normal.logpdf(beta, mean=0,
24                                                  cov=100**2))
25
26     def get_log(self, x, val=1e-300):       # avoid log(0) = nan
27         return np.log(np.clip(x, a_min=val, a_max=np.infty))
28
29     def log_likelihood(self, beta, X, y):   # ln L(beta)
30         pi = self.sigmoid(beta, X, y)
31         return np.sum(y*self.get_log(pi) + (1-y)*self.get_log(1-pi))
32
33     def log_target(self, beta, X, y):       # approx. ln Pi(beta| X, y)
34         return self.log_likelihood(beta, X, y) + self.log_prior(beta)
35
36     def fit(self, X, y):
37         np.random.seed(self.seed)
38         X_ = np.c_[np.ones(len(y)), X]       # add intercept
39         self.n_accept = 0
40         self.n_col = X_.shape[1]
41
42         Cov_nu = self.nu * np.linalg.inv(X_.T @ X_)
43         beta_hist = np.empty(shape=(self.num_it, self.n_col))
44         beta_hist[0] = self.proposal_rvs(mean=np.repeat(0,self.n_col),
45                                          cov=Cov_nu, size=1)
46         for i in trange(1, self.num_it):
47             beta_old = beta_hist[i-1]
48             beta_new = self.proposal_rvs(mean=beta_old, cov=Cov_nu,
49                                          size=1).reshape(-1, 1)
50
51             # Since the proposal distribution, normal, is symmetric,
52             # we have g(beta_old| beta_new) = g(beta_new| beta_old)
```

```
53              log_target_old = self.log_target(beta_old, X_, y)
54              log_target_new = self.log_target(beta_new, X_, y)
55              alpha = np.exp(log_target_new - log_target_old)
56              if np.random.rand() < min(1, alpha):
57                  beta_old = beta_new
58                  self.n_accept += 1
59
60              beta_hist[i] = beta_old.T
61
62          self.beta_dist = beta_hist
63
64      def get_coefficient(self, method="median"):
65          assert method in ["mean", "median"]
66          func = getattr(np, method)
67          return func(self.beta_dist[self.burn_in:], axis=0)
68
69      def predict(self, X, method="median"):
70          X_ = np.c_[np.ones(len(y)), X]
71          beta = self.get_coefficient(method)
72          pi = self.sigmoid(beta, X_, np.arange(len(X_)))
73          return (pi > 0.5).astype(np.int32)
74
75      def plot(self, save_name, method="median"):
76          beta_MCMC = self.get_coefficient(method)
77          for col in range(self.n_col):
78              fig, (ax1, ax2) = plt.subplots(ncols=2,
79                                             figsize=(20, 10))
80              ax1.plot(self.beta_dist[:,col], color="black",
81                       linewidth=2)
82              ax1.axvline(self.burn_in, color="b", linestyle="--",
83                          linewidth=4)
84              ax1.axhline(beta_MCMC[col], color="r", linestyle="--",
85                          linewidth=4)
86              ax1.set_xlabel("iterations")
87              ax1.set_ylabel(f"beta {col}")
88              ax1.set_title("MCMC path")
89
90              ax2.hist(self.beta_dist[self.burn_in:,col], alpha=0.5,
91                       color="gray", ec=*'black', bins="sturges")
92              ax2.axvline(beta_MCMC[col], color="r", linestyle="--",
93                          linewidth=4)
94              ax2.set_xlabel(f"beta {col}")
95              ax2.set_ylabel("frequencies")
96              ax2.set_title("MCMC histogram")
97
98              plt.tight_layout()
99              fig.savefig(save_name + f"_beta_{col}.png", dpi=200)
```

Programme 6.14 Applying the Metropolis–Hastings algorithm to parameter estimation of binary logistic regression via Python.

Here, the hyperparameters (i) `num_it` specifies the total number of iterations; (ii) `burn_in` stands for the number of the first batch of Markov chain samples $\boldsymbol{\beta}^{(0)}, \ldots, \boldsymbol{\beta}^{(\text{burn\_in}-1)}$ to be discarded; (iii) nu is the value of ν in the variance of the proposal transition density; and (iv) `seed` determines the random initial seed for the simulation for $\boldsymbol{\beta}$. Like most functions for machine learning in the `sklearn` library, we have the `fit()` and `predict()` functions for model fitting and predictions, respectively. The latter takes an extra argument `method`, which indicates the choice of estimation statistics for $\boldsymbol{\beta}$. Lastly, the `plot()` function is used to plot the simulation results.

In Section 10.4, the 1999–2002 HSI dataset will be used for demonstrations. Let us apply the Metropolis–Hastings algorithm to this dataset now; see Programme 6.15 in Python.

```
1  import pandas as pd
2
3  df = pd.read_csv("fin-ratio.csv")
4  X, y = df.drop(columns=["HSI"]).values, df.HSI.values
5
6  MCMC_model = MCMC_logistic(num_it=5e5, burn_in=1e5)
7  MCMC_model.fit(X, y)
8  y_hat_MCMC = MCMC_model.predict(X, method="median")
9  print(MCMC_model.n_accept / MCMC_model.num_it)
10 print("MCMC_median", MCMC_model.get_coefficient("median"))
11 print("MCMC_mean", MCMC_model.get_coefficient("mean"))
```

```
1  100% 499999/499999 [04:59<00:00, 1669.39it/s]
2  0.940952
3  MCMC_median [-5.99686791e+01 3.61588751e-01 -1.02498382e+00
4  6.38029105e+00 -2.20645963e-01 4.16788814e-02 -7.99494745e-02]
5  MCMC_mean [-6.24694061e+01 5.33396504e-01 -9.32235227e-01
6  6.65004031e+00 -2.31888242e-01 1.27580410e-02 -6.68279925e-02]
```

Programme 6.15 Applying Programme 6.14 to the 1999–2002 HSI dataset via Python.

From the outputs obtained in Programme 6.15, the acceptance probability is as high as 0.940952, and the parameter $\boldsymbol{\beta}$ is estimated by the median and mean, respectively. Note that different random seeds often lead to slightly different estimates in MCMC, especially when some of the feature variables are highly correlated. We shall defer its discussion in Programme 6.17 via Python. On the other hand, as shown in Figure 6.8, the estimated posterior distribution of each parameter component in the simulated Markov chain $\{\boldsymbol{\beta}^{(t)}\}$ is often skewed, so in practice we often adopt the median as the estimate instead of the mean in order to maintain the robustness of the estimation. Now, let us compute the MLE for $\boldsymbol{\beta}$, and compare the misclassification of the resulting fitted logistic regression models with those obtained from the Metropolis–Hastings algorithm; the Python codes and results are shown in the following Programme 6.16.

```
1  from sklearn.linear_model import LogisticRegression
2  sklearn_model = LogisticRegression(penalty=None)
3  sklearn_model.fit(X, y)
4  y_hat_MLE = sklearn_model.predict(X)
5  print("MLE", np.append(sklearn_model.intercept_,
6                         sklearn_model.coef_[0]))
7
8  print(f"MCMC_median: {np.mean(y == y_hat_MCMC)}")
9  print(np.where(y_hat_MCMC != y)[0])
10
11 print(f"MLE: {np.mean(y == y_hat_MLE)}")
12 print(np.where(y_hat_MLE != y)[0])
```

```
1  MLE [-55.49779339   1.10573882  -1.31293087   5.89156546  -0.15997237
2     0.14034441  -0.10858718]
3  MCMC_median: 0.9897058823529412
4  [255 301 453 500 506 534 570]
5  MLE: 0.9926470588235294
6  [255 301 506 534 570]
```

Programme 6.16 Comparing the fitted models from the MH algorithm and MLE using `sklearn` in Python.

The results obtained from Programme 6.16 indicate that both methods yield a nearly perfect fit, but they arrive with slightly different estimates for $\boldsymbol{\beta}$; indeed, the feature vectors are linearly dependent (see their correlation matrix in Programme 6.18 in Python), and this results in the instability of the estimation for $\boldsymbol{\beta}$. The accuracy of the MLE of 0.9926 is slightly higher than that of the MH median estimate of 0.9897. In Programme 6.17 (via Python), we plot the simulated paths for $\boldsymbol{\beta}$.

```
1  MCMC_model.plot("../Picture/MCMC", method="median")
```

Programme 6.17 Plotting the simulations and histograms for each parameter component of $\boldsymbol{\beta}$ via Python.

```
1 print(np.round(np.corrcoef(X, rowvar=False), 5))
```

```
1 [[ 1.         0.35195  0.16285  0.11988  0.04158  0.0218 ]
2  [ 0.35195  1.         0.24072  0.17285  0.11477  0.04746]
3  [ 0.16285  0.24072  1.         0.15756 -0.05649 -0.01242]
4  [ 0.11988  0.17285  0.15756  1.        -0.00837  0.01178]
5  [ 0.04158  0.11477 -0.05649 -0.00837  1.         0.06585]
6  [ 0.0218   0.04746 -0.01242  0.01178  0.06585  1.      ]]
```

Programme 6.18 Correlations among parameter components via Python.

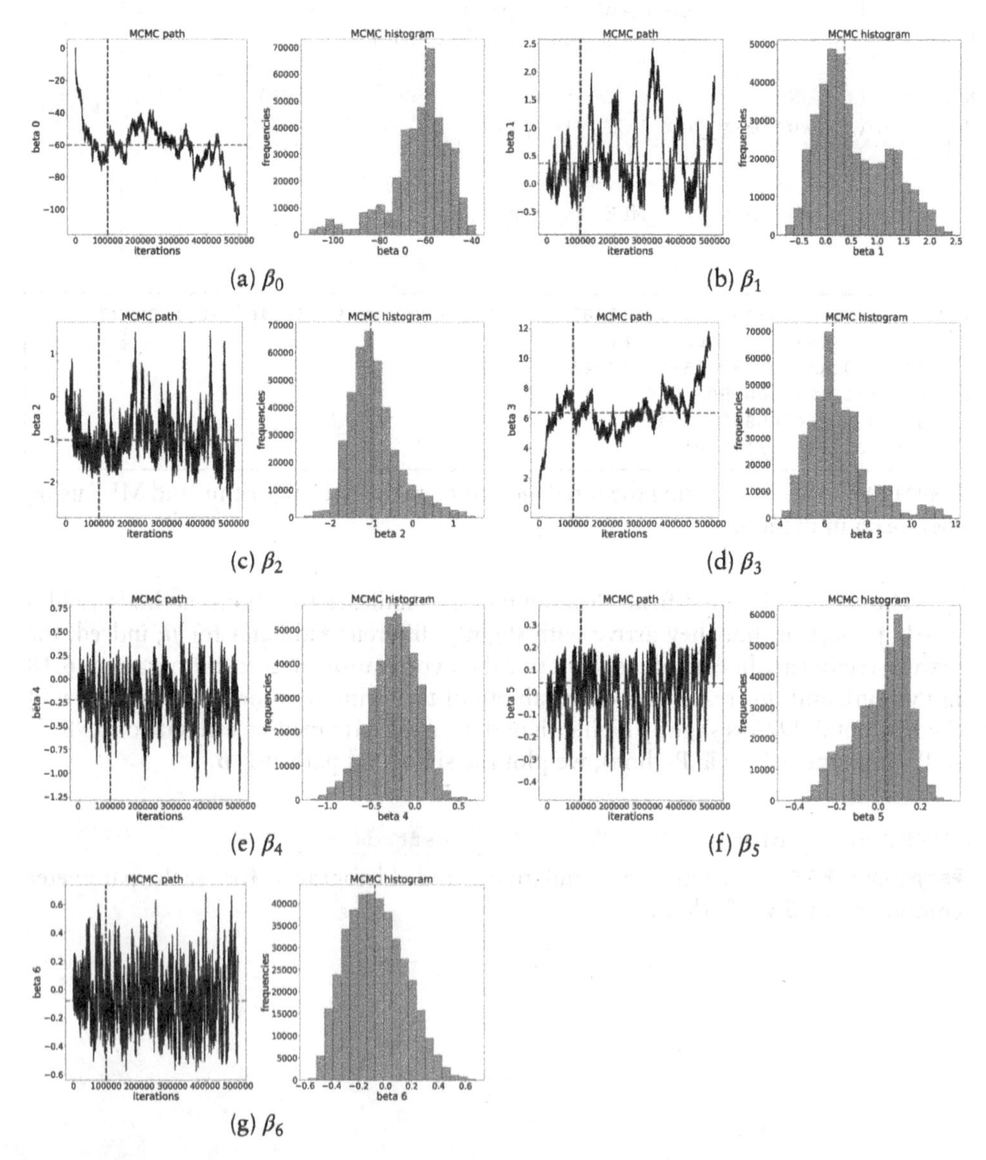

(a) β_0 (b) β_1 (c) β_2 (d) β_3 (e) β_4 (f) β_5 (g) β_6

FIGURE 6.8 Simulation paths and histograms for components of $\boldsymbol{\beta}$, generated in Programme 6.17 via Python.

The simulation paths and the corresponding histograms are shown in Figure 6.8 for each component of $\boldsymbol{\beta} = (\beta_0, \beta_1, \ldots, \beta_6)^\top$. In the simulation plots on the left-hand side, the blue vertical dashed lines represent the end of the burn-in period determined by burn_in, whereas the red horizontal dashed lines are the estimates of $\hat{\beta}_i$'s via the median statistics. In the histograms on the right-hand side, the red vertical dashed lines represent the $\hat{\beta}_i$ estimates obtained by the median. From both diagrams, we observe that the simulations for β_0, β_1 and β_3 give two modes in their respective histograms. Indeed, according to the correlation matrix shown in Programme 6.18, relatively large linear correlations were observed between features x_1 and x_2 with a correlation coefficient of 0.35195 and between features x_2 and x_3 with a correlation coefficient of 0.24072. In order to obtain a better estimate, one can also (i) vary the size of the hyperparameter ν; (ii) increase the number of iterations; (iii) burn in more samples; or (iv) adopt more random seeds.

Now, let us turn to the corresponding implementation in **R**. Again, we start with the construction of the Metropolis–Hastings algorithm; see the following Programme 6.19.

```
1  > library(mvtnorm)
2  >
3  > sigmoid <- function(beta, X, y=NA){
4  +   exp_eta <- exp(X%*%beta)
5  +   return(exp_eta / (1 + exp_eta))
6  + }
7  >
8  > MCMC_logistic <- function(X, y, num_it=5e5, burn_in=1e5, nu=0.5,
9  +                           seed=4002){
10 +   # ln Pi(beta): large cov to be a non-informative prior
11 +   log_prior <- function(beta) sum(dnorm(beta, 0, 100, log=TRUE))
12 +   get_log <- function(x, val=1e-300) log(pmax(x, val))
13 +
14 +   log_likelihood <- function(beta, X, y){    # ln L(beta)
15 +     pi <- sigmoid(beta, X, y)
16 +     return(sum(y * get_log(pi) + (1 - y) * get_log(1 - pi)))
17 +   }
18 +
19 +   log_target <- function(beta, X, y){         # approx. ln Pi(beta|X, y)
20 +     return(log_likelihood(beta, X, y) + log_prior(beta))
21 +   }
22 +
23 +   set.seed(seed)
24 +   X_ <- cbind(1, as.matrix(X))
25 +   n_accept <- 0
26 +
27 +   Cov_nu <- nu * solve(t(X_)%*%X_)
28 +   beta_hist <- matrix(NA, nrow=num_it, ncol=ncol(X_))
29 +   beta_hist[1,] <- rmvnorm(1, rep(0, ncol(X_)), Cov_nu)
30 +   prog_bar <- txtProgressBar(min=0, max=num_it, width=50, style=3)
31 +   for (i in 2:num_it){
32 +     beta_old <- beta_hist[i-1,]
33 +     beta_new <- rmvnorm(1, beta_old, Cov_nu)
34 +
```

```
35  +     # Since the proposal distribution, normal, is symmetric,
36  +     # we have g(beta_old| beta_new) = g(beta_new| beta_old)
37  +     log_target_old <- log_target(matrix(beta_old, ncol=1), X_, y)
38  +     log_target_new <- log_target(t(beta_new), X_, y)
39  +
40  +     alpha <- exp(log_target_new - log_target_old)
41  +     if (runif(1) < min(1, alpha)){
42  +       beta_old <- beta_new
43  +       n_accept <- n_accept + 1
44  +     }
45  +
46  +     beta_hist[i,] <- beta_old
47  +     setTxtProgressBar(prog_bar, i)
48  +   }
49  +   cat("\n")
50  +   return(list(beta_hist=beta_hist, n_accept=n_accept, num_it=num_it,
51  +               burn_in=burn_in))
52  + }
53  >
54  > get_coefficient <- function(MCMC_lst, method="median"){
55  +   beta_hist <- MCMC_lst$beta_hist
56  +   burn_in <- MCMC_lst$burn_in
57  +   func <- get(method)
58  +   return(apply(beta_hist[-c(1:burn_in),], MARGIN=2, FUN=func))
59  + }
60  >
61  > MCMC_predict <- function(X, MCMC_lst, method="median"){
62  +   X_ <- cbind(1, as.matrix(X))
63  +   beta <- get_coefficient(MCMC_lst, method)
64  +   pi <- sigmoid(beta, X_)
65  +   return(as.numeric(pi > 0.5))
66  + }
```

Programme 6.19 Applying the Metropolis-Hastings algorithm to parameter estimation of binary logistic regression via **R**.

As a remark, in the **R** environment we are unable to save and retrieve the parameters in the same way as Python. Therefore, we modify the main function `MCMC_logistic()` so that it returns a list containing (i) `beta_hist` for the historical estimate values of the coefficient parameter $\boldsymbol{\beta}$ for the simulation paths; (ii) `n_accept` for the total number of acceptance for the overall simulations; (iii) `num_it` for the total number of iterations; and (iv) `burn_in` for the number of the first batch of Markov chain samples to be discarded. Next, we apply Programme 6.19 to the HSI dataset, and compare the corresponding estimation with the MLE obtained from `glmnet()`; see the **R** codes and results in Programme 6.20.

```
1  > df <- read.csv("fin-ratio.csv")   # read in data file
2  > X <- subset(df, select=-c(HSI)); y <- df$HSI
3  >
4  > burn_in <- 1e5
5  > MCMC_lst <- MCMC_logistic(X, y, num_it=5e5, burn_in=burn_in)
6    |==================================================| 100%
7  > y_hat_MCMC <- MCMC_predict(X, MCMC_lst, method="median")
```

```
8   > MCMC_lst$n_accept / MCMC_lst$num_it
9   [1] 0.942432
10  > (MCMC_median <- get_coefficient(MCMC_lst, "median"))
11  [1] -62.773800393   0.229449009  -1.085862659   6.700959926
12  [5]  -0.268713003   0.002058367  -0.031104178
13  > (MCMC_mean <- get_coefficient(MCMC_lst, "mean"))
14  [1] -62.17471872   0.37468246  -1.02369069   6.64432446  -0.27659777
15  [6]  -0.04180551  -0.02923624
16  >
17  > library(glmnet)
18  > # 0 for penalty such that no regularization effect
19  > glm_model <- glmnet(X, y, family="binomial", lambda=c(1e-4, 0),
20  +                     standardize=FALSE)
21  > y_hat_MLE <- predict(glm_model, s=0, newx=as.matrix(X),
22  +                      type="response") > 0.5
23  > cat("MLE", coef(glm_model, s=0)@x, "\n")
24  MLE -55.26867 1.103012 -1.307148 5.867102 -0.159527 0.1397058 -0.1082043
25  >
26  > cat("MCMC_median:", mean(y == y_hat_MCMC), "\n")
27  MCMC_median: 0.9926471
28  > which(y_hat_MCMC != y)
29  [1] 256 302 501 535 571
30  >
31  > cat("MLE:", mean(y == y_hat_MLE), "\n")
32  MLE: 0.9926471
33  > which(y_hat_MLE != y)
34  [1] 256 302 507 535 571
```

Programme 6.20 Comparing the fitted models from the MH algorithm and MLE from `glmnet` in **R**.

In the `glmnet()` function in **R**, λ is set to 0 so that no penalty term is added to the objective function of the binary logistic regression, just like the argument `penalty=None` for `LogisticRegression()` in Programme 6.16 via Python. `glmnet` adopts a gradient descent algorithm to estimate the coefficient $\boldsymbol{\beta}$, yet under an unsatisfactory choice of initial point, the algorithm is subject to convergence error if $\boldsymbol{\beta}$ is directly estimated with λ set exactly to 0. To address this issue, we first estimate $\boldsymbol{\beta}$ with $\lambda = 10^{-4}$, and the resulting estimate, which should be quite close to the minimizer of the original objective function, is then used as the initial point for estimation, by then λ can be set to 0. Both the MLE method and MH algorithm with median estimates attain the same training accuracy of 0.9926471.

Lastly, we plot the simulation paths and histograms from using **R**; see the code in Programme 6.21 and the resulting plots in Figure 6.9.

```
1  > MCMC_median <- get_coefficient(MCMC_lst, "median", burn_in=burn_in)
2  > beta_median <- MCMC_lst$beta_hist
3  > par(mfrow=c(1, 2), cex.lab=1.5, cex.axis=1.5, cex.main=2)
4  > for (col in 1:(ncol(X)+1)){
5  +     plot(beta_median[,col], type="l", lwd=2, main="MCMC path",
6  +          xlab="iterations", ylab=paste("beta", col-1))
7  +     abline(v=burn_in, lty=2, lwd=3, col="blue")
8  +     abline(h=MCMC_median[col], lty=2, lwd=3, col="red")
```

```
9  +
10 +     hist(beta_median[-c(1:burn_in), col], main="MCMC histogram",
11 +          xlab=paste("beta", col-1))
12 +     abline(v=MCMC_median[col], lty=2, lwd=3, col="red")
13 + }
```

Programme 6.21 Plotting the simulation paths and histograms for each parameter component of $\boldsymbol{\beta}$ via R.

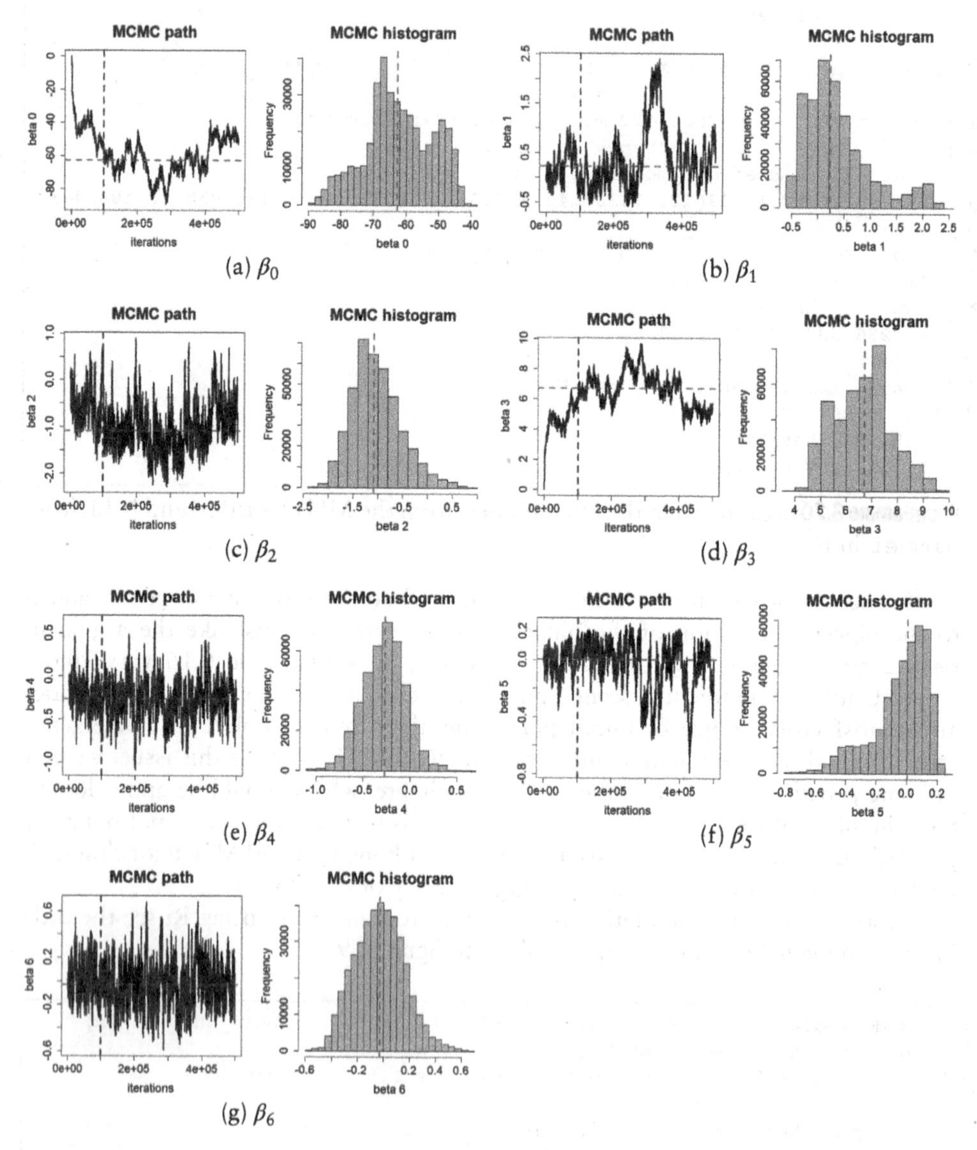

(a) β_0 (b) β_1 (c) β_2 (d) β_3 (e) β_4 (f) β_5 (g) β_6

FIGURE 6.9 Simulation paths and histograms for components of $\boldsymbol{\beta}$, generated in Programme 6.21 via R.

*6.4.2 Variants of the Metropolis–Hastings Algorithm

For sampling in Bayesian inference, for the ease of calculating the acceptance probability and simulating the Markov chain, we may sometimes generate the parameters one by one, rather than independently from those generated previously. The resulting joint transition density can still possess a dependence structure because of the use of the acceptance probability in the construction of the corresponding reversible Markov chain. To this point, for the parameter $\theta \in \mathbb{R}^r$, firstly the proposal transition density takes the simple form of a product of r univariate conditional densities, i.e., $q(\theta \mid \vartheta) = \prod_{i=1}^{r} q_i(\theta_i \mid \vartheta_{-i})$, where ϑ_{-i} stands for the vector with all components in the same order as in ϑ except the i-th component is removed. The collection of $q_i(\cdot \mid \cdot) : \mathbb{R} \times \mathbb{R}^{r-1} \to \mathbb{R}_+$, for $i = 1, \ldots, r$, are all pre-specified marginal conditional densities. In addition, we often simply take the q_i's to be the same. The resulting special algorithm is commonly called the *Single Component Adaptive Metropolis–Hastings algorithm* for Bayesian inference; see Algorithm 6.3.

Algorithm 6.3 Single Component Adaptive Metropolis–Hastings algorithm for posterior distribution

1: Initialize $\theta^{(0)} = (\theta_1^{(0)}, \theta_2^{(0)}, \ldots, \theta_r^{(0)})^\top$ with random values;
2: **for** $t = 0, 1, \ldots, T$ **do**
3: **for** $i = 1, \ldots, r$ **do**
4: Generate $\tilde{\theta}_i^{(t+1)}$ independently from $q_i(\theta_i \mid \theta_{-i}^{(t)})$
5: **end for**
6: $\tilde{\theta}^{(t+1)} = (\tilde{\theta}_1^{(t+1)}, \ldots, \tilde{\theta}_r^{(t+1)})^\top$;
7: Generate u from $U(0, 1)$, and calculate

$$\alpha = \min\left\{ \frac{\pi(\tilde{\theta}^{(t+1)}) L(x_1, \ldots, x_n \mid \tilde{\theta}^{(t+1)}) \prod_{i=1}^{r} q_i(\theta_i^{(t)} \mid \tilde{\theta}_{-i}^{(t+1)})}{\pi(\theta^{(t)}) L(x_1, \ldots, x_n \mid \theta^{(t)}) \prod_{i=1}^{r} q_i(\tilde{\theta}_i^{(t+1)} \mid \theta_{-i}^{(t)})}, 1 \right\}.$$

8: Set

$$\theta^{(t+1)} = \begin{cases} \tilde{\theta}^{(t+1)}, & \text{if } u < \alpha; \\ \theta^{(t)}, & \text{otherwise.} \end{cases}$$

9: **end for**

As shown in Algorithm 6.2, a vanilla Metropolis–Hastings algorithm generates the candidate sample values of all parameters in one go, which might not be convenient in practice. On the other hand, the single component variant in Algorithm 6.3 generates one parameter value at a time, making calculations much simpler, yet without identifying any inherent dependence relations among the parameters in advance.

Another natural variant is to decide the update of each single parameter individually, and to utilize the updated parameter values immediately to infer the update of the next parameter component. This alternative approach is the well-known *Gibbs*

Sampling; see [3, 26] for more detailed discussions. As before, we let $x_1, \ldots, x_n$ be n independent realizations from the same distribution with r unknown parameters, $\theta = (\theta_1, \ldots, \theta_r)^\top$. For each parameter θ_i, we specify a one-dimensional conjugate prior $\pi_i(\theta_i)$ together with the likelihood function $L(x_1, \ldots, x_n | \vartheta_1, \ldots, \vartheta_{i-1}, \theta_i, \vartheta_{i+1}, \ldots, \vartheta_r)$ (being conjugate to this prior), where $\vartheta_1, \ldots, \vartheta_{i-1}, \vartheta_{i+1}, \ldots, \vartheta_r$ are treated as given. We then lay down the marginal (also called *full*) conditional posterior $\pi_i(\theta_i \mid \vartheta_{-i}, x_1, \ldots, x_n)$ given by the Bayes' theorem, i.e.

$$\pi_i(\theta_i \mid \vartheta_{-i}, x_1, \ldots, x_n) = \frac{L(x_1, \ldots, x_n \mid \vartheta_1, \ldots, \theta_i, \ldots, \vartheta_r)\pi_1(\vartheta_1) \cdots \pi_i(\theta_i) \cdots \pi_r(\vartheta_r)}{\int L(x_1, \ldots, x_n \mid \vartheta_1, \ldots, \vartheta_i, \ldots, \vartheta_r)\pi_1(\vartheta_1) \cdots \pi_i(\vartheta_i) \cdots \pi_r(\vartheta_r) d\vartheta_i}.$$

These are the key quantities used for updating in Gibbs sampling, as shown in Algorithm 6.4. Being different from Algorithm 6.3 where all marginal densities q_i's can be chosen to be the same, each marginal posterior $\pi_i(\theta_i | \vartheta_{-i}, x_1, \ldots, x_n)$ in Gibbs sampling can likely be distinct from the other $\pi_j(\theta_j | \vartheta_{-j}, x_1, \ldots, x_n)$'s for the updating mechanism of the parameter θ_i, $i = 1, \ldots, r$, at the i-th step within one *cycle* at time t.

Algorithm 6.4 Gibbs sampling for posterior distribution

1: Initialize $\theta^{(0)} = (\theta_1^{(0)}, \theta_2^{(0)}, \ldots, \theta_r^{(0)})^\top$ with random values
2: **for** $t = 0, 1, \ldots, T-1$ **do**
3: Generate $\theta_1^{(t+1)}$ from $\pi_1(\theta_1 \mid \theta_2^{(t)}, \theta_3^{(t)}, \ldots, \theta_r^{(t)}, x_1, \ldots, x_n)$
4: Generate $\theta_2^{(t+1)}$ from $\pi_2(\theta_2 \mid \theta_1^{(t+1)}, \theta_3^{(t)}, \ldots, \theta_r^{(t)}, x_1, \ldots, x_n)$
⋮
5: Generate $\theta_r^{(t+1)}$ from $\pi_r(\theta_r \mid \theta_1^{(t+1)}, \theta_2^{(t+1)}, \ldots, \theta_{r-1}^{(t+1)}, x_1, \ldots, x_n)$
6: $\theta^{(t+1)} = (\theta_1^{(t+1)}, \ldots, \theta_r^{(t+1)})^\top$
7: **end for**

Despite the apparent absence of the acceptance probability and the seemingly different parameter updating mechanism in Gibbs sampling (compared with the Metropolis–Hastings algorithm or its single component variant), there is a close connection between them as each updating substep i in a cycle in the former is actually a step in the latter with the corresponding acceptance probability $\alpha_i \equiv 1$. To see this, we consider the i-th updating step in the cycle at the time t, where the joint distribution $\pi(\theta \mid x_1, \ldots, x_n)\big|_{\theta_{-i}=\theta_{-i}^{(t)}}$ is the target distribution, where $\theta_{-i}^{(t)} := (\theta_1^{(t+1)}, \ldots, \theta_{i-1}^{(t+1)}, \theta_{i+1}^{(t)}, \ldots, \theta_r^{(t)})^\top$. Denote the parameter vectors before and after updating by $\theta^{(i,t)} := (\theta_1^{(t+1)}, \ldots, \theta_{i-1}^{(t+1)}, \theta_i^{(t)}, \theta_{i+1}^{(t)}, \ldots, \theta_r^{(t)})^\top$ and $\tilde{\theta}^{(i,t)} := (\theta_1^{(t+1)}, \ldots, \theta_{i-1}^{(t+1)}, \theta_i^{(t+1)}, \theta_{i+1}^{(t)}, \ldots, \theta_r^{(t)})^\top$, respectively. At this i-th substep, by choosing the proposal transition density q to be $\pi_i(\theta_i \mid \theta_{-i}^{(t)}, x_1, \ldots, x_n)$, the acceptance probability (6.29) for this substep then becomes:

$$\alpha_i = \min\left\{\frac{\pi(\tilde{\theta}^{(i,t)}|x_1,\ldots,x_n)\cdot q(\theta_i^{(t)}|\tilde{\theta}_{-i}^{(i,t)})}{\pi(\theta^{(i,t)}|x_1,\ldots,x_n)\cdot q(\theta_i^{(t+1)}|\theta_{-i}^{(i,t)})}, 1\right\}$$

$$= \min\left\{\frac{\pi_i(\theta_i^{(t+1)}|\theta_{-i}^{(t)},x_1,\ldots,x_n)\pi(\theta_{-i}^{(t)}|x_1,\ldots,x_n)\cdot\pi_i(\theta_i^{(t)}|\theta_{-i}^{(t)},x_1,\ldots,x_n)}{\pi_i(\theta_i^{(t)}|\theta_{-i}^{(t)},x_1,\ldots,x_n)\pi(\theta_{-i}^{(t)}|x_1,\ldots,x_n)\cdot\pi_i(\theta_i^{(t+1)}|\theta_{-i}^{(t)},x_1,\ldots,x_n)}, 1\right\}$$

$$= \min\{1,1\} = 1,$$

where the second line holds since $\theta_{-i}^{(i,t)} = \tilde{\theta}_{-i}^{(i,t)} = \theta_{-i}^{(t)}$ by their very definitions.

To establish the convergence of the Gibbs sampler, it is noted that in general the joint posterior distribution $\pi(\theta \mid x_1,\ldots,x_n)$ is the stationary distribution of the underlying Markov chain constructed by Gibbs sampling in Algorithm 6.4. That is, we shall verify the following statement: for any t,

$$\int \pi(\theta^{(t)} \mid x_1,\ldots,x_n)K(\theta^{(t)},\theta^{(t+1)} \mid x_1,\ldots,x_n)d\theta^{(t)} = \pi(\theta^{(t+1)} \mid x_1,\ldots,x_n),$$

where

$$\begin{aligned}K(\theta^{(t)},\theta^{(t+1)} \mid x_1,\ldots,x_n) := {}& \pi_1(\theta_1^{(t+1)} \mid \theta_2^{(t)},\theta_3^{(t)},\ldots,\theta_r^{(t)},x_1,\ldots,x_n)\\ &\cdot\pi_2(\theta_2^{(t+1)} \mid \theta_1^{(t+1)},\theta_3^{(t)},\ldots,\theta_r^{(t)},x_1,\ldots,x_n)\\ &\ldots\\ &\cdot\pi_r(\theta_r^{(t+1)} \mid \theta_1^{(t+1)},\theta_2^{(t+1)},\ldots,\theta_{r-1}^{(t+1)},x_1,\ldots,x_n)\end{aligned}$$

is the *transition kernel* of the Gibbs sampler. To see this, when $r = 2$ (the general case $r > 2$ can be worked out similarly), we note that

$$\begin{aligned}&\int \pi(\theta^{(t)} \mid x_1,\ldots,x_n)K(\theta^{(t)},\theta^{(t+1)} \mid x_1,\ldots,x_n)d\theta^{(t)}\\ &= \int\int \pi(\theta_1^{(t)},\theta_2^{(t)} \mid x_1,\ldots,x_n)\pi_1(\theta_1^{(t+1)} \mid \theta_2^{(t)},x_1,\ldots,x_n)\\ &\qquad \pi_2(\theta_2^{(t+1)} \mid \theta_1^{(t+1)},x_1,\ldots,x_n)d\theta_1^{(t)}d\theta_2^{(t)}\\ &= \int \pi(\theta_2^{(t)} \mid x_1,\ldots,x_n)\pi_1(\theta_1^{(t+1)} \mid \theta_2^{(t)},x_1,\ldots,x_n)\pi_2(\theta_2^{(t+1)} \mid \theta_1^{(t+1)},x_1,\ldots,x_n)d\theta_2^{(t)}\\ &= \int \pi(\theta_1^{(t+1)},\theta_2^{(t)} \mid x_1,\ldots,x_n)\pi_2(\theta_2^{(t+1)} \mid \theta_1^{(t+1)},x_1,\ldots,x_n)d\theta_2^{(t)}\\ &= \pi(\theta_1^{(t+1)} \mid x_1,\ldots,x_n)\pi_2(\theta_2^{(t+1)} \mid \theta_1^{(t+1)},x_1,\ldots,x_n) = \pi(\theta^{(t+1)} \mid x_1,\ldots,x_n).\end{aligned}$$

As a result, according to Theorem 6.7, if the associated Markov chain of the Gibbs sampler is set to be irreducible and aperiodic[6], the convergence of the Gibbs sampler can be automatically warranted. As a remark, as long as the likelihood function is not symmetric in the parameters $\theta_1, \ldots, \theta_r$, the associated Markov chain is generally not time-reversible.

We have elaborated on the MCMC approach for estimating the posterior distribution in Bayesian inference. In Chapter 12, we shall provide more discussion on Bayesian learning in which more tailor-made tools for different *FinTech* and *InsurTech* problems will be discussed.

Another practical and relevant application of Gibbs sampling in finance is the parameter estimation of stochastic volatility (SV) models [13, 14, 17, 29, 30]. In financial time series analysis, volatility is a key ingredient, yet it is not directly observable. As an alternative to the ARCH/GARCH framework (to be introduced in Chapter 7), SV models treat volatility as a latent (hidden) variable, evolving over time in a stochastic manner. This leads to an immediate problem of parameter estimation in light of the intractable form of the likelihood, to which traditional MLE method may fail to apply.

Denote y_t as the logarithmic return of the stock price and h_t as the latent variable for volatility at time t. Then the canonical form of SV models[7] is defined as follows:

$$\begin{cases} y_t = \exp\left(\frac{h_t}{2}\right) u_t, \quad u_t \overset{iid}{\sim} \mathcal{N}(0,1); \\ h_{t+1} = \mu + \phi(h_t - \mu) + \sqrt{v_\eta}\eta_{t+1}, \\ \eta_t \overset{iid}{\sim} \mathcal{N}(0,1), \quad t = 1, \ldots, T, \end{cases} \tag{6.31}$$

where $h_1 \sim \mathcal{N}(\mu, v_\eta/(1-\phi^2))$, and u_t and η_t are independent standard normal random variables, for some unknown parameters $\theta = (\mu, \phi, v_\eta)^\top$ to be determined in the Bayesian framework. The hidden (one-half) log-volatility h_t is assumed to follow a stationary AR(1) process with the persistence parameter $|\phi| \le 1$; also refer to Chapter 7 for more discussions on AR processes as a special case of ARMA processes.

---

[6] According to the celebrated *Hammersley–Clifford Theorem* [11], the *positivity condition*, namely that for all θ_i, $i = 1, \ldots, r$, $\pi_i(\theta_i) > 0$, altogether implies that $\pi(\theta \mid x_1, \ldots, x_n) > 0$, and this ensures the Markov chain to be irreducible and aperiodic.

[7] Its continuous counterpart can be constructed below: from $y_t = \ln S_{t+1} - \ln S_t = \exp(h_t/2)u_t$ over one unit of time, as an analogy, after a scaling in space-time for a small δt, $\ln S_{t+\delta t} - \ln S_t = \exp(h_t/2)\sqrt{\delta t}\, u_t$, passing δt to dt, we then have $d \ln S_t = \exp(h_t/2) dw_t^S$, where w_t^S is a standard Brownian motion; applying Itô's lemma to the function $\exp(S_t)$ yields $dS_t = \frac{\exp(h_t)}{2} S_t dt + \exp(h_t/2) S_t dw_t^S$. Similarly, for h_t with $h_{t+1} - h_t = \mu(1-\phi) + h_t(\phi - 1) + \sqrt{v_\eta}\eta_{t+1}$ over one unit of time, after scaling in space-time for a small δt so that $h_{t+\delta t} - h_t = (\mu - h_t)(1 - \phi)\delta t + \sqrt{v_\eta}\eta_{t+\delta t}\sqrt{\delta t}$, and so $dh_t = (\mu - h_t)(1-\phi)dt + \sqrt{v_\eta} dw_t^h$, where w_t^h is another standard Brownian motion that may be correlated with w_t^S, and applying Itô's lemma to the function $v_t := \exp(h_t)$ yields $dv_t = \left(\mu(1-\phi) + \frac{1}{2}v_\eta - (1-\phi)\ln v_t\right) v_t dt + \sqrt{v_\eta} v_t dw_t^h$, meanwhile $dS_t = \frac{v_t}{2} S_t dt + \sqrt{v_t} S_t dw_t^S$.

Given observations $\{y_t\}_{t=1}^T$, our goal is to provide a Bayesian framework to model the conditional joint density of $\mathbf{h}$ and $\boldsymbol{\theta}$:

$$p(h, \theta|y) \propto p(y|h, \theta)p(h|\theta)\pi(\theta), \tag{6.32}$$

where π is a prior distribution of $\boldsymbol{\theta}$ and different $p(\cdot|\cdot)$'s are the corresponding conditional densities or parametrized densities. However, due to the high-dimensional integration resulting from a long time series of observations, jointly sampling $\mathbf{h}$ and $\boldsymbol{\theta}$ directly from (6.32) is highly inconvenient. Bayesian researchers [13, 14, 17] decomposed this sampling problem into two steps: (i) sampling h_t from $p(h_t|y, h_{-t}, \theta)$, for $t = 1, \dots, T$; and (ii) sampling θ_i from $p(\theta_i|y, h, \theta_{-i})$, for $i = 1, 2, 3$; the corresponding procedure is depicted in Algorithm 6.5.

Algorithm 6.5 Gibbs sampler for stochastic volatility model of (6.31) [29]

1: Initialize parameters $\mu^{(0)}, \phi^{(0)}, v_\eta^{(0)}$ and $h^{(0)}$
2: **for** $i = 1, \dots, n$ **do**
3: Sample v_η from $p(v_\eta|y, h^{(i-1)}, \mu^{(i-1)}, \phi^{(i-1)})$
4: Sample ϕ from $p(\phi|y, h^{(i-1)}, \mu^{(i-1)}, v_\eta^{(i)})$
5: Sample μ from $p(\mu|y, h^{(i-1)}, \phi^{(i)}, v_\eta^{(i)})$
6: **for** $t = 1, \dots, T$ **do**
7: Sample h_t from $p(h_t|y, h_{-t}^{(i)}, \mu^{(i)}, \phi^{(i)}, v_\eta^{(i)})$
8: **end for**
9: **end for**

For setting prior distributions for the unknown parameters μ, ϕ, and v_η, since the likelihood is a product of normal densities, we consider Gibbs sampling with the following conjugate priors: (I) $\mu \sim \mathcal{N}(\alpha_\mu, \beta_\mu^2)$, a normal distribution; (II) $\phi \sim \mathcal{N}(\alpha_\phi, \beta_\phi^2)\mathbb{1}_{\{-1\le\phi\le1\}}$, a truncated normal distribution; and (III) $v_\eta \sim \mathcal{IG}(\alpha_v, \beta_v)$, an inverse gamma distribution (see Chapter 1).

Let us establish the conditional distributions in Algorithm 6.5 one by one as follows. In the following, we augment the proportional sign with a subscript of a variable to indicate the treating of all variables as constant except the variable specified, so as to illustrate the development more clearly.

(I) For μ:

$$p(\mu|y, h, \phi, v_\eta) \propto_\mu \left(\prod_{t=1}^{T} p(h_t|h_{t-1}, y_t, \phi, v_\eta)\right) \cdot \pi(\mu)$$

$$\propto_\mu p\left(h_1|\mu, \phi, v_\eta\right) \cdot \left(\prod_{t=1}^{T-1} p\left(h_{t+1}|h_t, \mu, \phi, v_\eta\right)\right) \cdot p_{\mathcal{N}}(\mu|\alpha_\mu, \beta_\mu^2)$$

$$\propto_\mu \exp\left\{-\frac{\left(h_1-\mu\right)^2\left(1-\phi^2\right)}{2v_\eta} - \frac{\sum_{t=1}^{T-1}\left(h_{t+1}-\mu-\phi\left(h_t-\mu\right)\right)^2}{2v_\eta} - \frac{\left(\mu-\alpha_\mu\right)^2}{2\beta_\mu^2}\right\}$$

$$\propto_{\mu} \exp\left\{-\frac{1}{2}\left[\mu^2\left(\frac{1-\phi^2+(T-1)(1-\phi)^2}{v_\eta}+\frac{1}{\beta_\mu^2}\right)\right.\right.$$
$$\left.\left.-2\mu\left(\frac{h_1\left(1-\phi^2\right)+(1-\phi)\sum_{t=1}^{T-1}\left(h_{t+1}-\phi h_t\right)}{v_\eta}+\frac{\alpha_\mu}{\beta_\mu^2}\right)\right]\right\}.$$

Therefore, the posterior distribution for μ is still a normal distribution $\mu|y, h, \phi, v_\eta \sim \mathcal{N}(\hat{\alpha}_\mu, \hat{\beta}_\mu^2)$, where

$$\hat{\alpha}_\mu := \hat{\beta}_\mu^2\left(\frac{h_1\left(1-\phi^2\right)+(1-\phi)\sum_{t=1}^{T-1}\left(h_{t+1}-\phi h_t\right)}{v_\eta}+\frac{\alpha_\mu}{\beta_\mu^2}\right);$$

$$\hat{\beta}_\mu^2 := \left(\frac{1-\phi^2+(T-1)(1-\phi)^2}{v_\eta}+\frac{1}{\beta_\mu^2}\right)^{-1};$$

also, note that this conditional density depends on y through h.

(II) For ϕ:

$$p\left(\phi \mid y, h, \mu, v_\eta\right)$$

$$\propto_\phi p\left(h_1 \mid \mu, \phi, v_\eta\right)\cdot\left(\prod_{t=1}^{T-1} p\left(h_{t+1} \mid h_t, \mu, \phi, v_\eta\right)\right)\cdot p_{\mathcal{N}}(\phi|\alpha_\phi, \beta_\phi^2)\mathbb{1}_{\{-1\leq\phi\leq1\}}$$

$$\propto_\phi \exp\left\{-\frac{\left(h_1-\mu\right)^2\left(1-\phi^2\right)}{2v_\eta}-\frac{\sum_{t=1}^{T-1}\left(h_{t+1}-\mu-\phi\left(h_t-\mu\right)\right)^2}{2v_\eta}-\frac{\left(\phi-\alpha_\phi\right)^2}{2\beta_\phi^2}\right\}\mathbb{1}_{\{-1\leq\phi\leq1\}}$$

$$\propto_\phi \exp\left\{-\frac{1}{2}\left[\phi^2\left(\frac{-\left(h_1-\mu\right)^2+\sum_{t=1}^{T-1}\left(h_t-\mu\right)^2}{v_\eta}+\frac{1}{\beta_\phi^2}\right)\right.\right.$$
$$\left.\left.-2\phi\left(\frac{\sum_{t=1}^{T-1}\left(h_{t+1}-\mu\right)\left(h_t-\mu\right)}{v_\eta}+\frac{\alpha_\phi}{\beta_\phi^2}\right)\right]\right\}\mathbb{1}_{\{-1\leq\phi\leq1\}}.$$

As a result, the posterior distribution for ϕ is another truncated normal distribution on $[-1, 1]$, such that $\phi|y, h, \mu, v_\eta \sim \mathcal{N}(\hat{\alpha}_\phi, \hat{\beta}_\phi^2)\mathbb{1}_{\{-1\leq\phi\leq1\}}$, where

$$\hat{\alpha}_\phi := \hat{\beta}_\phi^2\left(\frac{\sum_{t=1}^{T-1}\left(h_{t+1}-\mu\right)\left(h_t-\mu\right)}{v_\eta}+\frac{\alpha_\phi}{\beta_\phi^2}\right);$$

$$\hat{\beta}_\phi^2 := \left(\frac{\sum_{t=2}^{T-1}\left(h_t-\mu\right)^2}{v_\eta}+\frac{1}{\beta_\phi^2}\right)^{-1}.$$

As before, the posterior is dependent on y through h.

(III) For v_η:

$$p\left(v_\eta \mid y, h, \mu, \phi\right)$$

$$\propto_{v_\eta} p\left(h_1 \mid \mu, \phi, v_\eta\right) \cdot \left(\prod_{t=1}^{T-1} p\left(h_{t+1} \mid h_t, \mu, \phi, v_\eta\right)\right) \cdot p_{\mathcal{IG}}(v_\eta | \alpha_v, \beta_v)$$

$$\propto_{v_\eta} \left(\frac{1}{v_\eta}\right)^{\frac{1+T-1}{2}} \exp\left(-\frac{\left(h_1-\mu\right)^2\left(1-\phi^2\right)}{2v_\eta} - \frac{\sum_{t=1}^{T-1}\left(h_{t+1}-\mu-\phi\left(h_t-\mu\right)\right)^2}{2v_\eta}\right)$$

$$\cdot \frac{\left(\beta_v\right)^{\alpha_v} e^{-\beta_v/v_\eta}}{\Gamma\left(\alpha_v\right)\left(v_\eta\right)^{\alpha_v+1}},$$

$$\propto_{v_\eta} \left(\frac{1}{v_\eta}\right)^{(\alpha_v+T/2)+1} \exp\left(-\frac{\beta_v + \frac{1}{2}\left(h_1-\mu\right)^2\left(1-\phi^2\right) + \frac{1}{2}\sum_{t=1}^{T-1}\left(h_{t+1}-\mu-\phi\left(h_t-\mu\right)\right)^2}{v_\eta}\right).$$

Hence, the posterior distribution for v_η is an alternative inverse gamma distribution $v_\eta | y, h, \mu, \phi \sim \mathcal{IG}(\hat{\alpha}_v, \hat{\beta}_v)$, where

$$\hat{\alpha}_v := \alpha_v + \frac{T}{2} \quad \text{and} \quad \hat{\beta}_v := \beta_v + \frac{1}{2}\left[\sum_{t=1}^{T-1}(h_{t+1}-\mu-\phi(h_t-\mu))^2 + (h_1-\mu)^2(1-\phi^2)\right].$$

Once again, the dependence of the posterior on the observation y is through the hidden variable h.

(IV) For h_t: by the iid nature of the u_t's, it is clear that the target density satisfies $p(h_t \mid y, h_{-t}, \theta) = \frac{p(h_t, y_t | h_{-t}, \theta)}{\int p(h_t, y_t | h_{-t}, \theta) dh_t}$, where $p(h_t, y_t \mid h_{-t}, \theta) = p(y_t \mid h_t, \theta) p(h_t \mid h_{-t}, \theta)$. First we consider the second term $p(h_t \mid h_{-t}, \theta)$ under each of the following cases:

1. For $t = 1$, h_1 depends on h_{-1} only through h_2,

$$p(h_1 \mid h_{-1}, \mu, \phi, v_\eta) = p(h_1 \mid h_2, \mu, \phi, v_\eta) \propto_{h_1} p(h_1 \mid \mu, \phi, v_\eta) \cdot p(h_2 \mid h_1, \mu, \phi, v_\eta)$$

$$\propto_{h_1} \exp\left\{-\frac{(h_1-\mu)^2(1-\phi^2)}{2v_\eta} - \frac{(h_2-\mu-\phi(h_1-\mu))^2}{2v_\eta}\right\}$$

$$\propto_{h_1} \exp\left\{-\frac{1}{2v_\eta}\left((1-\phi^2)h_1^2 - 2\mu(1-\phi^2)h_1 - 2(h_2-\mu)\phi h_1 + \phi^2 h_1^2 - 2\phi^2\mu h_1\right)\right\},$$

which is a normal density $p_{\mathcal{N}}$ with mean $\hat{\alpha}_1$ and variance $\hat{\beta}_1^2$, where

$$\hat{\alpha}_1 := \mu + \phi(h_2 - \mu) \quad \text{and} \quad \hat{\beta}_1^2 := v_\eta.$$

2. For $t = 2, \ldots, T-1$, h_t depends on and will affect $\boldsymbol{h}_{-t}$ only through h_{t-1} and h_{t+1},

$$
\begin{aligned}
& p(h_t \mid \boldsymbol{h}_{-t}, \mu, \phi, v_\eta) = p(h_t \mid h_{t-1}, h_{t+1}, \mu, \phi, v_\eta) \\
& \propto_{h_t} p(h_t \mid h_{t-1}, \mu, \phi, v_\eta) \cdot p(h_{t+1} \mid h_t, \mu, \phi, v_\eta) \\
& \propto_{h_t} \exp\left\{ -\frac{(h_t - \mu - \phi(h_{t-1} - \mu))^2}{2v_\eta} - \frac{(h_{t+1} - \mu - \phi(h_t - \mu))^2}{2v_\eta} \right\} \\
& \propto_{h_t} \exp\left\{ -\frac{1}{2v_\eta} \left((h_t - \mu)^2 - 2h_t\phi(h_{t-1} - \mu) + \phi^2(h_t - \mu)^2 - 2(h_{t+1} - \mu)\phi h_t\right) \right\} \\
& \propto_{h_t} \exp\left\{ -\frac{1}{2v_\eta} \left((1 + \phi^2)h_t^2 - 2\mu h_t - 2\mu\phi^2 h_t - 2\phi(h_{t+1} - \mu + h_{t-1} - \mu)h_t\right) \right\},
\end{aligned}
$$

which is a normal density $p_{\mathcal{N}}$ with mean $\hat{\alpha}_t$ and variance $\hat{\beta}_t^2$, where

$$
\hat{\alpha}_t := \mu + \frac{\phi((h_{t+1} - \mu) + (h_{t-1} - \mu))}{1 + \phi^2} \quad \text{and} \quad \hat{\beta}_t^2 := \frac{v_\eta}{1 + \phi^2}.
$$

3. For $t = T$, as h_T only depends on $\boldsymbol{h}_{-T}$ through h_{T-1},

$$
p(h_T \mid \boldsymbol{h}_{-T}, \mu, \phi, v_\eta) = p(h_T \mid h_{T-1}, \mu, \phi, v_\eta),
$$

which is a normal density $p_{\mathcal{N}}$ with mean $\hat{\alpha}_T$ and variance $\hat{\beta}_T^2$, where

$$
\hat{\alpha}_T := \mu + \phi(h_{T-1} - \mu) \quad \text{and} \quad \hat{\beta}_T^2 := v_\eta.
$$

To summarize, $p(h_t \mid \boldsymbol{h}_{-t}, \boldsymbol{\theta}) = \frac{1}{\sqrt{2\pi\hat{\beta}_t^2}} \exp\left(\frac{(h_t - \hat{\alpha}_t)^2}{2\hat{\beta}_t^2}\right)$, for $t = 1, \ldots, T$ for specific forms of $\hat{\alpha}_t$ and $\hat{\beta}_t$. Together with the conditional density of y_t, namely $p(y_t|h_t, \boldsymbol{\theta}) = \frac{1}{\sqrt{2\pi e^{h_t}}} \exp\left(-\frac{y_t^2}{2e^{h_t}}\right)$, we have

$$
p(h_t \mid \boldsymbol{y}, \boldsymbol{h}_{-t}, \boldsymbol{\theta}) = \frac{\exp\left(-\frac{y_t^2}{2}e^{-h_t} - \frac{h_t}{2} - \frac{(h_t - \hat{\alpha}_t)^2}{2\hat{\beta}_t^2}\right)}{\int \exp\left(-\frac{y_t^2}{2}e^{-h_t} - \frac{h_t}{2} - \frac{(h_t - \hat{\alpha}_t)^2}{2\hat{\beta}_t^2}\right) dh_t}.
$$

On the other hand, by its convexity, $\exp(x)$ always lies above its supporting line at $-\hat{\alpha}_t$, hence

$$
\exp(-h_t) \geq \exp(-\hat{\alpha}_t) + \exp(-\hat{\alpha}_t) \cdot (-h_t - (-\hat{\alpha}_t)). \tag{6.33}
$$

Thus, the logarithm of the conditional density $p(y_t|h_t, \theta)$ has the following upper bound:

$$\begin{aligned}
\ln p(y_t|h_t, \theta) &= -\frac{1}{2}h_t - \frac{y_t^2}{2}\exp(-h_t) - \frac{1}{2}\ln(2\pi) \\
&\leq -\frac{1}{2}h_t - \frac{y_t^2}{2}\left(\exp(-\hat{\alpha}_t) - \exp(-\hat{\alpha}_t)(h_t - \hat{\alpha}_t)\right) - \frac{1}{2}\ln(2\pi) \\
&\leq -\frac{y_t^2}{2}(1 + \hat{\alpha}_t)\exp(-\hat{\alpha}_t) + \frac{1}{2}h_t\left(y_t^2\exp(-\hat{\alpha}_t) - 1\right) - \frac{1}{2}\ln(2\pi) \\
&=: \ln g(y_t|h_t, \theta).
\end{aligned}$$

This motivates us to adopt the commonly used acceptance–rejection method (see Section 5.1) with the following candidate distribution:

$$\begin{aligned}
g(h_t \mid y, h_{-t}, \theta) &\propto_{h_t} g(y_t \mid h_t, \theta) \cdot p(h_t \mid h_{-t}, \theta) \\
&\propto_{h_t} \exp\left\{\frac{1}{2}\left(y_t^2\exp(-\hat{\alpha}_t) - 1\right)h_t - \frac{(h_t - \hat{\alpha}_t)^2}{2\hat{\beta}_t^2}\right\} \\
&\propto_{h_t} \exp\left(-\frac{(h_t - \tilde{\alpha}_t)^2}{2\hat{\beta}_t^2}\right),
\end{aligned}$$

which is the kernel of a normal density with mean $\tilde{\alpha}_t := \hat{\alpha}_t + \frac{\hat{\beta}_t^2}{2}\left(y_t^2\exp(-\hat{\alpha}_t) - 1\right)$ and variance $\hat{\beta}_t^2$. To verify that $g(h_t \mid y, h_{-t}, \theta)$ "envelopes" the target density $p(h_t \mid y, h_{-t}, \theta)$, we note that

$$\begin{aligned}
\frac{p(h_t \mid y, h_{-t}, \theta)}{g(h_t \mid y, h_{-t}, \theta)} &= \frac{\exp\left(-\frac{y_t^2}{2}e^{-h_t} - \frac{h_t}{2} - \frac{(h_t - \hat{\alpha}_t)^2}{2\hat{\beta}_t^2}\right)}{\frac{1}{\sqrt{2\pi\hat{\beta}_t^2}}\exp\left(-\frac{(h_t - \tilde{\alpha}_t)^2}{2\hat{\beta}_t^2}\right)\int \exp\left(-\frac{y_t^2}{2}e^{-h_t} - \frac{h_t}{2} - \frac{(h_t - \hat{\alpha}_t)^2}{2\hat{\beta}_t^2}\right)dh_t} \\
&= \frac{\exp\left(-\frac{y_t^2}{2}e^{-h_t} - \frac{h_t}{2} - \frac{(2h_t - \hat{\alpha}_t - \tilde{\alpha}_t)(\tilde{\alpha}_t - \hat{\alpha}_t)}{2\hat{\beta}_t^2}\right)}{\frac{1}{\sqrt{2\pi\hat{\beta}_t^2}}\int \exp\left(-\frac{y_t^2}{2}e^{-h_t} - \frac{h_t}{2} - \frac{(h_t - \hat{\alpha}_t)^2}{2\hat{\beta}_t^2}\right)dh_t} \\
&= \frac{\exp\left(-\frac{y_t^2}{2}e^{-h_t} - \frac{h_t}{2} - \frac{h_t}{\hat{\beta}_t^2}\frac{\hat{\beta}_t^2}{2}\left(y_t^2\exp(-\hat{\alpha}_t) - 1\right)\right)}{\frac{1}{\sqrt{2\pi\hat{\beta}_t^2}}\exp\left(-\frac{(\tilde{\alpha}_t + \hat{\alpha}_t)(\tilde{\alpha}_t - \hat{\alpha}_t)}{2\hat{\beta}_t^2}\right)\int \exp\left(-\frac{y_t^2}{2}e^{-h_t} - \frac{h_t}{2} - \frac{(h_t - \hat{\alpha}_t)^2}{2\hat{\beta}_t^2}\right)dh_t} \\
&=: \frac{\exp\left(-\frac{y_t^2}{2}e^{-h_t} - \frac{y_t^2}{2}e^{-\hat{\alpha}_t}h_t\right)}{C(\hat{\alpha}_t, \hat{\beta}_t, y_t)},
\end{aligned}$$

where

$$C(\hat{\alpha}_t, \hat{\beta}_t, y_t) := \frac{1}{\sqrt{2\pi\hat{\beta}_t^2}} \exp\left(-\frac{(\tilde{\alpha}_t + \hat{\alpha}_t)(\tilde{\alpha}_t - \hat{\alpha}_t)}{2\hat{\beta}_t^2}\right)$$

$$\cdot \int \exp\left(-\frac{y_t^2}{2}e^{-h_t} - \frac{h_t}{2} - \frac{(h_t - \hat{\alpha}_t)^2}{2\hat{\beta}_t^2}\right) dh_t,$$

whose value is independent of h_t and generally depends only on θ, y_t, h_{t-1}, h_{t+1}, through the expressions of $\hat{\alpha}_t$, $\tilde{\alpha}_t$ and $\hat{\beta}_t$. All of these are kept unchanged throughout the current updating substep of h_t. Then, by the convexity result (6.33), we see that

$$\frac{p(h_t \mid y, h_{-t}, \theta)}{g(h_t \mid y, h_{-t}, \theta)} \leq \frac{\exp\left(-\frac{y_t^2}{2}(1 + \hat{\alpha}_t)e^{-\hat{\alpha}_t}\right)}{C(\hat{\alpha}_t, \hat{\beta}_t, y_t)} =: M(\hat{\alpha}_t, \hat{\beta}_t, y_t) < \infty, \text{for any } h_t.$$

Hence our proposed normal density $g(h_t \mid y, h_{-t}, \theta)$ is indeed valid for the acceptance–rejection method, and the acceptance probability is given by:

$$\frac{p(h_t \mid y, h_{-t}, \theta)}{M(\hat{\alpha}_t, \hat{\beta}_t, y_t) g(h_t \mid y, h_{-t}, \theta)} = \exp\left(-\frac{y_t^2}{2}\left(e^{-h_t} + e^{-\hat{\alpha}_t}h_t - (1 + \hat{\alpha}_t)e^{-\hat{\alpha}_t}\right)\right).$$

Therefore, with the aid of the acceptance–rejection method, we implement the updating substep for h_t as follows:

1. Generate $\mathrm{h}_t \sim \mathcal{N}(\tilde{\alpha}_t, \hat{\beta}_t^2)$;
2. Generate $\mathrm{u} \sim U(0, 1)$;
3. If $\mathrm{u} \leq \exp\left(-\frac{y_t^2}{2}\left(e^{-\mathrm{h}_t} + e^{-\hat{\alpha}_t}\mathrm{h}_t - (1 + \hat{\alpha}_t)e^{-\hat{\alpha}_t}\right)\right)$, then accept h_t and finish the updating step; otherwise, reject h_t and return to Step 1).

To implement Algorithm 6.5 for this SV model (6.31), we use the ready function `svsample()` from the `stochvol` package[8] in **R**. Programme 6.22 via **R** illustrates the Gibbs sampler for the logarithmic returns (assuming they are of mean zero) of the stock prices of HSBC, CLP, and CK from 1999 to 2002. Here, `draws` plus `burnin` is the total number of iterations required, while the first `burnin` number of iterations will be discarded. Moreover, since the general package `stochvol` adopts a t-distribution with nu degrees of freedom for the proportional noise u_t, the argument `priornu=0` is used to ensure the degrees of freedom equal infinity so that the resulting limiting distribution resembles a standard normal distribution. Since u_t and η_t are independent, the argument `priorrho=NA` indicates a zero correlation between u_t and η_t. Despite their slow convergence, MH algorithms serve as a

[8] For further details of this package, one can refer to https://cran.r-project.org/web/packages/stochvol/stochvol.pdf.

first and most fundamental approach for these estimation problems, while obtaining alternative approaches with faster convergence has been a popular research topic.

```
1  > library(stochvol)
2  >
3  > df <- read.csv("stock_1999_2002.csv", row.names=1)
4  > y_HSBC <- diff(log(df$HSBC))
5  > y_CLP <- diff(log(df$CLP))
6  > y_CK <- diff(log(df$CK))
7  >
8  > set.seed(4002)
9  > params <- list(draws=50000, burnin=10000, priornu=0, priorrho=NA,
10 +                priormu=c(0, 100), priorphi=c(5, 1.5), priorsigma=1)
11 > sv_HSBC <- do.call(svsample, c(list(y=y_HSBC), params))
12 > sv_CLP <- do.call(svsample, c(list(y=y_CLP), params))
13 > sv_CK <- do.call(svsample, c(list(y=y_CK), params))
14 > summary(sv_HSBC$para)$statistics[,"Mean"]
15         mu         phi       sigma          nu         rho
16 -8.3843454   0.9338522   0.2192907         Inf   0.0000000
17 > summary(sv_CLP$para)$statistics[,"Mean"]
18         mu         phi       sigma          nu         rho
19 -9.1104018   0.9512893   0.2699840         Inf   0.0000000
20 > summary(sv_CK$para)$statistics[,"Mean"]
21         mu         phi       sigma          nu         rho
22 -7.7276158   0.9498354   0.1665525         Inf   0.0000000
```

Programme 6.22 Applying SV model (6.31) to logarithmic returns of HSBC, CLP, and CK from 1999 to 2002 via **R**.

After the three SV models are fitted, the *Ljung–Box test* with $K = 20$ is used to test against the standardized residuals $y_t/\exp(h_t/2)$.[9] From the output of Programme 6.23 via **R**, all p-values of the related Ljung–Box tests are greater than 0.05, so we do not reject the null hypothesis that the standardized residuals for HSBC, CLP, CK from 1999 to 2002 have no autocorrelation. Moreover, also in Programme 6.23, all p-values of the corresponding Shapiro–Wilk tests[10] are also greater than 0.05, and the

[9] Note that the asymptotic distribution of the corresponding test statistic for model diagnostics is a chi-square distribution with $K - J$ degrees of freedom, where $J = 3$ is the number of parameters estimated in the SV model; see Section 7.4 and footnotes therein for more detailed discussions.

[10] For a Shapiro–Wilk diagnostic test on fitted residuals, say $\hat{\epsilon}_i$'s for notational simplicity, the test statistic becomes $W = \frac{(\sum_{i=1}^n a_i \hat{\epsilon}_{(i)})^2}{\sum_{i=1}^n (\bar{\hat{\epsilon}} - \hat{\epsilon}_i)^2}$. Since $\hat{\epsilon}_i = \epsilon_i + O_p(\frac{1}{\sqrt{n}})$, where ϵ_i's are iid true random noises, and recalling that $\boldsymbol{a}$ is a unit vector, the numerator equals $(\sum_{i=1}^n a_i \epsilon_{(i)})^2 + 2\sum_{i \neq j} a_i a_j \epsilon_{(i)} O_p(\frac{1}{\sqrt{n}})$ + higher order terms, whose first term $(\sum_{i=1}^n a_i \epsilon_{(i)})^2 \sim \sum_{i=1}^n a_i^2 \cdot \sum_{i=1}^n \epsilon_i^2 = \sum_{i=1}^n \epsilon_i^2 \sim O_p(n)$; while by applying the Cauchy–Schwarz inequality on the square of the second cross-term, $(\sum_{i,j;i \neq j} a_i a_j \epsilon_{(i)} O_p(\frac{1}{\sqrt{n}}))^2 \leq (\sum_{j=1}^n a_j \sum_{i \neq j} a_i O_p(\frac{1}{\sqrt{n}}))^2 \leq \sum_{j=1}^n a_j^2 \cdot (\sum_{i \neq j} a_i O_p(\frac{1}{\sqrt{n}}))^2 \leq \sum_{j=1}^n a_j^2 \cdot \sum_{i=1}^n a_i^2 \cdot \sum_{i=1}^n O_p(\frac{1}{n}) = O_p(1)$, hence the first term $(\sum_{i=1}^n a_i \epsilon_{(i)})^2$ is the dominating one in

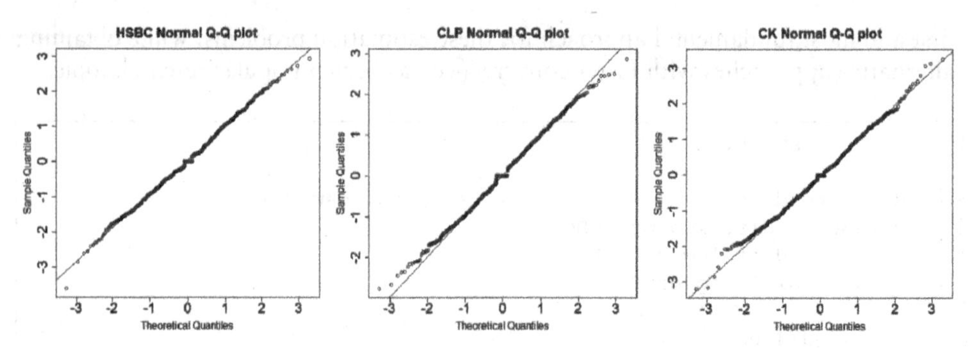

FIGURE 6.10 Normal Q-Q plots of the standardized residuals from the SV models (6.31) for HSBC, CLP, and CK via **R**, generated by Programme 6.22.

standardized residuals in the corresponding normal Q-Q plots in Figure 6.10 (via **R**) demonstrate quite a good fit with respect to its Q-Q reference line. It appears that the standardized residuals for HSBC, CLP, and CK likely do follow a normal distribution. The proposed SV models fit quite well with the real financial data.

```
1  > get_h <- function(sv) summary(sv$latent)[1]$statistics[,1]
2  > h_HSBC <- get_h(sv_HSBC); h_CLP <- get_h(sv_CLP);h_CK <- get_h(sv_CK)
3  >
4  > resid_HSBC <- y_HSBC / exp(h_HSBC/2)
5  > resid_CLP <- y_CLP / exp(h_CLP/2)
6  > resid_CK <- y_CK / exp(h_CK/2)
7  >
8  > # the degrees of freedom for the test:
9  > # K - (number of est. param.) = 20 - 3 = 17
10 > Box.test(resid_HSBC, lag=20, type="Ljung-Box", fitdf=3)
11     Box-Ljung test
12 data:  resid_HSBC
13 X-squared = 10.093, df = 17, p-value = 0.8997
14 > Box.test(resid_CLP, lag=20, type="Ljung-Box", fitdf=3)
15     Box-Ljung test
16 data:  resid_CLP
17 X-squared = 27.505, df = 17, p-value = 0.05107
18 > Box.test(resid_CK, lag=20, type="Ljung-Box", fitdf=3)
19     Box-Ljung test
20 data:  resid_CK
21 X-squared = 26.113, df = 17, p-value = 0.07242
22 >
23 > shapiro.test(resid_HSBC)
24     Shapiro-Wilk normality test
25 data:  resid_HSBC
26 W = 0.99781, p-value = 0.2223
27 > shapiro.test(resid_CLP)
```

---

the numerator. Likewise, the denominator $\sum_{i=1}^{n} (\bar{\hat{\epsilon}} - \hat{\epsilon}_i)^2$ is dominated by the term $\sum_{i=1}^{n} (\bar{\epsilon} - \epsilon_i)^2$ involving true random noises only. As a result, the direct plug-in of the original test statistic is justifiable.

```
28       Shapiro-Wilk normality test
29   data:   resid_CLP
30   W = 0.99706, p-value = 0.06793
31   > shapiro.test(resid_CK)
32       Shapiro-Wilk normality test
33   data:   resid_CK
34   W = 0.9969, p-value = 0.0526
35   >
36   > par(mfrow=c(1,3), cex.lab=1.5, cex.axis=2, cex.main=2)
37   > qqnorm(resid_HSBC, main="HSBC Normal Q-Q plot"); qqline(resid_HSBC)
38   > qqnorm(resid_CLP, main="CLP Normal Q-Q plot"); qqline(resid_CLP)
39   > qqnorm(resid_CK, main="CK Normal Q-Q plot"); qqline(resid_CK)
```

Programme 6.23 Model diagnostics of the fitted SV models (6.31) for HSBC, CLP, and CK from 1999 to 2002 via R.

*6.A APPENDIX

This appendix collects proofs of statements concerning Markov chains with uncountable state spaces as formulated in Section *6.3.

*6.A.1 Communication as Equivalence Relation

For any $s \in \mathcal{S}$, since $\mathcal{P}^0 f = f$ for all $f \in C_b^+(\mathcal{S})$, we have $s \leftrightarrow s$, and so the relation *communication* is reflexive. The symmetric property is also clear by its very definition. To show that communication is transitive, let $s_1, s_2, s_3 \in \mathcal{S}$ with $s_1 \leftrightarrow s_2$ and $s_2 \leftrightarrow s_3$. For any $f \in C_b^+(\mathcal{S})$ with $f(s_3) > 0$, using the fact that $s_2 \leftrightarrow s_3$, there exists $n_1 \in \mathbb{N}_0$ independent of f such that $\mathcal{P}^{n_1} f(s_2) > 0$. Let $g := \mathcal{P}^{n_1} f$ and using the fact that $s_1 \leftrightarrow s_2$, we infer the existence of $n_2 \in \mathbb{N}_0$ independent of f and n_1 such that

$$0 < \mathcal{P}^{n_2} g(s_1) = \mathcal{P}^{n_1+n_2} f(s_1).$$

This implies that $s_1 \to s_3$. The fact that $s_3 \to s_1$ can be shown similarly. Therefore, communication is indeed an equivalence relation.

*6.A.2 Proof of Proposition 6.1

We prove the three assertions in Proposition 6.1 as follows.

(i) Since $s_1 \to s_2$, there exists $m \in \mathbb{N}_0$ such that $\mathcal{P}^m f(s_1) > 0$ for any $f \in C_b^+(\mathcal{S})$ with $f(s_2) > 0$. Take $g := \mathcal{P}^m f$, and using the fact that $s_2 \to s_1$, we infer the existence of $n \in \mathbb{N}_0$, independent of f and m, such that $\mathcal{P}^n g(s_2) > 0$. Then

$$\mathcal{P}^{n+m} f(s_2) = \mathcal{P}^n g(s_2) > 0.$$

As this holds for any such f, so by the definition of $d(s_2)$ we deduce that $d(s_2) \mid n + m$. Next, for any $l \in I_{\mathcal{P}}(s_1)$, $\mathcal{P}^l h(s_1) > 0$ whenever $h(s_1) > 0$ in $C_b^+(\mathcal{S})$.

By considering the path $s_2 \to s_1 \to s_1 \to s_2$, for any $f \in C_b^+(S)$ with $f(s_2) > 0$, we have

$$\mathcal{P}^{n+l+m} f(s_2) = \mathcal{P}^n \left(\mathcal{P}^l(\mathcal{P}^m f)\right)(s_2) = \mathcal{P}^n \left(\mathcal{P}^l g\right)(s_2) > 0.$$

Therefore, we also have $d(s_2) \mid n+m+l$; together with $d(s_2) \mid n+m$, we infer that $d(s_2) \mid l$. Since $l \in I_{\mathcal{P}}(s_1)$ is arbitrary, we then have $d(s_2) \mid d(s_1)$ and so $d(s_1) \geq d(s_2)$. The reverse direction $d(s_1) \leq d(s_2)$ can be shown similarly by reflexivity.

(ii) This follows simply from Lemma 6.1 and its immediate remark by considering $\mathcal{A} = I_{\mathcal{P}}(s)$.

(iii) Since s_1 can access s_2 in m steps, for any $f \geq 0$ with $f(s_2) > 0$, it holds that $g(s_1) > 0$ with $g := \mathcal{P}^m f$. Using Statement (ii), we have $\mathcal{P}^{m+nd(s_1)} f(s_1) = \mathcal{P}^{nd(s_1)} g(s_1) > 0$ whenever n is sufficiently large.

*6.A.3 Proofs of Theorem 6.2 and Corollary 6.2

In this part, we denote the set $\mathbb{N}_\infty := \mathbb{N} \cup \{\infty\}$. We also take the convention that min $\emptyset = \infty$.

*6.A.3.1 Proof of Theorem 6.2

Define $N_1 : S \to \mathbb{N}_\infty$ by

$$N_1(s) := \min\{m : \mathcal{P}^n f(s) > 0 \text{ for any } n \geq m \text{ and } f \in C_b^+(S) \text{ with } f(s) > 0\}. \quad (6.34)$$

By the aperiodicity of the Markov chain and Proposition 6.1(ii), $N_1(s) < \infty$ for every $s \in S$.

Fix $s_0 \in S$ and let $N_1(s_0) = m$. We claim that there exists an open set $\mathcal{U}_{s_0} \subset S$ containing s_0 such that $N_1(s) \leq N_1(s_0)$ for any $s \in \mathcal{U}_{s_0}$. Indeed, by the definition of $N_1(s)$, for every $n \geq m$,

$$\mathcal{P}^n g(s_0) = \int_S \mu^{(n)}(s|s_0) g(s) ds > 0 \quad (6.35)$$

for any $g \in C_b^+(S)$ with $g(s_0) > 0$, where $\mu^{(n)}(\cdot|s_0) > 0$ is the n-step transition density of $\mathbf{x}_n$ given $\mathbf{x}_0 = s_0$. Since the function $g \in C_b^+(S)$ with $g(s_0) > 0$ is arbitrary, $\mu^{(n)}(s_0|s_0) > 0$, hence by the joint continuity of $\mu^{(n)}(\cdot|\cdot)$, (6.35) implies the existence of an open set $\mathcal{U}_{s_0}^{(n)}$ containing s_0 such that $\min_{s,s' \in \mathcal{U}_{s_0}^{(n)}} \mu^{(n)}(s'|s) > 0$.

Set $\mathcal{U}_{s_0} := \cap_{n=m}^{2m-1} \mathcal{U}_{s_0}^{(n)}$. Then, $\min_{s,s' \in \mathcal{U}_{s_0}} \mu^{(n)}(s'|s) > 0$ for any $n = m, \ldots, 2m-1$. For any $n \geq 2m$, $n = qm + r = (q-1)m + (m+r)$ with $q, r \in \mathbb{N}_0$, $q \geq 2$ and $0 \leq r < m$. Then we have

$$\mu^{(n)}(s'|s) \geq \int_{\mathcal{U}_{s_0}} \cdots \int_{\mathcal{U}_{s_0}} \mu^{(m)}(s^{(q-1)}|s^{(q-2)}) \mu^{(m)}(s^{(q-2)}|s^{(q-3)}) \cdots \mu^{(m)}(s^{(1)}|s)$$
$$\cdot \mu^{(m+r)}(s'|s^{(q-1)}) ds^{(q-1)} \ldots ds^{(1)} > 0$$

whenever $s, s' \in \mathcal{U}_{s_0}$. Now, take any $s \in \mathcal{U}_{s_0}$ and $f \in C_b^+(\mathcal{S})$ with $f(s) > 0$. By the continuity of f, there exists an open set $\mathcal{V}_s$ with $s \in \mathcal{V}_s \subseteq \mathcal{U}_{s_0}$ such that $f(s') > 0$ for any $s' \in \mathcal{V}_s$. Hence, we have

$$\mathcal{P}^n f(s) = \int_{\mathcal{S}} \mu^{(n)}(s'|s)f(s')ds' \geq \int_{\mathcal{V}_s} \mu^{(n)}(s'|s)f(s')ds' > 0$$

for any $n \geq m$. Since $f \in C_b^+(\mathcal{S})$ is arbitrary, we conclude by the definition of $N_1(\cdot)$ that $N_1(s) \leq N_1(s_0)$ for any $s \in \mathcal{U}_{s_0}$.

From the result of the above paragraph, for each $s \in \mathcal{S}$, there exists an open set $\mathcal{U}_s$ containing s on which the function $N_1(\cdot)$ is bounded above by $N_1(s)$. As $\mathcal{S}$ is compact, it can be covered by finitely many such open sets, that is, there exists a finite set $\mathcal{B} = \{s_1, \ldots, s_K\} \subseteq \mathcal{S}$ such that

$$\mathcal{S} \subset \bigcup_{s_j \in \mathcal{B}} \mathcal{U}_{s_j} \subset \bigcup_{j=1}^{K} \{s : N_1(s) \leq N_1(s_j)\},$$

which implies that $N_1(s) \leq N := \max_{j=1,\ldots,K} N_1(s_j) < \infty$ for all $s \in \mathcal{S}$.

*6.A.3.2 Proof of Corollary 6.2

Define $N_2 : \mathcal{S} \times \mathcal{S} \to \mathbb{N}_\infty$ by

$$N_2(s_1, s_2) := \min\{m : \mathcal{P}^n f(s_1) > 0 \text{ for any } n \geq m \text{ whenever } f \in C_b^+(\mathcal{S}) \text{ with } f(s_2) > 0\}.$$

By Proposition 6.1(iii), $N_2(s_1, s_2) < \infty$ for every couple of $(s_1, s_2) \in \mathcal{S} \times \mathcal{S}$. Following a similar continuity argument to that in the proof for Theorem 6.2, we can establish the existence of a finite collection of product open sets $\{\mathcal{U}_{s_i} \times \mathcal{U}_{s'_j} : (s_i, s'_j) \in \mathcal{U}_{s_i} \times \mathcal{U}_{s'_j}\}$ covering $\mathcal{S} \times \mathcal{S}$ such that $\min_{(s,s') \in \mathcal{U}_{s_i} \times \mathcal{U}_{s'_j}} \mu^{(N_2(s_i,s'_j))}(s|s') > 0$. Denote the set of all these pairs (s_i, s'_j) by $\mathcal{B} \subseteq \mathcal{S} \times \mathcal{S}$.

Take $N := \max_{(s_i,s'_j) \in \mathcal{B}} N_2(s_i, s'_j) + N_1$, where N_1 is the upper bound, we defined in the last line of the proof for Theorem 6.2, for the function $N_1(\cdot)$ defined in (6.34). For every $(s, s') \in \mathcal{S} \times \mathcal{S}$, it must lie inside one of the covering product open sets $\mathcal{U}_{s_i} \times \mathcal{U}_{s'_j}$, and for any $n \geq N$, we express $n = N_2(s_i, s'_j) + (n - N_2(s_i, s'_j))$, where $n - N_2(s_i, s'_j) \geq N_1$. By the proof of Theorem 6.2, we infer the existence of an open set $\mathcal{V}_{s'} \subseteq \mathcal{U}_{s'_j}$ such that $\min_{s'' \in \mathcal{V}_{s'}} \mu^{(n-N_2(s_i,s'_j))}(s''|s') > 0$. Therefore,

$$\begin{aligned}\mu^{(n)}(s|s') &= \int_{\mathcal{S}} \mu^{(N_2(s_i,s'_j))}(s|s'')\mu^{(n-N_2(s_i,s'_j))}(s''|s')ds'' \\ &\geq \int_{\mathcal{V}_{s'}} \mu^{(N_2(s_i,s'_j))}(s|s'')\mu^{(n-N_2(s_i,s'_j))}(s''|s')ds'' > 0.\end{aligned}$$

In other words, the n-step transition density for any two states is always positive whenever $n \geq N$. Finally, consider any $(s_1, s_2) \in S \times S$, and any $f \in C_b^+(S)$ with $f(s_2) > 0$. By continuity of f, there exists an open set $\mathcal{W}_{s_2} \subset S$ containing s_2 such that $\min_{s \in \mathcal{W}_{s_2}} f(s) > 0$. Hence, we conclude with

$$\mathcal{P}^n f(s_1) = \int_S f(s)\mu^{(n)}(s|s_1)ds \geq \int_{\mathcal{W}_{s_2}} f(s)\mu^{(n)}(s|s_1)ds > 0,$$

whenever $n \geq N$. The claim then follows.

*6.A.4 Proof of Theorem 6.6

Fix an arbitrary $s_0 \in S$. For $n \in \mathbb{N}$, define $\pi^{(n)} : C_b(S) \to \mathbb{R}$ by

$$\pi^{(n)}(f) := \frac{1}{n} \sum_{i=0}^{n-1} \mathcal{P}^i f(s_0), \quad f \in C_b(S).$$

By the *Stone–Weierstrass theorem*, $C_b(S)$ is separable, which allows us to choose a countable basis $\mathfrak{B} := \{f_j\}_{j \in \mathbb{N}} \subseteq C_b(S)$. It is clear that the sequence $\{\pi^{(n)}(f_j)\}_{n \in \mathbb{N}}$ is bounded by $\| f_j \|_\infty$ for each $j \in \mathbb{N}$. By the sequential compactness theorem and a diagonal argument, we infer the existence of a subsequence $\{n_l\}_{l \in \mathbb{N}}$ with $n_l \to \infty$ such that the subsequence $\{\pi^{(n_l)}(f_j)\}_{l \in \mathbb{N}}$ is convergent for every $j \in \mathbb{N}$. For every $j \in \mathbb{N}$, define

$$\pi(f_j) := \lim_{l \to \infty} \frac{1}{n_l} \sum_{l=0}^{n_l - 1} \mathcal{P}^l f_j(s_0) = \lim_{l \to \infty} \pi^{(n_l)}(f_j).$$

The definition of π can be naturally extended to span($\mathfrak{B}$): for any $\alpha, \beta \in \mathbb{R}$ and $f_i, f_j \in \mathfrak{B}$, we define

$$\begin{aligned}
\pi(\alpha f_i + \beta f_j) &:= \lim_{l \to \infty} \left(\pi^{(n_l)}\left(\alpha f_i + \beta f_j\right)\right) \\
&= \lim_{l \to \infty} \left(\pi^{(n_l)}(\alpha f_i) + \pi^{(n_l)}(\beta f_j)\right) \\
&= \alpha \lim_{l \to \infty} \pi^{(n_l)}(f_i) + \beta \lim_{l \to \infty} \pi^{(n_l)}(f_j) = \alpha\pi(f_i) + \beta\pi(f_j).
\end{aligned}$$

On the other hand, by the projection nature of $\mathcal{P}$, for any $f \in$ span($\mathfrak{B}$), we have $|\pi(f)| \leq \| f \|_\infty$. Hence, by the *Hahn–Banach theorem* [28], π can be extended to a bounded linear functional defined on $C_b(S)$. Notice that, for any $f_j \in \mathfrak{B}$,

$$\pi(\mathcal{P}f_j) = \lim_{l \to \infty} \frac{1}{n_l} \sum_{l=0}^{n_l - 1} \mathcal{P}^{l+1} f_j(s_0)$$

$$= \lim_{l \to \infty} \left(\frac{\mathcal{P}^{n_l} f_j(s_0) - f_j(s_0)}{n_l} + \boldsymbol{\pi}^{(n_l)}(f_j) \right)$$

$$= \lim_{l \to \infty} \boldsymbol{\pi}^{(n_l)}(f_j) = \boldsymbol{\pi}(f_j),$$

where the second last equality follows from the fact that $|\mathcal{P}^{n_l} f_j(s_0) - f_j(s_0)| \leq 2 \| f_j \|_\infty < \infty$. Hence, by extension, $\boldsymbol{\pi}(\mathcal{P}f) = \boldsymbol{\pi}(f)$ for any $f \in C_b(S)$. Finally, by the *Riesz representation theorem* [28] on Hausdorff spaces, there exists a probability measure $\mu_{\boldsymbol{\pi}}$, due to the positive nature of $\mathcal{P}$, such that

$$\boldsymbol{\pi}(f) = \int_S f(s) \mu_{\boldsymbol{\pi}}(ds)$$

for any $f \in C_b(S)$, which preserves the operator-invariant property $\boldsymbol{\pi}(\mathcal{P}f) = \boldsymbol{\pi}(f)$ by dominated convergence theorem. This $\boldsymbol{\mu}_{\boldsymbol{\pi}}$ serves as a stationary distribution.

*6.A.5 Proof of Theorem 6.7

This proof is based on generalizing the coupling argument from the discrete state setting. Again, we construct two independent Markov chains as follows. Let $\{\mathbf{u}_n\}_{n \in \mathbb{N}_0}$ be a sequence of independent uniform random vectors on $[0,1]^p$. By the method of inverse sampling, we can express $\{\mathbf{x}_n\}_{n \in \mathbb{N}_0}$ as

$$\mathbf{x}_0 := \psi_{\boldsymbol{\mu}^{(0)}}(\mathbf{u}_0) \sim \boldsymbol{\mu}^{(0)} \text{ and } \mathbf{x}_n := \phi(\mathbf{x}_{n-1}, \mathbf{u}_n) \sim \boldsymbol{\mu}(\mathbf{x}_n | \mathbf{x}_{n-1}), \text{ for } n \in \mathbb{N},$$

for some suitable functions[11] $\psi_{\boldsymbol{\mu}^{(0)}}$ and ϕ, respectively; this part is essentially the same as the proof of Theorem 6.4. The marginal distribution of $\mathbf{x}_n$ is $\boldsymbol{\mu}^{(n)}$. On the other hand, by utilizing another sequence of uniform random variables $\{\mathbf{u}'_n\}_{n \in \mathbb{N}_0}$ on $[0,1]^p$ which is independent of $\{\mathbf{u}_n\}_{n \in \mathbb{N}_0}$, we construct another Markov chain $\{\mathbf{x}'_n\}_{n \in \mathbb{N}_0}$ by

$$\mathbf{x}'_0 := \psi_{\boldsymbol{\pi}}(\mathbf{u}'_0) \sim \boldsymbol{\pi} \text{ and } \mathbf{x}'_n := \phi(\mathbf{x}'_{n-1}, \mathbf{u}'_n) \sim \boldsymbol{\mu}(\mathbf{x}'_n | \mathbf{x}'_{n-1}), \text{ for } n \in \mathbb{N},$$

where we again observe that the marginal distribution of every $\mathbf{x}'_n$ is the stationary distribution $\boldsymbol{\pi}$. By construction, $\{\mathbf{x}_n\}_{n \in \mathbb{N}_0}$ and $\{\mathbf{x}'_n\}_{n \in \mathbb{N}_0}$ are independent.

Next, for $\varepsilon > 0$, define the *ε-coupling time* as

$$\mathrm{T}_\varepsilon := \min\{n : \|\mathbf{x}_n - \mathbf{x}'_n\|_2 \leq \varepsilon\}.$$

[11] For instance, in the two-dimensional case, namely $\mathbf{x}_0 = (\mathrm{x}_{0,1}, \mathrm{x}_{0,2})$, we can generate the two elements in a sequential manner, by first simulating $\mathrm{x}_{0,1} \sim F^{-1}_{\mathrm{x}_1}(\mathrm{u}_1)$, and then simulating $\mathrm{x}_{0,2} | \mathrm{x}_{0,1} = x_{0,1} \sim F^{-1}_{\mathrm{x}_2|\mathrm{x}_1}(\mathrm{u}_2 | x_{0,1})$; the construction for ϕ is similar.

By Corollary 6.2, we can find an integer M such that

$$\mathcal{P}^n f(s) = \int_S f(s')\mu^{(n)}(s'|s)ds' > 0,$$

simultaneously for all $s \in S$ and for any $f \in C_b^+(S)$ that is not identically zero, whenever $n \geq M$. On the other hand, by the compactness of S, we have

$$\alpha := \min_{s,s' \in S} \mu^{(M)}(s'|s) > 0.$$

Clearly, by construction, the transition density of $\mathbf{x}'_n$ given $\mathbf{x}'_0 = s$ is $\mu^{(n)}(\cdot|s)$. Next, we consider:

$$\begin{aligned}
&\mathbb{P}(\mathrm{T}_\varepsilon \leq M) \\
&\geq \mathbb{P}(\|\mathbf{x}_M - \mathbf{x}'_M\|_2 \leq \varepsilon) \\
&= \int_S \int_S \int_{\{(s_M, s'_M) \in S \times S : \|s_M - s'_M\|_2 \leq \varepsilon\}} \mu^{(M)}(ds_M|s_0)\mu^{(M)}(ds'_M|s'_0)\mu^{(0)}(ds_0)\pi(ds'_0) \\
&\geq \alpha^2 \int_{\{(s_M, s'_M) \in S \times S : \|s_M - s'_M\|_2 \leq \varepsilon\}} ds_M ds'_M \int_S \mu^{(0)}(ds_0) \int_S \pi(ds'_0) \\
&= C_p \alpha^2 \varepsilon^{p-1} > 0
\end{aligned}$$

for some constant $C_p > 0$ depending on the dimension p and the geometric shape of S only. Hence, $\mathbb{P}(\mathrm{T}_\varepsilon > M) \leq 1 - C_p\alpha^2\varepsilon^{p-1}$, where the bound is clearly in $(0, 1)$ for a sufficiently small $\varepsilon > 0$.

Given that $\mathrm{T}_\varepsilon > M$, the Markov chains can be considered to start afresh no matter where it restarts, hence

$$\mathbb{P}\left(\|\mathbf{x}_{2M} - \mathbf{x}'_{2M}\|_2 > \varepsilon \mid \mathrm{T}_\varepsilon > M\right) \leq 1 - C_p\alpha^2\varepsilon^{p-1}.$$

Consequently,

$$\begin{aligned}
\mathbb{P}(\mathrm{T}_\varepsilon > 2M) &= \mathbb{P}(\mathrm{T}_\varepsilon > M) \cdot \mathbb{P}(\mathrm{T}_\varepsilon > 2M \mid \mathrm{T}_\varepsilon > M) \\
&\leq (1 - C_p\alpha^2\varepsilon^{p-1})\mathbb{P}(\|\mathbf{x}_{2M} - \mathbf{x}'_{2M}\|_2 > \varepsilon \mid \mathrm{T}_\varepsilon > M) \\
&\leq (1 - C_p\alpha^2\varepsilon^{p-1})^2.
\end{aligned}$$

By a recursive argument, we see that for any $l \in \mathbb{N}$,

$$\mathbb{P}(\mathrm{T}_\varepsilon > lM) \leq (1 - C_p\alpha^2\varepsilon^{p-1})^l \to 0 \text{ as } l \to \infty,$$

which implies that $\mathbb{P}(\mathrm{T}_\varepsilon = \infty) = 0$ for any $\varepsilon > 0$.

Now, we construct a third Markov chain $\{\mathbf{x}''_n\}_{n\in\mathbb{N}_0}$ as follows: set $\mathbf{x}''_0 := \mathbf{x}_0$ and for any $n \in \mathbb{N}_0$,

$$\mathbf{x}''_{n+1} := \begin{cases} \phi(\mathbf{x}''_n, \mathbf{u}_{n+1}), & \text{if } n < \mathrm{T}_\varepsilon; \\ \phi(\mathbf{x}''_n, \mathbf{u}'_{n+1}), & \text{if } n \geq \mathrm{T}_\varepsilon. \end{cases}$$

By construction, $\mathbf{x}''_n = \mathbf{x}_n$ for $n \leq T_\varepsilon$, and $\mathbf{x}''_n \sim \boldsymbol{\mu}^{(n)}$ for any $n \in \mathbb{N}_0$. On the other hand, unlike the case of discrete countable state spaces, the chains $\{\mathbf{x}''_n\}$ and $\{\mathbf{x}'_n\}$, albeit close to each other, do not necessarily coincide when $n > \mathrm{T}_\varepsilon$. To proceed, consider an arbitrary measurable subset $A \subset \mathcal{S}$. We can deduce the following coupling inequality:

$$\begin{aligned} |\boldsymbol{\mu}^{(n)}(A) - \boldsymbol{\pi}(A)| &= |\mathbb{P}(\mathbf{x}''_n \in A) - \mathbb{P}(\mathbf{x}'_n \in A)| \\ &\leq |\mathbb{P}(\mathbf{x}''_n \in A, n \leq \mathrm{T}_\varepsilon) - \mathbb{P}(\mathbf{x}'_n \in A,\ n \leq \mathrm{T}_\varepsilon)| \\ &\quad + |\mathbb{P}(\mathbf{x}''_n \in A, n > \mathrm{T}_\varepsilon) - \mathbb{P}(\mathbf{x}'_n \in A,\ n > \mathrm{T}_\varepsilon)| \\ &\leq \mathbb{P}(\mathrm{T}_\varepsilon \geq n) + |\mathbb{P}(\mathbf{x}''_n \in A, n > \mathrm{T}_\varepsilon) - \mathbb{P}(\mathbf{x}'_n \in A,\ n > \mathrm{T}_\varepsilon)|. \end{aligned}$$

Consider the second summand,

$$\begin{aligned} &\left|\mathbb{P}(\mathbf{x}''_n \in A, n > \mathrm{T}_\varepsilon) - \mathbb{P}(\mathbf{x}'_n \in A,\ n > \mathrm{T}_\varepsilon)\right| \\ &= \left|\sum_{m=0}^{n-1} \mathbb{P}(\mathrm{T}_\varepsilon = m)\left(\mathbb{P}(\mathbf{x}''_n \in A \mid \mathrm{T}_\varepsilon = m) - \mathbb{P}(\mathbf{x}'_n \in A \mid \mathrm{T}_\varepsilon = m)\right)\right| \\ &= \left|\sum_{m=0}^{n-1} \mathbb{P}(\mathrm{T}_\varepsilon = m) \iint\limits_{\|s''-s'\|_2\leq\epsilon} \left(\mathcal{P}^{n-m}\mathbb{1}_A(s'') - \mathcal{P}^{n-m}\mathbb{1}_A(s')\right) \boldsymbol{\mu}^{(m)}(ds'')\boldsymbol{\pi}(ds')\right|, \end{aligned}$$

where the last equality follows from the fact that once after T_ε, the chain $\{\mathbf{x}''_n\}$ shares the same transition distribution with $\{\mathbf{x}'_n\}$. Notice that

$$\begin{aligned} &\left|\iint\limits_{\|s''-s'\|_2\leq\epsilon} \left(\mathcal{P}^{n-m}\mathbb{1}_A(s'') - \mathcal{P}^{n-m}\mathbb{1}_A(s')\right) \boldsymbol{\mu}^{(m)}(ds'')\boldsymbol{\pi}(ds')\right| \\ &= \left|\iint\limits_{\|s''-s'\|_2\leq\epsilon} \left(\int_A \left(\boldsymbol{\mu}^{(n-m)}(y|s'') - \boldsymbol{\mu}^{(n-m)}(y|s')\right) dy\right) \boldsymbol{\mu}^{(m)}(ds'')\boldsymbol{\pi}(ds')\right| \\ &= \left|\iint\limits_{\|s''-s'\|_2\leq\epsilon} \left(\int_A\int_{\mathcal{S}} \boldsymbol{\mu}^{(n-m-1)}(y|z)\left(\boldsymbol{\mu}(z|s'') - \boldsymbol{\mu}(z|s')\right) dz dy\right) \boldsymbol{\mu}^{(m)}(ds'')\boldsymbol{\pi}(ds')\right| \\ &\leq \mathcal{M}(\mathcal{S})\Delta_\varepsilon, \end{aligned}$$

where $\mathcal{M}(\mathcal{S})$ is the Lebesgue measure of $\mathcal{S}$, and

$$\Delta_\varepsilon := \sup_{z\in\mathcal{S}} \sup_{s,s'\in\mathcal{S}:\|s-s'\|_2\leq\epsilon} |\boldsymbol{\mu}(z|s) - \boldsymbol{\mu}(z|s')|$$

converges to 0 as $\varepsilon \to 0$, thanks to the compactness of S and thus the uniform continuity of the transition density $\mu(\cdot|\cdot)$. Combining the above, we arrive at

$$\limsup_{n\to\infty} |\mu^{(n)}(A) - \pi(A)| \leq \lim_{n\to\infty} \mathbb{P}(T_\varepsilon \geq n) + \mathcal{M}(S)\Delta_\varepsilon = \mathcal{M}(S)\Delta_\varepsilon,$$

which is valid uniformly in A. Since $\varepsilon > 0$ is arbitrary, we conclude that $\mu^{(n)} \to \pi$ in total variation distance.

Remark 6.2 The proof of Theorem 6.7 provides an exponential rate of convergence to the stationary distribution up to a truncation error ($\mathcal{M}(S)\Delta_\varepsilon$), which can be arbitrary small. We refer interested readers to [24, 27] for more refined convergence rates.

*6.A.6 Proof of Theorem 6.9

Suppose that π is a time-reversible distribution. For any $f \in C_b(S)$,

$$\begin{aligned}
\pi(f) &= \int_S f(s_1)\pi(s_1)ds_1 \\
&= \int_S f(s_1)\pi(s_1)\left(\int_S \mu(s_2|s_1)ds_2\right) ds_1 \\
&= \int_S f(s_1)\int_S \pi(s_1)\mu(s_2|s_1)ds_2 ds_1 \\
&= \int_S f(s_1)\int_S \pi(s_2)\mu(s_1|s_2)ds_2 ds_1 \\
&= \int_S \pi(s_2)\left(\int_S f(s_1)\mu(s_1|s_2)ds_1\right) ds_2 \\
&= \int_S \pi(s_2)Pf(s_2)ds_2 \\
&= \pi(Pf).
\end{aligned}$$

This shows that π is a stationary distribution.

REFERENCES

1. Asmussen, S. (2003). *Applied Probability and Queues* (2nd ed.). Springer.
2. Box, G.E. (1979). Robustness in the strategy of scientific model building. In R.L. Launer and G.N. Wilkinson (Eds.), *Robustness in Statistics* (pp. 201–236). Academic Press.
3. Chen, M.H., Shao, Q.M., and Ibrahim, J.G. (2012). *Monte Carlo Methods in Bayesian Computation*. Springer Science & Business Media.
4. De Pierro, A.R. (1993). On the relation between the ISRA and the EM algorithm for positron emission tomography. *IEEE Transactions on Medical Imaging*, 12(2), 328–333.

5. Dempster, A.P., Laird, N.M., and Rubin, D.B. (1977). Maximum likelihood from incomplete data via the EM algorithm. *Journal of the Royal Statistical Society: Series B (Methodological)*, 39(1), 1–22.
6. Diaconis, P., and Freedman, D. (1997). On Markov chains with continuous state space. *Annals of Probability*, 27, 261–283.
7. Gamerman, D., and Lopes, H.F. (2006). *Markov Chain Monte Carlo: Stochastic Simulation for Bayesian Inference* (2nd ed.). Chapman & Hall/CRC.
8. Gilks, W.R., Richardson, S., and Spiegelhalter, D. (Eds.). (1995). *Markov Chain Monte Carlo in Practice*. CRC Press.
9. Grimmett, G., and Stirzaker, D. (2020). *Probability and Random Processes* (4th ed.). Oxford University Press.
10. Häggström, O. (2002). *Finite Markov Chains and Algorithmic Applications*. Cambridge University Press.
11. Hammersley, J.M., and Clifford, P. (1971). Markov fields on finite graphs and lattices. Unpublished manuscript. https://ora.ox.ac.uk/objects/uuid:4ea849da-1511-4578-bb88-6a8d02f457a6 (accessed 7 May 2024).
12. Hunter, D.R., and Lange, K. (2000). Quantile regression via an MM algorithm. *Journal of Computational and Graphical Statistics*, 9(1), 60–77.
13. Jacquier, E., Polson, N.G., and Rossi, P.E. (2002). Bayesian analysis of stochastic volatility models. *Journal of Business & Economic Statistics*, 20(1), 69–87.
14. Jacquier, E., Polson, N.G., and Rossi, P.E. (2004). Bayesian analysis of stochastic volatility models with fat-tails and correlated errors. *Journal of Econometrics*, 122(1), 185–212.
15. Kallenberg, O., and Kallenberg, O. (1997). *Foundations of Modern Probability* (Vol. 2). Springer.
16. Karlin, S., and Taylor, H.M. (1975). *A First Course in Stochastic Processes* (2nd ed.). Academic Press.
17. Kim, S., Shephard, N., and Chib, S. (1998). Stochastic volatility: likelihood inference and comparison with ARCH models. *The Review of Economic Studies*, 65(3), 361–393.
18. Lange, K. (2010). *Numerical Analysis for Statisticians* (2nd ed.). Springer.
19. Lange, K. (2013). *Optimization* (2nd ed.). Springer.
20. Lange, K. (2016). *MM Optimization Algorithms*. SIAM.
21. Levin, D.A., and Peres, Y. (2017). *Markov Chains and Mixing Times* (2nd ed.). American Mathematical Society.
22. McLachlan, G.J., and Krishnan, T. (2008). *The EM Algorithm and Extensions* (2nd ed.). Wiley.
23. Norris, J.R. (1998). *Markov Chains*. Cambridge University Press.
24. Persi, D., and Freedman, D. (1997). On Markov chains with continuous state space. Technical report.
25. Privault, N. (2013). *Understanding Markov Chains*. Springer.
26. Robert, C.P., and Casella, G. (2004). *Monte Carlo Statistical Methods* (2nd ed.). Springer.
27. Roberts, G., and Rosenthal, J. (2004). General state space Markov chains and MCMC algorithms. *Probability Surveys*, 1, 20–71.
28. Rudin, W. (1987). *Real and Complex Analysis* (3rd ed.). McGraw-Hill.
29. Tanner, M.A., and Wong, W.H. (1987). The calculation of posterior distributions by data augmentation. *Journal of the American Statistical Association*, 82(398), 528–540.
30. Taylor, S.J. (1982). Financial returns modelled by the product of two stochastic processes – a study of daily sugar prices 1961–79. In O.D. Anderson (Ed.), *Time Series Analysis: Theory and Practice* (Vol. 1, pp. 203–226). North-Holland.
31. Williams, D. (1991). *Probability with Martingales*. Cambridge University Press.

CHAPTER 7

Time-Varying Volatility Matrix and Kelly Fraction

Volatilities and correlations are fundamental concepts in quantitative finance, financial statistics and risk management. They serve as parameters in forecasting the trend of riskiness of stock prices, in pricing options and derivatives, in calculating the Value-at-Risk (V@R) of portfolios, in determining the beta factors in CAPM, and so on. In this chapter, first we explain how historical security price data can be used to estimate volatilities and correlations. In particular, we consider some common models like *exponentially weighted moving average* (EWMA), *autoregressive conditional heteroscedasticity* (ARCH) [11], and *generalized ARCH* (GARCH) [3]. The effectiveness of these models will be illustrated by real financial data. An important characteristic of these models is that volatilities and correlations are not constant, rather they change rapidly with time. A model that allows for this is closer to reality than one that assumes volatility is constant, like the celebrated Black–Scholes–Merton model [2]. In the second part of this chapter, we present some original research results of the authors, which are about the possible use of the *Kelly fraction* for selling (resp. buying) a stock at a favourable time over a planning horizon. Its implications for apparent calendar effects will also be discussed. Before we begin this journey, let us introduce some notations and definitions of some relevant technical concepts.

7.1 FLUCTUATION OF VOLATILITIES

Under the classical Black–Scholes–Merton model for a security price, the percentage change in the stock price over a short period of time is assumed to be normally distributed, namely

$$\delta S_t / S_t \sim \mathcal{N}(\mu \delta t, \sigma^2 \delta t), \tag{7.1}$$

where $\delta S_t := S_{t+\delta t} - S_t$ is the change in the stock price S_t over $[t, t+\delta t]$, $\mu \delta t$ is the mean percentage change and $\sigma\sqrt{\delta t}$ is the standard deviation of this percentage change.

To apply this model to real-life data, define the i-th daily percentage change of a stock as

$$u_i := (S_i - S_{i-1})/S_{i-1} \sim \mathcal{N}(\mu\tau, \sigma^2\tau), \quad i = 1, \ldots, n, \tag{7.2}$$

where $\tau = 1/252$, taking 252 trading days (roughly) in a year. Distribution (7.2) suggests that the usual sample variance of u_i's,

$$s^2 = \frac{1}{n-1}\sum_{i=1}^{n}(u_i - \overline{u})^2,$$

can serve as an unbiased estimate of $\sigma^2\tau$, and so we can estimate σ by $\hat{\sigma} = s/\sqrt{\tau}$. However, choosing an appropriate value for n is not easy. Generally speaking, the estimate is more accurate when n is large. Unfortunately, in reality σ does change in time. Including too many past observations may be useless or even harmful for the prediction of the spontaneous volatility. As a rule of thumb, the choice of n should reflect the length of time the volatility is used. For instance, if we want to estimate the value of a one-year option, the daily stock price for the last year should be taken. More sophisticated methods for estimating volatility will be discussed in later sections of this chapter. For the moment, we first illustrate some simple models with the stocks in Example 1.1 of Chapter 1. The file `stock_1999_2002.csv` contains adjusted daily closing prices for the stocks HSBC (0005)[1], CLP (0002)[2] and CK (0001)[3] from 1 Jan 1999 to 31 Dec 2002. Recall that in Programme 3.1 in Python (resp. 3.2 in **R**), we have demonstrated how to read in these stock prices and compute the u_i for each stock respectively according to (7.2), which we reproduce below in Programme 7.1 in Python (resp. Programme 7.2 in **R**) for readers' convenience.

```
1  import pandas as pd
2  import numpy as np
3  import matplotlib.pyplot as plt
4  import matplotlib.ticker as ticker
5
6  plt.rc('font', size=20); plt.rc('axes', titlesize=30, labelsize=30)
7  plt.rc('xtick', labelsize=25); plt.rc('ytick', labelsize=25)
8
9  d = pd.read_csv("stock_1999_2002.csv", index_col=0)
10 u = np.diff(d, axis=0) / d.iloc[:-1, :] # Arithmetic return
11 u.columns = d.columns.values + "_Return"
```

Programme 7.1 Excerpt from Programme 3.1: Computing arithmetic returns for HSBC, CLP, and CK in Python.

[1] *Hongkong and Shanghai Banking Corporation* (HSBC) is a British multinational universal bank founded in 1865 and has a ticker symbol of 0005 in Stock Exchange of Hong Kong (SEHK).

[2] *China Light and Power* (CLP) Hong Kong Ltd is a Hong Kong electricity company founded in 1901 and has a ticker symbol of 0002 in Stock Exchange of Hong Kong (SEHK).

[3] *Cheung Kong* (CK) is a Hong Kong multi-national conglomerate founded in 1972 and has a ticker symbol of 0001 in Stock Exchange of Hong Kong (SEHK).

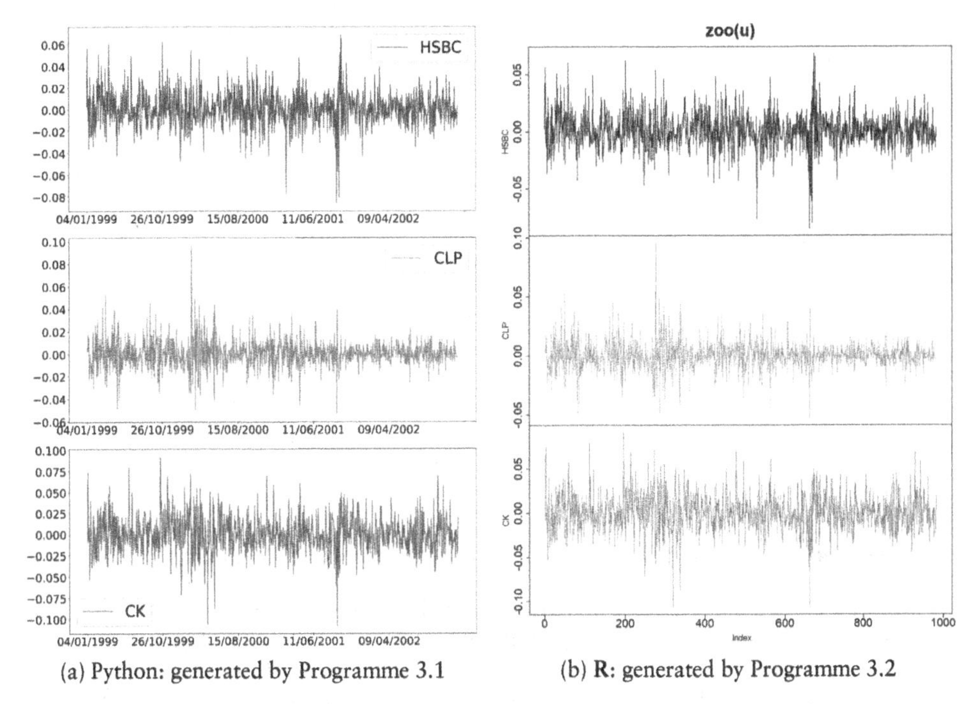

(a) Python: generated by Programme 3.1 (b) **R**: generated by Programme 3.2

FIGURE 7.1 Reproduced from Figures 3.1 and 3.3: Plots of returns of HSBC, CLP, and CK.

```
1 > d <- read.csv("stock_1999_2002.csv", row.names=1)
2 > date <- as.Date(rownames(d), format="%d/%m/%Y")
3 > d <- as.ts(d)
4 > u <- (lag(d) - d)/d
5 > colnames(u) <- paste0(colnames(d), "_Return")
```

Programme 7.2 Excerpt from Programme 3.2: Computing arithmetic returns for HSBC, CLP, and CK in **R**.

Based on the time series plots of the returns of these three stocks in Figure 7.1, we see that the u_i's are roughly fluctuating around zero, while the volatility amplitude is small in certain periods but large in others. Obviously the volatility varies with time rather than staying constant. Let us focus on computing the evolution in time of the standard deviation of u_i's based on a moving window of width w. First, we compute the standard deviation of the first w observations $u_1, \ldots, u_w$, and then that of $u_2, \ldots, u_{w+1}$, and so forth. This gives rise to a series of estimated standard deviations.

```
1 # compute 90-day moving sd and 180-day moving sd
2 u1, u2, u3 = u["HSBC_Return"], u["CLP_Return"], u["CK_Return"]
3
4 s_90, s_180 = u.rolling(90).std(ddof=1), u.rolling(180).std(ddof=1)
5
6 fig, ax = plt.subplots(figsize=(15,10))
```

```
 7  ax.plot(s_90["HSBC_Return"], "b-", label="s_90", linewidth=4)
 8  ax.plot(s_180["HSBC_Return"], "r--", label="s_180", linewidth=4)
 9  ax.set_title("Simple moving standard deviation of HSBC")
10  ax.xaxis.set_major_locator(ticker.MultipleLocator(200))
11  ax.legend(fontsize=25)
```

Programme 7.3 90-day and 180-day simple moving standard deviations in Python.

```
 1  > library(xts)
 2  >
 3  > u1 <- u[,"HSBC_Return"]; u2 <- u[,"CLP_Return"]; u3 <- u[,"CK_Return"]
 4  > s_HSBC_90 <- xts(rollapply(u1, 90, sd), order.by=date[-c(1:90)])
 5  > s_HSBC_180 <- xts(rollapply(u1, 180, sd), order.by=date[-c(1:180)])
 6  >
 7  > plot(s_HSBC_90, col="blue", lty=1, lwd=4, cex=1,
 8  +      main="Simple moving standard deviation of HSBC")
 9  > lines(s_HSBC_180, col="red", lty=2, lwd=4)
10  > addLegend("topright", legend.names=c("s_90", "s_180"),
11  +           col=c("blue", "red"), lty=1:2, lwd=4)
```

Programme 7.4 90-day and 180-day simple moving standard deviations in **R**.

Through Programme 7.3 in Python (resp. Programme 7.4 in **R**), we illustrate the time-varying nature of volatility estimates in Figure 7.2 (resp. Figure 7.3), and we also notice that different window widths have a very different effect on the estimates. For the 90-day version, the minimum and maximum values of HSBC's `s_90` for 1999–2002 are respectively 0.00984 and 0.02439, so the minimum and maximum annual volatilities are respectively $(\sqrt{252})(0.00984) = 15.62\%$ and $(\sqrt{252})(0.02439) = 38.72\%$. Similarly, with a window of 180 days, the minimum and maximum values of HSBC's `s_180` are 0.01172 and 0.02072 and the annual

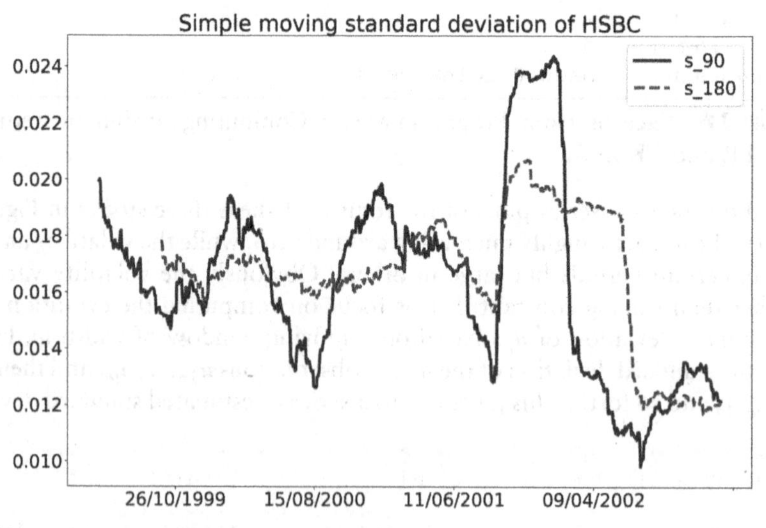

FIGURE 7.2 90-day and 180-day simple moving standard deviations of HSBC arithmetic returns by Python, generated by Programme 7.3.

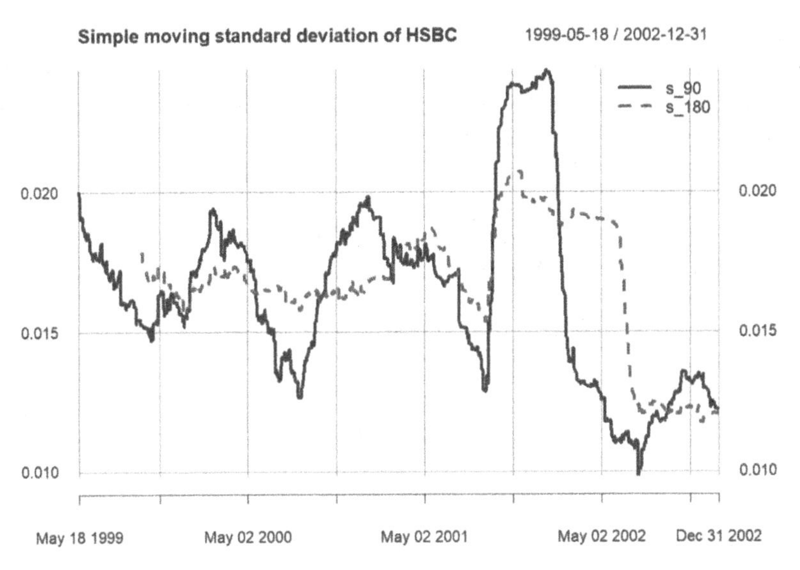

FIGURE 7.3 90-day and 180-day simple moving standard deviations of HSBC arithmetic returns by R, generated by Programme 7.4.

volatilities are 18.60% and 32.89%, respectively. Empirical studies often show that the annual volatility of a stock usually lies within the range from 10% to 50%.

7.2 EXPONENTIALLY WEIGHTED MOVING AVERAGE

In the previous section, the "moving" variance at time t was computed as

$$s_t^2 = \frac{1}{w-1} \sum_{i=t-w+1}^{t} (u_i - \overline{u})^2,$$

which is an average of the most recent w terms of $(u_i - \overline{u})^2$ and $\overline{u}$ is the moving average of $u_{t-w+1}, \ldots, u_t$. This estimate assigns equal weights to these w terms, even though the more recent terms are expected to have a greater relevance for the instantaneous volatility. On the other hand, an influential daily return from the distant past may still have an effect on the volatility today, but this distant past value is now completely excluded from the estimate as it lies outside the moving window. To address these two issues, we introduce the *Exponentially Weighted Moving Average* (EWMA) model that gives the highest weight to the most recent term and gradually down-weights as we move back in time. Before presenting the precise definition of this model, we state the following:

1. In estimating the volatility, instead of using $u_i = \ln(S_i/S_{i-1})$, we usually use the percentage change $u_i = (S_i - S_{i-1})/S_{i-1}$. In fact, the values obtained from these

two definitions are close if the quotient is small, which is normally the case in practice. This is especially the case when the time segment $[t, t + \delta t]$ is short.

2. The mean $\overline{u}$ of u_i's can be assumed to be zero for simplicity in the course of volatility estimation. In practice, the daily percentage change is small and fluctuates around zero. Assuming a zero mean of the random variables u_i's greatly simplifies the model as well.
3. By replacing $w - 1$ in the denominator of the definition of s_t^2 by w, we obtain the maximum likelihood estimate instead of an unbiased estimate.

To this end, we adopt the maximum likelihood estimation (MLE) approach from now on, by taking

$$s_t^2 = \frac{1}{w} \sum_{i=t-w+1}^{t} u_i^2. \tag{7.3}$$

The standard EWMA model for estimating the variance rate, σ_n^2 at the time $n\delta t$, is defined recursively by

$$\sigma_n^2 = \lambda\sigma_{n-1}^2 + (1 - \lambda)u_{n-1}^2, \tag{7.4}$$

where $0 < \lambda < 1$ is a parameter to be estimated in this model. To see why this model is called EWMA, let us substitute σ_{n-1}^2 in (7.4) to obtain

$$\begin{aligned}\sigma_n^2 &= \lambda\left(\lambda\sigma_{n-2}^2 + (1 - \lambda)u_{n-2}^2\right) + (1 - \lambda)u_{n-1}^2 \\ &= (1 - \lambda)(u_{n-1}^2 + \lambda u_{n-2}^2) + \lambda^2\sigma_{n-2}^2.\end{aligned}$$

Continuing in this manner by recursion, we obtain

$$\sigma_n^2 = (1 - \lambda)\sum_{i=1}^{m} \lambda^{i-1}u_{n-i}^2 + \lambda^m\sigma_{n-m}^2. \tag{7.5}$$

Letting m go to infinity, the last term of (7.5) vanishes under the assumption of uniform boundedness of $\sigma_0^2, \sigma_{-1}^2, \sigma_{-2}^2, \ldots,$ and the coefficient of the u_{n-i}^2 term is $(1 - \lambda)\lambda^{i-1}$. These weights decrease exponentially as we run backward in time, and they all sum up to 1. The *Riskmetrics* database developed by *JP Morgan & Chase* (1996) [18] uses this EWMA model with $\lambda = 0.94$ for updating the daily volatility estimate. Empirical studies found that this value of λ gives a reasonable volatility estimate for most common stocks [18].

Of course, we can also estimate λ from historical data by the MLE method. We defer the details of estimating λ this way to later in this chapter. In the next section, we shall introduce the celebrated ARIMA time series model, and then more importantly the GARCH(1, 1) model which includes the EWMA model as a special case. We shall then describe the MLE of the parameters in the GARCH model, which can also be applied to estimating λ for the EWMA model.

7.3 ARIMA TIME SERIES MODEL

ARIMA stands for "*Autoregressive Integrated Moving Average*", a model for time series data. *Autoregressive* refers to the dependence of the present observation on some predefined number of time-lagged observations. "*Integrated*" indicates that the model employs differencing to the raw observations, i.e. the subtraction of an observation from the one at previous time step, or even higher order of differencing, so as to make the time series stationary. *Moving Average* means that the model expresses the relationship between the residual errors incurred over the last few time lags of the observations and the model. An ARIMA model is characterized by three parameters: p, d and q, where p and q are the respective orders of AR and MA components, and d is the number of finite differencing required to ensure the stationary of the time series; if the time series is itself stationary, d is then taken to be 0. The ARIMA model for a time series $\{\mathrm{y}_t\}$ can be written in the following general form:

$$\mathrm{x}_t = \beta_1 \mathrm{x}_{t-1} + \beta_2 \mathrm{x}_{t-2} + \cdots + \beta_p \mathrm{x}_{t-p} + \epsilon_t + \phi_1 \epsilon_{t-1} + \phi_2 \epsilon_{t-2} + \cdots + \phi_q \epsilon_{t-q}; \tag{7.6}$$

here, $\mathrm{x}_t := (1 - B)^d \mathrm{y}_t$ where B is the backward operator so that $B\mathrm{y}_t := \mathrm{y}_{t-1}$ for every t. Note the intercept vanishes as x_t is now simplified to an ARMA(p, q) model with a constant mean of zero due to the finite differencing.

A commonly used approach in determining the values of p and q of the resulting time series $\{\mathrm{x}_t\}$ is via the use of the plots of sample *autocorrelation function* (ACF) and sample *partial autocorrelation functions* (PACF). These are simply the respective bar plots of the sample ACF values r_k and sample PACF values $\hat{\phi}_{kk}$, defined as follows, against time lag k. Recall that the ACF at lag k of a stationary time series $\{\mathrm{x}_t\}$ is $\rho_k := \mathrm{Corr}(\mathrm{x}_t, \mathrm{x}_{t-k})$, which is independent of t due to the stationarity of x_t. Its sample version, with the observed time series values $x_1, \ldots, x_n$, is defined as:

$$r_k := \frac{\sum_{i=1}^{n-k} (x_i - \overline{x})(x_{i+k} - \overline{x})}{\sum_{i=1}^{n-k} (x_i - \overline{x})^2}, \tag{7.7}$$

where $\overline{x} = \frac{1}{n} \sum_{i=1}^{n} x_i$. On the other hand, the PACF at lag k of a time series $\{\mathrm{x}_t\}$ is $\phi_{kk} := \mathrm{Corr}(\mathrm{x}_t, \mathrm{x}_{t-k} \mid \mathrm{x}_{t-1}, \ldots, \mathrm{x}_{t-k+1})$; it is the coefficient of x_{t-k} in the following AR(k) representation of x_t:

$$\mathrm{x}_t = \phi_{k1} \mathrm{x}_{t-1} + \cdots + \phi_{kk} \mathrm{x}_{t-k} + \epsilon'_t. \tag{7.8}$$

The validity of this representation can be established from the viewpoint of Hilbert space projection; it is the projection of x_t on $\mathrm{x}_{t-1}, \ldots, \mathrm{x}_{t-k}$, where the coefficients ϕ_{ki}'s are independent of t by joint stationarity, and ϵ'_t is orthogonal to $\mathrm{x}_{t-1}, \ldots, \mathrm{x}_{t-k}$. Note that the coefficients can be solved via the following *Yule–Walker equations*, which can be easily derived by considering Corr(x_t, x_{t-j}) for $j = 1, \ldots, k$ from (7.8), together with the symmetric property of ACF, namely $\rho_j = \rho_{-j}$ for all j:

$$\begin{pmatrix} \rho_1 \\ \vdots \\ \rho_k \end{pmatrix} = \begin{pmatrix} 1 & \rho_1 & \cdots & \rho_{k-1} \\ \rho_1 & 1 & \cdots & \rho_{k-2} \\ \vdots & \vdots & \ddots & \vdots \\ \rho_{k-1} & \rho_{k-2} & \cdots & 1 \end{pmatrix} \begin{pmatrix} \phi_{k1} \\ \vdots \\ \phi_{kk} \end{pmatrix}. \tag{7.9}$$

The sample PACF at lag k, $\hat{\phi}_{kk}$, is calculated by solving the sample version of (7.9), with ρ_k's replaced by r_k's.

In particular, for an AR(p) (resp. MA(q)) model, its ACF (resp. PACF) tails off / decays exponentially (can be sinusoidally) with the lag k, while its PACF (resp. ACF) cuts off / vanishes after lag $k > p$ (resp. $k > q$); this latter result is called the *cut-off property*. Their sample counterparts also demonstrate similar trends, see Figure 7.4 for a numerical illustration. In particular, for the AR(2) model, the sample PACF value does not deviate significantly from 0 for $k > 2$, which justifies the cut-off property of its PACF at lag 2, and the same argument applies to the ACF of the MA(2) model. The mathematical details of these properties have been thoroughly discussed in various classical textbooks on time series, including [15, 39] to name a couple. While the properties mentioned above are for pure AR and MA models, they can also be applied to the determination of the values of p and q in the general ARMA model, namely by observing the cut-off point of the sample PACF and ACF plots, respectively. See Table 7.1 for a summary of the properties of ACF and PACF for ARMA models.

ARIMA models are a useful tool commonly used in a wide range of disciplines, including but not limited to financial forecasting, economic analysis, operations research, and climate modelling. Here, let us illustrate their application to finance under the context of a recent trendy topic, namely the price prediction of *Bitcoin*. Bitcoin is a digital crypto-currency that enables individuals to transfer monetary value to one another without the requirement of the presence of a trusted third party like a bank or payment processor. It is the first decentralized peer-to-peer payment network solely powered by its own users without a central authority. The Bitcoin network monitors the movement of the cryptocurrency among digital wallets to prevent fraud and double spending. Bitcoin represents an innovative new form of digital money that is censorship-proof and accessible to anyone with internet access. New Bitcoins are issued through a process called "mining" in which computers solve

TABLE 7.1 Properties of ACF and PACF for ARMA models.

Plot	AR(p)	MA(q)	ARMA(p, q)
ACF	Tail off	Cut off after lag q	Tail off
PACF	Cut off after lag p	Tail off	Tail off

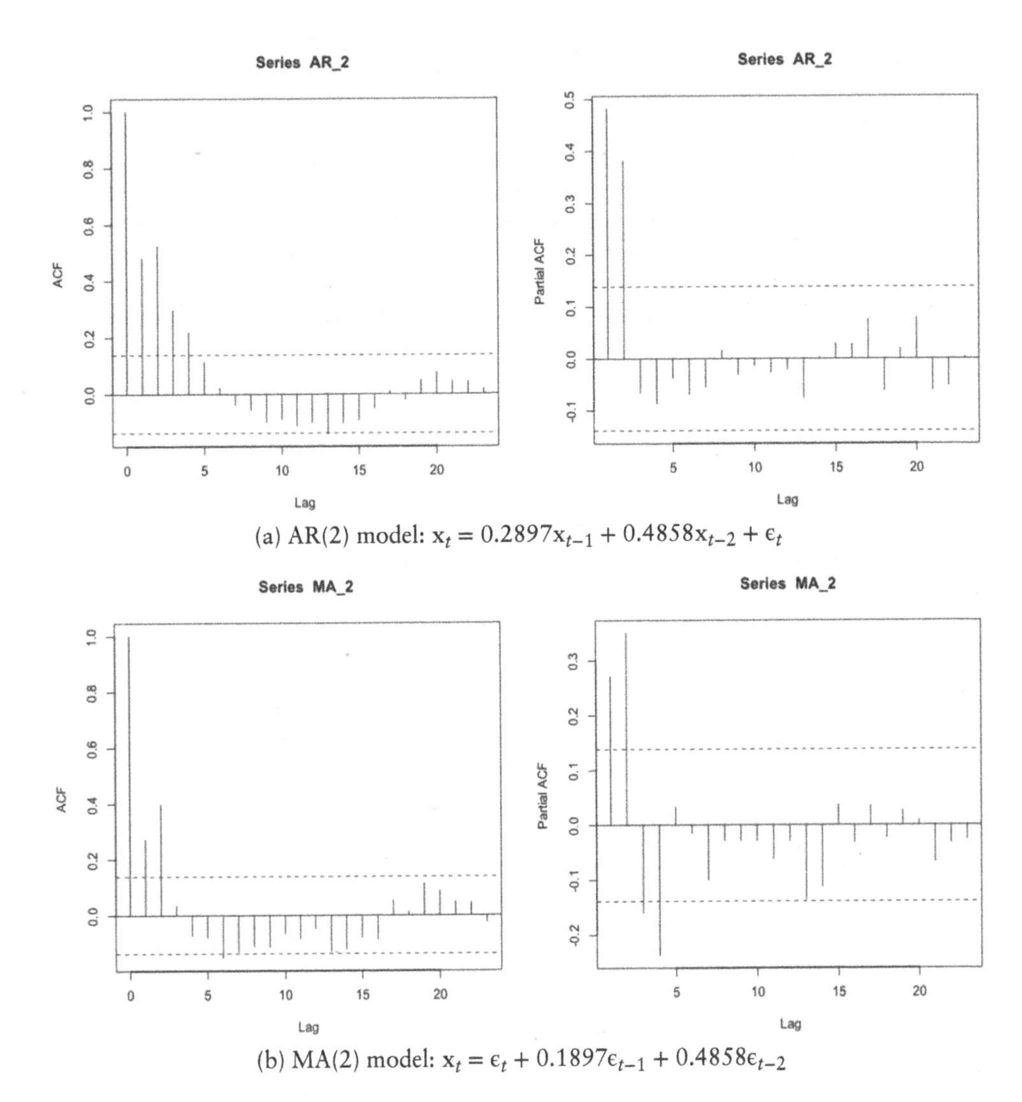

(a) AR(2) model: $x_t = 0.2897x_{t-1} + 0.4858x_{t-2} + \epsilon_t$

(b) MA(2) model: $x_t = \epsilon_t + 0.1897\epsilon_{t-1} + 0.4858\epsilon_{t-2}$

FIGURE 7.4 Respective sample ACF and PACF plots of AR(2) and MA(2) models.

complex math problems to discover a new block in the blockchain. Miners are then rewarded with new Bitcoins for verifying transactions and securing the network.

From the late 2010s onwards in particular, Bitcoin has been fiercely volatile and risky as an investment, yet it has gained popularity among those seeking decentralized digital currencies. The value of Bitcoin has increased significantly since it was first introduced in 2009, making many early investors wealthy. However, the high volatility and unregulated nature of Bitcoin make it too risky as a way to save money.

The `Bitstamp_BTCUSD_2018_minute.csv` dataset contains intraday minute data for Bitcoin in 2018. Let us train an ARIMA model using the observations of `Close` in the fourth quarter of 2018, save for the last day, 31 December. Indeed, we want to use our ARIMA model to predict the closing prices every minute that day. Note that in 2018, unlike the rapid rise in the previous years, the value of a Bitcoin dropped drastically from USD 13,841 to USD 3,689. The upper plot in Figure 7.5 via Python (resp. 7.6 via **R**) illustrates the movement of Bitcoin minute closing prices in the fourth quarter of 2018, and we note that there are many fierce jumps in the closing prices. These are likely to be regime-switching signals for parameters, and it would be reasonable to assume that the prices before and after the jumps should have different moving dynamics, and different statistical mean and volatility. Therefore, with this in mind, we choose only the time period after the most recent significant spike of standard deviation as the training period for our ARIMA model, that is after 00:21:00 on 30 Dec 2018 as indicated by the orange dashed line on the bottom plot in Figure 7.5 via Python (resp. 7.6 via **R**).

```
1  import numpy as np
2  import pandas as pd
3  import matplotlib.pyplot as plt
4  from scipy import stats
5  from statsmodels.graphics.tsaplots import plot_acf, plot_pacf
6  from statsmodels.tsa.arima.model import ARIMA
7  from statsmodels.stats.diagnostic import acorr_ljungbox
8
9  plt.rc('font', size=20); plt.rc('axes', titlesize=30, labelsize=25)
10 plt.rc('xtick', labelsize=20); plt.rc('ytick', labelsize=20)
11
12 df = pd.read_csv("Bitstamp_BTCUSD_2018_minute.csv", header=1)
13 df = df.iloc[::-1]              # Reverse the order of dates
14 df.index = df.date
15
16 # Select the last quarter as the training dataset
17 date_index = pd.to_datetime(df.index)
18 mask_train = pd.Series(date_index).between("2018-10-01", "2018-12-31",
19                                            inclusive="left")
20 train_close = df.close.loc[mask_train.values]
21
22 # Select the first day as the test dataset
23 mask_test = pd.Series(date_index).between("2018-12-31", "2019-01-01",
24                                           inclusive="left")
25 test_close = df.close[mask_test.values]
26
27 print(train_close.index[0], train_close.index[-1], test_close.index[0])
28
29 fig, (ax1, ax2) = plt.subplots(nrows=2, figsize=(20, 17))
30 ax1.plot(train_close, "b-", linewidth=3)
31 ax1.set_title("Bitcoin Price")
32 skip_minutes = len(train_close)//4
33 ax1.set_xticks(np.arange(0, len(train_close), skip_minutes),
34                train_close.index[::skip_minutes])
35
36 win_size = 30
37 s_win = train_close.tail(5000).rolling(win_size).std(ddof=1)
38 ax2.plot(s_win, "b-", linewidth=3)
39 ax2.set_title(f"{win_size}-minute simple moving standard deviation")
40 skip_minutes = len(s_win)//4
41 ax2.set_xticks(np.arange(0, len(s_win), skip_minutes),
42                s_win.index[::skip_minutes])
43
```

```
44 split_idx = np.where(s_win > 50)[0][-1] + 10     # add 10 to adjust
45 print(split_idx)     # len(s_win) = 5000
46 ax2.axvline(split_idx, linestyle="dashed", color="orange",
47             linewidth=3)
48 ax2.axvline(split_idx, ls="--", c="orange", linewidth=3)
49
50 # len(s_win) = 5000; -1 for Python indexing starting from 0
51 split_train = train_close.tail(5000 - split_idx - 1)
52 print(split_train.index[0])
53 print(len(split_train))
```

```
1 2018-10-01 00:00:00 2018-12-30 23:59:00 2018-12-31 00:00:00
2 3580
3 2018-12-30 00:21:00
4 1419
```

Programme 7.5 Plots of Bitcoin minute prices and standard deviations over 30-minute moving windows via Python.

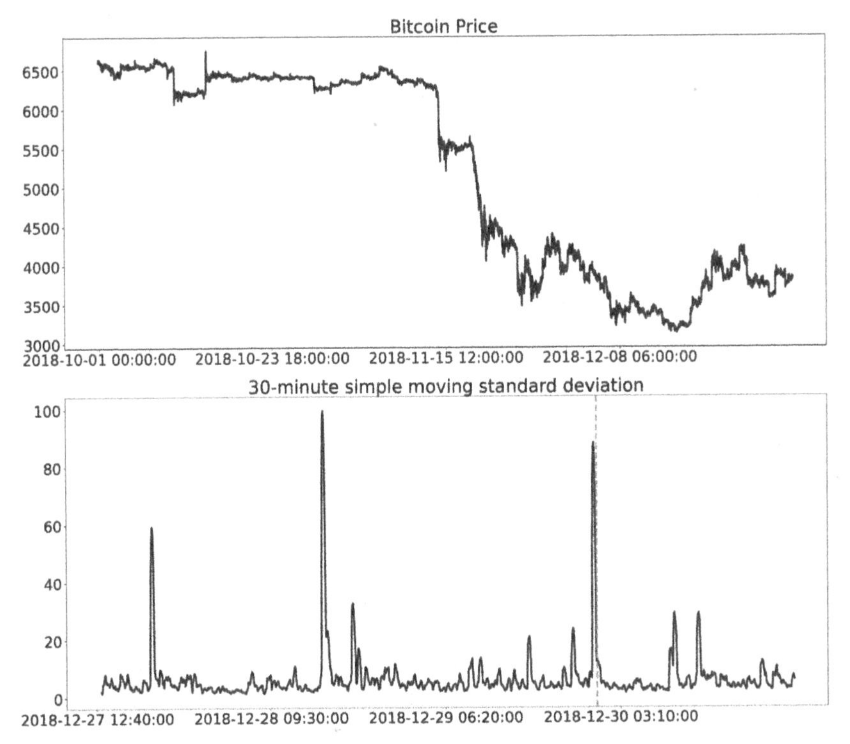

FIGURE 7.5 Plots of Bitcoin minute prices and standard deviations over 30-minute moving windows via Python, generated by Programme 7.5.

```
1  > library("tseries")
2  > library("xts")
3  > library("forecast")
4  >
5  > df <- read.csv("Bitstamp_BTCUSD_2018_minute.csv", skip=1)
6  >
7  > # Split prices into train (last quarter) and test (last day)
8  > train_idx <- "2018-10-01" < df$date & df$date < "2018-12-31"
9  > train_close <- rev(df$close[train_idx])
10 > train_date <- rev(df$date[train_idx])
11 > test_idx <- df$date > "2018-12-31"
12 > test_close <- rev(df$close[test_idx])
13 > test_date <- rev(df$date[test_idx])
14 > c(train_date[1], tail(train_date, 1), test_date[1])
15 [1] "2018-10-01 00:00:00" "2018-12-30 23:59:00" "2018-12-31 00:00:00"
16 > train_date <- strptime(train_date, format="%Y-%m-%d %H:%M:%S")
17 >
18 > par(mfrow=c(1, 1))
19 > plot(xts(train_close, order.by=train_date), type="l",
20 +      ylab="", main="Bitcoin Price", cex=1)
21 >
22 > win_size <- 30
23 > s_win <- xts(rollapply(tail(train_close, 5000), win_size, sd),
24 +              order.by=tail(train_date, 5000-win_size+1))
25 > plot(s_win, type="l", ylab="", cex.axis=1.1, cex=1,
26 +   main=paste0(win_size, "-minute simple moving standard deviation"))
27 >
28 > # length(s_win) + win_size - 1 = 5000
29 > (split_idx <- tail(which(s_win > 50), 1) + 10)  # add 10 to adjust
30 [1] 3552
31 > addEventLines(events=xts(x='', order.by=index(s_win)[split_idx]),
32 +               lty=2, col='orange', lwd=2, on=0)
33 >
34 > # length(s_win) + win_size - 1 = 5000
35 > split_train <- tail(train_close, 5000-(split_idx+win_size-1))
36 > tail(train_date, 5000-win_size+1-split_idx)[1]
37 [1] "2018-12-30 00:21:00 GMT"
38 > length(split_train)
39 [1] 1419
```

Programme 7.6 Plots of Bitcoin minute prices and standard deviations over 30-minute moving windows via **R**.

We then select an appropriate ARIMA model by determining the values of the model parameters p, d, q; see the respective Python codes in Programme 7.7 and **R** codes in Programme 7.8, and the corresponding sample ACF and PACF plots are shown in Figures 7.7 and 7.8, respectively. We observe that the sample ACF plot of the time series does not have a cut-off point, nor does it decay exponentially, indicating the non-stationarity of the time series and the need for differencing. After applying 1-lag differencing to the Bitcoin prices, i.e., $d = 1$, both sample ACF and PACF plots of the resulting time series roughly exhibit the cut-off property at $k = 2$. We therefore consider ARIMA(2, 1, 0), ARIMA(0, 1, 2), and ARIMA(2, 1, 2) as the candidate models.

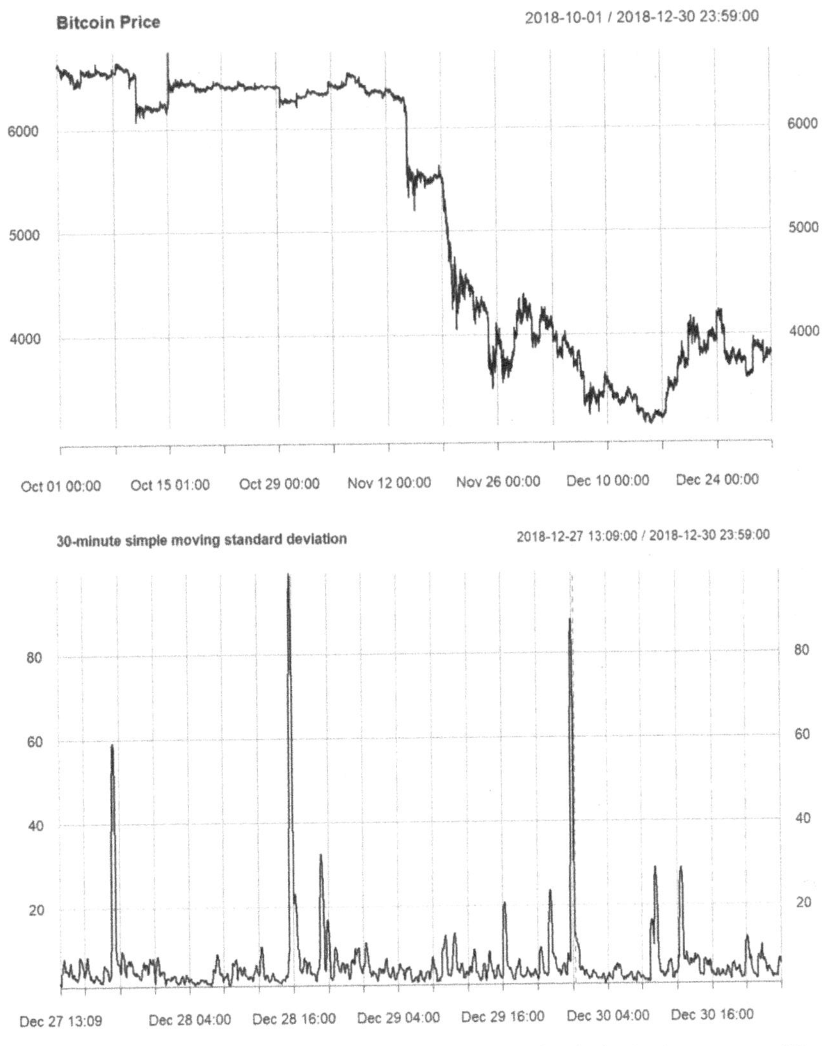

FIGURE 7.6 Plots of Bitcoin minute prices and standard deviations over 30-minute moving windows via **R**, generated by Programme 7.6.

For the selection between these three candidates, the arguably most prevalent thresholds are various *information criteria*, which depend on both the likelihood $\hat{L}$ of the fitted model and the number m of model parameters. Here we adopt the *Akaike Information Criterion* $\text{AIC} := -2\ln\hat{L} + 2m$ as the decision criterion of model selection; a smaller value of AIC indicates a better model; also see Section 10.6 in Chapter 10 for more discussions on AIC and BIC. We can simply fit the candidate models above and prompt for their respective AIC values, as shown in Programme 7.9 via Python and 7.10 via **R**. The AIC values of the fitted ARIMA$(2, 1, 0)$ and ARIMA$(0, 1, 2)$ are 8,319.95 and 8,316.41, respectively; both are higher than that of the fitted ARIMA$(2, 1, 2)$ model (less than 8,300), therefore ARIMA$(2, 1, 2)$ is our preferred model for this example.

```
alpha = 0.05
c_i = stats.norm.ppf(1-alpha/2) / np.sqrt(len(split_train))
fig, ax = plt.subplots(figsize=(10,10))
plot_acf(split_train.values, lags=30, ax=ax, alpha=None,
         color="blue", title="ACF Plot of Bitcoin Price")
ax.set_ylim((-0.15, 1.1))
ax.axhline(c_i, linestyle="--", c="purple")
ax.axhline(-c_i, linestyle="--", c="purple")

lag_price = split_train.diff().values[1:]
fig, axs = plt.subplots(ncols=2, figsize=(20,10))
plot_acf(lag_price, lags=30, ax=axs[0], alpha=None,
         color="blue", title="ACF Plot of 1-lagged Bitcoin Price")
axs[0].set_ylim((-0.2, 1.1))
axs[0].axhline(c_i, linestyle="--", c="purple")
axs[0].axhline(-c_i, linestyle="--", c="purple")

plot_pacf(lag_price, lags=30, ax=axs[1], alpha=None, zero=False,
          color="blue", title="PACF Plot of 1-lagged Bitcoin Price")
axs[1].set_ylim((-0.155, 0.055))
axs[1].axhline(c_i, linestyle="--", c="purple")
axs[1].axhline(-c_i, linestyle="--", c="purple")

# 1 lag difference; ACF and PACF: 2 significant lags and looks similar
print(ARIMA(split_train, order=(2, 1, 0)).fit().aic)
print(ARIMA(split_train, order=(0, 1, 2)).fit().aic)
print(ARIMA(split_train, order=(2, 1, 2)).fit().aic)
```

```
8319.95426608393
8316.40877058989
8299.159940396014
```

Programme 7.7 Sample ACF and PACF plots of Bitcoin minute prices in Python.

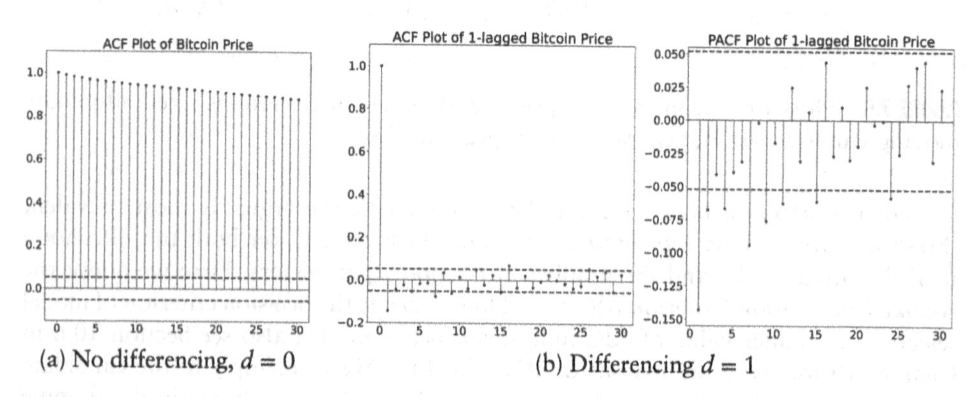

(a) No differencing, $d = 0$ (b) Differencing $d = 1$

FIGURE 7.7 Sample ACF and PACF plots of Bitcoin prices in Python, generated in Programme 7.7.

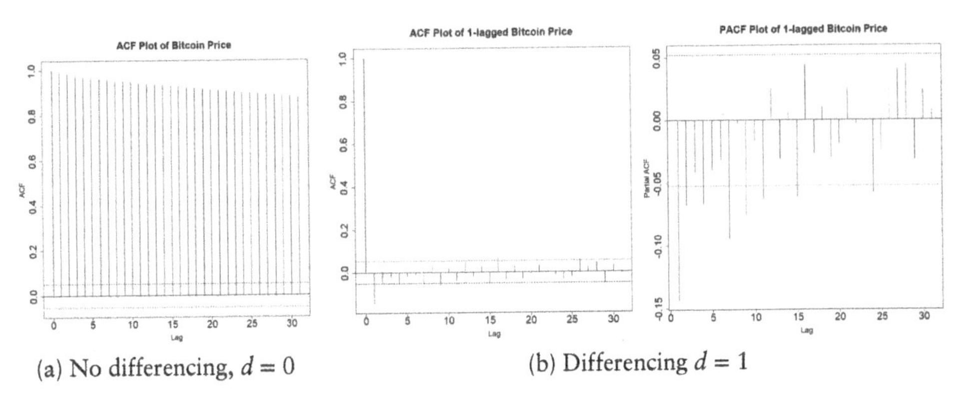

(a) No differencing, $d = 0$ (b) Differencing $d = 1$

FIGURE 7.8 Sample ACF and PACF plots of Bitcoin prices in **R**, generated in Programme 7.8.

```
1  > par(mfrow=c(1, 1), cex.lab=1.5, cex.axis=2, cex.main=2)
2  > acf(split_train, main=paste("ACF Plot of Bitcoin Price"))
3  >
4  > par(mfrow=c(1, 2))
5  > lag_price <- diff(split_train)
6  > acf(lag_price, main=paste("ACF Plot of 1-lagged Bitcoin Price"))
7  > pacf(lag_price, main=paste("PACF Plot of 1-lagged Bitcoin Price"))
8  >
9  > # 1 lag difference; ACF and PACF: 2 significant lags and looks similar
10 > AIC(Arima(ts(split_train), order=c(2, 1, 0)),
11 +     Arima(ts(split_train), order=c(0, 1, 2)),
12 +     Arima(ts(split_train), order=c(2, 1, 2)))
13                                            df      AIC
14 Arima(ts(split_train), order = c(2, 1, 0))  3 8319.952
15 Arima(ts(split_train), order = c(0, 1, 2))  3 8316.406
16 Arima(ts(split_train), order = c(2, 1, 2))  5 8294.059
```

Programme 7.8 Sample ACF and PACF plots of Bitcoin minute prices in **R**.

In fact, both Python and **R** already have various built-in functions that carry out the entire procedure of model fitting (as introduced above) in one go, such as the `auto_arima()` function in the `pmdarima` library for Python and the `auto.arima()` function in the `forecast` library for **R**. Specifically, given the user inputs of $p_{\max}, d_{\max}, q_{\max}$ and the information criterion selected, these functions fit all the ARIMA(p, d, q) models for $p = 0, \ldots, p_{\max}$, $d = 0, \ldots, d_{\max}$, $q = 0, \ldots, q_{\max}$, and return the best-fitted model in terms of the specified information criterion. Interestingly, with $p_{\max} = q_{\max} = 5$ and $d_{\max} = 2$, the output from Programme 7.10 via **R** indicates that the ARIMA(2, 1, 2) model is the best in terms of AIC, which agrees with our earlier conclusion. However, Programme 7.9 via Python determines that the best model in terms of AIC is ARIMA(2, 1, 3) with an AIC value of 8,295.22, lower than that of 8,299.16 for ARIMA(2, 1, 2) in Programme 7.7. In view of this, we shall conduct further analysis on both of these models as follows:

```
1 import pmdarima as pm
2
3 best_model = pm.auto_arima(split_train, max_p=5, max_q=5, max_d=2,
4                            information_criterion="aic", seasonal=False,
5                            stepwise=False, with_intercept=False)
6 print(best_model.summary())
```

```
1                                SARIMAX Results
2 ==============================================================================
3 Dep. Variable:                      y   No. Observations:                 1419
4 Model:               SARIMAX(2, 1, 3)   Log Likelihood               -4141.607
5 Date:                Fri, 19 Jan 2024   AIC                           8295.215
6 Time:                        15:50:37   BIC                           8326.757
7 Sample:                    12-30-2018   HQIC                          8306.998
8                          - 12-30-2018
9 Covariance Type:                  opg
10 ==============================================================================
11                  coef    std err          z      P>|z|      [0.025      0.975]
12 ------------------------------------------------------------------------------
13 ar.L1         -0.2013      0.058     -3.469      0.001      -0.315      -0.088
14 ar.L2          0.7451      0.067     11.097      0.000       0.614       0.877
15 ma.L1          0.0332      0.064      0.515      0.606      -0.093       0.159
16 ma.L2         -0.8416      0.058    -14.539      0.000      -0.955      -0.728
17 ma.L3          0.0330      0.037      0.892      0.372      -0.039       0.105
18 sigma2        20.1554      0.197    102.429      0.000      19.770      20.541
19 ==============================================================================
20 Ljung-Box (L1) (Q):                   0.00   Jarque-Bera (JB):            217026.60
21 Prob(Q):                              0.97   Prob(JB):                         0.00
22 Heteroskedasticity (H):               2.89   Skew:                            -1.21
23 Prob(H) (two-sided):                  0.00   Kurtosis:                        63.56
24 ==============================================================================
```

Programme 7.9 Use of `auto_arima()` function in determining the best set of hyperparameters for modelling Bitcoin prices in Python.

```
1 > (best_model <- auto.arima(split_train, max.p=5, max.q=5, max.d=2,
2 +                           seasonal=FALSE, ic='aic', allowdrift=FALSE))
3 Series: split_train
4 ARIMA(2,1,2)
5
6 Coefficients:
7           ar1     ar2     ma1      ma2
8       -0.2376  0.7079  0.0850  -0.8201
9 s.e.   0.0540  0.0485  0.0455   0.0394
10
11 sigma^2 = 20.23:  log likelihood = -4142.03
12 AIC=8294.06   AICc=8294.1   BIC=8320.34
```

Programme 7.10 Use of `auto.arima()` function in determining the best set of hyperparameters for modelling Bitcoin prices in **R**.

We evaluate the goodness-of-fit of the fitted models, namely ARIMA(2, 1, 3) via Python and ARIMA(2, 1, 2) via **R**, by using the Ljung–Box test, a kind of portmanteau test. Specifically, given an observed times series $\{x_t\}_{t=1}^{n}$, the null hypothesis H_0 of this test states that the data are independently distributed, while the alternative hypothesis

H_1 states that the data exhibits serial correlation. The test statistic is

$$\chi_K^2 := n(n+2)\sum_{k=1}^{K}\frac{r_k^2}{n-k},$$

where r_k is the sample ACF of $\{x_t\}$ of lag k. In general, as a means of hypothesis testing, the test statistic has an approximate chi-square distribution of K degrees of freedom; see Appendix *7.A.1 for more detailed discussions. In contrast, in our current context of diagnostics for fitted ARIMA models, the r_k's computed based on fitted residuals are not linearly independent; it can be shown that $p+q$ extra linear constraints are imposed accordingly in relation to the ARMA parameters β_i's and ϕ_j's, see equation (2.9) in [6]. To establish the asymptotic distribution of the corresponding test statistic under this scenario, we can work in parallel with the arguments in Appendix *7.A.1, which relies on the quadratic form expression, yet the sandwiched square matrix, originally with full rank, will have its rank reduced by the number of parameters estimated, namely $p+q$, hence showing that the asymptotic distribution of the corresponding test statistic in this case is still a chi-square distribution, yet with $K-(p+q)$ degrees of freedom[4]; see [6] for detailed discussions. From the results in Programmes 7.11 via Python and 7.12 via **R**, respectively, both models have p-values slightly greater than 0.05 in the Ljung–Box test with $K = 10$, and we fail to reject the null hypothesis that the residuals of the Bitcoin prices under the respective fitted models (ARIMA(2, 1, 3) via Python and ARIMA(2, 1, 2) via **R**) show no autocorrelation up to lag 10; see Figure 7.9 by Python and Figure 7.10 by **R** for the plots of serial residuals. When we use either of the fitted models to make predictions, we shall treat the results with extra caution, due to the fact that they only marginally pass the Ljung–Box test.

```
1  p, d, q = best_model.order
2  arima_name = f"ARIMA({p}, {d}, {q})"
3
4  resid = best_model.resid()[1:]           # remove the first residual
5  # Ljung-Box test on (non-standardized) residuals
6  print(acorr_ljungbox(resid, lags=[10], model_df=p+q))
7  # the degrees of freedom for the test are K - (p+q) = 10 - (2+3) = 5
8
9  fig, ax = plt.subplots(figsize=(13,8))
10 ax.plot(resid, "b-", linewidth=2)
11 skip_minutes = len(split_train)//4
12 ax.set_xticks(np.arange(0, len(split_train), skip_minutes),
13               split_train.index[::skip_minutes])
14 ax.tick_params(axis='x', which='major', labelsize=16)
15 ax.set_title(f"Fitted residuals of {arima_name}")
16 ax.set_ylabel("")
```

[4] Likewise, the same philosophy can also be seen for the sampling distribution of variance of a normal sample, namely $\sum_{i=1}^{n}\frac{(x_i-\overline{x})^2}{\sigma^2}\sim\chi_{n-1}^2$ (see Section 1.4), while $\sum_{i=1}^{n}\frac{(x_i-\mu)^2}{\sigma^2}\sim\chi_n^2$, due to the extra linear constraint that $\sum_{i=1}^{n}(x_i-\overline{x})=0$ imposed in replacing the theoretical mean μ by the sample mean $\overline{x}$.

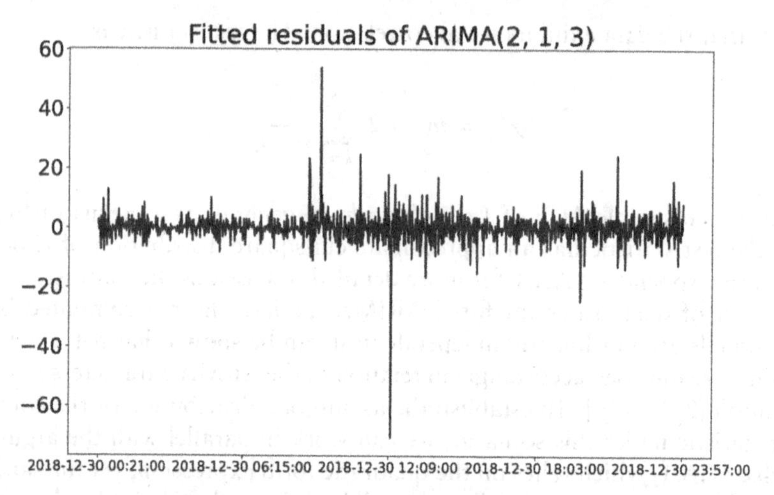

FIGURE 7.9 Plot of the residuals of the fitted ARIMA(2, 1, 3) model in Python, generated by Programme 7.11.

```
1        lb_stat  lb_pvalue
2  10   9.954177   0.076544
```

Programme 7.11 Fitting an ARIMA(2, 1, 3) model for Bitcoin prices in Python.

```
1  > order <- arimaorder(best_model)
2  > arima_name <- paste0("ARIMA(", paste(order, collapse=", "), ")")
3  >
4  > resid <- xts(best_model$residuals,
5  +              order.by=tail(train_date, 5000-win_size+1-split_idx))
6  > par(mfrow=c(1, 1))
7  > plot(resid, main=paste("Fitted residuals of", arima_name),
8  +      type="h", ylab="", cex=1)
9  >
10 > # Ljung-Box test on (non-standardized) residuals
11 > # the degrees of freedom for the test are K - (p+q) = 10 - (2+2) = 6
12 > Box.test(resid, lag=10, type="Ljung-Box", fitdf=order[1]+order[3])
13
14         Box-Ljung test
15
16 data:  resid
17 X-squared = 12.121, df.p = 6, p-value = 0.05932
```

Programme 7.12 Fitting an ARIMA(2, 1, 2) model to Bitcoin prices in R.

Finally, we apply the fitted ARIMA models to the test dataset and predict the Bitcoin prices on a minute basis on 31 Dec 2018. We also include a naïve prediction strategy for comparison, namely setting the closing price of the previous minute as the predicted closing price of the current minute; see Programmes 7.13 and 7.14 for the codes in Python and R, respectively. The testing mean squared error (MSE) of the naïve approach is 19.726, while the fitted ARIMA(2, 1, 3) model in Python performs

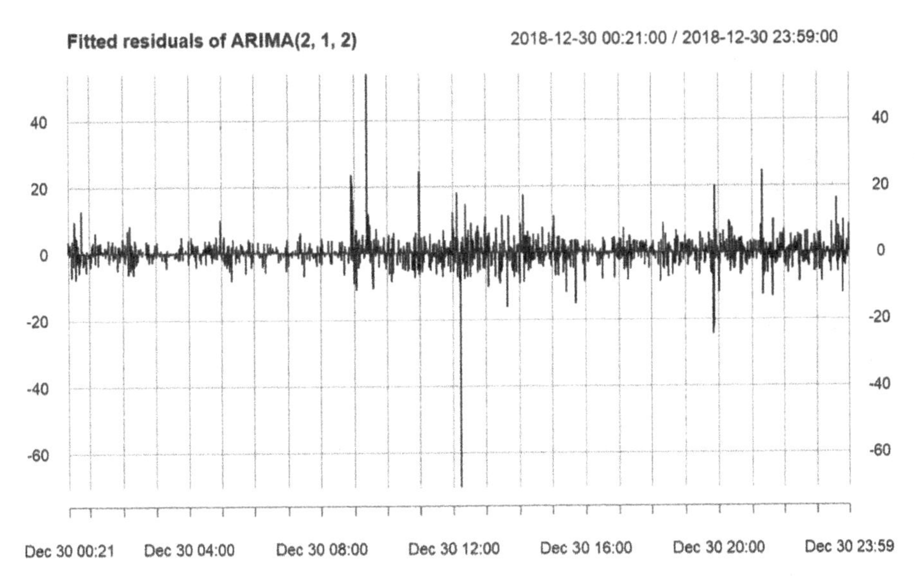

FIGURE 7.10 Plot of the residuals of the fitted ARIMA(2, 1, 2) model in **R**, generated by Programme 7.12.

better with a lower testing MSE of 19.249; the fitted ARIMA(2, 1, 2) model in **R** results in a slightly higher testing MSE (19.261), yet it still outperforms the naïve approach; see Figure 7.11 in Python and Figure 7.12 in **R** for the illustration of price predictions. In Chapter 15, we shall introduce a more sophisticated model, namely the *Long-Short Term Memory* (LSTM), to strive for a slightly better prediction of the Bitcoin price.

```
1  hist_data = train_close[-30:].tolist()
2  arima_pred = np.empty(test_close.shape)
3  for i, x_t in enumerate(test_close):
4      model = ARIMA(hist_data, order=best_model.order)
5      model_fit = model.smooth(best_model.params())
6      arima_pred[i] = model_fit.forecast()[0]
7      hist_data.append(x_t)
8
9  date_val = pd.to_datetime(test_close.index)
10 xticks = date_val.strftime('%H:%M')
11
12 fig, ax = plt.subplots(figsize=(13,8))
13 ax.plot(test_close, color='pink', label='Actual', linewidth=8)
14 ax.plot(arima_pred, 'b- ', linewidth=2, label=arima_name)
15 skip_minutes = len(test_close)//4
16 ax.set_xticks(np.arange(0, len(test_close), skip_minutes),
17               xticks[::skip_minutes])
18 ax.tick_params(axis='both', which='major', labelsize=17)
19 ax.set_title(f'Bitcoin Price Prediction on {date_val.date[0]}')
20 ax.set_xlabel('Date')
```

```
21 ax.set_ylabel('Price')
22 plt.legend(fontsize=20)
23
24 price = np.append(train_close[-1], test_close.values)
25 print(np.mean(np.diff(price)**2))
26 print(np.mean((test_close.values - arima_pred)**2))
```

```
1 19.725824513888924
2 19.248766796251108
```

Programme 7.13 Bitcoin price prediction by the fitted ARIMA(2, 1, 3) model in Python.

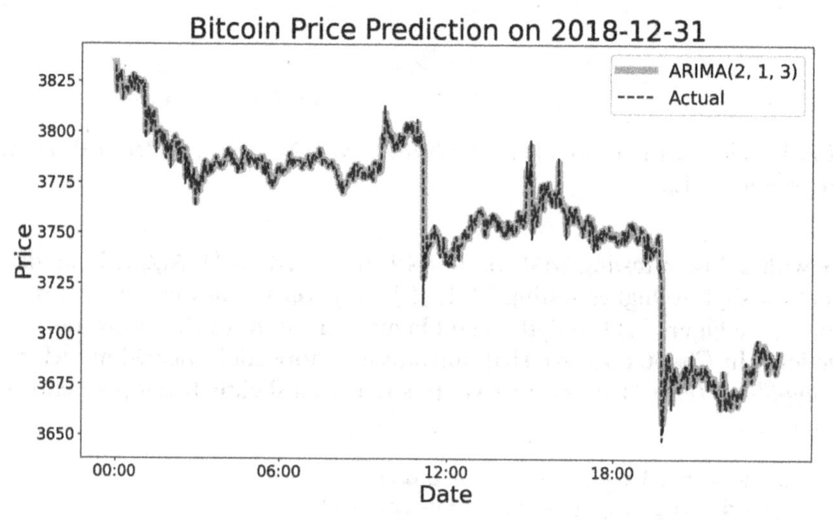

FIGURE 7.11 Bitcoin price prediction with the fitted ARIMA(2, 1, 3) model on 31 December 2018 in Python, generated by Programme 7.13.

```
1  > hist_data <- tail(split_train, 30)
2  > arima_pred <- array(NA, dim=length(test_close))
3  > for (i in 1:length(test_close)) {
4  +   # update the model with old estimates of ARIMA parameters
5  +   best_model <- Arima(hist_data, model=best_model)
6  +   arima_pred[i] <- predict(best_model, n.ahead=1)$pred[1]
7  +   hist_data <- c(hist_data, test_close[i])
8  + }
9  >
10 > ts_pred <- xts(cbind(test_close, arima_pred),
11 +               order.by=strptime(test_date, format="%Y-%m-%d %H:%M:%S"))
12 >
13 > plot(ts_pred, main="Bitcoin Price Prediction", lwd=c(2, 8), lty=2:1,
```

```
14 +        col=c("blue", "pink"), cex=1)
15 > addLegend("topright", lwd=c(2, 8), lty=2:1, ncol=1, bg="white",
16 +           bty="o", legend.names=c("Actual", arima_name))
17 >
18 > # using previous price to predict the next day price
19 > mean(diff(c(tail(train_close, 1), ts_pred$test_close))**2, na.rm=TRUE)
20 [1] 19.72582
21 > mean((ts_pred$test_close - ts_pred$arima_pred)**2)
22 [1] 19.26076
```

Programme 7.14 Bitcoin price prediction by the fitted ARIMA(2, 1, 2) model in R.

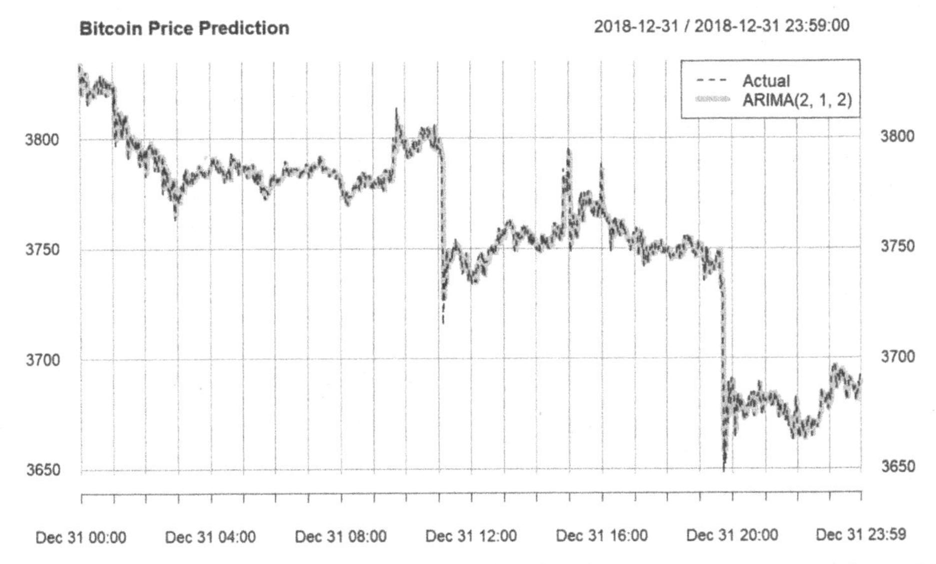

FIGURE 7.12 Bitcoin price prediction with the fitted ARIMA(2, 1, 2) model on 31 December 2018 in R, generated by Programme 7.14.

7.4 ARCH AND GARCH MODELS

While ARIMA models do take possible non-stationarity of the time series itself into consideration, they still require stationary behaviour for the time series' variance process. This is not always satisfied in reality, as we saw in the Bitcoin price modelling. Therefore, we have to proceed to the next level of sophistication that can decently explain those inhomogeneous volatility phenomena. In this regard, we first consider the *autoregressive conditional heteroscedasticity* model ARCH(m) for variance proposed by the Nobel-winning economist Robert F. Engle in 1982 [11]. This model assumes that the current variance rate σ_n^2 depends on the m most recent values of u_i^2 as well as a long-run level average variance rate V_L, so that

$$\sigma_n^2 = \gamma V_L + \alpha_1 u_{n-1}^2 + \cdots + \alpha_m u_{n-m}^2, \tag{7.10}$$

where the coefficients $\gamma > 0, \alpha_1, \ldots, \alpha_m \geq 0$ satisfy $\gamma + \alpha_1 + \cdots + \alpha_m = 1$. We further assume the following conditional independence of u_i's given the latest volatility:

$$\frac{\mathrm{u}_i}{\sigma_i}\bigg|(\sigma_i, \mathrm{u}_{i-1}, \mathrm{u}_{i-2}, \ldots, \mathrm{u}_1) \stackrel{d}{=} \frac{\mathrm{u}_i}{\sigma_i}\bigg|\sigma_i = \sigma_i \sim \mathcal{N}(0,1). \tag{7.11}$$

One advantage of this ARCH(m) model is that it introduces a long-run variance rate around which the variance rates fluctuate over time. The disadvantages are that there are too many parameters to be estimated especially if m is large, and that this model lacks a simple autoregressive recursive relationship as found in the EWMA model. However, by combining the ideas in the EWMA model (7.4) and the ARCH model (7.10), one arrives at the following celebrated GARCH model.

The *generalized autoregressive conditional heteroscedasticity* model GARCH(p, q) was first proposed in the doctoral thesis of the economist *Tim Bollerslev* in 1986 [3], under the supervision of Robert Engle. Recall that in the EWMA model (7.4), σ_n^2 is defined recursively in terms of both σ_{n-1}^2 and u_{n-1}^2; while in the ARCH(m) model (7.10), σ_n^2 is defined in terms of m most recent values of u_i^2 with a long-run variance rate V_L. As a hybrid, a GARCH(p, q) model takes the following general form:

$$\sigma_n^2 = \gamma V_L + \alpha_1 \mathrm{u}_{n-1}^2 + \cdots + \alpha_p \mathrm{u}_{n-p}^2 + \beta_1 \sigma_{n-1}^2 + \cdots + \beta_q \sigma_{n-q}^2,$$

where $\gamma > 0, \alpha_1, \ldots, \alpha_p, \beta_1, \ldots, \beta_q \geq 0$ satisfy $\gamma + \sum_{i=1}^p \alpha_i + \sum_{j=1}^q \beta_j = 1$, and V_L is the long-run average variance rate. We shall explain later why we call V_L the long-run average variance rate. In particular, the simplest case of $p = q = 1$ corresponds to the following GARCH(1, 1) model:

$$\sigma_n^2 = \gamma V_L + \alpha \mathrm{u}_{n-1}^2 + \beta \sigma_{n-1}^2, \tag{7.12}$$

which clearly reduces to the simpler EWMA model when $\gamma = 0$. In financial practice, we use the GARCH(1, 1) model much more often than other models for time-varying volatility estimation.

7.4.1 MLE for GARCH(1, 1) Model

We now turn to the estimation of the parameters of the GARCH(1, 1) model (7.12). There are four unknown positive parameters V_L, γ, α, β to be estimated, subject to one linear constraint: $\gamma + \alpha + \beta = 1$. We rewrite the GARCH(1, 1) model as

$$\sigma_n^2 = \omega + \alpha \mathrm{u}_{n-1}^2 + \beta \sigma_{n-1}^2, \tag{7.13}$$

where $\omega := \gamma V_L = (1 - \alpha - \beta) V_L$. By recalling the conditional distribution of $\mathrm{u}_i | \sigma_i = \sigma_i \sim \mathcal{N}(0, \sigma_i^2)$, and abbreviating $v_i := \sigma_i^2$, the likelihood function for $u := (u_1, \ldots, u_n)$ is

$$L(\omega, \alpha, \beta; u) = \prod_{i=1}^n \frac{1}{\sqrt{2\pi v_i}} \exp\left(-\frac{u_i^2}{2v_i}\right), \tag{7.14}$$

hence the log-likelihood function is

$$\ln L(\omega, \alpha, \beta; \boldsymbol{u}) = -\frac{n}{2}\ln(2\pi) + \frac{1}{2}\sum_{i=1}^{n}\left(-\ln(v_i) - \frac{u_i^2}{v_i}\right). \tag{7.15}$$

We aim to find the parameters ω, α, β such that the function

$$l(\omega, \alpha, \beta; \boldsymbol{u}) = \sum_{i=1}^{n}\left(-\ln(v_i) - \frac{u_i^2}{v_i}\right) \tag{7.16}$$

is maximized. Note that the parameters are entered implicitly into function (7.16) through v_i which satisfies the recursive relation of (7.13), so it is a rather complicated maximization problem. Let us illustrate numerically with the HSBC stock data, using $u_i = (S_i - S_{i-1})/S_{i-1}$, the percentage change in stock price, to fit a GARCH(1, 1) model and obtain the corresponding MLEs for ω, α, β; see Programmes 7.15 and 7.16 for Python and **R** codes, respectively.

```
1 from arch import arch_model
2
3 res_HSBC = arch_model(u1, mean='Zero', vol='GARCH', p=1, q=1).fit()
4 print(res_HSBC.params.values) # retrieve the MLE for ω,α,β
5 print(res_HSBC.loglikelihood) # compute log-likelihood value of (7.15)
```

```
1 [6.17171909e-06 4.99676502e-02 9.29302027e-01]
2 2659.992959126217
```

Programme 7.15 MLEs of ω, α, β, and the corresponding log-likelihood value in Python; also see Programme 7.17 for the outputs of HSBC.

```
 1 > library(fGarch) # load library "fGarch"
 2 >
 3 > # GARCH(1,1) on HSBC return
 4 > res_HSBC <- garchFit(garch(1, 1), data=u1, include.mean=FALSE)
 5 > round(coef(res_HSBC), 6)   # display coefficient in 6 digits
 6    omega    alpha1    beta1
 7 0.000015 0.062788 0.881989
 8 > res_HSBC@fit$llh            # compute negative log-likelihood value
 9 LogLikelihood
10     -2660.307
```

Programme 7.16 MLEs of ω, α, β using `fGarch` library, and the corresponding log-likelihood value in **R**; see Programme 7.18 for the HSBC output.

From the Python output of Programme 7.15, and the summary of results in Programme 7.17 by using the command `res_HSBC.summary()`, the MLEs of ω, α, β for HSBC over the period of 1999–2002 are respectively 6.172×10^{-6}, 4.997×10^{-2} and 9.293×10^{-1}. The maximum log-likelihood value in (7.15) is $2,659.993$ and the estimated long-run variance rate is $V_L = \omega/(1 - \alpha - \beta) = 0.0002977$. However, from the R output in Programme 7.16 and the summary in Programme 7.18, the MLEs of ω, α, β for the same set of 1999–2002 HSBC data are respectively 1.491×10^{-5}, 6.279×10^{-2}, and 8.820×10^{-1}, the corresponding maximum log-likelihood value of (7.15) is 2, 660.307, and the estimated long-run variance rate is 0.0002652.

```
1                          Zero Mean - GARCH Model Results
2  ==============================================================================
3  Dep. Variable:                HSBC_Return   R-squared:                       0.000
4  Mean Model:                     Zero Mean   Adj. R-squared:                  0.001
5  Vol Model:                          GARCH   Log-Likelihood:                2659.99
6  Distribution:                      Normal   AIC:                          -5313.99
7  Method:                Maximum Likelihood   BIC:                          -5299.31
8                                              No. Observations:                  983
9  Date:                    Wed, Jan 31 2024   Df Residuals:                      983
10 Time:                            16:33:42   Df Model:                            0
11                               Volatility Model
12 ============================================================================
13                 coef    std err          t       P>|t|      95.0% Conf. Int.
14 ----------------------------------------------------------------------------
15 omega      6.1717e-06  1.954e-09   3157.723       0.000 [6.168e-06,6.176e-06]
16 alpha[1]       0.0500  4.063e-02      1.230       0.219 [-2.966e-02,   0.130]
17 beta[1]        0.9293  3.407e-02     27.279 7.431e-164     [  0.863,   0.996]
18 ============================================================================
```

Programme 7.17 Output from fitting GARCH(1, 1) model to HSBC returns over 1999–2002 using the `arch_model` library in Python.

```
1 > res_HSBC@fit$matcoef
2              Estimate   Std. Error    t value      Pr(>|t|)
3 omega   1.490724e-05 5.230795e-06  2.849900 0.0043733045
4 alpha1  6.278833e-02 1.703009e-02  3.686906 0.0002269976
5 beta1   8.819895e-01 2.941547e-02 29.983860 0.0000000000
```

Programme 7.18 Summary statistics from fitting a GARCH(1, 1) model to HSBC returns over 1999–2002 in **R**.

The discrepancy between the outputs obtained from Python and **R** could be explained as follows. From the Python output of Programme 7.17, the parameter estimates for ω and β have small p-values below 0.05, thus these two coefficients are significantly different from zero. However, the p-value for α exceeds the threshold of 0.05, meaning that we cannot reject the null hypothesis that α is equal to zero. Therefore, based on the Python output, we suspect that the (arithmetic) return of HSBC does not follow a GARCH(1, 1) model; another possibility is that the MLEs obtained in the `arch` library of Python are not accurate enough.

In fact, we can change the optimization method used in Python to a two-step scheme, first using an extension of the *limited-memory Broyden–Fletcher–Goldfarb–Shanno* (L-BFGS) algorithm, called L-BFGS-B for short, followed by the *sequential least squares programming* (SLSQP). The first stage of using L-BFGS-B is often adopted in practice to solve an optimization problem. However, L-BFGS-B only supports bounded constraints for each parameter such as $\omega, \alpha, \beta > 0$ and $\alpha, \beta < 1$ as required by the GARCH$(1,1)$ model, while other constraints that involve multiple parameters at once, such as $\alpha + \beta < 1$, are not supported. We therefore require a second stage of optimization, namely by SLSQP. By the implementation as shown in Programme 7.19, the resulting MLEs are close to those obtained by **R** in Programme 7.16. This also justifies the accuracy of the results obtained by **R**; to this end, in the later development of the various GARCH models, while we still present results via both Python and **R**, we may primarily use the outputs from **R** for demonstration if the results from the two environments do not match.

```
1  from scipy.optimize import minimize
2  from scipy.optimize import Bounds
3
4  # set lower and upper bounds for omega, alpha, beta
5  bds = Bounds(lb=[1e-15,]*3, ub=[np.inf] + [1-1e-15,]*2)
6  # stationarity constraint: α+β < 1
7  cons = [{'type': 'ineq', 'fun': lambda x: 1-x[1]-x[2]}]
8
9  def GARCH_11(x, r):
10         omega, alpha, beta = x
11         nu = omega + alpha*np.mean(r**2) + beta*np.mean(r**2) # nu_1
12         log_like = -1/2*(np.log(2*np.pi) + np.log(nu) + r[0]**2/nu)
13         for i in range(1, len(r)):
14                 nu = omega + alpha*r[i-1]**2 + beta*nu
15                 log_like -= 1/2*(np.log(2*np.pi) + np.log(nu) + r[i]**2/nu)
16         return -log_like
17
18 def GARCH_11_MLE(r):
19     x0 = [0.05, 0.04, 0.9]
20
21     # two stage optimization: L-BFGS-B only support bounded constraints
22     model1 = minimize(GARCH_11, x0, args=(r), method='L-BFGS-B',
23                       bounds=bds, tol=1e-20)
24     model2 = minimize(GARCH_11, model1.x, args=(r), method='SLSQP',
25                       constraints=cons, bounds=bds, tol=1e-20)
26     return model2.x, model2.fun                  # negative log-likelihood
27
28 model_HSBC, llike_HSBC = GARCH_11_MLE(u1)
29 print(model_HSBC, llike_HSBC)
```

```
1  [1.48474391e-05 6.27356786e-02 8.82262810e-01] -2660.307068692025
```

Programme 7.19 MLEs of GARCH$(1,1)$ using L-BFGS-B and SLSQP optimization algorithms in Python.

To further justify the results from **R**, we also demonstrate the MLE seeking procedure in **R** without the use of the `fGARCH` package. In particular, in Programme 7.20 via **R**, we use the `constrOptim()` function to perform minimization subject to inequality constraints. The first argument is the decision variable; the second argument is the function to be minimized; `grad=NULL` means that no gradient function is provided; `u=u1` is the extra argument inserted in the `GARCH_11()` function; `method=c("Nelder-Mead")` indicates that the *Nelder–Mead* optimization algorithm proposed in [24] is adopted[5]; and finally `ui` and `ci` are the inputs for the inequality constraints. With ω, α, β being the decision variables, these inequality constraints take the form of

$$\theta_{11}\omega + \theta_{12}\alpha + \theta_{13}\beta \geq c_1,$$
$$\theta_{21}\omega + \theta_{22}\alpha + \theta_{23}\beta \geq c_2,$$
$$\vdots$$
$$\theta_{61}\omega + \theta_{62}\alpha + \theta_{63}\beta \geq c_6,$$

where the matrix $(\theta_{i,j})_{1\leq i\leq 6, 1\leq j\leq 3}$ is the input for `u_i` and the column vector $c := (c_1, \cdots, c_6)^{\mathsf{T}}$ is the input for `c_i`. In our current context, there are 6 inequalities constraints, namely (1) $\alpha + \beta < 1$; (2) $\omega > 0$; (3) $\alpha > 0$; (4) $\beta > 0$; (5) $\alpha < 1$; and (6) $\beta < 1$, we can construct the inputs `ui` and `ci` accordingly as `ui=rbind(c(0,-1,-1), diag(3), -cbind(0, diag(2)))` and `ci=c(-1, rep(0, 3), rep(-1, 2))` where `diag(3)` $= \begin{pmatrix} 1 & 0 & 0 \\ 0 & 1 & 0 \\ 0 & 0 & 1 \end{pmatrix}$, `cbind(0, diag(2))` $= \begin{pmatrix} 0 & 1 & 0 \\ 0 & 0 & 1 \end{pmatrix}$, `rep(0, 3)` $= \begin{pmatrix} 0 \\ 0 \\ 0 \end{pmatrix}$, and `rep(-1, 2)` $= \begin{pmatrix} -1 \\ -1 \end{pmatrix}$.

```
1  > GARCH_11 <- function(para, u){
2  +     u <- as.numeric(u)
3  +     omega0 <- para[1]
4  +     alpha <- para[2]
5  +     beta <- para[3]
6  +     nu <- var(u)
7  +     #nu <- omega0/(1-alpha-beta)
8  +     loglik <- dnorm(u[1], 0, sqrt(nu), log=TRUE)
9  +
10 +     for (i in 2:length(u)){
11 +         nu <- omega0 + alpha*u[i-1]^2 + beta*nu
12 +         loglik <- loglik + dnorm(u[i], 0, sqrt(nu), log=TRUE)
13 +
14 +     }
15 +     return(-loglik)
16 + }
17 >
18 > para <- c(0.3,0.1,0.4)
19 > self_model <- constrOptim(para, GARCH_11, grad=NULL, u=u1,
20 +   method=c("Nelder-Mead"),
```

[5] We adopt the Nelder–Mead method since it is the only option in `constrOptim()` which does not require the specification of gradient function `grad`.

```
21  +     ui=rbind(c(0,-1,-1), diag(3), -cbind(0, diag(2))),
22  +     ci=c(-1, rep(0, 3), rep(-1, 2))
23  + )
24  >
25  > self_model$par
26  [1] 1.489017e-05 6.260546e-02 8.821044e-01
27  > self_model$value        # negative log-likelihood
28  [1] -2660.314
```

Programme 7.20 MLEs of GARCH(1, 1) using the Nelder–Mead optimization algorithm in R.

Note that the parameters estimated by our own GARCH(1, 1) model are similar to those estimated by the `fGarch` library in Programme 7.16. We shall use the parameters estimated by the `fGarch` library for the discussion of model diagnostics in the next subsection.

7.4.2 Model Diagnostics for GARCH

In Programme 7.21 via Python, we can plot the fitted volatility, the sample ACF of HSBC's squared returns u_i^2, as well as the sample ACF of the squared standardized residuals $(u_i/\sigma_i)^2$ and the normal Q-Q plot of the standardized residuals.

Recall that $\mathrm{u}_i \sim \mathcal{N}(0, \sigma_i^2)$ given $\sigma_i = \sigma_i$. If GARCH(1, 1) is the true model, then each standardized residual $\mathrm{u}_i/\sigma_i \sim \mathcal{N}(0, 1)$ holds true for every i. The bottom right plot in Figure 7.13 is the normal Q-Q plot of the standardized residuals u_i/σ_i's, which indicates that the residuals are approximately normally distributed except at both extreme tails. More importantly, the sample ACF plot of the returns on the top right corner in Figure 7.13 clearly indicates that serial correlations exist in u_i^2, which can be removed by fitting a GARCH(1, 1) model, as demonstrated in the bottom left sample ACF plot of squared standardized residuals $(u_i/\sigma_i)^2$.

Similar to Section 7.3, we perform the Ljung–Box test to determine whether the observed time series $\{x_t\}$ exhibits autocorrelations. See the summary output in Programmes 7.21 for Python and 7.22 for **R**, where we compute the Ljung–Box test statistics with $K = 15$ for both u_i^2 and the squared residuals u_i^2/σ_i^2, respectively[6]. According to the results, the small p-value of the test for u_i^2 means that there exist

[6] Different from ARIMA model diagnostics, the sample ACF r_k's of fitted residuals here are subject to a set of possibly non-linear constraints, say $\psi_j(r_1, \ldots, r_T) = 0$ for $j = 1, \ldots, J$, where $J = p + q + 1$ is the number of estimated parameters in a GARCH(p, q) model. Nevertheless, r_k's are close to their respective true population values when the sample size n is large, then by Taylor expansion (a perturbation analytical approach, see [4, 33, 34]), these ψ_j's are dominated by their respective zeroth- and first-order terms, and hence we arrive at J approximate linear constraints similar to the ARIMA case. That is why the asymptotic distribution of the corresponding test statistic, when applied to fitted residuals, is again a chi-square distribution with $K - J$ degrees of freedom; readers may refer to [1, 5, 20, 21] for more discussions on this issue.

autocorrelations among u_i^2's, while the large p-value of that for u_i^2/σ_i^2 indicates no apparent autocorrelations among them. Altogether, this clearly indicates that the autocorrelation of u_i^2 is removed by the GARCH model and that σ_i^2 is a plausibly good estimate of the time-changing variance rate.

```
1  import matplotlib.pyplot as plt
2  from scipy import stats
3  from statsmodels.graphics.tsaplots import plot_acf
4
5  # Residuals u_i/sigma_i ~ N(0, 1)
6  omega, alpha, beta = model_HSBC
7  nu = [omega + alpha*np.mean(u1**2) + beta*np.mean(u1**2)]
8  for i in range(1, len(u1)):
9      nu.append(omega + alpha*u1[i-1]**2 + beta*nu[-1])
10
11 resid_HSBC = u1/np.sqrt(nu)
12
13 plt.rc('axes', titlesize=25, labelsize=25)
14 fig, axs = plt.subplots(2, 2, figsize=(15,15))
15 vol = pd.Series(np.sqrt(nu), index=d.index[1:])
16 axs[0,0].plot(vol, c="blue")
17 axs[0,0].set_title("Conditional SD")
18 axs[0,0].xaxis.set_major_locator(ticker.MultipleLocator(400))
19 plot_acf(u1**2, alpha=0.05, c="blue", ax=axs[0,1],
20          title="ACF Plot of Squared HSBC Return")
21 plot_acf(resid_HSBC**2, alpha=0.05, c="blue", ax=axs[1,0],
22          title="ACF Plot of Squared Residuals")
23 stats.probplot(resid_HSBC, dist="norm", plot=axs[1,1])
24 axs[1,1].set_title("QQ Plot of Standardized Residuals")
25
26 from statsmodels.stats.diagnostic import acorr_ljungbox
27
28 print(acorr_ljungbox(u1**2, lags=[15]))
29 # the degrees of freedom for the test are K - (p+q+1) = 15 - 3 = 12
30 print(acorr_ljungbox(np.array(resid_HSBC)**2, lags=[15], model_df=3))
```

```
1      lb_stat      lb_pvalue
2  15  185.250131   2.071455e-31
3        lb_stat  lb_pvalue
4  15  12.348686    0.418098
```

Programme 7.21 Calculating the standardized residuals of the GARCH(1, 1) model calibrated in Programme 7.19, and performing model diagnostics in Python.

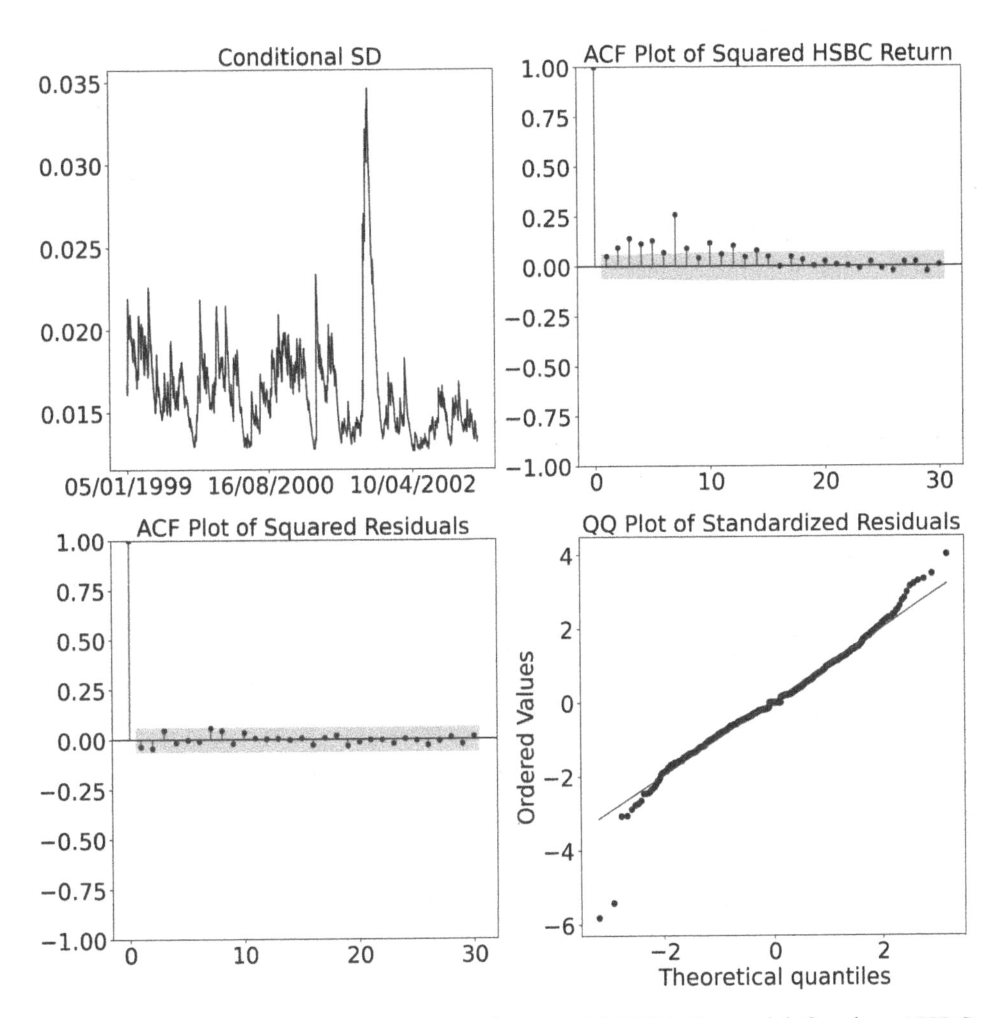

FIGURE 7.13 Plots of model diagnostics for a GARCH(1, 1) model fitted to HSBC returns in Python, generated by Programme 7.21.

```
1  > omega <- coef(res_HSBC)[1]
2  > alpha <- coef(res_HSBC)[2]
3  > beta <- coef(res_HSBC)[3]
4  > # initialize u_0^2 and nu_0 being mean of u_i^2
5  > nu <- omega + alpha*mean(u1**2) + beta*mean(u1**2) # nu_1
6  > for (i in 2:length(u1)){
7  +     nu <- c(nu, omega + alpha*u1[i-1]^2 + beta*nu[i-1])
8  + }
9  > all.equal(as.vector(nu), res_HSBC@h.t)
10 [1] TRUE
11 >
12 > resid_HSBC <- as.vector(u1/sqrt(nu))  # u1/sqrt(res_HSBC@h.t)
13 > all.equal(resid_HSBC, residuals(res_HSBC, standardize=TRUE))
```

```
14  [1] TRUE
15  >
16  > Box.test(u1^2, lag=15, type="Ljung")
17      Box-Ljung test
18  data:  u1^2
19  X-squared = 185.25, df = 15, p-value < 2.2e-16
20  > # the degrees of freedom for the test are K - (p+q+1) = 15 - 3 = 12
21  > Box.test(resid_HSBC^2, lag=15, type="Ljung", fitdf=3)
22      Box-Ljung test
23  data:  resid_HSBC^2
24  X-squared = 12.361, df = 12, p-value = 0.4171
25  >
26  > par(mfrow=c(2,2),mar=c(4,4,3,3),cex.lab=1.5,cex.axis=1.5,cex.main=1.5)
27  > plot(res_HSBC, which=c(2, 5, 11, 13))
```

Programme 7.22 Calculating the standardized residuals of the GARCH(1, 1) model calibrated in Programme 7.16, and performing model diagnostics in **R**.

Finally, using Programmes 7.23 and 7.24 via Python and **R** respectively, we plot the fitted values of volatilities using the GARCH(1, 1) model (7.13) (solid line in green) with the MLEs of ω, α, β, together with the 90-day and 180-day "moving" standard deviations (dashed line in blue and dotted line in red, respectively). As shown in both Figure 7.15 by Python and Figure 7.16 by **R**, some clustered and spiky patterns are present in the estimated volatilities from GARCH(1, 1), while the plots of the estimated volatilities using the "moving" standard deviation are smoother.

```
1   sig = pd.Series(np.sqrt(nu), index=d.index[1:])
2   df = pd.concat([sig, s_90["HSBC_Return"], s_180["HSBC_Return"]],
3                  axis=1)
4   df.columns = ["nu", "s_90", "s_180"]
5
6   fig, ax = plt.subplots(1, 1, figsize=(15,8))
7   df.plot(ax=ax, color=["green", "blue", "red"],
8           style=["-", "--", "."], linewidth=3)
9   ax.xaxis.set_major_locator(ticker.MultipleLocator(200))
10  ax.set_title("HSBC volatilities", fontsize=30)
```

Programme 7.23 Plotting GARCH(1, 1) fitted volatilities, 90-day and 180-day simple moving standard deviations for HSBC returns in Python.

```
1  > par(mfrow=c(1,1))
2  > vol_HSBC <- xts(sqrt(res_HSBC@h.t), order.by=date[-1])
3  > plot(vol_HSBC, col="green", lwd=3, ylab="", lty=1, cex=1,
4  +      ylim=c(0.01, 0.04), xaxt="n", main="HSBC volatilities")
5  > lines(s_HSBC_90, col="blue", lwd=3, lty=2)
6  > lines(s_HSBC_180, col="red", lwd=3, lty=3)
7  > addLegend("topleft", legend.names=c("nu", "s_90", "s_180"),
8  +           col=c("green", "blue", "red"), lwd=3, lty=1:3)
```

Programme 7.24 Plotting GARCH(1, 1) fitted volatilities, 90-day and 180-day simple moving standard deviations for HSBC returns in **R**.

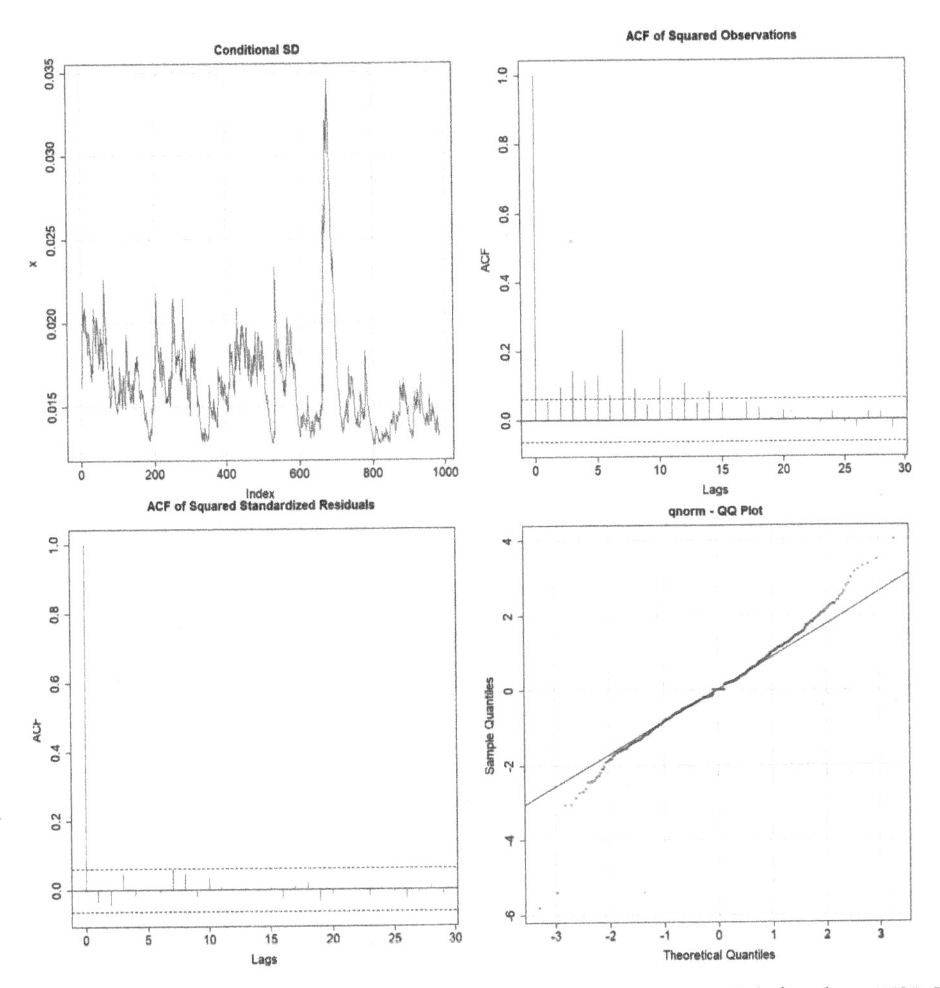

FIGURE 7.14 Plots of model diagnostics for a GARCH(1, 1) model fitted to HSBC returns in **R**, generated by Programme 7.22. 3 parameters

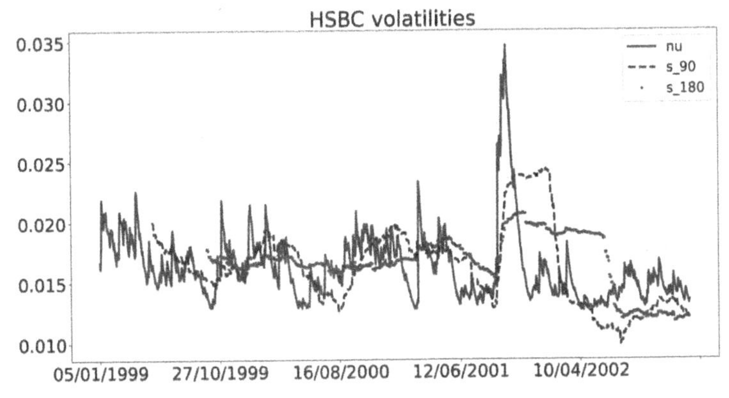

FIGURE 7.15 Plots of GARCH(1, 1) fitted volatilities, 90-day, and 180-day simple moving standard deviations for HSBC returns in Python, generated by Programme 7.23.

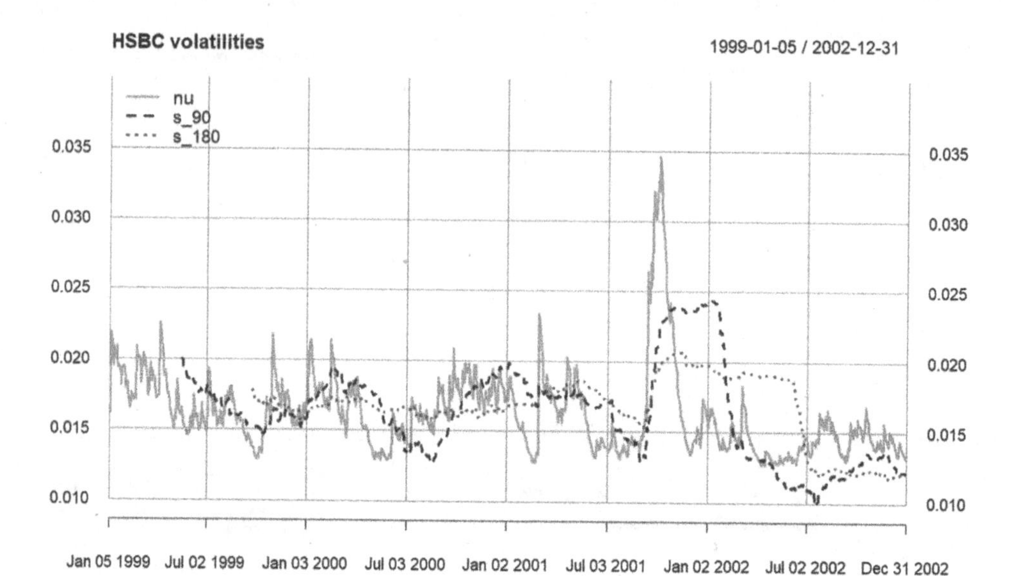

FIGURE 7.16 Plots of GARCH(1, 1) fitted volatilities, 90-day, and 180-day simple moving standard deviations for HSBC returns in **R**, generated by Programme 7.24.

7.4.3 Forecasting Future Volatilities

With the estimates of ω, α, β, we can use the fitted GARCH(1, 1) model to forecast future volatilities. Recall that in (7.13),

$$\sigma_n^2 = \omega + \alpha u_{n-1}^2 + \beta\sigma_{n-1}^2 = (1 - \alpha - \beta)V_L + \alpha u_{n-1}^2 + \beta\sigma_{n-1}^2,$$

so $\sigma_n^2 - V_L = \alpha\left(u_{n-1}^2 - V_L\right) + \beta\left(\sigma_{n-1}^2 - V_L\right)$. On day $n + k$ in the future, we still have the same form of recursive relation, but all variables involved are random:

$$\sigma_{n+k}^2 - V_L = \alpha\left(\mathrm{u}_{n+k-1}^2 - V_L\right) + \beta\left(\sigma_{n+k-1}^2 - V_L\right).$$

As $\mathbb{E}\left(\mathrm{u}_{n+k-1}^2 | \sigma_{n+k-1}^2\right) = \sigma_{n+k-1}^2$, from the assumption of (7.11), using tower property we have:

$$\begin{aligned}\mathbb{E}\left(\sigma_{n+k}^2 - V_L\right) &= (\alpha+\beta)\mathbb{E}\left(\sigma_{n+k-1}^2 - V_L\right) = (\alpha+\beta)^2\mathbb{E}\left(\sigma_{n+k-2}^2 - V_L\right)\\ &= \cdots = (\alpha+\beta)^k\,\mathbb{E}\left(\sigma_n^2 - V_L\right),\end{aligned}$$

which implies that

$$\mathbb{E}\left(\sigma_{n+k}^2\right) = V_L + (\alpha+\beta)^k\mathbb{E}\left(\sigma_n^2 - V_L\right). \tag{7.17}$$

We remark the following about (7.17):

1. Note that when k is large and $0 < \alpha + \beta < 1$, $\mathbb{E}\left(\sigma_n^2\right)$ is a convergent series in n, and so $\sup_n \mathbb{E}(\sigma_n^2)$ is bounded. Therefore, the expected variance rate tends to V_L as k goes to infinity. This is the reason why V_L is called the long-run variance rate.
2. The EWMA model can be considered a special case of GARCH(1, 1) with $\omega = 0$ and $\lambda = \beta$ so that $\alpha + \beta = 1$. In this case, from (7.17), the expected variance rate is constant over time. This is possibly an undesirable property of the EWMA model.
3. As shown in Appendix *7.A.2, the rescaled GARCH(1, 1) model, as the step size between two consecutive time instances approaches 0, tends to a continuous variance model $\mathrm{V}_t := \sigma_t^2$ satisfying the following stochastic differential equation (SDE):

$$dV_t = (1 - \alpha - \beta)(V_L - V_t)dt + \sqrt{2}\alpha V_t dz_t,$$

where $\{z_t\}$ is the standard Brownian motion, (see Section 5.3 of Chapter 5 for a quick review). This is also called the *diffusion limit* of the GARCH(1, 1) model, see [25]. Note that it has a similar form as the celebrated *Cox–Ingersoll–Ross* (CIR) interest rate model[7] in that they both possess a mean-reverting structure. However, the volatility of the diffusion limit is proportional to V_t, not to its square root.

Example 7.1 To illustrate how (7.17) can be used to forecast future volatilities, let us go back to our HSBC example. From Python's GARCH output in Programme 7.19, the MLEs of ω, α and β for the period 1999–2002 are 0.0000148474391, 0.0627356786, and 0.882262810 respectively. The current daily variance rate can be obtained from the last value of nu and is equal to 0.000181013, hence the current daily volatility rate is $\sigma_i = \sqrt{0.000181013} = 0.01345411$ and the long-run variance rate is $V_L = \omega/(1 - \alpha - \beta) = 0.000269946$.

Let $\mathrm{V}_t := \sigma_{n+t}^2$ denote the estimate of the instantaneous variance on the t-th day counting from the present day n, and $\theta := -\ln(\alpha + \beta)$. From Appendix *7.A.2 or Equation (7.17), we have

$$\mathbb{E}(V_t) = V_L + e^{-\theta t}(V_0 - V_L),$$

where $V_0 = \sigma_n^2$ is the initial condition. The average of the estimated variance from n to $n + T$ days is

$$\frac{1}{T}\int_0^T \mathbb{E}(V_t)\,dt = V_L + \frac{1 - e^{-\theta T}}{\theta T}(V_0 - V_L).$$

[7] The CIR model is popular for modelling term structures of interest rates. The dynamics of the model admits the form: $dr_t = \theta(\mu - r_t)\,dt + \sigma\sqrt{r_t}dz_t$, where θ, μ and σ are constants; μ is also called the *retention level* about which the interest rate fluctuates.

Suppose that we have a 10-day option for HSBC and we want to obtain a single estimate of the annual volatility to price this option via the Black–Scholes formula. To this end, we first estimate the average daily variance rate by

$$\frac{1}{10}\int_0^{10} \mathbb{E}(V_t)\,dt = V_L + \frac{1-e^{-10\theta}}{10\theta}(V_0 - V_L) = 0.000202144,$$

and then convert it into an estimate of annual volatility. The estimated daily volatility rate is $\sqrt{0.000202144} = 0.0142177$ and the estimated annual volatility rate is $\sqrt{252} \times 0.0142177 = 22.57\%$.

7.4.4 Correlation and GARCH

GARCH models are commonly used to model the dynamical movement of the volatility of one single asset. What about the volatility of a basket of securities? In addition to the evolution of an individual volatility, knowing the correlations among securities is also essential. In fact, correlation plays an important role in assessing the risk of a portfolio, being an important parameter for calculating Value-at-Risk (See Chapter 8), and in the active portfolio management. As variance varies over time, it is not surprising that correlation also varies over time. We can extend the EWMA model and GARCH(1, 1) model to estimate the inherent pairwise correlations among several securities.

Here, we illustrate the detailed procedure for a pair of stocks. Let

$$\mathrm{x}_i = \frac{\mathrm{S}_{1,i} - \mathrm{S}_{1,i-1}}{\mathrm{S}_{1,i-1}} \quad \text{and} \quad \mathrm{y}_i = \frac{\mathrm{S}_{2,i} - \mathrm{S}_{2,i-1}}{\mathrm{S}_{2,i-1}}$$

be the daily percentage changes of stocks S_1 and S_2, respectively. The estimated covariance between S_1 and S_2 based on the most recent m observations on day $n+t$, where n stands for the current day, is as follows:

$$\mathrm{cov}_{n+t-1} = \frac{1}{m}\sum_{i=1}^{m} \mathrm{x}_{n+t-i}\mathrm{y}_{n+t-i}, \qquad \text{for} \quad t = 0, 1, 2, \ldots. \tag{7.18}$$

The correlation between S_1 and S_2 on day $n+t$ can be naïvely estimated by

$$\rho_{n+t-1} = \frac{\mathrm{cov}_{n+t-1}}{\sigma_{\mathrm{x},n+t-1}\sigma_{\mathrm{y},n+t-1}}, \qquad \text{for} \quad t = 0, 1, 2, \ldots, \tag{7.19}$$

where

$$\sigma^2_{\mathrm{x},n+t-1} = \frac{1}{m}\sum_{i=1}^{m} \mathrm{x}^2_{n+t-i} \quad \text{and} \quad \sigma^2_{\mathrm{y},n+t-1} = \frac{1}{m}\sum_{i=1}^{m} \mathrm{y}^2_{n+t-i}.$$

However, this naïve approach faces the same inaccuracy issues as described for the estimation of variance rates in Sections 7.1 and 7.2. Following the discussion there, we shall first take the current estimates of cov_{n+1} and ρ_{n+1} on day n from (7.18) and (7.19) with a reasonable choice of m (neither too short nor too long). Then the estimates of cov_{n+t} and ρ_{n+t} can be updated daily, for $t = 0, 1, 2, \ldots$, by using an EWMA or a GARCH(1, 1) model as follows. For an EWMA model,

$$\text{cov}_{n+t} = \lambda \text{cov}_{n+t-1} + (1-\lambda)\text{x}_{n+t-1}\text{y}_{n+t-1} \text{and } \rho_{n+t} = \frac{\text{cov}_{n+t}}{\sigma_{\text{x},n+t}\sigma_{\text{y},n+t}}, \tag{7.20}$$

where $\sigma^2_{\text{x},n}$ and $\sigma^2_{\text{y},n}$ are updated according to the EWMA model by using (7.4) with the same value of λ throughout. For a GARCH(1, 1) model,

$$\text{cov}_{n+t} = \omega + \alpha \text{x}_{n+t-1}\text{y}_{n+t-1} + \beta \text{cov}_{n+t-1} \text{and } \rho_{n+t} = \frac{\text{cov}_{n+t}}{\sigma_{\text{x},n+t}\sigma_{\text{y},n+t}}, \tag{7.21}$$

where $\sigma^2_{\text{x},n}$ and $\sigma^2_{\text{y},n}$ are updated according to the GARCH(1, 1) model by using (7.13) with the same parameter values of ω, α and β throughout.

In the general case of p securities, a similar GARCH(1, 1) model can be built. In finance, a standardized version of the corresponding dynamical correlation movement analysis is considered, called the *Dynamic Conditional Correlation GARCH* (DCC-GARCH) model in the literature [13]. The implementation of DCC-GARCH(1, 1) will be deferred to Subsection 7.4.6.2.

```
1  print(u.iloc[-1, :])
2  print(np.corrcoef(u.iloc[-90:, :], rowvar=False))
3  print(np.cov(u.iloc[-90:, :], rowvar=False))
```

```
1  HSBC_Return     0.002941
2  CLP_Return     -0.003175
3  CK_Return       0.019076
4  Name: 30/12/2002, dtype: float64
5  [[ 1.          -0.00699531  0.60391534]
6   [-0.00699531  1.          -0.03430252]
7   [ 0.60391534 -0.03430252  1.        ]]
8  [[ 1.48238246e-04 -6.57818920e-07  1.44118250e-04]
9   [-6.57818920e-07  5.96539297e-05 -5.19288140e-06]
10  [ 1.44118250e-04 -5.19288140e-06  3.84171883e-04]]
```

Programme 7.25 Correlation matrix and covariance matrix for the last 90 days of returns of HSBC, CLP, CK in Python.

```
1  > u[dim(u)[1],]
2   HSBC_Return   CLP_Return     CK_Return
3   0.002941176 -0.003174603  0.019076305
4  > cor(u[(dim(u)[1]-89):dim(u)[1],])
5                 HSBC_Return   CLP_Return   CK_Return
6  HSBC_Return   1.000000000 -0.006995306  0.60391534
7  CLP_Return   -0.006995306  1.000000000 -0.03430252
```

```
8  CK_Return     0.603915335 -0.034302525  1.00000000
9  > var(u[(dim(u)[1]-89):dim(u)[1],])
10                 HSBC_Return    CLP_Return     CK_Return
11 HSBC_Return  1.482382e-04 -6.578189e-07  1.441182e-04
12 CLP_Return  -6.578189e-07  5.965393e-05 -5.192881e-06
13 CK_Return    1.441182e-04 -5.192881e-06  3.841719e-04
```

Programme 7.26 Correlation matrix and covariance matrix for the last 90 days of returns of HSBC, CLP, CK in **R**.

It is important to note that the estimated values of the parameters in (7.20) and (7.21) used in updating the variance rate and the covariance should be the same for the sake of convenience and tractability of the models. Let us illustrate this by the following example. First, we compute the correlation matrix of the daily percentage returns of HSBC, CLP and CK for 1999–2002; see Programmes 7.25 and 7.26 for implementation in Python and **R**, respectively.

From the outputs, we note that HSBC and CK are relatively more strongly correlated while CLP has relatively weaker correlations with HSBC and CK. Next, we consider the DCC-GARCH(1, 1) model without standardization. To this end, the current values of u_1, u_2 and u_3, i.e., $u_{i,0}$'s, and the entries in the covariance matrix will be used as the initial condition for the recursive relations (7.13) and (7.21). However, computing the common values of ω, α and β could be cumbersome. Instead, we may adopt the following convenient numerical scheme, which is still accurate enough for most practical situations. We first fit the GARCH(1, 1) models to `u1`, `u2` and `u3` separately, which returns three different sets of estimated parameters for ω, α and β, and then use the average of these three sets of estimated parameters as the combined estimates for ω, α and β. Then we update the variance rates and the correlations accordingly. This ensures that the correlation matrix obtained is positive definite; see Programme 7.27 via Python and Programme 7.28 via **R**.

```
1 model_CLP, llike_CLP = GARCH_11_MLE(u2)
2 model_CK, llike_CK = GARCH_11_MLE(u3)
3
4 coef = np.array([model_HSBC, model_CLP, model_CK])
5 print(coef)
6 print(np.mean(coef, axis=0))
```

```
1 [[1.48474391e-05 6.27356786e-02 8.82262810e-01]
2  [3.67580316e-06 1.25144123e-01 8.58713421e-01]
3  [1.32870834e-05 4.35618355e-02 9.29130022e-01]]
4 [1.06034419e-05 7.71472124e-02 8.90035417e-01]
```

Programme 7.27 MLEs of GARCH(1, 1) with HSBC, CLP, CK returns using `GARCH_11_MLE()` function defined in Programme 7.19 in Python.

Now we have all the numbers required to update the variance rates and the correlations according to (7.13) and (7.21), using the corresponding values for $u_{i,0}$'s, $\sigma_{i,0}^2$'s and $\sigma_{ij,0}$'s in Programme 7.27 via Python. They are:

$$\omega = 0.0000106034, \quad \alpha = 0.0771472, \quad \beta = 0.890035.$$

Therefore, by using the initial conditions, we have

$$\begin{aligned}
\sigma_{1,1}^2 &= \omega + \alpha(0.002941)^2 + \beta(0.000148238) = 0.00014321,\\
\sigma_{2,1}^2 &= \omega + \alpha(-0.003175)^2 + \beta(0.0000595393) = 0.00006448,\\
\sigma_{3,1}^2 &= \omega + \alpha(0.019076)^2 + \beta(0.000384172) = 0.00038060,\\
\text{cov}_{12,1} &= \omega + \alpha(0.002941)(-0.003175) + \beta(-0.00000066) = 0.00000930,\\
\text{cov}_{13,1} &= \omega + \alpha(0.002941)(0.019076) + \beta(0.00014412) = 0.00014320,\\
\text{cov}_{23,1} &= \omega + \alpha(-0.003175)(0.019076) + \beta(-0.000005193) = 0.00000131,\\
\rho_{12,1} &= \text{cov}_{12,1}/(\sigma_{1,1}\sigma_{2,1}) = 0.09675929,\\
\rho_{13,1} &= \text{cov}_{13,1}/(\sigma_{1,1}\sigma_{3,1}) = 0.61337951,\\
\rho_{23,1} &= \text{cov}_{23,1}/(\sigma_{2,1}\sigma_{3,1}) = 0.00835988.
\end{aligned}$$

```
1  > res_HSBC <- garchFit(~garch(1, 1), data=u1, include.mean=FALSE)
2  > res_CLP <- garchFit(~garch(1, 1), data=u2, include.mean=FALSE)
3  > res_CK <- garchFit(~garch(1, 1), data=u3, include.mean=FALSE)
4  > (coef <- rbind(coef(res_HSBC), coef(res_CLP), coef(res_CK)))
5                omega      alpha1     beta1
6  [1,] 1.490724e-05 0.06278833 0.8819895
7  [2,] 3.710697e-06 0.12535032 0.8582773
8  [3,] 1.338107e-05 0.04367150 0.9288332
9  > round(colMeans(coef), 6) # compute the column mean
10    omega    alpha1    beta1
11 0.000011 0.077270 0.889700
```

Programme 7.28 MLEs of GARCH(1, 1) with HSBC, CLP, CK returns using `fGarch` library in R.

7.4.5 Combining GARCH with Linear Models

One obvious application of the GARCH model is to address the constant volatility assumption of the Black–Scholes model (as introduced in Chapter 3). Recall that, for the Black–Scholes model,

$$\text{u}_t = \frac{\text{S}_t - \text{S}_{t-1}}{\text{S}_{t-1}} \sim \mathcal{N}(\mu, \sigma^2), \quad \text{or} \quad \frac{\text{u}_t}{\sigma} \sim \mathcal{N}(0, 1), \tag{7.22}$$

if we just assume $\mu = 0$ for simplicity. It can be immediately generalized by replacing σ by σ_t where σ_t^2 follows a GARCH(1, 1) model, i.e.,

$$\frac{u_t}{\sigma_t} \sim \mathcal{N}(0,1) \text{ subject to } \sigma_t^2 = \omega + \alpha u_{t-1}^2 + \beta\sigma_{t-1}^2. \tag{7.23}$$

This is known as the Normal-GARCH(1, 1) stock price model. Given ω, α, β and σ_t^2, we can simulate p sample paths for p different stocks with $u_{i,t+1}, \ldots, u_{i,t+K}$ for $i = 1, \ldots, p$ according to the following scheme:

1. Generate p iid standard normal random numbers $z_1, \ldots, z_p$;
2. For $i = 1, \ldots, p$, compute $u_{i,t+1} = \sigma_{i,t+1} z_i$ and $\sigma_{i,t+2}^2 = \omega + \alpha u_{i,t+1}^2 + \beta\sigma_{i,t+1}^2$;
3. Repeat Steps 1 and 2 from $t = t + 1$ to $t = t + K$.

Note that we have to generate a new set of p normal random numbers in Step 1 for every new value of t. Once we have p simulated sample paths of $u_{i,t+1}, \ldots, u_{i,t+K}$, we can evaluate the prices of different derivatives from these sample paths by taking an appropriate average of the corresponding payoffs, each of which is evaluated along one simulated sample path.

Other models can be simulated similarly. For example, we can combine the standardized t_ν distribution with the GARCH(1, 1) model: $u_i/\sigma_i \sim t_\nu$ and $\sigma_i^2 = \omega + \alpha u_{i-1}^2 + \beta\sigma_{i-1}^2$. This is known as the t-GARCH(1, 1) stock price model. Alternatively, one can also use the EWMA(λ) model, instead of the GARCH(1, 1) model, to assimilate the evolutional dynamics of volatilities.

When using regression models for financial time series data, the errors are usually autocorrelated. We can handle this problem by fitting an AR(m) model to several consecutive error terms in the regression model of the time series. For example,

$$y_t = x_t^\mathsf{T}\beta + v_t,$$
$$v_t = -\varphi_1 v_{t-1} - \varphi_2 v_{t-2} - \ldots - \varphi_m v_{t-m} + \epsilon_t,$$

where $\epsilon_t \overset{\text{iid}}{\sim} \mathcal{N}(0, \sigma^2)$. However, if the error variance is not constant but varies with time t, we can go one step further and fit a GARCH(p, q) model to the error term ϵ_t as follows:

$$y_t = x_t^\mathsf{T}\beta + v_t,$$
$$v_t = -\varphi_1 v_{t-1} - \varphi_2 v_{t-2} - \ldots - \varphi_m v_{t-m} + \epsilon_t,$$
$$\epsilon_t = \sqrt{h_t}e_t, \text{ where } e_t \overset{\text{iid}}{\sim} \mathcal{N}(0,1) \Rightarrow \epsilon_t \sim \mathcal{N}(0, h_t),$$
$$h_t = \omega + \sum_{i=1}^{p} \alpha_i \epsilon_{t-i}^2 + \sum_{j=1}^{q} \beta_j h_{t-j}.$$

This model is sometimes called the AR(m)-GARCH(p, q) model. In SAS, there is an AUTOREG procedure to fit this model, and the `rugarch` package in R can also serve this purpose. However, there is no well-established library in Python with the same function, hence we need to perform the fitting in the sequential manner described above. This procedure allows us to fit a GARCH(1, 1) model to the percentage change of stock price.

7.4.6 Extensions of the GARCH(1, 1) Model

In this subsection, we introduce two commonly used extensions of the GARCH(1, 1) model. We shall analyze the results obtained in **R**, while for Python, only the calibration procedure is provided.

7.4.6.1 Leverage GARCH(1, 1) (L-GARCH) The *Leverage GARCH* (L-GARCH) model is also known as the *Threshold GARCH* (T-GARCH) model, or the *Glosten, Jagannathan and Runkle GARCH* (GJR-GARCH) model [14]. To account for the asymmetric effect of negative movement of the stock price on the volatility, we extend the GARCH(1, 1) model to the following GJR-GARCH(1, 1) model:

$$\sigma_n^2 = \omega + \beta\sigma_{n-1}^2 + \alpha u_{n-1}^2 + \theta I_{n-1} u_{n-1}^2, \tag{7.24}$$

where $I_{n-1} = 1$ if $u_{n-1} < 0$ or $I_{n-1} = 0$ otherwise. The stationarity condition for the threshold GARCH(1,1) model is $\alpha + \beta + \gamma\theta < 1$, where γ is the proportion of the number of negative returns to the total number of returns; this is set at 0.5 in practice under the assumption of a zero-mean and symmetrically distributed returns. Note that if $\theta > 0$, a piece of bad news ($u_{n-1} < 0$) has an effect of $(\alpha + \theta)u_{n-1}^2$ on the variance rate, while a piece of good news ($u_{n-1} \geq 0$) has the relatively lower effect of αu_{n-1}^2.

```
1  > library(rugarch)
2  >
3  > gjr_mean_model <- list(armaOrder=c(0,0), include.mean=FALSE)
4  > gjr_var_model <- list(model="gjrGARCH", garchOrder=c(1,1))
5  > gjr_spec <- ugarchspec(mean.model=gjr_mean_model,
6  +                        variance.model=gjr_var_model,
7  +                        distribution.model="norm")
8  > gjr_HSBC <- ugarchfit(data=u1, spec=gjr_spec)
9  > gjr_CLP <- ugarchfit(data=u2, spec=gjr_spec)
10 > gjr_CK <- ugarchfit(data=u3, spec=gjr_spec)
11 >
12 > gjr_param <- rbind(coef(gjr_HSBC), coef(gjr_CLP), coef(gjr_CK))
13 > colnames(gjr_param) <- c("omega", "alpha", "beta", "theta")
14 > rownames(gjr_param) <- c("HSBC", "CLP", "CK")
15 > gjr_param
16             omega      alpha      beta      theta
17 HSBC 1.402690e-05 0.02570286 0.8902723 0.06947573
18 CLP  3.658758e-06 0.08756579 0.8624418 0.06955776
19 CK   9.597023e-06 0.01857886 0.9358707 0.05534282
```

Programme 7.29 L-GARCH(1, 1) using rugarch library for HSBC, CLP, CK returns in R.

We try to fit this GJR-GARCH(1, 1) model for HSBC, CLP and CK stocks in Programme 7.29 via **R**. We then analyze the fitted standardized residuals for the GJR-GARCH(1, 1) model in Programme 7.30 via **R**.

```
1  > omega <- coef(gjr_HSBC)[1]
2  > alpha <- coef(gjr_HSBC)[2]
3  > beta <- coef(gjr_HSBC)[3]
4  > theta <- coef(gjr_HSBC)[4]
5  > # initialize nu_0 being mean of u_i^2
6  > nu <- mean(u1^2) # nu_1
7  > for (i in 2:length(u1)){
8  +     nu <- c(nu, omega + alpha*u1[i-1]^2 + beta*nu[i-1] +
9  +                theta*u1[i-1]^2*(u1[i-1] < 0))
10 + }
11 > all.equal(sqrt(as.vector(nu)), gjr_HSBC@fit$sigma)
12 [1] TRUE
13 >
14 > resid_HSBC <- as.vector(u1/sqrt(nu)) # u1/gjr_HSBC@fit$sigma
15 > gjr_resid_HSBC <- as.numeric(residuals(gjr_HSBC, standardize=TRUE))
16 > all.equal(resid_HSBC, gjr_resid_HSBC)
17 [1] TRUE
18 >
19 > # the degrees of freedom for the test are K - (p+q+2) = 15 - 4 = 11
20 > Box.test(gjr_resid_HSBC^2, lag=15, type="Ljung", fitdf=4)
21    Box-Ljung test
22 data:  gjr_resid_HSBC^2
23 X-squared = 10.344, df = 11, p-value = 0.4998
24 > gjr_resid_CLP <- as.numeric(residuals(gjr_CLP, standardize=TRUE))
25 > Box.test(gjr_resid_CLP^2, lag=15, type="Ljung", fitdf=4)
26    Box-Ljung test
27 data:  gjr_resid_CLP^2
28 X-squared = 12.453, df = 11, p-value = 0.3306
29 > gjr_resid_CK <- as.numeric(residuals(gjr_CK, standardize=TRUE))
30 > Box.test(gjr_resid_CK^2, lag=15, type="Ljung", fitdf=4)
31    Box-Ljung test
32 data:  gjr_resid_CK^2
33 X-squared = 12.145, df = 11, p-value = 0.3528
34 >
35 > par(mfrow=c(2,2), mar=c(5,5,4,4))
36 > plot(gjr_HSBC@fit$sigma, type="l")
37 > acf(gjr_resid_HSBC)
38 > acf(gjr_resid_HSBC^2)
39 > plot(gjr_HSBC, which=9)
```

Programme 7.30 Model diagnostics and plotting the standardized residuals of L-GARCH(1, 1) model calibrated in Programme 7.29 in **R**.

The bottom right plot in Figure 7.17 is the normal Q-Q plot of the (standardized) residuals $u_{1,i}/\sigma_{1,i}$ of HSBC returns. The residuals are approximately normally distributed except at the two tails. More importantly, the sample ACF plots on the bottom left of Figure 7.17 clearly indicate that the autocorrelation of squared HSBC

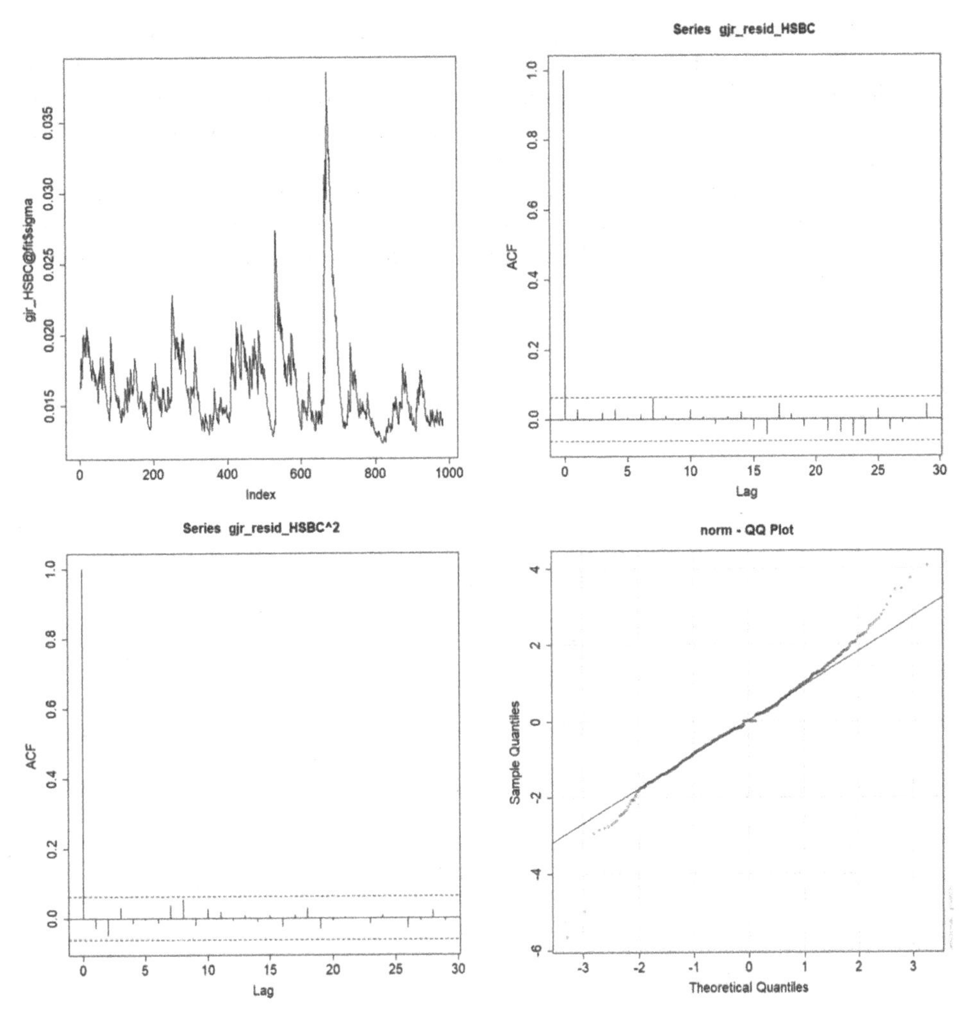

FIGURE 7.17 Plots of model diagnostics for L-GARCH(1, 1) fitted with HSBC returns in **R**, generated by Programme 7.30. 4 parameters

returns $u_{1,i}^2$ are successfully removed by fitting an L-GARCH(1, 1) model. In particular, the large p-values of the Ljung–Box tests, with $K = 15$,[8] for $u_{1,i}^2/\sigma_{1,i}^2$ in HSBC, $u_{2,i}^2/\sigma_{2,i}^2$ in CLP, and $u_{3,i}^2/\sigma_{3,i}^2$ in CK, mean that there are no autocorrelations in their respective $u_{j,i}^2/\sigma_{j,i}^2$. It is also clearly indicated that the autocorrelations of $u_{1,i}^2$, $u_{2,i}^2$, and $u_{3,i}^2$ are removed by the L-GARCH(1, 1) model and $\sigma_{1,i}^2$, $\sigma_{2,i}^2$, and $\sigma_{3,i}^2$ are plausibly good estimates of the variance rates of HSBC, CLP and CK stocks, respectively. Comparing

[8] The general argument remains valid, similar to the standard GARCH model, the test statistic for fitted residuals should follow a chi-square distribution with $K - 4 = 11$ degrees of freedom, as there are 4 parameters estimated.

the sample ACF plots of the standardized residuals of HSBC returns obtained from the GARCH(1, 1) and L-GARCH(1, 1) models, corresponding to the respective top right plots in Figures 7.14 and 7.17, the plot for GARCH(1, 1) shows that autocorrelation still exists in the standardized residuals $u_{1,i}/\sigma_{1,i}$, while that for L-GARCH(1, 1) exhibits no such autocorrelation. This illustrates a positive sign for its goodness-of-fit.

Similarly, we implement the same estimation in Programme 7.31 via Python, by first finding the MLEs of the GARCH(1, 1) models using the same two-stage optimization scheme in Programme 7.19, with $\theta = 0.005$ as the initial condition for the GJR-GARCH(1, 1) model.

```
1  L_bds = [(1e-15, np.inf), (1e-15, 1-1e-15), (1e-15, 1-1e-15),
2            (-1+1e-15, 1-1e-15)]
3
4  def L_GARCH_11(x, r):
5      omega, alpha, beta, theta = x
6      nu = np.mean(r**2)  # following rugarch in R
7      log_like = -1/2*(np.log(2*np.pi) + np.log(nu) + r[0]**2/nu)
8      for i in range(1, len(r)):
9          nu = omega+alpha*r[i-1]**2+beta*nu+theta*r[i-1]**2*(r[i-1]<0)
10         log_like -= 1/2*(np.log(2*np.pi) + np.log(nu) + r[i]**2/nu)
11     return -log_like
12
13 def L_GARCH_11_MLE(r):
14     # stationarity constraint: alpha + beta + theta*gamma < 1
15     fun = lambda x: 1-x[1]-x[2]-x[3]*np.mean(r < 0)
16     L_cons = [{'type': 'ineq', 'fun': fun}]
17
18     x0 = [0.05, 0.03, 0.9, 0.005]
19     model1 = minimize(L_GARCH_11, x0, args=(r), method='L-BFGS-B',
20                       bounds=L_bds, tol=1e-20)
21     model2 = minimize(L_GARCH_11, model1.x, args=(r), method='SLSQP',
22                       constraints=L_cons, bounds=L_bds, tol=1e-20)
23     return model2.x, model2.fun
24
25 print(L_GARCH_11_MLE(u1))  # HSBC
26 print(L_GARCH_11_MLE(u2))  # CLP
27 print(L_GARCH_11_MLE(u3))  # CK
```

```
1 (array([1.40231370e-05, 2.56912309e-02, 8.90288371e-01, 6.94781581e-02]),
      -2664.658549112674)
2 (array([3.62091310e-06, 8.75550255e-02, 8.62677416e-01, 6.97173832e-02]),
      -2985.3688307899356)
3 (array([9.59529009e-06, 1.85762151e-02, 9.35882596e-01, 5.53353153e-02]),
      -2366.7798745782147)
```

Programme 7.31 L-GARCH(1, 1) for HSBC, CLP, CK returns in Python.

7.4.6.2 Dynamic Conditional Correlation GARCH(1,1) The *Dynamic Conditional Correlation GARCH* (DCC-GARCH) model, proposed by Robert Engle [12], is a widely used multivariate volatility model that allows for time-varying conditional correlations among financial assets. This DCC-GARCH model combines the flexibility of the GARCH model for modelling the marginal evolution of each asset with the ability to capture dynamic correlations. In this model, the volatility of each asset is captured using a GARCH process, while the conditional correlation matrix, representing the interdependencies between assets, is modeled separately and updated dynamically. Also refer to [32] for more discussions on the model.

The DCC-GARCH model proposes to estimate the dynamic covariance matrix $\mathbf{H}_t$ by

$$\mathbf{H}_t = \mathbf{D}_t\mathbf{R}_t\mathbf{D}_t, \tag{7.25}$$

where $\mathbf{D}_t := \text{diag}(\sigma_{1t}, \dots, \sigma_{pt})$ and $\mathbf{R}_t$ is the conditional time-varying correlation matrix given σ_{it}'s. The variance σ_{it}^2 in $\mathbf{D}_t$ for each security $i = 1, 2, \dots, p$ can be modelled by an independent GARCH(1, 1) process. Denote the marginally standardized innovation vector by $\boldsymbol{\eta}_t = \mathbf{D}_t^{-1}\mathbf{r}_t$, where r_{it} is the return of security i at time t. The DCC-GARCH(1, 1) model incorporates time-varying conditional correlations through the evolutionary covariance matrix, denoted by $\mathbf{Q}_t$. This covariance matrix is recursively modeled as:

$$\mathbf{Q}_t = (1 - \theta_1 - \theta_2)\overline{\mathbf{Q}} + \theta_1\boldsymbol{\eta}_{t-1}\boldsymbol{\eta}_{t-1}^{\top} + \theta_2\mathbf{Q}_{t-1}, \quad \text{for } t = n, n+1, \dots \tag{7.26}$$

where $\overline{\mathbf{Q}}$ is the unconditional covariance matrix of $\boldsymbol{\eta}_t$. The positive coefficients θ_1 and θ_2 are constrained by $\theta_1 + \theta_2 < 1$, determining the impact of past innovations on the current covariance matrix. To obtain the evolutionary correlation matrix $\mathbf{R}_t$ at time t, we standardize the covariance matrix $\mathbf{Q}_t$ with the following formula:

$$\mathbf{R}_t = \text{diag}(\mathbf{Q}_t)^{-1/2}\mathbf{Q}_t\text{diag}(\mathbf{Q}_t)^{-1/2}. \tag{7.27}$$

After obtaining $\mathbf{R}_t$, we can estimate the conditional covariance matrix from (7.25).

Under the multivariate normal assumption $\mathbf{u}_t \sim \mathcal{N}_k(\mathbf{0}, \mathbf{H}_t)$ given $\mathbf{H}_t$, the logarithmic likelihood function is given by:

$$\begin{aligned}
&\ln L(\theta_1, \theta_2; \boldsymbol{\eta}_0, \boldsymbol{D}_0) \\
&= \sum_{t=1}^{n} \left(-\frac{p}{2}(2\pi) - \frac{1}{2}\ln|\boldsymbol{H}_t| - \frac{1}{2}\boldsymbol{u}_t^{\top}\boldsymbol{H}_t^{-1}\boldsymbol{u}_t \right) \\
&= \sum_{t=1}^{n} \left(-\frac{p}{2}(2\pi) - \ln|\boldsymbol{D}_t| - \frac{1}{2}\ln|\boldsymbol{R}_t| - \frac{1}{2}\boldsymbol{u}_t^{\top}\boldsymbol{D}_t^{-1}\boldsymbol{R}_t^{-1}\boldsymbol{D}_t^{-1}\boldsymbol{u}_t \right) \\
&= \sum_{t=1}^{n} \left(-\frac{p}{2}(2\pi) - \ln|\boldsymbol{D}_t| - \frac{1}{2}\ln|\boldsymbol{R}_t| - \frac{1}{2}\boldsymbol{\eta}_t^{\top}\boldsymbol{R}_t^{-1}\boldsymbol{\eta}_t \right).
\end{aligned}$$

Clearly, $\boldsymbol{D}_t$ is known, since the variance σ_{it}^2 for each security i is modelled by an independent GARCH(1, 1) process before the determination of θ_1 and θ_2. The MLEs for θ_1 and θ_2 are the minimizers of the function below:

$$\ell(\theta_1, \theta_2; \boldsymbol{\eta}_0) = \sum_{t=1}^{n} \left(\ln |\boldsymbol{R}_t| + \boldsymbol{\eta}_t^\top \boldsymbol{R}_t^{-1} \boldsymbol{\eta}_t \right). \tag{7.28}$$

Next, we shall fit the DCC-GARCH(1, 1) model to the HSBC, CLP, and CK stock returns with the `rmgarch` library, where `m` stands for multivariate. Indeed, the standardized residuals $\boldsymbol{\eta}_t$ for HSBC, CLP, and CK can be obtained by fitting the corresponding returns into independent GARCH(1, 1) models using the `rugarch` library, where `u` stands for univariate. After fitting the DCC-GARCH(1, 1) model in Programme 7.32 via **R**, the evolutionary covariance matrix $\boldsymbol{Q}_t$ and the evolutionary correlation matrix $\boldsymbol{R}_t$ for all time t can be retrieved through the codes `dcc_GARCH@mfit$Q` in Programme 7.34 and `dcc_GARCH@mfit$R` in Programme 7.35 via **R**, respectively.

```
1  > library(rmgarch)
2  >
3  > garch_mean_model <- list(armaOrder=c(0,0), include.mean=FALSE)
4  > garch_var_model <- list(model="sGARCH", garchOrder=c(1,1))
5  > garch_spec <- ugarchspec(mean.model=garch_mean_model,
6  +                          variance.model=garch_var_model,
7  +                          distribution.model="norm")
8  >
9  > dcc_mult_spec <- multispec(replicate(garch_spec, n=3))
10 > dcc_spec <- dccspec(uspec=dcc_mult_spec, dccOrder=c(1,1),
11 +                     distribution="mvnorm", model="DCC")
12 > dcc_GARCH <- dccfit(spec=dcc_spec, data=u)
13 > coef(dcc_GARCH)
14  [HSBC_Return].omega [HSBC_Return].alpha1  [HSBC_Return].beta1
15         1.482490e-05         6.252427e-02          8.824222e-01
16   [CLP_Return].omega   [CLP_Return].alpha1   [CLP_Return].beta1
17         3.680377e-06         1.253980e-01          8.584504e-01
18    [CK_Return].omega    [CK_Return].alpha1    [CK_Return].beta1
19         1.327619e-05         4.348044e-02          9.291816e-01
20        [Joint]dcca1          [Joint]dccb1
21         1.883122e-02         9.636941e-01
```

Programme 7.32 Fitting the DCC-GARCH(1, 1) model with the `rmgarch` library in R.

In Python in Programme 7.33, with the MLEs obtained in Programme 7.27, the MLEs for θ_1 and θ_2 of the DCC-GARCH(1, 1) model can be obtained as follows: note that each `eta[i,:]`, for $i = 1, \ldots, n$, is a one-dimensional numpy array, `eta[i,:] @ inv(R) @ eta[i,:]` will return a numerical value equivalent to $\boldsymbol{\eta}_t^\top \boldsymbol{R}_t \boldsymbol{\eta}_t$.

```
DCC_bds = Bounds(lb=[1e-15, 1e-15], ub=[1-1e-15, 1-1e-15])
# stationarity constraint: θ1 + θ2 < 1
DCC_cons = [{'type': 'ineq', 'fun': lambda x: 1-sum(x)}]
from numpy.linalg import det, inv

def DCC_GARCH_11(x, eta):
        theta_1, theta_2 = x
        bar_Sigma = np.cov(eta, rowvar=False)
        Q = bar_Sigma
        R = np.diag(np.diag(Q)**(-1/2)) @ Q @ np.diag(np.diag(Q)**(-1/2))
        func = np.log(det(R)) + eta[1,:] @ inv(R) @ eta[1,:]
        for i in range(1, np.shape(eta)[0]):
                z = eta[i-1,:].reshape(-1, 1)
                Q = (1-theta_1-theta_2)*bar_Sigma + theta_2*Q + theta_1*z @ z.T
                R = np.diag(np.diag(Q)**(-1/2))@Q@np.diag(np.diag(Q)**(-1/2))
                func += np.log(det(R)) + eta[i,:] @ inv(R) @ eta[i,:]

        return func

def resid(x, r):
        omega, alpha, beta = x
        nu = [omega + alpha*np.mean(r**2) + beta*np.mean(r**2)]
        for i in range(1, len(r)):
            nu.append(omega + alpha*r[i-1]**2 + beta*nu[-1])

        return r/np.sqrt(nu)

eta = np.array([resid(model_HSBC, u1), resid(model_CLP, u2),
                resid(model_CK, u3)]).T
x0 = [0.1, 0.4]
model1 = minimize(DCC_GARCH_11, x0, args=(eta), method='L-BFGS-B',
                  bounds=DCC_bds, tol=1e-20)
model2 = minimize(DCC_GARCH_11, model1.x, args=(eta), method='SLSQP',
                  constraints=DCC_cons, bounds=DCC_bds, tol=1e-20)
print(model2.x)
```

```
[0.01885065 0.9636796 ]
```

Programme 7.33 DCC-GARCH(1,1) model in Python, using the MLEs of GARCH(1,1) model in Programme 7.27.

```
> library(rugarch)
>
> garch_mean_model <- list(armaOrder=c(0,0), include.mean=FALSE)
> garch_var_model <- list(model="sGARCH", garchOrder=c(1,1))
> garch_spec <- ugarchspec(mean.model=garch_mean_model,
+                          variance.model=garch_var_model,
+                          distribution.model="norm")
> garch_HSBC <- ugarchfit(data=u1, spec=garch_spec)
> garch_CLP <- ugarchfit(data=u2, spec=garch_spec)
> garch_CK <- ugarchfit(data=u3, spec=garch_spec)
> resid_HSBC <- as.numeric(residuals(garch_HSBC, standardize=TRUE))
> resid_CLP <- as.numeric(residuals(garch_CLP, standardize=TRUE))
```

```
13  > resid_CK <- as.numeric(residuals(garch_CK, standardize=TRUE))
14  > eta <- cbind(resid_HSBC, resid_CLP, resid_CK)
15  >
16  > DCC_GARCH_11 <- function(para, Q, eta){
17  +     theta_1 <- para[1]
18  +     theta_2 <- para[2]
19  +
20  +     R <- diag(diag(Q)^(-1/2))%*%Q%*%diag(diag(Q)^(-1/2))
21  +
22  +     loglik <- -1/2*log(det(R)) - 1/2*eta[1,]%*%solve(R)%*%eta[1,]
23  +     for (i in 2:nrow(eta)){
24  +         Q <- ((1-theta_1-theta_2)*bar_Sigma + theta_2*Q
25  +               + theta_1*eta[i-1,]%*%t(eta[i-1,]))
26  +         R <- diag(diag(Q)^(-1/2))%*%Q%*%diag(diag(Q)^(-1/2))
27  +
28  +         loglik <- (loglik - 1/2*log(det(R))
29  +                    - 1/2*eta[i,]%*%solve(R)%*%eta[i,])
30  +     }
31  +     return (-loglik)
32  + }
33  >
34  > para <- c(0.3, 0.3)
35  > Q <- dcc_GARCH@mfit$Q[[1]]
36  > self_model <- constrOptim(para, DCC_GARCH_11, grad=NULL, Q=Q, eta=eta,
37  +                          method=c("Nelder-Mead"),
38  +                          ui=rbind(c(-1,-1), diag(2)),
39  +                          ci=c(-1, rep(0, 2)))
40  >
41  > self_model$par
42  [1] 0.01857228 0.96428835
```

Programme 7.34 Extracting the evolutionary covariance matrix Q_t DCC-GARCH(1, 1) model with the rugarch library in **R**.

Finally, we shall adopt the Mahalanobis distance and use the last 180 days of returns in 2002 to check whether the residuals obtained in the DCC-GARCH(1, 1) model follow a tri-variate normal distribution.

```
1   > n <- 180
2   > u_180 <- tail(u, n)
3   > mu_180 <- apply(u_180, 2, mean)
4   > S_180 <- cov(u_180)
5   >
6   > z_180 <- sweep(u_180, 2, mu_180)
7   > md2_180 <- rowSums((z_180%*%solve(S_180)) * z_180)
8   > smd2_180 <- sort(md2_180)      # sort md2 in ascendingly
9   > i <- ((1:n)-0.5)/n             # create percentile vector
10  > q <- qchisq(i,3)               # compute quantiles
11  >
12  > par(mfrow=c(1,2))
13  > qqplot(q, smd2_180, main="Chi2 Q-Q Plot")
14  > qqline(smd2_180, distribution=function(p) qchisq(p, df=3))
15  >
16  > md2_DCC <- c()
17  > for (i in 1:nrow(u)){
```

```
18 +     D <- diag(sigma[i,])
19 +     R <- dcc_GARCH@mfit$R[[i]]
20 +     H <- D%*%R%*%D
21 +     md2_DCC <- c(md2_DCC, u[i,]%*%solve(H)%*%u[i,])
22 + }
23 >
24 > smd2_180_DCC <- sort(tail(md2_DCC, n))
25 > qqplot(q, smd2_180_DCC, main="Chi2 Q-Q Plot under DCC",
26 +         ylim=c(0,14))
27 > qqline(smd2_180_DCC, distribution=function(p) qchisq(p, df=3))
```

Programme 7.35 Extracting evolutionary correlation matrix R_t and implementing model diagnostics of the DCC-GARCH(1, 1) model in **R**.

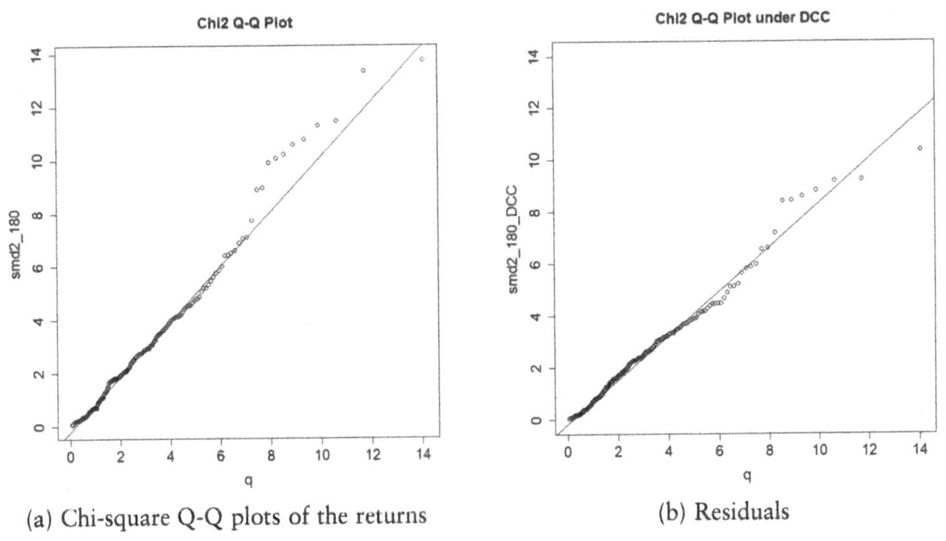

(a) Chi-square Q-Q plots of the returns (b) Residuals

FIGURE 7.18 Plots obtained in a DCC-GARCH(1, 1) model in **R**, generated by Programme 7.35.

From Figure 7.18, the residuals obtained from a DCC-GARCH(1, 1) model present a much better fit as a tri-variate normal distribution than the residuals computed from the sample mean and sample covariance of the last 180 days of returns of the three stocks.

*7.5 KELLY FRACTION

All stock trading strategies strive for a sole ultimate goal: buy at the lowest possible price and sell at the highest viable one so as to maximize the trading profit. However, without a crystal ball, the highest and the lowest stock prices over a fixed investment horizon can never be anticipated with certainty until the terminal time of the planning horizon is reached. To this end, a more reasonable and achievable objective is to find a trading price which, under some notion of difference, is as close as possible to the non-previsible target price. This problem formulation is classified as an optimal prediction problem [27, 28], a branch of *optimal stopping theory* in *Stochastic Analysis*. To facilitate our discussion, we assume that the underlying stock price process $\{S_t\}_{0\leq t\leq T}$ follows a Black–Scholes model with a constant rate of return μ and a constant volatility σ:

$$\frac{dS_t}{S_t} = \mu\, dt + \sigma\, dw_t, \qquad S_0 = x_0, \tag{7.29}$$

where $\{w_t\}_{0\leq t\leq T}$ is a standard Brownian motion. A reasonable objective proposed in [29] is the minimization of the expected relative error between the selling price S_t of a stock and its ultimate maximum price $M_T := \sup_{0\leq t\leq T} S_t$ over the investment horizon $[0, T]$, or equivalently maximizing the expected relative ratio of S_t to M_T. More formally, the problem can be formulated as follows:

$$\sup_{0\leq \tau\leq T} \mathbb{E}\left(\frac{S_\tau}{M_T}\right), \tag{7.30}$$

where the supremum is taken over all stopping times $\tau \leq T$[9]. Under this framework, it is shown in [29] that the index $\alpha := \mu/\sigma^2$ is relevant to the optimal selling time τ^*, given by:

$$\tau^* = \begin{cases} 0, & \text{if}\, \alpha < \frac{1}{2}; \\ T, & \text{if}\, \alpha \geq \frac{1}{2}. \end{cases}$$

This α is actually the well-known *Kelly fraction* ([30, 31]):

$$\kappa = \frac{\mu - r}{\sigma^2}, \tag{7.31}$$

with the risk-free interest rate r set to 0. This Kelly fraction also plays a major role in portfolio selection. Under the Black–Scholes model, κ and $1 - \kappa$ are the respective optimal constant proportions of a wealth to be invested in the stock and bond so as to maximize the expected logarithmic return of the portfolio. To see this, recall that the dynamics of the bond value is described by the following simple differential equation:

$$\frac{dB_t}{B_t} = r dt. \tag{7.32}$$

Consider a portfolio in which a constant portion of wealth ω is invested in the stock and the remaining portion $1 - \omega$ is invested in the bond. Let $\{X_t\}$ be the wealth process at time t. Then the number of stocks bought at time t is $\omega X_t/S_t$, while the units

[9] A stopping time τ is such that $\{\tau \leq t\} \in \mathcal{F}_t$, the filtration up to time t generated by the Brownian motion w_t, meaning that by making use of all historical information up to the present time t, we decide whether to stop.

of bond bought is $(1-\omega)X_t/B_t$. By (7.29) and (7.32), the wealth process $\{X_t\}$ then satisfies the dynamics $dX_t = \left(\frac{\omega X_t}{S_t}\right) dS_t + \left(\frac{(1-\omega)X_t}{B_t}\right) dB_t$. Thus $\{X_t\}$ satisfies the SDE:

$$\frac{dX_t}{X_t} = \omega\frac{dS_t}{S_t} + (1-\omega)\frac{dB_t}{B_t} = (\omega\mu + (1-\omega)r)dt + \omega\sigma dw_t,$$

whose solution, given an initial wealth X_0, is simply:

$$X_t = X_0 \exp\left(\left(\omega\mu + (1-\omega)r - \frac{\omega^2\sigma^2}{2}\right)t + \omega\sigma w_t\right).$$

Therefore, the (one period) expected logarithmic return of the portfolio wealth is equal to:

$$\mathbb{E}\left(\ln X_t - \ln X_0\right) = \left(\omega\mu + (1-\omega)r - \frac{\omega^2\sigma^2}{2}\right)t,$$

and the Kelly fraction is the value of ω that maximizes this latest quantity. In particular, the first order condition $\frac{d}{d\omega}\mathbb{E}\left(\ln X_t \mid X_0 = X_0\right) = \mu - r - \sigma^2\omega^* = 0$ gives $\omega^* = \frac{\mu-r}{\sigma^2}$, which is exactly the κ in (7.31).

Turning back to the optional selling problem (7.30), we can interpret the mathematical result as saying one should buy a stock and hold it until the end of the investment horizon at T when $\alpha \geq 1/2$; otherwise, one should sell it at once when $\alpha < 1/2$. Detailed proofs and variants of the problem can be found in [9, 10, 17, 29, 36–38]. Nonetheless, for this buy-and-hold or sell-at-once strategy, as it is dependent on a single index, the ultimate benefit can only be achieved on the average. Instead, in the following trading strategy, we focus on the optimal stopping time in a *probabilistic* manner (more along *pathwise sense*), which is more realistic and likely leads to a more profitable strategy. To this end, we first formally state both the corresponding optimal selling and buying problems as follows:

1. (*Selling problem*) For a fixed $p \in (0,1)$, find the optimal selling (stopping) time τ_S^* which maximizes the probability:

$$\sup_{0\leq\tau_S\leq T} \mathbb{P}(S_{\tau_S} \geq pM_T), \tag{7.33}$$

   where $M_t := \sup_{0\leq\tau\leq t} S_\tau$ is the running maximum stock price over $[0,t]$.

2. (*Buying problem*) For a fixed $q > 1$, find the optimal buying (stopping) time τ_B^* which maximizes the probability:

$$\sup_{0\leq\tau_B\leq T} \mathbb{P}(S_{\tau_B} \leq qm_T), \tag{7.34}$$

   where $m_t := \inf_{0\leq\tau\leq t} S_\tau$ is the running minimum stock price over $[0,t]$.

These problems were completely solved in one of our recent working papers [26]. It was found that their optimal solutions are determined by: the *goodness index* λ of the stock (a modified version of Kelly fraction with $r = 0$); the selling threshold ε_S; and the buying threshold ε_B; we define these as:

$$\lambda := \sigma\left(\alpha - \frac{1}{2}\right) = \frac{\mu}{\sigma} - \frac{\sigma}{2} = \frac{\mu - \sigma^2/2}{\sigma}, \quad \varepsilon_S := -\frac{\ln p}{\sigma}, \quad \text{and} \quad \varepsilon_B := \frac{\ln q}{\sigma}. \tag{7.35}$$

The corresponding optimal stopping times for the selling and the buying strategies can be obtained by solving for the root(s) of the following equation:

$$\varphi(t; \lambda, \varepsilon) := \eta(t; \lambda, \varepsilon) - \gamma(t; \lambda, \varepsilon), \tag{7.36}$$

where the function $\eta(t; \lambda, \varepsilon)$ is presented below up to the third order of its Taylor series expansion in λ (its more elaborate expression and more relevant discussions can be found in [26]):

$$\frac{\sqrt{2\pi t}}{2}\eta(t; \lambda, \varepsilon) = \frac{\varepsilon^4 - 4\varepsilon t^3\lambda^3 - 4\varepsilon^2 t(1 + \varepsilon\lambda) + 2\varepsilon\lambda t^2(4 + 3\varepsilon\lambda)}{8t^2} + \frac{\lambda t}{\varepsilon + \lambda t} - \frac{\lambda t^2}{(\varepsilon + \lambda t)^3} + \frac{3\lambda t^3}{(\varepsilon + \lambda t)^5} - 1 + \frac{1}{\lambda^2 t} - \frac{3}{\lambda^4 t^2} + \cdots. \tag{7.37}$$

On the other hand, there is no closed-form expression for the function γ. However, γ can be computed numerically by the *inverse Laplace transform* of another function $\mathcal{G}$ (also see [26]):

$$\mathcal{G}(s; \lambda, \varepsilon) = \int_0^{\infty} e^{-st}\gamma(t; \lambda, \varepsilon)\, dt = \frac{4\varepsilon\sqrt{\lambda^2 + 2s}(\varepsilon^2\sqrt{\lambda^2 + 2s} - \varepsilon - 2\lambda\varepsilon)}{2\varepsilon\sqrt{\lambda^2 + 2s}(1 + s\varepsilon^2) + \cdots}. \tag{7.38}$$

The following two propositions delineate the possible optimal times for selling and buying, that is, the solutions to Problems (7.33) and (7.34).

Proposition 7.1 (Refer to [26]) The optimal selling time τ_S^* of the stock for Problem (7.33) is given in a specific form in exactly one of the following four cases:

- Case 1: if $\varphi(t; \lambda, \varepsilon_S)$ admits no root, then $\tau_S^* := \inf\{t \leq T : S_t = pM_t\} \wedge T$. If λ is very negative in value, the stock simply drops without rising much throughout the planning horizon, so its initial value is likely the largest, and so it is almost optimal to sell-at-once. See the blue solid line in Figure 7.19;
- Case 2: if $\varphi(t; \lambda, \varepsilon_S)$ admits a unique (double) root s^* with $\varphi(s; \lambda, \varepsilon_S) < 0$ for any $s \neq s^*$, then $\tau_S^* := \inf\{t \leq T : S_t = pM_t\} \wedge T$. See the red dashed line in Figure 7.19;

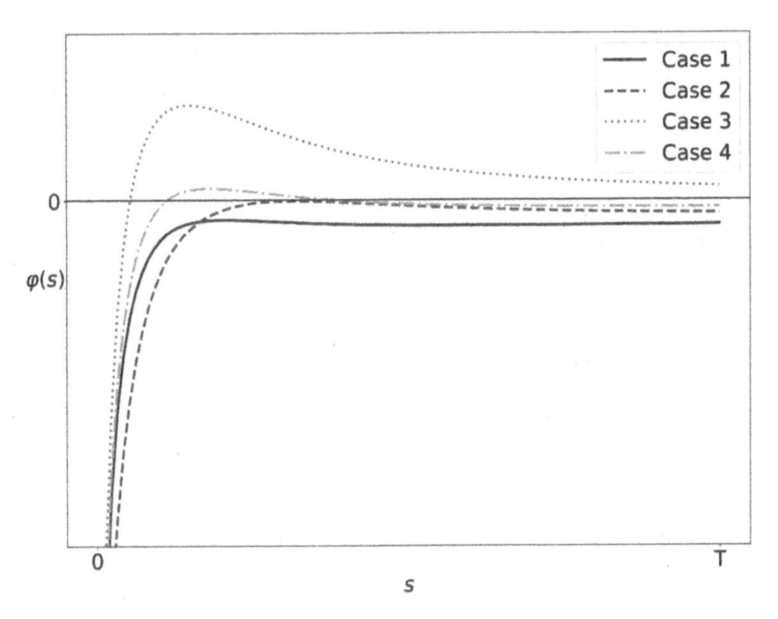

FIGURE 7.19 Illustration of the four cases of φ as in Propositions 7.1 and 7.2.

- Case 3: if $\varphi(t; \lambda, \varepsilon_S)$ admits a unique root s^* with $\varphi(s; \lambda, \varepsilon_S) < 0$ when $s < s^*$; and $\varphi(s; \lambda, \varepsilon_S) > 0$ when $s > s^*$, then by defining $t^* := (T - s^*)_+$, the optimal selling time is given by $\tau_S^* := \inf\{t \in [t^*, T] : S_t = pM_t\} \wedge T$. When λ is large, s^* will get close to 0, then t^* is close to T, and the resulting optimal selling is essentially no different from buy-and-hold. See the grey dotted line in Figure 7.19;
- Case 4: if $\varphi(t; \lambda, \varepsilon_S)$ admits two distinct roots $s_2^* < s_1^*$, then by defining $t_1^* := (T - s_1^*)_+$ and $t_2^* := (T - s_2^*)_+$, the optimal selling time is given by $\tau_S^* := \inf\{t \in [0, t_1^*] \cup [t_2^*, T] : S_t = pM_t\} \wedge T$. In particular, as λ gets more negative in value, t_1^* and t_2^* approach to each other. Ultimately, this closes the gap between these two numbers, making this case reduce to Case 1. See the yellow dash-dotted line in Figure 7.19.

We remark here that $\lambda \geq 0$ will only lead to Case 3, while all the other cases 1, 2 and 4 are possible when $\lambda < 0$. A similar characterization can be formulated for the optimal buying strategy, now with $\varphi(t; -\lambda, \varepsilon_B)$ in place of $\varphi(t; \lambda, \varepsilon_S)$.

Proposition 7.2 (Refer to [26]) The optimal buying time τ_B^* of the stock from Problem (7.34) is given in a specific form in exactly one of the following four cases:

- Case 1: if $\varphi(t; -\lambda, \varepsilon_B)$ admits no root, then $\tau_B^* := \inf\{t \leq T : S_t = qm_t\} \wedge T$;
- Case 2: if $\varphi(t; -\lambda, \varepsilon_B)$ admits a unique (double) root s^* with $\varphi(s; -\lambda, \varepsilon_B) < 0$ for any $s \neq s^*$, then $\tau_B^* := \inf\{t \leq T : S_t = qm_t\} \wedge T$;
- Case 3: if $\varphi(t; -\lambda, \varepsilon_B)$ admits a unique root s^* with $\varphi(s; -\lambda, \varepsilon_B) < 0$ when $s < s^*$; and $\varphi(s; -\lambda, \varepsilon_B) > 0$ when $s > s^*$, then by defining $t^* := (T - s^*)_+$, the optimal buying time is given by $\tau_B^* := \inf\{t \in [t^*, T] : S_t = qm_t\} \wedge T$;

- Case 4: if $\varphi(t; -\lambda, \varepsilon_B)$ admits two distinct roots $s_2^* < s_1^*$, then by defining $t_1^* := (T - s_1^*)_+$ and $t_2^* := (T - s_2^*)_+$, the optimal buying time is given by $\tau_B^* := \inf\{t \in [0, t_1^*] \cup [t_2^*, T] : S_t = q\mathrm{m}_t\} \wedge T$.

In Appendix *7.A.3, we present a table containing the root(s) s^* (s_1^* and s_2^*) of φ under various combinations of values of ε and λ. For any intermediate values at which φ is evaluated, the corresponding roots can then be interpolated.

*7.5.1 Trading Strategies Using the Goodness Index

Suppose that in a given calendar year, we would like to trade one or multiple stocks without any of their derivatives, so as to maximize our profit. This problem is time-inconsistent in nature, as our objective may change day-by-day. In the following, we shall introduce a class of trading strategies with different odds that utilize consecutive optimal selling and buying times, as introduced in Propositions 7.1 and 7.2.

Generally, while stock returns are expected to be non-stationary, their changes are usually mild within a shorter time frame. For the sake of illustration, here we set T to 4 so that each trading horizon contains $T + 1 = 5$ trading days, or one week. This motivates us to assume that μ and σ are constant throughout the trading horizon, and we denote their estimates of the k-th trading horizon spanning from day $t_{k,0}$ to day $t_{k,T} := t_{k,0} + T$ by $\mu_{t_{k,0}}$ and $\sigma_{t_{k,0}}$, respectively. We defer the discussion of parameter estimation to the subsequent subsections, while here we focus on discussing our proposed trading mechanism which is quite counterintuitive and not so immediately straightforward. However, it is inspired by and works consistently with our proposed theory.

The design of trading horizons can greatly influence trading strategies. Take a pre-commitment strategy as an example: a calendar year can be split into a number of relatively short trading horizons that are fixed in advance in a non-overlapping manner, namely the k-th trading horizon spans from day $t_{k,0} = (k-1)(T+1)$ to day $t_{k,T} = k(T+1) - 1$, then within each trading horizon, our decision on whether or not to trade on day $s = t_{k,0}, \ldots, t_{k,T}$ depends on the root(s) of the function $\varphi\left(s - t_{k,0}; \lambda_{t_{k,0}}, \varepsilon_{S,t_{k,0}}\right)$ in Proposition 7.1 for selling and the root(s) of the function $\varphi\left(s - t_{k,0}; -\lambda_{t_{k,0}}, \varepsilon_{B,t_{k,0}}\right)$ in Proposition 7.2 for buying. Note that we may hold the stock until the end of one of the trading horizons, and at that point we refresh our decision whether to sell or hold it for the next period.

However, by adopting the aforementioned strategy, at most one buying transaction or one selling transaction will take place within a trading horizon, and no action will be considered for the remaining portion of the horizon. As such, we may miss out the possible profit over this remaining period; indeed, this strategy's performance is probably not as good as simply holding the S&P500. This hints that pre-commitment strategies may not perform extremely well for any time-inconsistent problem. For example, consider a pre-commitment strategy of buy-and-holding the stock throughout a 10-day trading horizon. Say the stock keeps decreasing for the next five days due to randomness, then on day 5, we may undo our investment decision (made at time 0) by selling the stock immediately to avoid greater losses, and this can fail our initial goal of profit maximization at time 0. As an alternative, we can devise another trading strategy to cope with this time-inconsistent problem by defining a number of more frequent trading horizons as follows: once we perform a transaction on day s within the k-th trading horizon, we set the subsequent trading day as $t_{k+1,0} = s + 1$, the start of the subsequent $(k + 1)$-th trading horizon; otherwise, if no transaction is made in this trading horizon, then the subsequent trading day of its end day $t_{k,T}$ is set as the start of the $(k + 1)$-th trading horizon, meaning that $t_{k+1,0} = t_{k,T} + 1$.[10] However, the return of this alternative approach is still unsatisfactory in comparison with buy-and-holding S&P500. Therefore, we propose our unconventional and counterintuitive, yet potentially profit-making trading strategy as follows.

We construct an artificial "trading horizon" of length of $T + 1$ days starting from every trading day, and from now on, we call this artificial "trading horizon" an *examination window* to avoid ambiguity. Clearly, for a chosen trading day s, it is enclosed in a total of $T + 1$ different examination windows, such that the $(j + 1)$-th examination window spans from day $s - T + j$ to day $s + j$, for $j = 0, \dots, T$. In the following, we explain what it meant by the *action period* in each examination window. Consider an examination window from day $s - T + j$ to day $s + j$, with respective estimates of μ and σ denoted by μ_{s-T+j} and σ_{s-T+j}. According to the four cases in Proposition 7.1 for selling (resp. Proposition 7.2 for buying), and given s^* (s_1^* and s_2^*) as the root(s) of $\varphi(\cdot - (s - T + j); \lambda_{s-T+j}, \varepsilon_{S,s-T+j})$ for selling (resp. $\varphi(\cdot - (s - T + j); -\lambda_{s-T+j}, \varepsilon_{B,s-T+j})$ for buying), where λ_{s-T+j} and $\varepsilon_{S,s-T+j}$ (resp. $\varepsilon_{B,s-T+j}$) follow the definition of (7.35) but using the current estimates of μ_{s-T+j} and σ_{s-T+j}, and denoting $t^* := (T - s^*)_+$ ($t_1^* := (T - s_1^*)_+$ and $t_2^* := (T - s_2^*)_+$), the *action period* of this examination window takes one of the following forms:

$$\begin{aligned}
&\text{Case 1 and 2:} && [s - T + j, s + j];\\
&\text{Case 3:} && [s + t^* - T + j, s + j]; \qquad\qquad (7.39)\\
&\text{Case 4:} && [s - T + j, s + t_1^* - T + j] \cup [s + t_2^* - T + j, s + j].
\end{aligned}$$

Next, we apply our proposed theory within each examination window, so as to determine if the current day s *were* the optimal trading day over that window. While this would hold for at least one of the $T + 1$ examination windows, namely from day $s - T + j$ to day $s + j$, for some $j = 0, \dots, T$, we perform the corresponding trading action as follows: for fixed values of $p \in (0, 1)$ and $q > 1$, we have the following action:

- Recall the historical rolling maximum price from the beginning of the $(j + 1)$-th examination window $\tilde{M}_{s,j} := \max_{s-T+j \le \tau \le s} S_\tau, j = 0, \dots, T$. We proceed on selling on day s if there exists a $j = 0, \dots, T$ such that the following two conditions are simultaneously satisfied:

[10] In the former situation, we start the next trading horizon on the subsequent day after transaction but not on the current day, so as to avoid the possible double tradings on the same day; we also adopt the same practice in this latter situation.

(S1) Either $\{S_s > p\tilde{M}_{s,j}$ and $S_{s-1} < p\tilde{M}_{s,j}\}$ or $\{S_s < p\tilde{M}_{s,j}$ and $S_{s-1} > p\tilde{M}_{s,j}\}$, but not both, holds; and

(S2) s falls in the *action period* of this examination window corresponding to exactly one of the four possible cases in (7.39).

- Recall the historical rolling minimum price from the beginning of the $(j+1)$-th examination window $\tilde{m}_{s,j} := \min_{s-T+j\leq\tau\leq s} S_\tau$, $j = 0, \ldots, T$. We proceed on a buying on day s if there exists a $j = 0, \ldots, T$ such that the following two conditions are satisfied simultaneously:

(B1) Either $\{S_s > q\tilde{m}_{s,j}$ and $S_{s-1} < q\tilde{m}_{s,j}\}$ or $\{S_s < q\tilde{m}_{s,j}$ and $S_{s-1} > q\tilde{m}_{s,j}\}$, but not both, holds; and

(B2) s falls in the *action period* of this examination window corresponding to exactly one of the four possible cases in (7.39).

Below, with an initial capital of USD 100,000, we propose the following three Kelly fraction-based candidate strategies with different odds for each transaction:

1. Strategy K_{IC}: the same amount of capital (USD 100,000) is used for every transaction;
2. Strategy K_{AC}: all the accumulated capital is used for every transaction;
3. Strategy K_L: a leverage-adjusted accumulated capital is used for every transaction. Specifically, on day s, the leverage k_s is defined as:

$$k_s := \min\left\{\max\left\{\frac{\mu_s}{\sigma_s^2\sqrt{256}}, -4\right\}, 5\right\}, \tag{7.40}$$

and the amount used for each transaction is then $(0.8 + 0.2k_s)\Pi_s$, where Π_s is the accumulated capital on day s, while the additional amount of $\min\{0.2(k_s - 1)\Pi_s, 0\}$ to be borrowed is subject to an interest rate of 10% per annum, to be further explained below.

Intuitively, among the three candidate strategies, strategy K_L is the most aggressive when $k_s > 1$ and becomes the most conservative when k_s is close to -4; the aggressiveness of strategies K_{IC} and K_{AC} lies between these two boundary cases.

For a trade on current day s, we explain the transactions taken based on the current position of the portfolio as follows; assume that the previous trade was made δt calendar days ago[11].

- If we borrowed $n_{s-\delta t} > 0$ units of stock and short-sold them in the previous trade, so that we are currently in a short position of $n_{s-\delta t}$ units of stock, then we buy back these $n_{s-\delta t}$ shares from the market to *close the short-selling position*. Assuming a short-selling interest rate of 10% per annum[12], we pay back to the borrower

[11] Numerical evidence indicates that trades do not occur frequently and $\delta t \in (2T, 6T)$.
[12] The short-selling interest rate fluctuates daily, but it is usually less than 10% in practice.

a cash amount equivalent to $n_{s-\delta t}\left(1+\frac{0.1}{365}\right)^{\delta t}$ units of stock at the current trading price, that is $n_{s-\delta t}S_s\left(1+\frac{0.1}{365}\right)^{\delta t}$. After that, we further purchase $n_s > 0$ shares to *open the long position*, where n_s is computed according to which of strategies K_{IC}, K_{AC}, or K_L is being used.

- If we purchased $n_{s-\delta t} > 0$ stocks in the previous trade, so that we are currently in a long position of $n_{s-\delta t}$ units of stocks, then we sell these $n_{s-\delta t}$ shares to the market to *close the long position*, and receive the cash proceeds of $n_{s-\delta t}S_s$. Note that if a cash amount $\mathcal{K} \geq 0$ was borrowed when we opened the position in the previous trade, which occurs more often when one of the strategies K_{IC} or K_L is used, then assuming a borrowing interest rate of 10% per annum, we repay the loan of its accumulated value of $\mathcal{K}\left(1+\frac{0.1}{365}\right)^{\delta t}$, leading to a net proceed of $n_{s-\delta t}S_s - \mathcal{K}\left(1+\frac{0.1}{365}\right)^{\delta t}$ by closing the long position. After that, we *open the short-selling position* by borrowing $n_s > 0$ units of stock subject to a short-selling interest rate of 10%, and short-selling them back to the market. Once again, n_s is computed according to which of strategies K_{IC}, K_{AC}, or K_L is being used.

Furthermore, for each transaction that either opens or closes a position of $n_s > 0$ units of stocks on day s, a commission fee of USD $\max(0.013n_s, 2.05)$[13] is paid to the broker and US market authorities such as the *US Settlement Agency*[14]. Moreover, we also take dividends into consideration, by adopting unadjusted closing prices in the calculation of the arithmetic return on day s, namely $u_s = (S_s + D_s - S_{s-1})/S_{s-1}$, where $D_s \geq 0$ is the dividend amount paid on day s, and assume that the dividend D_s is reinvested in the stock on the same day s. We also note that no *Stop-Loss Orders* are adopted in the present proposed strategy, and so the loss can go beyond the initial capital. The trading strategy in the presence of a Stop-Loss Order is left for readers to explore.

*7.5.2 Backtesting

Recall that in determining whether the current day s is in the action period (7.39) within the $(j+1)$-th examination window from day $s-T+j$ to day $s+j$, the parameters $\lambda_{s-T+j}, \varepsilon_{S,s-T+j}, \varepsilon_{B,s-T+j}$ involve the estimates of μ_{s-T+j} and σ_{s-T+j} on day $s-T+j$, and the performance of this newly proposed trading strategy is certainly linked to the accuracy of these estimations. In this subsection, we assess its performance from the viewpoint of *backtesting*.

Regarding the estimation of μ, let's say for a given trading day t, being the beginning of some $(j+1)$-th examination window, we can obtain a better estimate by also using future returns as well as historical ones. To this end, the following *two-stage approach* is introduced: we obtain a preliminary estimate $\tilde{\mu}_t$ in the first step, by

---

[13] Source: https://www.futuhk.com/en/commissionnew; other brokers charge their commission fee at a similar level.

[14] The *Depository Trust Company* (DTC) https://www.dtcc.com/settlement-and-asset-services/settlement.

calculating the weighted geometric return over 3 consecutive trading days starting from the previous day $t-1$:

$$\tilde{\mu}_t = \prod_{i=-1}^{1} (1 + u_{t+i})^{\omega_i} - 1, \tag{7.41}$$

where $\sum_{i=-1}^{1} \omega_i = 1$; for the current backtesting, we rely more on the return of the following day, by taking $\omega_{-1} = 0.1$, $\omega_0 = 0.3$, and $\omega_1 = 0.6$.

In the second step, $\tilde{\mu}_t$ is adjusted further based on whether the stock has been in a superior, neutral or inferior state, with reference to a designated period of 11 trading days, namely from day $t-9$ to day $t+1$. Specifically, we compute the standard deviation using only the positive (resp. negative) returns among u_i's within the period, denoted by $\tilde{\sigma}_{+,t}$ (resp. $\tilde{\sigma}_{-,t}$), with the convention that $\tilde{\sigma}_{+,t} = 0$ (resp. $\tilde{\sigma}_{-,t} = 0$) whenever none of the trading days within the period has a positive (resp. negative) return. Using these, we obtain three candidate estimates, namely $\tilde{\mu}_t + \tilde{\sigma}_{+,t}/2$, $\tilde{\mu}_t$, and $\tilde{\mu}_t - \tilde{\sigma}_{-,t}/2$, corresponding to superior, neutral and inferior states for the stock, respectively. Finally, we set μ_t to the candidate estimate that is closest to the return u_t on day t. As a remark, the hyperparameters $p \in (0,1)$ and $q > 1$ should be chosen with care, usually based on some preliminary analysis on the historical performance of the stock, say via the goodness index λ_t. The actual tuning of these hyperparameters based on this fundamental principle is left as a project for readers. We also note that the aforementioned estimation procedure is different from knowing and using the exact price from the future, which requires much more information, as it even includes the noisy observations. Instead, for our proposed approach, it suffices for us to know the trend only, which we can also guess by analyzing the historical information contained in different sources, including volume trading as well as social medias and news, using tools such as text mining beyond the numerical approaches.

Regarding the estimation of σ mentioned above, say for an arbitrary trading day t, being the beginning of some $(j+1)$-th examination window, we use 61 days of historical returns $u_{t-60}, \cdots, u_t$ to fit an L-GARCH(1, 1) model as specified in (7.24)[15], in order to obtain an estimate of σ_t. Note that no future returns are used in the estimation of σ_t, since our numerical experiments show that backtesting on volatility by using the return on day $t+1$ does not notably improve the performance of our approach, therefore we stick to using historical returns only.

Furthermore, based on our observations, the stock prices also exhibit a one-day lagging effect, which can be justified since investors may need some time to digest and react to the information emerging in the market. Therefore, a possible approach to further improve the performances of our proposed investment strategy is to consider the following back-shifted rolling maximum and minimum, particularly for the $(j+1)$-th examination window, $j = 0, \ldots, T$:

$$\tilde{\mathrm{M}}_{s,j}^{(-1)} := \max_{s-T+j-1 \le \tau \le s-1} \mathrm{S}_\tau \quad \text{and} \quad \tilde{\mathrm{m}}_{s,j}^{(-1)} := \min_{t-T+j-1 \le \tau \le s-1} \mathrm{S}_\tau; \tag{7.42}$$

while the root(s) of the function φ remain unchanged. To sum up the discussions above, Algorithm 7.1 lays down an outline of our proposed trading strategy.

[15] Earlier in this chapter, recall that u_i's are assumed to have a zero mean. Still, numerical estimates for σ do not differ much, as pointed out before.

Algorithm 7.1 Trading strategy with Kelly fraction on an asset

Input: Hyperparameters $p \in (0,1)$ and $q > 1$; $60 + T$ daily historical arithmetic returns $u_{-(60+T)}, \cdots, u_{-1}$ computed from the closing prices, where the subscript 0 denotes the first trading day of the selected year.

Initialization: USD 10,000 initial capital as cash on hand.

1: **for** each trading day s in the specified year **do**
2: Compute the daily arithmetic return $u_s = (S_s + D_s - S_{s-1})/S_{s-1}$ and then the daily drift rate μ_s via the *two-step approach*
3: Fit an L-GARCH(1, 1) model with $u_{s-60}, \cdots, u_s$ and retrieve its estimated volatility σ_s
4: **if** we are not in a long position **AND** today s is not the last trading day of the specified year **then**
5: **for** $j = 0, \cdots, T$ of the $(j+1)$-th examination window **do**
6: Compute the back-shifted rolling minimum price $\tilde{m}^{(-1)}_{s,j}$ in (7.42)
7: Find the root(s) of $\varphi(\cdot - (s - T + j); -\lambda_{s-T+j}, \varepsilon_{B,s-T+j})$ with (7.35) using μ_{s-T+j} and σ_{s-T+j}
8: **if** both conditions (B1) and (B2) are fulfilled **then**
9: Close the short-selling position, if any, with a short-selling interest rate of 10%
10: Buy the stock at price S_s with a commission fee
11: Escape the current **for** loop of running index j
12: **end for**
13: **if** we are not in a short position **AND** today s is not the last trading day of the specified year **then**
14: **for** $j = 0, \cdots, T$ of the $(j+1)$-th examination window **do**
15: Compute the back-shifted rolling maximum price $\tilde{M}^{(-1)}_{s,j}$ in (7.42)
16: Find the root(s) of $\varphi(\cdot - (s - T + j); \lambda_{s-T+j}, \varepsilon_{S,s-T+j})$ with (7.35) using μ_{s-T+j} and σ_{s-T+j}
17: **if** both conditions (S1) and (S2) are fulfilled **then**
18: Close the long position, if any, with a borrowing interest rate of 10%
19: Short-sell the stock at price S_s with a commission fee
20: Escape the current **for** loop of running index j
21: **end for**
22: **if** today is the last trading day **AND** we are in a long (resp. short-selling) position **then**
23: Close the long (resp. short-selling) position with a commission fee and a borrowing (resp. short-selling) interest rate of 10%
24: **end for**

With an initial capital of USD 10,000 per stock available at the beginning of each year to trade the stocks listed in the S&P500 of that year[16], we implement the three candidate strategies with different odds, namely K_{IC}, K_{AC}, and K_L, throughout the year, and then start afresh with the same amount of initial capital of USD 10,000 for each stock at the beginning of the next year. The power of our proposed trading strategy and its robustness under various scenarios of the financial market are evident from the track record during the years 2003–2022, as shown in Figure 7.20; clearly,

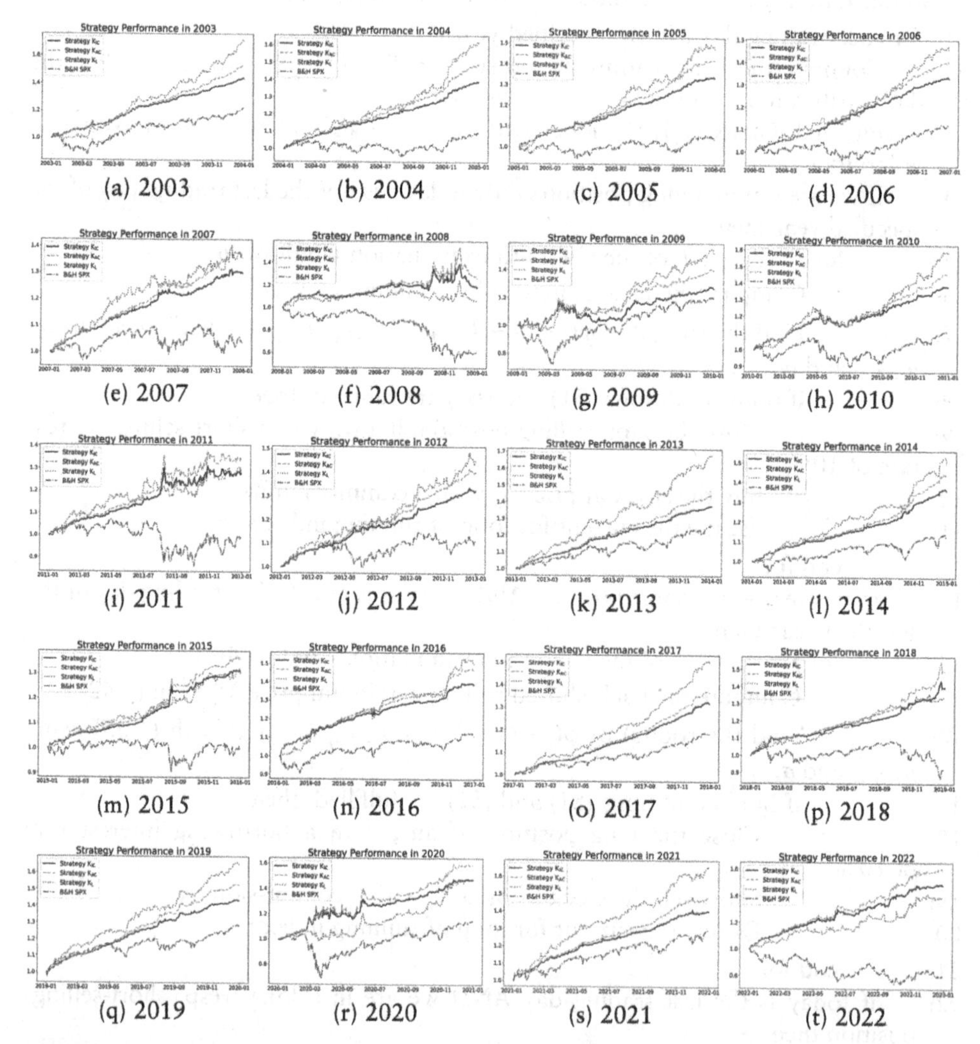

(a) 2003 (b) 2004 (c) 2005 (d) 2006
(e) 2007 (f) 2008 (g) 2009 (h) 2010
(i) 2011 (j) 2012 (k) 2013 (l) 2014
(m) 2015 (n) 2016 (o) 2017 (p) 2018
(q) 2019 (r) 2020 (s) 2021 (t) 2022

FIGURE 7.20 Changes of portfolio values in years 2003–2022, in terms of the average accumulated capital per S&P500 stock, using Kelly fraction-based trading strategies K_{IC} (solid line in blue), K_{AC} (dashed line in orange), and K_L (dotted line in red), each with an initial capital of USD 10, 000 per stock. The dash-dotted line in grey indicates the performance of the buy-and-hold strategy of S&P500 for the entire year.

[16] Not all stocks listed in the S&P500 of that year are traded since some companies are privatized or merged with other companies; they are no longer available in *Yahoo Finance*.

all three strategies significantly outperform the buy-and-hold strategy in all years. In particular, K_{IC} and K_{AC} achieve average annual returns of 30.74% and 36.55% from 2003 to 2022, respectively, and K_{AC} consistently outperforms K_{IC} all the time; overall, strategy K_L is the best-performing of the three candidates, achieving an average annual return of 38.55%, and it indeed outperforms the other two in most of the years, except in years 2008, 2011, 2018, 2020 and 2022 where bear markets are observed. This can be explained by the design of K_L, which makes it more conservative than the others in a bear market.

*7.5.3 Kelly Fraction in Practice

A key contributing factor for the high profitability of strategies K_{IC}, K_{AC}, and K_L is the unrealistically high accuracy of the estimate of μ_t on day t due to the use of the return on day $t+1$ in the calibration of the trend. In reality, these strategies can hardly be expected to perform as well with only past returns available. The accuracy of estimates of μ_t's based on past returns is relatively lower when compared to those of σ_t's. For example, consider the weighted geometric returns of the past 3 consecutive trading days:

$$\tilde{\mu}_t := \prod_{i=-2}^{0} (1 + u_{t+i})^{\omega_i} - 1, \tag{7.43}$$

where $\sum_{i=-2}^{0} \omega_i = 1$, say $\omega_{-2} = 0.1$, $\omega_{-1} = 0.1$, $\omega_0 = 0.8$, and adjust it further by $\tilde{\sigma}_{+,t}$ and $\tilde{\sigma}_{-,t}$, as discussed in Subsection *7.5.2 but now using returns from day $t-10$ to day t without an additional day in the future. Such an estimate for μ would lead to notable deviations in the root(s) of the function $\varphi(\cdot - (s-T+j); \lambda_{s-T+j}, \epsilon_{S,s-T+j})$ for selling (resp. $\varphi(\cdot - (s-T+j); -\lambda_{s-T+j}, \epsilon_{B,s-T+j})$ for buying) in (7.36), thereby distorting the decision on whether the current day s satisfies the selling (resp. buying) condition and is within the action period of the $(j+1)$-th examination window. It is possible that the resulting trading strategy may indicate that conditions (S1) and (S2) for selling (resp. (B1) and (B2) for buying) were satisfied on day s, but in reality they are not, hence leading to less profitable investment decisions. Worse still, for strategy K_L in particular, the inaccurate estimation also affects the leverage weight k_s, yielding an even more unstable performance. Therefore, we usually prefer the more moderate and conservative Strategy K_{AC} in practice.

As a remedy, we can mitigate the limitation of relatively insecure estimations of μ_t's by utilizing some common technical indicators, which are normally used by traders to predict future price trends and then make trading decisions. One of the commonly used technical indicators is the *Bollinger Bands*, first proposed by *John Bollinger* in the early 1980s; see [19, 22, 23] for details. For an arbitrary trading day t as discussed above, Bollinger Bands are computed from the simple moving average and simple moving standard deviation of the stock prices over the past d trading days. The bands are in the form of a stock price interval with the upper and lower endpoints given by

$$\text{BOLU}_t := \overline{S}_t + \kappa \cdot \sqrt{\frac{1}{d-1} \sum_{i=t-d+1}^{t} (S_i - \overline{S}_t)^2}; \tag{7.44}$$

$$\text{BOLL}_t := \overline{S}_t - \kappa \cdot \sqrt{\frac{1}{d-1} \sum_{i=t-d+1}^{t} (S_i - \overline{S}_t)^2}, \tag{7.45}$$

respectively, where $\overline{S}_t := \frac{1}{d}\sum_{i=t-d+1}^{t} S_i$. In practice, one usually chooses $d = 20$ and $\kappa = 2$.

Naturally, a stock price S_t above the upper band $BOLU_t$ (resp. below the lower band $BOLL_t$) indicates a selling (resp. buying) signal. Therefore, we can also take this into consideration when determining whether or not to trade on the current day s, so as to make up for our lack of firm grasp of the conditions (S1) and (S2) for selling (resp. (B1) and (B2) for buying) based on (7.36). To this end, we perform the selling action on day s if at least one of the following holds:

1. Both conditions (S1) and (S2) are satisfied;
2. $S_s > BOLU_s$;

while we perform the buying action on day s if at least one of the following holds:

1. Both conditions (B1) and (B2) are satisfied;
2. $S_s < BOLL_s$.

Recall that prior to the selling (resp. buying) action, we shall first close the existing long (resp. short-selling) position in the same manner as mentioned in Subsection *7.5.1. With this modified strategy, K_{AC} can yield an annual return of 10.4% on average from 2003 to 2022, still outperforming the buy-and-hold strategy of the S&P500 index which only yields an average annual return of 6.8% throughout the period.

As a closing remark, better trading strategies are not impossible, while they may require the use of different combinations of technical indicators, they ultimately relate to the Kelly fraction. We encourage readers to explore their own profitable trading strategies.

7.6 CALENDAR EFFECTS

Just as we commonly observe various business cycles in economies, seasonal effects are believed to be prevalent in stock price movements. These effects are also called *calendar effects* when they appear to be related to the calendar. It is undoubtedly crucial for profit-seeking investors to take full advantage of these effects by purchasing stocks at the start of periods of upward trends within cycles, and selling them just before periods of downward trends begin. For example, the well-known maxim of "*sell in May and go away*"[17] mentioned in [16] suggests that the performance of the stock market in general is significantly weaker on average from May onwards, as is commonly observed in Western markets. With this belief, investors should sell stocks at the beginning of May and hold their next major purchase until autumn, typically around *Halloween*[17,18], when stock prices tend to exhibit an upward trend in general. While many believe that the Hong Kong market should exhibit the same patterns as

[17] See descriptions and references in the link https://en.wikipedia.org/wiki/Sell_in_May.
[18] Also see the description of *Halloween effect* in the link https://www.investopedia.com/terms/h/halloween-strategy.asp.

western markets, based on our data analysis below, we find that different patterns were present at least before 2015. The market was relatively quiet in the few months after the *Lunar New Year* (usually between late January and early February), and its performance tended to improve only after the *Dragon Boat Festival* (usually within June). Analogous to a well-known Chinese adage that translates to "*before Dragon Boat, the weather can still be cold*" (未食五月粽，寒衣不敢送), we can adapt this traditional Chinese wisdom to describe the aforementioned trend in the Hong Kong stock market as "*before Dragon Boat, stock holdings should not be bold*" (未食五月粽，持股量勿重).

Despite the lack of consensus on the cause of such calendar effects, we can still examine their empirical validity using the data from stock markets. In particular, we adopt the *goodness index* λ_t in (7.35) to measure the performance of the market on any given day t:

$$\lambda_t = \frac{\hat{u}_t}{\hat{\sigma}_t} - \frac{\hat{\sigma}_t}{2};$$

here $\hat{u}_t$ and $\hat{\sigma}_t$ are the estimated daily drift and volatility of the stock on day t, respectively, now calculated by the following formulae:

$$\hat{u}_t := \frac{1}{2m+1} \sum_{s=t-m}^{t+m} r_s \quad \text{and} \quad \hat{\sigma}_t^2 := \frac{1}{2m+1} \sum_{s=t-m}^{t+m} (r_s - \overline{u}_t)^2, \tag{7.46}$$

where r_t is the (logarithmic) return on day t, and m is some positive integer (hyperparameter) so that $2m + 1$ is the width of the moving window. Note that in the current context of experiential study instead of real-life trading, we adopt moving windows including both historical and future returns for the sake of better estimation, so as to study the regular pattern of the markets concerned.

Following the procedure above, for the three selected stock indices, namely S&P500 Index in the US, *FTSE 100* Index in the UK and Hang Seng Index in Hong Kong, we compute the value of the goodness index λ_t for each stock index on every trading day between 2000 and 2015[19]. The best-fit values of m are 65 for both the US and Hong Kong markets, and 70 for the UK market, for which we shall provide further statistical justifications below. Then for each month from January to December, we calculate the aggregate proportion of days with a positive index $\lambda_t > 0$ and that of days with a negative index $\lambda_t < 0$ in the same month across all years[20]. Those proportions are represented in a bar chart by vertical bars colored in green at the bottom and red at the top, respectively; see Figures 7.21a, 7.21b, and 7.21c for the respective charts for S&P500 Index, FTSE 100 Index, and Hang Seng Index. We can clearly observe the downward trends starting in May and ending in October for both S&P500 and FTSE 100 indices, while that of Hang Seng Index mainly spans from February to June. This observation, mined from the actual data, serves

[19] The data were extracted from *Yahoo Finance*.

[20] So far we have not observed any trading day with $\lambda_t = 0$, which is as expected.

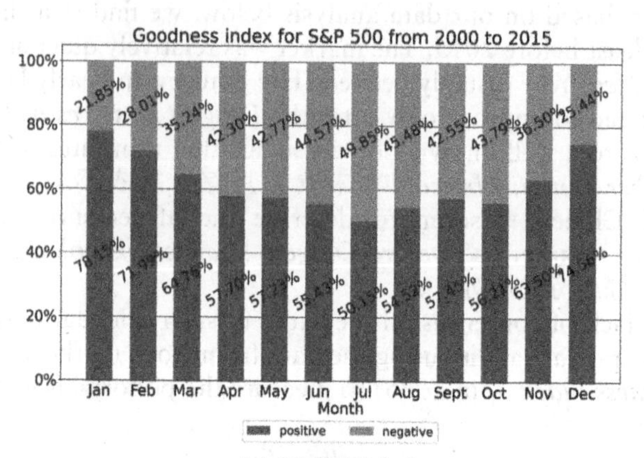

(a) S&P500 Index

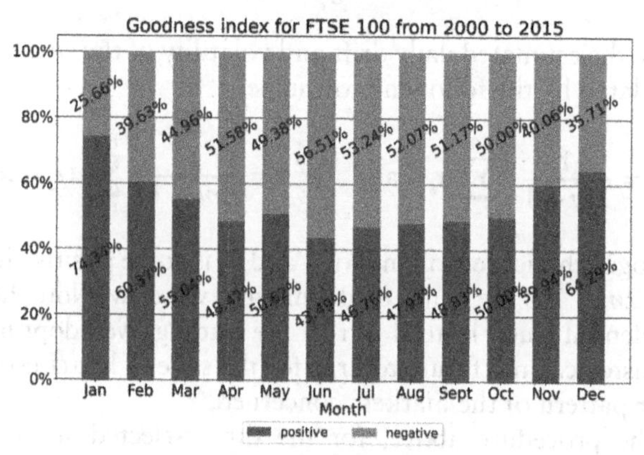

(b) FTSE 100 Index

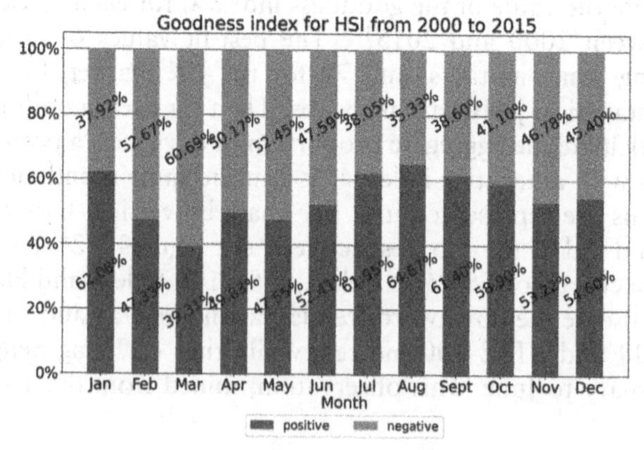

(c) Hang Seng Index

FIGURE 7.21 Bar chart of three stock indices based on the goodness index from 2000 to 2015.

as a solid empirical evidence of the commonly believed calendar effects, as well as the difference between the effects in the Hong Kong and Western stock markets. The Python codes are implemented in Programme 7.36, while readers may have a go at writing **R** programme themselves.

```
1  import pandas as pd
2  import numpy as np
3  import matplotlib.pyplot as plt
4  import datetime as dt
5  import yfinance as yf
6  from statsmodels.stats.diagnostic import acorr_ljungbox
7
8  def find_mu(returns, tail=5):
9      mu = returns.rolling(tail + 1 + tail).mean()
10     return mu.shift(-tail)
11
12 def find_sigma(returns, tail=15):
13     sig = returns.rolling(tail + 1 + tail).std(ddof=1)
14     return sig.shift(-tail)
15
16 INDICES = ["^SPX", "^FTSE", "^HSI"]
17 NAMES = ["S&P 500", "FTSE 100", "HSI"]
18
19 width = 0.6
20 xtick_labels = ["Jan", "Feb", "Mar", "Apr", "May", "Jun",
21                 "Jul", "Aug", "Sept", "Oct", "Nov", "Dec"]
22 ytick_labels = ["0%", "20%", "40%", "60%", "80%", "100%"]
23
24 def plot_monthly_index(daily_return, start, end, index_name, m=65):
25     date_start = dt.datetime(start, 1, 1)
26     date_end = dt.datetime(end, 12, 31)
27     date = daily_return.index
28     mask = (date >= date_start) * (date <= date_end)
29     title = f"Goodness index for {index_name} from {start} to {end}"
30
31     mu = find_mu(daily_return, tail=m)
32     sigma = find_sigma(daily_return, tail=m)
33     mu, sigma = mu[mask], sigma[mask]
34
35     stand_return = (daily_return - mu)/sigma
36     date = stand_return.index
37     lb_pvalue = np.empty(end - start + 1)
38     for i, year in enumerate(range(start, end+1)):
39         mask = (date >= dt.datetime(year,1,1))*(date <= dt.datetime(year,12,31))
40         lb_pvalue[i] = acorr_ljungbox(stand_return[mask],
41                                       lags=[10]).lb_pvalue.values[0]
42
43     goodness_index = mu/sigma - 0.5*sigma
44     monthly_goodness = {mon: None for mon in range(1, 13)}
45     date = goodness_index.index
46     for mon in range(1, 13):
47         lamb = goodness_index[date.month == mon]
48         total = np.sum(lamb < 0) + np.sum(lamb > 0)
```

```
49          monthly_goodness[mon] = {"neg": np.sum(lamb < 0)/total,
50                                   "pos":np.sum(lamb > 0)/total}
51
52      montly_pos = [monthly_goodness[mon]["pos"] for mon in range(1, 13)]
53      monthly_neg = [monthly_goodness[mon]["neg"] for mon in range(1, 13)]
54
55      xtick = np.arange(12)
56      fig, ax = plt.subplots(figsize=(15, 10))
57      rects1 = ax.bar(xtick+width, montly_pos, width=width, color="green",
58                      label="positive")
59      rects2 = ax.bar(xtick+width, monthly_neg, width=width, color="red",
60                      label="negative", bottom=montly_pos)
61
62      ax.set_xticks(xtick+width)
63      ax.set_yticks([0, 0.2, 0.4, 0.6, 0.8, 1])
64      ax.set_xticklabels(xtick_labels, fontsize=25)
65      ax.set_yticklabels(ytick_labels, fontsize=25)
66      ax.set_xlabel("Month", fontsize=25)
67      ax.yaxis.grid(color="b")
68      ax.set_axisbelow(True)
69      ax.set_title(title, fontsize=30)
70      ax.legend(loc="lower center", bbox_to_anchor=(0.5, -0.2),
71                prop={"size": 20}, ncol=2)
72      fig.subplots_adjust(bottom=0.2)
73
74      for (rect1, rect2) in zip(rects1, rects2):
75          xtick = rect1.get_x()
76          height1 = rect1.get_height()
77          height2 = rect2.get_height()
78          ax.text(xtick + width/2, 0.5*height1, f"{height1*100:.2f}%",
79                  rotation=30, ha="center", va="center", fontsize=25)
80          ax.text(xtick + width/2, 0.5*height2+height1, f"{height2*100:.2f}%",
81                  rotation=30, ha="center", va="center", fontsize=15)
82
83      return lb_pvalue
84
85  start, end = 2000, 2015
86  m_lst = [65, 70, 65]
87
88  df_lb = pd.DataFrame(index=range(start, end+1), columns=INDICES)
89  for index, index_name, m in zip(INDICES, NAMES, m_lst):
90      historical_data = yf.Ticker(index).history(period="max").tz_localize(None)
91      daily_return = pd.Series(np.diff(np.log(historical_data.Close)),
92                               index=historical_data.index[1:])
93
94      df_lb[index] = plot_monthly_index(daily_return, start, end, index_name, m=m)
95
96  print(df_lb)
```

```
1            ^SPX      ^FTSE       ^HSI
2  2000  0.169074  0.075855  0.819349
3  2001  0.664233  0.002878  0.054348
4  2002  0.951868  0.031194  0.982815
5  2003  0.014240  0.002365  0.296244
```

```
6  | 2004  0.209353  0.314925  0.881101
7  | 2005  0.924799  0.921799  0.848718
8  | 2006  0.163027  0.000366  0.853359
9  | 2007  0.044262  0.350629  0.316655
10 | 2008  0.422912  0.057794  0.041019
11 | 2009  0.631727  0.112146  0.829974
12 | 2010  0.658545  0.752840  0.980991
13 | 2011  0.053091  0.667483  0.580839
14 | 2012  0.198314  0.328899  0.643485
15 | 2013  0.534881  0.412555  0.098828
16 | 2014  0.833636  0.775693  0.883356
17 | 2015  0.256484  0.694237  0.706845
```

Programme 7.36 Calender effects of the three stock indices, S&P500, FTSE 100, and HSI, from 2000 to 2015 via Python.

Last but not least, we provide statistical justifications for our fitted models via the Ljung–Box test, so as to demonstrate the statistical reliability of the results obtained above. The output of Programme 7.36 via Python is a `dataframe` that records the p-values from the Ljung–Box test with lag[21] $K = 10$ for the standardized returns $(r_t - \hat{\mu}_t)/\hat{\sigma}_t$ (where $\hat{\mu}_t$ and $\hat{\sigma}_t$ are given in (7.46)), of each stock index, year by year. The p-values all exceed 0.05 except two years for the S&P500 (2003, 2007), three years for the FTSE 100 (2001, 2003 and 2006), and one year for the HSI (2008). Other than the years mentioned above, we do not reject the null hypothesis of independence among residuals. Furthermore, the standardized returns show no autocorrelation up to lag 10, meaning that $\hat{u}_t$'s and $\hat{\sigma}_t$'s are proper estimates of the running mean (instantaneous drift) and standard deviation (instantaneous volatility). Therefore, for each market, we have built a workable and justifiable stochastic diffusion process for market index dynamics S_t in the form $dS_t = S_t(u_t dt + \sigma_t dw_t)$, with u_t and σ_t given empirically as above; this is certainly far more complicated than a simple Black–Scholes model.

*7.A APPENDIX

*7.A.1 Asymptotic Distribution of the Ljung–Box Test Statistic

Let us prove that for a time series $\{a_i\}$ with mean zero and no serial autocorrelation, under some mild technical conditions as specified below, we have

$$Q(K) := \sum_{k=1}^{K} \frac{n(n+2)}{n-k} r_{k,n}^2 \sim \chi_K^2, \text{ approximately,}$$

[21] Note that no degree of freedom needs to be removed here, since this is a genuine hypothesis testing for model validation, rather than model diagnostics; indeed, we here do not specify any particular parametric model.

where $r_{k,n} = \frac{\sum_{i=1}^{n-k} a_i a_{i+k}}{\sum_{i=1}^{n} a_i^2}$ is the k-time-lag sample autocorrelation. More precisely, the chi-square distribution of χ_K^2 is the asymptotic distribution of $Q(K)$ as n goes to infinity. Readers may also refer to [5] for more discussions on this test statistic.

To establish this claim, it suffices to show that as n goes to infinity,

$$\sqrt{\frac{n(n+2)}{n-k}}\, r_{k,n} \overset{\text{iid}}{\sim} \mathcal{N}(0,1),$$

for $k = 1, 2, \ldots, K$. First, we assume the following conditions for the time series $\{a_i\}$:

1. There is no serial correlation in a_i such that for every i, $\mathbb{E}(a_i \mid \mathcal{F}_{i-1}) = 0$ and $\mathbb{E}\left(a_i^2 \mid \mathcal{F}_{i-1}\right) = \mathbb{E}\left(a_i^2\right) = \sigma^2$, where $\mathcal{F}_{i-1} = \sigma(a_{i-1}, a_{i-2}, \ldots)$ is the σ-field of information generated by $a_{i-1}, a_{i-2}, \ldots$. In other words, the time series $\{a_i\}$ is a *martingale difference*. The time series $\{a_i\}$ is also *weakly stationary* (constant up to second order moments).
2. For all i, $\mathbb{E}(a_i^4) = 3\left[\mathbb{E}\left(a_i^2\right)\right]^2$.[22]
3. $\sup_{i,j} \mathbb{E}\left(|a_i a_j|^{2+\delta}\right) < \infty$, for some $\delta > 0$.

We simplify the notation by writing:

$$r_{k,n} := \frac{x(k,n)}{y(n)},$$

where $y(n) := \sum_{i=1}^{n} a_i^2$ and $x(k,n) := \sum_{i=1}^{n-k} a_i a_{i+k}$. First we evaluate the moments of $y(n)$:

$$\mathbb{E}(y(n)) = \mathbb{E}\left(\sum_{i=1}^{n} a_i^2\right) = \sum_{i=1}^{n} \mathbb{E}(a_i^2) = n\sigma^2,$$

and observe that, for $i > j$, the tower property leads to $\mathbb{E}\left(a_i^2 a_j^2\right) = \mathbb{E}\left(\mathbb{E}\left(a_i^2 \mid \mathcal{F}_j\right) a_j^2\right) = \mathbb{E}(a_i^2)\mathbb{E}(a_j^2)$. Hence, we have

$$\begin{aligned}
\mathbb{E}[y^2(n)] &= \mathbb{E}\left(\sum_{i=1}^{n} a_i^2 \sum_{j=1}^{n} a_j^2\right) \\
&= \mathbb{E}\left(\sum_{i=1}^{n} a_i^4 + \sum_{i=1}^{n} \sum_{j=1, i\neq j}^{n} a_i^2 a_j^2\right) \\
&= \sum_{i=1}^{n} \mathbb{E}(a_i^4) + \sum_{i=1}^{n} \sum_{j=1, i\neq j}^{n} \mathbb{E}(a_i^2)\mathbb{E}(a_j^2).
\end{aligned}$$

[22] If the normalizing factor "3" is replaced by a generic number κ, the number "2" in the test statistic $Q(K)$ should be substituted by $\kappa - 1$. As an example, Condition 2 is fulfilled by a time series of normally distributed white noise.

Using Condition 2, we have

$$\begin{aligned}\mathbb{E}[y^2(n)] &= \sum_{i=1}^{n} 3[\mathbb{E}(a_i^2)]^2 + \sum_{i=1}^{n}\sum_{j=1, i\neq j}^{n} \mathbb{E}(a_i^2)\mathbb{E}(a_j^2)\\ &= 3n[\mathbb{E}(a_1^2)]^2 + n(n-1)[\mathbb{E}(a_1^2)]^2\\ &= n(n+2)\sigma^4. \end{aligned} \tag{7.47}$$

Hence, $\mathrm{Var}\,(y(n)) = 2n\sigma^4$. Therefore, $\mathbb{E}\left(\frac{y(n)}{n}\right) = \sigma^2$, and $\mathrm{Var}\left(\frac{y(n)}{n}\right) \to 0$ as $n \to \infty$, altogether implying that $\frac{y(n)}{n} \to \sigma^2$ in probability as $n \to \infty$. For the moments of $x(k,n) = \sum_{i=1}^{n-k} a_i a_{i+k}$, it holds that, for $k \geq 1$,

$$\mathbb{E}(x(k,n)) = \mathbb{E}\left(\sum_{i=1}^{n-k} a_i a_{i+k}\right) = \sum_{i=1}^{n-k} \mathbb{E}(a_i \mathbb{E}(a_{i+k} \mid \mathcal{F}_{i+k-1})) = 0, \tag{7.48}$$

and (to avoid complications in the following summations, we adopt the convention that $a_i = 0$ for $i = -1, -2, \ldots$)

$$\begin{aligned}\mathbb{E}\left(x^2(k,n)\right) &= \mathbb{E}\left(\sum_{i=1}^{n-k} a_i a_{i+k} \cdot \sum_{j=1}^{n-k} a_j a_{j+k}\right) = \mathbb{E}\left(\sum_{i=1}^{n-k}\sum_{j=1}^{n-k} a_i a_{i+k} a_j a_{j+k}\right)\\ &= \mathbb{E}\left(\sum_{i=1}^{n-k}\left(a_i a_{i+k} a_i a_{i+k} + a_i a_{i+k} a_{i-k} a_i + \sum_{\substack{j=1\\ j\neq i, i-k}}^{n-k} a_i a_{i+k} a_j a_{j+k}\right)\right)\\ &= \sum_{i=1}^{n-k}(\mathbb{E}(a_i^2)\mathbb{E}(a_{i+k}^2) + \mathbb{E}\left(\mathbb{E}(a_{i+k} \mid \mathcal{F}_{i+k-1}) a_i^2 a_{i-k}\right)\\ &\quad + \sum_{\substack{j=1\\ j\neq i, i-k}}^{n-k} \mathbb{E}\left(\mathbb{E}\left(a_{(i\vee j)+k} \mid \mathcal{F}_{(i\vee j)+k-1}\right) a_{i\vee j} a_{i\wedge j} a_{(i\wedge j)+k}\right))\\ &= (n-k)[\mathbb{E}(a_1^2)]^2 = (n-k)\sigma^4. \end{aligned} \tag{7.49}$$

Therefore, $\mathbb{E}\left(\left(\frac{x(k,n)}{n}\right)^2\right) \to 0$. On the other hand, for $k \neq l$, again adopting the convention that $a_i = 0$ for $i = -1, -2, \ldots$, we have:

$$\begin{aligned}&\mathrm{Cov}\left(x(k,n), x(l,n)\right)\\ &= \mathbb{E}\left(\sum_{i=1}^{n-k} a_i a_{i+k} \sum_{j=1}^{n-l} a_j a_{j+l}\right) = \mathbb{E}\left(\sum_{i=1}^{n-k}\sum_{j=1}^{n-l} a_i a_{i+k} a_j a_{j+l}\right)\end{aligned}$$

$$= \mathbb{E}\left(\sum_{i=1}^{n-l}\left(a_i a_{i+k} a_i a_{i+l} + a_i a_{i+k} a_{i+k} a_{i+k+l} + a_i a_{i+k} a_{i-l} a_i\right.\right.$$

$$\left.\left. + a_i a_{i+k} a_{i+k-l} a_{i+k} + \sum_{\substack{j=1 \\ j\neq i, i+k, i-l, i+k-l}}^{n-l} a_i a_{i+k} a_j a_{j+l}\right)\right)$$

$$= \sum_{i=1}^{n-l}\Big[\mathbb{E}\left(a_{i+(k\wedge l)} a_i^2 \mathbb{E}\left(a_{i+(k\vee l)} \mid \mathcal{F}_{i+(k\vee l)-1}\right)\right) + \mathbb{E}\left(a_i a_{i+k}^2 \mathbb{E}\left(a_{i+k+l} \mid \mathcal{F}_{i+k+l-1}\right)\right)$$

$$+ \mathbb{E}\left(a_i^2 a_{i-l} \mathbb{E}\left(a_{i+k} \mid \mathcal{F}_{i+k-1}\right)\right) + \mathbb{E}\left(a_{i+k}^2\right)\mathbb{E}\left(a_i a_{i+k-l}\right)$$

$$+ \sum_{\substack{j=1 \\ j\neq i, i+k, i-l, i+k-l}}^{n-l} \mathbb{E}\left(a_i a_j a_{(i+k)\wedge(j+l)} \mathbb{E}\left(a_{(i+k)\vee(j+l)} \mid \mathcal{F}_{(i+k)\vee(j+l)}\right)\right)\Big]$$

$$= 0. \tag{7.50}$$

Let $\epsilon > 0$, define a set $C_\epsilon(n) := \left\{\left|\frac{y(n)}{n} - \sigma^2\right| \leq \epsilon\right\}$; also denote $C_\epsilon^C(n)$ as the complement of the set $C_\epsilon(n)$. Then,

$$\mathbb{E}(r_{k,n}) = \mathbb{E}\left(\frac{\frac{1}{n}x(k,n)}{\frac{1}{n}y(n)}\right) = \mathbb{E}\left(\frac{\frac{1}{n}x(k,n)}{\frac{1}{n}y(n)}\left(\mathbb{1}_{C_\epsilon(n)} + \mathbb{1}_{C_\epsilon^C(n)}\right)\right)$$

$$= \mathbb{E}\left(\frac{\frac{1}{n}x(k,n)}{\frac{1}{n}y(n)}\mathbb{1}_{C_\epsilon(n)}\right) + \mathbb{E}\left(\frac{\frac{1}{n}x(k,n)}{\frac{1}{n}y(n)}\mathbb{1}_{C_\epsilon^C(n)}\right),$$

implying that $\left|\mathbb{E}(r_{k,n}) - \mathbb{E}\left(\frac{\frac{1}{n}x(k,n)}{\frac{1}{n}y(n)}\mathbb{1}_{C_\epsilon(n)}\right)\right| \leq 1 \cdot \mathbb{P}(C_\epsilon^C(n))$, which holds because $\left|\frac{\frac{1}{n}x(k,n)}{\frac{1}{n}y(n)}\right| \leq 1$ by the very definition of empirical correlation. Since $\mathbb{P}(C_\epsilon^C(n)) \to 0$ as demonstrated before, we only have to focus on $\mathbb{E}\left(\frac{\frac{1}{n}x(k,n)}{\frac{1}{n}y(n)}\mathbb{1}_{C_\epsilon(n)}\right)$. Indeed, by simple telescoping, we have

$$\mathbb{E}\left(\frac{\frac{1}{n}x(k,n)}{\frac{1}{n}y(n)}\mathbb{1}_{C_\epsilon(n)}\right) = \left\{\mathbb{E}\left(\frac{\frac{1}{n}x(k,n)}{\frac{1}{n}y(n)}\mathbb{1}_{C_\epsilon(n)}\right) - \mathbb{E}\left(\frac{\frac{1}{n}x(k,n)}{\sigma^2}\mathbb{1}_{C_\epsilon(n)}\right)\right\}$$

$$+ \mathbb{E}\left(\frac{\frac{1}{n}x(k,n)}{\sigma^2}\right) - \mathbb{E}\left(\frac{\frac{1}{n}x(k,n)}{\sigma^2}\mathbb{1}_{C_\epsilon^C(n)}\right).$$

We consider the first two terms in the last equality,

$$\left|\mathbb{E}\left(\frac{\frac{1}{n}x(k,n)}{\frac{1}{n}y(n)}\mathbb{1}_{C_\epsilon(n)}\right) - \mathbb{E}\left(\frac{\frac{1}{n}x(k,n)}{\sigma^2}\mathbb{1}_{C_\epsilon(n)}\right)\right| = \left|\mathbb{E}\left(\frac{\frac{x(k,n)}{n}}{\frac{y(n)}{n}}\frac{\sigma^2 - \frac{y(n)}{n}}{\sigma^2}\mathbb{1}_{C_\epsilon(n)}\right)\right| \leq \frac{\epsilon}{\sigma^2},$$

since $\left|\frac{y(n)}{n} - \sigma^2\right| \le \epsilon$ on $C_\epsilon(n)$ and $\left|\frac{\frac{x(k,n)}{n}}{\frac{y(n)}{n}}\right| \le 1$ by the definition of empirical correlation again. We also note that, by (7.48),

$$\mathbb{E}\left(\frac{\frac{1}{n}x(k,n)}{\sigma^2}\right) = \frac{\mathbb{E}(\frac{1}{n}x(k,n))}{\sigma^2} = 0. \tag{7.51}$$

It remains to calculate $\mathbb{E}\left(\frac{\frac{1}{n}x(k,n)}{\sigma^2}\mathbb{1}_{C_\epsilon^C(n)}\right)$. Denote $g(k,n) = \frac{\frac{x(k,n)}{n}}{\sigma^2}$ for simplicity, recall (7.49) that $\mathbb{E}(x^2(k,n)) = (n-k)\sigma^4$, so $\sup_n \mathbb{E}(g^2(k,n)) \le \sup_n \frac{1}{n} < \infty$. By the Cauchy–Schwarz inequality,

$$\mathbb{E}\left(\left|g(k,n)\mathbb{1}_{C_\epsilon^C(n)}\right|\right) \le \sqrt{\mathbb{E}(g^2(k,n))\mathbb{E}(\mathbb{1}_{C_\epsilon^C(n)})} = \sqrt{\mathbb{E}(g^2(k,n))}\sqrt{\mathbb{P}(C_\epsilon^C(n))},$$

which converges to 0 uniformly in k and n, as n goes to infinity. Since ϵ is arbitrary, combining all discussions above, we have $\lim_{n\to\infty}\mathbb{E}(r_{k,n}) = 0$.

On the other hand, we can express

$$\begin{aligned}\mathbb{E}(r_{k,n}r_{l,n}) = &\left\{\mathbb{E}\left(r_{k,n}r_{l,n}\mathbb{1}_{C_\epsilon(n)}\right) - \mathbb{E}\left(g(k,n)g(l,n)\mathbb{1}_{C_\epsilon(n)}\right)\right\} \\ &+ \mathbb{E}\left(g(k,n)g(l,n)\right) - \mathbb{E}\left(g(k,n)g(l,n)\mathbb{1}_{C_\epsilon^C(n)}\right) + \mathbb{E}\left(r_{k,n}r_{l,n}\mathbb{1}_{C_\epsilon^C(n)}\right).\end{aligned}$$

The difference of the first two terms on the right-hand side can be shown to converge to 0 as $n \to \infty$. We expand the difference:

$$\begin{aligned}&r_{k,n}r_{l,n}\mathbb{1}_{C_\epsilon(n)} - g(k,n)g(l,n)\mathbb{1}_{C_\epsilon(n)} \\ &= \mathbb{1}_{C_\epsilon(n)}\left[r_{k,n}\left(r_{l,n} - g(l,n)\right) + g(l,n)\left(r_{k,n} - g(k,n)\right)\right],\end{aligned}$$

where $|r_{k,n}|, |r_{l,n}| \le 1$ by their definitions, $|r_{i,n} - g(i,n)| = |r_{i,n}|\left|\frac{\sigma^2 - y(n)}{n}\right| \le 1 \cdot \frac{\epsilon}{\sigma^2}$ on $C_\epsilon(n)$ for $i = k, l$ by using the same argument as before, and by (7.49), $\mathbb{E}\left(|g(l,n)|\right) \le \frac{1}{\sigma^4}\mathbb{E}\left(\left(\frac{1}{n}x(l,n)\right)^2\right) \to 0$. Therefore, the claim of convergence to 0 follows. Besides, the definitions above also imply that $\mathbb{E}(|r_{k,n}r_{l,n}|\mathbb{1}_{C_\epsilon^C(n)}) \to 0$, while (7.48) and (7.50) tell us $\mathbb{E}\left(g(k,n)g(l,n)\right) = 0$. For the remaining fourth term, we first note that, by Condition 3,

$$\begin{aligned}\mathbb{E}\left(\left|\frac{x(k,n)}{n}\right|^{2+\delta}\right) &= \frac{1}{n^{2+\delta}}\mathbb{E}\left(\left|\sum_{i=1}^{n-k} a_i a_{i+k}\right|^{2+\delta}\right) \\ &\le \frac{1}{n^{2+\delta}}\mathbb{E}\left((n-k)^{2+\delta-1}\sum_{i=1}^{n-k}|a_i a_{i+k}|^{2+\delta}\right)\end{aligned}$$

$$\leq \frac{1}{n}\sum_{i=1}^{n-k}\mathbb{E}\left(|a_i a_{i+k}|^{2+\delta}\right)$$

$$\leq \frac{n-k}{n}\sup_{i,j}\mathbb{E}\left(|a_i a_j|^{2+\delta}\right) < \infty.$$

As a result, we have $\sup_n \mathbb{E}\left(|g(k,n)|^{2+\delta}\right) < \infty$ uniformly for all k. By *Hölder's inequality*,

$$\mathbb{E}(|g(k,n)g(l,n)\mathbb{1}_{C_\epsilon(n)}|) \leq (\mathbb{E}(|g(k,n)|^{2+\delta}))^{\frac{1}{2+\delta}}(\mathbb{E}(|g(l,n)|^{2+\delta}))^{\frac{1}{2+\delta}}\mathbb{E}(\mathbb{1}_{C_\epsilon(n)}^{\frac{2+\delta}{\delta}})^{\frac{\delta}{2+\delta}}$$

$$\leq \left(\sup_n \mathbb{E}\left(|g(k,n)|^{2+\delta}\right)\right)^{\frac{2}{2+\delta}}(\mathbb{P}(C_\epsilon(n)))^{\frac{\delta}{2+\delta}} \to 0.$$

In summary, we have $\mathbb{E}(r_{k,n}\cdot r_{l,n}) \to 0$, and so $\mathrm{Cov}(r_{k,n}, r_{l,n}) \to 0$ as $n \to \infty$. Consequently, $r_{k,n}$ and $r_{l,n}$ are asymptotically uncorrelated for $k \neq l$. Furthermore, if $r_{k,n}$ and $r_{l,n}$, after rescaling, are asymptotically normal, then they will also be independent asymptotically.

Next, we investigate the asymptotic distribution of the following term for an arbitrary $k = 1, \ldots, K$:

$$\frac{n(n+2)}{n-k}r_{k,n}^2 = \left(1+\frac{2}{n}\right)\frac{\frac{x^2(k,n)}{n-k}}{\left(\frac{y(n)}{n}\right)^2}.$$

Recall that $\frac{y_n}{n}$ converges to σ^2 in probability. As for the numerator, we note that $\mathbb{E}(a_i a_{i+k} \mid \mathcal{F}_i) = 0$ by a simple application of the tower property, hence the $a_i a_{i+k}$'s form another time series of martingale differences. Then, by recalling that $x(k,n) = \sum_{i=1}^{n-k} a_i a_{i+k}$ with mean 0 and variance $(n-k)\sigma^4$, an application of the martingale central limit theorem[23] implies that $\frac{x(k,n)}{\sigma^2\sqrt{n-k}}$ converges to $\mathcal{N}(0,1)$ as $n \to \infty$. Together with Slutsky's theorem, we further see that the asymptotic distribution of $\frac{n(n+2)}{n-k}r_{k,n}^2$ is a chi-square distribution χ_1^2 with 1 degree of freedom. Consequently, as $Q(K)$ is the sum of $\frac{n(n+2)}{n-k}r_{k,n}^2$'s for $k = 1, \ldots, K$, which are asymptotically independent and follow the same asymptotic distribution of χ_1^2, it is asymptotically chi-square distributed with K degrees of freedom as $n \to \infty$, by virtue of the property of chi-square distributions mentioned in Chapter 1. For a more rigorous treatment and detailed analysis on this part, we refer the reader to two recent works of the authors [7, 8].

[23] The martingale central limit theorem extends the classical central limit theorem in the sense that the sequence of sample means of a sequence of martingale differences converges in distribution to a normal distribution. We refer to Chapter 6 of [35] for detailed discussions.

*7.A.2 Differential Equation Analogue of GARCH(1, 1)

Recall that the GARCH(1, 1) process is

$$\sigma_{n+1}^2 = \gamma V_L + \alpha u_n^2 + \beta \sigma_n^2,$$

where $\alpha + \beta + \gamma = 1$. Hence, at the n-th time unit given $\sigma_n^2 = \sigma_n^2$,

$$\sigma_{n+1}^2 - \sigma_n^2 = \gamma(V_L - \sigma_n^2) + \alpha(u_n^2 - \sigma_n^2).$$

Using the iid random variables $z_i = u_i/\sigma_i \sim \mathcal{N}(0, 1)$ for all i, we have

$$\sigma_{n+1}^2 - \sigma_n^2 = \gamma(V_L - \sigma_n^2) + \alpha\sigma_n^2(z_n^2 - 1). \tag{7.52}$$

Note that $\mathbb{E}(z_n^2) = 1$ and $\mathrm{Var}(z_n^2) = 2$.

Now, we let δt be the uniform step size between two time instances. Define $V_t := \sigma_n^2$, such that $t = n \cdot \delta t$ is the present time at the n-th instance. Then (7.52) can be rewritten by taking care of the wait time being δt now,

$$V_{t+\delta t} - V_t = \gamma(V_L - V_t)\delta t + \alpha V_t \sqrt{2} w_t \sqrt{\delta t},$$

where w_t's are iid random variables with common mean 0 and common variance 1. As a result, the term $w_t\sqrt{\delta t}$ has mean 0 and variance δt. As we pass δt to dt, $w_t\sqrt{\delta t}$ tends to a normal distribution with mean 0 and variance dt by invoking the central limit theorem. Thus in the limit, $w_t\sqrt{\delta t}$ becomes an infinitesimal increment dz_t where $\{z_t\}$ is a standard Brownian motion. To this end, we arrive at the following SDE:

$$dV_t = \gamma(V_L - V_t)dt + \alpha V_t \sqrt{2} dz_t. \tag{7.53}$$

Indeed, one can verify that this is the diffusion limit of the GARCH(1, 1) model, also see [25].

Taking the expectation on both sides of (7.53), we obtain the following ordinary differential equation:

$$d\mathbb{E}(V_t) = \gamma(V_L - \mathbb{E}(V_t))dt,$$

whose solution is

$$\mathbb{E}(V_t) = V_L + (V_0 - V_L)e^{-\gamma t}.$$

Finally, by noting that $\ln(1 + x) \approx x$ for small x,

$$\ln(\alpha + \beta) = \ln(1 - \gamma) \approx -\gamma$$

for small enough γ, we arrive at the following approximation:

$$\mathbb{E}(V_t) \approx V_L + (V_0 - V_L)(\alpha + \beta)^t,$$

which agrees with (7.17).

*7.A.3 Roots Table for φ from the Optimal Buying / Selling Problem

$\varepsilon\backslash\lambda$	−2	−0.5	−0.1	−0.05	0.05	0.1	0.5	2	5	8	11	14	17	18
0.01	(0.0012394, 0.000077468)	(0.0048717, 0.000073706)	(0.024597, 0.000072784)	(0.049391, 0.000726706)	0.00007	0.00007	0.00007	0.00007	0.00006	0.00006	0.00006	∞	∞	1
0.025	(0.00342662 0.00054335)	(0.01222597, 0.00047202)	(0.06111955, 0.00045704)	(0.12280604, 0.00045525)	0.00045	0.00045	0.00044	0.00039	0.00034	0.00029	∞	∞	1	∞
0.05	(0.00696065, 0.00291224)	(0.02511279, 0.00197109)	(0.12179341, 0.00184264)	(0.24447822, 0.00182815)	0.0018	0.00179	0.00169	0.00141	0.00109	0.0009	∞	∞	∞	∞
0.1	∞	(0.05482597, 0.00869366)	(0.24385397, 0.00749038)	(0.48717365, 0.00737058)	0.00715	0.00704	0.00632	0.00469	0.00323	0.00251	0.00208	0.00178	0.00156	0.0015
0.5	∞	∞	(1.37064935, 0.21734143)	(2.51127947, 0.19710883)	0.16864	0.15798	0.10869	0.05509	0.02975	0.02087	0.01621	0.01331	0.01131	0.01078
2	∞	∞	∞	(11.13704537, 4.65957957)	2.25564	1.87774	0.8814	0.33386	0.15771	0.10477	0.07879	0.06327	0.05291	0.05018
5	∞	∞	∞	∞	10.86943	8.07636	2.9753	0.98571	0.43855	0.28437	0.21047	0.13514	0.10593	0.18644
8	∞	∞	∞	∞	22.97198	16.09064	5.34173	1.67626	0.72799	0.41484	0.48675	0.12947	0.40243	0.37404
11	∞	∞	∞	∞	37.26751	25.14277	7.84438	2.3835	1.01869	0.67756	0.75689	0.72077	0.70285	0.64736
14	∞	∞	∞	∞	53.11932	34.89896	10.43237	3.10017	1.05948	1.07615	0.28234	0.97823	1.01014	0.49655
17	∞	∞	∞	∞	70.15675	45.17692	13.07982	3.82293	1.66686	1.92159	0.17141	1.31603	0.12799	1.28471
21	∞	∞	∞	∞	94.31501	59.502	16.67658	4.79301	2.56704	2.72787	1.1773	0.07906	0	0
22	∞	∞	∞	∞	100.5711	63.17526	17.58526	5.03641	2.7011	2.69477	0.16385	1.5629	0	0

REFERENCES

1. Anderson, R.L. (1942). Distribution of the serial correlation coefficient. *The Annals of Mathematical Statistics*, 13(1), 1–13.
2. Black, F., and Scholes, M. (1973). The pricing of options and corporate liabilities. *The Journal of Political Economy*, 81(3), 637–654.
3. Bollerslev, T. (1986). Generalized autoregressive conditional heteroskedasticity. *Journal of Econometrics*, 31(3), 307–327.
4. Bonnans, J.F., and Shapiro, A. (2000). *Perturbation Analysis of Optimization Problems.* Springer.
5. Box, G.E., Jenkins, G.M., Reinsel, G.C., and Ljung, G.M. (2015). *Time Series Analysis: Forecasting and Control* (5th ed.). Wiley.
6. Box, G.E.P., and Pierce, D.A. (1970). Distribution of residual autocorrelations in autoregressive-integrated moving average time series models, *Journal of the American Statistical Association*, 65(332), 1509–1526.
7. Chan, K.C.G., Cheng, P.H., Ling, H.K., Or, M.K.B., and Yam, S.C.P. (2024+). Test for network independence with distance correlation. Working Paper.
8. Chan, K.C.G., Han, J., Kennedy, A.P., and Yam, S.C.P. (2022). Testing network autocorrelation without replicates. *PLoS ONE*, 17(11), e0275532.
9. Dai, M., Jin, H., Zhong, Y., and Zhou, X.Y. (2010). Buy low and sell high. In C. Chiarella and A. Novikov (Eds.), *Contemporary Quantitative Finance* (pp. 317–333). Springer.
10. Du Toit, J., and Peskir, G. (2009). Selling a stock at the ultimate maximum. *The Annals of Applied Probability*, 19(3), 983–1014.
11. Engle, R.F. (1982). Autoregressive conditional heteroscedasticity with estimates of the variance of United Kingdom inflation. *Econometrica*, 50(4), 987–1007.
12. Engle, R.F. (2002). Dynamic conditional correlation: a simple class of multivariate generalized autoregressive conditional heteroskedasticity models. *Journal of Business and Economic Statistics*, 20(3), 339–350.
13. Engle, R.F., and Sheppard, K. (2001). *Theoretical and Empirical Properties of Dynamic Conditional Correlation Multivariate GARCH* (No. w8554). National Bureau of Economic Research.
14. Glosten, L.R., Jagannathan, R., and Runkle, D.E. (1993). On the relation between expected value and volatility of the nominal excess return on stocks. *The Journal of Finance*, 48(50), 1779–1801.
15. Hamilton, J.D. (2020). *Time Series Analysis.* Princeton University Press.
16. Hui, E.C.M., Wright, J.A., and Yam, S.C.P. (2014). Calendar effects and real estate securities. *The Journal of Real Estate Finance and Economics*, 49(1), 91–115.
17. Hui, E., Yam, S.C.P., Wright, J., and Chan, K. (2014). Shall we buy and hold? Evidence from Asian real estate markets. *Journal of Property Investment & Finance*, 32(2), 168–186.
18. J.P. Morgan (1996). *Risk Metrics: Technical Document.* New York.
19. Kirkpatrick II, C.D., and Dahlquist, J.A. (2010). *Technical Analysis: The Complete Resource for Financial Market Technicians.* Financial Times/Prentice Hall.
20. Li, W.K., and Mak, T.K. (1994). On the squared residual autocorrelations in non-linear time series with conditional heteroskedasticity. *Journal of Time Series Analysis*, 15(6), 627–636.
21. Ling, S., and Li, W.K. (1997). On fractionally integrated autoregressive moving-average time series models with conditional heteroscedasticity. *Journal of the American Statistical Association*, 92(439), 1184–1194.
22. Lo, A.W., and Hasanhodzic, J. (2010). *The Evolution of Technical Analysis: Financial Prediction from Babylonian Tablets to Bloomberg Terminals.* Wiley.

23. Murphy, J.J. (1999). *Technical Analysis of the Financial Markets: A Comprehensive Guide to Trading Methods and Applications*. Penguin.
24. Nelder, J.A., and Mead, R. (1965). A simplex method for function minimization. *The Computer Journal*, 7(4), 308–313.
25. Nelson, D.B. (1990). ARCH models as diffusion approximations. *Journal of Econometrics*, 45(1–2), 7–38.
26. Ng, K.T.H., Wong, T.K., and Yam, S.C.P. (2024+). Selling high/buying low at a good chance. Working Paper.
27. Peskir, G., and Shiryaev, A. (2006). *Optimal Stopping and Free-Boundary Problems* (pp. 123–142). Birkhäuser.
28. Shiryaev, A.N. (2007). *Optimal Stopping Rules*. Springer Science & Business Media.
29. Shiryaev, A., Xu, Z., and Zhou, X.Y. (2008). Thou shalt buy and hold. *Quantitative Finance*, 8(8), 765–776.
30. Thorp, E.O. (2008). *The Kelly Criterion: Part I*. Wilmott Magazine.
31. Thorp, E.O. (2008). The Kelly criterion in blackjack sports betting, and the stock market. In *Handbook of Asset and Liability Management* (pp. 385–428). North-Holland.
32. Tse, Y.K., and Tsui, A.K.C. (2002). A multivariate generalized autoregressive conditional heteroscedasticity model with time-varying correlations. *Journal of Business & Economic Statistics*, 20(3), 351–362.
33. Van der Vaart, A.W., and Wellner, J.A. (1996). *Weak Convergence and Empirical Processes with Applications to Statistics*. Springer.
34. Van der Vaart, A.W. (2000). *Asymptotic Statistics* (Vol. 3). Cambridge University Press.
35. Varadhan, S.S. (2001). *Probability Theory* (No. 7). American Mathematical Society.
36. Yam, S.C.P., Yung, S.P., and Zhou, W. (2009). Two rationales behind the 'buy-and-hold or sell-at-once' strategy. *Journal of Applied Probability*, 46(3), 651–668.
37. Yam, S.C.P., Yung, S.P., and Zhou, W. (2012). Optimal selling time in stock market over a finite time horizon. *Acta Mathematicae Applicatae Sinica, English Series*, 28(3), 557–570.
38. Yam, S.C.P., Yung, S.P., and Zhou, W. (2013). A unified "Bang-Bang" principle with respect to R-invariant performance benchmarks. *Theory of Probability & Its Applications*, 57(2), 357–366.
39. Zhang, M. (2018). Time series: Autoregressive models – AR, MA, ARMA, ARIMA. University of Pittsburgh.

CHAPTER 8

Risk Measures, Extreme Values, and Copulae

Financial institutions usually manage hundreds, or even thousands, of portfolios of securities and derivatives every day. It is important to have a single number that summarizes the total risk in a portfolio of financial assets, so as to measure the overall potential contingencies to which the financial institution is exposed. *Value-at-Risk* (V@R) is an attempt to provide such an index (measurement). It was pioneered by J.P. Morgan [32] and has been adopted by the *Basel Committee* [7] in setting capital requirements for banks throughout the world. Closely related to V@R is *Expected Shortfall* (ES) which is utilized in accordance with *Basel III*[1]. We discuss these in this chapter along with other related issues such as backtesting, historical simulation, and Extreme Value theory, which provide a scientific approach on systematically studying very large tail values. Besides the conventional risk measures, more sophisticated topics on dependence modelling will be introduced here, including the use and properties of dependence copulae, as copulae are the most primitive mathematical object for describing the joint distributional behavior of multivariate risks.

8.1 VALUE-AT-RISK AND EXPECTED SHORTFALL

Consider a portfolio and let L be the loss in its value over a 1-day horizon. We say that the 1-day value-at-risk of such portfolio at the confidence level $p \in (0, 1)$, denoted by $\mathrm{V@R_L}(p)$, equals V if we are $100p\%$ certain that we shall not lose more than V dollars on the next day. Usually, the confidence level p is taken to be close to 1, such as $p = 0.99$. If we denote by F_L the cdf of the loss variable L, then $\mathrm{V@R_L}(p)$ is just the p-quantile of the loss distribution:

$$\mathrm{V@R_L}(p) = F_\mathrm{L}^{-1}(p) = \min\{l \in \mathbb{R} : F_\mathrm{L}(l) \geq p\}.$$

In particular, if F_L is continuous and strictly increasing, then $\mathrm{V@R_L}(p)$ is simply the unique solution to the equation $F_\mathrm{L}(l) = p$, also see Figure 8.1.

[1] Basel III: international regulatory framework for banks. Retrieved from https://www.bis.org/bcbs/basel3.htm.

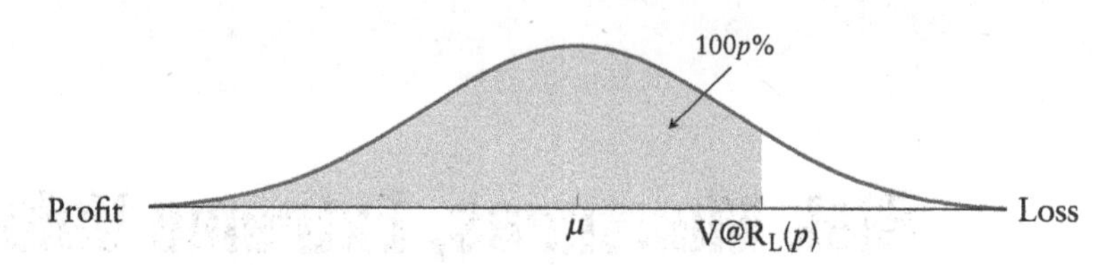

FIGURE 8.1 Illustration of $V@R_L(x)$.

Note that the N-day $V@R_L(x)$, denoted by $V@R_L(N, x)$, equals $\sqrt{N} \cdot V@R_L(x)$ assuming the daily changes are iid normal over the N-day period; this calculation also implicitly assumes that $\mathbb{E}(L) = 0$.

Example 8.1 Suppose that $L \sim \mathcal{N}(\mu, \sigma^2)$. By solving the equation $p = F_L(l) = \Phi(\frac{l-\mu}{\sigma}) = p$ for l, we obtain $V@R_L(p) = \mu + \sigma\Phi^{-1}(p)$.

Example 8.2 Assume that the daily loss distribution F_L of a portfolio is normally distributed with a mean of 0 and a standard deviation of \$20M. Then the 1-day $V@R_L(0.95) = (\$20M)(1.645) = 32.9M$ and the 1-day $V@R_L(0.99) = (\$20M)(2.326) = \$46.52M$. If we further assume that daily losses are independent, then the loss over an N-day period is normally distributed with mean 0 and standard deviation $\$20\sqrt{N}M$. Therefore, the 10-day $V@R_L(0.95)$ is $(\sqrt{10})(\$32.9M) = \$104.04M$ and the 252-day $V@R_L(0.99)$ is $(\sqrt{252})(\$46.52M) = \$738.48M$.

The V@R risk measure asks how bad the loss will be, while *Expected Shortfall* (ES) asks: if things do get bad, what is the expected loss? Mathematically speaking, the expected shortfall $ES_L(p)$ at confidence level p of the loss variable L is defined as

$$ES_L(p) := \frac{1}{1-p}\int_p^1 V@R_L(x)\,dx,$$

which is finite when L is integrable; see Figure 8.2 for an illustration. ES is also known as *conditional tail expectation* (CTE) for the following reason: if the loss variable L is continuous, then $ES_L(p)$ equals $\mathbb{E}(L \mid L > V@R(p))$; see also Section 2.3.4 of [33].

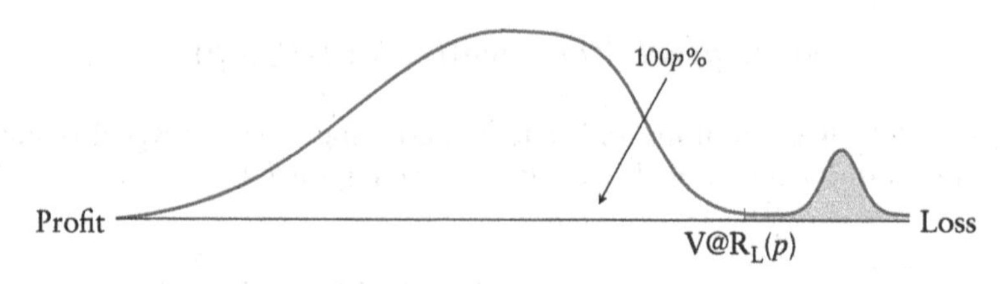

FIGURE 8.2 Illustration of $ES_L(p)$, which equals the area of the blue region.

Example 8.3 Consider a \$10 million 1-year loan which has a 1.25% chance of defaulting. If the loan defaults, the recovery of the loan principal is equally likely from 0% to 100%. Find the 1-year 99% V@R and the 1-year 99% ES, respectively.

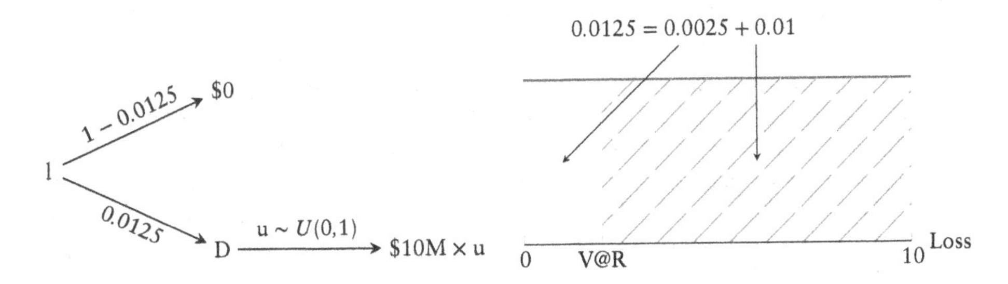

If the loan defaults (with the probability of 0.0125), let x be the percentage of loan principal recovered. Then $(0.0125)(x) = 0.01$, or $x = 0.8$, implies that 80% of \$10M = \$8M will be recovered with a probability of 0.01; or V@R(0.99) = \$10M − \$8M = \$2M. In other words, the probability of loss greater than \$2M is 80% of 1.25% = 1%.

Besides, the 1-year 99% ES is the expected loss given that the loss is greater than \$2M. Since the loss is uniformly distributed between \$2M to \$10M, the mean is \$6M. Note that a uniform random variable x conditional on the event that x > x still has a uniform distribution.

Example 8.4 Consider a portfolio consisting of two \$10 million 1-year loans as in Example 8.3. For simplicity, we further assume that if one loan defaults then it is certain that the other loan will not default. If a loan does not default, a profit of \$0.2 million is made. Find the 1-year 99% V@R and 1-year 99% ES of this portfolio.

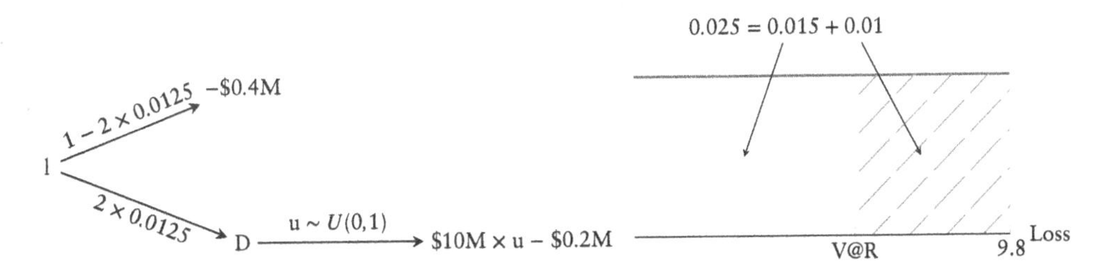

Each loan defaults with a probability of 1.25% and they never default together. Therefore a default occurs with a probability of 2.5%. Let x be the percentage of loan principal recovered. Then $(0.025)(x) = 0.01$, or $x = 0.4$, implies that 40% of \$10M = \$4M will be recovered with a probability of 0.01. However, a profit of \$0.2M is made on the other loan, hence the 1-year 99% V@R is \$10M − \$4M − \$0.2M = \$5.8M.

Moreover, the 1-year 99% ES of this portfolio is the expected loss given that the loss is greater than \$5.8M. Since the loss is uniformly distributed between \$5.8M and \$9.8M, the mean is \$7.8M.

8.2 BASEL ACCORDS AND RISK MEASURES

Risk measures is used for specifying capital requirements that must be added to a bank or a financial institution to provide a buffer for the underlying risk in order to satisfy regulators. In 1980, with the aim of monitoring and maintaining a healthy investment environment, supervisory authorities of Belgium, Canada, France, Germany, Italy, Japan, Luxembourg, Netherlands, Sweden, Switzerland, UK and US formed a committee for banking supervision, known as the *Bank for International Settlements* (BIS). The authorities of several other countries and territories have joined since. They meet regularly in the city of Basel in Switzerland (see the map in Figure 8.3), and in 1996 the Basel Committee issued an amendment to the 1988 Accord, which was then referred to as "BIS 88" [6] or "Basel I"[2]. In the amendment, banks are required to calculate the 10-day 99% V@R measure, and the Basel committee requires a bank to hold k times this V@R measure. This regulatory multiplier k is chosen on a bank-by-bank basis, and it must be at least 3. For a bank with an excellent well-tested V@R model, it is likely that $k = 3$; while for banks whose V@R model does not perform well during the last 250 days, k may be set as high as 4 [5].

Mathematically, there are some desirable properties that a risk measure $\rho(\cdot) : \mathcal{L}^2(\Omega; \mathbb{P}) \to \mathbb{R}$, defined on the space of random variables with finite second moments, should preferably have. In what follows, L_A and L_B are loss variables of portfolios A and B, respectively.

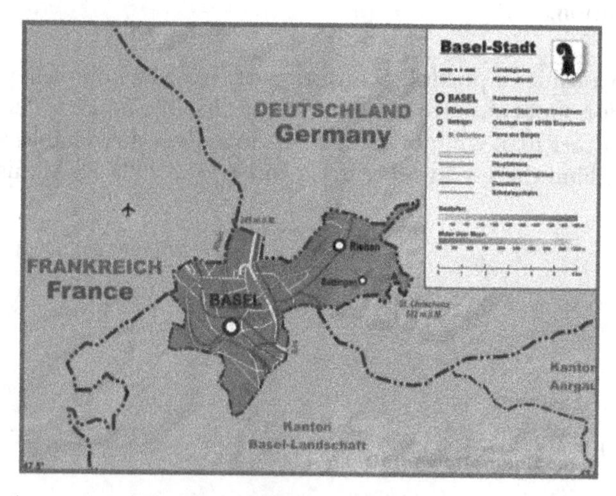

FIGURE 8.3 The location of Basel, Switzerland. Canton of Basel-Stadt map with cities and towns. Photo retrieved from https://ontheworldmap.com/switzerland/canton/basel-stadt/canton-of-basel-stadt-map-with-cities-and-towns.html.

[2] In June 2004, "Basel II" was released to replace "Basel I"; after the financial crisis in 2008, there is a need to update the "Basel II", and in December 2010, "Basel III" was released, which was implemented on 1 January 2023; see https://www.bis.org/bcbs/history.htm for more details.

1. *Monotonicity:* If the loss amount of a portfolio is less than that of another portfolio for every possible outcome from the sample space, its risk measure should be smaller than that of the second. $[L_A \le L_B \Rightarrow \rho(L_A) \le \rho(L_B)]$
2. *Translational invariance:* If a deterministic cash amount of K is added to a loss, its risk measure should go up by K. $[\rho(L_A + K) = \rho(L_A) + K]$
3. *Homogeneity:* Changing the size of a portfolio by a factor k should result in the risk measure being multiplied by k. $[\rho(kL_A) = k\rho(L_A)]$
4. *Subadditivity:* The risk measure of two portfolios after being merged should not be greater than the sum of the individual risk measures of the respective portfolio components. $[\rho(L_A + L_B) \le \rho(L_A) + \rho(L_B)]$

The first three conditions are natural enough while the fourth condition means that the diversification helps to reduce the inherent risks. Risk measures satisfying all four conditions are said to be *coherent*, see [3].

Recall that in Examples 8.3 and 8.4, the 1-year 99% V@R of the two loans separately is \$2M + \$2M = \$4M which is less than \$5.8M, the V@R of the portfolio. This implies that V@R does not always satisfy the subadditivity condition, yet it remains subadditive among *elliptical distributions*[3]. More precise results about asymptotic additivity will be discussed in Section 8.8. Meanwhile, the 1-year 99% ES of the two loans separately is \$6M + \$6M = \$12M, which is greater than \$7.8M, the ES of the portfolio. Indeed, ES generally fulfills the subadditivity condition [19].

Proposition 8.1 *(Subadditivity of ES)* Let x and y be two random variables, and let $ES_x(p)$ (resp. $ES_y(p)$) denote the expected shortfall of x (resp. y) at confidence level $p \in (0, 1)$. Then

$$ES_{x+y}(p) \le ES_x(p) + ES_y(p).$$

Proof. Note that $ES_x(p)$ admits the following expression (see [17]):

$$ES_x(p) = V@R_x(p) + \frac{1}{1-p}\mathbb{E}((x - V@R_x(p))_+).$$

By Lemma 3.2 of [19],

$$V@R_x(p) \in \underset{t \in \mathbb{R}}{arg\,min}\left\{t + \frac{1}{1-p}\mathbb{E}((x-t)_+)\right\}, \qquad p \in (0,1);$$

or to see this, assume for convenience that x is continuous with a density f, and consider the first-order condition of the function $g(t) := t + \frac{1}{1-p}\mathbb{E}((x-t)_+)$ as follows:

$$g'(t) = 1 + \frac{1}{1-p}\frac{d}{dt}\int_t^\infty \mathbb{P}(x > u)\,du = 1 - \frac{1}{1-p}\mathbb{P}(x > t) = 0,$$

[3] An elliptical distribution is any member of a broad family that generalizes the multivariate normal distribution, including the multivariate t-distribution, for example.

then the solution $t = \text{V@R}_\text{x}(p)$ is obvious. Based on this, we have

$$\text{ES}_\text{x}(p) = \min_{t\in\mathbb{R}} \left\{ t + \frac{1}{1-p}\mathbb{E}((\text{x}-t)_+) \right\}.$$

Denote $t_1 := \text{V@R}_\text{x}(p)$, $t_2 := \text{V@R}_\text{y}(p)$ and $t_0 := t_1 + t_2$. Using the inequality $(a+b)_+ \leq a_+ + b_+$ for any real numbers a and b, we have

$$\begin{aligned} \text{ES}_\text{x}(p) + \text{ES}_\text{y}(p) &= t_1 + t_2 + \frac{1}{1-p}\mathbb{E}((\text{x}-t_1)_+ + (\text{y}-t_2)_+) \\ &\geq t_0 + \frac{1}{1-p}\mathbb{E}((\text{x}+\text{y}-t_0)_+) \\ &\geq \min_{t\in\mathbb{R}} \left\{ t + \frac{1}{1-p}\mathbb{E}((\text{x}+\text{y}-t)_+) \right\} \\ &= \text{ES}_{\text{x+y}}(p). \end{aligned}$$

□

Although V@R is not coherent, it is easier to understand and backtest than ES. Therefore, V@R has become the most popular risk measure among regulators and risk managers, as suggested by Basel II. In fact, although coherent risk measures are a rather commonly used concept within academia, they were not as important in actual practice before Basel III. Now, in accordance with Basel III, the ES of the portfolio at a base liquidity horizon of 10 days has to be provided [8].

8.3 HISTORICAL SIMULATION (BOOTSTRAPPING)

In the previous (simplified) examples, we calculated V@R by assuming that the return of the portfolio follows a uniform or normal distribution for simplicity. Similar calculations can be performed for other distributions as long as the inverse of the distribution function can be computed analytically; but this is generally not viable in practice. There is another approach to calculating V@R based on historical data. Suppose that today is day n and we define v_i to be the value of a market variable (stock price or index) on day i, for $i = 0, 1, 2, \ldots, n$. The daily returns of the past n days provide us with n possible scenarios of the return on the day n. The predicted value $\hat{v}_{n+1}(i)$ of the value tomorrow v_{n+1}, based on the i-th scenario, is given by $v_n \times v_i / v_{i-1}$ for $i = 1, \ldots, n$. From these n estimates of v_{n+1}, we can then estimate its 1-day V@R as the empirical quantile of the set $\{\hat{v}_{n+1}(1), \ldots, \hat{v}_{n+1}(n)\}$. Let us illustrate this by the data in `stock_1999_2002.csv`. Suppose that we spend $40,000 for the purchase of stocks of HSBC, $30,000 on stocks of CLP and $30,000 on those of CK on 31 December 2002. The following programs compute the 1-day V@R of this portfolio using historical simulation, see Programmes 8.1 and 8.2 for Python and **R**, respectively.

```
1  import pandas as pd
2  import numpy as np
3
4  d = pd.read_csv("stock_1999_2002.csv", index_col=0)
5
6  x_n = d.iloc[-1,:]              # select the last obs
7  w = [40000, 30000, 30000]       # investment amount on each stock
8  p_0 = sum(w)                    # total investment amount
9  w_s = w/x_n                     # no. of shares bought at day n
10
11 h_sim = (d/d.shift(1) * x_n)[1:]
12 p_n = h_sim @ w_s               # portfolio value at day n
13
14 loss = p_0 - p_n                # loss
15 VaR_sim = np.quantile(loss, 0.99) # 1-day 99% V@R
16 print(VaR_sim)
```

```
1  3543.961475718493
```

Programme 8.1 Bootstrapping V@R of a portfolio consisting of HSBC, CLP, and CK in Python.

```
1  > d <- read.csv("stock_1999_2002.csv", row.names=1) # read in data file
2  > d <- as.ts(d)
3  >
4  > x_n <- as.vector(d[nrow(d),]) # select the last obs
5  > w <- c(40000, 30000, 30000)   # investment amount on each stock
6  > p_0 <- sum(w)                 # total investment amount
7  > w_s <- w/x_n                  # no. of shares bought at day n
8  >
9  > h_sim <- t(t(lag(d)/d) * x_n)
10 > p_n <- h_sim%*%w_s            # portfolio value at day n
11 >
12 > loss <- p_0 - p_n             # loss
13 > (VaR_sim <- quantile(loss, 0.99)) # 1-day 99% V@R
14      99%
15 3543.961
```

Programme 8.2 Bootstrapping V@R of a portfolio consisting of HSBC, CLP, and CK in R.

Note that the cost of the portfolio is \$100,000 based on the closing price on 31 December 2002. Then we compute the estimated stock prices under these n scenarios and save them in `h_sim`. Further, we compute the estimated portfolio value of `p_n` and the losses of these n scenarios, and finally the estimated 1-day 99% V@R is taken as the 99-th percentile, `VaR_sim = $3,543.961`.

The historical simulation (bootstrapping) method above puts equal weight to each historical observation. This method can be modified by putting different weights depending on the ratio of the $(n+1)$-th and i-th volatilities. To this end, we assume that the normalized daily returns $\frac{1}{\sigma_i}\frac{v_i - v_{i-1}}{v_{i-1}}$ follow a common distribution such as $\mathcal{N}(0,1)$ or t_ν. An estimate of the value of v_{n+1} can be obtained from the i-th scenario by matching $\frac{1}{\sigma_{n+1}}\frac{\hat{v}_{n+1}(i)-v_n}{v_n} = \frac{1}{\sigma_i}\frac{v_i - v_{i-1}}{v_{i-1}}$. This produces the following n estimates of v_{n+1}:

$$\hat{v}_{n+1}(i) = v_n \times \frac{v_{i-1} + (v_i - v_{i-1})\sigma_{n+1}/\sigma_i}{v_{i-1}} \qquad \text{with} \quad i = 1, \ldots, n, \tag{8.1}$$

where σ_i is the estimated volatility using the EWMA or GARCH(1, 1) model introduced in Chapter 7; see Programmes 8.3 and 8.4 for the corresponding calibrations in Python and **R**, respectively. Note that the estimated 1-day V@R under this approach, `VaR_GARCH` = `$2,538.037` (in Python), is much lower than `VaR_sim`; indeed, for these three major stocks in the Hong Kong market, their prices exhibited a similar downward trend to the Hang Seng Index near the end of 2002, while being less volatile compared with the earlier years, altogether this leads to a lower V@R estimate by the GARCH model compared with the bootstrapping approach. While the estimate can yield a better prediction, the underlying model may change completely within a short period, and by then the V@R calculated by the current model may become too low.

```
1  d = pd.read_csv("stock_1999_2002.csv", index_col=0)
2  u = np.diff(d, axis=0) / d.iloc[:-1, :] # Arithmetic return
3  x_n = d.iloc[-1,:]                 # select the last obs
4  w = [40000, 30000, 30000]          # investment amount on each stock
5  p_0 = sum(w)                       # total investment amount
6  w_s = w/x_n                        # no. of shares bought at day n
7
8  model_HSBC, _ = GARCH_11_MLE(u["HSBC"])
9  model_CLP, _ = GARCH_11_MLE(u["CLP"])
10 model_CK, _ = GARCH_11_MLE(u["CK"])
11
12 def Bootstrap_GARCH(model, u, t):
13     omega, alpha, beta = model
14     nu = [omega + alpha*np.mean(u**2) + beta*np.mean(u**2)]
15     for i in range(1, len(u)):
16         nu.append(omega + alpha*u[i-1]**2 + beta*nu[-1])
17
18     # Fitted variance on day n + 1
19     var_n_1 = omega + alpha*u[-1]**2 + beta*nu[-1]
20     t_i, t_1i, t_n = t[1:].values, t[:-1].values, t[-1]
21
22     return t_n*(t_1i+(t_i-t_1i)*np.sqrt(var_n_1/nu))/t_1i
23
```

```
24 h_sim1 = Bootstrap_GARCH(model_HSBC, u["HSBC"], d["HSBC"])
25 h_sim2 = Bootstrap_GARCH(model_CLP, u["CLP"], d["CLP"])
26 h_sim3 = Bootstrap_GARCH(model_CK, u["CK"], d["CK"])
27 h_sim = np.c_[h_sim1, h_sim2, h_sim3]
28 p_n = h_sim @ w_s                  # portfolio value at day n
29 loss_GARCH = p_0 - p_n             # loss
30 VaR_GARCH = np.quantile(loss_GARCH, 0.99)    # 1-day 99% VaR
31 print(VaR_GARCH)
```

```
1 2538.0371566229624
```

Programme 8.3 Unequal weighting bootstrapping V@R of a portfolio consisting of HSBC, CLP, and CK using GARCH(1,1) through the function `GARCH_11_MLE()` in Programme 7.19 via Python.

```
1  > library(fGarch)                  # load library "fGarch"
2  >
3  > d <- read.csv("stock_1999_2002.csv", row.names=1) # read in data file
4  > d <- as.ts(d)
5  > u <- (lag(d)-d)/d
6  > colnames(u) <- colnames(d)
7  > x_n <- as.vector(d[nrow(d),]) # select the last obs
8  > w <- c(40000, 30000, 30000)    # investment amount on each stock
9  > p_0 <- sum(w)                  # total investment amount
10 > w_s <- w/x_n                    # no. of shares bought at day n
11 >
12 > model_HSBC <- garchFit(~garch(1, 1), data=u[,"HSBC"], include.mean=F)
13 > model_CLP <- garchFit(~garch(1, 1), data=u[,"CLP"], include.mean=F)
14 > model_CK <- garchFit(~garch(1, 1), data=u[,"CK"], include.mean=F)
15 >
16 > Bootstrap_GARCH <- function(model, u, t){
17 +   var_n_1 <- coef(model)[1] + coef(model)[2]*u[length(u)]^2 +
18 +     coef(model)[3]*model@h.t[length(u)]
19 +   t_i <- t[-1]; t_1i <- t[-length(t)]; t_n <- t[length(t)]
20 +   t_n*(t_1i+(t_i-t_1i)*sqrt(var_n_1/model@h.t))/t_1i
21 + }
22 >
23 > h_sim1 <- Bootstrap_GARCH(model_HSBC, u[,"HSBC"], d[,"HSBC"])
24 > h_sim2 <- Bootstrap_GARCH(model_CLP, u[,"CLP"], d[,"CLP"])
25 > h_sim3 <- Bootstrap_GARCH(model_CK, u[,"CK"], d[,"CK"])
26 > h_sim <- cbind(h_sim1, h_sim2, h_sim3)
27 >
28 > p_n <- h_sim%*%w_s
29 > loss_GARCH <- p_0 - p_n      # loss
30 > (VaR_GARCH <- quantile(loss_GARCH, 0.99))    # 1-day 99% VaR
31      99%
32 2538.814
```

Programme 8.4 Unequal weighting bootstrapping V@R of a portfolio consisting of HSBC, CLP, and CK using GARCH(1,1) in **R**.

8.4 STATISTICAL MODEL BUILDING APPROACH

Another approach for calculating V@R is based on the assumption that the corresponding relative returns of the p market variables $\mathrm{u}_{i,j} = \frac{\mathrm{v}_{i,j}-\mathrm{v}_{i-1,j}}{\mathrm{v}_{i-1,j}}$ at time i, and for $j = 1, \ldots, p$, follow a p-variate independently and identically distributed normal random vector, i.e. $\mathbf{u}_i \overset{iid}{\sim} \mathcal{N}_p(\mathbf{0}, \mathbf{\Sigma})$. Assume that we have $\$\boldsymbol{\omega} = (\omega_1, \ldots, \omega_p)^\top$ invested on these p market variables. We bought, for $j = 1, \ldots, p$, $\frac{\omega_j}{\mathrm{v}_{0,j}}$ shares of the j-th market variable at time 0. When the market price changes from $\mathrm{v}_{0,j}$ to $\mathrm{v}_{1,j}$ at time 1, the market value of the newly acquired stock becomes $\omega_j \frac{\mathrm{v}_{1,j}}{\mathrm{v}_{0,j}}$ and the monetary change is $\omega_j \frac{\mathrm{v}_{1,j}}{\mathrm{v}_{0,j}} - \omega_j = \omega_j \frac{\mathrm{v}_{1,j}-\mathrm{v}_{0,j}}{\mathrm{v}_{0,j}} = \omega_j \mathrm{u}_{1,j}$. Hence the change in the portfolio value on the i-th period is $\Delta \mathrm{p}_i = \boldsymbol{\omega}^\top \mathbf{u}_i = \omega_1 \mathrm{u}_{i,1} + \ldots + \omega_p \mathrm{u}_{i,p}$ with mean and variance respectively given by

$$\mathbb{E}(\Delta \mathrm{p}_i) = \boldsymbol{\omega}^\top \mathbb{E}(\mathbf{u}_i) = 0 \quad \text{and} \quad \mathrm{Var}(\Delta \mathrm{p}_i) = \boldsymbol{\omega}^\top \mathrm{Var}(\mathbf{u}_i) \boldsymbol{\omega} = \boldsymbol{\omega}^\top \mathbf{\Sigma} \boldsymbol{\omega}.$$

Therefore, we estimate the standard deviation of the random change $\Delta \mathrm{p}$ by $\sqrt{\boldsymbol{\omega}^\top S \boldsymbol{\omega}}$, where S is the sample covariance matrix of $\mathbf{u} = (\mathrm{u}_1, \ldots, \mathrm{u}_p)^\top$. The loss variable is $\mathrm{L} = -\Delta \mathrm{p}$. Since $\mathrm{z} := \frac{\mathrm{L}-0}{\sqrt{\boldsymbol{\omega}^\top S \boldsymbol{\omega}}} \sim \mathcal{N}(0, 1)$, the 1-day 99%-V@R based on the normal model above, denoted by $\mathrm{V@R}_N(0.99)$, satisfies the equation:

$$\mathbb{P}(\mathrm{L} \leq \mathrm{V@R}_N(0.99)) = \mathbb{P}\left(\mathrm{z} \leq \frac{\mathrm{V@R}_N(0.99) - 0}{\sqrt{\boldsymbol{\omega}^\top S \boldsymbol{\omega}}}\right) = 0.99,$$

hence

$$\mathrm{V@R}_N(0.99) = \Phi^{-1}(0.99) \times \sqrt{\boldsymbol{\omega}^\top S \boldsymbol{\omega}} = 2.32635 \times \sqrt{\boldsymbol{\omega}^\top S \boldsymbol{\omega}}.$$

Note that $\sqrt{\boldsymbol{\omega}^\top S \boldsymbol{\omega}}$ can also be obtained by directly computing the sample standard deviation of $\Delta \mathrm{p}$. Let us illustrate this using the `stock_1999_2002.csv` example again with the Python and **R** codes in Programmes 8.5 and 8.6, respectively.

```
1  from scipy.stats import norm
2
3  d = pd.read_csv("stock_1999_2002.csv", row.names=1)
4  u = np.diff(d, axis=0) / d.iloc[:-1, :] # Arithmetic return
5  S = np.cov(u, rowvar=False)              # sample cov. matrix
6  w = np.array([40000, 30000, 30000]) # investment amount on each stock
7  delta_p = u @ w                          # Delta P
8  sd_p = np.std(delta_p, ddof=1)  # sample sd of portfolio (empirical)
9  VaR_N = norm.ppf(0.99)*sd_p              # 1-day 99% V@R with normal
10 print(VaR_N)
11 print(norm.ppf(0.99)*np.sqrt(w.T @ S @ w)) # z_0.99 × √(ω^T Sω)
```

```
1  3143.7528835999915
2  3143.752883599992
```

Programme 8.5 Normal modelling for V@R of a portfolio consisting of HSBC, CLP, and CK in Python.

```
1  > d <- read.csv("stock_1999_2002.csv", row.names=1)
2  > t <- as.ts(d)
3  > u <- (lag(t)-t)/t
4  > S <- var(u)                       # sample cov. matrix
5  > w <- c(40000, 30000, 30000)      # investment amount on each stock
6  > delta_p <- u%*%w                 # Delta P
7  > sd_p <- sd(delta_p)     # sample sd of portfolio (empirical)
8  > (VaR_N <- qnorm(0.99)*sd_p)      # 1-day 99% V@R with normal
9  [1] 3143.753
10 > qnorm(0.99)*sqrt(t(w)%*%S%*%w)  # z_{0.99} × √(ω^T S ω)
11          [,1]
12 [1,] 3143.753
```

Programme 8.6 Normal modelling for V@R of a portfolio consisting of HSBC, CLP, and CK in R.

The 1-day 99% V@R using the normal model is `VaR_N = $3,143.753` which is less than the historically simulated `VaR_sim = $3,543.961`. The normality assumption may not be valid since most returns have heavier tails than those specified by the normal distribution. Hence, the V@R under the normality assumption mostly falls small. Recall, from Section 3.2, that we can model the return by a Student's t-distribution t_ν with ν degrees of freedom. Notice that for $\mathrm{t} \sim t_\nu$ where $\nu > 2$, $\mathrm{Var(t)} = \frac{\nu}{\nu-2}$. Therefore, we may assume $\mathrm{t} := \frac{\mathrm{L}-0}{\sqrt{\omega^\top S\omega}} \times \sqrt{\frac{\nu}{\nu-2}} \sim t_\nu$, and the 1-day 99%-V@R, denoted as $\mathrm{V@R_t}$, satisfies the equation:

$$\mathbb{P}(\mathrm{L} \le \mathrm{V@R_t}) = \mathbb{P}\left(\mathrm{t} \le \frac{\mathrm{V@R_t} - 0}{\sqrt{\omega^\top S\omega}} \times \sqrt{\frac{\nu}{\nu-2}}\right) = 0.99,$$

which gives

$$\mathrm{V@R_t} = T_\nu^{-1}(0.99) \cdot \sqrt{\omega^\top S\omega} \cdot \sqrt{\frac{\nu-2}{\nu}},$$

where $T_\nu^{-1}(\cdot)$ is the inverse cdf of the standard Student's t-distribution with ν degrees of freedom.

Let us compute the sample excess kurtosis $\hat{\zeta}_2$ of Δp, as then we estimate the degrees of freedom by $\frac{6}{\hat{\zeta}_2+4}$, rounded to the nearest integer; see Programmes 8.7 in Python and 8.8 in R.

```
1 from scipy.stats import t
2 
3 ku = sum((delta_p/sd_p)**4)/len(delta_p)-3
4 nu = round(6/ku+4)
5 VaR_t = t.ppf(0.99, nu)*sd_p   # 1-day 99% V@R with t
6 print(VaR_t)
```

```
1 3424.0045816435672
```

Programme 8.7 Student's t-modelling for V@R of a portfolio consisting of HSBC, CLP, and CK in Python.

```
1 > ku <- sum((delta_p/sd_p)^4)/length(delta_p)-3
2 > nu <- round(6/ku+4)
3 > (VaR_t <- qt(0.99, nu)*sd_p*sqrt((nu-2)/nu)) # 1-day 99% V@R with t
4 [1] 3424.005
```

Programme 8.8 Student's t-modelling for V@R of a portfolio consisting of HSBC, CLP, and CK in R.

The 1-day 99% V@R using Student's t-model is `VaR_t = $3,424.005`, which is larger than the one obtained by using the normal model (`VaR_N = $3,143.753`), but smaller than the historically simulated value `VaR_sim = $3,543.961`.

8.5 USE OF EXTREME VALUE THEORY

In modelling empirical data, one useful law is the *Power Law*, first properly proposed by *Vilfredo Pareto* in 1897 [37]. This law is said to hold if the survival function of a random variable y is proportional to a power function: $\mathbb{P}(y > x) \propto x^{-\alpha}$ for some $\alpha > 0$; it also provides a good model for estimating the tail probability of rare events. Let F be the cdf of a random variable x, i.e., $\mathbb{P}(x \le u) = F(u)$, then given $x > u$, the conditional probability of $u < x \le u + y$ is

$$F_u(y) := \mathbb{P}(x - u \le y \mid x > u) = \frac{F(u+y) - F(u)}{1 - F(u)}.$$

It has been shown ([4, 23, 38]) that for a large class of distributions F, $F_u(y) \to G_{\xi,\beta}(y)$ uniformly in $y > 0$ as $u \to \infty$, where

$$G_{\xi,\beta}(y) = \begin{cases} 1 - \left(1 + \xi\frac{y}{\beta}\right)^{-1/\xi}, & \xi \neq 0, \\ 1 - \exp\left(-\frac{y}{\beta}\right), & \xi = 0, \end{cases}$$

and is called the *generalized Pareto distribution* with a shape parameter ξ and a scale parameter $\beta > 0$, in which $y \ge 0$ when $\xi \ge 0$, or $0 \le y \le -\frac{\beta}{\xi}$ when $\xi < 0$. For common financial data, $\xi > 0$ and it falls in the range from 0.1 to 0.4, see [27] for details. For more introduction to extreme value theory (EVT), we refer to [25, 40].

Suppose that we sort n financial observations $x_1, \ldots, x_n$ in descending order, and there are n_u observations greater than u. Assume that the previous theorem of *Gnedenko–Pickands–Balkema–de Haan* ("*Gnedenko's theorem*" in short hereafter) is applicable to the underlying distribution of this sample, then for a large enough u, we can estimate $\mathbb{P}(x > u + y \mid x > u)$ by $1 - G_{\xi,\beta}(y)$, and $1 - F(u)$ by $\frac{n_u}{n}$. As a result, the unconditional probability is approximately given by

$$\mathbb{P}(x > u + y) \approx (1 - F(u))\left(1 - G_{\xi,\beta}(y)\right) = \frac{n_u}{n}\left(1 + \xi\frac{y}{\beta}\right)^{-1/\xi}, \quad \text{for all } y > 0,$$

which implies that for large $x > u$,

$$\mathbb{P}(\mathrm{x} > x) \approx \frac{n_u}{n}\left(1 + \xi\frac{x-u}{\beta}\right)^{-1/\xi},$$

and so

$$\mathbb{P}(\mathrm{x} \leq x) \approx 1 - \frac{n_u}{n}\left(1 + \xi\frac{x-u}{\beta}\right)^{-1/\xi}. \tag{8.2}$$

In particular, if we take $u = \frac{\beta}{\xi}$, then $\mathbb{P}(\mathrm{x} > x) \approx \frac{n_u}{n}\left(\frac{\xi}{\beta}\right)^{-1/\xi} x^{-1/\xi}$, which satisfies the usual Power Law. If the random variable x represents the loss in value of a portfolio over a 1-day horizon, then from (8.2), we can calculate the 1-day V@R at the confidence level p by solving the following equation for x:

$$p = 1 - \frac{n_u}{n}\left(1 + \xi\frac{x-u}{\beta}\right)^{-\frac{1}{\xi}},$$

which gives

$$\mathrm{V@R}_{\mathrm{x}}(p) = u + \frac{\beta}{\xi}\left[\left(\frac{n(1-p)}{n_u}\right)^{-\xi} - 1\right].$$

Since the density of the generalized Pareto distribution is

$$g_{\xi,\beta}(y) = \frac{dG_{\xi,\beta}(y)}{dy} = \frac{1}{\beta}\left(1 + \frac{\xi y}{\beta}\right)^{-1/\xi-1},$$

the log-likelihood function of the sample $x_1 > \cdots > x_{n_u}$ is

$$\begin{aligned} l(\xi,\beta) &= \sum_{i=1}^{n_u}\left(-\left(\frac{1}{\xi}+1\right)\ln\left[1 + \frac{\xi(x_i-u)}{\beta}\right] - \ln\beta\right) \\ &= -n_u\ln\beta - \left(\frac{1}{\xi}+1\right)\sum_{i=1}^{n_u}\ln\left[1 + \frac{\xi(x_i-u)}{\beta}\right]. \end{aligned}$$

Accordingly, we can find the MLEs $\hat{\xi}$ and $\hat{\beta}$ by maximizing this log-likelihood function, or simply solving the following equations:

$$\frac{\partial l(\xi,\beta)}{\partial\xi} = \frac{1}{\xi^2}\sum_{i=1}^{n_u}\ln\left[1 + \frac{\xi(x_i-u)}{\beta}\right] - \left(\frac{1}{\xi}+1\right)\sum_{i=1}^{n_u}\frac{x_i-u}{\beta+\xi(x_i-u)} = 0,$$

$$\frac{\partial l(\xi,\beta)}{\partial\beta} = -\frac{n_u}{\beta} + \left(\frac{1}{\xi}+1\right)\sum_{i=1}^{n_u}\frac{\xi(x_i-u)}{\beta^2+\beta\xi(x_i-u)} = 0,$$

and then obtain an estimate of the 1-day V@R of x at the confidence level p given by

$$\text{V@R}_x(p) = u + \frac{\hat{\beta}}{\hat{\xi}}\left[\left(\frac{n(1-p)}{n_u}\right)^{-\hat{\xi}} - 1\right]. \tag{8.3}$$

The only remaining question is the choice of the threshold value u. In practice, we should first apply the standardized transformation $z_i = \frac{x_i - \bar{x}}{s}$, set u to be close to the 0.95-quantile of the empirical distribution of the standardized sample $z_1 > \ldots > z_n$[4], and proceed to estimate $\text{V@R}_z(p)$ using the equations described above. An estimate of $\text{V@R}_x(p)$ will then be given by reversing the transformation: $\bar{x} + s \cdot \text{V@R}_z(p)$.

Let us illustrate this by continuing the previous example. Note that we use the `minimize()` function in the `scipy` library in Python in Programme 8.9 or the **R** built-in function `optim()` in **R** in Programme 8.10 to minimize the negative of the log-likelihood function; also see `help(optim)` in **R** for details.

```
1  from scipy.optimize import minimize
2
3  u = 3.2                                        # threshold value
4  m = np.mean(loss)                              # mean loss
5  s = np.std(loss)                               # sd loss
6  z = (loss-m)/s                                 # standardize loss
7  z_u = z[z>u]                                   # select z>u
8  n_u = len(z_u)                                 # no. of z>u
9  print(n_u)
10
11 def n_log_lik(theta, y):                       # theta=(xi, beta)
12     xi, beta = theta
13     return(len(y)*np.log(beta)+(1/xi+1)*sum(np.log(1+xi*y/beta)))
14
15 theta_0 = [0.2, 0.01]                          # initial theta_0
16 # min -ve log_likelihood
17 res = minimize(n_log_lik, theta_0, method='Nelder-Mead', args=(z_u-u,))
18 xi, beta = res.x                               # MLE theta=(xi, beta)
19 print(res.x)
20 print(-res.fun)                                # max value
21 q = 0.99
22 VaR = u+(beta/xi)*((len(z)*(1-q)/n_u)**(-xi)-1)
23 print(VaR)
24 VaR_EVT = m+VaR*s                              # 1day 99% V@R by EVT
25 print(VaR_EVT)
```

```
1 5
2 [0.70874035 0.32496192]
3 -2.9234321245340515
4 3.0254626292904656
5 4045.81659088324
```

Programme 8.9 EVT approach to V@R of a portfolio consisting of HSBC, CLP, and CK in Python.

[4] A favorable practice is to choose a value for u between 2.7 and 3.2.

```
1  > u <- 3.2                                # threshold value
2  > m <- mean(loss)                         # mean loss
3  > s <- sd(loss)                           # sd loss
4  > z <- (loss-m)/s                         # standardize loss
5  > z_u <- z[z>u]                           # select z>u
6  > (n_u <- length(z_u))                    # no. of z>u
7  [1] 5
8  >
9  > n_log_lik <- function(p, y){           # p=(xi, beta)
10 +      length(y)*log(p[2])+(1/p[1]+1)*sum(log(1+p[1]*y/p[2]))
11 + }
12 >
13 > p0 <- c(0.2, 0.01)                      # initial p0
14 > # min -ve log_likelihood
15 > res <- optim(p0, n_log_lik, y=(z_u-u))
16 > (p <- res$par)                          # MLE p=(xi, beta)
17 [1] 0.7189946 0.3196722
18 > -res$value                              # max value
19 [1] -2.895636
20 > q <- 0.99
21 > (VaR <- u+(p[2]/p[1])*(((1-q)*length(z)/n_u)^(-p[1])-1))
22 [1] 3.02885
23 > (VaR_EVT <- m+VaR*s)                    # 1day 99% V@R by EVT
24 [1] 4052.475
```

Programme 8.10 EVT approach to V@R of a portfolio consisting of HSBC, CLP, and CK in R.

Recall that the 1-day 99% V@R using the normal-model is `VaR_N = $3,143.753`; using the t-model it is `VaR_t = $3,424.005`; and using extreme value theory it is `VaR_EVT = $4,045.817` (in Python). A natural question is: which V@R is more appropriate? In general, EVT is theoretically the most rigorous, and should be preferred if there are abundant data, as EVT uses only data at the tail of the loss distribution. For small datasets, the t-model is preferred against the normal-model as loss distributions are more likely to have heavy tails. We can actually test these V@Rs using historical data to see which one is more reasonable; this procedure is called *backtesting*.

8.6 BACKTESTING

Calculation of V@R is based on the knowledge of the loss distribution. It is important to test how accurate the V@R estimate really is. An important reality check is back-testing. Let the 1-day V@R at the confidence level p be $\$V$. We say that an exception occurs if the loss in the portfolio value is greater than $\$V$ on a given day. If the V@R model is accurate, the probability of the loss in the portfolio value greater than $\$V$ on any given day is $q = 1 - p$. Suppose that over an n-day horizon we observe that m exceptions occur in these n days, where the proportion $\frac{m}{n} > q$. Should we reject the model as it produces an underestimate of the V@R? This can be tested formally by using the following binomial test [34]:

> H_0 (null hypothesis): Probability of an exception on any given day is q;
> H_1 (alternative hypothesis): Probability of an exception on any given day is greater than q.

Note that this is a one-sided test as we now observe that $\frac{m}{n} > q$, and we place greater attention on the underestimation of V@R and are less concerned about any overestimation. Under H_0, the probability of having at least m exceptions out of n days is

$$P_0 = \sum_{k=m}^{n} \binom{n}{k} q^k (1-q)^{n-k}. \tag{8.4}$$

We reject H_0 at the significance level of α if $P_0 < \alpha$, as a smaller q tends to give a smaller P_0; typically, we take $\alpha = 0.05$. Let us set $n = 250$ days, $p = 0.99$ so that $q = 0.01$. We compute P_0 for $m = 0, \ldots, 10$, in Python in Programme 8.11 and in **R** in Programme 8.12.

```
1 from scipy.stats import binom
2 
3 m = np.arange(11)
4 print(np.round(1-binom.cdf(m, 250, 0.01), 4))
```

```
1 [9.189e-01 7.142e-01 4.568e-01 2.419e-01 1.078e-01 4.120e-02
2  1.370e-02 4.000e-03 1.100e-03 3.000e-04 1.000e-04]
```

Programme 8.11 Computing the probability of (8.4) with $n = 250$ and $q = 0.01$ in Python.

```
1 > m <- 0:10
2 > round(1-pbinom(m, 250, 0.01), 4)
3 [1] 0.9189 0.7142 0.4568 0.2419 0.1078 0.0412 0.0137 0.0040
4 [9] 0.0011 0.0003 0.0001
```

Programme 8.12 Computing the probability of (8.4) with $n = 250$ and $q = 0.01$ in R.

From the outputs, it is clear that $P_0 > 0.05$ if $m < 5$. In practice, the 1986 BIS Amendment requires V@R models to be backtested. Banks should look at the number of exceptions m during the previous 250 days. If $m < 5$, then the riskiness of the portfolio is considered in the green zone. The capital requirement is then computed by multiplying the V@R with the regulatory multiplier $k = 3$. If m is 5, 6, 7, 8, or 9, the riskiness of the portfolio is considered in the yellow zone, and k is set at $3.4, 3.5, 3.65, 3.75$ and 3.85, respectively. Finally, if $m \geq 10$, k will be set to 4, and the riskiness of the portfolio is considered in the red zone; also see [5] for further discussions, and a summary is provided in Table 8.1. Figure 8.4 shows that k possesses a linear relationship with m in the yellow zone, and the other values of k in this zone can be calculated by linear interpolation from the values of k for m in the range of 4 to 10.

TABLE 8.1 1986 BIS V@R Amendment from [5].

Zone	Number of exceptions	Increase in scaling factor	Cumulative probability	Regulatory multiplier
Green Zone	0	0.00	8.11 %	3.00
	1	0.00	28.58 %	3.00
	2	0.00	54.32 %	3.00
	3	0.00	75.81 %	3.00
	4	0.00	89.22 %	3.00
Yellow Zone	5	0.40	95.88 %	3.40
	6	0.50	98.63 %	3.50
	7	0.65	99.60 %	3.65
	8	0.75	99.89 %	3.75
	9	0.85	99.97 %	3.85
Red Zone	10 or more	1.00	99.99 %	4.00

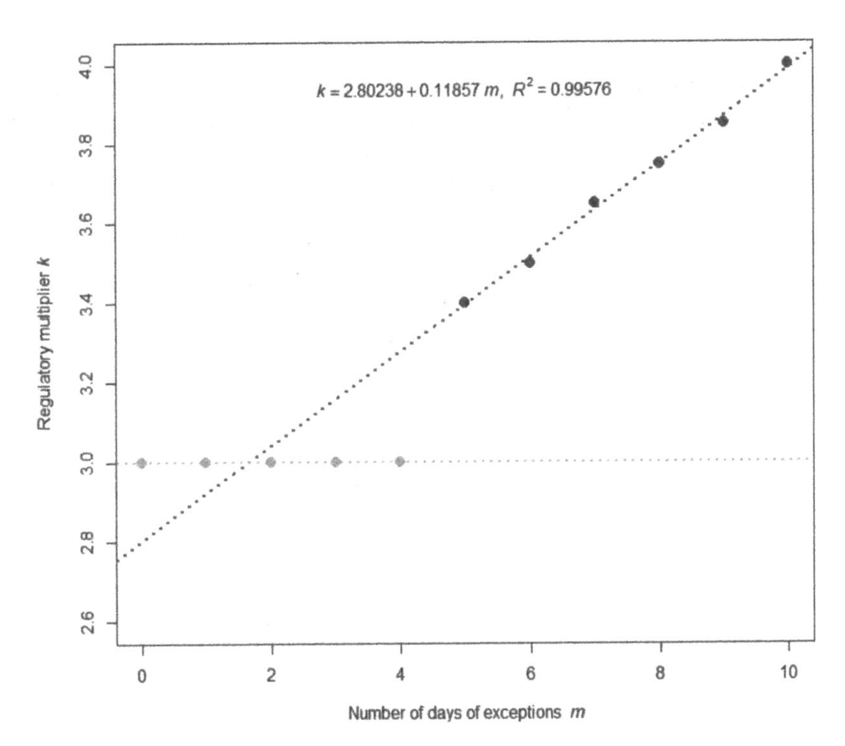

FIGURE 8.4 The fitted line of k against m using figures from Table 8.1.

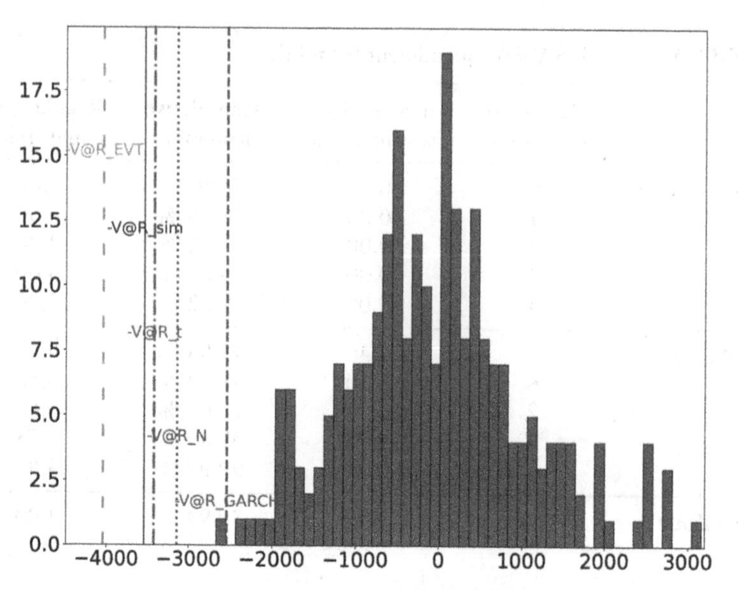

FIGURE 8.5 Histogram of profit and loss during the last 250 days in Python, generated in Programme 8.13.

Now let us continue with the data in `stock_1999_2002.csv` and backtest the 1-day 99% V@R using all the aforementioned V@R models. From the outputs in Programmes 8.13 (in Python) and 8.14 (in **R**), the numbers of exceptions in the past 250 days for all approaches are all less than 5, and as such, the regulatory multiplier should be set to 3 for each approach. In Figure 8.5 for Python and in Figure 8.6 for **R**, we depict the histogram of profit and loss over the previous 250-day period, and also the calibrated V@Rs under normal, *t*-, and bootstrapping with equal or unequal weighting models.

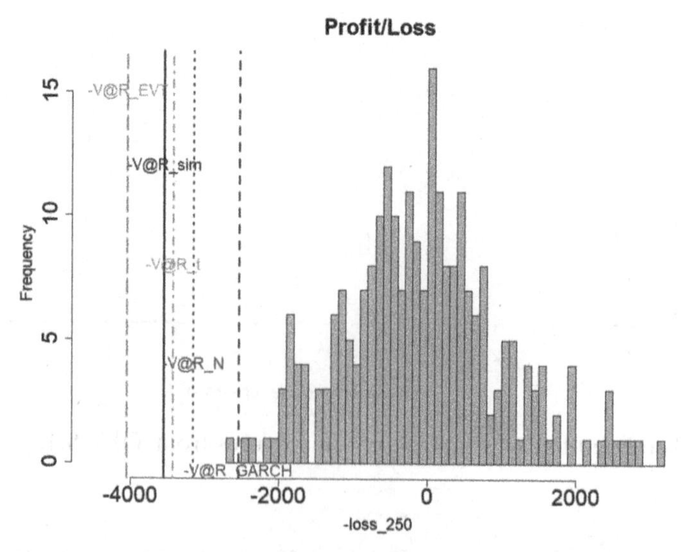

FIGURE 8.6 Histogram of profit and loss during the last 250 days in **R**, generated in Programme 8.14.

```
1  x_n = d.iloc[-1,:]                       # select the last obs
2  w = np.array([40000, 30000, 30000]) # investment amount on each stock
3  p_0 = sum(w)                             # total investment amount
4  w_s = w/x_n                              # no. of shares bought at day n
5
6  ns = 250                                 # 250 days
7  x_250 = d.iloc[-ns:,:]                   # recent 250 days
8  x_250 = d.iloc[n_250:n,:]                # recent 250 days
9  ps_250 = x_250 @ w_s.T                   # portfolio value
10 ps_250 = np.append(ps_250, p_0)          # add total amount
11 loss_250 = -np.diff(ps_250, axis=0)      # compute daily loss
12
13 print(sum(loss_250 > VaR_sim))           # no. of exceptions
14 print(sum(loss_250 > VaR_GARCH))
15 print(sum(loss_250 > VaR_N))
16 print(sum(loss_250 > VaR_t))
17 print(sum(loss_250 > VaR_EVT))
18
19 import matplotlib.pyplot as plt
20
21 plt.rcParams["figure.figsize"] = 13, 10
22 plt.rc('font', size=20); plt.rc('axes', titlesize=30, labelsize=30)
23 plt.rc('xtick', labelsize=25); plt.rc('ytick', labelsize=25)
24
25 plt.hist(-loss_250, ec='black', bins=50, alpha=0.8)
26 plt.xlim(-4500, 3200)
27 plt.axvline(x=-VaR_sim, c="blue")
28 plt.axvline(x=-VaR_GARCH, c="red", linestyle="--", linewidth=3)
29 plt.axvline(x=-VaR_N, c="green", linestyle="dotted", linewidth=3)
30 plt.axvline(x=-VaR_t, c="gray", linestyle="-.", linewidth=3)
31 plt.axvline(x=-VaR_EVT, c="orange", linestyle=(0, (5, 10)),
32             linewidth=3)
33
34 plt.text(-VaR_sim, 12, "-V@R_sim", ha='center', c="blue")
35 plt.text(-VaR_GARCH, 1.5, "-V@R_GARCH", ha='center', c="red")
36 plt.text(-VaR_N, 4, "-V@R_N", ha='center', c="green")
37 plt.text(-VaR_t, 8, "-V@R_t", ha='center', c="gray")
38 plt.text(-VaR_EVT, 15, "-V@R_EVT", ha='center', c="orange")
```

```
1  0
2  1
3  0
4  0
5  0
```

Programme 8.13 Summary of the numbers of exceptions using various V@R approaches in Python.

```
1  > x_n <- d[nrow(d),]                    # select the last obs
2  > w <- c(40000, 30000, 30000)           # investment amount on each stock
3  > p_0 <- sum(w)                         # total investment amount
4  > w_s <- w/x_n                          # no. of shares bought at day n
5  >
6  > ns <- 250                             # 250 days
7  > x_250 <- as.matrix(tail(d, ns))       # recent 250 days
8  > ps_250 <- x_250%*%t(w_s)              # portfolio value
9  > ps_250 <- c(ps_250, p_0)              # add total amount
10 > loss_250 <- ps_250[1:ns]-ps_250[2:(ns+1)] # compute daily loss
11 >
12 > sum(loss_250 > VaR_sim)               # no. of exceptions
13 [1] 0
14 > sum(loss_250 > VaR_GARCH)
15 [1] 1
16 > sum(loss_250 > VaR_N)
17 [1] 0
18 > sum(loss_250 > VaR_t)
19 [1] 0
20 > sum(loss_250 > VaR_EVT)
21 [1] 0
22 >
23 > par(mfrow=c(1,1), cex.lab=1.5, cex.axis=2, cex.main=2)
24 > hist(-loss_250, main="Profit/Loss", breaks=50, xlim=c(-4500, 3200))
25 > abline(v=-VaR_sim, col="blue", lwd=3, lty=1)
26 > abline(v=-VaR_GARCH, col="red", lwd=3, lty=2)
27 > abline(v=-VaR_N, col="green", lwd=3, lty=3)
28 > abline(v=-VaR_t, col="gray", lwd=3, lty=4)
29 > abline(v=-VaR_EVT, col="orange", lwd=3, lty=5)
30 >
31 > text(-VaR_sim, 12, "-V@R_sim", col="blue", cex=1.5)
32 > text(-VaR_GARCH, -0.35, "-V@R_GARCH", col="red", cex=1.5)
33 > text(-VaR_N, 4, "-V@R_N", col="green", cex=1.5)
34 > text(-VaR_t, 8, "-V@R_t", col="gray", cex=1.5)
35 > text(-VaR_EVT, 15, "-V@R_EVT", col="orange", cex=1.5)
```

Programme 8.14 Summary of the numbers of exceptions using various V@R approaches in R.

8.7 ESTIMATES OF EXPECTED SHORTFALL

Recall that when the loss random variable L is continuous, its expected shortfall at the confidence level p can be expressed as $\mathrm{ES_L}(p) = \mathbb{E}(\mathrm{L} \mid \mathrm{L} > \mathrm{V@R_L}(p))$. Again, let $f(x)$ and $F(x)$ be the density and distribution functions of L, respectively. Then we can express the expected shortfall as:

$$
\begin{aligned}
\mathrm{ES_L}(p) = \mathbb{E}(\mathrm{L} \mid \mathrm{L} > \mathrm{V@R_L}(p)) &= \frac{\mathbb{E}(\mathrm{L}; \mathrm{L} > \mathrm{V@R_L}(p))}{\mathbb{P}(\mathrm{L} > \mathrm{V@R_L}(p))} \\
&= \frac{1}{1 - F(\mathrm{V@R_L}(p))} \int_{\mathrm{V@R_L}(p)}^{\infty} x f(x)\, dx \\
&= \frac{1}{1-p} \int_{\mathrm{V@R_L}(p)}^{\infty} x f(x)\, dx.
\end{aligned}
$$

This expression can be calculated under different distributional assumptions mentioned before.

1. If $L \sim \mathcal{N}(\mu, \sigma^2)$, then

$$\mathrm{ES_L}(p) = \mu + \frac{\sigma}{1-p}\phi(\Phi^{-1}(p)), \tag{8.5}$$

where $\phi(\cdot)$ is the standard normal density function and $\Phi^{-1}(p)$ is the V@R of $\mathcal{N}(0,1)$ at the level p. To derive (8.5), we first consider $z \sim \mathcal{N}(0,1)$. The expected shortfall of z is

$$\begin{aligned}
\mathrm{ES_z}(p) &= \frac{1}{1-p}\int_{\mathrm{V@R_z}(p)}^{\infty} x \times \frac{1}{\sqrt{2\pi}} \exp\left(-\frac{x^2}{2}\right) dx \\
&= \frac{1}{(1-p)\sqrt{2\pi}}\int_{\Phi^{-1}(p)}^{\infty} d\left(-\exp\left(-\frac{x^2}{2}\right)\right) \\
&= -\frac{1}{(1-p)\sqrt{2\pi}} \exp\left(-\frac{x^2}{2}\right)\Bigg|_{\Phi^{-1}(p)}^{\infty} \\
&= \frac{1}{1-p}\phi(\Phi^{-1}(p)).
\end{aligned}$$

In general, when $L \sim \mathcal{N}(\mu, \sigma^2)$, $L = \mu + \sigma z$ with $z \sim \mathcal{N}(0,1)$. As expected shortfall is translationally invariant and positively homogeneous, $\mathrm{ES_L}(p) = \mu + \sigma \mathrm{ES_z}(p)$, therefore we arrive at the formula (8.5). For example, when $p = 0.99$, $\Phi^{-1}(p) \approx 2.326348$, and thus $\mathrm{ES_L}(0.99) \approx \mu + \frac{\sigma}{1-0.99}\phi(2.326348) = \mu + 2.665214\sigma$.

2. If $t := \frac{L-\mu}{\sigma} \times \sqrt{\frac{\nu}{\nu-2}} \sim t_\nu$ with $\nu > 2$, then letting $q_p := T_\nu^{-1}(p)$ be the p-quantile of t_ν, the expected shortfall of t is

$$\begin{aligned}
\mathrm{ES_t}(p) &= \frac{1}{1-p}\int_{q_p}^{\infty} t \cdot \frac{\Gamma\left(\frac{\nu+1}{2}\right)}{\sqrt{\nu\pi}\gamma\left(\frac{\nu}{2}\right)}\left(1+\frac{t^2}{\nu}\right)^{-(\nu+1)/2} dt \\
&= \frac{\nu}{2(1-p)}\int_{q_p}^{\infty} \frac{\Gamma\left(\frac{\nu+1}{2}\right)}{\sqrt{\nu\pi}\gamma\left(\frac{\nu}{2}\right)}\left(1+\frac{t^2}{\nu}\right)^{-(\nu+1)/2} d\left(1+\frac{t^2}{\nu}\right) \\
&= \frac{\nu}{(1-\nu)(1-p)}\left[\frac{\Gamma\left(\frac{\nu+1}{2}\right)}{\sqrt{\nu\pi}\gamma\left(\frac{\nu}{2}\right)}\left(1+\frac{t^2}{\nu}\right)^{(1-\nu)/2}\right]_{q_p}^{\infty}.
\end{aligned}$$

Recall that the standard Student's t-distribution with ν degrees of freedom has a finite mean because for $\nu > 1$, $\lim_{t\to\infty}\left(1+\frac{t^2}{\nu}\right)^{(1-\nu)/2} = 0$, hence

$$\mathrm{ES}_t(p) = 0 - \frac{\nu}{(1-\nu)(1-p)} \cdot \frac{\Gamma\left(\frac{\nu+1}{2}\right)}{\sqrt{\nu\pi\gamma}\left(\frac{\nu}{2}\right)}\left(1+\frac{q_p^2}{\nu}\right)^{-(\nu+1)/2}\left(1+\frac{q_p^2}{\nu}\right)$$

$$= \frac{\nu+q_p^2}{\nu-1} \cdot \frac{\tau_\nu(q_p)}{1-p},$$

where $\tau_\nu(\cdot)$ is the density function of the standard Student's t-distribution with ν degrees of freedom. Therefore, the expected shortfall on L is:

$$\mathrm{ES_L}(p) = \mu + \sigma\sqrt{\frac{\nu-2}{\nu}}\frac{\nu+q_p^2}{\nu-1} \cdot \frac{\tau_\nu(q_p)}{1-p}. \tag{8.6}$$

3. On the other hand, if there are enough data, the expected shortfall can be more accurately estimated by the generalized Pareto distribution under EVT. Recall that for any large value of $u > 0$, the generalized Pareto distribution function $G_{\xi,\beta}$ serves as an approximation of the conditional probability $\mathbb{P}(\mathrm{x} \le u + x \mid \mathrm{x} > u)$ by Gnedenko's theorem. Following [33], suppose that $0 < \xi < 1$ (usually appropriate for financial data), then

$$\begin{aligned}\mathbb{E}(\mathrm{x} \mid \mathrm{x} > u) &= u + \mathbb{E}(\mathrm{x} - u \mid \mathrm{x} > u)\\ &= u + \int_0^\infty (1 - G_{\xi,\beta}(y))dy\\ &= u + \int_0^\infty \left(1+\xi\frac{y}{\beta}\right)^{-1/\xi} dy\\ &= u + \left[\frac{\beta}{\xi} \cdot \frac{\left(1+\xi\frac{y}{\beta}\right)^{1-1/\xi}}{1-\frac{1}{\xi}}\right]_0^\infty;\end{aligned}$$

for $0 < \xi < 1$, $\lim_{y\to\infty}\left(1+\xi\frac{y}{\beta}\right)^{1-1/\xi} = 0$, hence

$$\mathbb{E}(\mathrm{x} \mid \mathrm{x} > u) = u + \frac{\beta}{1-\xi}.$$

Thus, we have the following:

(a) If the conditional distribution of $\mathrm{x} - \mathrm{V@R_x}(p)$ given that $\mathrm{x} > \mathrm{V@R_x}(p)$ is approximated by the generalized Pareto distribution $G_{\xi,\beta}$ with $0 < \xi < 1$, then the expected shortfall of x at level p can be approximated by:

$$\mathrm{ES_x}(p) = \mathbb{E}(\mathrm{x} \mid \mathrm{x} > \mathrm{V@R_x}(p)) \approx \mathrm{V@R_x}(p) + \frac{\beta}{1-\xi}.$$

Given plenty of data of losses greater than $V@R_x(p)$, the parameters ξ and β can then be accurately estimated. If we apply the approximation above to the standardized loss $x = \frac{L-\mu}{\sigma}$, we can obtain the following approximation of the expected shortfall of L:

$$ES_L(p) \approx \mu + \sigma\left(V@R_x(p) + \frac{\beta}{1-\xi}\right). \tag{8.7}$$

(b) In the more general case of a shifted Pareto distribution $x \sim GPD(u, \xi, \beta)$ where $u > 0$, with survival function $\left(1 + \frac{\xi}{\beta}(x-u)\right)^{-1/\xi}$, we note that $x - \theta \mid x > \theta \sim GPD(0, \xi, \beta + \xi(\theta - u))$, and the corresponding expected shortfall is given by:

$$ES_L(\epsilon) = \mu + \sigma\left(V@R_x(p) + \frac{\beta + \xi(V@R_x(p) - u)}{1-\xi}\right), \tag{8.8}$$

where $x := \frac{L-\mu}{\sigma}$. To see this, consider the conditional tail probability:

$$\begin{aligned}
\mathbb{P}(x - \theta > x \mid x > \theta) &= \frac{\mathbb{P}(x > x + \theta)}{\mathbb{P}(x > \theta)} \\
&= \left(1 + \frac{\xi}{\beta}(x + \theta - u)\right)^{-1/\xi} \Big/ \left(1 + \frac{\xi}{\beta}(\theta - u)\right)^{-1/\xi} \\
&= \left(\frac{\beta + \xi(x + \theta - u)}{\beta + \xi(\theta - u)}\right)^{-1/\xi} \\
&= \left(1 + \frac{\xi x}{\beta + \xi(\theta - u)}\right)^{-1/\xi},
\end{aligned}$$

which is another generalized Pareto distribution with $u^* = 0$, $\xi^* = \xi$, and $\beta^* = \beta + \xi(\theta - u)$. With $\theta = V@R_x(p)$ at the level of $100p\%$, we get the expected shortfall of x:

$$\begin{aligned}
ES_x(p) &= \mathbb{E}\left(x \mid x > V@R_x(p)\right) \\
&= V@R_x(p) + \mathbb{E}\left(x - V@R_x(p) \mid x > V@R_x(p)\right) \\
&= V@R_x(p) + \frac{\beta + \xi(V@R_x(p) - u)}{1-\xi};
\end{aligned}$$

with μ and σ, the expected shortfall of L can be obtained as above.

4. If we do not impose any distributional assumption on L, we can estimate $ES_L(p)$ by sorting the sample of observed losses: $L_{(1)}, \ldots, L_{(n)}$ in descending order. Assuming that there is no tie in the sample, then

$$ES_L(p) \approx \frac{1}{n(1-p)} \sum_{k=1}^{K-1} L_{(k)} + \left(1 - \frac{K-1}{n(1-p)}\right) L_{(K)}, \tag{8.9}$$

where $K = \lceil n(1-p) \rceil$ is the smallest integer greater than or equal to $n(1-p)$. We note that this formula is simply the expected shortfall of the empirical distribution of the sample at the confidence level p. It assigns equal weights for each ordered sample point $L_{(k)}$, with a continuity correction for the final least term.

Lastly, we shall compute the empirical expected shortfall for the portfolio consisting of HSBC, CLP, and CK under all the aforementioned V@R calibration methods.

```
1  print(np.mean(loss[loss > VaR_sim]))      # expected shortfall
2  print(np.mean(loss[loss > VaR_GARCH]))
3  # expected shortfall with losses bootstrapped from unequal weight model
4  print(np.mean(loss_GARCH[loss_GARCH > VaR_GARCH]))
5
6  mu = np.mean(loss)
7  sig = np.std(loss, ddof=1)
8  # normal from (8.5)
9  from scipy.stats import norm
10 print(mu + sig/0.01*norm.pdf(norm.ppf(0.01)))
11
12 # student t from (8.6)
13 from scipy.stats import t
14 Term1 = (nu + t.ppf(0.01, nu)**2)/(nu-1)
15 Term2 = t.pdf(t.ppf(0.01, nu), nu)/0.01
16 print(mu + sig*np.sqrt((nu-2)/nu)*Term1*Term2)
17
18 # extreme value theorem from (8.8)
19 EVT = VaR + (beta + xi*(VaR - u))/(1 - xi)
20 print(mu + sig*EVT)
21
22 # distribution free from (8.9)
23 n = len(loss)
24 K = int(np.floor(n*0.01))
25 sort_loss = np.sort(loss)[::-1]
26 Term1 = 1/(0.01*n)*sum(sort_loss[:(K-1)])
27 Term2 = (1-(K-1)/(0.01*n))*sort_loss[K-1]
28 print(Term1 + Term2)
```

```
1 4578.422449521
2 3599.8859488028506
3 3399.268338408097
4 3561.0684333505187
5 4265.071129783812
6 4981.691432110931
7 4611.053472844207
```

Programme 8.15 Expected shortfall using various V@R approaches in Python.

```
1  > # expected shortfall
2  > mean(loss[loss > VaR_sim])
3  [1] 4578.422
4  > mean(loss[loss > VaR_GARCH])
5  [1] 3599.886
6  > # expected shortfall with losses bootstrapped from unequal weight model
7  > mean(loss_GARCH[loss_GARCH > VaR_GARCH])
8  [1] 3400.294
9  >
10 > mu <- mean(loss)
11 > sig <- sd(loss)
12 > # normal
13 > mu + sig/0.01*dnorm(qnorm(0.01))
14 [1] 3561.068
15 >
16 > # student-t
17 > s <- sig*sqrt((nu-2)/nu)
18 > mu + s*(nu + qt(0.01, nu)^2)/(nu-1)*(dt(qt(0.01, nu), nu)/0.01)
19 [1] 4265.071
20 >
21 > # extreme value theorem
22 > EVT <- VaR + (p[2] + p[1]*(VaR - u))/(1 - p[1])
23 > mu + sig*EVT
24 [1] 4998.013
25 >
26 > # distribution free
27 > n <- length(loss)
28 > K <- floor(n*0.01)
29 > sort_loss <- sort(loss, decreasing=TRUE)
30 > 1/(0.01*n)*sum(sort_loss[1:(K-1)]) + (1-(K-1)/(0.01*n))*sort_loss[K]
31 [1] 4611.053
```

Programme 8.16 Expected shortfall using various V@R approaches in **R**.

From Programmes 8.15 via Python and 8.16 via **R**, the expected shortfalls computed from the GARCH(1, 1) non-parametric bootstrapping approach ($3,599.886) and the normal parametric model ($3,561.068) are the smallest; while the expected shortfalls computed from the Student's t-model ($4,265.071), the EVT approach in Python ($4,981.691) (resp. $4,998.013 in **R**) and the distribution-free approach (8.9) ($4,611.053) are larger. Since we know that the Student's t and the EVT are better at capturing the tail behaviour of the loss distribution, we tend to adopt the ES coming from either one of these two models.

8.8 DEPENDENCE MODELLING VIA COPULAE

In this section, we shall use d for the dimension of the underlying multivariate risks, and p for the critical level of the probability in context. Copulae are used to model and extract the dependence structure among marginal random variables in multivariate distribution functions. Mathematically speaking, a d-dimensional copula is simply a d-dimensional distribution function on the hyper-cube $[0,1]^d$ whose d marginal distributions follow a uniform distribution on $[0,1]$. The main reason why copulae are relevant is due to the celebrated *Sklar's theorem* [41], which asserts that for any d-dimensional joint distribution function F with marginal distributions $F_1,\dots,F_d$, there always exists a d-dimensional copula $C : [0,1]^d \to \mathbb{R}$ such that for any $x_1,\dots,x_d \in \mathbb{R}$,

$$F(x_1,\dots,x_d) = C(F_1(x_1),\dots,F_d(x_d)). \tag{8.10}$$

Such a copula is unique if all the marginal distributions are continuous; in this case, this unique copula in (8.10) can be expressed explicitly as

$$C(u_1,\dots,u_d) = F(F_1^{-1}(u_1),\dots,F_d^{-1}(u_d)), \quad \text{for } u_1,\dots,u_d \in [0,1]. \tag{8.11}$$

Equation (8.10) demonstrates that the joint distribution function can always be decomposed into the underlying marginal distributions linked by a copula function. Hence the copula C captures all the information about how the marginal random variables depend on one another. This fact, that the copula fully reflects the dependence structure, is manifested by its invariance property: the copula remains the same upon any strictly increasing transformations of the marginal distributions. Conversely, for any copula C with d arguments and marginal distributions $F_1,\dots,F_d$, the function on the right of (8.10) defines a joint distribution function whose marginal distributions are $F_1,\dots,F_d$, respectively. This fact gives us a useful way to construct multivariate distribution functions from given marginal distributions through various choices of copulae. Refer to [28, 31, 33, 35] for further discussions on the theoretical and practical aspects of copulae.

Apart from (8.10) which relates the joint lower tail probability to a copula, one may also consider the upper tail counterpart, which can be formulated as follows. For any random vector x,

$$\mathbb{P}(\mathrm{x}_1 > x_1,\dots,\mathrm{x}_d > x_d) = \widehat{C}(\overline{F}_1(x_1),\dots,\overline{F}_d(x_d)), \quad x \in \mathbb{R}^d,$$

where $\widehat{C}$ is called the *survival copula* of x, and $\overline{F}_i$ is the survival distribution of the random variable x_i for $i = 1,\dots,d$.

In finance, copulae have proven useful in various fields of finance, including but not limited to quantitative risk management, portfolio management and optimization, and pricing of financial derivatives. A typical example of its application is tests on the stress or robustness of companies, which played a particularly crucial role during the extreme downtime of economies in 2007–2008, when the global financial crisis struck. Besides, the estimation for loss distributions of various financial products, such as bonds and loans, also involve copulae as an important tool.

Many investors would exhibit the behavior known as *flight-to-quality* during bear markets, namely making a switch from the riskier equities or real estates to the "safer" cash or bonds for a refuge; as a result, they tend to get rid of most of the riskier assets within a short time, and that explains why dependencies across equities are stronger during a bear market than in a bull market, which could have disastrous effects on the overall economy; indeed, news on gigantic losses in millions in one day appears on financial headlines far more often than that on huge profits at the same level during a similar time span.

Next, we shall introduce several commonly used copulae.

(I) Independence Copula
The *independence copula* is arguably the simplest copula. As the name suggests, it corresponds to the situation where the marginal variables are independent of each other. This copula can be written as follows:

$$\Pi(\boldsymbol{u}) = \prod_{i=1}^{d} u_i. \tag{8.12}$$

It is an elementary fact that random variables are independent if and only if their copula is the independence copula. Later in this section, we shall see that independence copula is also an Archimedean copula with the generator function $\psi(x,\theta) = -\ln x$.

(II) Gaussian Copula
Suppose that F is a d-variate normal distribution function with a mean vector $\boldsymbol{\mu}$ and a covariance matrix $\boldsymbol{\Sigma}$. By the invariance property, its copula remains unchanged upon standardization, and so we may simply take $\boldsymbol{\mu} = \boldsymbol{0}$ and $\boldsymbol{\Sigma}$ as its own corresponding correlation matrix $\boldsymbol{P}_\rho$. The *Gaussian copula* is defined as the copula of such a d-variate normal distribution:

$$C_{\boldsymbol{P}_\rho}^{\text{Gauss}}(\boldsymbol{u}) = \Phi_{\boldsymbol{P}_\rho}(\Phi^{-1}(u_1), \ldots, \Phi^{-1}(u_d)), \quad \text{for } u_i \in [0,1], i = 1, \ldots, d,$$

where $\Phi_{\boldsymbol{P}_\rho}$ is the joint cdf of the d-variate normal distribution with $\boldsymbol{\mu} = \boldsymbol{0}$ and the covariance matrix $\boldsymbol{P}_\rho$, and Φ is the cdf of the standard normal distribution. The copula density can be found as follows:

$$c_{\text{Gauss}}(\boldsymbol{u}) = \frac{1}{|\boldsymbol{P}_\rho|^{1/2}} \exp\left\{-\frac{1}{2}\boldsymbol{z}^\top \left(\boldsymbol{P}_\rho^{-1} - \boldsymbol{I}_d\right)\boldsymbol{z}\right\}, \tag{8.13}$$

where $\boldsymbol{z} := (\Phi^{-1}(u_1), \ldots, \Phi^{-1}(u_d))^\top$ and $\boldsymbol{I}_d$ is the $d \times d$ identity matrix.

The Gaussian copula was widely used in performing stress tests or credit risk analyses before 2008, but as the normal distribution would typically understate the probability of extreme losses, it has been argued that it contributed to the 2007–2008 Credit Crunch. Let us apply the Gaussian copula with empirical marginals to our stock data from `stock_1999_2002.csv` as an illustration.

```
1  import pandas as pd
2  import numpy as np
3  import seaborn as sns
4  import matplotlib.pyplot as plt
5
6  plt.rc('font', size=20); plt.rc('axes', titlesize=30, labelsize=30)
7  plt.rc('xtick', labelsize=25); plt.rc('ytick', labelsize=25)
8
9  d = pd.read_csv("stock_1999_2002.csv", index_col=0)
10 returns = np.diff(d, axis=0) / d.iloc[:-1, :] # Arithmetic return
```

```
11 n_days, n_stocks = returns.shape
12 n_sim = int(1e5)
13
14 # compute pseudo observations
15 from scipy.stats import rankdata
16 pse_u = rankdata(returns, method='average', axis=0) / (n_days + 1)
17
18 # Q-Q plot for pseudo observations
19 col = ["blue", "orange", "green"]
20 q = (np.arange(1, n_days + 1) - 0.5) / n_days
21
22 for k in range(n_stocks):
23     pse_u_k = pse_u[:,k]
24     com = returns.columns[k]
25     b, w = np.linalg.lstsq(np.c_[np.ones(n_days), q],
26                            np.sort(pse_u_k), rcond=None)[0]
27
28     fig, ax = plt.subplots(figsize=(10, 10))
29     ax.scatter(q, np.sort(pse_u_k), color=col[k], facecolor="white",
30                marker="o", s=100)
31     ax.plot(q, w*q + b, color="blue", linewidth=3)
32     ax.set_xlabel("Theoretical quantiles")
33     ax.set_ylabel("Returns quantiles")
34     ax.set_title(f"Q-Q Plot with {com}'s returns")
35
36 def pseudo_quantile(p, samples):
37     p, samples = np.array(p), np.array(samples)
38     q = np.empty(p.shape)
39     for k in range(p.shape[1]):
40         q[:,k] = np.quantile(samples[:,k], p[:,k],
41                              method='interpolated_inverted_cdf')
42
43     return q
```

Programme 8.17 Data prepreparation in Python.

In Programme 8.17 using Python, we state that the library `copulae` will transform the data into *pseudo observations*, defined as $v_{i,j} := \text{rank}(x_{i,j})/(n+1)$, $i = 1, 2, \ldots, n$ and $j = 1, 2, \ldots, d$, before fitting the model. In this programme we demonstrate this transformation via the `rankdata()` function from the `scipy` library. After we sample observations $(v_{i,1}, \ldots, v_{i,d})$ from the fitted Gaussian copula in `copulae`, we compute the empirical inverse of each marginal cdf with `pseudo_quantile()` to obtain component-wise simulated returns of $\hat{F}_j^{-1}(v_{i,j})$ for $j = 1, \ldots, d$.

From Figure 8.7, the plots are fitted by the sample generated by pseudo observations. All three Q-Q plots should be a perfect fit against the theoretical quantiles except for ties, such as in the middle portion of the plots where the return is zero. The same implementation is carried out in R through Programme 8.18, with an illustration in Figure 8.8.

```
1  > d <- read.csv("stock_1999_2002.csv", row.names=1) # read in data file
2  > d <- as.ts(d)
3  > returns <- (lag(d) - d)/d
4  > colnames(returns) <- paste0(colnames(d), "_Return")
5  > n_sim <- 1e5
6  >
7  > # compute pseudo observations
8  > library(copula)  # Package for copula computation
9  > pse_u <- pobs(returns, ties.method="average")
10 >
11 > par(cex.lab=2, cex.axis=2, cex.main=2, mar=c(5,5,4,4))
12 > # Q-Q plot for pseudo observations
13 > col <- c("blue", "orange", "green")
14 > n_days <- nrow(returns)
15 > q <- ((1:n_days) - 0.5) / n_days
16 > for (k in 1:ncol(returns)){
17 +   qqplot(q, sort(pse_u[,k]), col=col[k],
18 +          xlab="Theoretical quantiles", ylab="Sample quantiles",
19 +          main=paste0("Q-Q Plot of ", colnames(returns)[k]))
20 +   abline(lsfit(q, sort(pse_u[,k])), lwd=2)
21 + }
22 >
23 > pseudo_quantile <- function(p, samples){
24 +   q <- matrix(NA, nrow=nrow(p), ncol=ncol(p))
25 +   for (k in 1:ncol(p)){
26 +     q[,k] <- quantile(samples[,k], probs=p[,k], type=4, names=FALSE)
27 +   }
28 +   return (q)
29 + }
```

Programme 8.18 Data prepreparation in R.

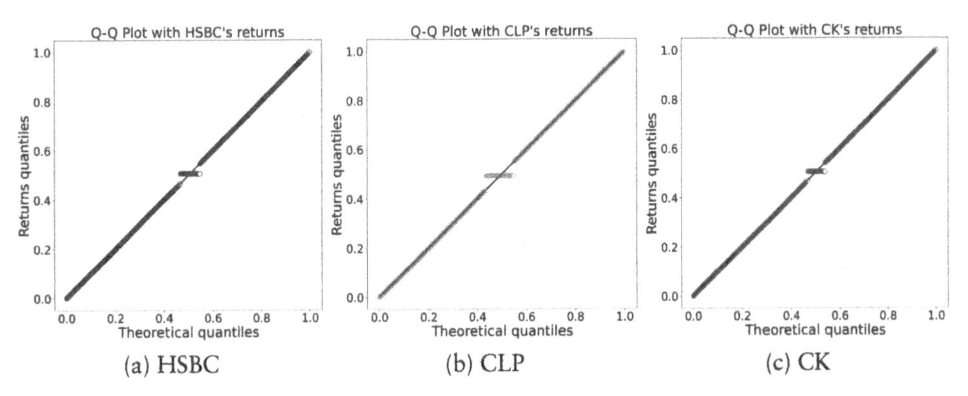

FIGURE 8.7 Q-Q plots of the sample generated by pseudo observations of HSBC, CLP, and CK via Python, generated by Programme 8.17.

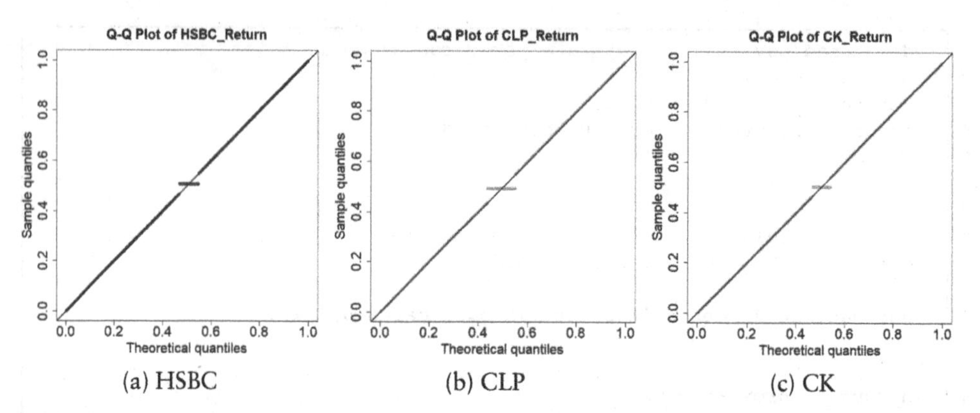

(a) HSBC (b) CLP (c) CK

FIGURE 8.8 Q-Q plots of the sample generated by pseudo observations of HSBC, CLP, and CK via **R**, generated by Programme 8.18.

Next, in Programmes 8.19 via Python and 8.20 via **R**, we aim to fit a Gaussian copula for the dependence structure for three security stocks from `stock_1999_2002.csv`, while keeping the marginal distributions as the individual empirical distribution of each stock's relative price changes. Technically, we fit a multivariate Gaussian distribution with a zero mean vector and covariance matrix $\boldsymbol{\Sigma} = P_\rho$ by using a sample of $(\Phi^{-1}(v_{i,1}), \ldots, \Phi^{-1}(v_{i,d}))$, for $i = 1, 2, \ldots, n$ and $d = 3$. We also depict the random samples $\boldsymbol{v}$'s in $[0,1]^3$ and the simulated returns $\boldsymbol{u}$'s generated by this fitted Gaussian copula dependence structure in Figures 8.9 via Python and 8.10 via **R**.

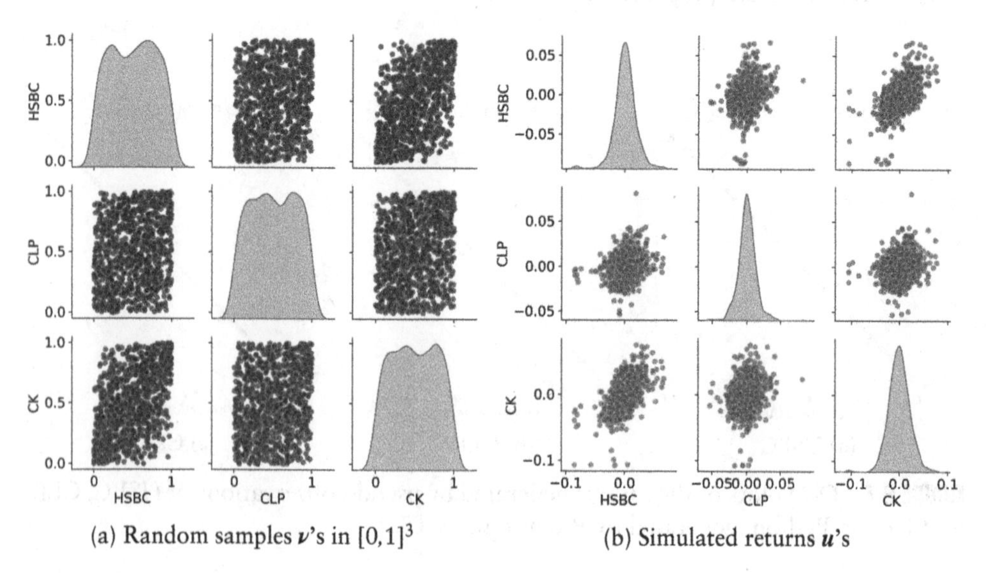

(a) Random samples $\boldsymbol{v}$'s in $[0,1]^3$ (b) Simulated returns $\boldsymbol{u}$'s

FIGURE 8.9 Plots from the fitted Gaussian copula in Python, generated by Programme 8.19.

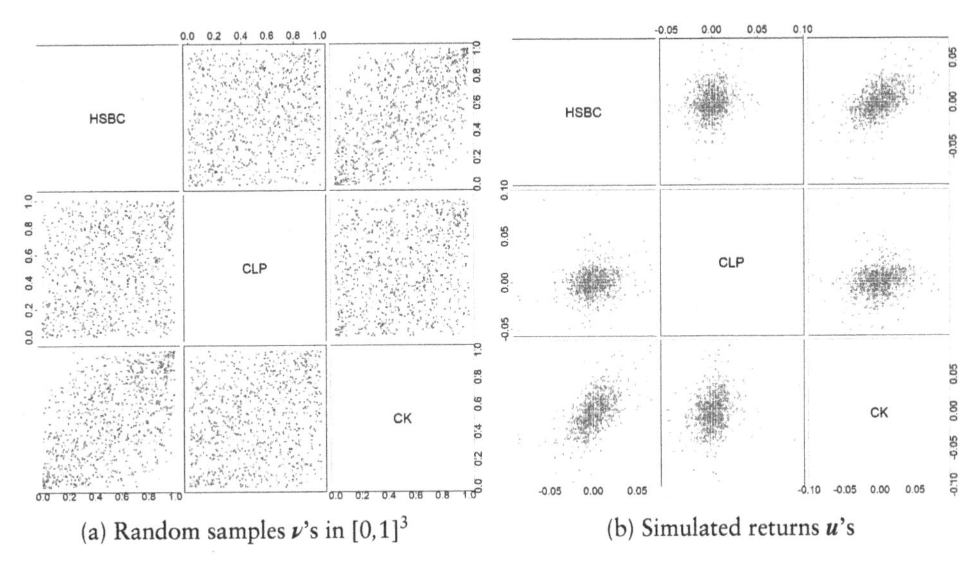

(a) Random samples $\boldsymbol{\nu}$'s in $[0,1]^3$ (b) Simulated returns $\boldsymbol{u}$'s

FIGURE 8.10 Plots from the fitted Gaussian copula in **R**, generated by Programme 8.20.

```
from copulae import NormalCopula

plt.rcParams.update(plt.rcParamsDefault)

N_cop_dist = NormalCopula(dim=len(d.columns))
N_cop_dist.fit(pse_u, verbose=0, to_pobs=False)
# Generate random samples based on gaussian copula
u_sim_N = pd.DataFrame(N_cop_dist.random(n_sim, seed=4002),
                       columns=returns.columns)
# only plot the first 1000 samples
sns.pairplot(u_sim_N[1:1000], diag_kind="kde",
             plot_kws={'alpha': 0.5, 'color': 'blue'})
print(np.corrcoef(u_sim_N, rowvar=False, ddof=1))
print(np.corrcoef(returns, rowvar=False, ddof=1))

# Get back returns based on the random samples
return_sim_N = pd.DataFrame(pseudo_quantile(u_sim_N, returns),
                            columns=returns.columns)
# only plot the first 1000 samples
sns.pairplot(return_sim_N[1:1000], diag_kind="kde",
             plot_kws={'alpha': 0.5, 'color': 'green'})
```

```
[[1.         0.23021469 0.55358563]
 [0.23021469 1.         0.23885371]
 [0.55358563 0.23885371 1.        ]]
[[1.         0.24613997 0.57718628]
 [0.24613997 1.         0.24520583]
 [0.57718628 0.24520583 1.        ]]
```

Programme 8.19 Fitting a Gaussian copula for HSBC, CLP, CK returns with empirical marginal distributions using Python.

```
1  > # Assume a normal-copula with ncol(d)=3
2  > # P2p: array of elements of upper triangular matrix
3  > N.cop <- normalCopula(dim=ncol(d), dispstr="un")
4  > fit <- fitCopula(N.cop, pse_u, "ml")
5  > (rho <- coef(fit))
6      rho.1     rho.2     rho.3
7  0.2427711 0.5746243 0.2481858
8  > N.cop_fit <- normalCopula(rho, dim=ncol(d), dispstr="un")
9  > set.seed(4002)
10 > # Generate random samples u~U(0, 1) from the fitted gaussian copula
11 > u_sim_N <- rCopula(n_sim, N.cop_fit)
12 > colnames(u_sim_N) <- colnames(d)
13 > # only show the first 1000
14 > pairs(u_sim_N[1:1e3,], col="blue", cex.axis=3, cex.labels=4)
15 > cor(u_sim_N)
16           HSBC       CLP        CK
17 HSBC 1.0000000 0.2361333 0.5533921
18 CLP  0.2361333 1.0000000 0.2410417
19 CK   0.5533921 0.2410417 1.0000000
20 > cor(returns)
21             HSBC_Return CLP_Return CK_Return
22 HSBC_Return   1.0000000  0.2461400 0.5771863
23 CLP_Return    0.2461400  1.0000000 0.2452058
24 CK_Return     0.5771863  0.2452058 1.0000000
25 >
26 > # Get back returns based on the random samples
27 > return_sim_N <- pseudo_quantile(u_sim_N, returns)
28 > colnames(return_sim_N) <- colnames(d)
29 > # only show the first 1000
30 > pairs(return_sim_N[1:1e3,], col="green", cex.axis=3, cex.labels=4)
```

Programme 8.20 Fitting a Gaussian copula for HSBC, CLP, CK returns with empirical marginal distributions using **R**.

Finally, we consider a Q-Q plot of the empirical returns against the theoretical (actually, simulated) returns from the joint distribution with the fitted Gaussian copula for each stock. In Programme 8.21 via Python, we lay down a function for drawing the Q-Q plot.

```
1  def Mahalanobis2(X):
2      X = np.array(X)
3      mu = np.mean(X, axis=0)
4      inv_Sig = np.linalg.inv(np.cov(X, rowvar=False))
5      X_minus_mu = X - mu
6      return np.sum((X_minus_mu @ inv_Sig) * X_minus_mu, axis=1)
7
8  def QQ_Plot(sim_data, raw_data, col="blue"):
9      n_days = len(raw_data)
10
11     i = (np.arange(1, n_days + 1) - 0.5) / n_days
12     q = np.quantile(sim_data, i, method='interpolated_inverted_cdf')
13
14     fig = plt.figure(figsize=(10, 10))
15     plt.scatter(q, np.sort(raw_data), color=col,
```

```
16                     facecolor="white", marker="o", s=100)
17      plt.axline([0, 0], [1, 1], color=col, linewidth=3)
18      plt.xlabel("Bootstrapped quantiles")
19      plt.ylabel("Sample quantiles")
20      plt.title("Copula Q-Q Plot")
21
22      return fig
23
24  returns_md2 = Mahalanobis2(returns)
```

Programme 8.21 Function for plotting Q-Q plots to simulate (bootstrap) the theoretical returns via Python.

Furthermore, we build an auxiliary function `Mahalanobis2()` for computing the squared Mahalanobis distance introduced in (3.3). The main plotting function `QQ_Plot()` aims to plot a Q-Q plot between the parametrically bootstrapped theoretical quantiles from `sim_data` and the actual observations from `raw_data`. Using Programme 8.22, we compare this Q-Q plot with the chi-square Q-Q plot depicted in Section 3.3, under the hypothesis that the returns follow a multivariate normal distribution.

```
1   plt.rc('font', size=20); plt.rc('axes', titlesize=30, labelsize=30)
2   plt.rc('xtick', labelsize=25); plt.rc('ytick', labelsize=25)
3
4   sim_N_md2 = Mahalanobis2(return_sim_N)
5   fig = QQ_Plot(sim_N_md2, returns_md2, col="blue")
6
7   from scipy import stats
8
9   i = (np.arange(1, n_days+1)-0.5)/n_days
10  q = stats.chi2.ppf(i, 3)
11
12  fig = plt.figure(figsize=(10, 10))
13  b, w = np.linalg.lstsq(np.vstack([np.ones(n_days), q]).T,
14                         np.sort(returns_md2), rcond=None)[0]
15  plt.scatter(q, np.sort(returns_md2), color="blue", marker="o", s=100)
16  plt.plot(q, w*q + b, color="blue", linewidth=3)
17  plt.title("Chi2 Q-Q Plot")
18
19  print(stats.kstest(np.sqrt(returns_md2), np.sqrt(sim_N_md2),
20                     method="asymp"))
21  print(stats.kstest(np.sqrt(returns_md2), stats.chi2.cdf,
22                     args=(3,), method="asymp"))
```

```
1  KstestResult(statistic=0.04023843336724314, pvalue=0.0833251702659672,
       statistic_location=0.9983424597718914, statistic_sign=1)
2  KstestResult(statistic=0.3697098718483404, pvalue=3.943732517218666e-117,
       statistic_location=2.171602746698503, statistic_sign=1)
```

Programme 8.22 Q-Q plots for a Gaussian copula and simple multivariate normal distribution fittings of HSBC, CLP, and CK via Python, using the `QQ_Plot()` function in Programme 8.21.

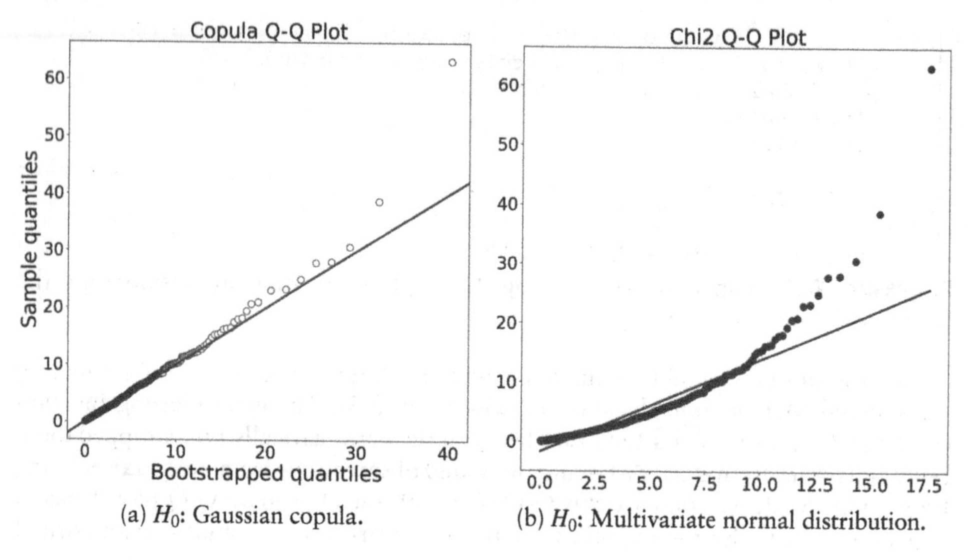

(a) H_0: Gaussian copula. (b) H_0: Multivariate normal distribution.

FIGURE 8.11 Q-Q plots generated in Programme 8.22.

From Figure 8.11, we can see how the Gaussian copula, which fits the dependence structure but does not force the marginals to be normal, performs much better. Indeed, except for a dozen or so outliers, the joint distribution fitted with a Gaussian copula and empirical marginals can explain the joint changing behavior of the three stocks well on most of the fair weather days. The Kolmogorov–Smirnov test does allow us to accept the goodness-of-fit of the Gaussian copula model but not the simple multivariate normality fitting. On the other hand, these outliers can be coped with by using the methods we have mentioned in the earlier sections of this chapter.

Similarly, we can carry out the whole procedure in **R** environment: firstly, the special function needed for drawing the Q-Q plot is built in Programme 8.23. Finally, the Q-Q plots of the squared Mahalanobis distances are illustrated in Figure 8.12. Note that the p-value of the KS test in the **R** output of Programme 8.24 is actually the same as that in the Python output of Programme 8.22, yet they display differently due to the mechanism in **R** that the minimum displayable positive number is `2.2e-16` (2.2×10^{-16}).

```
1  > Mahalanobis2 <- function(X){
2  +   mu <- apply(X, 2, mean)
3  +   inv_Sig <- solve(cov(X))
4  +   X_minus_mu <- sweep(X, 2, mu, FUN="-")
5  +   return (rowSums((X_minus_mu%*%inv_Sig) * X_minus_mu))
6  +
7  + }
8  >
9  > QQ_Plot <- function(sim_data, raw_data, col="blue"){
10 +   n_days <- length(raw_data)
11 +   i <- ((1:n_days) - 0.5) / n_days
```

```
12  +    q <- quantile(sim_data, probs=i, type=4, names=FALSE)
13  +
14  +    qqplot(q, sort(raw_data), col=col, cex=2,
15  +           xlab="Bootstrapped quantiles", ylab="Sample quantiles",
16  +           main="Copula Q-Q Plot")
17  +    abline(0, 1, lwd=2)
18  + }
19  >
20  > returns_md2 <- Mahalanobis2(returns)
```

Programme 8.23 Function for plotting Q-Q plots to simulate (bootstrap) the theoretical returns via R.

```
1  > sim_N_md2 <- Mahalanobis2(return_sim_N)
2  > QQ_Plot(sim_N_md2, returns_md2, col="blue")
3  >
4  > i <- ((1:n_days) - 0.5) / n_days
5  > q <- qchisq(i, 3)
6  > qqplot(q, sort(returns_md2), main="Chi2 Q-Q Plot", cex=2)
7  > abline(lsfit(q, sort(returns_md2)))
8  >
9  > ks.test(sqrt(returns_md2), sqrt(sim_N_md2))
10      Asymptotic two-sample Kolmogorov-Smirnov test
11 data:  sqrt(returns_md2) and sqrt(sim_N_md2)
12 D = 0.037998, p-value = 0.1203
13 alternative hypothesis: two-sided
14 > ks.test(sqrt(returns_md2), "pchisq", df=3)
15      Asymptotic one-sample Kolmogorov-Smirnov test
16 data:  sqrt(returns_md2)
17 D = 0.36971, p-value < 2.2e-16
18 alternative hypothesis: two-sided
```

Programme 8.24 Q-Q plots for Gaussian copula and simple multivariate normal distribution fittings of HSBC, CLP, and CK via **R**, using the `QQ_Plot()` function in Programme 8.23.

(III) *t*-copula

A number of contemporary articles such as [9, 43] have shown that *Student's t-copulae* can in general provide a better empirical fit for financial data than Gaussian copulae, mainly due to the domino effect when the underlying runs to extremes. Generally speaking, t-copulae can better capture the phenomenon of dependent extreme values which are often present in financial data such as returns. Recalling that the notation $t_\nu^{-1}(p)$ is the p-quantile of Student's t-distribution with ν degrees of freedom, the t-copula with ν degrees of freedom is defined as: for $u_i \in [0,1], i = 1, \ldots, d$,

$$C^t_{\nu,\boldsymbol{\Sigma}}(\boldsymbol{u}) = \int_{-\infty}^{t_\nu^{-1}(u_1)} \cdots \int_{-\infty}^{t_\nu^{-1}(u_d)} f^t_{\nu,\boldsymbol{\Sigma}}(\boldsymbol{x})\, d\boldsymbol{x}.$$

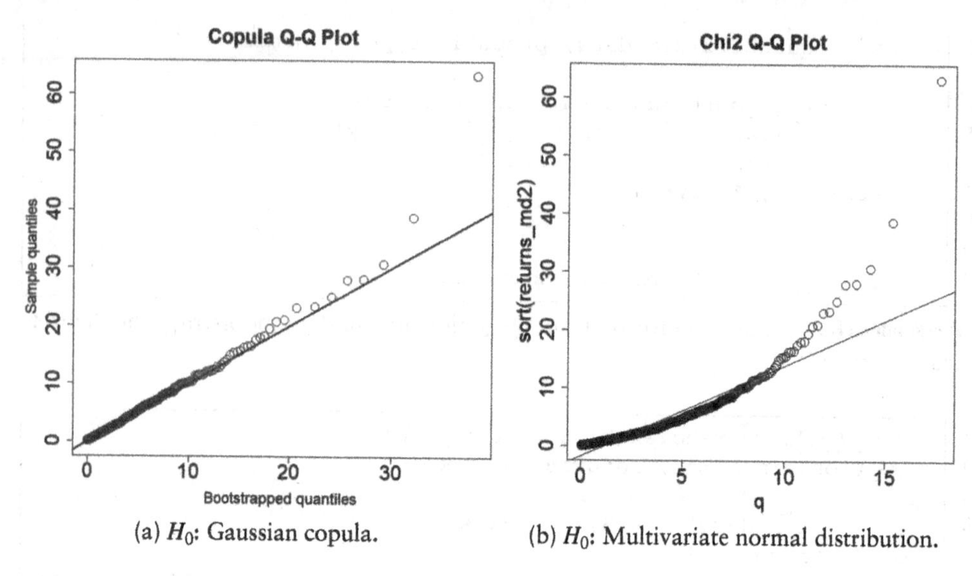

(a) H_0: Gaussian copula.

(b) H_0: Multivariate normal distribution.

FIGURE 8.12 Q-Q plots generated in Programme 8.24.

where $f^t_{\nu,\boldsymbol{\Sigma}}(\boldsymbol{x}) = \frac{\Gamma(\frac{\nu+d}{2})}{\Gamma(\frac{\nu}{2})\sqrt{(\pi\nu)^d|\boldsymbol{\Sigma}|}}\left(1 + \frac{\boldsymbol{x}^\top\boldsymbol{\Sigma}^{-1}\boldsymbol{x}}{\nu}\right)^{-\frac{\nu+d}{2}}$ is the density function of a multivariate t-distribution with ν degrees of freedom, and it possesses a mean vector $\boldsymbol{\mu} = \boldsymbol{0}$ and a dispersion matrix (positive definite and symmetric) $\boldsymbol{\Sigma}$. The copula density can be shown to be:

$$c^t_{\nu,\boldsymbol{\Sigma}}(\boldsymbol{u}) = \frac{f^t_{\nu,\boldsymbol{\Sigma}}(\boldsymbol{z})}{\Pi^d_{i=1} f^t_\nu(t^{-1}_\nu(u_i))},$$

where $\boldsymbol{z} := (t^{-1}_\nu(u_1), \ldots, t^{-1}_\nu(u_d))^\top$, and f^t_ν is the density function of the standard univariate Student's t-distribution with the same degrees of freedom ν.

Let us fit a t-copula to the financial data in `stock_1999_2002.csv` through Programme 8.25 via Python with the `copulae` library. The procedure of finding the parameters of the best-fit t-copula is essentially the same as that for the Gaussian copula, and so we leave it as an exercise for readers. Again we depict the random samples $\boldsymbol{v}$'s in $[0, 1]^3$ and the simulated returns $\boldsymbol{u}$'s, generated by the joint distribution with this fitted t-copula, in Figure 8.13.

```
1 from copulae import StudentCopula
2
3 plt.rcParams.update(plt.rcParamsDefault)
4
5 t_cop_dist = StudentCopula(dim=len(d.columns))
6 t_cop_dist.fit(pse_u, verbose=0, to_pobs=False)
7 print(t_cop_dist.params.rho)
```

```
8  print(t_cop_dist.params.df)
9
10 # Generate random samples based on t-copula
11 u_sim_t = pd.DataFrame(t_cop_dist.random(n_sim, seed=4002),
12                        columns=returns.columns)
13 # only plot the first 1000 samples
14 sns.pairplot(u_sim_t[1:1000], diag_kind="kde",
15              plot_kws={'alpha': 0.5, 'color': 'blue'})
16
17 # Get back returns based on the random samples
18 return_sim_t = pd.DataFrame(pseudo_quantile(u_sim_t, returns),
19                             columns=returns.columns)
20 # only plot the first 1000 samples
21 sns.pairplot(return_sim_t[1:1000], diag_kind="kde",
22              plot_kws={'alpha': 0.5, 'color': 'green'})
```

```
1 [0.23468915 0.57438888 0.24802204]
2 7.100775048482457
```

Programme 8.25 Fitting a t-copula with empirical marginals for HSBC, CLP, CK returns using Python.

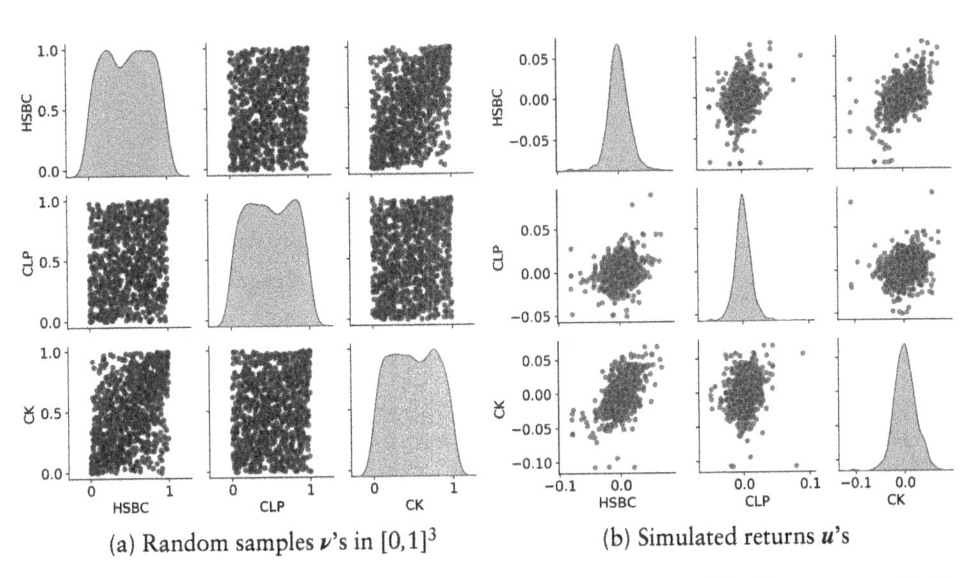

(a) Random samples ν's in $[0,1]^3$ (b) Simulated returns u's

FIGURE 8.13 Plots from the fitted t-copula in Python, generated by Programme 8.25.

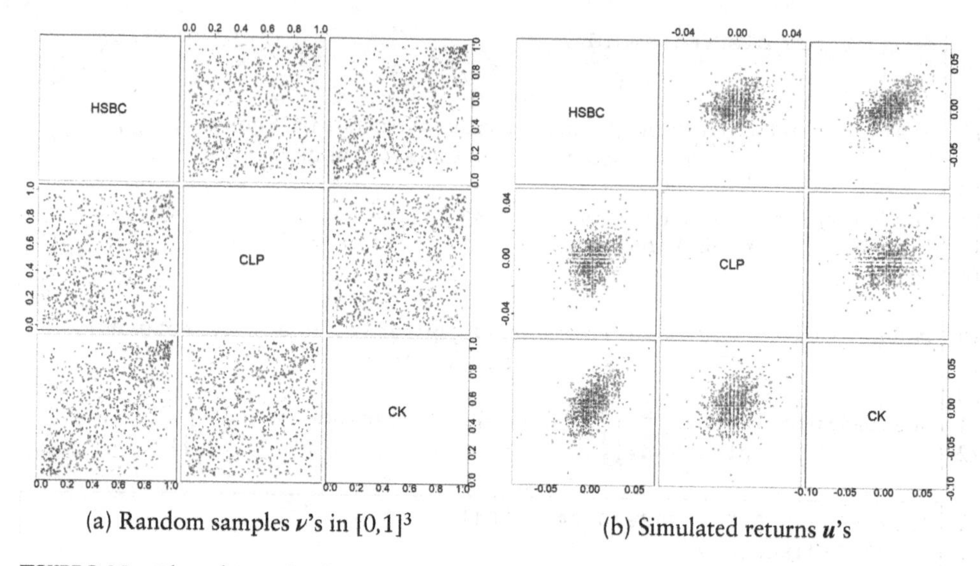

(a) Random samples $\boldsymbol{\nu}$'s in $[0,1]^3$ (b) Simulated returns $\boldsymbol{u}$'s

FIGURE 8.14 Plots from the fitted t-copula via **R**, generated in Programme 8.26.

Similarly, we implement the same procedure in Programme 8.26 via **R** with an illustration in Figure 8.14.

```
1  > # Assume a t-copula  with ncol(d)=3
2  > t.cop <- tCopula(dim=ncol(d), dispstr='un')
3  > fit <- fitCopula(t.cop, pse_u, "ml")
4  > (rho <- coef(fit)[1:ncol(d)])
5      rho.1     rho.2     rho.3
6  0.2346630 0.5743798 0.2479884
7  > (df <- coef(fit)[length(coef(fit))])
8        df
9  7.093516
10 > t.cop_fit <- tCopula(dim=ncol(d), rho, df=df, dispstr="un")
11 >
12 > # Generate random samples u~U(0, 1) from the fitted t copula
13 > u_sim_t <- rCopula(n_sim, t.cop_fit)
14 > colnames(u_sim_t) <- colnames(d)
15 > # only show the first 1000
16 > pairs(u_sim_t[1:1e3,], col="blue", cex.axis=3, cex.labels=4)
17 >
18 > # Get back returns based on the random samples
19 > return_sim_t <- pseudo_quantile(u_sim_t, returns)
20 > colnames(return_sim_t) <- colnames(d)
21 > # only show the first 1000
22 > pairs(return_sim_t[1:1e3,], col="green", cex.axis=3, cex.labels=4)
```

Programme 8.26 Fitting a t-copula with empirical marginals for HSBC, CLP, CK returns using **R**.

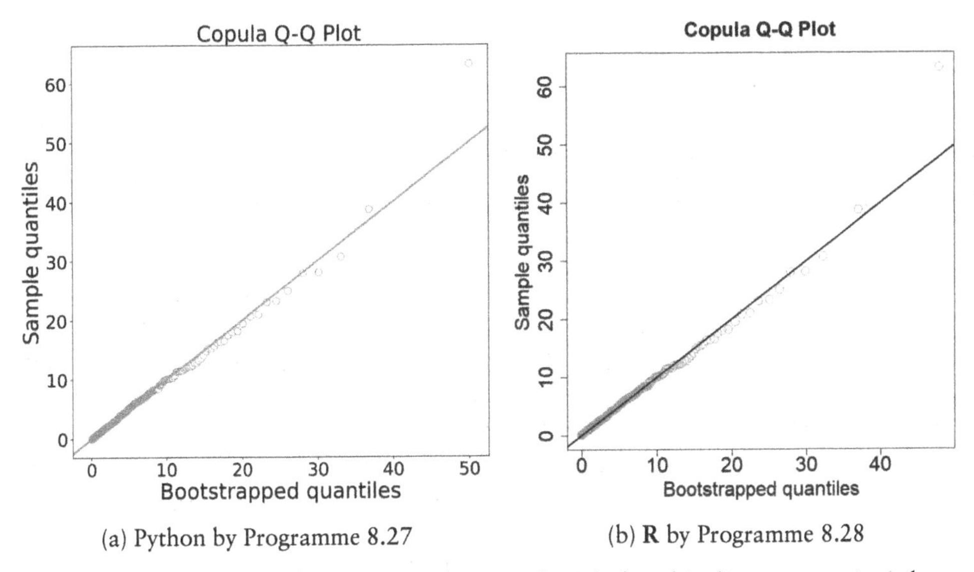

(a) Python by Programme 8.27 (b) **R** by Programme 8.28

FIGURE 8.15 Q-Q plots of the empirical squared Mahalanobis distances against those bootstrapped estimates from the fitted *t*-copula of HSBC, CLP, and CK in Python.

Next, we shall plot the Q-Q plot for the fitted *t*-copula, first by using Python in Programme 8.27 and then by using **R** in Programme 8.28. In Figure 8.15, we also depict the Q-Q plots of the empirical squared Mahalanobis distances against the corresponding bootstrapped quantile estimates from the joint distribution with the fitted *t*-copula. From the Kolmogorov–Smirnov test, we can see the goodness-of-fit of the *t*-copula model, noting it has a larger *p*-value than the Gaussian copula.

```
1 plt.rc('font', size=20); plt.rc('axes', titlesize=30, labelsize=30)
2 plt.rc('xtick', labelsize=25); plt.rc('ytick', labelsize=25)
3
4 sim_t_md2 = Mahalanobis2(return_sim_t)
5 fig = QQ_Plot(sim_t_md2, returns_md2, col="orange")
6
7 print(stats.kstest(np.sqrt(returns_md2), np.sqrt(sim_t_md2),
8                    method="asymp"))
```

```
1 KstestResult(statistic=0.01887967446592065, pvalue=0.8719373472493792,
      statistic_location=0.8987346602526389, statistic_sign=-1)
```

Programme 8.27 Q-Q plots for the fitted *t*-copula of HSBC, CLP, and CK via Python, using the `QQ_Plot()` function in Programme 8.21.

```
1 > sim_t_md2 <- Mahalanobis2(return_sim_t)
2 > QQ_Plot(sim_t_md2, returns_md2, col="orange")
3 >
4 > ks.test(sqrt(returns_md2), sqrt(sim_t_md2))
5
```

```
6        Asymptotic two-sample Kolmogorov-Smirnov test
7
8   data:  sqrt(returns_md2) and sqrt(sim_t_md2)
9   D = 0.01816, p-value = 0.9052
10  alternative hypothesis: two-sided
```

Programme 8.28 Q-Q plots for the fitted *t*-copula of HSBC, CLP, and CK via **R**, using the `QQ_Plot()` function in Programme 8.23.

Lastly, we inspect the goodness-of-fit of the Gaussian and *t*-copula models by using a Q-Q plot of the empirical squared Mahalanobis distances, and the plot of squared differences of the squared Mahalanobis distances off from the 45-degree reference line. For ease of comparison, we consider quantiles larger than 15 separately in the Q-Q plot and plot the largest 50 squared residuals, excluding the one with the largest value. We implement this comparison in Programmes 8.29 via Python and 8.30 via **R**.

```
1   n_days = len(returns)
2   i = (np.arange(1, n_days + 1) - 0.5) / n_days
3   q_N = np.quantile(sim_N_md2, i, method='interpolated_inverted_cdf')
4   q_t = np.quantile(sim_t_md2, i, method='interpolated_inverted_cdf')
5
6   a = 15
7   # find the common index where both coordinates > a
8   sort_returns_md2 = np.sort(returns_md2)
9   idx = set.intersection(set(np.where(q_N>a)[0]),
10                         set(np.where(q_t>a)[0]),
11                         set(np.where(sort_returns_md2>a)[0]))
12  idx_start = min(idx)
13  print(idx_start)
14
15  fig, ax = plt.subplots(figsize=(10, 10), dpi=200)
16  # skip the largest (last) entry
17  ax.scatter(q_N[idx_start:], sort_returns_md2[idx_start:],
18             color="blue", facecolor="white", marker="o", s=100)
19  ax.plot(q_t[idx_start:], sort_returns_md2[idx_start:],
20          "x", color="orange", markersize=15)
21  x = np.linspace(*ax.get_xlim())
22  ax.plot(x, x, color="black", linewidth=2)
23  ax.set_xlabel("Bootstrapped quantiles")
24  ax.set_ylabel("Sample quantiles")
25  ax.set_title("Copula Q-Q Plot")
26  ax.legend(["Gaussian copula", "t-copula"])
27
28  # residuals with respect to the 45-degree line
29  resid2_N = (sort_returns_md2 - q_N)**2
30  resid2_t = (sort_returns_md2 - q_t)**2
31
32  fig = plt.figure(figsize=(10, 10), dpi=200)
33  # plot the largest 50 and skip the largest (last) entry
34  plt.plot(np.sort(resid2_N)[-50:-1], "bo",
35           markerfacecolor="white", markersize=15)
36  plt.plot(np.sort(resid2_t)[-50:-1], "x", color="orange", markersize=15)
```

```
37 plt.xlabel("Index")
38 plt.ylabel("Squared residuals")
39 plt.legend(["Gaussian copula", "t-copula"])
```

```
1 966
```

Programme 8.29 Q-Q plots and the plot of squared differences for the fitted Gaussian copula and t-copula models via Python.

```
1  > n_days <- nrow(returns)
2  > i <- ((1:n_days) - 0.5) / n_days
3  > q_N <- quantile(sim_N_md2, probs=i, type=4, names=FALSE)
4  > q_t <- quantile(sim_t_md2, probs=i, type=4, names=FALSE)
5  >
6  > a <- 15
7  > leg <- c("Gaussian copula", "t-copula")
8  > # find the common index where both coordinates > a
9  > sort_returns_md2 <- sort(returns_md2)
10 > idx <- min(Reduce(intersect, list(which(q_N > a), which(q_t > a),
11 +                                   which(sort_returns_md2 > a))))
12 > (idx_start <- min(idx))
13 [1] 966
14 >
15 > plot(q_N[idx_start:n_days], sort_returns_md2[idx_start:n_days],
16 +      xlab="Bootstrapped quantiles", ylab="Sample quantiles",
17 +      main="Copula Q-Q Plot",
18 +      pch=1, cex=2, col="blue", xlim=c(a, max(q_N, q_t)))
19 > points(q_t[idx_start:n_days], sort_returns_md2[idx_start:n_days],
20 +        pch=4, cex=2, lwd=2, col="orange")
21 > legend("topleft", pch=c(1, 4), cex=2, lwd=c(1, 2),
22 +        col=c("blue", "orange"), lty=0, legend=leg)
23 > abline(0, 1, lwd=2)
24 >
25 > # residuals with respect to the 45-degree line
26 > sort_resid2_N <- sort((sort(returns_md2) - q_N)^2)
27 > sort_resid2_t <- sort((sort(returns_md2) - q_t)^2)
28 > # plot the largest 50 and skip the largest (last) entry
29 > idx_plots <- (n_days-50):(n_days-1)
30 > plot(sort_resid2_N[idx_plots], ylab="squared residuals",
31 +      ylim=c(0,max(sort_resid2_N[idx_plots],sort_resid2_t[idx_plots])),
32 +      pch=1, cex=2, col="blue")
33 > points(sort_resid2_t[idx_plots], pch=4, cex=2, lwd=2, col="orange")
34 > legend("topleft", pch=c(1, 4), cex=2, lwd=c(1, 2),
35 +        col=c("blue", "orange"), lty=0, legend=leg)
```

Programme 8.30 Q-Q plots and the plot of squared differences for the fitted Gaussian copula and t-copula models via R.

From Figures 8.16 via Python and 8.17 via **R**, the line of best fit for the t-copula model is closer to the 45-degree reference line than that for the Gaussian copula model, which indicates that the t-copula fits better with the dataset. On the other hand, the squared differences under the fitted t-copula model are mostly very slightly

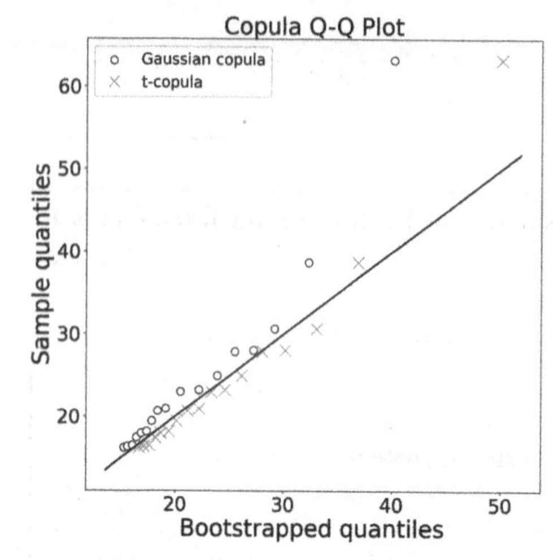

(a) Solid line in black is the 45-degree reference line

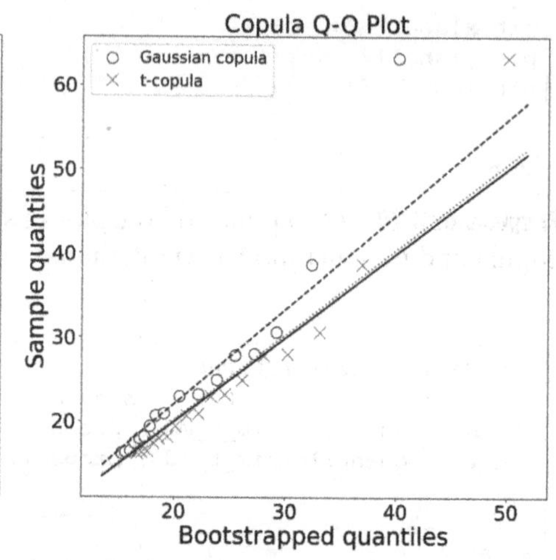

(b) Dashed line in blue (resp. dotted line in orange) is the line of best-fit under Gaussian copula (resp. t-copula) model

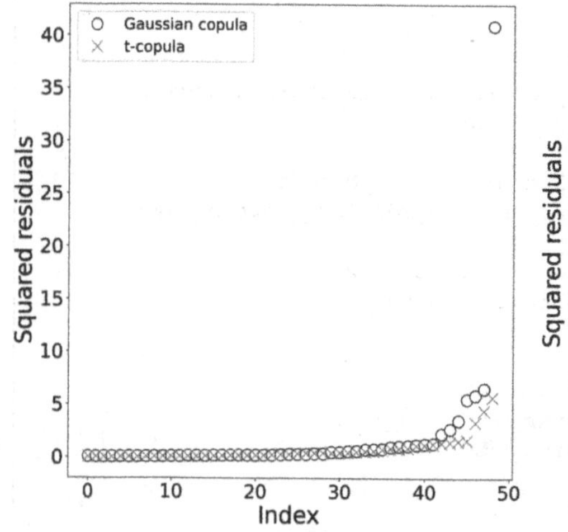

(c) Largest 50 squared differences excluding the last and largest entry (49 of them only)

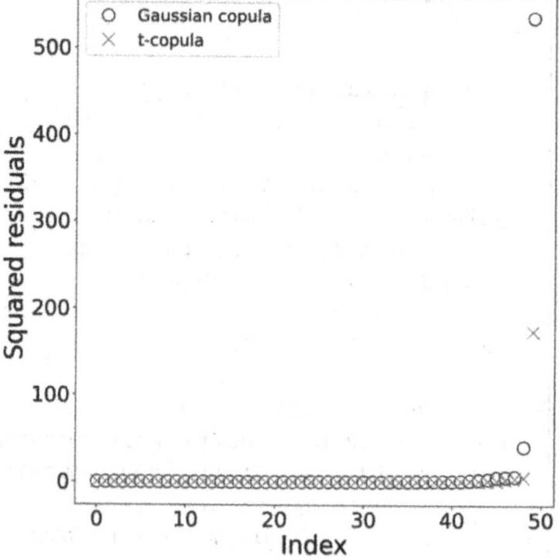

(d) Largest 50 squared differences

FIGURE 8.16 Q-Q plot and the plot of squared differences for the fitted Gaussian copula and t-copula models for the returns of HSBC, CLP, and CK over the period of 1999 to 2002 using Python, generated by Programme 8.29.

smaller than those obtained from the fitted Gaussian copula model. In particular, from both the Q-Q plots and the plots of squared differences, the largest quantile has the largest deviation from the 45-degree reference line; for the Gaussian copula this entry deviates more heavily than for the t-copula. This observation indicates that t-copula is indeed more capable of capturing extremal events than Gaussian copula, especially useful for periods like the 2007–2008 Credit Crunch.

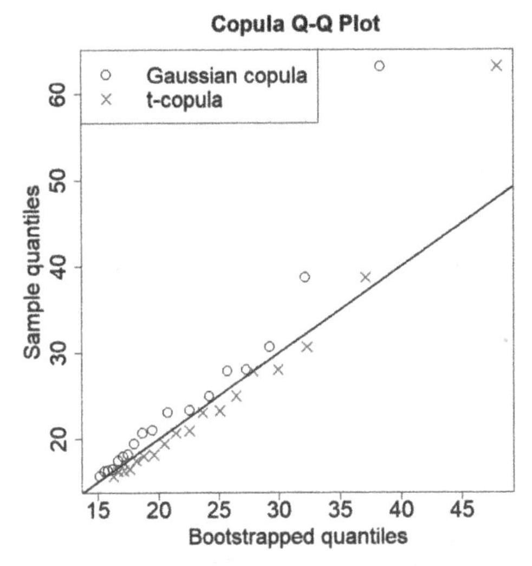

(a) Solid line in black is the 45-degree reference line

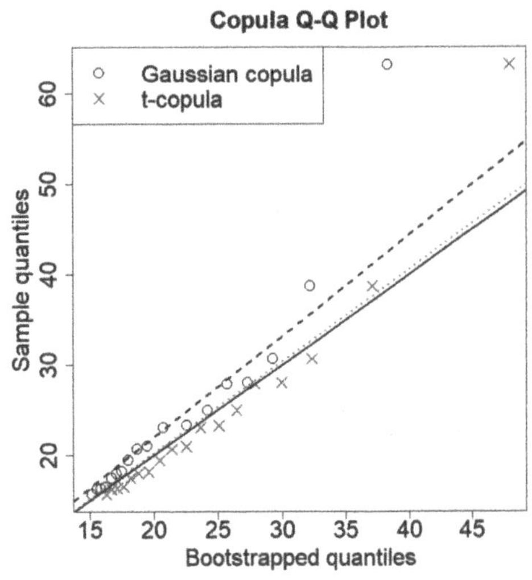

(b) Dashed line in blue (resp. dotted line in orange) is the line of best-fit under Gaussian copula (resp. t-copula) model

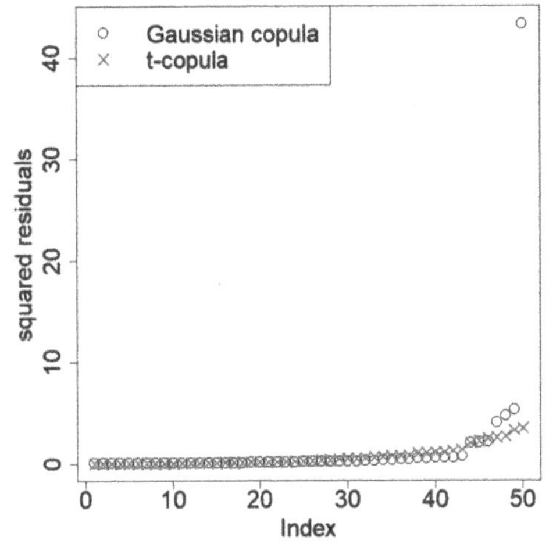

(c) Largest 50 squared differences excluding the last and largest entry (49 of them only)

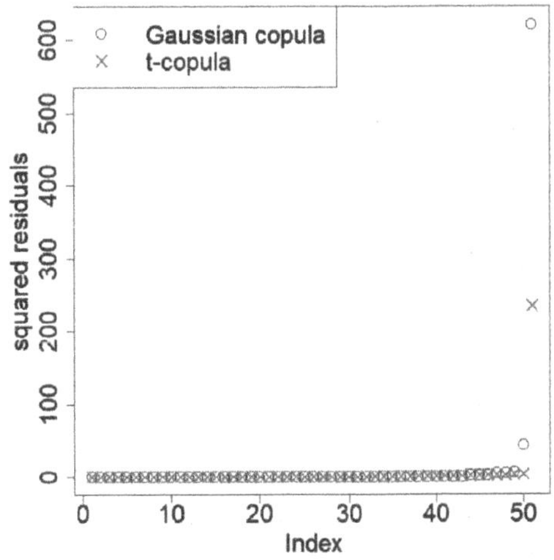

(d) Largest 50 squared differences

FIGURE 8.17 Q-Q plot and the plot of squared differences for the fitted Gaussian copula and t-copula models for the returns of HSBC, CLP, and CK over the period of 1999 to 2002 using **R**, generated by Programme 8.30.

Indeed, let us examine the goodness-of-fit of the Gaussian copula and the t-copula models during the 2007–2008 Credit Crunch period via the dataset `stock_2006_2009.csv`, which records the adjusted closing prices of HSBC, CLP, and CK from 2006 to 2009. From Figures 8.18 via Python and 8.19 via **R**, the largest 50 squared differences under the Gaussian copula model are generally slightly

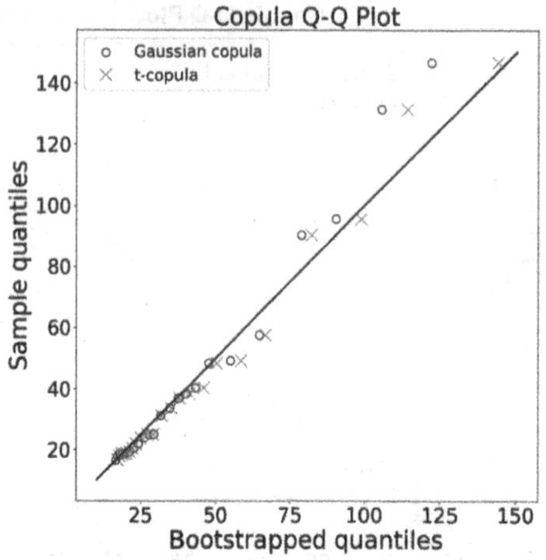

(a) Solid line in black is the 45-degree reference line

(b) Dashed line in blue (resp. dotted line in orange) is the line of best-fit under Gaussian copula (resp. t-copula) model

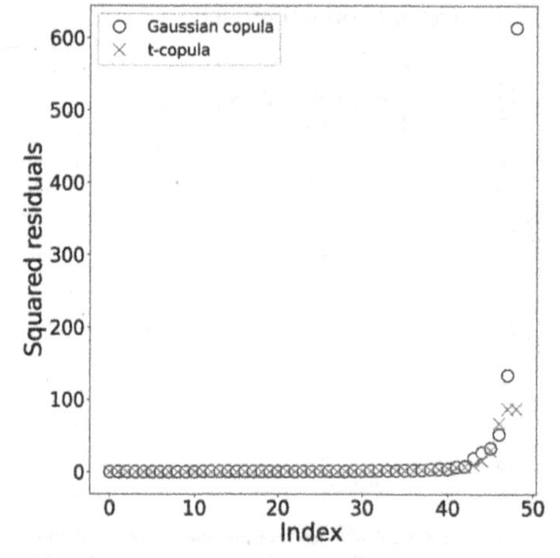

(c) Largest 50 squared differences excluding the last and largest entry (49 of them only)

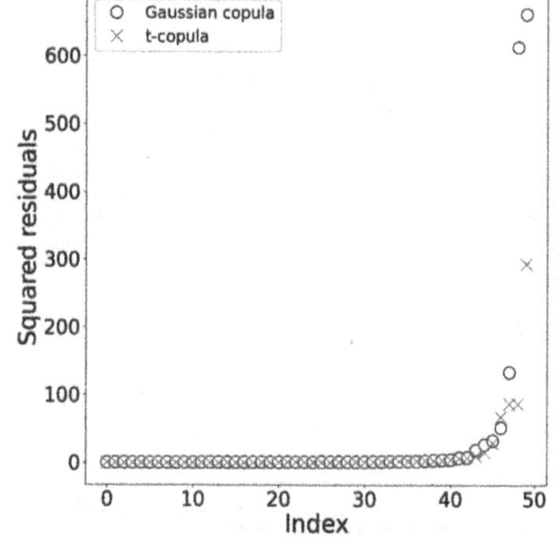

(d) Largest 50 squared differences

FIGURE 8.18 Q-Q plot and the plot of squared differences for the fitted Gaussian copula and t-copula models for the returns of HSBC, CLP, and CK over the period of 2006 to 2009 using Python, generated by Programme 8.29.

smaller than those under the t-copula model; however they also have much heavier tail sample values. In particular, as expected, on fair weather days, Gaussian copulae perform well for fitting financial return data, while t-copulae handle extreme values better. During the 2007–2008 Credit Crunch, the volatilities of the stock prices increased sharply and these prices plummeted drastically, leading to more extreme returns compared to those during normal periods.

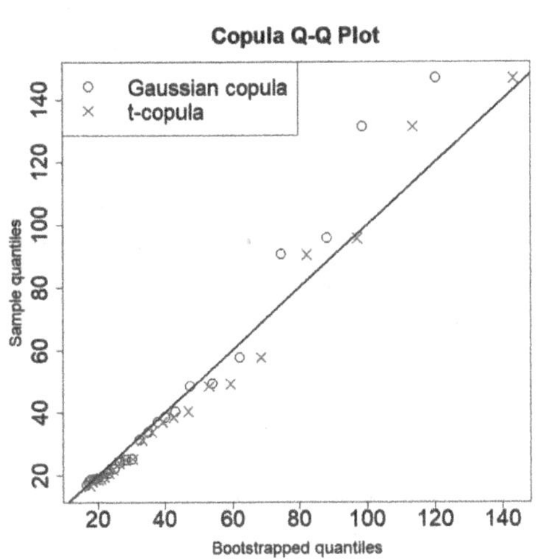

(a) Solid line in black is the 45-degree reference line

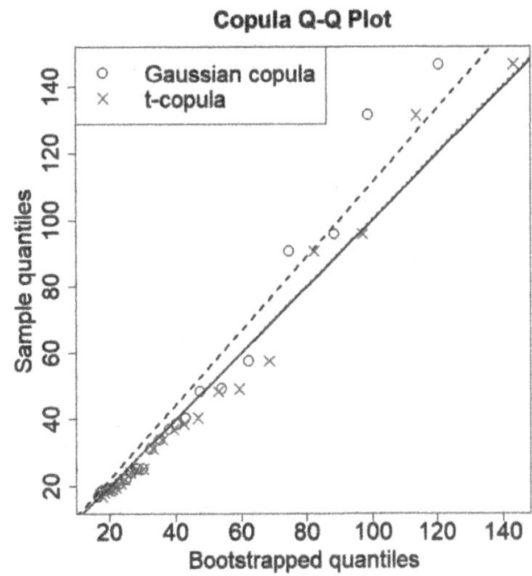

(b) Dashed line in blue (resp. dotted line in orange) is the line of best-fit under Gaussian copula (resp. t-copula) model

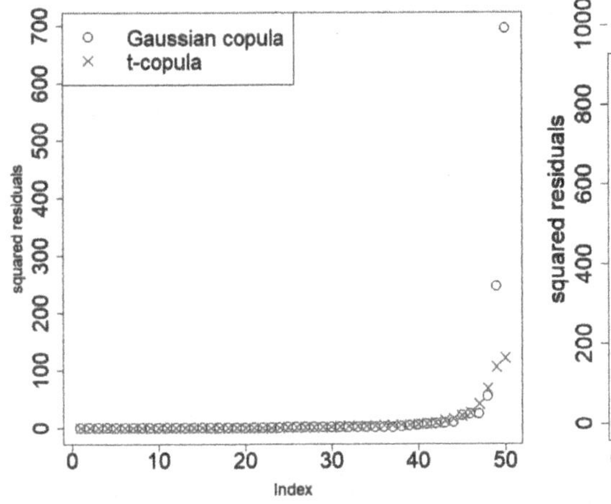

(c) Largest 50 squared differences excluding the last and largest entry (49 of them only)

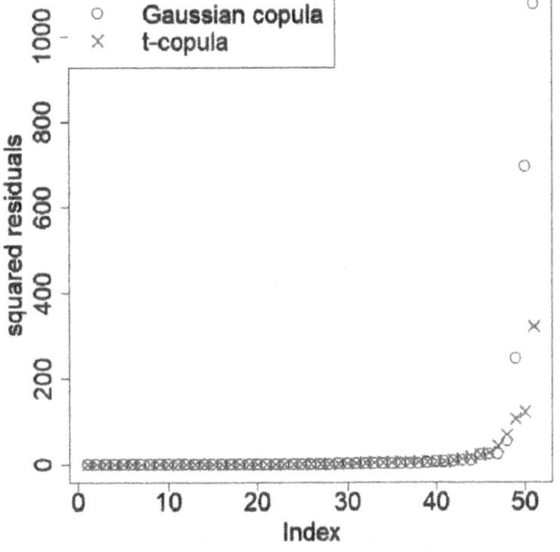

(d) Largest 50 squared differences

FIGURE 8.19 Q-Q plot and the plot of squared differences for the fitted Gaussian copula and t-copula models for the returns of HSBC, CLP, and CK over the period of 2006 to 2009 using **R**, generated by Programme 8.30.

(IV) Comonotonic Copula

The *comonotonic copula* is given by

$$M(u) := \min(u_1, \dots, u_d), \quad \text{for}\, u_1, \dots, u_d \in [0, 1]. \tag{8.14}$$

It corresponds to the case of extremal positive dependence (sometimes called *complete dependence*). The random variables $x_1, \dots, x_d$ have the comonotonic copula if and only if they are increasing transformations of a common source of randomness in the sense that $x_i = f_i(w)$, for every i, there is an increasing function $f_i : \mathbb{R} \to \mathbb{R}$, and w is a common random variable.

(V) Upper Tail Comonotonic Copula[5]

Despite its simplicity and nice analytical tractability, global comonotonicity over the whole domain of definition is too restrictive and has limited scope of application in quantitative finance and risk management. Nevertheless, comonotonicity happening only at the extreme upper tail is a far more realistic modelling choice and this leads us to consider the prevalent notion of *upper tail comonotonicity*, in which the survival copula $\widehat{C}$ satisfies:

$$\widehat{C}(\overline{F}_1(x_1), \dots, \overline{F}_d(x_d)) \approx \min\{\overline{F}_1(x_1), \dots, \overline{F}_d(x_d)\}$$

only for all sufficiently large values of x_i's. Copulae having this property regardless of the marginal distributions should behave in such a way that

$$\widehat{C}(u_1, \dots, u_d) \approx \min\{u_1, \dots, u_d\},$$

for all sufficiently small values of u_i's. In addition, to ensure mathematical tractability and facilitate its practical implementation, it is appealing to impose further regularity conditions on $\widehat{C}$. One reasonable requirement is scaling invariance, in the sense of a comparable rate of convergence with respect to u_i's, namely:

$$\widehat{C}(uw_1, \dots, uw_d) \sim \min\{uw_1, \dots, uw_d\}, \quad u \downarrow 0, \quad \text{for any } w_1, \dots, w_d \in [0, \infty).$$

These considerations lead us to define upper tail comonotonicity (cf. [26]) as follows. A random vector $\mathbf{x}$ with a copula C and a survival copula $\widehat{C}$ is said to be *upper tail comonotonic* if for any $w_1, \dots, w_d \in [0, \infty)$,

$$\lim_{u \downarrow 0} \frac{\widehat{C}(uw_1, \dots, uw_d)}{u} = \min\{w_1, \dots, w_d\}.$$

Such a copula $\widehat{C}$ is then said to be an *upper tail comonotonic copula*.

Upper tail comonotonicity is closely connected to the asymptotic additivity of Value-at-Risk. Consider a portfolio of d loss variables $x_1, \dots, x_d$. We aim to examine the Value-at-Risk of the aggregate loss $s = x_1 + \cdots + x_d$. Suppose that for some $1 \leq m \leq d$, $x_1, \dots, x_m \in \text{MDA}(\Phi_\alpha)$[6] for a common index α and they are upper tail comonotonic. Also, for $j = m + 1, \dots, d$, each x_j satisfies one of the following requirements:

1. $x_j \in \text{MDA}(\Phi_{\alpha'})$[6] with $\alpha' > \alpha$;
2. $x_j \in \text{MDA}(\Lambda)$[6];
3. x_j has a finite endpoint.

[5] From here on, in the rest of this chapter, the discussions are aligned with the joint work with our former postgraduate student, and part of the results is also included in her dissertation [44].

[6] We refer to Appendix *8.A.1 for more information about the meaning of *Maximum Domain of Attraction* (MDA) and related concepts from classical Extreme Value Theory.

Under the above setting, our recent work [11] establishes that Value-at-Risk is asymptotically additive for such portfolios:

$$\mathrm{V@R}_p(\mathrm{s}) \sim \sum_{i=1}^{d} \mathrm{V@R}_p(\mathrm{x}_i) \quad p \uparrow 1. \tag{8.15}$$

Moreover, there are two other scenarios where such asymptotic additivity of Value-at-Risk remains true:

1. The non-negative random variables $\mathrm{x}_1, \ldots, \mathrm{x}_m$ are in MDA(Λ)[6] with infinite endpoints and they are upper tail comonotonic, while other non-negative $\mathrm{x}_{m+1}, \ldots, \mathrm{x}_d$ have finite endpoints.
2. All the non-negative random variables $\mathrm{x}_1, \ldots, \mathrm{x}_d$ have finite endpoints and they are upper tail comonotonic.

(VI) Archimedean Copulae

Archimedean copulae form an associative class of copulae. Explicit formulae are readily available for most commonly used Archimedean copulae, while this is not the case for other classes of copulae such as the commonly used Gaussian copula. Besides, Archimedean copulae can be easily used to model dependent risks in an arbitrarily high dimension, which is another reason of their popularity in practice.

A copula $C_{Archimedean}$ is called *Archimedean* if it admits the following representation:

$$C_{Archimedean}(\boldsymbol{u};\theta) = \psi^{[-1]}(\psi(u_1;\theta) + \cdots + \psi(u_d;\theta);\theta), \tag{8.16}$$

where$[0,1] \times \Theta \ni (u,\theta) \mapsto \psi(u,\theta) \in [0,\infty)$ is a continuous, strictly decreasing and convex function in u such that $\psi(1;\theta) = 0$, θ is a parameter within some parameter space Θ governing the underlying dependence of multivariate risks; ψ is often called the *generator function* and $\psi^{[-1]}$ is its *pseudo-inverse* defined by

$$\psi^{[-1]}(t;\theta) := \begin{cases} \psi^{-1}(t;\theta), & \text{if } 0 \le t \le \psi(0;\theta); \\ 0, & \text{if } \psi(0;\theta) \le t \le \infty. \end{cases}$$

Moreover, the above formula (8.16) for $C_{Archimedean}$ yields a copula for $\psi^{[-1]}$ if and only if $\psi^{[-1]} = \psi^{-1}$ is d-monotone on $[0,\infty)$, meaning that if it is $d-2$ times differentiable and its derivatives satisfy the following inequalities (closely related to the notion of completely monotone functions):

$$(-1)^k \left(\psi^{-1}\right)^{(k)}(t;\theta) \ge 0,$$

for all $t \ge 0$ and $k = 0, 1, \ldots, d-2$, while $(-1)^{d-2}\left(\psi^{-1}\right)^{(d-2)}(t;\infty)$ is also non-increasing and convex.

The class of Archimedean copulae is broad and covers many commonly used copulae as special cases. Below is a partial list; more examples can be found in Table 4.1 of [35].

1. *Ali–Mikhail–Haq* family: introduced in [2], the generator of this family is given by

$$\psi(t,\theta) = \ln\left(\frac{1-\theta}{t} + \theta\right), \quad \theta \in [0,1).$$

2. *Clayton* family: introduced in [13], the generator of this family is given by

$$\psi(t,\theta) = \frac{1}{\theta}\left(t^{-\theta} - 1\right), \quad \theta \in [-1,\infty)\backslash\{0\}.$$

   The function inverse of this generator function is the Laplace transform of a Gamma distribution with a shape parameter $1/\theta$ and a scale parameter θ, thus the Clayton copula model has a *Gamma-distributed frailty representation* [42].
3. *Frank* family: first appeared in [20], the generator of this family is given by

$$\psi(t,\theta) = -\ln\left(\frac{e^{-\theta t} - 1}{e^{-\theta} - 1}\right), \quad \theta \in (0,\infty).$$

   In the bivariate case, it is the only Archimedean family which is *radially symmetric*, meaning that its copula and survival copula are the same.
4. *Gumbel* family: introduced in [24], the generator of this family is given by

$$\psi(t,\theta) = (-\ln(t))^{\theta}, \quad \theta \in [1,\infty).$$

5. *Joe* family: discussed in [29, 30], the generator of this family is given by

$$\psi(t,\theta) = -\ln\left(1 - (1-t)^{\theta}\right), \quad \theta \in [1,\infty).$$

Note that for all five families, a larger value of the parameter θ indicates a stronger dependence; all (except the Frank copula) reduce to the independence copula when θ takes the minimum value in the respective ranges, while for the Frank copula, the limiting case with θ approaching 0 corresponds to the independence copula. The scatter plots for the bivariate case of these five copula families, each based on 2,000 simulated points, are depicted in Figures 8.20 and 8.21.

Knowing all of the above, we examine the asymptotic behaviour of the Value-at-Risk, as the probability level approaches 1, of the aggregate loss of a portfolio of risks $\mathrm{x}_1, \ldots, \mathrm{x}_d$ whose dependence is given by an Archimedean copula. More precisely, we assume that

$$\mathbb{P}(\mathrm{x}_1 > x_1, \mathrm{x}_2 > x_2, \ldots, \mathrm{x}_d > x_d) = \psi^{[-1]}\left(\sum_{i=1}^{d} \psi(\overline{F}_i(x_i))\right),$$

where $\overline{F}_i$ is the survival function of x_i, $\psi : [0,1] \to [0,\infty)$ is a generator that is *regularly varying*[7] at 0 (or equivalently, $\psi(1/x)$ is regularly varying at infinity) with an index $-\gamma < 0$. Under this setup, [12] showed that when at least one of these d loss

[7] We refer to Appendix *8.A.1 for the definition of *regularly varying functions*.

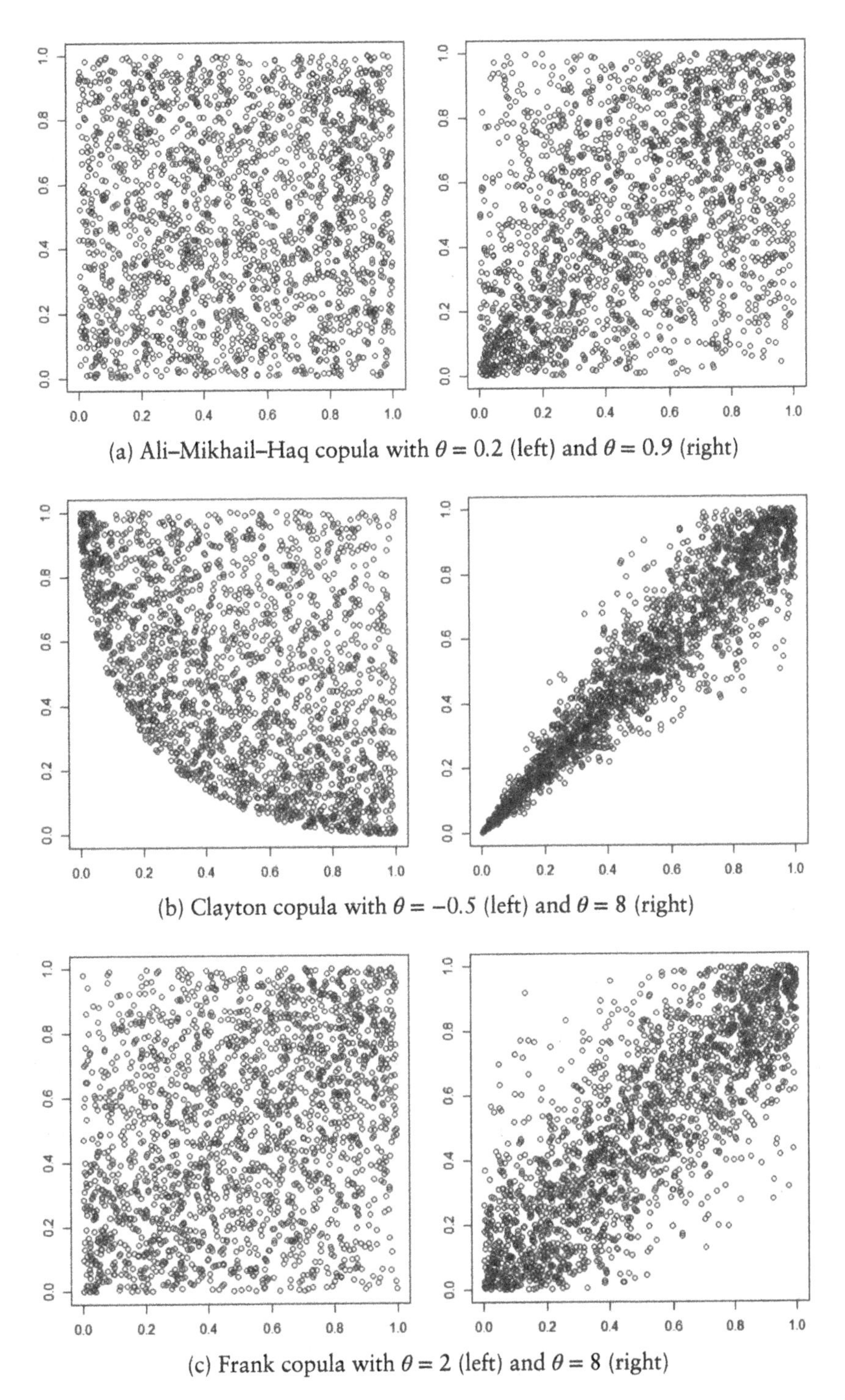

(a) Ali–Mikhail–Haq copula with $\theta = 0.2$ (left) and $\theta = 0.9$ (right)

(b) Clayton copula with $\theta = -0.5$ (left) and $\theta = 8$ (right)

(c) Frank copula with $\theta = 2$ (left) and $\theta = 8$ (right)

FIGURE 8.20 Scatter plots for bivariate risks by using different Archimedean copulae in R.

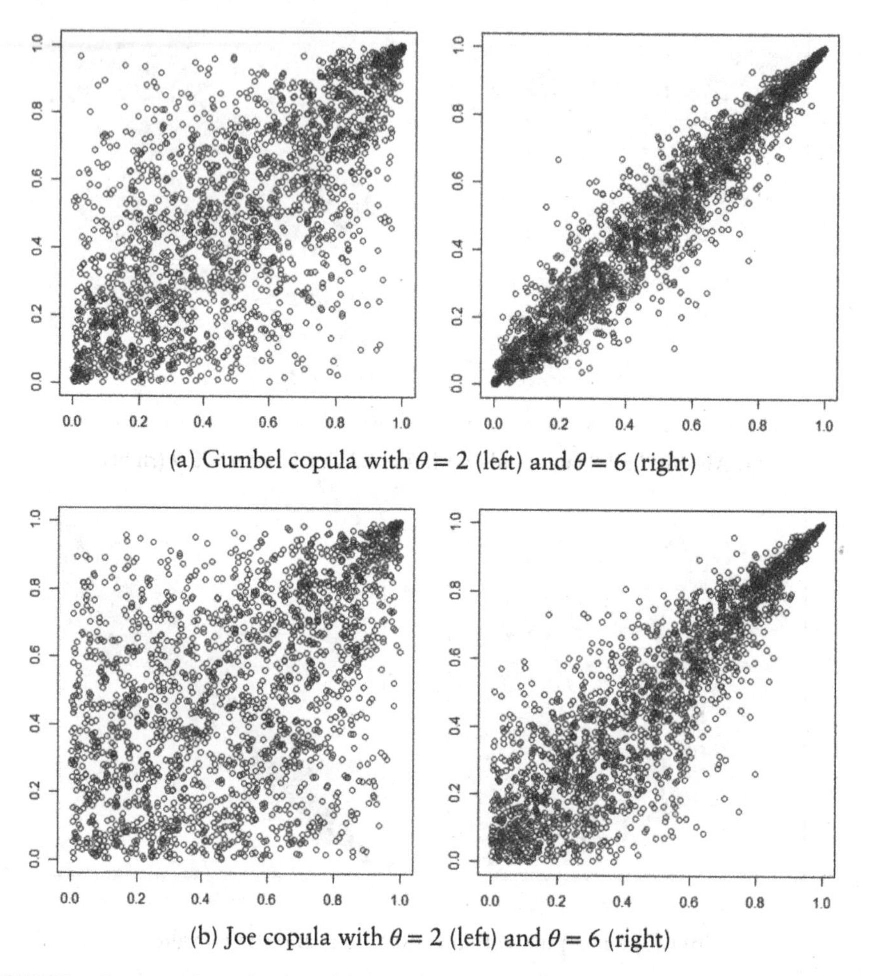

(a) Gumbel copula with $\theta = 2$ (left) and $\theta = 6$ (right)

(b) Joe copula with $\theta = 2$ (left) and $\theta = 6$ (right)

FIGURE 8.21 Scatter plots for bivariate risks by using different Archimedean copulae in R (continued).

variables is heavy-tailed so that it belongs to $\mathrm{MDA}(\Phi_\alpha)$ for some α, and the tails of all remaining variables are not as heavy as this chosen one, then asymptotically the Value-at-Risk of the aggregate loss $s = x_1 + \cdots + x_d$ can exhibit different behaviour, depending on whether α is greater than, equal to, or less than 1. More precisely, suppose that for some $1 \leq m \leq d$, $x_1, \ldots, x_m \in \mathrm{MDA}(\Phi_{\alpha^*})$ for a common index α^*, while for $i \in \{m+1, \ldots, d\}$, each x_i satisfies one of the following requirements: (i) $x_i \in \mathrm{MDA}(\Phi_{\alpha_i})$ with $\alpha_i > \alpha^*$; or (ii) $x_i \in \mathrm{MDA}(\Lambda)$; or (iii) x_i has a finite endpoint, then

1. when $\alpha^* > 1$, V@R is asymptotically subadditive:

$$\limsup_{p \to 1} \frac{\mathrm{V@R}_p(s)}{\sum_{j=1}^{d} \mathrm{V@R}_p(x_j)} \leq 1;$$

2. when $0 < \alpha^* < 1$, V@R is asymptotically superadditive:

$$\liminf_{p\to 1} \frac{\mathrm{V@R}_p(\mathrm{s})}{\sum_{j=1}^d \mathrm{V@R}_p(\mathrm{x}_j)} \geq 1;$$

3. when $\alpha^* = 1$, V@R is asymptotically additive:

$$\lim_{p\to 1} \frac{\mathrm{V@R}_p(\mathrm{s})}{\sum_{j=1}^d \mathrm{V@R}_p(\mathrm{x}_j)} = 1.$$

If none of the loss variables belongs to MDA(Φ_α) for whatever α, the following results hold:

1. If $\mathrm{x}_i \in \mathrm{MDA}(\Lambda)$ for $i = 1, \ldots, m$ and they all have infinite endpoints, while the remaining loss variables $\mathrm{x}_{m+1}, \cdots, \mathrm{x}_d$ have finite endpoints, then V@R is asymptotically subadditive.
2. If x_i has a finite endpoint x_{F_i} for all $i = 1, \ldots, d$, then V@R is asymptotically additive; indeed, we clearly have:

$$\lim_{p\to 1} \mathrm{V@R}_p\left(\sum_{i=1}^d \mathrm{x}_i\right) = \lim_{p\to 1} \sum_{i=1}^d \mathrm{V@R}_p\left(\mathrm{x}_i\right) = \sum_{i=1}^d x_{F_i}.$$

Let us present several numerical examples to illustrate these claims. We consider a Clayton copula with the parameter $\theta = 1$, whose generator is given by $\psi(t) = t^{-1} - 1$. We use the **R** package `copula` to simulate the Clayton copula with some given marginals. By generating 10^7 data points from the corresponding joint distributions, we plot the V@R ratios of $\frac{\mathrm{V@R}_p(\mathrm{s})}{\sum_{i=1}^d \mathrm{V@R}_p(\mathrm{x}_i)}$ in the y-axis against the probability level p in the x-axis, with $d = 2, 5$. Illustrations for various scenarios are depicted in Figures 8.22 to 8.24.

(VII) Extreme-value Copulae

Finally, we introduce the family of extreme-value copulae, whose first construction dates back to [15, 22]. A copula C is called an *extreme-value copula* if there exists a copula C_F such that

$$\left(C_F(u_1^{1/n}, \ldots, u_d^{1/n})\right)^n \to C(u_1, \ldots, u_d) \quad \text{as} \quad n \to \infty,$$

for all $(u_1, \ldots, u_d) \in [0,1]^d$. The copula C_F is said to be in the *domain of attraction* of the copula C. A closely related concept is *max-stability:* a d-dimensional copula C is *max-stable* if it satisfies

$$C(u_1, \ldots, u_d) = \left(C\left(u_1^{1/m}, \ldots, u_d^{1/m}\right)\right)^m,$$

for all integers $m \geq 1$ and all $(u_1, \ldots, u_d) \in [0,1]^d$. A max-stable copula is clearly in its own domain of attraction, and hence it is an extreme-value copula itself. Indeed, these two notions are equivalent: a copula is an extreme-value copula if and only if

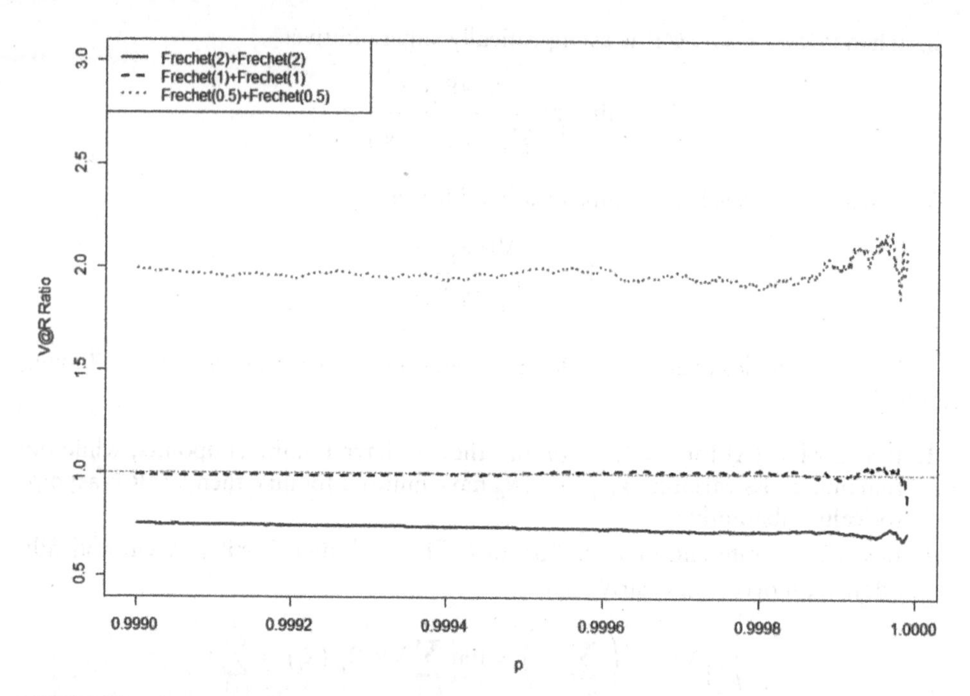

FIGURE 8.22 Bivariate Clayton copula: Fréchet + Fréchet marginals with $d = 2$.

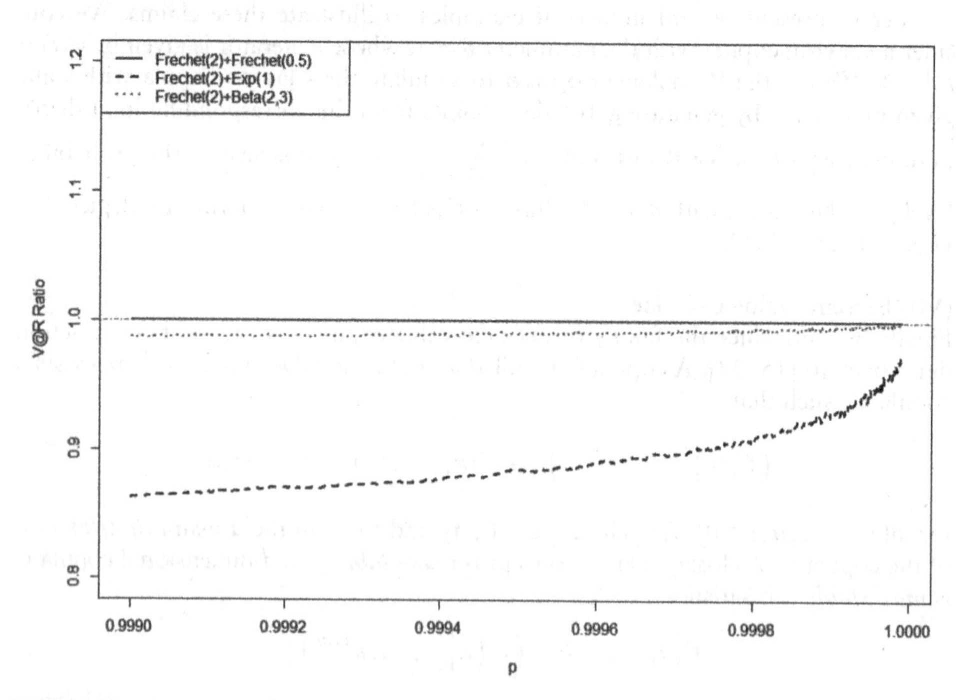

FIGURE 8.23 Bivariate Clayton copula: Fréchet + Gumbel marginals with $d = 2$.

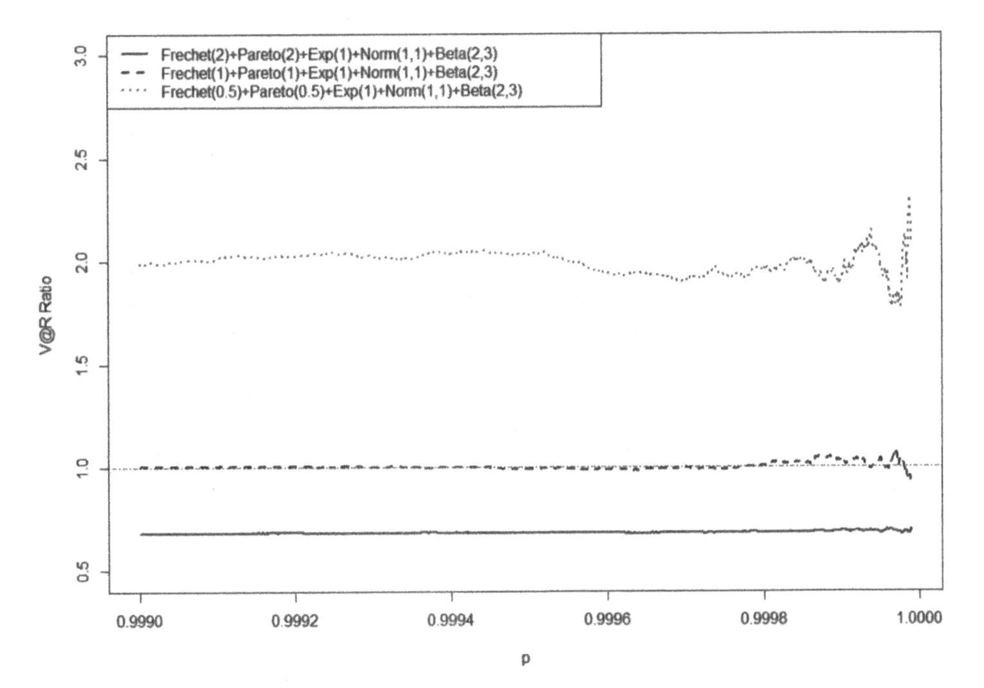

FIGURE 8.24 High-dimensional Clayton copula: Frechet + Gumbel + Weibull marginals with $d = 5$.

it is max-stable. To see that extreme-value copulae are also max-stable, let C be an extreme-value copula, and consider a fixed integer $m \geq 1$ and take $n = mk$. On the one hand, there is another copula C_F such that when $k \to \infty$ and so $n \to \infty$,

$$\left(C_F\left(u_1^{1/n},\ldots,u_d^{1/n}\right)\right)^n = \left[\left(C_F\left(\left(u_1^{1/m}\right)^{1/k},\ldots,\left(u_d^{1/m}\right)^{1/k}\right)\right)^k\right]^m \to \left(C\left(u_1^{1/m},\ldots,u_d^{1/m}\right)\right)^m;$$

on the other hand,

$$C_F^n\left(u_1^{1/n},\ldots,u_d^{1/n}\right) \to C(u_1,\ldots,u_d),$$

as $n \to \infty$. Therefore, C is max-stable.

We next introduce some common examples of extreme-value copulae.

I. *Gumbel–Hougaard* copula (Logistic Model): This copula is both an Archimedean copula, with a generator of $\psi(t,\theta) = (-\ln t)^\theta$, and an extreme-value copula. In general, it is defined as

$$C\left(u_1,\ldots,u_d;\alpha\right) = \exp\left[-\left(\left(-\ln u_1\right)^\alpha + \cdots + \left(-\ln u_d\right)^\alpha\right)^{1/\alpha}\right],$$

where $\alpha \geq 1$. The parameter α measures the degree of dependence, ranging from independence with $\alpha = 1$ to complete dependence (corresponding to the comonotonic compula) when $\alpha = \infty$.

II. *Galambos* copula (Negative Logistic Model): The Galambos copula is an extreme-value copula, defined as

$$C(u_1,\ldots,u_d;\alpha) = \exp\left[\sum_{I\subset\{1,\ldots,d\}} (-1)^{|I|}\left(\sum_{i\in I}\left|\ln(u_i)\right|^{-\alpha}\right)^{-1/\alpha}\right],$$

where $\alpha \in (0,\infty)$. When $\alpha \to \infty$, it refers to complete dependence, corresponding to the comonotonic copula; while when $\alpha \to 0$, it refers to independence. This multivariate extreme value copula was obtained by [21] as an extreme value limit, just as in (8.17) but for the multivariate case of the multivariate Pareto distribution; also see [31] for more discussions. In the extreme value literature, such as [14], this copula is sometimes called the *negative logistic model.*

III. *Hüsler–Reiss* copula: It is another extreme-value copula, and for the bivariate case, it is defined as

$$C(u_1,u_2;\alpha) = \exp\left(-\hat{u}_1\Phi\left[\frac{1}{\alpha}+\frac{\alpha}{2}\ln\left(\frac{\hat{u}_1}{\hat{u}_2}\right)\right] - \hat{u}_2\Phi\left[\frac{1}{\alpha}+\frac{\alpha}{2}\ln\left(\frac{\hat{u}_2}{\hat{u}_1}\right)\right]\right),$$

where $\hat{u}_1 = -\ln(u_1)$, $\hat{u}_2 = -\ln(u_2)$, $\Phi(\cdot)$ is the standard normal distribution function, and $\alpha \in (0,\infty)$. It corresponds to independence and complete dependence (again it degenerates to comonotonic copula) when α approaches 0 and infinity, respectively.

IV. The t-EV copula: The name t-EV was proposed in [16], where the bivariate t-EV copula is derived. The multivariate t-EV copula is derived in [36]. Let T_ν be the cumulative distribution function of the univariate t-distribution with $\nu > 0$ degrees of freedom. For the bivariate case, with a correlation coefficient $\rho \in [-1,1]$, the copula is defined as

$$C(u_1,u_2;\rho,\nu) = \exp\left[\ln(u_1u_2)B(\ln u_1/\ln(u_1u_2);\rho,\nu)\right],$$

where the function:

$$B(w;\rho,\nu) = wT_{\nu+1}\left(\frac{\sqrt{\nu+1}}{\sqrt{1-\rho^2}}\left[\left(\frac{w}{1-w}\right)^{1/\nu}-\rho\right]\right) + (1-w)T_{\nu+1}\left(\frac{\sqrt{\nu+1}}{\sqrt{1-\rho^2}}\left[\left(\frac{1-w}{w}\right)^{1/\nu}-\rho\right]\right).$$

V. *Tawn* copula: For the bivariate case, it is defined as

$$C(u_1,u_2) = u_1u_2\exp\left(-\alpha\frac{\ln(u_1)\ln(u_2)}{\ln(u_1u_2)}\right),$$

where $\alpha \in (0,1)$. When $\alpha = 0$, the independence copula can be achieved, but it is not possible to attain complete dependence.

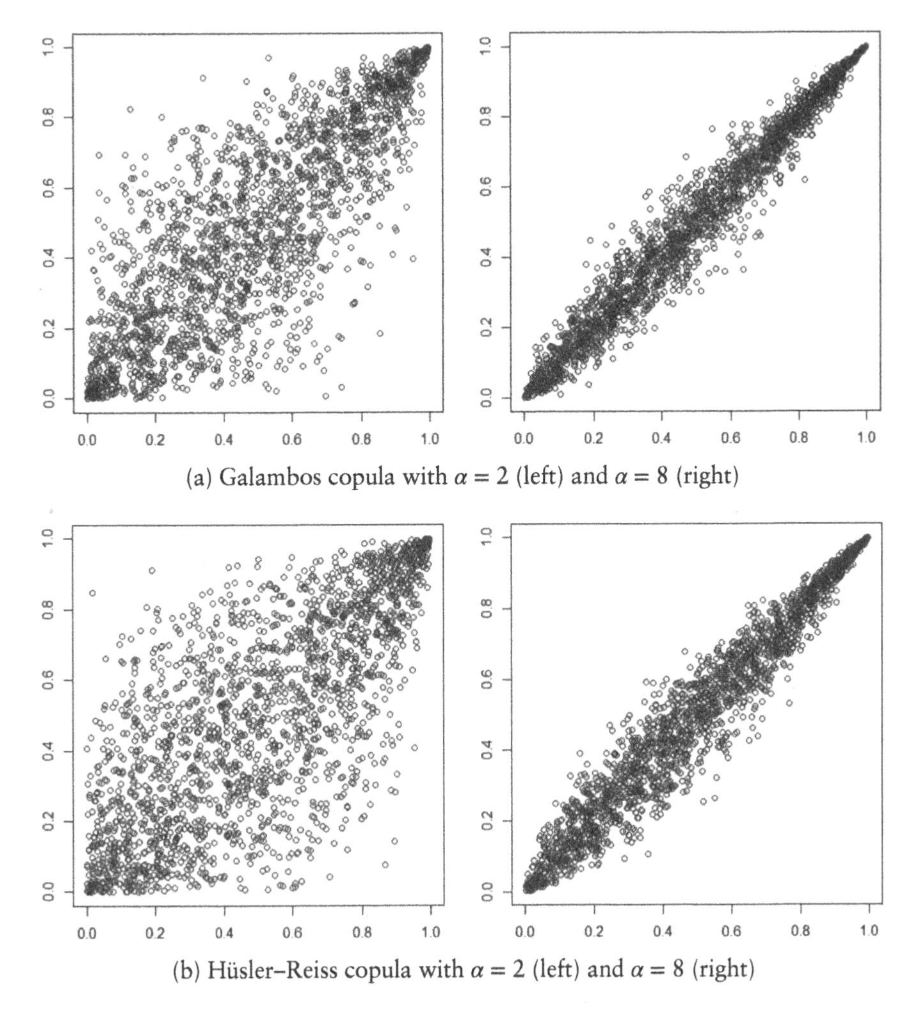

(a) Galambos copula with $\alpha = 2$ (left) and $\alpha = 8$ (right)

(b) Hüsler–Reiss copula with $\alpha = 2$ (left) and $\alpha = 8$ (right)

FIGURE 8.25 Scatter plots of some extreme-value copulae in R.

The scatter plots of 1,000 observation samples generated by these five copulae for selected parameter values are illustrated in Figures 8.25 and 8.26.

Finally, we have a similar asymptotic result for V@R under extreme-value copulae as that under Archimedean copulae. When the loss variables $x_1, \ldots, x_d$ have an *identical distribution* and their dependence is given by an extreme-value copula, [10] established the following results concerning the Value-at-Risk of the aggregate loss $s = x_1 + \cdots + x_d$ when the probability level approaches 1:

1. If $x_1 \in \text{MDA}(\Phi_\beta)$, then V@R is (i) asymptotically subadditive when $\beta > 1$; (ii) asymptotically superadditive when $\beta < 1$; and (iii) asymptotically additive when $\beta = 1$.
2. If $x_1 \in \text{MDA}(\Lambda)$ with an infinite right endpoint, then V@R is asymptotically subadditive.
3. If x_1 has a finite endpoint, then V@R is asymptotically additive.

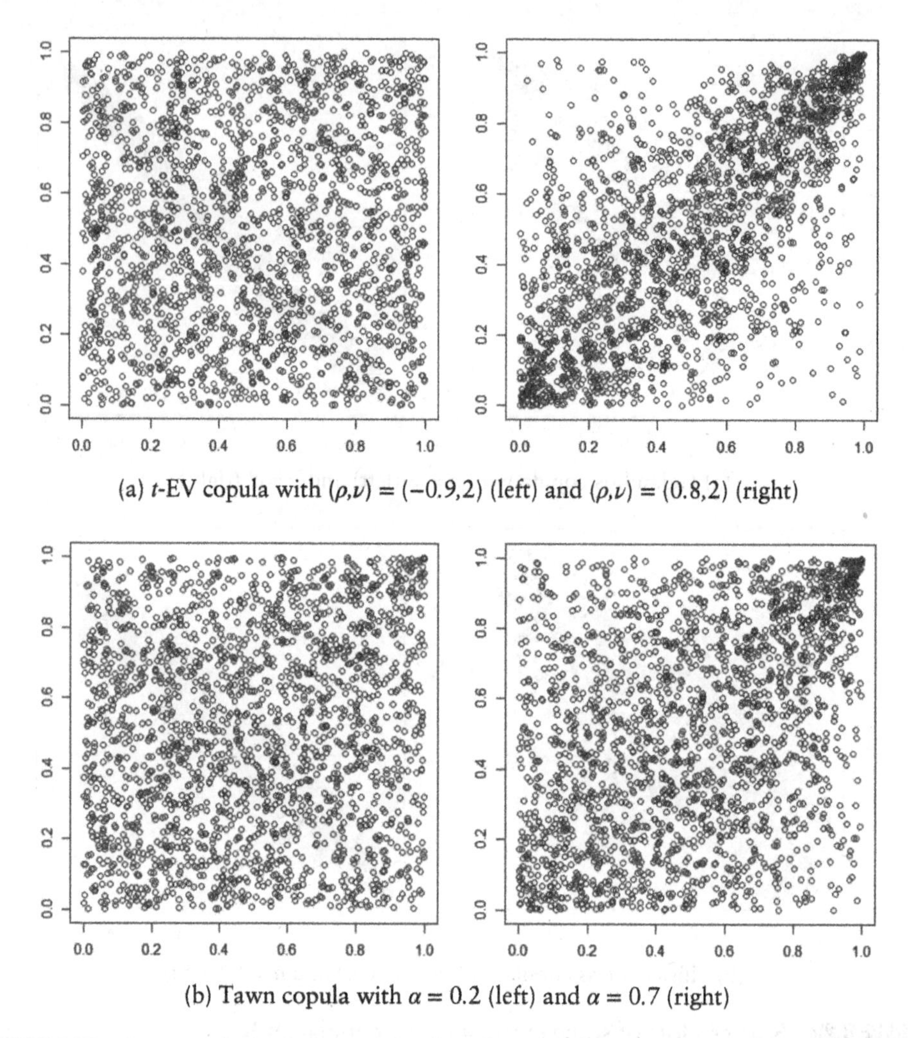

(a) t-EV copula with $(\rho,\nu) = (-0.9,2)$ (left) and $(\rho,\nu) = (0.8,2)$ (right)

(b) Tawn copula with $\alpha = 0.2$ (left) and $\alpha = 0.7$ (right)

FIGURE 8.26 Scatter plots of some extreme-value copulae in **R** (continued).

We use the **R** package evd to simulate the extreme-value copula with some given marginals. As an example, we generate 10^7 data points from the corresponding distributions, and plot the V@R ratios of $\frac{\text{V@R}_p(s)}{\sum_{i=1}^{d} \text{V@R}_p(x_i)}$ with given marginals under the Gumbel–Hougaard copula. In our simulation study, we simply set $\alpha = 1.5$. See Figures 8.27–8.29 for the illustrations.

It is believed that the above claims, 1) to 3), remain valid by and large, even for heterogeneous distributions and when the dependence structures among the variables are modelled by general copulae exhibiting certain kinds of positive dependence, and this would be an interesting future research project.

Last but not least, we give a brief introduction to the trendy *vine copula*, which has found extensive use in finance, such as for the effective modelling of tail risk in portfolio optimization applications (see [1] for detailed discussions). A *vine* is a graphical tool used to label constraints in high-dimensional probability distributions.

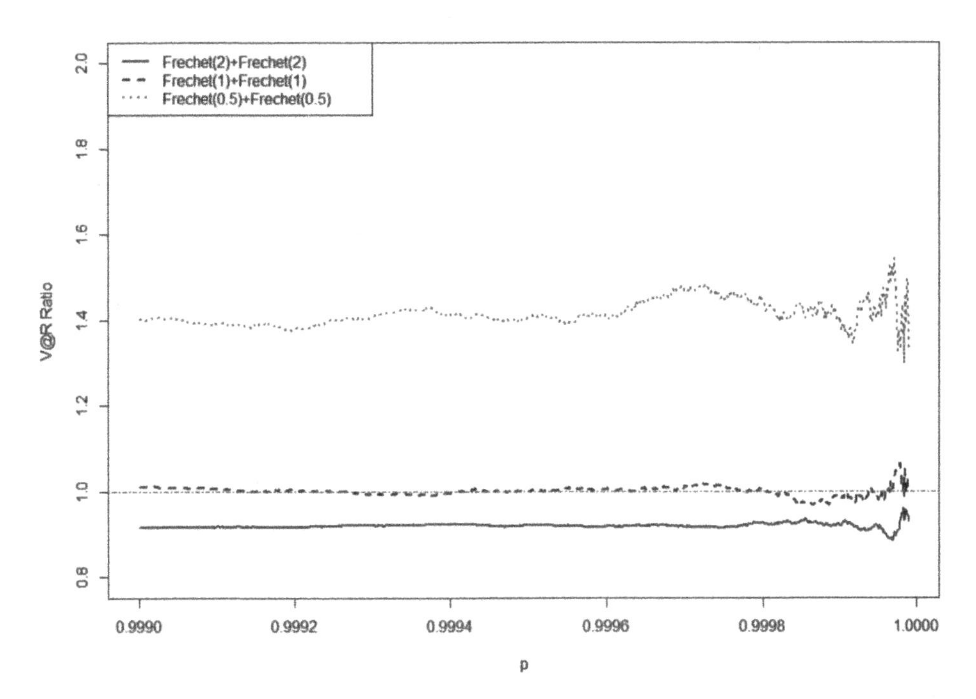

FIGURE 8.27 Gumbel–Hougaard copula: Bivariate Fréchet cases; also see [11].

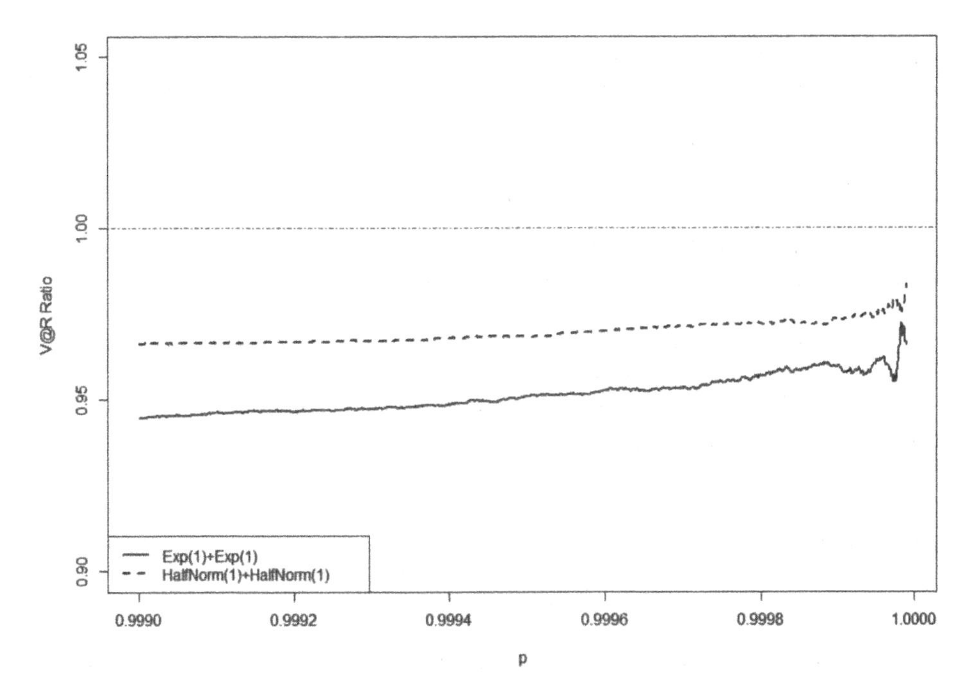

FIGURE 8.28 Gumbel–Hougaard copula: Gumbel marginals; also see [11].

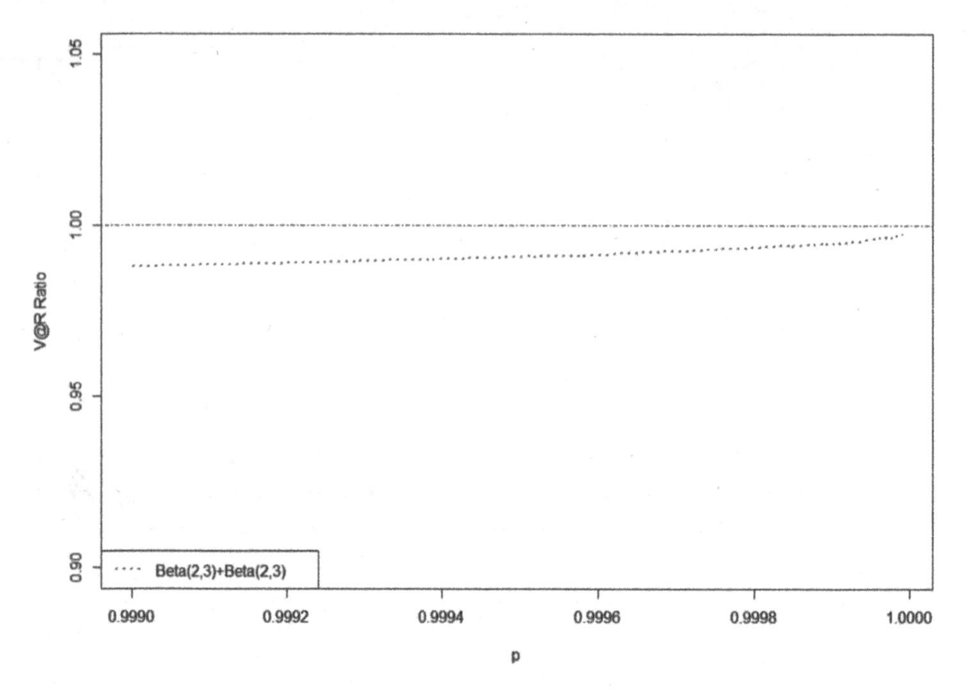

FIGURE 8.29 Gumbel–Hougaard copula: Weibull marginals; also see [11].

It is a nested set of connected trees, where the edges in one tree become the nodes of the next tree to be built. As a special case in which all constraints are two-dimensional or conditional two-dimensional, regular vines (a.k.a. R-vines) get more popular since the computation of joint probabilities is carried out in a sequential pairwise manner. Indeed, when the constraints are associated with conditional bivariate copulae, we impose the vine copula, such that any multivariate density can be represented as a product of marginal densities and conditional copula densities on an R-vine. One can interpret vine copulae as copulae that characterize the dependence structures among clusters of random variables. To this end, the *Comonotone-Independence Bayes classifier* (CIBer), which we shall introduce in Chapter 12, can be regarded as a special case of vine copula, in that the dependence structures there are handled in an all-or-nothing manner, namely either comonotonicity or independence. Due to space limitations, we shall not present its formulation and applications in detail.

*8.A APPENDIX

*8.A.1 A Quick Review of Extreme Value Theory

In Section 8.5, we introduced a basic fact in Extreme Value Theory (EVT) that the generalized Pareto distribution is the limit of the tail conditional distributions of a large class of distributions. In this appendix, we present a more comprehensive overview of some other related results in EVT. They are relevant to the discussion on the asymptotic behaviour of Value-at-Risk in Section 8.8. For a thorough discussion of EVT, one can also refer to [25, 40].

Let $\mathrm{x}_1, \ldots, \mathrm{x}_n$ be iid non-degenerate random variables and denote their maximum by $\mathrm{M}_n := \max\{\mathrm{x}_1, \ldots, \mathrm{x}_n\}$. Two distribution functions $F(x)$ and $G(x)$ are said to be of the *same type* if there exist $a > 0$ and $b \in \mathbb{R}$ such that $F(x) = G(ax + b)$ for all $x \in \mathbb{R}$. The *Fisher–Tippett theorem* [18], one of the cornerstone results of EVT, asserts that if there exist norming constants $a_n > 0$ and $b_n \in \mathbb{R}$, for $n = 1, 2, \ldots$ and some non-degenerate distribution function H such that

$$a_n^{-1}(\mathrm{M}_n - b_n) \xrightarrow{d} H, \quad \text{as } n \to \infty, \tag{8.17}$$

then H must belong to one of the classes of distributions being of the same type as one of the following three types of distribution functions: for $\alpha > 0$,

$$\text{(i) Fréchet:} \quad \Phi_\alpha(x) = \begin{cases} 0, & x \leq 0, \\ \exp\{-x^{-\alpha}\}, & x > 0; \end{cases}$$

$$\text{(ii) Gumbel:} \quad \Lambda(x) = \exp\{-e^{-x}\}, \quad x \in \mathbb{R};$$

$$\text{(iii) Weibull:} \quad \Psi_\alpha(x) = \begin{cases} \exp\{-(-x)^\alpha\}, & x < 0, \\ 1, & x \geq 0. \end{cases}$$

These three distribution functions are called *extreme value distributions*. We say that the random variable x belongs to the *Maximum Domain of Attraction* (MDA) of the extreme value distribution $H = \Phi_\alpha, \Lambda$, or Ψ_α, namely Fréchet, Gumbel, or Weibull distributions, respectively, written as $\mathrm{x} \in \mathrm{MDA}(H)$, if there exist constants $a_n > 0$ and $b_n \in \mathbb{R}$, for $n = 1, 2, \ldots$, such that the condition (8.17) holds.

These three MDAs cover most of the distributions that are of interest in finance, insurance, and quantitative risk management. For instance, the distributions in $\mathrm{MDA}(\Phi_\alpha)$ include Cauchy, Pareto, Burr, Stable with index $\alpha < 2$ and log-gamma distributions; the distributions in $\mathrm{MDA}(\Lambda)$ include exponential-like, Weibull-like, gamma, normal, lognormal, exponential behaviour at a finite point $x_F < \infty$, *Benktander-type I* and *Benktander-type II* distributions; the distributions in $\mathrm{MDA}(\Psi_\alpha)$ include uniform, power law behaviour at a finite point $x_F < \infty$ and beta distributions. Here, x_F stands for the finite endpoint of the Gumbel or Weibull distributions.

A measurable function $h : \mathbb{R}_+ \to \mathbb{R}_+$ is said to be *regularly varying* at ∞ with an index $\alpha \in \mathbb{R}\backslash\{0\}$, denoted as $h \in \mathrm{RV}_\alpha$, if for all $t > 0$,

$$\lim_{x \to \infty} \frac{h(tx)}{h(x)} = t^\alpha.$$

It is well-known that $\mathrm{x} \in \mathrm{MDA}(\Phi_\alpha)$ if and only if $\mathrm{x} \in \mathrm{RV}_{-\alpha}$; in addition, the right endpoint of x is infinity. On the other hand, if $\mathrm{x} \in \mathrm{MDA}(\Lambda)$, then $\mathrm{x} \in \mathrm{RV}_{-\infty}$ but the converse may not hold in general. To characterize distributions in MDA(Λ), we introduce the notion of *von Mises functions*. A von Mises function $F_{\#}$ with a right endpoint x_0 is a distribution function such that there exist $z_0 < x_0$ and $c > 0$ and

$$1 - F_{\#}(x) = c(x) \exp\left\{ - \int_{z_0}^{x} \frac{1}{f(u)} du \right\}, \quad \text{for all } z_0 < x < x_0,$$

where f is a positive absolutely continuous function on (z_0, x_0) with *Radon–Nikodym derivative* f' such that $\lim_{u \uparrow x_0} f'(u) = 0$; and $c(x)$ is a function such that $\lim_{x \to x_0} c(x) = c$. A distribution F with a right endpoint $x_F \leq \infty$ is in MDA(Λ) if and only if there exists a von Mises function $F_{\#}$, with x_F as its right endpoint, and a point $z_0 < x_F$ such that for all $x \in (z_0, x_F)$,

$$1 - F(x) = 1 - F_{\#}(x) = c(x) \exp\left(- \int_{z_0}^{x} \frac{1}{f(u)} du \right),$$

and $\lim_{x \to x_F} c(x) = c > 0$.

REFERENCES

1. Alcock, J., and Satchell, S. (Eds.). (2018). *Asymmetric Dependence in Finance: Diversification, Correlation and Portfolio Management in Market Downturns*. Wiley.
2. Ali, M., Mikhail, N., and Haq, M. (1978). A class of bivariate distributions including the bivariate logistic. *Journal of Multivariate Analysis*, 8(3), 405–412.
3. Artzner, P., Delbaen, F., Eber, J.M., and Heath, D. (1999). Coherent measures of risk. *Mathematical Finance*, 9(3), 203–228.
4. Balkema, A.A., and De Haan, L. (1974). Residual life time at great age. *The Annals of Probability*, 2(5), 792–804.
5. Basel Committee on Banking Supervision. (1996). Supervisory framework for the use of "backtesting" in conjunction with the internal models approach to market risk capital requirements. *Basel: Bank for International Settlements*. https://www.bis.org/publ/bcbs22.htm (accessed 3 May 2024).
6. Basel Committee on Banking Supervision. (1996). Amendment to the capital accord to incorporate market risks. *Basel: Bank for International Settlements*. https://www.bis.org/publ/bcbs24.htm (accessed 3 May 2024).
7. Basel Committee on Banking Supervision. (2016). Minimum capital requirements for market risk. *Basel: Bank for International Settlements*. https://www.bis.org/bcbs/publ/d352.htm (accessed 3 May 2024).
8. Basel Committee on Banking Supervision. (2019). Minimum capital requirements for market risk. *Basel: Bank for International Settlements*. https://www.bis.org/bcbs/publ/d457.htm (accessed 3 May 2024).
9. Breymann, W., Dias, A., and Embrechts, P. (2003). Dependence structures for multivariate high-frequency data in finance. *Quantitative Finance*, 3(1), 1.
10. Chen, Y., Cheung, K.C., Yam, S.C.P., Yuen, F.L., and Zeng, J. (2023). On the diversification effect in Solvency II for extremely dependent risks. *Risks*, 11, 143.
11. Cheung, K.C., Ling, H.K., Tang, Q., Yam, S.C.P., and Yuen, F.L. (2019). On additivity of tail comonotonic risks. *Scandinavian Actuarial Journal*, 2019(10), 837–866.

12. Cheung, K.C., Tang, Q., Yam, S.C.P., and Yuen, F.L. (2024+). When is Value-at-Risk subadditive or superadditive? Working Paper.
13. Clayton, D.G. (1978). A model for association in bivariate life tables and its application in epidemiological studies of familial tendency in chronic disease incidence. *Biometrika*, 65(1), 141–151.
14. Coles, S.G., and Tawn, J.A. (1991). Modelling extreme multivariate events. *Journal of the Royal Statistical Society: Series B*, 53(2), 377–392.
15. Deheuvels, P. (1984). Probabilistic aspects of multivariate extremes. In J. Tiago Oliveira (Ed.), *Statistical Extremes and Applications* (pp. 117–130). Springer.
16. Demarta, S., and McNeil, A.J. (2005). The *t* copula and related copulas. *International Statistical Review*, 73(1), 111–129.
17. Dhaene, J., Vanduffel, S., Goovaerts, M.J., Kaas, R., Tang, Q., and Vyncke, D. (2006). Risk measures and comonotonicity: a review. *Stochastic Models*, 22(4), 573–606.
18. Embrechts, P., Klüppelberg, C., and Mikosch, T. (2013). *Modelling Extremal Events: For Insurance and Finance* (Vol. 33). Springer Science & Business Media.
19. Embrechts, P., and Wang, R. (2015). Seven proofs for the subadditivity of expected shortfall. *Dependence Modelling*, 3(1), 126–140.
20. Frank, M.J. (1979). On the simultaneous association of $F(x, y)$ and $x + y - F(x, y)$. *Aequationes Mathematicae*, 19, 194–226.
21. Galambos, J. (1975). Order statistics of samples from multivariate distributions. *Journal of the American Statistical Association*, 70(351), 674–680.
22. Galambos, J. (1987). *The Asymptotic Theory of Extreme Order Statistics*. Wiley.
23. Gnedenko, B. (1943). Sur la distribution limite du terme maximum d'une série aléatoire. *Annals of Mathematics*, 44(3), 423.
24. Gumbel, E.J. (1960). Distributions des valeurs extrêmes en plusiers dimensions. *Publications de l'Institut de Statistique de l'Université de Paris*, 9, 171–173.
25. Haan, L. de, and Ferreira, A. (2011). *Extreme Value Theory: An Introduction*. Springer.
26. Hua, L., and Joe, H. (2012). Tail comonotonicity: properties, constructions, and asymptotic additivity of risk measures. *Insurance: Mathematics and Economics*, 51(2), 492–503.
27. Hull, J. (2012). *Risk Management and Financial Institutions*. Wiley.
28. Jaworski, P., Durante, F., Härdle, W.K., and Rychlik, T. (2010). *Copula Theory and Its Applications*. Springer.
29. Joe, H. (1993). Parametric families of multivariate distributions with given margins. *Journal of Multivariate Analysis*, 46(2), 262–282.
30. Joe, H. (1997). *Multivariate Models and Dependence Concepts*. Chapman & Hall.
31. Joe, H. (2014). *Dependence Modelling with Copulas*. CRC Press.
32. J.P. Morgan. (1996). *Risk Metrics: Technical Document*. New York.
33. McNeil, A.J., Frey, R., and Embrechts, P. (2015). *Quantitative Risk Management: Concepts, Techniques and Tools* (revised ed.). Princeton University Press.
34. Moreno-Montoya, J. (2020). Benford's law with small sample sizes: a new exact test useful in health sciences during epidemics. *Salud UIS*, 52(2), 161–163.
35. Nelsen, R.B. (2006). *An Introduction to Copulas* (2nd ed.). Springer.
36. Nikoloulopoulos, A.K., Joe, H., and Li, H. (2009). Extreme value properties of multivariate *t* copulas. *Extremes*, 12(2), 129–148.
37. Pareto, V. (1897). *Cours d'Économique Politique*, Vol. 2. Macmillan.
38. Pickands III, J. (1975). Statistical inference using extreme order statistics. *The Annals of Statistics*, 3(1), 119–131.
39. Pickands III, J. (1981). Multivariate extreme value distributions. In *Proceedings of the 43rd Session of the International Statistical Institute*, Vol. 2 (Buenos Aires, 1981), pp. 49, 859–878, 894–902.
40. Resnick, S.I. (2008). *Extreme Values, Regular Variation, and Point Processes*. Springer.

41. Sklar, M. (1959). Fonctions de répartition à n dimensions et leurs marges. *Annales de l'ISUP*, 8(3), 229–231.
42. Yeh, H. (2007). The frailty and the Archimedean structure of the general multivariate Pareto distributions. *Bulletin of the Institute of Mathematics Academia Sinica*, 2(3), 713.
43. Zeevi, A., and Mashal, R. (2002). Beyond correlation: extreme co-movements between financial assets. Available at SSRN 317122. https://papers.ssrn.com/sol3/papers.cfm?abstract_id=317122 (accessed 3 May 2024).
44. Zeng, J. (2022). Contributions to risk management and mean-field type problems (Doctoral dissertation, King's College London).

PART Two

Linear Models

CHAPTER 9

Principal Component Analysis and Recommender Systems

Principal Component Analysis (PCA) is a classical multivariate statistical technique for a dimension reduction of an underlying set of random variables. It is used to find linear combinations of original variables such that the information in the original data is "essentially" preserved even after the reduction. As a generalization, the collaborative filtering approach in recommender systems is used to preserve the essential information contained in a two-dimensional array of random numbers. Before we go into its details, we first look at an interesting application of PCA to identify the important factors explaining the term structure of interest rates extracted from the USA bonds market.

9.1 US ZERO-COUPON RATES

The file `us-rate.csv`, of sample size $n = 588$, contains US zero-coupon rates[1], measured in percentages[2], with maturities of `1m` (months) to `15y` (years), recorded monthly over the period from 1944 to 1992. To begin with, we shall read in the data file, then assign labels and compute the correlation matrix of these rates, see Programmes 9.1 for Python and 9.2 for **R**.

```
1  import numpy as np
2  import pandas as pd
3
4  d = pd.read_csv("us-rate.csv")  # Read in data
5
6  label = ["1m","3m","6m","9m","12m","18m","2y",
7           "3y","4y","5y","7y","10y","15y"]
8  d.columns = label                    # Apply labels
9  np.set_printoptions(precision=2)     # Display the number of 2 digits
10 print(np.corrcoef(d, rowvar=False)) # Compute correlation matrix
```

[1] As a usual practice, bond prices are typically calculated semiannually.
[2] Another commonly used unit of measurement in practice is *basis point* (BP); 100 BPs equal 1 percent.

```
[[1.   0.99 0.99 0.99 0.98 0.98 0.97 0.96 0.95 0.95 0.94 0.92 0.91]
 [0.99 1.   1.   0.99 0.99 0.99 0.98 0.97 0.96 0.96 0.95 0.94 0.92]
 [0.99 1.   1.   1.   1.   0.99 0.99 0.98 0.97 0.96 0.96 0.94 0.93]
 [0.99 0.99 1.   1.   1.   1.   0.99 0.99 0.98 0.97 0.96 0.95 0.94]
 [0.98 0.99 1.   1.   1.   1.   1.   0.99 0.98 0.98 0.97 0.96 0.94]
 [0.98 0.99 0.99 1.   1.   1.   1.   1.   0.99 0.99 0.98 0.97 0.96]
 [0.97 0.98 0.99 0.99 1.   1.   1.   1.   0.99 0.99 0.98 0.98 0.96]
 [0.96 0.97 0.98 0.99 0.99 1.   1.   1.   1.   1.   0.99 0.99 0.98]
 [0.95 0.96 0.97 0.98 0.98 0.99 0.99 1.   1.   1.   1.   0.99 0.98]
 [0.95 0.96 0.96 0.97 0.98 0.99 0.99 1.   1.   1.   1.   1.   0.99]
 [0.94 0.95 0.96 0.96 0.97 0.98 0.98 0.99 1.   1.   1.   1.   0.99]
 [0.92 0.94 0.94 0.95 0.96 0.97 0.98 0.99 0.99 1.   1.   1.   1.  ]
 [0.91 0.92 0.93 0.94 0.94 0.96 0.96 0.98 0.98 0.99 0.99 1.   1.  ]]
```

Programme 9.1 Correlation matrix of zero-coupon rates in Python.

```
> d <- read.csv("us-rate.csv")  # Read in data
>
> label <- c("1m","3m","6m","9m","12m","18m","2y",
+            "3y","4y","5y","7y","10y","15y")
> names(d) <- label             # Apply labels
> options(digits=2)             # Display the number of 2 digits
> cor(d)                        # Compute correlation matrix
      1m   3m   6m   9m  12m  18m   2y   3y   4y   5y   7y  10y  15y
1m  1.00 0.99 0.99 0.99 0.98 0.98 0.97 0.96 0.95 0.95 0.94 0.92 0.91
3m  0.99 1.00 1.00 0.99 0.99 0.99 0.98 0.97 0.96 0.96 0.95 0.94 0.92
6m  0.99 1.00 1.00 1.00 1.00 0.99 0.99 0.98 0.97 0.96 0.96 0.94 0.93
9m  0.99 0.99 1.00 1.00 1.00 1.00 0.99 0.99 0.98 0.97 0.96 0.95 0.94
12m 0.98 0.99 1.00 1.00 1.00 1.00 1.00 0.99 0.98 0.98 0.97 0.96 0.94
18m 0.98 0.99 0.99 1.00 1.00 1.00 1.00 1.00 0.99 0.99 0.98 0.97 0.96
2y  0.97 0.98 0.99 0.99 1.00 1.00 1.00 1.00 0.99 0.99 0.98 0.98 0.96
3y  0.96 0.97 0.98 0.99 0.99 1.00 1.00 1.00 1.00 1.00 0.99 0.99 0.98
4y  0.95 0.96 0.97 0.98 0.98 0.99 0.99 1.00 1.00 1.00 1.00 0.99 0.98
5y  0.95 0.96 0.96 0.97 0.98 0.99 0.99 1.00 1.00 1.00 1.00 1.00 0.99
7y  0.94 0.95 0.96 0.96 0.97 0.98 0.98 0.99 1.00 1.00 1.00 1.00 0.99
10y 0.92 0.94 0.94 0.95 0.96 0.97 0.98 0.99 0.99 1.00 1.00 1.00 1.00
15y 0.91 0.92 0.93 0.94 0.94 0.96 0.96 0.98 0.98 0.99 0.99 1.00 1.00
```

Programme 9.2 Correlation matrix of zero-coupon rates in R.

First note that these $p = 13$ variables are highly correlated. This suggests that it is very likely we can use only a few variables (say two or three) to represent the whole dataset without losing a significant amount of information, which is exactly the goal of dimension reduction. A natural question to ask is which variables should we use such that they can contain as much information in the dataset as possible? An immediate answer is not from any of these original variables but linear combinations of them; indeed, let the original variables be $\mathrm{x}_1, \ldots, \mathrm{x}_p$, we first look for a new variable

$$\mathrm{y}_1 = b_{11}\mathrm{x}_1 + \cdots + b_{p1}\mathrm{x}_p = b_1^\top \mathbf{x},$$

where $\boldsymbol{b}_1 = (b_{11}, \ldots, b_{p1})^\top$ and $\mathbf{x} = (x_1, \ldots, x_p)^\top$, such that its variance:

$$\mathrm{Var}(y_1) = \boldsymbol{b}_1^\top S \boldsymbol{b}_1$$

is maximized subject to the constraint:

$$b_{11}^2 + \cdots + b_{p1}^2 = \boldsymbol{b}_1^\top \boldsymbol{b}_1 = 1,$$

where S is the sample covariance matrix of the vector $\mathbf{x}$ of the original variables. The variable y_1 is called the first *principal component* (PC) and the $p \times 1$ vector $\boldsymbol{b}_1$ is called the *loadings* of this first PC.

We can continue to find the second principal component $y_2 = b_{12}x_1 + \ldots + b_{p2}x_p = \boldsymbol{b}_2^\top \mathbf{x}$, where $\boldsymbol{b}_2 = (b_{12}, \ldots, b_{p2})^\top$, such that its variance $\mathrm{Var}(y_2) = \boldsymbol{b}_2^\top S \boldsymbol{b}_2$ is maximized subject to (i) $\boldsymbol{b}_2^\top \boldsymbol{b}_2 = 1$; and (ii) $\boldsymbol{b}_2^\top \boldsymbol{b}_1 = 0$. The latter condition means that $\boldsymbol{b}_1$ and $\boldsymbol{b}_2$ are *orthogonal* to each other (or equivalently, y_1 and y_2 are statistically uncorrelated[3]).

The third PC can be found similarly, and the process can continue up to the last p-th PC. This gives the basic idea of PCA and the optimization procedure has a nice solution via linear algebra. In particular, it turns out that these loadings are the *normalized eigenvectors*, denoted by $\boldsymbol{b}_j = (b_{1j}, \ldots, b_{pj})^\top$, for $j = 1, 2, \ldots, p$, for the j-th PC, of the correlation matrix or the variance-covariance matrix of $\mathbf{x}$ and the variance of y_j, the j-th PC, is the corresponding *eigenvalue* $\lambda_j = \mathrm{Var}(y_j) = \boldsymbol{b}_j^\top S \boldsymbol{b}_j$. This claim is discussed in detail in Appendix *9.A.1, while readers may refer back to Section 1.2 of Chapter 1 for the basic theory of spectral decomposition.

The reason for maximizing the variance of the transformed variable y_j is that we want to preserve as much information in these $x_1, \ldots, x_p$ as possible. Conceptually, variation represents information, and so we want the variances of the linearly transformed variables to be as large as possible, also "as soon as possible", so that the original information in $\mathbf{x}$ is preserved. The j-th eigenvalue $\lambda_j = \mathrm{Var}(y_j)$ indicates the amount of information retained in the j-th PC.

9.2 PCA ALGORITHM

PCA is implemented in Python through the `PCA()` function in the `sklearn` library; while in **R** it is implemented by the built-in function `princomp()`. First let us continue the US zero-coupon rates example in Programmes 9.3 and 9.4 for Python and R, respectively. Throughout this section and the next, for the sake of illustration, we shall treat the standardized random variables $\tilde{x}_i := \frac{x_i - \mathbb{E}(x_i)}{\sqrt{\mathrm{Var}(x_i)}}$, $i = 1, \ldots, 13$ as the original variables, and denote $\tilde{\mathbf{x}} := (\tilde{x}_1, \ldots, \tilde{x}_{13})^\top$. Note that this standardization may not be necessary when observations of the variables share a similar order of magnitude.

[3] To see this, first note that, each $\boldsymbol{b}_k$ is actually an eigenvector of the symmetric matrix S with respect to the distinct eigenvalue λ_k. Therefore, by Property 23 in Section 1.2, $\mathrm{Cov}(y_i, y_j) = \boldsymbol{b}_i^\top S \boldsymbol{b}_j = \boldsymbol{b}_i^\top(\lambda_j \boldsymbol{b}_j) = \lambda_j(\boldsymbol{b}_i^\top \boldsymbol{b}_j) = 0$, for $i, j = 1, 2, \ldots, p$; also see Section 9.2.

```
1 from sklearn.decomposition import PCA
2
3 np.set_printoptions(precision=4)      # Display the number of 4 digits
4 pca = PCA()
5 standard_d = (d - np.mean(d, axis=0))/np.std(d, axis=0)
6 pca.fit(standard_d)
7 # components_: an ndarray of shape (n_components, n_features)
8 print(pca.components_[:6,:].T)
```

```
 1 [[ 0.2732  0.4043 -0.5976 -0.5659 -0.2719  0.0604]
 2  [ 0.2758  0.3446 -0.2823  0.2941  0.6122 -0.3492]
 3  [ 0.277   0.3027  0.0285  0.3677  0.1699  0.3833]
 4  [ 0.2783  0.2361  0.1771  0.2332 -0.2162  0.2511]
 5  [ 0.2786  0.2023  0.252   0.1651 -0.4103  0.1819]
 6  [ 0.2798  0.0915  0.2765 -0.0345 -0.2009 -0.3118]
 7  [ 0.2799  0.0353  0.2885 -0.1355 -0.0939 -0.5615]
 8  [ 0.2798 -0.0998  0.2226 -0.2212  0.1388 -0.1092]
 9  [ 0.2791 -0.1682  0.1881 -0.2654  0.2558  0.1215]
10  [ 0.2785 -0.2298  0.0836 -0.2061  0.2137  0.2002]
11  [ 0.2772 -0.3015 -0.0391 -0.1377  0.1623  0.2945]
12  [ 0.2754 -0.3754 -0.2187  0.1056 -0.0493  0.0824]
13  [ 0.2729 -0.4438 -0.4082  0.4062 -0.3146 -0.242 ]]
```

Programme 9.3 The first six principal components' loadings for the US rates example in Section 9.1 via Python.

```
 1 > options(digits=4)          # Display the number of 4 digits
 2 > pca <- princomp(d, cor=T)
 3 > pca$loadings[,1:6]  # Display first six loadings of PCAs
 4      Comp.1   Comp.2   Comp.3   Comp.4   Comp.5   Comp.6
 5 1m   0.2732  0.40434  0.59756  0.56592  0.27195  0.06041
 6 3m   0.2758  0.34462  0.28225 -0.29408 -0.61221 -0.34923
 7 6m   0.2770  0.30272 -0.02848 -0.36767 -0.16994  0.38335
 8 9m   0.2783  0.23612 -0.17710 -0.23317  0.21616  0.25111
 9 12m  0.2786  0.20227 -0.25195 -0.16505  0.41027  0.18186
10 18m  0.2798  0.09148 -0.27652  0.03453  0.20085 -0.31179
11 2y   0.2799  0.03527 -0.28849  0.13555  0.09394 -0.56147
12 3y   0.2798 -0.09978 -0.22264  0.22118 -0.13877 -0.10916
13 4y   0.2791 -0.16815 -0.18806  0.26537 -0.25584  0.12151
14 5y   0.2785 -0.22976 -0.08357  0.20610 -0.21373  0.20016
15 7y   0.2772 -0.30150  0.03910  0.13771 -0.16234  0.29454
16 10y  0.2754 -0.37538  0.21868 -0.10556  0.04935  0.08244
17 15y  0.2729 -0.44384  0.40821 -0.40615  0.31461 -0.24199
```

Programme 9.4 The first six principal components' loadings for the US rates example in Section 9.1 via R.

From the output, the first and second PCs are

$$y_1 = 0.2732\tilde{x}_1 + \cdots + 0.2729\tilde{x}_{13} = \boldsymbol{h}_1^\top \tilde{\mathbf{x}};$$
$$y_2 = 0.4043\tilde{x}_1 + \cdots - 0.4438\tilde{x}_{13} = \boldsymbol{h}_2^\top \tilde{\mathbf{x}},$$

respectively. However, for the third PC, we observe the loadings generated in Python are exactly the negative of those given in **R**. Notice that the unit norm eigenvectors $\boldsymbol{h}_i$'s are only confined by the conditions that $\boldsymbol{h}_i^\top \boldsymbol{h}_i = 1$ (of unit length) and $\boldsymbol{h}_i^\top \boldsymbol{h}_j = 0$ (being orthogonal) for $i \neq j$. Therefore, these loadings can be presented by their negative counterpart. It is easy to illustrate that $\boldsymbol{h}_1^\top \boldsymbol{h}_1 = \boldsymbol{h}_2^\top \boldsymbol{h}_2 = 1$ and $\boldsymbol{h}_1^\top \boldsymbol{h}_2 = 0$; also see Programmes 9.5 for Python and 9.6 for **R**.

```
1  pc1 = pca.components_[0,:]     # Save the loading of 1st PC
2  pc2 = pca.components_[1,:]     # Save the loading of 2nd PC
3
4  print(pc1 @ pc1)               # h1^T h1
5  print(pc2 @ pc2)               # h2^T h2
6  print(pc1 @ pc2)               # h1^T h2
```

```
1  1.0
2  0.9999999999999989
3  1.3877787807814457e-17
```

Programme 9.5 Justification of the unit norm nature and orthogonality of the first two PCs from the US rates example via Python.

```
1  > pc1 <- pca$loadings[,1]    # Save the loading of 1st PC
2  > pc2 <- pca$loadings[,2]    # Save the loading of 2nd PC
3  >
4  > pc1%*%pc1                  # h1^T h1
5        [,1]
6  [1,]    1
7  > pc2%*%pc2                  # h2^T h2
8        [,1]
9  [1,]    1
10 > pc1%*%pc2                  # h1^T h2
11              [,1]
12 [1,] 1.388e-17
```

Programme 9.6 Justification of the unit norm nature and orthogonality of the first two PCs from the US rates example via **R**.

PCA aims at dimension reduction in variables, i.e. we want to represent this `us-rate.csv` dataset with only a few variables. How many PCs should we use? Before answering this question, let us first look at the variance of these PCs.

```
1 s2 = pca.explained_variance_      # save the variance of all PC to s2
2 print(np.round(s2, 4))            # Display the variances of all PC's
3 print(sum(s2))
4 # Proportion of variance explained by PC's
5 print(np.round(pca.explained_variance_ratio_, 4))
6 # Cumulative sum of proportion of variance
7 print(np.cumsum(pca.explained_variance_ratio_))
```

```
1 [1.2746e+01 2.3810e-01 2.4900e-02 7.5000e-03 2.9000e-03 1.5000e-03
2  1.0000e-03 7.0000e-04 0.0000e+00 0.0000e+00 0.0000e+00 0.0000e+00
3  0.0000e+00]
4 13.022146507666104
5 [9.788e-01 1.830e-02 1.900e-03 6.000e-04 2.000e-04 1.000e-04 1.000e-04
6  1.000e-04 0.000e+00 0.000e+00 0.000e+00 0.000e+00 0.000e+00]
7 [0.9788 0.997  0.999  0.9995 0.9998 0.9999 0.9999 1.     1.     1.
8  1.     1.     1.    ]
```

Programme 9.7 Variances of all 13 PCs from the US rates example via Python.

```
1  > s <- pca$sdev     # Save the SD of all PC's to s
2  > round(s^2, 4)       # Display the variances of all PC"s
3   Comp.1  Comp.2  Comp.3  Comp.4  Comp.5  Comp.6  Comp.7  Comp.8
4  12.7239  0.2377  0.0249  0.0075  0.0029  0.0015  0.0010  0.0007
5   Comp.9 Comp.10 Comp.11 Comp.12 Comp.13
6   0.0000  0.0000  0.0000  0.0000  0.0000
7  > (t <- sum(s^2))    # Compute total variance (should equal 13)
8  [1] 13
9  > round(s^2/t, 4)    # Proportion of variance explained by PC's
10 Comp.1  Comp.2  Comp.3  Comp.4  Comp.5  Comp.6  Comp.7  Comp.8
11 0.9788  0.0183  0.0019  0.0006  0.0002  0.0001  0.0001  0.0001
12 Comp.9 Comp.10 Comp.11 Comp.12 Comp.13
13 0.0000  0.0000  0.0000  0.0000  0.0000
14 > cumsum(s^2/t)       # Cumulative sum of proportion of variance
15 Comp.1  Comp.2  Comp.3  Comp.4  Comp.5  Comp.6  Comp.7  Comp.8
16 0.9788  0.9970  0.9990  0.9995  0.9998  0.9999  0.9999  1.0000
17 Comp.9 Comp.10 Comp.11 Comp.12 Comp.13
18 1.0000  1.0000  1.0000  1.0000  1.0000
```

Programme 9.8 Variances of all 13 PCs from the US rates example via R.

From the output in Programme 9.8 via **R**, we know that the variance of the first PC is 12.7239, and so on; while in Programme 9.7 via Python, it computes the variance of the sample instead of sample variance, hence these variances are slightly different from those obtained in **R** by a factor of $(n-1)/n = 587/588$. By Property 25 in Section 1.2, we see that the total variance is equal to tr($\boldsymbol{S}$), the trace of $\boldsymbol{S}$. Note that in our example, all the diagonal elements of $\boldsymbol{S}$ are equal to 1 after standardization, hence their sum, the total variance, is simply equal to $p = 13$.

Based on the **R** output, the first PC explained 12.7239/13 = 97.88% of the total variance of 13. The second PC explained 1.83%, the third PC explained 0.19%, and so on. The first PC already captured almost all (≈ 97.88%) the information (of the total variance). Even if we only include the first three PCs, 99.90% of all the information is already captured! A plot of variance values (called a *Scree Plot*, "scree" meaning large loose broken "gems") can help us determine the suitable number of PCs to be used. This plot graphically represents the information retained in each PC graphically; see Figure 9.1 generated by Programmes 9.9 in Python and 9.10 in **R**.

```
1  import matplotlib.pyplot as plt
2
3  plt.rcParams["figure.figsize"] = 10, 10
4  plt.rcParams.update({'font.size': 25})
5
6  PC_comp = np.arange(pca.n_components_) + 1
7  plt.plot(PC_comp, pca.explained_variance_ratio_*sum(s2),
8          "bo-", linewidth=2)
9  plt.title('Scree Plot')
10 plt.xlabel('Principal Component')
11 plt.ylabel('Variance')
```

Programme 9.9 Plotting a scree plot of the US rates example in Python.

```
1  > par(cex.lab=2, cex.axis=2, cex.main=2, mar=c(5,5,4,4))
2  > screeplot(pca, type="lines")
```

Programme 9.10 Plotting a scree plot of the US rates example in **R**.

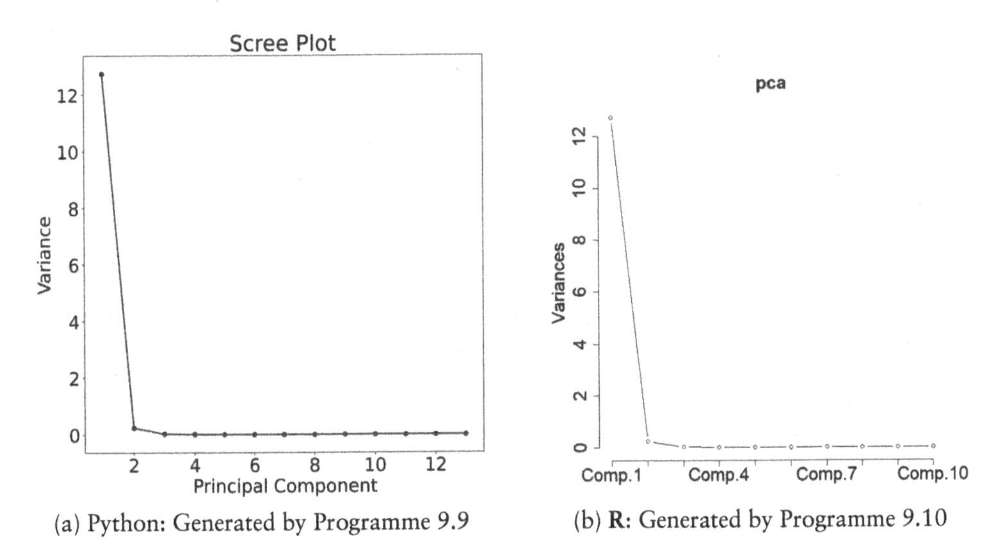

(a) Python: Generated by Programme 9.9

(b) **R**: Generated by Programme 9.10

FIGURE 9.1 Scree plots of the US rates example.

For most applications, we can look at the scree plot to determine the number of PCs to be used. In this US rate example, we can just use the first PC to represent all the 13 variables. However, the first three PCs altogether have more interesting interpretations so that we may want to use these to represent the term structure of the 13 US zero-coupon rates.

To further elucidate how much information we lose by only using the first few PCs, based on the general theory for PCA given in Appendix *9.A.1 we note that all y_i's are orthogonal to each other. Indeed, by spectral decomposition, $S = HDH^\top$ and $y = H^\top \tilde{x}$,

$$\mathrm{Cov}(y) = \mathrm{Cov}(H^\top \tilde{x}) = H^\top \mathrm{Cov}(\tilde{x}) H = H^\top HDH^\top H = D,$$

which is a diagonal matrix of eigenvalues, and so all off-diagonal elements vanished, i.e. $\mathrm{Cov}(y_i, y_j) = 0$, for $i \neq j$. Note that in Section 9.1, we defined $h_j = (h_{1j}, \ldots, h_{pj})^\top$ such that $H = (h_{ij})$ and $y = H^\top \tilde{x}$; therefore, $\tilde{x} = Hy$, or equivalently, $\tilde{x}_i = \sum_{j=1}^{p} h_{ij} y_j$, for $i = 1, \ldots, p$. Clearly, H is orthogonal such that $H^\top H = HH^\top = I_p$, hence $\sum_{j=1}^{p} h_{ij}^2 = 1$ with $h_{ij}^2 \leq 1$. The i-th residual information of $\tilde{x}_i$, resulting from using only the first m PCs is then:

$$\mathrm{Var}\left(\tilde{x}_i - \sum_{j=1}^{m} h_{ij} y_j\right) = \mathrm{Var}\left(\sum_{j=m+1}^{p} h_{ij} y_j\right) = \sum_{j=m+1}^{p} h_{ij}^2 \lambda_j \leq \sum_{j=m+1}^{p} \lambda_j.$$

For the US rates example, if we choose $m = 3$, the latter sum is even less than 0.01% of the total variance for each $i = 1, 2, \ldots, p$. Therefore, the total L^2-error[4], by using only the first m PCs, should be

$$\sum_{i=1}^{p} \mathrm{Var}\left(\tilde{x}_i - \sum_{j=1}^{m} h_{ij} y_j\right) = \sum_{i=1}^{p} \sum_{j=m+1}^{p} h_{ij}^2 \lambda_j = \sum_{j=m+1}^{p} \lambda_j \sum_{i=1}^{p} h_{ij}^2 = \sum_{j=m+1}^{p} \lambda_j,$$

where the last equality follows by the orthogonality of H, which implies $\sum_{i=1}^{p} h_{ij}^2 = 1$. Again, if we set $m = 3$, there is only less than 0.01% of the original total variance still remaining in the L^2-error in our example. In practice, we may only use the first few m PCs to approximate $\tilde{x}$, $\tilde{x}_i \approx h_{i1} y_1 + \cdots + h_{im} y_m$, for all $i = 1, \ldots, p$. This provides an interesting and useful application of PCA to approximate the Value-at-Risk of a portfolio and immunizes the portfolio from various risk factors arising in markets; see Sections 9.5, 9.6, and 9.7.

[4] Recall that $\tilde{x}_i$'s have been de-meaned (or centralized), such that $\mathbb{E}(\tilde{x}_i) = 0$ for all $i = 1, 2, \ldots, p$, so does $\mathbb{E}(y_j) = 0$ for all $j = 1, 2, \ldots, p$.

9.3 FINANCIAL INTERPRETATION OF PCS FOR US ZERO-COUPON RATES

Let us plot the first three PCs in Python with Programme 9.11 and in **R** with Programme 9.12.

```
pc1 = pca.components_[0,:]     # Save the loading of 1st PC
pc2 = -pca.components_[1,:]    # Save the loading of 2nd PC
pc3 = -pca.components_[2,:]    # Save the loading of 3rd PC

# Multi-frame for plotting
fig, axs = plt.subplots(3, 1, figsize=(10,17))
axs[0].plot(PC_comp, pc1, "bo-")    # blue-circle-line
axs[1].plot(PC_comp, pc2, "ro-")    # red-circle-line
axs[2].plot(PC_comp, pc3, "co-")    # cyan-circle-line

for i, ax in enumerate(axs):
    axs[i].set_ylim(-0.65, 0.65)
    axs[i].set_ylabel(f"PC{i+1}")
    axs[i].set_xlabel("Index", fontsize=20)
```

Programme 9.11 Python code for generating and plotting loadings of the first three principal components of the US rates example. Note that pc2 and pc3 are the negative of the originally generated PC values.

```
> pc1 <- pca$loadings[,1]    # Save the loading of 1st PC
> pc2 <- -pca$loadings[,2]   # Save the loading of 2nd PC
> pc3 <- pca$loadings[,3]    # Save the loading of 3rd PC
>
> par(mfrow=c(3,1), cex.lab=2.5, cex.axis=2.5, cex.main=2)
> plot(pc1, ylim=c(-0.6, 0.6), type="o")
> plot(pc2, ylim=c(-0.6, 0.6), type="o")
> plot(pc3, ylim=c(-0.6, 0.6), type="o")
```

Programme 9.12 R code for generating and plotting loadings of the first three principal components of the US rates example. Note that pc2 is the negative of the originally generated PC values.

Identifying these principal components is important for understanding the term structure of interest rates. Many investment instruments are very sensitive to the movement of interest rates, including stocks, bonds, forward contracts and currency swaps. Knowledge of these principal components is very useful in valuating these instruments.

Recall that the first PC is

$$y_1 = 0.2732\tilde{x}_1 + \cdots + 0.2729\tilde{x}_{13},$$

whose loadings are almost constant at the level of 0.27 for bond rates of all maturity times. In fact, these loadings are not unique. They are unique only up to a positive

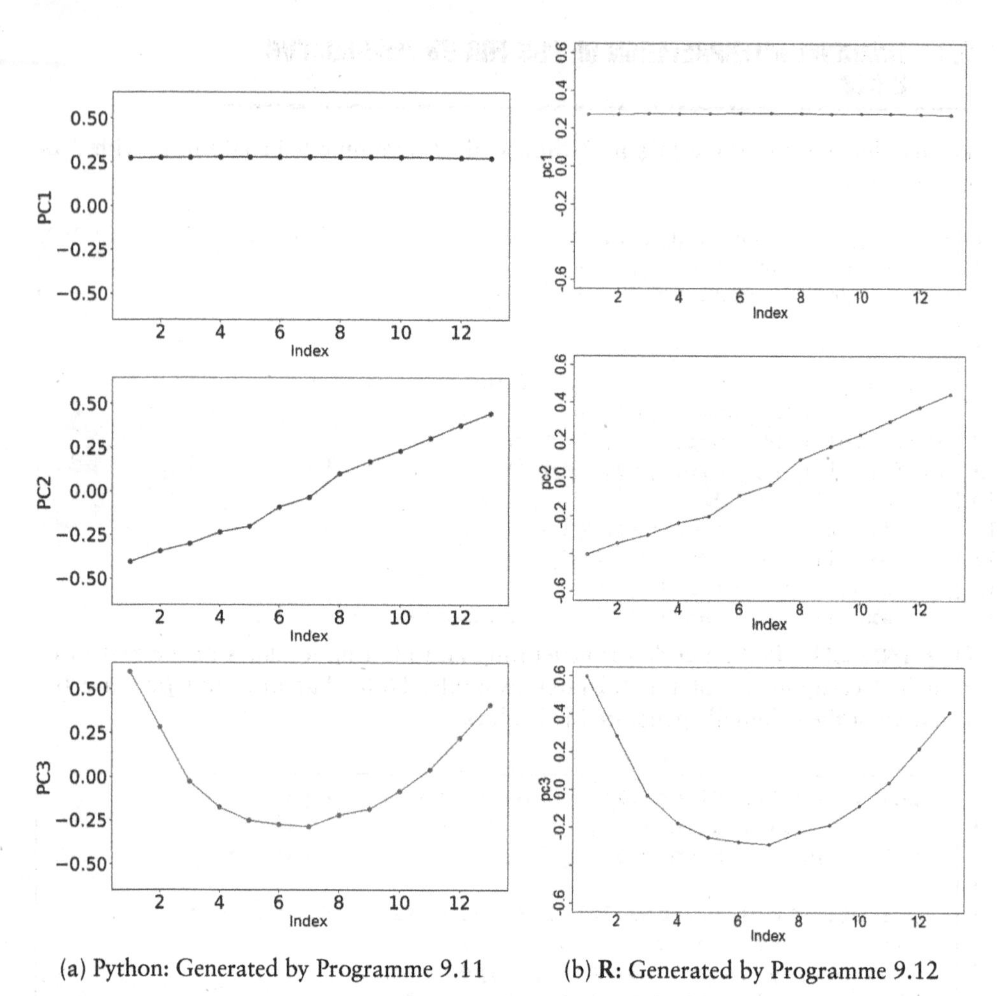

(a) Python: Generated by Programme 9.11 (b) **R**: Generated by Programme 9.12

FIGURE 9.2 Loadings of h_{ij}'s for the first three principal components of the US rates example.

or negative sign. That is, we can define a new first PC as $y_1 = -(0.2732\tilde{x}_1 + \cdots + 0.2729\tilde{x}_{13})$ whose variance is the same at that of the old one. This PC is interpreted as a *parallel shift* (caused by a "level" shock for example) of the yield curve, and explains the majority of the variances of all bonds. To explain this parallel shift, we refer to a shift in the economic environment caused by some major macroeconomic event, so that the rates of returns for all zero-coupon bonds with different maturities contribute equally to this principal component representing an average effect. On the other hand, $\tilde{x}_i \approx h_{i1}y_1$, so each bond rate is equally affected by the first PC.

The second PC is

$$y_2 = 0.4043\tilde{x}_1 + \cdots - 0.4438\tilde{x}_{13},$$

where the loadings are decreasing with maturity. Again, the loadings are not unique. In particular, the alternative second PC could be $y_2 = -0.4043\tilde{x}_1 - \cdots + 0.4438\tilde{x}_{13}$, which seems more reasonable since it agrees with *liquidity preference theory*. This theory states that investors prefer to preserve their liquidity and invest in investment vehicles that last for a shorter period of time, and so additional premiums are paid for holding securities with longer time of maturity. This second component is therefore termed as the *tilt* component of the yield curve.

The third PC is

$$y_3 = 0.5976\tilde{x}_1 + \cdots + 0.4082\tilde{x}_{13},$$

whose loadings decrease with maturity up to a certain point and then start increasing again with maturity. This can be interpreted as the *curvature* of the yield curve, which is caused by the higher demand on the relatively short-term and relatively long-term bonds. Indeed, the market for short-term bonds is popular with frequent traders in securities, derivatives and insurance markets, while bonds of longer terms are welcomed by long-term market participants including insurance and reinsurance companies. Short-term and long-term bonds can be used for immunization of interest rate risk for investment vehicles and liabilities of intermediate lifespans, contributing to the convexity in the loadings of this third PC. That said, its eigenvalue (and also variance) is much smaller than those of the first and second PCs, because of the relatively small proportion that insurance markets take up in the entire financial industry; we refer to Appendix *9.A.2 for a rationale behind this observation.

The original data can be expressed approximately in terms of y_1, y_2 and y_3 by using the loadings of PC1, PC2 and PC3 in light of the fact that $\mathbf{y} = H^{\top}\tilde{\mathbf{x}}$. These y_j's are also called the principal component scores with respect to PC1, PC2 and PC3, respectively. The geometric interpretation of PCA is illustrated by the scatter plot matrix of these scores, see Figures 9.3 via Python and 9.4 via **R**.

```
1 from seaborn import pairplot
2 
3 score = standard_d @ pca.components_.T  # save scores of all PC's
4 score.columns = [f"PC{i}" for i in PC_comp]
5 print(score.iloc[:, :3])
6 plots = pairplot(score.iloc[:, :3], diag_kind="hist", height=5)
7 plots.savefig("PCA first 3 scores pairplot.png", dpi=400)
```

Programme 9.13 Python code for scatter plots of the first three principal components.

```
1 > score <- pca$scores[,1:3]        # save scores of PC1-PC3
2 > pairs(score, cex.labels=4)       # scatterplot of scores
```

Programme 9.14 R code for scatter plots of the first three principal components.

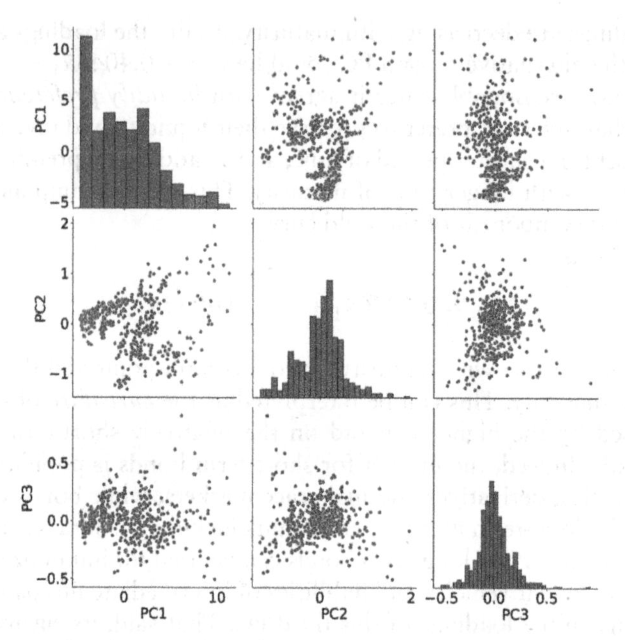

FIGURE 9.3 Scatter plot for the first three principal components from the US rates example. Generated in Programme 9.13 via Python, we expect to see that these components are uncorrelated.

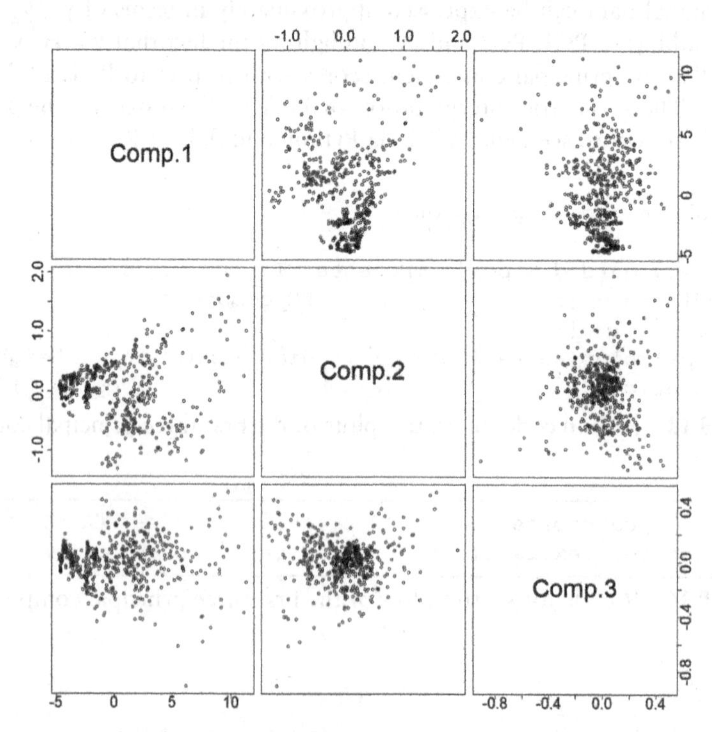

FIGURE 9.4 Scatter plot for the first three principal components from the US rates example. Generated in Programme 9.14 via **R**, we expect to see that these components are uncorrelated.

9.4 PCA AS AN EIGENVALUE PROBLEM

In general, let $X = (x_1, x_2, \dots, x_n) \in \mathbb{R}^{p\times n}$ be a data matrix of n sample points (assuming mean zero for simplicity) such that each $x_i \in \mathbb{R}^p$, for $i = 1, 2, \dots, n$. The main goal of PCA is to compress the information in X so that each sample point is effectively projected onto $\mathbb{R}^k$ with $k < p$, while the loss of information is minimized.

Let $W = (\omega_1, \dots, \omega_k) \in \mathbb{R}^{p\times k}$ be a projection matrix which satisfies

$$\|\omega_i\|_2 = 1,\ i = 1, \dots, k, \qquad \text{and} \qquad \omega_i^\top \omega_j = 0 \quad \text{for } i \neq j.$$

Then, the compressed data matrix Z takes the form $Z = W^\top X$. From this compressed data matrix, we can retrieve the original data matrix approximately by using $\hat{X} = WZ = WW^\top X$. The question now is how to determine W so that the induced approximation error can be minimized. In other words, we aim to solve the following optimization problem:

$$\begin{aligned} &\min_W \|X - \hat{X}\|_F^2 = \min_W \|X - WW^\top X\|_F^2, \\ &\text{subject to } W^\top W = I_k, \end{aligned} \tag{9.1}$$

where $\|\cdot\|_F$ is the Frobenius norm (see Section 1.3). We can first simplify the objective function of (9.1) as

$$\begin{aligned} \|X - WW^\top X\|_F^2 &= \operatorname{tr}\left((X - WW^\top X)^\top (X - WW^\top X)\right) \\ &= \operatorname{tr}\left(X^\top X - 2X^\top WW^\top X + X^\top WW^\top WW^\top X\right) \\ &= \operatorname{tr}\left(X^\top X - X^\top WW^\top X\right), \end{aligned}$$

where the last equality follows by using $W^\top W = I_k$. Since $X^\top X$ does not depend on W, we have

$$\begin{aligned} \arg\min_W \|X - WW^\top X\|_F^2 &= \arg\min_W \operatorname{tr}\left(-X^\top WW^\top X\right) \\ &= \arg\max_W \operatorname{tr}\left(X^\top WW^\top X\right). \end{aligned}$$

By the cyclic invariance of trace, the optimization problem can then be rewritten as

$$\begin{aligned} &\max_W \operatorname{tr}\left(W^\top XX^\top W\right), \\ &\text{subject to } W^\top W = I_k. \end{aligned} \tag{9.2}$$

One can solve (9.2) by using the Lagrange multiplier method. Consider

$$L(W, \Lambda) = \operatorname{tr}\left(W^\top XX^\top W\right) + \operatorname{tr}(\Lambda(I_k - W^\top W)), \quad \text{for } \Lambda \in \mathbb{R}^{k\times k}.$$

Referring to Section 1.2, we can equate the derivative $\nabla_W L(W, \Lambda)$ with zero and obtain the first order condition:

$$2(XX^\top)W = W(\Lambda + \Lambda^\top). \tag{9.3}$$

Let $\Xi := \frac{1}{2}(\Lambda + \Lambda^\top) \in \mathbb{R}^{k\times k}$, which is symmetric, and note that spectral decomposition gives $\Xi = \tilde{H}\tilde{D}\tilde{H}^\top$ for some orthogonal matrix $\tilde{H}$ and a diagonal matrix $\tilde{D}$. Therefore,

$$(XX^\top)(W\tilde{H}) = (W\tilde{H})\tilde{D}. \tag{9.4}$$

Also, letting $H := W\tilde{H}$, we observe that $\text{tr}(W^\top XX^\top W) = \text{tr}(H^\top XX^\top H)$ and $H^\top H = I_k$.

From (9.4), we see that H is composed of the eigenvectors of $XX^\top$, as $(XX^\top)H = H\tilde{D}$. Hence, the trace in (9.2) can be maximized by picking the k largest eigenvalues and then summing them up. From Appendix *9.A.1, we construct $H = (h_1, \dots, h_k)$, where h_j is the unit eigenvector corresponding to the j-th largest eigenvalue of $XX^\top$. In summary, we can now conclude with the following pseudo-code for the PCA algorithm.

Algorithm 9.1 PCA Algorithm.

1: **Input:** Data matrix $X = (x_1, \dots, x_n)$, $x_i \in \mathbb{R}^p$.
2: **Initialize:**
3: Sample mean $\mu \leftarrow 0$;
4: $x_i \leftarrow x_i - \frac{1}{n}\sum_{j=1}^n x_j$;
5: Sample variance $\sigma^2 \leftarrow 1$;
6: Assign $\sigma \leftarrow \sqrt{\frac{1}{n}\sum_{i=1}^n \|x_i\|_2^2}$, and then $x_i \leftarrow x_i/\sigma$;
7: Compute the covariance matrix of $\frac{1}{n-1}XX^\top$, or equivalently just $XX^\top$;
8: Spectrally decompose $XX^\top$ to obtain the eigenvalues and eigenvectors;
9: Sort the eigenvalues of $XX^\top$ in descending order of magnitude and the corresponding eigenvectors, label them as $h_1, \dots, h_k$;
10: **return:** $H = (h_1, \dots, h_k)$.

9.5 FACTOR MODELS VIA PCA

In finance, a reasonable model should provide effective and interpretable predictions with a tractable performance. Factor models are among those most commonly used in practice. Indeed, factor models play an important role in finance: for instance, they provide a decomposition to assist with portfolio hedging; see Section 9.7. A p-dimensional random vector $\mathbf{x}_t$ is said to follow an *m-linear factor model* if

$$\mathbf{x}_t = a + B\mathbf{f}_t + \epsilon_t, \quad t = 0, 1, 2, \dots, n, \tag{9.5}$$

where $\mathrm{f}_t := (\mathrm{f}_t^1, \ldots, \mathrm{f}_t^m)^\top$ is an m-dimensional random process, with $m < p$ (m is normally much less than p), having a positive definite variance-covariance matrix; $\boldsymbol{\epsilon}_t = (\epsilon_t^1, \ldots, \epsilon_t^p)^\top$ is a zero-mean p-dimensional random error; $\boldsymbol{a}$ and $\boldsymbol{B}$ are a p-dimensional constant vector and a $p \times m$ constant coefficient matrix respectively. It is further assumed that these factors are the major ones such that the residual information is uncorrelated, i.e. $\mathrm{Cov}(\mathrm{f}_t^i, \epsilon_t^j) = 0$ for all $i = 1, 2, \ldots, m$, $j = 1, 2, \ldots, p$ and $t \geq 0$. To determine the factors, one can incorporate PCA by analyzing the data matrix of $x_0, \ldots, x_n$, so as to represent a random vector $\mathbf{x}_t$ in terms of its first m principal components with a small error term. We remark that our discussions in this section lay the foundations for *Principal Component Regression* (PCR); we postpone delicate discussions on its mathematics and implementation to Section *10.7.

Suppose that $\mathbf{x}_t$ is a p-dimensional random vector with mean vector $\boldsymbol{\mu} = (\mu_1, \ldots, \mu_p)^\top$ and (homogeneous) covariance matrix $\boldsymbol{\Sigma}$ for all $t \geq 0$, and all $\mathbf{x}_t$'s were collected independently, so that they are supposed to follow a common distribution. We first standardize it to $z_t = (z_t^1, \ldots, z_t^p)^\top$ given by

$$z_t^i = \frac{x_t^i - \mu_i}{\sigma_i}, \qquad i = 1, \ldots, p, \quad t = 0, 1, 2, \ldots, n,$$

where $\sigma_i^2 = \mathrm{Var}(\mathrm{x}_t^i)$. One can apply PCA to the (sample) variance-covariance matrix of $\mathbf{z}_t$:

$$\mathrm{y}_t = H^\top \mathbf{z}_t,$$

where $\mathrm{y}_t = (\mathrm{y}_t^1, \ldots, \mathrm{y}_t^p)^\top$ with y_t^j being the j-th principal component of $\mathbf{z}_t$, and H is the collection of columns of the corresponding loadings of the principal components, and is independent of t. If the first m components are sufficient to explain a significant portion of the underlying total variance, one may partition the component vector y_t into $\mathrm{y}_t^{(1)}$ and $\mathrm{y}_t^{(2)}$ being m and $p - m$ dimensional random vectors respectively. Similarly, we can partition the corresponding loadings H into $H^{(1)} \in \mathbb{R}^{p \times m}$ and $H^{(2)} \in \mathbb{R}^{p \times (p-m)}$. In other words, we write

$$\mathrm{y}_t = \begin{pmatrix} \mathrm{y}_t^{(1)} \\ \hdashline \mathrm{y}_t^{(2)} \end{pmatrix} \quad \text{and} \quad H = \left(H^{(1)} \,\vdots\, H^{(2)} \right).$$

Hence with $HH^\top = I_p$, we have $\mathbf{z}_t = H\mathrm{y}_t = H^{(1)}\mathrm{y}_t^{(1)} + H^{(2)}\mathrm{y}_t^{(2)}$. With these partitions, we can rewrite $\mathbf{x}_t$ into the form of (9.5):

$$\mathbf{x}_t = \boldsymbol{\mu} + \mathrm{diag}(\sigma_1, \ldots, \sigma_p)\mathbf{z}_t = \boldsymbol{\mu} + \mathrm{diag}(\sigma_1, \ldots, \sigma_p)H^{(1)}\mathrm{y}_t^{(1)} + \boldsymbol{\epsilon}_t, \tag{9.6}$$

with $\boldsymbol{a} = \boldsymbol{\mu}$, $\boldsymbol{B} = \mathrm{diag}(\sigma_1, \ldots, \sigma_p)H^{(1)}$, $\mathrm{f}_t = \mathrm{y}_t^{(1)}$, and $\boldsymbol{\epsilon}_t = \mathrm{diag}(\sigma_1, \ldots, \sigma_p)H^{(2)}\mathrm{y}_t^{(2)}$.

9.6 VALUE-AT-RISK VIA PCA

Denote $\tilde{\mathbf{x}}$ as the normalized version of the coupon rate vector $\mathbf{x}$, where each x_i is measured in percentages as mentioned in Section 9.1, such that $\tilde{x}_i = \frac{x_i - \mu_i}{\sigma_i}$, where μ_i and σ_i are the mean and standard deviation of x_i, respectively. Suppose that we hold a bond portfolio for a year with an exposure (money invested) ω_i on the bond i with an annual yield-to-maturity x_i, then the future value of this bond portfolio Π in year one is $\sum_{i=1}^{p} \omega_i \left(1 + \frac{x_i}{100}\right)$, and hence the random annual profit and loss is given by $\Delta\Pi := \boldsymbol{\omega}^\top \left(\frac{1}{100}\mathbf{x}\right) = \omega_1 \frac{x_1}{100} + \ldots + \omega_p \frac{x_p}{100}$. From Chapter 8, assume that the daily changes are independently and identically normally distributed over the year with 365 calendar days[5], we know that the N-day 99% V@R of the portfolio is $\sqrt{N/365} z_{0.99} \sqrt{\mathrm{Var}(-\Delta\Pi)} = \sqrt{N/365} z_{0.99} \sqrt{\boldsymbol{\omega}^\top \boldsymbol{\Sigma} \boldsymbol{\omega}}/100$.

With the first m PCs, we have $\tilde{\mathbf{x}} \approx H^{(1)}\mathbf{y}^{(1)}$, and hence $\Delta\Pi$ can be approximated as follows:

$$\begin{aligned}
\Delta\Pi &= \frac{1}{100}\boldsymbol{\omega}^\top \mathbf{x} \\
&= \frac{1}{100}\boldsymbol{\omega}^\top(\mathrm{diag}(\sigma_1,\ldots,\sigma_p)\tilde{\mathbf{x}} + \boldsymbol{\mu}) \approx \frac{1}{100}(\boldsymbol{\omega}^\top \mathrm{diag}(\sigma_1,\ldots,\sigma_p)H^{(1)}\mathbf{y}^{(1)} + \boldsymbol{\omega}^\top\boldsymbol{\mu}) \\
&= \frac{1}{100}(\omega_1\sigma_1(h_{11}y_1 + \cdots + h_{1m}y_m) + \cdots + \omega_p\sigma_p(h_{p1}y_1 + \cdots + h_{pm}y_m) + \boldsymbol{\omega}^\top\boldsymbol{\mu}) \\
&= \frac{1}{100}((\omega_1\sigma_1 h_{11} + \cdots + \omega_p\sigma_p h_{p1})y_1 + \cdots + (\omega_1\sigma_1 h_{1m} + \cdots + \omega_p\sigma_p h_{pm})y_m + \zeta) \\
&= \frac{1}{100}(\delta_1 y_1 + \cdots + \delta_m y_m + \zeta),
\end{aligned} \tag{9.7}$$

where $\delta_j := \sum_{i=1}^{p} \omega_i\sigma_i h_{ij}$ for $j = 1, 2, \ldots, m$ and $\zeta := \boldsymbol{\omega}^\top\boldsymbol{\mu}$. Since all PCs are orthogonal to each other, the variance of $\Delta\Pi$ can be approximated using the first m PCs by:

$$\mathrm{Var}(\Delta\Pi) \approx \frac{\delta_1^2 \mathrm{Var}(y_1) + \cdots + \delta_m^2 \mathrm{Var}(y_m)}{10{,}000} = \frac{\delta_1^2\lambda_1 + \cdots + \delta_m^2\lambda_m}{10{,}000},$$

and these first m PCs maintain a large portion of the variation information (variance) contained in $\mathbf{x}$.

To illustrate this, let us continue with our example of US zero-coupon yield; their respective individual standard deviations are computed in Programme 9.15 via Python and Programme 9.16 via R. Suppose that we have a bond portfolio with the money exposure $\boldsymbol{\omega}$ shown in Table 9.1, and we are interested in the 10-day 99% V@R of this bond portfolio. As a benchmark from further calculations in Programmes 9.15 via Python and 9.16 via R, the actual standard deviation of $\Delta\Pi$ is $\sqrt{\boldsymbol{\omega}^\top\boldsymbol{\Sigma}\boldsymbol{\omega}}/100 \approx 0.07786\text{M} = 77{,}860$, by using the sample one to estimate the population covariance matrix, and the corresponding 10-day 99% V@R is $\sqrt{10/365} z_{0.99}(77{,}860) = 29{,}980$. Note that all computations hereafter are based on actual values from the programme outputs, and may be subject to rounding issues.

[5] Unlike the examples in the stock trading context where we assumed 252 trading days in a year, the interest is accrued on every calendar day.

TABLE 9.1 Money exposure of a bond portfolio.

Bond maturity	12m	2y	3y	4y	5y
Money exposure ω	+\$10M	+\$4M	−\$8M	−\$7M	+\$2M

```
1  Sig = np.cov(d, ddof=1, rowvar=False)
2  omega = np.array([0, 0, 0, 0, 10, 0, 4, -8, -7, 2,
3                    0, 0, 0])                # 1 M = 10^6
4
5  # compute the covariance matrix and extract the variances
6  var = np.diag(Sig)
7  print(np.sqrt(var))
8
9  # compute the standard deviation of the bond portfolio (in millions)
10 print(np.sqrt(omega.T @ Sig @ omega) / 100)
```

```
1  [3.2686 3.4637 3.4947 3.5063 3.5158 3.4966 3.492  3.4504 3.4363 3.3958
2   3.3566 3.3184 3.2921]
3  0.07786363554530
```

Programme 9.15 Computation of standard deviations of respective coupon rates and the bond portfolio in Python.

```
1  > omega <- c(0, 0, 0, 0, 10, 0, 4, -8, -7, 2, 0, 0, 0)  # 1 M = 10^6
2  > # compute the covariance matrix and extract the variances
3  > var <- diag(cov(d))
4  > sqrt(var)
5     1m    3m    6m    9m   12m   18m    2y    3y    4y
6  3.269 3.464 3.495 3.506 3.516 3.497 3.492 3.450 3.436
7     5y    7y   10y   15y
8  3.396 3.357 3.318 3.292
9  >
10 > # compute the standard deviation of the bond portfolio (in millions)
11 > (s <- sqrt(t(omega)%*%cov(d)%*%omega) / 100)
12         [,1]
13 [1,] 0.07786
```

Programme 9.16 Computation of standard deviations of respective coupon rates and the bond portfolio in R.

Suppose that there is an increase of 2 percent in the first PC, after changing the sign, (i.e. parallel shift in the yield curve). From Figures 9.2a and 9.2b, since $\mathbf{y} = H^{\mathsf{T}}\tilde{\mathbf{x}}$ so $\tilde{\mathbf{x}} = H\mathbf{y}$ and $\mathbf{x} = \text{diag}(\sigma_1, \ldots, \sigma_p)H\mathbf{y} + \boldsymbol{\mu}$, this change will contribute a $(2)(0.2732)(3.269) = 1.786$ percent increase in the `1m` interest rate, $(2)(0.2758)(3.464) = 1.911$ percent increase in the `3m` interest rate, and so on. Similar effects of the second and third PCs on any individual bond can be obtained in the same manner. We can also apply PCA to calculate the V@R of a portfolio of investment instruments depending on the interest rate.

Now assume that there is an increase of 1% in the interest rates of all bonds with different maturities, i.e., $\Delta x_i = 1$ for $i = 1, \ldots, 13$, then according to the entries in the first column for the first PC in Programmes 9.3 in Python and 9.4 in **R**, we can calculate the change in y_1, denoted by $\Delta y_1 = \sum_{i=1}^{13} h_{i1}\Delta\tilde{x}_i = \sum_{i=1}^{13} h_{i1}\frac{1}{\sigma_i}\Delta x_i = (\frac{0.2732}{3.269} + \frac{0.2752}{3.464} + \cdots + \frac{0.2729}{3.292}) \cdot 1 = 1.0547$; this value is close to 1, which echoes the interpretation of the first PC as a parallel shift. For a given portfolio of bonds as shown in Table 9.1 , by noting that $x_i \approx \sigma_i h_{i1} y_1 + \mu_i$, the exposure of the portfolio to the first PC (measured in \$ per one percent) can be calculated by differencing (9.7) when $m = 1$:

$$
\begin{aligned}
\text{Change in } \Delta\Pi &\approx \delta_1 \cdot \frac{1}{100}\Delta y_1 \\
&= (\omega_5\sigma_5 h_{51} + \omega_7\sigma_7 h_{71} + \omega_8\sigma_8 h_{81} + \omega_9\sigma_9 h_{91} + \omega_{10}\sigma_{10} h_{10,1}) \cdot \frac{1}{100}\Delta y_1 \\
&= [(10\text{M})(3.516)(0.2786) + (4\text{M})(3.492)(0.2799) \\
&\quad + (-8\text{M})(3.450)(0.2798) \\
&\quad + (-7\text{M})(3.436)(0.2791) + (2\text{M})(3.396)(0.2785)] \cdot 0.010547 \\
&= 1.161\text{M} \cdot 0.010547 = 0.012245\text{M}.
\end{aligned}
$$

That is, a change of 1 percent in interest rates will result in a \$12,245 increase in the value of the portfolio. Therefore, by using the first PC, the standard deviation of $\Delta\Pi$ is approximately $s \approx \sqrt{\text{Var}\left(\frac{(1.161\text{M})y_1}{100}\right)} = 11,610\sqrt{12.7239} = 11,610 \cdot 3.567 = 41,413$, and the estimated 10-day 99% V@R of this portfolio is $\sqrt{10/365}z_{0.99}s = (\sqrt{10/365})(2.33)(41,413) = 15,971$.

Similarly, from Programmes 9.3 via Python and 9.4 via **R**, the exposures to the second (the negative counterpart that results in a positive increase in liquidity premium with maturity) and third PCs are respectively

$$
\begin{aligned}
\delta_2 &= (10\text{M})(3.516)(-0.2023) + (4\text{M})(3.492)(-0.0353) + (-8\text{M})(3.450)(0.0998) \\
&\quad + (-7\text{M})(3.436)(0.1682) + (2\text{M})(3.396)(0.2298) \\
&= -12.845\text{M}, \\
\delta_3 &= (10\text{M})(3.516)(-0.2520) + (4\text{M})(3.492)(-0.2885) + (-8\text{M})(3.450)(-0.2226) \\
&\quad + (-7\text{M})(3.436)(-0.1881) + (2\text{M})(3.396)(-0.0836) \\
&= -2.790\text{M}.
\end{aligned}
$$

With $\text{Var}(y_1) = \lambda_1 = 3.567^2$, $\text{Var}(y_2) = \lambda_2 = 0.4876^2$, and $\text{Var}(y_3) = \lambda_3 = 0.1577^2$, the estimate of the standard deviation of the change in portfolio value by using only the first and second PCs is

$$
\begin{aligned}
s &\approx \sqrt{\text{Var}\left(\frac{(1.161\text{M})y_1}{100}\right) + \text{Var}\left(\frac{(-12.845\text{M})y_2}{100}\right)} \\
&= \sqrt{(11,610)^2(3.567)^2 + (-128,450)^2(0.4876)^2} = 75,085.
\end{aligned}
$$

The corresponding estimate of the 10-day 99% V@R is

$$\sqrt{10/365}z_{0.99}s = \sqrt{10/365}(2.33)(75,090) = 28,958.$$

Note that since the PCs are orthogonal or uncorrelated, we do not need to include any cross or covariance terms in estimating the variance of $\Delta\Pi$. Similarly, if we use all the first three PCs to estimate the standard deviation of $\Delta\Pi$, then it is

$$s \approx \sqrt{(11,610)^2(3.567)^2 + (-128,450)^2(0.4876)^2 + (-27,900)^2(0.1577)^2} = 75,214,$$

and the 10-day 99% V@R is approximately

$$\sqrt{10/365}z_{0.99}s = \sqrt{10/365}(2.33)(75,210) = 29,007.$$

In this example, it is clear that the first PC alone could only explain around 53% of the total variance, hence using only the first PC will severely underestimate the V@R, as the component change for each bond is quite different from that of another.

As a final remark, note that choosing a suitable number of PCs could be a quite subtle decision in practice. For an extreme scenario, consider a portfolio with weights $\omega_i = c\frac{h_{ip}}{\sigma_i}$ for all $i = 1, \ldots, p$ with a common constant c. Since the last PC takes the form $\mathrm{y}_p = \sum_i h_{ip}\tilde{\mathrm{x}}_i$, due to its lack of correlation with y_j's, we can see that $\delta_j = c\sum_i h_{ip}h_{ij} = c\boldsymbol{h}_p^\top \boldsymbol{h}_j = 0$ for all $j = 1, \ldots, p-1$; and so $\Delta\Pi \approx \zeta$ whenever the number m of PCs is less than p. Since the last PC is highly unlikely to appear in any truncated PC procedure, the change in the portfolio value by using any common PCA procedure will never be accurately estimated. However, in most financial applications, investors are only interested in securities, governed by and related to the major risk factors $\mathrm{y}_1, \ldots, \mathrm{y}_m$ with $m \ll p$, that contribute to the majority of impacts on the market. Those securities related to y_j for $j \geq m+1$ are barely considered since their effects are negligible, and therefore play a very limited role in the investment plans of common investors. Consequently, the corresponding $|\delta_j|$'s would be quite small. Unfortunately, concentrating only on a few major financial risk factors, hence a few investment vehicles, may result in increased systemic risk, which mainly consists of risk factors y_j for $j \geq m+1$.

9.7 PORTFOLIO IMMUNIZATION

This section only provides a schematic outline on how to hedge a portfolio using principal components against major risk factors. Consider a portfolio that depends on N fundamental assets, which are even non-tradable or highly illiquid, such as future investment projects, real estates and tangible properties, insurance contracts, or intangible assets such as brand names. Let r_t^k be the value[6] of the k-th fundamental

[6] The value of a fundamental asset may not necessarily refer to its quoting price or net worth, it can stand for income generated regularly by consuming this fundamental asset.

asset at time t, for $k = 1, 2, \ldots, N$. According to Section 9.5, one can extract the major risk components by using PCA. We model the change in value of the k-th fundamental asset $\Delta r_t^k := r_{t+1}^k - r_t^k$ by using a linear m-factor model:

$$\Delta r_t^k = \sum_{j=1}^{m} \omega_j^k y_t^j + \epsilon_t^k, \quad k = 1, 2, \ldots, N,$$

where y_t^j is the value of the j-th principal component or risk factor of these N fundamental assets at time t for $j = 1, 2, \ldots, m$; ω_j^k is the sensitivity in the value of the k-th fundamental asset with respect to the j-th risk factor; and ϵ_t^k is the residual, or miscellaneous, risk factor.

On one hand, it is common for companies to run several lines of business which plausibly depend on many non-tradable assets; on the other hand, they often like to hedge out their risky positions, especially after acquiring some notable profit during an up-and-down period, as by then they prefer to have a more stable income until the expiry of these projects. Mathematically, to achieve this goal, let Π_t be the value of the portfolio at time t, which depends on those N fundamental assets. By assumption, the change in value of Π can be mostly explained by using the m risk factors of y_t^j's. We expect to use (at least) m other tradable securities, also depending on these N fundamental assets, to hedge out most of the risks in Π. Let S_t^l, $l = 1, \ldots, m$, be the respective prices of those other m tradable securities at time t; we also hedge the portfolio by holding ϕ_t^l units of the l-th tradable security, $l = 1, \ldots, m$. Hence, the value of the combined portfolio at time t is given by

$$\overline{\Pi}_t = \Pi_t + \sum_{l=1}^{m} \phi_t^l S_t^l.$$

Now, denote the change in value of the combined portfolio from t to $t+1$ by $\Delta\overline{\Pi}_t$. We can approximate it by:

$$\begin{aligned}
\Delta\overline{\Pi}_t &\approx \sum_{k=1}^{N} \left(\frac{\partial \Pi_t}{\partial r_t^k} + \sum_{l=1}^{m} \phi_t^l \frac{\partial S_t^l}{\partial r_t^k} \right) \Delta r_t^k \\
&= \sum_{k=1}^{N} \left[\left(\frac{\partial \Pi_t}{\partial r_t^k} + \sum_{l=1}^{m} \phi_t^l \frac{\partial S_t^l}{\partial r_t^k} \right) \left(\sum_{j=1}^{m} \omega_j^k y_t^j \right) \right] + \text{miscellaneous error involving } \epsilon_t^k \\
&= \sum_{j=1}^{m} \left[\sum_{k=1}^{N} \omega_j^k \left(\frac{\partial \Pi_t}{\partial r_t^k} + \sum_{l=1}^{m} \phi_t^l \frac{\partial S_t^l}{\partial r_t^k} \right) \right] y_t^j + \text{miscellaneous error involving } \epsilon_t^k.
\end{aligned}$$

To (delta) hedge out the risky position, we want to ensure that all coefficients of y_t^j's vanish for all t. The positions ϕ_t^l, $l = 1,\ldots,m$, should be chosen so that they satisfy the following system of linear equations:

$$\sum_{k=1}^{N} \omega_j^k \left(\frac{\partial \Pi_t}{\partial \mathrm{r}_t^k} + \sum_{l=1}^{m} \phi_t^l \frac{\partial \mathrm{S}_t^l}{\partial \mathrm{r}_t^k} \right) = 0, \quad \text{for} \quad j = 1,\ldots,m,$$

or equivalently in matrix form,

$$\mathbf{P}_t + \mathbf{S}_t \mathbf{\Phi}_t = 0,$$

where

$$\mathbf{\Phi}_t = (\phi_t^1,\ldots,\phi_t^m)^\top,\ \mathbf{P}_t = \left(\sum_{k=1}^{N} \omega_1^k \frac{\partial \Pi_t}{\partial \mathrm{r}_t^k},\ldots,\sum_{k=1}^{N} \omega_m^k \frac{\partial \Pi_t}{\partial \mathrm{r}_t^k} \right)^\top,\ \mathbf{S}_t = \left(\sum_{k=1}^{N} \omega_j^k \frac{\partial \mathrm{S}_t^l}{\partial \mathrm{r}_t^k} \right)_{j,l}.$$

Therefore, by a suitable choice of those m other securities in the market such that S_t is invertible, the weight of the hedging portfolio is given by

$$\mathbf{\Phi}_t = -(\mathbf{S}_t)^{-1}\mathbf{P}_t. \tag{9.8}$$

When solving the system (9.8), companies need the performance data on their own lines of business against all these N fundamental assets, as well as the corresponding sensitivities of other securities publicly traded in the market. In principle, for the former, companies can determine these coefficients using their internal business information extracted from their own financial report figures. As a result, it is relatively easy to acquire all the information for $\mathbf{P}_t$, which is specifically the sensitivity of our portfolio Π_t to the value of each individual fundamental asset r_t^k, for $k = 1,\ldots,N$. For instance, as a simple case from a cash flow perspective, let r_τ^k be the average annual rate of return for the investment made at time τ on the k-th asset, with a cash flow of c_τ^k, and assume that the asset cannot be traded until an expiry date T. Then the time-t present value of the portfolio of all N assets is given by $\Pi_t = (1 + r_f)^{-(T-t)} \sum_{k=1}^{N} \sum_{\tau=t}^{T} c_\tau^k (1 + \mathrm{r}_\tau^k)^{T-\tau}$, where r_f is the deterministic risk-free rate. Indeed, this scenario is not uncommon in reality; consider the purchase of government bonds for example. Another example is the construction of public infrastructure, where capital injections are required every year in the construction stage, while the profit will not be realized in few years' time until the facilities are put in service and start to generate revenues. In this case, one can easily evaluate the sensitivity as $\frac{\partial \Pi_t}{\partial \mathrm{r}_t^k} = (1 + r_f)^{-(T-t)} c_t^k (T - t)(1 + \mathrm{r}_t^k)^{T-t-1}$ for $k = 1,\ldots,N$. In real practice, despite a possibly more complicated form of Π_t, investors should have a rather thorough understanding of their own projects, leading to a reasonable knowledge of the functional form of how the portfolio depends on the fundamental assets, such as the one above. By then they can readily estimate the sensitivity with respect to any argument variables.

One approach to approximate these gradients at r^k_{t0} by the *symmetric difference quotient* (a.k.a. *central difference*) method:

$$\frac{\partial \Pi_t(r^k_{t0})}{\partial r^k_t} \approx \frac{\Pi_t(r^k_{t0} + \Delta r^k_{t0}) - \Pi_t(r^k_{t0} - \Delta r^k_{t0})}{2\Delta r^k_{t0}},$$

for some small change Δr^k_{t0} in the value of the k-th fundamental asset in portfolio Π_t, whereas other r^j_{t0}'s, $j \neq k$, are held fixed.

However, the sensitivities of the hedging tradable securities S^l_t's to the value r^k_t of each individual fundamental asset, for $l = 1, \ldots, m$ and $k = 1, \ldots, N$, require much effort as the information of these tradable securities is neither under the control nor a regular entry of the company. One simple approach is to construct a regression model on each historical security price S^l_t against the historical values r^k_t's. The required sensitivity is just the tangent of the corresponding regression model, possibly extrapolated at r^k_{t0}. For more details on regression models, see Chapter 10. If there are only one or two major risk factors, the immunization strategy would be achievable in reality, just like CAPM; however, if there are quite a number of factors, we would have to rely on big data in order to implement this immunization strategy in reality. In particular, extensions of deep learning methods (to be discussed in Chapter 15) can be useful for future research endeavours on this topic.

9.8 FACIAL RECOGNITION VIA PCA

In general, PCA is also useful in identifying major factors behind a collection of very high-dimensional random variables. The same principle can be utilized for image recognition by selecting key features from a class of photos; see [2, 8, 10, 18, 21, 22] for more discussions. Certainly, image recognition plays vital roles in business such as providing more tailor-made customer services that facilitate the development of online private banking. For instance, facial recognition, as illustrated in Figure 9.5, can be used to verify a customer's identity, and henceforth his/her policy and background information can be retrieved and reviewed instantly. This enriches the potential use of online banking with the minimum possible manpower yet can still achieve high precision in the absence of human intervention. Facial recognition is generally much safer to use than the simple log-in via usernames and passwords. With this technology, future customers can simply capture photos of the relevant documents, and then submit them for services online. The image recognition engine converts the uploaded images to the desired information for further process, and then successful applications can be realized in seconds. In a nutshell, financial corporations and customers can attain a win-win situation: the management cost for the customers can be much reduced, while the companies can obtain more accurate information.

Let us explain the actual implementation of facial recognition via PCA. The dataset $\boldsymbol{X}$ is an $(h \cdot w) \times n$ matrix, where each observation $\boldsymbol{x}_i$ is a face image of a shape of height $h \times$ width w being flattened into a one-dimensional vector of a length

(a) IU, *Ji-eun LEE*, a contemporary popular South Korean singer–songwriter and actress. Source: Sipa USA.

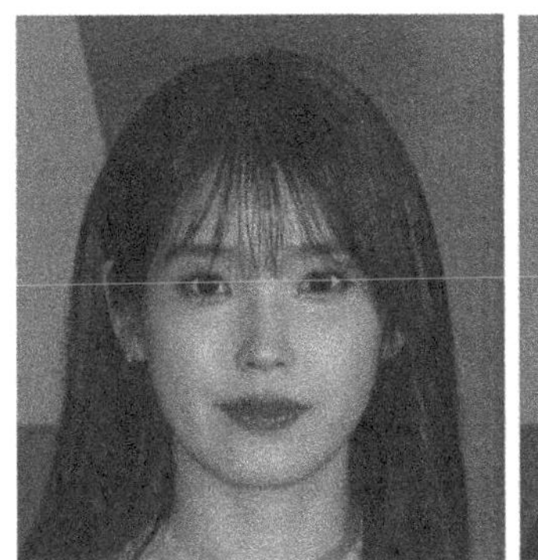

(b) Courtesy: `MediaPipe`

(c) Courtesy: `MediaPipe`

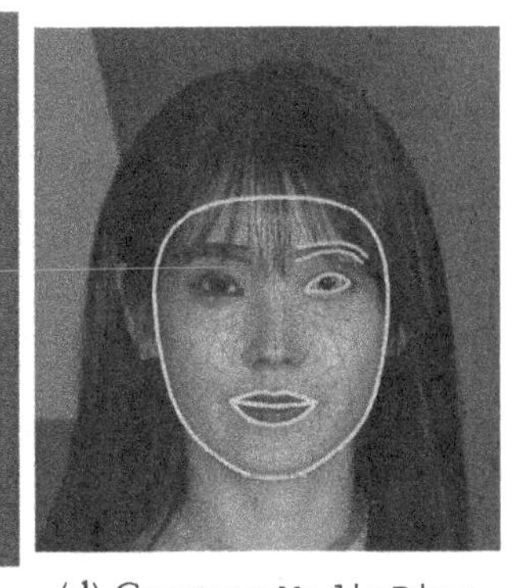

(d) Courtesy: `MediaPipe`

FIGURE 9.5 Facial recognition for the actress *IU*.

$h \cdot w$. The feature variables, the various pixels of the face image, are highly correlated locally, because the neighborhoods of a pixel often take similar values because they share similar color. Therefore, PCA can be adopted for dimension reduction. In the context of facial recognition, for classification, we compute the Euclidean distance of a query face image $\boldsymbol{x}_*$ against the database $\boldsymbol{x}_1, \ldots, \boldsymbol{x}_n$ with respective labels $y_1, \ldots, y_n$. Suppose that the first m PCs are selected with the corresponding loadings $\boldsymbol{H}^{(1)} = (\boldsymbol{h}_1, \ldots, \boldsymbol{h}_m) \in \mathbb{R}^{(h \cdot w) \times m}$, the predicted label of that query face image is given by $y_* = y_k$, where

$$k = \underset{i=1,\ldots,n}{\arg\min}\, d((\boldsymbol{H}^{(1)})^\top \boldsymbol{x}_i, (\boldsymbol{H}^{(1)})^\top \boldsymbol{x}_*) = \underset{i=1,\ldots,n}{\arg\min} \sum_{j=1}^{m} (\boldsymbol{h}_j^\top (\boldsymbol{x}_i - \boldsymbol{x}_*))^2.$$

This methodology actually shares the same philosophy as the *K-Nearest Neighbour* technique to be discussed in Chapter 14. The *AT&T Database of Faces*[7] [16] contains 10 face images of size $112 \times 92 (= h \times w)$ in different scenarios of various persons, each of these 10 face images comes from 40 individuals taken between April 1992 and April 1994. Examples of these 10 scenarios include varying lighting, having closed eyes, smiling, and wearing glasses. For the subfolders downloaded from *Kaggle*, `s1`, `s2`, ..., `s40`, each of them contains 10 face images of the i-th individual for $i = 1, \dots, 40$. In this example, the first image of each individual is collected together to form the test set, while the remaining 9 images of each individual are considered in the training set. We here lay down the Python codes in Programme 9.17, where the `cv2`, also known as *OpenCV*, is a well-known library for image processing and computer vision. Moreover, the training and test sets are first standardized by the training mean `mu_train` and standard deviation `std_train`.

```
1  # Operating System to link with the specified directory
2  import os
3  import numpy as np
4  import cv2       # OpenCV = OPEN source for Computer Vision library
5  # tqdm = taqadum, meaning the processing passed for each iteration
6  from tqdm import tqdm, trange
7
8  SUB_FOLDERS = [f"s{i+1}" for i in range(40)]
9  train_img, test_img = [], []
10 for sub_folder in SUB_FOLDERS:
11     # join() = JOIN the text strings by slash "/"
12     path = os.path.join("ATT", sub_folder)
13     # For all image filenames in the current working directory
14     for i, img_Name in enumerate(tqdm(os.listdir(path))):
15         img_Path = os.path.join(path, img_Name)
16         img_array = cv2.imread(img_Path, cv2.IMREAD_COLOR)
17         gray_array = cv2.cvtColor(img_array, cv2.COLOR_BGR2GRAY)
18         rescale_array = gray_array.flatten()
19         # take the first image in every sub_folder as test
20         if i == 0:
21             test_img.append(rescale_array)
22         else:
23             train_img.append(rescale_array)
24
25 train_img, test_img = np.array(train_img).T, np.array(test_img).T
26 mu_train = np.mean(train_img, axis=1).reshape(-1, 1)
27 std_train = np.std(train_img, axis=1).reshape(-1, 1)
28 rescale_train = (train_img - mu_train)/std_train
29 rescale_test = (test_img - mu_train)/std_train
```

Programme 9.17 Creating training and test datasets from the AT&T dataset via Python.

[7] The dataset can be readily downloaded from https://www.kaggle.com/datasets/kasikrit/att-database-of-faces.

In **R**, the `pixmap` (PIXel MAPs) library is used to process bitmapped images. In particular, the `read.pnm()` (Portable Any Map) function is adopted to read images; see the corresponding implementation in Programme 9.18 in the **R** environment.

```
1  > library(pixmap)
2  >
3  > SUB_FOLDERS <- paste0("s", 1:40)
4  > train_img <- c()
5  > test_img <- c()
6  >
7  > for (sub_folder in SUB_FOLDERS) {
8  +   path <- file.path("../Datasets/ATT", sub_folder)
9  +   files <- list.files(path)
10 +   for (i in 1:length(files)) {
11 +     img_array <- read.pnm(file=file.path(path, files[i]))
12 +     # matching python arrays
13 +     rescale_array <- as.vector(t(img_array@grey) * 255)
14 +
15 +     if (i == 1) {
16 +       test_img <- cbind(test_img, rescale_array)
17 +     } else {
18 +       train_img <- cbind(train_img, rescale_array)
19 +     }
20 +   }
21 + }
22 >
23 > n <- ncol(train_img)
24 > mu_train <- apply(train_img, 1, mean)
25 > std_train <- apply(train_img, 1, sd) * sqrt((n-1)/n)
26 > rescale_train <- (train_img - mu_train) / std_train
27 > rescale_test <- (test_img - mu_train) / std_train
```

Programme 9.18 Creating training and test datasets from the AT&T dataset via **R**.

Notice that the training set X is a $(112 \cdot 92) \times (9 \cdot 40) = 10,304 \times 360$ matrix, leading to a singular $10,304 \times 10,304$ matrix $XX^{\top}$. From Section 1.2 of Chapter 1, we know that the largest 360 (non-zero) eigenvalues of $XX^{\top}$ and $X^{\top}X$ are the same. For $i = 1, \ldots, 360$, denote by u_i the i-th eigenvector of $X^{\top}X$, then the i-th eigenvector of $XX^{\top}$ is given by $h_i = Xu_i$; meanwhile, the remaining eigenvalues $\lambda_{361}, \ldots, \lambda_{10,304}$ of the singular $XX^{\top}$ are simply zero. Taking advantage of this, we need only consider the spectral decomposition of $X^{\top}X$ instead. We implement this in Programme 9.19 via Python.

```
1 _, n = rescale_train.shape
2 lamb, U = np.linalg.eig(rescale_train.T @ rescale_train / n)
3 H = U.T @ rescale_train.T
4 # normalize each eigenvector
5 H = H / np.sqrt(np.sum(H**2, axis=1).reshape(-1, 1))
6 print(H.shape)                          # n x (hw) matrix
7 print(sum(lamb))
```

```
8
9  CumSum_s2 = np.cumsum(lamb)/sum(lamb)
10 idx = np.where(CumSum_s2 > 0.9)[0][0]
11 tilde_H = H[:idx, :]
12 print(tilde_H.shape)                    # m x (hw) matrix
```

```
1 (360, 10304)
2 10303.999999999995
3 (107, 10304)
```

Programme 9.19 Applying PCA to the AT&T training dataset via Python.

From the outputs of Programme 9.19 in Python, 107 PCs suffice to preserve 90% of all information contained in the total variance. Notice that the training set is standardized, so the total variance is then equal to $p = h \cdot w = 10,304$. A similar implementation via **R** can be found in Programme 9.20.

```
1  > eig <- eigen(t(rescale_train)%*%rescale_train / n)
2  > lamb <- eig$values; H <- t(eig$vectors)%*%t(rescale_train)
3  > H <- H / sqrt(rowSums(H^2))      # normalize each eigenvector
4  > dim(H)                           # n x (hw) matrix
5  [1]   360 10304
6  > sum(lamb)
7  [1] 10304
8  >
9  > CumSum_s2 <- cumsum(lamb) / sum(lamb)
10 > idx <- which(CumSum_s2 > 0.9)[1]
11 > tilde_H <- H[1:idx,]
12 > dim(tilde_H)                     # m x (hw) matrix
13 [1]   107 10304
```

Programme 9.20 Applying PCA to the AT&T training dataset via **R**.

In Python, the PCA in `sklearn` can also be used directly with the argument `n_components=0.9` to preserve 90% of information; also see Programmes 9.21 in Python and 9.22 in **R**. For **R**, since the training dataset has already been normalized, we call the `prcomp()` function with the two arguments `center=FALSE` and `scale=FALSE`.

```
1 from sklearn.decomposition import PCA
2 pca = PCA(n_components=0.9).fit(rescale_train.T)
3 # components_ : an ndarray of shape (n_components, n_features)
4 tilde_H = pca.components_
5 print(tilde_H.shape)                    # m x (hw) matrix
```

```
1 (107, 10304)
```

Programme 9.21 Applying PCA with `sklearn` to the AT&T training dataset via Python.

```
1  > pca <- prcomp(t(rescale_train), center=FALSE, scale=FALSE)
2  > idx <- which(cumsum(pca$sdev^2)/sum(pca$sdev^2) > 0.9)[1]
3  > tilde_H <- t(pca$rotation[,1:idx])
4  > dim(tilde_H)                          # m x (hw) matrix
5  [1]   107 10304
```

Programme 9.22 Applying PCA with `prcomp` to the AT&T training dataset via **R**.

Both the training and test datasets are transformed by projecting onto these 107 PCs. We then compare the Euclidean distances between each test observation against all in the training dataset. The observation from the training dataset that yields the shortest distance with the query face image is said to be a match with that test observation.

```
1  import matplotlib.pyplot as plt
2  
3  score_train = tilde_H @ rescale_train    # m x n matrix
4  score_test = tilde_H @ rescale_test      # m x n matrix
5  m, n_test = score_test.shape
6  ncol = 10
7  fig, axes = plt.subplots(nrows=8, ncols=ncol, figsize=(6*ncol,8*8))
8  pred_dist = []
9  for i in trange(n_test):
10     distance = np.linalg.norm(score_train -
11                          score_test[:,i].reshape(-1, 1), axis=0)
12     idx_train = np.argmin(distance)
13 
14     j = 2*i
15     ax1 = axes[j // ncol][j % ncol]
16     ax1.imshow(test_img[:,i].reshape(112, 92), cmap='gray')
17     ax1.axis('off')
18     ax1.set_title("Test image", fontsize=45)
19     ax2 = axes[(j+1) // ncol][(j+1) % ncol]
20     ax2.imshow(train_img[:,idx_train].reshape(112, 92),
21                cmap='gray')
22     ax2.axis('off')
23     ax2.set_title(f"Distance {distance[idx_train]:.3f}", fontsize=45)
24     pred_dist.append(distance[idx_train])
25 
26 col = ["blue"] * n_test
27 col[34] = "burlywood"
28 fig = plt.figure(figsize=(10, 7))
29 plt.bar(np.arange(n_test), pred_dist, ec="black", color=col, alpha=0.7)
30 plt.xlabel("Index", fontsize=20)
31 plt.ylabel("Distance", fontsize=20)
32 plt.axhline(65, ls="--", color="black", linewidth=3)
```

Programme 9.23 Testing of facial recognition with PCA for the AT&T dataset via Python.

From Figure 9.6, only the rightmost pair of images in the second last row returns a mismatch, as seen in the red box. The test image was of a lady but the programme returns a male with a distance of 78.546, which is the largest among all pairs in Figure 9.6. From Figure 9.7, there are three bars whose values are greater than 65, which are query faces No. 1, 16 and 35, indicating that we have a lower confidence on the matching of these query faces with the correct person. Therefore, in practice, if the distance is larger than some threshold $\tau > 0$, for example 65 as mentioned above, and as indicated in the horizontal dashed line in Figure 9.7, we regard the matching process of that query face image with the database as a failure. We also implement the same analysis in Programme 9.24 via **R**; see the corresponding testing results and the

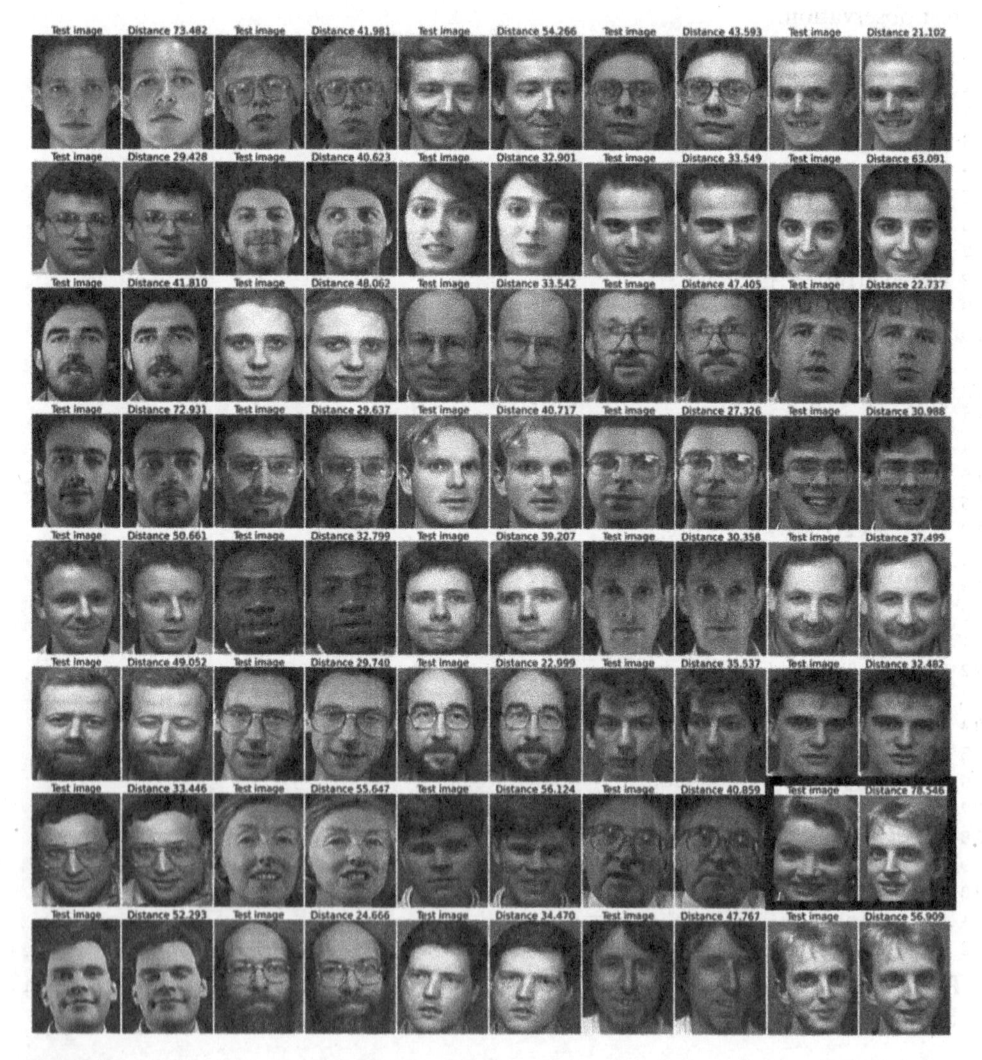

FIGURE 9.6 Testing results of matching faces, generated by Programme 9.23 in Python.

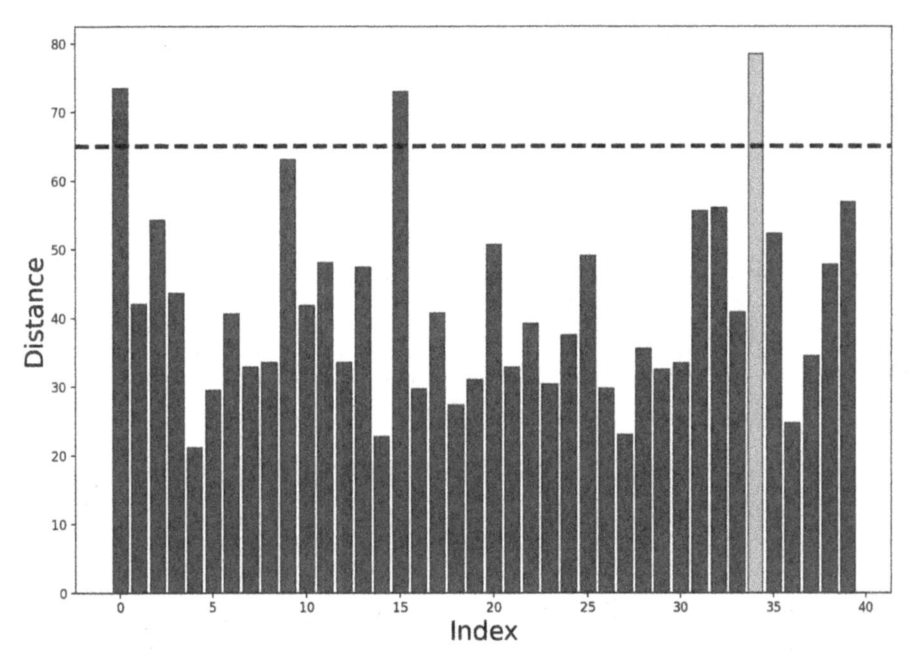

FIGURE 9.7 Bar chart of the Euclidean distances of the projected PCs between the training and test datasets, generated by Programme 9.23 in Python.

bar chart in Figures 9.8 and 9.9, which agree with the outcomes obtained via Python. Recently, a higher accuracy can be attained by using *Convolutional Neural Networks* (ConvNets or CNNs); more discussions can be found in our upcoming book [5].

```
1  > score_train <- tilde_H%*%rescale_train     # m x n matrix
2  > score_test <- tilde_H%*%rescale_test       # m x n matrix
3  > pred_dist <- c()
4  > n_test <- ncol(score_test)
5  >
6  > img_arr <- function(img){
7  +   t(apply(matrix(img, nrow=112, ncol=92, byrow=TRUE), 2, rev))
8  + }
9  >
10 > nrow <- 8; ncol <- 10
11 > opar <- par()
12 > par(mfrow=c(nrow, ncol), plt=c(0.05,0.95,0,0.7), oma=c(1,1,1,1))
13 > for (i in 1:n_test) {
14 +   distance <- sqrt(colSums((score_train - score_test[,i])^2))
15 +   idx_train <- which.min(distance)
16 +
17 +   image(img_arr(test_img[,i]),
18 +         col=grey(seq(0, 1, length=256)), xaxt='n', yaxt='n')
19 +   title("Test image", font.main=1 , line=1)
20 +   image(img_arr(train_img[,idx_train]),
21 +         col=grey(seq(0, 1, length=256)), xaxt='n', yaxt='n')
```

```
22 +   title(paste("Distance", round(distance[idx_train], 3)),
23 +         font.main=1, line=1)
24 +   pred_dist <- c(pred_dist, distance[idx_train])
25 + }
26 > par(opar)
27 > par(cex.lab=2, cex.axis=2, cex.main=2, mar=c(5,5,4,4))
28 > col <- rep("blue", n_test)
29 > col[35] <- "burlywood1"
30 > barplot(pred_dist, names=1:n_test, col=col, ylim=c(0, 80),
31 +         xlab="Index", ylab="Distance")
32 > abline(h=65, lty=2, lwd=3)
```

Programme 9.24 Testing of facial recognition with PCA for the AT&T dataset via R.

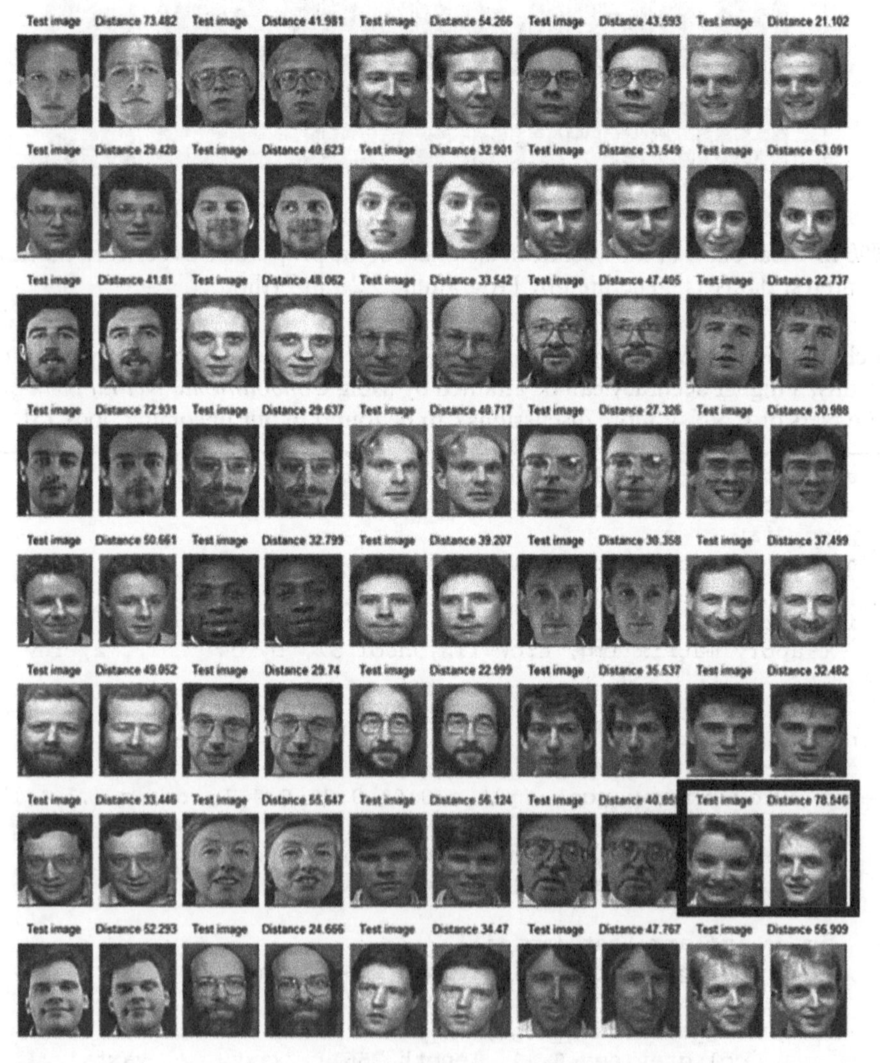

FIGURE 9.8 Testing results of matching faces, generated by Programme 9.24 in R.

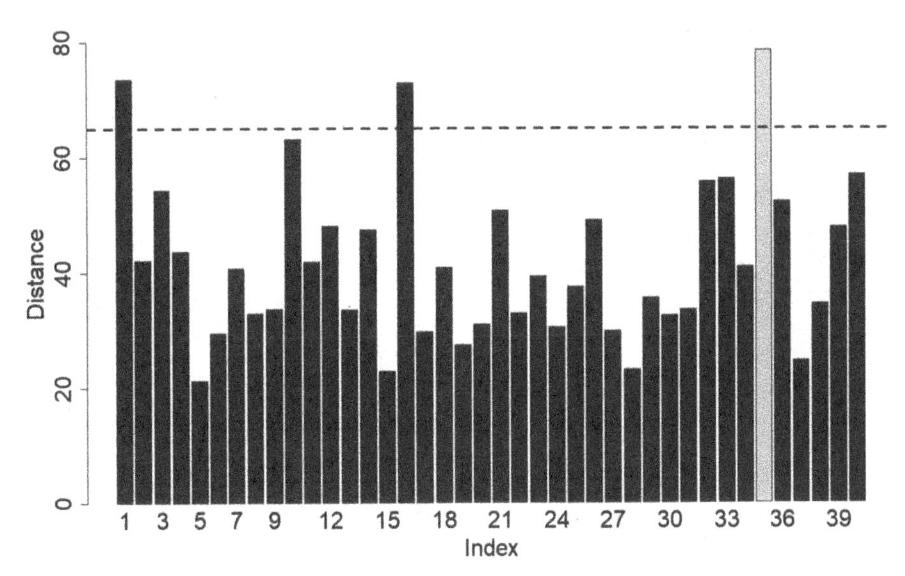

FIGURE 9.9 Bar chart of the Euclidean distances of the projected PCs between the training and test datasets, generated by Programme 9.24 in **R**.

9.9 NON-LIFE INSURANCE VIA PCA

Casualty insurance can also benefit from image recognition. With the plenty of images available on the Internet, one can identify key information that is crucial for a financial property to be insurable. For example, photos of a house roof can be captured by drones to better understand the condition of the house and to determine any collapse risk, as in Figure 9.10. In the old days, insurance companies could only obtain partial information from written historical records. With these pictures, the underwriting process can be implemented in a more comprehensive manner.

Health insurance is another field in which image recognition can help uncover any hidden risk of insureds, which could not be easily detected in the past. In modern medicine applications, image recognition is of great use in diagnosis and prognosis, especially in identifying cancer cells or tumors at a very early stage. Besides taking doctors' viewpoints into account, *Computed Tomography* (CT) images of cancer patients (see Figure 9.11) can also be utilized as another factor for insurance pricing, during steps of underwriting and determination of risk loading. These medical images may provide richer information than the record of diagnosis provided by a medical doctor, which often take a much longer time, and they can therefore help improve the precision of the pricing model and timely action to be taken. Nevertheless, biomedical images are usually corrupted by noisy signals which should be de-noised, as from these one can extract features and classify images to facilitate better diagnosis. PCA can be used as a digital image compression algorithm without causing too much loss in information. Indeed, PCA can be incorporated into image processing; for example, see Figure 9.12, which enhances data texture segmentation (depicting the outlines

(a) Good Condition
Risk Level: 1
Risk Loading: -4%
Source: Forbes Home

(b) Bad Condition
Risk Level: 6
Risk Loading: 12%
Source: Cambridgeshire Live

FIGURE 9.10 Property risk analysis through image processing.

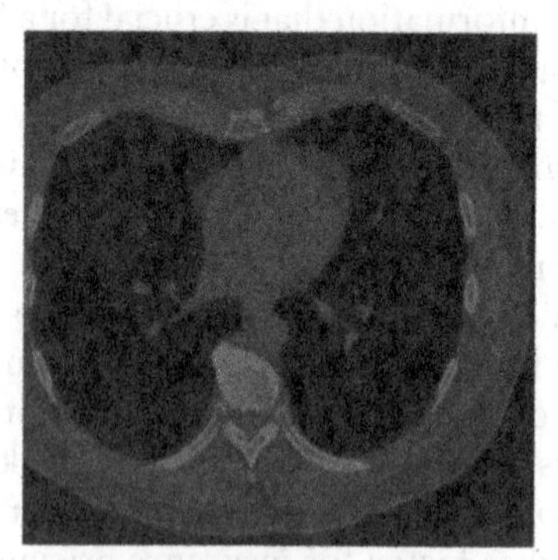

Healthy
Life Expectancy: 25 years
Risk Loading: 0%

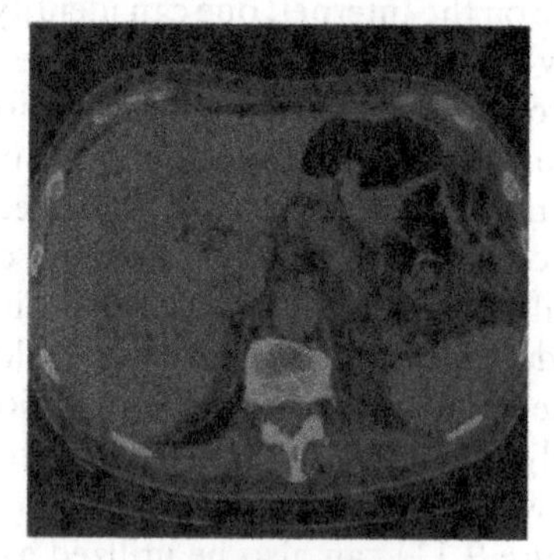

Lung Cancer Stage I
Life Expectancy: 4 years
Risk Loading: 350%

Photo credit: Joanna Shang, 2013.

FIGURE 9.11 Biomedical image recognition is commonly used to determine whether a patient has a tumor and how severe it is. Source: [19].

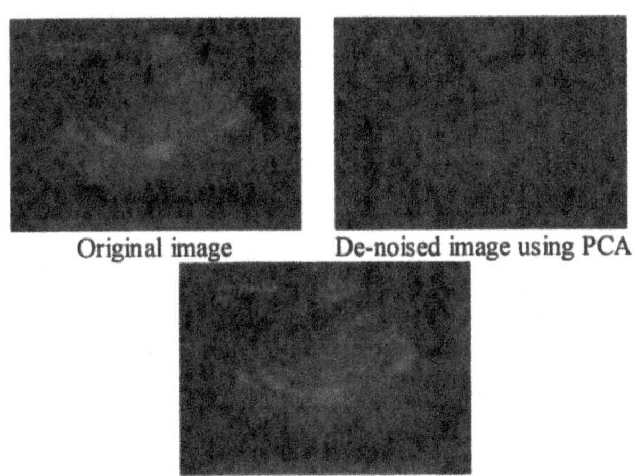

FIGURE 9.12 Enhancing visibility for biomedical images. Source: [4].

Number of insured household: 56
Estimated insurance claim: $ 12.3 M

FIGURE 9.13 Housing insurance analysis through satellite geographical images. Source: [19].

via PCs), compression, helps removal of redundant information (removing PCs with smaller eigenvalues), and enables noise removal and feature extraction.

Besides, image recognition techniques also prove helpful in monitoring natural catastrophes, such as typhoons, floods and forest fires, in a real-time manner; it can also assist in handling the aftermath of the incidents. For instance, upon detection of the trace of a typhoon or a flooded area from the satellite image, the severity of damage of the affected regions can be determined with the aid of image recognition. This gives light to the development of automatic systems for loss prediction, so as to facilitate a better management of foreseeable claims for insurers. Take the analysis on property insurance in [19] for example, where the grey cloud of tornado in the satellite photo (Figure 9.13) is captured to run from west to northeast, whereas the insured properties are denoted by black dots. By applying some image recognition algorithm to this photo, the identification of the tornado path leads to further judgements on the post-tornado status of the insured homes, namely whether and how severely are they damaged. It is also possible to feed latest photos to the algorithm, such that the estimation procedure can be carried out in a progressive manner, which leads to timely updates of the status of the buildings concerned.

9.10 INVESTMENT STRATEGIES USING PCA

Apart from image recognition, we can also use PCA to formulate investment strategies. During the first half of 2020, there was a global stock decline, largely attributed to the outbreak of COVID-19. Let us try to apply PCA to the US stock prices and see if this method can identify whether the key factors behind the decline really coincide with the virus outbreak. Once these key factors are unveiled, we can find those securities that align most favorably with these factors. This information can then be used to formulate a profitable investment strategy; indeed, we can simply pick up well-performing stocks from the national stock index to build a portfolio that outperforms the underlying market.

On 15 March 2020, the *Federal Reserve* in the US began to conduct its fourth round of *quantitative easing* (QE); the money flow figures are readily available in the public domain, and one naturally expected the stock market to respond positively. To this end, we shall use PCA to select some representative stocks to invest. Suppose that the current time is the end of March 2020, and we have observed the response from the market in the form of a rapid upward trend. Assume that we have an initial investment of USD 100,000 and we would like to invest evenly in 10 US stocks. In Programme 9.25 via Python (resp. Programme 9.26 via **R**), PCA is applied to the centered logarithmic returns computed from the opening prices from 2 January 2020 to 31 March 2020. The top 10 stocks with the largest loadings (likely positive in value) and the bottom 10 stocks with the smallest loadings (likely negative in value) with respect to the first PC y_1 are then selected as candidates[8]. Intuitively, this first PC is certainly viewed as the major market factor. The opening price difference between 2 January 2020 to 31 March 2020 for each selected candidate is computed to analyze its performance.

```
1  import pandas as pd
2  import numpy as np
3
4  df = pd.read_csv("us_Open_2020H.csv", index_col=0)
5  r = df.apply(np.log).diff()
6  r_mask_train = pd.to_datetime(r.index,format='%d/%m/%Y') < "2020-04-01"
```

[8] The loadings of the first PC generated from Python have the opposite sign to those generated in **R**, so the basket of 10 stocks with the smallest loadings and the basket of 10 stocks with the largest loadings are interchanged in Python and **R**. PCA can extract the factors, while the extraction itself does not take the orientation and direction into account.

```
 7 r_train = r.loc[r_mask_train].dropna()
 8
 9 from sklearn.decomposition import PCA
10 # PCA: Input data is centered but not scaled for each feature
11 pca = PCA(n_components=1).fit(r_train)
12 pc1 = pd.Series(index=r.columns, data=pca.components_[0])
13
14 # Select 10 equities to invest base on the loadings of pc1
15 n_stock = 10
16 smallest_n = pc1.nsmallest(n_stock).index
17 largest_n = pc1.nlargest(n_stock).index
18
19 df_train_last = sum(r_mask_train) - 1 # python starts idx from 0
20 print(df[smallest_n].iloc[[0, df_train_last],:].diff())
21 print(df[largest_n].iloc[[0, df_train_last],:].diff())
```

```
 1            VAL UN Equity  NBR UN Equity  APA UN Equity  RIG UN Equity  \
 2 Date
 3 2/1/2020             NaN            NaN            NaN            NaN
 4 31/3/2020          -6.41      -121.7018       -21.0311          -5.78
 5
 6            NBL UN Equity  FLR UN Equity  OXY UN Equity  DVN UN Equity  \
 7 Date
 8 2/1/2020             NaN            NaN            NaN            NaN
 9 31/3/2020       -19.2219       -12.7683        -28.869       -19.0759
10
11            MUR UN Equity  KBH UN Equity
12 Date
13 2/1/2020             NaN            NaN
14 31/3/2020       -20.4311       -16.0407
15            GILD UW Equity  CLX UN Equity  WBA UN Equity  BDX UN Equity  \
16 Date
17 2/1/2020              NaN            NaN            NaN            NaN
18 31/3/2020         10.8778         21.259       -13.4532        -48.971
19
20            CTXS UW Equity  LM UN Equity  CPB UN Equity  MKC UN Equity  \
21 Date
22 2/1/2020              NaN           NaN            NaN            NaN
23 31/3/2020         32.7623       12.5407         -2.536       -29.2857
24
25            LLY UN Equity  WMT UN Equity
26 Date
27 2/1/2020             NaN            NaN
28 31/3/2020          5.842        -4.0087
```

Programme 9.25 Python code for the PCA investment example.

```
1  > df <- read.csv("us_Open_2020H.csv", header=TRUE, row.names=1)
2  > # Compute daily logarithmic return for each equity
3  > r <- sapply(df, function(x) diff(log(x)))
4  > date <- as.Date(rownames(df), format="%d/%m/%Y")
5  > r_idx_train <- which(date[-1] < as.Date("2020/04/01"))
6  > r_train <- r[r_idx_train,]
7  >
8  > PCA <- prcomp(r_train, center=TRUE, scale=FALSE)
9  > pc1 <- PCA$rotation[,1]          # Save the loading of 1st PC
10 >
11 > # Select 10 equities to invest base on the loadings of pc1
12 > n_stock <- 10
13 > smallest_n <- order(pc1, decreasing=FALSE)[1:n_stock]
14 > largest_n <- order(pc1, decreasing=TRUE)[1:n_stock]
15 >
16 > # check which group performs the best with training data
17 > df_train_last <- tail(r_idx_train, 1) + 1     # add back 1 in r for d
18 > df[df_train_last, smallest_n] - df[1, smallest_n]
19           GILD.UW.Equity CLX.UN.Equity WBA.UN.Equity BDX.UN.Equity
20 31/3/2020        10.8778        21.259      -13.4532       -48.971
21           CTXS.UW.Equity LM.UN.Equity CPB.UN.Equity MKC.UN.Equity
22 31/3/2020        32.7623      12.5407        -2.536      -29.2857
23           LLY.UN.Equity WMT.UN.Equity
24 31/3/2020         5.842       -4.0087
25 > df[df_train_last, largest_n] - df[1, largest_n]
26           VAL.UN.Equity NBR.UN.Equity APA.UN.Equity RIG.UN.Equity
27 31/3/2020         -6.41     -121.7018      -21.0311         -5.78
28           NBL.UN.Equity FLR.UN.Equity OXY.UN.Equity DVN.UN.Equity
29 31/3/2020      -19.2219      -12.7683       -28.869      -19.0759
30           MUR.UN.Equity KBH.UN.Equity
31 31/3/2020      -20.4311      -16.0407
```

Programme 9.26 R code for the PCA investment example.

Table 9.2 summarizes the background of the 20 selected candidate stocks from the outputs of Programmes 9.25 via Python and Programme 9.26 via R. In the first basket, most of the 10 stocks are in the sectors of *Health Care* and *Consumer Staples*; in the second basket, all 10 stocks are in either *Energy* or *Construction* sector[9]. We notice all 10 stocks in the second basket suffered from a substantial market downturn due to COVID-19. This can be explained by the disruption of supply chains and the drastic drop in electricity and fuel demands due to the restriction of movement due to quarantine. On the other hand, stocks in the first basket remained strong in COVID-19 due to their relevance to consumer products and healthcare, which were still mandatory for citizens during the period. Therefore, after the QE on 15 March, we expect a more significant revival in the energy sector than in the household necessities.

[9] Also see *Stock Market Analytics with PCA*. Retrieved from https://web.archive.org/web/20210304165314/https://towardsdatascience.com/stock-market-analytics-with-pca-d1c2318e3f0e?gi=da1a691e72b.

TABLE 9.2 The 20 selected candidates that have the 10 smallest and 10 largest loadings in the 1st PC.

Basket	Ticker	Name	Sector
1	GILD	Gilead Sciences, Inc.	Health Care
	CLX	The Clorox Company	Consumer Staples
	WBA	Walgreens Boots Alliance, Inc.	Consumer Staples
	BDX	Becton Dickinson and Co.	Health Care
	CTXS	Citrix Systems, Inc.	Technology
	LM	Legg Mason, Inc.	Financials
	CPB	Campbell Soup Company	Consumer Staples
	MKC	McCormick & Company Inc.	Consumer Staples
	LLY	Eli Lilly and Company	Health Care
	WMT	Walmart Inc.	Consumer Staples
2	VAL	Valaris Limited	Energy
	NBR	Nabors Industries Ltd.	Energy
	APA	APA Corporation	Energy
	RIG	Transocean Ltd.	Energy
	NBL	Fluor Corporation	Energy
	FLR	Transocean Ltd.	Construction
	OXY	Occidental Petroleum	Energy
	DVN	Devon Energy Corporation	Energy
	MUR	Murphy Oil Corporation	Energy
	KBH	KB Home	Construction

As a comparison, let us formulate three portfolios, the performances of which are shown in Figure 9.14[10], namely (i) long the 10 securities in first basket; (ii) long the 10 securities in the second basket; and (iii) long the S&P index indicated by the black line. For fair comparison, each selected stock in strategies (i) and (ii) has an initial investment of USD 10,000, and strategy (iii) is set off with an initial investment of $10 \times \text{USD}\,10,000 = \text{USD}\,100,000$ in total. We here adopt the same commission fee, namely USD $\max(0.013 \times \text{number of stocks traded}, 2.05)$ for each transaction[11]. From the results in Figure 9.14, as of the end of June 2020, portfolio (ii) (dashed lines in blue) performs significantly better than (iii) (dotted lines in black); while the performance of the strategy (i) (solid lines in red) is inferior to that of (iii), it still ends up being profitable, thanks to the upward trend stimulated by the implementation of QE as we indicated above.

[10] As a remark, since the loading of the first PC changes with time, often the expression of the PC now is different from that obtained in the past. To put it in real practice, a close real-time monitoring is required to see if there is any substantial shift in the nature of the first few PCs.
[11] Source: https://www.futuhk.com/en/commissionnew.

```python
from matplotlib import pyplot as plt

plt.rcParams["figure.figsize"] = 15, 10
plt.rcParams.update({'font.size': 25})

investment = 100_000
def buy_and_hold(test_price, n_stock=n_stock):
    n_units = np.floor(investment/n_stock/test_price.iloc[0,:])
    commission = sum(n_units.apply(lambda x: max(2.05, 0.013*x)))
    stocks_amount = test_price.iloc[0,:] @ n_units
    price_path = test_price.iloc[1:,:] @ n_units
    remain = investment - 2*commission
    return remain + (price_path - stocks_amount)

spx = pd.read_csv("SPX_Open_2020H.csv", index_col=0)
equity_price = df.loc[np.invert(r_mask_train)]
index_price = spx.loc[np.invert(r_mask_train)]

pca_bnh_smallest = buy_and_hold(equity_price[smallest_n], n_stock)
pca_bnh_largest = buy_and_hold(equity_price[largest_n], n_stock)
market_bnh = buy_and_hold(index_price, 1)

compare = pd.concat([pca_bnh_smallest, pca_bnh_largest,
                     market_bnh], axis=1)
compare.plot(linewidth=3, color=["blue", "red", "black"],
             style=["--", "-", "."])
plt.legend(["PCA smallest", "PCA largest", "S&P500"])
plt.ylim(0.9*investment, 3*investment)
```

Programme 9.27 Buy-and-hold strategy for the two baskets of stocks and the S&P500 market index in Python.

```r
> library(zoo)
>
> investment <- 100000
> buy_and_hold <- function(test_price, test_date, n_stock=1){
+   n_units <- floor(investment/n_stock/test_price[1,])
+   commission <- sum(sapply(n_units, function(x) max(2.05, 0.013*x)))
+   stocks_amount <- as.numeric(test_price[1,]%*%n_units)
+   price_path <- test_price%*%n_units
+   remain <- investment - 2*commission
+   values <- zoo(remain + (price_path - stocks_amount))
+   index(values) <- test_date
+   return (values)
+ }
>
> spx <- read.csv("SPX_Open_2020H.csv", header=TRUE, row.names=1)
> equity_price <- as.matrix(df[(df_train_last+1):nrow(df),])
> index_price <- as.matrix(spx[(df_train_last+1):nrow(df),])
> test_date <- date[(df_train_last+1):nrow(df)]
>
> pca_bnh_smallest <- buy_and_hold(equity_price[,smallest_n], test_date,
+                                  n_stock)
```

```
22  > pca_bnh_largest <- buy_and_hold(equity_price[,largest_n], test_date,
23  +                                  n_stock)
24  > market_bnh <- buy_and_hold(index_price, test_date, 1)
25  >
26  > par(mfrow=c(1,1), cex.lab=2, cex.axis=1.5, cex.main=2)
27  > plot_date <- test_date[seq(1, length(test_date), 10)]
28  > plot(pca_bnh_smallest, ylim=c(0.9*investment, 3*investment),
29  +      col='red', lwd=3, xaxt="n", ylab="")
30  > lines(pca_bnh_largest, col='blue', lwd=3, lty=2)
31  > lines(market_bnh, col='black', lwd=3, lty=3)
32  > legend("topleft", c("PCA smallest", "PCA largest", "S&P500"),
33  +        col=c("red","blue","black"), lwd=3, lty=1:3, cex=1.5)
34  > axis(side=1, plot_date , format(plot_date , "%d-%m-%y"), cex.axis=1)
```

Programme 9.28 Buy-and-hold strategy for the two baskets of stocks and the S&P500 market index in **R**.

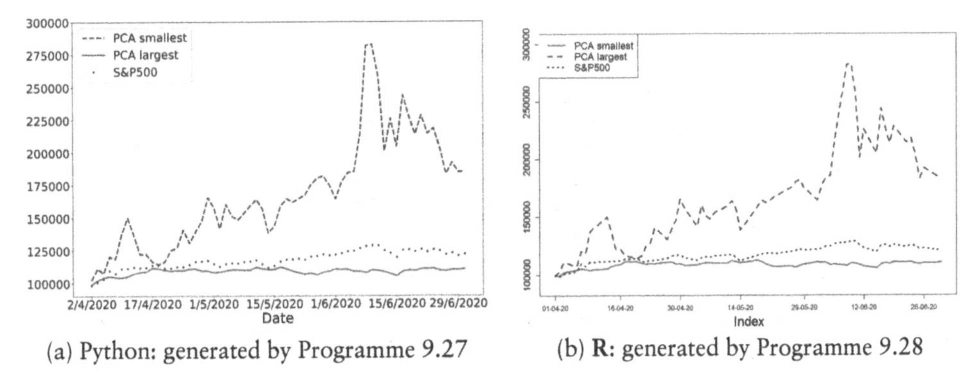

(a) Python: generated by Programme 9.27 (b) **R**: generated by Programme 9.28

FIGURE 9.14 Performance comparison of the three portfolios from April to June 2020.

*9.11 RECOMMENDER SYSTEM

A recommender system, commonly encountered in *Meta*, *Amazon*, and *Netflix*, helps predict the "rating" or "preference" that a subscriber would give to an item. It has become increasingly popular in recent years, and is utilized in a variety of areas including movies, music, news, books, research articles, search queries, social tags, and even groceries. Recommender systems can be used in marketing science research and business informatics to formulate a better business solution to maximize companies' profit. There are also recommender systems for entertainment enterprises, restaurants, garments, financial services, insurance products, romantic partner finding (online dating), and X (formerly *Twitter*) pages.

In the rest of this section, we shall provide the basic connection between the theory behind one of the commonly used recommender systems and PCA. To motivate this, imagine that we found a startup which provides online movie viewing. The company has attracted $i = 1, 2, \ldots, m$ subscribers, and has also acquired $j = 1, 2, \ldots, n$ movies.

Each subscriber is invited to rate a movie after viewing with $r = 1, 2, \ldots, 5$ (a continuum rating is certainly allowed) such that $r = 1$ means the worst and $r = 5$ means the best. These ratings are stored in an $m \times n$ sparse[12] matrix $\mathbf{R} = (r_{i,j})_{m\times n}$, containing records of each subscriber's ratings on all movies seen by him/her in our database; Figure 9.15 illustrates some real-life examples of $\mathbf{R}$ in the context of the *Netflix Prize.*

Note that the ratings remain at $r = 0$ for movies that the subscriber has not yet watched or rated. Based on the data, we want to build a recommender system to learn about the taste of each subscriber, so that we can make predictions on a particular subscriber's ratings of those movies that he/she has not watched, and also make recommendations of any future movies.

Here we describe a popular approach called the *latent-factor method* (matrix factorization) in *collaborative filtering*[13], which relies on the global structure of $\mathbf{R}$. Each movie is evaluated against certain general features, labeled as $k = 1, 2, \ldots, p$,[14] that describe movies. Typically, only a few features are needed and p is far less than n. These "hidden/latent" features, which arise in our implementation algorithm, depend

FIGURE 9.15 *Netflix Prize*, October 2006. The rating matrix $\mathbf{R}$ is large and sparse with $m = 480,189$ and $n = 17,770$. Source: Getty Images North America.

[12] Most of the entries are zero as there are too many good movies missed due to limited time! Usually only about 1% of all entries are non-zero.

[13] While *Netflix* makes use of collaborative filtering, *Amazon* adopts *content-based filtering* based on purchase and browsing history.

[14] Typical numbers for p for the *Netflix Prize* ranged from 10 to 200.

entirely on the data, and are not necessarily related to typical classifications such as comedy, tragedy, thriller, action, etc.. More precisely, for a particular movie j, we are to evaluate a sequence of scores $q_{j,k}$ against all hidden features $k = 1, 2, \dots, p$ that best describe this j-th movie's appeal. The collection of all such scores $q_{j,k}$ is to be stored in the $n \times p$ matrix $Q = (q_{j,k})_{n\times p}$, which is to be estimated.

On the other hand, these hidden features can also be used to describe the subscribers' preferences for different movies. For the subscriber i, we want to evaluate a sequence of scores $p_{i,k}$ against all hidden features $k = 1, 2, \dots, p$ that best describe the i-th subscriber's taste on movies. These scores are to be collectively stored in another $m \times p$ matrix $P = (p_{i,k})_{m\times p}$, which is also to be estimated.

For a fixed p, our task is to approximate the matrices P and Q, which contain a total of $(m+n) \times p$ unknown parameters. The product matrix, $PQ^\top$ of size $m \times n$ can be compared to the subscriber's ratings stored in R. While R is sparse (since each subscriber has only viewed or rated a small portion of the entire movie database), our algorithm aims to minimize the deviation of the average between the known ratings of $r_{i,j}$'s and the corresponding ratings predicted by using $\sum_{k=1}^{p} p_{i,k} q_{j,k}$. In other words, the objective is to minimize the following mean squared error (MSE):

$$J(p_{i,k}, q_{j,k}; 1 \le i \le m, 1 \le j \le n, 1 \le k \le p) := \sum_{i=1}^{m} \sum_{j=1}^{n} \left(r_{i,j} - \sum_{k=1}^{p} p_{i,k} q_{j,k} \right)^2$$

$$= \sum_{i=1}^{m} \sum_{j=1}^{n} (r_{i,j} - p_i^\top q_j)^2 = \operatorname{tr}\left((R - \hat{R})^\top (R - \hat{R}) \right) = \| R - \hat{R} \|_F^2, \tag{9.9}$$

where $\|\cdot\|_F$ is the Frobenius norm (see Section 1.3), $p_i = (p_{i,1}, \dots, p_{i,p})^\top$ and $q_j = (q_{j,1}, \dots, q_{j,p})^\top$, for $i = 1, 2, \dots, m$ and $j = 1, 2, \dots, n$; $\hat{R} := PQ^\top$ such that

$$P = \begin{pmatrix} p_1^\top \\ \vdots \\ p_m^\top \end{pmatrix} \quad \text{and} \quad Q^\top = \begin{pmatrix} q_1 & \cdots & q_n \end{pmatrix};$$

see Figure 9.16 for an illustration. Note the similarity of the format of (9.9) with that of eigenvalue problem associated with PCA in Section 9.4. We shall adopt the technique of *Singular Value Decomposition* (SVD) for general rectangular matrices, which can be considered as the generalization of spectral decomposition for positive definite square symmetric matrices.

An optimization scheme is used to find the minimizer by treating P and Q as decision variables. However, this J does not correspond to a standard convex optimization problem. To find the minimizer, we first rewrite J as:

$$J = \operatorname{tr}(R^\top R) - 2\operatorname{tr}(PQ^\top R^\top) + \operatorname{tr}(PQ^\top QP^\top). \tag{9.10}$$

By using formulae for matrix derivatives from Section 1.2 of Chapter 1, we have

$$\begin{cases} \nabla_P J = -2RQ + 2P(Q^\top Q), \\ \nabla_Q J = -2R^\top P + 2Q(P^\top P). \end{cases} \tag{9.11}$$

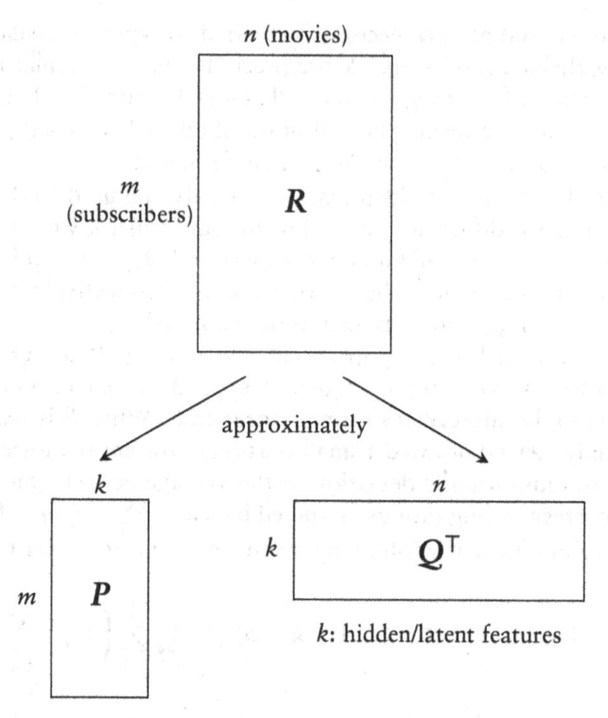

FIGURE 9.16 *Singular Value Decomposition* of the rating matrix $\boldsymbol{R}$.

By equating the gradients with zero, we obtain the first-order condition as

$$\begin{cases} \boldsymbol{R}\boldsymbol{Q} = \boldsymbol{P}(\boldsymbol{Q}^\top\boldsymbol{Q}) \quad (= \hat{\boldsymbol{R}}\boldsymbol{Q}), \\ \boldsymbol{R}^\top\boldsymbol{P} = \boldsymbol{Q}(\boldsymbol{P}^\top\boldsymbol{P}) \quad (= \hat{\boldsymbol{R}}^\top\boldsymbol{P}). \end{cases} \tag{9.12}$$

Since the original problem is not a convex one, (9.12) may contain many local optimal solutions. With a *singular value decomposition* of $\boldsymbol{R} = \boldsymbol{U}\boldsymbol{\Sigma}\boldsymbol{V}^\top$ where $\boldsymbol{U} \in \mathbb{R}^{m\times m}$ is the matrix of *left singular vectors*, $\boldsymbol{V} \in \mathbb{R}^{n\times n}$ is the matrix of *right singular vectors*, and $\boldsymbol{\Sigma} \in \mathbb{R}^{m\times n}$ is a matrix with all entries vanished except along the main diagonal, we propose a solution to the first order condition (9.12) as follows (more discussion on singular value decomposition of rectangular matrices is provided in Appendix *9.A.3). Take $\boldsymbol{P} = \boldsymbol{U}\boldsymbol{\Sigma}_1$, and $\boldsymbol{Q} = \boldsymbol{V}\boldsymbol{\Sigma}_2^\top$ where $\boldsymbol{\Sigma}_1 = \begin{pmatrix} \boldsymbol{D}_1 \\ \boldsymbol{0} \end{pmatrix} \in \mathbb{R}^{m\times p}$ and $\boldsymbol{\Sigma}_2^\top = \begin{pmatrix} \boldsymbol{D}_2 \\ \boldsymbol{0} \end{pmatrix} \in \mathbb{R}^{n\times p}$,[15] such that $\boldsymbol{D}_1$ and $\boldsymbol{D}_2$ are $p \times p$ diagonal matrices, and $\boldsymbol{D}_1\boldsymbol{D}_2^\top = \text{diag}(\sigma_1, \ldots, \sigma_p)$; here $\sigma_1, \ldots, \sigma_p$ are the first p largest *singular values* of $\boldsymbol{R}$; it is left as an exercise for readers to verify that these choices of $\boldsymbol{P}$ and $\boldsymbol{Q}$ satisfy both the equations of (9.12) simultaneously.

[15] Note that the range of entries in $\boldsymbol{P}$ and $\boldsymbol{Q}$ need not be the same as that for the entries of $\boldsymbol{R}$.

This guessed solution can indeed be shown correct by identifying the optimal solution to Problem (9.9) in accordance with the celebrated *Eckart–Young–Mirsky theorem* in the context of *Low-rank Approximation*, which has different versions in terms of either the *Frobenius norm* or the *spectral norm*. However, their respective proofs are not simple analogies to each other, and here we only present the Frobenius norm version of the theorem, which echoes the setting of Problem (9.9). Interested readers can refer to Appendix *9.A.3 for the spectral norm version.

Theorem 9.1 *(Eckart–Young–Mirsky theorem (Frobenius Norm)* [3, 13, 17]*)* Let R be an $m \times n$ matrix with rank r and with SVD given by $R = U\Sigma V^\top$, where $U = (u_1, \dots, u_m) \in \mathbb{R}^{m\times m}$ and $V = (v_1, \dots, v_n) \in \mathbb{R}^{n\times n}$ are orthogonal matrices, $\Sigma = \begin{pmatrix} \mathrm{diag}(\sigma_1, \dots, \sigma_r) & \mathbf{0} \\ \mathbf{0} & \mathbf{0} \end{pmatrix} \in \mathbb{R}^{m\times n}$ is a diagonal matrix with singular values $\sigma_1 \geq \sigma_2 \geq \cdots \geq \sigma_r > 0$. Fix any $p = 1, \dots, r-1$. Let $U_p = (u_1, \dots, u_p) \in \mathbb{R}^{m\times p}$, $V_p = (v_1, \dots, v_p) \in \mathbb{R}^{n\times p}$ and $\Sigma_p = \mathrm{diag}(\sigma_1, \dots, \sigma_p) \in \mathbb{R}^{p\times p}$. Then the matrix $R_p = U_p\Sigma_p V_p^\top = \sum_{i=1}^{p} \sigma_i u_i v_i^\top$ has a rank p, and it is the best rank-p approximation of R with respect to the Frobenius norm $\|\cdot\|_F$:

$$\|R - R_p\|_F \leq \|R - B\|_F \quad \text{for any rank-}p\text{ matrix } B \in \mathbb{R}^{m\times n}.$$

Moreover, $\|R - R_p\|_F = \sqrt{\sum_{i=p+1}^{r} \sigma_i^2}$.

Proof. For any matrix $B \in \mathbb{R}^{m\times n}$ of rank p, we have, by the unitary invariance of the Frobenius norm,

$$\|R - B\|_F = \|U\Sigma V^\top - B\|_F = \|U^\top(U\Sigma V^\top - B)V\|_F = \|\Sigma - U^\top BV\|_F.$$

Denote $C := U^\top BV \in \mathbb{R}^{m\times n}$, which is also of rank p. Then we have

$$\|\Sigma - C\|_F^2 = \sum_{i=1}^{r} (\sigma_i - c_{ii})^2 + \sum_{i>r} (0 - c_{ii})^2 + \sum_{i\neq j} (0 - c_{ij})^2 \geq \sum_{i=1}^{r} (\sigma_i - c_{ii})^2,$$

where the last inequality holds if and only if $c_{ij} = 0$ for all $i \neq j$ and $c_{ii} = 0$ for all $i > r$. Furthermore, with the constraint that rank(C) $= p$, for the minimizer, there can be exactly p non-zero c_{ii}'s and the remaining $r - p$ c_{ii}'s must vanish; denote these two groups by C_1 and C_2 respectively. Therefore, $\sum_{i=1}^{r} (\sigma_i - c_{ii})^2 = \sum_{C_1} (\sigma_i - c_{ii})^2 + \sum_{C_2} \sigma_i^2$ which attains a local minimum value of $\sum_{C_2} \sigma_i^2$ by taking $c_{ii} = \sigma_i$ on C_1. Hence, by the non-increasing nature of σ_i's, $\sum_{i=1}^{r} (\sigma_i - c_{ii})^2$ reaches the minimum value of $\sum_{i=p+1}^{r} \sigma_i^2$ when $c_{ii} = \sigma_i$ for $i = 1, \dots, p$, and $c_{ii} = 0$ for $i = p+1, \dots, r$. Altogether leads to the minimizer of $\|\Sigma - C\|_F$, that is $\tilde{C} = \begin{pmatrix} \mathrm{diag}(\sigma_1, \dots, \sigma_p) & \mathbf{0} \\ \mathbf{0} & \mathbf{0} \end{pmatrix} \in \mathbb{R}^{m\times n}$, with $\|\Sigma - \tilde{C}\|_F = \sqrt{\sum_{i=p+1}^{r} \sigma_i^2}$.

Finally, we can recover the minimizer $\tilde{B}$ of $\|R - B\|_F$ as follows:

$$\tilde{B} = U\check{C}V^{\mathsf{T}} = \sum_{i=1}^{p} \sigma_i u_i v_i^{\mathsf{T}} = R_p,$$

and $\|R - R_p\|_F = \|\Sigma - \tilde{C}\|_F = \sqrt{\sum_{i=p+1}^{r} \sigma_i^2}$, as desired. □

Therefore, our respective guesses of P and Q above are correct; indeed, the resulting $\hat{R} = PQ^{\mathsf{T}}$ is precisely the minimizer of (9.9).

Note that the optimal solution of Problem (9.9) suffers from *overfitting*; that is, it fits the known data in the training set so well that it loses the flexibility to adapt to a new dataset, e.g. the test dataset, thereby losing predictive power. To avoid overfitting, we employ common technique of *regularization*, by adding the sum of the squared matrix entries in P and Q together with some penalty parameters. That is, we add $\mathrm{tr}(\Lambda_1 PP^{\mathsf{T}}) + \mathrm{tr}(\Lambda_2 QQ^{\mathsf{T}})$ to the original objective function (9.10) for some diagonal matrices $\Lambda_1 \in \mathbb{R}^{m\times m}$ and $\Lambda_2 \in \mathbb{R}^{n\times n}$, which serve as Lagrange multipliers[16].

In general, to optimize J (equivalently, to solve (9.12), regularized or not), rather than make a guess through SVD as described before, we can instead hold P constant and then vary over Q, or vice versa. This reduces each iteration problem to the standard least squares problem, which is repeated until convergence is achieved. This approach is commonly called *alternating projections* in the literature based on the usual fixed point argument.

In addition to the latent-factor method as described above, there is another approach called the *neighborhood model* which is perhaps intuitively simpler. A similarity score for each pair of subscribers is computed, based on the cosine coefficient between the two vectors of ratings of each. The larger the score, the closer these two subscribers are in their taste of movies. We then aim to look for a group of subscribers who are close to a particular user of interest. In order to predict the rating of a movie by a subscriber of interest, we can use a weighted sum of the ratings of a reasonably large number of similar subscribers who have shared their views on that movie. More details on this method can be found in [11].

Going back to the latent-factor method, upon completion of the optimization procedure that minimizes the MSE J, both the subscribers' tastes on movies $\hat{p}_{i,k}$ and the nature of movies $\hat{q}_{j,k}$ can be estimated. The $m \times n$ matrix $\hat{R} = \hat{P}\hat{Q}^{\mathsf{T}}$ then stores the predicted ratings of all movies by all subscribers. For instance, the i-th rows of $\hat{R}$ are row vectors that list the predicted ratings on all movies $j = 1, 2, \ldots, n$ by the i-th subscriber. Notably, these predicted ratings of $\hat{R}$ should be non-zero for those unknown ratings $r_{i,j} = 0$; while for any $r_{i,j} \neq 0$, the corresponding estimate $\hat{r}_{i,j}$ should be roughly equal to that known rating $r_{i,j}$. By sorting these estimated

[16] In practice, the *multipliers* Λ_1 and Λ_2 are chosen in advance and are often selected via *cross-validation*, where the available data are divided into M parts. Particularly, in the i-th round of cross-validation, all parts except the i-th one out of these M portions serve as the training data, while the i-th part is used for testing.

ratings in descending order, our recommender system should be able to show the i-th subscriber top recommendations that best fit their taste. Furthermore, if we want to predict all ratings for the testing data, say for a newly joined user $m+1$, based on some known ratings $r_{m+1,j}$'s of his, then we solve for the $\hat{p}_{m+1} = (\hat{p}_{m+1,1}, \ldots, \hat{p}_{m+1,p})^{\mathsf{T}}$ that minimizes $\sum_{j=1}^{n} \left(r_{m+1,j} - \sum_{k=1}^{p} p_{m+1,k}\hat{q}_{j,k} \right)^2$. Here, $\hat{Q}$ has already been obtained using the training data from the existing users $1, \ldots, m$, and it is not affected by just one new user $m+1$ or a few more. In other words, the best approximant for the ratings of the user $m+1$, the column vector $\hat{r}_{m+1} = \hat{Q}\hat{p}_{m+1} := \left(\sum_{k=1}^{p} \hat{p}_{m+1,k}\hat{q}_{j,k} \right)_{j=1}^{n}$, is just the projection of $r_{m+1} := (r_{m+1,1}, \ldots, r_{m+1,n})^{\mathsf{T}}$ onto the column space of $\hat{Q}$. Certainly, if subscriber $m+1$ really just joins the service, normally some random recommendations will be popped up, so as to see what choices this new user would make, then more non-zero values of $r_{m+1,j}$'s can be learnt first. Afterwards, the procedure described above could then be implemented.

Next we discuss and illustrate, by using the **R** package `recommenderlab`, how recommender systems can serve as an important marketing analytic tool. For example, movie websites want to analyze the collective data (such as watching history and movie ratings) obtained from visitors, and then recommend new movies for them to watch based on the preference predictions acquired from the whole group. Generally, this **R** package provides an infrastructure for testing and developing recommender algorithms, and supports the rating (e.g. 1 to 5 stars) and unary (0 or 1) datasets. We first briefly describe the mathematics behind for training a recommender system using the package via the *SVD with column-mean imputation* method. This method is toggled by the argument `method="SVD"` in `recommenderlab`, where the rating matrix $\boldsymbol{R}$ is first normalized per subscriber (i.e., subtracting the mean and then rescaling by the standard deviation for each row) and the missing entries of the resulting normalized matrix are imputed by their corresponding column means, i.e., the movie-wise average of existing ratings. The programme then applies SVD to look for the p largest singular values $\sigma_1, \ldots, \sigma_p$ and their corresponding singular vectors $\boldsymbol{U}_p$ and $\boldsymbol{V}_p$. In the prediction stage, given a testing rating matrix $\boldsymbol{R}^{(\text{test})}$, with the same number of columns (movies) as $\boldsymbol{R}$ but possibly different number of rows (subscribers), and recalling that $\boldsymbol{R}_p = \boldsymbol{U}_p \boldsymbol{\Sigma}_p \boldsymbol{V}_p^{\mathsf{T}}$, we first expect $\boldsymbol{R}^{(\text{test})}$ to be close to $\boldsymbol{U}^{(\text{test})}\boldsymbol{\Sigma}_p\boldsymbol{V}_p^{\mathsf{T}}$. Hence we simply take $\boldsymbol{U}^{(\text{test})} = \boldsymbol{R}^{(\text{test})}\boldsymbol{V}_p(\boldsymbol{\Sigma}_p)^{-1}$, and then the subscriber-normalized predicted rating matrix is given by $\boldsymbol{R}_p^{(\text{test})} = \boldsymbol{U}^{(\text{test})}\boldsymbol{\Sigma}_p\boldsymbol{V}_p^{\mathsf{T}} = \boldsymbol{R}^{(\text{test})}\boldsymbol{V}_p(\boldsymbol{\Sigma}_p)^{-1}\boldsymbol{\Sigma}_p\boldsymbol{V}_p^{\mathsf{T}} = \boldsymbol{R}^{(\text{test})}\boldsymbol{V}_p\boldsymbol{V}_p^{\mathsf{T}}$. Finally, we denormalize $\boldsymbol{R}_p^{(\text{test})}$ to give the predictions.

In the meantime, since the actual implementation of SVD in `recommenderlab` adopts the *Lanczos bidiagonalization* algorithm (IRLBA) [1] to approximate the largest few singular values and the corresponding singular vectors of a matrix, where the initial random seed plays a crucial role, it is therefore necessary in practice to build multiple recommender systems and select the one with the best model performance metric, e.g., the lowest MSE, under the validation set; see the **R** codes in Programme 9.29. In particular, the function `evaluationScheme` in the `recommenderlab` package splits a dataset into training and validation, two separate datasets, at the ratio of 9:1 as indicated by the argument `train=0.9`, and the validation dataset is further partitioned into two subgroups called known and unknown. More specifically, given a validation dataset $\boldsymbol{X} \in \mathbb{R}^{n \times p}$, we generate a Bernoulli random matrix $\mathbf{B} = (\mathrm{b}_{ij}) \in \mathbb{R}^{n \times p}$ where each entry $\mathrm{b}_{ij} \overset{iid}{\sim} \mathrm{Ber}(q)$, for $i = 1, \ldots, n$ and $j = 1, \ldots, p$, with the success probability q. The known (resp. unknown) subgroup is then given by $\boldsymbol{X} \odot \mathbf{B}$ (resp. $\boldsymbol{X} \odot (\mathbf{1} - \mathbf{B})$), where $\odot$ denotes the Hadamard product. Next, all resulting entries of known and unknown with a value of 0 are regarded as missing. After building a recommender system based on the training dataset, we make predictions on the known subgroup of the validation dataset, including those that are not rated by a specific user at all and those that have been rated but whose ratings are assigned to the unknown subgroup since their corresponding b_{ij}'s take a value of 0. Then, for the sake of model evaluation, we only compare the non-missing actual ratings in the unknown subgroup of the validation dataset with their predicted counterparts in the known subgroup from the previous step.

```
> library(recommenderlab)
>
> recstat <- function(train, known, unknown, n_PC){
+   rec <- Recommender(train, method="SVD", parameter=list(k=n_PC))
+   pre <- predict(rec, newdata=known, type="ratings")
+   stat <- calcPredictionAccuracy(pre, unknown)
+   c(list(rec=rec, pre=pre), split(unname(stat), names(stat)))
+ }
>
> # Try Recommender(X, method="SVD", ...) several times and output the best trial
> # metric: any one of "MAE", "MSE", and "RMSE"
> # train: training data
> # known: validation data; unknown: validation data for model evaluation
> best_rec <- function(X, n_PC=50, trial=5, metric="MSE", seed=4002){
+   set.seed(seed)
+   data <- evaluationScheme(data=X, train=0.9, given=-1)
+   train <- getData(data, "train")
+   known <- getData(data, "known")
+   unknown <- getData(data, "unknown")
+   l0 <- Inf
+   for (i in 1:trial) {
+     res <- recstat(train, known, unknown, n_PC)  # new trial
+     if (res[[metric]] < l0) {       # update l0 if it is greater than l
+       l0 <- res[[metric]]; rec0 <- res$rec
+     }
+   }
+   print(paste0("K=", n_PC, "; ", metric, "=", l0))
+   rec0                                # best recommender
+ }
```

Programme 9.29 R codes for training a recommender system by using the SVD method with column-mean imputation.

We are now ready to demonstrate the use of this package by training a recommender system on the `MovieLense` dataset. The dataset collected 99,392 ratings (1 (worst), 2, ..., 5 (best)) from 943 users on 1,664 movies through the *MovieLens* website (movielens.umn.edu) during the seven-month period from 19 September 1997 to 22 April 1998. In this example, we only consider those data satisfying the following two conditions: (i) users that have rated over 100 movies, so as to ensure sufficient data for the sake of mean centering; and (ii) movies that have at least 30 ratings, so as to ensure sufficient data for the sake of column-mean imputation. Based on the reduced dataset with the ratings from 358 users on 792 movies, we choose the rating data of the first 300 users as the joint dataset for training and validation `train_valid`, which serves as the input of the `best_rec()` function in Programme 9.29, and the best recommender system is then the one that yields the lowest MSE among all `trial=5` trials. Meanwhile, the remaining 358 − 300 = 58 users form the test dataset `test`.

The output in Programme 9.30 via **R** illustrates the use of the trained recommender system by selecting the top 10 recommended movies tailor-made for two users, namely Users 798 and 804. Note that the `predict()` function in `recommenderlab` only makes predictions on the missing entries, and therefore only the movies that are not rated by the user will be recommended by the model. As an interesting observation, while the recommendations for the two users are fairly personalized, let say perhaps User 804 prefers older movies while User 798 prefers more recent ones, they also share several recommended items in common, including *Titanic* in particular. This record-breaking movie was released in December 1997, which was around the middle of the seven-month period of ratings collected, so these two users had not watched it at the time of data collection, although they might already have done so shortly afterwards. Anyhow, the high profile of this movie is already strongly evident from its appearance in the top 10 recommendations of both two users, whose preferences are actually quite different. What do you think? Which user do you find closer to your preference on movies?

```
1  > data("MovieLense")
2  > MovieLense
3  943 x 1664 rating matrix of class `realRatingMatrix' with 99392 ratings.
4  >
5  > # only counts rows (users) with more than 100 ratings (non-missing)
6  > row_idx <- rowCounts(MovieLense) > 100
7  > # only counts columns (movies) with more than 30 ratings (non-missing)
8  > col_idx <- colCounts(MovieLense) > 30
9  > # 358 users (792 movies) with more than 100 (30) ratings
10 > (data <- MovieLense[row_idx, col_idx])
11 358 x 792 rating matrix of class `realRatingMatrix' with 66594 ratings.
12 >
13 > # train: rating (training) data from the first 300 users
14 > # test: rating (test) data from the remaining 58 users
15 > (train_valid <- data[1:300])
16 300 x 792 rating matrix of class `realRatingMatrix' with 56586 ratings.
17 > (test <- data[301:nrow(data)])
18 58 x 792 rating matrix of class `realRatingMatrix' with 10008 ratings.
19 > # train the model with the first 50 PCs and choose the best model
```

```
20 > (rec <- best_rec(train_valid, n_PC=50))
21 [1] "K=50; MSE=0.688067687368031"
22 Recommender of type `SVD' for `realRatingMatrix' learned using 270 users.
23 > (pre <- predict(rec, test, data=train_valid, n=10))
24 Recommendations as `topNList' with n = 10 for 58 users.
25 >
26 > row_idx[row_idx][301:302]                      # Users 798 and 804
27  798  804
28 TRUE TRUE
29 > as(pre, "list")[c('0', '1')]
30 $`0`
31  [1] "Usual Suspects, The (1995)"          "Close Shave, A (1995)"
32  [3] "Braveheart (1995)"                   "Wrong Trousers, The (1993)"
33  [5] "Titanic (1997)"                      "Shawshank Redemption, The (1994)"
34  [7] "Godfather: Part II, The (1974)"      "Good Will Hunting (1997)"
35  [9] "Godfather, The (1972)"               "Babe (1995)"
36
37 $`1`
38  [1] "Titanic (1997)"
39  [2] "Wrong Trousers, The (1993)"
40  [3] "Close Shave, A (1995)"
41  [4] "Wallace & Gromit: The Best of Aardman Animation (1996)"
42  [5] "Duck Soup (1933)"
43  [6] "Usual Suspects, The (1995)"
44  [7] "Jean de Florette (1986)"
45  [8] "Maltese Falcon, The (1941)"
46  [9] "12 Angry Men (1957)"
47 [10] "Cyrano de Bergerac (1990)"
```

Programme 9.30 R codes for training a recommender system with the `MovieLense` dataset.

*9.A APPENDIX

*9.A.1 Spectral Decompositions of Square Matrices

The basic idea in PCA, as the first step, is to find a linear combination of the original variables x_i's,

$$y_1 = b_{11}x_1 + \ldots + b_{p1}x_p = b_1^{\mathrm{T}}x,$$

where $x = (x_1, \ldots, x_p)^{\mathrm{T}}$, so that $\mathrm{Var}(y_1) = b_1^{\mathrm{T}}\mathrm{Var}(x)b_1 \approx b_1^{\mathrm{T}}Sb_1$, and the latter is maximized subject to the constraint $b_1^{\mathrm{T}}b_1 = 1$. The solution of this optimization problem is related to a well-known and very useful result in matrix algebra called the *spectral decomposition* of symmetric matrices. For a better understanding of this technique, readers may revisit Section 1.2 of Chapter 1 for the relevant concepts of eigenvalues (a.k.a. latent roots), eigenvectors and normalized eigenvectors, as well as some of their important properties.

The following Lemma 9.1 shows that the maximum value of the variance of $y_1 = b_1^{\mathrm{T}}x$ is the largest eigenvalue of S and is achieved when b_1 is equal to the corresponding normalized eigenvector of S. Next, the second PC is the linear combination

$$y_2 = b_{12}x_1 + \ldots + b_{p2}x_p = b_2^{\mathrm{T}}x,$$

so that $\text{Var}(y_2) = \boldsymbol{h}_2^\top S \boldsymbol{h}_2$ is maximized subject to $\boldsymbol{h}_2^\top \boldsymbol{h}_2 = 1$ and $\boldsymbol{h}_1^\top \boldsymbol{h}_2 = 0$. It turns out that this maximum value of the variance of y_2 is the second largest eigenvalue of S and $\boldsymbol{h}_2$ is the corresponding normalized eigenvector of S. The same argument applies to the j-th PC and each of the corresponding variance being equal to the j-th largest eigenvalue, and the corresponding vector of loadings being equal to the respective eigenvector.

Lemma 9.1 Suppose that the $p \times p$ positive definite matrix $A \in \mathbb{R}^{p\times p}$ has eigenvalues $\lambda_1 \geq \cdots \geq \lambda_p > 0$, and the corresponding normalized eigenvectors are $\boldsymbol{h}_1, \ldots, \boldsymbol{h}_p$. Then,

1. $\max_{l\neq 0} \dfrac{\boldsymbol{l}^\top A \boldsymbol{l}}{\boldsymbol{l}^\top \boldsymbol{l}} = \lambda_1$, and this maximum is attained at $\boldsymbol{l} = \boldsymbol{h}_1$;
2. $\min_{l\neq 0} \dfrac{\boldsymbol{l}^\top A \boldsymbol{l}}{\boldsymbol{l}^\top \boldsymbol{l}} = \lambda_p$, and this minimum is attained at $\boldsymbol{l} = \boldsymbol{h}_p$;
3. In general, for $k = 1, 2, \ldots, p-1$, $\max_{\boldsymbol{l}\perp \boldsymbol{h}_1,\ldots,\boldsymbol{h}_k} \dfrac{\boldsymbol{l}^\top A \boldsymbol{l}}{\boldsymbol{l}^\top \boldsymbol{l}} = \lambda_{k+1}$, and this maximum is attained at $\boldsymbol{l} = \boldsymbol{h}_{k+1}$.

Proof. Let $A = HDH^\top$ be the spectral decomposition of A, where $H = (\boldsymbol{h}_1, \ldots, \boldsymbol{h}_p)$, $D = \text{diag}(\lambda_1, \ldots, \lambda_p)$, and also define $A^{1/2} = HD^{1/2}H^\top$ so that $A^{1/2}A^{1/2} = A$. Let $\boldsymbol{u}$, $\boldsymbol{l} \in \mathbb{R}^p$ be arbitrary vectors such that $\boldsymbol{u} = H^\top \boldsymbol{l}$. Then,

$$\frac{\boldsymbol{l}^\top A \boldsymbol{l}}{\boldsymbol{l}^\top \boldsymbol{l}} = \frac{\boldsymbol{l}^\top A^{1/2}A^{1/2}\boldsymbol{l}}{\boldsymbol{l}^\top HH^\top \boldsymbol{l}} = \frac{\boldsymbol{l}^\top HD^{1/2}H^\top HD^{1/2}H^\top \boldsymbol{l}}{\boldsymbol{u}^\top \boldsymbol{u}} = \frac{\boldsymbol{u}^\top D\boldsymbol{u}}{\boldsymbol{u}^\top \boldsymbol{u}} = \frac{\sum_{j=1}^p \lambda_j u_j^2}{\sum_{j=1}^p u_j^2}$$

$$\leq \lambda_1 \frac{\sum_{j=1}^p u_j^2}{\sum_{j=1}^p u_j^2} = \lambda_1.$$

When $\boldsymbol{l} = \boldsymbol{h}_1$, $\boldsymbol{u} = H^\top \boldsymbol{h}_1 = (1, 0, \ldots, 0)^\top$ and $\dfrac{\boldsymbol{h}_1^\top A\boldsymbol{h}_1}{\boldsymbol{h}_1^\top \boldsymbol{h}_1} = \dfrac{\boldsymbol{u}^\top D\boldsymbol{u}}{\boldsymbol{u}^\top \boldsymbol{u}} = \lambda_1$. Similarly, we can prove the second claim via a parallel argument. For the third claim, if $\boldsymbol{u} = H^\top \boldsymbol{l}$ then $\boldsymbol{l} = H\boldsymbol{u} = u_1\boldsymbol{h}_1 + \cdots + u_p\boldsymbol{h}_p$. Hence, for if $\boldsymbol{l} \perp \boldsymbol{h}_1, \ldots, \boldsymbol{h}_k$, this implies that $0 = \boldsymbol{h}_j^\top \boldsymbol{l} = u_1\boldsymbol{h}_j^\top \boldsymbol{h}_1 + \ldots, + u_p\boldsymbol{h}_j^\top \boldsymbol{h}_p = u_j$, for all $j = 1, 2, \ldots, k$. Therefore,

$$\frac{\boldsymbol{l}^\top A \boldsymbol{l}}{\boldsymbol{l}^\top \boldsymbol{l}} = \frac{\sum_{j=k+1}^p \lambda_j u_j^2}{\sum_{j=k+1}^p u_j^2} \leq \lambda_{k+1}.$$

The claim then follows by using the same argument as that for the first claim. □

*9.A.2 Immunization against Interest Rate Risks

Here we provide a quick review of immunization; more details on this topic can be found in [7, 9, 14, 15]. For the sake of simplicity, we only consider deterministic cash flows here.

Redington Immunization [15] is a well-known technique for reducing or even eliminating the influence of interest rate fluctuations on values of portfolios of cash flows. A typical portfolio consists of assets and liabilities. We first define some notations to be used below. The force of interest is denoted by $\delta = \ln(1+\imath)$, where $\imath$ is the annualized interest rate. The present value at time 0 of the assets evaluated at rate δ is

$$P_A(\delta;0) := \sum_i c_i^A e^{-\delta t_i},$$

where c_i^A and t_i are the cash flow amount and the time of the i-th cash flow of the assets, respectively.

The *Macaulay duration* $D_A(\delta)$ (see [12]) of the assets is defined as

$$D_A(\delta) := -\frac{dP_A/P_A}{d\delta} = -\frac{P'_A(\delta;0)}{P_A(\delta;0)} = \frac{\sum_i t_i c_i^A e^{-\delta t_i}}{\sum_i c_i^A e^{-\delta t_i}},$$

which is the average payment time of cash flows weighted by the corresponding discounted cash flows. It can also be interpreted as the percentage change of the price with respect to a change in the force of interest, and so it is regarded as the *sensitivity* of the price with respect to a change in the force of interest rate. Alternatively, this concept can also be interpreted as the effective lifespan of a project, i.e., the effective number of years before the initial cost for a project is recovered by the future cash flows it generates. The convexity of the assets is defined as

$$\frac{P''_A(\delta;0)}{P_A(\delta;0)} = \frac{\sum_i t_i^2 c_i^A e^{-\delta t_i}}{\sum_i c_i^A e^{-\delta t_i}}.$$

The present value at time 0 of the liabilities evaluated at the rate δ is

$$P_L(\delta;0) := \sum_i c_i^L e^{-\delta t_i},$$

where the notations are defined in the same way as those for the assets, except that the amount of the i-th cash flow is replaced by c_i^L. Definitions for the Macaulay duration $D_L(\delta)$ and the convexity of the liabilities follow suit.

At the force of interest δ, a fund is immunized against a small change ε in the force of interest if (i) $P_A(\delta;0) = P_L(\delta;0)$ and (ii) $P_A(\delta+\varepsilon;0) \geq P_L(\delta+\varepsilon;0)$ for ε sufficiently close to 0. Using the Taylor series expansion, these conditions translate to $P'_A(\delta;0) = P'_L(\delta;0)$ and $P''_A(\delta;0) \geq P''_L(\delta;0)$, i.e., the Macaulay duration (resp. convexity) of the assets should be equal to (resp. greater than) that of the liabilities. If we denote the difference of their time values at time t by $P(\delta;t) := P_A(\delta;t) - P_L(\delta;t)$, the conditions above can be rewritten as $P'(\delta;0) = 0$ and $P''(\delta;0) \geq 0$.

Let us provide an example of achieving the *full immunization* against changes of any magnitude in the interest rate. Consider a single liability cash outflow of L_T at the time T. Full immunization can be achieved by holding two assets providing cash inflows of A at time $0 \le T-a$ and B at time $T+b$ respectively, for some $a, b > 0$ and $a \le T$. Given any two out of the four constants A, B, a, b are known, we can obtain the full immunization strategy by solving the remaining two unknowns from the following two equations, using T rather than 0 as the comparison date:

$$\begin{cases} P(\delta; T) = Ae^{a\delta} + Be^{-b\delta} - L_T = 0, \\ P'(\delta; T) = aAe^{a\delta} - bBe^{-b\delta} = 0. \end{cases} \tag{9.13}$$

Assume that the two unknowns are solved so that the four constants satisfy (9.13). We can easily see that $Be^{-b\delta} = \frac{a}{b}Ae^{a\delta} = \frac{a}{a+b}L_T$ by solving the two equations of (9.13), hence

$$P''(\delta; T) = a^2Ae^{a\delta} + b^2Be^{-b\delta} = \left(ab + b^2\right)Be^{-b\delta} = b(a+b)\frac{a}{a+b}L_T = abL_T > 0.$$

By continuity, $P''(\delta + \varepsilon; T) > 0$ for sufficiently small $|\varepsilon|$. By the second-order mean value theorem, for ε sufficiently small (positive or negative) and for some δ^* between δ and $\delta + \varepsilon$, $P(\delta + \varepsilon; T) = P(\delta; T) + P'(\delta; T) \cdot \delta + \frac{1}{2}P''(\delta^*; T) \cdot \delta^2$, it follows from $P''(\delta^*; T) > 0$ that $P(\delta + \varepsilon; T) > P(\delta; T)$ holds as long as ε is sufficiently close to 0. This argument provides an intuition for full immunization locally around δ. It is interesting to note that this strategy indeed achieves full immunization for all possible values of $\varepsilon \in \mathbb{R}$. Consider an arbitrary $\varepsilon \neq 0$. The time value of the portfolio at time T is given by

$$\begin{aligned} P(\delta + \varepsilon; T) &= Ae^{a(\delta+\varepsilon)} + Be^{-b(\delta+\varepsilon)} - L_T \\ &= Ae^{a(\delta+\varepsilon)} + Be^{-b(\delta+\varepsilon)} - \left(Ae^{a\delta} + Be^{-b\delta}\right) \\ &= Ae^{a\delta}\left(e^{a\varepsilon} + \frac{a}{b}e^{-b\varepsilon} - \left(1 + \frac{a}{b}\right)\right). \end{aligned}$$

where the final two equalities follow from the equalities $Be^{-b\delta} = \frac{a}{b}Ae^{a\delta} = \frac{a}{a+b}L_T$ mentioned above. It then suffices to show that the function $f(\varepsilon) := e^{a\varepsilon} + \frac{a}{b}e^{-b\varepsilon} - \left(1 + \frac{a}{b}\right) > 0$ for all $\varepsilon \neq 0$. By differentiation, $f'(\varepsilon) = a\left(e^{a\varepsilon} - e^{-b\varepsilon}\right)$, which is (i) negative when $\varepsilon < 0$; (ii) positive when $\varepsilon > 0$; and (iii) 0 when $\varepsilon = 0$. Therefore, f takes its global minimum at $\varepsilon = 0$ with $f(0) = 0$, hence $f(\varepsilon) > 0$ for all $\varepsilon \neq 0$, from which we conclude that $P(\delta + \varepsilon; T) \ge P(\delta; T)$ for any $\varepsilon \in \mathbb{R}$, no matter positive or negative or how large it is, hence the full immunization is reached. This result may explain the high demand for short-term and long-term bonds in the markets: many insurance products involve random claim times which are usually between 1 and 10 years, and so short-term and long-term bonds are needed in order to immunize the interest rate risks. This phenomenon also justifies the convexity in the loadings of the third PC. While the insurance market is a notable component of the financial market, its proportion is still relatively small, which may in turn explain the significantly smaller eigenvalue of the third PC, compared with those of the first two.

*9.A.3 Singular Value Decomposition

Mathematically, *Singular Value Decomposition* (SVD) is a generalization of eigenvalue decompositions for square symmetric matrices, but now applied to general rectangular $m \times n$ matrices. It finds wide applications in machine learning, for instance, in recommender systems as commonly encountered in marketing sciences. SVD claims the following (see [6]): given an $m \times n$ matrix A, there exist matrices U, Σ and V such that

$$A = U\Sigma V^{\top}, \tag{9.14}$$

or equivalently

$$AV = U\Sigma, \tag{9.15}$$

where

1. $U = (u_1, \dots, u_m)$ is an $m \times m$ matrix, and each u_i is a unit eigenvector corresponding to the i-th largest eigenvalue of $AA^{\top}$, for $i = 1, 2, \dots, m$. The vectors u_i's are called the *left singular vectors* of A, and they are orthogonal to each other;
2. $V = (v_1, \dots, v_n)$ is an $n \times n$ matrix, and each v_i is a unit eigenvector corresponding to the i-th largest eigenvalue of $A^{\top}A$, for $i = 1, 2, \dots, n$. The vectors v_i's are called the *right singular vectors* of A, and they are orthogonal to each other;
3. $\Sigma = \begin{pmatrix} \text{diag}(\sigma_1, \dots, \sigma_r) & \mathbf{0} \\ \mathbf{0} & \mathbf{0} \end{pmatrix}$ is a diagonal $m \times n$ matrix, where $r = \text{rank}(A)$, and $\sigma_i = \sqrt{\lambda_i(A^{\top}A)} > 0$ is the square root of the i-th largest eigenvalue of $A^{\top}A$, which is called a *singular value* of A.

To establish the first claim of this SVD representation (9.14), recall Section 1.2 of Chapter 1. We first note that $AA^{\top}$ and $A^{\top}A$ share the same set of non-zero eigenvalues; indeed, for $AA^{\top}u_i = \lambda_i u_i$ and $A^{\top}Av_j = \theta_j v_j$, for $\lambda_i, \theta_j \neq 0$ (that is to say, u_i is an eigenvector for $AA^{\top}$ corresponding to the non-zero eigenvalue λ_i; similarly v_j is an eigenvector for $A^{\top}A$ corresponding to the non-zero eigenvalue θ_j), then $A^{\top}(AA^{\top})u_i = A^{\top}(\lambda_i u_i)$, hence

$$(A^{\top}A)(A^{\top}u_i) = \lambda_i(A^{\top}u_i).$$

Since we also have

$$\|A^{\top}u_i\|_2^2 = u_i^{\top}AA^{\top}u_i = \lambda_i \|u_i\|_2^2 > 0,$$

we conclude that $A^{\top}u_i$ is an eigenvector of $A^{\top}A$ with respect to eigenvalue λ_i. A similar argument leads to the claim that Av_j is an eigenvector of $AA^{\top}$ with respect to eigenvalue θ_j. Therefore, $AA^{\top}$ and $A^{\top}A$ possess the same set of non-zero eigenvalues.

For $r = \text{rank}(A)$, by the spectral decomposition of the square symmetric matrix $AA^{\top}$, there exist r positive constants $\sigma_1, \sigma_2, \dots, \sigma_r$, r orthogonal $m \times 1$ unit vectors $u_1, u_2, \dots, u_r$, and, based on the above discussion, r orthogonal $n \times 1$ unit vectors

$$v_1(= A^{\top}u_1), \quad v_2(= A^{\top}u_2), \quad \dots \quad v_r(= A^{\top}u_r), \tag{9.16}$$

such that by multiplying them with A on the right,

$$Av_1 = \sigma_1 u_1, \quad Av_2 = \sigma_2 u_2, \quad \dots \quad Av_r = \sigma_r u_r. \tag{9.17}$$

Note that for $i, j = 1, 2, \dots, r$,

$$\langle v_i, v_j \rangle = \langle A^\top u_i, A^\top u_j \rangle = u_i^\top AA^\top u_j = \sigma_j u_i^\top u_j = \sigma_j \langle u_i, u_j \rangle,$$

hence $v_i, \dots, v_r$ are orthogonal.

By defining $U_r = (u_1, u_2, \dots, u_r)$, $V_r = (v_1, v_2, \dots, v_r)$, and $\Sigma_r = \text{diag}(\sigma_1, \sigma_2, \dots, \sigma_r)$, we can rewrite (9.17) in a compact matrix form:

$$AV_r = U_r \Sigma_r. \tag{9.18}$$

Also note that the right singular vectors $v_1, v_2, \dots, v_r$ are in the row space of A by (9.16); while the left singular vectors $u_1, u_2, \dots, u_r$ are in the column space of A by (9.18).

To conclude the claim that $AV = U\Sigma$, we only have to pick $n - r$ more v_i's and $m - r$ more u_j's from the null spaces $\mathcal{N}(A)$ and $\mathcal{N}(A^\top)$, respectively. Each set of these can be chosen to be the respective orthonormal bases for those two null spaces. Note that by then, they would be orthogonal to the first r v_i's (being in the row space of A) and u_j's (being in the column space of A) respectively. For instance, for $v_{r+1}, \dots, v_n$ from $\mathcal{N}(A)$, $i = r + 1, \dots, n$, and $j = 1, \dots, r$, we have

$$\langle v_i, v_j \rangle = v_i^\top (A^\top u_j) = (Av_i)^\top u_j = 0,$$

by definition. Similarly,

$$\sigma_j \langle u_j, u_i \rangle = \langle Av_j, u_i \rangle = v_j^\top A^\top u_i = v_j^\top \mathbf{0} = 0.$$

Now we include all the newly constructed v_i's and u_j's in matrices V and U, by letting $V = (V_r; v_{r+1}, \dots, v_n)$ and $U = (U_r; u_{r+1}, \dots, u_m)$, and so V and U are both square matrices. Also form the $m \times n$ matrix Σ by adding to Σ_r with $m - r$ new zero rows and $n - r$ new zero columns. These definitions lead to $Av_i = \mathbf{0}$ for $i = r + 1, \dots, n$, and together with (9.18), we conclude that $AV = U\Sigma$. Finally, since V is orthogonal by construction, we also have $A = U\Sigma V^\top$.

To summarize, u's and v's form the bases for the following four fundamental spaces:

- $u_1, \dots, u_r$ form an orthonormal basis for the column space of A;
- $u_{r+1}, \dots, u_m$ form an orthonormal basis for the null space $\mathcal{N}(A^\top)$;
- $v_1, \dots, v_r$ form an orthonormal basis for the row space of A;
- $v_{r+1}, \dots, v_n$ form an orthonormal basis for the null space $\mathcal{N}(A)$.

To specify those singular values σ_i's, we consider

$$A^\top A = (U\Sigma V^\top)^\top (U\Sigma V^\top) = V\Sigma^\top U^\top U\Sigma V^\top = V\Sigma^\top \Sigma V^\top,$$

hence $\Sigma^\top\Sigma$ is the diagonal matrix of eigenvalues of $A^\top A$, which means that σ_i^2 are non-zero eigenvalues of $A^\top A$ (and also $AA^\top$ by the discussion above). Also recall that v_i's are the eigenvectors of $A^\top A$, while u_i's are the eigenvectors of $AA^\top$. Finally, we can also write (9.14) in a "reduced SVD" form as:

$$A = \sum_{i=1}^{r} \sigma_i u_i v_i^\top = U_r \Sigma_r V_r^\top.$$

For most applications in practice, we do not always need the exact SVD of the data matrix A. It often suffices to have a factorization which well approximates A so as to save storage space and to reduce future computational complexity. In matrix Σ, the singular values can be arranged in descending order. Quite often the singular values decrease at a relatively fast rate, and the sum of the first 10% or even 1% of the singular values (say p out of r, with $p \ll r$) already amounts to 99% of the total sum of all the singular values. We also note that both u_i's and v_j's have unit norm. Therefore, it is plausible to describe matrix A by using only the first p singular values via the decomposition:

$$A \approx U_p \Sigma_p V_p^\top,$$

where $U_p \in \mathbb{R}^{m\times p}, \Sigma_p \in \mathbb{R}^{p\times p}$ and $V_p \in \mathbb{R}^{n\times p}$. Here, U_p and V_p contain the corresponding first p left and right singular vectors, and $\Sigma_p = \operatorname{diag}(\sigma_1, \dots, \sigma_p)$. Usually, we require $p \ll r = \min\{m, n\}$. Hence, $m\times p + p\times p + p\times n = (m+n+p)\times p \ll m\times n$,[17] in other words, we can recover A from matrices U_p, V_p and Σ_p, which are much smaller in size. Finally, in Appendix *9.A.4, we point out some pitfalls when one uses **R** functions of SVD and SD.

We conclude this section of the appendix with an alternative version of the *Eckart–Young–Mirsky theorem* introduced earlier in Section *9.11 formulated in terms of the spectral norm instead of the Frobenius norm. Recall that the spectral norm of a matrix A is defined as $\|A\|_2 := \max_{\|x\|_2=1} \|Ax\|_2$, where $\|x\|_2 = \sqrt{x^\top x}$ is the usual Euclidean norm; indeed, it is clearly equal to the largest singular value of A by the *spectral decomposition theorem.*

Theorem 9.2 *(Eckart–Young–Mirsky theorem (Spectral Norm)* [3, 13, 17]*)* Let R be an $m\times n$ matrix with rank r and with SVD given by $R = U\Sigma V^\top$, where $U = (u_1, \dots, u_m) \in \mathbb{R}^{m\times m}$ and $V = (v_1, \dots, v_n) \in \mathbb{R}^{n\times n}$ are orthogonal matrices, $\Sigma = \begin{pmatrix} \operatorname{diag}(\sigma_1, \dots, \sigma_r) & \mathbf{0} \\ \mathbf{0} & \mathbf{0} \end{pmatrix} \in \mathbb{R}^{m\times n}$ is a diagonal matrix with $\sigma_1 \geq \sigma_2 \geq \cdots \geq \sigma_r > 0$.

[17] Note that p is often chosen so that the product $(m+n+p)\times p$ is about the same as the number of entries in the large and sparse data matrix A.

Fix any $p = 1, \dots, r-1$. Let $U_p = (u_1, \dots, u_p) \in \mathbb{R}^{m\times p}$, $V_p = (v_1, \dots, v_p) \in \mathbb{R}^{n\times p}$ and $\Sigma_p = \text{diag}(\sigma_1, \dots, \sigma_p) \in \mathbb{R}^{p\times p}$. Then the matrix $R_p = U_p\Sigma_p V_p^\top = \sum_{i=1}^{p} \sigma_i u_i v_i^\top$ has a rank p, and it is the best rank-p approximation of R with respect to the spectral norm $\|\cdot\|_2$:

$$\|R - R_p\|_2 \le \|R - B\|_2 \quad \text{for any rank-}p\text{ matrix } B \in \mathbb{R}^{m\times n}.$$

Moreover, $\|R - R_p\|_2 = \sigma_{p+1}$.

Proof. We first notice that

$$\|R - R_p\|_2 = \left\| \sum_{i=1}^{r} \sigma_i u_i v_i^\top - \sum_{i=1}^{p} \sigma_i u_i v_i^\top \right\|_2 = \left\| \sum_{i=p+1}^{r} \sigma_i u_i v_i^\top \right\|_2 = \sigma_{p+1}.$$

Let $B \in \mathbb{R}^{m\times n}$ be an arbitrary rank-p matrix. By the *rank nullity theorem* in [20], the dimension of $\mathcal{N}(B)$, the null space of B, is equal to $n - p$. Note that $v_1, \dots, v_{p+1}$ are orthogonal and therefore linearly independent: they generate a subspace S_p of dimension $p+1$ in $\mathbb{R}^n$. As $(p+1) + (n-p) > n$, $S_p \cap \mathcal{N}(B) \neq \emptyset$, one can always find a non-zero vector in the form $\omega = \gamma_1 v_1 + \cdots + \gamma_{p+1} v_{p+1}$ that also lies in $\mathcal{N}(B)$. Without loss of generality, we can scale ω so that the square of Euclidean norm $\|\omega\|_2^2 = \gamma_1^2 + \cdots + \gamma_{p+1}^2$ equals 1. Therefore,

$$\begin{aligned}
\|R - B\|_2^2 = \max_{\|x\|_2=1} \|(R-B)x\|_2^2 &\ge \|(R-B)\omega\|_2^2 \\
&= \|R\omega\|_2^2 \\
&= \omega^\top V\Sigma^2 V^\top \omega \\
&= \gamma_1^2\sigma_1^2 + \cdots + \gamma_{p+1}^2\sigma_{p+1}^2 \\
&\ge \sigma_{p+1}^2 = \|R - R_p\|_2^2.
\end{aligned}$$

The result follows by taking the square root of both sides of the last inequality. □

*9.A.4 Comparison between R functions of SVD and SD

For a given matrix A, the singular value decomposition (SVD) $A = U\Sigma V^\top$ always exists, regardless of whether A is square or rectangle, whereas the spectral decomposition (SD) $A = PDP^{-1}$ can only be valid when A is square and symmetric. Even among square matrices, the SD might not always exist. Notice that in Programme 9.31, the two R functions `prcomp()` and `princomp()` seem to yield the same results.

```
1  > d <- read.csv("fin-ratio.csv")  # read in dataset
2  > # standardize the data so the columns of large loadings
3  > # will not dominate the data
4  > d_scale <- scale(d[1:5])
5  > pca_svd <- prcomp(d_scale)      # run PCA with SVD
6  > pca_svd$rotation
7              PC1     PC2      PC3      PC4      PC5
8  EY     0.53261 -0.1429  0.40547  0.47331 -0.55450
9  CFTP   0.59671 -0.1891  0.15288  0.03542  0.76392
10 ln_MV  0.45438  0.3740  0.09225 -0.76483 -0.24535
11 DY     0.37958  0.2721 -0.83468  0.28201 -0.07519
12 BTME   0.09864 -0.8543 -0.32714 -0.33203 -0.20763
13 >
14 > s_svd <- pca_svd$sdev
15 > cumsum(s_svd^2/sum(s_svd^2))
16 [1] 0.3253 0.5372 0.7151 0.8762 1.0000
17 >
18 > pca_sd <- princomp(d_scale)   # run PCA with Spectral Decomposition
19 > pca_sd$loadings
20
21 Loadings:
22       Comp.1 Comp.2 Comp.3 Comp.4 Comp.5
23 EY     0.533  0.143  0.405  0.473  0.555
24 CFTP   0.597  0.189  0.153        -0.764
25 ln_MV  0.454 -0.374        -0.765  0.245
26 DY     0.380 -0.272 -0.835  0.282
27 BTME          0.854 -0.327 -0.332  0.208
28
29                Comp.1 Comp.2 Comp.3 Comp.4 Comp.5
30 SS loadings       1.0    1.0    1.0    1.0    1.0
31 Proportion Var    0.2    0.2    0.2    0.2    0.2
32 Cumulative Var    0.2    0.4    0.6    0.8    1.0
33 >
34 > s_sd <- pca_sd$sdev
35 > cumsum(s_sd^2/sum(s_sd^2))
36 Comp.1 Comp.2 Comp.3 Comp.4 Comp.5
37 0.3253 0.5372 0.7151 0.8762 1.0000
```

Programme 9.31 Comparison between SVD and SD

In fact, when the dataset contains more rows than columns, computing PCA via SVD using the `prcomp()` function provides the same result as that via SD using the `princomp()` function. For example, see the illustration in Programme 9.32 and the caption referred therein when the row number is less than the column number. In SVD, the entries returned in the diagonal matrix $\mathbf{\Sigma}$ are always real and nonnegative by construction, while in general the entries of $\boldsymbol{D}$ in SD could be a complex number. Nevertheless, in data analytics, PCA is implemented for the variance-covariance/correlation matrix of a dataset, and since those matrices are always in squared form, running PCA through both SVD and SD should provide the same result.

```
1 > d_scale <- t(scale(d[1:5]))
2 > pca_svd <- prcomp(d_scale)       # run PCA with SVD
3 > s_svd <- pca_svd$sdev
4 > cumsum(s_svd^2/sum(s_svd^2))
5 [1] 0.3175 0.5835 0.8177 1.0000 1.0000
6 >
7 > pca_sd <- princomp(d_scale)    # run PCA with Spectral Decomposition
8 Error in princomp.default(d_scale) :
9 'princomp' can only be used with more units than variables
```

Programme 9.32 Both SVD and SD work when the row number exceeds the column number for a dataset, yet SD does not work when the dataset is transposed so that the row number is less than the column number; on the other hand, SVD still works in this case, and produces similar singular values.

REFERENCES

1. Baglama, J., and Reichel, L. (2005). Augmented implicitly restarted Lanczos bidiagonalization methods. *SIAM Journal on Scientific Computing*, 27(1), 19–42.
2. Candès, E.J., Li, X., Ma, Y., and Wright, J. (2011). Robust principal component analysis? *Journal of the ACM*, 58(3), 1–37.
3. Eckart, C., and Young, G. (1936). The approximation of one matrix by another of lower rank. *Psychometrika*, 1(3), 211–218.
4. Emami, T., Janney, S.S., and Chakravarty, S. (2019). Elements of medical image processing. *Biomedical Engineering and its Applications in Healthcare*, 473–517.
5. Chen, Y., Fan, N.S., and Yam, S.C.P. (2024+). *Statistical Deep Learning with Python and R*. Preprint.
6. Golub, G.H., and Loan, C.F.V (2013). *Matrix Computations*. The Johns Hopkins University Press.
7. Kellison, S.G. (2006). *The Theory of Interest*. McGraw Hill.
8. Kim, K.I., Jung, K., and Kim, H.J. (2002). Face recognition using kernel principal component analysis. *IEEE Signal Processing Letters*, 9(2), 40–42.
9. Klugman, S.A., Beckley, J.A., Scahill, P.L., Varitek, M.C., and White, T.A. (2012). *Understanding Actuarial Practice*, Society of Actuaries.
10. Kong, H., Wang, L., Teoh, E.K., Li, X., Wang, J.G., and Venkateswarlu, R. (2005). Generalized 2D principal component analysis for face image representation and recognition. *Neural Networks*, 18(5–6), 585–594.
11. Koren, Y., and Bell, R. (2015). Advances in collaborative filtering. *Recommender Systems Handbook*, 77–118.
12. Macaulay, F.R. (1938). Some theoretical problems suggested by the movements of interest rates, bond yields and stock prices in the United States Since 1856. National Bureau of Economic Research, New York.
13. Mirsky, L. (1960). Symmetric gauge functions and unitarily invariant norms. *The Quarterly Journal of Mathematics*, 11(1), 50–59.
14. Panjer, H.H., Boyle, D.D., Cox, S.H., Dufresne, D., et al. (1998). *Financial Economics. With Applications to Investments, Insurance, and Pensions*. The Actuarial Foundation.

15. Redington, F.M. (1952). Review of the principles of life-office valuations. *Journal of the Institute of Actuaries (1886–1994)*, 78(3), 286–340.
16. Samaria, F.S., and Harter, A.C. (1994). Parameterisation of a stochastic model for human face identification. In *Proceedings of 1994 IEEE Workshop on Applications of Computer Vision* (pp. 138–142), IEEE.
17. Schmidt, E. (1908). Zur Theorie der linearen und nichtlinearen Integralgleichungen. III. *Teil. Mathematische Annalen*, 63, 433–476.
18. Schölkopf, B., Smola, A., and Müller, K.R. (1997). Kernel principal component analysis. In *International Conference on Artificial Neural Networks* (pp. 583–588). Springer.
19. Shang, K. (2018). Applying image recognition to insurance. Society of Actuaries. Retrieved from https://www.soa.org/globalassets/assets/Files/resources/research-report/2018/applying-image-recognition.pdf (accessed 21 April 2024).
20. Strang, G. (2012). *Linear Algebra and Its Applications* (4th ed.). Cengage Learning.
21. Wright, J., Ganesh, A., Rao, S., Peng, Y., and Ma, Y. (2009). Robust principal component analysis: Exact recovery of corrupted low-rank matrices via convex optimization. In *Advances in Neural Information Processing Systems* (pp. 2080–2088). https://proceedings.neurips.cc/paper/2009/file/c45147dee729311ef5b5c3003946c48f-Paper.pdf (accessed 7 May 2024).
22. Zhao, W., Krishnaswamy, A., Chellappa, R., Swets, D.L., and Weng, J. (1998). Discriminant analysis of principal components for face recognition. In H. Wechsler, P.J. Phillips, V. Bruce, F.F. Soulié, and T.S. Huang (Eds), *Face Recognition* (pp. 73–85). Springer.

CHAPTER 10

Regression Learning

Regression Learning (a.k.a. *Regression Analysis*) is undoubtedly one of the most widely used statistical tools, and its predictive power is mostly determined by appropriate choices of independent variables or covariates, perhaps after transformations. Its name of "regression" was first coined by British statistician *Francis Galton* in 1885 [7], and its context was further consolidated by his disciple *Karl Pearson* in 1903 [18]. Generally speaking, regression analysis estimates the relationship between the response (dependent) and independent variables, which is not completely known but subject to some random error caused by noisy data. To name its immediate application in finance, the celebrated Capital Asset Pricing Model (CAPM) by Markowitz in 1952 [16] is indeed a representative example. In this chapter, starting from this conventional model, we shall revisit the simple and multiple linear regression models, including the special case of polynomial regression, then we discuss several commonly used generalized linear models, namely *logistic regression* and *Poisson regression*, together with their applications in finance as well as the practical considerations for issues such as model selection and evaluation. In addition, echoing the PCA discussion in Chapter 9, we shall also introduce the closely related *principal component regression* and illustrate its use in practice.

10.1 SIMPLE AND MULTIPLE LINEAR REGRESSION MODELS AND BEYOND

CAPM, as mentioned above, is in fact a simple linear regression of an individual asset's return against the market rate of return, namely:

$$\mathrm{r}_i = r_f + \beta_i(\mathrm{r}_m - r_f) + \epsilon_i, \qquad i = 1, 2, \ldots, n, \tag{10.1}$$

so that

$$\mathbb{E}(\mathrm{r}_i) = r_f + \beta_i[\mathbb{E}(\mathrm{r}_m) - r_f], \tag{10.2}$$

where r_i is the relative return of the i-th asset, r_m is the relative return of the market of interest, r_f is the risk-free interest rate which is assumed to be the only one in the market; β_i is the β-factor or the sensitivity of the expected excess return of the i-th security to the expected excess market return. Also, n is the total number of different securities in this market sector. Finally, ϵ_i is the idiosyncratic noise component with the mean zero and a finite variance σ^2. We further assume that

1. $\epsilon_i \sim \mathcal{N}(0, \sigma^2)$ with $\sigma^2 < \infty$;
2. $\text{Cov}(\epsilon_i, \epsilon_j) = 0$ for $i \neq j$;
3. $\text{Cov}(\mathrm{r}_m, \epsilon_i) = 0$ for all i; that means the market return and the individual idiosyncratic noise component are uncorrelated.

CAPM is widely applied in modern finance. For instance, the linear relationship (10.2) between the β-factor and the expected return of the security in the CAPM, a.k.a. the *Security Market Line* (SML), is often used to determine whether an asset is properly valued. Usually, β_i stays roughly constant at the equilibrium level for almost every industrial sector, and if the sample mean of r_i over the latest period becomes too high (resp. low) for a stock, it then receives an overall downward (resp. upward) pressure that pushes its price back to the corresponding equilibrium level on the SML. In the real market, these kinds of adjustments often take place in a relatively short time. Moreover, *Jensen's index* [11] $J := \hat{\mu} - (r_f + \hat{\beta}(\hat{\mu}_m - r_f))$, where $\hat{\mu}_m$ is the empirical mean of the market index r_m, is a well-known performance evaluation reference index that detects any possible mispricing via the difference between the estimated return of the stock $\hat{\mu}$ observed from the market data and the "natural" return of $r_f + \hat{\beta}(\hat{\mu}_m - r_f)$ as predicted by the SML in CAPM. The *Sharpe ratio*, the empirical realization of β_i, $\hat{\beta}_i = \frac{\hat{\mu}_i - r_f}{\hat{\sigma}_i}$, measures the leverage effect of the i-th stock, i.e. its sensitivity to changes of the underlying market risk. If $\hat{\beta}_i$ of a stock is equal to (resp. less than) the market beta β_m (or its empirical counterpart $\hat{\beta}_m = \frac{\hat{\mu}_m - r_f}{\hat{\sigma}_m}$), then the stock is said to be *efficient* (resp. *inefficient*); likewise if $\hat{\beta}_i > \hat{\beta}_m$, then the stock is considered more efficient than the market portfolio. Last but not least, the certainty equivalent pricing formula is frequently adopted in evaluating (single-period) future projects for companies, and it is derived from CAPM as follows. To calculate the present value P of a project with a random terminal payoff Q, we apply CAPM to the expectation of its rate of return $\mathrm{r} = \frac{\mathrm{Q} - P}{P}$:

$$\mathbb{E}(\mathrm{r}) = \frac{\mathbb{E}(\mathrm{Q}) - P}{P} = r_f + \beta(\mu_m - r_f). \tag{10.3}$$

The *beta factor* of this project, $\beta = \frac{\text{Cov}(\mathrm{r}, \mathrm{r}_m)}{\sigma_m^2} = \frac{\text{Cov}(\mathrm{Q}, \mathrm{r}_m)}{P\sigma_m^2}$, is then substituted to (10.3) and after further algebraic rearrangement, we can obtain the following pricing formula:

$$P = \frac{1}{1 + r_f}\left[\mathbb{E}(\mathrm{Q}) - \frac{\text{Cov}(\mathrm{Q}, \mathrm{r}_m)(\mu_m - r_f)}{\sigma_m^2}\right].$$

That said, it is common practice to enhance the prediction power of the model by including more covariates. Simple linear regression models can then be extended to *multiple linear regression* models, resulting in the well-known multi-factor model in finance, a special, celebrated case of which was constructed by Fama and French in 1993 [5]. The general form of this model is

$$y_i = \beta_0 + \beta_1 x_{i1} + \cdots + \beta_p x_{ip} + \varepsilon_i, \qquad i = 1, 2, \ldots, n,$$

where $\beta_j \in \mathbb{R}, j = 0, \ldots, p$, and there are p different fundamental factors x_j, $j = 1, \ldots, p$; the idiosyncratic noises ε_i's possess similar structures as in the simple linear regression case. This general regression procedure is included in many statistical packages, such as `lm()` in **R**, and each will return the least squared estimate of $\boldsymbol{\beta}$,[1] which is $\hat{\boldsymbol{\beta}} = (\tilde{\boldsymbol{X}}^\top \tilde{\boldsymbol{X}})^{-1} \tilde{\boldsymbol{X}}^\top \boldsymbol{y}$, where

$$\tilde{\boldsymbol{X}} := (\mathbf{1}, \boldsymbol{X}) = \begin{pmatrix} 1 & x_{11} & \cdots & x_{1p} \\ 1 & \vdots & \ddots & \vdots \\ 1 & x_{n1} & \cdots & x_{np} \end{pmatrix} \in \mathbb{R}^{n \times (p+1)} \quad \text{and} \quad \boldsymbol{y} = \begin{pmatrix} y_1 \\ \vdots \\ y_n \end{pmatrix},$$

and the fitted values are:

$$\hat{y}_i = \hat{\beta}_0 + \hat{\beta}_1 x_{i1} + \ldots + \hat{\beta}_p x_{ip},$$

with the corresponding residual $e_i := y_i - \hat{y}_i$, for $i = 1, \ldots, n$.

The Fama–French three-factor model uses three feature variables (instead of only one, the market portfolio, in CAPM) to describe the return of a stock. These correspond to: the factor of market risk; the outperformance index of small companies against big ones; and the outperformance indicator of companies with a high book-to-market ratio. More precisely, this three-factor model for the i-th security is written as:

$$\mathbb{E}(\mathrm{r}_i) - r_f = \alpha_i + \beta_{i1}(\mathbb{E}(\mathrm{r}_m) - r_f) + \beta_{i2} \cdot \mathrm{SMB} + \beta_{i3} \cdot \mathrm{HML},$$

where the two new factors SMB and HML respectively stand for "Small Minus Big", the historical excess return of companies with smaller market capitalization over those with larger market capitalization, and "High Minus Low", the historical excess return of stocks with a higher book-to-market ratio over those with a lower one. Let us give a brief summary of the calculations of these two factors[2]. Given the evaluation day t, we first collect the data of returns, market capitalizations, and book-to-market ratios of all stocks in the market on that day[3]. We then sort the stocks by their market

---

[1] The estimate is obtained by minimizing the L^2-loss function $L(\boldsymbol{\beta}) := (\boldsymbol{y} - \tilde{\boldsymbol{X}}\boldsymbol{\beta})^\top (\boldsymbol{y} - \tilde{\boldsymbol{X}}\boldsymbol{\beta}) = \boldsymbol{y}^\top \boldsymbol{y} - \boldsymbol{y}^\top \tilde{\boldsymbol{X}}\boldsymbol{\beta} - \boldsymbol{\beta}^\top \tilde{\boldsymbol{X}}^\top \boldsymbol{y} + \boldsymbol{\beta}^\top \tilde{\boldsymbol{X}}^\top \tilde{\boldsymbol{X}}\boldsymbol{\beta}$, whose first order condition $\nabla_{\boldsymbol{\beta}} L = -2\tilde{\boldsymbol{X}}^\top \boldsymbol{y} + 2\tilde{\boldsymbol{X}}^\top \tilde{\boldsymbol{X}}\boldsymbol{\beta} = \boldsymbol{0}$ gives $\tilde{\boldsymbol{X}}^\top (\tilde{\boldsymbol{X}}\hat{\boldsymbol{\beta}}) = \tilde{\boldsymbol{X}}^\top \boldsymbol{y}$. Note that this is a regression without regularization especially when $\boldsymbol{X}$ is not so sparse. Otherwise, for a sparse $\boldsymbol{X}$, we add an additional regularizer of $\sum_{\{i: \boldsymbol{x}_i \text{ is sparse}\}} \lambda_i |\beta_i|$ to the objective function; further discussion can be found in Chapter 11.

[2] More details can be found in http://mba.tuck.dartmouth.edu/pages/faculty/ken.french/Data_Library/f-f_3developed.html.

[3] From *Bloomberg* and *Yahoo Finance*, say.

capitalizations in descending order; the top 90% are classified as "big-cap" (B) with larger market capitalizations, while the bottom 10% are classified as "small-cap" (S) with smaller market capitalizations. Next, for each of these two classes of stocks, we further sort them by the book-to-market ratios, also in descending order within each class; the top 30% are classified as "value" (V) stocks with higher book-to-market ratios, while the bottom 30% are classified as "growth" (G) stocks with lower book-to-market ratios, and the remaining 40% in the middle of the list are classified as "neutral" (N) stocks with moderate book-to-market ratios. To this end, the collection of "small-cap" and "big-cap" stocks are partitioned into six ($= 2 \times 3$) subgroups, namely SV, SN, SG, BV, BN, and BG, and we denote the respective average returns of the stocks within these six subgroups as $\overline{\text{SV}}, \overline{\text{SN}}, \overline{\text{SG}}, \overline{\text{BV}}, \overline{\text{BN}}, \overline{\text{BG}}$. The formulae for calculating SMB and HML are then given by:

$$\text{SMB} = \frac{1}{3}(\overline{\text{SV}} + \overline{\text{SN}} + \overline{\text{SG}}) - \frac{1}{3}(\overline{\text{BV}} + \overline{\text{BN}} + \overline{\text{BG}}),$$

$$\text{HML} = \frac{1}{2}(\overline{\text{SV}} + \overline{\text{BV}}) - \frac{1}{2}(\overline{\text{SG}} + \overline{\text{BG}}).$$

According to [4], the Fama–French three-factor model explains over 90% of the diversified portfolios returns at their time of study, while the figure is only 70% on the average for CAPM.

In general, given a random variable $\mathbf{x} \in \mathbb{R}^p$ and another response variable of interest $y \in \mathbb{R}$, regression analysis aims to learn the functional dependence of y on $\mathbf{x}$. That is, to find a reasonably simple function $f : \mathbb{R}^p \to \mathbb{R}$ so that $f(\mathbf{x})$ can well approximate y by minimizing a certain loss function, for instance, the mean squared error:

$$L(f) := \mathbb{E}((y - f(\mathbf{x}))^2).$$

We can easily deduce that the conditional expectation $m(x) := \mathbb{E}(y|\mathbf{x} = x)$ is the optimal answer, called the *regression function*; indeed, for any $f : \mathbb{R}^p \to \mathbb{R}$,

$$L(f) = \mathbb{E}((y - m(\mathbf{x}))^2) + \mathbb{E}((m(\mathbf{x}) - f(\mathbf{x}))^2) \geq \mathbb{E}((y - m(\mathbf{x}))^2),$$

hence $m(x) = \arg\min_f L(f)$ as claimed. In practice, we often demand more regularities from f, such as being a polynomial; a smooth enough function; or a linear combination of some regular enough base functions. Therefore, the general rationale behind regression analysis is to estimate the best regression functions (also called *regressors*) of this possibly highly nonlinear conditional expectation under different settings.

Practically, regression analysis consists of two common approaches: *parametric* and *non-parametric* regressions. A parametric model assumes a prior knowledge of the class of the functions upon certain parameters to be determined. All simple and multiple regressions, together with the non-linear regressions shown as follows, are typical examples. For non-linear regression, the regressor f takes the general form:

$$f(\mathbf{x}) = f(\mathbf{x}; \beta_0, \beta_1, \ldots \beta_p) \quad \text{and} \quad y = f(\mathbf{x}) + \epsilon,$$

where the random variable ϵ is the random error that varies independently of the pair $(\mathbf{x}, y)$. Each of the error terms corresponding to different samples therefore follows a common distribution; and these error terms are mostly assumed to be independent of each other.

In contrast, a non-parametric model does not assume any specific form for the regressor. To name two common methods, they are *Kernel Regression* (to be introduced in Section *14.3) and Principal Component Regression (to be introduced in Section *10.7). As a mixture of parametric and non-parametric models, for a *semi-parametric model*, the regressor takes the following separable form:

$$f(\mathbf{x}) = f_1(\mathbf{x}; \beta_0, \beta_1, \ldots, \beta_p) + f_2(\mathbf{x}) \qquad \text{and} \qquad y = f(\mathbf{x}) + \epsilon,$$

where $f_1(\mathbf{x}; \boldsymbol{\beta})$ is the parametric functional part involving several parameters $\boldsymbol{\beta} = (\beta_0, \beta_1, \ldots, \beta_p)^\top$ with a known functional form of f_1, and the unknown function $f_2(\cdot)$ is the non-parametric functional part.

In regard to finding estimates of the optimal parameters β_j's in the parametric functional part, here we introduce a generic iterative numerical scheme. Consider a dataset $S = \{(x_i, y_i)\}_{i=1}^n$, with which we empirically estimate the theoretical mean squared error, and the least square method corresponding to the former is to minimize the empirical mean squared error between the actual observation (label) y_i and the approximant $f(x_i; \boldsymbol{\beta})$ for $i = 1, 2, \ldots, n$:

$$S(\boldsymbol{\beta}) := \hat{\mathbb{E}}[(y - f(\mathbf{x}; \boldsymbol{\beta}))^2] = \frac{1}{n} \sum_{i=1}^{n} \left(y_i - f(x_i; \boldsymbol{\beta})\right)^2, \tag{10.4}$$

where the best estimate $\hat{\boldsymbol{\beta}} = \arg\min_{\boldsymbol{\beta}} \frac{1}{n} \sum_{i=1}^n \left(y_i - f(x_i; \boldsymbol{\beta})\right)^2 \in \mathbb{R}^{p+1}$ should satisfy the first-order condition which is a system of nonlinear equations:

$$\nabla_{\boldsymbol{\beta}} S \Big|_{\boldsymbol{\beta}=\hat{\boldsymbol{\beta}}} = \mathbf{0}. \tag{10.5}$$

Generally speaking, it is often difficult to obtain an analytic solution to the system (10.5). Instead, it is more relevant to solve it numerically. To this point, the *Gauss–Newton method* is one of the commonly used numerical algorithms, and we introduce a modified version here. Readers may refer to Appendix *10.A.1 for a formal derivation of the Gauss–Newton method.

Define n functions called residuals; $r_i(\boldsymbol{\beta}) := (y_i - f(x_i; \boldsymbol{\beta}))^2$, for $i = 1, \cdots, n$. The minimization of $S(\boldsymbol{\beta})$ is equivalent to minimizing the following sum of squares:

$$S(\boldsymbol{\beta}) \propto \frac{1}{2} \sum_{i=1}^{n} r_i^2(\boldsymbol{\beta}) = \frac{1}{2} \langle r(\boldsymbol{\beta}), r(\boldsymbol{\beta}) \rangle.$$

Denote by

$$J_r(\boldsymbol{\beta}) := \left(\frac{\partial r_i(\boldsymbol{\beta})}{\partial \beta_j} \right)_{1 \le i \le n, 0 \le j \le p}, \tag{10.6}$$

the Jacobian (rectangular) matrix of (partial derivatives of) $(r_1, \cdots, r_n)$ with respect to $\boldsymbol{\beta}$. The procedure of the *modified Gauss–Newton algorithm* is the following:

Algorithm 10.1 Modified Gauss–Newton Algorithm

1) Initialize $\boldsymbol{\beta}^{(0)}$ and compute $S(\boldsymbol{\beta}^{(0)})$. Label $k = 0$.
2) For $i = 1, \dots, n$ and $j = 0, \dots, p$, compute $\left(d_{ij}^{(k)}\right)_{1\le i\le n, 0\le j\le p} := \boldsymbol{J}_r(\boldsymbol{\beta}^{(k)}) = \left(\frac{\partial r_i(\boldsymbol{\beta}^{(k)})}{\partial \beta_j}\right)$, denoted by $\boldsymbol{J}_r^{(k)}$, and $\boldsymbol{b}^{(k)} = \left(b_j^{(k)}\right)_{0\le j\le p} = (\boldsymbol{J}_r^{(k)})^\top \boldsymbol{r}(\boldsymbol{\beta}^{(k)}) = \left(\sum_{i=1}^n r_i(\boldsymbol{\beta}^{(k)}) d_{ij}^{(k)}\right)$.
3) Solve $\boldsymbol{p}^{(k)}$ for the linear system $(\boldsymbol{J}_r^{(k)})^\top \boldsymbol{J}_r^{(k)} \boldsymbol{p}^{(k)} = -\boldsymbol{b}^{(k)}$. For if rank $\left((\boldsymbol{J}_r^{(k)})^\top \boldsymbol{J}_r^{(k)}\right) = n$, the system admits a unique solution $\boldsymbol{p}^{(k)} = -\left((\boldsymbol{J}_r^{(k)})^\top \boldsymbol{J}_r^{(k)}\right)^{-1} \boldsymbol{b}^{(k)}$; otherwise take $\boldsymbol{p}^{(k)} = -\boldsymbol{b}^{(k)}$.
4) Set $q(\lambda) := S\left(\boldsymbol{\beta}^{(k)} + \lambda \boldsymbol{p}^{(k)}\right)$. Find $\lambda^{(k)}$ such that $q(\lambda^{(k)}) = \min_\lambda q(\lambda)$ using the golden-section search method. Also set $\boldsymbol{\beta}^{(k+1)} = \boldsymbol{\beta}^{(k)} + \lambda^{(k)} \boldsymbol{p}^{(k)}$ and compute $S(\boldsymbol{\beta}^{(k+1)})$.
5) Repeat Steps (2) to (4) for $k \leftarrow k+1$ until the terminal condition is satisfied, i.e. $\left|S(\boldsymbol{\beta}^{(k+1)}) - S(\boldsymbol{\beta}^{(k)})\right| < \delta$ and $\left\|\boldsymbol{\beta}^{(k+1)} - \boldsymbol{\beta}^{(k)}\right\|_2 < \delta$ for some small enough prescribed δ.
6) Return $\hat{\boldsymbol{\beta}} = \boldsymbol{\beta}^{(k+1)}$ if the terminal condition is reached.

Note that the error $S(\boldsymbol{\beta})$ may not be decreasing for every iteration in the Gauss–Newton algorithm but could be oscillating instead. Yet for any sufficiently small $\lambda > 0$, it always holds that $S(\boldsymbol{\beta}^{(k)} + \lambda\boldsymbol{\Delta}) < S(\boldsymbol{\beta}^{(k)})$ in the neighbourhood of a minimum point of S, for $\hat{\boldsymbol{\Delta}} = \boldsymbol{\beta}^{(k+1)} - \boldsymbol{\beta}^{(k)}$. Thus the modified version of the Gauss–Newton algorithm incorporates a small enough factor λ (learning rate) in each consecutive update. For each iteration, to choose a suitable $\lambda \in (0, 1)$ which minimizes the error $S(\boldsymbol{\beta}^{(k)} + \lambda\hat{\boldsymbol{\Delta}})$, we can simply apply some popular search algorithms such as the *golden-section search method*[4] as adopted in Algorithm 10.1.

In large-scale optimization, we want to find an efficient way of computing $\boldsymbol{J}_r(\boldsymbol{\beta})^\top \boldsymbol{J}_r(\boldsymbol{\beta})\boldsymbol{p}$ for some known vector $\boldsymbol{p}$. Denote by $\mathcal{J}_i(\boldsymbol{\beta})$ the i-th row of $\boldsymbol{J}_r(\boldsymbol{\beta})$, then we can write

$$\boldsymbol{J}_r(\boldsymbol{\beta})^\top \boldsymbol{J}_r(\boldsymbol{\beta})\boldsymbol{p} = \sum_{i=1}^n (\mathcal{J}_i(\boldsymbol{\beta})\boldsymbol{p})\mathcal{J}_i(\boldsymbol{\beta})^\top, \tag{10.7}$$

where the i-th summand solely depends on the i-th row of $\boldsymbol{J}_r(\boldsymbol{\beta})$ which is the gradient of the i-th residual $r_i(\boldsymbol{\beta})$ that in turn is just the residual with respect to the i-th datum

---

[4] The golden-section search method initializes by calculating the values of $g(\lambda) := S(\boldsymbol{\beta}^{(k)} + \lambda\hat{\boldsymbol{\Delta}})$ at the two initial endpoints $a_0 = \lambda_0^{(1)} = 0$ and $b_0 = \lambda_0^{(2)} = 1$. For the i-th iteration, set $\lambda_i^{(1)} := \varphi a_{i-1} + (1-\varphi)b_{i-1}$ and $\lambda_i^{(2)} := (1-\varphi)a_{i-1} + \varphi b_{i-1}$ where $\varphi = \frac{\sqrt{5}-1}{2}$ is the golden ratio, and calculate $g(\lambda_i^{(1)})$ and $g(\lambda_i^{(2)})$. If $g(\lambda_i^{(1)}) < g(\lambda_i^{(2)})$, then set $a_i = a_{i-1}$ and $b_i = \lambda_i^{(2)}$; otherwise, set $a_i = \lambda_i^{(1)}$ and $b_i = b_{i-1}$. This procedure is repeated until the distance between the endpoints after the j-th iteration falls below a pre-specified threshold value ε, and the minimizer $\hat{\lambda}$ can then be taken as their midpoint $\frac{a_j+b_j}{2}$. This method is often used to minimize functions whose derivatives or gradients cannot be obtained explicitly, as they are actually not explicitly required in the algorithm.

only; so each of these $J_i(\beta)$'s contributes to $S(\beta)$ independently of the others. On the other hand, consider $\frac{1}{n}J_r(\beta)^\top J_r(\beta)p$ as an average: in light of the law of large numbers, it stabilizes as n is large, and one can save much calculation by only coping with a small proportion $c \in (0,1)$ of the overall n residuals. More precisely, one can approximate the average of the total sum by only summing over a mini-batch of the original sample as $\frac{1}{n}J_r(\beta)^\top J_r(\beta)p \approx \frac{1}{cn}\sum_{i=1}^{cn}(J_i(\beta)p)J_i(\beta)^\top$ for some $c \ll 1$. As an aside, we note the independence among the summands in the representation (10.7) makes parallel computing feasible.

In addition to ordinary linear regression models, there are many other useful variants to be introduced in the following sections; readers can also refer to [15] and references therein for more discussions.

10.2 POLYNOMIAL REGRESSION

Polynomial regression models a regressor by a polynomial function. Given a dataset $S = \{(x_i, y_i)\}_{i=1}^n$, a quadratic regression model takes the form:

$$y_i = \beta_0 + \beta_1 x_i + \beta_2 x_i^2 + \varepsilon_i, \quad i = 1, \ldots, n,$$

while a cubic model is:

$$y_i = \beta_0 + \beta_1 x_i + \beta_2 x_i^2 + \beta_3 x_i^3 + \varepsilon_i, \quad i = 1, \ldots, n,$$

where ε_i's are independent idiosyncratic noises as before. In practice, the degree of the polynomial regressor barely exceeds 5, otherwise the parameters are hard to interpret. Even worse, a high degree polynomial regressor becomes unstable, drastically affecting its performance as it leans towards to overfitting.

For more effective applications, we can include more explanatory variables in regression. For instance, a two-variable quadratic regression model is given by:

$$y_i = \beta_0 + \beta_1 x_{i1} + \beta_2 x_{i2} + \beta_{11} x_{i1}^2 + \beta_{22} x_{i2}^2 + \beta_{12} x_{i1} x_{i2} + \varepsilon_i, \quad i = 1, \ldots, n,$$

where $x_{i1}x_{i2}$ stands for the *interaction term* for the datum i.

In principle, a polynomial regression is essentially no different from ordinary linear regressions, except that the regressor is now in terms of powers of the same data input. This means that the same formula for the least squared estimate of $\hat{\beta}$ applies, but the data matrix, say for a regressor with a degree p, takes the form:

$$X = \begin{pmatrix} 1 & x_1 & x_1^2 & \ldots & x_1^p \\ \vdots & \vdots & \vdots & \ddots & \vdots \\ 1 & x_n & x_n^2 & \ldots & x_n^p \end{pmatrix},$$

which is a *Vandermonde* matrix (see [10]). Also, the statistical diagnostics is essentially the same as that for multiple linear regressions (via MANOVA).

We next present a real estate price example by using polynomial regression models in an **R** environment. The dataset used is the *Boston Housing Dataset*[5] containing information collected by the U.S. Census Service regarding housing in the Boston Mass area (Figure 10.1). This dataset has been used extensively in the statistics literature to evaluate different algorithms. It contains 506 entries and 13 feature variables[6] including one categorical feature variable `CHAS` which takes value 1 if the tract bounds a river, or 0 otherwise; one discrete feature variable `RAD`, representing the accessibility to radial highways; and 11 continuous feature variables. Among these, `CRIM` represents the per capita crime rate by town; `ZN` stands for the proportion of residential

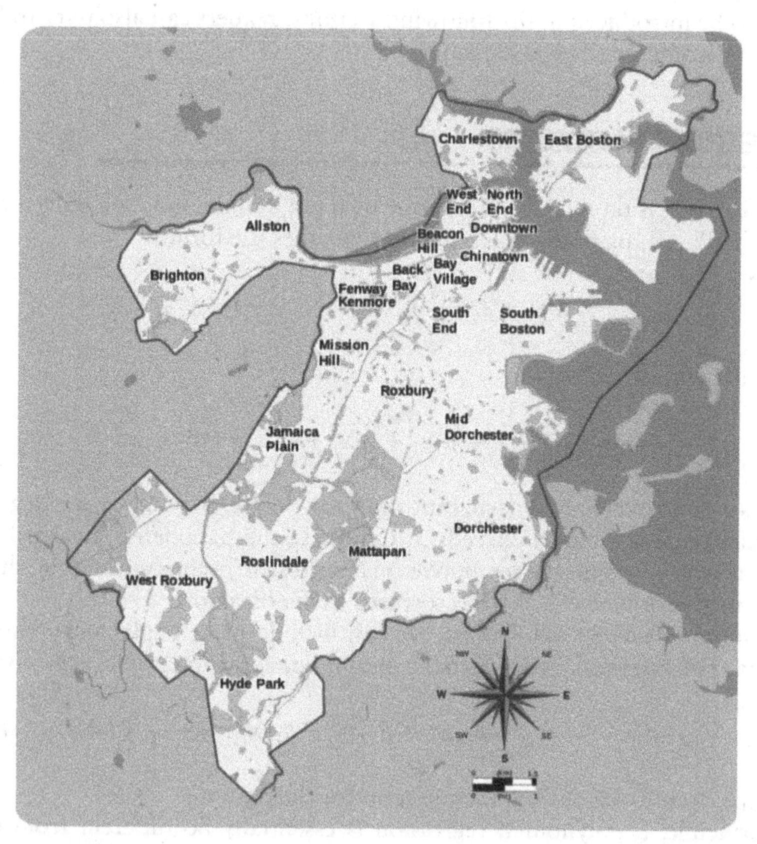

FIGURE 10.1 Neighborhoods of Boston (https://upload.wikimedia.org/wikipedia/commons/3/35/Boston_ONS_Neighborhoods.svg).

[5] The Boston Housing Dataset; retrieved from https://www.cs.toronto.edu/~delve/data/boston/bostonDetail.html.

[6] One of the feature variables is controversial in the modern era, i.e. `B - 1000 (Bk - 0.63)**2` where `Bk` is the proportion of black people by town. This may be regarded as an indicator of racism, and we sincerely hope that such feature will never become a contributor to any socioeconomic factors in our future generations as we strive for diversity and equality without reservation.

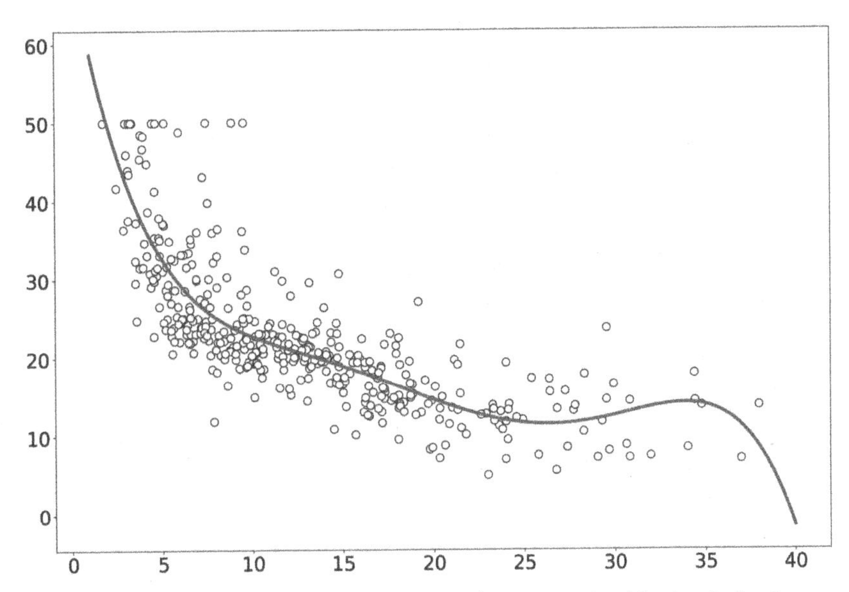

FIGURE 10.2 The output graph from Programme 10.1; the black circle dots are the training data points, and the red solid line represents the fitted regression.

land zoned for lots over 25,000 sq. ft.; and `INDUS` indicates the proportion of non-retail business acres per town. However, in this example, we shall only use the `LSTAT` continuous feature variable, which stores information related to the percentage of lower status of the population[7], to predict the median value of owner-occupied homes in the unit of USD 1,000 (`MEDV`). After splitting the dataset into training and testing data, we apply a polynomial regression with a degree $p = 5$ on the training data (as indicated by the Python function `poly = PolynomialFeatures(degree=5)` in the `sklearn.preprocessing` package and the **R** function `poly(lstat, 5, ...)` in the `stats` package, respectively), and then test the fitted model by calculating the *residual mean-squared error* (RMSE in this section) of the testing data. The polynomial regression lines are depicted in Figure 10.2 for Python and Figure 10.3 for **R**. From the Python and **R** codes, all the coefficients of polynomial and the intercepts have a small p-value < 0.05, indicating that they are significantly different from zero.

```
1  from sklearn.model_selection import train_test_split
2  from sklearn.preprocessing import PolynomialFeatures
3  import statsmodels.api as sm
4  import matplotlib.pyplot as plt
5  import numpy as np
6  import pandas as pd
7
8  plt.rc('xtick', labelsize=25); plt.rc('ytick', labelsize=25)
9
```

[7] According to [9], it is defined as "$\frac{1}{2}$(proportion of adults without some high school education and proportion of male workers classified as laborers)".

```
10  df = pd.read_csv("Boston.csv", index_col=0)
11  LSTAT, y = df.lstat.values, df.medv.values
12
13  # Split the data into training and test dataset in ratio of 8:2
14  LSTAT_train, LSTAT_test,
15  y_train, y_test = train_test_split(LSTAT, y, test_size=0.2,
16                                     random_state=4002)
17  # Build the model
18  poly = PolynomialFeatures(degree=5)
19  LSTAT_train_5 = poly.fit_transform(LSTAT_train.reshape(-1, 1))
20  LSTAT_test_5 = poly.fit_transform(LSTAT_test.reshape(-1, 1))
21
22  model = sm.OLS(y_train, LSTAT_train_5).fit()
23  print(model.summary())
24
25  # Model predictions
26  predict = model.predict(LSTAT_test_5)
27
28  # Model performance
29  from sklearn.metrics import r2_score, mean_squared_error
30  print(mean_squared_error(y_test, predict, squared=False))   # RMSE
31  print(r2_score(y_test, predict))    # R2
32
33  fig, ax = plt.subplots(figsize=(15, 10))
34  plt.scatter(LSTAT_train, y_train, edgecolors="black",
35              color="white", s=20)
36
37  line = np.linspace(1, 40, 100)
38  line_5 = poly.fit_transform(line.reshape(-1, 1))
39  plt.plot(line, model.predict(line_5), color="red")
```

```
1                            OLS Regression Results
2   ==============================================================================
3   Dep. Variable:                      y   R-squared:                       0.667
4   Model:                            OLS   Adj. R-squared:                  0.663
5   Method:                 Least Squares   F-statistic:                     159.4
6   Date:                Thu, 11 Nov 2021   Prob (F-statistic):           1.15e-92
7   Time:                        21:40:25   Log-Likelihood:                -1242.4
8   No. Observations:                 404   AIC:                             2497.
9   Df Residuals:                     398   BIC:                             2521.
10  Df Model:                           5
11  Covariance Type:            nonrobust
12  ==============================================================================
13                   coef    std err          t      P>|t|      [0.025      0.975]
14  ------------------------------------------------------------------------------
15  const         70.2003      4.515     15.549      0.000      61.324      79.076
16  x1           -12.8288      1.856     -6.913      0.000     -16.477      -9.180
17  x2             1.3957      0.266      5.246      0.000       0.873       1.919
18  x3            -0.0767      0.017     -4.542      0.000      -0.110      -0.043
19  x4             0.0020      0.000      4.109      0.000       0.001       0.003
20  x5         -1.923e-05   5.08e-06     -3.785      0.000   -2.92e-05   -9.24e-06
21  ==============================================================================
22  Omnibus:                      129.209   Durbin-Watson:                   2.066
23  Prob(Omnibus):                  0.000   Jarque-Bera (JB):              465.077
```

```
24  Skew:                         1.409   Prob(JB):                   1.02e-101
25  Kurtosis:                     7.437   Cond. No.                    1.62e+08
26  ==============================================================================
27
28  Notes:
29  [1] Standard Errors assume that the covariance matrix of the errors is
        correctly specified.
30  [2] The condition number is large, 1.62e+08. This might indicate that there are
31  strong multicollinearity or other numerical problems.
32  5.036160278727456
33  0.7250683211169218
```

Programme 10.1 Polynomial regression for the Boston housing dataset in Python.

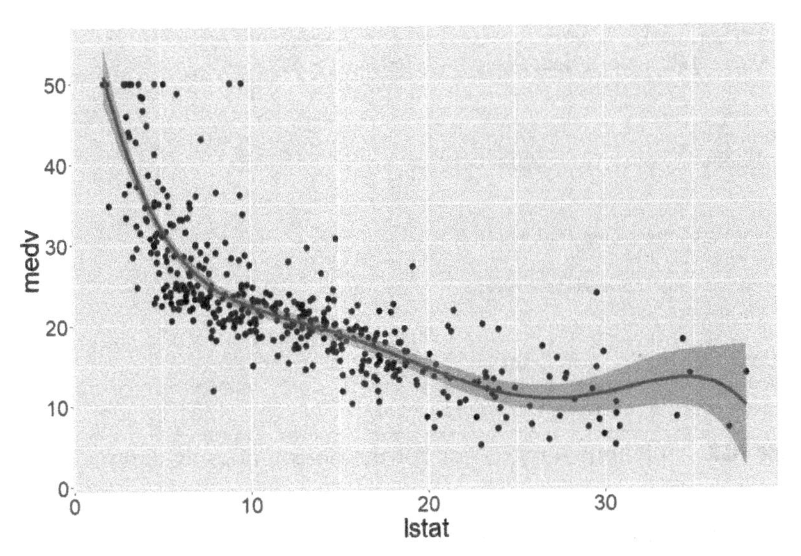

FIGURE 10.8 The output graph from Programme 10.2; the black dots are the training data points, and the blue solid line represents the fitted regression.

```
1   > library(caret)                        # Library for RMSE and R2
2   >
3   > data("Boston", package="MASS")        # Load the data
4   >
5   > # Split the data into training and test dataset in ratio of 8:2
6   > set.seed(123)
7   > training.samples <- sample(1:nrow(Boston), round(0.8*nrow(Boston)))
8   > train.data <- Boston[training.samples,]
9   > test.data <- Boston[-training.samples,]
10  >
11  > # Build the model
12  > model <- lm(medv ~ poly(lstat, degree=5, raw=TRUE), data=train.data)
13  > summary(model)
14
15  Call:
16  lm(formula = medv ~ poly(lstat, degree = 5, raw = TRUE), data = train.data)
```

```
17
18 Residuals:
19      Min       1Q   Median       3Q      Max
20 -13.9120  -2.9729  -0.6636   1.8563  27.2912
21
22 Coefficients:
23                                          Estimate Std. Error t value Pr(>|t|)
24 (Intercept)                              6.927e+01  3.864e+00  17.927  < 2e-16 ***
25 poly(lstat, degree = 5, raw = TRUE)1 -1.281e+01  1.635e+00  -7.837 4.24e-14 ***
26 poly(lstat, degree = 5, raw = TRUE)2  1.396e+00  2.394e-01   5.830 1.15e-08 ***
27 poly(lstat, degree = 5, raw = TRUE)3 -7.589e-02  1.544e-02  -4.914 1.30e-06 ***
28 poly(lstat, degree = 5, raw = TRUE)4  1.932e-03  4.473e-04   4.318 1.99e-05 ***
29 poly(lstat, degree = 5, raw = TRUE)5 -1.832e-05  4.738e-06  -3.867 0.000129 ***
30 ---
31 Signif. codes:  0 '***' 0.001 '**' 0.01 '*' 0.05 '.' 0.1 ' ' 1
32
33 Residual standard error: 5.109 on 399 degrees of freedom
34 Multiple R-squared:  0.6947,    Adjusted R-squared:  0.6909
35 F-statistic: 181.6 on 5 and 399 DF,  p-value: < 2.2e-16
36
37 >
38 > # Make predictions
39 > predictions <- predict(model, test.data)
40 >
41 > # Model performance
42 > RMSE(predictions, test.data$medv)
43 [1] 5.637608
44 > R2(predictions, test.data$medv)
45 [1] 0.6347266
46 >
47 > (ggplot(train.data, aes(lstat, medv)) + geom_point(cex=2)
48 +    + stat_smooth(method=lm, formula=y ~ poly(x, 5, raw=TRUE), cex=2)
49 +    + theme(text = element_text(size=20)))
```

Programme 10.2 Polynomial regression for the Boston housing dataset in R.

10.3 GENERALIZED LINEAR MODELS

In traditional linear regression models, the response variable y is assumed to be normally distributed, while the functional relationship between the predictors $x_1, \ldots, x_p$ and y is linear. However, many real-world applications require models to accommodate different types of response variables and nonlinear relationships. Enriching the potential applications of regression analysis, the vast class of *Generalized Linear Models* (GLMs) is prevalent in these scenarios. In these models, the response variables follow a distribution from the so-called *exponential dispersion family* (EDF), whose mean is connected, via the *link function*, with the *linear predictor* or *systematic component* $\eta = \tilde{x}^\top \beta$, which is a linear combination of the standard versions of the feature (explanatory) variables $x_1, \ldots, x_p$.

Let us go into more detail of the GLM. For simplicity, we consider a one-dimensional setting only. We first briefly introduce some fundamental properties of the EDF. The pdf (or pmf in the discrete case) of a distribution in the EDF takes the following general form:

$$f(y; \theta, \phi) = \exp\left(\frac{y\theta - b(\theta)}{\phi} + S(y, \phi)\right), \tag{10.8}$$

where θ is called the *natural parameter*, and $b(\theta) \in C^2$ and $S(y, \phi)$ are known functions. From (10.8), we can see that EDF is a sub-class of the exponential family where the sufficient statistic of θ equals y divided by the scale parameter ϕ. Many commonly used distributions in financial and insurance analytics belong to the EDF, including but not limited to the normal, gamma, Poisson and binomial distributions. From (10.8), we can readily derive the mgf of y:

$$
\begin{aligned}
M_{\mathrm{y}}(t) = \mathbb{E}(\exp(t\mathrm{y})) &= \int \exp(ty)\exp\left(\frac{y\theta - b(\theta)}{\phi} + S(y,\phi)\right) dy \\
&= \int \exp\left(\frac{y(t\phi+\theta) - b(\theta)}{\phi} + S(y,\phi)\right) dy \\
&= \exp\left(\frac{b(t\phi+\theta) - b(\theta)}{\phi}\right) \int \exp\left(\frac{y(t\phi+\theta) - b(t\phi+\theta)}{\phi} + S(y,\phi)\right) dy \\
&= \exp\left(\frac{b(t\phi+\theta) - b(\theta)}{\phi}\right);
\end{aligned}
$$

where the last equation holds because S only depends on y and ϕ so that the integrand is another EDF density with a new natural parameter $t\phi + \theta$. The mean and variance of y can then be expressed as functions of θ as follows:

$$
\mu = \mathbb{E}(\mathrm{y}) = \left.\frac{dM_{\mathrm{y}}(t)}{dt}\right|_{t=0} = b'(t\phi+\theta)\exp\left.\left(\frac{b(t\phi+\theta) - b(\theta)}{\phi}\right)\right|_{t=0} = b'(\theta); \tag{10.9}
$$

and by noting that

$$
\begin{aligned}
\mathbb{E}\left(\mathrm{y}^2\right) &= \left.\frac{d^2 M_{\mathrm{y}}(t)}{dt^2}\right|_{t=0} \\
&= \left((b'(t\phi+\theta))^2 \exp\left(\frac{b(t\phi+\theta) - b(\theta)}{\phi}\right)\right. \\
&\qquad \left.\left. + \phi b''(t\phi+\theta)\exp\left(\frac{b(t\phi+\theta) - b(\theta)}{\phi}\right)\right)\right|_{t=0} \\
&= (b'(\theta))^2 + \phi b''(\theta),
\end{aligned}
$$

therefore $\sigma^2 = \mathbb{E}\left(\mathrm{y}^2\right) - \mu^2 = \phi b''(\theta) =: \phi V(\mu)$. In particular, the function V is called the *variance function*, which is a function of μ by taking $\theta = (b')^{-1}(\mu)$, if the inverse of b' exists, from (10.9), and so $V(\mu) = b''((b')^{-1}(\mu))$.

Another property of the EDF that is essential for insurance data analytics is the closure of the distribution upon averaging iid variables. If the risk of each exposure unit y_i is in the EDF, and are iid with parameters $(\theta, \phi, b(\cdot))$, and there are a total of ω units of exposure, then the mgf of the average risk per unit of exposure $\overline{\mathrm{y}} = \frac{1}{\omega}\sum_{i=1}^{\omega} \mathrm{y}_i$ is given by

$$
M_{\overline{\mathrm{y}}}(t) = \mathbb{E}\left(\exp\left(\frac{t}{\omega}\sum_{i=1}^{\omega}\mathrm{y}_i\right)\right) = \left(M_{\mathrm{y}}\left(\frac{t}{\omega}\right)\right)^{\omega} = \exp\left(\frac{b\left(\frac{t\phi}{\omega}+\theta\right) - b(\theta)}{\phi/\omega}\right),
$$

implying that $\bar{y}$ is still in the EDF with the same natural parameter θ and function $b(\cdot)$, but the scale parameter changes from ϕ to $\frac{\phi}{\omega}$.

On the other hand, the link function $g(\cdot)$ is a one-to-one function which maps $\mathbb{E}(y|\tilde{x} = \tilde{x})$ to $\eta = \tilde{x}^{\top}\beta$, i.e. $g(\mathbb{E}(y|\tilde{x} = \tilde{x})) = \eta$, or equivalently,

$$\mathbb{E}(y|\tilde{x} = \tilde{x}) = g^{-1}(\eta) = g^{-1}(\tilde{x}^{\top}\beta).$$

Note that the selection of the link function may be restricted by the model of the response variable y, as it affects the range of $\mathbb{E}(y|\tilde{x} = \tilde{x})$. For example, to model a Bernoulli distributed response variable as a GLM, its link function should be a mapping of $[0, 1] \to \mathbb{R}$. Moreover, we also want the link function to be able to explain the effect of the explanatory variables on the response.

One common choice of the link function is the *canonical link* function that satisfies

$$g \circ b'(\theta) = \theta. \tag{10.10}$$

Using (10.9), where $b'(\theta)$ stands for the conditional mean $\mathbb{E}(y|\tilde{x} = \tilde{x})$, we see that the canonical link function equates the linear predictor η with the natural parameter θ. The canonical link function not only possesses appropriate domain and range, but also usually provides a direct interpretation. Unless otherwise specified, all link functions used in this book refer to the canonical link functions.

As a side note in terms of model assessment for GLMs, a commonly adopted measure is *deviance*, defined as twice the difference of the loglikelihood of the *saturated model* (whose number of parameters equals the sample size, but not necessarily the empirical distribution) subtracted by that of the fitted model. In particular, for candidate models under the same distributional family, their corresponding saturated model is the same, and comparing deviances of the candidate models reduces to comparing their likelihoods.

We next introduce a financial example with an inappropriate use of ordinary regression, and then explain how to rescue the situation by using GLMs. To this end, we consider the dataset `fin-ratio.csv` that contains financial ratios of 680 securities listed in the main board of Hong Kong Stock Exchange in 2002, in which there are six financial variables, namely: *Earning Yield* (`EY`); *Cash Flow to Price* (`CFTP`); *logarithm of Market Value* (`ln_MV`); *Dividend Yield* (`DY`); *Book to Market Equity* (`BTME`); *Debt to Equity Ratio* (`DTE`). The information of these financial variables are publicly available, via *Yahoo Finance* (https://finance.yahoo.com/) for instance. Among these companies, there were 32 Blue Chips which used to be the Hang Seng Index (HSI) Constituent Stocks. The last column `HSI` is a binary variable indicating whether a stock is a Blue Chip or not. We want to find any possible relationship between `HSI` and these six financial variables. In Programme 10.3, using the built-in function `lm()` in **R**, we fit an ordinary regression model for `HSI` against these six financial variables. From the output, the p-value of `d$DTE` is the largest, so the coefficient of `DTE` is not significantly different from 0. We then exclude it, and fit the remaining five independent financial variables in another regression as in Programme 10.4.

```
1  > d <- read.csv("fin-ratio.csv")
2  > summary(lm(HSI~EY+CFTP+ln_MV+DY+BTME+DTE, data=d))
3
4  Call:
5  lm(formula = HSI ~ EY + CFTP + ln_MV + DY + BTME + DTE, data = d)
6
7  Residuals:
8       Min       1Q   Median       3Q      Max
9  -0.32104 -0.08546 -0.01672  0.05592  0.73866
10
11 Coefficients:
12              Estimate Std. Error t value Pr(>|t|)
13 (Intercept) -0.4591209  0.0268310 -17.112  < 2e-16 ***
14 EY          -0.0017172  0.0016181  -1.061  0.28896
15 CFTP        -0.0103792  0.0037321  -2.781  0.00557 **
16 ln_MV        0.0810286  0.0040887  19.818  < 2e-16 ***
17 DY          -0.0027336  0.0017826  -1.534  0.12561
18 BTME         0.0004798  0.0007938   0.604  0.54575
19 DTE          0.0010610  0.0018035   0.588  0.55655
20 ---
21 Signif. codes:  0 '***' 0.001 '**' 0.01 '*' 0.05 '.1' 0.1 ' ' 1
22
23 Residual standard error: 0.1689 on 673 degrees of freedom
24 Multiple R-squared:  0.3708,    Adjusted R-squared:  0.3652
25 F-statistic: 66.09 on 6 and 673 DF,  p-value: < 2.2e-16
```

Programme 10.3 Ordinary least square regression of HSI against the six financial variables via **R**.

```
1  > summary(lm(HSI~EY+CFTP+ln_MV+DY+BTME, data=d))
2
3  Call:
4  lm(formula = HSI ~ EY + CFTP + ln_MV + DY + BTME, data = d)
5
6  Residuals:
7       Min       1Q   Median       3Q      Max
8  -0.32132 -0.08534 -0.01732  0.05568  0.73824
9
10 Coefficients:
11              Estimate Std. Error t value Pr(>|t|)
12 (Intercept) -0.4581051  0.0267624 -17.117   <2e-16 ***
13 EY          -0.0017108  0.0016173  -1.058   0.2905
14 CFTP        -0.0102957  0.0037275  -2.762   0.0059 **
15 ln_MV        0.0809789  0.0040859  19.819   <2e-16 ***
16 DY          -0.0027259  0.0017816  -1.530   0.1265
17 BTME         0.0005074  0.0007921   0.641   0.5220
18 ---
19 Signif. codes: 0 '***' 0.001 '****' 0.01 '***' 0.05 '.' 0.1 ' ' 1
20
21 Residual standard error: 0.1688 on 674 degrees of freedom
22 Multiple R-squared:  0.3704,    Adjusted R-squared:  0.3658
23 F-statistic: 79.32 on 5 and 674 DF,  p-value: < 2.2e-16
```

Programme 10.4 Ordinary least square regression of HSI against the remaining five financial variables (excluding DTE) via **R**.

We then see that the p-value of d$BTME remains large, and we can exclude this d$BTME to fit rest for another regression. We keep on excluding variables with the largest p-values step by step until all the listed p-values left behind are small (say, less than 0.1). This overall procedure is known as *backward elimination* for model selection. Finally, we arrive at the "ultimate" model in Programme 10.5.

```
1  > summary(lm(HSI~CFTP+ln_MV, data=d))
2
3  Call:
4  lm(formula = HSI ~ CFTP + ln_MV, data = d)
5
6  Residuals:
7       Min       1Q   Median       3Q      Max
8  -0.32409 -0.08559 -0.01729  0.05688  0.73488
9
10 Coefficients:
11              Estimate Std. Error t value Pr(>|t|)
12 (Intercept) -0.454781   0.026284 -17.303  < 2e-16 ***
13 CFTP        -0.012026   0.003475  -3.461 0.000573 ***
14 ln_MV        0.079630   0.004032  19.751  < 2e-16 ***
15 ---
16 Signif. codes:  0 '***' 0.001 '**' 0.01 '*' 0.05 '.' 0.1 ' ' 1
17
18 Residual standard error: 0.1689 on 677 degrees of freedom
19 Multiple R-squared:  0.3666,    Adjusted R-squared:  0.3648
20 F-statistic: 195.9 on 2 and 677 DF,  p-value: < 2.2e-16
```

Programme 10.5 The final regression model after backward elimination via R.

The least square regression model reads:

$$\texttt{HSI} = -0.454781 - 0.01026 \cdot \texttt{CFTP} + 0.07963 \cdot \texttt{ln\_MV}.$$

There are some serious concerns with this. Firstly, the predicted labels distribute continuously over a large range of values other than 0 or 1. Secondly, if the assumption of the normality of residuals were correct, the normal Q-Q plot should be close to a straight line, while all other residual plots should show a random pattern.

```
1  > reg <- lm(HSI~CFTP+ln_MV, data=d)        # save regression results
2  > names(reg)
3   [1] "coefficients"  "residuals"     "effects"       "rank"
4   [5] "fitted.values" "assign"        "qr"            "df.residual"
5   [9] "xlevels"       "call"          "terms"         "model"
6  >
7  > par(mfrow=c(3,2), cex.lab=2, cex.axis=2, cex.main=2, mar=c(5,5,4,4))
8  > hist(reg$fitted.values)                 # fitted values histogram
9  > hist(reg$residuals)                     # residuals histogram
10 > plot(reg$fitted.values, reg$residuals)# residuals vs fitted values
11 > qqnorm(reg$residuals)                   # qq-normal plot of residuals
12 > qqline(reg$residuals)                   # add reference line
13 > res <- as.ts(reg$residuals)             # change res to time series
14 > plot(res, lag(res), cex=1.5)            # residuals vs lag(residuals)
15 > plot(reg$residuals, cex=1.5)            # residuals vs index number
```

Programme 10.6 Analysis of the fitted residuals via R.

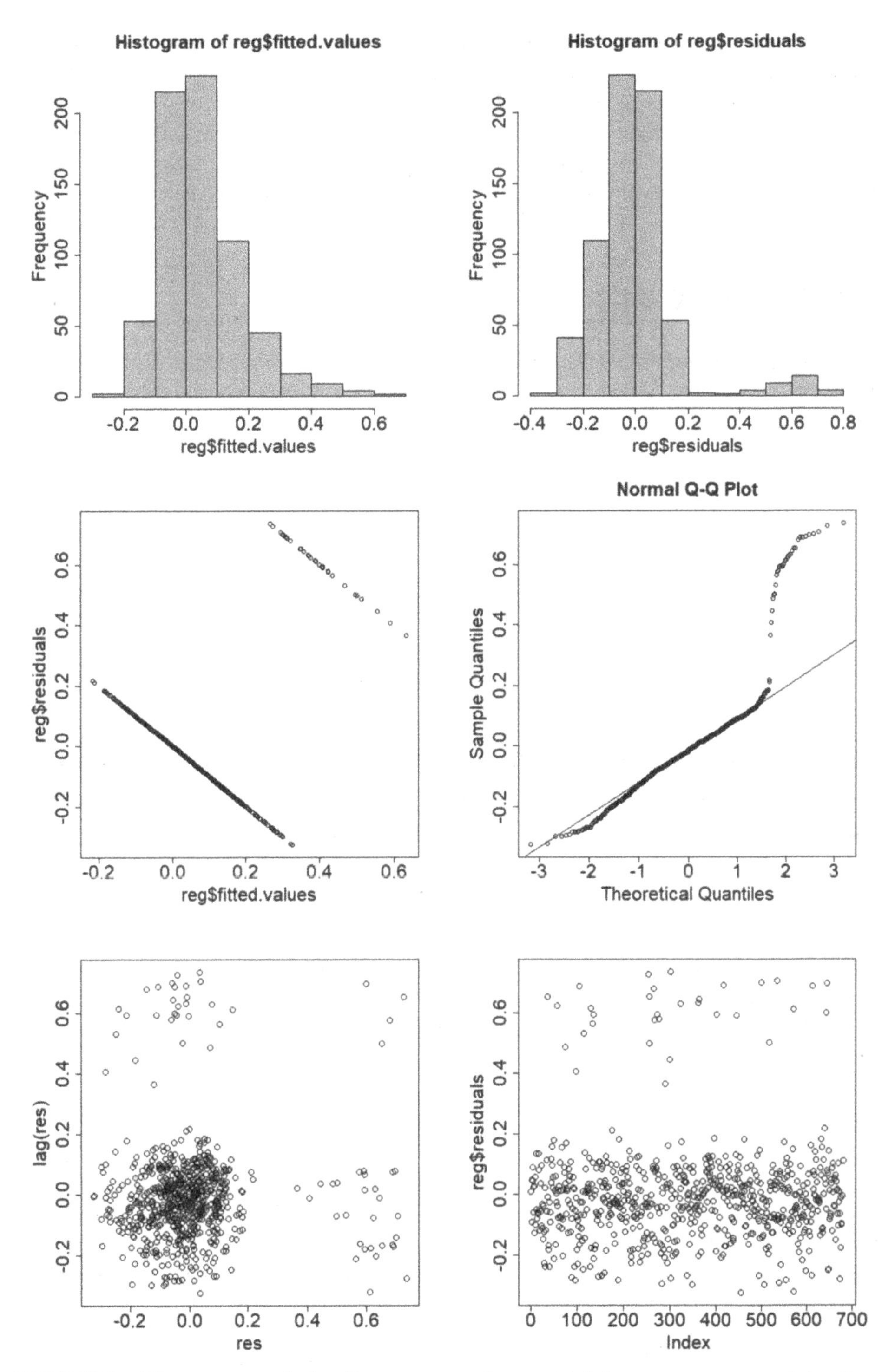

FIGURE 10.4 The output of the linear regression model for `HSI` generated by Programme 10.6.

From Figure 10.4, for the rightmost bottom plot, the residuals are mainly distributed around 0 and have another, secondary peak at larger positive values; the top left plot of residuals against fitted values indicates the secondary peak of large positive residual values is likely associated with those data with `HSI` = 1. In addition, the middle right Q-Q plot in Figure 10.4 also shows the heavy-tail nature of the fitted model. As a whole, the linear regression assumption is seriously violated as the low adjusted R^2 of 0.3648, as seen in the output in Programme 10.5, also implies.

10.4 LOGISTIC REGRESSION

The last example highlights the shortcoming of ordinary linear regression in the face of binary responses. A better approach is a commonly used special case of GLMs: *logistic regression*. We first verify that the Bernoulli response y_i with $\pi_i = \mathbb{P}(y_i = 1)$ is in the EDF; its pmf is:

$$\begin{aligned} f_{y_i}(y) &= \pi_i^y (1-\pi_i)^{1-y} \\ &= \exp\left(y \ln \pi_i + (1-y)\ln(1-\pi_i)\right) \\ &= \exp\left(y \ln\left(\frac{\pi_i}{1-\pi_i}\right) + \ln(1-\pi_i)\right) \\ &= \exp\left(y \ln\left(\frac{\pi_i}{1-\pi_i}\right) - \ln\left(1+\exp\left(\ln\left(\frac{\pi_i}{1-\pi_i}\right)\right)\right)\right). \end{aligned} \tag{10.11}$$

The pmf in (10.11) has the same form as (10.8), where the natural parameter $\theta = \ln\left(\frac{\pi_i}{1-\pi_i}\right)$, the scale parameter $\phi = 1$, and $b(\theta) = \ln(1+e^\theta)$. From (10.10), the canonical link function of a logistic regression can be derived, so that

$$g^{-1}(\theta) = b'(\theta) = \frac{e^\theta}{1+e^\theta}. \tag{10.12}$$

This canonical link function g in (10.12) is sometimes called the *log-odds ratio* of the probability of success π_i. By the properties of natural parameter $\theta = \eta = \tilde{x}_i^\top \beta$ and the Bernoulli distribution, (10.12) can be rewritten as

$$\pi_i = \mathbb{E}(y_i | \tilde{\mathbf{x}}_i = \tilde{x}_i) = \frac{\exp(\tilde{x}_i^\top \beta)}{1+\exp(\tilde{x}_i^\top \beta)}. \tag{10.13}$$

This function is also called the *logit transform* of $\tilde{x}_i^\top \beta$. Obviously, π_i always lies in $[0, 1]$, and this is consistent with its very nature as a probability. Clearly, (10.13) has a gradient of $\nabla_\beta \pi_i = \pi_i(1-\pi_i)\tilde{x}_i \in \mathbb{R}^{p+1}$. The likelihood function for the data (x_i, y_i), for $i = 1, \ldots, n$, is $L(\beta) = \prod_{i=1}^n \left(\pi_i^{y_i}(1-\pi_i)^{1-y_i}\right)$, hence the log-likelihood function is:

$$\ln L(\beta) = \sum_{i=1}^n \left(y_i \ln \pi_i + (1-y_i)\ln(1-\pi_i)\right). \tag{10.14}$$

Once the maximum likelihood estimate of $\boldsymbol{\beta}$ is obtained from (10.14), the predicted probability of success given $\boldsymbol{x}_i$ can be computed using (10.13); indeed, the gradient of $\ln L(\boldsymbol{\beta})$ in $\boldsymbol{\beta}$ is given by

$$\begin{aligned}
\nabla_{\boldsymbol{\beta}} \ln L(\boldsymbol{\beta}) &= \sum_{i=1}^{n} \left(y_i \frac{\nabla_{\boldsymbol{\beta}} \pi_i}{\pi_i} + (1 - y_i) \frac{(-\nabla_{\boldsymbol{\beta}} \pi_i)}{1 - \pi_i} \right) \\
&= \sum_{i=1}^{n} \left(y_i \frac{\pi_i(1 - \pi_i)}{\pi_i} \tilde{\boldsymbol{x}}_i + (1 - y_i) \left(\frac{-\pi_i(1 - \pi_i)}{1 - \pi_i} \tilde{\boldsymbol{x}}_i \right) \right) \\
&= \sum_{i=1}^{n} (y_i(1 - \pi_i)\tilde{\boldsymbol{x}}_i - (1 - y_i)\pi_i \tilde{\boldsymbol{x}}_i) = \sum_{i=1}^{n} (y_i - \pi_i)\tilde{\boldsymbol{x}}_i.
\end{aligned}$$

Generally, the root(s) of $\nabla_{\boldsymbol{\beta}} \ln L(\boldsymbol{\beta}) = \boldsymbol{0}$ cannot be solved explicitly due to the nonlinearity of the logit transform of $\tilde{\boldsymbol{x}}_i^{\top} \boldsymbol{\beta}$. Instead, we can efficiently find them numerically via the Gauss–Newton algorithm, see Appendix *10.A.1. Its detailed adaptation for logistic regression is put in Appendix *10.A.2. The derived method is called the *Iteratively Reweighted Least Squares* (IRLS); see formula (10.26) for the iteration scheme.

Next, we illustrate the effectiveness of logistic regression on the HSI example. **R** has a built-in function `glm()` (standing for generalized linear model) for performing logistic regression, whose output format is similar to that of `lm()`. The last parameter `binomial` specifies the *link function* of logistic regression in `glm()`. In Programme 10.7, the MLEs of the coefficients are generally large, this may be caused by the presence of many outliers.

```
1  > summary(glm(HSI~EY+CFTP+ln_MV+DY+BTME+DTE, data=d, family=binomial))
2
3  Call:
4  glm(formula = HSI ~ EY + CFTP + ln_MV + DY + BTME + DTE,
5       family = binomial, data = d)
6
7  Deviance Residuals:
8      Min       1Q   Median       3Q      Max
9    -8.49     0.00     0.00     0.00     8.49
10
11 Coefficients:
12                 Estimate Std. Error    z value Pr(>|z|)
13 (Intercept) -4.121e+15  1.066e+07 -386410689   <2e-16 ***
14 EY           1.516e+13  6.431e+05   23570644   <2e-16 ***
15 CFTP        -6.364e+13  1.483e+06  -42902735   <2e-16 ***
16 ln_MV        4.945e+14  1.625e+06  304287297   <2e-16 ***
17 DY          -1.144e+14  7.085e+05 -161536188   <2e-16 ***
18 BTME        -7.907e+12  3.155e+05  -25063060   <2e-16 ***
19 DTE          8.744e+12  7.168e+05   12198713   <2e-16 ***
20 ---
21 Signif. codes:  0 '***' 0.001 '**' 0.01 '*' 0.05 '.' 0.1 ' ' 1
22
23 (Dispersion parameter for binomial family taken to be 1)
```

```
24
25     Null deviance:  258.08  on 679  degrees of freedom
26 Residual deviance: 1153.40  on 673  degrees of freedom
27 AIC: 1167.4
28
29 Number of Fisher Scoring iterations: 18
```

Programme 10.7 Fitted coefficients of the logistic regression for HSI via R.

```
1 > lreg <- glm(HSI~EY+CFTP+ln_MV+DY+BTME+DTE, data=d, family=binomial)
2 > pr <- (lreg$fitted.values > 0.5)         # pr=TRUE if fitted > 0.5
3 > table(pr, d$HSI)
4
5 pr        0   1
6   FALSE 634   2
7   TRUE   14  30
```

Programme 10.8 Output and prediction of the logistic regression for HSI via R.

In Programme 10.8, the `glm()` output is saved in `lreg`, which contains many items. An important one is `fitted.values`, which are the estimated success probabilities $\hat{\pi}_i$ from (10.13) for these 680 stocks in 2002. The extremely large MLEs of the coefficients mean the fitted values are very close to 0 or 1. To obtain predicted labels, we can assign a value of `pr=True` if the fitted value is greater than or equal to 0.5; otherwise `pr=False` is placed. Finally, we generate a confusion matrix / cross tabulation table of `pr` versus `HSI` to see how well this logistic regression model predicts. We see there are $634 + 30 = 664$ correct classifications with $14 + 2 = 16$ misclassifications. There are 2 stocks with `HSI` $= 1$ being misclassified as `pr=False`, while 14 stocks with `HSI` $= 0$ are misclassified as `pr=True`. The correct classification rate is therefore $664/680 = 97.65\%$.

As mentioned before, regarding Programme 10.7, the large coefficients could be caused by outliers. To get a more robust model with smaller coefficients, we may use the Mahalanobis distance (defined in (3.3) in Section 3.3) to detect outliers, remove them, then fit a more reasonable model. In Programme 10.9, we introduce a function `mdist()` in **R** to compute the Mahalanobis distance. For the dataset of `fin-ratio.csv`, we separate the data into two groups, `d0` of those with `HSI` $= 0$ and `d1` of those with `HSI` $= 1$, as shown in Programme 10.10. We detect and throw away the outliers in `d0`, while leaving `d1` untouched as it contains too few cases.

```
1 > mdist <- function(x) {
2 +   t <- as.matrix(x)                    # transform x to a matrix
3 +   m <- apply(t, 2, mean)               # compute column mean
4 +   s <- var(t)                          # compute sample cov. matrix
5 +   mahalanobis(t, m, s)                 # built-in mahalanobis func.
6 + }
```

Programme 10.9 Code for `mdist()` to compute the Mahalanobis distances of all data in **R**.

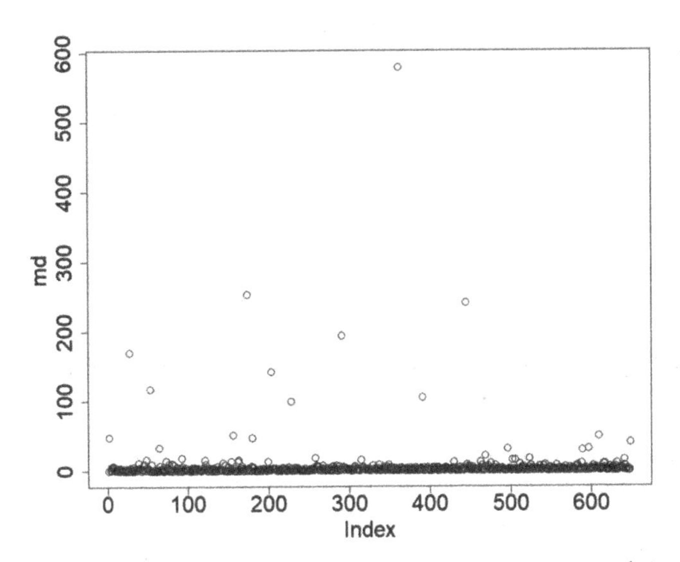

FIGURE 10.5 Scatter plot of the Mahalabonis distances, generated in Programme 10.10.

```
1  > d <- read.csv("fin-ratio.csv")          # read in dataset
2  > d0 <- d[d$HSI==0,]                       # select HSI=0
3  > d1 <- d[d$HSI==1,]                       # select HSI=1
4  > dim(d0)
5  [1] 648   7
6  > x <- d0[,1:6]                            # save d0 to x
7  > md <- mdist(x)                           # compute mdist
8  > plot(md, cex=1.5)
```

Programme 10.10 Computing Mahalabonis distances using `mdist()` in Programme 10.9 for all non-HSI constituent stocks via **R**.

In Figure 10.5, points with a large distance are potential outliers which will be discarded. But what is an appropriate cut-off value? Recall that for an iid sample of $x_1, \ldots, x_n$ following $\mathcal{N}_p(\mu, \Sigma)$, the Mahalanobis distance D_i for each x_i, for $i = 1, 2, \ldots, n$, satisfies $D_i^2 = (x_i - \overline{x})^\top S^{-1}(x_i - \overline{x}) \sim \chi_p^2$ approximately for large n. Moreover, these D_i^2's are also approximately independent of each other for large n. Therefore, based on this observation, we can remove the stocks in `d0` with Mahalanobis distances exceeding the top 1% percentile of the chi-square distribution with degrees of freedom $p = 6$; and then combine this resulting family with the data of `d1` to obtain the cleansed dataset `d3`; see the implementation in Programme 10.11.

```
1  > (c <- qchisq(0.99, df=6))               # p=6, type-I error = 0.01
2  [1] 16.81189
3  >
4  > d2 <- d0[md<c,]                         # select case in d0 with md<c
5  > dim(d2)                                 # throw away 648-626=22 cases
6  [1] 626   7
7  > d3 <- rbind(d1, d2)                     # combine d1 with d2
8  > dim(d3)
9  [1] 658   7
10 > # save the cleansed dataset
11 > write.csv(d3, file="fin-ratio_cleansed.csv", row.names=FALSE)
```

Programme 10.11 Cleansing the `fin-ratio.csv` dataset by dropping the top 1% outliers of `d0` via R.

Next, we fit a logistic regression to this cleansed dataset `d3`, and remove financial variables with coefficients having large p-values by backward eliminations. A final model can eventually be obtained following Programme 10.12. From the corresponding confusion matrix, the correct classification rate has increased to $653/658 = 99.24\%$ on this reduced training dataset. In particular, throwing away the outliers actually gives a simpler model with smaller coefficients.

```
1  > summary(glm(HSI~CFTP+ln_MV+BTME, data=d3, family=binomial))
2
3  Call:
4  glm(formula = HSI ~ CFTP + ln_MV + BTME, family = binomial, data = d3)
5
6  Deviance Residuals:
7        Min        1Q     Median        3Q       Max
8  -2.37704  -0.00019  -0.00001   0.00000   1.73796
9
10 Coefficients:
11              Estimate Std. Error z value Pr(>|z|)
12 (Intercept) -69.9309    21.3821  -3.271  0.00107 **
13 CFTP         -3.0376     1.2178  -2.494  0.01262 *
14 ln_MV         7.2561     2.2284   3.256  0.00113 **
15 BTME          1.3222     0.6418   2.060  0.03940 *
16 ---
17 Signif. codes:  0 '***' 0.001 '**' 0.01 '*' 0.05 '.' 0.1 ' ' 1
18
19 (Dispersion parameter for binomial family taken to be 1)
20
21     Null deviance: 255.920  on 657  degrees of freedom
22 Residual deviance:  23.087  on 654  degrees of freedom
23 AIC: 31.087
24
25 Number of Fisher Scoring iterations: 12
26
27 Warning message:
28 glm.fit: fitted probabilities numerically 0 or 1 occurred
29 >
30 > lreg <- glm(HSI~CFTP+ln_MV+BTME, data=d3, family=binomial)
```

```
31 Warning message:
32 glm.fit: fitted probabilities numerically 0 or 1 occurred
33 > pr <- (lreg$fitted.values > 0.5)      # pr=TRUE if fitted > 0.5
34 > table(pr, d3$HSI)
35
36 pr        0   1
37   FALSE 624   3
38   TRUE    2  29
```

Programme 10.12 A simpler logistic regression model for the cleansed dataset d3 via R.

10.4.1 Logistic Regression (Threshold Type) with Dummy Variables

Like ordinary regression, there could be categorical variables as independent (feature) variables in a logistic regression. In Chapter 13, by using a decision tree, we shall see that `ln_MV>9.4776` serves as an important indicator for classifying stocks with `HSI` = 0 or 1. Taking this as given, we propose to use a dummy (categorical) variable $g := 1 + \mathbb{1}_{\{\ln\_MV>9.4776\}}$, and fit a logistic regression with this g in addition to the other five financial variables, apart from MV, together with their interaction terms. This primitive threshold logistic regression model is shown in Programme 10.13.

```
1  > # create dummy variable g=2 if ln_MV>9.4776 and g=1 otherwise
2  > g <- (d3$ln_MV > 9.4776) + 1
3  > summary(glm(HSI~EY+CFTP+DY+BTME+DTE+g+EY*g+CFTP*g+DY*g+BTME*g+DTE*g,
4  +              data=d3, binomial))
5
6  Call:
7  glm(formula = HSI ~ EY + CFTP + DY + BTME + DTE + g + EY * g +
8      CFTP * g + DY * g + BTME * g + DTE * g, family = binomial,
9      data = d3)
10
11 Deviance Residuals:
12     Min       1Q    Median       3Q      Max
13 -1.9745  -0.1049  -0.0852  -0.0755   3.3153
14
15 Coefficients:
16             Estimate Std. Error z value Pr(>|z|)
17 (Intercept) -21.0469     5.6643  -3.716 0.000203 ***
18 EY           29.0955    38.2277   0.761 0.446592
19 CFTP         -0.1916     3.8931  -0.049 0.960745
20 DY            1.0979     0.8182   1.342 0.179637
21 BTME          1.4157     1.5673   0.903 0.366398
22 DTE          -0.3659     0.8731  -0.419 0.675166
23 g            15.2686     5.3888   2.833 0.004606 **
24 EY:g        -28.8939    38.2187  -0.756 0.449641
25 CFTP:g        0.1269     3.6225   0.035 0.972054
26 DY:g         -0.9840     0.7821  -1.258 0.208316
27 BTME:g       -1.3220     1.4948  -0.884 0.376499
28 DTE:g         0.2058     0.5665   0.363 0.716429
29 ---
```

```
30 Signif. codes:  0 '***' 0.001 '**' 0.01 '*' 0.05 '.' 0.1 ' ' 1
31
32 (Dispersion parameter for binomial family taken to be 1)
33
34     Null deviance: 255.920  on 657  degrees of freedom
35 Residual deviance:  48.391  on 646  degrees of freedom
36 AIC: 72.391
37
38 Number of Fisher Scoring iterations: 10
```

Programme 10.13 First primitive threshold model with the additional dummy variable $g = 1 + \mathbb{1}_{\{\ln\_MV>9.4776\}}$ via R.

The coefficient of `CFTP*g` has the largest p-value (0.9721), and that of `CFTP` has the second largest (0.9607). We can then exclude `CFTP` and `CFTP*g` in the next regression fitting; see Programme 10.14 for the revised model.

```
1  > summary(glm(HSI~EY+DY+BTME+DTE+g+EY*g+DY*g+BTME*g+DTE*g,
2  +             data=d3, binomial))
3
4  Call:
5  glm(formula = HSI ~ EY + DY + BTME + DTE + g + EY * g + DY *
6      g + BTME * g + DTE * g, family = binomial, data = d3)
7
8  Deviance Residuals:
9      Min       1Q   Median       3Q      Max
10 -1.9738  -0.1052  -0.0855  -0.0754   3.3138
11
12 Coefficients:
13             Estimate Std. Error z value Pr(>|z|)
14 (Intercept) -21.0169     5.5409  -3.793 0.000149 ***
15 EY           28.6141    27.2787   1.049 0.294200
16 DY            1.0968     0.7752   1.415 0.157089
17 BTME          1.4089     1.4991   0.940 0.347279
18 DTE          -0.3558     0.8474  -0.420 0.674536
19 g            15.2425     5.2592   2.898 0.003752 **
20 EY:g        -28.4191    27.2658  -1.042 0.297274
21 DY:g         -0.9858     0.7396  -1.333 0.182540
22 BTME:g       -1.3146     1.4225  -0.924 0.355404
23 DTE:g         0.2015     0.5513   0.366 0.714692
24 ---
25 Signif. codes:  0 '***' 0.001 '**' 0.01 '*' 0.05 '.' 0.1 ' ' 1
26
27 (Dispersion parameter for binomial family taken to be 1)
28
29     Null deviance: 255.920  on 657  degrees of freedom
30 Residual deviance:  48.398  on 648  degrees of freedom
31 AIC: 68.398
32
33 Number of Fisher Scoring iterations: 9
```

Programme 10.14 Threshold model with `CFTP*g` and `CFTP` removed via R.

In Programme 10.14, the coefficient DTE*g has the largest p-value (0.7147) and that of DTE itself also has the second largest (0.6745), so we exclude DTE and DTE*g in the next trial. We keep on excluding variables with large p-values, and after several rounds we finally obtain the ultimate model as shown in Programme 10.15. This threshold model is simpler than that in Programme 10.12, yet its classification power remains as good as before, and it reads:

$$\ln\left(\frac{\hat{\pi}}{1-\hat{\pi}}\right) = -18.6304 + 12.9236\text{g} + 1.4036\text{DY} - 1.2860(\text{g}^*\text{DY}),$$

or equivalently,

$$\ln\left(\frac{\hat{\pi}}{1-\hat{\pi}}\right) = \begin{cases} -5.6944 + 0.1176\text{DY}, & \text{for } \ln\_\text{MV} \leq 9.4776; \\ 7.2416 - 1.1684\text{DY}, & \text{for } \ln\_\text{MV} > 9.4776; \end{cases}$$

where π is the probability that HSI = 1 given the features. When ln_MV > 9.4776, the intercept is bigger than that if ln_MV ≤ 9.4776. Thus the probability that HSI = 1 looks higher when ln_MV is large, as it should.

```
1   > summary(glm(HSI~DY+g+DY*g, data=d3, binomial))
2
3   Call:
4   glm(formula = HSI ~ DY + g + DY * g, family = binomial, data = d3)
5
6   Deviance Residuals:
7       Min        1Q    Median       3Q       Max
8   -2.3141   -0.1020   -0.0815   -0.0815    3.3351
9
10  Coefficients:
11                Estimate Std. Error z value Pr(>|z|)
12  (Intercept) -18.6304     3.4117  -5.461 4.74e-08 ***
13  DY            1.4036     0.6485   2.165   0.0304 *
14  g            12.9236     3.1251   4.135 3.54e-05 ***
15  DY:g         -1.2860     0.6082  -2.115   0.0345 *
16  ---
17  Signif. codes:  0 '***' 0.001 '**' 0.01 '*' 0.05 '.' 0.1 ' ' 1
18
19  (Dispersion parameter for binomial family taken to be 1)
20
21      Null deviance: 255.92  on 657  degrees of freedom
22  Residual deviance:  51.17  on 654  degrees of freedom
23  AIC: 59.17
24
25  Number of Fisher Scoring iterations: 8
```

```
26
27  >
28  > threshold_lreg <- glm(HSI~DY+g+DY*g, data=d3, binomial)
29  > pr <- (threshold_lreg$fit>0.5)
30  > table(pr, d3$HSI)
31
32  pr        0   1
33    FALSE 624   3
34    TRUE    2  29
```

Programme 10.15 The final threshold logistic regression model via **R**.

10.4.2 Decision Threshold for Binary Logistic Regression

In logistic regression, casual users may take an observation, with a feature vector $\boldsymbol{x}$, and classify it to group c_1 if the predicted probability satisfies $\hat{\mathbb{P}}(y = c_1|\boldsymbol{x}) \geq 0.5$. However, this decision threshold of 0.5 is not always appropriate. For instance, in a medical diagnosis of a rare disease, a considerable number of infected patients may receive a negative diagnosis result that delays treatments if the decision threshold is set too high. On the other hand, the diagnosis becomes meaningless if the threshold is set too low. It is crucial to select an appropriate deciding probability u, so that a patient with a feature vector $\boldsymbol{x}$ is classified as positive when the predicted probability is greater than u; or negative otherwise. Another example, from finance, is the estimation of the probability of default of a client and the classification of his credit-worthiness via binary logistic regression. For a loan application, the decision threshold u can be chosen suitably as follows: first, consider the following table of potential losses incurred for lending:

Decision \ Borrower	Non-default (actual $y = 0$)	Default (actual $y = 1$)
Accept (predict $y = 0$)	correct	error (cost = \$$u_2$)
Reject (predict $y = 1$)	error (cost = \$$u_1$)	correct

There are two costs of misclassification errors u_1 and u_2, and they need not be equal. Let $p(\boldsymbol{x}) = \mathbb{P}(\text{Default}|\boldsymbol{x})$ and $1 - p(\boldsymbol{x}) = \mathbb{P}(\text{Non-default}|\boldsymbol{x})$. The expected loss of rejecting a non-default borrower is therefore given by $u_1(1 - p(\boldsymbol{x}))$; while the expected loss of accepting a default borrower is given by $u_2 p(\boldsymbol{x})$. On the average, it is sensible to reject an application if $u_1(1 - p(\boldsymbol{x})) < u_2 p(\boldsymbol{x})$, or equivalently $p(\boldsymbol{x}) > \frac{u_1}{u_1+u_2} =: u$, which is a desired *decision threshold*. In the special case when \$$u_1$ equals \$$u_2$, u is simply $\frac{1}{2}$.

10.4.3 Multinomial Logistic Regression for Categorical Response

Binary logistic regressions can be easily extended to *Multinomial Logistic Regression* in which the response y is categorical, with more than two categories. For instance, let y take k categories, namely $c_1, \cdots, c_k$, and we model the corresponding probabilities by:

$$\mathbb{P}(y_i = c_1|\boldsymbol{x}_i) = \frac{1}{\xi}\exp(\boldsymbol{x}_i^\top \boldsymbol{\beta}_1), \ldots, \mathbb{P}(y_i = c_k|\boldsymbol{x}_i) = \frac{1}{\xi}\exp(\boldsymbol{x}_i^\top \boldsymbol{\beta}_k),$$

where $\xi := \exp(\boldsymbol{x}_i^\top \boldsymbol{\beta}_1) + \cdots + \exp(\boldsymbol{x}_i^\top \boldsymbol{\beta}_k)$. These $\boldsymbol{\beta}_j$'s cannot be uniquely estimated unless one of them is standardized to be zero. Fortunately, it does not matter which $\boldsymbol{\beta}_j$ is chosen as zero. Setting $\boldsymbol{\beta}_1 = \mathbf{0}$, the model becomes:

$$\mathbb{P}(y_i = c_1|\boldsymbol{x}_i) = \frac{1}{\xi}, \ \mathbb{P}(y_i = c_2|\boldsymbol{x}_i) = \frac{\exp(\boldsymbol{x}_i^\top \boldsymbol{\beta}_2)}{\xi}, \ldots, \mathbb{P}(y_i = c_k|\boldsymbol{x}_i) = \frac{\exp(\boldsymbol{x}_i^\top \boldsymbol{\beta}_k)}{\xi}, \tag{10.15}$$

where $\xi = 1 + \exp(\boldsymbol{x}_i^\top \boldsymbol{\beta}_2) + \cdots + \exp(\boldsymbol{x}_i^\top \boldsymbol{\beta}_k)$, and the group $y = c_1$ serves as the baseline. Consider the likelihood function

$$L(\boldsymbol{\beta}_2, \boldsymbol{\beta}_3, \ldots, \boldsymbol{\beta}_k|\boldsymbol{x}_1, \boldsymbol{x}_2, \cdots, \boldsymbol{x}_n) := \prod_{i=1}^{n}\left(\xi^{-1}\prod_{j=2}^{k}\exp\left(\boldsymbol{x}_i^\top \boldsymbol{\beta}_j\right)^{\mathbb{1}_{\{y_i=c_j\}}}\right), \tag{10.16}$$

where for $j = 2, \ldots, k$,

$$\mathbb{1}_{\{y_i=c_j\}} = \begin{cases} 0, & \text{if } y_i \neq c_j; \\ 1, & \text{if } y_i = c_j. \end{cases}$$

These $\boldsymbol{\beta}_j$'s can be estimated by maximum likelihood estimation, numerically via the Gauss–Newton algorithm similar to those in Section 10.1 for binary responses. In particular, the IRLS in Appendix *10.A.2 can also be extended.

In **R**, there is a built-in function `multinom()`, in the `nnet` (standing for neural network) library, which estimates the parameters in a multinomial logistic regression model. Referring to Programme 10.16, this example uses Fisher's famous *Iris flower dataset*[8]. The built-in dataset `iris`[9] contains information about a sample of

[8] The Iris flower dataset was collected by the renowned British statistician and biologist *Ronald Fisher* in 1936 [6], during his study of iris flowers and their related species in the Gaspé Peninsula. The data consists of features such as the length and the width of the sepal and petal of each sample. Based on his understanding, he classified the samples into different species according to their features. Nowadays, the Iris data is often used to demonstrate the training and testing of a classifier.

[9] Iris function — R Documentation. Retrieved from https://www.rdocumentation.org/packages/datasets/versions/3.6.2/topics/iris.

FIGURE 10.6 Photo of an Iris virginica flower; original picture from https://en.m.wikipedia.org/wiki/File:Iris_virginica.jpg.

Iris flowers in five columns. The first four columns are the measurements of *Sepal Length*, *Sepal Width*, *Petal Length* and *Petal Width*, respectively; see Figure 10.6. The last column indicates which of the three different species of flower is (1 - `setosa`, 2 - `versicolor`, 3 - `virginica`). There are 150 observations with 50 observations from each species. We can estimate $\mathbb{P}(y_i = c_j|\boldsymbol{x}_i)$, for $j = 1, 2, 3$, according to (10.15) and predict y belongs to group c_j if $\hat{\mathbb{P}}(y_i = c_j|\boldsymbol{x}_i)$ is the maximum for this j. **R** has a built-in function `predict()` to obtain these predictions. Referring to Programme 10.16 for the cross tabulation table, we see there are only two misclassification cases, and the misclassification rate is just $2/150 = 1.33\%$.

```
1  > library(nnet)                                # load nnet
2  >
3  > names(iris)
4  [1] "Sepal.Length" "Sepal.Width"  "Petal.Length" "Petal.Width"
5  [5] "Species"
6  > mnl <- multinom(Species~., data=iris) # perform MNL
7  # weights:  18 (10 variable)
8  initial  value 164.791843
9  iter  10 value 16.177348
10 iter  20 value 7.111438
11 iter  30 value 6.182999
12 iter  40 value 5.984028
13 iter  50 value 5.961278
14 iter  60 value 5.954900
15 iter  70 value 5.951851
16 iter  80 value 5.950343
17 iter  90 value 5.949904
18 iter 100 value 5.949867
19 final  value 5.949867
20 stopped after 100 iterations
```

```
21  > summary(mnl)                                  # MNL summary
22  Call:
23  multinom(formula = Species ~ ., data = iris)
24
25  Coefficients:
26             (Intercept) Sepal.Length Sepal.Width Petal.Length Petal.Width
27  versicolor    18.69037    -5.458424   -8.707401     14.24477   -3.097684
28  virginica    -23.83628    -7.923634  -15.370769     23.65978   15.135301
29
30  Std. Errors:
31             (Intercept) Sepal.Length Sepal.Width Petal.Length Petal.Width
32  versicolor    34.97116     89.89215    157.0415     60.19170    45.48852
33  virginica     35.76649     89.91153    157.1196     60.46753    45.93406
34
35  Residual Deviance: 11.89973
36  AIC: 31.89973
37  >
38  > pred <- predict(mnl)                          # prediction
39  > table(pred, iris$Species)                     # tabulate results
40
41  pred         setosa versicolor virginica
42    setosa         50          0         0
43    versicolor      0         49         1
44    virginica       0          1        49
```

Programme 10.16 Multinomial logistic regression for the Iris flower dataset via **R**.

For our last example of logistic regression, we revisit the classification problem for the MNIST dataset mentioned in Subsection 6.1.2. To this, we first apply the multinomial logistic regression in Programme 10.17 via Python (resp. Programme 10.18 via **R**). There are $28 \times 28 = 784$ feature variables with $784 \times (10 - 1) = 7,056$ coefficient weights and $10 - 1 = 9$ biases.

```
1   import numpy as np
2   import tensorflow as tf
3   from sklearn.metrics import confusion_matrix
4   from datetime import datetime as dt
5
6   ((X_train, y_train),
7    (X_test, y_test)) = tf.keras.datasets.mnist.load_data()
8
9   N_train = len(X_train)                          # 60000 training samples
10  N_test = len(X_test)                            # 10000 test samples
11  X_train = X_train.reshape(N_train, -1).astype('float32') / 255.0
12  X_test = X_test.reshape(N_test, -1).astype('float32') / 255.0
13
14  start = dt.now()
15  from sklearn.linear_model import LogisticRegression
16
17  np.random.seed(4002)
18  clf = LogisticRegression(max_iter=10000)
19  clf.fit(X_train, y_train)
20  y_hat_logit = clf.predict(X_test)
```

```
21
22 # Table of predictions and prediction accuracy
23 tab = confusion_matrix(y_hat_logit, y_test)      # Confusion matrix
24 print(tab)
25 print(sum(tab.diagonal()) / len(y_test))          # Accuracy
26 print(dt.now() - start)                    # Logistic regression
```

```
1  [[ 955    0    6    4    1    9    8    1    9    9]
2   [   0 1110    9    1    3    2    3    7   11    8]
3   [   2    5  930   16    7    3    8   23    6    1]
4   [   4    2   14  925    3   35    2    7   22    9]
5   [   1    0   10    1  921   10    6    6    7   21]
6   [  10    2    3   23    0  777   16    1   29    7]
7   [   4    3   12    2    6   15  912    0   13    0]
8   [   3    2   10   10    5    6    2  947   10   21]
9   [   1   11   34   19    6   31    1    4  855    9]
10  [   0    0    4    9   30    4    0   32   12  924]]
11 0.9256
12 0:01:01.985001
```

Programme 10.17 Multinomial logistic regression for the MNIST dataset via Python.

From Programme 10.17, the accuracy on the test dataset is 92.56%, which is much greater than the 67.84% achieved by the EM algorithm. However, one should notice that the EM algorithm is an unsupervised learner, and it is reasonable to expect a lower accuracy. The difference in accuracies obtained via different programming languages is due to difference in values of initial seeds taken and variation in the numerical computation schemes used.

```
1  > library(keras)          # MNIST dataset
2  > library(nnet)
3  >
4  > # Load MNIST dataset
5  > mnist <- dataset_mnist()
6  > X_train <- mnist$train$x / 255
7  > y_train <- mnist$train$y / 255
8  > X_test <- mnist$test$x
9  > y_test <- mnist$test$y
10 >
11 > X_train <- matrix(X_train, nrow=length(y_train))
12 > X_test <- matrix(X_test, nrow=length(y_test))
13 >
14 > df_train <- data.frame(cbind(X_train, y_train))
15 > names(df_train) <- c(paste0('x_', 1:ncol(X_train)), 'y')
16 > df_test <- data.frame(X_test)
17 > names(df_test) <- paste0('x_', 1:ncol(X_train))
18 >
19 > start <- Sys.time()
20 > set.seed(4002)
21 > mnl <- multinom(y~., data=df_train, MaxNWts=1e6, maxit=200, trace=F)
```

```
22  > y_hat <- predict(mnl, newdata=df_test, type="class")
23  > table(y_hat, y_test)        # Confusion matrix
24       y_test
25  y_hat    0    1    2    3    4    5    6    7    8    9
26      0  952    0   10    4    3    8   10    3    8   11
27      1    0 1110   10    3    3    4    7   10   13    7
28      2    2    4  912   23    9    4   13   25    7    3
29      3    2    3   19  912    4   32    0    6   22   10
30      4    3    0   10    3  905   10   10    9    7   26
31      5    9    2    4   29    1  776   26    1   25   11
32      6    4    3    7    1    8   13  889    0   11    0
33      7    4    1   10   10    5    9    1  943   11   24
34      8    4   12   44   20   10   32    1    4  858    7
35      9    0    0    6    5   34    4    1   27   12  910
36  > mean(y_hat == y_test)       # Accuracy
37  [1] 0.9167
38  > Sys.time() - start          # Logistic regression with full training set
39  Time difference of 14.37146 mins
```

Programme 10.18 Multinomial logistic regression for the MNIST dataset via R.

10.5 POISSON REGRESSION

For insurance policies, a policyholder files claims to insurance companies once an insured contingent event occurs, so as to compensate his/her financial losses. It is essential for the insurance industry to model the random arrivals of claims within a specified period of time. Poisson regression is commonly used to model this type of count data by assuming the response variable y follows a Poisson distribution with a hazard rate, or mean arrival rate, of λ. As a representative example of GLMs, we note that a Poisson response y is in the EDF; indeed, the pmf of y can be written as:

$$f_{\mathrm{y}}(y) = e^{-\lambda}\frac{\lambda^y}{y!} = \exp(y\ln\lambda - \lambda - \ln(y!)),$$

so that the natural parameter is $\theta = \ln\lambda$, the scale parameter is $\phi = 1$, and $b(\theta) = e^{\theta}$. By using (10.10), the canonical link function satisfies

$$g^{-1}(\theta) = b'(\theta) = e^{\theta},$$

from which we obtain

$$\lambda = \mathbb{E}(\mathrm{y}|\mathbf{x} = \boldsymbol{x}) = \exp(\boldsymbol{x}^{\top}\boldsymbol{\beta}).$$

The likelihood function for a sample can be written as:

$$L(\boldsymbol{\beta}; \boldsymbol{x}_1, \cdots, \boldsymbol{x}_n, y_1, \ldots, y_n) = \prod_{i=1}^{n} \frac{\lambda_i^{y_i} e^{-\lambda_i}}{y_i!} = \prod_{i=1}^{n} \frac{\exp\left(y_i \boldsymbol{x}_i^{\top}\boldsymbol{\beta}\right)\exp\left(-e^{\boldsymbol{x}_i^{\top}\boldsymbol{\beta}}\right)}{y_i!},$$

while the log-likelihood is:

$$\ln L(\boldsymbol{\beta}; \boldsymbol{x}_1, \ldots, \boldsymbol{x}_n, y_1, \ldots, y_n) = \sum_{i=1}^{n} \left(y_i \boldsymbol{x}_i^\top \boldsymbol{\beta} - e^{\boldsymbol{x}_i^\top \boldsymbol{\beta}} - \ln(y_i!) \right).$$

In general, there is no closed-form solution to $\nabla_{\boldsymbol{\beta}} L(\boldsymbol{\beta}) = \mathbf{0}$, and we may find the numerical maximizer $\hat{\boldsymbol{\beta}}_{\text{MLE}}$ via the usual Gauss–Newton algorithm.

Next, we give an application of Poisson regression on predicting counts. The *1976 Panel Study of Income Dynamics Dataset*[10] contains the work and household characteristics of 753 individuals in the United States in 1976. Two of the recorded features are the number of children under 6 and the number of children between 6 and 18. We aim to figure out the relationship between the number of children under 18 and the rest of the factors. To this end, we build a Poisson regression model on the number of children under 18 against all other features in the following Programme 10.19.

```
1  > df <- read.csv('Mroz.csv')  # read the dataset
2  >
3  > # create a new feature: number of children under 18
4  > df$child <- df$child6 + df$child618
5  > # remove the '#','child6', 'child618' features
6  > df <- df[,-c(1,4,5)]
7  >
8  > # fit the 1st model with all features
9  > model_1 <- glm(formula=child~., family=poisson, data=df)
10 > summary(model_1)               # print summary of the 1st model
11
12 Call:
13 glm(formula = child ~ ., family = poisson, data = df)
14
15 Deviance Residuals:
16     Min       1Q   Median       3Q      Max
17 -2.6848  -1.0983  -0.2022   0.4994   3.3255
18
19 Coefficients:
20               Estimate Std. Error z value Pr(>|z|)
21 (Intercept)  3.411e+00  3.168e-01  10.767  < 2e-16 ***
22 workyes      1.962e-01  9.965e-02   1.969  0.04896 *
23 hoursw      -1.920e-04  5.983e-05  -3.208  0.00133 **
24 agew        -4.578e-02  8.406e-03  -5.446 5.15e-08 ***
25 educw       -1.814e-02  1.863e-02  -0.974  0.33020
26 hearnw      -1.047e-02  1.386e-02  -0.755  0.45013
27 wagew       -5.130e-03  1.916e-02  -0.268  0.78886
28 hoursh       1.349e-05  6.038e-05   0.223  0.82319
```

[10] The dataset can be downloaded from Kaggle: https://www.kaggle.com/datasets/utkarshx27/labor-supply-data.

```
29 ageh          -1.447e-02  7.904e-03  -1.830  0.06723 .
30 educh         -2.728e-03  1.333e-02  -0.205  0.83788
31 wageh         -1.797e-02  1.348e-02  -1.334  0.18235
32 income         8.473e-06  4.211e-06   2.012  0.04421 *
33 educwm        -3.053e-03  1.133e-02  -0.269  0.78757
34 educwf        -1.717e-02  1.044e-02  -1.646  0.09982 .
35 unemprate      1.689e-02  9.606e-03   1.758  0.07868 .
36 cityyes        3.827e-02  6.544e-02   0.585  0.55870
37 experience    -2.511e-02  5.318e-03  -4.723 2.33e-06 ***
38 ---
39 Signif. codes:  0 '***' 0.001 '**' 0.01 '*' 0.05 '.' 0.1 ' ' 1
40
41 (Dispersion parameter for poisson family taken to be 1)
42
43     Null deviance: 1185.45  on 752  degrees of freedom
44 Residual deviance:  838.91  on 736  degrees of freedom
45 AIC: 2246.5
46
47 Number of Fisher Scoring iterations: 5
```

Programme 10.19 The primitive Poisson regression model for the 1976 Panel Study via **R**.

In Programme 10.19, **R** reported that the relations between the number of children under 18 and most of the features are insignificant, so we first remove the insignificant features and then fit the model again. This backward elimination is repeated until all the remaining features are significant enough.

```
1  > # fit the 2nd model with the remaining features
2  > model_2 <- glm(formula=child~work+hoursw+agew+income+experience,
3  +                family=poisson, data=df)
4  >
5  > summary(model_2)              # print summary of the 2nd model
6
7  Call:
8  glm(formula = child ~ work + hoursw + agew + income + experience,
9      family = poisson, data = df)
10
11 Deviance Residuals:
12     Min       1Q   Median       3Q      Max
13 -2.6348  -1.0897  -0.2041   0.5075   3.3465
14
15 Coefficients:
16               Estimate Std. Error z value Pr(>|z|)
17 (Intercept)  2.948e+00  1.662e-01  17.735  < 2e-16 ***
18 workyes      1.178e-01  8.411e-02   1.401  0.16133
19 hoursw      -1.630e-04  5.445e-05  -2.993  0.00276 **
20 agew        -5.480e-02  4.169e-03 -13.144  < 2e-16 ***
21 income       1.313e-06  2.322e-06   0.566  0.57165
22 experience  -2.636e-02  5.212e-03  -5.058 4.24e-07 ***
23 ---
```

```
24 Signif. codes:  0 '***' 0.001 '**' 0.01 '*' 0.05 '.' 0.1 ' ' 1
25
26 (Dispersion parameter for poisson family taken to be 1)
27
28     Null deviance: 1185.45  on 752  degrees of freedom
29 Residual deviance:  858.99  on 747  degrees of freedom
30 AIC: 2244.6
31
32 Number of Fisher Scoring iterations: 5
33
34 >
35 > # fit the 3rd model with the remaining features
36 > model_3 <- glm(formula=child~hoursw+agew+experience,
37 +                family=poisson, data=df)
38 >
39 > summary(model_3)              # print summary of the 3rd model
40
41 Call:
42 glm(formula = child ~ hoursw + agew + experience, family = poisson,
43     data = df)
44
45 Deviance Residuals:
46     Min       1Q   Median       3Q      Max
47 -2.6873  -1.0950  -0.1949   0.5357   3.4549
48
49 Coefficients:
50               Estimate Std. Error z value Pr(>|z|)
51 (Intercept)  3.0141230  0.1585206  19.014  < 2e-16 ***
52 hoursw      -0.0001087  0.0000395  -2.752  0.00593 **
53 agew        -0.0550786  0.0041208 -13.366  < 2e-16 ***
54 experience  -0.0259452  0.0051628  -5.025 5.02e-07 ***
55 ---
56 Signif. codes:  0 '***' 0.001 '**' 0.01 '*' 0.05 '.' 0.1 ' ' 1
57
58 (Dispersion parameter for poisson family taken to be 1)
59
60     Null deviance: 1185.45  on 752  degrees of freedom
61 Residual deviance:  861.21  on 749  degrees of freedom
62 AIC: 2242.8
63
64 Number of Fisher Scoring iterations: 5
```

Programme 10.20 The final Poisson regression model for the 1976 Panel Study via R.

Programme 10.20 presents the ultimate model with remaining features `hoursw`, `agew`, and `experience`. Firstly, the feature `hoursw` stands for a wife's hours of work in 1975. Its coefficient is negative (–0.0001087), which implies that if the wife would work for longer hours, her expected number of children decreases. Secondly, the feature `experience` is the actual number of years of a wife's previous working experience in the labour market. Its negative coefficient (–0.0259452) indicates a negative correlation between the wife's working experience and her own expected number of children. Finally, we also observe a similar negative correlation between the response variable and the feature agew.

A reasonable explanation for the negative correlations between the count variable and the features is as follows: if a woman spent more time on work, she would have less time available for her family, and so she tends to have fewer kids. Finally, the feature agew represents the wife's age with its coefficient being negative (−0.0550786). This negativity could be attributed to the fact that as the mother got older, her children would eventually become adults, resulting in a decline in the number of children under 18; in addition, as the wife ages, she tends to not bear more children.

10.6 MODEL EVALUATION AND CONSIDERATIONS IN PRACTICE

10.6.1 Model Selection: Backward Elimination and Information Criterion

In ordinary linear regression or logistic regression, selecting the most relevant feature variables is a critical issue. One common procedure is backward elimination, a stepwise approach mentioned earlier, which starts with all the variables and then eliminates variables with the largest p-values one-by-one in each step. There are alternative methods for model selection. For instance, an *Information Criterion* (IC) can be applied to choose the suitable complexity and parameter estimation in one step of the regression model by minimizing the overall IC which is defined as $-2\ln L + k\theta$, where $\ln L$ is the log-likelihood function corresponding to the underlying statistical model, and θ is the effective degrees of freedom which is the number of free parameters in the model. In particular, for $k = 2$, the corresponding IC is called the *Akaike Information Criterion* (AIC) (see [1]). The motivation for AIC is that a decent model should have a good predictive power for the *out-sample* by using the likelihood approach based on the in-sample; and AIC serves as an asymptotic unbiased estimate for the expected predictive log-likelihood. For $k = \ln n$, it is called the *Bayesian Information Criterion* (BIC) (see [19]). For a finite set $\mathcal{M} = \{M\}$ of models, by using a Laplace integral approximation and Taylor expansion up to order 2, one can show that given a sample $x_1, \ldots, x_n$, the posterior probability of a model M, $\mathbb{P}(M|x_1, \ldots, x_n)$, is proportional to $\exp(-\text{BIC}/2)\mathbb{P}(M)$, asymptotically for a large sample size n. Therefore, the smaller the BIC, the more preferable the model; readers may refer to Appendix 12.A.2 in Chapter 12 for a more detailed introduction to BIC. In general, with more parameters included in the model, the log-likelihood function tends to increase in value and so $-2\ln L$ decreases, yet for a more complex model, it tends to overfit the in-sample and result in a larger $k\theta$. The IC penalizes an excessive inclusion of parameters, and the corresponding optimal model, with reasonable accuracy and model complexity, is indicated at the IC minimum.

Let us illustrate with the cleansed dataset of `fin-ratio_cleansed.csv`, which has had outliers removed by Programme 10.11, we choose the model with the smallest AIC. In Programme 10.21, the `step()` function in **R** implements the stepwise regression. Note that the ultimate model (with smallest AIC) is HSI = −69.931 − 3.038CFTP + 7.256ln_MV + 1.322BTME, which is the same as that determined by the backward elimination in Programme 10.12; and this agreement is generally the case.

```
1  > d3 <- read.csv("fin-ratio_cleansed.csv")
2  > lreg <- glm(HSI~., data=d3, binomial) # save the logistic reg
3  Warning message:
4  glm.fit: fitted probabilities numerically 0 or 1 occurred
5  > step(lreg)                          # perform stepwise selection
6  Start:  AIC=36.47
7  HSI ~ EY + CFTP + ln_MV + DY + BTME + DTE
8  
9          Df Deviance     AIC
10 - DTE    1   22.495  34.495
11 - DY     1   22.769  34.769
12 - EY     1   22.822  34.822
13 <none>       22.468  36.468
14 - BTME   1   27.628  39.628
15 - CFTP   1   30.586  42.586
16 - ln_MV  1  245.018 257.018
17 
18 Step:  AIC=34.5
19 HSI ~ EY + CFTP + ln_MV + DY + BTME
20 
21         Df Deviance     AIC
22 - DY     1   22.787  32.787
23 - EY     1   22.857  32.857
24 <none>       22.495  34.495
25 - BTME   1   27.756  37.756
26 - CFTP   1   30.589  40.589
27 - ln_MV  1  245.026 255.026
28 
29 Step:  AIC=32.79
30 HSI ~ EY + CFTP + ln_MV + BTME
31 
32         Df Deviance     AIC
33 - EY     1   23.087  31.087
34 <none>       22.787  32.787
35 - BTME   1   28.033  36.033
36 - CFTP   1   33.290  41.290
37 - ln_MV  1  245.776 253.776
38 
39 Step:  AIC=31.09
40 HSI ~ CFTP + ln_MV + BTME
41 
42         Df Deviance     AIC
43 <none>       23.087  31.087
44 - BTME   1   28.051  34.051
45 - CFTP   1   33.623  39.623
46 - ln_MV  1  246.700 252.700
47 
48 Call:  glm(formula = HSI ~ CFTP + ln_MV + BTME, family = binomial,
49     data = d3)
50 
```

```
51 Coefficients:
52 (Intercept)          CFTP         ln_MV        BTME
53     -69.931        -3.038         7.256       1.322
54
55 Degrees of Freedom: 657 Total (i.e. Null);  654 Residual
56 Null Deviance:        255.9
57 Residual Deviance: 23.09     AIC: 31.09
```

Programme 10.21 Model with the smallest AIC, using the cleansed HSI data via **R**.

10.6.2 Lift Chart for Assessing Performance

The fitted value (`lreg$fit`) in the logistic regression output gives the estimated probability of success of $\hat{\mathbb{P}}(y = 1|\boldsymbol{x})$. Sorting variables y's (`d3$HSI` in our example) according to the decreasing order of `lreg$fit` and assigning them with rankings, we expect most of the records with $y = 1$ to appear on the top of the list. However, due to the inaccuracy of the model, some with $y = 1$ may appear low in the list. We can then compute the cumulative percentage of $y = 1$ and plot them against their rankings, which gives a plot called a *lift chart* that indicates how well this logistic regression model predicts.

More precisely, suppose that the sample data are randomly arranged according to indexing subscript $i = 1, 2, \ldots, n$. We then sort them in descending order of $\hat{\pi}_i =$ `lreg$fit`, and label them with a new index (ranking) $i' = 1, 2, \ldots, n$. Programme 10.22 gives the lift chart of ranking i' versus $f_1(i') := \frac{1}{i'} \sum_{j=1}^{i'} \mathrm{HSI}_j$.

```
1  > ysort <- d3$HSI[order(lreg$fit, decreasing=T)] # sort y descending
2  > # ideal case of y
3  > yideal <- c(rep(1, sum(d3$HSI)), rep(0,length(d3$HSI)-sum(d3$HSI)))
4  >
5  > n <- length(ysort)                          # length of ysort
6  > perc1 <  cumsum(ysort)/(1:n)                # cum. percentage
7  > plot(perc1, type="l", col="blue", lwd=4)    # plot percentage
8  > abline(h=sum(d3$HSI)/n, lty=3, lwd=4)       # add horizontal baseline
9  > perc_ideal <- cumsum(yideal)/(1:n)          # ideal cum. percentage
10 > lines(perc_ideal, lty=2, col="red", lwd=4)# plot ideal case
11 > legend("topright", ncol=1, c("Actual", "Ideal", "Base"),
12 +        lty=1:3, lwd=4, col=c("blue", "red"), cex=1.5)
```

Programme 10.22 Plotting the lift chart of ranking versus the cumulative percentage using the cleansed dataset d3 from Programme 10.11 via **R**.

As a variant, instead of cumulative percentage, we can plot against the cumulative proportion, namely, i' versus $f_2(i') := \frac{\sum_{j=1}^{i'} \mathrm{HSI}_j}{\sum_{j=1}^{n} \mathrm{HSI}_j}$; see Programme 10.23.

```
1 > perc2 <- cumsum(ysort)/sum(ysort)                  # cum. percentage
2 > pop <- (1:n)/n                                     # x-coordinate
3 > plot(pop, perc2, type="l", col="blue", lwd=4) # plot percentage
4 > lines(pop, pop, lty=3, lwd=4)                      # add reference line
5 > # cumulative perc. of success for ideal case
```

```
6 > perc2_ideal <- cumsum(yideal)/sum(yideal)
7 > lines(pop, perc2_ideal, lty=2, col="red", lwd=4)# plot ideal case
8 > legend("bottomright", ncol=1, c("Actual", "Ideal", "Reference"),
9 +           lty=1:3, lwd=4,col=c("blue", "red", "black"), cex=1.5)
```

Programme 10.23 Plotting lift chart of ranking versus the cumulative proportion using the cleansed dataset d3 from Programme 10.11 via **R**.

For both resulting plots in Figure 10.7, we also plot the ideal cases (dashed lines in red), where all y = 1 records are ranked higher than y = 0 records against the sorted index (ranking) i'. Deviations of the actual cases (solid lines in blue) from the ideal cases reflect the inaccuracy of the logistic regression model; the more the discrepancy, the less reliable the model is.

10.6.3 Confusion Matrices and ROC Curve

When assessing the performance of a classifier, its accuracy as shown in the classification table is not a very robust measure, and can be sometimes misleading. To remedy its shortcomings, we can introduce other measurements such as *precision*, *recall* (a.k.a. *True Positive Rate* (TPR)), *False Positive Rate* (FPR) and the F_1 *score*[11]. For instance, in a medical test, we say the test result is positive if the subject is predicted to have a certain disease; see the possible outcomes in the confusion matrix (a.k.a. error matrix or cross-tabulation table, see [20]) in Table 10.1 as an example.

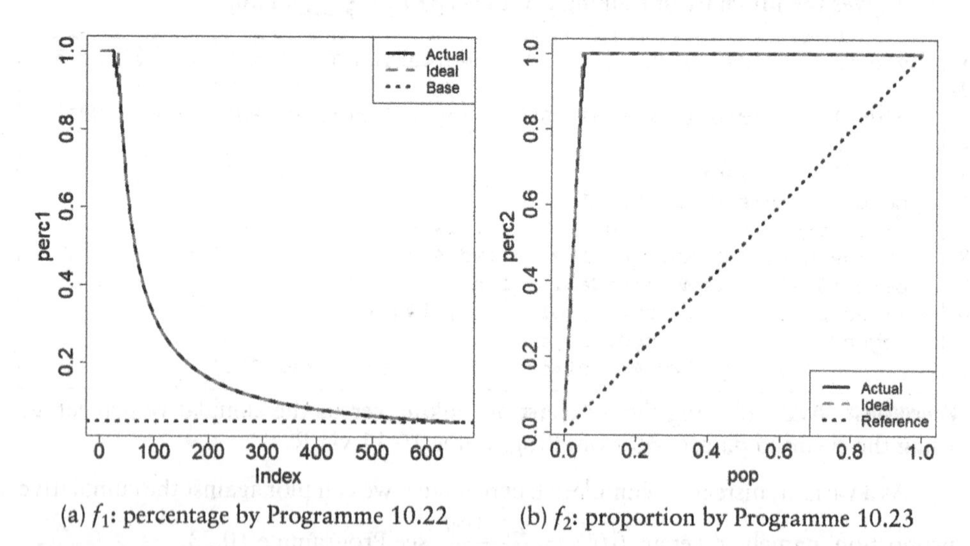

(a) f_1: percentage by Programme 10.22 (b) f_2: proportion by Programme 10.23

FIGURE 10.7 Using the cleansed dataset d3 from Programme 10.11, lift charts of ranking versus cumulative percentage and proportion, respectively.

[11] The name comes from the definition of the F-measure with a parameter $\beta \geq 0$ as $F_\beta = \frac{(\beta^2+1)\cdot\text{Precision}\cdot\text{Recall}}{\beta^2\cdot\text{Precision}+\text{Recall}}$ (see [3]); indeed, the F_1-score is obtained simply by setting $\beta = 1$.

TABLE 10.1 Confusion matrix for a medical test.

Test \ Truth	Disease	No Disease	Row Sum
Positive	True Positive (TP)	False Positive (FP)	TP + FP
Negative	False Negative (FN)	True Negative (TN)	FN + TN
Column Sum	TP + FN	FP + TN	Total Sum

A *true positive* (resp. *true negative*) is an outcome where the model correctly predicts the positive (resp. *negative*) class; while a *false positive* (resp. *false negative*) is an outcome where the model incorrectly predicts the positive (resp. negative) class but the subject actually belongs to the negative class.

Based on Table 10.1, the precision, recall (a.k.a. True Positive Rate, TPR), FPR (False Positive Rate), and F_1-score of the test are defined one-by-one as follows:

$$\text{Precision} := \frac{\text{TP}}{\text{TP} + \text{FP}}$$

is the proportion of correct test results among positive results;

$$\text{Recall (a.k.a. TPR)} := \frac{\text{TP}}{\text{TP} + \text{FN}}$$

is the proportion of observations correctly classified as positive among all observations which are actually positive;

$$\text{FPR} := \frac{\text{FP}}{\text{TN} + \text{FP}}$$

is defined as the proportion of observations that are incorrectly classified as positive among all negative observations (also see Figure 10.8 for an illustration of the definitions of TPR and FPR);

$$F_1 := \frac{2}{1/\text{Precision} + 1/\text{Recall}} = \frac{2 \times \text{Precision} \times \text{Recall}}{\text{Precision} + \text{Recall}}$$

is the harmonic mean of precision and recall. Its advantage can be seen by considering the test of a rare disease as in Table 10.2.

TABLE 10.2 A confusion matrix of a test of a rare disease.

Test \ Truth	Disease	No Disease	Row Sum
Positive	2	2	4
Negative	1	195	196
Column Sum	3	197	200

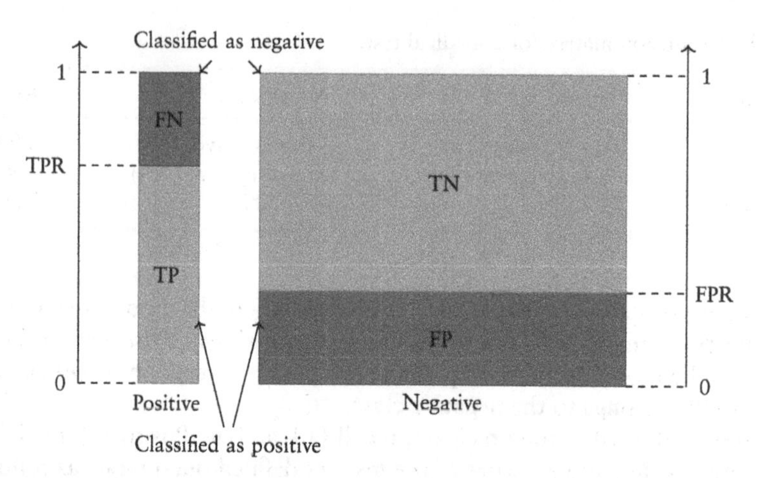

FIGURE 10.8 Illustration of TPR and FPR.

The overall accuracy of the test is $(2 + 195)/200 = 98.5\%$, which is high only due to the rareness of the disease. Indeed, most of the correct predictions come from the true negatives; yet the precision of the test is only $2/4 = 50\%$ while the recall (TPR) of the test is $2/3 = 66.7\%$. Together, these make the F_1-score become $4/7 = 57.14\%$; also, the FPR of the test is $2/197 = 1.02\%$.

Another use of TPR and FPR is to see whether a classification algorithm is effective enough. In particular, an efficient classifier should maximize the TPR as much as possible, while ideally at the same time it could reduce the FPR. This mutual relation can be better understood by using the *Receiver Operating Characteristic* (ROC) curve[12], which is a plot of TPR against FPR; one can check out [14] for a more thorough discussion.

Consider Table 10.3, which shows a training dataset of 20 observations (x_{i1}, x_{i2}) with their corresponding label y_i, for $i = 1, \ldots, 20$. Ten of the observations have $y_i = 1$ (in positive class) while the rest are $y_i = 0$ (in negative class). The last column shows the predicted probability $\hat{\pi}_i = \hat{\mathbb{P}}(\mathrm{y}_i = 1 | \mathrm{x}_{i1} = x_{i1}, \mathrm{x}_{i2} = x_{i2})$ given by a logistic regression in (x_{i1}, x_{i2}). The observations are arranged in descending order of $\hat{\pi}_i$.

Let the probability threshold u (Subsection 10.4.2) vary from positive infinity down to 0. We can observe the changes of TPR and FPR as functions of u. Note that a classification algorithm applied to a dataset gives one classifier for each value of the probability threshold u; in essence, it produces a collection of classifiers. For instance, for the extreme scenario of $u \geq 1$, all observations in Table 10.3 are simply classified as negative as clearly all $\hat{\pi}_i \leq 1 < u$. No observation is classified as positive, making both TPR = FPR = 0, which gives the following confusion matrix (Table 10.4).

[12] The ROC curve was first used during *World War II* for the analysis of radar signals, particularly after the *Pearl Harbor* incident in 1941, to increase the accuracy of detecting Japanese aircrafts from radar signals. The US army measured the effectiveness of radar receivers, which gives this curve its name.

TABLE 10.3 The dataset of 20 labeled samples with the respective predicted probabilities using binary logistic regression.

i	x_{i1}	x_{i2}	y_i	$\hat{\pi}_i$
1	24.43	6.95	1	0.98
2	8.84	11.92	1	0.91
3	18.69	−1.17	1	0.87
4	17.37	−0.07	1	0.86
5	4.77	11.66	1	0.85
6	0.83	10.74	0	0.73
7	1.57	8.51	1	0.69
8	10.07	−0.53	1	0.66
9	0.99	6.04	1	0.58
10	10.73	−4.88	0	0.53
11	11.16	−6.77	0	0.47
12	−11.21	14.64	0	0.46
13	−5.67	5.05	1	0.31
14	−0.06	−1.47	0	0.28
15	−9.25	6.74	0	0.26
16	1.05	−4.86	0	0.21
17	−12.35	5.61	0	0.16
18	−6.12	−2.41	1	0.12
19	−2.17	−14.40	0	0.04
20	−4.06	−15.70	0	0.02

TABLE 10.4 A confusion matrix of a test using the dataset in Table 10.3.

Test \ Truth	Disease	No Disease	Row Sum
Positive	0	0	0
Negative	10	10	20
Column Sum	10	10	20

Next, for $u = 0.95$, any observation is classified as positive whenever $\hat{\pi}_i > 0.95$. Then, only one of the 10 positive observations ($i = 1$) will be correctly classified, hence TPR $= 1/10 = 0.1$ as TP $= 1$ and FN $= 9$. On the other hand, there is no false positive, i.e. FP $= 0$, so FPR $= 0$.

From the above two particular values of u, we can see that, as the probability threshold u decreases, both TPR and FPR generally increase. The ROC curve, depicted in Figure 10.9, illustrates the change of TPR (y-axis) and FPR (x-axis) as the probability threshold u varies from 1 down to 0. In general, the ROC curve tends to be smoother as the sample size n increases. In fact, given the observations ranked in descending order of $\hat{\pi}_i$, a general procedure to draw the ROC curve is as follows: we start from the origin $(0, 0)$, and refer to the labels of the observations in order. If $\hat{\pi}_i = 1$ (resp. 0), then we move up by $\frac{1}{n_1}$ in size vertically (resp. by $\frac{1}{n_0}$ in size to the right

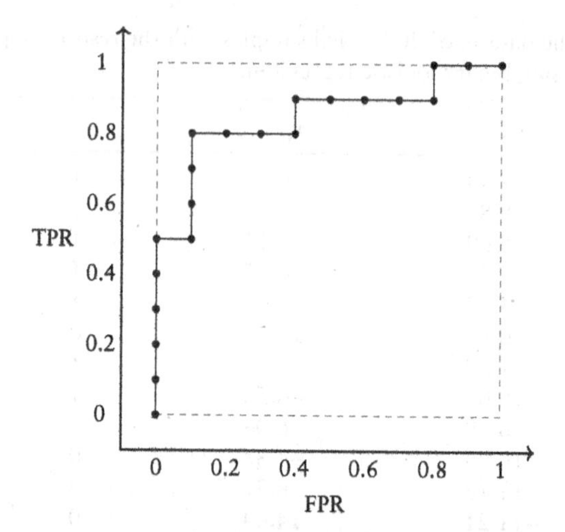

FIGURE 10.9 ROC curve for the data extracted from Table 10.3, where each point represents a value $u = \hat{\pi}_i$ of the threshold, for $i = 1, \ldots, 20$, along the curve.

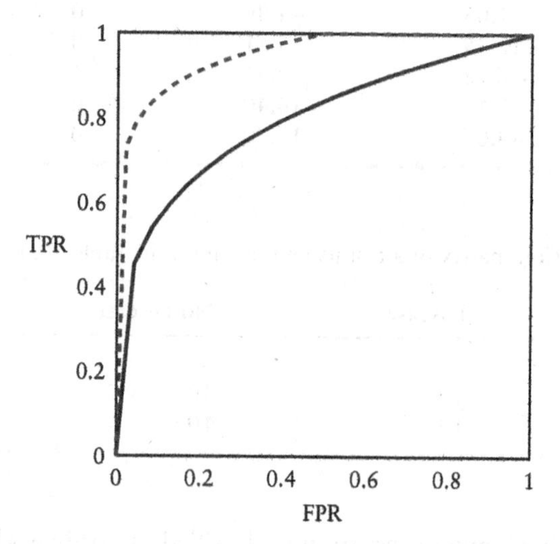

FIGURE 10.10 ROC curves of two binary logistic regressions deployed to two different datasets.

horizontally), where n_1 and n_0 are the number of observations in the classes 1 and 0, respectively. We ultimately end up at $(1, 1)$ when we repeat this for all observations. This procedure also indicates the concave nature of the curve.

Figure 10.10 shows two different ROC curves examining the performance of two binary logistic regressions for two respective datasets. The dashed line in red shows the ROC curve of a dataset which is easier to be separated into two groups, so that a higher value of TPR matches with a smaller value of FPR. In contrast, the solid ROC curve in blue corresponds to a classification which is relatively harder to be separated cleanly. Generally, the ROC curve of an easily separable dataset or from a very reliable classification algorithm tends to lean towards the top left-hand corner of the unit square. When all positive observations can be correctly classified (TPR = 1) while no negative observation is misclassified as positive (FPR = 0), no matter what values of u, then the corresponding ROC curve degenerates to the single point at the top left-hand corner of the unit square.

Besides, ROC curves can be used to compare the performances of different classification algorithms applied to the same dataset. One possible way is to compare the *Area Under Curve* (AUC) of the ROC curves, so that the greater the AUC, the better the performance of the algorithm is.

Just as the top left-hand corner of the unit square represents a perfect classifier, the other corners of the unit square also possess their own interpretations; see Figure 10.11. For instance, at the bottom left-hand corner, the classifier identifies all observations as negative; since no observation is classified as positive, we have TPR = FPR = 0. On the other hand, the top right-hand corner corresponds to another extreme scenario: the corresponding classifier identifies all observations as positive; now all positive observations are correctly classified while all negative observations are misclassified, and we have TPR = FPR = 1. The diagonal line joining the origin and the top right-hand corner has another important interpretation. Suppose that we randomly classify an observation as positive with a probability p and as negative with a probability $1 - p$. Let n_1 and n_2 be the total number of observations having a disease and having no disease, respectively. Since no matter the observation is positive or negative it will be classified as positive with a probability p, from Table 10.5, both ratios TPR and FPR approach p as both n_1 and n_2 tend to infinity. In other words, the point (p, p) on the main diagonal is associated with the algorithm which randomly classifies any observation as positive with a probability p. Therefore, if

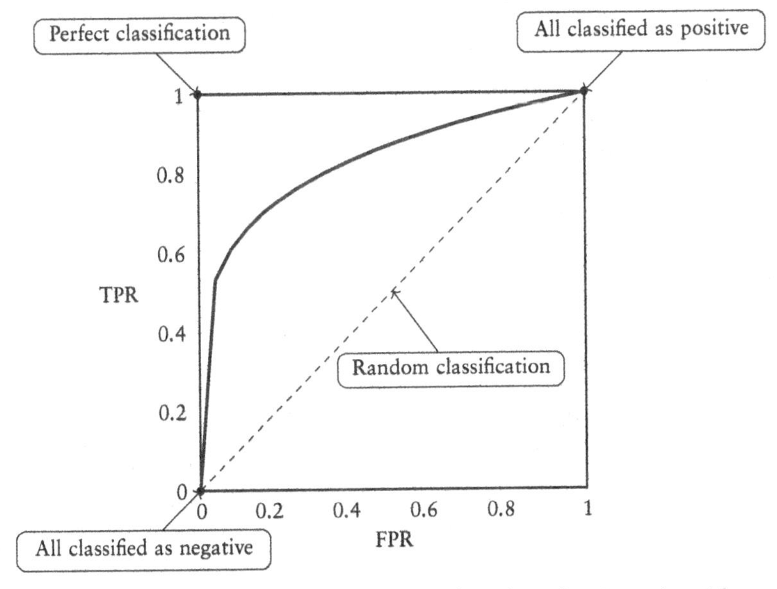

FIGURE 10.11 Evaluation of the performance of a classification algorithm using its ROC curve.

TABLE 10.5 Confusion matrix for a test obtained by a random classification rule; the values are taken approximately, assuming that n_1 and n_2 are sufficiently large.

Test \ Truth	Disease	No Disease	Row Sum
Positive	$\approx n_1 p$	$\approx n_2 p$	$\approx (n_1 + n_2)p$
Negative	$\approx n_1(1-p)$	$\approx n_2(1-p)$	$\approx (n_1 + n_2)(1-p)$
Column Sum	n_1	n_2	$n_1 + n_2$

the ROC curve of a classification algorithm lies below this main diagonal, we can interpret its performance even worse than unintentional random classification.

*10.7 PRINCIPAL COMPONENT REGRESSION

Given a dataset $S = \{(x_i, y_i) : x_i \in \mathbb{R}^p \text{ and} y_i \in \mathbb{R}^q\}_{i=1,\dots,n}$, *Principal Component Regression* (PCR for short, a.k.a. *Spectral Regression*) predicts y while also adopting dimension reduction on the family of input covariates $x_i := (x_{i1}, x_{i2}, \dots, x_{ip})$. One advantage of using PCR is to remove the multicollinearity in the set of input covariates, especially when two or more x_j's are close to collinear. Generally, rather than just the linear structure, there can be a functional relation between y and covariates **x** given by:

$$y_i = f(x_i) + \epsilon_i, \quad i = 1, 2, \dots, n,$$

where $f : \mathbb{R}^p \to \mathbb{R}^q$ is an unknown function to be estimated[13] and $\boldsymbol{\epsilon}_i$'s are iid q-dimensional random error terms. We further assume the statistical independence of $\mathbf{x}_i$ and $\boldsymbol{\epsilon}_i$ for the sake of simplicity. To avoid technical details, we simply take $q = 1$ here. To this end, we choose a set of basis functions $\{\tilde{f}_j(\cdot)\}_{j=1}^K$, where each $\tilde{f}_j$ is regarded as a feature element, and we aim to find a linear combination of these feature elements to approximate the unknown function f in a pointwise manner. We then take a step back by allowing certain minor additional errors and try to regress y_i in the following form (here we abuse notation slightly by using the same symbol ε_i for the enlarged error):

$$y_i = \sum_{j=1}^{K} \alpha_j \tilde{f}_j(x_i) + \varepsilon_i, \quad i = 1, \dots, n, \tag{10.17}$$

where α_j, to be determined, is the coefficient serving as a kind of "weight" associated with $\tilde{f}_j(\cdot)$. We essentially consider $(\tilde{f}_1(x), \dots, \tilde{f}_K(x))^\top$ as a K-dimensional random input vector. Generally speaking, K is mostly large, so we use a PCA approach to these $\tilde{f}_j(x_i)$'s to reduce the input data dimension, at the same time also removing the collinearity

[13] Usually f is nontrivial, and one useful method is to determine to which class the function f belongs by getting the first feeling through some rough sketch.

(if present). With a sample of size n, consider the following $K \times n$ data matrix:

$$\tilde{\mathbb{D}} := \left(\tilde{f}_{\cdot}(x_1)\ \ldots\ \tilde{f}_{\cdot}(x_n)\right),\ \text{where}\ \tilde{f}_{\cdot}(x_i) := \left(\tilde{f}_1(x_i)\ \ldots\ \tilde{f}_K(x_i)\right)^{\top},$$

with which the corresponding $K \times K$ sample covariance matrix $\tilde{S}$ is constructed; see Chapter 1 for how to build $\tilde{S}$. PCR applies PCA to extract principal components out of $\tilde{S}$, these being different linear combinations of these $\tilde{f}_j(x)$'s, and use them as the new covariates in a regression model for y. In particular, the spectral decomposition of $\tilde{S}$ is $H\Lambda H^{\top}$ where $H = (h_1, h_2, \ldots, h_K)$, and each $h_j = (h_{1j}, h_{2j}, \ldots, h_{Kj})^{\top}$ is a unit eigenvector corresponding to the j-th largest eigenvalue λ_j of $\tilde{S}$. The corresponding j-th principal component (function in x_i) $P_j(x_i)$, for $j = 1, 2, \ldots, K$, can be computed by:

$$\begin{pmatrix} P_1(x_i) \\ P_2(x_i) \\ \vdots \\ P_K(x_i) \end{pmatrix} = H^{\top} \begin{pmatrix} \tilde{f}_1(x_i) \\ \tilde{f}_2(x_i) \\ \vdots \\ \tilde{f}_K(x_i) \end{pmatrix} = \begin{pmatrix} h_1^{\top} \\ h_2^{\top} \\ \vdots \\ h_K^{\top} \end{pmatrix} \begin{pmatrix} \tilde{f}_1(x_i) \\ \tilde{f}_2(x_i) \\ \vdots \\ \tilde{f}_K(x_i) \end{pmatrix},\ \text{for}\ i = 1, 2, \cdots, n.$$

In other words, $P_j(x_i) = h_j^{\top}\tilde{f}_{\cdot}(x_i) = \sum_{k=1}^{K} h_{kj}\tilde{f}_k(x_i)$. By inverting $H^{\top}$, we can write $\tilde{f}_j(x_i) = \sum_{k=1}^{K} h_{jk}P_k(x_i)$, for $j = 1, 2, \cdots, K$. Therefore

$$\begin{aligned} y_i &= \sum_{l=1}^{K} \alpha_l \tilde{f}_l(x_i) + \varepsilon_i = \sum_{l=1}^{K} \alpha_l \sum_{j=1}^{K} h_{lj}P_j(x_i) + \varepsilon_i \\ &= \sum_{l=1}^{K}\sum_{j=1}^{K} \alpha_l h_{lj}P_j(x_i) + \varepsilon_i = \sum_{j=1}^{K} \beta_j P_j(x_i) + \varepsilon_i, \end{aligned}$$

where $\beta_j := \sum_{l=1}^{K} \alpha_l h_{lj}$, and more neatly $\alpha^{\top}H = (\beta_1, \ldots, \beta_K) =: \beta^{\top}$ which gives $\alpha = H\beta$. To achieve the dimension reduction, suppose that the first N ($\ll K$) principal components explain nearly all information in the sense that $\sum_{j=1}^{N} \lambda_j \geq (1-\delta)\sum_{j=1}^{K} \lambda_j$, for $\delta = 0.01$, say (see Chapter 9 for an explanation). These N number of PCs are crucial while the remaining $K - N$ ones are considered to be of negligible effect, meaning that the unknown sensitivity coefficients β_j's ($j = N+1, \ldots, K$) are all small such that $\sup_{j>N}|\beta_j| < \varepsilon$ for some $\varepsilon > 0$. Recall that

$$\alpha_i = \sum_{j=1}^{K} h_{ij}\beta_j = \sum_{j=1}^{N} h_{ij}\beta_j + \sum_{j=N+1}^{K} h_{ij}\beta_j,$$

where the second summation on the right is regarded as negligible, and we think of approximating $\alpha_i \approx \sum_{j=1}^{N} h_{ij}\beta_j := \alpha_i^{(N)}$. To see why this is justifiable, we note that

$$\sum_{j=1}^{K} \alpha_j^{(N)}\tilde{f}_j(x_i) = \sum_{j=1}^{K}\sum_{k=1}^{N} h_{jk}\beta_k\tilde{f}_j(x_i) = \sum_{k=1}^{N} \beta_k \sum_{j=1}^{K} h_{jk}\tilde{f}_j(x_i) = \sum_{k=1}^{N} \beta_k P_k(x_i).$$

Under the assumption that the sample size n is so large such that the random orthogonal matrix $\boldsymbol{H}$ is stabilized, we can then view $\boldsymbol{\beta} = \boldsymbol{H}^{\top}\boldsymbol{\alpha}$ as essentially constant. Then the (remaining) variance of the residual term $\hat{y} - \sum_{j=1}^{K} \alpha_j^{(N)}\tilde{f}_j(\boldsymbol{x})$ with $\hat{y} = \sum_{j=1}^{K} \beta_j P_j(\boldsymbol{x})$ by using only the truncated coefficients $\alpha_j^{(N)}$'s is:

$$\mathrm{Var}\left(\hat{y} - \sum_{j=1}^{K} \alpha_j^{(N)}\tilde{f}_j(\boldsymbol{x})\right) = \mathrm{Var}\left(\hat{y} - \sum_{j=1}^{N} \beta_j P_j(\boldsymbol{x})\right) = \mathrm{Var}\left(\sum_{j=N+1}^{K} \beta_j P_j(\boldsymbol{x})\right)$$

$$= \sum_{j=N+1}^{K} \beta_j^2 \mathrm{Var}(P_j(\boldsymbol{x})) \leq \varepsilon^2 \sum_{j=N+1}^{K} \mathrm{Var}\left(P_j(\boldsymbol{x})\right) \leq \varepsilon^2 \delta \sum_{j=1}^{K} \lambda_j.$$

Hence, as we expect ε to be not too large, one can just take $\delta = 0.01$ and we adjust the number N of PCs to be taken. Based on the discussion above, we can work backward, so the PCR can be carried out as follows:

1. Regress y_i against the first N PCs of $P_1, \ldots, P_N$ evaluated at $\boldsymbol{x}_i$ by establishing the least square estimator for the corresponding $\boldsymbol{\beta}$ coefficients using the new data matrix $\boldsymbol{P}^{(N)}$ of $(P_1(\boldsymbol{x}_i), \ldots, P_N(\boldsymbol{x}_i))^{\top}$, that is

$$\hat{\boldsymbol{\beta}} = \left[\left(\boldsymbol{P}^{(N)}\right)^{\top} \boldsymbol{P}^{(N)}\right]^{-1} \left(\boldsymbol{P}^{(N)}\right)^{\top} \boldsymbol{y};$$

2. Set $\hat{\alpha}_i^{(N)} = \sum_{j=1}^{N} h_{ij}\hat{\beta}_j$, and obtain the final regressor $\hat{y}_i = \sum_{j=1}^{K} \hat{\alpha}_j^{(N)}\tilde{f}_j(\boldsymbol{x}_i)$.

For the diagnostics, we essentially consider the dropped miscellaneous term $\sum_{j=N+1}^{K} \beta_j P_j(\boldsymbol{x}_i)$ to merge with the idiosyncratic noise term ε_i, meaning that they add up to become another error term $\tilde{\varepsilon}_i$. As long as ϵ_i is independent of $\mathbf{x}_i$, ϵ_i is uncorrelated with $P_k(\mathbf{x}_i)$ for all $k = 1, \ldots, K$. Therefore $\tilde{\epsilon}_i$ is also uncorrelated with $P_j(\mathbf{x}_i)$ for those $j = 1, \ldots, N$, and the same theory as MANOVA (Multivariate Analysis of Variances, see [12]) on the diagnostics of the significance of various coefficients can still be applied.

As an illustration of PCR, we again use the Boston housing dataset depicted in Subsection 10.2. We first split the housing dataset into training and test datasets with a ratio of 8 to 2, then PCA is applied to the correlation matrix of the training dataset. Programme 10.24 shows the application of PCA in Python. As discussed in Programme 9.3, we standardize each feature variable first before feeding the training dataset into the `PCA()` class in Python, and we shall also use the same in-sample sample means and sample standard deviations to standardize the corresponding feature variables in the test dataset.

```
1  from sklearn.model_selection import train_test_split
2  from sklearn.decomposition import PCA
3  from sklearn.linear_model import LinearRegression
4  import matplotlib.pyplot as plt
5  import numpy as np
6  import pandas as pd
7
8  plt.rcParams["figure.figsize"] = 10, 10
9  plt.rc('font', size=20); plt.rc('axes', titlesize=30, labelsize=30)
10 plt.rc('xtick', labelsize=25); plt.rc('ytick', labelsize=25)
11
12 df = pd.read_csv("Boston.csv", index_col=0)
13 X, y = df.drop(["medv"], axis=1).values, df.medv.values
14 # Split the data into training and test dataset in ratio of 8:2
15 (X_train, X_test, y_train,
16  y_test) = train_test_split(X, y, test_size=0.2, random_state=4012)
17
18 mu_train = np.mean(X_train, axis=0)
19 sigma_train = np.std(X_train, axis=0)
20 scale_X_train = (X_train - mu_train)/sigma_train
21 scale_X_test = (X_test - mu_train)/sigma_train
22
23 pca = PCA()
24 pca.fit(scale_X_train)
25 print(np.cumsum(pca.explained_variance_ratio_))
26
27 s2 = pca.explained_variance_   # save the variance of all PC to s2
28 PC_comp = np.arange(pca.n_components_) + 1
29 plt.plot(PC_comp, pca.explained_variance_ratio_*sum(s2),
30          "bo-", linewidth=2)
31 plt.title('Scree Plot')
32 plt.xlabel('Principal Component')
33 plt.ylabel('Variance')
```

```
1  [0.47826441 0.58202731 0.678612   0.7467532  0.80766512 0.85709552
2   0.8990518  0.92954972 0.95165464 0.96883299 0.98264471 0.99522195
3   1.        ]
```

Programme 10.24 Applying PCA to the Boston housing training dataset in Python.

From the output of Programme 10.24, we observe that the first 9 principal components can explain over 95% of the total variance; Figure 10.12 further justifies this claim and shows that the total variance decreases only slightly after the first 9 principal components. Therefore, the first 9 principal components can be adopted to fit a multiple linear regression model.

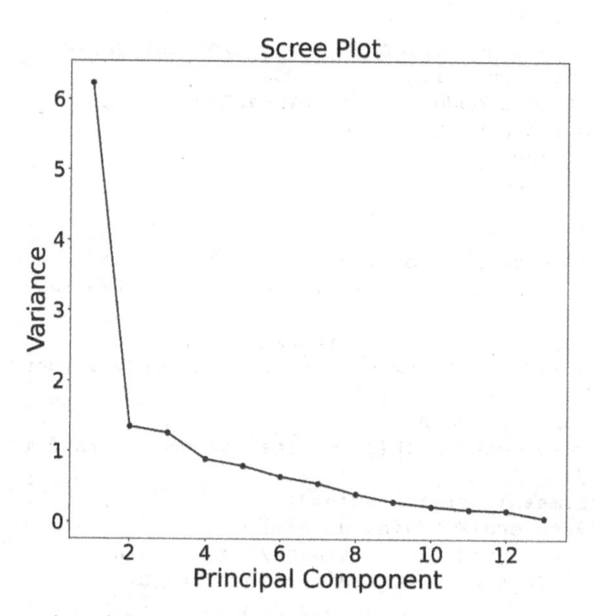

FIGURE 10.12 Scree plot of the Boston housing training dataset in Python, generated in Programme 10.24.

```
1  n_comp = 9 # for 95% explained variance
2  pcr_model = LinearRegression()
3  scores_X_train = pca.transform(scale_X_train)
4  pcr_model.fit(scores_X_train[:,:n_comp], y_train)
5
6  scores_X_test = pca.transform(scale_X_test)
7  y_hat_pcr = pcr_model.predict(scores_X_test[:,:n_comp])
```

Programme 10.25 PCR in Python fitted to Boston housing dataset.

In Programme 10.26, we compare the PCR model with the multiple linear regression model for full dataset via Python, by computing the corresponding root-mean-squared errors (RMSEs, different from the residual mean squared error defined before) based on the test dataset:

```
1  lr_model = LinearRegression()
2  lr_model.fit(X_train, y_train)
3  y_hat_lr = lr_model.predict(X_test)
4
5  print(np.sqrt(np.mean((y_test - y_hat_pcr)**2)))     # RMSE
6  print(np.sqrt(np.mean((y_test - y_hat_lr)**2)))      # RMSE
```

```
1  5.73465499391599
2  5.558104701368164
```

Programme 10.26 PCR versus linear regression for full dataset with Boston housing training dataset in Python.

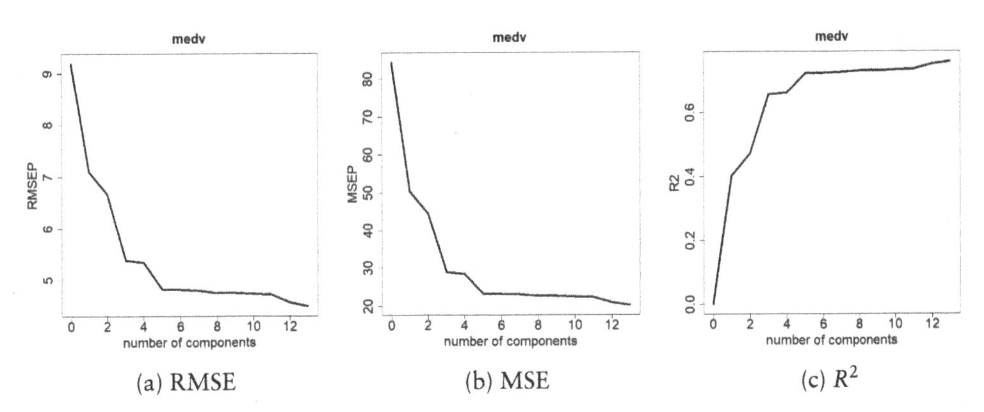

(a) RMSE (b) MSE (c) R^2

FIGURE 10.13 Scree plot of the Boston housing training dataset in **R**, generated in Programme 10.27.

As expected, the RMSE of PCR is slightly higher than that of the multiple linear regression model for full dataset, because in the end we did ignore some information in the original Boston housing dataset by selecting only 9 of the total of 13 PCs.

Analogously, the `pcr()` function in the `pls` (principal least square) package adopted in **R** can carry out PCR. From Figure 10.13, we observe that the fifth principal component is the kink point of the scree plot, suggesting that the first 5 PCs can be used for PCR. Comparing the respective outputs in Figures 10.12 (Python) and 10.13 (**R**), despite the small difference in the shape of curves due to the random subsamples of training datasets being drawn, some of their key characteristics, such as the locations of kink points are indeed very similar. Besides, the difference in the numbers of PCs selected by the two programmes is due to the difference of selection criteria between Python and **R**; while Python chooses the PCs that explain over 95% total variance, **R** simply sets the kink point as the maximum number of PCs chosen.

```
1  > data("Boston", package="MASS")        # Load the data
2  >
3  > # Split the data into training and test dataset in ratio of 8:2
4  > set.seed(4002)
5  > training.samples <- sample(1:nrow(Boston), round(0.8*nrow(Boston)))
6  > train.data <- Boston[training.samples,]
7  > test.data <- Boston[-training.samples,]
8  >
9  > # Principal Component Regression model
10 > library(pls)
11 > pcr_model <- pcr(medv~., data=train.data, scale=T, center=T,
12 +                  validation="none")
13 >
14 > RMSEP(pcr_model)
15 (Intercept)      1 comps      2 comps      3 comps      4 comps
16       9.187        7.098        6.672        5.383        5.340
17     5 comps      6 comps      7 comps      8 comps      9 comps
18       4.824        4.819        4.801        4.757        4.756
19    10 comps     11 comps     12 comps     13 comps
20       4.735        4.720        4.566        4.492
```

```
21  > MSEP(pcr_model)
22  (Intercept)      1 comps      2 comps      3 comps      4 comps
23        84.40        50.39        44.51        28.98        28.51
24      5 comps      6 comps      7 comps      8 comps      9 comps
25        23.27        23.22        23.05        22.63        22.62
26     10 comps     11 comps     12 comps     13 comps
27        22.42        22.28        20.85        20.18
28  > R2(pcr_model)
29  (Intercept)      1 comps      2 comps      3 comps      4 comps
30       0.0000       0.4030       0.4727       0.6566       0.6622
31      5 comps      6 comps      7 comps      8 comps      9 comps
32       0.7242       0.7249       0.7269       0.7319       0.7320
33     10 comps     11 comps     12 comps     13 comps
34       0.7344       0.7361       0.7530       0.7609
35  >
36  > par(cex.lab=2, cex.axis=2, cex.main=2, mar=c(5,5,4,4))
37  > validationplot(pcr_model, val.type="RMSE", lwd=4)
38  > validationplot(pcr_model, val.type="MSE", lwd=4)
39  > validationplot(pcr_model, val.type="R2", lwd=4)
40  >
41  > # Make prediction with 5 principal components
42  > y_hat_pcr <- predict(pcr_model, test.data, ncomp=5)
43  >
44  > # Linear regression model
45  > lr_model <- lm(medv~., data=train.data)
46  > y_hat_lr <- predict(lr_model, test.data)
47  >
48  > # Model comparison with RMSE
49  > sqrt(mean((test.data$medv - y_hat_pcr)^2))
50  [1] 5.832351
51  > sqrt(mean((test.data$medv - y_hat_lr)^2))
52  [1] 5.441909
```

Programme 10.27 PCR on the Boston housing training dataset in R.

Recall that in the MNIST example in Section 10.4, fitted with a generic multinomial logistic regression model, there were $28 \times 28 = 784$ feature variables, leading to a huge 7,056 coefficient weights and 9 biases. Let us take a leap by extending the dimension reduction technique via PCA from linear regression to multinomial logistic regression. To this end, as before, we choose the number of principal components that explain 95% of the total variance. Since the pixel feature variables are of the same gray scale between 0 and 255, we only standardize each feature variable in the training feature data matrix by centering them with their respective feature mean; we do not rescale each feature variable with their respective standard deviations. This process is directly performed via `PCA()` in Programme 10.28 via Python.

```
1  start = dt.now()
2  from sklearn.decomposition import PCA
3  from sklearn.linear_model import LogisticRegression
4
5  pca = PCA(n_components=.95)
6  X_train_pca = pca.fit_transform(X_train)
7  X_test_pca = pca.transform(X_test)
8  print(f'Total number of components: {pca.n_components_}')
9
10 np.random.seed(4002)
11 clf_pca = LogisticRegression(max_iter=10000)
12 clf_pca.fit(X_train_pca, y_train)
13 y_hat_pca = clf_pca.predict(X_test_pca)
14
15 # Table of predictions and prediction accuracy
16 tab = confusion_matrix(y_hat_pca, y_test)        # Confusion matrix
17 print(tab)
18 print(sum(tab.diagonal()) / len(y_test))          # Accuracy
19 print(dt.now() - start)                   # PCA + Logistic regression
```

```
1  Total number of components: 154
2  [[ 955    0    5    4    1   10   11    1    7   11]
3   [   0 1110    8    1    2    3    3    6    9    8]
4   [   1    4  929   23    5    5    9   24    7    1]
5   [   3    2   18  915    2   33    1    7   26    9]
6   [   0    0    7    1  915    9    7    8    7   33]
7   [   7    2    5   26    1  775   12    1   25    6]
8   [   7    4   12    2    8   14  910    0   12    0]
9   [   4    2    8   10    4    5    3  947    9   21]
10  [   1   11   34   19    9   32    2    1  864    7]
11  [   2    0    6    9   35    6    0   33    8  913]]
12 0.9233
13 0:00:08.095001
```

Programme 10.28 PC multinomial logistic regression for the MNIST dataset via Python.

From the results obtained in Programme 10.28, 154 principal components are selected and the resulting accuracy for the test dataset is 92.33%, which is similar to Programme 10.17 with the use of full set of feature variables, but the number of parameters is reduced to $(154+1)\times(10-1)=1,395$. Moreover, for a common desktop computer, the time required for running the codes in Python is greatly reduced from 1 minute to 8 seconds.

For the **R** codes in Programme 10.29, we plug the MNIST training data into the `princomp()` function with the two arguments `center=TRUE` and `scale=FALSE`, as we only perform the centering for each feature variable. Comparing the results obtained in Programmes 10.18 (without PCA) and 10.29 (with PCA) in **R**, the accuracy for the test dataset slightly increases from 91.67% to 92.19% with the use of PCA. Moreover, **R** generally executes much slower than Python.

```
1  > start <- Sys.time()
2  > pca <- prcomp(X_train, center=TRUE, scale=FALSE)
3  > s <- pca$sdev
4  > t <- sum(s^2)              # Compute total variance
5  > cumvar <- cumsum(s^2/t) # Cumulative sum of proportion of variance
6  > (idx <- which(cumvar >= 0.95)[1])
7  [1] 154
8  > X_train_pca <- predict(pca, newdata=X_train)[,1:idx]
9  > X_test_pca <- predict(pca, newdata=X_test)[,1:idx]
10 > df_train_pca <- data.frame(cbind(X_train_pca, y_train))
11 > names(df_train_pca) <- c(paste0('x_', 1:idx), 'y')
12 > df_test_pca <- data.frame(X_test_pca)
13 > names(df_test_pca) <- paste0('x_', 1:idx)
14 >
15 > set.seed(4002)
16 > mnl_pca <- multinom(y~., data=df_train_pca, MaxNWts=1e6, maxit=200,
17 +                     trace=F)
18 > y_hat_pca <- predict(mnl_pca, newdata=df_test_pca, type="class")
19 > table(y_hat_pca, y_test)     # Confusion matrix
20          y_test
21 y_hat_pca    0    1    2    3    4    5    6    7    8    9
22         0  959    0    8    6    2    9   12    1    8   10
23         1    0 1107    6    2    3    2    5    9    8    9
24         2    1    3  929   20    4    4    9   19    8    1
25         3    1    2   14  911    3   29    0    8   21    9
26         4    1    1    8    1  913   13    9    6   10   35
27         5    4    3    6   29    0  773   14    2   27    8
28         6    5    5   10    2    9   15  905    1   10    0
29         7    5    2   10   12    3   10    2  947    8   21
30         8    1   12   33   19    8   29    2    1  866    7
31         9    3    0    8    8   37    8    0   34    8  909
32 > mean(y_hat_pca == y_test)     # Accuracy
33 [1] 0.9219
34 > Sys.time() - start              # PCA + Logistic regression
35 Time difference of 5.145588 mins
```

Programme 10.29 PC multinomial logistic regression for the MNIST dataset via R.

*10.A APPENDIX

*10.A.1 Derivation of Gauss–Newton Algorithm

The Gauss–Newton algorithm is a modification of Newton's method, used to numerically evaluate the (local) minimum point of a broad class of functions. It helps to solve many practical least square problems subject to regularization, commonly found in the context of semi-parametric regressions. The algorithm makes use of a linear approximation for the residual function $r_i(\boldsymbol{\beta})$ (see below) so that it is not necessary to compute its second or higher-order derivatives. The formalization of the algorithm first appeared in [8] written by Carl Friedrich Gauss in 1809.

More precisely, consider the parameter vector $\boldsymbol{\beta} = (\beta_1, \ldots, \beta_p)^\top$ and the residual function $\boldsymbol{r}(\boldsymbol{\beta}) = (r_1(\boldsymbol{\beta}), \ldots, r_n(\boldsymbol{\beta}))^\top$, with $p \leq n$, where $r_i(\boldsymbol{\beta}) = y_i - f(\boldsymbol{x}_i; \boldsymbol{\beta})$ for some parameter function f, and the sum of squares $S(\boldsymbol{\beta}) := \sum_{i=1}^{n} r_i^2(\boldsymbol{\beta})$. Algorithm 10.1 can be derived as follows. Given an initial seed $\boldsymbol{\beta}^{(0)}$, assume that it is close to the unknown minimum point $\hat{\boldsymbol{\beta}} := \arg\min_{\boldsymbol{\beta}} \frac{1}{n} S(\boldsymbol{\beta}) = \arg\min_{\boldsymbol{\beta}} \frac{1}{n} \sum_{i=1}^{n} \left(y_i - f(\boldsymbol{x}_i; \boldsymbol{\beta})\right)^2$. By expanding $\boldsymbol{r}$ using its linear approximation, we have

$$\boldsymbol{r}(\boldsymbol{\beta}) \approx \boldsymbol{r}(\boldsymbol{\beta}^{(0)}) + \boldsymbol{J}_r(\boldsymbol{\beta}^{(0)})(\boldsymbol{\beta} - \boldsymbol{\beta}^{(0)}),$$

where $\boldsymbol{J}_r(\boldsymbol{\beta}) = \left(\frac{\partial r_i(\boldsymbol{\beta})}{\partial \beta_j}\right)_{1 \le i \le n, 0 \le j \le p}$ is the Jacobian matrix (10.6), and the minimization of $S(\boldsymbol{\beta})$ is "nearly" equivalent to

$$\min_{\boldsymbol{\beta}} \left\| \boldsymbol{r}(\boldsymbol{\beta}^{(0)}) + \boldsymbol{J}_r(\boldsymbol{\beta}^{(0)})(\boldsymbol{\beta} - \boldsymbol{\beta}^{(0)}) \right\|_2^2, \tag{10.18}$$

where $\|\cdot\|_2$ is the Euclidean norm as defined in Section 1.3. The minimization problem of (10.18) is no different from the usual linear least square problem in terms of $\boldsymbol{\Delta} := \boldsymbol{\beta} - \boldsymbol{\beta}^{(0)}$. Recalling the multiple regression formula (also see the footnote after (10.3)) or otherwise the first order Euler condition, to find the minimizer of (10.18), $\hat{\boldsymbol{\Delta}} = \hat{\boldsymbol{\beta}} - \boldsymbol{\beta}^{(0)}$ should satisfy

$$\boldsymbol{J}_r(\boldsymbol{\beta}^{(0)})^\top \left(\boldsymbol{r}(\boldsymbol{\beta}^{(0)}) + \boldsymbol{J}_r(\boldsymbol{\beta}^{(0)}) \hat{\boldsymbol{\Delta}} \right) = \mathbf{0}, \tag{10.19}$$

which indicates

$$\hat{\boldsymbol{\beta}} \approx \boldsymbol{\beta}^{(0)} - \left(\boldsymbol{J}_r(\boldsymbol{\beta}^{(0)})^\top \boldsymbol{J}_r(\boldsymbol{\beta}^{(0)})\right)^{-1} \boldsymbol{J}_r(\boldsymbol{\beta}^{(0)})^\top \boldsymbol{r}(\boldsymbol{\beta}^{(0)}), \tag{10.20}$$

which suggests the following iteration scheme:

$$\boldsymbol{\beta}^{(k+1)} = \boldsymbol{\beta}^{(k)} - \left(\boldsymbol{J}_r(\boldsymbol{\beta}^{(k)})^\top \boldsymbol{J}_r(\boldsymbol{\beta}^{(k)})\right)^{-1} \boldsymbol{J}_r(\boldsymbol{\beta}^{(k)})^\top \boldsymbol{r}(\boldsymbol{\beta}^{(k)}), \tag{10.21}$$

which can plausibly be solved as $\mathrm{rank}(\boldsymbol{J}_r(\boldsymbol{\beta}^{(k)})) = p + 1$ almost certainly in practice. Clearly,

$$\frac{\partial r_i}{\partial \beta_j} = -\frac{\partial f}{\partial \beta_j}(\boldsymbol{x}_i; \boldsymbol{\beta}), \quad \text{for } j = 0, 1, \ldots, p, \text{ and } i = 1, 2, \cdots, n.$$

so the iteration scheme (10.21) can be rewritten as:

$$\boldsymbol{\beta}^{(k+1)} = \boldsymbol{\beta}^{(k)} + \left(\boldsymbol{J}_f(\boldsymbol{\beta}^{(k)})^\top \boldsymbol{J}_f(\boldsymbol{\beta}^{(k)})\right)^{-1} \boldsymbol{J}_f(\boldsymbol{\beta}^{(k)})^\top \boldsymbol{r}(\boldsymbol{\beta}^{(k)}), \tag{10.22}$$

where $\boldsymbol{J}_f(\boldsymbol{\beta}) = \left(\frac{\partial f}{\partial \beta_j}(\boldsymbol{x}_i; \boldsymbol{\beta})\right)_{1 \le i \le n, 0 \le j \le p}$. For each iteration step, the system of equations (10.21) (or (10.22)), in terms of $\hat{\boldsymbol{\Delta}}^{(k)} = \boldsymbol{\beta}^{(k+1)} - \boldsymbol{\beta}^{(k)}$ can be solved by applying Cholesky decomposition (see Section 1.2) or general *QR factorization* (see for instance [2]) to the symmetric square matrix $\boldsymbol{J}_r(\boldsymbol{\beta}^{(k)})^\top \boldsymbol{J}_r(\boldsymbol{\beta}^{(k)})$.

Alternatively, this Gauss–Newton algorithm can also be argued from the celebrated Newton's method; indeed, to seek for the root(s) (critical point(s)) of $\nabla_{\boldsymbol{\beta}} S(\boldsymbol{\beta}) = \mathbf{0}$, using Newton's method, we see that

$$\nabla_{\boldsymbol{\beta}} S(\boldsymbol{\beta}^{(0)}) + (\nabla_{\boldsymbol{\beta}} \nabla_{\boldsymbol{\beta}}^{\mathsf{T}}) S(\boldsymbol{\beta}^{(0)})(\boldsymbol{\beta} - \boldsymbol{\beta}^{(0)}) + O(\|\boldsymbol{\beta} - \boldsymbol{\beta}^{(0)}\|_2^2) = \mathbf{0},$$

based on which the recursive formula for the update can be defined via

$$\boldsymbol{\beta}^{(k+1)} = \boldsymbol{\beta}^{(k)} - \left(\boldsymbol{H}(\boldsymbol{\beta}^{(k)})\right)^{-1} \nabla_{\boldsymbol{\beta}} S(\boldsymbol{\beta}^{(k)}), \tag{10.23}$$

where $\boldsymbol{H}(\boldsymbol{\beta})$ is the Hessian matrix of $S(\boldsymbol{\beta})$ given by $\left(\frac{\partial^2 S}{\partial \beta_i \partial \beta_j}(\boldsymbol{\beta})\right)_{0 \le i \le p, 0 \le j \le p}$. Since $S(\boldsymbol{\beta}) = \sum_{k=1}^{n} r_k^2(\boldsymbol{\beta})$, a simply calculation gives:

$$\nabla_{\boldsymbol{\beta}} S(\boldsymbol{\beta}) = \left(2 \sum_{k=1}^{n} r_k(\boldsymbol{\beta}) \frac{\partial r_k(\boldsymbol{\beta})}{\partial \beta_0}, \ldots, 2 \sum_{k=1}^{n} r_k(\boldsymbol{\beta}) \frac{\partial r_k(\boldsymbol{\beta})}{\partial \beta_p}\right)^{\mathsf{T}},$$

which can be further differentiated to give its Hessian $\boldsymbol{H}(\boldsymbol{\beta}) = (H_{ij}(\boldsymbol{\beta}))_{0 \le i \le p, 0 \le j \le p}$, where

$$H_{ij}(\boldsymbol{\beta}) = 2 \sum_{k=1}^{n} \left(\frac{\partial r_k(\boldsymbol{\beta})}{\partial \beta_i} \frac{\partial r_k(\boldsymbol{\beta})}{\partial \beta_j} + r_k(\boldsymbol{\beta}) \frac{\partial^2 r_k(\boldsymbol{\beta})}{\partial \beta_i \partial \beta_j}\right), \; i, j = 0, 1, \ldots, p. \tag{10.24}$$

If one ignores the second-order terms in (10.24) (see below for a justification), the Hessian is approximately equal to:

$$H_{ij}(\boldsymbol{\beta}) \approx 2 \sum_{k=1}^{n} (\boldsymbol{J}_r(\boldsymbol{\beta}))_{ki} (\boldsymbol{J}_r(\boldsymbol{\beta}))_{kj},$$

where $(\boldsymbol{J}_r(\boldsymbol{\beta}))_{kl}$ is just the (k, l)-entry of the Jacobian matrix $\boldsymbol{J}_r(\boldsymbol{\beta})$ defined before. Writing in matrix form, we have $\nabla_{\boldsymbol{\beta}} S(\boldsymbol{\beta}) = 2(\boldsymbol{J}_r(\boldsymbol{\beta}))^{\mathsf{T}} \boldsymbol{r}(\boldsymbol{\beta})$ and $\boldsymbol{H}(\boldsymbol{\beta}) \approx 2\boldsymbol{J}_r(\boldsymbol{\beta})^{\mathsf{T}} \boldsymbol{J}_r(\boldsymbol{\beta})$, which can be substituted back into (10.23), and (10.21) can then be obtained as desired.

The effectiveness of the convergence of the Gauss–Newton algorithm can be much enhanced, especially when the magnitude of the second-order cross-term is relatively small, for instance, in the sense that

$$\left| r_k(\boldsymbol{\beta}) \frac{\partial^2 r_k(\boldsymbol{\beta})}{\partial \beta_i \partial \beta_j} \right| \ll \left| \frac{\partial r_k(\boldsymbol{\beta})}{\partial \beta_i} \frac{\partial r_k(\boldsymbol{\beta})}{\partial \beta_j} \right| \quad \text{for all } i, j, k,$$

which is plausible particularly when (i) $r_k(\boldsymbol{\beta})$'s are small and stable in value; and (ii) these $r_k(\boldsymbol{\beta})$'s are nearly linear, for $\boldsymbol{\beta}$ lying around the minimum point $\hat{\boldsymbol{\beta}}$. In general, the convergence rate of the algorithm is linear; recently, there has been work [13] which claims that by rectifying and modifying the tail of the square loss function in almost linear growth, a quadratic convergence rate can still be attained.

*10.A.2 Iteratively Reweighted Least Squares (IRLS)

Let us adopt Newton's method, similar to (10.23), to the finding of MLEs for logistic regression. Denote $-\ln L(\boldsymbol{\beta})$ (as defined in (10.14)) by $-l(\boldsymbol{\beta})$ for the sake of notational simplicity. With an initial seed $\boldsymbol{\beta}^{(0)}$, we construct iteratively the sequence of approximants, with $k = 0, 1, \ldots$:

$$\begin{aligned}\boldsymbol{\beta}^{(k+1)} &= \boldsymbol{\beta}^{(k)} - [\nabla_{\boldsymbol{\beta}}\nabla_{\boldsymbol{\beta}}^{\top}(-l(\boldsymbol{\beta}^{(k)}))]^{-1}\nabla_{\boldsymbol{\beta}}(-l(\boldsymbol{\beta}^{(k)})) \\ &= \boldsymbol{\beta}^{(k)} - (H(\boldsymbol{\beta}^{(k)}))^{-1}\nabla_{\boldsymbol{\beta}}l(\boldsymbol{\beta}^{(k)}),\end{aligned} \tag{10.25}$$

where $H(\boldsymbol{\beta})$ is the Hessian matrix of $l(\boldsymbol{\beta})$, which is computed explicitly in the following. Recall that the logit function $\text{logit}(t) := \frac{e^t}{1+e^t}$ satisfies the simple relation $\frac{d}{dt}\text{logit}(t) = \text{logit}(t)(1 - \text{logit}(t))$ as pointed out in Section 10.4, hence for $\mu, \nu = 0, 1, \ldots, p$ and $i = 1, \ldots, n$, we have

$$\left.\frac{\partial \pi_i}{\partial \beta_\mu}\right|_{x_i} = \pi_i(1-\pi_i)x_{i\mu},$$

$$\left.\frac{\partial l}{\partial \beta_\mu}\right|_{x_\mu} = \sum_{i=1}^{n}\left(\frac{y_i}{\pi_i} - \frac{1-y_i}{1-\pi_i}\right)\left.\frac{\partial \pi_i}{\partial \beta_\mu}\right|_{x_i} = \sum_{i=1}^{n}(y_i - \pi_i)x_{i\mu},$$

$$\left.\frac{\partial^2 l}{\partial \beta_\mu \partial \beta_\nu}\right|_{x_i} = -\sum_{i=1}^{n}\pi_i(1-\pi_i)x_{i\mu}x_{i\nu},$$

where $\pi_i = \text{logit}(\boldsymbol{x}_i^{\top}\boldsymbol{\beta})$ and $x_{i0} = 1$ for all $i = 1, \ldots, n$. We then rewrite these equations in matrix form:

$$\nabla_{\boldsymbol{\beta}}l(\boldsymbol{\beta}) = \mathbf{X}^{\top}(\boldsymbol{y} - \boldsymbol{\pi}) \quad \text{and} \quad H(\boldsymbol{\beta}) = -\mathbf{X}^{\top}\boldsymbol{R}\mathbf{X},$$

where the data matrix $\mathbf{X} = (\boldsymbol{x}_1, \ldots, \boldsymbol{x}_n)^{\top}$ is $n \times (p+1)$, $\boldsymbol{R} = \text{diag}(\pi_1(1-\pi_1), \ldots, \pi_n(1-\pi_n))$, and both $\boldsymbol{y} = (y_1, \ldots, y_n)^{\top}, \boldsymbol{\pi} = (\pi_1, \ldots, \pi_n)^{\top} \in \mathbb{R}^n$. Therefore, the iterative formula (10.25) can be rewritten neatly as:

$$\boldsymbol{\beta}^{(k+1)} = \boldsymbol{\beta}^{(k)} + (\mathbf{X}^{\top}\boldsymbol{R}^{(k)}\mathbf{X})^{-1}\mathbf{X}^{\top}(\boldsymbol{y} - \boldsymbol{\pi}^{(k)}), \tag{10.26}$$

where $\boldsymbol{R}^{(k)}$ and $\boldsymbol{\pi}^{(k)}$ are respectively $\boldsymbol{R}$ and $\boldsymbol{\pi}$ evaluated at $\boldsymbol{\beta} = \boldsymbol{\beta}^{(k)}$. We remark that since $\pi_i = \frac{\exp(\boldsymbol{x}_i^{\top}\boldsymbol{\beta})}{1+\exp(\boldsymbol{x}_i^{\top}\boldsymbol{\beta})} \in (0, 1)$, for all $i = 1, \ldots, n$, the matrix $\mathbf{X}^{\top}\boldsymbol{R}\mathbf{X}$ is positive definite, hence the Hessian H is invertible. The procedure of finding $\boldsymbol{\beta}$ via (10.26) is commonly called *Iteratively Reweighted Least Squares* (IRLS). As the changes in the values of $\boldsymbol{\beta}^{(k)}$'s tend to reduce in order as the log-likelihood l (as well as the likelihood L) approaches its global maximum, we may terminate the iteration when the percentage change of $\boldsymbol{\beta}^{(k)}$ is smaller than a prescribed threshold $\varepsilon > 0$, i.e., stopping at the iteration step $k + 1$ when

$$\frac{\|\boldsymbol{\beta}^{(k+1)} - \boldsymbol{\beta}^{(k)}\|_2}{\|\boldsymbol{\beta}^{(k)}\|_2} < \varepsilon. \tag{10.27}$$

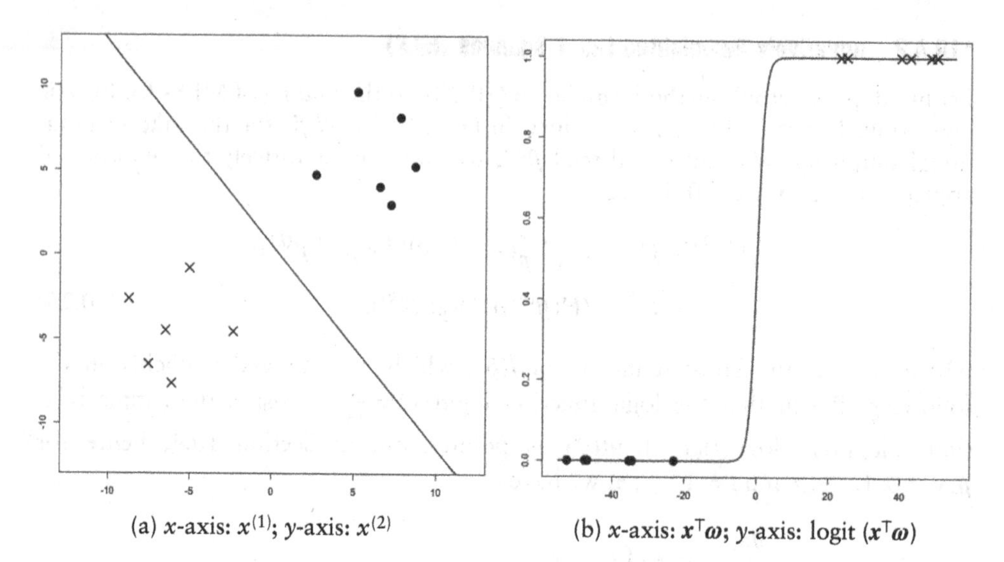

(a) x-axis: $x^{(1)}$; y-axis: $x^{(2)}$ (b) x-axis: $\boldsymbol{x}^{\top}\boldsymbol{\omega}$; y-axis: logit $(\boldsymbol{x}^{\top}\boldsymbol{\omega})$

FIGURE 10.14 Causing singularity of H for logistic regression as $\pi_i^{(t)}$ approaching to 0 or 1.

For if both p and n are very large, computing the inverse of Hessian $\mathbf{X}^{\top}\boldsymbol{R}^{(k)}\mathbf{X}$ for each iterative step becomes costly. Supposing that the initial seed $\boldsymbol{\beta}^{(0)}$ is already quite close to the maximum point to be sought, we can replace $\mathbf{X}^{\top}\boldsymbol{R}^{(k)}\mathbf{X}$ by an approximated value, for instance, replacing $H(\boldsymbol{\beta}^{(k)}) = \mathbf{X}^{\top}\boldsymbol{R}^{(k)}\mathbf{X}$ simply by $H(\boldsymbol{\beta}^{(m)}) = \mathbf{X}^{\top}\boldsymbol{R}^{(m)}\mathbf{X}$ for some $m \ll k$, meaning that we only update the Hessian matrix once after quite a number of steps. Such a shortcut approach is a special case of the *quasi-Newton method* (see [17]). This method can be useful yet it can lead to some limitations, namely as the iteration proceeds, some of the $\pi_i^{(k)}$'s approach rapidly to 0 or 1, and their values may then be incorrectly recorded by the computer program as 0 or 1 due to a limited number of floating digits used. At that point, the Hessian $H(\boldsymbol{\beta}^{(k)})$ becomes singular and the iteration (10.26) turns to be ill-conditioned; also see Figure 10.14.

REFERENCES

1. Akaike, H. (1974). A new look at the statistical model identification. *IEEE Transactions on Automatic Control*, 19(6), 716–723.
2. Atkinson, K.E. (2008). *An Introduction to Numerical Analysis*. Wiley.
3. Chinchor, N. (1992). MUC-4 evaluation metrics. *Proceedings of the Fourth Message Understanding Conference*, pp. 22–29.
4. Fama, E., and French, K. (1992). The cross-section of expected stock returns. *The Journal of Finance*, 47(2), 427–465.
5. Fama, E.F., and French, K.R. (1993). Common risk factors in the returns on stocks and bonds. *Journal of Financial Economics*, 33(1), 3–56.
6. Fisher, R.A. (1936). The use of multiple measurements in taxonomic problems. *Annals of Eugenics*, 7(2), 179–188.

7. Galton, F. (1885). Presidential address, section H, anthropology. *British Association Reports*, 55, 1206–1214.
8. Gauss, C.F. (1809). *Theoria Motus Corporum Coelestium in Sectionibus Conicis Solem Ambientium Auctore Carolo Friderico Gauss. Sumtibus Frid.* Perthes et IH Besser.
9. Harrison Jr, D., and Rubinfeld, D.L. (1978). Hedonic housing prices and the demand for clean air. *Journal of Environmental Economics and Management*, 5(1), 81–102.
10. Horn, R.A., and Johnson, C.R. (2012). *Matrix Analysis*. Cambridge University Press.
11. Jensen, M.C. (1968). The performance of mutual funds in the period 1945–1964. *The Journal of Finance*, 23(2), 389–416.
12. Johnson, R.A., and Wichern, D.W. (2007). *Applied Multivariate Statistical Analysis* (6th ed.). Prentice Hall.
13. Karimireddy, S.P., Stich, S.U., and Jaggi, M. (2018). Global linear convergence of Newton's method without strong-convexity or Lipschitz gradients. *ArXiv preprint* arXiv:1806.00413.
14. Krzanowski, W.J., and Hand, D.J. (2009). *ROC Curves for Continuous Data*. CRC Press.
15. Lee, Y., Nelder, J.A., and Pawitan, Y. (2018). *Generalized Linear Models with Random Effects: Unified Analysis via H-Likelihood*. CRC Press.
16. Markowitz, H. (1952). Portfolio selection. *The Journal of Finance*, 7(1), 77–99.
17. Nocedal, J., and Wright, S. (2006). *Numerical Optimization*. Springer.
18. Pearson, K. (1903). The law of ancestral heredity. *Biometrika*, 2(2), 211–228.
19. Schwarz, G. (1978). Estimating the dimension of a model. *The Annals of Statistics*, 6(2), 461–464.
20. Stehman, S.V. (1997). Selecting and interpreting measures of thematic classification accuracy. *Remote Sensing of Environment*, 62(1), 77–89.

CHAPTER 11

Linear Classifiers

A linear classifier aims to find a $(p-1)$-dimensional hyperplane to separate a given collection of data points into two categories. Mathematically, suppose that each data point is represented by a p-dimensional feature vector $\boldsymbol{x} = (x_1, \ldots, x_p)^\top \in \mathbb{R}^p$ together with a label $y \in \{-1, +1\}$, representing to which category it belongs. A linear classifier is a hyperplane defined by the equation $f(\boldsymbol{x}) := \boldsymbol{x}^\top \boldsymbol{\omega} + b = 0$, where $\boldsymbol{\omega} \in \mathbb{R}^p$ is the weight vector and $b \in \mathbb{R}$ is the bias. A data point $(\boldsymbol{x}_i, y_i)$ is classified as belonging to the category of "+1" if $f(\boldsymbol{x}_i) > 0$; and is classified as belonging to the category of "−1" if $f(\boldsymbol{x}_i) < 0$. The data point is correctly classified by this hyperplane if $f(\boldsymbol{x}_i) \cdot y_i > 0$; but it is misclassified if $f(\boldsymbol{x}_i) \cdot y_i < 0$.

Figure 11.1 illustrates a dataset $\{(x_{i1}, x_{i2}, y_i)\}_{i=1}^n$ of two-dimensional feature vectors, which is linearly separated by the line:

$$f(\boldsymbol{x}) = b + \omega_1 x_1 + \omega_2 x_2 = 0. \tag{11.1}$$

Given its feature vector $\boldsymbol{x}$, an observation is classified as the category +1 (resp. −1) if it lies above (resp. below) this separating line of (11.1). In this chapter, we shall introduce two traditional binary linear classifiers, namely the *perceptron* and *support vector machine* (SVM).

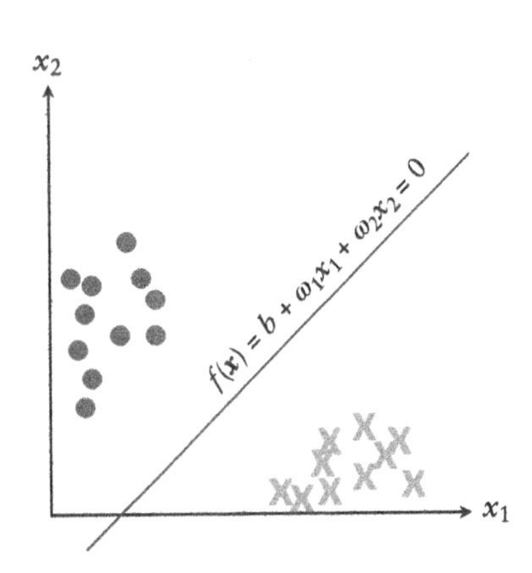

FIGURE 11.1 An example of a linear classifier for two-dimensional feature vectors with blue dots (resp. orange crosses) having a label +1 (resp. −1).

Both methods are processes of reducing the number of misclassified data. Intuitively, the algorithms improve the separating hyperplane by leaning it forward to the misclassified data in order to reduce their distances to the hyperplane. During the training, a *loss function* is chosen as a measure of misclassification. A greater value of the loss function indicates a lower accuracy in the prediction. Perceptron and SVM differ from the choice of the loss function. In addition, SVM can perform non-linear classification while the perceptron is mostly restricted to linear classification.

11.1 PERCEPTRON

Rosenblatt [20], inspired by the mechanism of the neurons in our brain to process information, invented the very first version of the perceptron. It is one of the earliest linear binary classifiers, and is generally regarded as an ancestor of recently developed *Neural Networks* and *Deep Learning*. In layman terms, each layer of neuron is connected to other layer of neurons, where information is passed from one layer to the latter. For the connection of one neuron to another, variation in the strength of impulse sent from the first to the latter, as neuroscientists believe, is crucial for our brain to process information; more discussions can be found in Chapter 15. In the simplest case of an abstract perceptron model, the strength of connection can be modelled by assigning different coefficient weights. After the neuron sums up the information received from the few neurons attached, if the total sum reaches a certain threshold, the neuron fires to the next neuron with output value 1; or with output value 0 otherwise.

Rosenblatt also proposed that perceptrons could perform visual tasks such as facial and object recognition. For instance, in recognizing the handwritten digit (as introduced in Chapter 6), a perceptron model can be used to differentiate whether a given handwritten digit is a certain number.

Minsky and Papert (1969) [18] criticized the effectiveness of perceptrons by showing that it can only solve limited types of questions perfectly and gave their mathematical evidence in a book titled *Perceptrons: an introduction to computational geometry*. For instance, no perceptron can handle the XOR (eXclusive OR) logic function, which receives two binary inputs. In Figure 11.2, the yellow dots represent the situation that exactly either one takes value 0 and the other one takes value 1; while the red crosses represent the situation that both take value 0 or both

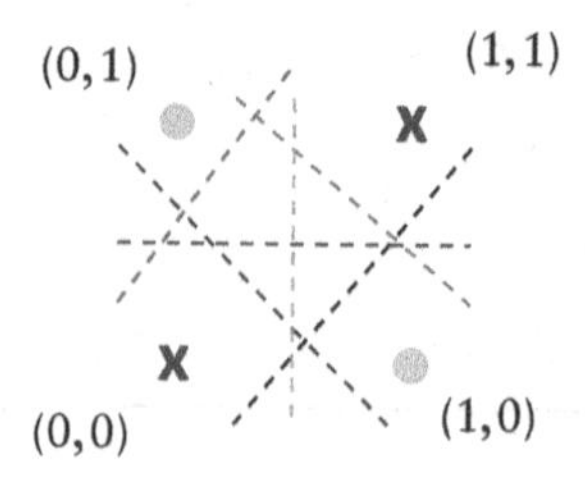

FIGURE 11.2 Two yellow dots and two red crosses which cannot be classified by any separating line.

take value 1. In the same book, Minsky and Papert suggested that this matter can be addressed by adding an additional "layer" of neurons such that the resulting multilayer neural network can deal with an arbitrary Boolean function. However, tuning a suitable multilayer neural network model was not broadly studied during that time; now this can commonly be accomplished by using *backpropagation* as we shall discuss in Chapter 15.

To understand how perceptrons work, we first lay down some notation. The i-th datum in a dataset is represented by $(\boldsymbol{x}_i, y_i)$, in which $\boldsymbol{x}_i = (x_{i0}, x_{i1}, \ldots, x_{ip})^\top$, where $x_{i0} = 1$ and $(x_{i1}, \ldots, x_{ip})^\top \in \mathbb{R}^p$ is a p-dimensional feature vector; and $y_i = +1$ or -1, which represents the label of this datum. The perceptron model is to learn a linear function $f : \mathbb{R}^{p+1} \to \mathbb{R}$, which classifies $\boldsymbol{x}_i$ in the category $+1$ (predicted label $\hat{y}_i = 1$) if $f(\boldsymbol{x}_i) > 0$ or in the category -1 (predicted label $\hat{y}_i = -1$) if $f(\boldsymbol{x}_i) < 0$. If $f(\boldsymbol{x}_i) = 0$, we consider it as undetermined, which barely happens in practice. In other words, the learner is $f(\boldsymbol{x}) = \boldsymbol{\omega}^\top \boldsymbol{x}$, where $\boldsymbol{\omega}$ is a vector of parameters to be estimated. A datum $(\boldsymbol{x}_i, y_i)$ is correctly classified if $f(\boldsymbol{x}_i) > 0$ with $y_i = 1$ or $f(\boldsymbol{x}_i) < 0$ with $y_i = -1$. In both cases, $f(\boldsymbol{x}_i) \cdot y_i > 0$. Similarly, a datum $(\boldsymbol{x}_i, y_i)$ is misclassified if $f(\boldsymbol{x}_i) \cdot y_i < 0$.

In the following, we illustrate with a planar example, in which a perceptron for two-dimensional feature vectors with a sample $\{(x_{i1}, x_{i2}, y_i)\}_{i=1}^n$ learns a linear function

$$f(\boldsymbol{x}) = \boldsymbol{\omega}^\top \boldsymbol{x} = \omega_0 + \omega_1 x_1 + \omega_2 x_2,$$

where $\boldsymbol{\omega} = (\omega_0, \omega_1, \omega_2)^\top$ and $\boldsymbol{x} = (1, x_1, x_2)^\top$. Geometrically, (x_{i1}, x_{i2}) is classified as category $+1$ (resp. -1) if it lies above (resp. below) the separation line $f(\boldsymbol{x}) = f(x_1, x_2) = 0$.

To design a criterion for obtaining suitable parameter $\boldsymbol{\omega}$, and hence obtaining the model, we observe from the separation line $f(x_1, x_2) = 0$ drawn in Figure 11.3 that the value $|f(x_{i1}, x_{i2})|$ increases if the point (x_{i1}, x_{i2}) lies further apart from the line, so it is reasonable to think of the modulus of $f(x_1, x_2)$ as an individual loss penalty corresponding to $\boldsymbol{x}$. This motivates us to define the loss function as

$$L(\boldsymbol{\omega}) := \sum_{i \in I_n} |f(\boldsymbol{x}_i)|,$$

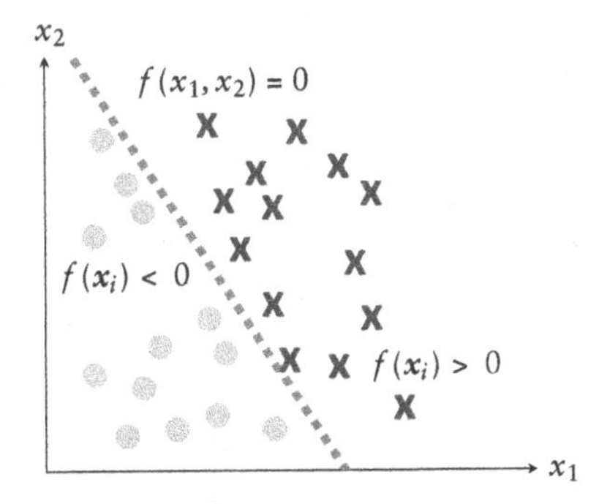

FIGURE 11.3 Perceptron: Category 1 (red crosses); Category 2 (yellow dots)

where $I_n := \{i : f(x_i) \cdot y_i < 0\}$ is the set of indexes of all misclassified data. An equivalent expression can be given by:

$$L(\boldsymbol{\omega}) = -\sum_{i \in I_n} y_i \boldsymbol{\omega}^\top \boldsymbol{x}_i = \sum_{i=1}^{n} \max(0, -y_i \boldsymbol{\omega}^\top \boldsymbol{x}_i), \tag{11.2}$$

and our goal now is to find the parameter $\boldsymbol{\omega} = (\omega_0, \omega_1, \ldots, \omega_p)^\top$, that minimizes the loss function L. One of the difficulties involved in this optimization problem is the implicit dependence of L on the parameter $\boldsymbol{\omega}$ through the index set I_n. We shall employ the *gradient descent* method, which can effectively refine the parameter values iteratively; also see Appendix *11.A.1 for an overview of the gradient descent method.

To this end, we first initialize $\boldsymbol{\omega}^{(0)}$, and then a sequence of estimates $\{\boldsymbol{\omega}^{(t)}\}_{t=1}^{\infty}$ is constructed as follows. For $t \in \mathbb{N}$, we define $I_n^{(t)}$ to be the set of indexes of misclassified data given the current linear function $f^{(t)}(\boldsymbol{x}) = \boldsymbol{x}^\top \boldsymbol{\omega}^{(t)}$ based on $\boldsymbol{\omega}^{(t)}$, and we also recursively define

$$L^{(t)}(\boldsymbol{\omega}) := \sum_{i \in I_n^{(t)}} |\boldsymbol{\omega}^\top \boldsymbol{x}_i| = -\sum_{i \in I_n^{(t)}} y_i \boldsymbol{\omega}^\top \boldsymbol{x}_i,$$

such that the new iterated estimate is given by

$$\boldsymbol{\omega}^{(t+1)} := \boldsymbol{\omega}^{(t)} - \eta \nabla_{\boldsymbol{\omega}} L^{(t)}(\boldsymbol{\omega}^{(t)}) = \boldsymbol{\omega}^{(t)} + \eta \sum_{i \in I_n^{(t)}} y_i \boldsymbol{x}_i,$$

where, $\eta > 0$ is a hyperparameter called the *learning rate*. The idea behind gradient descent is that the value of a function g decreases most rapidly in the direction $-\nabla g$. The iteration scheme above will be terminated when $\|\boldsymbol{\omega}^{(t+1)} - \boldsymbol{\omega}^{(t)}\|_2$ is less than some prescribed threshold $\varepsilon > 0$, or when $|L^{(t+1)}(\boldsymbol{\omega}^{(t+1)}) - L^{(t)}(\boldsymbol{\omega}^{(t)})| < \varepsilon$. Geometrically, as seen in Figure 11.4, we update the separation line $f(\boldsymbol{x}) = 0$ by correcting its intercept and the normal vector $\boldsymbol{\omega}_{-0} := (\omega_1, \omega_2)^\top$ step by step.

The gradient descent method described above is not recommended when the sample size n is large, as this results in a high demand of computational effort for the sum $\sum_{i \in I_n^{(t)}} y_i \boldsymbol{x}_i$ in each iteration. As a popular alternative, the *stochastic gradient descent* (SGD) method can be used; also see Appendix *11.A.1 for a brief introduction. In SGD, the parameter is updated by considering only one single datum in each iteration; and we repeat the procedures, aiming to obtain the ultimate estimate for $\boldsymbol{\omega}$. If the data happen to be truly linearly separable, that is, there exists $\boldsymbol{\omega}^* \in \mathbb{R}^{p+1}$ such that $\operatorname{sgn}((\boldsymbol{\omega}^*)^\top \boldsymbol{x}_i) = y_i$ for all i, then the convergence to optimal weights of $\boldsymbol{\omega}$ via stochastic gradient descent method can be guaranteed. To this end, by comparing $\boldsymbol{\omega}^{(t)}$ and $\boldsymbol{\omega}^{(t+1)}$, the respective parameters before and after each update, the loss function dealing with SGD at the $(t+1)$-th step is actually reduced. At each step, there are two scenarios to consider:

(1) if $(\boldsymbol{x}_{i_{t+1}}, y_{i_{t+1}})$ chosen is already correctly classified, then $\boldsymbol{\omega}^{(t+1)} = \boldsymbol{\omega}^{(t)}$; otherwise,

(2) the loss function to be minimized is $\ell_{i_{t+1}}^{(t+1)}(\boldsymbol{\omega}) = -y_{i_{t+1}} \boldsymbol{\omega}^\top \boldsymbol{x}_{i_{t+1}}$, where $(\boldsymbol{x}_{i_{t+1}}, y_{i_{t+1}})$ is the selected datum. Together with a learning rate η, the parameter $\boldsymbol{\omega}^{(t)}$ is updated

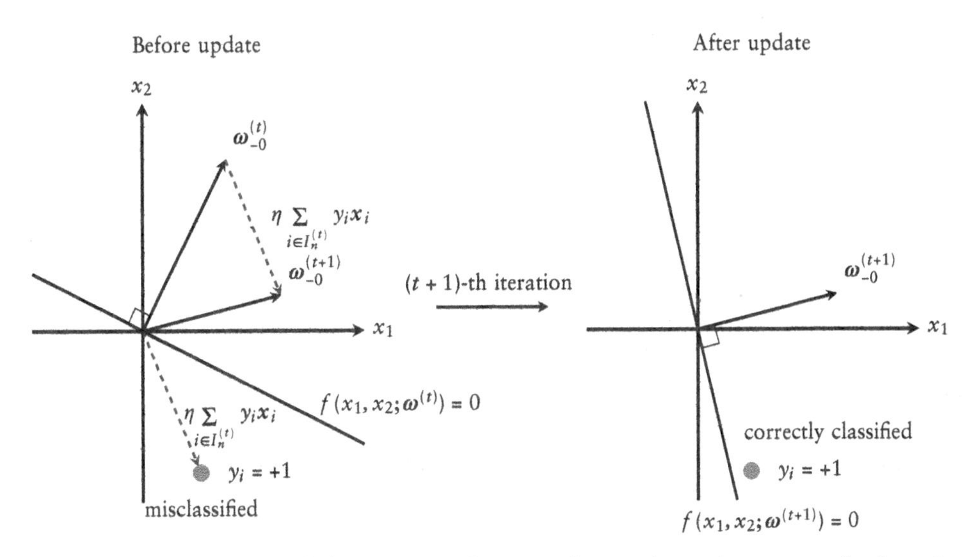

FIGURE 11.4 An update of the estimate for ω under gradient descent method at the $(t + 1)$-th iteration.

to $\omega^{(t+1)} = \omega^{(t)} + \eta y_{i_{t+1}} x_{i_{t+1}}$. Since $(x_{i_{t+1}}, y_{i_{t+1}})$ is misclassified before the update, the loss function $\ell_{i_{t+1}}^{(t+1)}(\omega^{(t)})$ is positive. Therefore,

$$\begin{aligned}
\ell_{i_{t+1}}^{(t+1)}(\omega^{(t+1)}) &= -y_{i_{t+1}}(\omega^{(t+1)})^{\top} x_{i_{t+1}} \\
&= -y_{i_{t+1}}(\omega^{(t)} + \eta y_{i_{t+1}} x_{i_{t+1}})^{\top} x_{i_{t+1}} \\
&= \ell_{i_{t+1}}^{(t+1)}(\omega^{(t)}) - \underbrace{\eta x_{i_{t+1}}^{\top} x_{i_{t+1}}}_{\text{strictly} > 0},
\end{aligned}$$

and the update corrects the weight vector in the direction of making the $(t + 1)$-th individual loss reduced.

Let us discuss the convergence of the SGD scheme to an optimal perceptron when the data truly are linearly separable, say by $\omega^* \in \mathbb{R}^{p+1}$ which is not unique in general (see Figure 11.5). In this case, there exists a strictly positive constant γ such that $y_i(\omega^*)^{\top} x_i \geq \gamma$ for all i. Let $R > 0$ be any constant such that $\|x_i\|_2^2 \leq R^2$ for all i; such an R can always be found as we only have finitely many data. For simplicity, we initialize the weights at 0, i.e. $\omega^{(0)} = \mathbf{0}$.

After a certain number of iterations, if there are no more misclassified points, the numerical scheme stops, and by definition, a perceptron classifier that correctly classifies all data is attained. Otherwise, if there are still misclassified points, modification of weights will still be carried out, and we now want to demonstrate that there is a ceiling for such a number of iterations, and so the optimal classifier should ultimately be obtained. The number of iterations we mean here is the effective number of

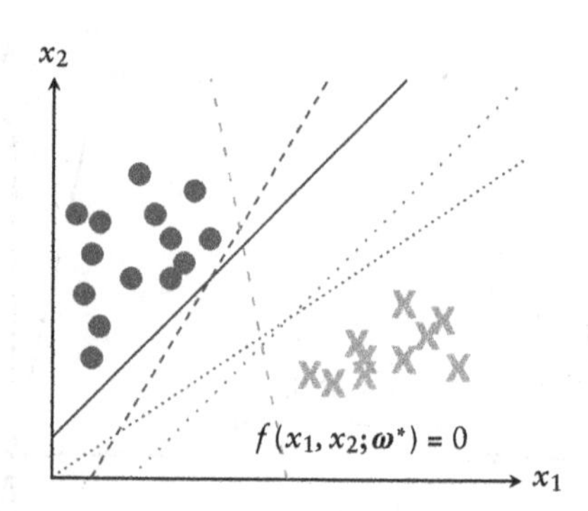

FIGURE 11.5 Some possible ω^*'s all of which satisfy $\text{sgn}((\omega^*)^\top x_i) = y_i$ for all i.

attempts so that $\omega^{(t+1)} \neq \omega^{(t)}$ for every step. We first observe that, at the t-th iteration step,

$$\begin{aligned}
(\omega^*)^\top \omega^{(t)} &= (\omega^*)^\top (\omega^{(t-1)} + \eta y_{i_t} x_{i_t}) \\
&\geq (\omega^*)^\top \omega^{(t-1)} + \eta\gamma \\
&\geq (\omega^*)^\top \omega^{(t-2)} + 2\eta\gamma \\
&\vdots \\
&\geq (\omega^*)^\top \omega^{(0)} + t\eta\gamma = t\eta\gamma > 0, \qquad (\text{since } \omega^{(0)} = \mathbf{0}).
\end{aligned} \tag{11.3}$$

On the other hand, if (x_{i_t}, y_{i_t}) is wrongly classified by $\omega^{(t-1)}$ so that an update is needed in the t-th iteration, then

$$\begin{aligned}
\|\omega^{(t)}\|_2^2 &= \|\omega^{(t-1)} + \eta y_{i_t} x_{i_t}\|_2^2 \\
&= \|\omega^{(t-1)}\|_2^2 + 2\eta y_{i_t} (\omega^{(t-1)})^\top x_{i_t} + \eta^2 \|x_{i_t}\|_2^2 \\
&\leq \|\omega^{(t-1)}\|_2^2 + \eta^2 \|x_{i_t}\|_2^2 \\
&\leq \|\omega^{(t-1)}\|_2^2 + \eta^2 R^2 \\
&\vdots \\
&\leq \eta^2 t R^2.
\end{aligned} \tag{11.4}$$

By the Cauchy–Schwarz inequality, $\|\omega^*\|_2 \cdot \|\omega^{(t)}\|_2 \geq (\omega^*)^\top \omega^{(t)} > 0$ because of (11.3). Furthermore, combining (11.3) with (11.4), we have the upper bound for the number of "effective" steps t^* (excluding those unchanged trivial steps) for identifying misclassified data points:

$$\|\omega^*\|_2 \cdot \sqrt{\eta^2 t R^2} \geq t\eta\gamma \qquad \text{which implies} \qquad t \leq t^* := \frac{R^2 \|\omega^*\|_2^2}{\gamma^2}, \tag{11.5}$$

which means that the effective number of updates required is at most t^*. In other words, there can be no more misclassified points beyond t^* steps, which holds regardless of the value of the learning rate η.

Logistic regression is another commonly used method for classification problems with binary responses, as we introduced in Section 10.4. A perceptron is essentially logistic regression by replacing the logit transform by the sign function (a step function) which can be considered as the derivative of the *Rectified Linear Unit* (ReLU) function. A generic numerical comparison between these two tools is given in Figure 11.6. The two illustrations on the top show two datasets that are linearly separable and the other two illustrations at the bottom show inseparable situations[1]. The advantage of using logistic regression over a perceptron is clear in classifying linearly

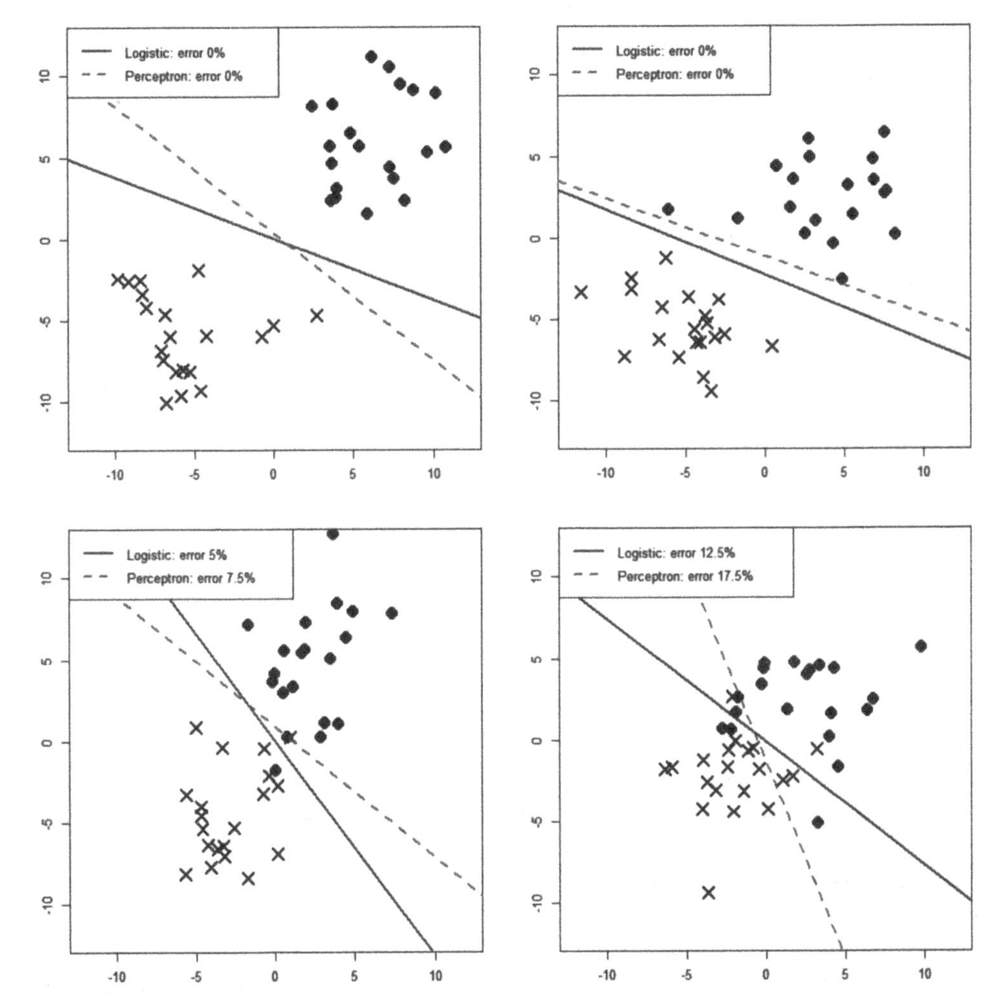

FIGURE 11.6 Performance of logistic regression represented by solid lines, and perceptron represented by dashed lines.

[1] 30 iterations are performed with a prior termination condition for logistic regression if (10.27) is satisfied for $\varepsilon = 0.01\%$.

separable data: though both approaches correctly classify two datasets, separating lines generated by logistic regression tends to separate the two groups more evenly, but those produced by a perceptron are slightly off-center. It is because the stochastic gradient descent used in perceptron ceases to update parameters once all the data are correctly classified, while IRLS for logistic regression picks the best possible one, in principle.

Finally, we illustrate the implementation of perceptron in Programme 11.1 via Python (resp. Programme 11.2 via **R**).

```
1  import numpy as np
2
3  def perceptron(X, y, eta=1e-3, n_iter=1e5):
4      X, y = np.array(X), np.array(y)
5      X_1 = np.c_[np.ones(len(y)), X]
6      n, p = X_1.shape
7      omega = np.repeat(1/p, p)              # initialize weight vector
8      for it in range(int(n_iter)):
9          i = it % n                         # Python index starts from 0
10         if i == 0:
11             new_idx = np.random.choice(np.arange(n), n, replace=False)
12
13         j = new_idx[i]
14         if y[j] * omega @ X_1[j] < 0:      # misclassified
15             omega += eta * y[j] * X_1[j]
16
17     return omega, np.sign(X_1 @ omega)
```

Programme 11.1 Perceptron function in Python.

```
1  > perceptron <- function(X, y, eta=1e-3, n_iter=1e5){
2  +   X <- as.matrix(X); y <- as.numeric(y)
3  +   X_1 <- cbind(1, X)
4  +   n <- nrow(X_1); p <- ncol(X_1)
5  +   omega <- rep(1/p, p)                  # initialize weight vector
6  +   for (it in 0:(n_iter-1)){
7  +     i <- it %% n + 1                    # R index starts from 1
8  +     if (i == 1) new_idx <- sample(1:n, n, replace=FALSE)
9  +     j <- new_idx[i]
10 +     if(y[j] * omega%*%X_1[j,] < 0) {    # misclassified
11 +       omega <- omega + eta * y[j] * X_1[j,]
12 +     }
13 +   }
14 +   return (list(omega=omega, y_pred=sign(X_1%*%omega)))
15 + }
```

Programme 11.2 Perceptron function in R.

Let us apply this `perceptron()` function in Programme 11.1 (resp. Programme 11.2) to the HSI financial ratio dataset in Programme 11.3 via Python (resp. Programme 11.4 via **R**). From the output of Programme 11.3 (resp. Programme 11.4), we see that the accuracy rate is $(646 + 23)/680 = 98.38\%$ (resp. $(646 + 24)/680 = 98.53\%$).

```
1  import pandas as pd
2  from sklearn.metrics import confusion_matrix
3
4  df = pd.read_csv("fin-ratio.csv")
5  X, y = df.drop(columns=["HSI"]), df["HSI"]
6  y = y.replace(0, -1)                          # transform the output to {-1, 1}
7
8  np.random.seed(4002)
9  omega, y_pred = perceptron(X, y, eta=0.01, n_iter=1e6)
10 conf = confusion_matrix(y_pred, y)          # confusion matrix
11 print(conf)
12 print(sum(np.diag(conf))/len(y))
13 print(omega)
```

```
1  [[646    9]
2   [  2   23]]
3  0.9838235294117647
4  [-17.04714286    0.53385014    0.46897014    1.79111614   -0.17228286
5    -0.11764886    0.05075914]
```

Programme 11.3 Perceptron using Programme 11.1 for the HSI financial ratio dataset in Python.

```
1  > df <- read.csv("fin-ratio.csv")
2  > X <- df[-ncol(df)]
3  > y <- df$HSI
4  > y[y==0] <- -1                           # transform the output to {-1, 1}
5  >
6  > set.seed(4002)
7  > model <- perceptron(X, y, eta=0.01, n_iter=1e6)
8  > (conf <- table(model$y_pred, y))       # confusion matrix
9      y
10         -1    1
11     -1 646    8
12     1    2   24
13 > sum(diag(conf))/length(y)
14 [1] 0.9852941
15 > model$omega
16                           EY           CFTP            ln_MV
17 -17.08714286   0.47466614   0.83761614   1.85155314
18            DY          BTME           DTE
19  -0.23556286  -0.34777686   0.02343714
```

Programme 11.4 Perceptron using Programme 11.2 for the HSI financial ratio dataset in R.

11.2 SUPPORT VECTOR MACHINE

Support vector machines (SVMs) choose a separating hyperplane which maximizes the margin between the two classes. In other words, the distance from the hyperplane to the nearest data point from each category is maximized. There are a number of monographs that discuss this tool in detail, such as [1, 10, 11, 23].

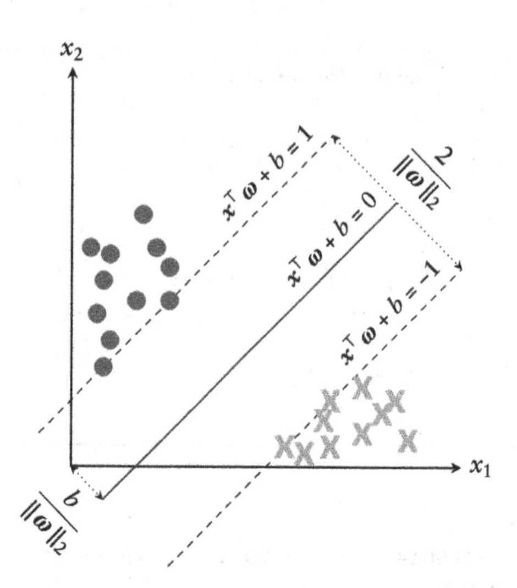

FIGURE 11.7 An example of an SVM model for two-dimensional feature vectors.

For choosing the hyperplane, SVM does not make use of all the available data. Instead, only the points, called *support vectors*, that are closest to the hyperplane are relevant; see Figure 11.7. Support vectors are used to define the hyperplane, and at the same time, they are the most difficult to be classified correctly; they carry the most important information for the classification problem.

Let x_0 be an arbitrary point lying on the plane. For any $x \in \mathbb{R}^p$, standard coordinate geometry tells us that the distance $d(x)$ between x and the hyperplane $x^\top \omega + b = 0$[2] is given by

$$d(x) = \|\text{proj}_\omega(x_0 - x)\|_2 = \left\| \frac{x_0^\top \omega - x^\top \omega}{\|\omega\|_2^2} \omega \right\|_2 = \frac{|x^\top \omega + b|}{\|\omega\|_2}. \tag{11.6}$$

If x is correctly classified by the plane, (11.6) can be further written as

$$d(x) = \frac{|x^\top \omega + b|}{\|\omega\|_2} = y \cdot \frac{x^\top \omega + b}{\|\omega\|_2},$$

where $y \in \{-1, 1\}$ is the actual label of x. Now, for a given dataset $\{(x_i, y_i)\}_{i=1}^n$, define

$$\gamma_i := \begin{cases} d(x_i), & \text{if } x_i \text{ is correctly classified;} \\ -d(x_i), & \text{if } x_i \text{ is misclassified.} \end{cases}$$

The *geometric margin* is then defined by

$$\gamma := \min_{i=1,\ldots,n} \gamma_i.$$

[2] Note that for any $c \in \mathbb{R}$, $x^\top \omega + b = 0$ whenever $x^\top (c\omega) + (cb) = 0$. This means the hyperplane remains unchanged after parameters' rescaling. More technically, the hyperplane is parametrized with a parameter vector lying on a *projective line*.

Suppose that the dataset is linearly separable, then there exists (ω, b) such that all the data are correctly classified by the hyperplane $x^\top \omega + b = 0$. Our goal is to look for (ω, b) such that this separating hyperplane is as far as possible from the two groups of points. In other words, we aim to find the optimal hyperplane that solves the following optimization problem:

$$\begin{aligned} \underset{\omega,b}{\text{maximize}} \quad & \gamma, \\ \text{subject to} \quad & y_i(x_i^\top \omega + b) \geq 0, \quad \forall i = 1, \ldots, n. \end{aligned} \tag{11.7}$$

Note that for any feasible solution (ω, b), we can always find an alternative separating hyperplane that is parallel to the former with parameters (ω, b') and sits in the middle between the two groups. Simple geometry tells us that the corresponding geometric margin of the alternative separating hyperplane is greater than that of the original feasible hyperplane. Finally, as pointed out in the footnote, the definition of a separating hyperplane is invariant with rescaling of parameter vector, we can assume that $y_i(x_i^\top \omega + b) \geq 1$ for all $i = 1, 2, \ldots, n$, so that the equality holds for the two points closest to the separating hyperplane, hence $\gamma = 1/\|\omega\|_2$. Now, Problem 11.7 can be converted as:

$$\begin{aligned} \underset{\omega,b}{\text{maximize}} \quad & \frac{1}{\|\omega\|_2}, \\ \text{subject to} \quad & y_i(x_i^\top \omega + b) \geq 1, \quad \forall i = 1, \ldots, n. \end{aligned} \tag{11.8}$$

The maximization problem of Problem (11.8) can be seen equivalent to:

$$\min_{\omega,b} \frac{1}{2} \|\omega\|_2^2,$$

subject to the same boundary conditions as before. Here we square the term $\|\omega\|_2$ to ensure the smoothness of this new objective function, while the fraction $1/2$ is introduced for the ease of computation. In summary, Problem (11.7) is equivalent to the following problem:

$$\begin{aligned} \underset{\omega,b}{\text{minimize}} \quad & \frac{1}{2} \omega^\top \omega, \\ \text{subject to} \quad & y_i(x_i^\top \omega + b) \geq 1, \quad \text{for all } i = 1, \ldots, n. \end{aligned} \tag{11.9}$$

11.2.1 Primal and Dual Problems in SVM

One can observe that (11.9) is indeed a *Quadratic Programming* (QP) problem. To solve this (primal) problem of finding the optimal parameters ω^* and b^* in (11.9), we can adopt the duality approach for QP problems by maximizing the *dual Lagrangian* (the corresponding *dual problem*); see Appendix *11.A.2.3 for more details. To this end, we first write down the Lagrangian:

$$\mathcal{L}(\boldsymbol{\omega}, b, \lambda_1, \ldots, \lambda_n) = \frac{1}{2}\boldsymbol{\omega}^\top\boldsymbol{\omega} - \sum_{i=1}^{n} \lambda_i\{y_i(\boldsymbol{x}_i^\top\boldsymbol{\omega} + b) - 1\}, \quad \lambda_i \geq 0, i = 1, \ldots, n.$$

The partial derivative of this Lagrangian with respect to $\boldsymbol{\omega}$ and b are respectively given by:

$$\nabla_{\boldsymbol{\omega}}\mathcal{L}(\boldsymbol{\omega}, b, \lambda_1, \ldots, \lambda_n) = \boldsymbol{\omega} - \sum_{i=1}^{n} \lambda_i y_i \boldsymbol{x}_i;$$

$$\frac{\partial \mathcal{L}(\boldsymbol{\omega}, b, \lambda_1, \ldots, \lambda_n)}{\partial b} = -\sum_{i=1}^{n} \lambda_i y_i.$$

When we equate these partial derivatives to zero, we obtain

$$\boldsymbol{\omega} = \sum_{i=1}^{n} \lambda_i y_i \boldsymbol{x}_i \quad \text{and} \quad \sum_{i=1}^{n} \lambda_i y_i = 0. \tag{11.10}$$

Notice that the first equation of (11.10) states that the optimal value of $\boldsymbol{\omega}$ is a linear combination of $\boldsymbol{x}_i$'s. Finally, substituting these expressions back to the Lagrangian so as to eliminate $\boldsymbol{\omega}$ and b, we derive the dual Lagrangian:

$$\begin{aligned}
D(\lambda_1, \cdots, \lambda_n) &= \frac{1}{2}\boldsymbol{\omega}^\top\boldsymbol{\omega} - \sum_{i=1}^{n} \lambda_i\{y_i(\boldsymbol{x}_i^\top\boldsymbol{\omega} + b) - 1\} \\
&= \frac{1}{2}\boldsymbol{\omega}^\top\boldsymbol{\omega} - \sum_{i=1}^{n} \lambda_i y_i \boldsymbol{x}_i^\top\boldsymbol{\omega} - b\sum_{i=1}^{n} \lambda_i y_i + \sum_{i=1}^{n} \lambda_i \\
&= \frac{1}{2}\boldsymbol{\omega}^\top\boldsymbol{\omega} - \boldsymbol{\omega}^\top\boldsymbol{\omega} + \sum_{i=1}^{n} \lambda_i = \sum_{i=1}^{n} \lambda_i - \frac{1}{2}\boldsymbol{\omega}^\top\boldsymbol{\omega} \\
&= \sum_{i=1}^{n} \lambda_i - \frac{1}{2}\sum_{i=1}^{n}\sum_{\ell=1}^{n} \lambda_i\lambda_\ell y_i y_\ell \boldsymbol{x}_i^\top\boldsymbol{x}_\ell.
\end{aligned}$$

Consequently, the dual problem is written as:

$$\begin{aligned}
&\underset{\lambda_1, \ldots, \lambda_n}{\text{maximize}} && \sum_{i=1}^{n} \lambda_i - \frac{1}{2}\sum_{i=1}^{n}\sum_{\ell=1}^{n} \lambda_i\lambda_\ell y_i y_\ell \boldsymbol{x}_i^\top\boldsymbol{x}_\ell, \\
&\text{subject to} && \text{i)} \ \sum_{i=1}^{n} \lambda_i y_i = 0; \\
& && \text{ii)} \ \lambda_1, \ldots, \lambda_n \geq 0.
\end{aligned} \tag{11.11}$$

More precisely, with strong duality [5, 21, 24] in place, the minimum value attained by Problem (11.9) and the maximum value attained by (11.11) coincide, and their respective solutions are related by the *complementary slackness condition*:

$$\lambda_i\{y_i(\boldsymbol{x}_i^\top\boldsymbol{\omega}+b)-1\}=0, \quad \text{for all } i=1,\ldots,n. \tag{11.12}$$

To obtain the optimal parameters $\boldsymbol{\omega}^*$ and b^* in (11.9), one can first solve the dual problem (11.11) to obtain λ_i^*'s. It turns out that in most cases, most of these λ_i^*'s are equal to zero. A feature vector $\boldsymbol{x}_i$ is called a *support vector* if the corresponding multiplier λ_i^* is strictly positive, then $y_i\boldsymbol{x}_i^\top\boldsymbol{\omega}+b=1$. After deriving the values of λ_i^*'s from the dual problem, $\boldsymbol{\omega}^*$ can then be obtained from (11.10), which is a linear combination of support vectors. On the other hand, from the complementary slackness condition (11.12), the optimal b^* can be obtained from any support vector $\boldsymbol{x}_i$ (with $\lambda_i^*>0$) as

$$b^*=\frac{1}{y_i}-\boldsymbol{x}_i^\top\boldsymbol{\omega}=y_i-\boldsymbol{x}_i^\top\boldsymbol{\omega}=y_i-\sum_{k=1}^{n}\lambda_k y_k\boldsymbol{x}_i^\top\boldsymbol{x}_k.$$

Furthermore, to obtain a more robust result, we may apply the simple trick of averaging the b^*'s obtained from all support vectors:

$$b^*=\frac{1}{|\{i:\lambda_i^*>0\}|}\sum_{\{i:\lambda_i^*>0\}}\left\{y_i-\sum_{k=1}^{n}\lambda_k y_k\boldsymbol{x}_i^\top\boldsymbol{x}_k\right\}.$$

11.2.2 Kernel Trick for Coping with Non-Linearity

As mentioned before, some datasets are only non-linearly separable in nature, such as the dataset shown in Figure 11.8a. However, if we manage to transform the original space into a higher dimensional space, there is a chance that the data points might become linearly separable in the new transformed space. This transformation method is called the *kernel trick*.

Figure 11.8 illustrates a dataset with a two-dimensional feature vector $\boldsymbol{x}=(x_1,x_2)^\top$, labeled in blue dots and green crosses. The original data points as shown in Figure 11.8a cannot by separated by any linear line. However, after adding one more dimension given by $K(x_1,x_2)=x_1^2+x_2^2$ to the feature vector as shown in Figure 11.8b, a plane can now be used to separate the blue dots and green crosses perfectly.

Alternatively, one can also think of the transformation $\boldsymbol{\phi}:(x_1,x_2)^\top\mapsto\left(x_1^2,\sqrt{2}x_1x_2,x_2^2\right)^\top$; also see Figure 11.9. This choice is intimately connected with the *quadratic kernel transformation* in the sense that, for any $\boldsymbol{x}_i,\boldsymbol{x}_\ell\in\mathbb{R}^2$,

$$\boldsymbol{\phi}(\boldsymbol{x}_i)^\top\boldsymbol{\phi}(\boldsymbol{x}_\ell)=\left(\boldsymbol{x}_i^\top\boldsymbol{x}_\ell\right)^2,$$

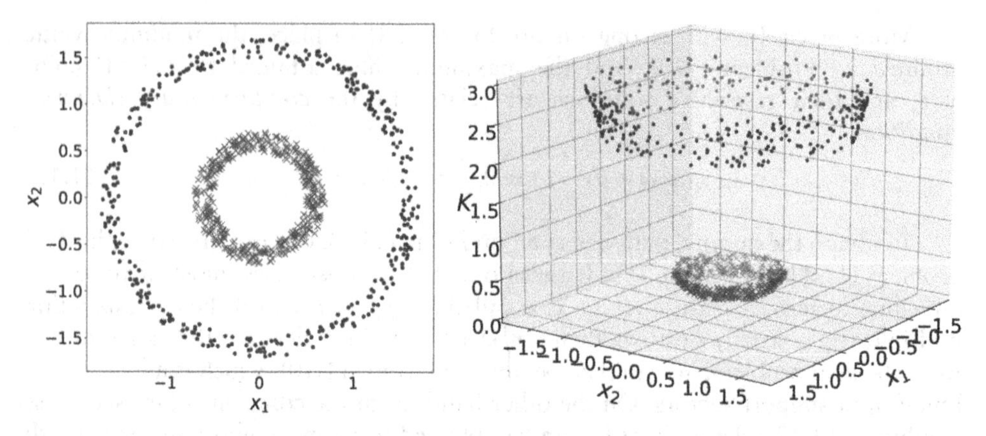

(a) Non-linearly separable. (b) Linearly separable after transformation.

FIGURE 11.8 Kernel trick: 2D to 3D transformation.

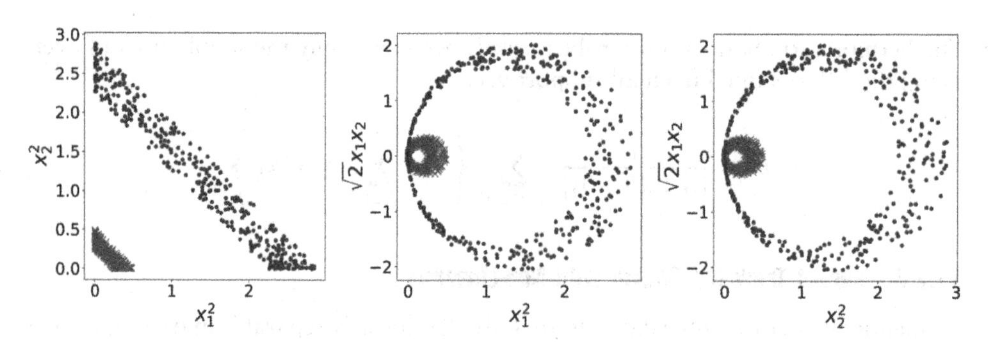

FIGURE 11.9 Applying quadratic kernel transformation to the data points generated in Figure 11.8a.

which can be seen as follows:

$$\begin{aligned}
\boldsymbol{\phi}(\boldsymbol{x}_i)^\top \boldsymbol{\phi}(\boldsymbol{x}_\ell) &= \begin{pmatrix} x_{1,i}^2 & \sqrt{2}x_{1,i}x_{2,i} & x_{2,i}^2 \end{pmatrix} \begin{pmatrix} x_{1,\ell}^2 \\ \sqrt{2}x_{1,\ell}x_{2,\ell} \\ x_{2,\ell}^2 \end{pmatrix} \\
&= \left(x_{1,i}x_{1,\ell}\right)^2 + 2x_{1,i}x_{2,i}x_{1,\ell}x_{2,\ell} + \left(x_{2,i}x_{2,\ell}\right)^2 \\
&= \left(x_{1,i}x_{1,\ell} + x_{2,i}x_{2,\ell}\right)^2 = \left(\boldsymbol{x}_i^\top \boldsymbol{x}_\ell\right)^2.
\end{aligned}$$

In most applications, instead of specifying the form of transformation to be used, the *kernel* of the transformation is proposed. Assume the required transformation function is $\boldsymbol{\phi} : \boldsymbol{x} \mapsto \boldsymbol{\phi}(\boldsymbol{x})$. In the dual problem of SVM in (11.11), the transformation

$\boldsymbol{\phi}$ is applied to $\boldsymbol{x}_i$ and $\boldsymbol{x}_\ell$ separately, that is,

$$\max_{\lambda_1,\ldots,\lambda_n} \left(\sum_{i=1}^{n} \lambda_i - \frac{1}{2} \sum_{i=1}^{n} \sum_{\ell=1}^{n} \lambda_i \lambda_\ell y_i y_\ell \boldsymbol{\phi}(\boldsymbol{x}_i)^\top \boldsymbol{\phi}(\boldsymbol{x}_\ell) \right), \tag{11.13}$$

which is computationally inefficient. It is more convenient to pick a *kernel* K such that, for any $\boldsymbol{x}_i, \boldsymbol{x}_\ell \in \mathbb{R}^p$,

$$K(\boldsymbol{x}_i, \boldsymbol{x}_\ell) = \boldsymbol{\phi}(\boldsymbol{x}_i)^\top \boldsymbol{\phi}(\boldsymbol{x}_\ell), \tag{11.14}$$

which is symmetric in its two arguments. One of the most widely used kernel functions is the *Radial Basis Function* (RBF) kernel: for any $\boldsymbol{x}_i, \boldsymbol{x}_\ell \in \mathbb{R}^p$,

$$K_{\mathrm{RBF}}(\boldsymbol{x}_i, \boldsymbol{x}_\ell) = \exp\left(-\frac{\|\boldsymbol{x}_i - \boldsymbol{x}_\ell\|_2^2}{2\sigma^2} \right),$$

where σ is a hyperparameter that determines the regularity and connectedness of the separating boundary hypersurface in the original space of $\boldsymbol{x}$. For instance, Figure 11.10 depicts the boundaries of the separating hypersurfaces obtained in SVM with the use of RBF kernels with different σ's. The original space is two-dimensional and the data points are transformed into infinite-dimensional ones via the RBF kernel. When σ is small, say $\sigma = 0.3$ as in Figure 11.10a, the zooming-in effect of the corresponding kernel is strong, allowing the two classes to be separated easily. The resulting separation boundary in the original two-dimensional space tends to be clear and sharp, yet it is more prone to overfitting. When σ is large, say $\sigma = 1$ as in Figure 11.10b, the zooming-out effect of the resulting kernel comes into play

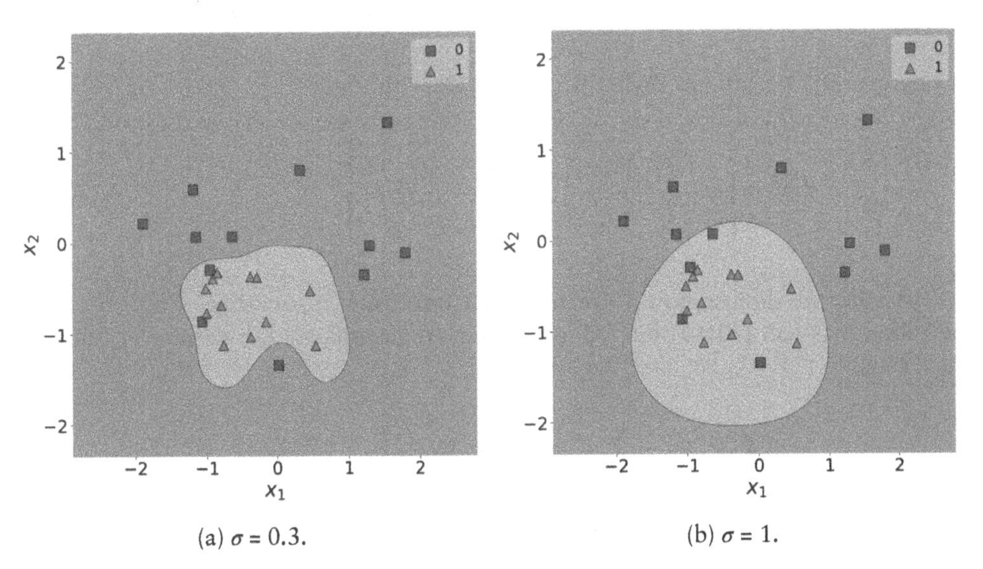

(a) $\sigma = 0.3$. (b) $\sigma = 1$.

FIGURE 11.10 SVM with RBF kernel resulting in different boundaries of the separating hypersurfaces shown in the original two-dimensional space.

and the separation boundary in the original two-dimensional space tends to be smoother and more connected in one piece. Though it often makes some occasional misclassifications, the predictions tend to be more robust.

For the RBF kernel, we can actually identify the corresponding transformation $\boldsymbol{\phi}$ as follows. This $\boldsymbol{\phi}$ actually lifts a finite-dimensional datum from $\mathbb{R}^p$ to the infinite-dimensional space of $\mathbb{R}^\infty$. To see this, we observe that:

$$\begin{aligned} K_{\text{RBF}}(\boldsymbol{x}_i, \boldsymbol{x}_\ell) &= \exp\left(-\frac{\|\boldsymbol{x}_i\|_2^2}{2\sigma^2}\right) \cdot \exp\left(\frac{\boldsymbol{x}_i^\top \boldsymbol{x}_\ell}{\sigma^2}\right) \cdot \exp\left(-\frac{\|\boldsymbol{x}_\ell\|_2^2}{2\sigma^2}\right) \\ &= \sum_{j=0}^{\infty} \frac{(\boldsymbol{x}_i^\top \boldsymbol{x}_\ell)^j}{j!\sigma^{2j}} \cdot \exp\left(-\frac{\|\boldsymbol{x}_i\|_2^2}{2\sigma^2}\right) \cdot \exp\left(-\frac{\|\boldsymbol{x}_\ell\|_2^2}{2\sigma^2}\right) \\ &= \sum_{j=0}^{\infty} \sum_{r_1+\cdots+r_p=j} \exp\left(-\frac{\|\boldsymbol{x}_i\|_2^2}{2\sigma^2}\right) \cdot \frac{(x_{i,1})^{r_1}\cdots(x_{i,p})^{r_p}}{\sigma^j\sqrt{r_1!\cdots r_p!}} \\ &\qquad \cdot \exp\left(-\frac{\|\boldsymbol{x}_\ell\|_2^2}{2\sigma^2}\right) \cdot \frac{(x_{\ell,1})^{r_1}\cdots(x_{\ell,p})^{r_p}}{\sigma^j\sqrt{r_1!\cdots r_p!}}, \end{aligned}$$

where the second line follows from the Taylor expansion $e^x = \sum_{j=0}^{\infty} \frac{x^j}{j!}$; and the last line follows from the multinomial expansion:

$$(\boldsymbol{x}_i^\top \boldsymbol{x}_\ell)^j = (x_{i,1}x_{\ell,1} + \cdots + x_{i,p}x_{\ell,p})^j = \sum_{r_1+\cdots+r_p=j} \frac{j!}{r_1!r_2!\cdots r_p!} \prod_{k=1}^{p} (x_{i,k}x_{\ell,k})^{r_k}.$$

Consequently, we can just define $\boldsymbol{\phi} : \mathbb{R}^p \to \mathbb{R}^\infty$ as follows:

$$\boldsymbol{\phi}(\boldsymbol{x}) = \left(\exp\left(-\frac{\|\boldsymbol{x}\|_2^2}{2\sigma^2}\right) \cdot \frac{(x_1)^{r_1}\cdots(x_p)^{r_p}}{\sigma^j\sqrt{r_1!\cdots r_p!}} \right)^\top_{j=1,2,\ldots;r_1+\cdots+r_p=j}.$$

Finally, we discuss how to recover the separating hypersurface once the optimization problem (11.13) has been resolved. In general, with a kernel function K, according to (11.10), we have the relation:

$$\boldsymbol{\phi}(\boldsymbol{x})^\top \boldsymbol{\omega}^* = \sum_{i=1}^{n} \lambda_i^* y_i \boldsymbol{\phi}(\boldsymbol{x})^\top \boldsymbol{\phi}(\boldsymbol{x}_i) = \sum_{i=1}^{n} \lambda_i^* y_i K(\boldsymbol{x}_i, \boldsymbol{x}).$$

Therefore, the equation of the separating hypersurface is given by:

$$f_{\boldsymbol{\omega},b}(\boldsymbol{x}) = \text{sgn}(\boldsymbol{\phi}(\boldsymbol{x})^\top \boldsymbol{\omega}^* + b^*) = \text{sgn}\left(\sum_{i=1}^{n} \lambda_i^* y_i K(\boldsymbol{x}_i, \boldsymbol{x}) + b^*\right) = 0.$$

11.2.3 Soft-Margin SVM for Dealing with Noise

In the previous subsection, we focused on the case where the data points are not linearly separable in their original space, but they can be separated after a transformation, such as a higher-degree polynomial. In this subsection, we focus on dataset with noise, as in Figure 11.11.

When dealing with noise, we impose some penalties to data points lying on the wrong side of the separation boundary. To this end, we introduce a *hinge loss function* for SVM:

$$\max\left\{0, 1 - y_i\left(x_i^{\mathsf{T}}\omega + b\right)\right\}.$$

In other words, if the i-th data point lies on the correct side of the separating hyperplane, i.e. $y_i(x_i^{\mathsf{T}}\omega + b) \geq 1$, the individual hinge loss is zero. On the other hand, if the i-th data point lies on the wrong side of the separation boundary, $y_i(x_i^{\mathsf{T}}\omega + b) < 1$, and the individual hinge loss is almost proportional to its distance from the separating hyperplane. The objective function for SVM with noise is the following cost function[3], which will be minimized:

$$C\|\omega\|_2^2 + \frac{1}{n}\sum_{i=1}^{n}\max\left\{0, 1 - y_i\left(x_i^{\mathsf{T}}\omega + b\right)\right\}, \tag{11.15}$$

where C is the hyperparameter that determines the tradeoff between increasing the geometric margin and ensuring that most of the x_i's lie on the correct side of the separating hyperplane. For sufficiently large values of C, the hinge loss function becomes negligible, so SVM tries to find the optimal solution with the greatest geometric margin by essentially ignoring misclassification. As we reduce the value of C, making classification errors becomes more costly, so SVM tries to make fewer mistakes by

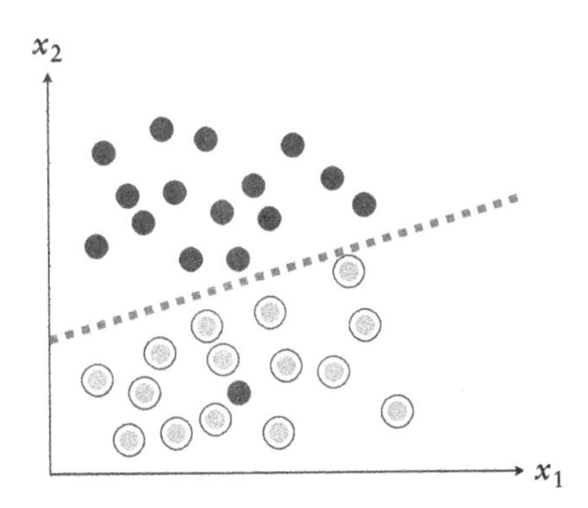

FIGURE 11.11 Example of the existence of some outliers or noise in a dataset.

[3] The SVM that optimizes the hinge loss is called a *soft-margin SVM*, while the original formulation (11.9) is referred to as a *hard-margin SVM*.

sacrificing the geometric margin size. However, a larger margin is better for generalization. Therefore, C regulates the balance between classifying the training data well (minimizing empirical risk) and classifying future examples well (generalization) with better out-sample fitting.

11.2.4 Application to Credit Scoring

Banks earn interest from loans, and it is crucial for banks to examine the credit scores of applicants for loans and approve applications accordingly. Let us use the credit scoring dataset [9][4] to illustrate the effect of σ for the RBF kernel on the generalization power of SVM. The dataset includes seven feature variables, such as `age`, `DebtRatio`, and `MonthlyIncome`, and one binary response variable `SeriousDlqin2yrs` representing whether an individual experienced serious delinquency (≥ 90 days past due of payment) in the last two years. For this example, we split the dataset randomly into a ratio of 8:2 for the training and test datasets. See the Python codes in Programme 11.5 and R codes in Programme 11.6, respectively.

```
1  import pandas as pd
2  import numpy as np
3  from tqdm import tqdm
4  from sklearn.model_selection import train_test_split
5  from sklearn.metrics import accuracy_score, confusion_matrix
6  from sklearn.svm import SVC
7
8  df = pd.read_csv("credit_scoring_sample.csv")
9  df = df.dropna()
10
11 y = df.SeriousDlqin2yrs.values
12 X = df.drop(columns="SeriousDlqin2yrs").to_numpy()
13 (X_train, X_test,
14  y_train, y_test) = train_test_split(X, y, test_size=0.2,
15                                      random_state=4002)
16
17 # exp(-gamma \norm{x-x'}**2); gamma = 1/(2*sigma2)
18 gamma = lambda sig: 1/(2*sig**2)
19 sigma = [0.5, 1, 5, 10, 50, 100]
20
21 train_acc, test_acc = list(), list()
22 for sig in tqdm(sigma, total=len(sigma)):
23     svm_rbf = SVC(kernel="rbf", gamma=gamma(sig))
24     svm_rbf.fit(X_train, y_train)
25     ypred_train = svm_rbf.predict(X_train)
26     ypred_test = svm_rbf.predict(X_test)
27     train_acc.append(accuracy_score(ypred_train, y_train))
28     test_acc.append(accuracy_score(ypred_test, y_test))
29
```

[4] The dataset can be downloaded from https://www.kaggle.com/code/kashnitsky/topic-1-exploratory-data-analysis-with-pandas/data?select=credit_scoring_sample.csv.

```
30 colnames = ["sigma", "gamma", "train-accuarcy", "test-accuracy"]
31 print(pd.DataFrame(dict(zip(colnames, [sigma, gamma(np.array(sigma)),
32                                          train_acc, test_acc]))))
33
34 import matplotlib.pyplot as plt
35
36 plt.rc('font', size=20); plt.rc('axes', labelsize=25)
37
38 fig = plt.figure(figsize=(10, 8))
39 plt.plot(sigma, test_acc, c="orange", linewidth=4, label="Test")
40 plt.plot(sigma, train_acc, "b--", linewidth=4, label="Train")
41 plt.ylim(0.75, 1)
42 plt.xlabel("sigma")
43 plt.ylabel("accuracy")
44 plt.legend(fontsize=20)
45
46 fig = plt.figure(figsize=(10, 8))
47 plt.plot(sigma, test_acc, c="orange", linewidth=4)
48 plt.xlabel("sigma")
49 plt.ylabel("accuracy")
```

```
1      sigma    gamma  train-accuarcy  test-accuracy
2 0      0.5  2.00000        0.984452       0.758100
3 1      1.0  0.50000        0.957063       0.760983
4 2      5.0  0.02000        0.826744       0.764003
5 3     10.0  0.00500        0.790809       0.764141
6 4     50.0  0.00020        0.775021       0.766063
7 5    100.0  0.00005        0.774677       0.766337
```

Programme 11.5 Application of SVM with the RBF kernel for different values of σ in Python.

For Python in Programme 11.5, the SVM model is implemented via the `SVC` function provided in the `sklearn` library. In particular, two arguments are required in `SVC`, namely `gamma` defined as $\gamma = 1/(2\sigma^2)$, and `kernel="rbf"` for using the radial basis function defined by $\exp(-\gamma\|\boldsymbol{x}_i - \boldsymbol{x}_\ell\|_2^2)$ as the kernel function. Based on the results obtained, a small (resp. large) value of σ yields a higher (resp. lower) training accuracy and a lower (resp. higher) testing accuracy. Indeed, the SVM model with a larger σ is less likely to result in overfitting, and is therefore better at generalizing. Figure 11.12a shows the training and testing accuracies of the SVM models against different values of σ for the RBF kernel. Clearly, the training accuracy levels as σ increases beyond 50, hence for this particular dataset, we may choose this kink point at the value for σ. This choice can be further justified by the leveling-out behavior of the test accuracy at $\sigma = 50$, as shown in Figure 11.12b.

Similarly, in Programme 11.6 via **R**, the SVM model is implemented via the `svm()` function provided in the library `e1071`. In particular, we adopt the RBF kernel via the argument `kernel="radial"` with `gamma`, γ parameter defined equivalently as in `SVC` in Python. The outcomes in Figure 11.13 agree with those in Python.

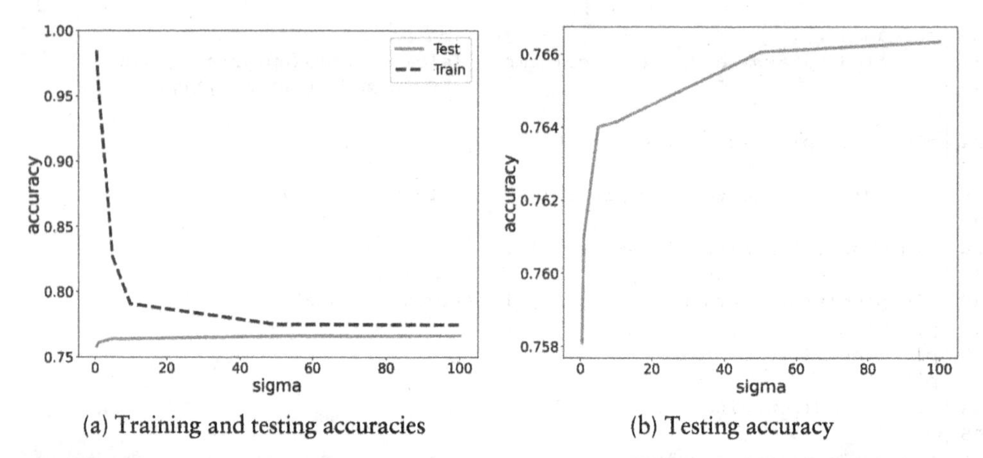

(a) Training and testing accuracies (b) Testing accuracy

FIGURE 11.12 Plot of training and testing accuracies of the SVM models with different σ's of the RBF kernel for the credit scoring dataset, based on Programme 11.5 via Python.

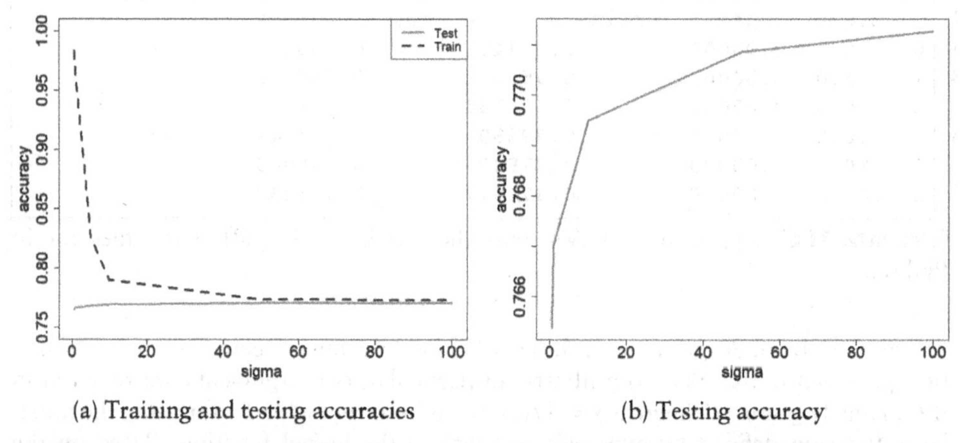

(a) Training and testing accuracies (b) Testing accuracy

FIGURE 11.13 Plot of training and testing accuracies of the SVM models with different σ's of the RBF kernel for the credit scoring dataset, based on Programme 11.6 via **R**.

```
1  > library(e1071)
2  > library(caret)
3  >
4  > df <- read.csv("credit_scoring_sample.csv")
5  > df <- na.omit(df)
6  >
7  > # split df into training and testing sets with a ratio of 8:2
8  > set.seed(4002)
9  > train_idx <- sample(seq_len(nrow(df)), size=floor(0.8*nrow(df)))
10 > df_train <- df[train_idx,]
11 > df_test <- df[-train_idx,]
12 >
13 > gamma <- function(sig) 1 / (2 * sig^2)
14 > sigma <- c(0.5, 1, 5, 10, 50, 100)
15 >
```

```
16 > train_acc <- numeric(length(sigma))
17 > test_acc <- numeric(length(sigma))
18 > for (i in seq_along(sigma)) {
19 +   svm_rbf <- svm(SeriousDlqin2yrs~., data=df_train,
20 +                  type="C-classification", scale=FALSE,
21 +                  kernel="radial", gamma=gamma(sigma[i]))
22 +   ypred_train <- predict(svm_rbf, newdata=df_train)
23 +   ypred_test <- predict(svm_rbf, newdata=df_test)
24 +   train_acc[i] <- mean(ypred_train == df_train$SeriousDlqin2yrs)
25 +   test_acc[i] <- mean(ypred_test == df_test$SeriousDlqin2yrs)
26 + }
27 >
28 > data.frame(sigma=sigma, gamma=gamma(sigma),
29 +            train_accuracy=train_acc, test_accuracy=test_acc)
30   sigma gamma train_accuracy test_accuracy
31 1   0.5 2e+00      0.9834226     0.7653762
32 2   1.0 5e-01      0.9569948     0.7670236
33 3   5.0 2e-02      0.8253707     0.7681219
34 4  10.0 5e-03      0.7903281     0.7694948
35 5  50.0 2e-04      0.7737850     0.7708677
36 6 100.0 5e-05      0.7734075     0.7712795
37 >
38 > par(cex.lab=2, cex.axis=2, cex.main=2, mar=c(5,5,4,4))
39 > plot(sigma, test_acc, col="orange", type="l", lwd=4, lty=1,
40 +      ylab="accuracy", ylim=c(0.75, 1))
41 > lines(sigma, train_acc, col="blue", type="l", lwd=4, lty=2)
42 > legend("topright", ncol=1, c("Test", "Train"), lty=1:2, lwd=4,
43 +        col=c("orange", "blue"), cex=1.5)
44 >
45 > plot(sigma, test_acc, col="orange", type="l", lwd=4, ylab="accuracy")
```

Programme 11.6 Application of SVM with the RBF kernel for different values of σ in R.

*11.A APPENDIX

*11.A.1 (Stochastic) Gradient Descent Methods

***11.A.1.1 Gradient Descent Method** In machine learning, one usually obtains a desired model by choosing optimal parameter values to minimize a certain loss function, which often takes the form of a sum or an average of a large number of individual losses. Gradient descent, and its stochastic version called stochastic gradient descent (SGD), are the two most frequently used iterative optimization algorithms for searching local minima. We shall briefly discuss these algorithms in this appendix, while we refer readers to our another treatise [8], as well as other textbooks such as [14, 17, 22], for more detailed discussions. The general steps for the gradient descent algorithm are:

1. start with a random seed; then
2. modify the current parameter estimate by an amount proportional to the negative of the gradient of the loss function evaluated at the current point.

Very often, the loss function to be minimized may possess multiple local minima. This is especially the case when one is confronted with training various deep neural networks. The choice of initial guess is therefore crucial for seeking and locating the ultimate local minimum.

In Step 2 of the gradient descent algorithm, the negative of the gradient is pointing in the direction towards the next guess, while the step size for the modification will be determined by a hyperparameter η, called the learning rate. With a learning rate η and loss function $L(\boldsymbol{\theta}) : \mathbb{R}^q \to \mathbb{R}$ in terms of the parameter $\boldsymbol{\theta} \in \boldsymbol{\Theta} \subset \mathbb{R}^q$, the gradient descent method updates the value of $\boldsymbol{\theta}$ as follows:

$$\begin{aligned} \text{Step Size (Signed)} &= \eta \cdot \nabla_{\boldsymbol{\theta}} L(\boldsymbol{\theta}); \\ \text{Next Move} &= \text{Last Move} - \text{Step Size}. \end{aligned} \tag{11.16}$$

Mathematically, we have

$$\boldsymbol{\theta}^{(t+1)} = \boldsymbol{\theta}^{(t)} - \eta \cdot \nabla_{\boldsymbol{\theta}} L(\boldsymbol{\theta})\Big|_{\boldsymbol{\theta}=\boldsymbol{\theta}^{(t)}}. \tag{11.17}$$

If the learning rate η is set too large, it is often not possible to reach the targeted nearby optimal point, as it may jump to one far away from the initial seed (see Figure 11.14a). On the other hand, if the learning rate is set too small, more iterations are needed to get to the optimal point (see Figure 11.14b). In practice, a learning rate of $\eta = 0.001$ is often used. Generally, it is more desirable to adopt a sequence of learning rates $\{\eta_t\}$ decreasing to zero. There are several techniques suggested in the literature on how these learning rates can be adjusted dynamically to achieve the best results. For the sake of simplicity, we adopt a constant learning rate in the rest of this discussion.

The popularity of the gradient descent method stems from its easiness to implement and its nice convergence property. In fact, one can prove that if the loss function $L(\boldsymbol{\theta}) : \mathbb{R}^q \to \mathbb{R}$ is convex and differentiable with a K-Lipschitz gradient, that is, $\|\nabla L(\boldsymbol{\theta}_1) - \nabla L(\boldsymbol{\theta}_2)\|_2 \leq K\|\boldsymbol{\theta}_1 - \boldsymbol{\theta}_2\|_2$ for any $\boldsymbol{\theta}_1, \boldsymbol{\theta}_2 \in \mathbb{R}^q$, then with a fixed learning rate $\eta \leq 1/K$, one has

$$L(\boldsymbol{\theta}^{(t)}) - L(\boldsymbol{\theta}^*) \leq \frac{\|\boldsymbol{\theta}^{(0)} - \boldsymbol{\theta}^*\|_2^2}{2\eta t}, \quad t = 1, 2, \ldots, \tag{11.18}$$

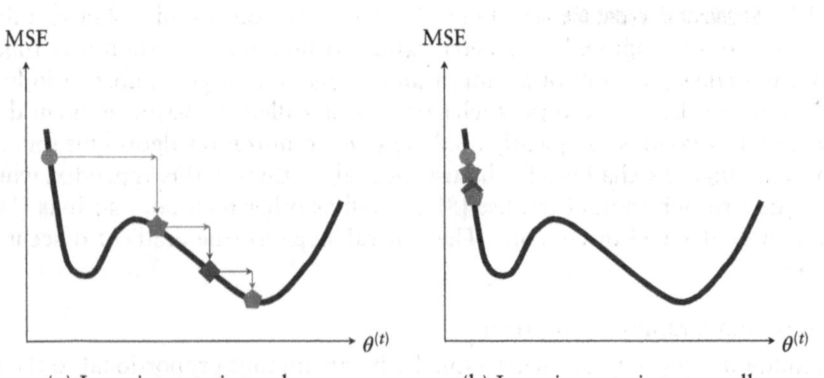

(a) Learning rate is too large. (b) Learning rate is too small.

FIGURE 11.14 The MSE plot against parameter $\theta^{(t)}$ of the first three iterations of gradient descent method.

where θ^* is the minimum. This means that the gradient descent method can guarantee convergence with a rate $O(1/t)$, where t is the number of iterations. The convergence rate can be improven if one further assumes that K satisfies stronger regularity conditions, such as strong convexity.

***11.A.1.2 Stochastic Gradient Descent Method** Recall that in a perceptron, the loss function to be minimized is (as given in (11.2)):

$$L(\omega) = -\sum_{i\in I_n} y_i\omega^\top x_i = \sum_{i=1}^{n} \max(0, -y_i\omega^\top x_i),$$

in which n is typically quite large. Computing the gradient of L is computationally intensive. To minimize L using gradient descent, it is more reasonable to update the model parameters by using only a portion of the dataset S (or training set), and the corresponding method is often referred to as *mini-batch stochastic gradient descent* (mini-batch sgd). To explain this, we consider a more general setting: given a training set $S = \{(x_i, y_i)\}_{i=1}^n$, and a particular learning method parameterized by $\theta \in \mathbb{R}^q$, consider the loss function

$$L_n(\theta) = \frac{1}{n}\sum_{i=1}^{n} \ell(\theta; y_i, x_i),$$

for some $\ell : \mathbb{R}^q \times \mathbb{R} \times \mathbb{R}^p \to \mathbb{R}$, where, for simplicity, we assume the true labels y_i's are real-valued, the feature vectors x_i are p-dimensional, and the model parameter θ is q-dimensional. The i-th individual loss is $\ell(\theta; y_i, x_i)$, where y_i is the true label of x_i. By the *Law of Large Numbers*, the gradient of the loss function $L_n(\theta)$ can be approximated by:

$$\nabla_\theta L_n(\theta) = \frac{1}{n}\sum_{i=1}^{n} \nabla_\theta \ell(\theta; y_i, x_i) \approx \mathbb{E}[\nabla_\theta \ell(\theta; \mathrm{y}, \mathbf{x})],$$

when the sample size n is large. Here, it is assumed that the data $(x_1, y_1), \ldots, (x_n, y_n)$ are iid realizations of $(\mathbf{x}, \mathrm{y})$. By the Law of Large Numbers again, with a sufficiently large $M \ll n$, the theoretical expectation $\mathbb{E}(\nabla_\theta \ell(\theta; \mathrm{y}, \mathbf{x}))$ can be approximated instead by a small portion of size M of S, that is,

$$\mathbb{E}[\nabla_\theta \ell(\theta; \mathrm{y}, \mathbf{x})] \approx \frac{1}{M}\sum_{m=1}^{M} \nabla_\theta \ell(\theta; y_m, x_m).$$

Therefore, with a very high probability, with both large M and n,

$$\nabla_\theta L_M(\theta) \approx \frac{1}{M}\sum_{m=1}^{M} \nabla_\theta \ell(\theta; y_m, x_m),$$

and this is the main philosophy of mini-batch SGD: replacing the gradient of the overall loss by only a portion average of gradient individual losses.

Instead of using the same batch of the first M data in S to approximate the gradient of the loss function for every iteration, it is more common to shuffle the data first. This can help to prevent the learned model from overfitting to specific patterns of the first M data and therefore having poor generalization. The batches used have to be selected randomly from S. One way to achieve this is by *random shuffle*. In this approach, we first shuffle the data randomly to obtain $\tilde{S} = \{(\tilde{x}_i, \tilde{y}_i)\}_{i=1}^n$, where $(\tilde{x}_i, \tilde{y}_i) := (x_{\pi(n)}, y_{\pi(n)})$ for some random permutation π on $\{1, 2, \ldots, n\}$ to itself. We then split $\tilde{S}$ into B batches, each of equal size of M, so that $B = \lceil n/M \rceil$, and for $j = 1, \ldots, B$, the j-th batch is given by $\{(\tilde{x}_i, \tilde{y}_i)\}_{i=(j-1)M+1}^{jM}$. These B batches will be used to approximate the gradient of the loss function as described above in the next B iteration steps. After all B batches have been used up, the whole process is then repeated.

In practice, a commonly used batch size is $M = 2^5 = 32 > 30$, thirty being the rule-of-thumb number for the law of large numbers to yield good enough approximation results. There are two particular cases of mini-batch SGD worth mentioning:

1. the common stochastic gradient descent (SGD) is a special case when $M = 1$, in which a randomly picked sample point is used in each iteration for an update of the parameter estimate;
2. the (whole) batch gradient descent is also a special case of mini-batch SGD but now the batch size $M = n$, so that all training data are used at a time.

A comparison of SGD, mini-batch SGD, and whole batch SGD methods for parameter updates is illustrated in Figure 11.15.

The red spot in the middle represents the ultimate local minimum point we are looking for. The solid trace in blue represents the sequence of updates under the whole batch gradient descent. It seems to be the fastest possible geographical path, going directly to the minimum point, and it involves the least number of steps. However, the computational cost is much higher in each step than the other two methods.

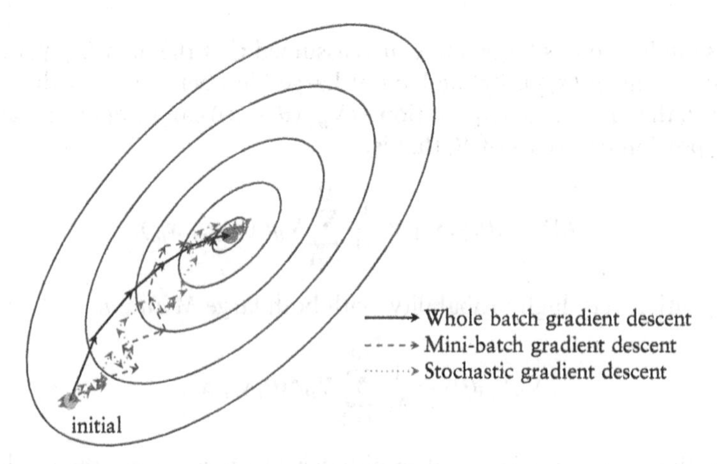

FIGURE 11.15 Gradient descent for SGD (dotted trace in grey), mini-batch SGD (dashed trace in green), and whole batch SGD (solid trace in blue).

Even worse, in the computation of the overall average loss $L(\boldsymbol{\theta})$, we might encounter numeric overflow or underflow due to a huge number of terms being dealt with at one time. The dotted trace in grey in Figure 11.15 corresponds to the stochastic gradient descent when $M = 1$. Though its path is of zigzag shape, the mean direction of the path is still pointing towards the ultimate minimum point. It requires more iterations in comparison with batch gradient descent to get sufficiently close to the ultimate minimum point. Interestingly, SGD only requires the computation of a single gradient $\nabla_{\boldsymbol{\theta}}\ell$ of an individual loss at a randomly selected point in each iteration, so that generally speaking, SGD requires far fewer computational steps (= number of zigzag steps $\sim$[5] $O_p\left(\frac{\theta_\alpha(\varepsilon)}{\eta}\right)$) to get sufficiently close to the ultimate minimum point than batch gradient descent (= number of steps[6] $\times$ number of calculations for each step $= O\left(\frac{1}{\eta\varepsilon}\right) \times O(n) = O\left(\frac{n}{\eta\varepsilon}\right)$) in order to achieve the same accuracy ε. Finally, for the mini-batch gradient descent represented by the dashed trace in green, it takes far fewer iterations than the stochastic gradient descent to get close to the minimum point, but with a computational time of $O\left(\frac{M}{\eta\varepsilon}\right)$. More discussions can be found in [13].

***11.A.1.3 Convergence of SGD** In this section, we briefly introduce the well-known *Robbins–Monro theorem* [19] on the convergence of SGD; more discussions can be found in [8]. To this, we write the individual loss as $\ell(\boldsymbol{\theta}; \mathbf{x})$ where $\mathbf{x} \in \mathbb{R}^p$ is a random variable, and $\boldsymbol{\theta} \in \mathbb{R}^q$ is the model parameter that we want to tune so as to minimize the sum (or average) of the individual losses. We assume that

(A1) (*Lipschitz continuity*) The individual loss function $\ell : \mathbb{R}^q \times \mathbb{R}^p \to \mathbb{R}$ is continuously differentiable and its gradient function $\nabla_{\boldsymbol{\theta}}\ell(\boldsymbol{\theta}; \boldsymbol{x})$ is Lipschitz continuous with a positive Lipschitz constant C: for any $\boldsymbol{\theta}, \boldsymbol{\theta}' \in \mathbb{R}^q$,

$$\|\nabla_{\boldsymbol{\theta}}\ell(\boldsymbol{\theta}; \boldsymbol{x}) - \nabla_{\boldsymbol{\theta}}\ell(\boldsymbol{\theta}'; \boldsymbol{x})\|_2 \le C\|\boldsymbol{\theta} - \boldsymbol{\theta}'\|_2; \tag{11.19}$$

(A2) ($\mathcal{L}^2$-*boundedness*)

$$\mathbb{E}\left[\|\nabla_{\boldsymbol{\theta}}\ell(\boldsymbol{\theta}; \mathbf{x})\|_2^2\right] < \infty, \qquad \text{for every} \quad \boldsymbol{\theta} \in \mathbb{R}^q; \tag{11.20}$$

(A3) (*Uniform (strong) convexity*) For some $\mu > 0$,

$$(\boldsymbol{\theta} - \boldsymbol{\theta}')^{\mathsf{T}}(\nabla_{\boldsymbol{\theta}}L(\boldsymbol{\theta}) - \nabla_{\boldsymbol{\theta}}L(\boldsymbol{\theta}')) \ge \mu\|\boldsymbol{\theta} - \boldsymbol{\theta}'\|_2^2, \text{for any } \boldsymbol{\theta}, \boldsymbol{\theta}' \in \mathbb{R}^q, \tag{11.21}$$

where $L(\boldsymbol{\theta}) := \mathbb{E}[\ell(\boldsymbol{\theta}; \mathbf{x})]$.

---

[5] The function θ_α is related to the (varying) learning rate at step t, $\eta_t = \eta \cdot \frac{1}{t^\alpha}$; particularly for $\alpha = 0$, with a constant learning rate $\eta_t = \eta$, we have $\theta_\alpha(\varepsilon) = O(1/\varepsilon)$ (yes, deterministic!).

[6] By (11.18), the number of steps t and the accuracy ε together satisfy $\varepsilon \le O(1/(\eta t))$, hence $t \le O(1/(\eta\varepsilon))$.

Assumption (**A3**) warrants that the function $L(\boldsymbol{\theta})$ has a unique minimizer $\boldsymbol{\theta}^*$ such that $\nabla_{\boldsymbol{\theta}} L(\boldsymbol{\theta}^*) = \mathbf{0}$. Assumptions (**A1**) and (**A3**) implicitly assume that $\mu < C$. Under these three assumptions, *Robbins–Monro theorem* states that when the learning rates $\{\eta_t\}$ are chosen in such a way that

$$\sum_{t=0}^{\infty} \eta_t = \infty \quad \text{and} \quad \sum_{t=0}^{\infty} \eta_t^2 < \infty, \tag{11.22}$$

then as $t \to \infty$, $\boldsymbol{\theta}^{(t)} \to \boldsymbol{\theta}^*$ almost surely using SGD. A more general result, known as *Kiefer–Wolfowitz theorem* [15], asserts that the same convergence holds true when the gradient $\nabla_{\boldsymbol{\theta}} L(\boldsymbol{\theta})$ in the updating rule of (11.17) is numerically approximated by the central finite difference method, with step sizes of the finite differencing $\{\alpha_t\}$ chosen in a way that $\sum_{t=0}^{\infty} \eta_t \alpha_t < \infty$.

*11.A.2 Quadratic Programming

***11.A.2.1 Formulation** Since the work of Frank and Wolfe [12], *Quadratic Programming* (QP) has become an indispensable tool to solve a broad class of quadratic optimization problems. Conventionally, a QP problem is formulated as follows:

$$\min_{\boldsymbol{x} \in \mathbb{R}^p} \frac{1}{2} \boldsymbol{x}^\top \boldsymbol{P} \boldsymbol{x} + \boldsymbol{q}^\top \boldsymbol{x} + r, \tag{11.23}$$

$$\text{subject to} \quad \boldsymbol{G}\boldsymbol{x} \preceq \boldsymbol{h} \text{ and } \boldsymbol{A}\boldsymbol{x} = \boldsymbol{b}, \tag{11.24}$$

where $\boldsymbol{P} \in \mathbb{R}^{p \times p}$ is positive definite matrix, $\boldsymbol{G} \in \mathbb{R}^{K \times p}, \boldsymbol{A} \in \mathbb{R}^{M \times p}, \boldsymbol{q} \in \mathbb{R}^p$ and $\boldsymbol{b} \in \mathbb{R}^M$; and the symbol $\preceq$ here means that every entry of vector $\boldsymbol{G}\boldsymbol{x}$ is less than or equal to the corresponding entry of the right-hand side vector $\boldsymbol{h} \in \mathbb{R}^K$. Note that the objective function in (11.23) is convex in $\boldsymbol{x}$, and $\boldsymbol{P} := (p_{ij})_{i,j=1,\dots,p}$ is usually, without loss of generality, assumed to be symmetric, since

$$\boldsymbol{x}^\top \boldsymbol{P} \boldsymbol{x} = (\boldsymbol{x}^\top \boldsymbol{P} \boldsymbol{x})^\top = \boldsymbol{x}^\top \boldsymbol{P}^\top \boldsymbol{x} = \boldsymbol{x}^\top \left(\frac{\boldsymbol{P} + \boldsymbol{P}^\top}{2} \right) \boldsymbol{x}$$

is a scalar, hence we can always replace any asymmetric matrix $\boldsymbol{P}$ with the symmetric one $\tilde{\boldsymbol{P}} := (\boldsymbol{P} + \boldsymbol{P}^\top)/2$ without changing the value of the objective function:

$$\boldsymbol{x}^\top \tilde{\boldsymbol{P}} \boldsymbol{x} := \frac{1}{2} \boldsymbol{x}^\top (\boldsymbol{P} + \boldsymbol{P}^\top) \boldsymbol{x} = \frac{1}{2} (\boldsymbol{x}^\top \boldsymbol{P} \boldsymbol{x} + \boldsymbol{x}^\top \boldsymbol{P}^\top \boldsymbol{x}) = \boldsymbol{x}^\top \boldsymbol{P} \boldsymbol{x}.$$

Observe that the feasible set $\mathcal{P}$ specified in (11.24), being an intersection of finitely many affine inequalities and equalities, is a polyhedron or its boundary.

As an example, consider the following QP problem:

$$\underset{x_1, x_2 \in \mathbb{R}}{\text{minimize}} \quad \frac{1}{2} (x_1 \; x_2) \begin{pmatrix} 2 & 1 \\ 1 & 6 \end{pmatrix} \begin{pmatrix} x_1 \\ x_2 \end{pmatrix} + (7 \; 3) \begin{pmatrix} x_1 \\ x_2 \end{pmatrix}, \tag{11.25}$$

$$\text{subject to} \quad \begin{pmatrix} 1 & 0 \\ -1 & 0 \\ 0 & 1 \\ 0 & -1 \end{pmatrix} \begin{pmatrix} x_1 \\ x_2 \end{pmatrix} \leq \begin{pmatrix} 1 \\ 1 \\ 1 \\ 1 \end{pmatrix}.$$

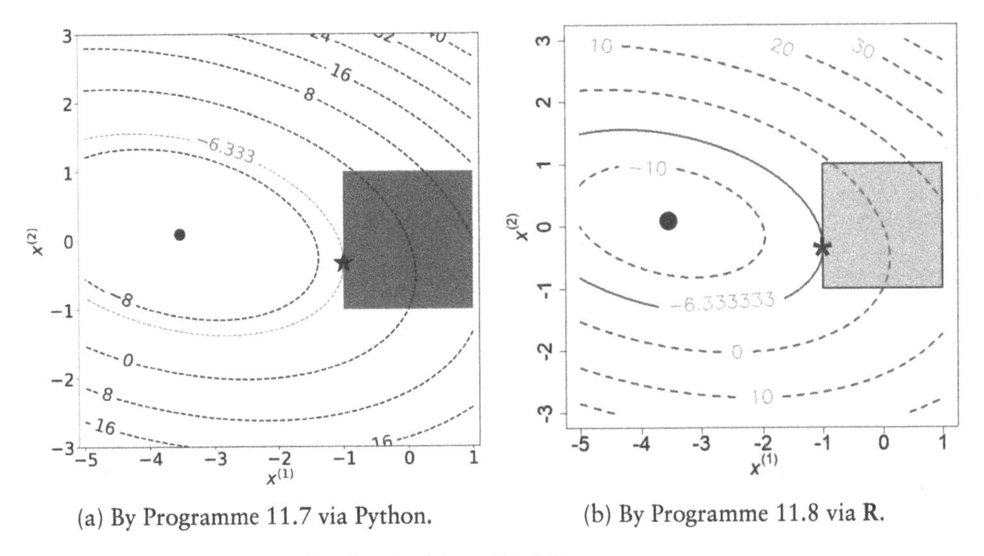

(a) By Programme 11.7 via Python.

(b) By Programme 11.8 via R.

FIGURE 11.16 A contour plot for Problem (11.25).

The contours of the objective function are given by the green ellipses as shown in Figure 11.16. The unconstrained problem has the global minimum attained at the red dot with coordinates $(x_1, x_2) = (-3.545, 0.0909)$.

Note that the constraints are equivalent to $|x_1| \leq 1, |x_2| \leq 1$, which is represented by the square shown in Figure 11.16. Within the square, the optimal point indicated by the star symbol is the contact point touched by the purple elliptical contour. We can use the `solve_qp()` function in the `qpsolvers` library of Python (Programme 11.7) or the `solve.QP()` function in the `quadprog` package of **R** (Programme 11.8) to solve and plot out the corresponding solution of this quadratic programming problem.

In Programme 11.7 via Python, the `np.meshgrid()` function returns a coordinated matrix for the two input arrays, and the `solve_qp()` function aims to solve a quadratic programming problem as formulated in (11.23) and (11.24).

```
1  import numpy as np
2  from qpsolvers import solve_qp
3  import matplotlib.pyplot as plt
4
5  plt.rc('font', size=25); plt.rc('axes', labelsize=25)
6
7  P = np.array([[2, 1], [1, 6]])
8  q = np.array([7, 3])
9  G = np.array([[1, 0], [-1, 0], [0, 1], [0, -1]])
10 h = np.array([1, ]*4)
11
12 def QPex(x_1, x_2):
13     # x: A 3-D tensor of shape 2 x len(x_1) x len(x_2)
14     x = np.stack([x_1, x_2], axis=0)
15     return np.sum(1/2 * x.T @ P * x.T, axis=-1) + x.T @ q
16
```

```
17 x1 = np.linspace(-4, 2, 1000)
18 x2 = np.linspace(-3, 3, 1000)
19 X1, X2 = np.meshgrid(x1, x2)
20 Z = QPex(X1, X2)
21
22 plt.figure(figsize=(10, 10))
23 # Need transpose for Z
24 fig = plt.contour(X1, X2, Z.T, 10, colors="green", linewidths=2,
25                   linestyles="dashed")
26 plt.xlim(-5.1, 1.1)
27 plt.clabel(fig, inline=1)
28
29 # solve_qp: minimize (1/2 x^T P x + q^T x) with constraints (G x <= h)
30 constr_ans = solve_qp(P, q, G, h, solver="ecos")
31 real_ans = solve_qp(P, q, solver="ecos")
32
33 plt.plot(*constr_ans, "r*", markersize=30)
34 plt.plot(*real_ans, "r.", markersize=30)
35 # Need transpose for Z
36 constr_fig = plt.contour(X1, X2, Z.T, levels=[QPex(*constr_ans)],
37                          colors="purple")
38 plt.clabel(constr_fig, inline=1)
39 plt.fill([-1,1,1,-1], [-1,-1,1,1], "gray", alpha=0.9)
40 plt.xlabel(r"$x^{(1)}$"); plt.ylabel(r"$x^{(2)}$")
```

Programme 11.7 Programme codes for Problem (11.25) in Python.

In Programme 11.8 via **R**, the `outer()` function takes three main arguments: (i) the first input array (vector) $\boldsymbol{x} \in \mathbb{R}^Q$ as `X`; (ii) the second input array (vector) $\boldsymbol{y} \in \mathbb{R}^K$ as `Y`; and (iii) the function $f : \mathbb{R} \times \mathbb{R} \to \mathbb{R}$ to be adopted in `outer()` as `FUN`. The output of this function is a two-dimensional grid matrix $\boldsymbol{O} = (o_{ij}) \in \mathbb{R}^{Q\times K}$ with entries $o_{ij} = f(x_i, y_j)$ for $i = 1, \cdots, Q$ and $j = 1, \cdots, K$. On the other hand, also in **R**, the `solve.QP()` function aims to solve the QP problem in the following form:

$$\min_{\boldsymbol{x}\in\mathbb{R}^p} \frac{1}{2}\boldsymbol{x}^\top \boldsymbol{P}\boldsymbol{x} - \boldsymbol{q}^\top \boldsymbol{x} \quad \text{subject to } \boldsymbol{G}^\top \boldsymbol{x} \geq \boldsymbol{h},$$

where the first `meq` constraints are treated as equality constraints, i.e. $\boldsymbol{G}_1^\top \boldsymbol{x} = h_1, \ldots, \boldsymbol{G}_{\text{meq}}^\top \boldsymbol{x} = h_{\text{meq}}$; while the rest are treated as inequality constraints, i.e. $\boldsymbol{G}_{\text{meq}+1}^\top \boldsymbol{x} \geq h_{\text{meq}+1}, \ldots, \boldsymbol{G}_K^\top \boldsymbol{x} \geq h_K$.

```
1  > library(graphics)
2  > library(quadprog)
3  >
4  > x1 <- seq(-5, 1, 0.01)
5  > x2 <- seq(-3, 3, 0.01)
6  > P <- matrix(c(2, 1, 1, 6), ncol=2)
7  > q <- c(7, 3)
8  > G.T <- matrix(c(1, 0, -1, 0, 0, 1, 0, -1), ncol=2, byrow=T)
9  > h <- c(1, 1, 1, 1)
10 >
11 > QPex <- function(x_1, x_2){
```

```
12 +     # x: A 2-D matrix of shape 2 x total number of grid
13 +     x <- matrix(c(x_1, x_2), nrow=2, byrow=TRUE)
14 +     as.numeric(rowSums(t(x)%*%P * t(x))/2 + t(q)%*%x)
15 + }
16 >
17 > par(cex.lab=2, cex.axis=2, cex.main=2, mar=c(5,5,4,4))
18 > z <- outer(x1, x2, FUN=QPex)
19 > contour(x1, x2, z, nlevels=8, col="green", lty=2, lwd=3, labcex=2)
20 > polygon(c(-1, -1, 1, 1, -1), c(-1, 1, 1, -1, -1),
21 +         col=adjustcolor("gray", alpha.f=0.5), border="black", lwd=2)
22 >
23 > # minimize (1/2 x∧T P x + q∧T x) with constraints (G∧T x <= h)
24 > # solve.QP: (1/2 x∧T P x - (-q∧T) x) with constraints (-(G∧T)∧T x>=-h)
25 > sol <- solve.QP(P, -q, -t(G.T), bvec=-h)
26 > (real_ans <- sol$unconstrained.solution)
27 [1] -3.54545455  0.09090909
28 > (const_ans <- sol$solution)
29 [1] -1.0000000 -0.3333333
30 >
31 > contour(x1, x2, z, levels=sol$value, col="purple",
32 +         lty=1, lwd=3, labcex=2, add=TRUE)
33 > points(real_ans[1], real_ans[2], pch=19, col="red", cex=4)
34 > points(const_ans[1], const_ans[2], pch="*", col="red", cex=6)
35 > title(xlab=expression(italic("x")∧"(1)"),
36 +       ylab=expression(italic("x")∧"(2)"))
```

Programme 11.8 Programme codes for Problem (11.25) in R.

***11.A.2.2 Minimax Theorem and Duality** In this subsection, we first recall the celebrated *von Neumann's Minimax theorem*, from which we derive general results concerning the duality of constrained minimization problems. The results will then be applied to solve the special case of quadratic programming problems in the next subsection. Readers may also refer to textbooks on convex analysis for more discussions, such as [2, 4].

A set $\mathcal{X} \subset \mathbb{R}^p$ is said to be *convex* if for any two points $x_1, x_2 \in \mathcal{X}$, the line segment joining them is still contained in $\mathcal{X}$, that is, $\alpha x_1 + (1-\alpha)x_2 \in \mathcal{X}$ for any $\alpha \in (0,1)$. A set $\mathcal{X} \subset \mathbb{R}^p$ is said to be *compact* if $\mathcal{X}$ is closed and bounded.

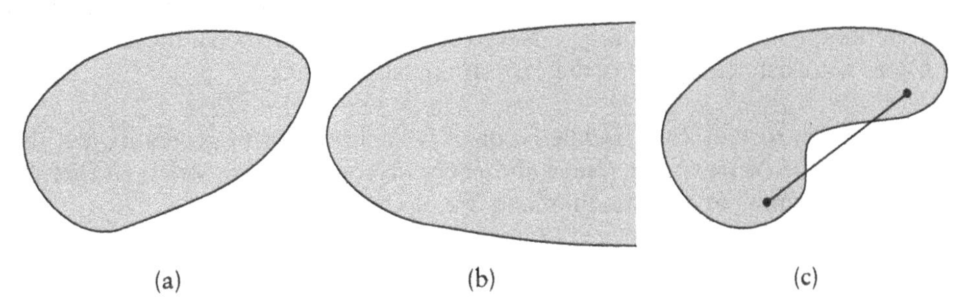

FIGURE 11.17 (a) A compact convex set; (b) A convex, but non-compact, set; (c) A compact non-convex set.

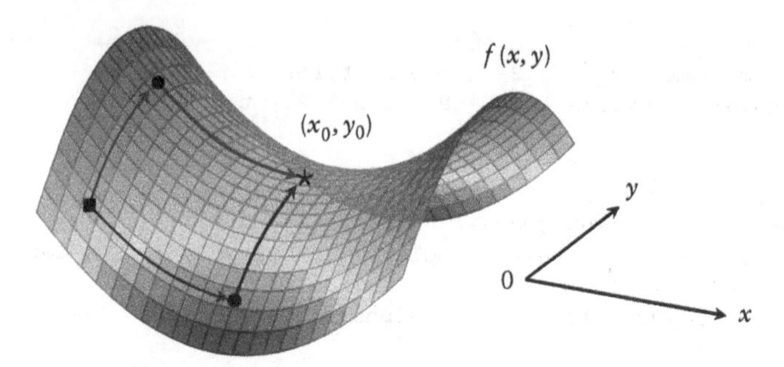

FIGURE 11.18 The two paths of optimizations taken with respect to von Neumann's Minimax Theorem.

Theorem 11.1 ***(von Neumann's Minimax theorem*** [25]*)* Let $\mathcal{X} \subset \mathbb{R}^p$ and $\mathcal{Y} \subset \mathbb{R}^q$ be compact convex sets. If $f : \mathcal{X} \times \mathcal{Y} \to \mathbb{R}$ is a continuous function such that

(i) for each $y \in \mathcal{Y}$, $f(\cdot, y) : \mathcal{X} \to \mathbb{R}$ is convex in $x \in \mathcal{X}$; and
(ii) for each $x \in \mathcal{X}$, $f(x, \cdot) : \mathcal{Y} \to \mathbb{R}$ is concave in $y \in \mathcal{Y}$.

Then, we have

$$\min_{x\in\mathcal{X}} \max_{y\in\mathcal{Y}} f(x,y) = \max_{y\in\mathcal{Y}} \min_{x\in\mathcal{X}} f(x,y). \tag{11.26}$$

In the proof below, we demonstrate that under the assumptions of Theorem 11.1, a saddle point (x_0, y_0) of f always exists, which by definition is a point that satisfies

$$\max_{y\in\mathcal{Y}} f(x_0,y) = f(x_0,y_0) = \min_{x\in\mathcal{X}} f(x,y_0);$$

moreover, the existence of a saddle point in fact implies the validity of (11.26):

$$\max_{y\in\mathcal{Y}} \min_{x\in\mathcal{X}} f(x,y) = \max_{y\in\mathcal{Y}} f(x_0,y) = f(x_0,y_0) = \min_{x\in\mathcal{X}} f(x,y_0) = \min_{x\in\mathcal{X}} \max_{y\in\mathcal{Y}} f(x,y).$$

Here, the leftmost expression corresponds to the lower path in blue in Figure 11.18, while the rightmost term corresponds to the upper path in red in the same figure. Both paths settle at the same point (x_0, y_0). Nevertheless, in general, even if the saddle point of f does not exist, the minimax theorem still remains valid.

Proof. We aim to look for a saddle point of f. Without loss of generality, we also assume that the convexity of f in x and concavity of f in y are strict, so that the following maximum and minimum points are uniquely defined:

$$g : x \in \mathcal{X} \mapsto \arg\max_{y\in\mathcal{Y}} f(x,y) =: y_x \in \mathcal{Y};$$

$$h : y \in \mathcal{Y} \mapsto \arg\min_{x\in\mathcal{X}} f(x,y) =: x_y \in \mathcal{X}.$$

Consider the composite map $h \circ g : \mathcal{X} \to \mathcal{X}$. Due to their corresponding (strict) convexity and concavity, and together with the continuity of f, both g and h are continuous and so is $h \circ g$. Since $\mathcal{X}$ is compact and convex, one can apply *Brouwer's fixed point theorem*[7] to $h \circ g$ and obtain $x_0 \in \mathcal{X}$ such that for $y_0 := y_{x_0}$, we have $x_{y_0} = x_0$. In particular, this implies that (x_0, y_0) is a saddle point of f. To see this, firstly notice:

$$\begin{aligned}\min_{x\in\mathcal{X}}\max_{y\in\mathcal{Y}} f(x,y) &\le \max_{y\in\mathcal{Y}} f(x_0,y)\\ &= f(x_0,y_0)\\ &= f(x_{y_0},y_0)\\ &= \min_{x\in\mathcal{X}} f(x,y_0) \le \max_{y\in\mathcal{Y}}\min_{x\in\mathcal{X}} f(x,y). \end{aligned} \tag{11.27}$$

On the other hand, for any $x' \in \mathcal{X}$ and $y \in \mathcal{Y}$,

$$\min_{x\in\mathcal{X}} f(x,y) \le f(x',y);$$

and taking the maximum over $y \in \mathcal{Y}$,

$$\max_{y\in\mathcal{Y}}\min_{x\in\mathcal{X}} f(x,y) \le \max_{y\in\mathcal{Y}} f(x',y).$$

As this inequality is valid for all $x' \in \mathcal{X}$, we arrive at the *max-min inequality*

$$\max_{y\in\mathcal{Y}}\min_{x\in\mathcal{X}} f(x,y) \le \min_{x'\in\mathcal{X}}\max_{y\in\mathcal{Y}} f(x',y). \tag{11.28}$$

Combining inequalities (11.27) and (11.28) yields the desired result. □

Notice that the max-min inequality (11.28) holds for any function f regardless of whether f possesses any convexity or concavity property, and regardless of the topological structure of $\mathcal{X}$ and $\mathcal{Y}$.

As a side discussion, von Neumann's minimax theorem is actually closely related to the vitally important concept of the *Nash equilibrium* in Economics, which is central to *noncooperative game theory*. Define a joint satisfaction function when playing a game between two players by $J(x,y)$, where $x \in \mathcal{X}$ and $y \in \mathcal{Y}$ are the respective strategies independently adopted by Players 1 and 2. We say a pair of strategies $(x^*, y^*) \in \mathcal{X} \times \mathcal{Y}$ is a *Nash equilibrium* if

$$\max_{x\in\mathcal{X}} J(x,y^*) = J(x^*,y^*) \quad \text{and} \quad \max_{y\in\mathcal{Y}} J(x^*,y) = J(x^*,y^*).$$

By definition, the Nash equilibrium represents a stable solution of the game in that if the players play in this way, either player has no incentive to unilaterally change

[7] *Brouwer's fixed point theorem* [6] asserts that any continuous function f mapping a nonempty compact convex set to itself has a fixed point, that is, there is a point x^* such that $f(x^*) = x^*$.

to another strategy because each player's strategy is the best response to the strategy adopted by his opponent. A fundamental result in game theory states that Nash equilibrium always exists for every such game, under the following assumptions:

(**A1**) both $\mathcal{X}$ and $\mathcal{Y}$ are nonempty compact and convex subsets of finite dimensional spaces;

(**A2**) $J(x, y)$ is concave in x for each $y \in \mathcal{Y}$, and is also concave in y for each $x \in \mathcal{X}$;

(**A3**) $J(x, y)$ is jointly continuous.

To establish its validity, the following argument is essentially the same as discussed above for von Neumann's minimax theorem. To ease the discussion here, we first assume that J is indeed strictly concave in both x and y, such that both maps below are well-defined and continuous:

$$h : \mathcal{X} \ni x \mapsto \arg\max_{y \in \mathcal{Y}} J(x, y) =: y_x \in \mathcal{Y},$$

$$g : \mathcal{Y} \ni y \mapsto \arg\max_{x \in \mathcal{X}} J(x, y) =: x_y \in \mathcal{X}.$$

Applying Brouwer's fixed point theorem to the composite continuous map $g \circ h : \mathcal{X} \to \mathcal{X}$ gives a fixed point $x^* \in \mathcal{X}$ of $g \circ h$ such that $g(h(x^*)) = x^*$. Denote $y^* := y_{x^*}$. The fixed point nature of x^* translates to $x_{y^*} = x^*$, and so

$$\max_{y \in \mathcal{Y}} J(x^*, y) = J(x^*, y_{x^*}) = J(x^*, y^*) = J(x_{y^*}, y^*) = \max_{x \in \mathcal{X}} J(x, y^*).$$

This shows that (x^*, y^*) is in fact a Nash equilibrium.

Next, we return to the study of the duality of a general constrained minimization problem:

$$\min_{x \in \mathbb{R}^p} f(x) \quad \text{subject to} \quad g_k(x) \le 0, \qquad k = 1, \ldots, K, \tag{11.29}$$

where $f, g_k : \mathbb{R}^p \to \mathbb{R}$ are convex functions, and we denote $\mathbf{g}(x) := (g_1(x), \cdots, g_K(x))^\top$. By the Lagrangian multiplier approach, the Lagrangian with a multiplier λ is given by:

$$\mathcal{L}(x, \lambda) := f(x) + \lambda^\top \mathbf{g}(x).$$

We set $\mathcal{X} := \{x \in \mathbb{R}^p : g_i(x) \le 0 \text{ for all } i\}$. Of course, we assume that $\mathcal{X} \neq \emptyset$. Note that

$$\max_{\lambda \ge \mathbf{0}} \left(f(x) + \lambda^\top \mathbf{g}(x) \right) = \begin{cases} f(x) (< \infty), & x \in \mathcal{X}; \\ \infty, & x \notin \mathcal{X}. \end{cases}$$

In particular, with $x \in \mathcal{X}$, the maximum is attained at $\lambda = \mathbf{0}$, while for $x \notin \mathcal{X}, g_i(x) > 0$ for some i, such that in general $\lambda^\top \mathbf{g}(x)$ is increasing in λ_i without bound. Hence, we have

$$m := \min_{x \in \mathcal{X}} f(x) = \min_{x \in \mathbb{R}^p} \max_{\lambda \ge \mathbf{0}} \left(f(x) + \lambda^\top \mathbf{g}(x) \right) = \min_{x \in \mathbb{R}^p} \max_{\lambda \ge \mathbf{0}} \mathcal{L}(x, \lambda). \tag{11.30}$$

By using the max-min inequality (11.28), which holds true for unbounded domains, we see that

$$\min_{x\in\mathbb{R}^p}\max_{\lambda\geq 0}\left(f(x)+\lambda^\top \mathbf{g}(x)\right)\geq \max_{\lambda\geq 0}\min_{x\in\mathbb{R}^p}\left(f(x)+\lambda^\top \mathbf{g}(x)\right).$$

If we define the *dual Lagrangian*

$$\mathcal{D}(\lambda) := \min_{x\in\mathbb{R}^p}\left(f(x)+\lambda^\top \mathbf{g}(x)\right),$$

then we arrive at the so-called *weak duality*:

$$m=\min_{x\in\mathcal{X}} f(x)\geq \max_{\lambda\geq 0}\mathcal{D}(\lambda). \tag{11.31}$$

The minimization on the left, our original problem, is called the *primal problem*, and the maximization on the right is called the *dual problem*. This result shows that the solution to the dual problem provides a lower bound to the solution of the primal problem. The difference between the optimal values of primal and dual is called the *duality gap*; see the illustration in Figure 11.19. We conclude this subsection by closing the gap via the *Fenchel–Rockafellar duality* (see Theorem 1.9 in [24]) of convex optimization problems.

We say that *strong duality* for (11.29) holds if the duality gap is equal to zero:

$$\min_{x\in\mathcal{X}} f(x)=\max_{\lambda\geq 0}\mathcal{D}(\lambda). \tag{11.32}$$

When Problem (11.29) is convex, that is, the objective function f and all constraint functions g_k's are convex, a classical condition that guarantees the validity of strong duality is the *Slater condition*: there exists $x\in\mathbb{R}^p$ such that $g_k(x)<0$ for all k; see [5] for a more comprehensive discussion. In what follows, we provide an elementary proof of strong duality under the following alternative assumptions on f and g_k's on top of their convexity:

Assumption 1: (f is coercive) $\lim_{\|x\|_2\to\infty} f(x)=\infty$.

Assumption 2: For all $1\leq k\leq K$, $\lim_{\|x\|_2\to\infty}\frac{g_k(x)}{f(x)}=0$.

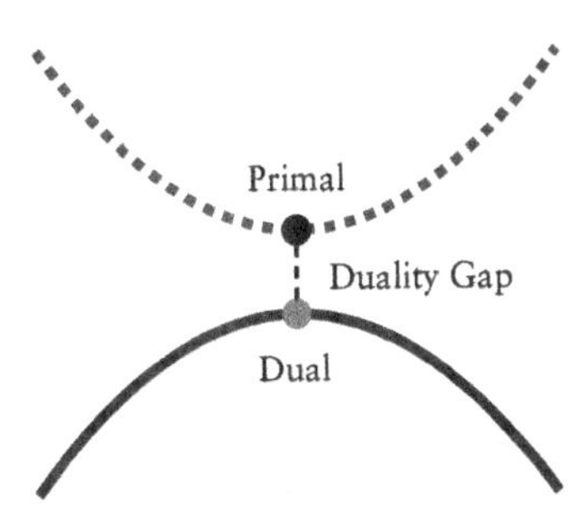

FIGURE 11.19 Duality Gap

As a remark, Assumption 2 is fulfilled if f is a function of quadratic growth, while all g_k's are of linear growth. To this end, let $m := \min_{x \in \mathcal{X}} f(x)$. For any $\epsilon > 0$, we denote by

$$\mathcal{X}_\epsilon := \{x \in \mathbb{R}^p : g_i(x) \leq \epsilon \text{ for all } i\}.$$

Also under Slater condition, since each g_k is convex (and hence continuous), $\mathcal{X}_\epsilon$ is a nonempty compact and convex subset of $\mathbb{R}^p$. We denote by $m_f := \min_{x \in \mathbb{R}^p} f(x)$, and $m_\epsilon := \min_{x \in \mathcal{X}_\epsilon} f(x)$ for any $\epsilon > 0$. Since f is continuous and for any $\epsilon > 0$, the set $\mathcal{X}_\epsilon$ is compact, m_ϵ is well-defined and finite. It is obvious that

$$m \geq m_{\epsilon'} \geq m_\epsilon \geq m_f, \quad 0 < \epsilon' < \epsilon, \tag{11.33}$$

and since f is continuous,

$$\lim_{\epsilon \to 0} m_\epsilon = m. \tag{11.34}$$

For $\epsilon > 0$, we define

$$\Lambda_\epsilon := \left\{ \lambda \in \mathbb{R}^K : \lambda \geq \mathbf{0}, \|\lambda\|_\infty \leq \frac{m - m_f}{\epsilon} \right\},$$

where $\|\cdot\|_\infty$ denotes the (uniform) ∞-norm. Note that Λ_ϵ is convex and compact; and it shortly becomes clear with this definition.

Lemma 11.1 For any fixed $\epsilon > 0$, we have

$$\min_{x \in \mathbb{R}^p} \max_{\lambda \in \Lambda_\epsilon} \left(f(x) + \lambda^\top \mathbf{g}(x) \right) \geq m_\epsilon.$$

Proof. We consider the following two cases.

Case 1. For $x \in \mathbb{R}^p \backslash \mathcal{X}_\epsilon$, from the definition of $\mathcal{X}_\epsilon$, there exists some k_0 such that $g_{k_0}(x) > \epsilon$. We choose $\lambda_0 := (0, \ldots, (m - m_f)/\epsilon, \ldots, 0)^\top \in \Lambda_\epsilon$, where the only non-zero term is in the k_0-th coordinate. Then,

$$f(x) + \lambda_0^\top \mathbf{g}(x) \geq m_f + \frac{m - m_f}{\epsilon}\epsilon = m.$$

Therefore, we have

$$\max_{\lambda \in \Lambda_\epsilon} \left(f(x) + \lambda^\top \mathbf{g}(x) \right) \geq f(x) + \lambda_0^\top \mathbf{g}(x) \geq m;$$

as a result, we also have

$$\min_{x \in \mathbb{R}^p \backslash \mathcal{X}_\epsilon} \max_{\lambda \in \Lambda_\epsilon} \left(f(x) + \lambda^\top \mathbf{g}(x) \right) \geq m \geq m_\epsilon.$$

Case 2. For $x \in \mathcal{X}_\epsilon$, since $\mathbf{0} \in \Lambda_\epsilon$, we have

$$\max_{\lambda \in \Lambda_\epsilon} \left(f(x) + \lambda^\top \mathbf{g}(x) \right) \geq f(x) + \mathbf{0}^\top \mathbf{g}(x) = f(x).$$

From the definition of m_ϵ, we then have

$$\min_{x\in\mathcal{X}_\epsilon}\max_{\lambda\in\Lambda_\epsilon}\left(f(x)+\lambda^\top g(x)\right)\geq\min_{x\in\mathcal{X}_\epsilon} f(x)=m_\epsilon.$$

Combining the conclusions from these two cases proves the lemma. □

For any $N>0$, we let $\mathcal{M}_N := \{x\in\mathbb{R}^p : \|x\|_2\leq N\}$, which is convex and compact. As an immediate consequence of Lemma 11.1,

$$\min_{x\in\mathcal{M}_N}\max_{\lambda\in\Lambda_\epsilon}\left(f(x)+\lambda^\top g(x)\right)\geq\min_{x\in\mathbb{R}^p}\max_{\lambda\in\Lambda_\epsilon}\left(f(x)+\lambda^\top g(x)\right)\geq m_\epsilon.$$

As both $\mathcal{M}_N$ and Λ_ϵ are nonempty, convex and compact, and the map

$$\mathcal{M}_N\times\Lambda_\epsilon\ni(x,\lambda)\mapsto f(x)+\lambda^\top g(x)\in\mathbb{R}$$

is convex in x and concave in λ, applying the minimax theorem yields

$$\max_{\lambda\in\Lambda_\epsilon}\min_{x\in\mathcal{M}_N}\left(f(x)+\lambda^\top g(x)\right)\geq m_\epsilon. \tag{11.35}$$

We also note that the right-hand side of the last inequality is independent of N. We then have the following result:

Lemma 11.2 For any fixed $\epsilon>0$ and any $\lambda\in\Lambda_\epsilon$, there exists a constant N_ϵ (which may depend on ϵ), such that

$$\min_{x\in\mathbb{R}^p}\left(f(x)+\lambda^\top g(x)\right)=\min_{x\in\mathcal{M}_{N_\epsilon}}\left(f(x)+\lambda^\top g(x)\right). \tag{11.36}$$

Proof. We first show that the left-hand side of (11.36) is well-defined. Actually, from the Cauchy–Schwarz inequality, for any $(x,\lambda)\in\mathbb{R}^p\times\Lambda_\epsilon$, we have

$$\begin{aligned}f(x)+\lambda^\top g(x)&\geq f(x)\left(1-\|\lambda\|_2\left\|\frac{g(x)}{f(x)}\right\|_2\right)\\&\geq f(x)\left(1-\frac{\sqrt{K}(m-m_f)}{\epsilon}\left\|\frac{g(x)}{f(x)}\right\|_2\right).\end{aligned} \tag{11.37}$$

Together with Assumption 2, this shows that for any $\lambda\in\Lambda_\epsilon$, the map $\mathbb{R}^p\ni x\mapsto f(x)+\lambda^\top g(x)$ is convex and coercive, and thus has an interior minimum point.

We now prove (11.36). Recall that $\mathcal{X}$ is compact, so there is a constant N_0 such that $\mathcal{X}\subset\mathcal{M}_{N_0}$. For any $x_0\in\mathcal{X}$ and $\lambda\in\Lambda_\epsilon$,

$$f(x_0)+\lambda^\top g(x_0)\leq f(x_0). \tag{11.38}$$

From Assumption 2, we know that there exists $N_{1,\epsilon} > 0$, such that whenever $\|x\|_2 \geq N_{1,\epsilon}$, we have

$$\frac{\sqrt{K}(m - m_f)}{\epsilon}\left\|\frac{g(x)}{f(x)}\right\|_2 \leq \frac{1}{2}.$$

It then follows from (11.37) that for $\|x\|_2 \geq N_{1,\epsilon}$,

$$f(x) + \lambda^\top g(x) \geq f(x)\left(1 - \frac{\sqrt{K}(m - m_f)}{\epsilon}\left\|\frac{g(x)}{f(x)}\right\|_2\right) \geq \frac{1}{2}f(x). \tag{11.39}$$

From Assumption 1, there exists $N_{2,\epsilon} > 0$ such that for $\|x\|_2 \geq N_{2,\epsilon}$,

$$f(x) \geq 2f(x_0) + 2. \tag{11.40}$$

Now we put $N_\epsilon := \max\{N_0, N_{1,\epsilon}, N_{2,\epsilon}\}$. Then for $\|x\|_2 \geq N_\epsilon$, from (11.39) and (11.40), we have

$$f(x) + \lambda^\top g(x) \geq f(x_0) + 1. \tag{11.41}$$

Since $x_0 \in \mathcal{X} \subset \mathcal{M}_{N_0}$, it follows from (11.38) and (11.41) that

$$\begin{aligned}\inf_{x\in\mathbb{R}^p\setminus\mathcal{M}_{N_\epsilon}} \left(f(x) + \lambda^\top g(x)\right) &\geq f(x_0) + 1 > f(x_0) \geq f(x_0) + \lambda^\top g(x_0)\\ &\geq \min_{x\in\mathcal{M}_{N_\epsilon}} \left(f(x) + \lambda^\top g(x)\right),\end{aligned}$$

and hence

$$\min_{x\in\mathbb{R}^p} \left(f(x) + \lambda^\top g(x)\right) \geq \min_{x\in\mathcal{M}_{N_\epsilon}} \left(f(x) + \lambda^\top g(x)\right).$$

From the last inequality, we obtain (11.36), and the lemma is proven. □

With all these preparations, we are now ready to establish the strong duality. For $\epsilon > 0$, from Lemma 11.2, there exists a constant N_ϵ (which may depend on ϵ) such that

$$\begin{aligned}\max_{\lambda\geq 0} \mathcal{D}(\lambda) &= \max_{\lambda\geq 0}\min_{x\in\mathbb{R}^p} \left(f(x) + \lambda^\top g(x)\right)\\ &\geq \max_{\lambda\in\Lambda_\epsilon}\min_{x\in\mathbb{R}^p} \left(f(x) + \lambda^\top g(x)\right)\\ &= \max_{\lambda\in\Lambda_\epsilon}\min_{x\in\mathcal{M}_{N_\epsilon}} \left(f(x) + \lambda^\top g(x)\right).\end{aligned}$$

Then, using (11.35), we have $\max_{\lambda\geq 0}\mathcal{D}(\lambda)\geq m_\epsilon$. As ϵ is arbitrary, applying (11.34), gives $\max_{\lambda\geq 0}\mathcal{D}(\lambda)\geq \lim_{\epsilon\to 0} m_\epsilon = m$. Together with the weak duality (11.31), under Slater condition together with Assumptions 1 and 2, we obtain the strong duality (11.32).

When strong duality holds, there is a close relationship between the optimal solutions of the primal problem and the dual problem. To this end, suppose that $x^*\in\mathcal{X}$ is optimal for the primal problem and $\lambda^*\geq \mathbf{0}$ is optimal for the dual problem. Then we have

$$\mathcal{D}(\lambda^*) = \min_{x\in\mathbb{R}^p}\left(f(x)+\lambda^{*\top}g(x)\right)\leq f(x^*)+\lambda^{*\top}g(x^*)\leq f(x^*). \tag{11.42}$$

Since the strong duality $f(x^*)=\mathcal{D}(\lambda^*)$ holds, all inequalities of (11.42) are satisfied as equalities. Two results then follow:

(i) We have $\lambda^{*\top}g(x^*)=\sum_{k=1}^{K}\lambda_k^* g_k(x^*)=0$, and the facts that $\lambda_k^*\geq 0$ and $g_k(x^*)\leq 0$ then imply

$$\lambda_k^* g_k(x^*)=0,\quad \text{for all } k=1,\ldots,K,$$

the so-called *complementary slackness condition.*

(ii) By (11.30) and (11.31), the point x^* is not only optimal for the primal problem, but also optimal for the Lagrangian problem:

$$\min_{x\in\mathbb{R}^p}\left(f(x)+\lambda^{*\top}g(x)\right).$$

***11.A.2.3 Primal and Dual Formulations of QP Problems** By using the duality result introduced in Subsection *11.A.2.2, here we reformulate the QP problem of (11.23) with the constraint of (11.24) by using the Lagrangian multiplier approach; see [3] for further discussions on this approach. The Lagrangian $\mathcal{L}$, with $\lambda_1\in\mathbb{R}^K, \lambda_2\in\mathbb{R}^M$, is given by:

$$\begin{aligned}\mathcal{L}(x,\lambda_1,\lambda_2) &= \frac{1}{2}x^\top Px+q^\top x+r+\lambda_1^\top(Gx-h)+\lambda_2^\top(Ax-b)\\ &=\frac{1}{2}x^\top Px+(q+G^\top\lambda_1+A^\top\lambda_2)^\top x-\lambda_1^\top h-\lambda_2^\top b+r,\end{aligned} \tag{11.43}$$

where $\lambda_1\geq\mathbf{0}$ and $\lambda_2\in\mathbb{R}^M$.[8] Being convex in x, the global minimum of $x\mapsto\mathcal{L}(x,\lambda_1,\lambda_2)$ can be located by equating the derivative of $\mathcal{L}(x,\lambda_1,\lambda_2)$, with

[8] To see why components of λ_2 can take any real values and not just positive ones, we actually consider the equality constraint $Ax=b$ as two inequality constraints in conjunction: $Ax\leq b$ and $-Ax\leq -b$, which are respectively augmented by positive multipliers $\lambda_2^{(1)}$ and $\lambda_2^{(2)}$. The corresponding Lagrangian penality term now involves $(\lambda_2^{(1)})^\top(Ax-b)+(\lambda_2^{(2)})^\top(-Ax+b)=(\lambda_2^{(1)}-\lambda_2^{(2)})^\top(Ax-b)$, and we simply take $\lambda_2=\lambda_2^{(1)}-\lambda_2^{(2)}$.

respect to x, with zero:

$$\nabla_x \mathcal{L}(x, \lambda_1, \lambda_2) = Px + (q + G^\top \lambda_1 + A^\top \lambda_2) = \mathbf{0},$$

which yields $x = -P^{-1}(q + G^\top \lambda_1 + A^\top \lambda_2)$. Substituting this back to (11.43), we get the dual Lagrangian:

$$\mathcal{D}(\lambda_1, \lambda_2) = -\frac{1}{2}(q + G^\top \lambda_1 + A^\top \lambda_2)^\top P^{-1}(q + G^\top \lambda_1 + A^\top \lambda_2) - \lambda_1^\top h - \lambda_2^\top b + r.$$

Therefore, the dual problem is given by

$$\max_{\lambda_1 \geq 0, \lambda_2 \in \mathbb{R}^M} \left\{ -\frac{1}{2}(q + G^\top \lambda_1 + A^\top \lambda_2)^\top P^{-1}(q + G^\top \lambda_1 + A^\top \lambda_2) - \lambda_1^\top h - \lambda_2^\top b + r \right\}.$$

***11.A.2.4 Quadratic Programming in Markowitz Portfolio Selection** Consider a financial market with p financial assets, and let their daily logarithmic returns on day t be $\mathbf{r}_t$. These returns are often assumed to be iid random vectors following a p-variate Gaussian distribution with a mean vector $\boldsymbol{\mu} \in \mathbb{R}^p$ and a covariance matrix $\boldsymbol{\Sigma} \in \mathbb{R}^{p \times p}$:

$$\mathbf{r}_t \overset{iid}{\sim} \mathcal{N}_p(\boldsymbol{\mu}, \boldsymbol{\Sigma}), \qquad \text{for } t = 1, 2, \ldots.$$

Let us assume short-selling has been prohibited and investors are risk averse whenever two investments yield the same expected return, they prefer the less risky one. In these circumstances, a portfolio is simply a combination of the p assets with a non-negative weight vector $\boldsymbol{\omega} \in [0, 1]^p$ that sums to 1 (no short-selling allowed), and the investors have to pick their desired portfolio to have an expected portfolio return of at least μ_P. Equivalently, the portfolio should satisfy $\mu_P \leq \boldsymbol{\mu}^\top \boldsymbol{\omega}$ but with the lowest possible level of risk, as measured by the variance of the portfolio return. Markowitz [16] formulated the optimal portfolio selection problem as a QP problem:

$$\underset{\boldsymbol{\omega} \in [0,1]^p}{\text{minimize}} \quad \frac{1}{2}\boldsymbol{\omega}^\top \boldsymbol{\Sigma} \boldsymbol{\omega}, \tag{11.44}$$

$$\text{subject to} \quad \text{i) } \boldsymbol{\mu}^\top \boldsymbol{\omega} \geq \mu_P; \quad \text{ii) } \mathbf{1}^\top \boldsymbol{\omega} = 1; \quad \text{iii) } \boldsymbol{\omega} \geq \mathbf{0}, \tag{11.45}$$

where $\mathbf{1} \in \mathbb{R}^p$ is a column vector of value 1 and $\mathbf{0} \in \mathbb{R}^p$ is a column vector of value 0. Readers can also refer to [7] for more discussions on portfolio selection.

For this problem, the Lagrangian $\mathcal{L}(\boldsymbol{\omega}, \lambda_1, \lambda_2, \boldsymbol{\lambda}_3)$ with Lagrange multipliers $\lambda_1 \geq 0$, $\lambda_2 \in \mathbb{R}$, and $\boldsymbol{\lambda}_3 \geq \mathbf{0}$ is given by

$$\mathcal{L}(\boldsymbol{\omega}, \lambda_1, \lambda_2, \boldsymbol{\lambda}_3) = \frac{1}{2}\boldsymbol{\omega}^\top \boldsymbol{\Sigma}\boldsymbol{\omega} + \lambda_1(\mu_P - \boldsymbol{\mu}^\top \boldsymbol{\omega}) + \lambda_2(\mathbf{1}^\top \boldsymbol{\omega} - 1) + \boldsymbol{\lambda}_3^\top(\mathbf{0} - \boldsymbol{\omega}).$$

Differentiating the Lagrangian with respect to $\boldsymbol{\omega}$ gives:

$$\nabla_{\boldsymbol{\omega}}\mathcal{L}(\boldsymbol{\omega}, \lambda_1, \lambda_2, \boldsymbol{\lambda}_3) = \frac{1}{2}(\boldsymbol{\Sigma} + \boldsymbol{\Sigma}^\top)\boldsymbol{\omega} - \lambda_1\boldsymbol{\mu} + \lambda_2\mathbf{1} - \boldsymbol{\lambda}_3 = \boldsymbol{\Sigma}\boldsymbol{\omega} - \lambda_1\boldsymbol{\mu} + \lambda_2\mathbf{1} - \boldsymbol{\lambda}_3.$$

Provided that the covariance matrix $\boldsymbol{\Sigma}$ is symmetric and invertible, setting this derivative to zero yields:

$$\boldsymbol{\omega} = -\boldsymbol{\Sigma}^{-1}(-\lambda_1\boldsymbol{\mu} + \lambda_2\mathbf{1} - \boldsymbol{\lambda}_3).$$

Finally, substituting the solution back to the Lagrangian gives the dual Lagrangian:

$$\mathcal{D}(\lambda_1, \lambda_2, \boldsymbol{\lambda}_3) = -\frac{1}{2}(-\lambda_1\boldsymbol{\mu} + \lambda_2\mathbf{1} - \boldsymbol{\lambda}_3)^\top \boldsymbol{\Sigma}^{-1}(-\lambda_1\boldsymbol{\mu} + \lambda_2\mathbf{1} - \boldsymbol{\lambda}_3) + \lambda_1\mu_P - \lambda_2.$$

Therefore, the dual optimization problem is written as the following:

$$\underset{\lambda_1, \lambda_2, \boldsymbol{\lambda}_3}{\text{maximize}} \quad -\frac{1}{2}(-\lambda_1\boldsymbol{\mu} + \lambda_2\mathbf{1} - \boldsymbol{\lambda}_3)^\top \boldsymbol{\Sigma}^{-1}(-\lambda_1\boldsymbol{\mu} + \lambda_2\mathbf{1} - \boldsymbol{\lambda}_3) + \lambda_1\mu_P - \lambda_2, \tag{11.46}$$

$$\begin{aligned} \text{subject to} \quad & \text{i)} \quad \lambda_1 \geq 0; \\ & \text{ii)} \quad \lambda_2 \in \mathbb{R}; \\ & \text{iii)} \quad \boldsymbol{\lambda}_3 \geq \mathbf{0}. \end{aligned} \tag{11.47}$$

Let us illustrate Markowitz Portfolio Selection with real data. The dataset `S&P500_Open_hourly_2021.csv` contains the (unadjusted) hourly *Opening Prices* of 375 securities[9] which are listed in the S&P500, from 1 Jan 2021 to 31 Jan 2022. The dataset `SPX_Open_hourly_2021.csv` contains the hourly *Opening Prices* of the S&P500 index within the same period. For simplicity, in the following analysis, we shall ignore transaction costs, market friction, dividends, etc. Moreover, we treat the hourly security data from the first trading hour on 1 Jan 2021 to the first trading hour on 31 Dec 2021 as the training dataset, while the first trading hour on 1 Jan 2022 to the last trading hour on 7 Jan 2022 becomes the test dataset. In reality, the hourly logarithmic returns are not necessarily iid, but in this implementation, we continue to assume that these random returns are iid and follow a multivariate Gaussian distribution.

[9] The list of securities in S&P500 was taken in 2020. Some securities are privatized and so we are no longer able to locate them in Yahoo Finance; while some securities were discarded for having many missing hourly returns. Therefore, the actual number of securities in this dataset is much lower than the full list.

```
1  import pandas as pd
2  import numpy as np
3  import datetime as dt
4
5  year = 2021
6  df_price = pd.read_csv(f"S&P500_Open_hourly_{year}.csv", index_col=0)
7  df_price.index = pd.to_datetime(df_price.index.values)
8  train_end = dt.datetime(year=year+1, month=1, day=1, hour=1)
9  test_end = train_end + dt.timedelta(days=7)
10 print(train_end, test_end) # 1-Jan, 2022 01:00:00; 8-Jan, 2022 01:00:00
11 mask_train = (df_price.index < train_end)
12 mask_test = (train_end < df_price.index) * (df_price.index <= test_end)
13 df_train, df_test = df_price.loc[mask_train], df_price.loc[mask_test]
14 n_stocks = len(df_price.columns)
```

```
1  2022-01-01 01:00:00 2022-01-08 01:00:00
```

Programme 11.8 Data preprocessing of the security data in Python.

The hourly logarithmic returns are first computed in Programme 11.9 via Python. Those in the training set are used to compute the sample covariance matrix $\boldsymbol{P}$ and estimate the mean vector $\boldsymbol{\mu}$ via EWMA (as seen in Section 7.2) with the mean vector defined as $\boldsymbol{\mu}_t = \lambda \boldsymbol{r}_t + (1-\lambda)\boldsymbol{\mu}_{t-1}$, for $t = 2, \ldots, n$, with $\lambda = 0.94$ and $\boldsymbol{\mu}_1 = \boldsymbol{r}_1$ by convention. Suppose that the investor would like to achieve an annual return of at least $\mu_P = 15\%$[10] (roughly 0.00837% per hour), and he has an initial investment amount of USD 100,000. We aim to solve the QP problem in (11.44) and (11.45). After the optimal weights are found, those weights having values less than 10^{-2} will be set to 0, meaning that we do not include the corresponding securities in the optimal portfolio.

```
1  from qpsolvers import solve_qp
2
3  PRINCIPAL = 100_000        # Initial Investment of 100,000 USD
4  mu_annual = 0.15           # 15% minimum desired portfolio annual return
5  mu_P = mu_annual / 7 / 256  # 7 hrs per day and 256 trading days per yr
6  print(mu_P)                # minimum desired portfolio hourly return
7
8  r = np.log(1 + df_train.pct_change()).iloc[1:, :]
9  mu = r.ewm(alpha=0.94, adjust=False).mean().iloc[-1, :]
10 P = r.cov(ddof=1)          # Sample covariance matrix
11 q = np.zeros(n_stocks)
12 G = np.vstack([-mu, -np.eye(n_stocks)])
13 h = np.append([-mu_P], np.zeros(n_stocks))
14 A = np.ones(n_stocks)
15 b = np.array([1]).astype(np.float32)
```

[10] This is roughly the rate of return that a well-performing fund manager in investment banks can warrant.

```
16
17 w = solve_qp(P, q, G, h, A, b, solver="ecos")
18
19 w[w < 1e-2] = 0            # Treat weights less than 0.01 as 0
20 w = w/np.sum(w)            # Rescale the weights
21 n_portfolio = PRINCIPAL * w / df_test.iloc[0,:].values
22 Portfolio_value = df_test.values @ n_portfolio
23 Markowitz_portfolio = df_test.columns.values[w > 0]
24 print(dict(zip(Markowitz_portfolio, np.round(w[w > 0], 5))))
```

```
1 8.370535714285714e-05
2 {'AMZN': 0.03464, 'BRK-B': 0.01985, 'PG': 0.06622, 'BAC': 0.03358,
3  'CRM': 0.01545, 'VZ': 0.19365, 'TMO': 0.03363, 'WMT': 0.04517,
4  'MRK': 0.02694, 'CME': 0.01299, 'BDX': 0.04375, 'CL': 0.02775,
5  'EQIX': 0.01515, 'PSA': 0.05983, 'KMB': 0.04709, 'ORLY': 0.03217,
6  'EA': 0.01878, 'GIS': 0.02036, 'BLL': 0.01241, 'ED': 0.0504,
7  'PPL': 0.01252, 'VRSN': 0.04365, 'DPZ': 0.04786, 'HRL': 0.03627,
8  'HST': 0.01517, 'RHI': 0.0159, 'CPB': 0.01883}
```

Programme 11.10 Solving the QP Problem depicted in (11.44) and (11.45) via Python.

From the output of Programme 11.10 in Python, 27 stocks are selected to be included in the optimal portfolio. In contrast, the weights for the other 375 – 27 = 348 securities are less than the threshold of 10^{-2}, so that they are totally discarded.

Finally, as a comparison, we consider the buy-and-hold strategy on the S&P500 portfolio with the same initial investment amount of USD 100,000, see Programme 11.11.

```
1 import matplotlib.pyplot as plt
2 import matplotlib.ticker as ticker
3 import matplotlib.dates as mdates
4
5 plt.rc('font', size=25); plt.rc('axes', labelsize=25)
6 plt.rc('xtick', labelsize=17); plt.rc('ytick', labelsize=17)
7
8 df_index = pd.read_csv(f"SPX_Open_hourly_{year}.csv", index_col=0)
9 df_test_index = df_index.loc[mask_test]
10 n_index = PRINCIPAL / df_test_index.iloc[0].values
11 Index_value = df_test_index.values.reshape(-1) * n_index
12
13 fig, ax = plt.subplots(figsize=(15, 10))
14 line1, = ax.plot(df_test.index, Portfolio_value, "r-", linewidth=3,
15                  label="MPT")
16 line2, = ax.plot(df_test.index, Index_value, "b--", linewidth=3,
17                  label="S&P 500")
18 ax.legend(loc="upper right", handles=[line1, line2])
19 plt.xlabel("Time")
20 plt.ylabel("Price")
21 ax.xaxis.set_major_locator(ticker.MaxNLocator(6))
```

```
22  ax.xaxis.set_major_formatter(
23      mdates.DateFormatter("%Y-%m-%d \n%H:%M:%S")
24  )
25
26  print(Portfolio_value[-1] - PRINCIPAL, Index_value[-1] - PRINCIPAL)
```

```
1  106.56726526544662 -1936.5332700568833
```

Programme 11.11 Plotting the optimal portfolio value of (11.46) and (11.47), and the buy-and-hold portfolio of S&P500 in Python.

From Figure 11.20, the optimal portfolio performs better than the S&P500 portfolio. However, even without counting in transaction costs and market friction, the optimal portfolio only yields a marginally positive return of USD 106.567 at the end of the test period when the market closes on 7 Jan 2022. In general, the performance of portfolio selection via quadratic programming, though certainly not the worst, is mostly mediocre. That is why we do not expect a very promising return for those utilizing this approach in actual trading.

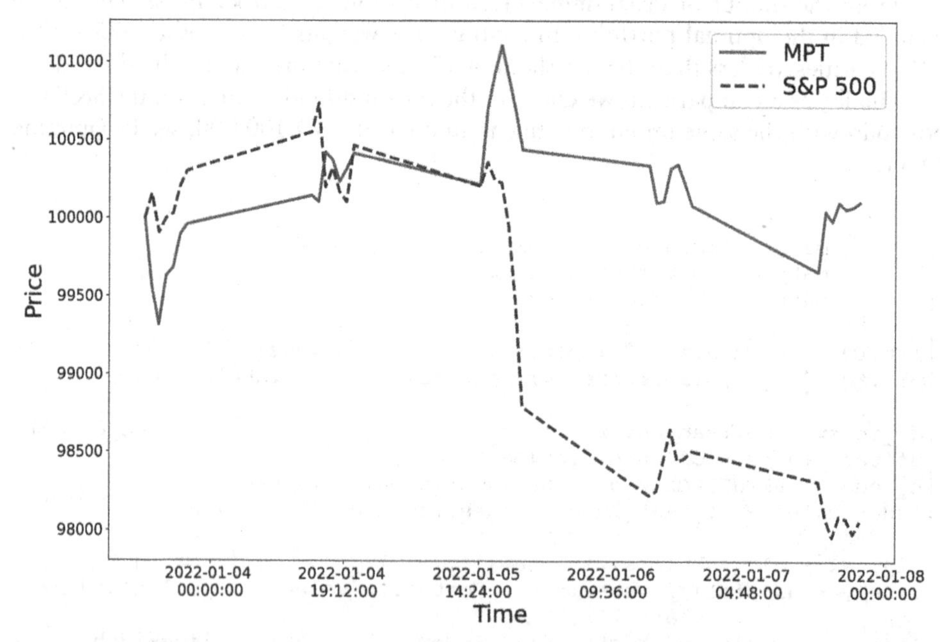

FIGURE 11.20 Value of the optimal portfolio and the S&P500 index in the test dataset, generated by Programme 11.11.

REFERENCES

1. Abe, S. (2005). *Support Vector Machines for Pattern Classification* (Vol. 2, p. 44). Springer.
2. Aubin, J.P. (2013). *Optima and Equilibria: An Introduction to Nonlinear Analysis*. Springer.
3. Berkovitz, L.D. (2003). *Convexity and Optimization in* $\mathbb{R}^n$. Wiley.
4. Bertsekas, D., Nedic, A., and Ozdaglar, A. (2003). *Convex Analysis and Optimization*. Athena Scientific.
5. Boyd, S.P., and Vandenberghe, L. (2004). *Convex Optimization*. Cambridge University Press.
6. Brouwer, L.E.J. (1912). Uber Abbildung von Mannigfaltigkeiten. *Mathematische Annalen*, 71, 97–115.
7. Capiński, M.J., and Kopp, E. (2014). *Portfolio Theory and Risk Management*. Cambridge University Press.
8. Chen, Y., Fan, N.S., and Yam, S.C.P. (2024+). *Statistical Deep Learning with Python and* R. Preprint.
9. Credit Fusion, Will, C. (2011). Give Me Some Credit. Kaggle. https://kaggle.com/competitions/GiveMeSomeCredit (accessed 7 May 2024).
10. Cristianini, N., and Shawe-Taylor, J. (2000). *An Introduction to Support Vector Machines and Other Kernel-Based Learning Methods*. Cambridge University Press.
11. Deng, N., Tian, Y., and Zhang, C. (2013). *Support Vector Machines: Optimization Based Theory, Algorithms, and Extensions*. Chapman & Hall/CRC.
12. Frank, M., and Wolfe, P. (1956). An algorithm for quadratic programming. *Naval Research Logistics Quarterly*, 3(1–2), 95–110.
13. Garrigos, G., and Gower, R.M. (2023). Handbook of Convergence Theorems for (Stochastic) Gradient Methods. *ArXiv preprint* arXiv:2301.11235.
14. Goodfellow, I., Bengio, Y., and Courville, A. (2016). *Deep Learning*. MIT Press.
15. Kiefer, J., and Wolfowitz, J. (1952). Stochastic estimation of the maximum of a regression function. *The Annals of Mathematical Statistics*, 23(3), 462–466.
16. Markowitz, H. (1952). Portfolio selection. *The Journal of Finance*, 7(1), 77–91.
17. Mehlig, B. (2021). *Machine Learning with Neural Networks: An Introduction for Scientists and Engineers*. Cambridge University Press.
18. Minsky, M., and Papert, S. (1969). *Perceptrons: An Introduction to Computational Geometry*. MIT Press.
19. Robbins, H., and Monro, S. (1951). A stochastic approximation method. *The Annals of Mathematical Statistics*, 22(3), 400–407.
20. Rosenblatt, F. (1958). The perceptron: a probabilistic model for information storage and organization in the brain. *Psychological Review*, 65(6), 386.
21. Ruszczynski, A. (2011). *Nonlinear Optimization*. Princeton University Press.
22. Shalev-Shwartz, S., and Ben-David, S. (2014). *Understanding Machine Learning: From Theory to Algorithms*. Cambridge University Press.
23. Steinwart, I., and Christmann, A. (2008). *Support Vector Machines*. Springer Science & Business Media.
24. Villani, C. (2021). *Topics in Optimal Transportation*. American Mathematical Society.
25. von Neumann, J. (1928). Zur theorie der gesellschaftsspiele. *Mathematische Annalen*, 100(1), 295–320.

PART Three

Nonlinear Models

CHAPTER 12

Bayesian Learning

In this chapter, we introduce another important class of semiparametric model-based learning, called *Bayesian inference*, or *Bayesian learning* especially if we focus on the model calibrations. Whereas Bayesian inference follows many of the conventional steps of semiparametric modelling, including model specification, criterion selection for evaluation, and parameter estimation, it differs from other approaches in terms of the criterion adopted. Indeed, the criterion used for Bayesian inference is the probability of attaining various parameter values, which serves as an alternative to the previously mentioned criteria such as the minimization of loss functions and maximum likelihood estimations. We shall highlight a few useful Bayesian-based tools for data analysis in *FinTech* and *InsurTech*.

12.1 SIMPLE CREDIBILITY THEORY

Credibility theory, as a linear Bayesian model, has long been a core topic in *actuarial science*. It involves the pricing of insurance products using the historical information from both the policyholder herself and the risk class she belongs to. Readers may refer to [2, 11] for a detailed and extensive discussion, as well as [4] and references therein for more recent developments in the field. More precisely, let $\mathrm{x}_i, i = 1, 2, \ldots$ be the loss amount of a policyholder at time i, whose risky position is governed by an unknown parameter $\theta \in \chi$; θ is also known as the *risk profile* of the policyholder, and χ is the collection of all possible risk profiles in the community. From the Bayesian perspective, θ is regarded as a realization of the random variable Θ, which describes the uncertainty due to the limit of knowledge. As a primitive model, we also assume that $\mathrm{x}_1, \mathrm{x}_2, \ldots$ are conditionally iid given $\Theta = \theta$.

The key question in credibility theory is to determine the fair premium to be collected in the next time period, say $n+1$, given the history of losses $\mathrm{x}_1, \ldots, \mathrm{x}_n$ of the policyholder. This would be straightforward with the knowledge of her risk profile θ and the distribution of x_i given $\Theta = \theta$; indeed, the fair premium should be $\mathbb{E}\left(\mathrm{x}_{n+1} \mid \Theta = \theta\right) =: \mu(\theta)$. This quantity is also called the *hypothetical mean*, and we note it does not depend on n because of the conditional iid assumption. However, as neither θ nor the conditional distribution is known in reality, we may adopt the Bayesian approach to estimate this quantity by using the whole loss history of the policyholder.

To this end, let $t(\mathbf{x})$ be an estimator of $\mu(\theta)$, a statistic of the historical loss vector $\mathbf{x} = (\mathrm{x}_1, \mathrm{x}_2, \ldots, \mathrm{x}_n)^{\mathsf{T}}$. For any given value of θ, we define the *risk function* to be the expected squared loss of $t(\mathbf{x})$ across all possible values of $\mathbf{x}$:

$$R_t(\theta) := \mathbb{E}\left((\mu(\theta) - t(\mathbf{x}))^2\right) = \int_{\mathbb{R}^n} (\mu(\theta) - t(x))^2 f_{\mathbf{x}|\Theta}(x \mid \theta)dx.$$

This allows us to define the *Bayes risk* of the estimator, by taking into account the randomness of Θ, as follows:

$$\begin{aligned} R(t(\mathbf{x})) &:= \mathbb{E}\left(R_t(\Theta)\right) \\ &= \int_{-\infty}^{\infty} \mathbb{E}(\mu(\theta) - t(\mathbf{x}))^2 \pi_\Theta(\theta)d\theta \\ &= \int_{-\infty}^{\infty}\int_{\mathbb{R}^n} (\mu(\theta) - t(x))^2 f_{\mathbf{x}|\Theta}(x \mid \theta)\pi_\Theta(\theta)dxd\theta \\ &= \int_{\mathbb{R}^n}\int_{-\infty}^{\infty} (\mu(\theta) - t(x))^2 \pi_{\Theta|\mathbf{x}}(\theta \mid x)d\theta f_{\mathbf{x}}(x)dx, \end{aligned} \tag{12.1}$$

where $\pi_\Theta(\theta)$ is the *prior distribution* of Θ. Naturally, we shall select the estimator that minimizes the Bayes risk, and it suffices to look for one that minimizes the inner integral in the last line of (12.1). Due to the squared loss function, it can be shown by standard arguments that the resulting minimizer is given by the posterior mean:

$$\widehat{t(\mathbf{x})} := \mathbb{E}\left(\mu(\Theta) \mid \mathbf{x}\right) = \arg\min_{t(\cdot)} R(t(\mathbf{x})), \tag{12.2}$$

which is also known as the *Bayesian premium*.

Next we illustrate the calculation of the Bayesian premium via a simple example. Suppose that $\mathrm{x}_i \mid \Theta = \theta \overset{iid}{\sim} \mathcal{N}\left(\theta, \sigma^2\right)$ for $i = 1, 2, \ldots$, and the prior distribution of Θ is $\mathcal{N}\left(\mu, \tau^2\right)$, where σ^2, μ and τ^2 are known constants. To determine the Bayesian premium for the $(n+1)$-th period based on the historical losses $\mathbf{x} = (\mathrm{x}_1, \ldots, \mathrm{x}_n)^{\mathsf{T}}$, we need to derive the posterior distribution of Θ given $\mathbf{x}$. By precisely the same derivations as in Section 1.5, we have:

$$\Theta \mid \mathbf{x} = x \sim \mathcal{N}\left(\mu_n, \beta_n^{-1}\right),$$

where $\mu_n = \frac{\frac{1}{\sigma^2}\sum_{i=1}^n x_i + \frac{1}{\tau^2}\mu}{\frac{n}{\sigma^2} + \frac{1}{\tau^2}} = \frac{n\tau^2\overline{x} + \sigma^2\mu}{n\tau^2 + \sigma^2}$ and $\beta_n = \frac{n}{\sigma^2} + \frac{1}{\tau^2}$. Therefore, together with the fact that $\mu(\theta) = \mathbb{E}(\mathrm{x} \mid \Theta = \theta) = \theta$, the Bayesian premium is simply equal to:

$$\widehat{t(\mathbf{x})} = \mathbb{E}(\mu(\Theta) \mid \mathbf{x}) = \mathbb{E}(\Theta \mid \mathbf{x}) = \frac{n\tau^2\overline{\mathrm{x}} + \sigma^2\mu}{n\tau^2 + \sigma^2} = z\overline{\mathrm{x}} + (1 - z)\mu, \tag{12.3}$$

where $z := \frac{n\tau^2}{n\tau^2 + \sigma^2}$ is called the *credibility factor* as it represents the extent to which, when determining the premium, we believe in the historical claims instead of our prior belief from the prior distribution. Moreover, this credibility factor tends to 1 as the sample size $n \to \infty$, indicating that we have a stronger belief in the policyholder's claim history over time.

The special linear form of the credibility estimator in (12.3) is not a coincidence. In fact, if the likelihood function belongs to the *exponential dispersion family* (EDF) introduced in Section 10.3 and the prior distribution is the conjugate prior with respect to the likelihood, the Bayesian estimator always takes the form of a convex combination of the sample mean and the theoretical expectation under the prior with weights that are the same as those in the example above. In this case the Bayesian credibility is said to be an *exact credibility*; its proof is given in [10].

*12.2 BAYESIAN ASYMPTOTIC INFERENCE

As we have seen in the previous section, though we illuminate the Bayesian nature of the credibility estimator, too many restrictions (including its simple linearity) limit its use in a broader scope. We want a more generalized *Bayesian decision theory* that can allow for (i) a larger family of loss functions, not just the squared loss; (ii) an explicit approximation of the corresponding Bayesian estimator, especially when the prior is not the conjugate prior of the likelihood; and (iii) an approximated posterior distribution for further statistical inference like hypothesis testing.

As before, we interpret the parameter $\theta \in \chi$ as an outcome of the random variable Θ following the prior distribution π, and the data $\mathbf{x}$ follows a distribution with a joint density function $f(x;\theta)$ given θ, that is the likelihood. The objective of *average risk optimality* is sought for:

$$\delta_\pi := \arg\min_{\delta(\cdot)} r(\pi,\delta) := \arg\min_{\delta(\cdot)} \mathbb{E}[L(\Theta,\delta(\mathbf{x}))] = \arg\min_{\delta(\cdot)} \int_\chi R(\theta,\delta)\pi(\theta)d\theta,$$

where $L(\theta,\delta(\mathbf{x}))$ is the loss function when θ is the true value of the parameter and $\delta(\mathbf{x})$ is used to estimate it,

$$R(\theta,\delta) = \mathbb{E}_{\mathbf{x}|\Theta=\theta} L(\theta,\delta(\mathbf{x})) = \int_{\mathbb{R}^n} L(\theta,\delta(x)) f_{\mathbf{x}|\Theta}(x|\theta)dx$$

is the risk function, and $r(\pi,\delta)$ is called *average risk*. The minimizer $\delta_\pi(\mathbf{x})$ is called the *Bayesian estimator* with respect to θ under the prior distribution π.

A useful existing result for the Bayesian estimator states that if there exists an estimator δ with a finite risk $R(\theta,\delta)$ for almost every θ, and there exists a statistic δ^* as a value that minimizes $\mathbb{E}[L(\Theta,\delta^*(\mathbf{x}))|\mathbf{x}=x]$ for almost all x, then the statistic $\delta^*(x)$ as a function is the Bayesian estimator. The proof can be found in many standard textbooks on statistical estimation, such as Chapter 4 in [16]. An example of this claim is that the posterior mean of (12.2) is the Bayesian estimator when the loss function is the squared loss. When the loss function is an absolute difference, by a simple combinatorial argument, the corresponding Bayesian estimator is given by the posterior median. Generally speaking, for any differentiable loss function satisfying enough regularity conditions, its Bayesian estimator satisfies the first order condition:

$$\mathbb{E}\left[\frac{\partial}{\partial\delta}L(\Theta,\delta_\pi(\mathbf{x}))\bigg|\,\mathbf{x}=x\right]=0. \tag{12.4}$$

The corresponding integrals (conditional expectations) are usually intractable, especially when the prior is not a conjugate of the likelihood, and so we lack the explicit expression of the Bayesian estimator. Nevertheless, if the sample is large, we can obtain an approximation by the *Laplace method*. Considering general likelihoods and priors, without loss of generality, we focus only on the one-dimensional case here. Most often, the corresponding Bayesian estimator can be written in terms of several conditional expectations of the following form:

$$\mathbb{E}(g(\Theta)\mid\mathbf{x}=x)=\frac{\int_\chi g(\theta)\pi(\theta)\exp\{l(\theta;x)\}d\theta}{\int_\chi \pi(\theta)\exp\{l(\theta;x)\}d\theta}=\frac{\int_\chi \exp\{n(\bar{l}(\theta;x)+q(\theta)/n)\}d\theta}{\int_\chi \exp\{n(\bar{l}(\theta;x)+p(\theta)/n)\}d\theta},$$

for some function g, where l stands for the conditional log-likelihood given θ, and we also use the notation $\bar{l}(\theta):=l(\theta)/n$ for the average log-likelihood, $q(\theta):=\ln(g(\theta)\pi(\theta))$ and $p(\theta):=\ln(\pi(\theta))$. Applying *Laplace's approximation* (refer to Appendix 12.A.1) when n is large, we can approximate:

$$\mathbb{E}(g(\Theta)\mid\mathbf{x}=x)\approx\frac{e^{n\bar{l}(\theta^*)+q(\theta^*)}}{e^{n\bar{l}(\bar{\theta})+p(\bar{\theta})}}\frac{\left\{-n\bar{l}''(\bar{\theta})-p''(\bar{\theta})\right\}^{1/2}}{\left\{-n\bar{l}''(\theta^*)-q''(\theta^*)\right\}^{1/2}}, \tag{12.5}$$

where θ^* maximizes the function $n\bar{l}(\theta)+\ln g(\theta)+\ln\pi(\theta)$, and $\bar{\theta}$ maximizes the function $n\bar{l}(\theta)+\ln\pi(\theta)$.

We have a key asymptotic result for the posterior distribution, known as the *Bayesian Central Limit Theorem* (Bayesian CLT), which asserts, in the general p-dimensional case, that

$$\pi_n(\theta|x)=\frac{f(x;\theta)\pi(\theta)}{\int_\chi f(x;\theta)\pi(\theta)d\theta}\to\mathcal{N}_p(\hat{\theta}_\pi,i_\pi^{-1}(\hat{\theta}_\pi)), \tag{12.6}$$

as n becomes large. To see the validity of (12.6), let $\hat{\theta}_\pi$ be the mode of $\pi_n(\theta\mid x)$, or equivalently $f(x;\theta)\pi(\theta)$ takes its maximum value at $\hat{\theta}_\pi$, when x is fixed. Then, consider

$$\begin{aligned}\ln\left(\frac{f(x;\theta)\pi(\theta)}{f(x;\hat{\theta}_\pi)\pi(\hat{\theta}_\pi)}\right)&\approx\ln\left(\frac{f(x;\hat{\theta}_\pi)\pi(\hat{\theta}_\pi)}{f(x;\hat{\theta}_\pi)\pi(\hat{\theta}_\pi)}\right)+\left(\nabla_\theta\ln\left(\frac{f(x;\theta)\pi(\theta)}{f(x;\hat{\theta}_\pi)\pi(\hat{\theta}_\pi)}\right)\bigg|_{\hat{\theta}_\pi}\right)^{\mathsf{T}}(\theta-\hat{\theta}_\pi)\\&\quad+\frac{1}{2}(\theta-\hat{\theta}_\pi)^{\mathsf{T}}\left(\nabla_\theta\nabla_\theta^{\mathsf{T}}\ln\left(\frac{f(x;\theta)\pi(\theta)}{f(x;\hat{\theta}_\pi)\pi(\hat{\theta}_\pi)}\right)\bigg|_{\hat{\theta}_\pi}\right)(\theta-\hat{\theta}_\pi)\\&=-\frac{1}{2}(\theta-\hat{\theta}_\pi)^{\mathsf{T}}i_\pi(\hat{\theta}_\pi)(\theta-\hat{\theta}_\pi),\end{aligned} \tag{12.7}$$

where $i_\pi(\hat{\theta}_\pi) = -\nabla_\theta \nabla_\theta^\top \ln\left(f(x;\theta)\pi(\theta)\right)\Big|_{\hat{\theta}_\pi}$ is the well-known empirical aggregated Fisher information. The approximation of (12.7) holds by Taylor expansion at $\hat{\theta}_\pi$ and omitting higher order terms, which is justifiable as they are collectively of order $o_p(1)$. The last equation holds because the gradient at the maximum point of the function vanishes, i.e. $\nabla_\theta \ln\left(f(x;\theta)\pi(\theta)\right)\Big|_{\hat{\theta}_\pi} = \mathbf{0}$. Finally we note that the leading constant term is simply $\ln 1 = 0$. As a result, we deduce that

$$f(x;\theta)\pi(\theta) \approx f(x;\hat{\theta}_\pi)\pi(\hat{\theta}_\pi)\exp\left(\frac{1}{2}\left(-(\theta-\hat{\theta}_\pi)^\top i_\pi(\hat{\theta}_\pi)(\theta-\hat{\theta}_\pi)\right)\right),$$

from which (12.6) holds.

Alternatively, using Laplace's approximation to approximate the integral in the denominator of (12.6), we can estimate the posterior density as follows. Firstly, we write

$$\int_\chi f(x;\theta)\pi(\theta)d\theta = \int_\chi \exp(n\bar{l}(\theta))\pi(\theta)d\theta,$$

and then apply (12.21) in 12.A.1 to obtain

$$\int_\chi f(x;\theta)\pi(\theta)d\theta \approx (2\pi)^{\frac{d}{2}}\left|i_\pi(\hat{\theta}_\pi)\right|^{-\frac{1}{2}} \exp\left(l(\hat{\theta}_\pi)\right)\pi(\hat{\theta}_\pi),$$

where i_π is the empirical aggregated Fisher information as defined immediately after (12.7). Therefore, the posterior density function can be approximated as the following when n is large,

$$\begin{aligned}\pi_n(\theta|x) &\approx (2\pi)^{-\frac{d}{2}}\left|i_\pi(\hat{\theta}_\pi)\right|^{\frac{1}{2}} \exp\left\{l(\theta) - l(\hat{\theta}_\pi)\right\}\frac{\pi(\theta)}{\pi(\hat{\theta}_\pi)}.\\ &= (2\pi)^{-\frac{d}{2}}\left|i_\pi(\hat{\theta}_\pi)\right|^{\frac{1}{2}} \cdot \frac{f(x;\theta)\pi(\theta)}{f(x;\hat{\theta}_\pi)\pi(\hat{\theta}_\pi)}. \qquad (12.8)\end{aligned}$$

Using Laplace's approximation, we can also motivate the setting behind the *Bayesian Information Criterion* (the BIC as mentioned in Section 10.6) commonly used in practice. We discuss this topic as a supplement in Appendix 12.A.2.

As a final remark, in the event of a relatively small finite sample, we usually use Markov Chain Monte Carlo (MCMC) methods to evaluate the posterior distribution or the relevant posterior statistics; we refer readers to Section *6.4 of Chapter 6 for more detailed discussions.

12.3 REVISITING POLYNOMIAL REGRESSION

We briefly introduced polynomial regression in Section 10.2. Let us revisit this class of models from the Bayesian learning perspective. We take the simplest univariate polynomial model as an illustration:

$$y_i = \sum_{m=0}^{M} \beta_m x_i^m + \epsilon_i =: g(x_i) + \epsilon_i, \quad i = 1, \ldots, n,$$

where $\epsilon_i \overset{iid}{\sim} \mathcal{N}(0, \sigma^2)$, and $\boldsymbol{\beta} = (\beta_0, \beta_1, \ldots, \beta_M)^\top$ are coefficients to be estimated. We assume that $\lambda := \sigma^{-2}$ is known. In the Bayesian framework, we view $\boldsymbol{\beta}$ as a random vector, and the prior distribution of $\boldsymbol{\beta}$ is given by

$$\pi(\boldsymbol{\beta}) = \left(\frac{\alpha}{2\pi}\right)^{\frac{M+1}{2}} \exp\left(-\frac{\alpha}{2}\boldsymbol{\beta}^\top\boldsymbol{\beta}\right), \tag{12.9}$$

where $\alpha > 0$ is a constant. In other words, $\boldsymbol{\beta}$ follows a multivariate normal distribution $\mathcal{N}_{M+1}(\mathbf{0}_{M+1}, \alpha^{-1}\boldsymbol{I}_{M+1})$, where $\mathbf{0}_{M+1}$ is the $(M+1)$-dimensional zero vector and $\boldsymbol{I}_{M+1}$ is the $(M+1)$-dimensional identity matrix. As $y_1, \ldots, y_n$ are independently normally distributed with mean $g(x_i)$ and variance λ^{-1},

$$f(\mathbf{y}|\boldsymbol{\beta}) = \prod_{i=1}^{n} f(y_i|g(x_i), \lambda) = \left(\frac{\lambda}{2\pi}\right)^{\frac{n}{2}} \exp\left(-\frac{\lambda}{2}\sum_{i=1}^{n}(y_i - g(x_i))^2\right). \tag{12.10}$$

Using (12.9) and (12.10), the posterior density of $\boldsymbol{\beta}$ is given by

$$f(\boldsymbol{\beta}|\mathbf{y}) \propto f(\mathbf{y}|\boldsymbol{\beta})\pi(\boldsymbol{\beta}) \propto \exp\left(-\frac{\lambda}{2}\sum_{i=1}^{n}(g(x_i) - y_i)^2 - \frac{\alpha}{2}\boldsymbol{\beta}^\top\boldsymbol{\beta}\right). \tag{12.11}$$

Instead of considering the posterior mean of $\boldsymbol{\beta}$, which is commonly used in Bayesian methods, we focus on the *maximum a posteriori probability* (MAP) estimate, which is defined as the maximizer of the posterior $f(\boldsymbol{\beta}|\mathbf{y})$, or equivalently, the minimizer of the following error function:

$$\mathcal{E} := \frac{\lambda}{2}\sum_{i=1}^{n}(g(x_i) - y_i)^2 + \frac{\alpha}{2}\boldsymbol{\beta}^\top\boldsymbol{\beta}. \tag{12.12}$$

Note that as $\alpha \to 0$, $\mathcal{E} \to \frac{\lambda}{2}\sum_{i=1}^{n}(g(x_i) - y_i)^2$, hence the MAP estimate converges to the least square estimate. Furthermore, as $\alpha > 0$, the second term of $\mathcal{E}$ in (12.12) increases with the norm of $\boldsymbol{\beta}$, thus it can be regarded as a regularization term, meaning that MAP estimation can be regarded as a regularization on the least square estimation. The resulting $\boldsymbol{\beta}$ tends to $\mathbf{0}$ for a larger value of α; while the prior distribution becomes closer to a uniform distribution as α decreases, echoing a weaker regularization in $\mathcal{E}$. As a remark, the selection of the parameter α may vary according to the data structure. For instance, when all x_i's lie between 0 and 1 while the responses y's have a larger magnitude, the estimate of $\boldsymbol{\beta}$ may inevitably be large. To this end we may choose a smaller α to strike a balance.

Apart from regularization, the penalty term in the expression of $\mathcal{E}$ in (12.12) can also help to prevent overfitting, as a simpler model is sought. Models with a larger degree M are usually subject to a larger penalty, and by comparing the values of $\mathcal{E}$ across models with various values of M, we can select an appropriate degree M of the estimated polynomial to avoid overfitting and to enhance out-sample predictions. This can be considered as an alternative to the well-known *Bayesian Information Criterion* (BIC), defined as BIC $:= -2\ln \hat{L} + k \ln n$, where $\hat{L}$ is the maximized likelihood, k is the number of parameters to be estimated, and n is the sample size; also see Appendix 12.A.2 for a motivation of BIC via Laplace's approximation.

Straightforward calculations of (12.11) show that the posterior is indeed another multivariate normal distribution:

$$\boldsymbol{\beta} | \mathrm{y} = y \sim \mathcal{N}_{M+1}\left(\lambda S \sum_{i=1}^{n} y_i \boldsymbol{\phi}(x_i), S \right),$$

where $S^{-1} = \alpha \boldsymbol{I}_{M+1} + \lambda \sum_{i=1}^{n} \boldsymbol{\phi}(x_i)\boldsymbol{\phi}^\top(x_i)$, with $\boldsymbol{\phi}(x) = (1, x, x^2, \ldots, x^M)^\top$. With a new observation x, the marginal density of the unknown response y is:

$$\begin{aligned} f_{\mathrm{y}}(y) &= \int_{\mathbb{R}^{M+1}} f(y | g(x), \lambda) f(\boldsymbol{\beta} | \mathbf{y}) d\boldsymbol{\beta} \\ &= \int_{\mathbb{R}^{M+1}} \varphi(y | g(x), \lambda^{-1}) \varphi\left(\boldsymbol{\beta} \,\middle|\, \lambda S \sum_{i=1}^{n} y_i \boldsymbol{\phi}(x_i), S \right) d\boldsymbol{\beta}, \end{aligned}$$

where $g(x) = \boldsymbol{\phi}^\top(x)\boldsymbol{\beta}$ and $\varphi(x | \boldsymbol{\mu}, \boldsymbol{\Sigma})$ is the density of $\mathcal{N}_{M+1}(\boldsymbol{\mu}, \boldsymbol{\Sigma})$. By using the identity:

$$\int_{\mathbb{R}^{M+1}} \varphi(y | \boldsymbol{a}^\top \boldsymbol{\beta}, \lambda^{-1}) \varphi(\boldsymbol{\beta} | \boldsymbol{\mu}, \boldsymbol{\Sigma}) d\boldsymbol{\beta} = \varphi(y | \boldsymbol{a}^\top \boldsymbol{\mu}, \lambda^{-1} + \boldsymbol{a}^\top \boldsymbol{\Sigma} \boldsymbol{a}),$$

with $\boldsymbol{\Sigma} = S$, $\boldsymbol{\mu} = \lambda S \sum_{i=1}^{n} y_i \boldsymbol{\phi}(x_i)$, $\boldsymbol{a} = \boldsymbol{\phi}(x)$, we have $f_{\mathrm{y}}(y) = \varphi(y | m(x), s(x))$, i.e., $\mathrm{y} \sim \mathcal{N}(m(x), s(x))$, where

$$m(x) = \lambda \boldsymbol{\phi}^\top(x) S \sum_{i=1}^{n} y_i \boldsymbol{\phi}(x_i) \quad \text{and} \quad s(x) = \lambda^{-1} + \boldsymbol{\phi}^\top(x) S \boldsymbol{\phi}(x).$$

Let us illustrate the Bayesian approach in polynomial regression using the following simulation experiment. We generate four random samples of sizes $n = 4, 5, 10$ and 100 respectively from the model $\mathrm{y} = \sin(2\pi x) + \epsilon$, where $x \in (0, 1)$ and $\epsilon \sim \mathcal{N}(0, 0.3^2)$. For each random sample, we use a polynomial with degree $M = 9$ to fit the model, with the prior $\mathcal{N}_{10}(\boldsymbol{0}_{10}, \frac{1}{10000}\boldsymbol{I}_{10})$ for the coefficients. Figure 12.1 shows the plots of the predicted mean $m(x)$ (solid line in blue), and its upper and lower one-standard deviation prediction limits $m(x) + \sqrt{s(x)}$ (dotted line in red above the solid line) and $m(x) - \sqrt{s(x)}$ (dotted lines in green below the solid line), respectively, and the true underlying function $\sin(2\pi x)$ (dashed line in black) for each

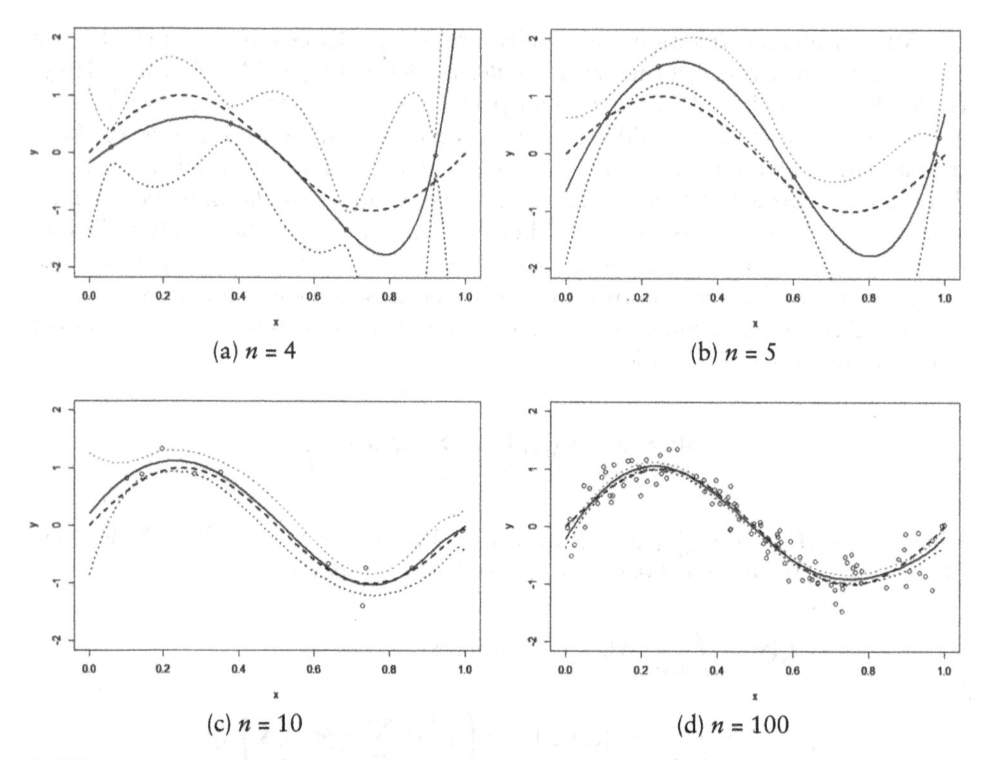

(a) $n = 4$ (b) $n = 5$

(c) $n = 10$ (d) $n = 100$

FIGURE 12.1 Polynomial regression using Bayesian approach with various sample sizes.

sample, while the observations (x_i, y_i), for $i = 1, \dots, n$, are also represented as hollow dots on the plots.

As observed from the plots, when n is small, the prediction interval band fluctuates rather heavily. The coefficient estimates $\hat{\boldsymbol{\beta}}$ possess a higher (resp. lower) confidence for values of x near (resp. far away from) the observed x_n's, resulting in a narrower (resp. wider) prediction interval. Meanwhile, as the sample size n increases, the interval band narrows down to the true underlying value uniformly over $x \in (0, 1)$.

12.4 BAYESIAN CLASSIFIERS

So far, we have only mentioned linear Bayesian models. Now we turn to non-linear Bayesian models. Bayesian methods play a vital role in many classification algorithms in supervised learning, and the classification threshold is the conditional probability of a certain label given a feature vector taking the observed values, calculated using the Bayes' theorem (1.10). It has found extensive applications in various fields, such as text document classification and spam filters.

Mathematically, suppose we are given the following training dataset with n samples:

$$S := \{(\boldsymbol{x}_1, c(\boldsymbol{x}_1)), (\boldsymbol{x}_2, c(\boldsymbol{x}_2)), \dots, (\boldsymbol{x}_n, c(\boldsymbol{x}_n))\},$$

where each datum is represented by its feature vector $\boldsymbol{x}_i = (x_{i1}, \dots, x_{ip})^\top \in \mathbb{R}^p$ and its actual label $c(\boldsymbol{x}_i)$. Take the detection of a list of p keywords in a collection of n text documents as an example. We set $x_{ij} = 1$ if the j-th keyword appears in the i-th document, or $x_{ij} = 0$ otherwise. In addition, we use $\mathcal{Y} := \{c_1, \dots, c_M\}$ to denote the set of all possible classification labels. In the case of spam filters, we usually have only two labels, $\mathcal{Y} = \{\text{"spam"}, \text{"not spam"}\}$.

Given a new feature vector, the Bayesian classifier first assigns probabilities to the feature vector belonging to each class in $\mathcal{Y}$ based on Bayes' Theorem. It then assigns the class with the largest probability as the final prediction. Mathematically, for a new observation with feature vector $\boldsymbol{x}_* := (x_{*1}, \dots, x_{*p})^\top$, Bayesian classifiers in general estimate the conditional probability $\mathbb{P}(c(\mathbf{x}) = c_k | \mathbf{x} = \boldsymbol{x}_*)$, the probability its classification label is c_k, for $k = 1, \dots, M$, by the following conditional empirical probability based on the training dataset $\mathcal{S}$:

$$p_k(\boldsymbol{x}_*) := \hat{\mathbb{P}}(c(\mathbf{x}) = c_k | \mathbf{x} = \boldsymbol{x}_*) := \frac{|\{i : \boldsymbol{x}_i = \boldsymbol{x}_*, c(\boldsymbol{x}_i) = c_k\}|}{|\{i : \boldsymbol{x}_i = \boldsymbol{x}_*\}|}, \quad k = 1, \dots, M.$$

This new observation is then classified with label $c_{\hat{k}}$, where

$$\hat{k} := \arg\max_{k=1,\dots,M} p_k(\boldsymbol{x}_*).$$

Note that these empirical (joint) probabilities would only be accurate enough when the sample size of $\mathcal{S}$ is large for different labels and subsets with various features. Meanwhile, when a large number of new observations need to be classified, it is more efficient to calculate these probabilities in advance. However, they would take up an enormous amount of time and memory as 2^p number of probabilities have to be calculated and stored for every classification label. In order to be more efficient in the training process while attaining a similar performance with a smaller training dataset, we usually consider a *naïve* model, with the extra assumption that all p feature variables $\mathrm{x}_1, \dots, \mathrm{x}_p$ are conditionally independent given any classification label. This special model is known as the *Naïve Bayes* classifier. Treating the new observation $\boldsymbol{x}_* = (x_{*1}, \dots, x_{*p})^\top$ as a realization of the random vector $\mathbf{x} = (\mathrm{x}_1, \dots, \mathrm{x}_p)^\top$, we have:

$$\begin{aligned}\mathbb{P}(c(\mathbf{x}) = c_k \mid \mathbf{x} = \boldsymbol{x}_*) &= \frac{\prod_{j=1}^{p} \mathbb{P}(\mathrm{x}_j = x_{*j} \mid c(\mathbf{x}) = c_k)\mathbb{P}(c(\mathbf{x}) = c_k)}{\mathbb{P}(\mathbf{x} = \boldsymbol{x}_*)} \\ &\propto \prod_{j=1}^{p} \mathbb{P}(\mathrm{x}_j = x_{*j} \mid c(\mathbf{x}) = c_k)\mathbb{P}(c(\mathbf{x}) = c_k). \end{aligned} \tag{12.13}$$

Hence, with a finite number of possible values for the i-th feature variable, we only need to compute $O(pM \max_{j=1,\dots,p} M_j)$ (not more than M_i) numbers of probabilities in advance.

The procedure of training the model and making classification predictions using the resulting *Naïve Bayes* classifier is as follows:

1. For each label c_k, compute the empirical unconditional probability $\hat{\mathbb{P}}(c(\mathbf{x}) = c_k) = |\{i : c(\boldsymbol{x}_i) = c_k\}|/n$.
2. For each feature variable x_j and a feature value x_{*j}, compute the empirical conditional probability for each class c_k:

$$\hat{\mathbb{P}}(\mathrm{x}_j = x_{*j} \mid c(\mathbf{x}) = c_k) = \frac{|\{i : x_{ij} = x_{*j}, c(\boldsymbol{x}_i) = c_k\}|}{|\{i : c(\boldsymbol{x}_i) = c_k\}|}.$$

3. The new observation $\boldsymbol{x}_*$ is classified as belonging to the class $c_{\hat{k}}$, where

$$\hat{k} = \underset{k=1,\dots,M}{\arg\max} \left(\prod_{j=1}^{p} \hat{\mathbb{P}}(\mathrm{x}_j = x_{*j} \mid c(\mathbf{x}) = c_k) \hat{\mathbb{P}}(c(\mathbf{x}) = c_k) \right). \tag{12.14}$$

Upon putting this procedure into code, the issue of arithmetic underflow may emerge when the products in (12.13) are computed. We usually deal with this by calculating their logarithms. Besides, while the conditional independence assumption yields effective computations of the probabilities, it often deviates from reality when the feature variables are dependent. For instance, feature variables in real data rarely exhibit conditional independence. To illustrate this drawback, consider a classification problem involving $p = 2$ binary feature variables x_1 and x_2. A new observation (x_1, x_2) is classified as class A when $x_1 = x_2$, or class B otherwise. Suppose that for a given training dataset, the number of samples taking every combination of values (x_1, x_2) is the same, so that each class has half of the observations, while the values of x_1 and x_2 are also evenly distributed in each class. By simple calculations, the product to be maximized on the right-hand side of (12.14) is $\frac{1}{2} \times \frac{1}{2} \times \frac{1}{2} = \frac{1}{8}$ for all combinations of (x_1, x_2), implying that the classifier is no better than flipping a coin. Due to this limitation and other ones like it, we propose in the next section how to enhance the Bayes classifiers by incorporating a well-known notion from actuarial science, namely tail comonotonicity for dependence modelling.

12.5 COMONOTONE-INDEPENDENCE BAYES CLASSIFIER (CIBER)

To remedy the drawbacks of the Naïve Bayes discussed in the previous section, we introduce a variant [5] that incorporates comonotonicity, as introduced briefly in Section 8.8 of Chapter 8. Our aim is to better recover and model the dependence structure among feature variables without complicating the calculations of conditional joint probabilities too much. This variant is called the *Comonotone-Independence Bayes Classifier*, or *CIBer* for short. This new classifier can effectively deal with discrete feature variables with numerical values, also known as "discrete features". However, in financial and insurance contexts, the presence of all three types of feature variable, namely discrete, continuous and categorical, is common. To that end, we introduce a unified feature engineering scheme: by binning continuous feature variables and employing the method of *Joint Encoding* for categorical ones, as we shall discuss in Sections 12.5.2 and 12.5.3, we convert all feature variables into discrete features, as the inputs of the new classifier.

Let us first clarify some terms and notation to be adopted in this section. As explained above, all feature variables are first transformed into discrete feature variables. We denote the sample space of each such discrete feature x_j, as an input of CIBer, by $\mathcal{X}_j := \{s_{j1}, s_{j2}, \ldots, s_{j,M_j}\}$, where $s_{j1} < s_{j2} < \cdots < s_{j,M_j}$ and $M_j := |\mathcal{X}_j| \in \mathbb{Z}^+$.

The notion of comonotonicity is central to CIBer. In addition to what we have introduced in Chapter 8, let us discuss some further properties. As a concept widely used in actuarial science, finance, and economics, comonotonicity describes the strong possible positive dependence among several random variables: whenever the realization of one random variable changes, all the rest follow suit along the same direction. Mathematically, a random vector $\mathbf{x} = (\mathrm{x}_1, \ldots, \mathrm{x}_p)^\top$ is said to be *comonotonic* if there is a *comonotonic subset* C of $\mathbb{R}^p$ such that $\mathbb{P}(\mathbf{x} \in C) = 1$. Here, C is comonotonic if it is totally ordered so that either $x_i \leq y_i$ or $x_i \geq y_i$ for all i, for any $x, y \in \mathbb{R}^p$. One can prove that the following characterizations of comonotonicity are equivalent:

1. the random vector $\mathbf{x} = (\mathrm{x}_1, \ldots, \mathrm{x}_p)^\top$ is comonotonic;
2. there exist a random variable w and increasing functions $h_1, \ldots, h_p$ such that $(\mathrm{x}_1, \ldots, \mathrm{x}_p) \overset{d}{=} (h_1(\mathrm{w}), \ldots, h_p(\mathrm{w}))$;
3. $(\mathrm{x}_1, \ldots, \mathrm{x}_p) \overset{d}{=} (F_{\mathrm{x}_1}^{-1}(\mathrm{u}), \ldots, F_{\mathrm{x}_p}^{-1}(\mathrm{u}))$ for any $\mathrm{u} \sim U(0,1)$, where the generalized inverse $F_{\mathrm{x}_j}^{-1}(u) := \inf\{x_j \in \mathbb{R} : F_{\mathrm{x}_j}(x_j) \geq u\}$ for $j = 1, 2, \ldots, p$;
4. $F_{\mathbf{x}}(x) = \mathbb{P}(\mathrm{x}_1 \leq x_1, \ldots, \mathrm{x}_p \leq x_p) = \min_{j=1,\ldots,p} F_{\mathrm{x}_j}(x_j)$ for any $x \in \mathbb{R}^p$.

Refer to [6, 7] for more detailed discussions.

The definition and properties of (unconditional) comonotonicity mentioned above could also be applied analogously to the conditional sense, resulting in the *conditional comonotonicity* that we shall find more useful in the context of Bayesian classification, as we are ultimately interested in computing the conditional probability. To this end, suppose that the *label function* $c(\mathbf{x})$ (indicating the actual label of $\mathbf{x}$) only takes finitely many values, say $c_1, \ldots, c_M$. The random vector $\mathbf{x}$ is said to be comonotonic conditional on $c(\mathbf{x})$ if the following holds for $k = 1, \ldots, M$:

$$F_{\mathbf{x}|c(\mathbf{x})}(x|c_k) := \mathbb{P}(\mathrm{x}_1 \leq x_1, \ldots, \mathrm{x}_p \leq x_p | c(\mathbf{x}) = c_k) = \min_{j=1,\ldots,p} F_{\mathrm{x}_j|c(\mathbf{x})}(x_j|c_k).$$

Analogously, the conditional comonotonicity of $\mathbf{x}$ given $c(\mathbf{x})$ is equivalent to the following: for any $\mathrm{u} \sim U(0,1)$ and $k = 1, \ldots, M$,

$$(\mathrm{x}_1, \ldots, \mathrm{x}_p) | c(\mathbf{x}) = c_k \overset{d}{=} (F_{\mathrm{x}_1|c(\mathbf{x})}^{-1}(\mathrm{u}|c_k), \ldots, F_{\mathrm{x}_p|c(\mathbf{x})}^{-1}(\mathrm{u}|c_k)),$$

where $F_{\mathrm{x}_j|c(\mathbf{x})}(\cdot|c_k)$ is the conditional cdf of x_j given $c(\mathbf{x}) = c_k$, and $F_{\mathrm{x}_j|c(\mathbf{x})}^{-1}(u|c_k) := \inf\{x_j \in \mathbb{R} : F_{\mathrm{x}_j|c(\mathbf{x})}(x_j|c_k) \geq u\}$.

Let us now compute the joint probability mass function of a discrete conditionally comonotonic random vector. Suppose that $\mathbf{x}$ is conditionally comonotonic given $c(\mathbf{x})$, where $\mathrm{x}_j \in \mathcal{X}_j$ for each j. Let $s = (s_{1,m_1}, \ldots, s_{p,m_p})^\mathsf{T}$, and $s_{j,m_j} \in \mathcal{X}_j$ for $j = 1, \ldots, p$, be the observed value of the feature vector $\mathbf{x}$. By the properties mentioned above, the conditional joint pmf of $\mathbf{x}$ given the label c_k, denoted by $\mathbb{P}_{\text{como}}(\mathbf{x} = s|c(\mathbf{x}) = c_k)$, can be conveniently expressed as follows:

$$
\begin{aligned}
\mathbb{P}_{\text{como}}(\mathbf{x} = s|c(\mathbf{x}) = c_k) &= \mathbb{P}(\mathrm{x}_1 = s_{1,m_1}, \ldots, \mathrm{x}_p = s_{p,m_p}|c(\mathbf{x}) = c_k) \\
&= \mathbb{P}_{\text{como}}(s_{1,m_1-1} < \mathrm{x}_1 \le s_{1,m_1}, \ldots, s_{p,m_p-1} < \mathrm{x}_p \le s_{p,m_p}|c(\mathbf{x}) = c_k) \\
&= \mathbb{P}(s_{1,m_1-1} < F^{-1}_{\mathrm{x}_1|c(\mathbf{x})}(\mathrm{u}|c_k) \le s_{1,m_1}, \ldots, s_{p,m_p-1} < F^{-1}_{\mathrm{x}_p|c(\mathbf{x})}(\mathrm{u}|c_k) \le s_{p,m_p}) \\
&= \mathbb{P}\left(F_{\mathrm{x}_j|c(\mathbf{x})}(s_{j,m_j-1}|c_k) < \mathrm{u} \le F_{\mathrm{x}_j|c(\mathbf{x})}(s_{j,m_1}|c_k); j = 1, \ldots, p\right) \\
&= \left(\min_j F_{\mathrm{x}_j|c(\mathbf{x})}(s_{j,m_j}|c_k) - \max_j F_{\mathrm{x}_j|c(\mathbf{x})}(s_{j,m_j-1}|c_k)\right)_+,
\end{aligned} \tag{12.15}
$$

where $F_{\mathrm{x}_j|c(\mathbf{x})}(s_{j0}|c_k) = 0$, for all j and k, by convention. If $\{(x_i, c(x_i))\}_{i=1}^n$ is a sample of observed values of $\mathbf{x}$ together with their labels, then $F_{\mathrm{x}_j|c(\mathbf{x})}$ can be estimated empirically by:

$$
\hat{F}_{\mathrm{x}_j|c(\mathbf{x})}(s_*|c_k) = \hat{\mathbb{P}}(\mathrm{x}_j \le s_*|c(\mathbf{x}) = c_k) = \frac{|\{i : x_{ij} \le s_*, c(i) = c_k\}|}{|\{i : c(i) = c_k\}|}, \tag{12.16}
$$

for $s_* \in \mathcal{X}_j$, $k = 1, \ldots, M$ and $j = 1, \ldots, p$. For actual implementation, in the case of sparse data, we shall refine (12.16) by using *Laplace shrinkage* to cope with rare events.

12.5.1 Clustered Comonotonicity

As in reality not all feature variables are necessarily comonotonic, a more appropriate and realistic approach in classification problems is to first partition the feature variables into several subgroups, and then treat the feature variables within each group as comonotonic while the totality of features in a group is independent of those in another group. However, the exhaustive searching time for the optimal partition grows rapidly with the number of features involved. We therefore have to deploy an efficient heuristic search to approximate this optimal partition without traversing all possibilities. The solution turns out to be a method of clustering. In principle, two features with a small value of some pre-specified "distance" are regarded as comonotonic and should be clustered into the same group. Some popular distance metrics for the purpose of measuring comonotonic dependence include the *Normalized Mutual Information U*, *Pearson's r*, *Kendall's τ_b*, and *Spearman's ρ*. We introduce another one based on Kendall's τ_b, specifically built for CIBer. It should be noted that the selection of the distance metric varies case by case, and we always aim to choose the one leading to the best performance. Let us list out these metrics in their natural empirical form, where x_i and x_j are any two distinct features with observed pairs $(x_{1i}, x_{1j}), \ldots, (x_{ni}, x_{nj})$:

1. **Normalized Mutual Information (NMI)** $\hat{U}$ [15, 21]: it quantifies the dependence between two features using entropy from information theory:

$$\hat{U} := 2 \cdot \frac{\hat{I}(\mathrm{x}_i; \mathrm{x}_j)}{\hat{H}(\mathrm{x}_i) + \hat{H}(\mathrm{x}_j)}.$$

   Here $\hat{I}(\mathrm{x}_i; \mathrm{x}_j)$ is the empirical mutual information between x_i and x_j defined by

$$\hat{I}(\mathrm{x}_i; \mathrm{x}_j) := \sum_{u=1}^{n} \sum_{v=1}^{n} \hat{\mathbb{P}}_{\mathrm{x}_i,\mathrm{x}_j}(x_{ui}, x_{vj}) \cdot \ln \frac{\hat{\mathbb{P}}_{\mathrm{x}_i,\mathrm{x}_j}(x_{ui}, x_{vj})}{\hat{\mathbb{P}}_{\mathrm{x}_i}(x_{ui}) \cdot \hat{\mathbb{P}}_{\mathrm{x}_j}(x_{vj})},$$

   where $\hat{\mathbb{P}}$ stands for the corresponding empirical pmf, and $\hat{H}(\mathrm{x}_i)$ and $\hat{H}(\mathrm{x}_j)$ are the empirical entropies for x_i and x_j, respectively:

$$\hat{H}(\mathrm{x}_i) := -\sum_{u=1}^{n} \hat{\mathbb{P}}_{\mathrm{x}_i}(x_{ui}) \cdot \ln(\hat{\mathbb{P}}_{\mathrm{x}_i}(x_{ui})),$$

$$\hat{H}(\mathrm{x}_j) := -\sum_{v=1}^{n} \hat{\mathbb{P}}_{\mathrm{x}_j}(x_{vj}) \cdot \ln(\hat{\mathbb{P}}_{\mathrm{x}_j}(x_{vj})).$$

2. **Pearson's** $\hat{r}$ [17]: it measures the magnitude and direction of linear association between two continuous features jointly distributed as bivariate normal:

$$\hat{r} := \frac{\sum_{u=1}^{n}(x_{ui} - \overline{x}_i)(x_{uj} - \overline{x}_j)}{\sqrt{\sum_{u=1}^{n}(x_{ui} - \overline{x}_i)^2}\sqrt{\sum_{v=1}^{n}(x_{vj} - \overline{x}_j)^2}},$$

   where $\overline{x}_i = \sum_{u=1}^{n} x_{ui}/n$ and $\overline{x}_j = \sum_{v=1}^{n} x_{vj}/n$ are the respective sample means of x_i and x_j.
3. **Kendall's** $\hat{\tau}_b$ [13]: this non-parametric metric quantifies the dependence between two features based on the concept of *concordant pairs*. A pair of observations (x_{ui}, x_{uj}) and (x_{vi}, x_{vj}), $u \neq v$, is said to be *concordant* if $(x_{ui} - x_{vi})(x_{uj} - x_{vj}) > 0$; *discordant* if $(x_{ui} - x_{vi})(x_{uj} - x_{vj}) < 0$; and tied if $(x_{ui} - x_{vi})(x_{uj} - x_{vj}) = 0$. The Kendall's $\hat{\tau}_b$ coefficient between x_i and x_j is computed by:

$$\hat{\tau}_b := \frac{C - D}{\sqrt{(n_0 - n_i)(n_0 - n_j)}},$$

where C and D denote the respective total numbers of concordant and discordant pairs, $n_0 := \binom{n}{2}$ is the total number of possible pairs of observations, and $n_k := \sum_{m:s_{k,m}\in\mathcal{X}_k} t_{km}(t_{km}-1)/2$ with t_{km} being the total number of tied pairs having the value $s_{k,m}$ for the feature x_k.

4. **Spearman's $\hat{\rho}$** [19]: it measures the rank correlation between two feature variables. Mathematically, let $(r_{1i}, r_{1j}), \ldots, (r_{n1}, r_{nj})$ denote the rankings[1] of the observed pairs, then Spearman's $\hat{\rho}$ is defined as:

$$\hat{\rho} := \frac{\sum_{u=1}^{n}(r_{ui}-\bar{r}_i)(r_{uj}-\bar{r}_j)}{\sqrt{\sum_{u=1}^{n}(r_{ui}-\bar{r}_i)^2}\sqrt{\sum_{v=1}^{n}(r_{vj}-\bar{r}_j)^2}},$$

where $\bar{r}_i = \sum_{u=1}^{n} r_{ui}/n$ and $\bar{r}_j = \sum_{v=1}^{n} r_{vj}/n$ are the respective average rankings of x_i and x_j. Note that the value of Spearman's $\hat{\rho}$ between x_i and x_j is equal to that of Pearson's $\hat{r}$ between r_i and r_j, the corresponding rank variables of x_i and x_j.

5. **Totally-ordered Kendall's $\hat{\tau}_{\text{como}}$**: this metric is based on Kendall's $\hat{\tau}_b$, and is specifically built for CIBer to utilize the totally ordered set as much as possible. Instead of focusing on concordant and discordant pairs as in Kendall's $\hat{\tau}_b$, we combine concordant and tied pairs, and the number of the resulting *concordant-tied* pairs $\overline{C}$ is used in calculating the metric:

$$\hat{\tau}_{\text{como}} := \frac{\overline{C}-D}{\binom{n}{2}}, \tag{12.17}$$

where $D = \binom{n}{2} - \overline{C}$ is the number of discordant pairs as defined before.

All metrics discussed above are measurements for statistical association or "similarity", hence they are also called *association measures*. However, this is inconsistent with the notion of "dissimilarity" generally used in clustering. Therefore, we also introduce the following notion of a random distance matrix: given a p-dimensional random vector $\mathbf{x} = (\mathrm{x}_1, \ldots, \mathrm{x}_p)^\top$, a random matrix $\mathbf{A} := (\mathrm{a}_{ij}) \in \mathbb{R}^{p\times p}$ is said to be a distance matrix if its entries satisfy:

$$\mathrm{a}_{ij} := 1 - |\mathrm{asso}(\mathrm{x}_i, \mathrm{x}_j)|, \qquad \text{for} \quad i,j = 1, \ldots, p,$$

where $\mathrm{asso}(\mathrm{x}_i, \mathrm{x}_j)$ is a chosen association measure between x_i and x_j. Its empirical counterpart is $A := (a_{ij}) \in \mathbb{R}^{p\times p}$, where

$$a_{ij} := 1 - |\widehat{\mathrm{asso}}(x_i, x_j)|, \qquad \text{for} \quad i,j = 1, \ldots, p,$$

[1] We first sort the values in decreasing order and then index the sorted values with numbers $1, \ldots, n$ as their primary ranking. For each group of ties, i.e. members have the same value, the rankings of all members in that group are equal to the average primary rankings of those values.

where $\widehat{\text{asso}}(x_i, x_j)$ is the corresponding sample association measure between x_i and x_j, such as those mentioned above.

This distance matrix plays a vital role in the following hierarchical clustering algorithm, called *Agglomerative Nesting* (AGNES), first proposed in [12], which we shall employ to cluster the features. The implementation of this algorithm relies on an additional notion of a distance measure between two disjoint clusters of features, say with respective index sets[2] $\mathcal{G}_1 := \{i_1, i_2, \ldots, i_{\ell_1}\}$, $\mathcal{G}_2 := \{j_1, j_2, \ldots, j_{\ell_2}\} \subset \{1, \ldots, p\}$ such that $\mathcal{G}_1 \cap \mathcal{G}_2 = \emptyset$, where ℓ_1 and ℓ_2 are the number of features in $\mathcal{G}_1$ and $\mathcal{G}_2$, respectively. Among various choices of distance measures between clusters, we adopt the *Complete Linkage* here, defined by the largest distance among all possible cross-cluster pairs of features:

$$d_{\text{complete}}(\mathcal{G}_1, \mathcal{G}_2) := \max_{i \in \mathcal{G}_1, j \in \mathcal{G}_2} a_{ij}.$$

As a bottom-up approach for hierarchical clustering, the AGNES algorithm begins with all p features as singletons, so that the initialized clusters are $\{1\}, \ldots, \{p\}$. We then iteratively merge two clusters with the smallest complete linkage into one, which reduces the total number of clusters by one in each iteration, until the complete linkage between any two clusters becomes greater than a preset threshold $\delta \in [0, 1]$, or only one cluster remains (which implies all features are regarded as comonotonic). The pseudocode of AGNES is shown in the following Algorithm 12.1.

Algorithm 12.1 Agglomerative Nesting (AGNES) algorithm

Input: An empirical distance matrix A; Maximum cluster-wise distance $\delta \in [0, 1]$.

1: $\mathcal{Y} \leftarrow \{\{1\}, \{2\}, \ldots, \{p\}\}$
2: $d_{\min} \leftarrow \min_{\mathcal{G}_1, \mathcal{G}_2 \in \mathcal{Y}} d_{\text{complete}}(\mathcal{G}_1, \mathcal{G}_2)$
3: **while** $d_{\min} \leq \delta$ **do**
4: $\quad \mathcal{G}_1^*, \mathcal{G}_2^* \leftarrow \arg\min_{\mathcal{G}_1, \mathcal{G}_2 \in \mathcal{Y}} d_{\text{complete}}(\mathcal{G}_1, \mathcal{G}_2)$
5: $\quad \mathcal{Y} \leftarrow \mathcal{Y} \setminus \{\mathcal{G}_1^*, \mathcal{G}_2^*\}$
6: $\quad \mathcal{Y} \leftarrow \mathcal{Y} \cup \{\mathcal{G}_1^* \cup \mathcal{G}_2^*\}$
7: $\quad d_{\min} \leftarrow \min_{\mathcal{G}_1, \mathcal{G}_2 \in \mathcal{Y}} d_{\text{complete}}(\mathcal{G}_1, \mathcal{G}_2)$
8: **end while**
9: **return** $\mathcal{Y}$

After partitioning the feature variables into different comonotonic clusters using AGNES, we compute the conditional joint pmf given the class label by adopting the philosophy of clustered comonotonicity. Mathematically, suppose that given the class label c_k, the feature variables $\mathrm{x}_1, \ldots, \mathrm{x}_p$ are partitioned into n_c clusters,

[2] Without cause of ambiguity, we shall use the index sets to represent the corresponding sets of features, e.g., $\{1, 2, 3\}$ stands for $\{\mathrm{x}_1, \mathrm{x}_2, \mathrm{x}_3\}$.

say $\mathcal{G}_{1,k}, \ldots, \mathcal{G}_{n_c,k}$, then the conditional joint pmf of $\mathbf{x}$ given the class labels is calculated by:

$$\mathbb{P}(\mathrm{x}_1 = s_{1,m_1}, \ldots, \mathrm{x}_p = s_{p,m_p} \mid c(\mathbf{x}) = c_k) = \prod_{v=1}^{n_c} \mathbb{P}_{\text{como}}(\mathrm{x}_j = s_{j,m_j} \text{ for } j \in \mathcal{G}_{v,k} \mid c(\mathbf{x}) = c_k).$$

12.5.2 Binning Continuous Features

Intuitively, individuals in the same class c_k should share similar characteristics, and the values of their respective continuous features are also expected to cluster in a certain range. It suffices to predict an individual's label based on whether her continuous features lie in a characterizing range, rather than on their exact values. To this point, it is natural to apply discretization to continuous variables for the purpose of classification. As an alternative to the commonly used approaches such as equal-width binning, we consider another discretization method proposed in [3, 9], based on the concept of entropy.

The goal of discretization with respect to a given continuous feature x_i is to find the "optimal" partition of $\mathcal{X}_i$ into a number of disjoint subsets $\mathcal{X}_1^{(i)}, \ldots, \mathcal{X}_{k_i}^{(i)}$ so that $\bigsqcup_{j=1}^{k_i} \mathcal{X}_j^{(i)} = \mathcal{X}_i$, where k_i can be arbitrarily chosen, such that the following *class information entropy* is minimized:

$$\mathcal{E}(\mathrm{x}_i, \mathcal{X}_i | \mathcal{P}_{k_i}^{(i)}) := \sum_{j=1}^{k_i} \frac{|\mathcal{X}_j^{(i)}|}{|\mathcal{X}_i|} H(\mathcal{X}_j^{(i)}), \tag{12.18}$$

where $\mathcal{P}_{k_i}^{(i)}$ stands for the partition $\{\mathcal{X}_1^{(i)}, \ldots, \mathcal{X}_{k_i}^{(i)}\}$ of $\mathcal{X}_i$, and H is the entropy of the set involved, on which we will elaborate in Section *13.2. Nevertheless, the arbitrariness of the choice of k_i and the lack of restriction on the partition render this optimization problem *NP-hard*. Hence, we resort to a binary search method to obtain the partitions instead.

As proposed in [8], we first seek for the "optimal" splitting boundary $t_1^{(i)}$, usually selected among midpoints of consecutive values in $\mathcal{X}_i$, such that the corresponding binary partition of $\mathcal{X}_i$, say $\mathcal{X}_{1,1}^{(i)}$ and $\mathcal{X}_{1,2}^{(i)}$, minimizes the class information entropy (12.18) when $k_i = 2$. We then repeat the same procedure to seek for the "optimal" splitting boundaries $t_{1,1}^{(i)}$ and $t_{1,2}^{(i)}$ for splitting $\mathcal{X}_{1,1}^{(i)}$ and $\mathcal{X}_{1,2}^{(i)}$, respectively, again by minimizing (12.18). Say the "optimal" binary partition of $\mathcal{X}_{1,1}^{(i)}$ is $\mathcal{X}_{2,1}^{(i)}$ and $\mathcal{X}_{2,2}^{(i)}$; while those for $\mathcal{X}_{1,2}^{(i)}$ is the pair of $\mathcal{X}_{2,3}^{(i)}$ and $\mathcal{X}_{2,4}^{(i)}$. This procedure is carried out iteratively until some pre-set stopping condition is fulfilled. The k_i number of subsets in the resulting partition $\mathcal{P}_{k_i}^{(i)}$ correspond to $k_i - 1$ splitting boundaries being specified throughout the procedure, which are sorted in ascending order $t_{(1)}^{(i)} < \cdots < t_{(k_i-1)}^{(i)}$. Then for $b = 1, \ldots, k_i$, the set of observed sample data points for x_i in the b-th bin can be characterized as:

$$\mathcal{X}_b^{(i)} := \begin{cases} \left\{s \in \mathcal{X}_i : s < t_{(1)}^{(i)}\right\}, & b = 1; \\ \left\{s \in \mathcal{X}_i : t_{(b-1)}^{(i)} \leq s < t_{(b)}^{(i)}\right\}, & b = 2, \ldots, k_i - 1; \\ \left\{s \in \mathcal{X}_i : s \geq t_{(k_i-1)}^{(i)}\right\}, & b = k_i. \end{cases}$$

In particular, a commonly used stopping criterion is the *Minimal Description Length Principle* (MDLP) proposed in [9]. Consider a subset of observations $\mathcal{X}_{l,1}^{(i)} \subseteq \mathcal{X}_i$ at the l-th level of the binary partition, and its "optimal" binary partition results in the subsets $\mathcal{X}_{l+1,1}^{(i)}$ and $\mathcal{X}_{l+1,2}^{(i)}$, where $\mathcal{P}_{l+1,2}^{(i)} := \{\mathcal{X}_{l+1,1}^{(i)}, \mathcal{X}_{l+1,2}^{(i)}\}$. We let $m_{l,1}^{(i)}$, $m_{l+1,1}^{(i)}$, and $m_{l+1,2}^{(i)}$ denote the number of actual class labels c_k's present among observations in $\mathcal{X}_{l,1}^{(i)}$, $\mathcal{X}_{l+1,1}^{(i)}$, and $\mathcal{X}_{l+1,2}^{(i)}$, respectively. Under the MDLP, no further splitting will be carried out for $\mathcal{X}_{l,1}^{(i)}$ if

$$H(\mathcal{X}_{l,1}^{(i)}) - \mathcal{E}(x_i, \mathcal{X}_{l,1}^{(i)} | \mathcal{P}_{l+2,2}^{(i)}) < \frac{\log_2(|\mathcal{X}_{l,1}^{(i)}| - 1)}{|\mathcal{X}_{l,1}^{(i)}|} + \frac{\Delta(x_i; \mathcal{X}_{l,1}^{(i)})}{|\mathcal{X}_{l,1}^{(i)}|},$$

where $\Delta(x_i; \mathcal{X}_{l,1}^{(i)}) : = \log_2(g_{l,1}^{(i)}) - [m_{l,1}^{(i)} \cdot H(\mathcal{X}_{l,1}^{(i)}) - m_{l+1,1}^{(i)} \cdot H(\mathcal{X}_{l+1,1}^{(i)}) - m_{l+1,2}^{(i)} \cdot H(\mathcal{X}_{l+1,2}^{(i)})]$, and $g_{l,1}^{(i)} := 3^{m_{l,1}^{(i)}} - 2$ is the total number of possible scenarios in terms of the distribution of the $m_{l,1}^{(i)}$ class labels into the subsets $\mathcal{X}_{l+1,1}^{(i)}$ and $\mathcal{X}_{l+1,2}^{(i)}$.[3]

[3] This total number can be shown by a standard combinatorial argument: we first focus on observations in $\mathcal{X}_{l,1}^{(i)}$ with an arbitrary class label c_k, then any classification would result in one of the following three scenarios:

(i) All of these observations are classified into $\mathcal{X}_{l+1,1}^{(i)}$, so that the class c_k is only present in $\mathcal{X}_{l+1,1}^{(i)}$;

(ii) All of these observations are classified into $\mathcal{X}_{l+1,2}^{(i)}$, so that the class c_k is only present in $\mathcal{X}_{l+1,2}^{(i)}$; and

(iii) Some of the observations are classified into $\mathcal{X}_{l+1,1}^{(i)}$, while others fall into $\mathcal{X}_{l+1,2}^{(i)}$, so that the class c_k is present in both subsets.

Then with a total of $m_{l,1}^{(i)}$ classes, we have $3^{m_{l,1}^{(i)}}$ possible scenarios altogether, regarding their distribution in the two subsets. However, the two scenarios where (i) all the $m_{l,1}^{(i)}$ classes are

12.5.3 Joint Encoding for Categorical Features

Categorical features are more difficult to cope with in general, in view of their pure distinguishable nature and lack of natural ordering. As a result, association measures cannot be directly applied to them, nor can the notion of comonotonicity be applied immediately. To overcome this issue, being inspired by *optimal transport* theory, we introduce a novel encoding scheme for them that can "boost up" the inherent dependence among categorical features by making use of the frequencies of various grouped feature values. This encoding scheme, capable of handling several categorical features in one go, is named *Joint Encoding*.

Let us illustrate the procedure and effectiveness of joint encoding via a simple example. Consider the contingency table of a sample dataset with each datum possessing two categorical variables x_1 and x_2. The corresponding contingency table is shown in Table 12.1.

To determine the magnitude of dependence between x_1 and x_2, we plan to assign numerical values to the possible feature values they can take, and then calculate the association measures. By the nature of this dataset, namely a large number of repeated observations (s_{1i}, s_{2j})'s, Pearson's $\hat{r}$ is a more suitable measure than others. Besides, it is also a useful association measure being invariant with respect to any affine transformation of coding. We first consider an arbitrary encoding scheme for x_1 and x_2, namely, for the values of x_1, we encode s_{1i} as i for $i = 1, 2, 3$; while for the values of x_2, we encode s_{2j} as j for $j = 1, 2$. Under this encoding scheme, we have

$$\overline{x}_1 = \frac{1\times 240 + 2\times 300 + 3\times 180}{720} = 1.91667 \text{ and } \overline{x}_2 = \frac{1\times 320 + 2\times 400}{720} = 1.55556,$$

TABLE 12.1 The contingency table of a sample dataset with two categorical feature variables x_1 and x_2.

$\mathrm{x}_2\backslash\mathrm{x}_1$	s_{11}	s_{12}	s_{13}	Total
s_{21}	40	250	30	320
s_{22}	200	50	150	400
Total	240	300	180	720

present only in $\mathcal{X}^{(i)}_{l+1,1}$; and (ii) all the $m^{(i)}_{l,1}$ classes are present only in $\mathcal{X}^{(i)}_{l+1,2}$, should not count as no classification is actually done. Specifically one of the subsets $\mathcal{X}^{(i)}_{l+1,1}$ or $\mathcal{X}^{(i)}_{l+2,1}$ is empty while the other simply equals $\mathcal{X}^{(i)}_{l,1}$. Therefore, we have the total number of possible scenarios equal to $3^{m^{(i)}_{l,1}} - 2$ as claimed.

yielding

$$\sum_{i=1}^{720}(x_{1i}-\bar{x}_1)(x_{2i}-\bar{x}_2) = -16.66667,$$

$$\sum_{i=1}^{720}(x_{1i}-\bar{x}_1)^2 = 415, \quad \sum_{i=1}^{720}(x_{2i}-\bar{x}_2)^2 = 177.77778.$$

We calculate the empirical Pearson's $\hat{r}$ as follows:

$$\hat{r} = \frac{-16.66667}{\sqrt{415 \times 177.77778}} = -0.06136,$$

which is fairly close to 0, and thus gives no hint of a possible (linear) relation between x_1 and x_2.

Nevertheless, we can indeed observe from Table 12.1 that the cells (s_{11}, s_{22}), (s_{12}, s_{21}) and (s_{13}, s_{22}) have higher frequencies than the others. Such non-uniform distribution of data points indicates some hidden association between x_1 and x_2, which could not be accurately reflected by the arbitrary encoding scheme as proposed above. This suggests that a more systematic encoding approach has to be adopted to reveal the hidden connection between x_1 and x_2 as much as possible. To this end, we propose the following heuristic scheme:

1. Find the combination of values of x_1 and x_2 with the highest frequency, and encode the corresponding feature values as 1;
2. Find the combination with the highest frequency within those of un-encoded values of x_1 and x_2, and encode the corresponding values by the next unassigned integer;
3. Repeat Step 2 until a feature variable has precisely one value left that is not encoded, then automatically assign to it the next unassigned integer. If this is the case for both feature variables, then we assign the same integer to the values of both feature variables. Otherwise, if the remaining feature variable has two or more values that are not encoded, then we encode its remaining values sequentially in descending order of their corresponding marginal frequencies.

Following this scheme, we shall first encode s_{12} and s_{21} as 1 by noting that (s_{12}, s_{21}) has the largest frequency of 250. As s_{22} now becomes the only value of x_2 that is not yet encoded, we encode it as 2. As x_1 now still has two un-encoded values, namely s_{11} and s_{13}, we encode them in descending order according to their marginal frequencies listed at the bottom row of Table 12.1, so that s_{11} is encoded as 2 and s_{13} as 3, respectively. Under this encoding scheme, we can compute that:

$$\bar{x}_1 = \frac{1 \times 300 + 2 \times 240 + 3 \times 180}{720} = 1.83333 \quad \text{and} \quad \bar{x}_2 = \frac{1 \times 320 + 2 \times 400}{720} = 1.55556,$$

which together yield

$$\sum_{i=1}^{720}(x_{1,i}-\bar{x}_1)(x_{2,i}-\bar{x}_2) = 166.66667,$$

$$\sum_{i=1}^{720}(x_{1,i}-\bar{x}_1)^2 = 460, \quad \sum_{i=1}^{720}(x_{2,i}-\bar{x}_2)^2 = 177.77778.$$

Finally the empirical Pearson's $\hat{r}$ is:

$$\hat{r} = \frac{166.66667}{\sqrt{460 \times 177.77778}} = 0.58282.$$

This significant increase in value in Pearson's $\hat{r}$ compared to the previous scheme is mainly due to the vastly increased empirical covariance in the numerator of $\hat{r}$.

The idea and procedure above can be extended to the high-dimensional case of p_c categorical feature variables $\mathrm{x}_1, \dots, \mathrm{x}_{p_c}$ as follows. Recall that the finite sample space of x_j is denoted by $\mathcal{X}_j$ and $M_j := |\mathcal{X}_j|$. To facilitate the implementation, we assume that $M_1 \geq M_2 \geq \cdots \geq M_{p_c}$. Now, we consider a p_c-dimensional hyperrectangle $\mathcal{N}$ of size $M_1 \times \cdots \times M_{p_c}$, in which each cell has a "coordinate" represented by a combination of the feature values that $\mathrm{x}_1, \dots, \mathrm{x}_{p_c}$ can take, i.e., $(s_{1,m_1}, s_{2,m_2}, \dots, s_{p_c,m_{p_c}})$ for $m_j = 1, \dots, M_j$ and $j = 1, \dots, p_c$. The value of the cell $(s_{1,m_1}, s_{2,m_2}, \dots, s_{p_c,m_{p_c}})$ in this hyperrectangle, denoted by $\mathcal{N}[s_{1,m_1}, s_{2,m_2}, \dots, s_{p_c,m_{p_c}}]$, is the observed frequency of the corresponding combination in the training dataset. All cells in this hyperrectangle are initially marked as `available`, indicating that they are free to be selected. The pseudocode of joint encoding for this high-dimensional setting is shown in Algorithm 12.2.

12.5.4 Computation of Empirical Probabilities in CIBer

Let us see how CIBer compares with a Naïve Bayes classifier via a numerical example. Suppose that there are six features, five of them continuous and one discrete, taking values 1, 2, or 3. We train a Naïve Bayes classifier and CIBer after discretization, respectively, for classifying the dummy label y with 2,614 of them taking value 0 and 1,386 of them taking value 1 in a training dataset. For the sake of illustration, we adopt the naïve equal-width binning approach for discretization, so that each continuous feature is discretized into 10 bins. The resulting empirical proportions for $y = 0$ and $y = 1$ are shown in Tables 12.2 and 12.3, respectively. Notice that some rows may not sum up exactly to 1 due to rounding. The frequencies of the sixth categorial variable are listed in Table 12.4.

Denote by $\mathrm{b}_j, j = 1, \dots, 6$, the corresponding binned variable after the discretization of the j-th feature variable, especially for the first five continuous ones. Suppose that a new datum falls into the bin $\boldsymbol{b} = (b_1, b_2, b_3, b_4, b_5, b_6)^\top = (5, 5, 6, 5, 5, 1)^\top$. We are interested in classifying this new datum by comparing the values of $\mathbb{P}(c(\mathbf{b}) = 0 | \mathbf{b} = \boldsymbol{b})$ and $\mathbb{P}(c(\mathbf{b}) = 1 | \mathbf{b} = \boldsymbol{b})$, via the Naïve Bayes classifier and CIBer, respectively. Note that the calculations below may be subject to rounding errors.

Algorithm 12.2 Joint Encoding for Categorical Features

Input: Hyperrectagle $\mathcal{N}$ with dimension $M_1 \times \cdots \times M_{p_c}$ such that $M_1 \geq \cdots \geq M_{p_c}$.

1: $p_{\max} \leftarrow p_c$, indicating the index of the first feature variable to be fully encoded
2: Mark all cells of $\mathcal{N}$ as `available`
3: **for** $j = 1$ to p_c **do**
4: Set $r_j(x) \leftarrow 0$ for any $x \in \mathcal{X}_j$, to initialize the mapping (as a function)
5: **end for**
6: $k \leftarrow 1$
7: **while** $p_{\max} > 0$ **do**
8: Among all `available` cells in $\mathcal{N}$, find the one with the largest value, let say with the coordinate $(s_{1,m_1}, s_{2,m_2}, \ldots, s_{p_{\max},m_{p_{\max}}})$
9: **for** $j = 1$ to $p_{\max}$ **do**
10: $r_j(s_{j,m_j}) \leftarrow k$
11: **end for**
12: $k \leftarrow k + 1$
13: Mark all cells in $\mathcal{N}$ with coordinates $(x_1, \ldots, x_{p_{\max}})$ such that $x_j = s_{j,m_j}$ for some $j = 1, \ldots, p_{\max}$ as `unavailable`
14: **while** there exists a unique $x^* \in \mathcal{X}_{p_{\max}}$ such that $r_{p_{\max}}(x^*) = 0$ **do**
15: $r_{p_{\max}}(x^*) \leftarrow k$
16: **if** $p_{\max} > 1$ **then**
17: Construct a hyperrectangle $\mathcal{N}^*$ with dimension $k_1 \times \cdots \times k_{p_{\max}-1}$, such that $\mathcal{N}^*[s_{1,m_1}, \ldots, s_{p_{\max}-1,m_{p_{\max}-1}}] = \sum_{x \in \mathcal{X}_{p_{\max}}} \mathcal{N}[s_{1,m_1}, \ldots, s_{p_{\max}-1,m_{p_{\max}-1}}, x]$
18: $\mathcal{N} \leftarrow \mathcal{N}^*$, and mark the jointly un-encoded cells in $\mathcal{N}$ as `available`, that means those with coordinates $(x_1, \ldots, x_{p_{\max}-1})$ such that $\sum_{j=1}^{p_{\max}-1} r_j(x_j) = 0$
19: $p_{\max} \leftarrow p_{\max} - 1$
20: **end while**
21: **end while**
22: **Return:** one-to-one mappings (as functions) $r_j : \mathcal{X}_j \to \{1, 2, \ldots, M_j\}$, for $j = 1, \ldots, p_c$.

TABLE 12.2 Proportions of the first five continuous features in the training dataset with label $y = 0$.

	Bins									
Feature #	1	2	3	4	5	6	7	8	9	10
1	0.0023	0.0199	0.0742	0.1818	0.2710	0.2403	0.1370	0.0566	0.0134	0.0034
2	0.0084	0.0279	0.0919	0.1853	0.2580	0.2167	0.1286	0.0597	0.0203	0.0031
3	0.0011	0.0138	0.0532	0.1328	0.2514	0.2648	0.1737	0.0792	0.0237	0.0061
4	0.0027	0.0203	0.0785	0.1833	0.2997	0.2415	0.1217	0.0440	0.0077	0.0008
5	0.0038	0.0191	0.0739	0.1684	0.2755	0.2491	0.1435	0.0528	0.0107	0.0031

TABLE 12.3 Proportions of the first five continuous features in the training dataset with label $y = 1$.

	Bins									
Feature #	1	2	3	4	5	6	7	8	9	10
1	0.0000	0.0000	0.0014	0.0267	0.1804	0.3889	0.3110	0.0837	0.0079	0.0000
2	0.0000	0.0014	0.0115	0.0642	0.1984	0.3196	0.2626	0.1169	0.0202	0.0051
3	0.0000	0.0000	0.0007	0.0101	0.1140	0.3362	0.3759	0.1364	0.0238	0.0029
4	0.0000	0.0000	0.0310	0.2496	0.4603	0.2244	0.0339	0.0007	0.0000	0.0000
5	0.0000	0.0043	0.0794	0.3146	0.4336	0.1515	0.0159	0.0007	0.0000	0.0000

TABLE 12.4 Frequencies of the sixth categorical variable in the training dataset with label $y = 0$ and $y = 1$, respectively.

	Values (Bins)		
Feature #	1	2	3
6 (with label $y = 0$)	423	1251	940
6 (with label $y = 1$)	227	641	518

(I) Naïve Bayes

By the conditional independence assumption, we have:

(i) From Table 12.2 and Table 12.4:

$$\widehat{\mathbb{P}}_{NB}(\mathbf{b} = b|c(\mathbf{b}) = 0) = \prod_{j=1}^{6} \widehat{\mathbb{P}}(\mathbf{b}_j = b_j|c(\mathbf{b}) = 0)$$

$$= 0.2710 \cdot 0.2580 \cdot 0.2648 \cdot 0.2997 \cdot 0.2755 \cdot \frac{423}{2614}$$

$$= 0.0002474;$$

(ii) From Table 12.3 and Table 12.4:

$$\widehat{\mathbb{P}}_{NB}(\mathbf{b} = b|c(\mathbf{b}) = 1) = \prod_{j=1}^{6} \widehat{\mathbb{P}}(\mathbf{b}_j = b_j|c(\mathbf{b}) = 1)$$

$$= 0.1804 \cdot 0.1984 \cdot 0.3362 \cdot 0.4603 \cdot 0.4336 \cdot \frac{227}{1386}$$

$$= 0.0003933;$$

hence

$$\hat{\mathbb{P}}_{\text{NB}}(c(\mathbf{b})=0\mid \mathbf{b}=b)=\frac{\hat{\mathbb{P}}_{\text{NB}}(\mathbf{b}=b\mid c(\mathbf{b})=0)\cdot\hat{\mathbb{P}}(c(\mathbf{b})=0)}{\hat{\mathbb{P}}_{\text{NB}}(\mathbf{b}=b\mid c(\mathbf{b})=0)\cdot\hat{\mathbb{P}}(c(\mathbf{b})=0)+\hat{\mathbb{P}}_{\text{NB}}(\mathbf{b}=b\mid c(\mathbf{b})=1)\cdot\hat{\mathbb{P}}(c(\mathbf{b})=1)}$$

$$=\frac{0.0002474\cdot\dfrac{2614}{2614+1386}}{0.0002474\cdot\dfrac{2614}{2614+1386}+0.0003933\cdot\dfrac{1386}{2614+1386}}=0.5426,$$

$$\hat{\mathbb{P}}_{\text{NB}}(c(\mathbf{b})=1\mid \mathbf{b}=b)=1-\hat{\mathbb{P}}_{\text{NB}}(c(\mathbf{b})=0\mid \mathbf{b}=b)=0.4574.$$

Since $\hat{\mathbb{P}}_{\text{NB}}(c(\mathbf{b})=0\mid \mathbf{b}=b)=0.5426$ is greater than $\hat{\mathbb{P}}_{\text{NB}}(c(\mathbf{b})=1\mid \mathbf{b}=b)=0.4574$, the new datum is classified with label 0.

(II) CIBer Using totally-ordered Kendall's $\hat{\tau}_{\text{como}}$, we computed the distance matrices A_0 and A_1 for labels 0 and 1 respectively:

$$A_0=\begin{pmatrix} 0.0000 & 0.0317 & 0.0028 & 0.6298 & 0.6393 & 0.5121\\ 0.0317 & 0.0000 & 0.0416 & 0.6451 & 0.6536 & 0.5240\\ 0.0028 & 0.0416 & 0.0000 & 0.6364 & 0.6472 & 0.5165\\ 0.6298 & 0.6451 & 0.6364 & 0.0000 & 0.0061 & 0.4825\\ 0.6393 & 0.6536 & 0.6472 & 0.0061 & 0.0000 & 0.4909\\ 0.5121 & 0.5240 & 0.5165 & 0.4825 & 0.4909 & 0.0000 \end{pmatrix},$$

$$A_1=\begin{pmatrix} 0.0000 & 0.0778 & 0.0051 & 0.4700 & 0.4684 & 0.4469\\ 0.0778 & 0.0000 & 0.0935 & 0.5286 & 0.5230 & 0.4722\\ 0.0051 & 0.0935 & 0.0000 & 0.4772 & 0.4766 & 0.4497\\ 0.4700 & 0.5286 & 0.4772 & 0.0000 & 0.0000 & 0.4248\\ 0.4684 & 0.5230 & 0.4767 & 0.0000 & 0.0000 & 0.4166\\ 0.4469 & 0.4722 & 0.4497 & 0.4248 & 0.4166 & 0.0000 \end{pmatrix}.$$

Let us elaborate on the formation of comonotonic clusters by applying AGNES Algorithm 12.1 with $\delta=0.1$. First, we consider the case of label 0, while the case for label 1 is similar. The algorithm begins with a collection of singletons as individual clusters $\mathcal{Y}_0=\{\{1\},\dots,\{6\}\}$. Since the pair of Features 1 and 3 has the smallest complete linkage among all cluster pairs, with $d_{\text{complete}}(\{1\},\{3\})=0.0028<\delta$, they are merged first leading to the revised $\mathcal{Y}_0=\{\{1,3\},\{2\},\{4\},\{5\},\{6\}\}$. In the next step, the pair of Features 4 and 5 has the smallest complete linkage of $d_{\text{complete}}(\{4\},\{5\})=0.0061<\delta$, hence they are merged, resulting in the second revised $\mathcal{Y}_0=\{\{1,3\},\{2\},\{4,5\},\{6\}\}$. Among these clusters, the pair of $\{1,3\}$ and Feature 2 has the smallest complete linkage of $d_{\text{complete}}(\{1,3\},\{2\})=\max\{0.0317,0.0416\}=0.0416<\delta$; they are then merged, and then the third revised $\mathcal{Y}_0$ becomes $\{\{1,2,3\},\{4,5\},\{6\}\}$. As the complete linkage of any pair of clusters is greater than $\delta=0.1$, the AGNES algorithm stops. By the same procedure, the set of comonotonic clusters $\mathcal{Y}_1$ with the label $y = 1$ can be shown the same as $\mathcal{Y}_0$ under $\delta = 0.1$. Next we use the empirical version of (12.15) to calculate the conditional joint pmf of each comonotonic cluster as follows:

- Cluster $\{1, 2, 3\}$, conditional on the label 0:

$$\begin{aligned}
&\widehat{\mathbb{P}}_{\text{Como}}(\text{b}_1 = 5, \text{b}_2 = 5, \text{b}_3 = 6 | c(\mathbf{b}) = 0) \\
&= \left(\min\left\{\hat{F}_{\text{b}_1|c(\text{b})}(5|0), \hat{F}_{\text{b}_2|c(\text{b})}(5|0), \hat{F}_{\text{b}_3|c(\text{b})}(6|0)\right\}\right. \\
&\quad \left.- \max\left\{\hat{F}_{\text{b}_1|c(\text{b})}(5-1|0), \hat{F}_{\text{b}_2|c(\text{b})}(5-1|0), \hat{F}_{\text{b}_3|c(\text{b})}(6-1|0)\right\}\right)_+ \\
&= \left(\min\{0.5492, 0.5716, 0.7172\} - \max\{0.2782, 0.3136, 0.4524\}\right)_+ = 0.0968,
\end{aligned}$$

where for instance $\hat{F}_{\text{b}_1|c(\text{b})}(5|0) = 0.0023 + 0.0199 + 0.0742 + 0.1818 + 0.2710 = 0.5492$, by using the figures in Table 12.2. All other estimates can be obtained similarly.

- Cluster $\{1, 2, 3\}$, conditional on the label 1:

$$\begin{aligned}
&\widehat{\mathbb{P}}_{\text{Como}}(\text{b}_1 = 5, \text{b}_2 = 5, \text{b}_3 = 6 | c(\mathbf{b}) = 1) \\
&= \left(\min\left\{\hat{F}_{\text{b}_1|c(\text{b})}(5|1), \hat{F}_{\text{b}_2|c(\text{b})}(5|1), \hat{F}_{\text{b}_3|c(\text{b})}(6|1)\right\}\right. \\
&\quad \left.- \max\left\{\hat{F}_{\text{b}_1|c(\text{b})}(5-1|1), \hat{F}_{\text{b}_2|c(\text{b})}(5-1|1), \hat{F}_{\text{b}_3|c(\text{b})}(6-1|1)\right\}\right)_+ \\
&= \left(\min\{0.2085, 0.2756, 0.4610\} - \max\{0.0281, 0.0772, 0.1248\}\right)_+ = 0.0837.
\end{aligned}$$

- Cluster $\{4, 5\}$, conditional on the label 0:

$$\begin{aligned}
&\widehat{\mathbb{P}}_{\text{Como}}(\text{b}_4 = 5, \text{b}_5 = 5 | c(\mathbf{b}) = 0) \\
&= \left(\min\left\{\hat{F}_{\text{b}_4|c(\text{b})}(5|0), \hat{F}_{\text{b}_5|c(\text{b})}(5|0)\right\} - \max\left\{\hat{F}_{\text{b}_4|c(\text{b})}(5-1|0), \hat{F}_{\text{b}_5|c(\text{b})}(5-1|0)\right\}\right)_+ \\
&= (\min\{0.5844, 0.5408\} - \max\{0.2847, 0.2652\})_+ = 0.2560.
\end{aligned}$$

- Cluster $\{4, 5\}$, conditional on the label 1:

$$\begin{aligned}
&\widehat{\mathbb{P}}_{\text{Como}}(\text{b}_4 = 5, \text{b}_5 = 5 | c(\mathbf{b}) = 1) \\
&= \left(\min\left\{\hat{F}_{\text{b}_4|c(\text{b})}(5|1), \hat{F}_{\text{b}_5|c(\text{b})}(5|1)\right\} - \max\left\{\hat{F}_{\text{b}_4|c(\text{b})}(5-1|1), \hat{F}_{\text{b}_5|c(\text{b})}(5-1|1)\right\}\right)_+ \\
&= \left(\min\{0.7410, 0.8319\} - \max\{0.2807, 0.3983\}\right)_+ = 0.3427.
\end{aligned}$$

- Cluster {6}, conditional on the label 0: $\hat{\mathbb{P}}(b_6 = 1|c(\mathbf{b}) = 0) = \frac{423}{2614}$.
- Cluster {6}, conditional on the label 1: $\hat{\mathbb{P}}(b_6 = 1|c(\mathbf{b}) = 1) = \frac{227}{1386}$.

With these quantities, we can calculate the conditional joint pmf of $\mathbf{b}$ given each label as follows:

$$\begin{aligned}
&\hat{\mathbb{P}}_{\text{CIBer}}(\mathbf{b}=\boldsymbol{b}|c(\mathbf{b})=0)\\
&=\hat{\mathbb{P}}_{\text{Como}}(b_1=5, b_2=5, b_3=6|c(\mathbf{b})=0)\cdot\hat{\mathbb{P}}_{\text{Como}}(b_4=5, b_5=5|c(\mathbf{b})=0)\cdot\hat{\mathbb{P}}(b_6=1|c(\mathbf{b})=0)\\
&=0.09682\cdot 0.2560\cdot\frac{423}{2614}=0.004011,\\
&\mathbb{P}_{\text{CIBer}}(\mathbf{b}=\boldsymbol{b}|c(\mathbf{b})=1)\\
&=\hat{\mathbb{P}}_{\text{Como}}(b_1=5, b_2=5, b_3=6|c(\mathbf{b})=1)\cdot\hat{\mathbb{P}}_{\text{Como}}(b_4=5, b_5=5|c(\mathbf{b})=1)\cdot\hat{\mathbb{P}}(b_6=1|c(\mathbf{b})=1)\\
&=0.08369\cdot 0.3427\cdot\frac{227}{1386}=0.004697,
\end{aligned}$$

which allows us to evaluate the following posterior probabilities of the labels given the observed bins:

$$\begin{aligned}
\hat{\mathbb{P}}_{\text{CIBer}}(c(\mathbf{b}) = 0|\mathbf{b} = \boldsymbol{b}) &= \frac{\hat{\mathbb{P}}_{\text{CIBer}}(\mathbf{b} = \boldsymbol{b}|c(\mathbf{b}) = 0)\cdot\hat{\mathbb{P}}(c(\mathbf{b}) = 0)}{\hat{\mathbb{P}}_{\text{CIBer}}(\mathbf{b} = \boldsymbol{b}|c(\mathbf{b}) = 0)\cdot\hat{\mathbb{P}}(c(\mathbf{b}) = 0) + \hat{\mathbb{P}}_{\text{CIBer}}(\mathbf{b} = \boldsymbol{b}|c(\mathbf{b}) = 1)\cdot\hat{\mathbb{P}}(c(\mathbf{b}) = 1)}\\
&= \frac{0.004011\cdot\frac{2614}{2614+1386}}{0.004011\cdot\frac{2614}{2614+1386} + 0.004697\cdot\frac{1386}{2614+1386}} = 0.6169,\\
\mathbb{P}_{\text{CIBer}}(c(\mathbf{b}) = 1|\mathbf{b} = \boldsymbol{b}) &= 1 - \mathbb{P}_{\text{CIBer}}(c(\mathbf{b}) = 0|\mathbf{b} = \boldsymbol{b}) = 0.3831.
\end{aligned}$$

Since $\hat{\mathbb{P}}_{\text{CIBer}}(c(\mathbf{b}) = 0|\mathbf{b} = \boldsymbol{b}) = 0.6169$ is greater than $\hat{\mathbb{P}}_{\text{CIBer}}(c(\mathbf{b}) = 1|\mathbf{b} = \boldsymbol{b}) = 0.3831$, the new datum should be classified with label 0. We note that while the result is consistent with that from the Naïve Bayes classifier, it is more clear-cut as the posterior probability of 0.6169 for label 0 using CIBer is much larger than 0.5426 of the Naïve Bayes classifier.

12.5.5 Robustness of Comonotonic Copula

As discussed in Chapter 8, copulae can be used to model the general dependence structure of underlying feature variables. While many copulae are readily available for use, CIBer focuses on utilizing the comonotonic copula. A natural question for readers to raise at this point would be: why choose comonotonic copula over other alternatives? To address this, let us provide a comprehensive simulation study to justify the use of comonotonic copula as first choice.

We investigate the robustness of comonotonicity as a modelling copula in CIBer, by comparing CIBer against three other Archimedean copula-based classifiers, namely

Gumbel, Frank, and Clayton copulae. To mimic the variety of real-world datasets, for each feature variable x_i we randomly pick a distribution from a pre-selected list of common distributions. For simplicity, on the one hand we consider only one cluster where all five feature variables in the class label c_1 have a strong dependence. On the other hand, those feature variables in the class label c_2 have a weak dependence, but the same type of copulae govern both dependence structures. Although this setting favors the three Archimedean copula-based classifiers, we shall demonstrate that CIBer performs comparatively well against them. The simulation procedure is depicted in Algorithm 12.3.

Algorithm 12.3 Simulation procedure of copula-based classifiers

1: **for** $t = 1, \ldots, 100$ **do**
2: Generate 10,000 iid uniform $U(0,1)$ random samples, $u_1, \ldots, u_{10,000} \in \mathbb{R}^5$, through a randomly chosen Archimedean copula with a parameter $\theta \sim U(5,30)$ governing the strength of dependence;
3: Generate another set of 10,000 iid uniform $U(0,1)$ random samples $u_{10,001}, \ldots, u_{20,000} \in \mathbb{R}^5$, through the selected Archimedean copula with $\theta = 1.5$ for weak dependence;
4: **for** $i = 1, \ldots, 5$ **do**
5: Randomly pick a distribution $F_{x_i,\phi}$ with parameter ϕ, either continuous or discrete, and compute the quantile $x_{j,i} := F^{-1}_{x_i,\phi}(u_{j,i})$ for $j = 1, \ldots, 10,000$;
6: Compute the quantile $x_{j,i} := F^{-1}_{x_i,\psi}(u_{j,i})$ with another parameter ψ for $j = 10,001, \ldots, 20,000$;
7: **end for**
8: Randomly select the cut points $a_{(0)} := \min x_5 - 10^{-6}, a_{(1)}, \ldots, a_{(n_B^{(5)})}, a_{(n_B^{(5)}+1)} := \max x_5$, where $n_B^{(5)} \sim U[5,15]$ and $a_k \sim U(\min x_5, \max x_5)$ for $k = 1, \ldots, n_B^{(5)}$;
9: For each $x_{j,5}$ falling in the interval $(a_{(k)}, a_{(k+1)}]$, assign $x_{j,5} \leftarrow (a_{(k)} + a_{(k+1)})/2$ as the mid-point value of that interval;
10: Label $x_1, \ldots, x_{10,000}$ as class c_1 and $x_{10,001}, \ldots, x_{20,000}$ as class c_2;
11: Randomly split $x_1, \ldots, x_{20,000}$ into the training dataset $\mathcal{T}_{\text{train}}$ and the test dataset $\mathcal{T}_{\text{test}}$ with a ratio of 8:2;
12: Compute the training empirical cdf for the training dataset $\mathcal{T}_{\text{train}}$, and then use $\mathcal{T}_{\text{train}}$ to compute the testing empirical cdf for the test dataset $\mathcal{T}_{\text{test}}$;
13: Build CIBer with $\delta_{\max} = 0.1$ and $\mathcal{T}_{\text{train}}$, whereas the training empirical cdf is used to fit the three Archimedean copula-based classifiers and estimate their respective parameters θ's governing the strength of dependence;
14: Obtain the predicted labels of CIBer with the test dataset $\mathcal{T}_{\text{test}}$ and other copula-based classifiers with the testing empirical cdf;
15: Compute the test F_1-score and accuracy for each of the models.
16: **end for**
17: Plot a boxplot from 100 records for each of the selected models and model assessment methods.

By construction, CIBer with $\delta_{\max} = 0.1$ captures only one comonotone cluster for the class label c_1, i.e. $\mathcal{G}_{1,0} = \{1, 2, 3, 4, 5\}$, and five singleton clusters for the class label c_2, i.e. $\mathcal{G}_{i,1} = \{i\}$ for $i = 1, \ldots, 5$. From Figure 12.2, boxplots of F_1-scores show that both `CIBer (normal)` and `CIBer (auto)` generally perform better than all other copula-based classifiers, including ones that were originally used to simulate the data. In particular, `CIBer (normal)` discretizes each continuous feature variable x_i by the quantiles $\hat{F}_{\mathrm{x}_i}^{-1}(1/k_i), \hat{F}_{\mathrm{x}_i}^{-1}(2/k_i), \ldots, \hat{F}_{\mathrm{x}_i}^{-1}((k_i - 1)/k_i)$ of the corresponding fitted normal distribution with a cdf $\hat{F}_{\mathrm{x}_i}$. Likewise, instead of only considering the normal distribution, `CIBer (auto)` chooses the best distribution (within a list of candidate distributions) by using the Kolmogorov–Smirnov test (introduced in Chapter 3), in which the fitted distribution from the list having the smallest value of KS statistic is selected. Similar findings regarding the accuracies are shown in Figure 12.3.

While copulae have been widely used in recent years for modelling dependence structures, using a copula requires parameter estimation and is usually suitable only for scenarios where all feature variables are of the same type. If only continuous feature variables are present, those three Archimedean copula-based classifiers naturally perform better than CIBer. With the presence of x_5, a randomly discretized feature variable, which happens normally in practice (perhaps due to censoring and personal privacy, for example), CIBer steadily performs better than those three copula-based classifiers. Instead of merely focusing on datasets with all feature variables of the same type, CIBer is designed to handle datasets with multiple feature variable types at the same time. In the real world, discrete, continuous, and categorical feature variables do appear in a dataset. Moreover, independence and comonotonicity are two extreme and universal dependence structures that are entirely non-parametric. By making use of only these two dependence structures, CIBer can avoid complex computations, in contrast to classifiers based on specific parametric copulae, which also risk overfitting on the training dataset. CIBer is not prone to this error, and it demonstrates a promising performance on the test dataset. To echo the end of Chapter 8, CIBer can be regarded as a special case of the highly versatile vine copula, with an all-or-nothing treatment of the dependence among feature variables which does not lose much generality.

12.5.6 CIBer in Action in Finance and Insurance

A Python package has been created and is available for download on PyPI *(The Python Package Index)*. It can be installed by the command `pip install CIBer`. As discussed in Section 2.3, we shall first open a command terminal such as Anaconda Prompt, and then type the command in. The package will be downloaded automatically, as seen in Figure 12.4, and an installation guide is available through the QR code in Figure 12.5.

In this subsection, we demonstrate the effectiveness of CIBer versus some representative machine learners through the evaluation of several model assessment methods by analyzing two real datasets from finance and insurance:

1. **Bank Churners** [20]: An important source of income for banks is the annual fees and charges paid by credit card holders. As a result, customer attrition is

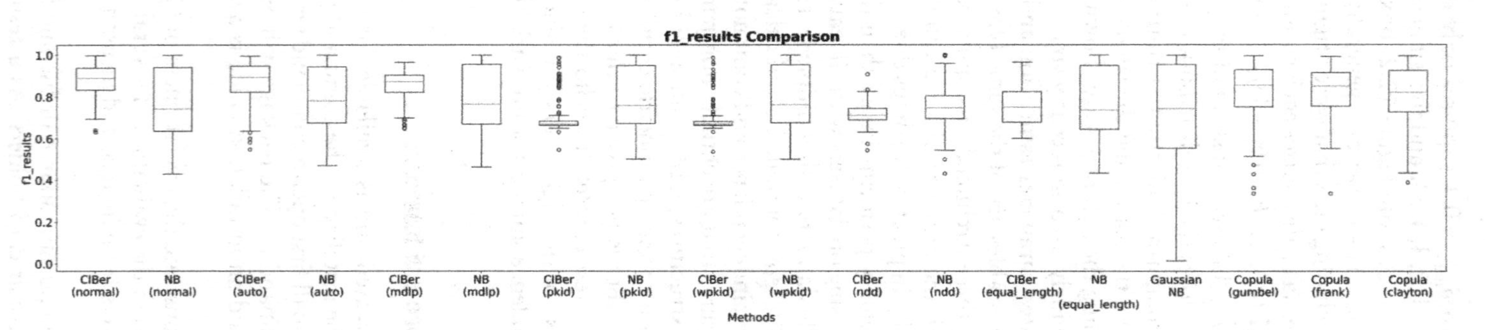

FIGURE 12.2 Boxplot of F_1-scores for all classifiers in the simulation study.

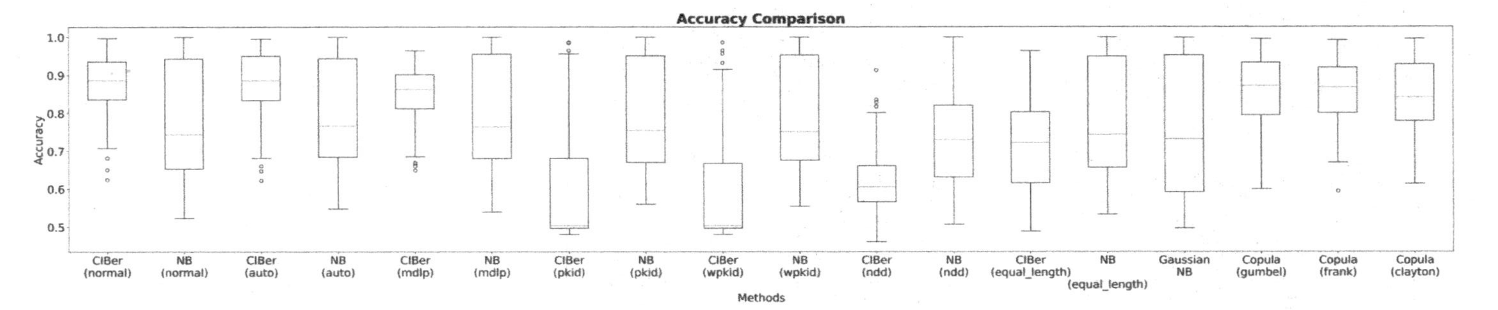

FIGURE 12.3 Boxplot of accuracies for all classifiers in the simulation study.

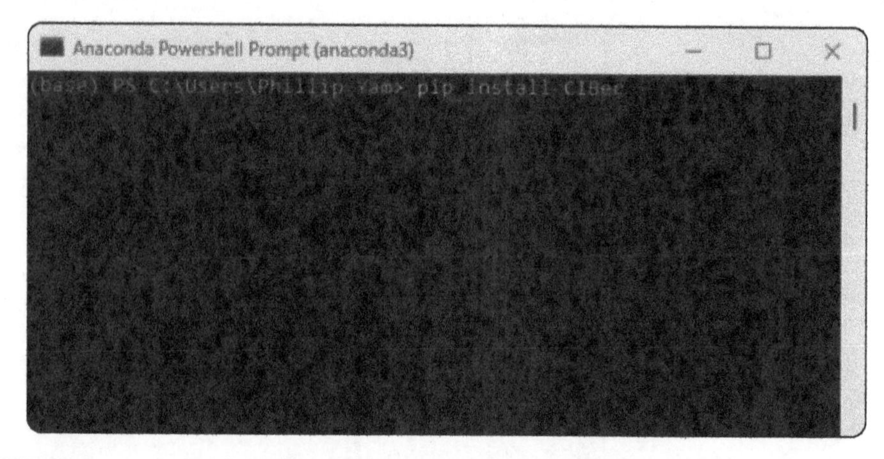

FIGURE 12.4 Downloading CIBer package in Python via terminal.

FIGURE 12.5 GitHub for CIBer.

one of the major issues of concern. It is crucial for banks to predict whether a cardholder is likely to withdraw their credit card services because this can better facilitate their analysis on the attributing factors, helping them to improve their services in a timely manner.

2. **Default Premium Payment** [18]: The on-time collection of insurance premiums is crucial to insurance companies as this is a major component of their cash flow. However, policyholders may defer or even terminate premium payments due to various reasons. Therefore, it is important for insurance companies to be able to predict the emergence of such unfavourable events and alert their company agents to take timely action in advance.

Both datasets are highly imbalanced in nature. For example, a balanced dataset with roughly half of their customers in default would obviously bankrupt the company. We also note that there are some missing entries in both datasets. For simplicity,

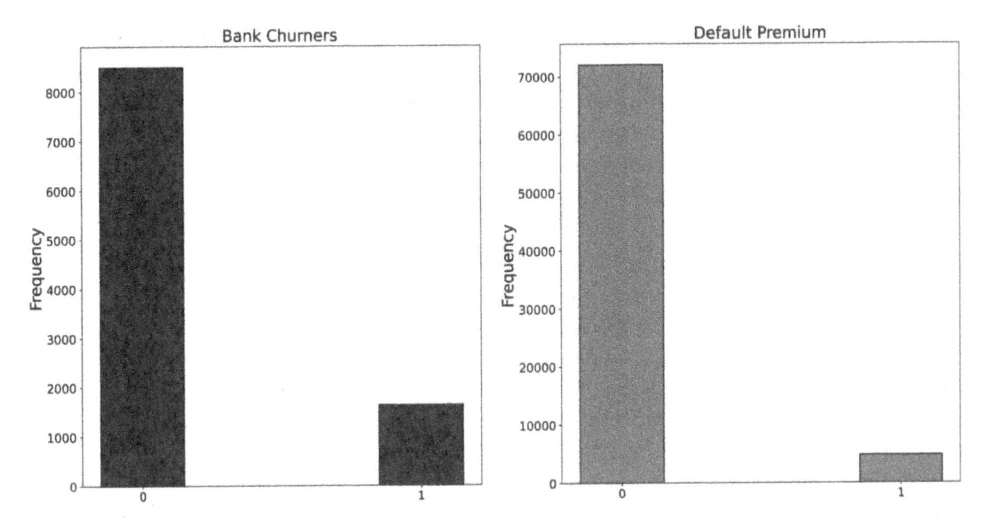

FIGURE 12.6 Class label distributions of bank churners (left) and default premium payment (right) datasets after removing missing entries.

we assume the missing mechanism of *missing completely at random* and remove observations containing any missing entry. Figure 12.6 shows the distribution of the labels in the bank churners (left) and default premium payment (right) datasets after missing entries are removed.

Next, we describe the experimental procedure for each of the two datasets. Since each dataset contains thousands of records, it suffices for us to (under-)sample a small, yet representative enough, portion of the original dataset as our training dataset. We denote the training sample size by n_{train}. Next, we shall randomly pick n_{test} data points from each label of the remaining dataset as our test dataset, resulting in a testing size of $2n_{\text{test}}$. The actual values of n_{train} and n_{test} for each dataset are subject to their original sizes of each binary label and hence their corresponding degrees of imbalance, which we shall state explicitly in the course of the following discussion. Algorithm 12.4 presents a summary of the experimental procedure.

Algorithm 12.4 Procedure of experimental studies

for $t = 1, \cdots, 100$ **do**

Randomly select n_{train} data points in the dataset without replacement as the training dataset;

From the remaining data points, randomly select n_{test} data points without replacement from each label, altogether they form the test dataset;

Build CIBer and other competitive models with the training dataset;

Obtain the predicted labels from each model for the test dataset;

Compute the test F_1-score, accuracy, and recall rate for each model.

end for

Plot a boxplot from 100 records for each of the selected models and model assessment methods.

As before, in implementing CIBer, we shall first discretize the continuous features by binning (referring to Section 12.5.2), and then perform a joint encoding scheme for categorical features. Comonotonic clusters are then identified using the AGNES algorithm with the totally-ordered Kendall's $\hat{\tau}_{como}$ (12.17) as the underlying association measure. The value of the maximum cluster-wise distance δ is determined by grid search at the values of $0, 0.05, 0.1, 0.15, \ldots, 1$, with the F_1-score as the assessment metric.

We shall compare the results of CIBer with those from other commonly used alternatives, namely *Composite Naïve Bayes* (CNB), *Gaussian Naïve Bayes* (GNB), Support Vector Machine (SVM) with RBF kernel, Logistic regression (LR), *Multi-layer Perceptron* (MLP), *Linear Discriminant Analysis* (LDA), *Decision Tree* (DT), and *K-Nearest Neighbour* (*K*-NN). Note that we choose only the primitive machine learners but not their boosted versions which involve various ensembling tools, and compare CIBer only with the primitive models (we could as well apply the same ensembling tools to CIBer and compare it with the ensembled variants of the others). In particular, CNB builds from two Naïve Bayes classifiers, namely GNB for modelling continuous and discrete features and the Categorical Naïve Bayes for categorical ones. Moreover, we also consider a variant of LDA, in which we first adopt PCA (see Section 9.1) to transform all features into continuously and uncorrelatedly distributed ones, and then perform the usual LDA. Last but not least, an additional CIBer with $\delta = 0$ is adopted as a base comparison without incorporating comonotonicity, similar to an empirical Naïve Bayes classifiers.

12.5.6.1 Bank Churners in Finance The Bank Churners dataset[4] records 19 features from both existing customers and attrited customers. Five of the features are categorical, namely `Gender` (2), `Education_Level` (7), `Marital_Status` (4), `Income_Category` (6), and `Card_Category` (4), where the numbers inside the brackets are the total numbers of categories of the features. The binary response variable `Attrition_Flag` takes a value of 1 if the cardholder terminates the services, or 0 otherwise. Since the dataset contains roughly 10,000 records, we consider: $n_{\text{train}} = 7,000$ and $n_{\text{test}} = 200$, and we set $\delta = 0.05$ after a grid search.

CIBer shares similar capabilities and functions for fitting[5] as other common machine learning libraries such as `scikit-learn`. In the following programmes, we illustrate how to fit CIBer to the Bank Churners dataset in Python. First, we import all necessary libraries and the dataset through Programme 12.1.

```
1 import numpy as np
2 import pandas as pd
3 from sklearn.metrics import roc_auc_score, precision_score,
     recall_score, f1_score, accuracy_score
4 import matplotlib.pyplot as plt
```

[4] The dataset can be downloaded from https://www.kaggle.com/code/thomaskonstantin/bank-churn-data-exploration-and-churn-prediction/data.

[5] For further details of its functionality and codes of examples, please refer to the project GitHub with link https://github.com/kaiser1999/CIBer or via the QR code in Figure 12.5.

```
5  #import Class CIBer from CIBer package
6  from CIBer import CIBer
7
8  #import dataset
9  df = pd.read_csv("BankChurners.csv")
```

Programme 12.1 Loading in all required libraries including CIBer package in Python.

Before fitting the model, the dataset needs to be prepared. In Programm 12.2 via Python, `pd.factorize` is a `pandas` function for jointly encoding categorical features' values into discrete ones.

```
1  # Drop unused columns
2
3  # Encode categorical feature values into discrete ones
4  df['Attrition_Flag'] = pd.factorize(df['Attrition_Flag'])[0] + 0
5  df['Gender'] = pd.factorize(df['Gender'])[0] + 1
6  df['Education_Level'] = pd.factorize(df['Education_Level'])[0] + 1
7  df['Marital_Status'] = pd.factorize(df['Marital_Status'])[0] + 1
8  df['Income_Category'] = pd.factorize(df['Income_Category'])[0] + 1
9  df['Card_Category'] = pd.factorize(df['Card_Category'])[0] + 1
10
11 # Move label y to the last column of df
12 cols = df.columns.to_list()
13 cols.append(cols[0])
14 cols = cols[1:]
15 df = df[cols]
```

Programme 12.2 Dataset preprocessing for CIBer in Python.

Then in Programme 12.3 via Python, we split the dataset into the training and testing sets. Since this dataset is highly imbalanced, we sample 7,000 data points as the training dataset and then pick 200 data points in each class as the test dataset.

```
1  # Dataset preparation
2  label_name = "Attrition_Flag"
3  df_bank_y = df[label_name]
4
5  n_sample = 7000
6  n_test = 200
7
8  np.random.seed(4012)
9  idx_train = np.random.choice(np.arange(len(df)), n_sample,
10                               replace=False)
11 X_train = df.iloc[idx_train,:-1].to_numpy()
12 y_train = df.iloc[idx_train,-1].to_numpy()
13
14 # obtain test. 1500 in EACH class
15 samples_per_group_dict = {0:n_test, 1:n_test}
16 df_test = df[~df.index.isin(idx_train)]
```

```
17 df_test = df_test.groupby(label_name).apply(lambda group: group.
       sample(samples_per_group_dict[group.name])).reset_index(drop=
       True)
18 X_test = df_test.iloc[:,:-1].to_numpy()
19 y_test = df_test.iloc[:,-1].to_numpy()
```

Programme 12.3 Sampling and Train-test split for CIBer in Python.

In Programm 12.4 in Python, `cont_col` is a list of indices for the continuous feature variables and `min_asso` is the minimum threshold adopted in measuring the linear association for all pairs of feature variables, which can be found by a grid search. To fit a CIBer, we create a CIBer object called `CIBer_clf`, then fit the model using the training dataset `X_train` and its label set `y_train`. Finally, we get the predictions and its corresponding probabilities from CIBer by calling the functions `predict()` and `predict_porba()`, respectively.

```
1  # parameters to be added in CIBer
2  cont_col = [7, 11, 12, 13, 14, 15, 16, 17, 18]
3  min_asso = 0.95
4
5  # Fit CIBer
6  CIBer_clf = CIBer(cont_col=cont_col, asso_method='modified',
7                    min_asso=min_asso, disc_method="auto", n_bins=50,
8                    joint_encode=True)
9  CIBer_clf.fit(X_train, y_train)
10 CIBer_predict = CIBer_clf.predict(X_test)
11 CIBer_proba = CIBer_clf.predict_proba(X_test)
```

Programme 12.4 Fitting a CIBer and making predictions in Python.

Figure 12.7 shows the boxplot of the F_1-scores for each classifier. We immediately observe that the F_1-score of CIBer with $\delta = 0.05$, denoted by `CIBer (Normal)`, is slightly higher than that of CIBer with $\delta = 0$, denoted by `CIBer (No CM)`, which does not involve any comonotonic structures. Besides, the ranges of both boxplots of CIBer are smaller than those of composite Naïve Bayes and Gaussian Naïve Bayes classifiers, and their positions are also higher than those of the latter two, indicating that the feature engineering techniques applied in CIBer improve both the robustness and effectiveness of the Naïve Bayes classifier.

The results show that both versions of CIBer are second only to the multilayer perceptron (MLP) and classification (decision) tree (DT). As a remark, we note that the results obtained by both MLP and classification tree are often unexplainable, whereas in the case of CIBer, we are able to justify the rationale behind the comonotonic clusters obtained. Taking this dataset as an example, we obtain one comonotonic cluster in the majority class with the actual label `0`, namely `Card_Category` and `Credit_Limit`. In particular, `Card_Category` is a categorical feature with four

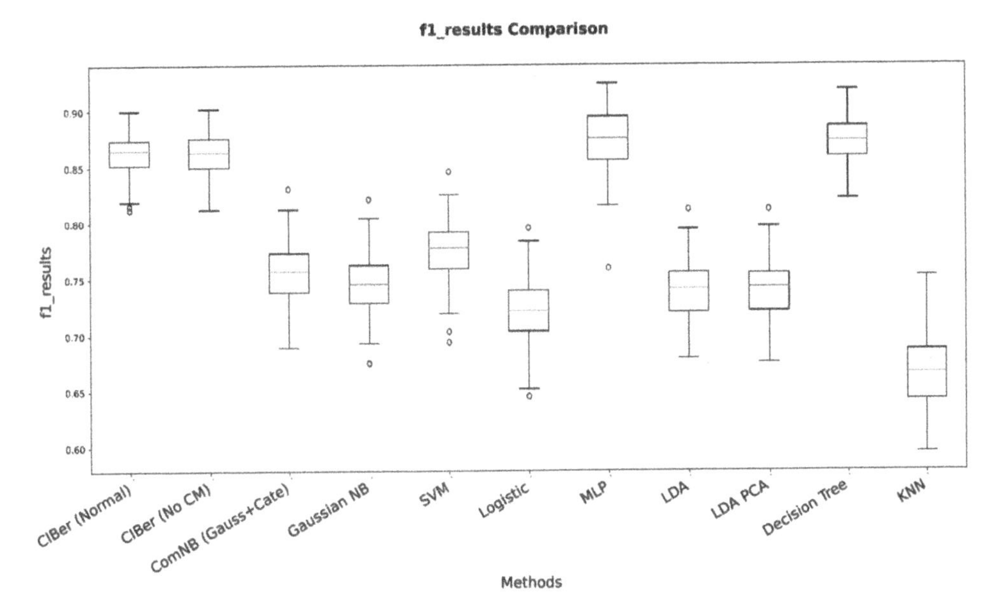

FIGURE 12.7 Boxplot of F_1-score for each classifier in the bank churners dataset.

categories, including `Blue`, `Silver`, `Gold`, and `Platinum`, representing the status level of a credit card holder in an ascending order. For an existing cardholder with actual label 0, it is natural that the higher status level of a credit card comes with a higher credit limit, which explains why the two features are comonotonic. On the other hand, we also reveal a comonotonic cluster with actual label 1, namely (`Card_Category`, `Avg_Open_To_Buy`). Attrited cardholders in the minority class with actual label 1 tend to use their credit cards less often, resulting in a larger value of average open-to-buy. As a higher credit card level is subject to higher fees and charges, cardholders with a large average open-to-buy are expected to be more likely to terminate the services. Identifying these comonotonic structures does bring some improvement to the performance of CIBer, as shown in the slight increase in the F_1-scores from `CIBer (No CM)` to `CIBer (Normal)`. The F_1-scores of the other classifiers are roughly the same, while K-NN performs the worst. Similar observations are made in the boxplots of the corresponding accuracies for all classifiers as illustrated in Figure 12.8.

The recall rate is important when examining the model performance on an imbalanced dataset, especially when one places more emphasis on the minority class than the majority. In this dataset, banks aim to solve the problem of customer attrition, hence they focus on making correct predictions for clients from the minority class. Figure 12.9 shows the boxplots of recall rates for all classifiers. It is observed that the recall rates of the two versions of CIBer, as well as those of MLP and classification tree, are roughly 0.8, which are significantly higher than those of all other competitors, indicating that these classifiers perform well in detecting customer attrition.

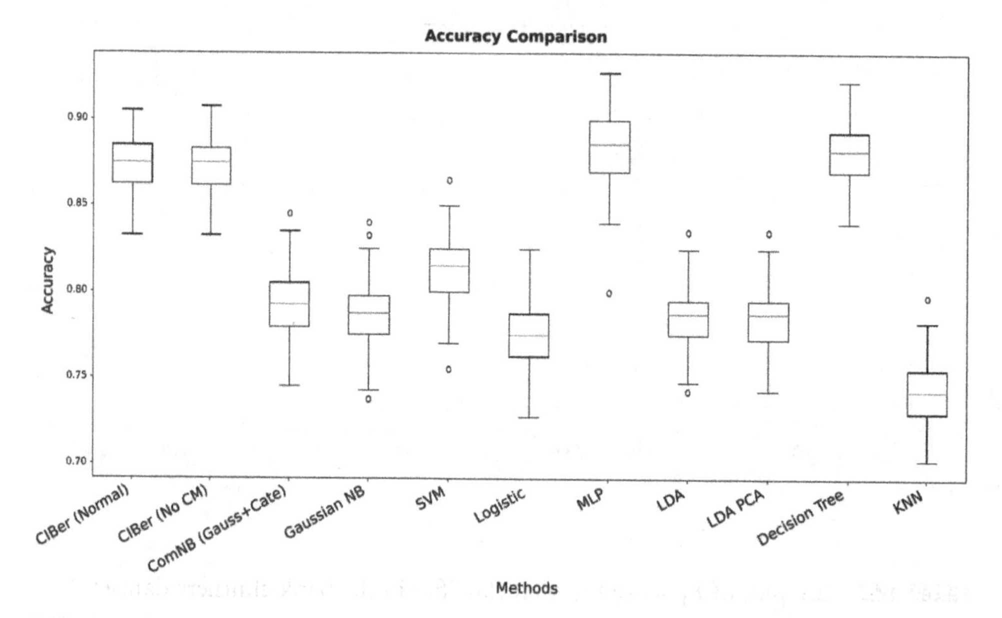

FIGURE 12.8 Boxplots of accuracies for all classifiers in the bank churners dataset.

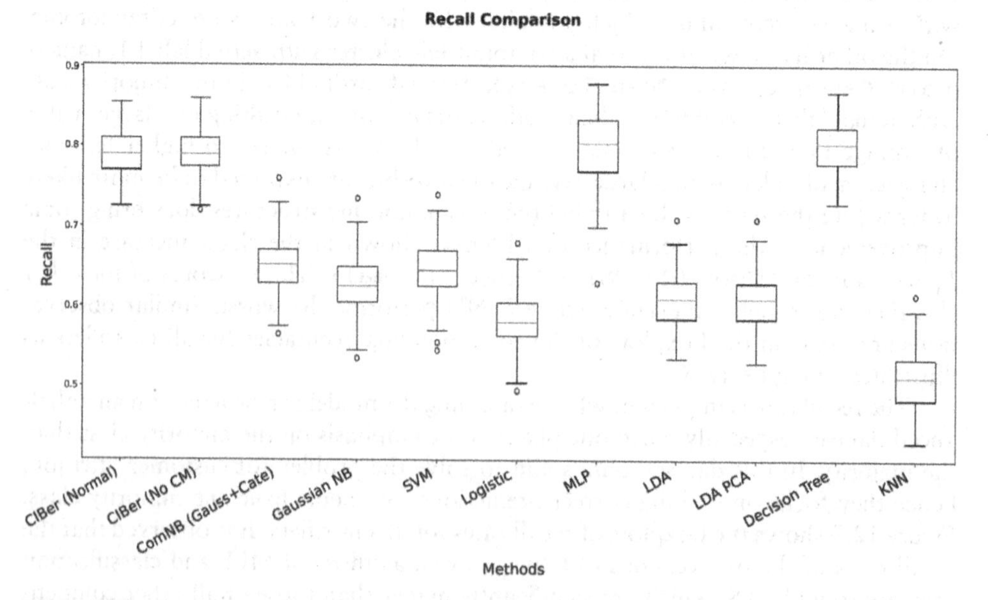

FIGURE 12.9 Boxplots of recall rates for all classifiers in the bank churners dataset.

12.5.6.2 Default Insurance Premium Payment This insurance dataset[6], about default in insurance premium payments, records 10 feature variables on the personal profiles and premium payment history of 79, 853 policyholders. There are two categorical feature variables, namely `sourcing` (5) and `residence:area_type` (2), where the numbers inside the brackets are the total numbers of categories. The binary response variable `Target` takes a value of 1 if the policyholder defaults on the premium payments, or 0 otherwise. For this dataset, we set $n_{\text{train}} = 30,000$ and $n_{\text{test}} = 1,000$, and we choose $\delta = 0.05$ after a grid search.

Figure 12.10 shows the boxplot of the F_1-score for each classifier. Similar to the bank churners dataset, the incorporation of comonotonicity is beneficial, as the F_1-score of CIBer with $\delta = 0.05$ (denoted as `CIBer (Normal)`) is significantly higher than that of CIBer with $\delta = 0$ (denoted as `CIBer (No CM)`). Unexpectedly, the F_1-scores for the Gaussian Naïve Bayes are 0 in most trials, indicating that it predicts all policyholders in the test dataset to not default. By using the composite Naïve Bayes classifier, the F_1-scores shoot up and even exceed those of CIBer with $\delta = 0$, but its performance still a bit inferior to CIBer with $\delta = 0.05$. This again indicates that the feature engineering techniques applied in CIBer improve both the robustness and effectiveness of a Naïve Bayes classifier.

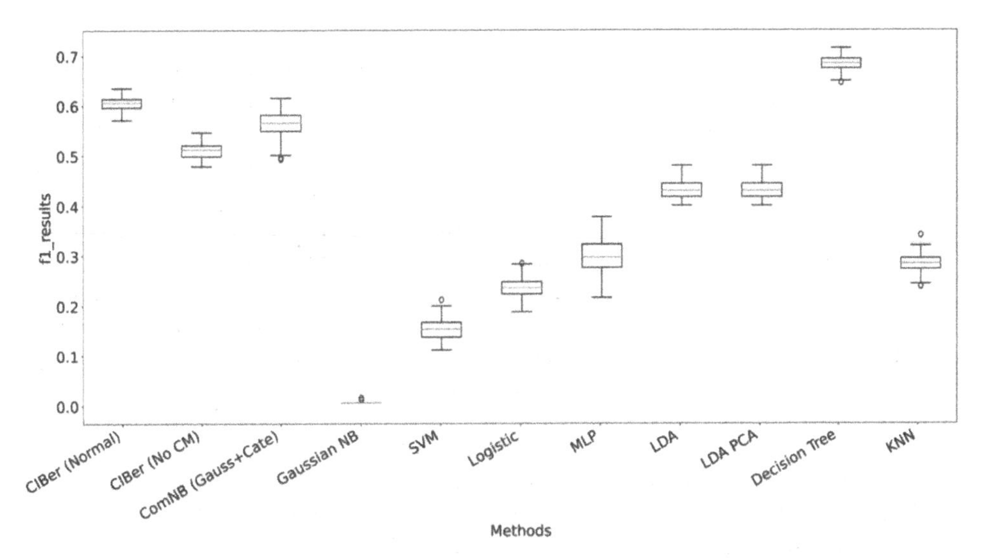

FIGURE 12.10 Boxplots of F_1-scores for all classifiers in the default premium payment dataset.

[6] The dataset can be downloaded from https://www.kaggle.com/datasets/prakharrathi25/insurance-company-dataset.

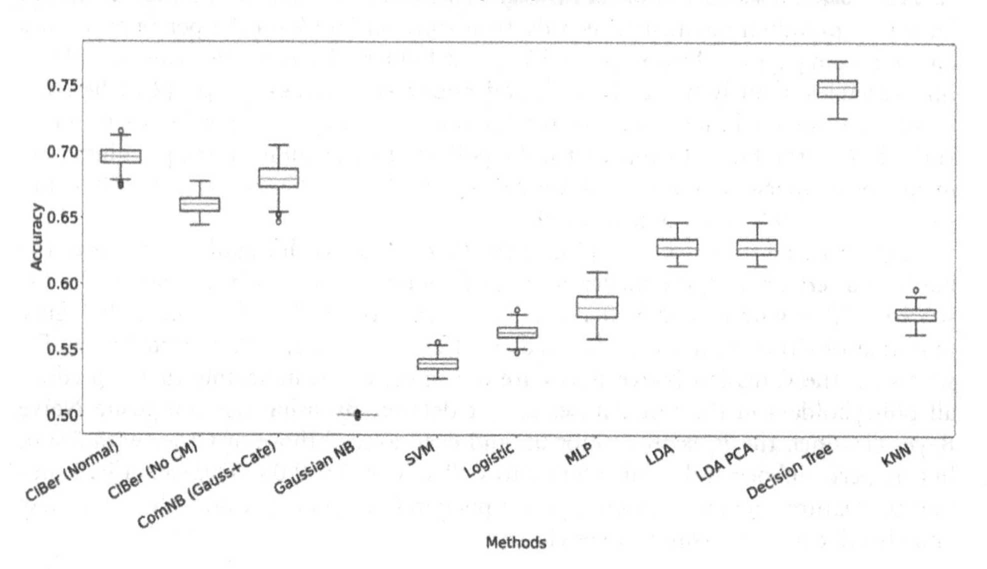

FIGURE 12.11 Boxplots of accuracies for all classifiers in the default premium payment dataset.

Let us provide some economic justifications for the comonotonic clusters obtained in CIBer with $\delta = 0.05$. For actual label 0 (policyholders do not default on premium payments), one comonotonic cluster is obtained, namely (`Count 3-6`, `Count 6-12`, `Count over 12`). Here, `Count xx` refers to the count of premium payments late by `xx` months. The strong dependence among these three variables is obvious: a client with a good record without a default on premium payments is also more likely to not delay future premium payments. This is also the reason why we do not observe such a cluster in the minority class with actual label 1 (policyholders default on premium payments), since those customers are more likely to defer premium payments by whatever number of months. This comonotonic relationship helps `CIBer (Normal)` perform significantly better than `CIBer (No CM)`. Similar observations can be made in comparison study on the accuracies and recall rates of all classifiers as shown in Figures 12.11 and 12.12.

In conclusion, CIBer has demonstrated its superiority and robustness over most traditional machine learners and is seen to be one of the best methods in the two real-data applications discussed above. Additionally, CIBer can be extended into a regression context, by using similar arguments to those related to regression trees (to be introduced in Chapter 13). Due to space limitations, details regarding this extension will not be discussed in this book; we invite interested readers to view our ongoing and future research works.

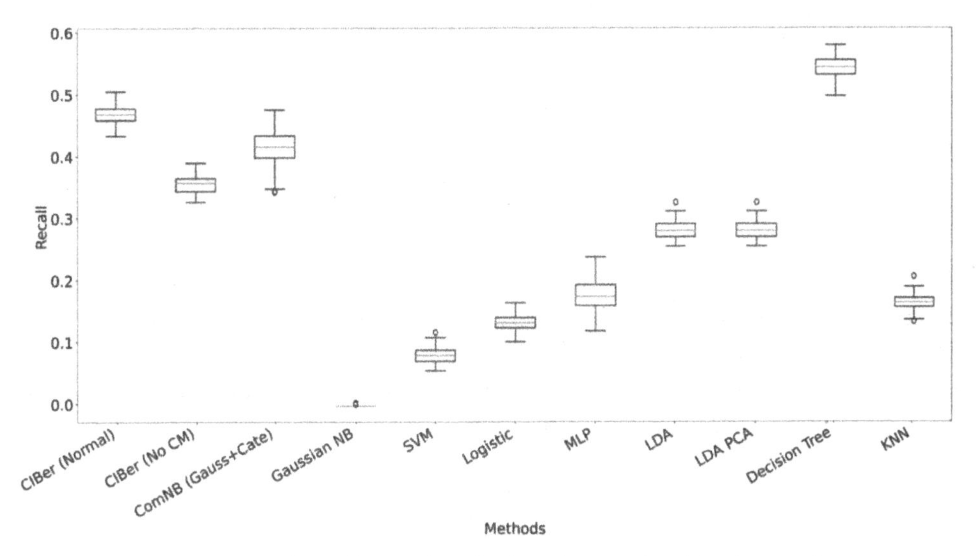

FIGURE 12.12 Boxplots of recall rates for all classifiers in the default premium payment dataset.

12.A APPENDIX

12.A.1 Laplace's Approximation for Integrals

To avoid unnecessary technical details, we mainly focus on the one-dimensional setting here. Results for higher dimensional framework follow analogously, and we leave these for readers. Laplace's approximation, or Laplace's method, is a commonly used approach for evaluating integrals of the following form:

$$G_n := \int_a^b e^{ng(y)} dy. \tag{12.19}$$

It is motivated by the observation that the integrand decays exponentially as y deviates away from $\hat{y}$, the global maximum of g, hence the major contributions to the integral mainly come from the values of y in a close neighborhood of $\hat{y}$.

Mathematically, assuming twice differentiablity of g, we consider the Taylor series expansion of g at $\hat{y}$ by recalling that $g'(\hat{y}) = 0$ and $g''(\hat{y}) < 0$:

$$g(y) = g(\hat{y}) + \frac{1}{2}(y - \hat{y})^2 g''(\hat{y}) + \text{higher order terms.}$$

Substituting the expression into (12.19), and dropping the higher order terms (which is justifiable, as becomes clear below), we have:

$$G_n \approx \int_a^b e^{n\left(g(\hat{y})+\frac{1}{2}(y-\hat{y})^2 g''(\hat{y})\right)} dy$$
$$\approx e^{ng(\hat{y})} \int_{-\infty}^{+\infty} e^{-\frac{n}{2}(y-\hat{y})^2(-g''(\hat{y}))} dy,$$

where we extend the domain of integration from (a, b) to the whole real line. This can be justified by noting that the integrand is proportional to the density kernel of $\mathcal{N}(\hat{y}, -\frac{1}{ng''(\hat{y})})$. It also validates the dropping of higher order terms as they are collectively of $o_p(1)$ order under this normal density. Therefore, we obtain the following approximation for G_n for a large value of n:

$$G_n \approx e^{ng(\hat{y})} \sqrt{-\frac{2\pi}{ng''(\hat{y})}}.$$

In fact, for a more general class of integrals of the form:

$$H_n := \int_a^b h(y) e^{ng_n(y)} dy, \tag{12.20}$$

we can also deduce the following result by a similar analysis:

$$H_n = e^{ng_n(\tilde{y})} h(\tilde{y}) \sqrt{-\frac{2\pi}{nq_n''(\tilde{y})}} \left(1 + O\left(\frac{1}{n}\right)\right),$$

where $q_n(y) := g_n(y) + \frac{1}{n} \ln h(y)$, and $\tilde{y}$ is the global maximizer of q_n. Analogously, the multi-dimensional version of Laplace's approximation takes the following form: as $n \to \infty$,

$$\int_D h(y) e^{ng_n(y)} dy \approx \left(\frac{2\pi}{n}\right)^{\frac{d}{2}} \frac{h(\tilde{y}) e^{ng_n(\tilde{y})}}{|-\nabla_y \nabla_y^\top q_n(\tilde{y})|^{\frac{1}{2}}}, \tag{12.21}$$

where the integral is one defined on some region $D \subseteq \mathbb{R}^d$, and $\nabla_y \nabla_y^\top q_n$ is the Hessian matrix of q_n as introduced in Section 6.2, which is required to be negative definite and the determinant of which is taken in the quotient of (12.21). Readers may refer to the existing literature for more detailed discussions of the method, such as Chapter 3 of [1] and Section 9.7 of [22].

12.A.2 Bayesian Information Criterion (BIC)

Next, let us see how Laplace's approximation can motivate BIC. Consider a model $\mathcal{M}$ involving k parameters to be estimated, which are represented by a vector $\boldsymbol{\theta} \in \chi \subseteq \mathbb{R}^k$. Then for the observations $\boldsymbol{x} := (x_1, x_2, \ldots, x_n)^\top$, we can compute their marginal likelihood under the model as follows:

$$f_{\mathcal{M}}(\boldsymbol{x}) = \int_{\chi} f_{\mathcal{M}}(\boldsymbol{x}|\boldsymbol{\theta})\pi_{\mathcal{M}}(\boldsymbol{\theta})d\boldsymbol{\theta},$$

where $f_{\mathcal{M}}(\boldsymbol{x}|\boldsymbol{\theta})$ is the joint likelihood of the observations, and $\pi_{\mathcal{M}}(\boldsymbol{\theta})$ is the prior distribution of the parameters. This can be expressed in the form of (12.20) with $h(\boldsymbol{\theta}) := \pi_{\mathcal{M}}(\boldsymbol{\theta})$ and $g_n(\boldsymbol{\theta}) := \frac{1}{n}\ln f_{\mathcal{M}}(\boldsymbol{x}|\boldsymbol{\theta})$. Putting this together leads to $q_n(\boldsymbol{\theta}) = \frac{1}{n}\ln f_{\mathcal{M}}(\boldsymbol{x}|\boldsymbol{\theta}) + \frac{1}{n}\ln \pi_{\mathcal{M}}(\boldsymbol{\theta})$, whose maximizer $\hat{\boldsymbol{\theta}}_{\mathcal{M}}$ is exactly the maximum a posteriori probability (MAP) estimate as it also maximizes

$$f_{\mathcal{M}}(\boldsymbol{\theta}|\boldsymbol{x}) \propto f_{\mathcal{M}}(\boldsymbol{x}|\boldsymbol{\theta})\pi_{\mathcal{M}}(\boldsymbol{\theta}) = e^{nq_n(\boldsymbol{\theta})}.$$

As $n \to \infty$, $q_n(\boldsymbol{\theta})$ is dominated by the first term corresponding to the likelihood since the prior distribution is independent of the sample size n, hence $\hat{\boldsymbol{\theta}}_{\mathcal{M}}$ approaches the MLE $\hat{\boldsymbol{\theta}}_{\mathrm{MLE}}$ in general. This allows us to replace $\hat{\boldsymbol{\theta}}_{\mathcal{M}}$ by the more easily computable quantity $\hat{\boldsymbol{\theta}}_{\mathrm{MLE}}$ in Laplace's approximation, and still obtain meaningful estimates. Now, by applying (12.21) and replacing $\hat{\boldsymbol{\theta}}_{\mathcal{M}}$ by $\hat{\boldsymbol{\theta}}_{\mathrm{MLE}}$, we have:

$$\begin{aligned} f_{\mathcal{M}}(\boldsymbol{x}) = \int h(\boldsymbol{\theta})e^{ng_n(\boldsymbol{\theta})}d\boldsymbol{\theta} &\approx \left(\frac{2\pi}{n}\right)^{\frac{k}{2}} \frac{h(\hat{\boldsymbol{\theta}}_{\mathrm{MLE}})e^{ng_n(\hat{\boldsymbol{\theta}}_{\mathrm{MLE}})}}{|-\nabla_{\boldsymbol{\theta}}\nabla_{\boldsymbol{\theta}}^\top q_n(\hat{\boldsymbol{\theta}}_{\mathrm{MLE}})|^{\frac{1}{2}}} \\ &\approx \hat{L}\left(\frac{2\pi}{n}\right)^{\frac{k}{2}} \pi_{\mathcal{M}}(\hat{\boldsymbol{\theta}}_{\mathrm{MLE}})|-\nabla_{\boldsymbol{\theta}}\nabla_{\boldsymbol{\theta}}^\top q_n(\hat{\boldsymbol{\theta}}_{\mathrm{MLE}})|^{-\frac{1}{2}}, \end{aligned}$$

where $\hat{L} := e^{ng_n(\hat{\boldsymbol{\theta}}_{\mathrm{MLE}})} = f_{\mathcal{M}}(\boldsymbol{x}|\hat{\boldsymbol{\theta}}_{\mathrm{MLE}})$ is the maximized likelihood. Therefore, the comparison among the candidate models $\mathcal{M}$'s reduces to comparing the marginal likelihood $f_{\mathcal{M}}(\boldsymbol{x})$ of the observed data $\boldsymbol{x}$ under each of the models. We should select the one with the largest marginal likelihood, or equivalently the largest log-marginal likelihood $\ln f_{\mathcal{M}}(\boldsymbol{x})$, which is approximately:

$$\begin{aligned} \ln f_{\mathcal{M}}(\boldsymbol{x}) &\approx \ln \hat{L} - \frac{k}{2}\ln n + \left(\frac{k}{2}\ln(2\pi) + \ln \pi_{\mathcal{M}}(\hat{\boldsymbol{\theta}}_{\mathrm{MLE}}) - \frac{1}{2}\ln|-\nabla_{\boldsymbol{\theta}}\nabla_{\boldsymbol{\theta}}^\top q_n(\hat{\boldsymbol{\theta}}_{\mathrm{MLE}})|\right) \\ &= -\frac{1}{2}(-2\ln \hat{L} + k\ln n) + O_p(1). \end{aligned}$$

This explains why $\mathrm{BIC} := -2\ln \hat{L} + k\ln n$ is used as a criterion for model selection under the Bayesian framework; indeed, the model with the smallest BIC has the largest log-marginal likelihood, and so should be favourably selected. For a more detailed discussion, we refer to [14] and references therein.

REFERENCES

1. Barndorff-Nielsen, O.E., and Cox, D.R. (1989). *Asymptotic Techniques for Use in Statistics*. Chapman & Hall.
2. Bühlmann, H., and Gisler, A. (2005). *A Course in Credibility Theory and its Applications*. Springer.
3. Catlett, J. (1991). On changing continuous attributes into ordered discrete attributes. In *European Working Session on Learning* (pp. 164–178). Springer.
4. Chen, Y., Cheung, K.C., Choi, H.M.C., and Yam, S.C.P. (2020). Evolutionary credibility risk premium. *Insurance: Mathematics and Economics*, 93, 216–229.
5. Chen, Y., Cheung, K.C., Fan, N.S., Sethi, S.P., and Yam, S.C.P. (2024+). Comonotone-independence Bayes classifier with applications in finance and insurance. Working Paper.
6. Dhaene, J., Denuit, M., Goovaerts, M.J., Kaas, R., and Vyncke, D. (2002). The concept of comonotonicity in actuarial science and finance: theory. *Insurance: Mathematics and Economics*, 31(1), 3–33.
7. Dhaene, J., Vanduffel, S., Goovaerts, M.J., Kaas, R., Tang, Q., and Vyncke, D. (2006). Risk measures and comonotonicity: a review. *Stochastic Models*, 22(4), 573–606.
8. Dougherty, J., Kohavi, R., and Sahami, M. (1995). Supervised and unsupervised discretization of continuous features. In *Machine Learning Proceedings 1995* (pp. 194–202). Morgan Kaufmann.
9. Fayyad, U., and Irani, K. (1993). Multi-interval discretization of continuous-valued attributes for classification learning. In *13th International Joint Conference on Artificial Intelligence*, 93(2), 1022–1029.
10. Jewell, W. (1974). Credible means are exact Bayesian for exponential families. *ASTIN Bulletin*, 8(1), 77–90.
11. Kaas, R., Goovaerts, M., Dhaene, J., and Denuit, M. (2008). *Modern Actuarial Risk Theory: Using R* (2nd ed.). Springer.
12. Kaufman, L., and Rousseeuw, P. J. (1990). *Finding Groups in Data: An Introduction to Cluster Analysis*. Wiley.
13. Kendall, M.G. (1938). A new measure of rank correlation. *Biometrika*, 30(1/2), 81–93.
14. Konishi, S., and Kitagawa, G. (2008). *Information Criteria and Statistical Modelling*. Springer Science & Business Media.
15. Kullback, S. (1997). *Information Theory and Statistics*. Courier Corporation.
16. Lehmann, E.L., and Casella, G. *Theory of Point Estimation* (2nd ed.). Springer.
17. Pearson, K. (1895). Notes on regression and inheritance in the case of two parents. *Proceedings of the Royal Society of London*, 58, 240–242.
18. Rathi, P., and Arpan, M. (2019). Insurance company dataset. https://www.kaggle.com/datasets/prakharrathi25/insurance-company-dataset (accessed 15 May 2022).
19. Spearman, C. (1904). The proof and measurement of association between two things. *The American Journal of Psychology*, 15(1), 72–101.
20. Thomas, K. (2021). Bank churn data exploration and churn prediction. https://www.kaggle.com/code/thomaskonstantin/bank-churn-data-exploration-and-churn-prediction/data (accessed 15 May 2022).
21. Witten, I.H., Frank, E., Hall, M.A., and Pal, C.J. (2016). *Data Mining: Practical Machine Learning Tools and Techniques* (4th ed.). Morgan Kaufmann.
22. Young, G.A., and Smith, R.L. (2005). *Essentials of Statistical Inference*. Cambridge University Press.

CHAPTER 13

Classification and Regression Trees, and Random Forests

13.1 CLASSIFICATION (DECISION) TREES

The binary logistic regression and the multinomial logit model introduced in Chapter 10 aim to classify observations based on linear combinations of feature variables. To common practitioners, especially for regulatory purposes, these models may be difficult to interpret, and it is desirable to have more transparent yet simple classification rules. *Classification (Decision) Tree* is a greedy algorithmic method to generate a set of simple rules that mechanically classify observations into different categories. It is extremely useful in data mining and has many potential financial applications. For example, it can be used to derive simple rules for credit approval decisions and determining financial stresses. Being one of the most popular methods in supervised learning, classification trees were first developed by Breiman, Friedman, Olshen, and Stone in their celebrated book [1]; from 1980 onwards, it has been an active research area, and some resulting developments can be found in [8]. It has been even more widely applied to various disciplines since the beginning of the 21st century, including but not limited to financial and insurance contexts; see [6, 7, 12, 15, 18] for example.

Let $\mathcal{S} = \{(x_i, y_i)\}_{i=1}^n$ be a dataset, in which $x_i = (x_i^{(1)}, x_i^{(2)}, \ldots, x_i^{(p)})^\top$ contains p feature or input variables and y_i is the corresponding label out from the set $\mathcal{Y} = \{c_1, \ldots, c_M\}$. We implicitly assume that whenever $x_i^{(j)} \equiv x_{i'}^{(j)}$ for all $j = 1, \ldots, p$, then $y_i = y_{i'}$. Furthermore, we denote the space of feature vectors by $\mathcal{D} := \prod_{j=1}^p \mathcal{R}(\mathrm{x}^{(j)})$, where $\mathcal{R}(\mathrm{x}^{(j)})$ is the range of all possible values taken by the feature variable $\mathrm{x}^{(j)}$. Mathematically, a classification tree partitions the space of feature vectors $\mathcal{D}$ into M disjoint subsets $\mathcal{D}_1, \ldots, \mathcal{D}_M$, which then induces the corresponding partition of $\mathcal{S}$ into $\mathcal{S}_1, \ldots, \mathcal{S}_M$, where

$$\mathcal{S}_k := \{(x, y) \in \mathcal{S} : x \in \mathcal{D}_k\}. \tag{13.1}$$

A classification tree is an acyclic graph, in which each internal node represents an attribute, which is characterized by certain quantitative relations specified by some feature components of $\mathbf{x}$, and the branches coming out from a node indicate the decision rule outcomes. Specifically, for simplicity, with a binary classification tree, each branching node of the tree examines a specific feature component of $\mathbf{x}$, say $x^{(j)}$: if the value of $x^{(j)}$ falls below a certain threshold $t^{(j)}$, then the left branch route is taken; otherwise, the right branch is taken. Going down the tree, as this procedure continues, the dataset is partitioned into smaller and smaller subsets. The model stops building any new branch, that is, it stops at a certain leaf node, if all the labels y_i's in that leaf are essentially in the same category. Such a leaf (terminal) node is then identified as giving a "pure" class label c_k in $\mathcal{Y}$. The route from the root to the leaf node indicates a classification rule. The depth of a tree is the maximum length of a path, that is, the maximum number of branches, from the root node to one leaf node. Naturally, one might ask the questions:

1. Which feature should be examined at each node, and based on what criterion should we choose that feature?
2. For a binary classification tree, what is the selection threshold value $t^{(j)}$ of each step? In general, if an attribute can take more than two values, or even a continuum, how many branches should we build?

These two issues will be addressed in this chapter; some of the discussions here are also extracted to our article [2] as a separate user guide.

Figure 13.1 shows a binary classification tree for a lady choosing her *Prince Charming*. The labels "Yes!" and "No" respectively indicate whether the lady would or would not consider that gentleman. The deciding attributes include the man's age, appearance, income and occupation. For instance, the lady would consider a man who is younger than 35, good looking and with high income.

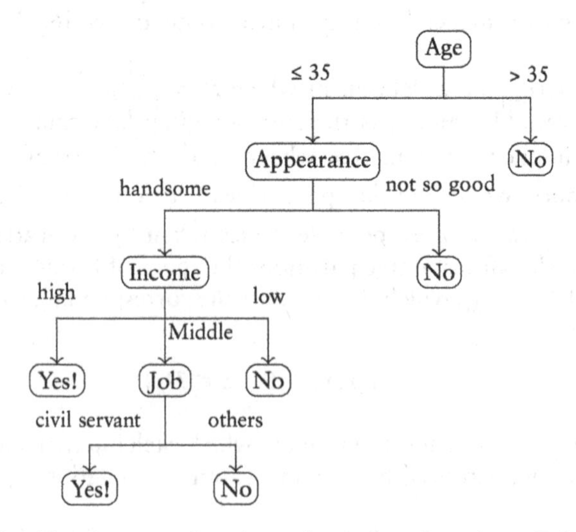

FIGURE 13.1 A classification tree showing criteria for choosing a Prince Charming.

*13.2 CONCEPTS OF ENTROPIES

To find classification rules that are highly accurate and have strong predictive power, we need to specify an appropriate measure for impurity, with which we select the suitable features for splitting at each node of the tree. In particular, we want to maximize the information gain after each splitting. So how to quantify information gain? Guided by the greedy algorithm, one approach is to maximize the reduction of entropy of the data after each splitting. The smaller the entropy value (on the average) after each splitting, the more the information gain is. Before diving into the details, we first introduce some commonly used information-theoretic quantities.

(I) Shannon and Differential Entropies

In 1948, Claude Shannon (see [13, 14]) introduced the concept of entropy from *thermodynamics* into *information theory*. In honour of him, the usual entropy used in the context of information is often called the *Shannon entropy* and is a common choice of measuring the concentration biasedness of a distribution other than a uniform distribution. In information theory, Shannon entropy is essential in defining the capacity of a communication channel, which measures the efficiency of the channel in information transmission.

Let x be a discrete random variable taking values in $\mathcal{X}$ with probability masses $p(x_i) = \mathbb{P}(\mathrm{x} = x_i)$ for finitely many i. Its (unconditional) Shannon entropy, denoted by $H(\mathrm{x})$, is defined as:

$$H(\mathrm{x}) := -\mathbb{E}(\log_2 p(\mathrm{x})) = -\sum_{x\in\mathcal{X}} p(x)\log_2 p(x), \tag{13.2}$$

where we adopt the convention of $0\log_2 0 = 0$. While the Shannon entropy is defined for random variables, it only depends on them through their distributions. Therefore, with a slight abuse of notation, we also talk about the entropy of a distribution. Empirically, given a dataset S, $p(x)$ can be estimated by the empirical frequency

$$p_S(x) = \frac{\text{number of data points in } S \text{ that are equal to } x}{\text{total number of data points in } S},$$

with which we can empirically estimate the entropy by $-\sum_{x\in\mathcal{X}} p_S(x)\log_2 p_S(x)$.

Clearly, if x is Bernoulli with $p = \mathbb{P}(\mathrm{x} = 1)$, its entropy is given by

$$H(\mathrm{x}) = -p\log_2 p - (1-p)\log_2(1-p). \tag{13.3}$$

Consider a sample S with 20 observations in which 16 of them belong to the first class ($\mathrm{x} = 0$) and the remaining 4 belong to the second class ($\mathrm{x} = 1$). With the empirical probability masses $p_S(0) = 16/20$ and $p_S(4) = 4/20$, the empirical entropy of x is given by:

$$-\frac{16}{20}\log_2\left(\frac{16}{20}\right) - \frac{4}{20}\log_2\left(\frac{4}{20}\right) = 0.7219.$$

The unit of entropy taken here is a *bit*, i.e. 0 or 1. As entropy increases, the amount of useful distinguishable information reduces[1], and the randomness and chaotic information both increase. Note that these two types of information, namely chaotic information which quantifies randomness and useful information, are in vast contrast to each other. In this book, the term "information" will be frequently used, and we shall clarify its meaning in case of any possible ambiguity.

As a remark, the concept of Shannon entropy has an intimate connection with Fisherian statistical inference. Given a random sample $x_1, \ldots, x_n$ of size n from a discrete random variable x, the likelihood is $\prod_{i=1}^{n} p(x_i)$, hence the average negative of log-likelihood of the sample is $-\frac{1}{n}\sum_{i=1}^{n} \ln p(x_i)$. By the weak law of large numbers, as the sample size n goes to infinity, the limit of this average is $-\mathbb{E}(\ln p(\mathrm{x})) = -\sum_{x\in\mathcal{X}} p(x)\ln p(x)$, which is exactly $H(\mathrm{x})\cdot\ln 2$, thus the likelihood is asymptotically equal to $\left(\frac{1}{2^{H(\mathrm{x})(1+o_p(1))}}\right)^n$. In addition, in information theory there is a strong motivation to use entropy for data compression. To compress the information stored in an n-dimensional random sample $(x_1, \ldots, x_n) \in \mathcal{X}^n$ of a common random variable x by using binary codes, on average with n large enough, it suffices to use $nH(\mathrm{x})$ bits. See Appendix 13.A.1 for more details.

From the definition, it is always clear that $H(\mathrm{x}) \geq 0$. If x is degenerate, then $H(\mathrm{x}) = 0$, and there is no ambiguity in x. If x is Bernoulli taking values 0 or 1, we have $H(\mathrm{x}) \in [0, 1]$, and $H(\mathrm{x}) = 1$ if and only if it is equally likely for x to take 0 and 1. In general, if x is a discrete random variable taking n different values, with the corresponding probabilities p_i for $i = 1, 2, \ldots, n$, the Shannon entropy is maximized when $p_i \equiv \frac{1}{n}$ for all i, and the corresponding entropy $H(\mathrm{x})$ is $\log_2 n$. This claim follows by an application of Jensen's inequality[2].

Next, if $\mathrm{x} \in \mathcal{X}$ is a continuous random variable with a continuous density function $f(x)$, the corresponding Shannon entropy is refined as the so-called *differential entropy*[3] (see [13]):

$$h(\mathrm{x}) := -\mathbb{E}(\ln f(\mathrm{x})) = -\int_{\mathcal{X}} f(x)\ln f(x)dx. \tag{13.4}$$

---

[1] In fields such as communication and signal processing, the term "information" quantifies the level of uncertainty, and the (chaotic) information increases with the entropy. For instance, by simple calculus, (13.3) attains its maximum value of 1 when $p = \frac{1}{2}$. Nevertheless, in the financial context, it is more natural to define the term otherwise, i.e., the (useful) information increases as the level of randomness becomes lower; indeed, it is least accurate to predict the outcome from tossing a fair coin, yet the more biased the coin is, the more certain one can predict, i.e. the more useful information in predicting the outcome.

[2] In particular, Jensen's inequality states that for a convex function ϕ and a random variable y, $\phi(\mathbb{E}(\mathrm{y})) \leq \mathbb{E}(\phi(\mathrm{y}))$. If we set $\phi(x) = -\log_2 x$, which is convex, and $\mathrm{y} = \frac{1}{p(\mathrm{x})}$ for a discrete random variable x, then $\mathbb{E}(\mathrm{y}) = \sum_{i=1}^{n} p_i \cdot \frac{1}{p_i} = n$, and $H(\mathrm{x}) = \mathbb{E}\left(\log_2\left(\frac{1}{p(\mathrm{x})}\right)\right) \leq \log_2\left(\mathbb{E}\left(\frac{1}{p(\mathrm{x})}\right)\right) = \log_2 n$, where the equality holds if and only if $p_i \equiv \frac{1}{n}$ for all i.

[3] As a common practice, the differential entropy is defined in terms of natural logarithm instead of binary logarithm as used in the Shannon entropy for the discrete cases.

However, differential entropy lacks a number of properties that Shannon entropy possesses, such as non-negativity and scale invariance, hence differential entropy is not a direct generalization of Shannon entropy. For instance, consider $\mathrm{x} \sim \mathcal{N}(\mu, \sigma^2)$, then

$$
\begin{aligned}
h(\mathrm{x}) &= -\int_{-\infty}^{\infty} \frac{1}{\sqrt{2\pi\sigma^2}} e^{-\frac{(x-\mu)^2}{2\sigma^2}} \left(-\frac{1}{2}\ln(2\pi\sigma^2) - \frac{(x-\mu)^2}{2\sigma^2}\right) dx \\
&= \frac{1}{2}\ln(2\pi\sigma^2) + \frac{\mathbb{E}((\mathrm{x}-\mu)^2)}{2\sigma^2} = \frac{1}{2}[1 + \ln(2\pi\sigma^2)],
\end{aligned}
\tag{13.5}
$$

which is clearly negative if $\sigma^2 < \frac{1}{2\pi e}$. Besides, when $\sigma^2 \to 0$, x becomes degenerate at μ, while $h(\mathrm{x})$ goes to $-\infty$ but not 0 as in the case of Shannon entropy. For some distributions, differential entropy may even not exist: for example, for $k > 1$, consider the density function $f(x) = \ln(k)/(x(\ln x)^2)$ for $x > k$, or 0 otherwise. The differential entropy for this density is not finite at all:

$$
\begin{aligned}
h(\mathrm{x}) &= -\int_{k}^{\infty} \frac{\ln k}{x(\ln x)^2} \ln \frac{\ln k}{x(\ln x)^2} dx \\
&= -\ln(\ln k) + \int_{k}^{\infty} \ln x \cdot \frac{\ln k}{x(\ln x)^2} dx - \int_{k}^{\infty} \frac{\ln k}{x(\ln x)^2} \ln \frac{1}{(\ln x)^2} dx \\
&= -\ln(\ln k) + \int_{\ln k}^{\infty} t \cdot \frac{\ln k}{t^2} dt - \int_{\ln k}^{\infty} \frac{\ln k}{t^2} \ln \frac{1}{t^2} dt \\
&= -\ln(\ln k) + [\ln k \cdot \ln t]_{\ln k}^{\infty} + 2\ln(\ln k) + 2 = \infty.
\end{aligned}
$$

To illustrate the power of making educated guesses via entropy, in Appendix 13.A.2 we apply the concept to solving a *Wordle* puzzle.

(II) Conditional Entropy

While unconditional entropy can be of use when there is no additional knowledge of x, it is actually more common in practice that we often possess some more knowledge beforehand from other source variables. Intuitively, we expect that the level of randomness reduces with this extra knowledge, and so does the entropy. For instance, in a linguistic model to "predict" the appearance of the next text passages, the possible range of next words is much shrunk once the current words are known. This concept can be made formal as follows. Let us consider the simplest case with only a pair of discrete random variables (x, y) with a joint probability mass function $\mathbb{P}(\mathrm{x} = x, \mathrm{y} = y) =: p(x, y)$ for $x \in \mathcal{X}$ and $y \in \mathcal{Y}$. We want to examine how knowing a piece of knowledge of x affects the randomness of y, which in turn changes the entropy of y. Using the notion of conditional probability of $p(y|x) = \mathbb{P}(\mathrm{y} = y|\mathrm{x} = x)$, we define the *conditional entropy* of y given x as:

$$
H(\mathrm{y}|\mathrm{x}) := -\sum_{x\in\mathcal{X}} \sum_{y\in\mathcal{Y}} p(x, y) \log_2 p(y|x).
$$

For the special case when y is completely determined by x, that is, $y = f(x)$ for some known function f, $p(y|x)$ takes the value 1 when $y = f(x)$; or 0 otherwise. As a result, $H(y|x) = -\sum_{x\in\mathcal{X}}\sum_{y=f(x)} p(x,y)\log_2 p(y|x) = -\sum_{x\in\mathcal{X}} 0 = 0$. On the other hand, from the definition of $H(y|x)$, we can express $p(x,y)$ in terms of the conditional probability $p(y|x)$ and then interchange the order of summations to obtain:

$$H(y|x) = -\sum_{x\in\mathcal{X}}\sum_{y\in\mathcal{Y}} p(x,y)\log_2 p(y|x)$$

$$= -\sum_{y\in\mathcal{Y}}\left(\sum_{x\in\mathcal{X}} p(x)p(y|x)\log_2 p(y|x)\right).$$

Denote $\phi(x) := x\log_2 x$ for $x > 0$. The inner summation can also be expressed as $\mathbb{E}(\phi(p(y|x)))$ for each y. Since $\phi'(x) = \frac{\ln x+1}{\ln 2}$ and $\phi''(x) = \frac{1}{x\ln 2} > 0$ for all $x > 0$, $\phi(x)$ is convex, which allows us to apply Jensen's inequality to conclude that:

$$H(y|x) = -\sum_{y\in\mathcal{Y}}\mathbb{E}(\phi(p(y|x))) \le -\sum_{y\in\mathcal{Y}}\phi(\mathbb{E}(p(y|x)))$$

$$= \sum_{y\in\mathcal{Y}} p(y)\log_2 p(y) = H(y), \tag{13.6}$$

which echoes our intuition that any extra information will reduce the randomness or entropy of the original random variable.

Similar to the manner above, if y is a continuous variable while x is another one of any type with a distribution function $F_x(x)$, we can use the conditional density $f(y|x)$ of y given x to define the *conditional differential entropy* of y given x as:

$$h(y|x) := -\int_{\mathcal{X}}\int_{\mathcal{Y}} f(y|x)\ln f(y|x)dydF_x(x).^4$$

For example, consider the bivariate normal pair x, y so that each has the common marginal distribution $\mathcal{N}(\mu,\sigma^2)$, and they also have a correlation coefficient $\rho \in (-1,1)$; clearly, $y|x = x \sim \mathcal{N}(\mu(1-\rho)+\rho x, \sigma^2(1-\rho^2))$, then the conditional differential entropy $h(y|x)$ can be obtained from (13.5) as:

$$h(y|x) = \frac{1}{2}\left[1+\ln(2\pi\sigma^2(1-\rho^2))\right],$$

which is less than $h(y) = \frac{1}{2}\left[1+\ln(2\pi\sigma^2)\right]$. Again, this demonstrates a reduction in entropy.

[4] For the sake of consistency with the concept of the unconditional differential entropy, we also use the natural logarithm instead of binary logarithm in this definition.

To further motivate the concept of conditional entropy, we consider two examples that relate to mixed distributions. Firstly, consider $\mathrm{x}|\mu \sim \mathcal{N}(\mu, \sigma^2)$ and $\mu \sim \mathcal{N}(\mu_0, \sigma_0^2)$, then the conditional distribution (posterior) of $\mu|\mathrm{x} = x$ is given by

$$\begin{aligned}
\pi(\mu|x) &\propto f(x|\mu) \times \pi(\mu) \\
&= \frac{1}{\sqrt{2\pi\sigma^2}} \exp\left(-\frac{(x-\mu)^2}{2\sigma^2}\right) \cdot \frac{1}{\sqrt{2\pi\sigma_0^2}} \exp\left(-\frac{(\mu-\mu_0)^2}{2\sigma_0^2}\right) \\
&\propto \exp\left(-\frac{(\sigma_0^2+\sigma^2)(\mu - (\sigma_0^2 x + \sigma^2\mu_0)/(\sigma_0^2+\sigma^2))^2}{2\sigma_0^2\sigma^2}\right).
\end{aligned}$$

Therefore, $\mu|\mathrm{x} = x$ also follows a normal distribution with mean and variance respectively given by:

$$\mathbb{E}(\mu|\mathrm{x} = x) = \frac{\sigma_0^2 x + \sigma^2\mu_0}{\sigma_0^2 + \sigma^2} \qquad \text{and} \qquad \mathrm{Var}(\mu|\mathrm{x} = x) = \frac{\sigma_0^2\sigma^2}{\sigma_0^2+\sigma^2}.$$

Hence, by (13.5) again, the conditional differential entropy is equal to:

$$h(\mu|\mathrm{x}) = \frac{1}{2}\left[1 + \ln\left(\frac{2\pi\sigma_0^2\sigma^2}{\sigma_0^2+\sigma^2}\right)\right].$$

Secondly, consider $\mathrm{N}|\lambda \sim \mathrm{Poi}(\lambda)$ and $\lambda \sim \mathrm{Exp}(\theta)$, then the posterior density of $\lambda|\mathrm{N} = n$ is given by

$$\pi(\lambda|n) \propto f(n|\lambda)\pi(\lambda) = \frac{e^{-\lambda}\lambda^n}{n!} \cdot \theta e^{-\theta\lambda} \propto e^{-\lambda(1+\theta)}\lambda^n;$$

so $\lambda|\mathrm{N} = n$ follows a Gamma distribution with a shape parameter $\alpha = n+1$ and a rate $\beta = 1+\theta$. On the other hand, in general, if $\mathrm{x} \sim \mathrm{Gamma}(\alpha, \beta)$ with the density $f_\Gamma(x) = \frac{\beta^\alpha}{\Gamma(\alpha)}x^{\alpha-1}e^{-\beta x}$ for $x > 0$, its entropy is given by:

$$\begin{aligned}
H(\mathrm{x}) &= -\int_0^\infty f(x) \ln\left\{\frac{\beta^\alpha}{\Gamma(\alpha)}x^{\alpha-1}e^{-\beta x}\right\} dx \\
&= -\ln\frac{\beta^\alpha}{\Gamma(\alpha)} - (\alpha-1)\int_0^\infty \ln x \cdot f(x)dx + \beta\int_0^\infty xf(x)dx \\
&= \ln\Gamma(\alpha) - \ln\beta + (1-\alpha)\psi(\alpha) + \alpha,
\end{aligned}$$

where $\psi(\alpha) := \frac{\Gamma'(\alpha)}{\Gamma(\alpha)}$ is the *Digamma function*, using the fact that:

$$\int_0^\infty \ln x \cdot f(x)dx = \int_0^\infty \ln x \cdot \frac{\beta^\alpha}{\Gamma(\alpha)} x^{\alpha-1} e^{-\beta x} dx = \int_{-\infty}^\infty y \cdot \frac{\beta^\alpha}{\Gamma(\alpha)} e^{\alpha y} e^{-\beta e^y} dy$$
$$= \frac{\beta^\alpha}{\Gamma(\alpha)} \frac{\partial}{\partial \alpha} \left(\int_{-\infty}^\infty e^{\alpha y - \beta e^y} dy \right) = \frac{\beta^\alpha}{\Gamma(\alpha)} \frac{\partial}{\partial \alpha} \left(\frac{\Gamma(\alpha)}{\beta^\alpha} \right)$$
$$= \frac{\beta^\alpha}{\Gamma(\alpha)} \frac{\beta^\alpha \Gamma'(\alpha) - \Gamma(\alpha)\beta^\alpha \ln \beta}{\beta^{2\alpha}} = \psi(\alpha) - \ln \beta.$$

Therefore, as n is an integer, the conditional differential entropy of $\lambda|N$ is:

$$h(\lambda|N) = -\ln \frac{(1+\theta)^{n+1}}{\Gamma(n+1)} - n(\ln(1+\theta) + \psi(n+1)) + n + 1$$
$$= -\ln \frac{(1+\theta)^{n+1}}{n!} - n(\ln(1+\theta) + H_n - \gamma) + n + 1,$$

where $H_n := \sum_{k=1}^n \frac{1}{k}$ is the truncated harmonic series and γ is *Euler's constant*, and the last equality holds by utilizing the infinite product representation of the gamma function and considering its logarithmic derivative.

Finally, we note that the conditional entropy $H(\mathrm{y}|\mathrm{x})$ can be expressed as $H((\mathrm{y}, \mathrm{x})) - H(\mathrm{x})$. Here we only give the proof for the discrete case, leaving the continuous case to readers:

$$H(\mathrm{y}|\mathrm{x}) = -\sum_{x\in\mathcal{X}} \sum_{y\in\mathcal{Y}} p(x,y) \log_2 p(y|x) = -\sum_{x\in\mathcal{X}} \sum_{y\in\mathcal{Y}} p(x,y) \log_2 \frac{p(x,y)}{p(x)}$$
$$= -\sum_{x\in\mathcal{X}} \sum_{y\in\mathcal{Y}} p(x,y) \log_2 p(x,y) + \sum_{x\in\mathcal{X}} \sum_{y\in\mathcal{Y}} p(x,y) \log_2 p(x)$$
$$= -\sum_{x\in\mathcal{X}} \sum_{y\in\mathcal{Y}} p(x,y) \log_2 p(x,y) + \sum_{x\in\mathcal{X}} p(x) \log_2 p(x)$$
$$= H((\mathrm{y}, \mathrm{x})) - H(\mathrm{x}). \tag{13.7}$$

(III) Mutual Information

Mutual information has a standalone position in information theory as it measures the mutual dependence between two distributions represented by two random variables of x and y. It is defined as $I(\mathrm{x}, \mathrm{y}) := H(\mathrm{y}) - H(\mathrm{y}|\mathrm{x})$, yet it is symmetric, i.e. $I(\mathrm{x}, \mathrm{y}) = H(\mathrm{x}) - H(\mathrm{x}|\mathrm{y}) = I(\mathrm{y}, \mathrm{x})$; indeed, from (13.7), we have $I(\mathrm{x}, \mathrm{y}) := H(\mathrm{y}) - H(\mathrm{y}|\mathrm{x}) = H(\mathrm{y}) - H((\mathrm{x}, \mathrm{y})) + H(\mathrm{x}) = H(\mathrm{x}) - H((\mathrm{x}, \mathrm{y})) + H(\mathrm{y}) = H(\mathrm{x}) - H(\mathrm{x}|\mathrm{y}) =: I(\mathrm{y}, \mathrm{x})$.

Based on the properties of entropy and conditional entropy, we can obtain the upper and lower bounds for mutual information, namely $0 \le I(\mathrm{x}, \mathrm{y}) \le H(\mathrm{y})$ since

$H(y) \geq H(y|x) \geq 0$ as shown in (13.6). In particular, the upper bound of $I(x, y) = H(y)$ can be achieved when y is completely determined by x, by then $H(y|x) = 0$; while the lower bound of $I(x, y) = 0$ is achieved when x and y are independent, giving $H(y|x) = H(y)$.

(IV) Relative Entropy and Cross Entropy

Relative entropy and *cross entropy* are common measures of difference between two probability distributions; however, they are not metric, though they still preserve some of the metric properties, for example nonnegativity[5].

i) Relative entropy:

Let P and Q be two discrete distributions (also identify them as their own probability mass functions) defined on a common discrete support $\mathcal{X}$,[6] the *relative entropy* (a.k.a. *Kullback–Leibler divergence*; see [9]) from Q to P is defined as:

$$\mathbb{D}(P\|Q) := -\mathbb{E}^P\left(\ln\left(\frac{Q(x)}{P(x)}\right)\right) = -\sum_{x\in\mathcal{X}} P(x)\ln\left(\frac{Q(x)}{P(x)}\right),$$

where $\mathbb{E}^P$ denotes the expectation with respect to P.

As an example, consider P and Q following Poisson distributions with rates λ and θ, respectively. The relative entropy from Q to P is

$$\begin{aligned}\mathbb{D}(P\|Q) &= -\sum_{x=0}^{\infty} P(x)\ln\left(\frac{\theta^x e^{-\theta}}{x!}\cdot\frac{x!}{\lambda^x e^{-\lambda}}\right)\\ &= -\sum_{x=0}^{\infty} P(x)\ln(e^{-(\theta-\lambda)}(\theta/\lambda)^x)\\ &= (\theta-\lambda) - \ln(\theta/\lambda)\sum_{x=0}^{\infty} xP(x)\\ &= \theta - \lambda - \lambda\ln(\theta/\lambda).\end{aligned}$$

[5] To see $\mathbb{D}(P\|Q) \geq 0$, we only need a simple application of Jensen's inequality to the convex function $-\ln x$:

$$\mathbb{D}(P\|Q) = \mathbb{E}^P\left(-\ln\left(\frac{Q(x)}{P(x)}\right)\right) \geq -\ln\left(\mathbb{E}^P\left(\frac{Q(x)}{P(x)}\right)\right) = -\ln\left(\sum_{x\in\mathcal{X}} Q(x)\right) = 0.$$

[6] If the supports of P and Q are $\mathcal{X}_P$ and $\mathcal{X}_Q$, respectively, we can simply set $\mathcal{X} := \mathcal{X}_P \cup \mathcal{X}_Q$.

As another example, for $P = \text{Bin}(n, \alpha)$ and $Q = \text{Bin}(n, \beta)$, the relative entropy from Q to P is

$$\begin{aligned}\mathbb{D}(P\|Q) &= -\sum_{x=0}^{n} P(x)\ln\left(\frac{n!}{x!(n-x)!}\cdot\frac{x!(n-x)!}{n!}\cdot\frac{\beta^x(1-\beta)^{n-x}}{\alpha^x(1-\alpha)^{n-x}}\right)\\ &= -\ln\left(\frac{\beta}{\alpha}\right)\cdot\sum_{x=0}^{n} xP(x) + \ln\left(\frac{1-\beta}{1-\alpha}\right)\sum_{x=0}^{n}(x-n)P(x)\\ &= n\alpha\ln\left(\frac{\alpha}{\beta}\right) + n(\alpha-1)\ln\left(\frac{1-\beta}{1-\alpha}\right).\end{aligned}$$

From these examples, we clearly see that the definition of relative entropy is generally asymmetric, violating the usual commutative property of a metric. Nevertheless, when $\alpha = 1 - \beta$ in the Binomial case, $\mathbb{D}(P\|Q) = \mathbb{D}(Q\|P)$ can hold.

Analogously, if P and Q are two continuous distributions (again, we identify their densities with the same symbols) defined on a common support $\mathcal{X}$, the *relative entropy (Kullback–Leibler divergence)* from Q to P is defined as, by replacing summation by integration,

$$\mathbb{D}(P\|Q) := -\mathbb{E}^P\left(\ln\left(\frac{Q(x)}{P(x)}\right)\right) = -\int_{x\in\mathcal{X}} P(x)\ln\left(\frac{Q(x)}{P(x)}\right)dx.$$

For the first example, let P follow a d-variate normal distribution with a mean vector $\boldsymbol{\mu}_P$ and a covariance matrix $\boldsymbol{\Sigma}_P$, while Q has the same Gaussian distribution but with a different mean vector $\boldsymbol{\mu}_Q$ and covariance matrix $\boldsymbol{\Sigma}_Q$. The relative entropy from Q to P is

$$\begin{aligned}\mathbb{D}(P\|Q) &= -\int_{-\infty}^{\infty} P(\boldsymbol{x})\ln\left[\frac{1}{(2\pi)^{d/2}|\boldsymbol{\Sigma}_Q|^{1/2}}\exp\left\{-\frac{1}{2}(\boldsymbol{x}-\boldsymbol{\mu}_Q)^\top\boldsymbol{\Sigma}_Q^{-1}(\boldsymbol{x}-\boldsymbol{\mu}_Q)\right\}\right]dx\\ &\quad+\int_{-\infty}^{\infty} P(\boldsymbol{x})\ln\left[\frac{1}{(2\pi)^{d/2}|\boldsymbol{\Sigma}_P|^{1/2}}\exp\left\{-\frac{1}{2}(\boldsymbol{x}-\boldsymbol{\mu}_P)^\top\boldsymbol{\Sigma}_P^{-1}(\boldsymbol{x}-\boldsymbol{\mu}_P)\right\}\right]dx\\ &= \frac{1}{2}\left(\ln\frac{|\boldsymbol{\Sigma}_Q|}{|\boldsymbol{\Sigma}_P|} - d + \text{tr}(\boldsymbol{\Sigma}_Q^{-1}\boldsymbol{\Sigma}_P) + (\boldsymbol{\mu}_P-\boldsymbol{\mu}_Q)^\top\boldsymbol{\Sigma}_Q^{-1}(\boldsymbol{\mu}_P-\boldsymbol{\mu}_Q)\right).\end{aligned}$$

In particular, when $\boldsymbol{\Sigma}_P = \boldsymbol{\Sigma}_Q$, the relative entropy from Q to P becomes larger when $\boldsymbol{\mu}_P$ and $\boldsymbol{\mu}_Q$ are farther apart. On the other hand, when $\boldsymbol{\mu}_P = \boldsymbol{\mu}_Q$, the direction of change whenever $\boldsymbol{\Sigma}_P \neq \boldsymbol{\Sigma}_Q$ is not as straightforward. Consider the one-dimensional case with variances σ_P^2 and σ_Q^2, and an arbitrary correlation coefficient ρ. The relative entropy from Q to P is then reduced to:

$$\mathbb{D}(P\|Q) = \frac{1}{2}\left(\frac{\sigma_P^2}{\sigma_Q^2} - \ln\frac{\sigma_P^2}{\sigma_Q^2} - 1\right),$$

which becomes larger if the ratio σ_P^2/σ_Q^2 is farther away from 1.

ii) Cross Entropy:
The *cross entropy* between two discrete distributions P and Q is defined as

$$H(P,Q) := H(P) + \mathbb{D}(P\|Q), \tag{13.8}$$

where $H(P) := -\sum_{x\in\mathcal{X}} P(x)\ln P(x)$ denotes the rescaled entropy of P. By simplifying (13.8), an equivalent definition is:

$$H(P,Q) = -\sum_{x\in\mathcal{X}} P(x)\ln Q(x).$$

As long as P is fixed, $H(P)$ is a constant independent of Q, so there is an equivalence between relative entropy and cross entropy. A similar notion of entropy can be defined for continuous distributions by replacing summation by integration. In real applications, cross entropy is often served as an objective function in deep learning to measure the similarities between the sample distribution of labels and the predicted distribution.

13.3 INFORMATION GAIN

Information gain plays a crucial role in selecting suitable attributes for a classification tree whenever it branches out. The entropy of the empirical distribution of observation is calculated at each decision node, and one decides whether the node should be split further by considering whether its information gain is sufficiently large enough, i.e., if the reduction of randomness (measured by entropy) after splitting is significant or not. At each node, among the p features, one $\mathrm{x}^{(j)}$ will be selected if the splitting causes the biggest possible information gain, which is actually the mutual information between the label y of the subsample at the node and this testing attribute $\mathrm{x}^{(j)}$. More precisely, let $\mathrm{x}^{(j_i)}$ be the chosen attribute at a node i with a subsample $\mathcal{S}^{(i)}$ attaching there. The information gain $\mathrm{IG}(\mathcal{S}^{(i)}, \mathrm{x}^{(j_i)})$ (as a special example of a goodness measure to be defined in (13.13)) for $\mathrm{x}^{(j_i)}$, considered as the difference of entropies on average before and after splitting $\mathcal{S}^{(i)}$ with respect to $\mathrm{x}^{(j_i)}$, is defined as:

$$\begin{aligned}\mathrm{IG}(\mathcal{S}^{(i)}, \mathrm{x}^{(j_i)}) = I(\mathrm{y}^{(i)}, \mathrm{x}^{(j_i)}) &= H(\mathrm{y}^{(i)}) - H(\mathrm{y}^{(i)}|\mathrm{x}^{(j_i)})\\ &= H(\mathrm{y}^{(i)}) - \sum_{v\in\mathcal{V}(\mathrm{x}^{(j_i)})} p_{\mathrm{x}^{(j_i)}}(v) H(\mathrm{y}^{(i,v)})\\ &= H(\mathrm{y}^{(i)}) - \sum_{v\in\mathcal{V}(\mathrm{x}^{(j_i)})} \frac{|\mathcal{S}^{(i,v)}|}{|\mathcal{S}^{(i)}|} H(\mathrm{y}^{(i,v)}),\end{aligned} \tag{13.9}$$

where $\mathcal{V}(\mathrm{x}^{(j_i)})$ is the set of all possible values taken by the attribute $\mathrm{x}^{(j_i)}$, $\mathcal{S}^{(i,v)}$ is the collection of sample points from $\mathcal{S}^{(i)}$ with the feature $\mathrm{x}^{(j_i)}$ taking value $v \in \mathcal{V}(\mathrm{x}^{(j_i)})$; while $\mathrm{y}^{(i)}$ and $\mathrm{y}^{(i,v)}$ are the respective labels of the subsamples $\mathcal{S}^{(i)}$ and $\mathcal{S}^{(i,v)}$. Note that the unconditional entropy $H(\mathrm{y}^{(i,v)})$ in (13.9) is computed by using the conditional probabilities of

$$\hat{\mathbb{P}}\left(\mathrm{y}^{(i)} = u | \mathrm{x}^{(j'_n)} = v, \mathrm{x}^{(j'_{n-1})} = v_{j'_{n-1}}, \ldots, \mathrm{x}^{(j'_1)} = v_{j'_1}\right)$$

$$= \frac{\left|\left\{(x, y^{(i)}) \in \mathcal{S} : y^{(i)} = u, x^{(j'_n)} = v, x^{(j'_{n-1})} = v_{j'_{n-1}}, \ldots, x^{(j'_1)} = v_{j'_1}\right\}\right|}{\left|\left\{(x, y) \in \mathcal{S} : x^{(j'_n)} = v, x^{(j'_{n-1})} = v_{j'_{n-1}}, \ldots, x^{(j'_1)} = v_{j'_1}\right\}\right|}, \quad (13.10)$$

for all $u \in \mathcal{Y}$ and for every $v \in \mathcal{V}(\mathrm{x}^{(j'_n)})$, where $j'_n = j_i$ is the current node, $j'_1, j'_2, \ldots, j'_{n-1}$ are the nodes traversed before, namely from the root of j'_1 along all corresponding branches until to the present node j'_n, and $v_{j'_j}$ is the corresponding value of the attribute $\mathrm{x}^{(j'_j)}$ for $j = 1, \ldots, n-1$, taken at the j-th traversed node; we refer to Figure 13.2 for a pictorial illustration. In the following, to simplify notations, we may use an abused symbol of $H(\mathcal{S}^{(i,v)})$ in place of the more proper one of $H(\mathrm{y}^{(i,v)})$, meaning that we indicate only the subsample but not the corresponding labels without any cause for ambiguity, as this may facilitate comparisons among entropies of different subsamples.

As an example to illustrate the concept of information gain, we consider a node with a sample of size 15, $\mathcal{S}^{(1)} = \{(x_i^{(1)} = \mathrm{R}, x_i^{(2)} = \mathrm{S}), y_i \in \{\mathrm{N}, \mathrm{Y}\}\}_{i=1}^{15}$, of a credit default dataset, in which 9 of them did not default (N) on their loan, and the remaining 6 did (Y). The attribute $\mathrm{x}^{(1)}$ refers to the "Risk level" (R), which can be either "High risk" (H) or "Low risk" (L). The second attribute $\mathrm{x}^{(2)}$ is for "Sex" (S), which can be either "Female" (F) or "Male" (M). Among the 15 sample points, 6 sample points from Class N and 3 from Class Y have attribute $\mathrm{x}^{(1)} = \mathrm{L}$, and the remaining sample points fall into the high risk level category with $\mathrm{x}^{(1)} = \mathrm{H}$; on the other hand, 5 sample points from Class N and 4 from Class Y have attribute $\mathrm{x}^{(2)} = \mathrm{F}$, and the remaining sample points fall into the male category with $\mathrm{x}^{(2)} = \mathrm{M}$; see Table 13.1 for the labels and attributes of each of the 15 samples.

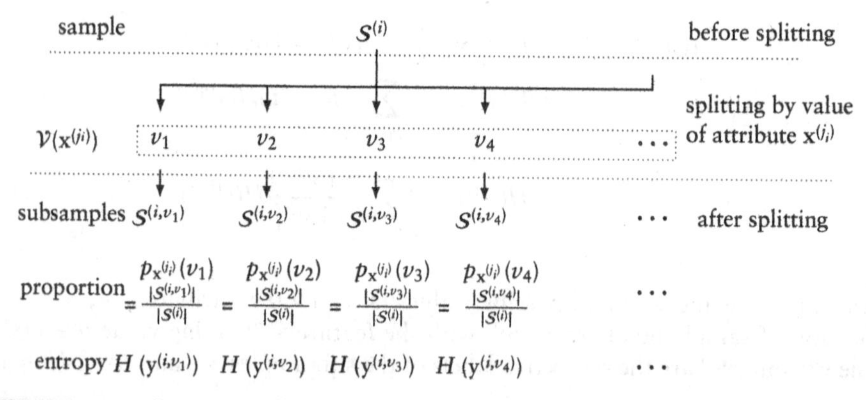

FIGURE 13.2 An illustration for splitting tree with a sample $\mathcal{S}$ at node i by using an attribute of $\mathrm{x}^{(j_i)}$.

TABLE 13.1 15 credit default samples with Class N or Y as their label y; attributes are "Risk level" $x^{(1)} = R$ and "Sex" $x^{(2)} = S$.

Index	Default	Risk level	Sex
1	N	L	F
2	N	L	F
3	N	L	M
4	N	L	M
5	N	L	M
6	N	L	M
7	N	H	F
8	N	H	F
9	N	H	F
10	Y	L	F
11	Y	L	M
12	Y	L	M
13	Y	H	F
14	Y	H	F
15	Y	H	F

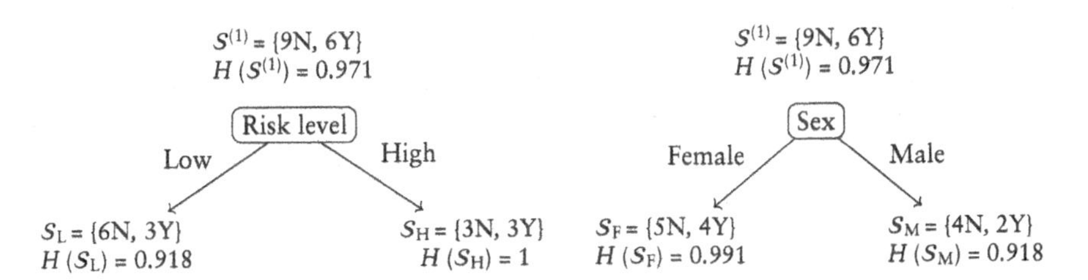

FIGURE 13.3 Selection of attribute $x^{(1)}$ or $x^{(2)}$ based on information gain.

Using the notations in Figure 13.3, the Shannon entropies are computed as follows:

$$H(S^{(1)}) = -\frac{9}{15}\log_2\left(\frac{9}{15}\right) - \frac{6}{15}\log_2\left(\frac{6}{15}\right) = 0.971;$$

$$H(S_{\mathrm{L}}) = -\frac{6}{9}\log_2\left(\frac{6}{9}\right) - \frac{3}{9}\log_2\left(\frac{3}{9}\right) = 0.918;$$

$$H(S_{\mathrm{H}}) = -\frac{3}{6}\log_2\left(\frac{3}{6}\right) - \frac{3}{6}\log_2\left(\frac{3}{6}\right) = 1;$$

$$H(S_{\mathrm{F}}) = -\frac{5}{9}\log_2\left(\frac{5}{9}\right) - \frac{4}{9}\log_2\left(\frac{4}{9}\right) = 0.991;$$

$$H(S_{\mathrm{M}}) = -\frac{4}{6}\log_2\left(\frac{4}{6}\right) - \frac{2}{6}\log_2\left(\frac{2}{6}\right) = 0.918,$$

and the information gains for two attributes are given by

$$\mathrm{IG}(S^{(1)}, \mathrm{R}) = H(S^{(1)}) - \frac{9}{15}H(S_{\mathrm{L}}) - \frac{6}{15}H(S_{\mathrm{H}}) = 0.020;$$

$$\mathrm{IG}(S^{(1)}, \mathrm{S}) = H(S^{(1)}) - \frac{9}{15}H(S_{\mathrm{F}}) - \frac{6}{15}H(S_{\mathrm{M}}) = 0.009.$$

Since $\mathrm{IG}(S^{(1)}, \mathrm{R}) > \mathrm{IG}(S^{(1)}, \mathrm{S})$, we conclude that "Risk level" is a better attribute for classifying the data into Class N or Y at this node. Interestingly, if we compare the entropies of the corresponding child nodes individually, we can see that $H(S_{\mathrm{H}}) > H(S_{\mathrm{F}})$ and $H(S_{\mathrm{L}}) = H(S_{\mathrm{M}})$. Based on these comparisons at child nodes, one might be tempted to conclude that the resulting entropy for the "Sex" attribute is lower than that of the "Risk level" attribute, suggesting the former can bring a higher information gain, which should be used for splitting. However, since a larger weight of $\frac{9}{15} = 0.6$ is assigned to S_{L} (with a relatively lower entropy) for the "Risk level" attribute, while the same larger weight is also assigned to S_{F} (with a relatively higher entropy) for the "Sex" attribute, it turns out that the actual average entropy for "Risk level" is lower than that of "Sex", and so "Risk level" is eventually selected due to a higher information gain. We now observe larger individual entropy values of two child nodes induced from "Risk level" than those of the corresponding ones induced from "Sex", yet after taking into account the proportion contribution of each child node, the combined effect (here the average entropy) for the "Risk level" becomes smaller than that of "Sex". That is a typical example of *Simpson's Paradox*, in which a trend appears in several different groups of data, but disappears or reverses when these groups are combined; also see [19].

13.4 OTHER IMPURITY MEASURES FOR INFORMATION

In addition to the various entropies and mutual information introduced before, here we introduce two additional commonly used measures of impurity.

1. *Gini-Index*
   Let x be a discrete random variable with a probability mass function $p(x)$ for $x \in \mathcal{X}$, the Gini-index G of x is defined as:

$$\mathrm{G(x)} := 1 - \sum_{x \in \mathcal{X}} p^2(x). \tag{13.11}$$

2. *Misclassification Error*
   The misclassification error of a discrete random variable x is defined as:

$$\text{Misclassification Error(x)} := 1 - \max_{x \in \mathcal{X}} p(x). \tag{13.12}$$

Figure 13.4 illustrates the respective behavior of entropy, Gini-index and misclassification error when x is a Bernoulli random variable with a success probability p. Simple calculus tells us that for all these impurity measures, the maxima are attained at $p = 1 - p = \frac{1}{2}$, when both outcomes are equally likely. At that point the corresponding node of a binary tree is at the "most impure" state.

As in the case for entropy, for a discrete random variable x now taking n possible values, all impurity measures attain their respective maximum when all the probability masses $p(x) \equiv \frac{1}{n}$, meaning that all outcomes are equally likely; the corresponding maximum values for various impurity measures are: (1) for entropy, it is $H(\mathrm{x}) = \log_2 n$; (2) for the Gini-index, it is $G(\mathrm{x}) = 1 - \frac{1}{n}$;[7] and (3) for misclassification error, it is $1 - \frac{1}{n}$ since $(\max_i p_i) \cdot n \geq \sum_{i=1}^{n} p_i = 1$.

As with maximizing information gain, to choose the best attribute for splitting a node, we generally compare the impurity of the parent node (before splitting) with the average of those of the child nodes (after splitting). Again we use $S^{(i)}$ to denote the subsample at the present node i. For an attribute $\mathrm{x}^{(j_i)}$, let $\mathcal{V}(\mathrm{x}^{(j_i)})$ be the set of all attribute values for $\mathrm{x}^{(j_i)}$, and $S^{(i,v)}$ be the subset of $S^{(i)}$ with the attribute $\mathrm{x}^{(j_i)}$ taking a value of $v \in \mathcal{V}(\mathrm{x}^{(j_i)})$. As for the case with entropy, we define the *goodness*[8]

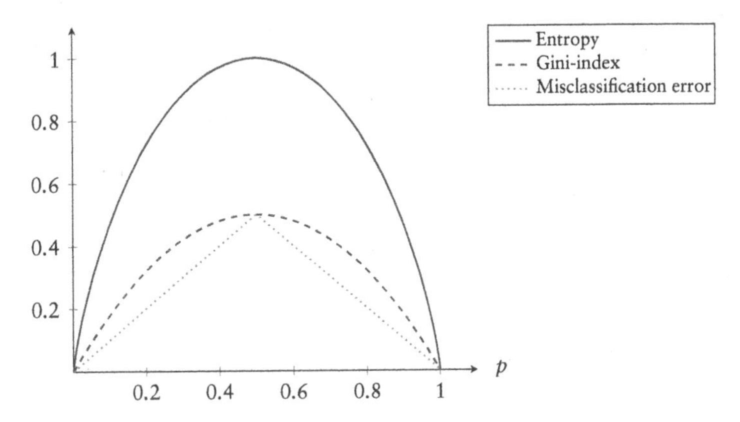

FIGURE 13.4 A graph of entropy, Gini-index and misclassification error as p varies.

[7] This can be shown using Lagrange multipliers: consider the Lagrangian $\overline{G}_\lambda = 1 - \sum_{i=1}^{n} p_i^2 - \lambda\left(\sum_{i=1}^{n} p_i - 1\right)$, then by the first-order condition of $\frac{\partial \overline{G}_\lambda}{\partial p_i} = -2p_i - \lambda = 0$ for $i = 1, \ldots, n$, we have $p_i \equiv \frac{1}{n}$ for all i by the constraint that $\sum_{i=1}^{n} p_i = 1$. Furthermore, since $\frac{\partial^2 \overline{G}_\lambda}{\partial p_i^2} = -2$ and $\frac{\partial^2 \overline{G}_\lambda}{\partial p_i \partial p_j} = 0$ for any $i, j = 1, \ldots, n$ and $i \neq j$, $\overline{G}_\lambda$ (as well as G without the Lagrangian penalty term) is strictly concave, hence $p_i \equiv \frac{1}{n}$ is the global maximizer for G.

[8] In fact, (13.13) is a version of a *Laplacian operator* Δ acting over the function of impurity measure Im. The Laplacian also plays a key role in the fundamental equation (for *Riemannian*

of an attribute $x^{(j_i)}$ by

$$\Delta \mathrm{Im}(x^{(j_i)}) = \mathrm{Im}(S^{(i)}) - \sum_{v \in \mathcal{V}(x^{(j_i)})} \frac{|S^{(i,v)}|}{|S^{(i)}|} \mathrm{Im}(S^{(i,v)}), \qquad (13.13)$$

where the function Im(·) is a pre-assigned impurity measure, such as one of the three introduced above. Again, each impurity measure is defined on the probability distributions of the labels, and the probability distribution in question here is the conditional one quoted in (13.10). Without causing much ambiguity, we slightly abuse the notation of $\mathrm{Im}(S^{(i,v)})$ to stand for the impurity measure of this conditional probability distribution of the labels from the subsample $S^{(i,v)}$. Furthermore, for the sake of convenience, we just identify the name of a node with the name of the subsample at the same node.

Once the impurity measure is chosen, it is consistently used throughout the construction of the whole tree (or even the whole random forest, see Section 13.9). We aim to choose an attribute $x^{(j_i)}$ such that $\Delta \mathrm{Im}(x^{(j_i)})$ is the largest at each parent node, or equivalently the attribute that reduces the impurity measure the most after splitting.

We now consider a simple example to illustrate the calculations involving these alternative impurity measures. We are given a dataset of size 12, in which 6 of them belong to the Group 1 and the remaining 6 belong to Group 2. We choose between two binary attributes A and B to split the root note into two child nodes based on the information in Table 13.2.

We first use Gini-index as the impurity measure. At the parent node,

$$G(\text{parent}) = 1 - \left(\frac{6}{12}\right)^2 - \left(\frac{6}{12}\right)^2 = 0.5.$$

At the child nodes, using the attribute A, we get:

$$G(N_1) = 1 - \left(\frac{4}{7}\right)^2 - \left(\frac{3}{7}\right)^2 = 0.4898, \quad G(N_2) = 1 - \left(\frac{2}{5}\right)^2 - \left(\frac{3}{5}\right)^2 = 0.48,$$

TABLE 13.2 Number of elements in Groups 1 and 2 after splitting by either attribute A or B.

	parent node (6,6)	parent node (6,6)
child node N_1	A = 0 (4,3) *	B = 0 (1,4) *
child node N_2	A = 1 (2,3) *	B = 1 (5,2) *

metric) modeling the flow (*Ricci flow*) of hot lava. Analogous to the lava flow that stops and solidifies when the change of its temperature becomes too small, for a classification tree, the branching procedure stops when the change in impurity measure induced by the subsequent splitting falls below a pre-determined tolerance level.

and hence the goodness of A is

$$\Delta G(A) = G(\text{parent}) - \frac{4+3}{12}G(N_1) - \frac{2+3}{12}G(N_2) = 0.014.$$

Similarly, for the attribute B, the Gini-indices at the two child nodes are respectively:

$$G(N_1) = 1 - \left(\frac{1}{5}\right)^2 - \left(\frac{4}{5}\right)^2 = 0.32, \quad G(N_2) = 1 - \left(\frac{5}{7}\right)^2 - \left(\frac{2}{7}\right)^2 = 0.408;$$

and so the goodness of B is

$$\Delta G(B) = G(\text{parent}) - \frac{1+4}{12}G(N_1) - \frac{5+2}{12}G(N_2) = 0.1285.$$

As a result, the attribute B is preferred to A since $\Delta G(B) = 0.1285 > 0.014 = \Delta G(A)$. If we repeat the calculation using entropy as the impurity measure, we get $\Delta G(A) = 0.0207$ and $\Delta G(B) = 0.1957$; again, the attribute B is preferable. The same conclusion can also be obtained by using misclassification error as the impurity measure.

13.5 SPLITTING AGAINST CONTINUOUS ATTRIBUTES

The calculation of $\Delta(\cdot)$ through a specific choice of impurity measure $\text{Im}(\cdot)$ can be easily extended to find the best splitting condition for a continuous attribute $x^{(j)}$, where we basically divide the feature space of $x^{(j)}$, its range of values, into a number of disjoint and consecutive intervals. Then, based on the distribution (now still probability mass) on all these intervals, every impurity measure at each of the different child nodes can be calibrated. As an example, consider Table 13.3, where values of taxable income are sorted, and by setting a fictitious initial value of 50 and terminal value of 240, the mid-points between two consecutive values are used as the splitting boundary points, and we denote them by m_k in order. Then we try to compute the goodness for each attribute of $\mathbb{1}_{\{\text{Taxable income}<m_k\}}$ for every boundary mid-point m_k; particularly, one can check that the splitting condition of Taxable Income ≤ 97 provides the smallest impurity measure. Hence the best attribute for this parent node can be chosen by comparing the magnitudes of impurity measures evaluated at other discrete and categorical attributes, and those of other indicators in the same form as above but now based on alternative continuous attributes. The attribute with the smallest impurity measure among all will be selected.

From Table 13.3, the detailed calculations concerning the splitting condition of Taxable Income ≤ 97 are shown below.

1. Child node with Taxable Income ≤ 97, $(Y, N) = (3, 3)$

$$\text{Gini-index} = 1 - \left(\frac{3}{6}\right)^2 - \left(\frac{3}{6}\right)^2 = \frac{1}{2};$$

$$\text{Entropy} = -\frac{3}{6}\log_2\frac{3}{6} - \frac{3}{6}\log_2\frac{3}{6} = 1;$$

$$\text{Misclassification Error} = 1 - \max\left(\frac{3}{6}, \frac{3}{6}\right) = \frac{1}{2}.$$

TABLE 13.3 Choosing the threshold value for the attribute "Taxable Income" with the smallest impurity measure (Average Gini-index, entropy, misspecification error).

Real label		No		No			No	Yes		Yes		Yes		No		No		No	No			
											Taxable Income											
		60		70			75	85		90		95		100		120		125	220			
Rules	55		65		72		80		87		92		97		110		122		172		230	
	≤	>	≤	>	≤	>	≤	>	≤	>	≤	>	≤	>	≤	>	≤	>	≤	>	≤	>
Yes	0	3	0	3	0	3	0	3	1	2	2	1	3	0	3	0	3	0	3	0	3	0
No	0	7	1	6	2	5	3	4	3	4	3	4	3	4	4	3	5	2	6	1	7	0
Gini	0.420		0.400		0.375		0.343		0.417		0.400		**0.300**		0.343		0.375		0.400		0.420	
Entropy	0.881		0.826		0.764		0.690		0.875		0.846		**0.600**		0.690		0.764		0.826		0.881	
Misspecification	0.300		0.300		0.300		0.300		**0.200**		0.300		0.300		0.300		0.300		0.300		0.300	

2. Child node with Taxable Income > 97, (Y, N) = (0, 4)

$$\text{Gini-index} = 1 - \left(\frac{0}{4}\right)^2 - \left(\frac{4}{4}\right)^2 = 0;$$

$$\text{Entropy} = -\frac{0}{4}\log_2\frac{0}{4} - \frac{4}{4}\log_2\frac{4}{4} = 0;$$

$$\text{Misclassification Error} = 1 - \max\left(\frac{0}{4}, \frac{4}{4}\right) = 0.$$

Consequently, the averages of the three impurity measures (among child nodes), based on "Taxable Income ≤ 97", after splitting are given by:

$$\text{Average of Gini-index}(\cdot) = \left(\frac{3+3}{10}\right)\left(\frac{1}{2}\right) + \left(\frac{0+4}{10}\right)(0) = 0.3;$$

$$\text{Average of Entropy}(\cdot) = \left(\frac{3+3}{10}\right)(1) + \left(\frac{0+4}{10}\right)(0) = 0.6;$$

$$\text{Average of Misclassification Error}(\cdot) = \left(\frac{3+3}{10}\right)\left(\frac{1}{2}\right) + \left(\frac{0+4}{10}\right)(0) = 0.3,$$

We observe that the average Gini-index and the average entropy are the smallest among all those of the other indicator attributes of "Taxable Income $\leq m_k$". All these values of average impurity measures can be calculated in the same manner, and they are listed at the bottom rows of Table 13.3. Other similar discussions can be found in various standard textbooks such as [16].

13.6 OVERFITTING IN CLASSIFICATION TREE

Recall that a classification tree corresponds to a partition of the space of feature vectors $\mathcal{D}$ into $\mathcal{D}_1, \ldots, \mathcal{D}_M$, corresponding to the predicted labels $c_1, \ldots, c_M$. This then induces the partition of the dataset $\mathcal{S}$ into $\mathcal{S}_1, \ldots, \mathcal{S}_M$, where $\mathcal{S}_k = \{(x,y) \in \mathcal{S} : x \in \mathcal{D}_k\}$. Generally speaking, for a given sample $\mathcal{S} = \{(x_i, y_i)\}_{i=1}^n$, we aim to construct a classification tree $\mathcal{T}$, whose number of terminal leaf nodes is denoted by T (see Figure 13.5 for an illustration), that minimizes the classification error:

$$R(\mathcal{T}) := \frac{1}{n} \sum_{\ell=1}^{T} \sum_{(x_i,y_i)\in\mathcal{T}_\ell} \mathbb{1}_{\{y_i \neq \bar{y}_{\mathcal{T}_\ell}\}}, \tag{13.14}$$

where $\mathcal{T}_\ell$ is the set of data points at the ℓ-th terminal node, for $\ell = 1, \ldots, T$, and $\bar{y}_{\mathcal{T}_\ell}$ is the majority vote for a particular label using all data points in $\mathcal{T}_\ell$:

$$\bar{y}_{\mathcal{T}_\ell} := \arg\max_{c\in\mathcal{Y}} \sum_{(x_i,y_i)\in\mathcal{T}_\ell} \mathbb{1}_{\{y_i = c\}}, \tag{13.15}$$

from which the partition of $\mathcal{S}$ can be obtained from some of these $\mathcal{T}_\ell$'s; that is $\mathcal{S}_k = \bigsqcup_{\ell:\bar{y}_{\mathcal{T}_\ell}=c_k} \mathcal{T}_\ell$, which is equivalent to our earlier definition of (13.1).

While the classification tree is often an effective classifier, it sometimes results in overfitting, especially when its depth is too large. We need to strive for a balance between its complexity and its training classification error $R(\mathcal{T})$ given in (13.14). For a more detailed discussion in this regard, for instance, using the chi-square test to determine whether the splitting should be performed one step further, we refer the readers to our user guide [2].

The first approach to reduce the complexity of a tree is *pre-pruning*. It is intuitive at the first glance: if the value of the chosen impurity measure exceeds a certain threshold level, one expects that subsequent splittings cannot reduce the training error much, while the resulting tree becomes more complex as the tree size grows, so it is better to stop at the current node. Nevertheless, pre-pruning is greedy in nature: the tree growing may stop at a seemingly bad splitting, even though a subsequent partition might be extremely valuable.

Another approach is called *post-pruning*, in which we first construct the classification tree $\mathcal{T}$, and then remove any unnecessary intermediate nodes, which prevents us from being misguided by seemingly worthless splittings. Intuitively, a post-pruning procedure purposely removes branches which give little reduction in the

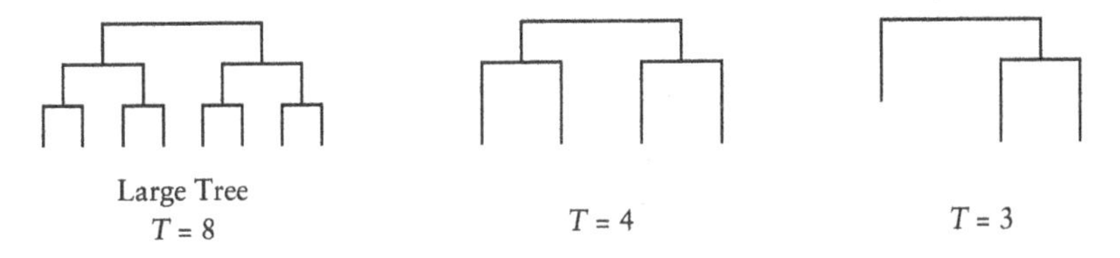

FIGURE 13.5 An illustration of trees with different numbers of terminal nodes.

training error. There are two major types of approach that depend on where the procedure is kicked off:

1. *bottom-up pruning* starts at any one of the terminal nodes, and one recursively traces upwards along the chosen path to determine whether the splitting at each corresponding parent node along this backward traversing path is "valuable" under a pre-determined measure; if a node is not "valuable", for instance, the reduction in the training error is below a pre-specified level, the splitting at this parent node is removed, which consequently becomes the immediate terminal node.
2. *top-down pruning* starts at the root of the tree. It checks the "value" of the splitting at each node, and removes the sub-tree under a node whose "contributed reduction" is below a certain threshold. This approach is seemingly similar to the pre-pruning, yet a significant difference between them is that pre-pruning stops building classification trees in the training stage, while top-down pruning starts to work only after building the whole tree.

A representative bottom-up post-pruning approach is called *(Minimal) Cost Complexity Pruning.* Let $\mathcal{T}^{(0)} = \mathcal{T}$ be the original tree and $\mathcal{T}^{(K)}$ be the tree which consists only of the root. For $k = 0, \ldots, K$, we aim to construct a sequence of trees $\mathcal{T}^{(0)} \supset \cdots \supset \mathcal{T}^{(K)}$, where $\mathcal{T}^{(k+1)}$ is formed by removing a subtree from $\mathcal{T}^{(k)}$. For example, consider the classification tree in Figure 13.6.

There are four possible trees that can be formed from the tree in Figure 13.6. Figure 13.7 shows three of them, and the last tree $\mathcal{T}^{(K)}$ is formed by removing $\mathcal{T}_{N_1}$ from $\mathcal{T}$, denoted by $\mathcal{T}\backslash\mathcal{T}_{N_1}$, leading to a single node A, the root.

In cost complexity pruning, one avoids overfitting by focusing not only on the original classification error $R(\mathcal{T})$ but also the tree complexity. This is achieved by adding a "penalty" term to the original classification error $R(\mathcal{T})$ so that the objective now is to construct a tree that minimizes

$$R_\alpha(\mathcal{T}) = R(\mathcal{T}) + \alpha T := \frac{1}{n}\sum_{\ell=1}^{T}\sum_{(x_i,y_i)\in\mathcal{T}_\ell} \mathbb{1}_{\{y_i \neq \bar{y}_{\mathcal{T}_\ell}\}} + \alpha T. \tag{13.16}$$

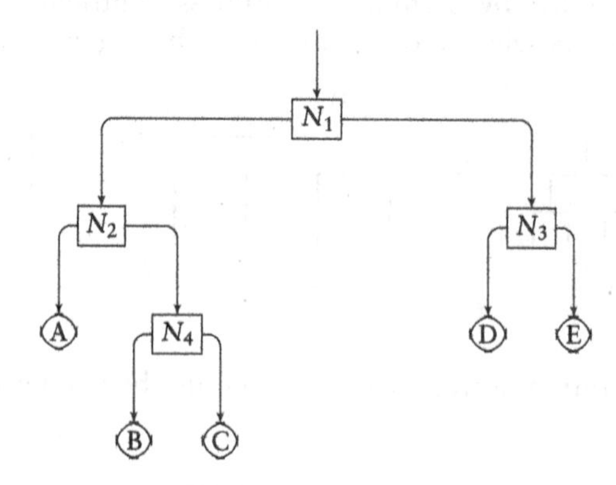

FIGURE 13.6 Classification tree with five leaf nodes and four decision nodes.

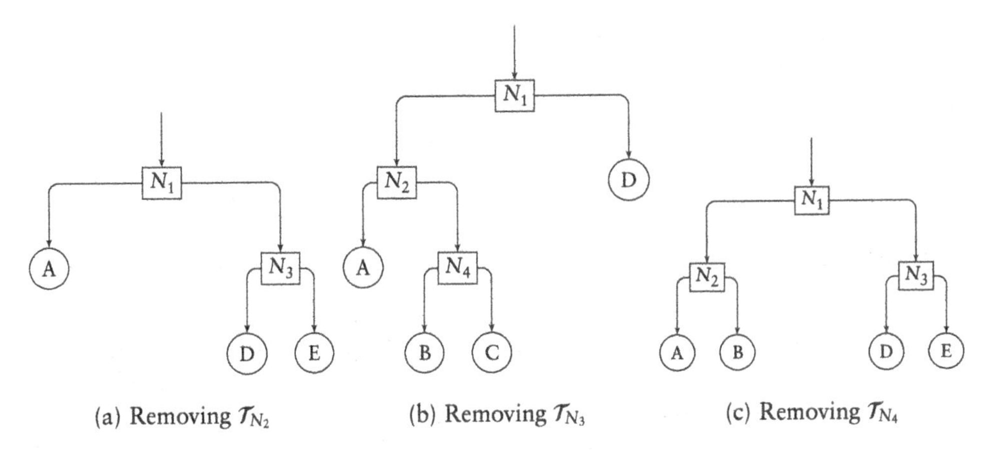

FIGURE 13.7 Three possible trees $\mathcal{T}^{(k+1)}$ formed from $\mathcal{T}^{(k)}$ in Figure 13.6.

Here, α is a hyperparameter which controls the effect of the model complexity. Given $\alpha \geq 0$, we aim to find a subtree $\mathcal{T}(\alpha)$ of $\mathcal{T}$, denoted as $\mathcal{T}(\alpha) \subseteq \mathcal{T}$, that minimizes $R_\alpha(\mathcal{T})$, namely:

$$\mathcal{T}(\alpha) := \arg\min_{\tilde{\mathcal{T}} \subseteq \mathcal{T}} R_\alpha(\tilde{\mathcal{T}}) = \arg\min_{\tilde{\mathcal{T}} \subseteq \mathcal{T}} (R(\tilde{\mathcal{T}}) + \alpha \tilde{T}). \tag{13.17}$$

If $\alpha = 0$, then no pruning is needed as there is no penalty coming from the number of the terminal leaves $\tilde{T}$, so $\mathcal{T} = \tilde{\mathcal{T}}$. As $\alpha \to \infty$, the tree size outweighs the classification error $R(\tilde{\mathcal{T}})$, and so we shall prune $\mathcal{T}$ until the root node is reached, yielding a single-node tree of size $\tilde{T} = 1$.

For more discussions on pre-pruning and post-pruning, readers can refer to various articles such as [4, 5, 11].

13.7 CLASSIFICATION TREES IN PYTHON AND R

In general, for both Python and R, a classification tree is built based on the binary splitting of feature variables $\mathrm{x}^{(j)}$, one at a time, for some $j = 1, 2, \ldots, p$. Since this method will search for all possible splittings caused by all feature variables, building a tree is computationally intensive and time-consuming. Usually, once the classification tree is built, given a carefully chosen hyperparameter α (for pruning), the subtree that minimizes (13.16) is chosen to be the final tree, and a set of simple classification rules can then be easily read off.

In Python, classification trees are implemented by `DecisionTreeClassifier` in the well-known `sklearn` package under the Python class `tree`; see Programme 13.1.

```
1 import pandas as pd
2 import matplotlib.pyplot as plt
3 from matplotlib.colors import ListedColormap
4 from sklearn.tree import DecisionTreeClassifier, plot_tree, export_text
5 from sklearn.metrics import confusion_matrix
6
7 plt.rc('font', size=20); plt.rc('axes', titlesize=30, labelsize=25)
8 plt.rc('xtick', labelsize=25); plt.rc('ytick', labelsize=25)
```

Programme 13.1 Loading all required libraries for building a classification tree in Python.

Here the `plot_tree()` is a function that helps to plot the resulting tree obtained by `DecisionTreeClassifier`, and `export_text()` is another function that returns a text report about its classification rules. Although the Python class `DecisionTreeClassifier` is quite easy to use, there are many options hidden in this class. In particular, `ccp_alpha` is the hyperparameter α used for cost complexity pruning, which is defaulted to be `0`; `criterion` picks the impurity measure function used and it is defaulted as `gini` for the Gini-index in (13.11), other choices include `entropy` for the Shannon entropy in (13.2), and `log_loss` for the differential entropy in (13.4).

In **R**, classification trees are implemented by `rpart()` function inside the built-in library `rpart`[9] (stands for *Recursive Partitioning and Regression Trees*); also see Programme 13.2.

```
1 > library(rpart)                                    # load rpart library
2 > library(rpart.plot)                               # plot rpart object
3 > par(cex.lab=2, cex.axis=2, cex.main=2, mar=c(5,5,4,4))
```

Programme 13.2 Loading `rpart` and `rpart.plot` libraries in **R**.

Again, the `rpart()` function contains several options: for example, differential entropy can be adopted by plugging in the argument `parms=list(split="information")`, while `"gini"` for the Gini-index is used as the default mode. Moreover, while the default value $\alpha = 0.01$ is used for the corresponding cost complexity pruning parameter, users can pick any non-negative value of α, say $\alpha = 0.05$, by taking the argument: `control=rpart.control(cp=0.05)`. There is one further important option which needs mentioning: the method option in `rpart` can be one of "`class`", "`anova`", "`poisson`", or "`exp`". For a classification problem, the target variable is a categorical one, therefore `method="class"` is used; meanwhile for a regular regression tree, we use `method="anova"` which attempts to minimize the mean squared errors summed at all terminal nodes. It is also worth mentioning that the "`poisson`" method is adopted for a Poisson regression. Last but not least, the "`exp`" method is adopted for building regression trees with exponential scaling, for example when the y_i's are data in survival analysis. Such trees are more commonly known as *survival trees*, which are a nonparametric alternative to the celebrated semi-parametric *Cox proportional hazards model*.

[9] See more details in https://cran.r-project.org/web/packages/rpart/rpart.pdf; see [17].

I. HSI dataset:

Let us first illustrate using our `fin-ratio.csv` file, which contains the stock data in 2002 and has not gone through outlier detection, for HSI classification. See Programme 13.3 for the Python version and Programme 13.4 for the **R** version.

```
1  df = pd.read_csv("fin-ratio.csv")
2  X = df.drop(columns="HSI")
3  y = df["HSI"]
4  ctree = DecisionTreeClassifier(ccp_alpha=0.01)
5  ctree.fit(X, y)
6  print(export_text(ctree, feature_names=list(X.columns),
7        show_weights=True))
8
9  fig, ax = plt.subplots(1, 1, figsize=(20, 15))
10 plot_tree(ctree, feature_names=X.columns, filled=False)
```

```
1  |--- ln_MV <= 9.48
2  |   |--- weights: [644.00, 3.00] class: 0
3  |--- ln_MV >  9.48
4  |   |--- weights: [4.00, 29.00] class: 1
```

Programme 13.3 Building a classification tree for the 2002 financial data via Python.

```
1  > df <- read.csv("fin-ratio.csv")              # read in data in csv format
2  > ctree <- rpart(HSI~., data=df, method="class")
3  > print(ctree)                                # print detailed information
4  n= 680
5
6  node), split, n, loss, yval, (yprob)
7        * denotes terminal node
8
9  1) root 680 32 0 (0.952941176 0.047058824)
10   2) ln_MV< 9.4776 647  3 0 (0.995363215 0.004636785) *
11   3) ln_MV>=9.4776 33  4 1 (0.121212121 0.878787879) *
12 > rpart.rules(ctree, nn=TRUE)                  # print classification rules
13  nn  HSI
14   2 0.00 when ln_MV <  9.5
15   3 0.88 when ln_MV >= 9.5
16 > rpart.plot(ctree, extra=1, cex=2.5, digits=4, nn=TRUE) # plot ctree
```

Programme 13.4 Building a classification tree for the 2002 financial data via **R**.

The classification in Figure 13.8 is very simple. The dataset is split into two classes of `0` and `1` according to the value of the variable `ln_MV`. The classification rules and the corresponding quality measures are shown as below:

R1: If `ln_MV` < 9.478, then return as class = `0` (not HSI) (644/3).

R2: If `ln_MV` ≥ 9.478, then return as class = `1` (HSI) (4/29).

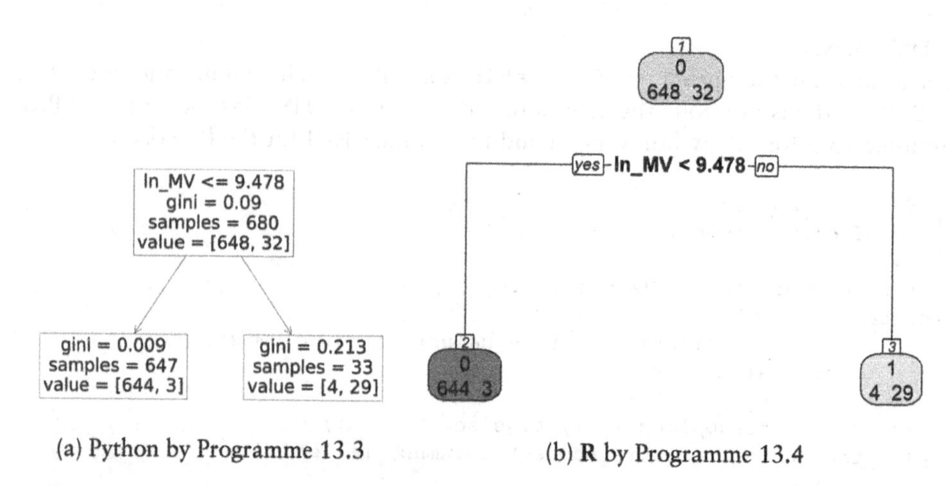

(a) Python by Programme 13.3 (b) **R** by Programme 13.4

FIGURE 13.8 Classification trees for the 2002 data without removing outliers.

The numbers in the terminal nodes represent the number of cases. For example, in the group with `ln_MV<9.478`, there are 644 "zeroes" and 3 "ones"; while in the group `ln_MV>=9.478`, there are 4 "zeroes" and 29 "ones". The outputs in Programmes 13.3 and 13.4 show the details at each node:

1. At node 1 (root), there are altogether $n = 680$ cases. Among them, 32 belong to group 1 (regarded as a loss). Since the majority is from group 0, the group label is classified as `yval=0` and the percentages of groups 0 and 1 are given in parentheses: they are $0.952941176 = 648/680$ and $0.047058824 = 32/680$, respectively.
2. At node 2 (the left branch), the split condition is `d$ln_MV<9.478`. There are $n = 647$ cases, 3 originating from group 1 of the root (node 1), and the group label is `yval=0`; again, the respective percentages are $0.995363215 = 644/647$ and $0.004636785 = 3/647$.
3. At node 3 (the right branch), $n = 33$. There are 4 cases originating from the group 0 of the root, and the group label is `yval=1`; here, the percentages of the respective groups are $0.121212121 = 4/33$ and $0.878787879 = 29/33$.

This output actually implies that we can predict or classify whether a stock is a Blue Chip based on only one simple condition of `ln_MV` < 9.478. Since the classification tree in this example is very simple, we can actually plot the observations: see the full plot in Figure 13.9.

```
1 fig, ax = plt.subplots(1, 1, figsize=(9, 8))
2 ax.scatter(y[y==0], X.ln_MV[y==0], c="blue", marker=".", s=1000,
3            label="Non-HSI")
4 ax.scatter(y[y==1], X.ln_MV[y==1], c="red", marker="^", s=700,
5            label="HSI")
6 ax.legend(fontsize=25)
```

```
7 ax.axhline(y=9.478)
8 ax.set_xlabel("HSI")
9 ax.set_ylabel("ln MV")
```

Programme 13.5 Plotting `ln_MV` versus `HSI` via Python.

```
1 > plot(df$HSI, df$ln_MV, pch=c(21,24)[df$HSI+1],
2 +      bg=c("blue","red")[df$HSI+1], cex=3)
3 > legend(0, 14, legend=c("Non-HSI", "HSI"), pch=c(19,17),
4 +         col=c("blue","red"), cex=2)
5 > abline(h=9.478, lwd=4)          # add a horizontal line at y=9.478
```

Programme 13.6 Plotting `ln_MV` versus `HSI` via **R**.

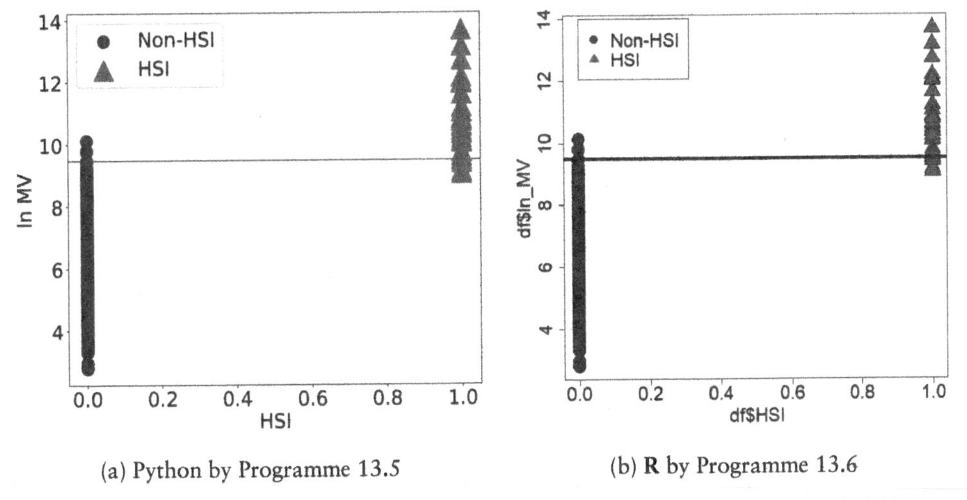

(a) Python by Programme 13.5 (b) **R** by Programme 13.6

FIGURE 13.9 Plot of `ln_MV` versus `HSI`, with the horizontal line `ln_MV` = 9.478 representing the classification rule.

A cross tabulation table for this classification tree can easily be found; see Programme 13.7 (Python) and Programme 13.8 (**R**). From the output, 3 Blue Chips are misclassified as non-Blue Chips while 4 non-Blue Chips are misclassified as Blue Chips.

```
1 y_hat = ctree.predict(X)
2 print(confusion_matrix(y_hat, y))
```

```
1 [[644   3]
2  [  4  29]]
```

Programme 13.7 Cross tabulation table for the classification tree in Programme 13.3 using the `fin-ratio` dataset via Python.

```
1 > prob <- predict(ctree)         # 2 columns of probabilities for 0 or 1
2 > y_hat <- colnames(prob)[max.col(prob)]
3 > table(y_hat, df$HSI)                                 # confusion matrix
4
5 y_hat   0   1
6     0 644   3
7     1   4  29
```

Programme 13.8 Cross tabulation table for the classification tree in Programme 13.4 using the `fin-ratio` dataset via **R**.

II. Iris flower dataset:

Recall the Iris flower dataset from Section 10.4.3. On this occasion, we build classification trees to categorize species of the Iris flowers; see Programme 13.9 (Python) and in Programme 13.10 (**R**).

```
1  from sklearn.datasets import load_iris
2
3  iris = load_iris()
4  X, y = iris["data"], iris['target']
5  ctree = DecisionTreeClassifier(ccp_alpha=0.01, random_state=4012)
6  ctree.fit(X, y)
7
8  fig, ax = plt.subplots(1, 1, figsize=(20, 15))
9  plot_tree(ctree, feature_names=iris['feature_names'], filled=False)
10
11 print(export_text(ctree, feature_names=iris['feature_names'],
12                   show_weights=True))
```

```
1  |--- petal length (cm) <= 2.45
2  |   |--- weights: [50.00, 0.00, 0.00] class: 0
3  |--- petal length (cm) >  2.45
4  |   |--- petal width (cm) <= 1.75
5  |   |   |--- petal length (cm) <= 4.95
6  |   |   |   |--- petal width (cm) <= 1.65
7  |   |   |   |   |--- weights: [0.00, 47.00, 0.00] class: 1
8  |   |   |   |--- petal width (cm) >  1.65
9  |   |   |   |   |--- weights: [0.00, 0.00, 1.00] class: 2
10 |   |   |--- petal length (cm) >  4.95
11 |   |   |   |--- weights: [0.00, 2.00, 4.00] class: 2
12 |   |--- petal width (cm) >  1.75
13 |   |   |--- weights: [0.00, 1.00, 45.00] class: 2
```

Programme 13.9 Building a classification tree for the Iris flower dataset via Python.

```
1  > data("iris")                  # load the built-in iris flower dataset
2  > df <- iris
3  > ctree <- rpart(Species~., data=df, method="class")
4  > print(ctree)                             # print detailed information
5  n= 150
6
7  node), split, n, loss, yval, (yprob)
8        * denotes terminal node
9
10 1) root 150 100 setosa (0.33333333 0.33333333 0.33333333)
11   2) Petal.Length< 2.45 50   0 setosa (1.00000000 0.00000000 0.00000000) *
12   3) Petal.Length>=2.45 100  50 versicolor (0.00000000 0.50000000 0.50000000)
13     6) Petal.Width< 1.75 54   5 versicolor (0.00000000 0.90740741 0.09259259) *
14     7) Petal.Width>=1.75 46   1 virginica (0.00000000 0.02173913 0.97826087) *
15 > rpart.rules(ctree, nn=TRUE)               # print classification rules
16  nn     Species  seto vers virg
17   2      setosa [1.00  .00  .00] when Petal.Length <  2.5
18   6 versicolor [ .00  .91  .09] when Petal.Length >= 2.5 & Petal.Width <  1.8
19   7  virginica [ .00  .02  .98] when Petal.Length >= 2.5 & Petal.Width >= 1.8
20 > rpart.plot(ctree, extra=1, cex=2, digits=4, nn=TRUE) # plot ctree
```

Programme 13.10 Building a classification tree for the Iris flower dataset via R.

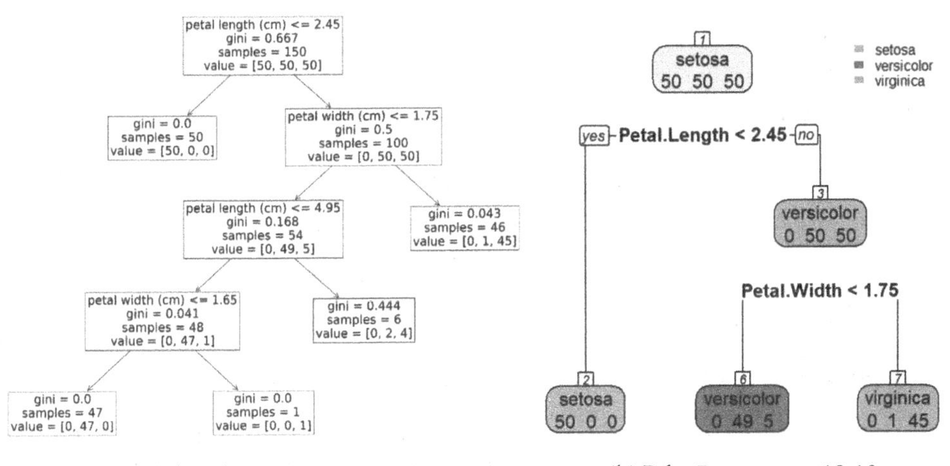

(a) Python by Programme 13.9 (b) **R** by Programme 13.10

FIGURE 13.10 Classification tree for the iris flower dataset.

From the outputs shown in Figure 13.10, the classification rules and their corresponding quality measures are:

R1: If `Petal.Length` < 2.45, then species = `setosa` (50/0/0).

R2: If `Petal.Length` ≥ 2.45 and `Petal.width` < 1.75, then species = `versicolor` (0/49/5);

R3: If `Petal.Length` ≥ 2.45 and `Petal.width` ≥ 1.75, then species = `virginica` (0/1/45).

Although the two classification trees produced by Python and **R** look similar, Python constructed a much larger tree. In fact, there are other hyperparameters involved in building a classification tree, including `minsplit`, which specifies the minimum number of observations that must exist in a node in order for a split to be attempted, and `maxdepth`, which sets the maximum depth of the final tree with the root node counted as at depth 0.

Since both classification trees are relatively simple and the rules also depend mainly on two variables, we can again plot the observations for illustration. See Programme 13.11 (Python) and Programme 13.12 (**R**).

```
1  fig, ax = plt.subplots(1, 1, figsize=(9, 8))
2  ax.scatter(X[y==0,2], X[y==0,3], c="red", marker=".", s=1000,
3             label=iris['target_names'][0])
4  ax.scatter(X[y==1,2], X[y==1,3], c="blue", marker="s", s=500,
5             label=iris['target_names'][1])
6  ax.scatter(X[y==2,2], X[y==2,3], c="green", marker="^", s=500,
7             label=iris['target_names'][2])
8  ax.legend(fontsize=25)
9  plt.axhline(y=1.75, linewidth=4)
10 plt.axvline(x=2.45, linewidth=4)
11 plt.axhline(y=1.65, linewidth=4)
12 plt.axvline(x=4.95, linewidth=4)
13 ax.set_xlabel("Petal length")
14 ax.set_ylabel("Petal width")
```

Programme 13.11 Plotting `Petal.Length` versus `Petal.width` via Python.

```
1 > plot(df$Petal.Length, df$Petal.Width, pch=c(21, 22, 24)[df$Species],
2 +       bg=c("red","blue","green")[df$Species], cex=3)
3 > legend(1, 2.5, legend=unique(df$Species), pch=c(21, 22, 24),
4 +        pt.bg=c("red", "blue", "green"), cex=2)
5 > abline(h=1.75, lwd=4, lty=2)                  # add a horizontal line
6 > abline(v=2.45, lwd=4, lty=2)                  # add a vertical line
```

Programme 13.12 Plotting `Petal.Length` versus `Petal.width` via **R**.

From Figure 13.11, we can clearly see how these classification rules work. Finally, we can produce cross tabulation via Python (Programme 13.13) and **R** (Programme 13.14).

```
1 y_hat = ctree.predict(X)
2 print(confusion_matrix(y_hat, y))
```

```
1 [[50  0  0]
2  [ 0 47  0]
3  [ 0  3 50]]
```

Programme 13.13 Cross tabulation table for the classification tree in Programme 13.9 using the Iris flower dataset via Python.

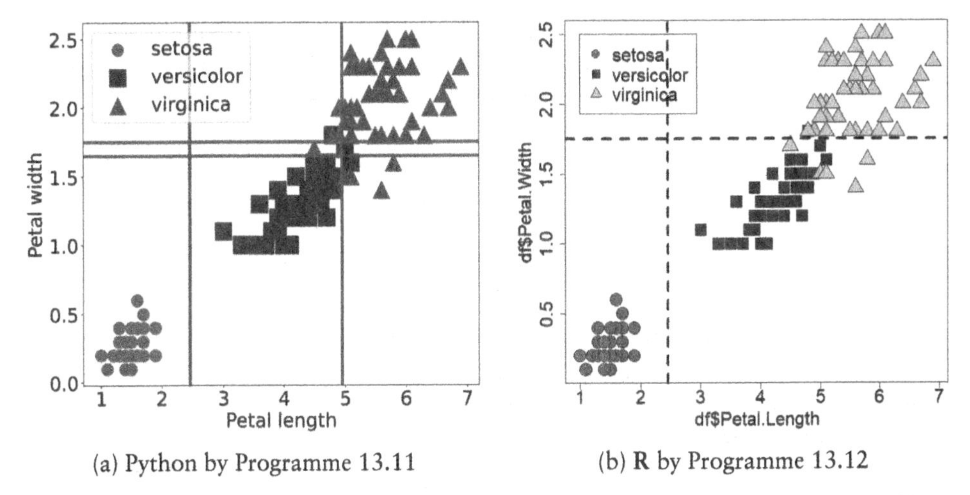

(a) Python by Programme 13.11 (b) **R** by Programme 13.12

FIGURE 13.11 Plot of `Petal.Length` versus `Petal.width`, with the horizontal lines `Petal.width` = 1.65 and 1.75, and the vertical lines `Petal.Length` = 2.45 and 4.95 representing the classification rules.

```
1  > prob <- predict(ctree)  # 3 columns of probabilities for 3 species
2  > y_hat <- colnames(prob)[max.col(prob)]
3  > table(y_hat, df$Species)                    # confusion matrix
4
5  y_hat        setosa versicolor virginica
6    setosa         50          0         0
7    versicolor      0         49         5
8    virginica       0          1        45
```

Programme 13.14 Cross tabulation table for the classification tree in Programme 13.10 using the Iris flower dataset via **R**.

From the table produced by Programme 13.13 via Python, there are 3 versicolor species being misclassified as virginica species, and the error rate is 3/150 = 2%. On the other hand, the table produced by Programme 13.14 via **R** shows an error rate of (1+5)/150 = 4%, with 5 virginica species being misclassified as versicolor species and 1 versicolor species being misclassified as virginica species. The difference in error rates between the uses of Python and **R** is due to the depth of their corresponding trees.

13.8 REGRESSION TREES

A *regression tree* is similar to a classification tree, except that while the label variable y must be categorical for a classification tree, for a regression tree it takes a continuum of values. Recall that the main inspiration behind classification trees is segmenting the space of feature vectors $\mathcal{D}$ into M simpler regions, namely $\mathcal{D}_1, \ldots, \mathcal{D}_M$. The predictor function $\hat{f}$ (to predict the label in this case) is written as:

$$\hat{f}(x) = \sum_{k=1}^{M} c_k \mathbb{1}_{\{x \in \mathcal{D}_k\}}. \tag{13.18}$$

Building a classification tree $\mathcal{T}$ is essentially finding a set of terminal leaf nodes $\{\mathcal{T}_1, \ldots, \mathcal{T}_T\}$ that minimizes the misclassification error $R(\mathcal{T})$ in (13.14) (or (13.16) if cost complexity pruning is adopted). For building a regression tree, we shall replace the 0-1 loss of $\mathbb{1}_{\{y_i \neq \hat{y}_{\mathcal{T}_\ell}\}}$ by the squared loss function:

$$R(\mathcal{T}) := \frac{1}{n} \sum_{\ell=1}^{T} \sum_{(x_i, y_i) \in \mathcal{T}_\ell} (y_i - \hat{y}_{\mathcal{T}_\ell})^2, \tag{13.19}$$

and the resulting tree built from minimizing (13.19) is commonly called a regression tree with $M = T$ being the total number of partitions. Certainly, we can simply assign a single but continuous label value for each $\mathcal{T}_\ell$, for example the mean for the subsample at each terminal node:

$$\hat{y}_{\mathcal{T}_\ell} = \frac{1}{|\mathcal{T}_\ell|} \sum_{(x_i, y_i) \in \mathcal{T}_\ell} y_i, \qquad \text{for } \ell = 1, \ldots, T. \tag{13.20}$$

On the other hand, we can also rewrite (13.19) as:

$$R(\mathcal{T}) := \frac{1}{n} \sum_{i=1}^{n} (y_i - \hat{f}(x_i))^2,$$

where the predictor function $\hat{f}(x)$ takes the same form as (13.18) where c_ℓ can now take any continuous value. With this in mind, a regression tree is just a special case of the threshold regression model, whose predictor function takes the more generic form:

$$\hat{f}(x) = \sum_{\ell=1}^{T} \hat{f}_\ell(x) \mathbb{1}_{\{x \in \mathcal{D}_\ell\}},$$

where $\hat{f}_\ell$ is a regression function with a domain of $\mathcal{D}_\ell$, for $\ell = 1, \ldots, T$.

Similar to classification trees, we build a regression tree by searching greedily via a top-down approach. Starting from the root node, we determine the best splitting attribute that minimizes the squared loss function, and we iterate the search followed by a child node. In the following, we only focus on the most common scenario of binary splitting, yet this is not necessary for regression trees, and the treatment below can easily be adapted for the general case when a parent node is split into three or more child nodes.

Let $\mathcal{S} = \{(x_i, y_i)\}_{i=1}^{n}$ be the dataset at this parent node. For a feature variable $\mathrm{x}^{(j)}$, and attribute value $t^{(j)}$ to be determined below, we aim to split the dataset into two groups:

$$\mathcal{S}_-^{(j)} = \{(x_i, y_i) \in \mathcal{S} : x_i^{(j)} < t^{(j)}\} \quad \text{and} \quad \mathcal{S}_+^{(j)} = \{(x_i, y_i) \in \mathcal{S} : x_i^{(j)} \geq t^{(j)}\},$$

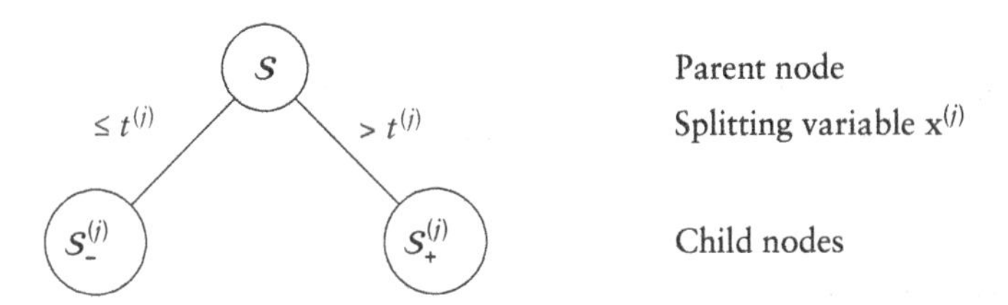

FIGURE 13.12 An illustration of splitting at a parent node with a dataset S.

see Figure 13.12 for a pictorial illustration. We then define the averages of the labels of the subsamples at two child nodes as

$$\bar{y}_{S_-^{(j)}} = \frac{\sum_{(x_i,y_i)\in S_-^{(j)}} y_i}{|S_-^{(j)}|} \quad \text{and} \quad \bar{y}_{S_+^{(j)}} = \frac{\sum_{(x_i,y_i)\in S_+^{(j)}} y_i}{|S_+^{(j)}|}.$$

The quality of the splitting is evaluated by the mean squared error:

$$\frac{1}{|S|}\left(\sum_{(x_i,y_i)\in S_-^{(j)}} (y_i - \bar{y}_{S_-^{(j)}})^2 + \sum_{(x_i,y_i)\in S_+^{(j)}} (y_i - \bar{y}_{S_+^{(j)}})^2 \right). \tag{13.21}$$

We try to determine the optimal pair of attribute variable $x^{(j)}$ and its threshold value $t^{(j)}$ that makes (13.21) as small as possible. To achieve this, for each attribute $x^{(j)}$, we first determine the optimal threshold $t^{(j)}$ by minimizing (13.21), using the boundary points obtained from a certain discretization method to serve as the set of all possible thresholds $t^{(j)}$; also see Section 13.5. Next, we compare the obtained minimum values of mean squared errors arising from all finitely many attributes, and pick the one with the smallest value. The same process continues until a stopping criterion is attained at a terminal node $\mathcal{T}_\ell$; by then no more splitting will be carried out. Typical stopping criteria include the following:

(i) the number of samples in the present node falls below a pre-specified threshold n_0:

$$|\mathcal{T}_\ell| < n_0; \quad \text{or}$$

(ii) the sum of squared errors at the present node falls below a pre-determined threshold ϵ:

$$\sum_{(x_i,y_i)\in\mathcal{T}_\ell} (y_i - \hat{y}_{\mathcal{T}_\ell})^2 < \epsilon; \text{ or}$$

(iii) the reduction in the mean squared error (13.21) by making an extra splitting of the present node S into $S_-^{(j)}$ and $S_+^{(j)}$, by using whatever feature variable $x^{(j)}$, is still smaller than a pre-specified threshold ϵ, i.e.:

$$\max_j \left(\frac{1}{|S|} \left(\sum_{(x_i,y_i)\in S} (y_i - \hat{y}_S)^2 - \left(\sum_{(x_i,y_i)\in S_-^{(j)}} (y_i - \overline{y}_{S_-^{(j)}})^2 + \sum_{(x_i,y_i)\in S_+^{(j)}} (y_i - \overline{y}_{S_+^{(j)}})^2 \right) \right) \right) < \epsilon.$$

After building the regression tree, the prediction for a given test observation is the mean of the training observations in the region $\mathcal{D}_\ell$ to which that test observation belongs.

As an example, Figure 13.13 shows a dataset containing four categories, with only one feature variable x and one label variable y, both real-valued. At the root node, the dataset is split into two groups according to whether $x < t_1$ or $x \geq t_1$. The averages $\hat{c}_1$ and $\hat{c}_2$ for the two groups are then calculated. Each subgroup is further divided into two smaller groups with respect to the splitting thresholds t_2 and t_3, respectively, so that we finally have that each cluster corresponds to its own label.

While regression trees are commonly used, one should be award of their pitfalls. Specifically, the tree-building process is a top-down, greedy binary search with the following characteristics:

1. Each splitting is the best given the past splittings in the previous nodes; it is only valid locally, but not necessarily the best one from the global viewpoint.
2. Some partitions of the space $\mathcal{D}$ cannot be obtained in a regression tree. For example, suppose that there are only two feature variables, then a regression tree can possibly partition $\mathcal{D}$ in the way of Figure 13.14(a), but the space $\mathcal{D}$ cannot be partitioned as in Figure 13.14(b). It is because there is no splitting in (b) that can serve as an initial one, while the leftmost vertical line at $x^{(1)} = s_{11}$ inside the big square in (a) can, since it splits this into two smaller rectangles. Therefore, the configuration depicted in Figure 13.14(b) cannot arise from a regression tree.

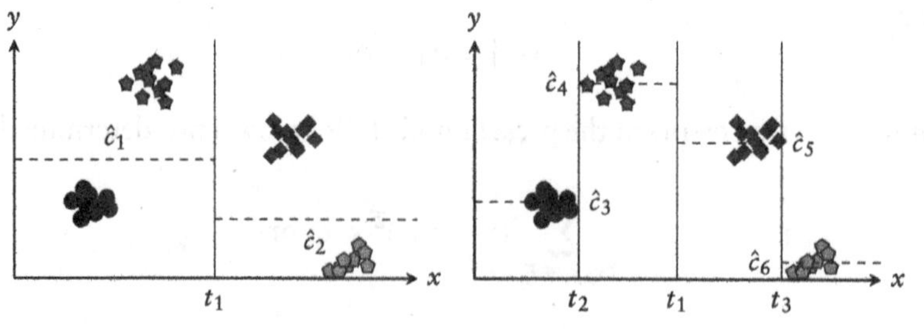

(a) Splitting at the root node. (b) Each subgroup is further split into 2 groups.

FIGURE 13.13 Regression tree splitting algorithm for one attribute.

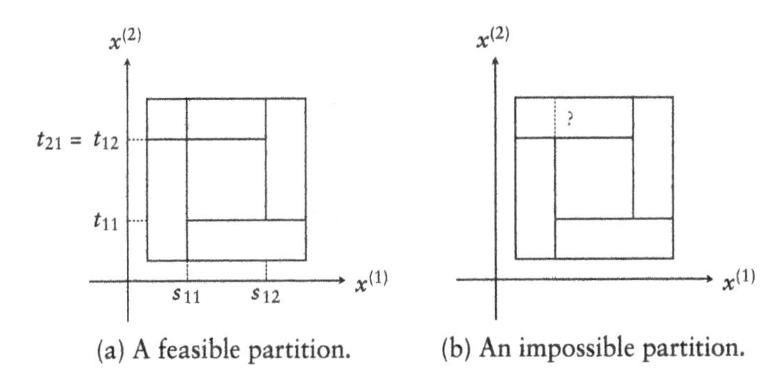

(a) A feasible partition. (b) An impossible partition.

FIGURE 13.14 Illustrations of a feasible partition of a space of feature vectors caused by a regression tree, and an impossible one.

3. Sometimes, this tree-building process may overfit the training dataset, leading to poor test performance. This could be remedied by constructing a smaller tree with fewer splits that may lead to higher variance but a better interpretation.

13.8.1 A Medical Insurance Example

We now consider another example, where the goal is to predict the `Premium Price` charged by a medical insurance company with two feature variables of customers, namely their `Age` and `Weight`.[10]

```
 1  import pandas as pd
 2  from sklearn.tree import DecisionTreeRegressor, plot_tree, export_text
 3  import matplotlib.pyplot as plt
 4  from matplotlib import collections  as mc
 5
 6  # Prepare the dataset
 7  df = pd.read_csv("Medicalpremium.csv")
 8  X = df[['Age', 'Weight']]
 9  y = df['PremiumPrice']
10
11  # Fit and plot the regression tree model
12  rtree = DecisionTreeRegressor(ccp_alpha=0.01, max_depth=3)
13  rtree.fit(X, y)
14  print(export_text(rtree, feature_names=list(X.columns),
15                    show_weights=True))
16
17  plt.figure(figsize=(27, 10))
18  plot_tree(rtree, feature_names=list(X.columns), filled=False,
19            fontsize=15, precision=1)
20
21  # Plot dataset with segments and text annotations
```

[10] The dataset can be downloaded from https://www.kaggle.com/datasets/tejashvi14/medical-insurance-premium-prediction.

```
22 fig, ax = plt.subplots(figsize=(16, 12))
23 s = ax.scatter(df['Age', df['Weight'], c=y, cmap='gray', s=200)
24
25 # Add vertical lines and text annotations
26 ax.axvline(x=29.5, color='black', linewidth=4)
27 ax.axvline(x=46.5, color='black', linewidth=4)
28 ax.axvline(x=38.5, color='black', linewidth=4)
29 lines = [[(0, 119), (29.5, 119)], [(24.5, 0), (24.5, 119)],
30          [(23.0, 119), (23.0, 140)], [(46.5, 94.5), (70, 94.5)]]
31 lc = mc.LineCollection(lines, colors="black", linewidths=4)
32 ax.add_collection(lc)
33 plt.text(17.7, 127, "R1", fontsize=40, weight='bold', c="red")
34 plt.text(24.3, 127, "R2", fontsize=40, weight='bold', c="red")
35 plt.text(18.5, 85, "R3", fontsize=40, weight='bold', c="red")
36 plt.text(25, 85, "R4", fontsize=40, weight='bold', c="red")
37 plt.text(32, 90, "R5", fontsize=40, weight='bold', c="red")
38 plt.text(40.7, 90, "R6", fontsize=40, weight='bold', c="red")
39 plt.text(55.5, 115, "R7", fontsize=40, weight='bold', c="red")
40 plt.text(55.5, 71, "R8", fontsize=40, weight='bold', c="red")
41
42 plt.xlabel('Age')
43 plt.ylabel('Weight')
44 fig.colorbar(s, label='Premium Price')
```

```
1  |--- Age <= 29.50
2  |   |--- Weight <= 119.00
3  |   |   |--- Age <= 24.50
4  |   |   |   |--- value: [15816.79]
5  |   |   |--- Age >  24.50
6  |   |   |   |--- value: [16990.20]
7  |   |--- Weight >  119.00
8  |   |   |--- Age <= 23.00
9  |   |   |   |--- value: [15000.00]
10 |   |   |--- Age >  23.00
11 |   |   |   |--- value: [32500.00]
12 |--- Age >  29.50
13 |   |--- Age <= 46.50
14 |   |   |--- Age <= 38.50
15 |   |   |   |--- value: [24081.08]
16 |   |   |--- Age >  38.50
17 |   |   |   |--- value: [25954.80]
18 |   |--- Age >  46.50
19 |   |   |--- Weight <= 94.50
20 |   |   |   |--- value: [28000.00]
21 |   |   |--- Weight >  94.50
22 |   |   |   |--- value: [32647.06]
```

Programme 13.15 Predicting medical insurance premium prices with a regression tree in Python.

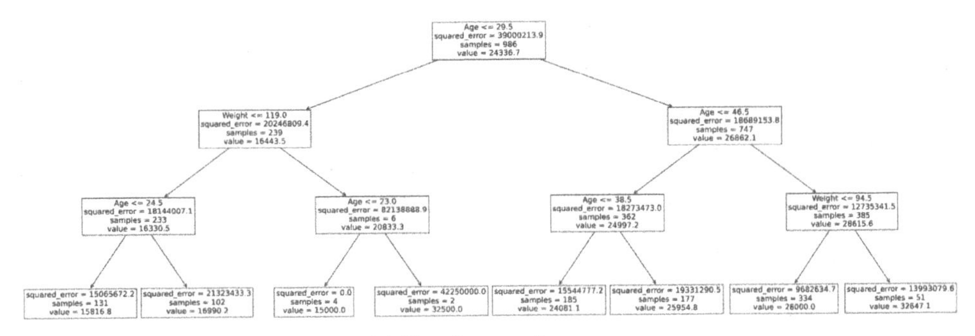

(a) Fitted regression tree

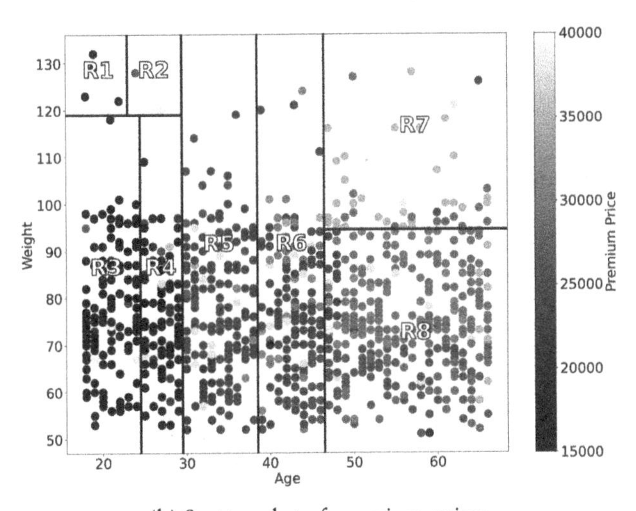

(b) Scatter plot of premium prices

FIGURE 13.15 A regression tree for the medical premium data via Python, generated in Programme 13.15.

```
> library(rpart)                                    # load rpart library
> library(rpart.plot)                               # plot rpart object
>
> # Prepare the dataset
> df <- read.csv("Medicalpremium.csv")
>
> # Fit and plot the regression tree model
> rtree <- rpart(PremiumPrice~Age+Weight, data=df, method="anova")
> rpart.rules(rtree, nn=TRUE)
 nn PremiumPrice
  2        16444 when Age <  30
  6        24997 when Age is 30 to 47
 28        26537 when Age >=       47 & Weight <  70
 29        28853 when Age >=       47 & Weight is 70 to 95
 15        32647 when Age >=       47 & Weight >=       95
> rpart.plot(rtree, extra=0, cex=2, digits=2, type=0, nn=TRUE)
>
> # Plot dataset with segments and text annotations
```

```
19 > layout(t(1:2), widths=c(6,1))
20 > par(mar=c(5,5,2,0))
21 > n_col <- 20
22 > y <- df$PremiumPrice
23 > y_scale <- round((y - min(y))/diff(range(y))*(n_col-1)) + 1
24 > plot(df$Age, df$Weight, pch=20, xlab="Age", ylab="Weight",
25 +      col=gray.colors(n_col)[y_scale], cex=3)
26 > abline(v=30, lwd=2)
27 > abline(v=47, lwd=2)
28 > segments(47, 70, 70, 70, lwd=2)
29 > segments(47, 95, 70, 95, lwd=2)
30 > text(23, 90, substitute(paste(bold("R1"))), cex=2.5)
31 > text(38, 90, substitute(paste(bold("R2"))), cex=2.5)
32 > text(57, 60, substitute(paste(bold("R3"))), cex=2.5)
33 > text(57, 83, substitute(paste(bold("R4"))), cex=2.5)
34 > text(57, 115, substitute(paste(bold("R5"))), cex=2.5)
35 > image(y=1:n_col, z=t(1:n_col), col=gray.colors(n_col), axes=FALSE,
36 +       ylab="Premium")
```

Programme 13.16 Predicting medical insurance premium prices with a regression tree in **R**.

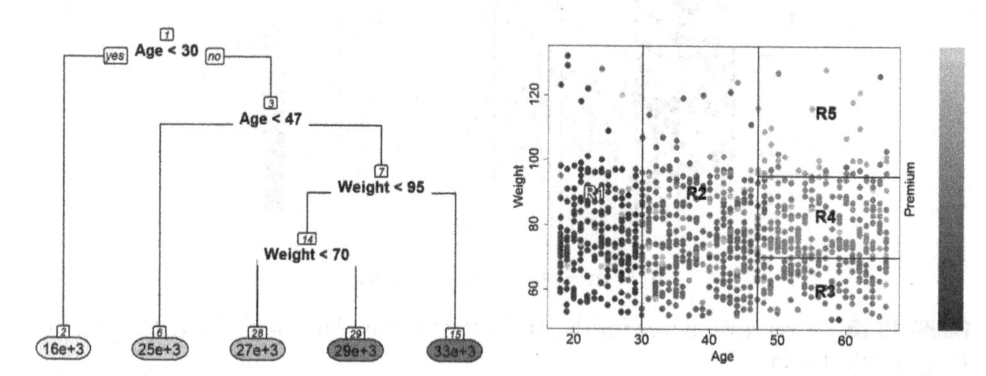

FIGURE 13.16 A regression tree for the medical premium data via **R**, generated in Programme 13.16.

Comparing the results in Figures 13.15 and 13.16, respectively from Programmes 13.15 via Python and 13.16 via **R**, the decision rules found by `sklearn` in Python and `rpart` in **R** are slightly different because of their choice of the splitting criterion, in which `sklearn` chooses the split that minimizes the squared error while `rpart` chooses the split that maximizes the between-groups sum of squares in ANOVA. Based on these two regression tree models, we obtain the following interpretations:

1. `Age` is the most relevant factor in determining the `Premium Price` of a customer. Customers under the age of 29.5 would be generally charged a lower amount of premium. Customers aged between 29.5 and 46.5 would be charged a medium amount of premium, and those customers above age 46.5 would be charged a higher amount of premium.

2. Given that a customer is either below the age of 29.5 or of age between 29.5 and 46.5, his or her `Weight` barely plays a role in the premium charged.
3. Given that a customer is of age above 46.5, his or her weight would affect his `Premium Price`. Customers heavier than 94.5 kg would be charged a higher amount of premium.

13.9 RANDOM FOREST

Random forests are based on the idea of *bagging*. Given a training set S, we create B (a hyperparameter) random samples $S_1, \ldots, S_B$ of S, on which we build B classification and regression tree models with respective labels $\hat{f}_1, \ldots, \hat{f}_B$. To sample each S_b for some $b = 1, 2, \ldots, B$, we perform sampling with replacement from S until $|S_b| = n = |S|$. Besides these bootstrapping, to reduce the computational complexity especially when the number of features p is large, one may confine the candidate features to $m(\ll p)$ randomly selected from the p originals when a specific tree is built. Alternatively, the random selection of m features can be carried out at every node, and the splitting attribute for the node is the most effective one out of the m selected features. After training, we obtain B different trees. In the regression tree case, the prediction for a new input x is obtained from the average of the B predictions:

$$\hat{f}_{rf}(x) = \frac{1}{B} \sum_{b=1}^{B} \hat{f}_b(x),$$

while in the classification case the prediction is the majority vote.

As an illustration, we apply the random forest algorithm to the 2002 financial dataset using Python and **R**, and we compare the results to those obtained from the (single) classification tree.

```
1 from sklearn.ensemble import RandomForestClassifier
2
3 df = pd.read_csv("fin-ratio.csv")
4 X = df.drop(columns="HSI")
5 y = df["HSI"]
6 rf_clf = RandomForestClassifier(max_features=2, random_state=4002)
7 rf_clf.fit(X, y)
8 y_hat = rf_clf.predict(X)
9 print(confusion_matrix(y_hat, y))
```

```
1 [[648   0]
2  [  0  32]]
```

Programme 13.17 A random forest for the 2002 financial dataset via Python.

The misclassification rate is 0%, which is much better than that of the classification tree in Programme 13.7 via Python (resp. Programme 13.8 via **R**).

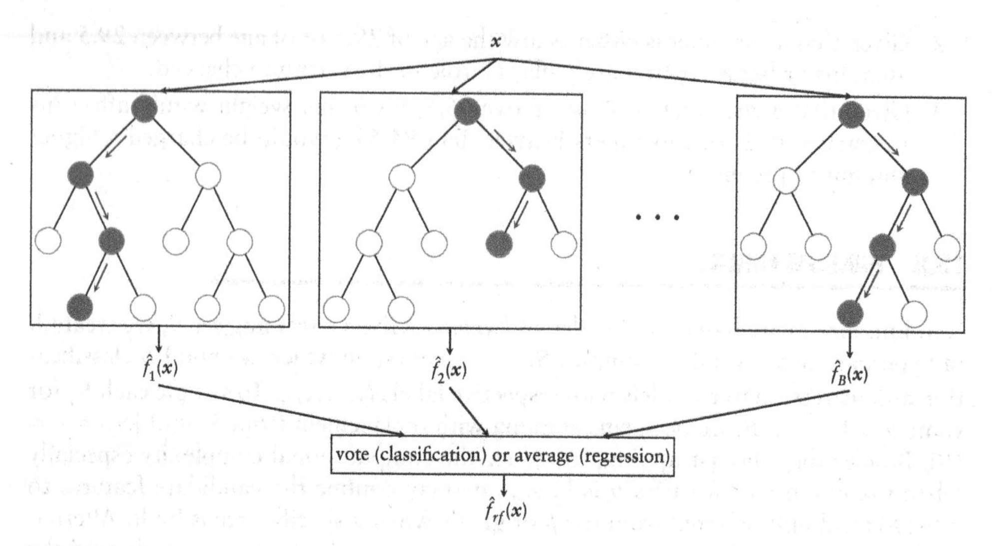

FIGURE 13.17 A graphical illustration of the random forest method.

```
1  > library(randomForest)
2  >
3  > set.seed(4002)
4  > df <- read.csv("fin-ratio.csv")          # read in data in csv format
5  > df$HSI <- as.factor(df$HSI) # change label into factor for classification
6  > rf_clf <- randomForest(HSI~., data=df, ntree=10, mtry=2, importance=TRUE)
7  > y_hat <- predict(rf_clf, newdata=df)
8  > table(y_hat, df$HSI)
9
10 y_hat   0   1
11     0 648   0
12     1   0  32
```

Programme 13.18 A random forest for the 2002 financial dataset via R.

One can observe from this example that at each splitting in the tree building process, only a random selection of $m = 2$ (`max_features=2` in Python and `mtry=2` in R) features from the original $p = 6$ features are inspected; this is the only difference between random forest and the vanilla bagging when $m = p$; as now $2 = m < p = 6$, this allows the possibility that a random forest can perform less favourably than an ordinary classification tree. The above value of m (hyperparameter) agrees with the common rules of thumb: choose $m = \lfloor\sqrt{p}\rfloor$ for classification and $m = \lfloor\frac{p}{3}\rfloor$ for regression.

The reason for the random selection of subsets of features is to de-correlate different trees that are built. This reduces the variance of the final model further from bagging. If one or a few features are very strong predictors for the label, these features will be selected to split the samples in many trees. This would result in many correlated trees in our "forest". Correlated predictors cannot help in improving the accuracy of prediction by reducing the variance. The main reason behind a better performance of model ensembling is that good models will likely agree on the same prediction, while bad models will likely disagree on different ones, so combining them can diversify the error and hence reduce the variance. Correlation makes bad models more likely to agree, possibly hampering the majority vote or the average.

13.9.1 Credit Card Default Prediction

In a credit card rating system, data are collected from potential customers to predict whether future customers will default or not. As an illustration, we consider a dataset which contains information on default payments, demographics, credit data, payment history, and bill statements of credit card clients in Taiwan from April 2005 to September 2005 [10]. Each feature vector contains 26 features, such as "Amount of credit", "Gender", "Education Level", "Marital status", and "Age". The label takes a value of 1 if the client defaults next month, or 0 otherwise. In Programmes 13.19 (Python) and 13.20 (**R**), classification trees and random forests are built for prediction.

```
1  from sklearn.model_selection import train_test_split
2  from sklearn.metrics import classification_report
3  
4  df = pd.read_csv("credit default.csv")
5  X = df.drop(columns=["default payment next month"])
6  y = df["default payment next month"]
7  
8  (X_train, X_test, y_train,
9   y_test) = train_test_split(X, y, train_size=0.8, random_state=4012)
10 ctree = DecisionTreeClassifier(ccp_alpha=0.01, random_state=4012)
11 ctree.fit(X_train, y_train)
12 y_hat_dt = ctree.predict(X_test)
13 print(confusion_matrix(y_hat_dt, y_test))
14 print(classification_report(y_test, y_hat_dt))
15 
16 fig, ax = plt.subplots(1, 1, figsize=(20, 15))
17 plot_tree(ctree, feature_names=X.columns, filled=False)
18 
19 rf_clf = RandomForestClassifier(random_state=4012)
20 rf_clf.fit(X_train, y_train)
21 y_hat_rf = rf_clf.predict(X_test)
22 print(confusion_matrix(y_hat_rf, y_test))
23 print(classification_report(y_test, y_hat_rf))
```

```
1  [[4511  870]              # [[TN, FP],
2   [ 198  421]]             #  [FN, TP]]
3                precision    recall  f1-score   support
4  
5             0       0.84      0.96      0.89      4709
6             1       0.68      0.33      0.44      1291
7  
8      accuracy                           0.82      6000
9     macro avg       0.76      0.64      0.67      6000
10 weighted avg       0.80      0.82      0.80      6000
```

```
11
12  [[4434  800]              # [[TN, FP],
13   [ 275  491]]             #  [FN, TP]]
14                precision    recall  f1-score   support
15
16             0       0.85      0.94      0.89      4709
17             1       0.64      0.38      0.48      1291
18
19      accuracy                           0.82      6000
20     macro avg       0.74      0.66      0.68      6000
21  weighted avg       0.80      0.82      0.80      6000
```

Programme 13.19 Applying CART and the random forest to the credit default dataset via Python.

```
1   > library("caret")                          # confusionMatrix
2   >
3   > set.seed(4002)                                    # set random seed
4   > df <- read.csv("credit default.csv")     # read in data in csv format
5   > df$default.payment.next.month <- as.factor(
6   +   df$default.payment.next.month
7   + )                       # change label into factor for classification
8   >
9   > train_idx <- sample(1:nrow(df), size=floor(nrow(df)*0.8))
10  > df_train <- df[train_idx,]                      # training dataset
11  > df_test <- df[-train_idx,]                      # test dataset
12  >
13  > ctree <- rpart(default.payment.next.month~., data=df_train,
14  +                method="class")
15  > rpart.plot(ctree, extra=1, cex=2, digits=4, nn=TRUE) # plot ctree
16  > prob <- predict(ctree, newdata=df_test)
17  > y_hat_dt <- colnames(prob)[max.col(prob)]
18  > # confusionMatrix(y_test, y_true, ...)
19  > dt_result <- confusionMatrix(as.factor(y_hat_dt),
20  +                              df_test$default.payment.next.month,
21  +                              mode="prec_recall", positive="1")
22  > dt_result$table              # Confusion matrix
23            Reference
24  Prediction    0    1
25           0 4440  920
26           1  196  444
27  > dt_result$byClass[c("Precision", "Recall")]
28  Precision    Recall
29     0.6937    0.3255
30  >
31  > rf_clf <- randomForest(default.payment.next.month~., data=df_train,
32  +                        ntree=10, importance=TRUE)
33  > y_hat_rf <- predict(rf_clf, newdata=df_test)
34  > rf_result <- confusionMatrix(as.factor(y_hat_rf),
35  +                              df_test$default.payment.next.month,
36  +                              mode="prec_recall", positive="1")
37  > rf_result$table                # Confusion matrix
```

```
38             Reference
39 Prediction      0     1
40             0 4324   872
41             1  312   492
42 > rf_result$byClass[c("Precision", "Recall")]
43 Precision      Recall
44    0.6119      0.3607
```

Programme 13.20 Applying CART and the random forest to the credit default dataset via R.

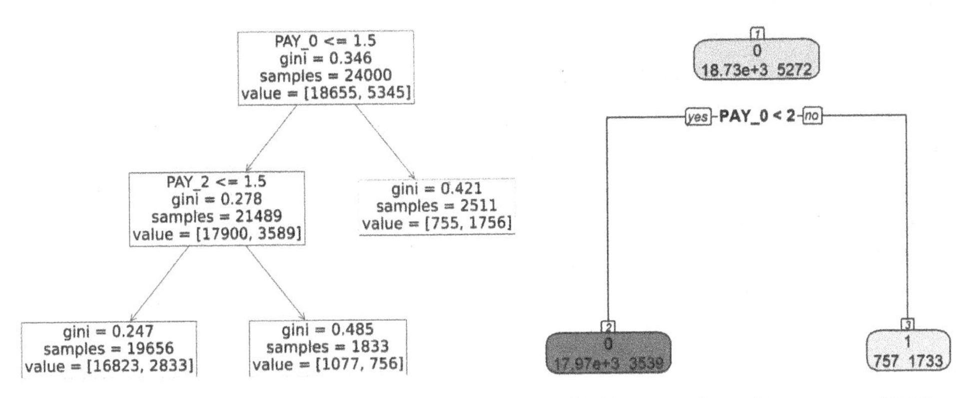

(a) Python output from Programme 13.19 (b) **R** output from Programme 13.20

FIGURE 13.18 Classification trees for the Taiwan credit default dataset.

From the output of Programme 13.19, various measures of the classification tree built in Python such as precision, recall, F_1-score, and accuracy are:

$$\text{Precision} = \frac{421}{421 + 198} = 0.680, \qquad \text{Recall} = \frac{421}{421 + 870} = 0.326,$$

$$f_1\text{-score} = \frac{2}{1/0.680 + 1/0.326} = 0.441, \qquad \text{Accuracy} = \frac{421 + 4511}{6000} = 0.822.$$

Similarly, the precision, recall, F_1-score, and accuracy for the random forest built computed in Python are:

$$\text{Precision} = \frac{491}{491 + 275} = 0.641, \qquad \text{Recall} = \frac{491}{491 + 800} = 0.380,$$

$$f_1\text{-score} = \frac{2}{1/0.641 + 1/0.380} = 0.477, \qquad \text{Accuracy} = \frac{491 + 4434}{6000} = 0.821.$$

From Figure 13.18a, Python splits the dataset according to only 2 out of 26 feature variables, namely `PAY_0` and `PAY_2`, which represent the repayment status in September of 2005 and August of 2005, respectively. They record the number of months that the client delayed his/her payments, with `-1` indicating no delay and a maximum value of nine months. It is true that a customer will likely default if his/her payments have already been delayed for several months, and the magic number found by Python is 2 (the smallest integer that is larger than 1.5) found by Python. Similarly, from Figure 13.18b, **R** splits the dataset according to `PAY_0` with the same magic number of 2. We here adopt the `confusionMatrix()` function from the `caret` library to compute various performance metrics in **R**. Readers can also refer to our user guide [2] for more discussions.

13.A APPENDIX

13.A.1 Entropy in Information Theory

We here motivate the definition of entropy introduced in Section *13.2 with its information-theoretic importance. Readers can refer to [3] for more details. Let $\mathrm{x}_1, \ldots, \mathrm{x}_n \in \mathcal{X}$ be iid discrete random variables with a common probability mass function $p(x)$. An application of the weak law of larger numbers readily implies the *Asymptotic Equipartition Property* (AEP), namely:

$$-\frac{1}{n}\log_2 p(\mathrm{x}_1, \ldots, \mathrm{x}_n) = -\frac{1}{n}\sum_{i=1}^{n} \log_2 p(\mathrm{x}_i) \to -\mathbb{E}(\log_2 p(\mathrm{x})) = H(\mathrm{x}),$$

in probability, as n goes to infinity. Fix an $\varepsilon > 0$ and $n \in \mathbb{N}$, consider a *typical set* $A_\varepsilon^{(n)}$ which is a subset of all sequences $(x_1, \ldots, x_n) \in \mathcal{X}^n$ that satisfy

$$2^{-n(H(\mathrm{x})+\varepsilon)} \leq p(x_1, \ldots, x_n) \leq 2^{-n(H(\mathrm{x})-\varepsilon)}, \tag{13.22}$$

or equivalently,

$$H(\mathrm{x}) - \varepsilon \leq -\frac{1}{n}\log_2 p(x_1, \ldots, x_n) \leq H(\mathrm{x}) + \varepsilon.$$

Since

$$\begin{aligned} 1 \geq \mathbb{P}\left((\mathrm{x}_1, \ldots, \mathrm{x}_n) \in A_\varepsilon^{(n)}\right) &= \sum_{(x_1, \ldots, x_n) \in A_\varepsilon^{(n)}} p(x_1, \ldots, x_n) \\ &\geq |A_\varepsilon^{(n)}| \cdot 2^{-n(H(\mathrm{x})+\varepsilon)}, \end{aligned}$$

the cardinality satisfies $\left|A_\varepsilon^{(n)}\right| \leq 2^{n(H(\mathrm{x})+\varepsilon)}$, and by virtue of the AEP, $A_\varepsilon^{(n)}$ should satisfy $\lim_{n\to\infty} \mathbb{P}\left((\mathrm{x}_1, \ldots, \mathrm{x}_n) \in A_\varepsilon^{(n)}\right) = 1$. With these in mind, for large n, the typical set $A_\varepsilon^{(n)}$ captures most of the sequences in $\mathcal{X}^n$, while the elements in $A_\varepsilon^{(n)}$ are almost equiprobable due to (13.22), and this intuition gives the central idea of data compression.

To this end, suppose that we aim to compress a message $x^{(n)} = (x_1, \ldots, x_n)$ with n alphabets from $\mathcal{X}$ into a binary code. We want to examine, on average, the number (length) of bits $l(x^{(n)})$ required to code an arbitrary message of $x^{(n)} \in \mathcal{X}^n$. By assuming the alphabets $\mathrm{x}_1, \ldots, \mathrm{x}_n$ arrive identically and independently, the problem can be reformulated as computing $\mathbb{E}(l(\mathrm{x}^{(n)}))$, where $\mathrm{x}^{(n)} = (\mathrm{x}_1, \ldots, \mathrm{x}_n)$ is the random message. We then decompose $\mathcal{X}^n$ into the typical set $A_\varepsilon^{(n)}$ and its complement $\overline{A_\varepsilon^{(n)}}$. From the discussion above, it suffices to use at most $n(H(\mathrm{x}) + \varepsilon) + 1$ bits to represent all sequences of alphabets $x^{(n)} \in A_\varepsilon^{(n)}$. On the other hand, by direct counting, we can certainly use at most $n \log_2 |\mathcal{X}| + 1$ bits to code all sequences in $\overline{A_\varepsilon^{(n)}}$. Together with the AEP and the fact that $\mathbb{P}(\mathrm{x}^{(n)} \in A_\varepsilon^{(n)}) \geq 1 - o_n(1)$, where $o_n(1) \to 0$ as $n \to \infty$, we can derive the following:

$$
\begin{aligned}
\mathbb{E}[l(\mathrm{x}^{(n)})] &= \sum_{x^{(n)} \in \mathcal{X}^n} p(x^{(n)}) l(x^{(n)}) \\
&= \sum_{x^{(n)} \in A_\varepsilon^{(n)}} p(x^{(n)}) l(x^{(n)}) + \sum_{x^{(n)} \notin A_\varepsilon^{(n)}} p(x^{(n)}) l(x^{(n)}) \\
&\leq \mathbb{P}(\mathrm{x}^{(n)} \in A_\varepsilon^{(n)})(n(H(\mathrm{x}) + \varepsilon) + 1) + \mathbb{P}(\mathrm{x}^{(n)} \notin A_\varepsilon^{(n)})(n \log_2 |\mathcal{X}| + 1) \\
&\leq n(H(\mathrm{x}) + \epsilon) + 1 + o_n(1)(n \log_2 |\mathcal{X}| + 1) \\
&= n(H(\mathrm{x}) + \varepsilon'),
\end{aligned}
$$

where $\varepsilon' = \varepsilon + \frac{1+o_n(1)}{n} + o_n(1) \log_2 |\mathcal{X}|$. Therefore,

$$
\mathbb{E}\left(\frac{1}{n} l(\mathrm{x}^{(n)})\right) \leq H(\mathrm{x}) + \varepsilon.
$$

By using similar arguments or otherwise, we can easily show that $\liminf_{n\to\infty} \mathbb{E}\left(\frac{1}{n} l(\mathrm{x}^{(n)})\right) \geq H(\mathrm{x})$, and hence $\lim_{n\to\infty} \mathbb{E}\left(\frac{1}{n} l(\mathrm{x}^{(n)})\right) = H(\mathrm{x})$; as a result, one only needs $nH(\mathrm{x})$ bits on average to code sequences in $\mathcal{X}^n$ when n is sufficiently large.

13.A.2 Solving Wordle with Entropy

Wordle is a web-based word game developed by *Josh Wardle*, released regularly by the *New York Times* as a word puzzle for its readers. Players have six attempts to guess a five-letter word, with feedback given for each guess: each guessed letter is marked as either green, yellow or gray. Green indicates that the corresponding letter appears in the correct position of the target word; yellow color indicates that the corresponding letter is in the target word but not positioned correctly; while gray color means that the corresponding letter is not in the target word. As a special rule, in the event of multiple copies of a specific letter in the guess, let say m of them, appear in a guess, such as the A's and R's in RADAR (where $m = 2$ for both letters) or the M's in MUMMY (where $m = 3$), those in the correct positions, let say m_0 of them, will first be colored as green, then the leftmost $(m - m_0)_+$ among the remaining copies will be colored yellow, while the rest of them are displayed as gray; see Figure 13.19 for an illustration, where readers can try to recover the color of each position in every attempt based on the aforementioned rules[11]. Furthermore, while players are allowed to make any guess as long as the guess is a valid five-letter word, the game also has a "hard mode" option, which requires players to include letters marked as green and yellow in subsequent guesses. Upon the knowledge of all valid five-letter words, we here design an algorithm that attempts to narrow down the list of possible words in each guess, in the hope of arriving at the target word in the smallest possible number of guesses on average. By the very nature of our proposed algorithm, it also works in the hard mode.

To illustrate the idea, suppose that the word bank contains the totality of n five-letter words, and that each word in the bank is equally probable to be selected as the target word. Furthermore, we represent the 26 letters from A to Z by numerals 1 to 26 in order, respectively. In this way, each five-letter word in the word bank corresponds to a 5-tuple $\boldsymbol{\omega}_i = (\omega_{i1}, \omega_{i2}, \omega_{i3}, \omega_{i4}, \omega_{i5})^\mathsf{T}$ where $\omega_{ij} \in \{1, \ldots, 26\}$, for $i = 1, \ldots, n$ and $j = 1, \ldots, 5$. We further denote the word bank of all $\boldsymbol{\omega}_i$'s, $i = 1, \ldots, n$, as $\mathcal{W}$, and the target word as $\boldsymbol{\omega}^* = (\omega_1^*, \omega_2^*, \omega_3^*, \omega_4^*, \omega_5^*)^\mathsf{T}$. Our goal is to make a guess $\tilde{\boldsymbol{\omega}} = (\tilde{\omega}_1, \tilde{\omega}_2, \tilde{\omega}_3, \tilde{\omega}_4, \tilde{\omega}_5)^\mathsf{T}$ such that $\tilde{\omega}_j = \omega_j^*$ for $j = 1, \ldots, 5$.

Given the target word $\boldsymbol{\omega}^*$ and the guess $\tilde{\boldsymbol{\omega}} = (\tilde{\omega}_1, \ldots, \tilde{\omega}_5)^\mathsf{T}$, we can obtain the realization of the color vector $\boldsymbol{c} = \boldsymbol{c}(\tilde{\boldsymbol{\omega}}) = (c_1, \ldots, c_5)^\mathsf{T}$, where $c_j \in \{\text{Green, Yellow, Gray}\}$, for $j = 1, \ldots, 5$, such that the color of each component is taken by:

1. Green, if $\tilde{\omega}_j = \omega_j^*$;
2. Yellow, if (i) $\tilde{\omega}_j \neq \omega_j^*$, (ii) $\prod_{\ell=1}^{5} (\tilde{\omega}_j - \omega_\ell^*) = 0$, and (iii) $\sum_{\ell=1}^{j-1} \mathbb{1}_{\{\tilde{\omega}_j = \tilde{\omega}_\ell\}} \leq \sum_{\ell=1}^{5} \mathbb{1}_{\{\tilde{\omega}_j = \omega_\ell^*, \tilde{\omega}_\ell \neq \omega_\ell^*\}} = m - m_0$;
3. Gray, otherwise.

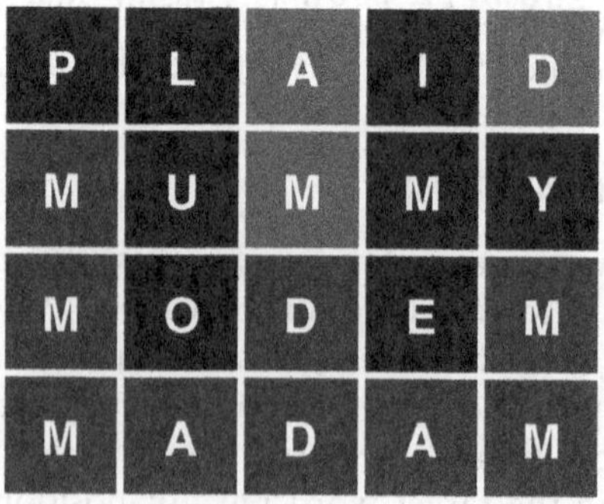

P	L	A	I	D
M	U	M	M	Y
M	O	D	E	M
M	A	D	A	M

FIGURE 13.19 An illustration of *Wordle* gameplay under normal mode. The target word MADAM is revealed in four attempts.

[11] Refer to the e-book online for a clear colored version.

There are at most $3^5 = 243$ possible realizations of $\boldsymbol{c}$. When a guess is made, the color vector is revealed, based on which one can narrow down the range of possible five-letter words. The key iteration step in the algorithm is to find the "best" guess in the sense that it can narrow down the range of candidate words for the next step as much as possible on average.

To explain the algorithm, we denote by $\mathcal{W}^{(0)} := \mathcal{W}$ the initial word bank, and $\mathcal{W}^{(k)}, k = 1, 2, \ldots$, the shortlisted collection of words after the k-th guess and the color information revealed so far, also denote $n^{(k)} := |\mathcal{W}^{(k)}|$. Let us now mathematically define the action of "narrowing down the range of possible words" in a recursive manner. Suppose that the k-th guess $\tilde{\boldsymbol{\omega}}^{(k)} = (\tilde{\omega}_1^{(k)}, \ldots, \tilde{\omega}_5^{(k)})^\top \in \mathcal{W}^{(k-1)}$ returns the color vector $\boldsymbol{c}^{(k)} = (c_1^{(k)}, \ldots, c_5^{(k)})^\top$. Then,

$$\begin{aligned}\mathcal{W}^{(k)} = \Big\{ \boldsymbol{\omega} = (\omega_1, \ldots, \omega_5)^\top \in \mathcal{W}^{(k-1)} \Big| \; & \omega_j = \tilde{\omega}_j^{(k)} \text{ if } c_j^{(k)} = \text{Green};\\ & \text{Conditions (i), (ii) and (iii) are fulfilled if } c_j^{(k)} = \text{Yellow};\\ & \prod_{\ell=1}^{5} (\omega_\ell - \tilde{\omega}_j^{(k)}) \neq 0 \text{ if } c_j^{(k)} = \text{Gray for all } j = 1, \ldots, 5 \Big\}.\end{aligned} \tag{13.23}$$

At the stage before making the k-th guess, for each specific guess, we can calculate the proportion of five-letter words in the word bank that lead to each possible realization of the color vector assuming that they are the correct word. More rigorously, we consider the random color vector $\boldsymbol{c}(\tilde{\boldsymbol{\omega}}^{(k)})$ that clearly depends on the guess $\tilde{\boldsymbol{\omega}}^{(k)}$, and randomness comes from the unknown target word. The probability mass function of $\boldsymbol{c}(\tilde{\boldsymbol{\omega}}^{(k)})$ is then the proportion of the words in the word bank $\mathcal{W}^{(k-1)}$ that, were they to be target words, would lead to a certain realization of the color vector. For example, the color vector (Green, Green, Green, Green, Green)$^\top$ can only appear when the target word happens to be the same as the guess, therefore we have

$$\mathbb{P}\left(\boldsymbol{c}(\tilde{\boldsymbol{\omega}}^{(k)}) = (\text{Green, Green, Green, Green, Green})^\top \right) = \frac{1}{n^{(k-1)}} =: p_1(\tilde{\boldsymbol{\omega}}^{(k)}),$$

where we systematically assign an index from 1 to 243 for each realization of the color vector for the sake of convenience. The entropy of this random variable can then be calculated as:

$$H\left(\boldsymbol{c}(\tilde{\boldsymbol{\omega}}^{(k)}) \right) := -\sum_{i=1}^{243} p_i(\tilde{\boldsymbol{\omega}}^{(k)}) \log_2 p_i(\tilde{\boldsymbol{\omega}}^{(k)}).$$

In order to find a multipurpose word that can provide as much information for subsequent guesses as possible via the color signals, the best word for the k-th guess is the one that maximizes the entropy, as opposed to maximizing "information gain" in the classification tree algorithm[12]:

$$\tilde{\omega}_*^{(k)} = \arg\max_{\omega \in \mathcal{W}^{(k-1)}} H(c(\omega)) = \arg\min_{\omega \in \mathcal{W}^{(k-1)}} \sum_{i=1}^{243} p_i(\omega) \log_2 p_i(\omega). \tag{13.24}$$

This Wordle algorithm can be formally stated as follows:

Algorithm 13.1 Wordle algorithm

1: Initialization: $\mathcal{W}^{(0)} := \mathcal{W}$ and $c^{(0)} = (\text{Gray, Gray, Gray, Gray, Gray})^\top$;
2: **while** $c^{(k)} \neq (\text{Green, Green, Green, Green, Green})^\top$ **do**
3: Search for the "optimal" guess $\tilde{\omega}_*^{(k)} \in \mathcal{W}^{(k-1)}$ by using (13.24);
4: Make the guess and obtain $c^{(k)}$, the realization of the color vector;
5: Based on $c^{(k)}$, update the word bank from $\mathcal{W}^{(k-1)}$ to $\mathcal{W}^{(k)}$ by (13.23);
6: **end while**

The implementation of this algorithm via **R** can be found on the book's GitHub repository with the file `Wordle.R` and the file `Wordle.csv`[13] for the alphabetically sorted word bank. According to Algorithm 13.1, the initial optimal guess $\tilde{\omega}_*^{(0)} \in \mathcal{W}^{(0)}$ is found to be the word `tares`. However, from Table 13.4, it is found that the word `alter` provides a greater count of success within six attempts and we shall use this word as the initial guess[14]. While the word `tares` performs better in early stages, it is optimal only for the initial trial but sub-optimal for the game as a whole.

Finally, by using the proposed algorithm, we can mostly solve the Wordle puzzle in four to five attempts, which is well within the limit.

TABLE 13.4 Total counts on the number of trials required for Algorithm 13.1.

# trials	1	2	3	4	5	6	7	8	9	10	11	12	13	14	15	16+
tares	1	213	2572	5632	3506	1524	678	328	187	109	54	27	13	8	2	1
alter	1	168	2510	5854	3607	1434	606	314	173	95	44	24	14	7	1	3

[12] Similar ideas have been around on the internet and in the public domain (see for instance https://towardsdatascience.com/information-theory-applied-to-wordle-b63b34a6538e). This entropy-based approach indeed has been commonly used in the context of speech recognition.
[13] This dataset (randomly shuffled) is retrieved from https://github.com/tabatkins/wordle-list.
[14] Readers can try other words as the initial guess but the programme runs for one day.

REFERENCES

1. Breiman, L., Friedman, J., Stone, C.J., and Olshen, R.A. (1984). *Classification and Regression Trees*. CRC Press.
2. Chen, Y., Cheung, K.C., Sun, R.Z., and Yam, S.C.P. (2024). A user guide of CART and random forests with applications in FinTech and InsurTech. Invited article in *Special Issue on Risk and Statistics in Actuarial Science* of *Japanese Journal of Statistics and Data Science*.
3. Cover, T.M., and Thomas, J.A. (2006). *Elements of Information Theory* (2nd ed.). Wiley.
4. Esposito, F., Malerba, D., Semeraro, G., and Kay, J. (1997). A comparative analysis of methods for pruning decision trees. *IEEE Transactions on Pattern Analysis and Machine Intelligence*, 19(5), 476–491.
5. Fürnkranz, J. (1997). Pruning algorithms for rule learning. *Machine Learning*, 27, 139–172.
6. Gepp, A., Kumar, K., and Bhattacharya, S. (2010). Business failure prediction using decision trees. *Journal of Forecasting*, 29(6), 536–555.
7. Gepp, A., Wilson, J.H., Kumar, K., and Bhattacharya, S. (2012). A comparative analysis of decision trees vis-à-vis other computational data mining techniques in automotive insurance fraud detection. *Journal of Data Science*, 10(3), 537–561.
8. Gordon, A.D. (1999). *Classification*. CRC Press.
9. Kullback, S. (1997). *Information Theory and Statistics*. Courier Corporation.
10. Lichman, M. (2013). UCI Machine Learning Repository, University of California, School of Information and Computer Science. http://archive.ics.uci.edu/ml (accessed 7 May 2024).
11. Mingers, J. (1989). An empirical comparison of pruning methods for decision tree induction. *Machine Learning*, 4, 227–243.
12. Quan, Z., and Valdez, E.A. (2018). Predictive analytics of insurance claims using multivariate decision trees. *Dependence Modeling*, 6(1), 377–407.
13. Shannon, C.E. (1948). A mathematical theory of communication. *The Bell System Technical Journal*, 27(3), 379–423.
14. Shannon, C.E. (1949). *The Mathematical Theory of Communication*. The University of Illinois Press.
15. Smith, K.A., Willis, R.J., and Brooks, M. (2000). An analysis of customer retention and insurance claim patterns using data mining: a case study. *Journal of the Operational Research Society*, 51, 532–541.
16. Tan, P.-N., Steinbach, M., Karpatne, A., and Kumar, V. (2018). *Introduction to Data Mining* (2nd ed.). Pearson.
17. Therneau, T., Atkinson, B., Ripley, B., and Ripley, M.B. (2015). Package 'rpart'. http://cran.ma.ic.ac.uk/web/packages/rpart/rpart.pdf (accessed 20 April 2016).
18. Viaene, S., Derrig, R.A., Baesens, B., and Dedene, G. (2002). A comparison of state-of-the-art classification techniques for expert automobile insurance claim fraud detection. *Journal of Risk and Insurance*, 69(3), 373–421.
19. Wagner, C.H. (1982). Simpson's paradox in real life. *The American Statistician*, 36(1), 46–48.

CHAPTER 14

Cluster Analysis

Most of the methods mentioned in this book, such as logistic regression in Chapter 10, classification and regression trees and random forest in Chapter 13, as well as *neural networks* (to be discussed in Chapter 15), require the presence of a response variable or a classification label for each observation. This allows us to monitor or "supervise" how well the methods perform by comparing their predicted results with the actual labels; as such, we say they are examples of *supervised learning*. For instance, for the 2002 HSI dataset of `fin-ratio.csv`, all stocks have the labels of `HSI` = 0 or 1. Other examples are: customer transaction records with bad debt or not; patient health records with a certain disease or not. However, real-life datasets that do not contain label information are prevalent. As a result, we may not know how many groups or labels actually exist in the dataset. Take marketing research for example. We want to classify the customers into several subgroups so that we can adopt different marketing strategies to promote tailor-made products to each of them, while the actual number of subgroups is mostly unknown to us with limited data. The methods used in such scenarios have no objective reference points for us to monitor their performance, and are thus named *unsupervised learning*. The problems in unsupervised learning are generally much more difficult to cope with than those in supervised learning. For instance, in the previous marketing example, selecting an appropriate number of subgroups is already a challenge; too few subgroups may yield too general strategies to address specific needs of customers, and hence cannot fulfill the taste of and so retain customers; while too many subgroups would add to the cost of promotion, which is undesirable for the companies.

14.1 *K*-MEANS CLUSTERING

As another example of an intrinsic difficulty in unsupervised learning, namely the absence of a unique objective criterion for grouping data, consider a deck of 52 playing cards. While normally we can group the cards according to their color (red or black), suits (spades, heart, club and diamond); or face values (J, Q, K, A or 1 to 10), we do not have clear objectives in unsupervised learning for picking an obvious choice of *measure of similarity* or *dissimilarity*. Nevertheless, as long as a measure of similarity (or dissimilarity) is fixed, we can classify the data into clusters such that observations within each cluster are as homogeneous as possible, meanwhile observations belonging to different clusters are as heterogeneous as possible. This approach is often referred to as *Cluster Analysis*, which is a branch of unsupervised learning. There are two major types of methods in cluster analysis, namely *hierarchical* and *non-hierarchical* clustering. Hierarchical clustering methods merge (resp. split) observations based on the outcomes of the previous step in an iterative manner, which corresponds to the branch of *agglomerative* (resp. *divisive*) methods. One typical example is the AGNES algorithm used in CIBer in Section 12.5. In general, this procedure formulates a tree-hierarchy structure with lines joining all the observations. However, it may not be feasible when the number of observations is large. To overcome the hurdle of handling very large datasets, a very popular non-hierarchical clustering method can be employed: *K-means clustering*.

(I) Lloyd's Algorithm

K-means clustering is fast, efficient and suitable for *data mining* very large datasets. While the term "K-means" was first coined by *James MacQueen* in [17], the standard algorithm for the clustering was in fact first proposed by *Stuart Lloyd* earlier in 1957 as a technique for *pulse-code modulation* (PCM)[1], but his paper remained unpublished until 1982 [15].

K-means clustering is an approach based on partitioning. Let $S = \{x_i\}_{i=1}^n$ be the dataset of observations, where $x_i \in \mathbb{R}^p$ for $i = 1, \ldots, n$. Suppose that they are partitioned into K regions (clusters), $S_1, \ldots, S_K$, such that each region S_k is represented generically by a vector c_k, which is usually the *geometric centroid* of S_k. With the number of clusters K specified in advance, a function q mapping an observation x to the centroid of the corresponding cluster is given by:

$$q(x) := \sum_{k=1}^{K} \mathbb{1}_{\{x \in S_k\}} c_k. \tag{14.1}$$

In K-means clustering, the centroids $c_1, \ldots, c_K$, and the map q are constructed in such a way that the mean-squared error between the observations and their assigned or associated centroids is minimized. Mathematically, the goal of K-means clustering is to partition S into a disjoint union $S = \bigsqcup_{k=1}^{K} S_k$ so that the following error sum of squares is minimized:

$$\sum_{k=1}^{K} \sum_{x \in S_k} \|x - c_k\|_2^2, \tag{14.2}$$

where $c_k = \frac{1}{|S_k|} \sum_{x \in S_k} x$ is the centroid of S_k. Indeed, when the dimension p and number of clusters K are fixed, [13] showed that solving for a globally optimal partition $\{S_1^*, \ldots, S_K^*\}$ of the objective function (14.2) can be done in time $O(n^{pK+1})$. However, when K is unknown, the problem becomes *NP-hard*[2]. Among the various heuristic

[1] PCM is a method in signal processing used to digitally represent analog signals, where an analog signal is first sampled, and then each sample is discretized by a predefined set of bins. Finally, the discretized samples are encoded into binary codes.

[2] In *computational complexity theory*, an NP-hard (non-deterministic polynomial-time hard) problem is characterized as follows: should an algorithm in polynomial time be developed for an NP-hard problem, for all other problems where we can check the solution in polynomial time, we could also develop algorithms in polynomial time to solve the problem. A typical example of an NP-hard problem is the *Travelling Salesman problem*.

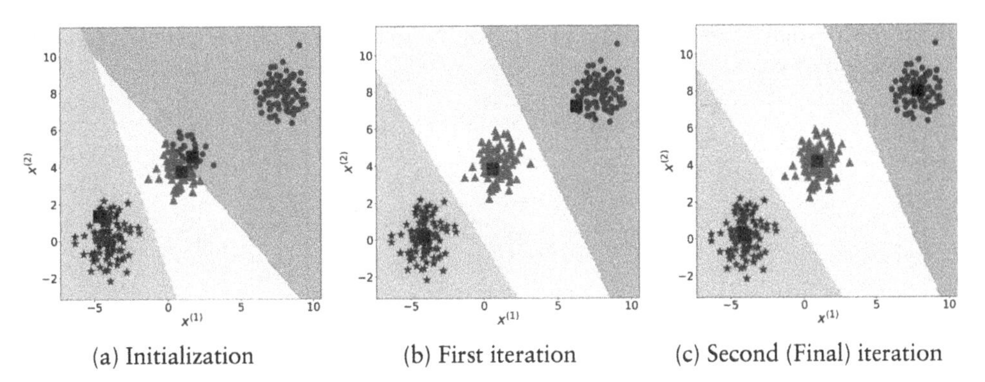

(a) Initialization (b) First iteration (c) Second (Final) iteration

FIGURE 14.1 The illustration for the iterations with $p = 2$ and $K = 3$; the *bisectors* of cluster boundaries are shown, and centroids of clusters are indicated by solid squares in black.

algorithms in place, the classical K-means algorithm first introduced by Stuart Lloyd (often referred to as *Lloyd's algorithm*) can effectively solve a local minimum of (14.2). Its procedure is described in the following Algorithm 14.1.

Algorithm 14.1 Lloyd's Algorithm

1: Randomly choose K data points in $\mathbb{R}^p$, say $\boldsymbol{c}_1^{(0)}, \dots, \boldsymbol{c}_K^{(0)}$, as initial seeds. They are the "centroids" of K initial clusters; see Figure 14.1a.

2: At each time step $t = 1, 2, \dots$, for each observation $\boldsymbol{x}_i$, calculate the distance of $\boldsymbol{x}_i$ to the current centroids $\boldsymbol{c}_1^{(t-1)}, \dots, \boldsymbol{c}_K^{(t-1)}$, and assign it to cluster k if it is closest to the k-th centroid $\boldsymbol{c}_k^{(t-1)}$, i.e.,

$$S_k^{(t)} = \left\{ \boldsymbol{x}_i : \left\| \boldsymbol{x}_i - \boldsymbol{c}_k^{(t-1)} \right\|_2^2 \leq \min_{r=1,\dots,K} \left\| \boldsymbol{x}_i - \boldsymbol{c}_r^{(t-1)} \right\|_2^2 \right\}. \tag{14.3}$$

3: Update the centroids for this time step t by using the sample mean of points from the cluster $S_k^{(t)}$ obtained in Step 2, also see Figures 14.1b and 14.1c:

$$\boldsymbol{c}_k^{(t)} = \frac{1}{\left|S_k^{(t)}\right|} \sum_{\boldsymbol{x}_i \in S_k^{(t)}} \boldsymbol{x}_i, \text{ for } k = 1, 2, \dots, K.$$

4: Repeat Step 2 and Step 3 until convergence, or stop when

$$\max_{k=1,\dots,K} \left\| \boldsymbol{c}_k^{(t)} - \boldsymbol{c}_k^{(t-1)} \right\|_2^2 < \varepsilon,$$

for a small enough pre-specified threshold $\varepsilon > 0$.

We show the local optimality of Lloyd's algorithm as follows. For any time step t, we denote the value of the objective function (14.2) after Steps 2 and 3 in Algorithm 14.1 by $L_{t,1}$ and $L_{t,2}$, respectively. By the nature of Step 2, we clearly have $L_{t,1} \leq L_{t-1,2}$. For Step 3 of the centroid updating, we fix an arbitrary cluster $S_k^{(t)}$, and focus on the mean squared error for this cluster only. Note that the following sum of squares is a function of c_k,

$$\sum_{x \in S_k^{(t)}} \|x - c_k\|_2^2 = \sum_{x \in S_k^{(t)}} (x - c_k)^\top (x - c_k)$$
$$= \sum_{x \in S_k^{(t)}} (x^\top x - 2x^\top c_k + c_k^\top c_k),$$

whose first order condition implies that

$$\sum_{x \in S_k^{(t)}} (-2x + 2c_k) = \mathbf{0} \iff c_k^* = \frac{\sum_{x \in S_k^{(t)}} x}{|S_k^{(t)}|} = c_k^{(t)}.$$

Furthermore, by the quadratic nature of the mean squared error, $c_k^{(t)}$ is the unique minimum of $\sum_{x \in S_k^{(t)}} \|x - c_k\|_2^2$. By the arbitrariness of k, this result holds for all clusters, which then yields $L_{t,2} \leq L_{t,1}$ by summing up across clusters.

Now, since $L_{t,2} \leq L_{t,1} \leq L_{t-1,2}$ for all t, and $L_{t,2} \geq 0$ by definition, we can conclude by the *monotone convergence theorem* that the objective function converges. Since there are only finitely many sample points, S is therefore bounded, and the *Heine–Borel theorem* ensures the limiting point exists and it is a local minimizer. Furthermore, using the fact that the sample size is finite again, then so is the total number of possible clustering scenarios. Recall that by the nature of Lloyd's algorithm, each update of clusters and centroids results in an objective function not greater in value than that in the previous step, hence we can usually achieve the optimal clustering after a finite number of steps and remain there. How many steps are needed to attain the optimal clustering depends on the choice of initial seeds.

(II) Illustrations for *K*-means clustering

The *K*-means clustering method is implemented in Python using the `KMeans()` function provided in the `sklearn` library, and in **R** using the built-in function `kmeans()`. We first illustrate its use by using Fisher's Iris data in Programme 14.1 via Python and Programme 14.2 via **R**.

```
1  import pandas as pd
2  import numpy as np
3  from sklearn.cluster import KMeans
4  import matplotlib.pyplot as plt
5  import seaborn as sns
6  from sklearn import datasets
7
8  plt.rc('font', size=20); plt.rc('axes', titlesize=30, labelsize=18)
9  plt.rc('xtick', labelsize=20); plt.rc('ytick', labelsize=20)
10
11 iris = datasets.load_iris(return_X_y=False)
12 X_iris, y_iris = iris["data"], iris["target"]
13 df_iris = pd.DataFrame(np.c_[X_iris, y_iris],
14                        columns=iris["feature_names"] + ["Species"])
15 df_iris["Species"].replace(dict(zip([0,1,2],iris["target_names"])), inplace=True)
```

```
16 fig = plt.figure(figsize=(10, 10))                   # Plot observations with color
17 sizes = np.array([600, 400, 300])
18 ax = sns.pairplot(df_iris, hue="Species", diag_kind=None,
19                   markers=[".", "*", "^"], plot_kws={'s': sizes[y_iris]},
20                   palette=dict(zip(iris["target_names"],
21                                   ["black", "red", "green"])))
22 for s, lh in zip(sizes, ax._legend.legendHandles): lh._sizes = [s]
23
24 np.random.seed(4002)
25 km_iris = KMeans(n_clusters=3, n_init="auto")   # K-means clustering with K=3
26 km_iris.fit(X_iris)
27
28 def get_bcss(km, X):
29     ng = np.bincount(km.labels_)
30     return sum(ng*np.sum((np.mean(X, axis=0) - km.cluster_centers_)**2, axis=1))
31
32 print(km_iris.cluster_centers_)                    # Mean of each cluster
33 print(km_iris.labels_)                             # Predicted cluster labels
34 print(np.bincount(km_iris.labels_))                # The size of each cluster
35 print("Between group sum of squares:", get_bcss(km_iris, X_iris))
36 print("Within group sum of squares:", km_iris.inertia_)
```

```
1  [[5.9016129  2.7483871  4.39354839 1.43387097]
2   [5.006      3.428      1.462      0.246     ]
3   [6.85       3.07368421 5.74210526 2.07105263]]
4  [1 1 1 1 1 1 1 1 1 1 1 1 1 1 1 1 1 1 1 1 1 1 1 1 1 1 1 1 1 1 1 1 1 1
5   1 1 1 1 1 1 1 1 1 1 1 1 1 1 1 1 0 0 2 0 0 0 0 0 0 0 0 0 0 0 0 0 0 0
6   0 0 0 0 0 0 0 2 0 0 0 0 0 0 0 0 0 0 0 0 0 0 0 0 0 0 0 0 0 0 2 0 2 2 2
7   2 0 2 2 2 2 2 2 0 0 2 2 2 2 0 2 0 2 0 2 2 0 0 2 2 2 2 2 0 2 2 2 2 0 2
8   2 2 0 2 2 2 0 2 2 0]
9  [62 50 38]
10 Between group sum of squares: 602.5191585738537
11 Within group sum of squares: 78.85144142614601
```

Programme 14.1 *K*-means clustering for the Iris flower dataset via Python.

```
1  > X_iris <- iris[,-5]                  # remove species label from iris
2  > par(cex.lab=2, cex.axis=2, cex.main=2, mar=c(5,5,4,4))
3  > plot(X_iris, col=iris[,5], pch=c(0,1,2)[y_iris],
4  +      cex=2, cex.labels=2.5)           # plot observations with color
5  >
6  > set.seed(4002)
7  > (km_iris <- kmeans(X_iris, 3))        # K-means clustering with K=3
8  K-means clustering with 3 clusters of sizes 50, 38, 62
9
10 Cluster means:
11   Sepal.Length Sepal.Width Petal.Length Petal.Width
12 1     5.006000    3.428000     1.462000    0.246000
13 2     6.850000    3.073684     5.742105    2.071053
14 3     5.901613    2.748387     4.393548    1.433871
15
16 Clustering vector:
17   [1] 1 1 1 1 1 1 1 1 1 1 1 1 1 1 1 1 1 1 1 1 1 1 1 1 1 1 1 1 1 1 1 1 1
18  [34] 1 1 1 1 1 1 1 1 1 1 1 1 1 1 1 1 1 3 3 2 3 3 3 3 3 3 3 3 3 3 3 3 3
19  [67] 3 3 3 3 3 3 3 3 3 3 3 2 3 3 3 3 3 3 3 3 3 3 3 3 3 3 3 3 3 3 3 3 3
```

```
20 [100] 3 2 3 2 2 2 2 3 2 2 2 2 2 2 3 3 2 2 2 2 3 2 3 2 3 2 2 3 3 2 2 2 2
21 [133] 2 3 2 2 2 2 3 2 2 2 3 2 2 2 3 2 2 3
22
23 Within cluster sum of squares by cluster:
24 [1] 15.15100 23.87947 39.82097
25  (between_SS / total_SS =  88.4 %)
26
27 Available components:
28
29 [1] "cluster"      "centers"      "totss"        "withinss"
30 [5] "tot.withinss" "betweenss"    "size"         "iter"
31 [9] "ifault"
32 > km_iris$betweenss                  # between group sum of squares
33 [1] 602.5192
34 > km_iris$tot.withinss               # within group sum of squares
35 [1] 78.85144
```

Programme 14.2 *K*-means clustering for the Iris flower dataset via **R**.

For the Iris data, both outputs from Programme 14.1 via Python and Programme 14.2 via **R** give 3 clusters of sizes 38, 50, and 62. Other items given in `km_iris` include:

1. `cluster_centers_` in Python and `centers` in **R**: the mean vectors for all 4 measurements of these 3 groups;
2. `labels_` in Python and `cluster` in **R**: the assigned cluster label for each datum based on the *K*-means clustering; note that the numbering of these group labels is arbitrary, e.g., the label 1 may represent any one of these 3 groups;
3. `inertia_` in Python and `tot.withinss` in **R**: the total within-cluster sum of squares; see more details and discussions in the next item (III) and Appendix *14.A.1.

Finally, we can produce the classification table of the labels from *K*-means clustering with the underlying "true" species in Programme 14.3 via Python and Programme 14.4 via **R**.

```
1 print(pd.crosstab(km_iris.labels_, y_iris))      # Classification table
```

```
1 col_0   0   1   2
2 row_0
3 0       0  48  14
4 1      50   0   0
5 2       0   2  36
```

Programme 14.3 Classification table with the "true" species via Python.

```
1 > table(km_iris$cluster, iris[,5])  # Classification table
2
3     setosa versicolor virginica
4   1     50          0         0
5   2      0          2        36
6   3      0         48        14
```

Programme 14.4 Classification table with the "true" species via **R**.

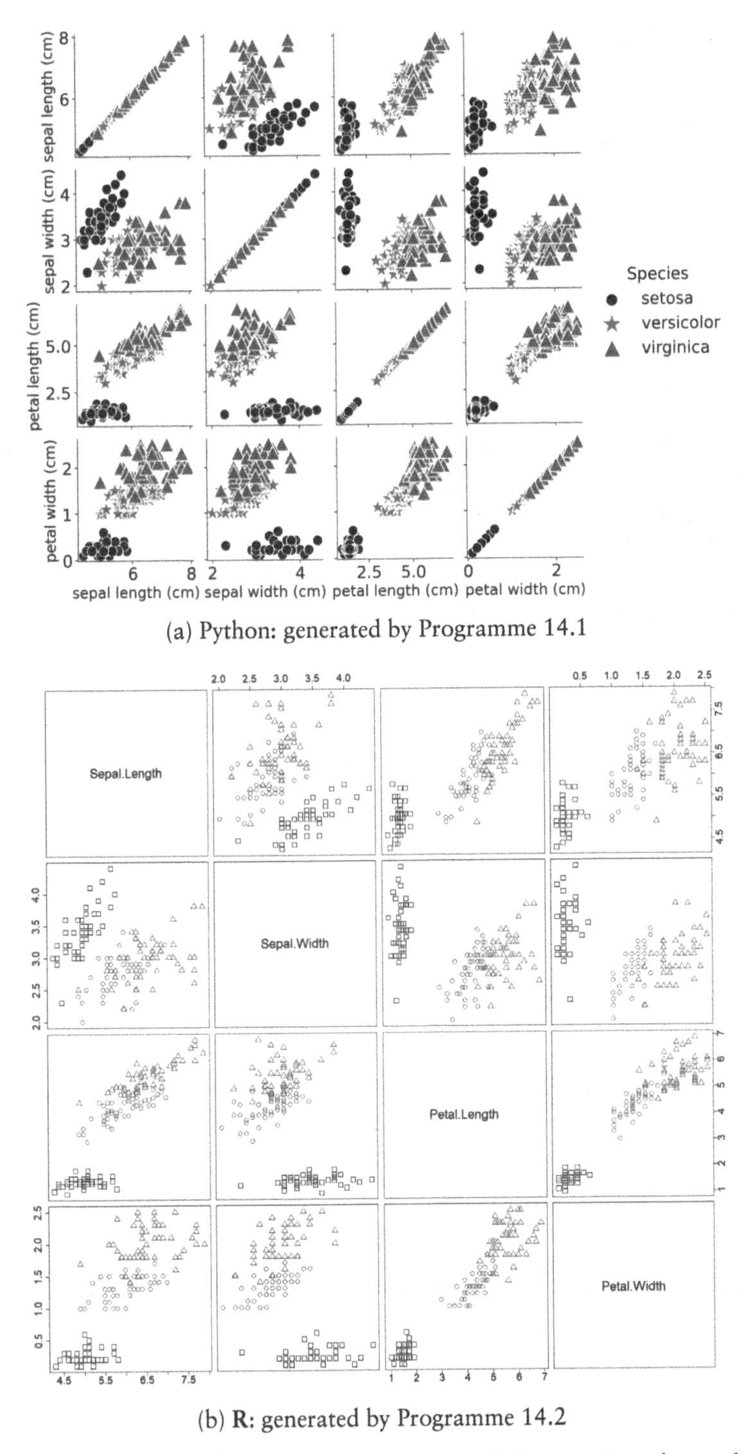

(a) Python: generated by Programme 14.1

(b) R: generated by Programme 14.2

FIGURE 14.2 Scatter plots of pairwise components of observations for each group in the Iris flower dataset.

TABLE 14.1 The correspondence between the original groups and the labels assigned for the Iris dataset.

Label assigned		Original group		
Python	R	setosa	virginica	versicolor
1	1	50	0	0
2	2	0	48	14
0	3	0	2	36

Again, the numbering of the label is arbitrary: in Python, the labels 0, 1, and 2 correspond to the original species of `versicolor`, `setosa`, and `virginica`, respectively; whereas in **R**, the labels 1, 2, and 3 correspond to the original species of `setosa`, `virginica`, and `versicolor` groups respectively; see Table 14.1. Therefore, the actual misclassification rates in Python and **R** are only $(14 + 2)/150 = 10.67\%$.

Next, we apply the *K*-means clustering to the 2002 HSI example, via Python in Programme 14.5 and via **R** in Programme 14.6.

```
1  plt.rc('font', size=20); plt.rc('axes', titlesize=30, labelsize=25)
2  plt.rc('xtick', labelsize=25); plt.rc('ytick', labelsize=25)
3
4  df_HSI = pd.read_csv("fin-ratio.csv")
5  X_HSI, y_HSI = df_HSI.iloc[:, :-1].values, df_HSI.HSI
6  fig = plt.figure(figsize=(15, 15))                # Plot observations with color
7  sizes = np.array([600, 300])
8  ax = sns.pairplot(df_HSI, hue="HSI", diag_kind=None, markers=[".", "^"],
9                    palette=dict(zip([0, 1], ["black", "red"])),
10                   plot_kws={'s': sizes[y_HSI]})
11 for s, lh in zip(sizes, ax._legend.legendHandles): lh._sizes = [s]
12
13 np.random.seed(4002)
14 km_HSI = KMeans(n_clusters=2, n_init="auto")     # K-means clustering with K=2
15 km_HSI.fit(X_HSI)
16
17 print(pd.crosstab(km_HSI.labels_, y_HSI))         # Classification table
```

```
1  HSI       0    1
2  row_0
3  0       159   11
4  1       489   21
```

Programme 14.5 *K*-means clustering for the 2002 HSI dataset via Python.

```
1  > df_HSI <- read.csv("fin-ratio.csv")     # read in HSI dataset
2  > X_HSI <- df_HSI[,-7]                    # remove HSI label
3  > plot(X_HSI, col=df_HSI[,7]+1, pch=c(1,2)[df_HSI[,7]+1],
4  +       cex=2, cex.labels=2.5)            # plot observations with color
5  >
6  > set.seed(4002)
7  > km_HSI <- kmeans(X_HSI, 2)              # K-means clustering with K=2
8  > table(km_HSI$cluster, df_HSI[,7]) # Classification table
9
10        0   1
11    1 496  24
12    2 152   8
```

Programme 14.6 *K*-means clustering for the 2002 HSI dataset via **R**.

From Figure 14.3, `ln_MV` is visually the most important and useful decision variable in the *K*-means clustering. In the third column, with `ln_MV` as the variable for the horizontal axis, the boundary between constituent and non-constituent stocks (the red triangles and black circles, respectively) seems the most clear compared to those using other variables for the horizontal axis in other columns. However, upon closer inspection, we can see that the two types of stock are not as separated as the different species of Iris flowers in the previous example, as quite a few red triangles are actually hidden in the cloud of black circles. This explains the relatively poor classification performance of *K*-means clustering in this case, with the misclassification rate as high as $(159 + 21)/680 = 26.47\%$ for Python and $(152 + 24)/680 = 25.88\%$ for **R**. Nevertheless, this result is not surprising; indeed, as an unsupervised learning approach, *K*-means clustering is more likely to give inferior results when the data with different class labels overlap.

(III) On the suitable choice of *K*

An immediate question is the choice of an appropriate number K of clusters in the *K*-means clustering. Fortunately, there is a useful statistic available to help us choosing a suitable K. Recall that our objective is to find clearly separable groups in the dataset, such that the records within the same group should be as homogeneous as possible while those between different groups are as different as possible.

The within-group and between-group variations can be obtained in Python via the `inertia_` component and the `get_bcss()` function of the model trained by `KMeans()` respectively. They are available in **R** as the respective components `tot.withinss` and `betweenss` of the model trained by `kmeans()`. Intuitively though, recalling the objective of clustering, these two variations can be combined to give a useful statistic. First, we revisit the Iris dataset and illustrate how these two statistics are computed; Programme 14.7 in Python (resp. Programme 14.8 in **R**) demonstrates how to reconcile the corresponding numerical results in Programme 14.1 (resp. Programme 14.2).

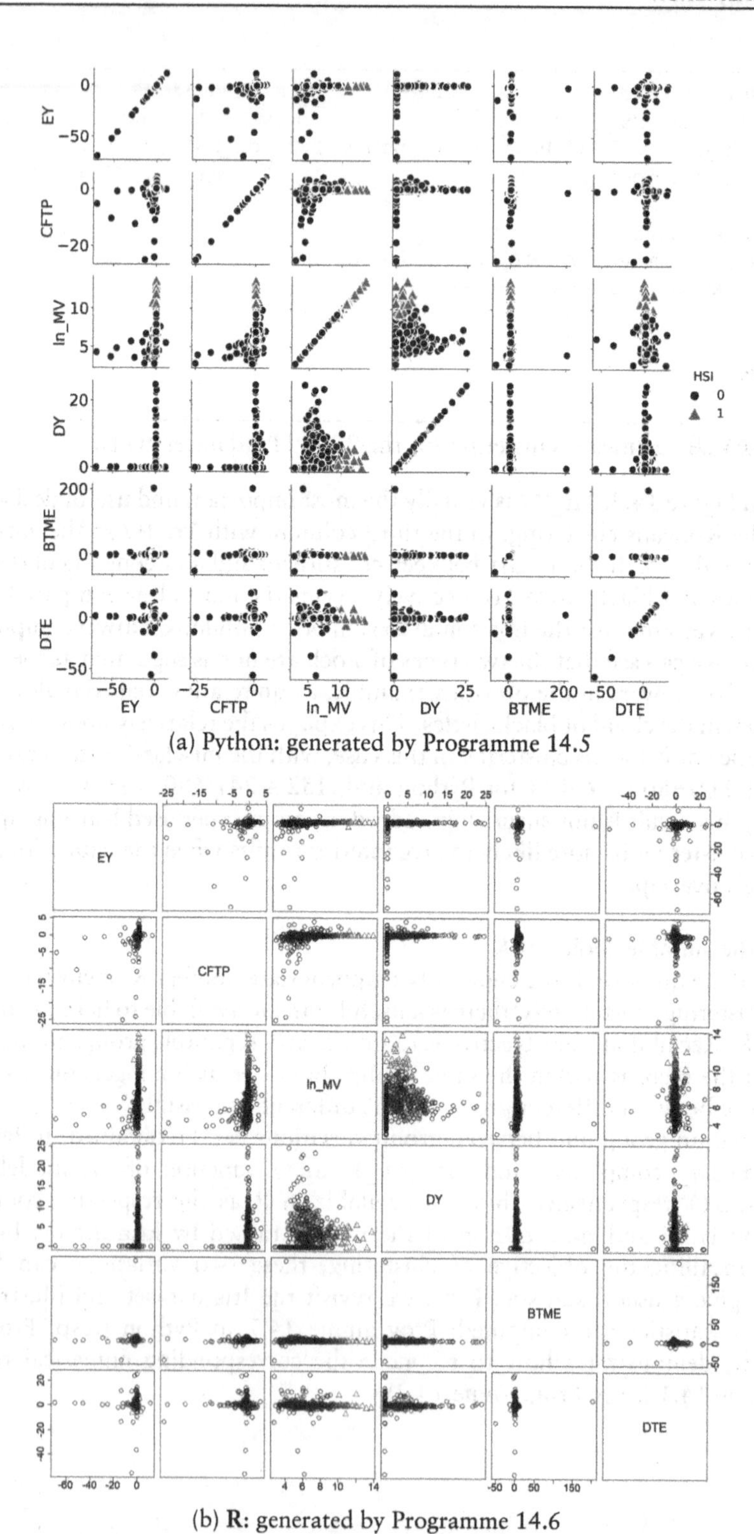

(a) Python: generated by Programme 14.5

(b) R: generated by Programme 14.6

FIGURE 14.3 Scatter plot matrix colored corresponding to the output from *K*-means clustering for the 2002 HSI data.

```
1  X1 = X_iris[km_iris.labels_ == 0]    # select group by cluster label
2  X2 = X_iris[km_iris.labels_ == 1]
3  X3 = X_iris[km_iris.labels_ == 2]
4  print("Cluster sizes:", len(X1), len(X2), len(X3))
5
6  # Cluster means
7  mean_X1 = np.mean(X1, axis=0)
8  mean_X2 = np.mean(X2, axis=0)
9  mean_X3 = np.mean(X3, axis=0)
10 print("Cluster 1 mean:", mean_X1)
11 print("Cluster 2 mean:", mean_X2)
12 print("Cluster 3 mean:", mean_X3)
13
14 # Within group sum of squares
15 wcss_X1 = np.sum((X1 - mean_X1)**2)
16 wcss_X2 = np.sum((X2 - mean_X2)**2)
17 wcss_X3 = np.sum((X3 - mean_X3)**2)
18 print("Within group sum of squares for Cluster 1:", wcss_X1)
19 print("Within group sum of squares for Cluster 2:", wcss_X2)
20 print("Within group sum of squares for Cluster 3:", wcss_X3)
21 print(wcss_X1 + wcss_X2 + wcss_X3)
```

```
1 Cluster sizes: 62 50 38
2 Cluster 1 mean: [5.9016129  2.7483871  4.39354839 1.43387097]
3 Cluster 2 mean: [5.006 3.428 1.462 0.246]
4 Cluster 3 mean: [6.85       3.07368421 5.74210526 2.07105263]
5 Within group sum of squares for Cluster 1: 39.82096774193548
6 Within group sum of squares for Cluster 2: 15.151000000000002
7 Within group sum of squares for Cluster 3: 23.879473684210527
8 78.85144142614601
```

Programme 14.7 Reconciling the sums of squares obtained in Programme 14.1 via Python.

```
1  > X1 <- X_iris[km_iris$cluster==1,] # select group by cluster label
2  > X2 <- X_iris[km_iris$cluster==2,]
3  > X3 <- X_iris[km_iris$cluster==3,]
4  > (n1 <- nrow(X1)); (n2 <- nrow(X2)); (n3 <- nrow(X3))  # cluster size
5  [1] 50
6  [1] 38
7  [1] 62
8  >
9  > apply(X1, 2, mean)                  # cluster mean
10 Sepal.Length  Sepal.Width Petal.Length  Petal.Width
11        5.006        3.428        1.462        0.246
12 > apply(X2, 2, mean)
13 Sepal.Length  Sepal.Width Petal.Length  Petal.Width
14     6.850000     3.073684     5.742105     2.071053
15 > apply(X3, 2, mean)
16 Sepal.Length  Sepal.Width Petal.Length  Petal.Width
17     5.901613     2.748387     4.393548     1.433871
18 > sum(diag((n1-1)*var(X1)))           # tr(SSCP)
19 [1] 15.151
20 > sum(diag((n2-1)*var(X2)))           # within group sum of squares
```

```
21  [1] 23.87947
22  > sum(diag((n3-1)*var(X3)))
23  [1] 39.82097
24  >
25  > m <- apply(X_iris, 2, mean)          # overall mean
26  > # (group mean - overall mean)^2
27  > dm <- sweep(km_iris$centers, 2, m, FUN="-")^2
28  > sum(km_iris$size*rowSums(dm))        # between group sum of squares
29  [1] 602.5192
```

Programme 14.8 Reconciling the sums of squares obtained in Programme 14.2 via R.

To better explain Programmes 14.7 and 14.8, we need to define some notations for the sake of clarity. Let $\boldsymbol{x}_1^{(k)},\ldots,\boldsymbol{x}_{n_k}^{(k)} \in \mathbb{R}^p$ be the data points classified with a predicted label $k = 1,2,\ldots,K$. Let $\boldsymbol{x}_i^{(k)} = (x_{i1}^{(k)},\ldots,x_{ip}^{(k)})^\top$, for $i = 1,\ldots,n_k$, and denote the sample mean components among those with the predicted label k as $\overline{x}_{\cdot j}^{(k)} := \frac{1}{n_k}\sum_{i=1}^{n_k} x_{ij}^{(k)}$ for $j = 1,\ldots,p$, and write $\overline{\boldsymbol{x}}^{(k)} := (\overline{x}_1^{(k)},\ldots,\overline{x}_p^{(k)})^\top$.

The *within-cluster sum of squares and cross products* (SSCP) matrix for the predicted label k is defined as:

$$\mathrm{SSCP}^{(k)} := \left[\left(\boldsymbol{x}_1^{(k)} - \overline{\boldsymbol{x}}^{(k)}\right) \cdots \left(\boldsymbol{x}_{n_k}^{(k)} - \overline{\boldsymbol{x}}^{(k)}\right)\right] \begin{bmatrix} \left(\boldsymbol{x}_1^{(k)} - \overline{\boldsymbol{x}}^{(k)}\right)^\top \\ \vdots \\ \left(\boldsymbol{x}_{n_k}^{(k)} - \overline{\boldsymbol{x}}^{(k)}\right)^\top \end{bmatrix},$$

and its trace is

$$\mathrm{tr}(\mathrm{SSCP}^{(k)}) = \sum_{i=1}^{n_k} \left(\left(\boldsymbol{x}_i^{(k)} - \overline{\boldsymbol{x}}^{(k)}\right)^\top \left(\boldsymbol{x}_i^{(k)} - \overline{\boldsymbol{x}}^{(k)}\right)\right),$$

which is the within-cluster sum of squares for the sample points with the predicted label k. A total of K different traces $\mathrm{tr}(\mathrm{SSCP}^{(k)})$ for $k = 1,\ldots,K$, are all stored in `km$withinss` in **R**. Summing them up over all the predicted labels k gives the total *within-cluster sum of squares* (WCSS), that is $\mathrm{WCSS} := \sum_{k=1}^{K} \mathrm{tr}(\mathrm{SSCP}^{(k)})$. This value is stored in `km_iris$inertia_` in Python (Programme 14.1) and `km_iris$tot.withinss` in **R** (Programme 14.2). Meanwhile, we also define the following *between-cluster* SSCP matrix:

$$\mathrm{BCSSCP} := \left[\left(\overline{\boldsymbol{x}}^{(1)} - \overline{\boldsymbol{x}}\right)\mathbf{1}_{n_1}^\top \cdots \left(\overline{\boldsymbol{x}}^{(K)} - \overline{\boldsymbol{x}}\right)\mathbf{1}_{n_K}^\top\right] \begin{bmatrix} \mathbf{1}_{n_1}\left(\overline{\boldsymbol{x}}^{(1)} - \overline{\boldsymbol{x}}\right)^\top \\ \vdots \\ \mathbf{1}_{n_K}\left(\overline{\boldsymbol{x}}^{(K)} - \overline{\boldsymbol{x}}\right)^\top \end{bmatrix},$$

where $\mathbf{1}_{n_k}$ is an n_k-dimensional vector with all elements equal to 1, $\overline{\boldsymbol{x}} = \sum_{k=1}^{K} \frac{n_k}{n}\overline{\boldsymbol{x}}^{(k)}$ is the cluster-size-weighted sample mean vector for the whole dataset, and

$n = n_1 + \cdots + n_K$ is the total sample size. Then the *between-cluster sum of squares* (BCSS) is defined and can be written as:

$$\begin{aligned}
\text{BCSS} := \text{tr(BCSSCP)} &= \text{tr}\left(\sum_{k=1}^{K}\left(\overline{\boldsymbol{x}}^{(k)} - \overline{\boldsymbol{x}}\right)\mathbf{1}_{n_k}^{\top}\mathbf{1}_{n_k}\left(\overline{\boldsymbol{x}}^{(k)} - \overline{\boldsymbol{x}}\right)^{\top}\right) \\
&= \text{tr}\left(\sum_{k=1}^{K}\left(\mathbf{1}_{n_k}\left(\overline{\boldsymbol{x}}^{(k)} - \overline{\boldsymbol{x}}\right)^{\top}\left(\overline{\boldsymbol{x}}^{(k)} - \overline{\boldsymbol{x}}\right)\mathbf{1}_{n_k}^{\top}\right)\right) \\
&= \sum_{k=1}^{K}\left(\text{tr}\left(\mathbf{1}_{n_k}\left(\overline{\boldsymbol{x}}^{(k)} - \overline{\boldsymbol{x}}\right)^{\top}\left(\overline{\boldsymbol{x}}^{(k)} - \overline{\boldsymbol{x}}\right)\mathbf{1}_{n_k}^{\top}\right)\right) \\
&= \sum_{k=1}^{K}\left(\left(\overline{\boldsymbol{x}}^{(k)} - \overline{\boldsymbol{x}}\right)^{\top}\left(\overline{\boldsymbol{x}}^{(k)} - \overline{\boldsymbol{x}}\right)\text{tr}\left(\mathbf{1}_{n_k}\mathbf{1}_{n_k}^{\top}\right)\right) \\
&= \sum_{k=1}^{K} n_k\left(\overline{\boldsymbol{x}}^{(k)} - \overline{\boldsymbol{x}}\right)^{\top}\left(\overline{\boldsymbol{x}}^{(k)} - \overline{\boldsymbol{x}}\right).
\end{aligned}$$

In Python (Programme 14.1); this sum of squares can be called out by the function `bcss`; while in **R** (Programme 14.2), this value is stored in `km_iris$betweenss`.

With all these quantities in mind, we introduce the following *Calinski–Harabasz index* [4]:

$$R(K) := \frac{n-K}{K-1}\frac{\text{BCSS}}{\text{WCSS}},$$

where both BCSS and WCSS are regarded as functions in K. Clearly, BCSS describes the dispersion of different clusters while WCSS describes the dispersion of points within each cluster. A good clustering should be one for which the former is large but the latter is small. Notice that as a ratio of the between-group and within-group sums of squares, divided by their respective degrees of freedom, the index is analogous to the F-test statistic in univariate analysis. We aim to seek for a value of K that simultaneously makes WCSS small and BCSS large as much as possible, and so make $R(K)$ as large as possible. In Appendix *14.A.1, we shall motivate why $R(L)$ will indeed attain its maximum at its true cluster number, i.e., when $L = K$, in most practical scenarios, and also discuss about pitfalls of using it under some special cases.

In Programme 14.9 via Python (resp. Programme 14.10 via **R**), we illustrate how to compute $R(K)$ using the self-defined function `kmstat()` in both Python and **R**. This function is designed to return the optimal choice of K, with the corresponding maximized $R(K)$ statistic, and the corresponding cluster model. We also introduce another function `best_km()` in both Python and **R** that performs K-means clustering several times for different values of K, and generates the same two outputs, namely the best possible trial with the largest $R(K)$ statistic. See Programmes 14.9 via Python and 14.10 via **R**, respectively.

```
1  # Try several values of K, choose K so that stat. is maximized
2  def kmstat(X, K):
3      km = KMeans(n_clusters=K, n_init="auto").fit(X)
4      n = X.shape[0]                          # sample size
5      wcss = km.inertia_                      # within group ss
6      bcss = get_bcss(km, X)                  # between group ss
7      return ((n - K) * bcss) / ((K - 1) * wcss), km
8
9  # Try KMeans() several times and output the best trial
10 def best_km(X, K, trial=5, seed=4002):
11     np.random.seed(seed)
12     r0 = 0
13     for i in range(trial):
14         r, km = kmstat(X, K)
15         if r > r0:                  # update r0 if it is less than r
16             r0, km0 = r, km
17     print(f"K = {K}; stat = {r0}")
18     return r0, km0
```

Programme 14.9 Python codes for finding the best possible clustering trial based on the statistic $R(K)$.

```
1  > # Try several values of K, choose K so that stat. is maximized
2  > kmstat <- function(X, K){
3  +   km <- kmeans(X, K)                  # K-means clustering
4  +   n <- nrow(X)                        # sample size
5  +   wcss <- sum(km$withinss)            # within group ss
6  +   bcss <- km$betweenss                # between group ss
7  +   # km$cluster: the cluster to which each point is allocated
8  +   list(stat=(n-K)*bcss/((K-1)*wcss), km=km)
9  + }
10 >
11 > # Try kmeans(X, K) several times and output the best trial
12 > best_km <- function(X, K, trial=5, seed=4002) {
13 +   set.seed(seed)
14 +   r0 <- 0
15 +   for (i in 1:trial) {
16 +     res <- kmstat(X, K)               # new trial
17 +     if (res$stat > r0) {              # update r0 if it is less than r
18 +       r0 <- res$stat; km0 <- res$km
19 +     }
20 +   }
21 +   print(paste0("K=", K, "; stat=", r0))
22 +   km0                                 # best cluster
23 + }
```

Programme 14.10 R codes for finding the best possible clustering trial based on the statistic $R(K)$.

Furthermore, using the Iris dataset once again, we shall try several values of K in Programme 14.11 via Python (resp. Programme 14.12 via **R**) and determine the specific value of K at which the function $R(K)$ is maximized.

```
1  _, km_iris2 = best_km(X_iris, 2)                    # try K=2
2  _, km_iris3 = best_km(X_iris, 3)                    # try K=3
3  _, km_iris4 = best_km(X_iris, 4)                    # try K=4
4  _, km_iris5 = best_km(X_iris, 5)                    # try K=5
```

```
1  K = 2; stat = 513.9245459802767
2  K = 3; stat = 561.6277566296197
3  K = 4; stat = 530.4871420421676
4  K = 5; stat = 459.5058152437512
```

Programme 14.11 Outputs of `best_km()` function for $K = 2, 3, 4, 5$ with the Iris flower data via Python.

```
1  > km_iris2 <- best_km(X_iris, 2)          # try K=2
2  [1] "K=2; stat=513.924545980277"
3  > km_iris3 <- best_km(X_iris, 3)          # try K=3
4  [1] "K=3; stat=561.62775662962"
5  > km_iris4 <- best_km(X_iris, 4)          # try K=4
6  [1] "K=4; stat=530.765808187285"
7  > km_iris5 <- best_km(X_iris, 5)          # try K=5
8  [1] "K=5; stat=459.505815243751"
```

Programme 14.12 Outputs of `best_km()` function for $K = 2, 3, 4, 5$ with the Iris flower data via R.

In Programme 14.11 via Python (resp. Programme 14.12 via **R**), we enter a few values of K in the `best_km()` function, and see that $K = 3$ gives the maximum value of 561.6278 for $R(K)$, so $K = 3$ could be a suitable choice. The cluster label is also saved in `km_iris3`, and the cross-tabulation table is exactly the one shown by Programme 14.3 via Python (resp. Programme 14.4 via **R**).

As another illustration, we also apply `best_km()` to the cleansed 2002 HSI dataset `fin-ratio_cleansed.csv` (produced in Programme 10.11 via **R**), with the output as shown in Programme 14.13 via Python (resp. Programme 14.14 via **R**).

```
1  # read in cleansed HSI dataset
2  df_cHSI = pd. read_csv ("fin - ratio_cleansed .csv")
3  X_cHSI, y_cHSI = df_cHSI.iloc[:, :-1].values, df_cHSI.HSI
4  _, km_cHSI2 = best_km(X_cHSI, 2)                  # try K=2
5  _, km_cHSI3 = best_km(X_cHSI, 3)                  # try K=3
6  _, km_cHSI4 = best_km(X_cHSI, 4)                  # try K=4
7  _, km_cHSI5 = best_km(X_cHSI, 5)                  # try K=5
```

```
1  K = 2; stat = 330.75352042884634
2  K = 3; stat = 247.14194830775617
3  K = 4; stat = 234.16770929263015
4  K = 5; stat = 239.88638277478742
```

Programme 14.13 Outputs of `best_km()` function for $K = 2, 3, 4, 5$ with the cleansed 2002 HSI dataset via Python.

```
1  > df_cHSI <- read.csv("fin-ratio_cleansed.csv")  # read in cleansed HSI
2  > X_cHSI <- df_cHSI[,-7]                          # remove HSI label
3  > km_cHSI2 <- best_km(X_cHSI, 2)                  # try K=2
4  [1] "K=2; stat=330.753520428843"
5  > km_cHSI3 <- best_km(X_cHSI, 3)                  # try K=3
6  [1] "K=3; stat=247.508632474202"
7  > km_cHSI4 <- best_km(X_cHSI, 4)                  # try K=4
8  [1] "K=4; stat=233.430694568438"
9  > km_cHSI5 <- best_km(X_cHSI, 5)                  # try K=5
10 [1] "K=5; stat=239.907986172307"
```

Programme 14.14 Outputs of `best_km()` function for $K = 2, 3, 4, 5$ with the cleansed 2002 HSI dataset via **R**.

As expected, $K = 2$ gives the maximum value of $R(K)$ at 330.7535, which seems consistent with the number of possible actual labels of the dataset. Despite the same number of categories, the classification results may not be in accordance with the stock categories in HSI. To find out the characteristics of these clusters, we use Programme 14.15 via Python (resp. Programme 14.16 via **R**). The resulting boxplots in Figure 14.4 are an effective graphical illustration of the characteristics of each cluster.

```
1  print(pd.crosstab(km_cHSI2.labels_, y_cHSI))      # classification table
2
3  df_cHSI_km2 = df_cHSI.copy()
4  df_cHSI_km2.HSI = km_cHSI2.labels_
5  fig, axs = plt.subplots(nrows=2, ncols=3, figsize=(15, 10))
6  # boxplots for each variable
7  for i in range(len(df_cHSI_km2.columns) - 1):
8      j, k = i // 3, i % 3
9      sns.boxplot(data=df_cHSI_km2, ax=axs[j,k],
10                 x="HSI", y=df_cHSI_km2.columns[i])
```

```
1  HSI      0    1
2  row_0
3  0      454   19
4  1      172   13
```

Programme 14.15 Plotting boxplots for each feature variable in the cleansed 2002 HSI dataset in Python.

```
1  > table(km_cHSI2$cluster, df_cHSI[,7])  # classification table
2
3         0   1
4     1 454  19
5     2 172  13
6  >
7  > par(mfrow=c(2,3))                      # 2x3 multi-frame graphic
8  > # boxplots for each variable
9  > for (i in 1:ncol(X_cHSI)){
10 +    boxplot(X_cHSI[,i]~km_cHSI2$cluster,
11 +            xlab="HSI", ylab=names(X_cHSI)[i])
12 + }
```

Programme 14.16 Plotting boxplots for each feature variable in the cleansed 2002 HSI dataset in R.

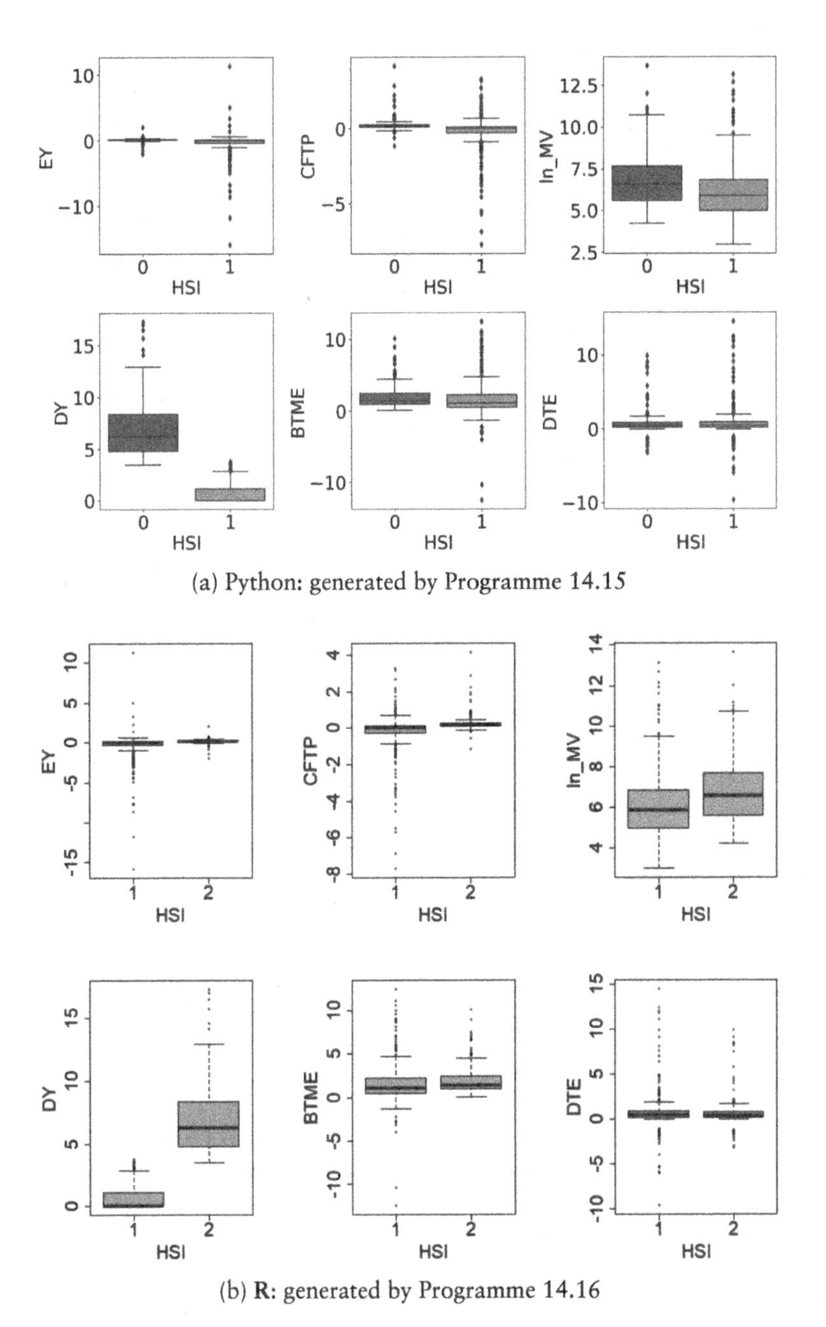

(a) Python: generated by Programme 14.15

(b) R: generated by Programme 14.16

FIGURE 14.4 Boxplots for each feature variable in the cleansed 2002 HSI dataset.

From the boxplots in Figure 14.4, the one for DY shows that the second cluster of points lies higher than those of the first one, while no significant difference is present in the plots of the other five variables. Based on this, we can simply describe these two clusters as the collection of stocks with lower values of DY as the first cluster and higher values of DY as the second.

14.1.1 Variants of *K*-Means Clustering

(I) Different Distance / Dis-similarity Metrics

K-means clustering as well as other clustering methods is based on the metric of distance (or *dis-similarity*) between observations. It is thus important to properly define the metric between any two observations. In view of the different types of variables in our dataset, these metrics should also be defined in a tailor-made manner. In the following, for the sake of convenience, first suppose that we have p variables of the same type in our dataset with size n, and the i-th observation $x_i = (x_{i1}, \ldots, x_{ip})^\top$, $i = 1, \ldots, n$. In the case of mixed variable types, we just combine different metrics for the various types.

Continuous variables

If all the variables are continuous, it is common to use the following q-norm as the distance measure:

$$d_q(x_i, x_j) := \|x_i - x_j\|_q = \left(\sum_{k=1}^{p} (x_{ik} - x_{jk})^q \right)^{1/q}.$$

This includes the well-known *Euclidean distance* ($q = 2$), *Manhattan distance* ($q = 1$), and *Minkowski distance* ($1 \le q \le \infty$) as special cases[3].

Ordinal variables

If all the variables are ordinal, we can rank the n observations of each variable, respectively, and then rescale the observations to $[0, 1]$ as follows:

$$z_{ik} = \frac{r_{ik} - 1}{M_k - 1}, \quad i = 1, \ldots, n, \quad k = 1, \ldots, p, \tag{14.4}$$

where r_{ik} is the ranking of the k-th component variable for the i-th observation, and M_k is the number of possible values the k-th variable can take, and the additional term of -1 in the numerator is included to ensure that the observations with the lowest rank of 1 can be rescaled to 0. After the rescaling transformation, we can then treat the component variables as if they were continuous variables, and measure their distance using the p-norm mentioned above.

[3] When $q = \infty$, $d_\infty(x_i, x_j) := \max_{k=1,\ldots,p} |x_{ik} - x_{jk}|$.

Binary variables
If all the component variables are binary, then for each pair of observations $\boldsymbol{x}_i$ and $\boldsymbol{x}_j$, we can count the number of variables under each possible combination. In particular, we let $m^{ij}_{l_1 l_2} := |\{k \in \{1, \ldots, p\} : x_{ik} = l_1, x_{jk} = l_2\}|$ denote the number of component variables that have a value of l_1 in $\boldsymbol{x}_i$ and l_2 in $\boldsymbol{x}_j$, respectively, for $l_1, l_2 = 0, 1$. Clearly $m^{ij}_{00} + m^{ij}_{01} + m^{ij}_{10} + m^{ij}_{11} = p$. Now we can use either of the following measures (for simplicity we omit the superscripts of i and j):

- Simple matching coefficient:

$$d(\boldsymbol{x}_i, \boldsymbol{x}_j) := \frac{m_{00} + m_{11}}{m_{00} + m_{01} + m_{10} + m_{11}} = \frac{m_{00} + m_{11}}{p};$$

- *Jaccard coefficient*:

$$d(\boldsymbol{x}_i, \boldsymbol{x}_j) := \frac{m_{10} + m_{01}}{m_{00} + m_{01} + m_{10}}.$$

Nominal or categorical variables
If all the component variables are nominal, then we can use the dis-similarity measure $d(\boldsymbol{x}_i, \boldsymbol{x}_j) := \frac{p-m}{p}$, where $m := |\{k \in \{1, \ldots, p\} : x_{ik} = x_{jk}\}|$ is the number of matches among these p component variables.

Mixed types, missing values and scaling of data
In practice, datasets usually contain missing values and variables of mixed types. The easiest way to handle the missing data is to delete all incomplete observations. The Python function `dropna()` and **R** function `complete.cases()` can be utilized to select all the complete observations and leave out the others. However, using only complete observations will potentially give up lots of hidden information and, more seriously, lead to the biased sampling of sub-samples, especially when the missing mechanism is *missing not at random*. A better but more complicated way of handling them is to proceed with the calculation of the distance, but only using the variables available in both observations, or we can impute the missing observations.

Another issue is that the ranges of the variables may differ a lot, and using the original scale inevitably assigns more weight to those with a large magnitude. Therefore, we should rescale variables to the $[0, 1]$ interval before performing cluster analysis.

Altogether, combining discussions above, we are led to the following distance metric:

$$d(\boldsymbol{x}_i, \boldsymbol{x}_j) := \frac{\sum_{k=1}^{p} w_{ij}^{(k)} d_{ij}^{(k)}}{\sum_{k=1}^{p} w_{ij}^{(k)}},$$

where $w_{ij}^{(k)}$ is 1 if both x_{ik} and x_{jk} are present, or equal to 0 otherwise; and $d_{ij}^{(k)}$ is calculated depending on the type of the k-th variable as follows:

$$d_{ij}^{(k)} := \begin{cases} \dfrac{|x_{ik} - x_{jk}|}{\max_u x_{uk} - \min_u x_{uk}}, & \\ |z_{ik} - z_{jk}| \text{ (defined in (14.4))}, & \text{if the } k\text{-th variable is } \begin{cases} \text{continuous;} \\ \text{ordinal;} \\ \text{binary or nominal.} \end{cases} \\ \mathbb{1}_{\{x_{ik}=x_{jk}\}}, & \end{cases}$$

(II) Enhancing Lloyd's Algorithm

Let us discuss further the computational nature and some drawbacks of Lloyd's algorithm [7, 28], and then incorporate certain ways to improve the overall performance of the algorithm.

From (14.3), the computational complexity of Step 2 in Algorithm 14.1 of Lloyd's algorithm is $O(Knp)$; to see this, note that for each of the n observations, its distances to all K centroids need to be calculated, and calculating a distance requires $O(p)$ units of time to complete since $\boldsymbol{x}_i \in \mathbb{R}^p$. When we update the centroids in Step 3, the computational complexity is $O(np)$ since each addition of $\boldsymbol{x}_i$ consumes $O(p)$ units of time, and there are essentially $O(n)$ additions to be done in total. If the number of iterations is T, then the total complexity of the algorithm becomes $O(KnpT)$; therefore, Lloyd's algorithm for K-means clustering is a linear-in-time algorithm.

As we hinted earlier, Lloyd's algorithm is rather sensitive to initialization, as different initial seeds might result in a quite substantially different clustering result, since the algorithm only yields a local minimum to (14.2) instead of the global one. To warrant a more robust performance, we can adopt a Monte Carlo-like approach, namely repeating the clustering procedure using random initial seeds for a number of times. We then compare the results obtained from different initial seeds, and then select the set of seeds with the smallest sum of squares. As a tradeoff, this approach certainly requires a much longer computational time.

In [3], *David Arthur* and *Sergei Vassilvitskii* introduced a simple yet effective alternative numerical approach, called *K-means++*, by enhancing the original Lloyd's algorithm with an additional "initialization algorithm". The algorithm guarantees the solution found is *$O(\ln K)$-competitive* to the optimal solution: let $\{S_1^{++}, \ldots, S_K^{++}\}$ be the resulting partition obtained by the K-means++ algorithm, with $\boldsymbol{c}_k^{++}$ being the centroid of the cluster $k = 1, \ldots, K$, and $\{S_1^*, \ldots, S_K^*\}$ be the optimal partition with cluster centroids $\boldsymbol{c}_k^*$ for (14.2), then

$$\mathbb{E}\left(\sum_{k=1}^{K} \sum_{x \in S_k^{++}} (x - c_k^{++})^2\right) \leq 8(\ln K + 2) \sum_{k=1}^{K} \sum_{x \in S_k^*} (x - c_k^*)^2. \tag{14.5}$$

The "initialization algorithm" is summarized in Algorithm 14.2. K-means++ has an extra computational complexity of order $O(K^2np)$ induced by Algorithm 14.2;

Algorithm 14.2 Initial seed selection of K-means++

1. Initialize a centroid $C = \{c_1\}$ randomly.

for each observation x_i **do**

2. Calculate the distance of x_i to the nearest centroid in C:

$$d(x_i, C) = \min_{c \in C} \|x_i - c\|_2 \, ;$$

3. Assign each x_i with probability $p(x_i)$ such that:

$$p(x_i) := \frac{d^2(x_i, C)}{\sum_{j=1}^{n} d^2(x_j, C)};$$

end for

4. a point x^* would be selected with probability $p(x^*)$, which is augmented in C:

$$C \leftarrow C \cup \{x^*\}.$$

5. Steps 2–4 are repeated until K centroids are selected.

indeed, when there are k centroids in C in Algorithm 14.2, a total of k calculations will be needed at its Step 2 for each of the n observations, while every such calculation takes $O(p)$ units of time. Therefore, the time needed in this iteration step is of the order $O(knp)$, and summing up for $k = 1, \ldots, K$ leads to a total complexity of $O(K^2np)$. Clearly, this newly revised algorithm still results in a polynomial (quadratic) computational time, yet on average it can guarantee the eventual clusters produced are effectively close to the ultimate optimal solution as indicated by the error bound (14.5).

Asymmetric Samples

Recall that the classical K-means algorithm makes use of the square of Euclidean distance as the measure of dis-similarity, which is symmetric in all component variables by assigning equal weights. It is therefore not surprising that the performance of K-means algorithm is relatively poor when handling samples that do not follow symmetric distributions. For instance, Figure 14.5 shows the performance of K-means clustering when the two samples actually follow asymmetric distributions. To eliminate the restrictive distribution assumptions, variants of the K-means algorithm for more general metric spaces are also available, including the *Kernel K-means clustering* and *Spectral Clustering.*

The Kernel K-means algorithm maps each observation $x_i \in \mathbb{R}^p$ to a high-dimensional space $\Phi(x_i) \in \mathbb{R}^m$ with $m > p$, via the *transformation* $x \mapsto \Phi(x)$; see more details in Section 11.2 on radial basis functions for SVM.

On the other hand, for spectral clustering, the collection of sample points in $\mathbb{R}^p$ is viewed as a *complete graph*, where each sample feature vector represents a vertex, and an edge between any two vertices is represented by a weight. The weights

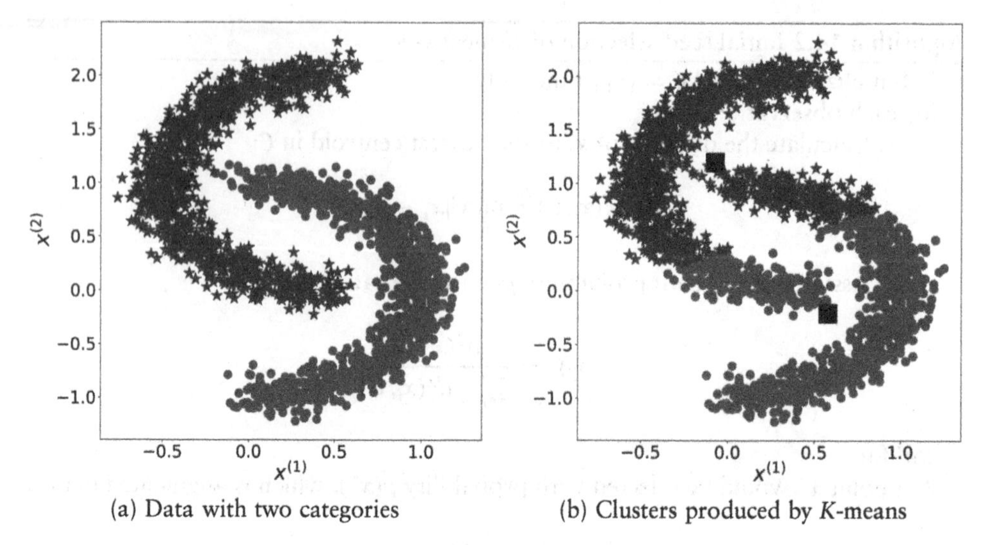

(a) Data with two categories (b) Clusters produced by K-means

FIGURE 14.5 An example of a two-dimensional asymmetric sample.

computed, based on a chosen similarity measure, form a similarity matrix, which is certainly symmetric due to the symmetry of the similarity measure. In addition, this similarity matrix is mostly positive semi-definite [20]; for example, if the similarity measure (signed) between two observed feature vectors x_i and x_j is given by the cosine of their inscribed angle, i.e. $\frac{\langle x_i, x_j \rangle}{\|x_i\|_2 \cdot \|x_j\|_2}$, then the corresponding similarity matrix is clearly positive semi-definite. By spectral decomposition (recall Section 1.2 for an introduction), we can obtain the first m ($m \gg K$) eigenvectors corresponding to the m largest eigenvalues (all eigenvalues are positive) of this similarity matrix[4]. Then the problem reduces to clustering the sample points by using the rows of the $p \times m$ matrix of these m $\mathbb{R}^p$-eigenvectors, such that the i-th row stands for compressed coordinates in terms of the first m principal components for the i-th data point. In this context, the classical K-means algorithm can now be utilized.

Bisecting K-means

We introduce one more variant of K-means, but in the realm of hierarchical clustering, called *bisecting K-means*, where the desired number of clusters K is pre-determined as usual. The whole sample is first considered as a single cluster, which is then divided into two clusters by using the classical K-means method with $K = 2$. In each subsequent iterative step, the same method of bisection is carried out to the cluster with the

[4] Actually, instead of the similarity matrix A, the problem in its general form is formulated as finding the smallest eigenvalues of its corresponding Laplacian matrix $L := D - A$, where D is the degree matrix. In the present case where the collection of sample points is viewed as a complete graph, we have $D = p \cdot I$, hence the general problem reduces to finding the largest eigenvalues of A. By considering L directly, the general problem is indeed connected with the renowned *Laplace spring-mass system* in the field of *Lagrangian mechanics*.

largest sum of squared error (SSE) from its corresponding centroid, hence the number of clusters increases by 1 after every new step. The procedure continues until the apparent stopping criterion that there are already K clusters.

(III) Applications of K-means Clustering

K-means clustering, especially Lloyd's algorithm, is very easily applicable to large datasets. Besides its immediate use in customer classification in marketing research as mentioned in the beginning of this chapter, we shall also discuss some of its interesting applications in other fields, such as *image processing* in computer science and *dictionary learning* in feature learning.

Here, we first illustrate the use of K-means clustering in customer credential classification using the dataset `Credit Card Customer Data.csv`[5]. In Programme 14.17 via Python, the two features `Sl_No` and `Customer Key` are unique identifiers for each individual customer, hence we first drop these two features from the dataset. Furthermore, since most feature variables are discrete except `Avg_Credit_Limit`, we apply standardization with respect to minimum and maximum values for each feature variable, instead of normalization by mean and standard deviation. The function `best_km` from Programme 14.9 via Python is used to find the best choice of the hyperparameter K for the subsequent K-means clustering.

```
1  import pandas as pd
2  import matplotlib.pyplot as plt
3  from sklearn.preprocessing import MinMaxScaler
4  from KMeansCluster import best_km
5
6  df_Cred = pd.read_csv('Credit Card Customer Data.csv')
7  df_Cred = df_Cred.drop(['Sl_No', 'Customer Key'], axis=1)
8  X = df_Cred.values
9  scaler = MinMaxScaler()
10 X = scaler.fit_transform(X)
11
12 _ , km_cCred2 = best_km(X, 2) # try K = 2
13 _ , km_cCred3 = best_km(X, 3) # try K = 3
14 _ , km_cCred4 = best_km(X, 4) # try K = 4
15 _ , km_cCred5 = best_km(X, 5) # try K = 4
```

```
1  K = 2; stat = 524.158745625086
2  K = 3; stat = 747.5389632707601
3  K = 4; stat = 732.9852790957698
4  K = 5; stat = 652.6946638008177
```

Programme 14.17 Selection of K using `best_km()` function based on $R(K)$ at $K = 2, 3, 4, 5$ with the credit card dataset via Python.

[5] The dataset can be downloaded from https://www.kaggle.com/datasets/aryashah2k/credit-card-customer-data.

From the output result of Programme 14.17 via Python, the $R(K)$ statistic reaches its largest value of 747.539 at $K = 3$, indicating that `km_cCred3` is the most appropriate K-means clustering scheme. Figures 14.6a and 14.6b shows the resulting scatter plot for the total number of credit cards against the average credit card limit, and boxplots for each feature variable generated by Programme 14.18 via Python. In particular, for the scatter plot, each orange star represents the centroids of the corresponding clusters. For the boxplots, one extra feature variable `Total_interactions` for the total number of times the customer made interactions with the bank is added.

```
1  plt_cmap = sns.color_palette("viridis", n_colors=3)
2  centroids = km_cCred3.cluster_centers_
3
4  fig, ax = plt.subplots(figsize=(10, 7))
5  # scatter plot w.r.t first two feature variables
6  markers = [".", "D", "^"]; s=[600, 200, 300]
7  for i in range(len(np.unique(km_cCred3.labels_))):
8      idx = np.where(km_cCred3.labels_ == i)[0]
9      ax.scatter(X[idx,0], X[idx,1], color=plt_cmap[i], s=s[i],
10                label=f'Cluster {i+1}', marker=markers[i])
11 plt.scatter(centroids[:, 0], centroids[:, 1], marker='*', c='orange',
12             s=500)
13 plt.legend(fontsize=20)
14 plt.xlabel(df_Cred.columns[0])
15 plt.ylabel(df_Cred.columns[1])
16 plt.title('Results of K-means Clustering')
17
18 df_Cred_km3 = df_Cred.copy()
19 df_Cred_km3["Total_interactions"] = (df_Cred["Total_visits_bank"] +
20                                      df_Cred["Total_visits_online"] +
21                                      df_Cred["Total_calls_made"])
22 df_Cred_km3["clust"] = km_cCred3.predict(X) + 1
23 fig, axs = plt.subplots(nrows=2, ncols=3, figsize=(15, 10))
24 # boxplots for each variable
25 for i in range(len(df_Cred_km3.columns) - 1):
26     j, k = i // 3, i % 3
27     sns.boxplot(data=df_Cred_km3, ax=axs[j,k], palette="viridis",
28                 x="clust", y=df_Cred_km3.columns[i])
```

Programme 14.18 K-means Clustering for the credit card user classification dataset via Python.

From Figure 14.6b, the three clusters can be interpreted as follows:

1. Cluster 1: Customers who interact with the bank the least, and make their interactions mostly through in-person visits. They have a moderate average credit limit and they hold a moderate number of credit cards;
2. Cluster 2: Customers who interact with the bank the most, and make their interactions mostly online. They have the highest average credit limit and they hold the greatest number of credit cards;
3. Cluster 3: Customers who interact with the bank slightly less often than the customers in Cluster 2, and make their interactions mostly by calling. They have the lowest average credit limit and they hold the least number of credit cards.

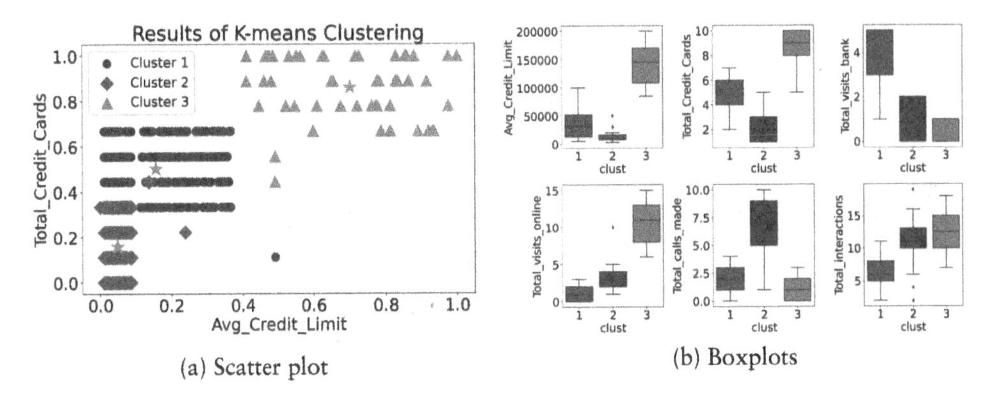

(a) Scatter plot (b) Boxplots

FIGURE 14.6 Scatter plot for two variables (left), and boxplots for each feature variable (right), for the credit card customer dataset via Python, generated by Programme 14.18.

Based on the analysis above, many suggestions can be formulated for the banking industry. For example, as high-net worth individuals tend to cherish the value of time more than others, they would rather save time by contacting the bank online directly, rather than queue up in the branch or make lengthy phone calls. It is then suggested that banks should promote and improve their online credit card services to attract more prestigious credit card users.

Image Segmentation

A typical example of image processing is *image segmentation*, which means partitioning an image into K different segments to make subsequent data processing easier, like using photo-editing software such as *Photoshop*. To implement K-means in this context, we usually adopt the *CIE 1976 Lab color space* (a.k.a. L*a*b* channel) instead of the usual RGB channel. This color space also expresses a pixel in a 3-tuple, namely the lightness L^*, the green-red component a^* and the blue-yellow component b^*. Compared with RGB setting, a numerical change in these three coordinates can better reflect the magnitude of color change perceived by a human.

To perform the segmentation according to colors, we can neglect the lightness component L^*, while the remaining ones a^* and b^* are extracted from each pixel, forming a sample of two-dimensional points, with a size equal to the total number of pixels in the picture. This sample is then clustered using the K-means algorithm or its variants as mentioned before, and pixels ending up in the same cluster are assigned the same values of a^* and b^* of the corresponding centroid. Figure 14.7 shows an illustration of image segmentation using K-means clustering with $K = 5$.[6] By making use of the information provided by the colors, this method can help to effectively identify various objects present in an image, such as the sky, the clouds, the sea, the mountain, and buildings, effectively.

[6] Refer to the e-book online for a clear colored version.

(a) Original image

(b) Output image when $K = 5$.

FIGURE 14.7 Image segmentation using K-means clustering.

Photo Denoising and Dictionary Learning

Another example that we may already benefit from is *photo denoising*, a mechanism that helps photos get rid of random noises on brightness or color caused by electrons, and so make the photo more aesthetically pleasant. It is so prevalent nowadays that we barely perceive its existence, thanks to the technique of *dictionary learning* that speeds up the procedure.

Dictionary learning is similar to PCA: given a dataset we wish to extract a number of basic elements called *atoms* such that almost all feature vectors can be represented in terms of these fundamental building blocks. The collection of such atoms is called a *dictionary*. It is at this stage that K-means clustering or its extension, *K-Singular Value Decomposition* (K-SVD) [2], is implemented.

Here, we briefly summarize K-SVD: let $\boldsymbol{Y} \in \mathbb{R}^{p\times n}$ be the matrix of n observed image signals so that each has p-dimensional features, and $\boldsymbol{D} \in \mathbb{R}^{p\times K}$ be the *dictionary* in which the k-th column $\boldsymbol{d}_k$ is the k-th atom to be learnt, for $k = 1, \ldots, K$.[7] Mathematically, we aim to solve the following optimization problem, for a fixed $T > 0$,

$$\begin{aligned} &\min_{\boldsymbol{D},\boldsymbol{X}} \quad \|\boldsymbol{Y} - \boldsymbol{D}\boldsymbol{X}\|_F^2, \\ &\text{subject to } \|\boldsymbol{x}_i\|_0 \leq T \text{ for all } i = 1, \ldots, n, \end{aligned}$$

where $\boldsymbol{X} = (\boldsymbol{x}_1, \ldots, \boldsymbol{x}_n)$ is a $K \times n$ matrix of coefficient columns with respect to dictionary $\boldsymbol{D}$ for each datum $i = 1, \ldots, n$, while $\|\boldsymbol{x}_i\|_0$ counts the number of non-zero components. Therefore, we aim to describe the key message contained in $\boldsymbol{y}_i$ by not more than T key terms from the dictionary of $\boldsymbol{d}_1, \boldsymbol{d}_2, \ldots, \boldsymbol{d}_K$. K-SVD can be regarded as a generalized K-means clustering problem because the problem degenerates to the latter if $T = 1$, or more precisely, $\|\boldsymbol{x}_i\|_0 = 1$; indeed, in K-means clustering, (i) the

[7] Here, for the sake of clarity in this context, we follow the notations from the literature such as [2] and deviate from the convention stated in Chapter 1, namely rows represent sample observations and columns represent feature variables.

observed matrix $Y \in \mathbb{R}^{p\times n}$ is the transpose of the data matrix (with features); (ii) the dictionary matrix $D = (c_1, \ldots, c_K) \in \mathbb{R}^{p\times K}$, where $c_k \in \mathbb{R}^p$ is the k-th centroid; and (iii) the sparse coefficient matrix $X = (x_1, \ldots, x_n) \in \mathbb{R}^{K\times n}$, where x_{ij} takes the value of 1 if the predicted label of y_i is Class j or is equal to zero otherwise, for $i = 1, \ldots, n$ and $j = 1, \ldots, K$, such that $\|x_i\|_0 = 1$.

Just like K-means clustering, it is in general not convenient to find the explicit solution in a direct manner, hence we adopt the following numerical method instead; particularly, we first solve for the sparse coefficient matrix X to locate its non-zero elements given a dictionary matrix D, and this step is called *sparse coding*. Noting that $\|Y - DX\|_F^2 = \sum_{i=1}^n \|y_i - Dx_i\|_2^2$, this leads to solving n sub-problems: for each $i = 1, \ldots, n$,

$$\begin{aligned} &\min_{x_i} && \|y_i - Dx_i\|_2^2, \\ &\text{subject to} && \|x_i\|_0 \leq T. \end{aligned} \tag{14.6}$$

Problem (14.6) can be solved through a chosen *matching pursuit algorithm*; see more discussions in [11, 18]. In the next step known as the *codebook update*, we update the k-th atom d_k and the k-th row of X denoted by x^k while fixing the rest of the others d_{j_1} and x^{j_2}:

$$\begin{aligned} \|Y - DX\|_F^2 &= \left\| Y - \sum_{j=1}^{K} d_j x^j \right\|_F^2 \\ &= \left\| \left(Y - \sum_{j=1,\, j\neq k}^{K} d_j x^j \right) - d_k x^k \right\|_F^2 \\ &=: \left\| E_k - d_k x^k \right\|_F^2. \end{aligned} \tag{14.7}$$

To fulfill the sparsity constraint, we define an ordered set $\omega_k := \{i : (x^k)_i \neq 0\}$, and denote $\omega_k(i)$ as the i-th element ordered in magnitude in ω_k. Also define a *compression matrix* $\Omega_k \in \mathbb{R}^{|\omega_k|\times n}$ whose $(\omega_k(i), i)$-th entries are 1, while all others are 0, so that $x^k \Omega_k^\top$ is a $1 \times |\omega_k|$ row vector containing all non-zero components (weights) in x^k. We note that

$$\begin{aligned} \left\| E_k - d_k x^k \right\|_F^2 &= \sum_{i,j} \left(\left(E_k - d_k x^k \right)_{i,j} \right)^2 \\ &= \sum_{i,j\in\omega_k} \left(\left(E_k - d_k x^k \right)_{i,j} \right)^2 + \sum_{i,j\notin\omega_k} \left(\left(E_k - d_k x^k \right)_{i,j} \right)^2 \\ &= \sum_{i,j\in\omega_k} \left((E_k)_{i,j} \right)^2 + \sum_{i,j\notin\omega_k} \left(\left(E_k - d_k x^k \right)_{i,j} \right)^2 \\ &= \sum_{i,j\in\omega_k} \left((E_k)_{i,j} \right)^2 + \left\| E_k \Omega_k^\top - d_k x^k \Omega_k^\top \right\|_F^2, \end{aligned} \tag{14.8}$$

where the third equality holds because $(d_k x^k)_{i,j} = 0$ for those $(x^k)_j = 0$. By noting that only the second summand involves d_k and x^k, we then consider the subproblem of the minimization:

$$\min_{d_k, x^k} \left\| E_k - d_k x^k \right\|_F^2 = \min_{d_k, x^k} \left\| E_k \Omega_k^\top - d_k x^k \Omega_k^\top \right\|_F^2.$$

Since d_k and $\left(x^k \Omega_k^\top\right)^\top = \Omega_k (x^k)^\top$ are both vectors, $d_k x^k \Omega_k^\top$ is a rank-1 matrix, by virtue of the Eckart–Young–Mirsky Theorem for the Frobenius norm (Theorem 9.1), we can apply SVD to $E_k \Omega_k^\top$ to get the closest rank-1 matrix (in Frobenius norm) that approximates $E_k \Omega_k^\top$. Therefore the solution d_k can be taken as the first column of U, and $x^k \Omega_k^\top$ is the scalar product of σ_1 (the largest singular value in Σ) and the first column of V, where $E_k \Omega_k^\top = U \Sigma V^\top$. The steps above are then implemented iteratively for k from 1 to K in this codebook update stage. Then we move to the next epoch of the sparse coding step and then the codebook update step again and again until convergence. Notice that convergence is guaranteed by the fact that MSE is monotonically non-increasing at each iteration.

Unlike PCA where the basic elements have to be mutually orthogonal, this is not a must for atoms; in fact, dictionary matrices are usually of a larger size than the data matrix we want to represent. To examine a new image, it suffices for us to only call up the dictionary along with the so-called *representation*, in the form of a linear combination of the atoms, which allows a fast and robust processing of the photos. As a simple example, consider a collection of ℓ grayscale (training) pictures where each picture is of size $h \times w$. It is suggested in [2] that partitioning each picture into m non-overlapping blocks of smaller images, each of which has p number of pixels such that $h \times w = m \times p$. Each of these blocks is then flattened into a one-dimensional vector $y_i^{(\text{train})} \in \mathbb{R}^p$, for $i = 1, \ldots, n = \ell \times m$, then the observed training matrix is given by $Y^{(\text{train})} = (y_1^{(\text{train})}, \ldots, y_{\ell \times m}^{(\text{train})})$; with this in mind, suppose that $D^{(\text{train})} \in \mathbb{R}^{p \times K}$, where $K > p$ for an over-complete dictionary, is the trained dictionary of K-SVD. A new grayscale test image partitioned into m blocks in the same manner as in the training stage yields the observed test matrix $Y^{(\text{test})} = (y_1^{(\text{test})}, \ldots, y_m^{(\text{test})}) \in \mathbb{R}^{p \times m}$. We aim to find a low-rank approximation $\hat{Y} \in \mathbb{R}^{p \times m}$ of the matrix $Y^{(\text{test})}$ of observed test image signals through K-SVD. We then adopt a chosen matching pursuit algorithm in the sparse coding step to find the sparse coefficient matrix $X^{(\text{test})}$ given the trained dictionary $D^{(\text{train})}$. The low-rank approximation is then $\hat{Y} = D^{(\text{train})} X^{(\text{test})}$, from which the reconstructed image can be built by reversing the block-partitioning steps in the training stage. This approach bears a high resemblance to the prediction of movie ratings for recommender systems discussed in Section *9.11, and is particularly useful for filling in missing pixels in images.

14.1.2 Application of PCA in *K*-means Clustering

In Subsection 10.4.3, we fitted a multinomial logistic regression model using the MNIST dataset before and after the dimension reduction via PCA. Here we shall illustrate the application of PCA in the context of K-means clustering using the same dataset. As a benchmark, we first perform a standard and direct K-means clustering on the dataset, with a training size of 60,000 and testing size of 10,000; see the Python codes in Programme 14.19 and **R** codes in Programme 14.20. Note that we set $K = 10$ in this illustration in view of the 10 different digits from 0 to 9 in the dataset. In fact, readers can verify by themselves that under the reasonable constraint of $K \geq 10$ inspired by the nature of the dataset, $R(K)$ indeed reaches its maximum at $K = 10$. Furthermore, since K-means clustering is supposed to be an unsupervised learning algorithm, the predicted clusters obtained should not be connected with the digits. However, for the sake of performance assessment in the present example, we still make such a connection via the majority vote approach, namely the digit with the highest frequency among the training data points in a cluster is selected as the label of the cluster. The labels assigned to the trained clusters are exhibited in `pos_dict`; for example, `4: 6` in the output means that the predicted label of the $(4 + 1 =)5$th cluster is the digit 6.

```
1  import numpy as np
2  import tensorflow as tf
3  from sklearn.metrics import confusion_matrix
4  from datetime import datetime as dt
5
6  ((X_train, y_train),
7   (X_test, y_test)) = tf.keras.datasets.mnist.load_data()
8
9  N_train = len(X_train)                         # 60000 training samples
10 N_test = len(X_test)                           # 10000 test samples
11 X_train = X_train.reshape(N_train, -1).astype('float32') / 255.0
12 X_test = X_test.reshape(N_test, -1).astype('float32') / 255.0
13 K = 10
14
15 start = dt.now()
16 _, km = best_km(X_train, K)                            # K-means with K=10
17 y_hat_km = km.predict(X_test)
18 tab = confusion_matrix(y_hat_km, y_test)               # Confusion matrix
19 print(tab)
20
21 pos_dict = dict(zip(np.arange(K), np.argmax(tab, axis=1)))
22 print(pos_dict)
23 y_pred = [pos_dict[k] for k in y_hat_km]
24 tab = confusion_matrix(y_pred, y_test)                 # Confusion matrix
25 print(tab)
26 print(sum(tab.diagonal()) / len(y_test))               # Accuracy
27 print(dt.now() - start)                                # K-means
```

```
1  K = 10; stat = 2296.59714881051
2  [[452   0   3   1   1   9  20   1   6   4]
3   [  0 661  60  73  29  23  27  59  34  29]
4   [  2   1 707  40   5   4  18  13   7   3]
5   [ 48   2  64 695   0 287   3   0 206   7]
6   [ 30   2  25   7  36  19 795   1  10   4]
7   [  4   0  30  15 558  55  21 292  31 541]
```

```
8   [  3    0   11    7 311   69    1 602   36 387]
9   [  5 468   84    7  37 107   31   58   46  11]
10  [419    0   20   20   1   41   24    1   12  10]
11  [ 17    1   28  145   4 278   18    1 586  13]]
12  {0: 0, 1: 1, 2: 2, 3: 3, 4: 6, 5: 4, 6: 7, 7: 1, 8: 0, 9: 8}
13  [[ 871    0   23   21    2   50   44    2   18   14]
14   [   5 1129  144   80   66  130   58  117   80   40]
15   [   2    1  707   40    5    4   18   13    7    3]
16   [  48    2   64  695    0  287    3    0  206    7]
17   [   4    0   30   15  558   55   21  292   31  541]
18   [   0    0    0    0    0    0    0    0    0    0]
19   [  30    2   25    7   36   19  795    1   10    4]
20   [   3    0   11    7  311   69    1  602   36  387]
21   [  17    1   28  145    4  278   18    1  586   13]
22   [   0    0    0    0    0    0    0    0    0    0]]
23  0.5943
24  0:00:15.002514
```

Programme 14.19 K-means clustering for the MNIST dataset with $K = 10$ via Python.

In the outputs obtained in Programme 14.19 via Python, the i-th row of the first "confusion matrix" obtained is actually the distribution of actual digits among the testing pictures falling into the corresponding i-th trained cluster, and for each testing picture, its predicted digit is set to be the label assigned to the cluster it belongs. In the resulting second and final confusion matrix, the entries in the sixth and tenth rows are all zero, indicating that no clusters are assigned with a predicted label of 5 or 9, hence no testing images are identified as either of these two digits. An explanation is that the digit 5 looks similar to digits 3 and 8, while digit 9 may be easily mistaken as either 4 or 7, hence the large entries in the sixth and tenth columns of the confusion matrix, respectively. The testing accuracy of K-means clustering with $K = 10$ is 0.5943, lower than the 0.6784 obtained by the EM algorithm in Programme 6.9 via Python.

In **R**, since the built-in function `kmeans()` for K-means clustering does not support the `predict()` function for computing the predicted labels given the new feature matrix, we need to define an additional function `predict.kmeans()` for this purpose. This self-defined function, as well as the codes and outputs for the standard K-means clustering, are shown in Programme 14.20. We also remark that the actual digits as columns in the confusion matrix are arranged in the order of $1, \ldots, 9, 0$, which is different from the order in Python (as $0, 1, \ldots, 9$). Lastly, the zero rows in the matrix are omitted in **R**. The prediction accuracy of 0.5848 is close to that obtained in Python.

```
1  > library(keras)           # MNIST dataset
2  > library(nnet)
3  >
4  > # Load MNIST dataset
5  > mnist <- dataset_mnist()
6  > X_train <- mnist$train$x / 255
7  > y_train <- mnist$train$y
8  > X_test <- mnist$test$x / 255
```

```
 9  > y_test <- mnist$test$y
10  >
11  > X_train <- matrix(X_train, nrow=length(y_train))
12  > X_test <- matrix(X_test, nrow=length(y_test))
13  > K <- 10
14  >
15  > predict.kmeans <- function(object, newdata) {
16  +   # Squared Euclidean distance of each sample to each cluster center
17  +   sd_by_center <- apply(object$centers, 1, function(x) {
18  +     colSums((t(newdata) - x)^2)
19  +   })
20  +   apply(sd_by_center, 1, which.min)
21  + }
22  >
23  > y_test[y_test == 0] <- K
24  >
25  > start <- Sys.time()
26  > km <- best_km(X_train, K)              # K-means with K=10
27  [1] "K=10; stat=2295.16643451436"
28  > y_hat_kmeans <- predict.kmeans(km, newdata=X_test)
29  > (tab <- table(y_hat_kmeans, y_test))  # Confusion matrix
30              y_test
31  y_hat_kmeans   1   2   3   4   5   6   7   8   9  10
32             1   0 709  38   3   4   8  11   6   2   4
33             2   2  20   9  23  23 735   1  10   2  42
34             3   0   9  18 298  50   2 427  44 493   2
35             4   0  34   7 363  29 127  74  18 222  14
36             5   0  11   8 251  57   0 429  34 241   2
37             6 660  50  58  18  20  25  36  34  15   0
38             7 469  86   5  24  91  18  48  46   7   5
39             8   3  76 685   0 297   1   0 250   7  57
40             9   0  13   4   0   6  16   1   7   7 780
41            10   1  24 178   2 315  26   1 525  13  74
42  > (pos_dict <- apply(tab, 1, which.max))
43   1  2  3  4  5  6  7  8  9 10
44   2  6  9  4  7  1  1  3 10  8
45  > y_pred <- pos_dict[as.character(y_hat_kmeans)]
46  > (tab <- table(y_pred, y_test))         # Confusion matrix
47        y_test
48  y_pred    1    2    3    4    5    6    7    8    9   10
49      1  1129  136   63   42  111   43   84   80   22    5
50      2     0  709   38    3    4    8   11    6    2    4
51      3     3   76  685    0  297    1    0  250    7   57
52      4     0   34    7  363   29  127   74   18  222   14
53      6     2   20    9   23   23  735    1   10    2   42
54      7     0   11    8  251   57    0  429   34  241    2
55      8     1   24  178    2  315   26    1  525   13   74
56      9     0    9   18  298   50    2  427   44  493    2
57     10     0   13    4    0    6   16    1    7    7  780
58  > mean(y_pred == y_test)                 # Accuracy
59  [1] 0.5848
60  > Sys.time() - start                     # K-means
61  Time difference of 1.313517 mins
```

Programme 14.20 *K*-means clustering for the MNIST dataset with $K = 10$ via R.

Next, we perform PCA on the training dataset and choose the number of PCs that explain 95% of the total variance, that is, the first 154 PCs (referring to Programme 10.28 via Python and Programme 10.29 via **R**), then compute the corresponding PCs for the testing data points, which serve as the feature variables for *K*-means clustering. From the outputs from Python in Programme 14.21 and from **R** in Programme 14.22, we observe that the resulting prediction accuracy remains at the same level of around 0.585. Furthermore, the implementation of this hybrid method takes less time than the standard *K*-means clustering on Python, while on **R**, it is more time-consuming.

In fact, if all the PCs are used in the example above, then the prediction accuracies of *K*-means clustering with and without PCA should be exactly the same under an appropriate choice of initial seeds. This result can be verified using Python and **R** by slightly modifying the respective codes, which is left as a programming exercise for readers. To see this mathematically, suppose that the initial seeds in the direct *K*-means clustering (with input data points $\boldsymbol{x}_1, \ldots, \boldsymbol{x}_n$) are $\boldsymbol{c}_1^{(0)}, \ldots, \boldsymbol{c}_K^{(0)}$. Then for the PCA-incorporated version of *K*-means clustering, using $\boldsymbol{y}_i = \boldsymbol{H}^\top \boldsymbol{x}_i$, $i = 1, \ldots, n$ as the input data, where $\boldsymbol{H}$ is a $p \times p$ orthogonal matrix of loading vectors for PCs (also see Chapter 9), we set $\boldsymbol{H}^\top \boldsymbol{c}_1^{(0)}, \ldots, \boldsymbol{H}^\top \boldsymbol{c}_K^{(0)}$ as the initial seeds. The distance metric $\|\boldsymbol{x}_i - \boldsymbol{c}_k^{(t-1)}\|_2^2$ involved in the updating step (14.3) of Lloyd's algorithm becomes $\|\boldsymbol{y}_i - \boldsymbol{H}^\top \boldsymbol{c}_k^{(t-1)}\|_2^2$; yet these two distances are in fact equal for all $i = 1, \ldots, n$, $k = 1, \ldots, K$, $t = 1, 2, \ldots$:

$$\begin{aligned}
\left\|\boldsymbol{y}_i - \boldsymbol{H}^\top \boldsymbol{c}_k^{(t-1)}\right\|_2^2 &= \left(\boldsymbol{H}^\top \boldsymbol{x}_i - \boldsymbol{H}^\top \boldsymbol{c}_k^{(t-1)}\right)^\top \left(\boldsymbol{H}^\top \boldsymbol{x}_i - \boldsymbol{H}^\top \boldsymbol{c}_k^{(t-1)}\right) \\
&= \left(\boldsymbol{x}_i - \boldsymbol{c}_k^{(t-1)}\right)^\top \boldsymbol{H}\boldsymbol{H}^\top \left(\boldsymbol{x}_i - \boldsymbol{c}_k^{(t-1)}\right) \\
&= \left(\boldsymbol{x}_i - \boldsymbol{c}_k^{(t-1)}\right)^\top \left(\boldsymbol{x}_i - \boldsymbol{c}_k^{(t-1)}\right) = \left\|\boldsymbol{x}_i - \boldsymbol{c}_k^{(t-1)}\right\|_2^2,
\end{aligned}$$

where the second last equality holds by the orthogonality of $\boldsymbol{H}$ such that $\boldsymbol{H}\boldsymbol{H}^\top = \boldsymbol{H}^\top \boldsymbol{H} = \boldsymbol{I}_p$. Therefore, *K*-means clustering with PCA is equivalent to the original version of *K*-means clustering, and should therefore have exactly the same predictive power.

```
1  from sklearn.decomposition import PCA
2
3  start = dt.now()
4  pca = PCA(n_components=.95)
5  X_train_pca = pca.fit_transform(X_train)
6  X_test_pca = pca.transform(X_test)
7  print(f'Total number of components: {pca.n_components_}')
8
9  _, km_pca = best_km(X_train_pca, K)                # K-means with K=10
10 y_hat_pca = km_pca.predict(X_test_pca)
11 tab = confusion_matrix(y_hat_pca, y_test)          # Confusion matrix
12 print(tab)
13
14 pos_dict = dict(zip(np.arange(K), np.argmax(tab, axis=1)))
15 print(pos_dict)
```

```
16 y_pred = [pos_dict[k] for k in y_hat_pca]
17 tab = confusion_matrix(y_pred, y_test)              # Confusion matrix
18 print(tab)
19 print(sum(tab.diagonal()) / len(y_test))            # Accuracy
20 print(dt.now() - start)                             # PCA + K-means
```

```
1  Total number of components: 154
2  K = 10; stat = 2459.6092505193346
3  [[ 41   2  20   9  23  23 736   1  10   2]
4   [ 57   3  76 686   0 297   1   0 251   7]
5   [ 74   1  24 177   2 316  26   1 524  13]
6   [  2   0   9  17 297  50   2 428  44 489]
7   [  0 660  50  59  18  20  24  36  33  16]
8   [ 14   0  34   7 366  29 126  76  18 225]
9   [  4   0 709  38   3   4   8  11   6   2]
10  [  2   0  11   8 249  55   0 426  34 241]
11  [781   0  13   4   0   6  16   1   7   7]
12  [  5 469  86   5  24  92  19  48  47   7]]
13 {0: 6, 1: 3, 2: 8, 3: 9, 4: 1, 5: 4, 6: 2, 7: 7, 8: 0, 9: 1}
14 [[ 781    0   13    4    0    6   16    1    7    7]
15  [   5 1129  136   64   42  112   43   84   80   23]
16  [   4    0  709   38    3    4    8   11    6    2]
17  [  57    3   76  686    0  297    1    0  251    7]
18  [  14    0   34    7  366   29  126   76   18  225]
19  [   0    0    0    0    0    0    0    0    0    0]
20  [  41    2   20    9   23   23  736    1   10    2]
21  [   2    0   11    8  249   55    0  426   34  241]
22  [  74    1   24  177    2  316   26    1  524   13]
23  [   2    0    9   17  297   50    2  428   44  489]]
24 0.5846
25 0:00:05.892999
```

Programme 14.21 K-means clustering with $K = 10$ for the MNIST dataset with PCA via Python.

```
1  > start <- Sys.time()
2  > pca <- prcomp(X_train, center=TRUE, scale=FALSE)
3  > s <- pca$sdev
4  > t <- sum(s^2)              # Compute total variance
5  > cumvar <- cumsum(s^2/t) # Cumulative sum of proportion of variance
6  > (idx <- which(cumvar >= 0.95)[1])
7  [1] 154
8  >
9  > X_train_pca <- predict(pca, newdata=X_train)[,1:idx]
10 > X_test_pca <- predict(pca, newdata=X_test)[,1:idx]
11 > km_pca <- best_km(X_train_pca, K)      # K-means with K=10
12 [1] "K=10; stat=2459.61669347933"
13 > y_hat_pca <- predict.kmeans(km_pca, newdata=X_test_pca)
14 > (tab <- table(y_hat_pca, y_test))      # Confusion matrix
15          y_test
```

```
16  y_hat_pca    1    2    3    4    5    6    7    8    9   10
17          1    0  709   38    3    4    8   11    6    2    4
18          2    2   20    9   23   23  735    1   10    2   42
19          3    0    9   18  297   50    2  429   44  492    2
20          4    0   34    7  364   29  127   74   18  223   14
21          5    0   11    8  251   57    0  427   34  241    2
22          6  660   50   59   18   20   24   36   33   15    0
23          7  469   87    5   24   91   19   48   47    7    5
24          8    3   75  684    0  297    1    0  250    7   57
25          9    0   13    4    0    6   16    1    7    7  780
26         10    1   24  178    2  315   26    1  525   13   74
27  > (pos_dict <- apply(tab, 1, which.max))
28   1  2  3  4  5  6  7  8  9 10
29   2  6  9  4  7  1  1  3 10  8
30  > y_pred <- pos_dict[as.character(y_hat_pca)]
31  > (tab <- table(y_pred, y_test))          # Confusion matrix
32        y_test
33  y_pred    1    2    3    4    5    6    7    8    9   10
34       1 1129  137   64   42  111   43   84   80   22    5
35       2    0  709   38    3    4    8   11    6    2    4
36       3    3   75  684    0  297    1    0  250    7   57
37       4    0   34    7  364   29  127   74   18  223   14
38       6    2   20    9   23   23  735    1   10    2   42
39       7    0   11    8  251   57    0  427   34  241    2
40       8    1   24  178    2  315   26    1  525   13   74
41       9    0    9   18  297   50    2  429   44  492    2
42      10    0   13    4    0    6   16    1    7    7  780
43  > mean(y_pred == y_test)                  # Accuracy
44  [1] 0.5845
45  > Sys.time() - start                      # PCA + K-means
46  Time difference of 2.622091 mins
```

Programme 14.22 K-means clustering with $K = 10$ for the MNIST dataset with PCA via R.

14.2 *K*-NEAREST NEIGHBOUR

Having obtained clustering results, the next stage is the cluster prediction of new observations. One of the most popular approaches is the *K-Nearest Neighbour algorithm* (K-NN), an example of *lazy-learning*[8]. It was first proposed in the technical report "*Discriminatory analysis: Nonparametric discrimination: small sample performance*" by *Evelyn Fix* and *Joseph Hodges* of University of California, Berkeley in 1951 [10]; while its name was coined in the paper "*Nearest neighbour pattern classification*" by *Thomas Cover* and *Peter Hart* of Stanford University in 1967 [8].

[8] Lazy learning methods wait until the query arrival to start processing the input. Indeed, as discussed below, each query of K-NN only requires a small neighborhood of observations, namely the K closest ones to the newly added observation. Readers may refer to [1] for more details.

(I) K-NN classification

In the K-NN algorithm, the distance metric (or dis-similarity) between the unlabelled observation and each existing one in the dataset is computed. The labels of the nearest (or most similar) K data points are then collected and counted, and the resulting class with the highest frequency is used to label the observation in question; see Figure 14.8 for an illustration.

In particular, in view of the majority voting nature of K-NN classification, the parameter K is usually chosen as an odd number to avoid ties, yet the classification result may differ with K. Take Figure 14.8 as an example, namely classifying the unlabelled observation (the green spot at the center) as either a blue square or red diamond. When $K = 3$, the closest three observations involve two blue squares and one red diamond, hence the observation will be classified as a blue square. However, when $K = 9$, the classification result becomes red diamond instead as there are 4 blue squares and 5 red diamonds among the 9 closest observations.

Notationally, we assume a dataset with n data points partitioned into c classes, $\{x_i^{(k)}\}$, $k = 1, \ldots, c$ and $i = 1, 2 \ldots, n_k$, where n_k is the number of data points in the k-th class. For general values of K, we classify x by the most frequently occurring class label among the K nearest data points. For $k = 1, \ldots, c$ and $i = 1, 2 \ldots, n_k$, define $r_x\left(x_i^{(k)}\right)$ as the ordered ranking of all Euclidean distances between $x_i^{(k)}$'s and x, namely $\|x - x_i^{(k)}\|_2$, in ascending order. Then among the K nearest data points, the number of observations in class k is $K_k := |\mathcal{K}_k|$, where $\mathcal{K}_k := \{i : r_x\left(x_i^{(k)}\right) \leq K\}$ represents the collection of these points. The classification result m_x for x is then the majority class:

$$m_x := \underset{k=1,\ldots,c}{\arg\max}\, K_k.$$

In the special simple case of $K = 1$, x is classified using the nearest data point and we have:

$$m_x = \underset{k=1,\ldots,c}{\arg\min} \left(\min_{i=1,\ldots,n_k} \left\| x - x_i^{(k)} \right\|_2 \right).$$

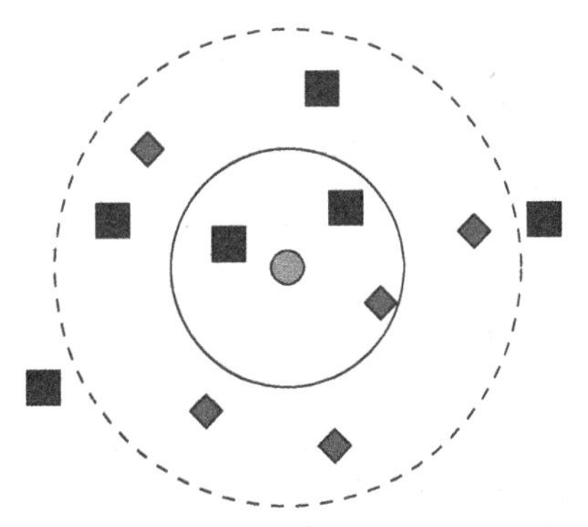

FIGURE 14.8 Illustration of K-NN classification, the newly added point is identified as blue square if $K = 3$, but red diamond if $K = 9$.

When the sample size is small, the performance of K-NN is very sensitive to the choice of K and the distance metric. In comparison with K-means clustering, which identifies the regions of different classes in the feature space, K-NN relies on the labels of the neighbouring data points of a new observation. This explains its suitability for handling out-samples with overlapping class regions. On the other hand, the straightforward design of K-NN also leads to several drawbacks. For one thing, the ranking of distance metric from each $x_i^{(k)}$ to x may be computationally expensive when the sample size is large. We can mitigate this issue by taking data points less significant to classification out of consideration at the preprocessing step. Another issue is that in the event of an imbalanced sample, a new observation from the minority class is more likely to be misclassified. Motivated by the intuition that minority observations often stay close to one another, we may tackle this issue by assigning weights to each of the K nearest points which are disproportional to their actual distance to the data point in question.

(II) K-NN Regression

Besides classification, K-NN can also be applied in a regression context, like the approach via CART; for example, see [14]. Recall that we can assign a larger weight to data points closer to a new observation x so as to reduce the misclassification rate for one from the minority class due to an imbalanced sample.

To predict the label of a new observation, we again focus on its K nearest data points, and the predicted label is then the average label of these K data points:

$$\hat{m} = \frac{1}{K} \sum_{r_x(x_i) \leq K} t_i,$$

where t_i is the label of the i-th data point x_i in the dataset. Despite the simplicity of the prediction formula, the performance of the fitted K-NN model may be severely affected by the choice of the hyperparameter K, which is fairly arbitrary in practice. Furthermore, an equal weight is assigned to each of the K nearest data points, while those observations close to the new one should be more important in the prediction task. It is more reasonable for them to receive higher weights compared with others farther away from the new observation. Besides the possible modifications on the K-NN algorithm to be introduced below, another common resolution to address these drawbacks is simply taking all data points into consideration, corresponding to the case of $K = \infty$, and introducing an appropriate weight function, also known as a *kernel function* Kernel(x, y), that decreases with the distance metric between x and y. The resulting prediction formula for the label of the new observation then becomes:

$$\hat{m} = \sum_i \frac{\text{Kernel}(x, x_i)}{\sum_j \text{Kernel}(x, x_j)} t_i.$$

In the context of a more generic tool of *Kernel Regression* (to be introduced in the next section), we shall see that this formula has the same form as the renowned *Nadaraya–Watson estimator* for a special scenario in kernel regression. To this end, while kernel regression does not include K-NN regression as a special case, it does try to tackle the aforementioned drawbacks of the latter.

(III) Enhancing the Algorithm

Another characteristic of K-NN is its sensitivity to noise. Isolated points in the sample may cause a drastic impact on the result of classification and regression. These outliers can be eliminated by an *a priori* filtering and selection procedure. Meanwhile, the choice of K may also significantly affect the outcome, as illustrated in Figure 14.8. To remedy these shortcomings, let us introduce some modifications to the plain K-NN algorithm.

Ensemble (Learning) Methods

In general, ensemble methods aim to improve the performance of a learning model by considering several independent models. The output is usually inferred from the results of voting from these collections of models. Depending on the weights assigned to each model, the voting is called *uniform voting* if all models share the same weight, or *weighted voting* otherwise. This general concept can also be applied to K-NN for either classification or regression. We first randomly select multiple subsets of features and fit K-NN models against the newly added observation accordingly. The predicted label of the classification (resp. regression) is then determined by voting (resp. taking a weighted average) among the outputs from these models. This is also known as *Random Subspace Ensemble K-NN*; also see [12] for details.

Kernel Methods

The kernel methods here (see [25]) refer to the kernel trick of radial basis functions discussed in Section 11.2, and should be distinguished from the kernel functions in K-NN regression or kernel regression. Recall that under the kernel trick, each sample $\boldsymbol{x} \in \mathbb{R}^p$ is mapped to a higher dimensional feature space, say $\mathbb{R}^q$ with $q > p$, to untangle the overlapping clusters in the hope of achieving linear separability of the samples in $\mathbb{R}^q$.

Here, we briefly describe the implementation of kernel methods in K-NN. Let the transformation be $\boldsymbol{\Phi} : \mathbb{R}^p \to \mathbb{R}^q$, so that the sample $\{\boldsymbol{x}_1, \dots, \boldsymbol{x}_n\}$ is mapped to $\{\boldsymbol{\Phi}(\boldsymbol{x}_1), \dots, \boldsymbol{\Phi}(\boldsymbol{x}_n)\}$. Let $K(\cdot,\cdot)$ be the kernel function, then for any $\boldsymbol{x}_i$ and $\boldsymbol{x}_j$, it is defined in such a way that

$$d(\boldsymbol{\Phi}(\boldsymbol{x}_i), \boldsymbol{\Phi}(\boldsymbol{x}_j)) = K(\boldsymbol{x}_i, \boldsymbol{x}_j),$$

where $d(\cdot,\cdot)$ is a distance function defined on $\mathbb{R}^q$. As we introduced in Section 11.2, the kernel $K(\cdot,\cdot)$ is usually proposed directly rather than the transformation $\boldsymbol{\Phi}$. The implementation of K-NN under this framework can be shown to greatly improve the performance, especially in a low-dimensional setting in which the features are not linearly separable.

Prior Clustering

A sample can be clustered first (e.g., by K-means) before implementing the K-NN algorithm; see [26]. Roughly speaking, given a clustered sample and a new observation $\boldsymbol{x}$, we first calculate its distance to the centroid of each cluster, and accordingly denote $S_{(i)}$, as the i-th nearest cluster. Taking these clusters into account, the K nearest neighbors can then be selected as follows: calculate the distance between the newly added observation $\boldsymbol{x}$ and each of the points in $S_{(1)}$, and one-by-one include those points, each of which has a distance from $\boldsymbol{x}$ smaller than a pre-set threshold, into the set of neighbouring points of x, in ascending order of the distances. The procedure is repeated for $i = 2, 3, \ldots$ until there are K points in the set of neighboring points of x. Then x attains its predicted label i_0 such that $S_{(i_0)}$ has the largest number of elements in this K-point list.

(IV) Application of K-NN Algorithm

While K-NN can be used for general classification or regression tasks, it is widely applied in facial recognition. Typically, given an observation of a human face, we aim to infer certain characteristics of the person, such as age and ethnicity, using the information of the K most similar faces in the training dataset. A general procedure is described as follows, assuming that there are M possible class labels in total:

1. For the i-th human face in the training dataset of size n, extract its values of pre-specified list of p features, possibly determined by principal components or more generally dictionary learning as discussed in Section 14.1, and store them as the feature vector $\boldsymbol{x}_i = (x_{i1}, \ldots, x_{ip})^\top$ for this face, for $i = 1, \ldots, n$.
2. Identify the feature vector $\boldsymbol{x}_{n+1}$ of the newly added human face.
3. Identify the K nearest faces in the training set using the following (signed) similarity measure of cosine as the criterion:

$$\text{Sim}(\boldsymbol{x}_i, \boldsymbol{x}_j) := \frac{\sum_{l=1}^{p} x_{il}x_{jl}}{\sqrt{\left(\sum_{l=1}^{p} x_{il}^2\right)\left(\sum_{l=1}^{p} x_{jl}^2\right)}} = \frac{\boldsymbol{x}_i^\top \boldsymbol{x}_j}{\|\boldsymbol{x}_i\|_2 \|\boldsymbol{x}_j\|_2},$$

   where this similarity measure is precisely $\cos\theta_{ij}$, where θ_{ij} is the angle between $\boldsymbol{x}_i$ and $\boldsymbol{x}_j$. In other words, we determine the K feature vectors that are closest in direction to $\boldsymbol{x}_{n+1}$, and store them in a set $\mathcal{P}_K$.
4. For each class label c_k, $k = 1, \ldots, M$, we calculate the following weight for the new photo with the feature vector $\boldsymbol{x}_{n+1}$:

$$w_k(\boldsymbol{x}_{n+1}) := \sum_{\boldsymbol{x} \in \mathcal{P}_K, c(\boldsymbol{x}) = c_k} \frac{1}{1 - \text{Sim}(\boldsymbol{x}_{n+1}, \boldsymbol{x})}. \tag{14.9}$$

5. The new photo is then classified as the label that maximizes the weight (14.9), i.e., $\hat{k} = \arg\max_k w_k(\boldsymbol{x}_{n+1})$.

Let us illustrate the idea in Programme 14.23 via Python using the same dataset for facial recognition as used in Programme 9.23. Note that the Python function `KNeighborsClassifier()` from the `sklearn.neighbors` library is used to perform K-NN, and the function requires three input arguments, namely `n_neighbors=K` for the value of the hyperparameter K, the number of nearest neighbours considered; `metric="cosine"` for using the cosine similarity metric described in Item 3 above; and `weight="distance"` for weighting the points by the inverse of their distance as described in (14.9). Recall that the photos in Programme 9.23 are not shuffled, and are arranged such that each individual has 9 faces and they are sorted in sequence, and stored in `y_train`. Programme 14.23

demonstrates the implementation of K-NN with a generally adopted choice of hyperparameter $K = 3$, using the 107 PCs obtained in Section 9.8 as the features. It achieves a high accuracy of $\frac{39}{40} = 97.5\%$ and there is only one wrong recognition for the 34th individual. Indeed, the output weight vector for this individual contains only two positive weights, namely 0.3079 and 0.6921 on the 34th and 39th position, respectively, and the predicted label is determined to be 39 as it has the greater weight.

```
1  y_train = np.array([np.repeat(x, 9) for x in range(40)]).reshape(-1)
2  K = 3
3
4  from sklearn.neighbors import KNeighborsClassifier
5
6  model = KNeighborsClassifier(n_neighbors=K, metric="cosine",
7                               weights="distance")
8  model.fit(score_train.T, y_train)
9  distances, indices = model.kneighbors(score_test.T)
10 indices = model.predict(score_test.T)
11 print(indices)
12 proba = model.predict_proba(score_test.T)
13 print(proba[34])                  # 34 is wrong
14
15 y_hat, y_weight = [], []
16 norm_train = np.linalg.norm(score_train, axis=0)
17 norm_test  = np.linalg.norm(score_test, axis=0)
18 for i in range(len(score_test.T)):
19     similarity = (score_train.T @ score_test[:,i] /
20                   (norm_train * norm_test[i]))
21     idx = np.argpartition(similarity, -K)          # cloest to theta=0
22     P_K = similarity[idx[-K:]]
23     weight = np.zeros(len(np.unique(y_train)))
24     for m in range(len(np.unique(y_train))):
25         weight[m] = np.sum(1 / (1 - P_K[y_train[idx[-K:]] == m]))
26
27     idx_train = np.argmax(weight)
28     y_hat.append(idx_train)
29     y_weight.append(weight)
30
31 print(y_hat)
32 print(y_weight[34] / np.sum(y_weight[34]))        # 34 is wrong
```

```
1  [ 0  1  2  3  4  5  6  7  8  9 10 11 12 13 14 15 16 17 18 19 20 21 22
2   23 24 25 26 27 28 29 30 31 32 33 39 35 36 37 38 39]
3  [0.         0.         0.         0.         0.         0.
4   0.         0.         0.         0.         0.         0.
5   0.         0.         0.         0.         0.         0.
6   0.         0.         0.         0.         0.         0.
7   0.         0.         0.         0.         0.         0.
8   0.         0.         0.         0.         0.30785606 0.
9   0.         0.         0.         0.69214394]
10 [ 0  1  2  3  4  5  6  7  8  9 10 11 12 13 14 15 16 17 18 19 20 21 22
11  23 24 25 26 27 28 29 30 31 32 33 39 35 36 37 38 39]
12 [0.         0.         0.         0.         0.         0.
```

```
13  0.         0.         0.         0.         0.         0.
14  0.         0.         0.         0.         0.         0.
15  0.         0.         0.         0.         0.         0.
16  0.         0.         0.         0.         0.         0.
17  0.         0.         0.         0.         0.30785606 0.
18  0.         0.         0.         0.69214394]
```

Programme 14.23 Continuation of Programme 9.23: Applying the `KNeighbors Classifier()` function to facial recognition in Python.

We also revisit the MNIST dataset, this time using the K-NN algorithm. For simplicity, we select $K = 3$ as in the previous example. As before, we first execute a standard direct K-NN algorithm as a benchmark. From the outputs obtained in Programme 14.24 via Python, the prediction accuracy of the K-NN algorithm with $K = 3$ on the test dataset is 0.9705, which is higher than the 0.9256 obtained in the logistic regression model coded in Programme 10.17 via Python. Furthermore, due to the fact that K-NN is a lazy algorithm, namely it does not process the training dataset until a prediction needs to be made, the running time for the Python programme is just about 6 seconds, much faster than the logistic regression (1 minute) in Programme 10.17 and the K-means clustering (15 seconds) in Programme 14.19. While the implementation via **R** in Programme 14.25, with the `knn()` function in the `class` library, yields a similar testing accuracy of 0.9716, it is much more time-consuming with the running time of over an hour[9].

```
1  from sklearn.neighbors import KNeighborsClassifier
2
3  start = dt.now()
4  knn = KNeighborsClassifier(n_neighbors=3)          # KNN with K=3
5  knn.fit(X_train, y_train)
6  y_hat_knn = knn.predict(X_test)
7  tab = confusion_matrix(y_hat_knn, y_test)          # Confusion matrix
8  print(tab)
9  print(sum(tab.diagonal()) / len(y_test))           # Accuracy
10 print(dt.now() - start)                            # KNN
```

```
1  [[ 974    0   10    0    1    6    5    0    8    4]
2   [    1 1133    9    2    6    1    3   21    2    5]
3   [    1    2  996    4    0    0    0    5    4    2]
4   [    0    0    2  976    0   11    0    0   16    8]
5   [    0    0    0    1  950    2    3    1    8    9]
```

[9] To the best of our knowledge, perhaps the Python function `KNeighborsClassifier()` encloses some advanced methods that adjust the data structure, so as to reduce the number of calculations required to find the nearest neighbors, while the **R** function `knn()` favours a brute-force search, so that the distances from the testing data point to all existing data points have to be calculated.

```
6  [   1    0    0   13    0  859    3    0   11    2]
7  [   2    0    0    1    4    5  944    0    3    1]
8  [   1    0   13    7    2    1    0  991    4    8]
9  [   0    0    2    3    0    3    0    0  914    2]
10 [   0    0    0    3   19    4    0   10    4  968]]
11 0.9705
12 0:00:05.816000
```

Programme 14.24 *K*-NN algorithm for the MNIST dataset with $K = 3$ via Python.

```
1  > library(class)
2  > y_test[y_test == K] <- 0
3  >
4  > start <- Sys.time()
5  > y_hat_knn <- knn(train=X_train, test=X_test,
6  +                  cl=factor(y_train), k=3) # KNN with K=3
7  > (tab <- table(y_hat_knn, y_test))      # Confusion matrix
8           y_test
9  y_hat_knn    0    1    2    3    4    5    6    7    8    9
10         0  974    0    9    0    0    4    4    0    5    3
11         1    1 1132    8    1    5    1    3   18    0    4
12         2    1    3  997    4    0    0    0    4    3    1
13         3    0    0    2  976    0   11    0    0   15    7
14         4    0    0    0    1  949    2    4    2    7    9
15         5    1    0    0   12    0  860    3    0   11    2
16         6    2    0    0    1    4    5  944    0    3    1
17         7    1    0   14    7    2    1    0  994    5    9
18         8    0    0    2    4    1    4    0    0  921    4
19         9    0    0    0    4   21    4    0   10    4  969
20 > mean(y_hat_knn == y_test)                # Accuracy
21 [1] 0.9716
22 > Sys.time() - start                      # KNN
23 Time difference of 1.051061 hours
```

Programme 14.25 *K*-NN algorithm for the MNIST dataset with $K = 3$ via **R**.

As in Section 14.1, we also investigate the effect of PCA on the *K*-NN algorithm, by reusing the PCs selected in the example of *K*-means clustering as the feature variable inputs for *K*-NN. From the outputs obtained in Programme 14.26 via Python and Programme 14.27 via **R**, the testing accuracy of *K*-NN still remains at around 0.97 without notable improvements, yet the running time for **R** on a common desktop computer greatly reduces from over an hour to just roughly 2.5 minutes thanks to the dimension reduction. The unimproven accuracy comes as no surprise; once again, if all PCs are involved, then the Euclidean distances will not change upon applying the linear transform $H^{\top}$ to the original data points, as shown in (14.8), and the resulting *K*-NN with PCA is then equivalent to the original version of *K*-NN, with exactly the same prediction accuracy.

We conclude this example by remarking that, although the testing accuracy of around 97% is already satisfactory, we can improve on this with various new tools from Deep Learning, e.g. the *Multilayer Perceptron* (an accuracy of 97.83%) or a *Convolutional Neural Network* (an accuracy of 99.02%). We shall introduce the former in Chapter 15, while the latter is beyond the scope of this book, and it will be discussed in detail in [6].

```
start = dt.now()
knn_pca = KNeighborsClassifier(n_neighbors=3)     # KNN with K=3
knn_pca.fit(X_train_pca, y_train)
y_hat_pca = knn_pca.predict(X_test_pca)
tab = confusion_matrix(y_hat_pca, y_test)         # Confusion matrix
print(tab)
print(sum(tab.diagonal()) / len(y_test))          # Accuracy
print(dt.now() - start)                           # PCA + KNN
```

```
[[ 974    0    9    1    1    6    6    0    7    4]
 [   1 1132    8    2    5    1    3   17    2    4]
 [   1    2  997    3    0    0    0    5    4    2]
 [   0    0    2  978    0   11    0    0   16    9]
 [   0    0    1    1  952    1    2    1    7   10]
 [   1    0    0   12    0  860    2    0    9    2]
 [   2    1    0    0    4    5  945    0    3    1]
 [   1    0   14    7    2    1    0  997    3    9]
 [   0    0    1    3    0    3    0    0  920    1]
 [   0    0    0    3   18    4    0    8    3  967]]
0.9722
0:00:01.149000
```

Programme 14.26 *K*-NN algorithm with $K = 3$ for the MNIST dataset with PCA via Python.

```
> start <- Sys.time()
> y_hat_pca <- knn(train=X_train_pca, test=X_test_pca,
+                  cl=factor(y_train), k=3) # KNN with K=3
> (tab <- table(y_hat_pca, y_test))         # Confusion matrix
         y_test
y_hat_pca    0    1    2    3    4    5    6    7    8    9
        0  974    0    8    0    0    3    4    0    5    3
        1    1 1132    6    2    4    0    3   16    1    4
        2    1    2  997    2    0    0    0    5    3    2
        3    0    0    2  975    0    9    0    0   12    6
        4    0    0    1    1  952    1    2    1    4   10
        5    1    0    0   15    0  862    4    0    9    3
        6    2    1    0    0    4    9  945    0    5    1
        7    1    0   13    7    2    1    0  998    3    8
        8    0    0    4    3    0    3    0    0  928    2
        9    0    0    1    5   20    4    0    8    4  970
> mean(y_hat_pca == y_test)                 # Accuracy
[1] 0.9733
> Sys.time() - start                        # PCA + KNN
Time difference of 2.512995 mins
```

Programme 14.27 *K*-NN algorithm with $K = 3$ for the MNIST dataset with PCA via R.

*14.3 KERNEL REGRESSION

While K-means clustering mainly focuses on separating out distinguishable clusters, there is still the puzzle of a prediction for the label of a new datum, which is the ultimate goal of clustering analysis. This is what K-NN regression does in the previous section. One apparent drawback of K-NN regression is the lack of clear guidelines for choosing K. A commonly adopted alternative approach is to take all data points into consideration, corresponding to the case of $K = \infty$. In light of this, we shall revisit a versatile tool of kernel regression in functional regression analysis, where we investigate a special case of a particular regressor called the *Nadaraya–Watson estimator* (see [19, 27]), which resembles the formula for K-NN regression with $K = \infty$. This greatly enriches the content of K-NN and its use in practice.

Recall that the fundamental task in regression analysis is the prediction of the response vector $\mathrm{y} \in \mathbb{R}^q$ using the feature vector $\boldsymbol{x} \in \mathbb{R}^p$. A key quantity of interest in this scenario is the conditional mean $\mathbb{E}(\mathrm{y} \mid \mathbf{x} = x)$. For the sake of simplicity, in this section we discuss the asymptotic theory for the simplest case of $p = q = 1$ only, while the argument for the general case is essentially the same but involves more tedious notations. The general regression framework takes the following form:

$$\mathrm{y} = m(\mathrm{x}) + \epsilon,$$

where $m(x)$ is an unknown function of x, ϵ and x are uncorrelated, and $\mathbb{E}(\epsilon \mid \mathrm{x} = x) = 0$. We define $\sigma^2(x) := \mathrm{Var}(\epsilon \mid \mathrm{x} = x) = \mathrm{Var}(\mathrm{y} \mid \mathrm{x} = x)$. We often assume that $m(x)$ is piecewise smooth, such that it is of class C^k for the interior points and C^{k-1} at the knots (boundary-splitting points); here C^k indicates that the derivatives up to the k-th order exist and are continuous. Then for a given point $x_0 \in \mathbb{R}$, we expect that $m(x_i)$ should be close to $m(x_0)$ for any observed value x_i in the proximity of x_0. Therefore, a reasonable estimate for $m(x_0)$ would be an average of the observed responses y_i's corresponding to those x_i's.

By Taylor expansion of $m(x)$ up to order k at x_0, in the proximity of x_0, we have:

$$\begin{aligned} m(x) &\approx m(x_0) + m^{(1)}(x_0)(x - x_0) + \frac{1}{2!}m^{(2)}(x_0)(x - x_0)^2 + \cdots + \frac{1}{k!}m^{(k)}(x_0)(x - x_0)^k \\ &=: \beta_0 + \beta_1(x - x_0) + \beta_2(x - x_0)^2 + \cdots + \beta_k(x - x_0)^k, \end{aligned}$$

which resembles a parametric model, with parameters $\{\beta_j := m^{(j)}(x_0)/j!\}_{j=0}^k$. While these can be estimated by parametric regression techniques using the observed data x_i's scattered around x_0, the results of this apparently local approach may deviate significantly if different datasets in closer or farther neighborhoods of x_0 are used. Worse still, in the event of a sparse dataset, the limited number of observations near x_0 further reduces the accuracy of estimates.

To remedy this shortcoming, we instead adopt a weighted least square approach by considering the following optimization problem which involves the full dataset $\{(x_i, y_i)\}_{i=1}^n$:

$$\min_{\beta_0,\ldots,\beta_k} \sum_{i=1}^n \left(y_i - \beta_0 - \beta_1(x_i - x_0) - \cdots - \beta_k(x_i - x_0)^k\right)^2 K_{b_n}(x_0 - x_i),$$

where the weighting function K_{h_n} is called a *kernel function*, and h_n is its bandwidth, the subscript of which indicates that its choice usually depends on the sample size n. This method is hence named *Kernel Regression*. The role of K_{h_n} is to control the amount of effect brought by each sample point (x_i, y_i) so that its contribution to the estimation of β_j's depends on the distance between x_0 and its own feature value x_i. Naturally, we expect a non-increasing contribution as the distance increases; indeed, the following common choice of K_{h_n} reflects this desired trend: $K_{h_n}(x_0 - x_i) = \frac{1}{h_n}K\left(\frac{x_0-x_i}{h_n}\right)$, where K is independent of h_n, satisfying the following technical requirements:

1. $K(\cdot)$ is a positive function, and $\int_{\mathbb{R}} K(u)du = 1$;
2. $K(u) = K(-u)$, i.e., K is symmetric about the origin; and
3. $K(u)$ attains its global maximum at $u = 0$, indicating that the weights are larger for the observations x_i's that are close to x_0.

Several commonly adopted choices of kernel functions are the following:

1. Uniform kernel (a.k.a. Box kernel): $K(u) = \frac{1}{2}\mathbb{1}_{\{|u|\leq 1\}}$;
2. Gaussian kernel: $K(u) = \frac{1}{\sqrt{2\pi}}\exp\left(-\frac{u^2}{2}\right)$;
3. *Epanechnikov* kernel: $K(u) = \frac{3}{4}(1-u^2)\mathbb{1}_{\{|u|\leq 1\}}$.

See Figure 14.9 for the illustrations of the plots of these functions. We notice that despite the discrepancy in numerical results using different kernels, in practice they yield comparable estimates with carefully selected bandwidths h_n's.

Next, we illustrate the construction of an estimate $\hat{m}(x)$ for the simplest case of $k = 1$, when $m(x)$ is approximated by a constant $\beta_0 = m(x_0)$, for a pre-specified $x_0 \in \mathbb{R}$. As we shall show in the following, the estimate is a weighted average of the responses mainly contributed by those observations whose feature variables x_i's lie in

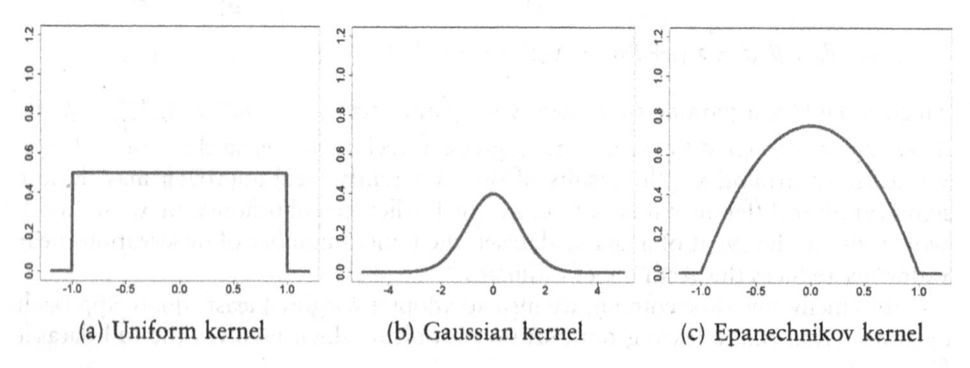

(a) Uniform kernel (b) Gaussian kernel (c) Epanechnikov kernel

FIGURE 14.9 Illustrations of three commonly adopted kernel functions.

a close proximity of x_0. This resulting optimization problem is also known as *local constant regression* (also see [16]), with the following problem specification:

$$\min_{\beta_0} \sum_{i=1}^{n} \frac{1}{h_n} K\left(\frac{x_0 - x_i}{h_n}\right)(y_i - \beta_0)^2, \qquad x_0 \in \mathbb{R}. \tag{14.10}$$

Differentiating the objective function in (14.10) with respect to β_0 and equating it with 0, we arrive at the following first order condition:

$$-\frac{2}{h_n}\sum_{i=1}^{n} K\left(\frac{x_0 - x_i}{h_n}\right)(y_i - \beta_0) = 0,$$

resulting in the following estimate:

$$\hat{\beta}_0 = \frac{\sum_{i=1}^{n} K\left(\frac{x_0 - x_i}{h_n}\right) y_i}{\sum_{i=1}^{n} K\left(\frac{x_0 - x_i}{h_n}\right)}. \tag{14.11}$$

Its estimator version, with (x_i, y_i) replaced by $(\mathrm{x}_i, \mathrm{y}_i)$, is called the *Nadaraya–Watson estimator* $\hat{\mathrm{m}}(x)$ (see [19, 27]) for $m(x)$ with $x = x_0$:

$$\hat{\mathrm{m}}(x_0) = \frac{\sum_{i=1}^{n} K\left(\frac{x_0 - \mathrm{x}_i}{h_n}\right) \mathrm{y}_i}{\sum_{i=1}^{n} K\left(\frac{x_0 - \mathrm{x}_i}{h_n}\right)}.$$

It can readily be extended to the case of general dimension p as follows, for $\boldsymbol{x}_0 \in \mathbb{R}^p$:

$$\hat{\mathrm{m}}(\boldsymbol{x}_0) = \frac{\sum_{i=1}^{n} K\left(\frac{\|\boldsymbol{x}_0 - \mathbf{x}_i\|_2}{h_n}\right) \mathrm{y}_i}{\sum_{i=1}^{n} K\left(\frac{\|\boldsymbol{x}_0 - \mathbf{x}_i\|_2}{h_n}\right)},$$

where $\|\cdot\|_2$ denotes the Euclidean norm. Again, echoing Section 14.2, the K-NN regression where $K = \infty$ resembles the Nadaraya–Watson estimator.

Under some regularity conditions (to be introduced later in this section), the bias and variance of $\hat{\mathrm{m}}(x_0)$ possess the following boundedness properties, where $C_1, C_2 > 0$ are some constants:

- $\mathrm{Bias}(\hat{\mathrm{m}}(x_0)) = \mathbb{E}(\hat{\mathrm{m}}(x_0)) - m(x_0)$ has a magnitude less than $C_1 h_n^2$;
- $\mathrm{Var}(\hat{\mathrm{m}}(x_0)) \leq \frac{C_2}{nh_n}$.

Clearly, as the bandwidth h_n decreases, the estimator $\hat{\mathrm{m}}(x_0)$ tends to be less biased but more volatile. Thus, a larger h_n is more likely to result in a smoother estimate

$\hat{m}(x_0)$, while a smaller h_n may lead to a more oscillatory $\hat{m}(x_0)$. Altogether, this phenomenon is sometimes called the *bias-variance tradeoff* in the context of statistical machine learning. The optimal bandwidth that minimizes the sum of (bias)^2+ variance is then of the order of $n^{-1/5}$.

Alternatively, the Nadaraya–Watson estimator can also be motivated using the kernel density estimator of an unknown density function $f_{\mathrm{x}}(x)$, also known as the *Rosenblatt–Parzen kernel density estimator* (see [21, 24]):

$$\hat{\mathrm{f}}_{\mathrm{x}}(x) = \frac{1}{nh_n}\sum_{i=1}^{n} K\left(\frac{x-\mathrm{x}_i}{h_n}\right). \tag{14.12}$$

The construction of the Rosenblatt–Parzen kernel density estimator is motivated by the original empirical distribution. To see why, we recall that the empirical cdf of an iid sample $\{\mathrm{x}_i\}_{i=1}^n$, is defined by $\hat{\mathrm{F}}(x) = \frac{1}{n}\sum_{i=1}^n \mathbb{1}_{\{\mathrm{x}_i\le x\}}$. Because $\mathbb{1}_{\{\mathrm{x}_i\le x\}}$ can be thought of as changing infinitely rapidly from 0 to 1 at x while staying unchanged everywhere else, we can interpret its formal derivative as a Dirac delta function (formally, as a generalized function, Dirac delta $\delta_z(u) = +\infty$ at $u = z$ and equals 0 elsewhere, such that $\int_{\mathbb{R}}\delta_z(u)du = 1$). Therefore, as a formal "derivative", the empirical density function can be taken to be $\frac{1}{n}\sum_{i=1}^n \delta_{\mathrm{x}_i}(x)$. As a kernel, $\frac{1}{h_n}K\left(\frac{x-\mathrm{x}_i}{h_n}\right)$ mimics the behaviour of $\delta_{\mathrm{x}_i}(x)$ asymptotically, and so it is reasonable to estimate the density by using (14.12). Indeed, $\hat{\mathrm{f}}_{\mathrm{x}}(x)$ is a valid density as $\int_{\mathbb{R}}\hat{\mathrm{f}}_{\mathrm{x}}(x)dx = 1$ by the properties of $K(\cdot)$. Furthermore, for each x, the estimator converges to the true density $f_{\mathrm{x}}(x)$ in probability under some regularity conditions. To see this, we first calculate the expectation of $\hat{\mathrm{f}}_{\mathrm{x}}(x)$:

$$\begin{aligned}
\mathbb{E}(\hat{\mathrm{f}}_{\mathrm{x}}(x)) &= \frac{1}{nh_n}\sum_{i=1}^{n}\mathbb{E}\left(K\left(\frac{x-\mathrm{x}_i}{h_n}\right)\right)\\
&= \frac{1}{h_n}\mathbb{E}\left(K\left(\frac{x-\mathrm{x}}{h_n}\right)\right)\\
&= \int_{\mathbb{R}}\frac{1}{h_n}K\left(\frac{x-u}{h_n}\right)f_{\mathrm{x}}(u)du\\
&= \int_{\mathbb{R}}K(z)f_{\mathrm{x}}(x+h_nz)dz\\
&= \int_{\mathbb{R}}K(z)f_{\mathrm{x}}(x)dz + o(1)\\
&= f_{\mathrm{x}}(x) + o(1),
\end{aligned} \tag{14.13}$$

where to obtain the second last equality, the integral $\int_{\mathbb{R}}K(z)f_{\mathrm{x}}(x+h_nz)dz$ has an integrand such that $f_{\mathrm{x}}(x+h_nz) = f_{\mathrm{x}}(x) + o(1)$ locally uniformly as $h_n \to 0$ if we assume

(i) f_{x} is globally Lipschitz, and (ii) $\int_{\mathbb{R}}|z|K(z)dz$ is finite, the integral is certainly equal to $\int K(z)f_{\mathrm{x}}(x)dz + o(1)$.

Meanwhile, the variance of $\hat{\mathrm{f}}_{\mathrm{x}}(x)$ is:

$$\begin{aligned}
\mathrm{Var}(\hat{\mathrm{f}}_{\mathrm{x}}(x)) &= \frac{1}{nh_n^2}\mathrm{Var}\left(K\left(\frac{x-\mathrm{x}}{h_n}\right)\right)\\
&= \frac{1}{nh_n^2}\left(\mathbb{E}\left(K^2\left(\frac{x-\mathrm{x}}{h_n}\right)\right) - O(h_n^2)\right)\\
&= \frac{1}{nh_n^2}\left(\int_{\mathbb{R}} K^2(z)f_{\mathrm{x}}(x+h_nz)h_n dz - O(h_n^2)\right)\\
&= \frac{1}{nh_n}\left(\int_{\mathbb{R}} K^2(z)f_{\mathrm{x}}(x)dz + O\left(\int_{\mathbb{R}} K^2(z)zdz\cdot h_n\right) - O(h_n)\right)\\
&= \frac{1}{nh_n}f_{\mathrm{x}}(x)\int_{\mathbb{R}} K^2(z)dz + O\left(\frac{1}{n}\right),
\end{aligned}$$

where the second line holds since $\mathbb{E}\left(K(\frac{x-\mathrm{x}}{h_n})\right) = O(h_n)$ as a direct consequence of (14.13). The second last line follows if both conditions (i) and (ii) mentioned above are satisfied. Therefore for all $\delta > 0$, by Chebyshev's inequality, $\mathbb{P}\left(|\hat{\mathrm{f}}_{\mathrm{x}}(x) - \mathbb{E}(\hat{\mathrm{f}}_{\mathrm{x}}(x))| \geq \delta\right) \leq \frac{\mathrm{Var}(\hat{\mathrm{f}}_{\mathrm{x}}(x))}{\delta^2}$, which tends to 0 when $h_n \to 0$ but $nh_n \to \infty$ as $n \to \infty$. Altogether, these lead to the weak convergence of $\hat{\mathrm{f}}_{\mathrm{x}}(x)$ to $f_{\mathrm{x}}(x)$ for every $x \in \mathbb{R}$, and so proves the consistency of this approach.

Now, for a random sample $\{(\mathrm{x}_i, \mathrm{y}_i)\}_{i=1}^n$, by virtue of the *Plug-in Principle*[10], we can similarly estimate the joint density of (x, y) by the product kernel:

$$\hat{\mathrm{f}}_{\mathrm{x,y}}(x,y) = \frac{1}{nh_n^2}\sum_{i=1}^{n} K\left(\frac{x-\mathrm{x}_i}{h_n}\right)K\left(\frac{y-\mathrm{y}_i}{h_n}\right), \quad (x,y)\in\mathbb{R}^2.$$

Recalling that for a fixed $x_0 \in \mathbb{R}$, by the model setting,

$$m(x_0) = \mathbb{E}(\mathrm{y} \mid \mathrm{x} = x_0) = \int_{\mathbb{R}} y f_{\mathrm{y}|\mathrm{x}}(y \mid x_0)dy = \frac{\int_{\mathbb{R}} y f_{\mathrm{x,y}}(x_0, y)dy}{f_{\mathrm{x}}(x_0)},$$

[10] This is a technique commonly adopted in Statistics, namely a characteristic of an unknown distribution, such as expected value or variance, can be approximated by its empirical counterpart based on a sample from the unknown distribution, such as sample mean and sample variance; see [23] for a detailed discussion on the principle.

where the denominator can be estimated by $\hat{f}_x(x_0)$ according to (14.12), and the integrand in the numerator can be estimated by:

$$\begin{aligned}
&\int_{\mathbb{R}} y\hat{f}_{x,y}(x_0, y)dy \\
&= \int_{\mathbb{R}} y\frac{1}{nh_n^2}\sum_{i=1}^{n} K\left(\frac{x_0 - x_i}{h_n}\right) K\left(\frac{y - y_i}{h_n}\right) dy \\
&= \frac{1}{nh_n}\sum_{i=1}^{n} K\left(\frac{x_0 - x_i}{h_n}\right) \int_{\mathbb{R}} \frac{y}{h_n} K\left(\frac{y - y_i}{h_n}\right) dy \\
&= \frac{1}{nh_n}\sum_{i=1}^{n} K\left(\frac{x_0 - x_i}{h_n}\right) \left(\int_{\mathbb{R}} \frac{y - y_i}{h_n} K\left(\frac{y - y_i}{h_n}\right) dy + y_i \int_{\mathbb{R}} \frac{1}{h_n} K\left(\frac{y - y_i}{h_n}\right) dy\right) \\
&= \frac{1}{nh_n}\sum_{i=1}^{n} K\left(\frac{x_0 - x_i}{h_n}\right) y_i,
\end{aligned}$$

where the second last line follows due to the fact that the first integral in the bracket becomes 0 by the symmetric property of $K(\cdot)$, while the second term can be rewritten as $y_i \int_{\mathbb{R}} K\left(\frac{y-y_i}{h_n}\right) d\left(\frac{y-y_i}{h_n}\right)$, which equals y_i by the very definition of the kernel function $K(\cdot)$. Hence, we can obtain an estimate for $m(x)$ as follows: for a fixed $x_0 \in \mathbb{R}$,

$$\hat{m}(x_0) = \frac{\int_{\mathbb{R}} y\hat{f}_{x,y}(x_0, y)dy}{\hat{f}_x(x_0)} = \frac{\frac{1}{nh_n}\sum_{i=1}^{n} K\left(\frac{x_0-x_i}{h_n}\right) y_i}{\frac{1}{nh_n}\sum_{i=1}^{n} K\left(\frac{x_0-x_i}{h_n}\right)} = \frac{\sum_{i=1}^{n} K\left(\frac{x_0-x_i}{h_n}\right) y_i}{\sum_{i=1}^{n} K\left(\frac{x_0-x_i}{h_n}\right)},$$

which is precisely the Nadaraya–Watson estimate (14.11).

Meanwhile, recall the case for a general value of order k, $m(x)$ is approximated by a k-th order polynomial in $x - x_0$ for a fixed $x_0 \in \mathbb{R}$, and the resulting estimator is called the *local polynomial kernel regression estimator*; see [9] for a detailed discussion. Indeed, for a general value of k, the corresponding minimization problem takes the following form: for a pre-specified $x_0 \in \mathbb{R}$,

$$\begin{aligned}
&\min_{\beta_0,\beta_1,\ldots,\beta_k} \sum_{i=1}^{n} \left(y_i - \sum_{j=0}^{k} \beta_j (x_i - x_0)^j\right)^2 \cdot \frac{1}{h_n} K\left(\frac{x_0 - x_i}{h_n}\right) \\
&= \min_{\beta} \frac{1}{h_n} (\boldsymbol{y} - \boldsymbol{X\beta})^\top \boldsymbol{K} (\boldsymbol{y} - \boldsymbol{X\beta}),
\end{aligned}$$

where $\boldsymbol{\beta} := (\beta_0, \beta_1, \ldots, \beta_k)^\top$, $\boldsymbol{y} := (y_1, \ldots, y_n)^\top$,

$$\boldsymbol{K} := \mathrm{diag}\left(K\left(\frac{x_0 - x_1}{h_n}\right), \ldots, K\left(\frac{x_0 - x_n}{h_n}\right)\right), \text{and}$$

$$\boldsymbol{X} := \begin{pmatrix} 1 & x_1 - x_0 & \ldots & (x_1 - x_0)^k \\ \vdots & \vdots & \ddots & \vdots \\ 1 & x_n - x_0 & \ldots & (x_n - x_0)^k \end{pmatrix} \text{is an } n \times (k+1) \text{ data matrix.}$$

The solution to this problem is the following weighted least squares estimate:

$$\hat{\beta} = \begin{pmatrix} \hat{\beta}_0 \\ \vdots \\ \hat{\beta}_k \end{pmatrix} = (X^\top KX)^{-1}X^\top Ky.$$

Let us illustrate the use of kernel regression through the following simulation. In Programme 14.28 via Python and Programme 14.29 via **R**, we generate a random sample of size $n = 500$ from $\mathrm{x} \sim \mathcal{N}(2, 16)$. Then for each observed $x_i, i = 1, \ldots, 500$, we generate the corresponding response y_i by the following model:

$$\mathrm{y}_i = m(x_i) + \epsilon_i,$$

where $m(x) := x^2 \cos x$ and $\epsilon_i \overset{iid}{\sim} \mathcal{N}(0, 4)$. The Nadaraya–Watson estimates $\hat{m}(x)$ for values of x from -15 to 15 with an increment of 0.3 are then calculated with the bandwidth $h_n = 0.5$, and they are plotted together with the true underlying function $m(x)$ as defined above; see Figure 14.10. It can be observed from the plots that the estimates are fairly close to (resp. deviate from) the true values for those x's close to (resp. far from) 0, which can be justified by the sample generation for x_i's. Indeed, most of the observed values lie within two standard deviations from the mean, that is, in the interval $[-6, 10]$, hence the estimates tend to be more accurate when x also falls within this interval.

```
1  import numpy as np
2  from scipy.stats import norm
3  import matplotlib.pyplot as plt
4
5  plt.rc('font', size=20); plt.rc('axes', titlesize=30, labelsize=30)
6  plt.rc('xtick', labelsize=25); plt.rc('ytick', labelsize=25)
7
8  def NW_estimator(x0, x, y, h, kernel=norm.pdf):
9          x0, x = np.asarray(x0)[:, None], np.asarray(x)
10         f_x = kernel((x0 - x)/h)/h
11         return f_x @ y / np.sum(f_x, axis=1)
12
13 np.random.seed(4002)
14 n = 500
15 eps = np.random.normal(loc=0, scale=2, size=n)
16 m = lambda x: x**2 * np.cos(x)
17 x = np.random.normal(loc=2, scale=4, size=n)
18 y = m(x) + eps
19 h = 0.5    # Bandwidth
20
21 XGrid = np.linspace(start=-15, stop=15, num=100)
22
23 fig, ax = plt.subplots(1, 1, figsize=(12, 8))
24 plt.scatter(x, y, edgecolors="black", color="white", s=200)
25 p1, = plt.plot(XGrid, m(XGrid), color='b', linewidth=3)
26 p2, = plt.plot(XGrid, NW_estimator(x0=XGrid, x=x, y=y, h=h), 'r-',
27                linewidth=3)
28
29 plt.xlabel("x0")
30 plt.ylabel("y")
31 plt.legend(handles=[p1, p2], loc="upper center",
32                        labels=["True regression", "Nadaraya-Watson"])
```

Programme 14.28 Implementation of kernel regression in Python.

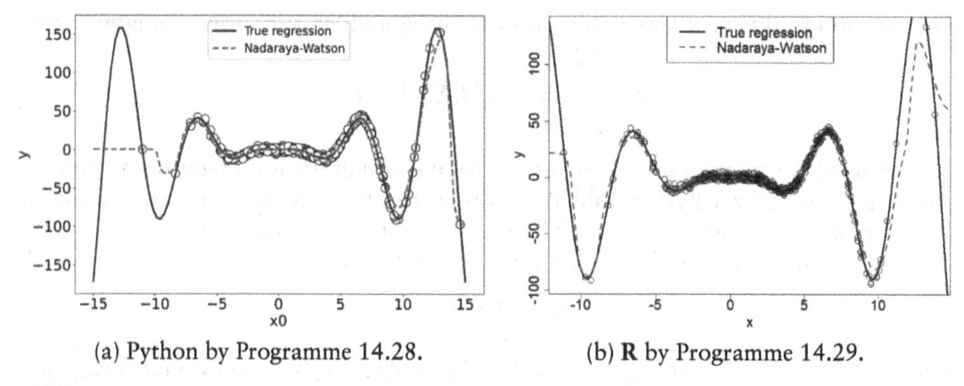

(a) Python by Programme 14.28. (b) **R** by Programme 14.29.

FIGURE 14.10 Comparison of Nadaraya–Watson estimates and the true underlying function $m(x) = x^2 \cos x$.

```
1  > NW_estimator <- function(x0, x, y, h, kernel=dnorm) {
2  +     f_x <- sapply(x, function(xi) kernel((x0 - xi)/h) / h)
3  +     f_x%*%y / rowSums(f_x)
4  + }
5  >
6  > set.seed(4002)
7  > n <- 500
8  > eps <- rnorm(n, sd=2)
9  > m <- function(x) x^2 * cos(x)
10 > x <- rnorm(n, mean=2, sd=4)
11 > y <- m(x) + eps
12 > h <- 0.5     # Bandwidth
13 >
14 > xGrid <- seq(-15, 15, l=100)
15 >
16 > # Plot data
17 > par(mfrow=c(1,1), cex.lab=2, cex.axis=2, cex.main=2, mar=c(5,5,4,4))
18 > plot(x, y, cex=2)
19 > #rug(x, side=1); rug(y, side=2)
20 > lines(xGrid, m(xGrid), col=1, lwd=3)
21 > lines(xGrid, NW_estimator(x0=xGrid,x=x,y=y,h=h), col=2, lwd=3, lty=2)
22 > legend("top", legend=c("True regression", "Nadaraya-Watson"),
23 +        lwd=3, col=1:2, cex=2, lty=1:2)
```

Programme 14.29 Implemetation of kernel regression in R.

Bias and Variance of $\hat{m}$

Next, we present a rigorous analysis regarding the bias and variance of the Nadaraya–Watson estimator $\hat{m}(x_0)$ at any pre-specified point $x_0 \in \mathbb{R}$. To this end, we lay down the following regularity conditions:

Condition 14.1 (Regularity Conditions). The following conditions are assumed:

1. $m(x)$ is twice continuously differentiable in x such that $|m''(x)|$ and $|m'(x)|$ are uniformly bounded in x on its domain of definition;
2. $\sigma^2(x) := \mathrm{Var}(\epsilon \mid \mathrm{x} = x)$ is continuous and uniformly positive (i.e. uniformly bounded away from zero), and globally Lipschitz in x;
3. The marginal density f_{x} is twice differentiable such that all its derivatives from zeroth up to the first order are globally bounded in magnitude, and f_{x} is also bounded away from zero at this chosen point x_0;
4. The kernel $K(\cdot)$ and its square have a finite fifth moment, meaning that both $\int_{\mathbb{R}} K(z)|z|^5 dz < \infty$, and also $\int_{\mathbb{R}} K^2(z)|z|^5 dz < \infty$;
5. $\{h_n\}$ is a deterministic sequence such that $nh_n \to \infty$ and $h_n \to 0$ as $n \to \infty$, i.e. $\frac{1}{h_n} = o(n)$.

We aim to establish the asymptotic normality of $\hat{\mathrm{m}}(x_0)$ and the following properties, as claimed earlier in this section:

(i) $\mathrm{Bias}(\hat{\mathrm{m}}(x_0))$ has a magnitude less than $C_1 h_n^2$;
(ii) $\mathrm{Var}(\hat{\mathrm{m}}(x_0)) \leq \frac{C_2}{nh_n}$,

where C_1 and C_2 are some positive constants. To this point, we first note that

$$
\begin{aligned}
&\frac{1}{nh_n}\sum_{i=1}^{n} K\left(\frac{x_0 - \mathrm{x}_i}{h_n}\right)\mathrm{y_i} \\
&= \frac{1}{nh_n}\sum_{i=1}^{n} K\left(\frac{x_0 - \mathrm{x}_i}{h_n}\right)\left(m(x_0) + (m(\mathrm{x}_i) - m(x_0)) + \epsilon_i\right) \\
&= \hat{\mathrm{f}}_{\mathrm{x}}(x_0)m(x_0) + \left(\frac{1}{nh_n}\sum_{i=1}^{n} K\left(\frac{x_0 - \mathrm{x}_i}{h_n}\right)\left(m(\mathrm{x}_i) - m(x_0)\right)\right) \\
&\quad + \left(\frac{1}{nh_n}\sum_{i=1}^{n} K\left(\frac{x_0 - \mathrm{x}_i}{h_n}\right)\epsilon_i\right) \\
&=: \hat{\mathrm{f}}_{\mathrm{x}}(x_0)m(x_0) + \hat{\mathrm{m}}_1(x_0) + \hat{\mathrm{m}}_2(x_0).
\end{aligned}
$$

Dividing both sides by $\hat{\mathrm{f}}_{\mathrm{x}}(x_0) = \frac{1}{nh_n}\sum_{i=1}^{n} K\left(\frac{x_0 - \mathrm{x}_i}{h_n}\right)$, which is legitimate for a large sample size n as $\hat{\mathrm{f}}_{\mathrm{x}}(x_0) \approx f_{\mathrm{x}}(x_0)$, which is different from 0, gives

$$
\hat{\mathrm{m}}(x_0) = m(x_0) + \frac{\hat{\mathrm{m}}_1(x_0)}{\hat{\mathrm{f}}_{\mathrm{x}}(x_0)} + \frac{\hat{\mathrm{m}}_2(x_0)}{\hat{\mathrm{f}}_{\mathrm{x}}(x_0)}. \tag{14.14}
$$

We next study the asymptotics of the last two terms. Firstly, we analyze those of $\frac{\hat{\mathrm{m}}_1(x_0)}{\hat{\mathrm{f}}_{\mathrm{x}}(x_0)}$:

$$\begin{aligned}
\mathbb{E}(\hat{\mathrm{m}}_1(x_0)) &= \frac{1}{nh_n}\mathbb{E}\left(nK\left(\frac{x_0-\mathrm{x}_1}{h_n}\right)(m(\mathrm{x}_1)-m(x_0))\right)\\
&= \frac{1}{h_n}\mathbb{E}\left(K\left(\frac{x_0-\mathrm{x}_1}{h_n}\right)(m(\mathrm{x}_1)-m(x_0))\right)\\
&= \int_{\mathbb{R}} K\left(\frac{x_0-x}{h_n}\right)(m(x)-m(x_0))f_{\mathrm{x}}(x)\frac{1}{h_n}dx\\
&= \int_{\mathbb{R}} K(z)(m(x_0+h_nz)-m(x_0))f_{\mathrm{x}}(x_0+h_nz)dz\\
&= \int_{\mathbb{R}} K(z)\left(m'(x_0)h_nz+m''(x_0)\frac{h_n^2z^2}{2}\right)(f_{\mathrm{x}}(x_0)+f_{\mathrm{x}}'(x_0)h_nz)dz+o(h_n^2)\\
&= h_nm'(x_0)f_{\mathrm{x}}(x_0)\int_{\mathbb{R}}K(z)zdz\\
&\quad +h_n^2\left(\frac{1}{2}m''(x_0)f_{\mathrm{x}}(x_0)+m'(x_0)f_{\mathrm{x}}'(x_0)\right)\int_{\mathbb{R}}K(z)z^2dz+o(h_n^2)\\
&= h_n^2\left(\frac{1}{2}m''(x_0)+m'(x_0)\frac{f_{\mathrm{x}}'(x_0)}{f_{\mathrm{x}}(x_0)}\right)f_{\mathrm{x}}(x_0)\int_{\mathbb{R}}K(z)z^2dz+o(h_n^2), \qquad (14.15)
\end{aligned}$$

where the third last equality holds by using the above arguments in light of Condition 14.1: 1), 3) and 4); the last equality follows by the symmetric property of $K(\cdot)$ around zero. We next consider the variance of $\hat{\mathrm{m}}_1(x_0)$,

$$\begin{aligned}
&\mathrm{Var}(\hat{\mathrm{m}}_1(x_0))\\
&= \frac{1}{nh_n^2}\mathrm{Var}\left(K\left(\frac{x_0-\mathrm{x}_1}{h_n}\right)(m(\mathrm{x}_1)-m(x_0))\right)\\
&= \frac{1}{nh_n^2}\left[\mathbb{E}\left(K^2\left(\frac{x_0-\mathrm{x}_1}{h_n}\right)(m(\mathrm{x}_1)-m(x_0))^2\right)-O(h_n^6)\right]\\
&= \frac{1}{nh_n^2}\left[\int_{\mathbb{R}}K^2(z)(m(x_0+h_nz)-m(x_0))^2f_{\mathrm{x}}(x_0+h_nz)h_ndz-O(h_n^6)\right]\\
&= \frac{1}{nh_n}\left[\int_{\mathbb{R}}K^2(z)(m'(x_0)h_nz+o(h_n))^2(f_{\mathrm{x}}(x_0)+O(h_n))dz-O(h_n^5)\right]\\
&= \frac{1}{nh_n}\left[h_n^2(m'(x_0))^2f_{\mathrm{x}}(x_0)\int_{\mathbb{R}}K^2(z)z^2dz+o(h_n^2)\right]\\
&= \frac{h_n}{n}(m'(x_0))^2f_{\mathrm{x}}(x_0)\int_{\mathbb{R}}K^2(z)z^2dz+o\left(\frac{h_n}{n}\right),
\end{aligned}$$

where the second equality follows by the expression for the expectation of $\hat{m}_1(x_0)$ obtained in (14.15), and the next third to fifth equalities also hold again in accordance with Condition 14.1: 1), 3) and 4). Since $O\left(\frac{h_n}{n}\right)$ is clearly of a smaller order than that of $O\left(\frac{1}{nh_n}\right)$ by Condition 14.1: 5), as a random sum in light of weak law of large numbers, due to $\mathrm{Var}(\sqrt{nh_n}\hat{m}_1(x_0)) = O(h_n^2)$, we have

$$\sqrt{nh_n}\left(\hat{m}_1(x_0) - h_n^2\left(\frac{1}{2}m''(x_0) + m'(x_0)\frac{f'_x(x_0)}{f_x(x_0)}\right)f_x(x_0)\int_{\mathbb{R}} K(z)z^2dz - o(h_n^2)\right) \xrightarrow{p} 0.$$

On the other hand, $\hat{f}_x(x_0) \xrightarrow{p} f_x(x_0)$ as shown earlier in this section and by Slutsky's theorem, we can conclude with:

$$\sqrt{nh_n}\left(\frac{\hat{m}_1(x_0)}{\hat{f}_x(x_0)} - h_n^2\left(\frac{1}{2}m''(x_0) + m'(x_0)\frac{f'_x(x_0)}{f_x(x_0)}\right)\int_{\mathbb{R}} K(z)z^2dz - o(h_n^2)\right) \xrightarrow{p} 0. \tag{14.16}$$

Next, we consider the asymptotic properties of $\frac{\hat{m}_2(x_0)}{\hat{f}_x(x_0)}$. Since $\mathbb{E}\left(\epsilon_1|x_1\right) = 0$,

$$\begin{aligned}
\mathbb{E}(\hat{m}_2(x_0)) &= \frac{1}{h_n}\mathbb{E}\left(K\left(\frac{x_0 - x_1}{h_n}\right)\epsilon_1\right) = \frac{1}{h_n}\mathbb{E}\left(\mathbb{E}\left(K\left(\frac{x_0 - x_1}{h_n}\right)\epsilon_1\middle| x_1\right)\right) \\
&= \frac{1}{h_n}\mathbb{E}\left(K\left(\frac{x_0 - x_1}{h_n}\right)\mathbb{E}\left(\epsilon_1|x_1\right)\right) = 0.
\end{aligned}$$

Combining this with Condition 14.1: 2), 3), 4) and 5), we can derive the variance of $\hat{m}_2$, by an application of the tower property:

$$\begin{aligned}
\mathrm{Var}(\hat{m}_2(x_0)) &= \frac{1}{(nh_n)^2}\sum_{i=1}^{n}\mathbb{E}\left(\left(K\left(\frac{x_0 - x_1}{h_n}\right)\epsilon_1\right)^2\right) \\
&= \frac{1}{nh_n^2}\mathbb{E}\left(K^2\left(\frac{x_0 - x_1}{h_n}\right)\sigma^2(x_1)\right) \\
&= \frac{1}{nh_n^2}\int_{\mathbb{R}} K^2\left(\frac{x - x_0}{h_n}\right)\sigma^2(x)f_x(x)dx \\
&= \frac{1}{nh_n}\int_{\mathbb{R}} K^2(z)\sigma^2(x_0 + h_n z)f_x(x_0 + h_n z)dz \\
&= \frac{1}{nh_n}\left(\int_{\mathbb{R}} K^2(z)\sigma^2(x_0)f_x(x_0)dz + O\left(h_n\right)\right) \\
&= \frac{1}{nh_n}\sigma^2(x_0)f_x(x_0)\int_{\mathbb{R}} K^2(z)dz + o\left(\frac{1}{nh_n}\right),
\end{aligned}$$

where the second last equality follows from Condition 14.1: 2), 3) and 4); and the last equality holds due to $O\left(\frac{1}{n}\right) = o\left(\frac{1}{nh_n}\right)$. Hence, as a random average, we can apply the Central Limit Theorem to deduce that

$$\sqrt{nh_n}\hat{\mathrm{m}}_2(x_0) \xrightarrow{d} \mathcal{N}\left(0, \sigma^2(x_0) f_{\mathrm{x}}(x_0) \int_{\mathbb{R}} K^2(z)dz\right),$$

and by another application of Slutsky's theorem, we conclude that

$$\sqrt{nh_n}\frac{\hat{\mathrm{m}}_2(x_0)}{\hat{\mathrm{f}}_{\mathrm{x}}(x_0)} \xrightarrow{d} \mathcal{N}\left(0, \frac{\sigma^2(x_0)}{f_{\mathrm{x}}(x_0)} \int_{\mathbb{R}} K^2(z)dz\right). \tag{14.17}$$

In summary, combining (14.16) and (14.17), we have:

$$\sqrt{nh_n}\left(\hat{\mathrm{m}}(x_0) - m(x_0) - h_n^2\left(\frac{1}{2}m''(x_0) + m'(x_0)\frac{f_{\mathrm{x}}'(x_0)}{f_{\mathrm{x}}(x_0)}\right)\int_{\mathbb{R}} K(z)z^2dz - o(h_n^2)\right)$$
$$\xrightarrow{d} \mathcal{N}\left(0, \frac{\sigma^2(x_0)}{f_{\mathrm{x}}(x_0)} \int_{\mathbb{R}} K^2(z)dz\right).$$

This result can be utilized to guide the optimal choice of h_n to minimize the mean squared error of the corresponding estimator $\hat{\mathrm{m}}(x_0)$; recall that

$$\mathrm{MSE}(\hat{\mathrm{m}}(x_0)) = \left(\mathrm{Bias}(\hat{\mathrm{m}}(x_0))\right)^2 + \mathrm{Var}(\hat{\mathrm{m}}(x_0)) \leq C_1^2 h_n^4 + \frac{C_2}{nh_n},$$

and we aim to determine the order of h_n that minimizes this upper bound as a function of h_n. By the first order condition, $4C_1^2h_n^3 - \frac{C_2}{nh_n^2} = 0$, which gives $h_n^* = \left(\frac{C_2}{4nC_1^2}\right)^{\frac{1}{5}} = O(n^{-1/5})$. Together with the convex nature of this upper bound, as its second derivative satisfies $12C_1^2h_n^2 + \frac{2C_2}{nh_n^3} > 0$ for all $h_n > 0$, we can conclude that choosing h_n to be of order $O(n^{-1/5})$ can reduce and bound the MSE of the model as much as possible, as mentioned before.

*14.A APPENDIX

*14.A.1 Asymptotic Optimality of the Calinski–Harabasz Index

The Calinski–Harabasz (CH) index is an evaluation assessment tool for clustering algorithms, defined as

$$R(L) := \frac{\mathrm{BCSS}(L)/(L-1)}{\mathrm{WCSS}(L)/(n-L)},$$

where L is the number of clusters, n is the size of the overall sample, and

$$\mathrm{WCSS}(L) := \sum_{l=1}^{L} \sum_{i=1}^{n_l} \|x_i^{(l)} - \overline{x}^{(l)}\|_2^2, \qquad \mathrm{BCSS}(L) := \sum_{l=1}^{L} n_l \|\overline{x}^{(l)} - \overline{x}\|_2^2.$$

Here, $x_i^{(l)} \in \mathbb{R}^p$ is a typical sample point classified in *artificial cluster* $l = 1, 2, \ldots, L$, which contains a total of n_l points. Moreover, $\overline{x}^{(l)}$ and $\overline{x}$ are the averages within the cluster l and of the whole dataset, respectively. In the following, we aim to derive the relative asymptotic value of $r_n(L) := \frac{R(L)}{n}$ for any L, based on some practically relevant assumptions. We shall claim that asymptotically the CH index attains its maximum value at $L = K$, the true cluster number. According to the classical reference [4], "*there is no satisfactory probabilistic foundation to support the use of CH index*", and there have not been any further significant breakthroughs along this direction in the literature until now, as echoed in the *Wikipedia* entry for CH index[11]. This appendix serves as a motivation for why the optimality of the CH index should be attained at the true value K of the number of clusters in many practical situations. We aim to lay down a thorough understanding of this issue under a rigorous mathematical framework in our future research article [5]. Firstly, we shall derive the asymptotic expressions of both BCSS and WCSS under a general framework, which not only provides an insight about the very nature of the CH index but also simplifies the later discussion about the CH index's optimality.

Let us state our assumptions and notations. We impose a hierarchical structure with latent information on the data points, that is, let $\mathbf{S}_i := (\mathbf{x}_i, \mathbf{U}_i)$, $i = 1, \ldots, n$, be the points of interest in which $\mathbf{x}_i$ is the observable random component of the i-th data point, and $\mathbf{U}_i$ is the latent (hidden) random component which together with $\mathbf{x}_i$ can completely determine the true cluster of $\mathbf{S}_i$ through some (not necessarily known) function $G(\mathbf{S}_i) = 1, \ldots, K$. We denote, under the iid assumption, $\mu_{|k\cdot} := \mathbb{E}[\mathbf{x}_i | G(\mathbf{S}_i) = k]$, $\mu_{j|k\cdot} := \mathbb{E}[\mathrm{x}_{ji} | G(\mathbf{S}_i) = k]$, and $\sigma^2_{j|k\cdot} := \mathrm{Var}[\mathrm{x}_{ji} | G(\mathbf{S}_i) = k]$, for $i = 1, \ldots, n$. We define another function $F(\mathbf{x}_i) = 1, \ldots, L$ to indicate the artificial cluster assigned to point i by the clustering algorithm, which should only depend on the observable component $\mathbf{x}_i$ as $\mathbf{U}_i$ can never be observed in reality. We denote $p_{k\cdot} := \mathbb{P}[G(\mathbf{S}_i) = k] > 0$, $p_{\cdot l} := \mathbb{P}[F(\mathbf{x}_i) = l] > 0$,[12] and $p_{kl} := \mathbb{P}[(G(\mathbf{S}_i) = k) \cap (F(\mathbf{x}_i) = l)] \geq 0$, for all $i = 1, \ldots, n$. The following two technical assumptions help to facilitate our asymptotic analysis; moreover, they are commonly satisfied in situations where clustering is appropriate.

Condition 14.2

(i) $\frac{1}{n}\sum_{i=1}^{n} \mathbb{1}_{\{G(\mathbf{S}_i)=k\}} \xrightarrow{p} p_{k\cdot}$, $\frac{1}{n}\sum_{i=1}^{n} \mathbb{1}_{\{F(\mathbf{x}_i)=l\}} \xrightarrow{p} p_{\cdot l}$, and $\frac{1}{n}\sum_{i=1}^{n} \mathbb{1}_{\{G(\mathbf{S}_i)=k \cap F(\mathbf{x}_i)=l\}} \xrightarrow{p} p_{kl}$;

(ii) the random elements $\mathbf{S}_i = (\mathbf{x}_i, \mathbf{U}_i)$, for $i = 1, \ldots, n$, are iid.

[11] Source: https://en.wikipedia.org/wiki/Calinski-Harabasz_index.

[12] For a fixed L, when the sample size is large enough, the optimal clustering cloud obtained from the algorithm will become stable in shape, so that any points belonging to the cluster l do so with a strictly positive probability $p_{\cdot l}$; see [22].

Let us rewrite WCSS(L) in our notations:

$$\begin{aligned}\frac{\mathrm{WCSS}(L)}{n-L} &= \frac{1}{n-L}\sum_{l=1}^{L}\sum_{i=1}^{n}\|\mathbf{x}_i-\mathbf{m}_{\cdot l}\|_2^2\cdot\mathbb{1}_{\{F(\mathbf{x}_i)=l\}}\\
&=\sum_{j=1}^{p}\sum_{l=1}^{L}\left(\frac{\sum_{i'=1}^{n}\mathbb{1}_{\{F(\mathbf{x}_{i'})=l\}}}{n-L}\right)\left(\frac{1}{\sum_{i'=1}^{n}\mathbb{1}_{\{F(\mathbf{x}_{i'})=l\}}}\right)\\
&\quad\cdot\sum_{i=1}^{n}\left(\mathrm{x}_{ji}-\mathrm{m}_{j\cdot l}\right)^2\mathbb{1}_{\{F(\mathbf{x}_i)=l\}}\\
&\xrightarrow{p}\sum_{j=1}^{p}\sum_{l=1}^{L}p_{\cdot l}\cdot\mathrm{Var}[\mathrm{x}_{ji}|F(\mathbf{x}_i)=l]\\
&=:\sum_{j=1}^{p}\sum_{l=1}^{L}p_{\cdot l}\cdot\sigma_{j|\cdot l}^2=\sum_{j=1}^{p}\mathbb{E}[\mathrm{Var}[\mathrm{x}_{ji}|F(\mathbf{x}_i)]],\end{aligned}$$

where $\mathbf{m}_{\cdot l} := \sum_{i=1}^{n}\left(\mathbf{x}_i\mathbb{1}_{\{F(\mathbf{x}_i)=l\}}\right)/\sum_{i=1}^{n}\mathbb{1}_{\{F(\mathbf{x}_i)=l\}}$ is the artificial *within-cluster* mean and $\mathrm{m}_{j\cdot l}$ is its j-th component, and the last line of convergence in probability is based on the two assumptions of Condition 14.2 by the law of large numbers and Slutsky's theorem. Note that the total sum of squares can be decomposed into WCSS(L) and BCSS(L); indeed, denoting $\mathbf{m}$ as the overall empirical mean of $\mathbf{x}_i$ over the whole dataset, we have, for any $i = 1, \ldots, n$,

$$\begin{aligned}\|\mathbf{x}_i-\mathbf{m}\|_2^2 &= \sum_{j=1}^{p}\sum_{l=1}^{L}\sum_{F(\mathbf{x}_i)=l}(\mathrm{x}_{ji}-\mathrm{m}_j)^2\\
&=\sum_{j=1}^{p}\sum_{l=1}^{L}\sum_{F(\mathbf{x}_i)=l}(\mathrm{x}_{ji}-\mathrm{m}_{j\cdot l}+\mathrm{m}_{j\cdot l}-\mathrm{m}_j)^2\\
&=\sum_{j=1}^{p}\sum_{l=1}^{L}\sum_{F(\mathbf{x}_i)=l}\left((\mathrm{x}_{ji}-\mathrm{m}_{j\cdot l})^2+(\mathrm{m}_{j\cdot l}-\mathrm{m}_j)^2+2\,(\mathrm{x}_{ji}-\mathrm{m}_{j\cdot l})\,(\mathrm{m}_{j\cdot l}-\mathrm{m}_j)\right)\\
&=\sum_{l=1}^{L}\sum_{F(\mathbf{x}_i)=l}\|\mathbf{x}_i-\mathbf{m}_{\cdot l}\|_2^2+\sum_{l=1}^{L}\sum_{F(\mathbf{x}_i)=l}\|\mathbf{m}_{\cdot l}-\mathbf{m}\|_2^2\\
&\quad+2\sum_{j=1}^{p}\sum_{l=1}^{L}(\mathrm{m}_{j\cdot l}-\mathrm{m}_j)\sum_{F(\mathbf{x}_i)=l}(\mathrm{x}_{ji}-\mathrm{m}_{j\cdot l})\\
&=\mathrm{WCSS}(L)+\mathrm{BCSS}(L),\end{aligned}$$

where the last equality holds because $\sum_{F(\mathbf{x}_i)=l}(\mathrm{x}_{ji} - \mathrm{m}_{j\cdot l}) = 0$. Since $\frac{1}{n}\|\mathbf{x}_i - \mathbf{m}\|_2^2 \xrightarrow{p} \sum_{j=1}^{p} \mathrm{Var}[\mathrm{x}_{ji}]$ under Condition 14.3, by applying the law of total variance, we also have

$$\frac{\mathrm{BCSS}(L)}{n} = \frac{1}{n}\|\mathbf{x}_i - \mathbf{m}\|_2^2 - \frac{\mathrm{WCSS}(L)}{n} \xrightarrow{p} \sum_{j=1}^{p} \mathrm{Var}[\mathrm{x}_{ji}] - \sum_{j=1}^{p} \mathbb{E}[\mathrm{Var}[\mathrm{x}_{ji}|F(\mathbf{x}_i)]]$$

$$= \sum_{j=1}^{p} \mathrm{Var}[\mathbb{E}[\mathrm{x}_{ji}|F(\mathbf{x}_i)]] =: \sum_{j=1}^{p}\sum_{l=1}^{L} p_{\cdot l} \cdot (\mu_{j|\cdot l} - \mu_j)^2 = \sum_{l=1}^{L} p_{\cdot l} \cdot \|\boldsymbol{\mu}_{j|\cdot l} - \boldsymbol{\mu}_j\|_2^2.$$

Therefore, we have the asymptotic value of $r_n(L)$, as $n \to \infty$,

$$\lim_{n\to\infty} r_n(L) =: r_\infty(L) \xrightarrow{p} \frac{1}{L-1} \frac{\mathrm{Var}[\mathbb{E}[\mathbf{x}_i|F(\mathbf{x}_i)]]}{\sum_{j=1}^{p} \mathbb{E}[\mathrm{Var}[\mathrm{x}_{ji}|F(\mathbf{x}_i)]]}. \tag{14.18}$$

Although the CH index as a clustering metric to evaluate the appropriateness of the hyperparameter K shows good performances in many practical examples such as those in this chapter, it still fails in an obvious manner in some cases; this issue is also briefly mentioned in [4], while we shall provide a mathematical discussion as follows. For instance, consider the data points generated by uniform distribution with the support $[0, 1]$, and we assume that the true clusters are $\left[0, \frac{1}{K}\right)$, $\left[\frac{1}{K}, \frac{2}{K}\right)$, $\cdots$, $\left[\frac{K-1}{K}, 1\right]$ with points generated uniformly from each bucket for some positive integer K. In the limiting case of $n \to \infty$, suppose that an L-means clustering is applied to the dataset, then by symmetry, the resulting L clusters are the L equal-length subintervals of $[0, 1]$, namely $\left[0, \frac{1}{L}\right)$, $\left[\frac{1}{L}, \frac{2}{L}\right)$, $\cdots$, $\left[\frac{L-1}{L}, 1\right]$. Clearly, each point has an equal probability of $\frac{1}{L}$ of falling into each of the L clusters, i.e. $p_{\cdot l} = \frac{1}{L}$ for all $l = 1, \ldots, L$, and given that the point is in the l-th cluster $\left[\frac{l-1}{L}, \frac{l}{L}\right)$, the conditional distribution of its value is another uniform distribution $U\left(\frac{l-1}{L}, \frac{l}{L}\right)$ with the mean $\frac{l-0.5}{L}$ and variance $\frac{1}{12L^2}$. To this end, we obtain the following asymptotic estimates for WCSS and BCSS, respectively:

$$\frac{\mathrm{WCSS}(L)}{n-L} \xrightarrow{p} \mathbb{E}[\mathrm{Var}[\mathrm{x}_i|F(\mathbf{x}_i)]] = \sum_{l=1}^{L} p_{\cdot l}\mathrm{Var}[\mathrm{x}_i|F(\mathbf{x}_i) = l] = L \cdot \frac{1}{L}\frac{1}{12L^2} = \frac{1}{12L^2},$$

$$\frac{\mathrm{BCSS}(L)}{n} \xrightarrow{p} \mathrm{Var}[\mathbb{E}[\mathrm{x}_i|F(\mathbf{x}_i)]] = \mathrm{Var}[\mathrm{x}_i] - \mathbb{E}[\mathrm{Var}[\mathrm{x}_i|F(\mathbf{x}_i)]] = \frac{1}{12} - \frac{1}{12L^2},$$

based on which we have

$$r_n(L) \xrightarrow{p} \frac{1}{L-1} \frac{\frac{1}{12} - \frac{1}{12L^2}}{\frac{1}{12L^2}} = L + 1 \quad \text{as } n \to \infty,$$

so $r_n(L)$ is asymptotically affinely increasing with L. In this case, the CH index loses its power as an evaluating metric for clustering; indeed, without the presence of natural clusters, no point can be more representative of its neighborhood than the others, hence as argued in [4], "*no reasonably better partition of the points exists than that into individuals*". Instead, when a *concentration effect* emerges so that natural clusters are clearly observed (for example, if the data points in each true cluster are normally distributed), and the distances among true cluster means are significantly greater than the standard deviation within any true cluster, the CH index can reach a maximum at the true value of K. In the following, we shall make an attempt to provide a set of sufficient conditions that warrant the optimality of $r_\infty(L)$ at $L = K$.

To this end, we first introduce different possible groupings to the collection of all artificial clusters: let us denote $B_k := B_{a\delta}(\boldsymbol{\mu}_{|k\cdot})$ and $F_l = \{\mathbf{x}_i | F(\mathbf{x}_i) = l\}$. Using these, we define

1. $\mathcal{A}_1$: for every $l \in \mathcal{A}_1$, the artificial cluster F_l contains all the points in one ball but no more points from any other ball, i.e., there is exactly one $k = 1, \ldots, K$ such that $B_k \subseteq F_l$ but $B_{k'} \cap F_l = \emptyset$ for any $k' \neq k$.
2. $\mathcal{A}_2$: for every $l \in \mathcal{A}_2$, the artificial cluster F_l contains the points from one ball but not exactly all its points, and it contains no points from any other ball, i.e. there is exactly one $k = 1, \ldots, K$ such that $B_k \cap F_l = \emptyset$ but $B_k \nsubseteq F_l$, while $B_{k'} \cap F_l = \emptyset$ for all $k' \neq k$.
3. $\mathcal{A}_3$: for every $l \in \mathcal{A}_3$, the artificial cluster F_l contains points from more than one ball B_k.
4. $\mathcal{A}_4$: for every $l \in \mathcal{A}_4$, the artificial cluster F_l is disjoint from all balls B_k's.[13]

Figure 14.11 illustrates two possible scenarios. On the left, $K = 4$ and $L = 3$, $\mathcal{A}_1 = \{1, 2\}$, $\mathcal{A}_2 = \emptyset$ and $\mathcal{A}_3 = \{3\}$. On the right, there are four true clusters and five artificial clusters such that $\mathcal{A}_1 = \{3, 4\}$, $\mathcal{A}_2 = \{1, 2\}$ and $\mathcal{A}_3 = \{5\}$. Next, we state the

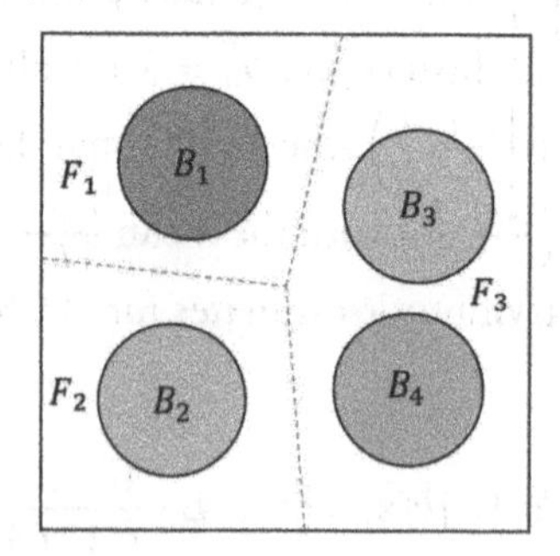

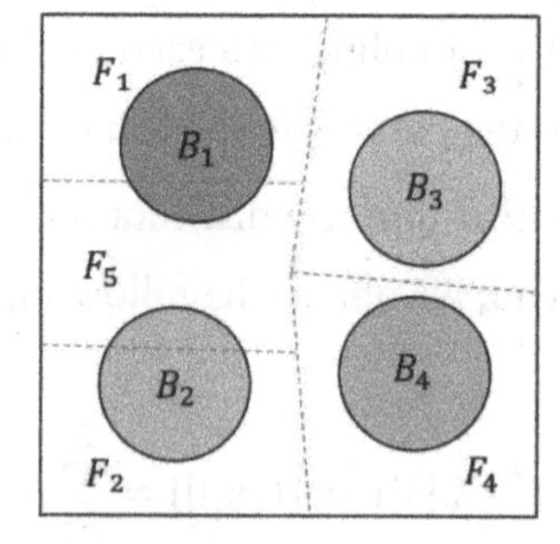

FIGURE 14.11 Illustrations for B_k, $k = 1, \ldots, K$ (colored circle) and the artificial clusters (divided by boundaries in dashed lines).

---

[13] In K-means clustering, we usually have $\mathcal{A}_4 = \emptyset$ due to its nature of minimizing MSE; indeed, partitions should be performed at those locations where data points are more agglomerated, rather than at locations with no points at all, so as to reduce the MSE as much as possible. However, this may not be the case for other clustering algorithms.

assumption that confines the sizes of variances and the distances between true cluster means. Combined, these hint at why the optimality of CH index is attained at the true cluster number K.

Condition 14.3

(i) The distance between any two true clusters' means of, say the k-th and k'-th ones, is bounded below by a common constant $C > 0$: $\|\boldsymbol{\mu}_{|k\cdot} - \boldsymbol{\mu}_{|k'\cdot}\|_2 \geq C$;
(ii) the true clusters' component variances are bounded above by $\sum_{j=1}^{p} \sigma_{j|k\cdot}^2 \leq \delta^2$ for every k; and
(iii) $\mathcal{A}_3 \neq \emptyset$ so that for each $l \in \mathcal{A}_3$, there is a pair of k and k' such that $\tilde{p}_{kl} := \mathbb{P}(F(\mathbf{x}_i) = l, \mathbf{x}_i \in B_k), \tilde{p}_{k'l} := \mathbb{P}(F(\mathbf{x}_i) = l, \mathbf{x}_i \in B_{k'}) \geq q > 0$ for some real q whose value is common for all $l \in \mathcal{A}_3$.[14]

In the following, we aim to show that for any $a > 0$ such that

$$\left(1 - \left(\frac{p}{a}\right)^{-p} \exp\left(-\frac{a^2 - p^2}{2p}\right)\right)^{\frac{1}{2}} \left(\frac{C}{\delta} - 2a\right) \geq \frac{\sqrt{3K}}{q} \sqrt{\frac{\sum_{j=1}^{p} \operatorname{Var}(x_{ji})}{\sum_{k=1}^{K} p_{k\cdot} \|\boldsymbol{\mu}_{|k\cdot} - \boldsymbol{\mu}\|_2^2}}, \tag{14.19}$$

then we have $r_\infty(L) < r_\infty(K)$.

Let us show by contradiction that $\mathcal{A}_3$ contains at least one artificial cluster when $L < K$. If $\mathcal{A}_3 = \emptyset$, we clearly have $|\mathcal{A}_2| \geq K - |\mathcal{A}_1|$ by the definition of $\mathcal{A}_2$. This, together with the fact that $\mathcal{A}_4$ contains no true clusters at all, shows that the number of artificial clusters is more than K. This is absurd if $L < K$. When $L > K$, we can sometimes see an artificial cluster containing points from at least two true clusters. In that case, $\mathcal{A}_3 \neq \emptyset$ still holds, and Condition 14.3 and the validity of (14.19) essentially demand that the cluster variances be small and the distances between the true clusters be large, and the following argument works well particularly when $\mathcal{A}_3 \neq \emptyset$. If $\mathcal{A}_3 = \emptyset$, especially when $L > K$, the clustering algorithm tends to carry out further splittings within true clusters; by combining Condition 14.3 with an additional condition regarding upper and lower bounds for all the variances of different components of $\mathbf{x}_i$, we can still establish that $r_\infty(L) < r_\infty(K)$ holds for any $L > K$ too. For details of the overall result, we refer to our article [5].

Next, we take a closer look at the respective bounds for the numerator and denominator of (14.18) under Condition 14.3 and the validity of (14.19). To this

[14] By the design of K-means clustering, the locations of the artificial clusters are driven by those of the balls B_k's. Indeed, when $L < K$, by the pigeonhole principle, there is at least one artificial cluster that encloses parts of more than one ball, justifying the non-emptyness of $\mathcal{A}_3$. Furthermore, such an artificial cluster resulting from K-means clustering usually contains a decent proportion of both balls, that is why $\tilde{p}_{kl}$ and $\tilde{p}_{k'l}$ are assumed to have a lower bound with a probability approaching 1 as $n \to \infty$.

end, we first establish an upper bound for the expectation of the conditional variance given the true clusters:

$$\sum_{j=1}^{p} \mathbb{E}[\text{Var}[x_{ji}|G(\mathbf{S}_i)]] = \sum_{j=1}^{p}\sum_{k=1}^{K} p_{k\cdot}\sigma^2_{j|k\cdot} \leq \delta^2,$$

due to Condition 14.3. Next, we derive a lower bound for the expectation of the conditional variance but now given the artificial clusters. Consider the probability

$$\mathbb{P}\left(\mathbf{x}_i \in B_{G(\mathbf{S}_i)}\right) = \mathbb{E}\left[\mathbb{P}\left(\mathbf{x}_i \in B_{G(\mathbf{S}_i)} \mid G(\mathbf{S}_i)\right)\right] = \sum_{k=1}^{K} p_{k\cdot}\mathbb{P}\left(\mathbf{x}_i \in B_k \mid G(\mathbf{S}_i) = k\right), \tag{14.20}$$

and the sub-term:

$$\begin{aligned}
&\mathbb{P}\left(\mathbf{x}_i \in B_k \mid G(\mathbf{S}_i) = k\right)\\
&= 1 - \mathbb{P}\left(\|\mathbf{x}_i - \boldsymbol{\mu}_{|k\cdot}\|_2^2 \geq a^2\delta^2 \middle| G(\mathbf{S}_i) = k\right)\\
&= 1 - \mathbb{P}\left(\frac{1}{\delta^2}\|\mathbf{x}_i - \boldsymbol{\mu}_{|k\cdot}\|_2^2 \geq a^2 \middle| G(\mathbf{S}_i) = k\right)\\
&\geq 1 - \mathbb{P}\left(\sum_{j=1}^{p} \frac{(x_{ji} - \mu_{j|k\cdot})^2}{\sigma^2_{j|k}} \mathbb{1}_{\{G(\mathbf{S}_i)=k\}} \geq a^2 \,|G(\mathbf{S}_i) = k\right),
\end{aligned}$$

where the last inequality follows by the upper bound assumption in Condition 14.3(ii) for the variance $\sum_{j=1}^{p} \sigma^2_{j|k\cdot} \leq \delta^2$:

$$\begin{aligned}
\frac{1}{\delta^2}\|\mathbf{x}_i - \boldsymbol{\mu}_{|k\cdot}\|_2^2 \mathbb{1}_{\{G(\mathbf{S}_i)=k\}} &\leq \sum_{j=1}^{p} \frac{(x_{ji} - \mu_{j|k\cdot})^2}{\sum_{j=1}^{p}\sigma^2_{j|k\cdot}} \mathbb{1}_{\{G(\mathbf{S}_i)=k\}}\\
&\leq \sum_{j=1}^{p} \frac{(x_{ji} - \mu_{j|k\cdot})^2}{\sigma^2_{j|k\cdot}} \mathbb{1}_{\{G(\mathbf{S}_i)=k\}}.
\end{aligned}$$

Therefore, by a simple use of the *Chernoff inequality*, (14.20) becomes

$$\begin{aligned}
\mathbb{P}\left(\mathbf{x}_i \in B_{G(\mathbf{S}_i)}\right) &\geq \sum_{k=1}^{K} p_{k\cdot}\left(1 - \mathbb{P}\left(\sum_{j=1}^{p} \frac{(x_{ji} - \mu_{j|k\cdot})^2}{\sigma^2_{j|k}} \mathbb{1}_{\{G(\mathbf{S}_i)=k\}} \geq a^2|G(\mathbf{S}_i) = k\right)\right)\\
&= 1 - \sum_{k=1}^{K} p_{k\cdot}\mathbb{P}\left(\sum_{j=1}^{p} \frac{(x_{ji} - \mu_{j|k\cdot})^2}{\sigma^2_{j|k}} \mathbb{1}_{\{G(\mathbf{S}_i)=k\}} \geq a^2|G(\mathbf{S}_i) = k\right)\\
&= 1 - \sum_{k=1}^{K} p_{k\cdot}\mathbb{P}(y_{ik} \geq a^2|G(\mathbf{S}_i) = k) \geq 1 - \sum_{k=1}^{K} p_{k\cdot}M_{y_{ik}}(t)\exp(-a^2 t),
\end{aligned} \tag{14.21}$$

for $t > 0$, where $\mathrm{y}_{ik} := \sum_{j=1}^{p} \frac{(\mathrm{x}_{ji}-\mu_{j|k\cdot})^2}{\sigma_{j|k}^2} \mathbb{1}_{\{G(\mathbf{S}_i)=k\}}$, which shows how concentrated the conditional distribution of x_{ji} in the k-th true cluster is. For some distributions, y_{ik} may not depend on k. For instance, in the Gaussian case, the conditional distribution of $\boldsymbol{\omega}_{ik} := \frac{\mathbf{x}_i-\boldsymbol{\mu}_{|k\cdot}}{\boldsymbol{\sigma}_{|k\cdot}} \mathbb{1}_{\{G(\mathbf{S}_i)=k\}} = \boldsymbol{A}_{ik}\mathbf{z}$, given $G(\mathbf{S}_i) = k$, follows a normal distribution $\mathcal{N}_p(\mathbf{0}_p, \boldsymbol{V}_{ik})$ with correlation matrix $\boldsymbol{V}_{ik}$, and $\mathbf{z} \sim \mathcal{N}_p(\mathbf{0}_p, \boldsymbol{I}_p)$ such that $\boldsymbol{A}_{ik}\boldsymbol{A}_{ik}^\top = \boldsymbol{V}_{ik}$. By the Spectral Decomposition Theorem, we have $\boldsymbol{V}_{ik} = \boldsymbol{H}\boldsymbol{\Lambda}\boldsymbol{H}^\top$ where the columns of the $p \times p$ orthogonal matrix $\boldsymbol{H}$ are the eigenvectors of $\boldsymbol{V}_{ik}$ and $\boldsymbol{\Lambda}$ is a $p \times p$ diagonal matrix of the corresponding eigenvalues. Let us denote the largest eigenvalue by $\lambda_{\max} := \max_{i=1,\dots,p} \lambda_i$, which is obviously smaller than p. Clearly, $\boldsymbol{A}_{ik}$ can be chosen as $\boldsymbol{H}\boldsymbol{\Lambda}^{1/2}\boldsymbol{H}^\top$, which is symmetric and positive definite, and satisfies $\boldsymbol{A}_{ik}\boldsymbol{A}_{ik}^\top = \boldsymbol{A}_{ik}^\top\boldsymbol{A}_{ik} = \boldsymbol{V}_{ik}$. Therefore, we have

$$
\begin{aligned}
\mathrm{y}_{ik} = \boldsymbol{\omega}_{ik}^\top \boldsymbol{\omega}_{ik} &= \mathbf{z}^\top \boldsymbol{A}_{ik}^\top \boldsymbol{A}_{ik}\mathbf{z} = \mathbf{z}^\top \boldsymbol{V}_{ik}\mathbf{z} = \mathbf{z}^\top \boldsymbol{H}\boldsymbol{\Lambda}\boldsymbol{H}^\top \mathbf{z} \\
&= (\boldsymbol{H}^\top \mathbf{z})^\top \boldsymbol{\Lambda}(\boldsymbol{H}^\top \mathbf{z}) =: (\mathbf{z}^H)^\top \boldsymbol{\Lambda}(\mathbf{z}^H),
\end{aligned}
$$

where $\mathbf{z}^H := (\mathrm{z}_1^H, \dots, \mathrm{z}_p^H)^\top = \boldsymbol{H}^\top \mathbf{z} \sim \mathcal{N}_p(\mathbf{0}_p, \boldsymbol{I}_p)$ by the orthogonality of $\boldsymbol{H}$. Thus $\mathrm{y}_{ik} = \sum_{i=1}^{p} \lambda_i (\mathrm{z}_i^H)^2$, where $\mathrm{z}_i^H \overset{iid}{\sim} \mathcal{N}(0,1)$ implies $(\mathrm{z}_i^H)^2 \overset{iid}{\sim} \chi_1^2$ with the mgf of $(1-2t)^{-1/2}$. The mgf of y_{ik} is therefore $M_{\mathrm{y}_{ik}}(t) = \prod_{i=1}^{p} (1-2t\lambda_i)^{-\frac{1}{2}}$ for $t < \frac{1}{2\lambda_{\max}}$, and (14.21) reduces to

$$
\mathbb{P}\left(\mathbf{x}_i \in B_{G(\mathbf{S}_i)}\right) \geq 1 - \exp\left(-a^2 t - \frac{1}{2}\sum_{i=1}^{p} \ln(1-2t\lambda_i)\right).
$$

First, we bound below the right-hand side of the inequality by another function of t, by noting that for all $i = 1, \dots, p$, $1 - 2t\lambda_i \geq 1 - 2t\lambda_{\max}$ for $0 \leq t < \frac{1}{2\lambda_{\max}}$:

$$
\begin{aligned}
&1 - \exp\left(-a^2 t - \frac{1}{2}\sum_{i=1}^{p} \ln(1-2t\lambda_i)\right) \\
&\geq 1 - e^{h(t)} := 1 - \exp\left(-a^2 t - \frac{p}{2}\ln(1-2t\lambda_{\max})\right).
\end{aligned}
$$

It then suffices to find a lower bound for the function $h(t)$. The first-order condition of $h'(t) = -a^2 + \frac{p\lambda_{\max}}{1-2t\lambda_{\max}} = 0$ gives $t = t_1 := \frac{1}{2\lambda_{\max}}\left(1 - \frac{p\lambda_{\max}}{a^2}\right) < \frac{1}{2\lambda_{\max}}$. Furthermore, since $h(t)$ is strictly convex (as $h''(t) = \frac{2p\lambda_{\max}^2}{(1-2t\lambda_{\max})^2} > 0$), we see that $h(t)$ attains its minimum over $\left[0, \frac{1}{2\lambda_{\max}}\right)$ at $t_1^* := \max(t_1, 0)$. Hence, $h_{\min}(t) = h(t_1^*)$, whose explicit expression depends on the magnitude of a^2:

(1) When $a^2 > p\lambda_{\max}$, we have $t_1 > 0$, hence $t_1^* = t_1$, and $h(t)$ takes the minimum of $h(t_1) = -\frac{p}{2}\ln\frac{p\lambda_{\max}}{a^2} - \frac{a^2 - p\lambda_{\max}}{2\lambda_{\max}}$;

(2) When $a^2 \leq p\lambda_{\max}$, we have $t_1 \leq 0$, thus $t_1^* = 0$, and $h(t)$ takes the minimum of $h(0) = 0$.

Combining the scenarios above, we see that $h(t) \geq -\frac{p}{2}\ln\frac{p\lambda_{\max}}{a^2} - \frac{a^2-p\lambda_{\max}}{2\lambda_{\max}}$ for all $0 \leq t < \frac{1}{2\lambda_{\max}}$, which leads to the following lower bound for (14.21): for all $i = 1, \ldots, n$,

$$\begin{aligned}
\mathbb{P}\left(\mathbf{x}_i \in B_{G(\mathbf{S}_i)}\right) &\geq 1 - e^{h(t)} \\
&\geq 1 - \left(\frac{p\lambda_{\max}}{a^2}\right)^{-\frac{p}{2}} \exp\left(-\frac{a^2 - p\lambda_{\max}}{2\lambda_{\max}}\right) \\
&\geq 1 - \left(\frac{p}{a}\right)^{-p} \exp\left(-\frac{a^2 - p^2}{2p}\right),
\end{aligned}$$

where the final inequality follows from the fact that $\lambda_{\max} \leq p$ by the property of the correlation matrix V_{ik}.

Finally, for artificial clusters, we have

$$\begin{aligned}
\sum_{j=1}^{p} \mathbb{E}[\mathrm{Var}[\mathrm{x}_{ji}|F(\mathbf{x}_i)]] &= \sum_{j=1}^{p} \Big(\mathbb{P}(\mathbf{x}_i \in B_{G(\mathbf{S}_i)})\mathbb{E}\left[\mathrm{Var}[\mathrm{x}_{ji}|F(\mathbf{x}_i)]\middle| \mathbf{x}_i \in B_{G(\mathbf{S}_i)}\right] \\
&\quad + \mathbb{P}\left(\mathbf{x}_i \notin B_{G(\mathbf{S}_i)}\right)\mathbb{E}\left[\mathrm{Var}[\mathrm{x}_{ji}|F(\mathbf{x}_i)]\middle| \mathbf{x}_i \notin B_{G(\mathbf{S}_i)}\right]\Big) \\
&\geq \sum_{j=1}^{p} \mathbb{P}\left(\mathbf{x}_i \in B_{G(\mathbf{S}_i)}\right)\mathbb{E}\left[\mathrm{Var}[\mathrm{x}_{ji}|F(\mathbf{x}_i)]\middle| \mathbf{x}_i \in B_{G(\mathbf{S}_i)}\right]. \qquad (14.22)
\end{aligned}$$

By another application of the variance decomposition formula, the expectation term in (14.22) can be bounded below:

$$\begin{aligned}
&\mathbb{E}\left(\mathrm{Var}[\mathrm{x}_{ji}|F(\mathbf{x}_i)]\middle| \mathrm{x}_{ji} \in B_{G(\mathbf{S}_i)}\right) \\
&= \mathbb{E}\left(\mathrm{Var}\left[\mathbb{E}\left(\mathrm{x}_{ji}|F(\mathbf{x}_i), G(\mathbf{S}_i)\right)\middle| F(\mathbf{x}_i)\right]\middle| \mathbf{x}_i \in B_{G(\mathbf{S}_i)}\right) \\
&\quad + \mathbb{E}\left(\mathbb{E}\left[\mathrm{Var}\left(\mathrm{x}_{ji}|F(\mathbf{x}_i), G(\mathbf{S}_i)\right)\middle| F(\mathbf{x}_i)\right]\middle| \mathbf{x}_i \in B_{G(\mathbf{S}_i)}\right) \\
&\geq \mathbb{E}\left(\mathrm{Var}\left[\mathbb{E}\left(\mathrm{x}_{ji}|F(\mathbf{x}_i), G(\mathbf{S}_i)\right)\middle| F(\mathbf{x}_i)\right]\middle| \mathbf{x}_i \in B_{G(\mathbf{S}_i)}\right) \\
&= \sum_{l\in A_3}\sum_{k=1}^{K} \tilde{p}_{kl}(\tilde{\mu}_{j|kl} - \tilde{\mu}_{j|\cdot l})^2,
\end{aligned}$$

where $\tilde{\mu}_{|\cdot l} := \mathbb{E}(\mathbf{x}_i|F(\mathbf{x}_i) = l, \mathbf{x}_i \in B_{G(\mathbf{S}_i)})$ and $\tilde{\mu}_{|kl} := \mathbb{E}(\mathbf{x}_i|F(\mathbf{x}_i) = l, \mathbf{x}_i \in B_k)$. Note that we only consider those points $\mathbf{x}_i$'s in the ball $B_{G(\mathbf{S}_i)}$ and B_k, which justifies the final equality above. More detailed discussions are available in [5]. In addition, we note that

$$\sum_{k'\neq k} \tilde{p}_{kl}\tilde{p}_{k'l}\|\tilde{\boldsymbol{\mu}}_{|kl} - \tilde{\boldsymbol{\mu}}_{|k'l}\|_2^2 = \sum_{k'\neq k} \tilde{p}_{kl}\tilde{p}_{k'l}\left\|\left(\tilde{\boldsymbol{\mu}}_{|kl} - \tilde{\boldsymbol{\mu}}_{|\cdot l}\right) - \left(\tilde{\boldsymbol{\mu}}_{|k'l} - \tilde{\boldsymbol{\mu}}_{|\cdot l}\right)\right\|_2^2$$

$$= \sum_{k'\neq k} \tilde{p}_{kl}\tilde{p}_{k'l}\left(\|\tilde{\boldsymbol{\mu}}_{|kl} - \tilde{\boldsymbol{\mu}}_{|\cdot l}\|_2^2 + \|\tilde{\boldsymbol{\mu}}_{|k'l} - \tilde{\boldsymbol{\mu}}_{|\cdot l}\|_2^2\right) - 2\sum_{k'\neq k} \tilde{p}_{kl}\tilde{p}_{k'l}\left(\tilde{\boldsymbol{\mu}}_{|kl} - \tilde{\boldsymbol{\mu}}_{|\cdot l}\right)^\top\left(\tilde{\boldsymbol{\mu}}_{|k'l} - \tilde{\boldsymbol{\mu}}_{|\cdot l}\right)$$

$$= \sum_{k=1}^{K-1}\sum_{k'=k+1}^{K} \tilde{p}_{k'l}\tilde{p}_{kl}\|\tilde{\boldsymbol{\mu}}_{|kl} - \tilde{\boldsymbol{\mu}}_{|\cdot l}\|_2^2 + \sum_{k'=2}^{K}\tilde{p}_{k'l}\sum_{k=1}^{k'-1}\tilde{p}_{kl}\|\tilde{\boldsymbol{\mu}}_{|k'l} - \tilde{\boldsymbol{\mu}}_{|\cdot l}\|_2^2$$
$$- 2\sum_{k'\neq k} \tilde{p}_{kl}\tilde{p}_{k'l}\left(\tilde{\boldsymbol{\mu}}_{|kl} - \tilde{\boldsymbol{\mu}}_{|\cdot l}\right)^\top\left(\tilde{\boldsymbol{\mu}}_{|k'l} - \tilde{\boldsymbol{\mu}}_{|\cdot l}\right)$$

$$\leq \sum_{k=1}^{K-1} 1\cdot\tilde{p}_{kl}\|\tilde{\boldsymbol{\mu}}_{|kl} - \tilde{\boldsymbol{\mu}}_{|\cdot l}\|_2^2 + \sum_{k'=2}^{K} 1\cdot\tilde{p}_{k'l}\|\tilde{\boldsymbol{\mu}}_{|k'l} - \tilde{\boldsymbol{\mu}}_{|\cdot l}\|_2^2$$
$$- 2\sum_{k'\neq k} \tilde{p}_{kl}\tilde{p}_{k'l}\left(\tilde{\boldsymbol{\mu}}_{|kl} - \tilde{\boldsymbol{\mu}}_{|\cdot l}\right)^\top\left(\tilde{\boldsymbol{\mu}}_{|k'l} - \tilde{\boldsymbol{\mu}}_{|\cdot l}\right)$$

$$\leq \sum_{k=1}^{K} \tilde{p}_{kl}\|\tilde{\boldsymbol{\mu}}_{|kl} - \tilde{\boldsymbol{\mu}}_{|\cdot l}\|_2^2 + \sum_{k'=1}^{K} \tilde{p}_{k'l}\|\tilde{\boldsymbol{\mu}}_{|k'l} - \tilde{\boldsymbol{\mu}}_{|\cdot l}\|_2^2$$
$$- 2\sum_{k'\neq k} \tilde{p}_{kl}\tilde{p}_{k'l}\left(\tilde{\boldsymbol{\mu}}_{|kl} - \tilde{\boldsymbol{\mu}}_{|\cdot l}\right)^\top\left(\tilde{\boldsymbol{\mu}}_{|k'l} - \tilde{\boldsymbol{\mu}}_{|\cdot l}\right)$$

$$= 2\sum_{k=1}^{K} \tilde{p}_{kl}\|\tilde{\boldsymbol{\mu}}_{|kl} - \tilde{\boldsymbol{\mu}}_{|\cdot l}\|_2^2 - 2\sum_{k'\neq k} \tilde{p}_{kl}\tilde{p}_{k'l}\left(\tilde{\boldsymbol{\mu}}_{|kl} - \tilde{\boldsymbol{\mu}}_{|\cdot l}\right)^\top\left(\tilde{\boldsymbol{\mu}}_{|k'l} - \tilde{\boldsymbol{\mu}}_{|\cdot l}\right)$$

$$\leq 2\sum_{k=1}^{K} \tilde{p}_{kl}\|\tilde{\boldsymbol{\mu}}_{|kl} - \tilde{\boldsymbol{\mu}}_{|\cdot l}\|_2^2 + \sum_{k=1}^{K} \tilde{p}_{kl}\|\tilde{\boldsymbol{\mu}}_{|kl} - \tilde{\boldsymbol{\mu}}_{|\cdot l}\|_2^2 - \sum_{k=1}^{K} \|\tilde{p}_{kl}\left(\tilde{\boldsymbol{\mu}}_{|kl} - \tilde{\boldsymbol{\mu}}_{||\cdot l}\right)\|_2^2$$
$$- 2\sum_{k'\neq k} \tilde{p}_{kl}\tilde{p}_{k'l}\left(\tilde{\boldsymbol{\mu}}_{|kl} - \tilde{\boldsymbol{\mu}}_{|\cdot l}\right)^\top\left(\tilde{\boldsymbol{\mu}}_{|k'l} - \tilde{\boldsymbol{\mu}}_{|\cdot l}\right)$$

$$= 3\sum_{k=1}^{K} \tilde{p}_{kl}\|\tilde{\boldsymbol{\mu}}_{|kl} - \tilde{\boldsymbol{\mu}}_{|\cdot l}\|_2^2 - \left(\sum_{k=1}^{K} \|\tilde{p}_{kl}\left(\tilde{\boldsymbol{\mu}}_{||kl} - \tilde{\boldsymbol{\mu}}_{|\cdot l}\right)\|_2^2\right.$$
$$\left. + 2\sum_{k'\neq k}\left(\tilde{p}_{kl}\left(\tilde{\boldsymbol{\mu}}_{|kl} - \tilde{\boldsymbol{\mu}}_{|\cdot l}\right)\right)^\top\left(\tilde{p}_{k'l}\left(\tilde{\boldsymbol{\mu}}_{|k'l} - \tilde{\boldsymbol{\mu}}_{|\cdot l}\right)\right)\right)$$

$$= 3\sum_{k=1}^{K} \tilde{p}_{kl}\|\tilde{\boldsymbol{\mu}}_{|kl} - \tilde{\boldsymbol{\mu}}_{|\cdot l}\|_2^2 - \left\|\sum_{k=1}^{K}\tilde{p}_{kl}\left(\tilde{\boldsymbol{\mu}}_{|kl} - \tilde{\boldsymbol{\mu}}_{|\cdot l}\right)\right\|_2^2$$

$$= 3\sum_{k=1}^{K} \tilde{p}_{kl}\|\tilde{\boldsymbol{\mu}}_{|kl} - \tilde{\boldsymbol{\mu}}_{|\cdot l}\|_2^2,$$

where all the inequalities above hold for $0 \le \tilde{p}_{kl}, \tilde{p}_{k'l} \le 1$ for all k and k', and the last equality holds since $\sum_{k=1}^{K} \tilde{p}_{kl} \left(\tilde{\boldsymbol{\mu}}_{|kl} - \tilde{\boldsymbol{\mu}}_{|\cdot l}\right) = \mathbf{0}$. Therefore as $d(B_k, B_{k'}) \le \|\tilde{\boldsymbol{\mu}}_{|kl} - \tilde{\boldsymbol{\mu}}_{|k'l}\|_2$ by convexity,

$$\sum_{j=1}^{p} \mathbb{E}\left(\mathrm{Var}[\mathrm{x}_{ji}|F(\mathbf{x}_i)]\middle| \mathbf{x}_i \in B_{G(\mathbf{S}_i)}\right) \ge \sum_{l \in A_3} \sum_{k' \neq k} \frac{\tilde{p}_{kl}\tilde{p}_{k'l}}{3} \|\tilde{\boldsymbol{\mu}}_{|kl} - \tilde{\boldsymbol{\mu}}_{|k'l}\|_2^2$$

$$\ge \sum_{l \in A_3} \sum_{k' \neq k} \frac{\tilde{p}_{kl}\tilde{p}_{k'l}}{3} d^2(B_k, B_{k'})$$

$$\ge \sum_{l \in A_3} \frac{q^2}{3}(C - 2a\delta)^2 = \frac{|A_3|q^2}{3}(C - 2a\delta)^2 .$$

Finally, in the context of the Gaussian case, (14.22) simplifies to:

$$\sum_{j=1}^{p} \mathbb{E}[\mathrm{Var}[\mathrm{x}_{ji}|F(\mathbf{x}_i)]] \ge \mathbb{P}\left(\mathbf{x}_i \in B_{G(\mathbf{S}_i)}\right) \frac{|A_3|q^2}{3}(C - 2a\delta)^2$$

$$\ge \left(1 - \left(\frac{p}{a}\right)^{-p} \exp\left(-\frac{a^2 - p^2}{2p}\right)\right) \frac{|A_3|q^2}{3}(C - 2a\delta)^2.$$

Therefore, if we choose an appropriate value of a so that (14.19) is satisfied, then simple algebra gives

$$\frac{\sum_{j=1}^{p} \mathbb{E}[\mathrm{Var}[\mathrm{x}_{ji}|F(\mathbf{x}_i)]]}{\sum_{j=1}^{p} \mathbb{E}[\mathrm{Var}[\mathrm{x}_{ji}|G(\mathbf{S}_i)]]} \ge \frac{\left(1 - \left(\frac{p}{a}\right)^{-p} \exp\left(-\frac{a^2-p^2}{2p}\right)\right) \frac{|A_3|q^2}{3}(C - 2a\delta)^2}{\delta^2}$$

$$\ge \frac{|A_3|q^2}{3} \frac{3K}{q^2} \frac{\sum_{j=1}^{p} \mathrm{Var}(\mathrm{x}_{ji})}{\sum_{k=1}^{K} p_{k\cdot}\|\boldsymbol{\mu}_{|k\cdot} - \boldsymbol{\mu}\|_2^2}$$

$$\ge \frac{K-1}{L-1} \frac{\sum_{l=1}^{L} p_{\cdot l}\|\boldsymbol{\mu}_{|\cdot l} - \boldsymbol{\mu}\|_2^2}{\sum_{k=1}^{K} p_{k\cdot}\|\boldsymbol{\mu}_{|k\cdot} - \boldsymbol{\mu}\|_2^2},$$

which implies that

$$r_\infty(L) = \frac{1}{L-1} \frac{\sum_{l=1}^{L} p_{\cdot l}\|\boldsymbol{\mu}_{|\cdot l} - \boldsymbol{\mu}\|_2^2}{\sum_{j=1}^{p} \mathbb{E}[\mathrm{Var}[\mathrm{x}_{ji}|F(\mathbf{x}_i)]]}$$

$$\le \frac{1}{K-1} \frac{\sum_{k=1}^{K} p_{k\cdot}\|\boldsymbol{\mu}_{|k\cdot} - \boldsymbol{\mu}\|_2^2}{\sum_{j=1}^{p} \mathbb{E}[\mathrm{Var}[\mathrm{x}_{ji}|G(\mathbf{S}_i)]]} = r_\infty(K).$$

REFERENCES

1. Aha, D.W. (1997). *Lazy Learning*, pp. 7–10. Springer.
2. Aharon, M., Elad, M., and Bruckstein, A. (2006). K-SVD: an algorithm for designing overcomplete dictionaries for sparse representation. *IEEE Transactions on Signal Processing*, 54(11), 4311–4322.
3. Arthur, D., and Vassilvitskii, S. (2006). k-means++: the advantages of careful seeding. *SODA '07: Proceedings of the Eighteenth Annual ACM-SIAM Symposium on Discrete Algorithms*, pp. 1027–1035. Stanford, CA.
4. Caliński, T., and Harabasz, J. (1974). A dendrite method for cluster analysis. *Communications in Statistics – Theory and Methods*, 3(1), 1–27.
5. Chan, K.C.G., Li, J., and Yam, S.C.P. (2024+). On asymptotic optimality of Calinski–Harabasz index and its statistical properties. Working Paper.
6. Chen, Y., Fan, N.S., and Yam, S.C.P. (2024+). *Statistical Deep Learning with Python and R*. Preprint.
7. Chiang, M.C., Tsai, C.W., and Yang, C.S. (2011). A time-efficient pattern reduction algorithm for k-means clustering. *Information Sciences*, 181(4), 716–731.
8. Cover, T., and Hart, P. (1967). Nearest neighbor pattern classification. *IEEE Transactions on Information Theory*, 13(1), 21–27.
9. Fan, J., and Gijbels, I. (1996). *Local Polynomial Modelling and its Applications*. Chapman & Hall.
10. Fix, E., and Hodges, J.L. (1951). Discriminatory Analysis. Nonparametric Discrimination: Consistency Properties (Report). USAF School of Aviation Medicine, Randolph Field, Texas.
11. Foucart, S., and Rauhut, H. (2013). *A Mathematical Introduction to Compressive Sensing*. Birkhäuser.
12. Ho, T.K. (1998). Nearest neighbors in random subspaces. In *Advances in Pattern Recognition: Joint IAPR International Workshops SSPR'98 and SPR'98 Sydney, Australia, August 11–13, 1998 Proceedings* (pp. 640–648). Springer.
13. Inaba, M., Katoh, N., and Imai, H. (1994). Applications of weighted Voronoi diagrams and randomization to variance-based k-clustering. In *Proceedings of the Tenth Annual Symposium on Computational Geometry* (pp. 332–339).
14. James, G., Witten, D., Hastie, T., and Tibshirani, R. (2013). *An Introduction to Statistical Learning*. Springer.
15. Lloyd, S.P. (1957). Least square quantization in PCM. Bell Telephone Laboratories Paper. Published later: Lloyd, S.P.: Least squares quantization in PCM. *IEEE Transactions on Information Theory* (1957/1982), 18.
16. Loader, C. (2006). *Local Regression and Likelihood*. Springer Science & Business Media.
17. MacQueen, J. (1967). Some methods for classification and analysis of multivariate observations. In *Proceedings of the Fifth Berkeley Symposium on Mathematical Statistics and Probability*, 1(4), 281–297.
18. Mallat, S.G., and Zhang, Z. (1993). Matching pursuits with time-frequency dictionaries. *IEEE Transactions on Signal Processing*, 41(12), 3397–3415.
19. Nadaraya, E.A. (1964). On estimating regression. *Theory of Probability & Its Applications*, 9(1), 141–142.
20. Nader, R., Bretto, A., Mourad, B., and Abbas, H. (2019). On the positive semi-definite property of similarity matrices. *Theoretical Computer Science*, 755, 13–28.
21. Parzen, E. (1962). On estimation of a probability density function and mode. *The Annals of Mathematical Statistics*, 33(3), 1065–1076.

22. Pollard, D. (1981). Strong consistency of K-means clustering. *The Annals of Statistics*, 9(1), 135–140.
23. Rice, J.A. (2006). *Mathematical Statistics and Data Analysis*. Nelson Education.
24. Rosenblatt, M. (1956). Remarks on some nonparametric estimates of a density function. *Annals of Mathematical Statistics*, 27(3), 832–837.
25. Terrell, G.R., and Scott, D.W. (1992). Variable kernel density estimation. *The Annals of Statistics*, 20(3), 1236–1265.
26. Wang, X. (2011). A fast exact k-nearest neighbors algorithm for high dimensional search using k-means clustering and triangle inequality. In *The 2011 International Joint Conference on Neural Networks* (pp. 1293–1299). IEEE.
27. Watson, G.S. (1964). Smooth regression analysis. *Sankhyā: The Indian Journal of Statistics, Series A*, 26(4), 359–372.
28. Xu, R., and Wunsch, D. (2005). Survey of clustering algorithms. *IEEE Transactions on Neural Networks*, 16(3), 645–678.

CHAPTER 15

Applications of Deep Learning in Finance

Binary logistic regression, multinomial logits and classification trees as introduced in previous chapters are common statistical classifiers. A completely different approach is based on Neural Networks (NNs) which aim to mimic the functioning mechanism of human brains. Although how a human brain memorizes, recognizes patterns and generalizes concepts is still not thoroughly understood, some of its aspects are well-known to neuroscientists, which helps to build a workable model for our brains as a NN.

In 1943, McCulloch and Pits [18] postulated a simple mathematical model to explain how biological neurons work. From then there was not much progress until 1970 when the modern computer was first invented. In the 1970s, *Linnainmaa Seppo* invented the *backpropagation* algorithm to train neural networks so as to match as much as possible the fitted output with the observed response [13, 14]. There followed many successful examples and applications in the laboratories. In the 1980s, research moved from the labs to the commercial world, with typical applications being the detection of fraudulent credit card transactions, real estate appraisal, and data mining. Before we dive into NNs, we introduce some typical structures of our brains.

15.1 HUMAN BRAINS AND ARTIFICIAL NEURONS

Our brains consist of approximately 86 billion special cells known as neurons. Each of these neurons is typically connected with thousands of other neurons. Most importantly, unlike other cells, these neurons will not regenerate. It is widely accepted that these neurons are responsible for our ability for memorizing, learning, generalizing concepts and thinking. Within each neuron, there are four basic components: *dendrites*, the *cell body*, the *axon* and *synapses*; see Figure 15.1. These neurons are connected to form a huge network. Though the exact function of these neurons is still a mystery, a simple mathematical model that mimics these neurons provides a surprisingly good performance in pattern recognition, classification and prediction problems in image processing and text mining.

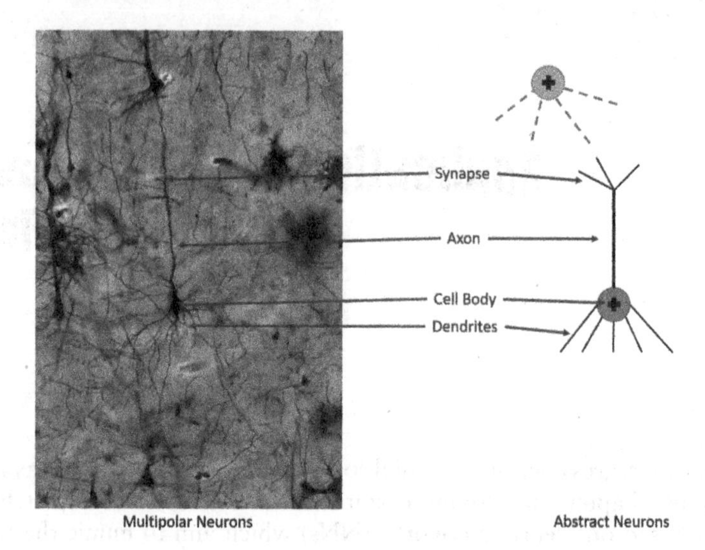

FIGURE 15.1 An illustration of the analogy between actual neurons and abstract ones. (Source: https://stock.adobe.com/in/images/cerebral-cortex-pyramidal-neuron/653658834)

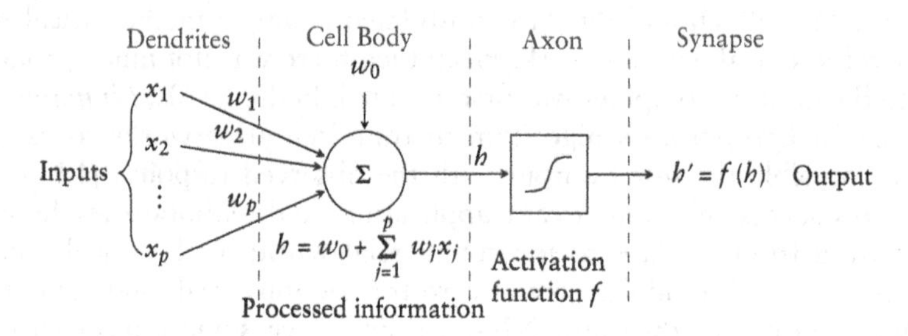

FIGURE 15.2 An artificial neuron.

Dendrites are responsible for receiving information from other neurons; a cell body is for processing information collected; an axon amplifies the signal carried by the processed information to other neurons; and a synapse is the junction between axon end and those dendrites of other neurons. To mimic the functioning of a neuron, we refer to the artificial neuron as illustrated in Figure 15.2.

Here $x_1, \ldots, x_p$ are the inputs received from other neurons or the outside environment. The total input h is formed from a linear combination of these inputs with weights $w_1, \ldots, w_p$ together with w_0 serving as a *bias* term. The *transfer function* (or *activation function*) converts the input h of x to the output $h' = f(h)$. In turn, this output h' may go to other neurons as input. Some commonly used transfer functions for the axon are shown in Figure 15.3.

Very often, a logistic transfer function is used; that is why artificial neural network is considered as a significant extension of logistic regression.

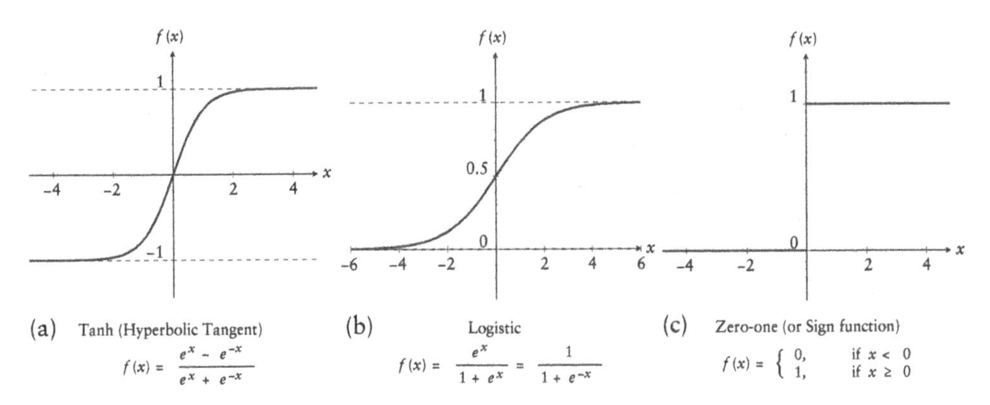

FIGURE 15.3 Commonly used activation functions for an axon.

15.2 FEEDFORWARD NETWORK

These artificial neurons are connected from one layer to another to form a network.

Referring to Figure 15.4, there are three layers: input, hidden and output layers. The numbers of neurons in the input, hidden and output layers are 4, 3 and 2 respectively; and so it is known as a 4-3-2 Artificial Neural Network (ANN).

1. For a standard ANN, there is only one hidden layer. With more than one hidden layer, it is called a Multilayer Perceptron (MLP); see Section 15.7 for a more detailed discussion on the latter.
2. The numbers of neurons in the input and output layers are determined by the nature of the problem, while the number of neurons in the hidden layer is user-defined.
3. Within each layer, neurons are not connected with each other. Neurons in one layer are connected only with neurons in the next layer in the forward direction; this mechanism is called *feeding forward*.

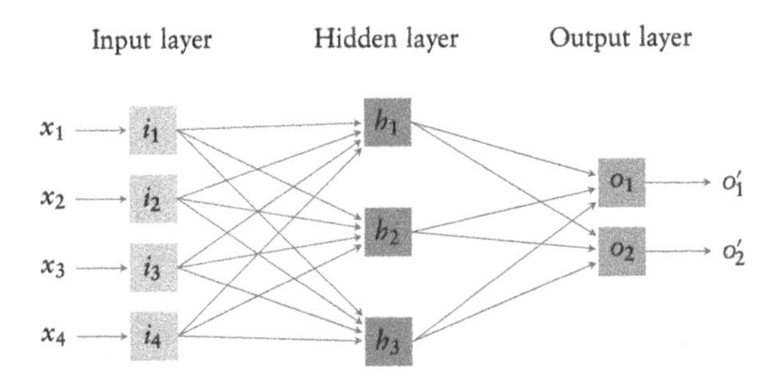

FIGURE 15.4 Illustration of a 4-3-2 ANN.

4. Each line joining two neurons, from the i-th neuron to the j-th neuron in the next layer, is associated with a weight w_{ji}. These weights are unknown parameters to be estimated from a training dataset.

15.3 ANN WITH LINEAR OUTPUTS

Given the input x and the output y in a training dataset, we aim to find the weights w_{ji}'s such that the sum of squared error:

$$E := \frac{1}{2}(y - o')^{\top}(y - o')$$

is minimized, where o' is the predicted label from the ANN; in particular, $o' = o$, the directed output without an additional activation, if a linear output is generated. In general, the objective function E has many local minima, and usually the backpropagation algorithm, an iterative form of the usual stochastic gradient descent method, is used to find the "almost global" minimum. In Python, the `MLPRegressor()` (*Multiple Layer Perceptron Regressor*) function in the `sklearn` library can implement single-layer or even multi-layer ANN, as specified by the parameter `hidden_layer_sizes`. In particular, the logistic activation function is used in the hidden layer with the argument `activation ="logistic"`, and 2 hidden neurons are used in the hidden layer with the parameter `hidden_layer_sizes=2` (If a 4-3-2-1 model is used, we can set `hidden_layer_sizes=(3,2)`, however, the activation function used is uniformly the same across all hidden layers). In R, ANN can be implemented using the built-in function `nnet()` inside the library `nnet`. We next illustrate the construction of an ANN with the Iris flower dataset in Programmes 15.1 and 15.2 via Python and R, respectively.

```
1  import numpy as np
2  from sklearn.datasets import load_iris
3
4  iris = load_iris()
5  X, y = iris.data, iris.target
6
7  from sklearn.neural_network import MLPRegressor
8  model = MLPRegressor(hidden_layer_sizes=2, max_iter=1000,
9                       solver="lbfgs", activation="logistic",
10                      random_state=1999)
11 model.fit(X, y)
12
13 W1, W2 = model.coefs_
14 b1, b2 = model.intercepts_
15 W1, W2 = W1.T, W2.T
16 print(W1)
17 print(W2)
18 print(b1)
19 print(b2)
```

```
1  [[-10.20043201 -10.14724349  16.40700302  14.79742642]
2   [ -3.44019264 -17.52913235  36.80899205  18.2599342 ]]
3  [[0.9987193   0.99304692]]
4  [-12.74264368  -4.04813158]
5  [3.1241105e-07]
```

Programme 15.1 4-2-1 ANN with the Iris flower dataset via Python.

```
1  > library(nnet)                    # load library nnet
2  > X <- iris[,1:4]; y <- as.numeric(iris[,5])
3  > # ANN with single hidden layer and linear output unit
4  > set.seed(1999)
5  > iris.nn <- nnet(X,y,size=2,linout=T) # linear output
6  # weights:  13
7  initial  value 540.881197
8  iter  10 value 10.640284
9  iter  20 value 4.863638
10 iter  30 value 4.847061
11 iter  40 value 4.846900
12 iter  50 value 4.824295
13 iter  60 value 4.602546
14 iter  70 value 3.515220
15 iter  80 value 2.001238
16 iter  90 value 1.911515
17 iter 100 value 1.812915
18 final  value 1.812915
19 stopped after 100 iterations
20 > summary(iris.nn)                  # summary of output
21 a 4-2-1 network with 13 weights
22 options were - linear output units
23  b->h1 i1->h1 i2->h1 i3->h1 i4->h1
24  -7.15   0.34  -2.17   4.23   4.63
25  b->h2 i1->h2 i2->h2 i3->h2 i4->h2
26 -30.20  -6.23  -7.05  13.65  13.32
27   b->o  h1->o  h2->o
28   1.00   1.00   1.00
```

Programme 15.2 4-2-1 ANN with the Iris flower dataset via R.

The above ANN is represented as the following Figure 15.5:

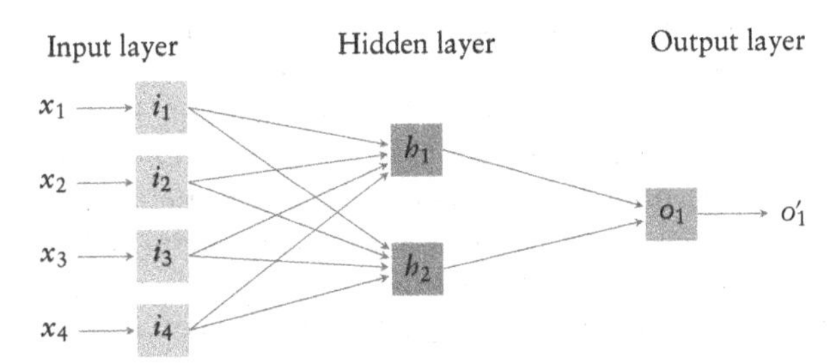

FIGURE 15.5 Illustration of a 4-2-1 ANN.

Here in Programme 15.1 via Python,

$$\begin{aligned} h_1 &= -12.74 - 10.20x_1 - 10.15x_2 + 16.41x_3 + 14.80x_4, \\ h_2 &= -4.05 - 3.44x_1 - 17.53x_2 + 36.81x_3 + 18.26x_4, \\ h'_1 &= \frac{\exp(h_1)}{1+\exp(h_1)}, \quad h'_2 = \frac{\exp(h_2)}{1+\exp(h_2)}, \\ o'_1 &= o_1 = 0 + h'_1 + 0.99h'_2; \end{aligned} \tag{15.1}$$

while in Programme 15.2 via **R**,

$$\begin{aligned} h_1 &= -7.15 + 0.34x_1 - 2.17x_2 + 4.23x_3 + 4.63x_4, \\ h_2 &= -30.2 - 6.23x_1 - 7.05x_2 + 13.65x_3 + 13.32x_4, \\ h'_1 &= \frac{\exp(h_1)}{1+\exp(h_1)}, \quad h'_2 = \frac{\exp(h_2)}{1+\exp(h_2)}, \\ o'_1 &= o_1 = 1 + h'_1 + h'_2. \end{aligned} \tag{15.2}$$

Let us use the 1st, 51st and 101st observations from the Iris dataset to illustrate how this ANN makes prediction. The 1st observation is $\boldsymbol{x} = (5.1, 3.5, 1.4, 0.2)^\top$. According to the formulae of (15.2),

$$\begin{aligned} h_1 &= -7.15 + (0.34)(5.1) + (-2.17)(3.5) + (4.23)(1.4) + (4.63)(0.2) \\ &= -6.163, \\ h_2 &= -30.2 + (-6.23)(5.1) + (-7.05)(3.5) + (13.65)(1.4) + (13.32)(0.2) \\ &= -64.874, \\ h'_1 &= \frac{\exp(h_1)}{1+\exp(h_1)} = 0.0021, \quad h'_2 = \frac{\exp(h_2)}{1+\exp(h_2)} = 0.0000, \\ o'_1 &= o_1 = 1 + h'_1 + h'_2 = 1.0021. \end{aligned}$$

Similarly, for the 51st observation, $\boldsymbol{x} = (7, 3.2, 4.7, 1.4)^\top$, $h_1 = 14.649$, $h_2 = -13.567$, $h'_1 = 1.0000$, $h'_2 = 0.0000$, and $o'_1 = 2.0000$; and the 101st observation, $\boldsymbol{x} = (6.3, 3.3, 6, 2.5)^\top$, $h_1 = 24.786$, $h_2 = 22.486$, $h'_1 = 1.0000$, $h'_2 = 1.0000$ and $o'_1 = 3.0000$.

As a remark, the coefficients of model trained by Python usually possess a much smaller magnitude than their counterparts trained by **R**; this is due to the presence of various underlying training mechanisms in Python, including but not limited to *batch normalization* (see [24]) and *dropout* (see [9]) procedures, which helps to improve the robustness of the trained model. Nevertheless, in this particular example, the coefficients in the two models exhibit similar orders of magnitude, and hence we illustrate only the calculation for the model trained in **R** due to the page limit.

In Python, o' can be obtained via the `predict()` function, and we can produce the confusion matrix as shown in Programme 15.3.

```
1  import pandas as pd
2
3  y_pred = model.predict(X)
4  y_pred = np.round(y_pred)
5
6  print(pd.crosstab(y_pred, y))   # confusion matrix
```

```
1  col_0  0.0  1.0  2.0
2  row_0
3  0       50    0    0
4  1        0   48    2
5  2        0    0   50
```

Programme 15.3 Output and prediction of the 4-2-1 ANN for the Iris flower dataset in Python.

In **R**, o' is stored in `iris.nn$fitted.values`, and we can produce the confusion matrix as in Programme 15.4.

```
1  > pred <- round(iris.nn$fit)     # round the fitted values
2  > table(y, pred)                  # confusion matrix
3      pred
4    y  1  2  3
5    1 50  0  0
6    2  0 49  1
7    3  0  0 50
```

Programme 15.4 Output and prediction of the 4-2-1 ANN for the Iris flower dataset in **R**.

We next try the HSI example in Programme 15.5 via Python and Programme 15.6 via **R**.

```
1  import pandas as pd
2
3  df = pd.read_csv("fin-ratio.csv")
4  X = df.drop(columns="HSI")
5  y = df["HSI"]
6  print(X.columns.values)
7
8  from sklearn.neural_network import MLPRegressor
9  model = MLPRegressor(hidden_layer_sizes=3, max_iter=1000,
10                      solver="lbfgs", activation="logistic",
11                      random_state=4002)
12
13 model.fit(X, y)
14 y_pred = model.predict(X)
15 y_pred = y_pred > 0.5
16 print(pd.crosstab(y_pred, y))   # confusion matrix
```

```
1 ['EY' 'CFTP' 'ln_MV' 'DY' 'BTME' 'DTE']
2 HSI       0   1
3 row_0
4 False   647   8
5 True      1  24
```

Programme 15.5 6-3-1 ANN with the financial HSI dataset via Python.

```
1  > d <- read.csv("fin-ratio.csv")
2  > names(d)
3  [1] "EY"     "CFTP"    "ln_MV" "DY"     "BTME"   "DTE"    "HSI"
4  > X <- subset(d, select=-c(HSI)); y <- d$HSI
5  > set.seed(4002)
6  > fin.nn <- nnet(X,y,size=3,linout=T,maxit=200)
7  # weights:  25
8  initial  value 109.269725
9  iter  10 value 26.766139
10 iter  20 value 17.495062
11 iter  30 value 15.160260
12 iter  40 value 14.932151
13 iter  50 value 14.749326
14 iter  60 value 14.605999
15 iter  70 value 14.053005
16 iter  80 value 9.114023
17 iter  90 value 5.960670
18 iter 100 value 4.793639
19 iter 110 value 4.051921
20 iter 120 value 3.717146
21 iter 130 value 3.664704
22 iter 140 value 3.522684
23 iter 150 value 3.409379
24 iter 160 value 3.224694
25 iter 170 value 2.833937
26 iter 180 value 2.706538
27 iter 190 value 2.663023
28 iter 200 value 2.637450
29 final  value 2.637450
30 stopped after 200 iterations
31 > # set max. number of iterations to 200
32 > pred <- round(fin.nn$fit)    # round the fitted values
33 > table(y, pred)               # confusion matrix
34    pred
35   y   0  1
36   0 645  3
37   1   0 32
```

Programme 15.6 6-3-1 ANN with the financial HSI dataset via R.

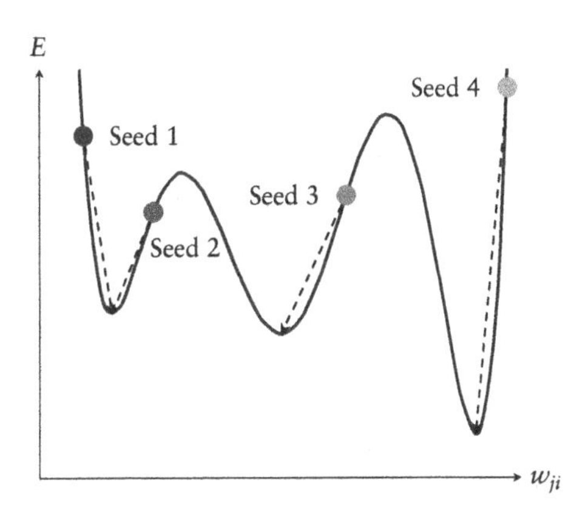

FIGURE 15.6 The three solutions of four different initial seeds in the parameter space.

A major problem of ANNs is that the solution of weights depends on the initial seeds. When we run `nnet()` each time, we may obtain different results. This is illustrated in Figure 15.6, where the solid curve represents the loss function.

The error function E, while depending on the weights $(w_{ji})_{i,j}$ of the ANN, contains a lot of local minima in the multi-dimensional spaces in w_{ji}'s. As `nnet()` randomly assigns an initial set of parameters w_{ji} (*Xavier initialization* [5]) and uses backpropagation for minimizing the error function, the estimated w_{ji}'s may only be a local minimum of the error function. Therefore, we usually run `MLPRegressor()` in Python (resp. `nnet()` in **R**) several times with different sets of initial seeds to attempt to obtain the "globally optimal" weights. See Programme 15.7 via Python (resp. Programme 15.8 via **R**) for an improven version of the function `MLPRegressor()` in Python (resp. `nnet()` in **R**).

```
1  from sklearn.neural_network import MLPRegressor, MLPClassifier
2
3  def ANNet(X, y, size, linout=False, max_iter=1e4, trial=5):
4      kwargs = {"hidden_layer_sizes": size, "max_iter": max_iter,
5                "solver": "lbfgs", "activation": "logistic"}
6      Best_ANN = MLPRegressor(**kwargs) if linout else MLPClassifier(**kwargs)
7      Best_ANN.fit(X, y)
8      Best_score = Best_ANN.score(X, y)
9      for i in range(1, trial):
10         model = MLPRegressor(**kwargs) if linout else MLPClassifier(**kwargs)
11         model.fit(X, y)
12         if model.score(X, y) > Best_score:      # check if improven
13             Best_score = model.score(X, y)      # save the best score
14             Best_ANN = model                    # save the best model
15     return Best_ANN
```

Programme 15.7 Return the best ANN model in Python, saved in a file with the name `ANNet.py` for subsequent use.

```
1  > # x is the matrix of input variable
2  > # y is the dependent value; which must be factor if linout=F
3  > library(nnet)
4  > ann <- function(x,y,size,maxit=100,linout=F,try=5) {
5  +     best <- nnet(y~.,data=x,size=size,maxit=maxit,linout=linout)
6  +     for (i in 2:try) {
7  +         ann <- nnet(y~.,data=x,size=size,maxit=maxit,linout=linout)
8  +         if (ann$value < best$value) best <- ann # save best ann
9  +     }
10 +     return (best)                         # return the results
11 + }
```

Programme 15.8 Return the best ANN model in **R**, saved in a file with the name `ANNet.R` for subsequent use.

This newly built function can specify the number of trials and save the best result; see Programme 15.9 via Python (resp. Programme 15.10 via **R**) for example.

```
1  from ANNet import ANNet
2  import pandas as pd
3  import numpy as np
4  np.random.seed(4002)
5  
6  df = pd.read_csv("fin-ratio.csv")
7  X = df.drop(columns="HSI")
8  y = df["HSI"]
9  
10 model = ANNet(X, y, size=3, linout=True, max_iter=1000, trial=10)
11 y_pred = model.predict(X)
12 y_pred = y_pred > 0.5
13 
14 print(pd.crosstab(y_pred, y))                  # confusion matrix
15 W1, W2 = model.coefs_
16 b1, b2 = model.intercepts_
17 W1, W2 = W1.T, W2.T
18 print(W1)
19 print(W2)
20 print(b1)
21 print(b2)
```

```
1  HSI       0    1
2  row_0
3  False   646    2
4  True      2   30
5  [[  1.33356741  -1.48179693   7.27142077  -0.33428612
6      0.15386847   0.04860259]
7   [  1.22921501  -1.51896643   7.97306514  -0.37603425
8     -0.04483474   0.0473285 ]
9   [ -1.77113521   2.92941111 -27.16425305   0.96350378
```

```
10      -0.39576395    1.83023616]]
11  [[ 9.22959561 -8.24136984 19.77134544]]
12  [-66.41224532 -72.38174725   -3.76514104]
13  [-0.00053662]
```

Programme 15.9 Best 6-3-1 ANN with the financial HSI dataset among 10 trials via Python.

```
1   > source("ANNet.R")
2   > d <- read.csv("fin-ratio.csv")
3   > X <- subset(d, select=-c(HSI)); y <- d$HSI
4   > set.seed(4002)
5   > fin.nn <- ANNet(X,y,size=3,linout=T,try=10) # 10 trials
6   > fin.nn$value                    # the best result
7   [1] 3.203018
8   > summary(fin.nn)                 # the best weights
9   a 6-3-1 network with 25 weights
10  options were - linear output units
11    b->h1  i1->h1  i2->h1  i3->h1  i4->h1  i5->h1  i6->h1
12   -50.70   -5.49   -7.22    2.95   -1.08    1.69    0.98
13    b->h2  i1->h2  i2->h2  i3->h2  i4->h2  i5->h2  i6->h2
14  -103.76   -7.43   -9.87   10.73    0.60    1.88    0.27
15    b->h3  i1->h3  i2->h3  i3->h3  i4->h3  i5->h3  i6->h3
16    -9.41   -0.26   -2.90    4.27   -2.50    0.75   -7.54
17     b->o   h1->o   h2->o   h3->o
18     0.00   -0.97    0.98    0.00
19  > pred <- round(fin.nn$fit)    # round the fitted values
20  > table(y, pred)               # confusion matrix
21      pred
22     y   0   1
23     0 645   3
24     1   2  30
```

Programme 15.10 Best 6-3-1 ANN with the financial HSI dataset among 10 trials via R.

15.4 ANN WITH LOGISTIC OUTPUTS

In the last section, the output is a linear function of the values from the neurons in the hidden layer, and it is a real number. When using this in a classification problem, like the Iris flower and HSI examples, we have to round the output to its nearest integer. However, we may use the default option `linout=False` in Python (resp. `linout=FALSE` in R) (but not `linout=True` now), then r different logistic functions are used to connect the neurons of the output layers and the final output. This can be interpreted as the probability of an observation with certain features belonging to a particular one out of r groups. Let us look at how to use this logistic output option in Programme 15.11 via Python (resp. in Programme 15.12 via **R**).

```
from sklearn.datasets import load_iris

iris = load_iris()
X, y = iris.data, iris.target
model = ANNet(X, y, size=2, linout=False, max_iter=1000, trial=10)
y_pred = model.predict(X)

print(pd.crosstab(y_pred, y))                    # confusion matrix
W1, W2 = model.coefs_
b1, b2 = model.intercepts_
W1, W2 = W1.T, W2.T
print(W1)
print(W2)
print(b1)
print(b2)
```

```
col_0    0   1   2
row_0
0       50   0   0
1        0  49   0
2        0   1  50
[[ -0.1927346    -1.47272604    2.76131431    1.87754883]
 [ 14.77507666  19.05618184 -41.40836685 -38.66317667]]
[[-22.35143221  10.65584082]
 [ 14.85940112   3.21161209]
 [  7.50009821 -14.79935427]]
[ -2.17768877 119.97401451]
[ 0.38269997 -5.78215224   5.55450809]
```

Programme 15.11 Best fitted 4-2-3 ANN with logistic output for the Iris flower dataset among 10 trials via Python.

```
> X <- iris[,1:4]; y <- as.factor(iris[,5])
> iris.nn <- ANNet(X,y,size=2,maxit=200,try=10)  # 10 trials, logistic
> iris.nn$value                          # best value
[1] 0.5648058
> summary(iris.nn)                       # display weights
a 4-2-3 network with 19 weights
options were - softmax modelling
  b->h1  i1->h1  i2->h1  i3->h1  i4->h1
   4.55   -0.15    0.22   -1.38    1.15
  b->h2  i1->h2  i2->h2  i3->h2  i4->h2
 111.04    2.55   37.50   -7.48 -112.19
  b->o1  h1->o1  h2->o1
   0.96   91.74  -55.32
  b->o2  h1->o2  h2->o2
-110.98   14.28  122.25
  b->o3  h1->o3  h2->o3
 110.83 -106.66  -67.85
```

```
18 > pred <- max.col(iris.nn$fit)# find column name with max. fitted values
19 > table(y, pred)                # confusion matrix
20      pred
21     y setosa versicolor virginica
22     1     50          0         0
23     2      0         50         0
24     3      0          0        50
```

Programme 15.12 Best fitted 4-2-3 ANN with logistic output for the Iris flower dataset among 10 trials via **R**.

Note that the fitted values in the output have r columns, and the probability (final output) of the i-th observation, of a sample of size n, belonging to the j-th group is

$$o'_{ij} = \frac{\exp(o_{ij})}{\sum_{k=1}^{r} \exp(o_{ik})}, \quad i = 1, 2, \ldots, n, \quad j = 1, 2, \ldots, r.$$

Instead of looking for the least squares estimates for the weights and biases as in the linear output case in Section 15.3, we here aim to minimize the so-called *categorical cross-entropy loss function*:

$$E := -\frac{1}{n} \sum_{i=1}^{n} \sum_{j=1}^{r} \mathbb{1}_{\{y_i = c_j\}} \ln o'_{ij}.$$

Normally speaking, we predict that the i-th observation belongs to group j if o'_{ij} is maximum in the i-th row of $(o'_{i1}, o'_{i2}, \ldots, o'_{ir})$. It is clear that o'_{ij} is the maximum if and only if o_{ij} is the maximum. For **R** in particular, we assign the label to the `pred` column according to the maximum o'_{ij} among all $j = 1, \ldots, r$ in each row of `iris.nn$fit`, and notably, the resulting model yields no error out of 150 cases and the error rate is 0%!

Since in the Iris flower example, the output has three levels, the diagram for this ANN with logistic output is shown in Figure 15.7:

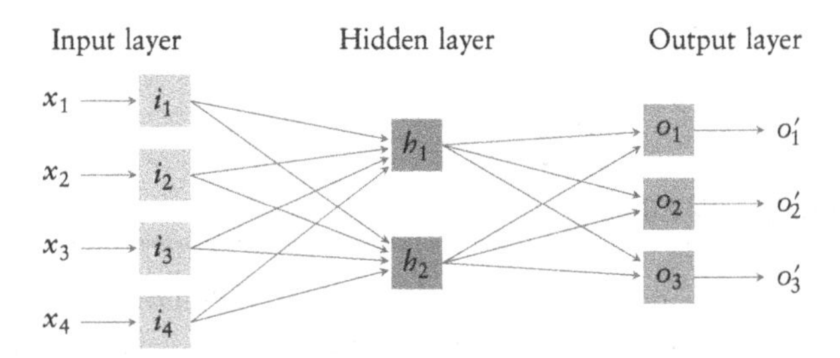

FIGURE 15.7 Illustration of a 4-2-3 ANN.

Note that for logistic output, the output of the i-th record is

$$o'_{ij} = \frac{e^{o_{ij}}}{1 + e^{o_{ij}}} = \mathbb{P}(y_i = c_j), \quad j = 1, 2, \ldots, r.$$

We next illustrate with the HSI example. For this example, the output variable is binary ($r = 2$). The fitted value in the output can only have one column, say the probability of belonging to Group 1, then the probability of belonging to Group 2 is the one minus the former.

```
1  > d <- read.csv("fin-ratio.csv")
2  > X <- subset(d, select=-c(HSI)); y <- d$HSI
3  > y <- as.factor(d$HSI)          # y as factor
4  > set.seed(4002)
5  > fin.nn <- ANNet(X,y,size=3,maxit=200,try=10)
6  > fin.nn$value                   # best value
7  [1] 4.713665
8  > summary(fin.nn)
9  a 6-3-1 network with 25 weights
10 options were - entropy fitting
11    b->h1   i1->h1   i2->h1   i3->h1   i4->h1   i5->h1   i6->h1
12     3.06     7.01    -0.10    10.08     2.38     0.20   -12.14
13    b->h2   i1->h2   i2->h2   i3->h2   i4->h2   i5->h2   i6->h2
14   212.98    87.59    52.83   -31.03    17.78     9.90   -12.35
15    b->h3   i1->h3   i2->h3   i3->h3   i4->h3   i5->h3   i6->h3
16  -204.12     3.30    -3.69    20.42     2.15     6.20     0.51
17     b->o    h1->o    h2->o    h3->o
18   -49.26     5.47   -70.07   112.74
19 > pred <- fin.nn$fit > 0.5      # check if it belongs to group 1
20 > table(y, pred)                 # confusion matrix
21        pred
22   y       0   1
23   FALSE 648   2
24   TRUE    0  30
```

Programme 15.13 Best fitted 6-3-1 ANN with logistic output for the financial HSI dataset among 10 trials via R.

We assign the label to the `pred` column according to the probability given in `fin.nn$fit`. Note that there are only 2 error cases out of a total of 680 cases and the error rate is only 0.29%!

15.5 ADAPTIVE LEARNING RATE

In the old-fashioned stochastic gradient descent (SGD) algorithm, the learning rate is fixed and uniform for the updates of all parameters. This is far from perfect, especially since the updates of less frequently occurring parameters are relatively slow. A better approach therefore is to take the current values of parameters into account and tune the learning rate accordingly. Roughly speaking, the learning rate for infrequently encountered parameters should be greater than that of frequently occurring ones. With this in mind, we introduce three commonly used variants of SGD in this section.

We first suppose that $E(w^{(t)})$ is a loss function with respect to the parameter $w^{(t)} = (w_1^{(t)}, \ldots, w_p^{(t)})^\top$ after t updates. For $i = 1, \ldots, p$, also denote by $g_i^{(t)} := \frac{\partial E}{\partial w_i}(w^{(t)})$, and let η be the starting learning rate.

I. AdaGrad

The *adaptive gradient method* (*AdaGrad*) adjusts the learning rate by dividing its starting value η by the root of the sum of squares of the gradients. Mathematically, the update from $w_i^{(t)}$ to $w_i^{(t+1)}$ is given by

$$w_i^{(t+1)} = w_i^{(t)} - \frac{\eta}{\sqrt{\sum_{s \le t} (g_i^{(s)})^2 + \epsilon}} g_i^{(t)}, \quad i = 1, 2, \ldots, p, \tag{15.3}$$

where ϵ is a small number to avoid division by zero.

This method is particularly suitable for handling sparse datasets in which the rates of update for both frequently and infrequently encountered parameters become comparable; indeed, as the gradient terms accumulate, the learning rate diminishes and by then the parameters cease to be updated.

II. RMSprop

To avoid the possibility of a vanishing learning rate (caused by truncation), *Geoffrey Hinton* suggested replacing the sum of square of historical gradients in (15.3) by a moving average [8]. This revised method is called the *root mean square propagation* (*RMSprop*); it starts with defining the second moment of gradient at time t:

$$v_i^{(t)} := \gamma v_i^{(t-1)} + (1 - \gamma)(g_i^{(t)})^2,$$

where $\gamma \in [0, 1]$. Then, the update of the parameter is given by

$$w_i^{(t+1)} = w_i^{(t)} - \frac{\eta}{\sqrt{v_i^{(t)} + \epsilon}} g_i^{(t)}. \tag{15.4}$$

Moreover, Hinton suggested that $\gamma = 0.9$ and $\eta = 0.001$ could be good candidates for an efficient algorithm.

III. Adam

The *adaptive moment estimation* (*Adam*) is another variant of SGD first introduced by *Diederik Kingma* and *Jimmy Ba* in 2015 [12]. It incorporates both the first and second moments of the gradients for parameter updates. Let $m_i^{(t)}$ and $v_i^{(t)}$ respectively be the average of historical gradients and their squares of the i-th weight parameter with respective decay rates $\beta_1, \beta_2 \in (0, 1)$; more precisely,

$$m_i^{(t)} = \beta_1 m_i^{(t-1)} + (1 - \beta_1) g_i^{(t)},$$
$$v_i^{(t)} = \beta_2 v_i^{(t-1)} + (1 - \beta_2)(g_i^{(t)})^2,$$

with initial zero values, i.e. $m_i^{(0)} = v_i^{(0)} = 0$. Then, the update at time t reads:

$$w_i^{(t+1)} = w_i^{(t)} - \frac{\eta}{\sqrt{v_i^{(t)}} + \epsilon} m_i^{(t)}, \tag{15.5}$$

where ϵ is a small number to avoid division by zero.

Empirical evidence in [12] argues that the performance is the best when both β_1 and β_2 are close enough to 1. To tackle the possibility that $m_i^{(t)}$ and $v_i^{(t)}$ are biased towards 0 due to zero values initially set, the update in (15.5) can be replaced by

$$w_i^{(t+1)} = w_i^{(t)} - \frac{\eta}{\sqrt{\hat{v}_i^{(t)}} + \epsilon} \hat{m}_i^{(t)}, \tag{15.6}$$

where

$$\hat{m}_i^{(t)} = \frac{m_i^{(t)}}{1 - \beta_1} \text{ and } \hat{v}_i^{(t)} = \frac{v_i^{(t)}}{1 - \beta_2}.$$

15.6 TRAINING NEURAL NETWORKS VIA BACKPROPAGATION

As seen in Section 15.3, training a neural network is equivalent to finding the appropriate weight parameter $\boldsymbol{w}$ such that the loss function $E = \frac{1}{2}(y - o')^\top (y - o')$ is minimized, and this is achieved by using a backpropagation algorithm. Let us illustrate this approach by a simple 2-2-1 ANN with a linear output; also see Figure 15.8 for an illustration.

Suppose the four sample input vectors, the target vector of the labels and the initial weights are as follows:

$$X = \begin{pmatrix} 0.4 & 0.7 \\ 0.8 & 0.9 \\ 1.3 & 1.8 \\ -1.3 & -0.9 \end{pmatrix}, \quad y = \begin{pmatrix} 0 \\ 0 \\ 1 \\ 0 \end{pmatrix},$$

$$W_1 = \begin{pmatrix} b_{11} & w_{11} & w_{12} \\ b_{12} & w_{21} & w_{22} \end{pmatrix} = \begin{pmatrix} 0.1 & -0.2 & 0.1 \\ 0.4 & 0.2 & 0.9 \end{pmatrix},$$

$$W_2 = (b_{21} \quad v_{11} \quad v_{12}) = (0.2 \quad -0.5 \quad 0.1).$$

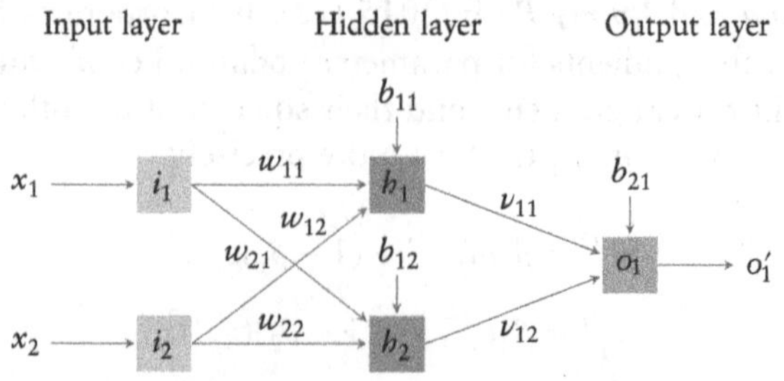

FIGURE 15.8 Illustration of a 2-2-1 ANN.

First, we compute the output from the ANN using the initial weights in Programme 15.14 via Python (resp. Programme 15.15 via **R**) as follows.

```
1  import numpy as np
2  logistic = lambda x: 1/(1 + np.exp(-x))
3  
4  X = np.array([[0.4,0.7], [0.8, 0.9], [1.3, 1.8], [-1.3, -0.9]])
5  y = np.array([0, 0, 1, 0])              # target value
6  
7  # hidden layer bias and weights
8  W1 = np.array([[0.1,-0.2,0.1], [0.4,0.2,0.9]])
9  # output layer bias and weights
10 W2 = np.array([[0.2,-0.5,0.1]])
11 
12 # transpose to fit the input format of ANN
13 X1 = np.c_[np.ones(len(y)), X]
14 h = logistic(W1 @ X1.T)                 # logistic hidden h'
15 h = np.c_[np.ones(len(y)), h.T].T
16 o = W2 @ h                              # linear output o'
17 err = y - o
18 print(err)                              # output error
19 print(np.mean(err**2))                  # mean SSE
```

```
1  [[-0.0139705  -0.0259883   0.96177921  0.04969677]]
2  0.23208989694608873
```

Programme 15.14 Forward propagation for 2-2-1 ANN via Python.

```
1  > logistic <- function(x) 1/(1+exp(-x))
2  >
3  > X <- matrix(c(0.4,0.7,0.8,0.9,1.3,1.8,-1.3,-0.9),ncol=2,byrow=T)
4  > y <- c(0, 0, 1, 0)                # target value
5  > # hidden layer bias and weights
6  > W1 <- matrix(c(0.1,-0.2,0.1,0.4,0.2,0.9),nrow=2,byrow=T)
7  > # output layer bias and weights
8  > W2 <- matrix(c(0.2,-0.5,0.1),nrow=1)
9  >
10 > X1 <- cbind(1, X)
11 > h <- logistic(W1%*%t(X1))         # logistic hidden h'
12 > h <- rbind(1, h)
13 > o <- W2%*%h                       # linear output o'
14 > (err <- y - o)                    # output error
15             [,1]       [,2]      [,3]       [,4]
16 [1,] -0.0139705 -0.0259883 0.9617792 0.04969677
17 > (mean_sse <- mean(err^2))         # mean SSE
18 [1] 0.2320899
```

Programme 15.15 Forward propagation for 2-2-1 ANN via **R**.

Next, we update the output layer weights `W2` and hidden layer weights `W1` in Programme 15.16 via Python (resp. Programme 15.17 via **R**) as follows.

```
1  lr = 0.5                               # learning rate: η
2  n = len(y)
3  del2 = -2*err                          # output layer δ2
4  Delta_W2 = -lr*del2 @ h.T              # ΔW2 = -ηδ2(h')ᵀ
5  new_W2 = W2 + Delta_W2 / n             # new output weights: W2 = W2 + ΔW2
6
7  del1 = (W2.T @ del2)*h*(1-h)           # hidden layer δ1
8  del1 = del1[1:,]                       # remove from X1
9  Delta_W1 = -lr*del1 @ X1               # ΔW1 = -ηδ1xᵀ
10 new_W1 = W1 + Delta_W1 / n             # new hidden weights: W1 = W1 + ΔW1
11
12 new_h = logistic(new_W1 @ X1.T)
13 new_h = np.c_[np.ones(len(y)), new_h.T].T
14 new_o = new_W2 @ new_h
15 new_err = y - new_o
16 print(np.mean(new_err**2))             # new mean SSE
```

```
1  0.20274498672045063
```

Programme 15.16 Backpropagation for 2-2-1 ANN via Python.

```
1  > lr <- 0.5                              # learning rate: η
2  > n <- length(y)
3  > del2 <- -2*err                         # output layer δ2
4  > Delta_W2 <- -lr*del2%*%t(h)            # ΔW2 = -ηδ2(h')ᵀ
5  > new_W2 <- W2 + Delta_W2 / n            # new output weights: W2 = W2 + ΔW2
6  >
7  > del1 <- (t(W2)%*%del2)*h*(1-h)         # hidden layer δ1
8  > del1 <- del1[-1,]                      # remove from X1
9  > Delta_W1 <- -lr*del1%*%X1              # ΔW1 = -ηδ1xᵀ
10 > new_W1 <- W1 + Delta_W1 / n            # new hidden weights: W1 = W1 + ΔW1
11 >
12 > new_h <- logistic(new_W1%*%t(X1))
13 > new_h <- rbind(1, new_h)
14 > new_o <- new_W2%*%new_h
15 > new_err <- y - new_o
16 > (new_mean_sse <- mean(new_err^2)) # new mean SSE
17 [1] 0.202745
```

Programme 15.17 Backpropagation for 2-2-1 ANN via R.

Note that the new weights `new_W1` and `new_W2` are computed and the mean SSE decreases from 0.232 to 0.203. We can update the weights `W1` by `new_W1` and `W2` by `new_W2` based on the above programmes, and iterate them until the SSE is small enough or the maximum number of iteration is exceeded. Note that we can scale up the learning rate (`lr = 1.1*lr`) in the next iteration if the mean SSE decreases, or we can scale it down (`lr = 0.5*lr`) if the mean SSE increases.

Let us justify the above two programmes by providing the mathematical details of the backpropagation algorithm. Consider a general *p-q-r* ANN with logistic transfer function f_h (so $f_h' = f_h(1 - f_h)$) and linear output f_o. The outputs of the hidden layer from the input variables are:

$$h'_j = f_h(h_j) = f_h\left(w_{j0} + \sum_{i=1}^{p} w_{ji}x_i\right), \qquad j = 1, \ldots, q;$$

and from the hidden layer to the final linear output, they are:

$$o'_k = f_o(o_k) = o_k = v_{k0} + \sum_{j=1}^{q} v_{kj}h'_j, \qquad k = 1, \ldots, r. \tag{15.7}$$

The backpropagation algorithm is the typical gradient descent method applied to the network by working from the final output backward in time to the layer of input variables for minimizing the error function:

$$E = \frac{1}{2}(y - o')^{\top}(y - o') = \frac{1}{2}\sum_{k=1}^{r}(y_k - o'_k)^2.$$

Based on the law of total derivative, we first have, by viewing the matrix $\mathbf{W}_1$ (resp. $\mathbf{W}_2$) as a vector after standard vectorization in machine learning,

$$\Delta E = \left(\frac{\partial E}{\partial \mathbf{W}_1}\right)^{\top} \Delta \mathbf{W}_1 + \left(\frac{\partial E}{\partial \mathbf{W}_2}\right)^{\top} \Delta \mathbf{W}_2.$$

The direction of $(\Delta \mathbf{W}_1, \Delta \mathbf{W}_2)$ which can maximize the reduction of ΔE by a unit change is along the negative direction of $\begin{pmatrix} \frac{\partial E}{\partial \mathbf{W}_1} \\ \frac{\partial E}{\partial \mathbf{W}_2} \end{pmatrix}$. The gradient descent algorithm for updating $\mathbf{W}$ reads as:

$$\mathbf{W}^{(t+1)} = \mathbf{W}^{(t)} + \Delta \mathbf{W}^{(t)} = \mathbf{W}^{(t)} - \eta \left.\frac{\partial E}{\partial \mathbf{W}}\right|_{\mathbf{W}^{(t)}},$$

where η is the learning rate, mostly kept at a low level. The negative sign means going inward while the vector $\begin{pmatrix} \frac{\partial E}{\partial \mathbf{W}_1} \\ \frac{\partial E}{\partial \mathbf{W}_2} \end{pmatrix}$ is pointing outward in the direction of positive increment.

We first compute the gradient for the weights $\mathbf{W}_2 = (v_{kj})_{k=1,\ldots,r;j=0,\ldots,q}$ (from hidden layer to output layer): for every $k = 1, \ldots, r$, when $j = 1, \ldots, q$,

$$\frac{\partial E}{\partial v_{kj}} = \frac{\partial E}{\partial o_k}\frac{\partial o_k}{\partial v_{kj}} = \frac{\partial E}{\partial o'_k}\frac{\partial o'_k}{\partial o_k}\frac{\partial o_k}{\partial v_{kj}} = -(y_k - o'_k) \cdot 1 \cdot h'_j = -(y_k - o'_k) \cdot h'_j,$$

and when $j = 0$,

$$\frac{\partial E}{\partial v_{k0}} = \frac{\partial E}{\partial o_k}\frac{\partial o_k}{\partial v_{k0}} = \frac{\partial E}{\partial o'_k}\frac{\partial o'_k}{\partial o_k}\frac{\partial o_k}{\partial v_{k0}} = -(y_k - o'_k).$$

Therefore, in matrix form:

$$W_2^{(t+1)} - W_2^{(t)} = -\eta \frac{\partial E}{\partial W_2} = -\eta \left(-(y - o')\right) \left[1 \ (h')^\top\right] =: -\eta \delta_2 \left[1 \ (h')^\top\right],$$

where $\delta_2 := -(y - o') = \partial E / \partial o$ denotes the error rate with respect to the output weighted sum $o = (o_1, o_2, \dots, o_r)^\top$, with o_k's given by (15.7), and $h' := (h'_1, \dots, h'_q)^\top$ is the vector of output values from the hidden layer.

Finally, we compute the gradient with respect to the weights $W_1 = (w_{ji})_{j=1,\dots,q;\, i=0,\dots,p}$ (from input layer to hidden layer): for every $j = 1, \dots, q$, when $i = 1, \dots, p$,

$$\frac{\partial E}{\partial w_{ji}} = \frac{\partial E}{\partial h'_j} \frac{\partial h'_j}{\partial h_j} \frac{\partial h_j}{\partial w_{ji}} = \sum_{k=1}^{r} \frac{\partial E}{\partial o_k} \frac{\partial o_k}{\partial h'_j} \frac{\partial h'_j}{\partial h_j} \frac{\partial h_j}{\partial w_{ji}} = \sum_{k=1}^{r} (\delta_2)_k \cdot v_{kj} \cdot h'_j (1 - h'_j) \cdot x_i,$$

and when $i = 0$,

$$\frac{\partial E}{\partial w_{j0}} = \sum_{k=1}^{r} \frac{\partial E}{\partial o_k} \frac{\partial o_k}{\partial h'_j} \frac{\partial h'_j}{\partial h_j} \frac{\partial h_j}{\partial w_{j0}} = \sum_{k=1}^{r} (\delta_2)_k \cdot v_{kj} \cdot h'_j (1 - h'_j);$$

therefore, in matrix form:

$$W_1^{(t+1)} - W_1^{(t)} = -\eta \frac{\partial E}{\partial W_1} = -\eta \left((W_2^\top \delta_2) \odot h' \odot (\mathbf{1} - h') \right) [1 \ x^\top] =: -\eta \delta_1 [1 \ x^\top],$$

where $\odot$ stands for the Hadamard product for vectors.

15.7 MULTILAYER PERCEPTRON

A multilayer perceptron (MLP) (see Figure 15.9) is essentially a deep[1] neural network (DNN) consisting of layers of perceptrons with non-linear activation functions. The input to each neuron in the intermediate layers is a linear combination of the outputs from the neurons of the previous layer. As a DNN consists solely of layers of linear transformations, without any nonlinear activation functions, it can be reduced to a simple two-layer input–output model, the presence of non-linear activation functions distinguishes a MLP from a single linear perceptron.

We next illustrate the application of a MLP to stock price prediction. Unlike traditional time series models such as ARIMA and GARCH ones, some restrictive and unrealistic assumptions such as stationarity of the stock movements are not necessary for MLPs.

[1] "Deep" means more than one hidden layer; even those with two hidden layers are called "deep".

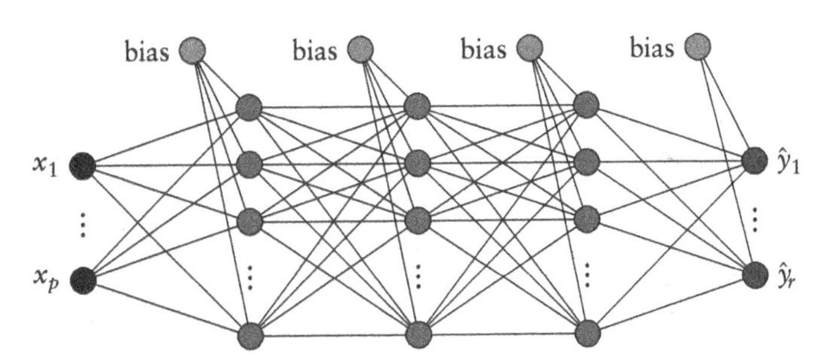

FIGURE 15.9 Illustration of a multilayer perceptron. Each node is connected with the preceding (hidden) layer, over which a non-linear activation function is applied. A linear transformation is carried within each node, with a (common) bias stored in a node disconnected from the preceding layers.

(I) Data source
Stock price data are readily available at *Yahoo Finance* (https://finance.yahoo.com) or *Bloomberg* (https://www.bloomberg.com/markets/stocks). Note that the prices quoted at Yahoo Finance are adjusted for share splits and dividend payments.

(II) Data pre-processing
We use the past 900 price quotes, starting from 15 March 2016 to 9 October 2019, as the training dataset, and the remaining 358 price quotes, starting from 10 October 2019 to 12 March 2021, as the test dataset. For each security, every set of consecutive ten trading days of price quotes is used as the observation input $x = (x_1, \cdots, x_{10})$, where x_i stands for the price quote on the first i-th day of these ten consecutive days, and the immediate following 11th-day price quote is considered as the label, i.e. $y = x_{11}$. By then, the size of the training dataset is $900 - 10 + 1 - 1 = 890$, and the size of the test dataset is $358 - 10 + 1 - 1 = 348$, where the last observation only serves as the label. The training dataset is first shuffled and then fed into the MLP model to be specified below.

(III) Model implementation
For each stock, we use the prices over a period, now 10 consecutive days, as the input features to predict the next day's price, as a label, after the censored period, i.e. the 11th-day price. We apply an MLP with 2 hidden layers each with 100 neurons. The activation function for each neuron is chosen to be the ReLU function, i.e. $(z)_+$. Figures 15.10 and 15.11 illustrate the proposed MLP model.

(IV) Backpropagation for parameter training
As before, we consider the mean squared error as the individual loss function:

$$L = \frac{(y - \hat{y})^2}{2}, \tag{15.8}$$

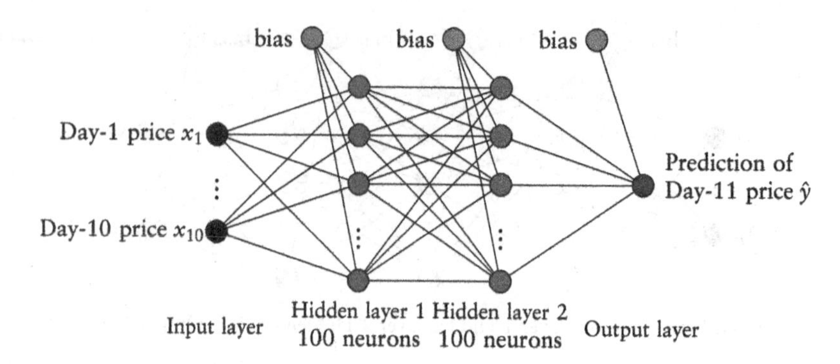

FIGURE 15.10 A 10-100-100-1 MLP model for stock price prediction.

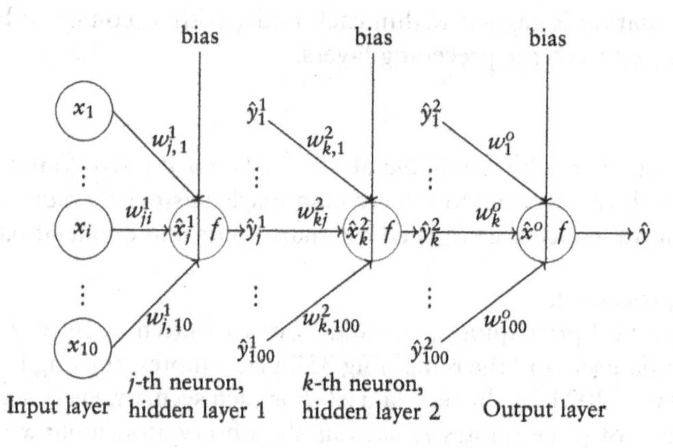

FIGURE 15.11 Input and output for each neuron with weights assigned.

and train the model via backpropagation by also incorporating the Adam method (see Section 15.5). Refer to Figure 15.11 for the different notations. Suppose that the algorithm is at the stage of t-th iteration. At the output layer, $k = 1, 2, \ldots, 100$,

$$\frac{\partial L}{\partial w^o_k} = -(y - \hat{y}_t)\frac{\partial \hat{y}_t}{\partial w^o_k} = -(y - \hat{y}_t)\frac{\partial}{\partial w^o_k}\text{ReLU}(\hat{x}^o) = -(y - \hat{y}_t)\mathbb{1}_{(0,\infty)}(\hat{x}^o_t)\hat{y}^2_{k,t},$$

where the indicator function $\mathbb{1}_{(0,\infty)}(x) = 1$ when $x \geq 0$, otherwise equals 0 when $x < 0$, is the sub-differential of the ReLU function. The first and second moments of the gradients are respectively: for $k = 1, 2, \ldots, 100$,

$$m^o_{k,t} = \beta_1 m^o_{k,t-1} - (1 - \beta_1)(y - \hat{y}_t)\mathbb{1}_{(0,\infty)}(\hat{x}^o_t)\hat{y}^2_{k,t},$$
$$v^o_{k,t} = \beta_2 v^o_{k,t-1} + (1 - \beta_2)(y - \hat{y}_t)^2\mathbb{1}_{(0,\infty)}(\hat{x}^o_t)(\hat{y}^2_{k,t})^2.$$

The update for the weights at the output layer is therefore given by, for $k = 1, 2, \ldots, 100$, with a learning rate η,

$$w^o_{k,t} = w^o_{k,t-1} - \frac{\eta}{\sqrt{v^o_{k,t}} + \epsilon} m^o_{k,t}. \tag{15.9}$$

At the second hidden layer, we can work out using the chain rule that, at the t-th iteration,

$$\frac{\partial L}{\partial w_{kj}^2} = \frac{\partial L}{\partial \hat{y}} \frac{\partial \hat{y}}{\partial \hat{x}^o} \frac{\partial \hat{x}^o}{\partial \hat{y}_k^2} \frac{\partial \hat{y}_k^2}{\partial \hat{x}_k^2} \frac{\partial \hat{x}_k^2}{\partial w_{kj}^2}$$

$$= -(y - \hat{y}_t)\mathbb{1}_{(0,\infty)}(\hat{x}_t^o) w_{k,t-1}^o \mathbb{1}_{(0,\infty)}(\hat{x}_{k,t}^2)\hat{y}_{j,t}^1.$$

Similarly, the partial derivative of the loss function with respect to the weight associated with the input variables connecting to the first hidden layer can be computed by, at the t-th iteration,

$$\frac{\partial L}{\partial w_{ji}^1} = \sum_{k=1}^{100} \frac{\partial L}{\partial \hat{y}} \frac{\partial \hat{y}}{\partial \hat{x}^o} \frac{\partial \hat{x}^o}{\partial \hat{y}_k^2} \frac{\partial \hat{y}_k^2}{\partial \hat{x}_k^2} \frac{\partial \hat{x}_k^2}{\partial \hat{y}_j^1} \frac{\partial \hat{y}_j^1}{\partial \hat{x}_j^1} \frac{\partial \hat{x}_j^1}{\partial w_{ji}^1}$$

$$= -(y - \hat{y}_t)\mathbb{1}_{(0,\infty)}(\hat{x}_t^o) x_i \mathbb{1}_{(0,\infty)}(\hat{x}_{j,t}^1) \sum_{k=1}^{100} w_{k,t-1}^o w_{kj,t-1}^2 \mathbb{1}_{(0,\infty)}(\hat{x}_{k,t}^2).$$

The partial derivative with respect to the bias parameter is left as an exercise for the readers; it can be tackled by similar arguments illustrated at the end of Section 15.6. The updates for both w_{kj}^2 and w_{ji}^1 can now be computed using formula (15.5), just like (15.9).

(V) Programming code in Python

We here enclose the detailed programming code in Python for the present MLP model of stock price prediction; see Programme 15.18.

```
1  import pandas as pd
2  import numpy as np
3  import tensorflow as tf
4  import matplotlib.pyplot as plt
5
6  def generate_dataset(price, seq_len):
7      X_list, y_list = [], []
8      for i in range(len(price) - seq_len):
9          X = np.array(price[i:i+seq_len])
10         y = np.array([price[i+seq_len]])
11         X_list.append(X)
12         y_list.append(y)
13     return np.array(X_list), np.array(y_list)
14
15 # self: a syntex referring to the class object itself, i.e. MLP_stock
16 class MLP_stock:
17     def build_model(self):
18         model = tf.keras.models.Sequential()
19         model.add(tf.keras.layers.Dense(100, activation=tf.nn.relu))
20         model.add(tf.keras.layers.Dense(100, activation=tf.nn.relu))
21         model.add(tf.keras.layers.Dense(1, activation=tf.nn.relu))
22         optimizer = tf.keras.optimizers.Adam(lr=0.01)
23         model.compile(optimizer=optimizer, loss="mse")
24         return model
```

```
25
26      def train(self, X_train, y_train, bs=32, ntry=5, n_epochs=50):
27          self.best_model = self.build_model()
28          self.best_model.fit(X_train, y_train, batch_size=bs,
29                              epochs=n_epochs, shuffle=True)
30          eval_data = (X_train[-50:], y_train[-50:])
31          best_loss = self.best_model.evaluate(*eval_data)
32          for i in range(1, ntry):
33              model = self.build_model()
34              model.fit(X_train, y_train, batch_size=bs,
35                        epochs=n_epochs, shuffle=True)
36              if model.evaluate(*eval_data) < best_loss:
37                  self.best_model = model
38                  best_loss = model.evaluate(*eval_data)
39
40      def predict(self, X_test):
41          return self.best_model.predict(X_test)
42
43  tf.random.set_seed(4002)
44  STOCKS = ["AAL", "GS", "FB", "MS"]
45  train_len = 900  # 900 trading days approximately 3.5 of calender years
46  seq_len = 10
47  for stock in STOCKS:
48      df = pd.read_csv(f"{stock}.csv")
49      stock_train = df["Adj Close"].iloc[:train_len].values
50      stock_test = df["Adj Close"].iloc[train_len:].values
51
52      X_train, y_train = generate_dataset(stock_train, seq_len)
53      X_test, y_test = generate_dataset(stock_test, seq_len)
54
55      MLP = MLP_stock()
56      MLP.train(X_train, y_train)
57      y_pred = np.squeeze(MLP.predict(X_test))
58
59      plt.figure(figsize=(15, 10))
60      # setting font size to be 20 for all the text in the plot
61      plt.rcParams['font.size'] = "20"
62      test_date = df.Date[-len(y_test):]
63      plt.plot(test_date,y_test,color=*'pink',linewidth=6,label="true")
64      plt.plot(test_date, y_pred, "r--", linewidth=3, label="predict")
65      plt.xticks(test_date[::50])
66
67      plt.title(f"{stock} prediction from {df.Date[train_len]}",
68                fontsize=35)
69      plt.ylabel("price", fontsize=25)
70      plt.xlabel("trading days", fontsize=25)
71      plt.legend(loc='lower right', fontsize=25)
```

Programme 15.18 A 10-100-100-1 MLP stock price prediction model in Python.

Before we finish this section, we make some remarks on some relevant syntax and coding practice.

The `tensorflow` library has been widely used for deep learning in Python and R. It was first developed by the *Google Brain* team for their own internal use [1], and then it was released for public use in 2015. However, `tensorflow` itself is built from Python, C++, and CUDA (Compute Unified Device Architecture)[2]. Even if one uses **R**, it will just call up the same set of syntax and commands. Therefore we shall not refer back to **R** for most of the applications related to `tensorflow`. Without loss of generality, from now on, we shall mainly focus on the Python coding for deep neural networks. The `keras` library is an open-source deep learning *Application Programming Interface* (API) written under Python. It aimed to provide an easy and fast implementation of deep neural networks, which can be run on top of TensorFlow. Keras was later integrated under TensorFlow by mid 2017, yet it can still be imported independently.

The `generate_dataset()` function receives two parameters, the `price` parameter takes in the training dataset as well as the test dataset, while the `seq_len` accepts an integer indicating the moving window width of input features, which is 10 in our stock price prediction example. This function returns two `numpy` arrays, the observations x and the label y for the dataset.

Next, the `MLP_stock` class object builds our MLP model. Under `tf.keras.models`, the `Sequential()` function builds a sequential/hierarchical model, which is composed of a plain stack of layers, where each layer has exactly one input tensor and one output tensor. For instance, our 10-100-100-1 model is a sequential model, where all layers are stacked transversely along a horizontal axis. The `Dense()` function under `tensorflow.keras.layers` builds a *densely* or fully-connected layer with the number of neurons being the first argument; the activation function is specified by the second argument `activation`, and here we adopted ReLU as the activation function, which can be imported by using `tf.nn.relu`, where `nn` stands for Neural Network, as expected. Finally, we configure our MLP model with `compile function`, where the Adam algorithm is chosen as the optimizer with a learning rate $\eta = 0.01$ being specified by the command `lr=0.01` and MSE is used as the loss function with a specification through the command `loss="mse"`. We feed in the `train()` function with four parameters, namely (i) the observation of the training set `X_train`; (ii) the label of the training set `y_train`; (iii) minibatch size M, by default set as `bs=32`; and (iv) the total number of models built, defaulted to be `ntry=5`. The purpose of this `train()` function is the same as the `ANNet()` function in Programme 15.8 in **R**. Different initial seeds lead to different set of optimal weights and biases, and we need to try several times in order to obtain the best possible model. Since the most recent stock prices are more relevant to our next-day prediction, the past $50 + 10 = 60$ days are enough for evaluating the loss of our model. Finally, the `predict()` function accepts only one parameter, the observation of the test dataset as `X_test`. Lastly, we plot both the test label and the predicted label for each stock. The resulting plots for the prediction performance for different companies are shown in Figure 15.12, from which the effectiveness of MLP is reasonably convincing.

---

[2] CUDA is an Application Programming Interface (API) created by *NVIDIA* that allows general purpose processing in a CUDA-enabled Graphics Processing Unit (GPU). It accelerates our applications by using the CUDA built-in functions or any customized functions written in C, C++, Fortran, and Python.

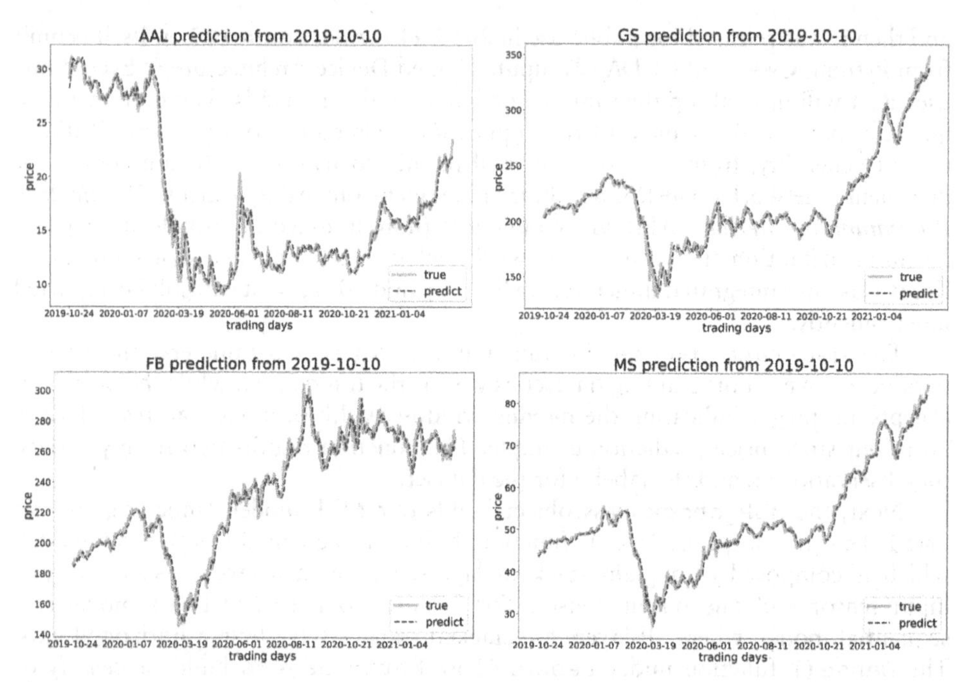

FIGURE 15.12 Predicting the day-11 price for *Anglo American* (AAL), *Goldman Sachs* (GS), *Meta* (FB), and *Morgan Stanley* (MS) by using Programme 15.18 in Python.

15.8 UNIVERSAL APPROXIMATION THEOREM

As the previous discussions have shown, the input to each neuron not in the input layer is a weighted average of the output from the neurons in the preceding layer, which is in turn fed into a non-linear activation function as its corresponding output. One common example of an activation function is a *sigmoid function*, which is an S-shaped increasing function defined on all real numbers, such as the logistic function $s(x) := 1/(1 + e^{-x})$ with a range of $(0, 1)$, or the hyperbolic tangent function $\tanh(x) := (e^x - e^{-x})/(e^x + e^{-x})$ with a range of $(-1, 1)$; see Figure 15.13.

In practice, we often have to cope with different sophisticated functional relations between feature variables and response variables. An ANN can approximate these complex functions by using simple building blocks, namely the sigmoid neurons. Likewise, a deep neural network with a certain number of hidden layers can also achieve the same goal while also allowing for more flexibility for self-adjustments. All of these are guaranteed by the celebrated *Universal Approximation Theorem.*

Theorem 15.1 ***(Universal Approximation)*** A feedforward network with a single hidden layer, in addition to the input and output layers, containing a finite number of neurons can approximate continuous functions on compact subsets of $\mathbb{R}$, if the activation functions are bounded, continuous and non-constant.

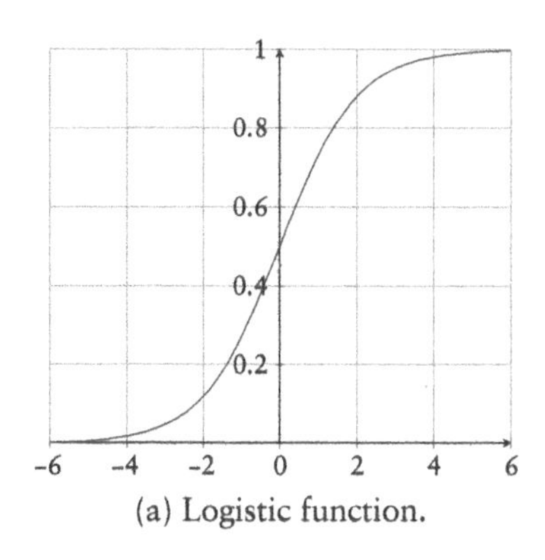

(a) Logistic function.

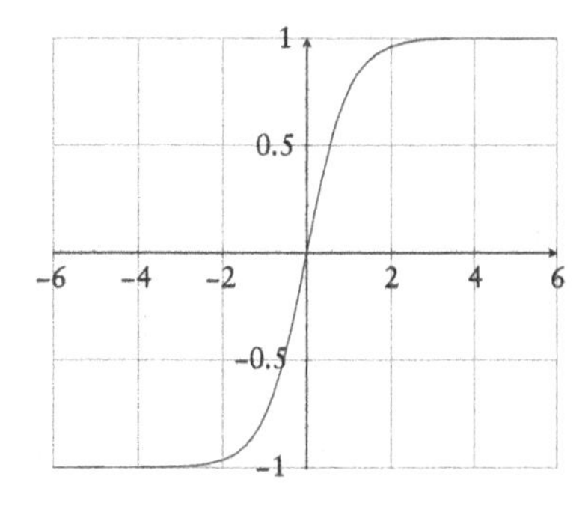

(b) Hyperbolic tangent function.

FIGURE 15.13 Examples of sigmoid functions.

The precise formulation of the theorem and its proof requires more advanced knowledge in *real analysis*, which we shall not discuss here. Nevertheless, we still illustrate the main idea behind its proof.

Consider the simple case when both input x and output y are in $\mathbb{R}$, and the functional relation between x and y is $y = f(x)$, such as in Figure 15.14a, for example. We approximate this function by a step function (see Figure 15.14b), such that each "step" can be considered as a simple function of the form $\mathbb{1}_{[x_i, x_{i+1}]}$, for some $x_i < x_{i+1}$. The linear combination of these simple functions clearly can approximate f arbitrarily well by reducing the mesh size of the underlying partition of the real line.

Consider two sigmoid functions h_1 and h_2 with very steep slopes such that for a small $\delta > 0$, h_1 stays at 0 for any $x \leq -\delta$, but reaches 1 linearly at $x = -\delta/2$, and it remains constant thereafter. Similarly, h_2 stays at 0 for any $x \leq \delta/2$, but reaches 1 at

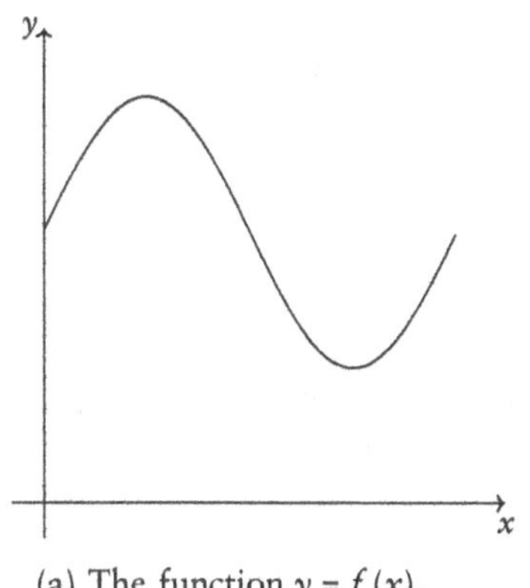

(a) The function $y = f(x)$.

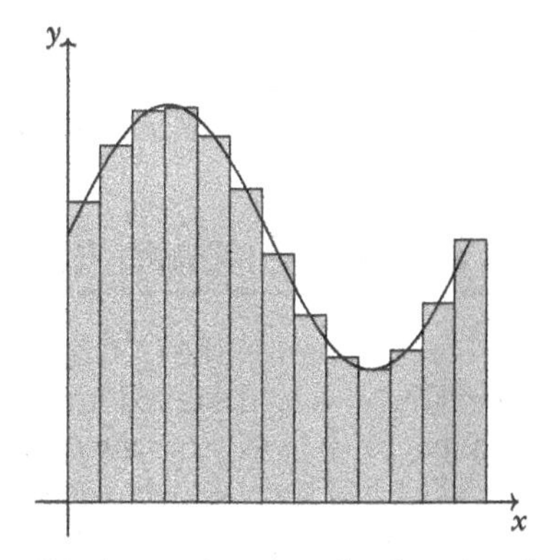

(b) Approximating the function f by step functions.

FIGURE 15.14 Illustration of approximating a function by simple functions.

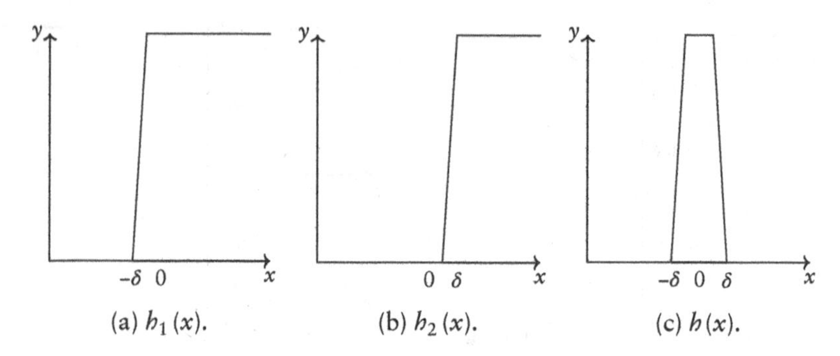

FIGURE 15.15 Construction of a simple function.

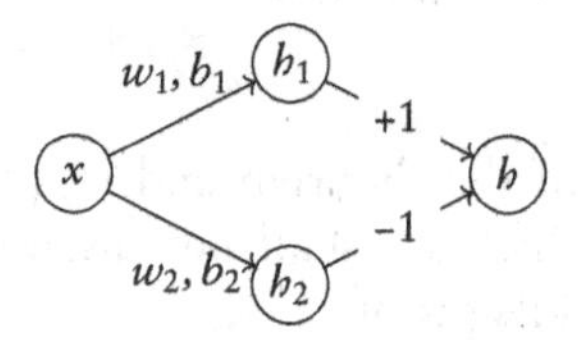

FIGURE 15.16 A component in an ANN used to construct a rectangular partition.

$x = \delta$ and remains constant thereafter, as seen in Figures 15.15a and 15.15b. Then the function $h := h_1 - h_2$ can approximately give us any one of the simple functions in Figure 15.14b as desired, as Figure 15.15c shows.

Hence, for an input x being passed into two sigmoid neurons with respective activation functions h_1 and h_2, if we apply the weights $+1$ and -1 to them in the neuron in the next layer, we can obtain the desired approximation for one simple function. By combining several similar components, as in Figure 15.16, we can approximate an arbitrary function f. Finally, note that by properly scaling the domain for the logistic or hyperbolic tangent function, either one can also well approximate h_1 or h_2. Therefore, the claim of the theorem follows, and this in some sense justifies the effective *representational power of deep neural networks* in general.

15.9 LONG SHORT-TERM MEMORY (LSTM)

Long Short-Term Memory (LSTM) model is a kind of *recurrent neural network* (RNN). This DNN was designed to learn and store long-term dependencies in temporal sequences. It was first introduced by Hochreiter and Schmidhuber [10]. Since then, LSTM has been widely used in various sequence-to-sequence tasks, such as natural language processing, speech recognition, and time series prediction; also see [6, 11, 15, 22].

The key ingredient in LSTM is the use of memory cells which enable the network to learn and remember long-term dependencies. Each memory cell consists of three main components: an input gate, a forget gate, and an output gate; see Figure 15.17 and more information in [10]. These gates control the flow of information within

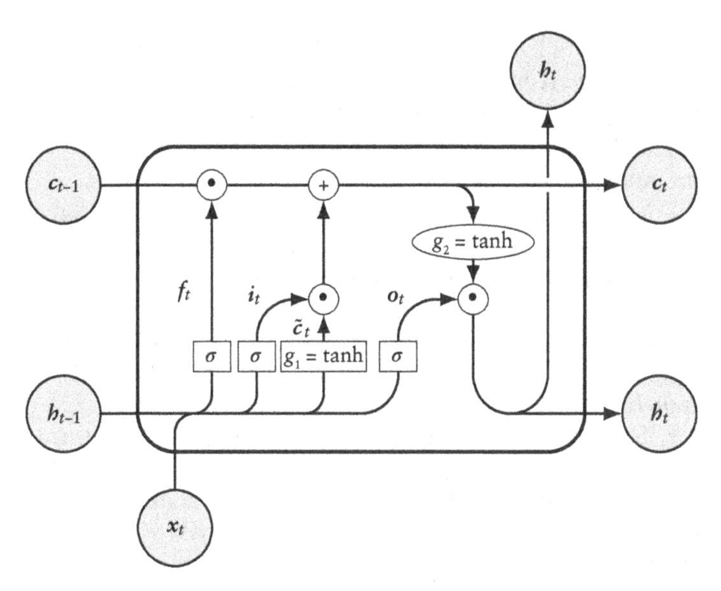

FIGURE 15.17 A standard LSTM memory cell; also see [20].

the cell, allowing the cell to selectively store, update, and retrieve information from its internal state. This gating mechanism helps to mitigate the *vanishing gradient problem*, which is a common issue in traditional RNNs [2]. The input gate determines the extent to which new input data is stored in the cell state. The forget gate decides which parts of the cell state should be forgotten or retained, allowing the network to learn when to discard irrelevant information. Finally, the output gate controls the output of the cell based on its current state; more discussions can be found in [7, 10, 23].

In the following, we define all gates properly:

$$\text{(Forget gate)} \quad f_t = \sigma(W_f x_t + V_f h_{t-1} + b_f); \tag{15.10a}$$

$$\text{(Input gate)} \quad i_t = \sigma(W_i x_t + V_i h_{t-1} + b_i); \tag{15.10b}$$

$$\text{(Cell input)} \quad \tilde{c}_t = g_1(W_{\tilde{c}} x_t + V_{\tilde{c}} h_{t-1} + b_{\tilde{c}}); \tag{15.10c}$$

$$\text{(Output gate)} \quad o_t = \sigma(W_o x_t + V_o h_{t-1} + b_o); \tag{15.10d}$$

$$\text{(Cell state)} \quad c_t = i_t \odot \tilde{c}_t + f_t \odot c_{t-1}; \tag{15.10e}$$

$$\text{(Cell output)} \quad h_t = o_t \odot g_2(c_t); \tag{15.10f}$$

for $t = 1, \cdots, T$, where $\odot$ stands for the pointwise (Hadamard) product for components of vectors as before; also

1. for this memory cell, $x_t \in \mathbb{R}^p$ stands for the input feature vector at the present timestep t;
2. matrices $W_f, W_i, W_{\tilde{c}}, W_o \in \mathbb{R}^{q \times p}$ correspond to the weights of inputs x_t used for the forget gate, input gate, cell input and output gate, respectively;

3. matrices $V_f, V_i, V_{\tilde{c}}, V_o \in \mathbb{R}^{q \times p}$ correspond to the weights of the cell output activated value h_{t-1}, fed from the one-step delay of itself for the forget gate, input gate, cell input and output gate, respectively;
4. vectors $b_f, b_i, b_{\tilde{c}}, b_o \in \mathbb{R}^q$ correspond to the biases of the forget gate, input gate, cell input, and output gate, respectively;
5. activation functions $g_1, g_2, \sigma \in \mathbb{R}^q$ are used for the cell input, cell output and all other gates, respectively. Typically, hyperbolic tangent function *tanh* is chosen for g_1 and g_2; while the logistic function can be adopted for σ.

In an LSTM model, the cells are connected in a sequential manner, forming a recurrent neural network. A single cell receives the output of the cell state $c_{t-1} \in \mathbb{R}^q$ and cell output $h_{t-1} \in \mathbb{R}^q$ from the previous time step as input, together with the input vector x_t from the current time. This connection allows information to flow from one cell to the next, enabling the model to capture long-term dependencies in sequential data. To initialize the LSTM model, the cell state c_0 and the cell output h_0 are often initialized with independent standard normals as a common practice of Xavier initialization. Figure 15.18 illustrates this architecture.

LSTM has been further improven upon by the introduction of different variants, such as the *Gated Recurrent Unit* (GRU) [4] which simplifies the LSTM architecture by merging the input and forget gates. Another notable variant is the *Bidirectional LSTM* (BiLSTM) [21] which processes input sequences in both forward (from the beginning) and backward (from the end) directions, capturing information from both top-down and bottom-up contexts.

Lastly, we shall use Figure 15.19 to motivate the idea of backpropagation in a multilayer LSTM model. For simplicity, consider the two-layer LSTM model with the first layer containing q LSTM memory cells (each depicted by a square with a round arrow inside, standing for their recurrent nature), each outputting a real value, as shown in the left diagram of Figure 15.19 in a compressed form; the input is a sequence of $T = 5$ temporal data each containing $p = 4$ features (these memory cells altogether form a q tensor of outputs). For the second layer, we just take one single memory cell with the input of (again) a sequence of $T = 5$ temporal data each containing q features from the first layer output with one real output by itself.

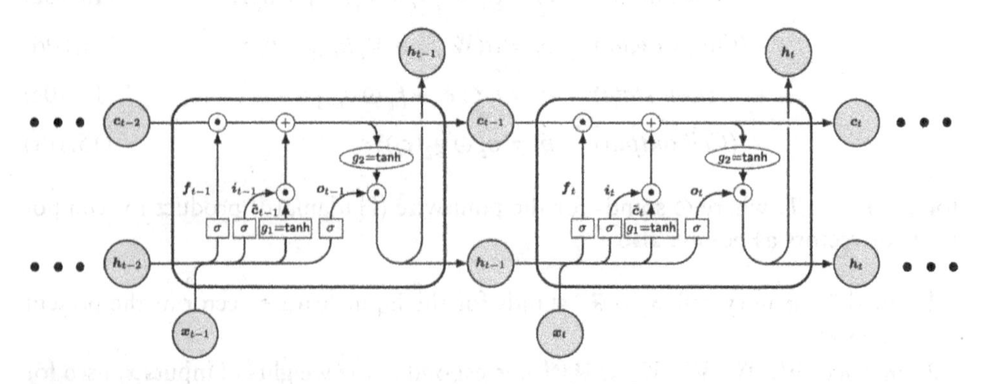

FIGURE 15.18 Connections of cells in a LSTM model.

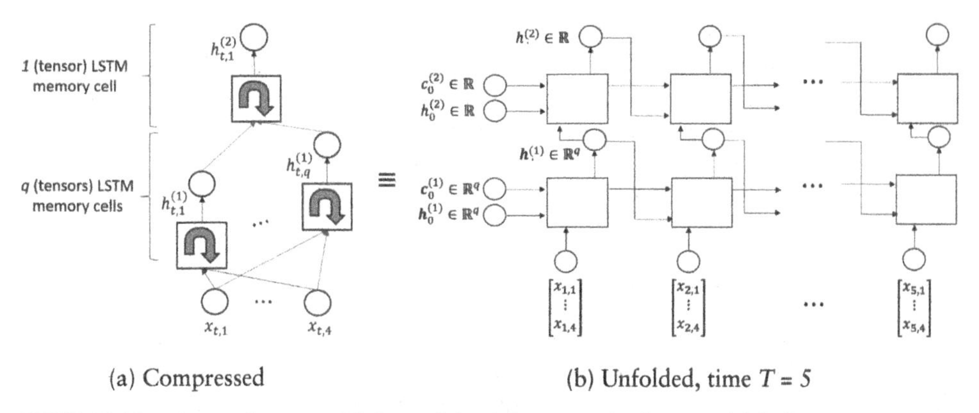

FIGURE 15.19 A two-layer LSTM model with an equivalent *unfolded representation* for backpropagation.

To consider the error propagation for this model, it is equivalent to consider the right diagram of Figure 15.19 an *unfolded form*, so that the cells are connected in a more elaborated manner than those in Figure 15.18. Therefore, schematically, training the right model of Figure 15.19 is equivalent to coping with the error flows in Figure 15.20. Particularly, to train an LSTM model, unlike a traditional DNN introduced before, the errors of the LSTM model can be backpropagated through both time (along horizontal direction) and layers (along vertical direction as for a DNN), where E_t is the individual loss function for each cell output $h_t^{(2)}$. Indeed, the dashed arrows represent the backpropagation process through layers, while the dash-dotted arrows correspond to the backpropagation process through time. This means

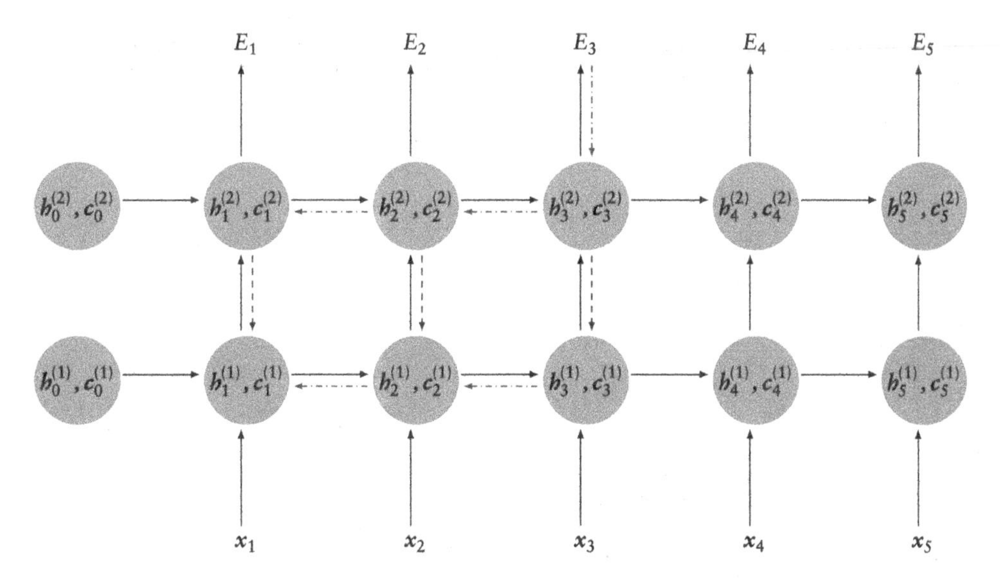

FIGURE 15.20 LSTM backpropagation for the two-layer model.

that during parameter training, the errors can flow through the LSTM model in both the layer and temporal directions. As the intricate calculus chain rules and formulas involved in the backpropagation process are beyond the scope of this book, we omit the details here but the underlying philosophy is actually the same as that of DNNs, the only difference is that there are two dimensions of error backpropagation for training LSTM compared to only one for DNN. We shall provide the details in our future book [3].

15.9.1 Bitcoin Price Prediction

In Section 7.3, we fitted an ARIMA model for predicting the price of Bitcoin for the last quarter of the year 2018, except for the very last day 31 December 2018. We now illustrate an application of LSTM in using intraday minute data with `Open`, `High`, `Low`, `Close`, and `Volume`. In fitting a deep learning model, especially for RNN, the range of each feature will first be limited to a narrow space, so as to prevent the issue of the gradient exploding or vanishing in the implementation of the mini-batch stochastic gradient descent method. Regarding this, we apply 1-lag differencing to the prices of `Open`, `High`, `Low` and `Close`, and apply a natural logarithmic transformation to (1+`Volume BTC`). Since the price process may exhibit rather different behaviours in the higher and lower price regions, an application of the 1-lag differencing "stabilizes all price level components". Then we shall add one additional feature variable of the *relative ratio of* the closing price in the current minute to that in the first minute of the training period at 00:00:00 on 01 Oct 2018. Note that while one usually uses the initial time points of the moving windows as the denominator of the aforementioned relative ratio, a simplified treatment by setting it to the starting point of the entire period will not cause much extra trouble thanks to the stability around a certain threshold of the price process. In fact it actually yields even better prediction results in this example. In our model, we propose using six feature variables ($p = 6$) to predict the closing prices per minute on the last day of 2018. We first lay down the codes for data cleansing in the following Programme 15.19 via Python:

```
1  import pandas as pd
2  import numpy as np
3
4  df = pd.read_csv("Bitstamp_BTCUSD_2018_minute.csv", header=1)
5  df.index = df.date
6  df.drop(["unix", "date", "symbol", "Volume USD"], axis=1, inplace=True)
7  df = df.iloc[::-1]              # Reverse the order of dates
8
9  BTC_vol = df["Volume BTC"].values
10 df_diff = df.diff()
11 df_diff["Volume BTC"] = np.log(1 + BTC_vol)
12
13 # Select the last quarter as the training dataset
14 date_index = pd.to_datetime(df_diff.index)
15 mask_train = pd.Series(date_index).between("2018-10-01", "2018-12-31",
16                                            inclusive="left")
17 df_train = df_diff.loc[mask_train.values].copy(deep=True)
18 y_close_train = df.close.loc[mask_train.values]
19 train_close = df.loc[mask_train.values].close.values
```

```
20 df_train["Relative_Close"] = train_close / train_close[0]
21
22 # Select the first day as the test dataset
23 mask_test = pd.Series(date_index).between("2018-12-31", "2019-01-01",
24                                           inclusive="left")
25 df_test = df_diff.loc[mask_test.values].copy(deep=True)
26 test_close = df.loc[mask_test.values].close.values
27 df_test["Relative_Close"] = test_close / train_close[0]
28 y_close_test = df.close[mask_test.values]
29
30 print(df_train.columns.values)
31
32 def generate_dataset(df, seq_len):
33     X_list, y_list = [], []
34     for i in range(len(df.index) - seq_len):
35         X_list.append(np.array(df.iloc[i:i+seq_len,:]))
36         y_list.append(df.close.values[i+seq_len])
37
38     return np.array(X_list), np.array(y_list)
```

```
1 ['open' 'high' 'low' 'close' 'Volume BTC' 'Relative_Close']
2 2018-10-01 00:00:00 2018-12-30 23:59:00 2018-12-31 00:00:00
```

Programme 15.19 Data preprocessing of Bitcoin dataset in Python.

In the data cleansing process, we first remove all redundant feature variables, such as `unix` referring to the unit timestamp, `symbol` representing the cryptocurrency Bitcoin (BTC), and the underlying currency USD with entries `BTC/USD`, and `Volume USD` standing for the total trading volume in USD, which is simply the product of `close` (closing price) and `Volume BTC` (volume of transactions). Secondly, the intraday data are arranged in backward order in time, which means that the very last minute of 2018 intraday data serves as the first observation; we perform this step by using the command `::-1`. As mentioned before, the price variables are large in magnitude, so we perform 1-lag differencing that subtracts the previously observed price from the current observed one, while we also put ln(1+`Volume BTC`) in place. Lastly, the parameter `seq_len` in the `generate_dataset()` function denotes the length of each input sequence block, i.e., T minutes of intraday data are used to predict the closing price in the next minute $T + 1$.

Dropout, first proposed in [9], is a regularization technique, which results in essentially the same effect as regularization for regression, as mentioned in Chapter 11. It is commonly used in deep neural networks to prevent overfitting and improve generalization to out-samples. During the training phase, a fraction of the units are randomly selected to halt their functioning by setting their corresponding outputs to zero; in other words, they are temporarily "dropped out" from the processing procedure. This introduces stochastics into the system yet reduces the interdependencies among the LSTM units, henceforth forcing the network to learn more robust and generalized representations. By randomly dropping units, dropout prevents individual units from relying too heavily on specific input patterns or co-adapting with other units. This encourages the network to learn more diverse and independent features, enhancing its ability to generalize unseen data. Loosely speaking, to understand the concept of dropout heuristically, one can imagine a factory with many workers. They work collaboratively on a given "massive" production procedure, so that each of the small duties for each worker is relatively simple while each worker needs not just work on one single service but could work on few; nevertheless, each worker is still responsible for one single simple duty. This massive production therefore is composed of a huge number of small tasks, and a few workers are responsible for each portion of the whole production. Certainly, different workers have different strengths and weaknesses, and they will affect the overall quality of the final output. For instance, a careless worker may make mistakes at an early stage of the production, and other workers in the subsequent stages may worsen the situation and fail to rescue the overall quality of the final output. Therefore, as a rule of thumb, in order to raise the quality of production, a successful factory should have workers that are able to collaborate with any others despite their relative weakness at the time, and the latter can be trained up by learning from the other capable workers. Similarly, in LSTM, if one LSTM unit outputs a "wrong" value, due to the highly connected network structure, all subsequent LSTM units cannot avoid making poorer predictions. That means different units have different prediction powers; as in the factory analogy above, a model can become more stable if most of its weaker units can learn from those more capable ones by a smaller grouped training. With the dropout layer mechanism, each unit has a chance to learn to work closely with a portion, around $1 - d$ where d is the dropout ratio, of other units in different layers. This normally increases the overall performance of each unit, and breaks the co-adaptation issue so that the units can now be less dependent on each other, and therefore reduce any possible redundancy of each unit by encouraging them to exercise in full power.

We shall build a two-layer LSTM model with the first and second layers containing p and q one-tensor LSTM memory cells, respectively, for the predictions of Bitcoin price. In particular, the number of one-tensor memory cells p and q in the two layers are known as hyperparameters. There are a large number of hyperparameters to be determined for the LSTM model. Other hyperparameters include, for example, the length of the input sequence T, and the dropout ratio d for each LSTM layer. In order to pick the correct set of hyperparameters, the only option available is to try them one by one. However, this approach can be extremely time-consuming and inefficient. In practice, we can limit the set of possible candidates and obtain a sub-optimal set among them by comparing the corresponding loss computed from a validation dataset. The validation dataset should be independent of the training dataset, and must not be used in training. In this example, we consider the following choices of hyperparameters: (i) $T = 5, 10, 15$; (ii) $p, q = 8, 16, 32$, corresponding to the number of memory cells for each LSTM layer; and (iii) $d_1, d_2 = 0, 0.2, 0.4$, corresponding to the dropout ratio in the first and second LSTM layer, respectively. We shall randomly draw 10% of the training data as validation data (In Python, this is done by stating `validation_split=0.1` when calling `fit()`), the candidate ($T = 10$, $q = 16$, $r = 32$, $d_1 = 0$, and $d_2 = 0.4$) achieves the lowest MSE on the validation dataset. Therefore, we shall employ this set of hyperparameters for our LSTM model based on the MSE of the validation data. Finally, we shall train our LSTM model with the full training dataset using `tensorflow`.

```
1  LAG = 5
2  # Add LAG number of observations in training dataset to test dataset
3  df_test_LAG = pd.concat((df_train.iloc[-LAG:,:], df_test), axis=0)
4
5  X_train, y_train = generate_dataset(df_train, seq_len=LAG)
6  X_test, y_test = generate_dataset(df_test_LAG, seq_len=LAG)
7
8  print(np.mean(y_train))
9  print(np.mean((y_train - np.mean(y_train))**2))
10 print(np.mean((y_test - np.mean(y_train))**2))
11
12 import tensorflow as tf
13 from tensorflow.keras import Sequential
14 from tensorflow.keras.layers import LSTM, Dense
15 from tensorflow.keras.optimizers import Adam
16
17 tf.keras.utils.set_random_seed(4002)
18
19 model = Sequential()
20 model.add(LSTM(8, return_sequences=True))
21 model.add(LSTM(8, dropout=0.2))
22
23 model.add(Dense(1))
24 model.compile(optimizer=Adam(learning_rate=0.001), loss='mse')
25 model.fit(X_train, y_train, batch_size=64, epochs=30, shuffle=True)
26
27 LSTM_pred = np.squeeze(model.predict(X_test))
28 print(np.mean((LSTM_pred - y_test)**2))
```

```
1  -0.021161063837905905
2  25.972813589924407
3  19.722084471197423
4  Epoch 1/30 2048/2048 - 10s 3ms/step - loss: 25.9267
5  Epoch 2/30 2048/2048 - 6s 3ms/step - loss: 25.8062
6  Epoch 3/30 2048/2048 - 7s 3ms/step - loss: 25.7302
7  Epoch 4/30 2048/2048 - 7s 3ms/step - loss: 25.7011
8  Epoch 5/30 2048/2048 - 7s 3ms/step - loss: 25.6776
9  Epoch 6/30 2048/2048 - 6s 3ms/step - loss: 25.6703
10 Epoch 7/30 2048/2048 - 6s 3ms/step - loss: 25.6399
11 Epoch 8/30 2048/2048 - 6s 3ms/step - loss: 25.6474
12 Epoch 9/30 2048/2048 - 6s 3ms/step - loss: 25.6345
13 Epoch 10/30 2048/2048 - 6s 3ms/step - loss: 25.6169
14 Epoch 11/30 2048/2048 - 6s 3ms/step - loss: 25.6173
15 Epoch 12/30 2048/2048 - 6s 3ms/step - loss: 25.5934
16 Epoch 13/30 2048/2048 - 6s 3ms/step - loss: 25.5890
17 Epoch 14/30 2048/2048 - 6s 3ms/step - loss: 25.5751
18 Epoch 15/30 2048/2048 - 6s 3ms/step - loss: 25.5380
19 Epoch 16/30 2048/2048 - 6s 3ms/step - loss: 25.5643
20 Epoch 17/30 2048/2048 - 6s 3ms/step - loss: 25.5573
```

```
21 Epoch 18/30 2048/2048 - 6s 3ms/step - loss: 25.5531
22 Epoch 19/30 2048/2048 - 6s 3ms/step - loss: 25.5353
23 Epoch 20/30 2048/2048 - 6s 3ms/step - loss: 25.5272
24 Epoch 21/30 2048/2048 - 6s 3ms/step - loss: 25.5391
25 Epoch 22/30 2048/2048 - 6s 3ms/step - loss: 25.5177
26 Epoch 23/30 2048/2048 - 6s 3ms/step - loss: 25.5215
27 Epoch 24/30 2048/2048 - 6s 3ms/step - loss: 25.4943
28 Epoch 25/30 2048/2048 - 6s 3ms/step - loss: 25.5079
29 Epoch 26/30 2048/2048 - 6s 3ms/step - loss: 25.5050
30 Epoch 27/30 2048/2048 - 6s 3ms/step - loss: 25.4789
31 Epoch 28/30 2048/2048 - 6s 3ms/step - loss: 25.4735
32 Epoch 29/30 2048/2048 - 6s 3ms/step - loss: 25.4495
33 Epoch 30/30 2048/2048 - 6s 3ms/step - loss: 25.4913
34 45/45 - 1s 1ms/step
35 19.171343590487826
```

Programme 15.20 A 6-8-8-1 LSTM model for Bitcoin price predictions in Python.

If we naïvely use the average of Bitcoin prices from the training dataset to predict prices in the test dataset, the mean squared error (MSE) of the out-sample predictions is 19.722. In contrast, when we employ a 6-8-8-1 LSTM model with the architecture as shown in Figure 15.21, the MSE decreases to 19.171, indicating that the LSTM model can achieve better prediction performance compared to a simple constant value, and even slightly outperform the fitted ARIMA(2, 1, 3) model with an MSE of 19.249 using Python and the fitted ARIMA(2, 1, 2) model with an MSE of 19.261 using **R**. On the other hand, the MSE of the in-sample predictions based on the

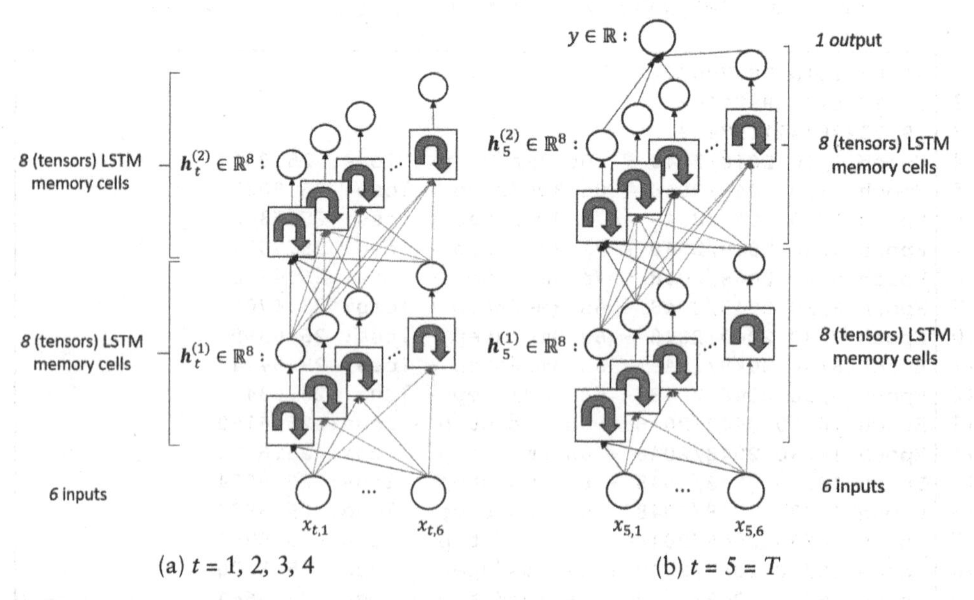

FIGURE 15.21 A 6-8-8-1 LSTM model for Bitcoin price predictions.

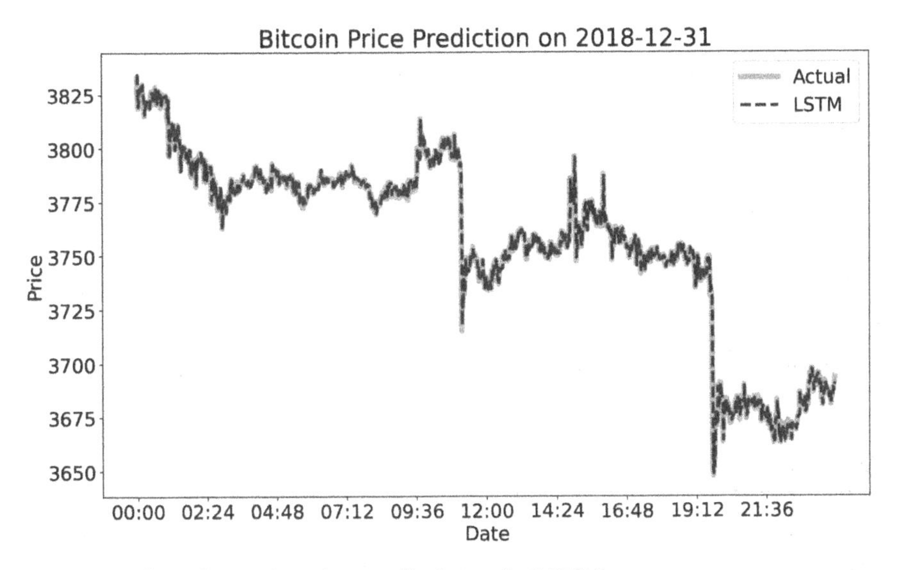

FIGURE 15.22 Plot of Bitcoin price predictions via LSTM.

average of Bitcoin prices from the training dataset is 25.927. When we evaluate the in-sample predictions of the LSTM model, the MSE is reduced to 25.491, which has a similar reduction compared to the corresponding out-sample prediction. Therefore, the representation power of this network can be improven by applying dropouts in the second LSTM layer. To visualize the price predictions over time for the test dataset, the following Programme 15.21 (via Python) plots the predictions in Figure 15.22 obtained from fitting the LSTM model in Programme 15.20 (via Python).

As a concluding remark, the example shows the best performance occurs in the presence of dropout, and LSTM models without a dropout layer may not always be optimal for prediction tasks due to the potential for overfitting. Nevertheless, they are still effective in detecting *market anomalies* or capturing exceptional patterns within financial numerical data by involving more convoluted information contained in a long temporal series from the present back to the past. This possible success relies typically on employing an effective and robust feature extraction procedure, such as signature methods in the *theory of rough paths*; see [16, 17, 19] for more discussion, and we may explore more in our future book [3].

```
1 import matplotlib.pyplot as plt
2
3 plt.rc('xtick', labelsize=20); plt.rc('ytick', labelsize=20)
4
5 date_val = pd.to_datetime(y_close_test.index)
6 xticks = date_val.strftime('%H:%M')
7
8 LSTM_close = y_close_test + (LSTM_pred - y_test)
9 fig = plt.figure(figsize=(13,8))
```

```
10 plt.plot(y_close_test, color='pink', linewidth=6, label='Actual')
11 plt.plot(LSTM_close, color='blue', linestyle='dashed',
12         linewidth=3, label="LSTM")
13 skip_minute = len(y_close_test)//10
14 plt.xticks(np.arange(0, len(y_close_test), skip_minute),
15           xticks[::skip_minute])
16 plt.title(f*'Bitcoin Price Prediction on {date_val.date[0]}',
17          fontsize=25)
18 plt.xlabel('Date', fontsize=20)
19 plt.ylabel('Price', fontsize=20)
20 plt.legend(fontsize=20)
```

Programme 15.21 Plotting the Bitcoin price predictions for LSTM via Python.

REFERENCES

1. Abadi, M., Barham, P., Chen, J., et al. (2016). TensorFlow: a system for large-scale machine learning. In *12th USENIX Symposium on Operating Systems Design and Implementation (OSDI 16)* (pp. 265–283).
2. Bengio, Y., Simard, P., and Frasconi, P. (1994). Learning long-term dependencies with gradient descent is difficult. *IEEE Transactions on Neural Networks*, 5(2), 157–166.
3. Chen, Y., Fan, N.S., and Yam, S.C.P. (2024+). *Statistical Deep Learning with Python and R*. Preprint.
4. Cho, K., Van Merriënboer, B., Gulcehre, C., Bahdanau, D., Bougares, F., Schwenk, H., and Bengio, Y. (2014). Learning phrase representations using RNN encoder–decoder for statistical machine translation. *ArXiv preprint* arXiv:1406.1078.
5. Glorot, X., and Bengio, Y. (2010). Understanding the difficulty of training deep feedforward neural networks. In *Proceedings of the Thirteenth International Conference on Artificial Intelligence and Statistics* (pp. 249–256). JMLR Workshop and Conference Proceedings.
6. Graves, A., Jaitly, N., and Mohamed, A.R. (2013). Hybrid speech recognition with deep bidirectional LSTM. In *2013 IEEE Workshop on Automatic Speech Recognition and Understanding* (pp. 273–278). IEEE.
7. Greff, K., Srivastava, R.K., Koutník, J., Steunebrink, B.R., and Schmidhuber, J. (2016). LSTM: a search space odyssey. *IEEE Transactions on Neural Networks and Learning Systems*, 28(10), 2222–2232.
8. Hinton, G., Nitish, S., and Swersky, K. (2012). Divide the gradient by a running average of its recent magnitude. *Coursera: Neural Networks for Machine Learning*. Technical Report 31.
9. Hinton, G.E., Srivastava, N., Krizhevsky, A., Sutskever, I., and Salakhutdinov, R.R. (2012). Improving neural networks by preventing co-adaptation of feature detectors. *ArXiv preprint* arXiv:1207.0580.
10. Hochreiter, S., and Schmidhuber, J. (1997). Long short-term memory. *Neural Computation*, 9(8), 1735–1780.
11. Hua, Y., Zhao, Z., Li, R., Chen, X., Liu, Z., and Zhang, H. (2019). Deep learning with long short-term memory for time series prediction. *IEEE Communications Magazine*, 57(6), 114–119.
12. Kingma, D.P., and Ba, J. (2014). Adam: a method for stochastic optimization. *ArXiv preprint* arXiv:1412.6980.

13. Linnainmaa, S. (1970). The representation of the cumulative rounding error of an algorithm as a Taylor expansion of the local rounding errors (Doctoral dissertation, Master's thesis (in Finnish), University of Helsinki).
14. Linnainmaa, S. (1976). Taylor expansion of the accumulated rounding error. *BIT Numerical Mathematics*, 16(2), 146–160.
15. Lipton, Z.C., Berkowitz, J., and Elkan, C. (2015). A critical review of recurrent neural networks for sequence learning. *ArXiv preprint* arXiv:1506.00019.
16. Lyons, T., and McLeod, A.D. (2022). Signature methods in machine learning. *ArXiv preprint* arXiv:2206.14674.
17. Lyons, T., Ni, H., and Oberhauser, H. (2014). A feature set for streams and an application to high-frequency financial tick data. In *Proceedings of the 2014 International Conference on Big Data Science and Computing* (pp. 1–8).
18. McCulloch, W.S., and Pitts, W. (1943). A logical calculus of the ideas immanent in nervous activity. *The Bulletin of Mathematical Biophysics*, 5, 115–133.
19. Morrill, J., Fermanian, A., Kidger, P., and Lyons, T. (2020). A generalised signature method for multivariate time series feature extraction. *ArXiv preprint* arXiv:2006.00873.
20. Olah, C. (2015). Understanding LSTM networks. https://colah.github.io/posts/2015-08-Understanding-LSTMs/ (accessed 7 May 2024).
21. Schuster, M., and Paliwal, K.K. (1997). Bidirectional recurrent neural networks. *IEEE Transactions on Signal Processing*, 45(11), 2673–2681.
22. Wang, S., and Jiang, J. (2015). Learning natural language inference with LSTM. *ArXiv preprint* arXiv:1512.08849.
23. Yu, Y., Si, X., Hu, C., and Zhang, J. (2019). A review of recurrent neural networks: LSTM cells and network architectures. *Neural Computation*, 31(7), 1235–1270.
24. Loffe, S. and Szegedy, C. (2015). Batch normalization: Accelerating deep network training by reducing internal covariate shift. In *International Conference on Machine Learning*, 448–456. PMLR.

Postlude

We have introduced different useful tools in financial data analytics in a rather independent way throughout different chapters. In reality, there is no single "silver bullet" that works best for every aspect of any single question, and all methods have their own pros and cons. To this end, it is important to understand the similarities and differences among them, and choose the approach best tailor-made for a specific context. Take the estimation of the drift μ in the calibration of the Kelly fraction as discussed in Chapter 7: while the trading strategy with the aid of Bollinger bands does lead to a decent performance that remedies the less accurate real-life estimation of μ, it can still be further improven by using the regressor version of CIBer in Chapter 12. In particular, as CIBer can handle the categorical data in financial reports in addition to the numerical information from various security prices and control indices, it can yield an even better estimate of μ, which in turn allows the formulation of more profit-making trading strategies even without the assistance of Bollinger bands.

We have reached the end of this book, but certainly not the end of our journey of exploration into financial data analytics. We hope that this book can nurture you, our readers, with the mindset to establish connections among all the approaches, and analyze each problem in a multifaceted manner via the use of a wide spectrum of methods, which will undoubtedly involve more advanced techniques to be formulated by future generations. Nevertheless, the core ingredients inside this rapid evolution, including the importance of mathematics, statistics, and metaprogramming, will stay unchanged over time.

Index

C

T

U

V